An Introduction to System Modeling and Control

An Introduction to System Modeling and Control

John Chiasson

Boise State University
Boise, Idaho
United States

WILEY

This edition first published 2022
© 2022 John Wiley & Sons, Inc.

All rights reserved. No part of this publication may be reproduced, stored in a retrieval system, or transmitted, in any form or by any means, electronic, mechanical, photocopying, recording or otherwise, except as permitted by law. Advice on how to obtain permission to reuse material from this title is available at http://www.wiley.com/go/permissions.

The right of John Chiasson to be identified as the author of this work has been asserted in accordance with law.

Registered Office
John Wiley & Sons, Inc., 111 River Street, Hoboken, NJ 07030, USA

Editorial Office
111 River Street, Hoboken, NJ 07030, USA

For details of our global editorial offices, customer services, and more information about Wiley products visit us at www.wiley.com.

Wiley also publishes its books in a variety of electronic formats and by print-on-demand. Some content that appears in standard print versions of this book may not be available in other formats.

Limit of Liability/Disclaimer of Warranty
The contents of this work are intended to further general scientific research, understanding, and discussion only and are not intended and should not be relied upon as recommending or promoting scientific method, diagnosis, or treatment by physicians for any particular patient. In view of ongoing research, equipment modifications, changes in governmental regulations, and the constant flow of information relating to the use of medicines, equipment, and devices, the reader is urged to review and evaluate the information provided in the package insert or instructions for each medicine, equipment, or device for, among other things, any changes in the instructions or indication of usage and for added warnings and precautions. While the publisher and authors have used their best efforts in preparing this work, they make no representations or warranties with respect to the accuracy or completeness of the contents of this work and specifically disclaim all warranties, including without limitation any implied warranties of merchantability or fitness for a particular purpose. No warranty may be created or extended by sales representatives, written sales materials or promotional statements for this work. The fact that an organization, website, or product is referred to in this work as a citation and/or potential source of further information does not mean that the publisher and authors endorse the information or services the organization, website, or product may provide or recommendations it may make. This work is sold with the understanding that the publisher is not engaged in rendering professional services. The advice and strategies contained herein may not be suitable for your situation. You should consult with a specialist where appropriate. Further, readers should be aware that websites listed in this work may have changed or disappeared between when this work was written and when it is read. Neither the publisher nor authors shall be liable for any loss of profit or any other commercial damages, including but not limited to special, incidental, consequential, or other damages.

Library of Congress Cataloging-in-Publication Data

Names: Chiasson, John, author.
Title: An introduction to system modeling and control / John Chiasson.
Description: First edition. | Hoboken, NJ : John Wiley & Sons, Inc., 2022.
 | Includes bibliographical references and index.
Identifiers: LCCN 2021030670 (print) | LCCN 2021030671 (ebook) | ISBN
 9781119842897 (cloth) | ISBN 9781119842903 (adobe pdf) | ISBN
 9781119842910 (epub)
Subjects: LCSH: Systems engineering–Mathematics. | Mathematical models.
Classification: LCC TA168 .C55 2022 (print) | LCC TA168 (ebook) | DDC
 620.001/171–dc23
LC record available at https://lccn.loc.gov/2021030670
LC ebook record available at https://lccn.loc.gov/2021030671

Cover Design: Wiley
Cover Image: © By Walter Barie

Set in 10/12pt Computer Modern Roman by Straive, Chennai, India

SKY10032480_011822

Contents

Preface		ix
About the Companion Website		xiv
1	**Introduction**	**1**
	1.1 Aircraft	1
	1.2 Quadrotors	7
	1.3 Inverted Pendulum	12
	1.4 Magnetic Levitation	13
	1.5 General Control Problem	15
2	**Laplace Transforms**	**17**
	2.1 Laplace Transform Properties	20
	2.2 Partial Fraction Expansion	24
	2.3 Poles and Zeros	35
	2.4 Poles and Partial Fractions	36
	Appendix: Exponential Function	39
	Problems	43
3	**Differential Equations and Stability**	**49**
	3.1 Differential Equations	49
	3.2 Phasor Method of Solution	52
	3.3 Final Value Theorem	57
	3.4 Stable Transfer Functions	62
	3.5 Routh–Hurwitz Stability Test	65
	Problems	77
4	**Mass–Spring–Damper Systems**	**89**
	4.1 Mechanical Work	89
	4.2 Modeling Mass–Spring–Damper Systems	90
	4.3 Simulation	96
	Problems	100
5	**Rigid Body Rotational Dynamics**	**111**
	5.1 Moment of Inertia	111
	5.2 Newton's Law of Rotational Motion	112
	5.3 Gears	120
	5.4 Rolling Cylinder	127
	Problems	135
6	**The Physics of the DC Motor**	**149**
	6.1 Magnetic Force	149

vi   Contents

6.2	Single-Loop Motor	151
6.3	Faraday's Law	155
6.4	Dynamic Equations of the DC Motor	163
6.5	Optical Encoder Model	165
6.6	Tachometer for a DC Machine*	168
6.7	The Multiloop DC Motor*	170
	Problems	175

7 Block Diagrams **185**
7.1	Block Diagram for a DC Motor	185
7.2	Block Diagram Reduction	187
	Problems	197

8 System Responses **203**
8.1	First-Order System Response	203
8.2	Second-Order System Response	205
8.3	Second-Order Systems with Zeros	217
8.4	Third-Order Systems	222
	Appendix: Root Locus Matlab File	224
	Problems	224

9 Tracking and Disturbance Rejection **233**
9.1	Servomechanism	233
9.2	Control of a DC Servo Motor	239
9.3	Theory of Tracking and Disturbance Rejection	252
9.4	Internal Model Principle	256
9.5	Design Example: PI-D Control of Aircraft Pitch	258
9.6	Model Uncertainty and Feedback*	265
	Problems	273

10 Pole Placement, 2 DOF Controllers, and Internal Stability **285**
10.1	Output Pole Placement	285
10.2	Two Degrees of Freedom Controllers	298
10.3	Internal Stability	308
10.4	Design Example: 2 DOF Control of Aircraft Pitch	316
10.5	Design Example: Satellite with Solar Panels (Collocated Case)	321
	Appendix: Output Pole Placement	324
	Appendix: Multinomial Expansions	328
	Appendix: Overshoot	329
	Appendix: Unstable Pole-Zero Cancellation	335
	Appendix: Undershoot	336
	Problems	339

11 Frequency Response Methods **361**
11.1	Bode Diagrams	361
11.2	Nyquist Theory	383
11.3	Relative Stability: Gain and Phase Margins	402
11.4	Closed-Loop Bandwidth	409
11.5	Lead and Lag Compensation	414

Contents vii

11.6	Double Integrator Control via Lead-Lag Compensation	419
11.7	Inverted Pendulum with Output $Y(s) = X(s) + \left(\ell + \dfrac{J}{m\ell}\right)\theta(s)$	426
	Appendix: Bode and Nyquist Plots in Matlab	427
	Problems	428

12 Root Locus 447

12.1	Angle Condition and Root Locus Rules	449
12.2	Asymptotes and Their Real Axis Intersection	457
12.3	Angles of Departure	463
12.4	Effect of Open-Loop Poles on the Root Locus	481
12.5	Effect of Open-Loop Zeros on the Root Locus	482
12.6	Breakaway Points and the Root Locus	483
12.7	Design Example: Satellite with Solar Panels (Noncollocated)	484
	Problems	488

13 Inverted Pendulum, Magnetic Levitation, and Cart on a Track 497

13.1	Inverted Pendulum	497
13.2	Linearization of Nonlinear Models	506
13.3	Magnetic Levitation	510
13.4	Cart on a Track System	516
	Problems	521

14 State Variables 537

14.1	Statespace Form	537
14.2	Transfer Function to Statespace	539
14.3	Laplace Transform of the Statespace Equations	551
14.4	Fundamental Matrix Φ	554
14.5	Solution of the Statespace Equation*	558
14.6	Discretization of a Statespace Model*	561
	Problems	563

15 State Feedback 569

15.1	Two Examples	569
15.2	General State Feedback Trajectory Tracking	578
15.3	Matrix Inverses and the Cayley–Hamilton Theorem	579
15.4	Stabilization and State Feedback	584
15.5	State Feedback and Disturbance Rejection	589
15.6	Similarity Transformations	593
15.7	Pole Placement	598
15.8	Asymptotic Tracking of Equilibrium Points	603
15.9	Tracking Step Inputs via State Feedback	605
15.10	Inverted Pendulum on an Inclined Track*	612
15.11	Feedback Linearization Control*	618
	Appendix: Disturbance Rejection in the Statespace	623
	Problems	626

16 State Estimators and Parameter Identification 643

16.1	State Estimators	643

	16.2	State Feedback and State Estimation in the Laplace Domain*	660
	16.3	Multi-Output Observer Design for the Inverted Pendulum*	663
	16.4	Properties of Matrix Transpose and Inverse	665
	16.5	Duality*	668
	16.6	Parameter Identification	669
		Problems	677

17 Robustness and Sensitivity of Feedback — 693

	17.1	Inverted Pendulum with Output x	694
	17.2	Inverted Pendulum with Output $y(t) = x(t) + \left(\ell + \dfrac{J}{m\ell}\right)\theta(t)$	708
	17.3	Inverted Pendulum with State Feedback	711
	17.4	Inverted Pendulum with an Integrator and State Feedback	715
	17.5	Inverted Pendulum with State Feedback via State Estimation	717
		Problems	720

*Sections marked with an asterisk may be skipped without loss of continuity.

References — 727

Index — 731

Preface

Quite honestly, I never thought I would write a book for a first controls course. So, what happened? Well, in my teaching career I have taught out of the textbooks by Ogata [1], Kuo [2], Franklin et al. [3], and Phillips and Harbor [4] with lecture notes taken from Qiu and Zhou [5] and Goodwin et al. [6]. I did find many great ideas in these texts that I do use.[1] However, I was also disappointed that the modeling was not done using first principles of physics, but rather the model was simply stated without a derivation. To address this I made up my own lecture notes on rigid body dynamics (Chapter 5), DC motors (Chapter 6), and the inverted pendulum and magnetic levitation systems (Chapter 13). I realize that my emphasis on detailed modeling may be a "bug" to some (rather than a "feature") as students find it difficult and it takes away from doing control "stuff". However, as a colleague of mine put it, students develop important insight when they understand where the dynamic models come from and any linearization, approximation, or simplification used to obtain them.

It seemed to me early in my career that teaching the first controls course seemed to be more about its techniques, i.e., manipulating block diagrams, drawing root locus, Bode and Nyquist plots, doing the Routh–Hurwitz test, etc. Yet I think the course should be about making some physical system do what you want it to do such as having a robot arm rotate 30° despite the weight of the object in its end effector, or ensure a magnetic bearing maintains an air gap despite various loads on it, keeping a pendulum rod pointing straight up, etc. I recall a colleague commenting on a lecture he gave in which he referred to the standard unity feedback controller block diagram (Figure 1) and told the class that the controller $G_c(s)$ was to be designed so that $C(s) \approx R(s)$. A student then simply asked why not just get rid of the blocks and set $C(s) = R(s)$! It seems that in teaching the first controls course we end up manipulating block diagrams so much and so easily that students get lost in the abstraction not understanding what they represent.

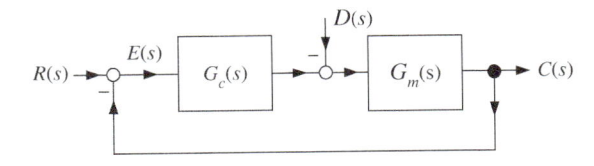

Figure 1. Control system in standard block diagram form.

Figures such as Figure 2 on the next page are included to help the students remember what a block diagrams represents.

Spending the time to do detailed derivations of a few models helps the student to understand and remember what the transfer function models represent. A similar confusion arises in the modeling of disturbances. When a disturbance is shown on the block diagram model, it is typically placed as input to the physical system. For example, one might have a load torque on a motor, but in the block diagram this disturbance $D(s)$ is modeled coming into the motor input, which is a voltage (Figure 1). I explain how this load torque is modeled as an equivalent voltage disturbance which has the same effect on

---

[1] To paraphrase Picasso: *Good ideas are inspired, great ideas are stolen!*

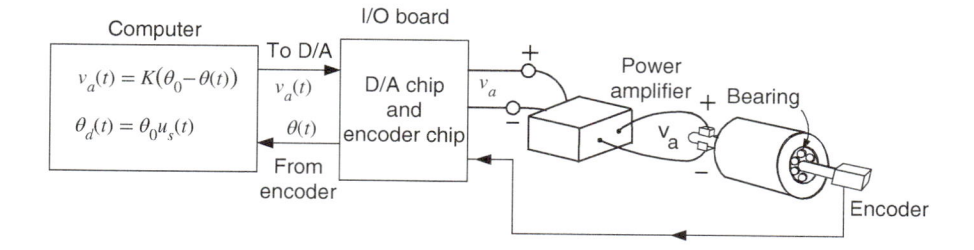

Figure 2. Position control of a DC motor.

the position/speed of the rotor as the actual load torque. This sort of understanding seems to be lost in the standard manipulations of converting a differential equation model into a block diagram model. Of course, some good laboratory work can really help to clarify these ideas as well.

Chapter 1 presents a qualitative description of the operation of an aircraft, a quadrotor, an inverted pendulum, and a magnetically levitated steel ball to motivate the need for modeling and control.

Chapter 2 is a standard presentation of the Laplace transform theory with an emphasis on partial fractions as a way to connect the time domain to the Laplace domain.

Chapter 3 on differential equations introduces stability by giving special attention to the final value theorem (FVT). This is important as it is used over and over again to determine asymptotic tracking and/or disturbance rejection of step inputs by showing the error $e(t) \to 0$ via the FVT $\lim_{s \to 0} sE(s) = 0$. This chapter also explains how to check a differential equation for stability using the Routh–Hurwitz test.

Chapters 4–6 are modeling chapters. Chapter 4 develops mass–spring–damper systems and uses them to introduce simulation using SIMULINK. Chapter 5 presents rigid body dynamics applied to gears and rolling motion. Chapter 6 on DC motors uses the first principles of physics to develop the equations that model a DC motor and explains how both an optical encoder and a tachometer work.

Chapter 7 on block diagrams is pretty standard. It is emphasized that a block diagram is simply a *graphical* representation of the relationships between the various (Laplace transformed) variables of a physical system. It is shown how to rearrange and simplify them using block diagram reduction which provides a straightforward and simple way to manipulate all the block diagrams considered in the textbook.

Chapter 8 on system responses is also pretty standard. The objective here is to make connections between the s-domain and the time domain.

Chapter 9 covers PID control that is explained using the internal model principle. It provides the understanding of why the controller must have an integrator to reject constant disturbances with the P and D parts (typically) needed to make the closed-loop system stable. It is very important to leave the student with an understanding of why PID control works in so many applications.

After covering Chapters 1–9 one typically covers root locus, Bode, and Nyquist to finish the term. The students seemed to catch on to root locus just fine, but the approach didn't seem that helpful in *designing* a controller. That is, one has to propose some form for a controller (seemingly out of thin air) and then vary a single gain to see if the closed-loop poles can be moved to a location that results in a good response. A typical example is to have an open-loop model of the form $G(s) = \dfrac{b}{s(s+a)}$, propose a controller of the form

$G_c(s) = K\dfrac{s+z}{s+p}$, and then do a root locus to choose the gain K so the two "dominant" poles of the closed-loop system result in a desired response. However, using this controller one can actually place all three closed-loop poles arbitrarily! The root locus method is then usually followed by teaching Bode diagrams, Nyquist plots, and Nyquist stability. In my experience Nyquist theory has always been harder for the students to grasp. After all that hard work of understanding Nyquist theory the design of a controller using the Bode diagram (lead, lag, lead–lag, etc.) to obtain desired gain and phase margins is still a trial and error approach. Further, the connection between gain and phase margins to system performance (settling time, overshoot, etc.) is not so direct. This leaves the students without much confidence in designing controllers.

Rather than proceeding directly to root locus, Bode, and Nyquist after PID control, my approach is to first cover Chapter 10 on output pole placement and two degree of freedom (2DOF) controllers. Starting with a transfer function model, these methods provide a straightforward systematic way to design an output feedback controller to achieve tracking and disturbance rejection objectives while often being able to also eliminate overshoot in step responses.

Before I added the material of Chapter 10 (pole placement and 2DOF controllers), the first semester course ended by covering Bode and Nyquist (Chapter 11) and root locus (Chapter 12). After covering Chapter 10 there is typically not enough time in one semester to also cover Chapters 11 and 12. Consequently, I now cover Chapter 11 at the beginning of the second course. Nyquist theory is fundamental to understanding robustness and sensitivity of control systems. Using the unity output feedback controller for the inverted pendulum designed in Chapter 10 a Nyquist analysis is presented to show that the resulting stability margins are too small for this control system to work in practice. The point here is show that there is more to control than just making the closed-loop system stable. Chapter 17 elaborates on these ideas.

Chapter 12 is a standard presentation of root locus. It ends with an example showing how to design a notch filter that cancels out stable open-loop poles close to the $j\omega$ axis and is also robust with respect to small perturbations in the location of these poles.

Chapter 13 derives the differential equation model of an inverted pendulum, a magnetically levitated steel ball, and a cart on a track using first principles of physics. It is shown how to obtain linear models of the inverted pendulum and the magnetically levitated steel ball, as well as how to control them using the methods of Chapters 9 and 10.

Chapter 14 on state variables gives the elementary linear (matrix) algebra background needed for the state feedback theory of Chapter 15.

Chapter 15 on state feedback starts out by giving a detailed derivation of a state trajectory tracking controller for the cart on the track system. This is followed a general approach to state feedback for linear statespace models including the development of a statespace pole placement algorithm. Also, based on the internal model principle, a detailed presentation of a disturbance rejection statespace controller for a servo system (DC motor) is given. An important goal here is to show that trajectory tracking accomplished by trajectory generation, pole placement, and state estimation (Chapter 16) provides a systematic procedure for designing feedback controllers in the statespace. This is to be compared with the systematic procedure for pole placement in the Laplace domain given in Chapter 10.

Chapter 16 is on state and parameter estimation. State estimation (state observer) is presented as "what to do" when one doesn't have full state measurements. An observer is presented that provides a smooth estimate of the speed of a motor using the output of an optical encoder in comparison to the noisy estimate found by differentiation of the encoder's output. Parameter estimation via least-squares is explained by going through the detailed calculations to estimate the model parameters of a DC motor.

Chapter 17 discusses robustness and sensitivity. Nyquist theory and the Bode sensitivity integrals are used to make the reader aware of the fundamental problems of controlling systems whose open-loop models have right half-plane poles. Specifically, the robustness and sensitivity of four different stabilizing controllers for the inverted pendulum are considered: (i) output feedback of the cart position, (ii) output feedback of a linear combination of cart position and pendulum rod angle, (iii) full state feedback, and (iv) output feedback of the cart position followed by state estimation and state feedback.

Supplementary Materials

There is a *solutions manual* for the end of chapter problems and a set of *slides* for each chapter. Further, there is also a complete set of MATLAB/SIMULINK *files* for the examples and problems that require them. These are all available on the book's companion website provided by Wiley.

Prerequisites

The prerequisites for this book are elementary differential equations, Laplace transforms, and a freshman/sophomore course in (calculus-based) physics. Chapters 14, 15, and 16 assume some familiarity with matrix algebra (matrix multiplication, determinants, inverses, etc.).

Typical coverage in the first semester course

Chapter 1, All sections

Chapter 2, All sections

Chapter 3, All sections except subsections 3.51 and 3.52

Chapter 4, All sections

Chapter 5, Sections 5.1 through 5.3

Chapter 6, Sections 6.1 through 6.4

Chapter 7, All sections

Chapter 8, All sections

Chapter 9, Sections 9.1 through 9.4

Chapter 10, Sections 10.1 through 10.3

Possible coverage in a second semester course

Chapter 11, All sections

Chapter 12, All sections

Chapter 13, Sections 13.1, 13.2, and either 13.3 or 13.4

Chapter 14, Sections 14.1 through 14.4

Chapter 15, Sections 15.1 through 15.8

Chapter 16, Section 16.1

Chapter 17, Sections 17.1 through 17.3

Logical Dependence of the Chapters

The logical dependence of the chapters is given in the figure below.

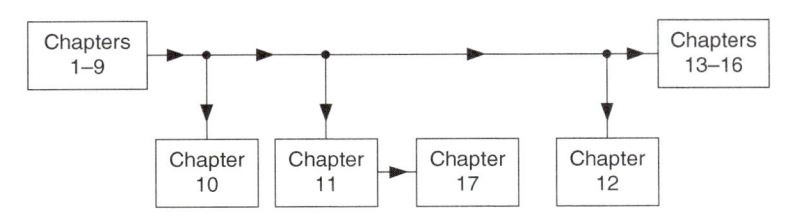

```
Chapters 1–9 → → → → Chapters 13–16
   ↓        ↓        ↓
Chapter 10  Chapter 11 → Chapter 17  Chapter 12
```

Acknowledgments

I would like to take this opportunity to thank Dr. Uri Rogers for making several valuable suggestions to improve the manuscript as well as for pointing out typos in its earlier versions. This was greatly appreciated by the author. I also want to thank Dr. Marc Bodson for his helpful comments and introducing me to parameter identification in our work on electric machines. Early in my career Professor Edward Kamen was very supportive of me teaching the introductory controls course for which I am very grateful. I am indebted to the late Professor Bruce Francis for his comments and kind remarks on an early version of this manuscript. I am also grateful to Dr. Aykut Satici for his crucial help in the final stages of this work. I want to express my gratitude to the anonymous reviewers of the manuscript who provided many constructive comments. John Wiley & Sons is gratefully acknowledged for allowing me to reuse material from Chapter 1 of my book *Modeling and High-Performance Control of Electric Machines* [7]. My Wiley editor Brett Kurzman and his production team of Sarah Lemore and Devi Ignasi have turned my manuscript into a published book. Thank you! Finally, I would like to thank the many students I have taught for putting up with the various versions of this textbook in manuscript form.

Any comments, criticisms, and corrections are most welcome and may be sent to the author at *chiasson@ieee.org*.

John Chiasson

About the Companion Website

This book is accompanied by a companion website.

www.wiley.com/go/chiasson/anintroductiontosystemmodelingandcontrol

This website contains:

MATLAB/SIMULINK files for the instructor

MATLAB/SIMULINK files for the students

Slides for each chapter

Solutions manual

1
Introduction

In this chapter we simply want to present some examples of physical control systems. The intent here is to show why feedback control is necessary and to give the "big picture" of what must be done to control physical systems.

1.1 Aircraft

Figure 1.1 is a drawing of a simple propeller powered airplane. There are four forces on the airplane: *lift* which is primarily provided by the wings, *drag* which is basically wind resistance, *thrust* which is provided by the propeller, and *gravity*.

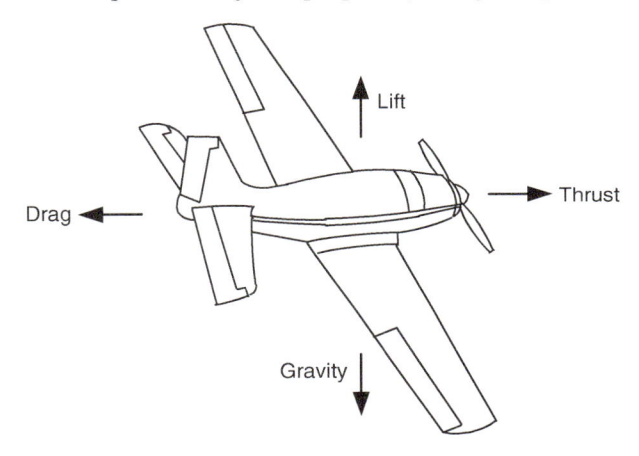

Figure 1.1. The four forces on an aircraft.

    Lift is due to the difference in air pressure between the top and bottom of an airfoil. The airfoil is a generic term that here refers to the wings, horizontal tail or the vertical tail of the aircraft. Figure 1.2 on the next page shows streamlines of air moving past an airfoil (wing). The air going over the top travels faster (as it goes a greater distance) than the air on the bottom of the wing. Because of this speed difference, it turns out that the air pressure on the top of the wing is then less than the air pressure on the bottom of the wing. The resulting upward force is what we call *lift*. We call lift an *aerodynamic* force. Figure 1.3 on the next page shows the air flow streamlines across an airfoil in a wind tunnel.

    The horizontal cylindrical glass tube (Venturi tube) in Figure 1.4 on the next page is used to experimentally demonstrate that air pressure decreases as the air speed increases. The original flow of air from the right in the Venturi tube is forced to flow through a bottleneck into a reduced diameter tube. Although not obvious, at speeds below supersonic, air is

---

An Introduction to System Modeling and Control, First Edition. John Chiasson.
© 2022 John Wiley & Sons, Inc. Published 2022 by John Wiley & Sons, Inc.
Companion website: www.wiley.com/go/chiasson/anintroductiontosystemmodelingandcontrol

2   1 Introduction

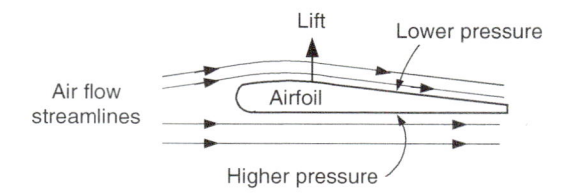

Figure 1.2. A lift force is due to the shape of the airfoil that results in the pressure above the airfoil being less than the pressure below it.

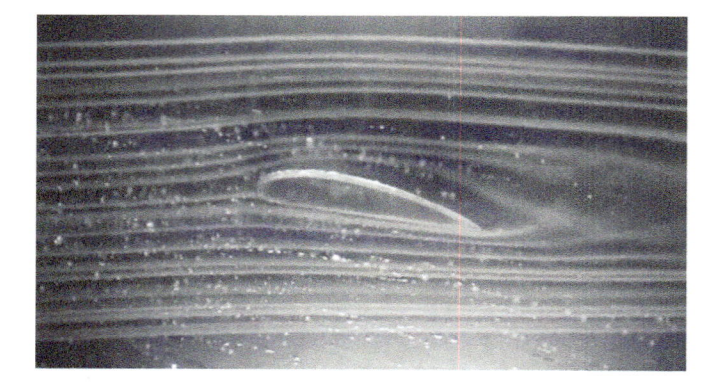

Figure 1.3. Airflow streamlines on a wing in a wind tunnel. Source: Screenshot from http://www.decodedscience.com/how-does-an-airplane-fly-lift-weight-thrust-and-drag-in-action/5200.

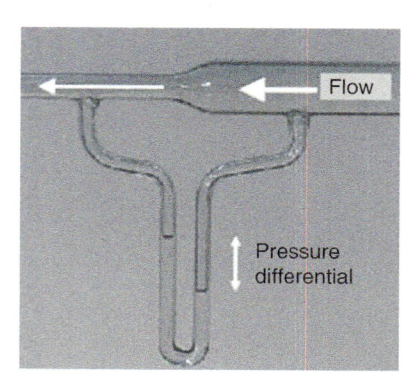

Figure 1.4. Venturi tube showing air flow through a bottleneck reducing the cross-section of the tube. In this situation the air is essentially incompressible. So the air speeds up going through the bottleneck as the mass flow is constant. The U-shaped tube below the Venturi tube (called a manometer) is filled with water to show the pressure on the left-side is reduced compared to the pressure on the right-side. Source: ComputerGeezer and Geof [8]. "Venturiflow", https://commons.wikimedia.org/wiki/File:Venturi-Flow.png, 2010, Licensed under CC BY-SA 3.0.

essentially incompressible so here its density is the same on both sides of the bottleneck.[1] The bottleneck results in increasing the air speed as it goes through it. That is, the same mass of air per unit time is flowing in both tubes (because the air does not compress) so it must speed up on the left side as the tube cross section is smaller there. The ∪-shaped columns below the Venturi tube are filled with water (called a manometer) and show the experimental fact that the air pressure on the left side is *less* than the air pressure on the right side. As air speeds up its pressure drops! This is referred to as Bernoulli's principle. The key point for airfoils is that the shape of the wing causes the air speed on the top of the wing to increase resulting in lower pressure on the top of the wing and therefore producing lift.

The left side of Figure 1.5 shows the centerline of the wing aligned with the air speed v. The air speed is simply the speed of the aircraft with respect to the air. The right side of Figure 1.2 shows the centerline of the wing at an angle α with respect to the air speed. As long as α is not too large (8 to 20 degrees depending on the plane), the lift force increases with α. We refer to α as the *angle of attack*.

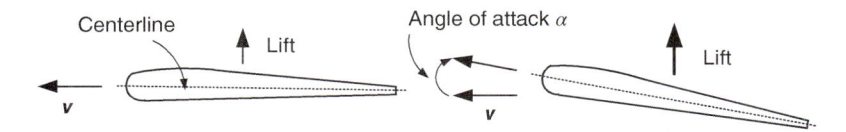

Figure 1.5. Angle of attack.

Let's go back to the airplane as shown in Figure 1.6. The *wings*, *horizontal tail*, and *vertical tail* are generically referred to as *airfoils*. The *ailerons* on the wings, the *elevators* on the horizontal tail, and the *rudder* on the vertical tail are referred to as *control surfaces*.

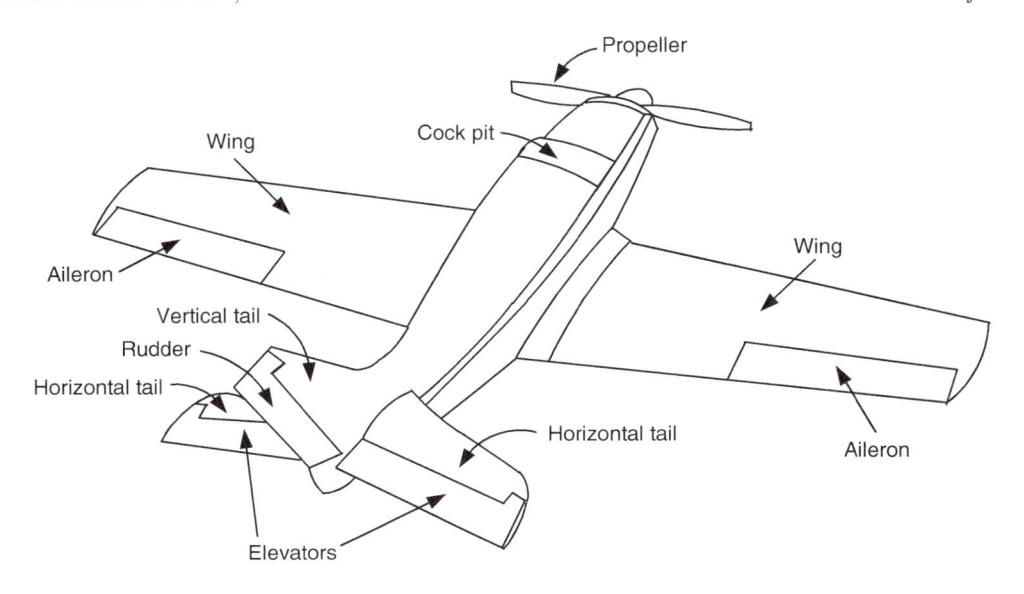

Figure 1.6. The airfoils and the control surfaces on the aircraft.

---

[1] Another way to say this is that the air is *not* compressed going through the bottleneck.

4   1   Introduction

As indicated in Figure 1.7, the control surface is connected by a hinge to the airfoil. The pilot can rotate the control surface about the hinge. As shown in Figure 1.7, if the control surface is rotated down (relative to the centerline) then the airfoil will pitch down due to the (aerodynamic) force on the control surface. Conversely, if the control surface is rotated up, it will cause the airfoil to pitch up.

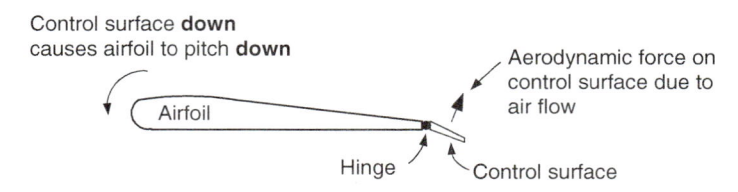

Figure 1.7. Control surface used to pitch the airfoil up or down. With the control surface down, the airfoil pitches down.

Control of Pitch Using the Elevators

As a first example of how control surfaces are used, we consider the elevators on the horizontal tail. Figure 1.8 shows the elevator (control surface) on the horizontal wing rotated up, which results in the aircraft being pitched up about its center of mass. By moving the elevator up or down, the pilot can cause the aircraft to pitch up or down, respectively.

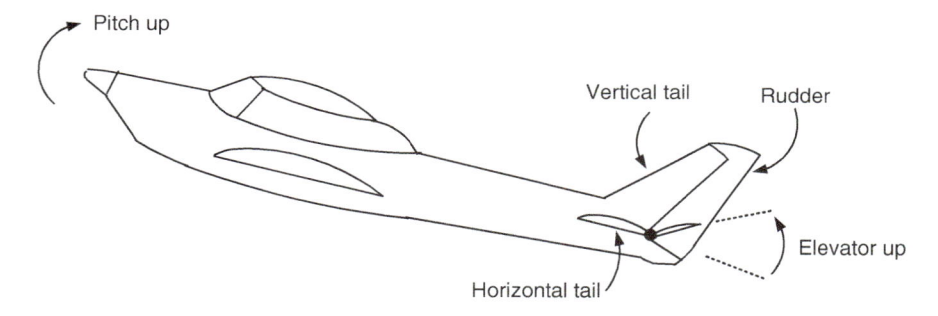

Figure 1.8. Using the elevators to pitch the aircraft up.

Control of Roll Using the Ailerons

Figure 1.9 shows the aircraft with its left aileron down and its right aileron is up. The aerodynamic forces on the two ailerons then cause the airplane to roll to the pilot's right. By adjusting the angle of the ailerons, the pilot can roll right or left.

Control of Yaw Using the Rudder

The final control surface is the rudder on the vertical tail. As indicated in Figure 1.10, if the pilot rotates the rudder to the right, then the plane will yaw (rotate) to the right and vice versa.

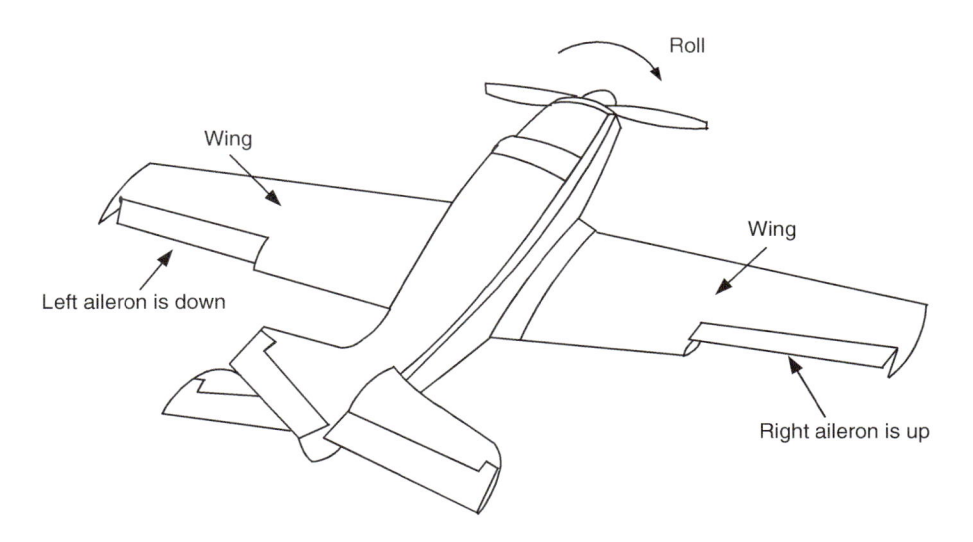

Figure 1.9. Using the ailerons to roll the aircraft.

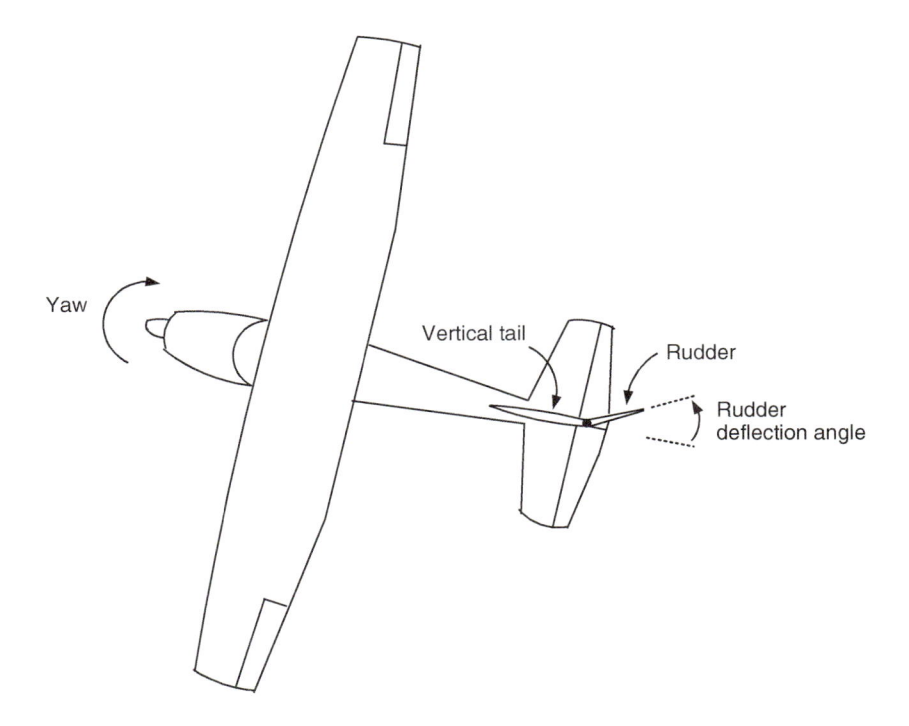

Figure 1.10. Using the rudder to change the heading of the aircraft.

Controlling the Airplane

The pilot has the four input controls:

(1) The elevator to control pitch.

(2) The ailerons to control roll.

(3) The rudder to control yaw.

(4) An engine connected to a propeller to control forward thrust.

Figure 1.11 depicts how a pilot in a small airplane controls the elevator, ailerons, and rudder. In Figure 1.11 the pilot's hands are shown on the *yoke* or *center stick* (looks like a bit like a car's steering wheel). If the pilot pulls the yoke toward himself, the plane will pitch up while alternatively pushing the yoke away from himself the plane will pitch down. See the YouTube animation at [9].

Figure 1.11. Pilot using a yoke, pedals, and throttle to control the aircraft. Source: Courtesy of Steve Karp [9]. "How It Works Flight Controls", October 12, 2013, https://www.youtube.com/watch?v=AiTk5r-4coc.

If the pilot turns the yoke to the right, the plane will roll to the right and conversely if the pilot turns the yoke to the left the plane will rotate to the left.

Figure 1.11 shows the pilot with each foot on a pedal. Pushing on the right pedal the plane will yaw to the right while pushing on the left pedal the plane will yaw to the left.

Not shown in Figure 1.11 is a throttle[2] that the pilot can use to control the engine speed and therefore the propeller thrust.

The angular position of the airplane as specified by its pitch, roll, and yaw is referred to as the *attitude* of the aircraft. The pilot uses the thrust input (propeller) to maintain air flow over the wings. On take off, the propeller must accelerate the airplane to a speed such that the flow of the air across the wings is fast enough to produce the necessary lift. The pilot will then adjust the pitch angle (using the elevators) to obtain an angle of attack that provides the necessary lift to get to the desired altitude. During the climb the pilot may bank (roll) and yaw using the ailerons and rudder, respectively to head off in

---
[2]A throttle controls the amount of fuel to the engine.

the direction of the destination. At a cruising speed at some fixed altitude the propeller is primarily producing enough thrust to cancel out the drag force so the plane can maintain air speed (and therefore the lift force). The first aircraft required a lot of physical effort on the part of the pilot to control them using mechanical linkages to move the control surfaces. However, using these inputs the pilot can make the airplane take off, cruise, and land safely even though there are cross winds, air gusts, etc. disrupting the plane's motion.

Automatic Control (Autopilot)

The basic idea of *automatic control* is to replace the human pilot with an autopilot. The autopilot is simply a computer with inputs to it being the air speed, the aircraft acceleration, and the aircraft attitude (roll, pitch, and yaw angles). The autopilot obtains the airspeed measurement from a *Pitot tube* mounted on the outside of the plane, the acceleration from a three-axis accelerometer set mounted on the airplane and the angular rates from a three-axis gyroscope set also mounted on the plane. Using these measurements, the autopilot continuously (typically every millisecond) determines what angles the rudder, elevators and ailerons should be at in order to keep the plane at a proper attitude as well as the throttle value so the engine thrust is able to maintain the required air speed. The autopilot sends out the required control surface positions to the control surfaces where motors move these surfaces to these commanded values. The autopilot also sends the required throttle value to the engine where another motor maintains the throttle at this value. Such an autopilot can maintain level flight despite wind gusts acting on the plane.

1.2 Quadrotors

Figure 1.12 is a picture of a quadrotor, which basically consists of four propellers all in the same plane and 90 degrees apart connected to a main body. By adjusting the speed of each propeller one can control the thrust (aerodynamic force) produced by each propeller. The propeller speeds are managed by an on board microprocessor that controls the electric motor of each propeller.

Figure 1.12. Photo of a PARROT quadrotor in hover. Source: Courtesy of Professor Aykut Satici.

8   1   Introduction

First let's look at a single propeller connected to an electric motor as indicated in Figure 1.13. For the purposes of this discussion think of the propeller/motor system as being on a frictionless surface. Thus the bottom of the motor is free to slide or rotate without being stopped by any friction from the table.

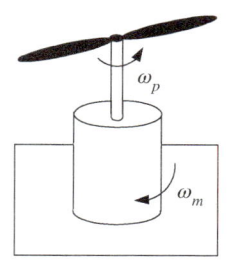

Figure 1.13. Conservation of angular momentum requires $J_p\omega_p = J_m\omega_m$.

Let the propeller have a moment of inertia J_p about the spin axis and the motor have a moment of inertial J_m about same axis. As indicated by the curved arrows in Figure 1.13, if the propeller rotates counterclockwise then $\omega_p > 0$, while if the motor rotates clockwise then $\omega_m > 0$. The angular velocity of the propeller is denoted by ω_p and it has angular momentum $L_p = J_p\omega_p$. The angular velocity of the motor is denoted by ω_m and it has angular momentum $L_m = -J_m\omega_m$. With no external forces acting on the motor–propeller system of Figure 1.13, the principle of *conservation on angular momentum* tells us that its total angular momentum is constant, i.e. $L_p + L_m =$ constant. At startup the propeller is off and the motor is stationary so this constant is zero. After turning on the propeller we must continue to have $L_p + L_m = 0$ or

$$J_p\omega_p - J_m\omega_m = 0. \tag{1.1}$$

Consequently, if the propeller is spinning counterclockwise the motor must spin clockwise. Typically $J_p \ll J_m$ so $\omega_p \gg \omega_m$.

Figure 1.14 is a schematic representation of a quadrotor in flight. The unit orthogonal vectors $\hat{x}_e, \hat{y}_e, \hat{z}_e$ represent earth (inertial) coordinates fixed to the earth.[3] $\hat{z}_e$ points straight up from the earth while $\hat{x}_e, \hat{y}_e$ are tangent to the earth and orthogonal to each other. We have $\hat{z}_e = \hat{x}_e \times \hat{y}_e$ so $\hat{x}_e, \hat{y}_e, \hat{z}_e$ form a right-handed coordinate system. Let $\hat{x}_b, \hat{y}_b, \hat{z}_b$ be unit orthogonal vectors attached to the quadrotor (rigid body) with their origin at the center of mass of the quadrotor. The unit vector $\hat{x}_b$ points to the front of the quadrotor, the unit vector $\hat{y}_b$ points to the left side of the quadrotor, and the unit vector $\hat{z}_b$ points up (from the point of view of the quadrotor). Note that $\hat{z}_b = \hat{x}_b \times \hat{y}_b$ so that they also form a right-handed coordinate system. $\vec{r}_{cm}$ is a vector from the origin of the $\hat{x}_e, \hat{y}_e, \hat{z}_e$ coordinate system to the center of mass of the quadrotor, which is the origin of the $\hat{x}_b, \hat{y}_b, \hat{z}_b$ coordinate system. Both sets of vectors span Euclidean 3-*space*, which we denote as $\mathbf{E}^3$.

Figure 1.15 is a schematic of a quadrotor in hover, i.e., $\hat{z}_b = \hat{z}_e$ with all the propellers rotating at the same angular speed ω_p and thus each producing the same upward thrust (force) f where $f = mg$ with m the mass of the quadrator. Note in Figure 1.15 that the front and rear propellers rotate counterclockwise, while the left and right propellers rotate

---

[3] The "hat" over the vectors is to denote them as *unit* vectors.

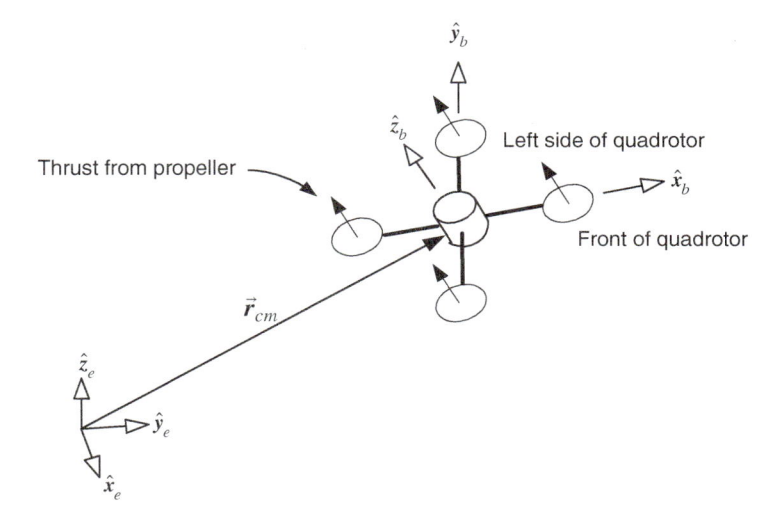

Figure 1.14. Quadrotor with the inertial and body coordinate systems shown.

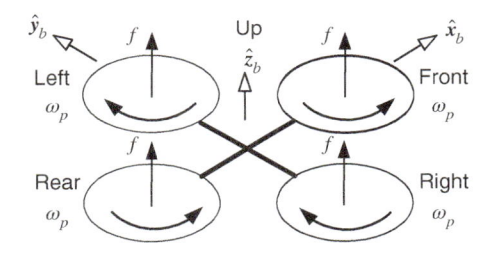

Figure 1.15. A hovering quadrotor. The front and back propellers spin counter clockwise, while the left and right propellers spin clockwise.

clockwise. As a result the total angular momentum L_{z_b} of the four propellers about the $\hat{z}_b$ axis is zero as

$$L_{z_b} = L_{z_b\_left} + L_{z_b\_right} + L_{z_b\_front} + L_{z_b\_rear}$$
$$= -J_{z_b}\omega_p - J_{z_b}\omega_p + J_{z_b}\omega_p + J_{z_b}\omega_p$$
$$= 0 \qquad (1.2)$$

where J_{z_b} is the moment of inertia of a *propeller*. As $L_{z_b} = 0$ in hover the body of the quadrotor must also have zero angular momentum and thus it does not rotate about the $\hat{z}_b$ axis.

Due to the construction of the quadrotor its thrust is always in the direction of the $\hat{z}_b$ axis. As a consequence, the quadrotor body must be rotated to have a component of its thrust point in the direction of travel. This is done by roll, pitch, and yaw motions.

Figure 1.16 on the next page shows how the quadrotor performs a *roll* motion, that is, rotates about its $\hat{x}_b$ axis. In Figure 1.16 p denotes the angular velocity of the quadrotor body about the $\hat{x}_b$ axis. For the quadrotor to roll to its right the propeller speed on the right side is decreased by $\Delta\omega_p$ with a consequential decrease of Δf in its thrust, while the left propeller has its rotor speed increased by $\Delta\omega_p$ with an increase in its thrust of Δf.

10   1   Introduction

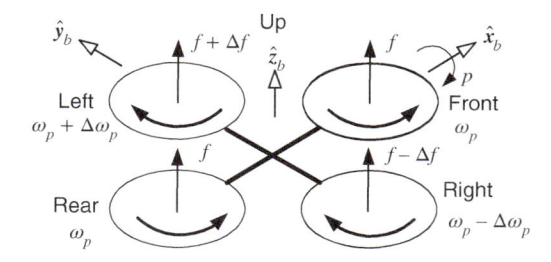

Figure 1.16. The quadrotor executing a roll motion.

With this approach for rolling the quadrotor the total angular momentum L_{z_b} of the four propellers about the $\hat{z}_b$ axis is zero as

$$L_{z_b} = L_{z_b\_left} + L_{z_b\_right} + L_{z_b\_front} + L_{z_b\_rear}$$
$$= -J_{z_b}(\omega_p + \Delta\omega_p) - J_{z_b}(\omega_p - \Delta\omega_p) + J_{z_b}\omega_p + J_{z_b}\omega_p$$
$$= 0. \quad (1.3)$$

By conservation of angular momentum the body of the quadrotor does not rotate about the $\hat{z}_b$ axis during this roll motion.

Figure 1.17 shows how the quadrotor performs a *pitch* motion, which is a rotation about its $\hat{y}_b$ axis. In Figure 1.17 q denotes the angular velocity of the quadrotor body about the $\hat{y}_b$ axis. For the quadrotor to pitch down the propeller speed of the rear propeller is increased by $\Delta\omega_p$ with a consequential increase in its thrust of Δf, while the front propeller has its rotor speed decreased by $\Delta\omega_p$ with an decrease in its thrust of Δf. With this way of pitching the quadrotor the total angular momentum of the four propellors about the $\hat{z}_b$ axis is zero as

$$L_{z_b} = L_{z_b\_left} + L_{z_b\_right} + L_{z_b\_front} + L_{z_b\_rear}$$
$$= +J_{z_b}\omega_p + J_{z_b}\omega_p - J_{z_b}(\omega_p - \Delta\omega_p) - J_{z_b}(\omega_p + \Delta\omega_p)$$
$$= 0. \quad (1.4)$$

As in the case of a roll motion, during a pitch maneuver the body of the quadrotor does not rotate about the $\hat{z}_b$ axis.

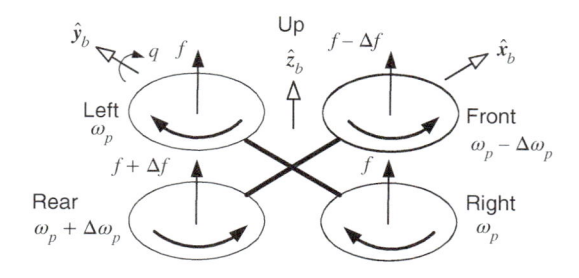

Figure 1.17. The quadrotor executing a pitch motion.

Finally Figure 1.18 shows how to get the quadrotor body to rotate about its $\hat{z}_b$ axis, which is referred to as a *yaw* motion. In Figure 1.18 r denotes the angular velocity of the quadrotor body about the $\hat{z}_b$ axis. For the quadrotor to yaw to the left (counterclockwise

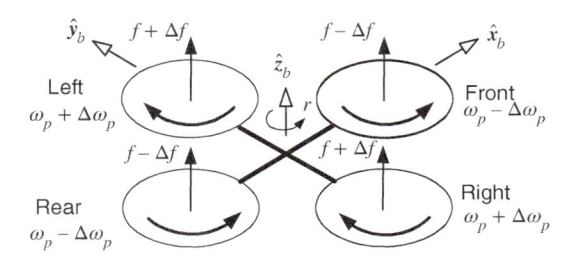

Figure 1.18. The quadrotor executing a yaw motion.

from above), the front and rear propeller speeds are decreased by $\Delta\omega_p$ with a consequential decrease in each of their thrusts by Δf, while the left and right propellers have their speeds increased by $\Delta\omega_p$ with an increase in their thrusts of Δf each. In contrast to the pitch and roll motions, the total angular momentum about the $\hat{z}_b$ axis of the four propellers is not zero. Specifically, their total angular momentum is

$$\begin{aligned} L_{z_b} &= L_{z_b\_left} + L_{z_b\_right} + L_{z_b\_front} + L_{z_b\_rear} \\ &= -J_{z_b}(\omega_p + \Delta\omega_p) - J_{z_b}(\omega_p + \Delta\omega_p) + J_{z_b}(\omega_p - \Delta\omega_p) + J_{z_b}(\omega_p - \Delta\omega_p) \\ &= -4J_{z_b}(\Delta\omega_p). \end{aligned} \quad (1.5)$$

L_{z_b} negative tells us that the net angular momentum of the propellers is in the clockwise direction. As angular momentum is conserved, the body of the quadrotor will rotate counterclockwise (left). With J_b the moment of inertia of the quadrotor *body* about the $\hat{z}_b$ axis its angular momentum must be $L_b = J_b r = 4J_{z_b}(\Delta\omega_p)$ so that $L_b + L_{z_b} = 4J_{z_b}(\Delta\omega_p) - 4J_{z_b}(\Delta\omega_p) = 0$. The angular velocity of the quadrotor body is then

$$r = (4J_{z_b}/J_b)\Delta\omega_p. \quad (1.6)$$

Automatic Control

We can consider the inputs to the quadrotor to be the four propeller speeds.[4] Typically a quadrotor has an inertial measurement unit (IMU), which consists of a three axis accelerometer, a three axis rate gyro, and a three axis magnetometer. The three accelerometers provide the inertial acceleration (i.e., the acceleration with respect to the earth) of the quadrotor in the $\hat{x}_b, \hat{y}_b, \hat{z}_b$ directions. The three axis rate gyro gives angular velocity vector (p, q, r) of the quadrotor in the $\hat{x}_b, \hat{y}_b, \hat{z}_b$ coordinate system. The three axis magnetometer gives the ambient magnetic field[5] in the $\hat{x}_b, \hat{y}_b, \hat{z}_b$ directions. Using an IMU there are methods to estimate the orientation of the quadrotor relative to the earth coordinate system (orientation of $\hat{x}_b, \hat{y}_b, \hat{z}_b$ with respect to $\hat{x}_e, \hat{y}_e, \hat{z}_e$) as well as the velocity vector of the center of mass of quadrotor with respect to earth. Based on the quadrotor's orientation, velocity, and position, a feedback controller (an algorithm on a microprocessor) figures out how fast each propeller should spin to make the quadrotor follow a reference trajectory.

---

[4] There is an "inner" controller that controls the propeller speeds by controlling the voltage of the motor connected to the propeller.
[5] Same as the earth's magnetic field if there are not power lines close by.

1.3 Inverted Pendulum

A classic control problem in academia is the inverted pendulum on a cart. It consists of a rod of length ℓ which is free to rotate about a pivot as shown in Figure 1.19. The idea is to apply the appropriate force u to the cart to keep it under the pendulum rod, i.e., to the keep the rod vertical. Specifically, one can get a measurement of θ and x (say every millisecond) and wants to determine the value of u (also every millisecond) that keeps the pendulum at a fixed position in x and keeps $\theta = 0$. In Chapter 13 we will derive the nonlinear differential equations of motion of this pendulum. For now we just state them as follows:

$$(J + m\ell^2)\frac{d^2\theta}{dt^2} = mg\ell \sin(\theta) - m\ell \frac{d^2x}{dt^2} \cos(\theta) \quad (1.7)$$

$$(M + m)\frac{d^2x}{dt^2} + m\ell\left(\frac{d^2\theta}{dt^2}\cos(\theta) - \left(\frac{d\theta}{dt}\right)^2 \sin(\theta)\right) = u(t). \quad (1.8)$$

The point here is that for any input $u(t)$ the solution to these differential equations gives position of the cart $x(t)$ and the angle $\theta(t)$ of the pendulum rod.

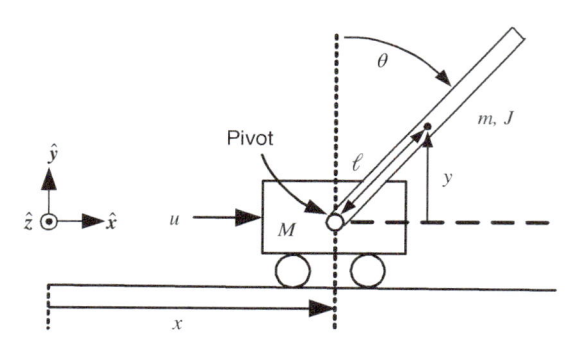

Figure 1.19. Inverted pendulum.

These equations may look complicated, but that is only because they are! To simplify them consider the situation for which $|\theta|$ and $|\dot\theta| = |d\theta/dt|$ remain small. Then $\sin(\theta) \approx \theta$, $\cos(\theta) \approx 1$, and $\left(\frac{d\theta}{dt}\right)^2 \sin(\theta) \approx (\dot\theta)^2 \theta \approx 0$ so that Eqs. (1.7) and (1.8) can be approximated by the *linear* differential equations

$$(J + m\ell^2)\frac{d^2\theta}{dt^2} = mg\ell\theta - m\ell\frac{d^2x}{dt^2} \quad (1.9)$$

$$(M + m)\frac{d^2x}{dt^2} + m\ell\frac{d^2\theta}{dt^2} = u(t). \quad (1.10)$$

In your first differential equations course you primarily studied *linear* differential equations. The reason for this is that we know how to solve them! As we will show in later chapters, it is also known how to *control* a physical system with a linear differential equation model. By the control of the inverted pendulum we mean that given the measurements $x(t), \theta(t)$ the input value $u(t)$ can be chosen to keep the pendulum rod vertical. If this controller based on the linear model keeps $|\theta|$ and $|\dot\theta|$ small, then the linear model is a valid approximation to the actual pendulum and the controller will also work on the actual pendulum.

We can use Laplace transforms to convert linear differential equations into *algebraic* equations. In Chapters 2 and 3 we will do a comprehensive review of Laplace transforms. For now we just note that the Laplace transfer function model of the inverted pendulum is

$$X(s) = \frac{1}{M m \ell^2 + J(M+m)} \frac{(J+m\ell^2)s^2 - mg\ell}{s^2(s^2 - \alpha^2)} U(s) \quad (1.11)$$

$$\theta(s) = -\frac{1}{M m \ell^2 + J(M+m)} \frac{m\ell}{s^2 - \alpha^2} U(s) \quad (1.12)$$

$$\alpha^2 = \frac{mg\ell(M+m)}{M m \ell^2 + J(M+m)}. \quad (1.13)$$

Control Problem

The control problem is to use the measured values of $x(t)$ and $\theta(t)$ to determine the force $u(t)$ applied to the cart so that the pendulum rod does not fall. The control algorithm is simply a function that gives the value of $u(t)$ given the values of $x(t)$ and $\theta(t)$. A controller can be designed using the above transfer function model or the above differential equation model. In the first half of the book we base the control design on the transfer function model, while in later chapters of the book a differential equation model is used.

1.4 Magnetic Levitation

In this example we consider a magnetic levitation system. The schematic given in Figure 1.20 shows a wire wrapped around a cylindrical core of iron making an electromagnet.

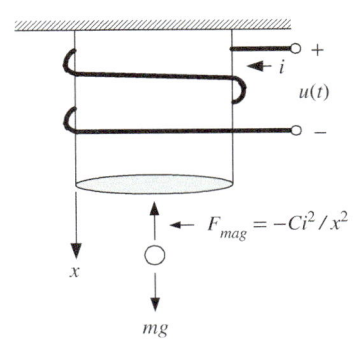

Figure 1.20. A simple magnetic levitation system.

Applying a voltage $u(t)$ to the coil results in a current $i(t)$. This current along with the iron core produces a magnetic field that extends below the electromagnet. This magnetic field then magnetizes the steel ball, i.e., makes it into a magnet. It turns out that the magnetic force of attraction between the steel ball and the electromagnet is upward and proportional to the square of the current and inversely proportional to the square of the distance. Mathematically, we write that $F_{mag} = -Ci^2/x^2$ where C is a constant. Note in the figure that positive x is downward and $x = 0$ corresponds to the bottom of the iron core.

Mathematical Model

There are current-command amplifiers that allow one to have the amplifier put out a desired current. Using such an amplifier we can consider the input to the magnetic levitation system to be the current $i(t)$. The differential equation model of this magnetic levitation system is then (see Chapter 13)

$$\frac{dx}{dt} = v \tag{1.14}$$

$$m\frac{dv}{dt} = -C\frac{i^2}{x^2} + mg. \tag{1.15}$$

Here C is the force constant, m is the mass of the steel ball, and g is the acceleration of gravity. This differential equation model is important because for any given input current $i(t)$ to the coil, the solution of these differential equations then gives the response of the position $x(t)$ and velocity $v(t) = \dot{x}(t)$ of the steel ball. However this model is nonlinear. As in the case of the inverted pendulum, we can obtain an approximate linear model. Specifically, let the desired (constant) position of the steel ball be x_0 and choose the current i_0 to make the ball's acceleration zero, i.e.,

$$0 = -C\frac{i_0^2}{x_0^2} + mg \quad \text{or} \quad i_0 = x_0\sqrt{\frac{mg}{C}}. \tag{1.16}$$

With $\Delta x \triangleq x - x_0, \Delta \dot{x} \triangleq \dot{x} - \dot{x}_0 = \dot{x}, u \triangleq i - i_0$ we will show in Chapter 13 that

$$m\Delta\ddot{x} = \frac{2g}{x_0}\Delta x - \frac{2g}{i_0}u \tag{1.17}$$

is a valid *linear* model for $\Delta x, \Delta \dot{x}$, and u small. After studying Chapter 3 you will be able to show that the corresponding Laplace transform model is

$$\Delta X(s) = -\frac{2g/i_0}{s^2 - 2g/x_0}U(s). \tag{1.18}$$

Figure 1.21 is a magnetic levitation system built by W. Barie [10]. It shows light being shone across the top of the ball to a photodetector sensor. Using this setup the position of the ball beneath the electromagnet can be found.

Control Problem

Based on measurements of $x(t)$ and $v(t)$, one wants to choose the input voltage $u(t)$ so the ball maintains a fixed position x_0 below the electromagnet, i.e., it does not fall nor gets sucked into the magnet. A procedure to choose $u(t)$ based on the values of $x(t)$ and $v(t)$ is called the control algorithm or controller. The control algorithm is found using either the differential equation model or the Laplace transform model.

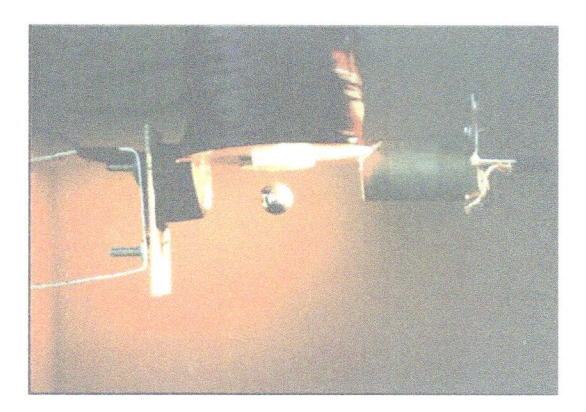

Figure 1.21. A laboratory magnetic levitation system. Source: Barie and Chiasson [10]. "Linear and nonlinear state-space controllers for magnetic levitation", International Journal of Systems Science, vol. 27, no. 11, pp. 1153–1163, November 1996. DOI-https://doi.org/10.1080/00207729608929322.

1.5 General Control Problem

In general there is some physical system that needs to be controlled. The steps involved are:

- Determine a mathematical model of this system (typically a set of differential equations). The model describes how the *output variables* behave for any given *input*.
- Design the control algorithm. The feedback control algorithm processes measurements of the output through a differential equation to produce the inputs to the physical system that force the output to a desired value.
- Simulate the combined controller and mathematical model to check if the system outputs are controlled as desired. If not, modify or (if necessary) redesign the controller. If the controller doesn't work in simulation, then it won't work in practice!
- The final step is to implement the controller on the actual physical system.

2
Laplace Transforms

In Chapter 1 we gave examples of physical systems modeled by differential equations. For each such physical system we will use the differential equation model to design algorithms to control the system. Furthermore, it turns out that these algorithms will also be described by differential equations. To ease the difficulty of working with differential equations, classical control uses the Laplace transform. This transform converts a differential equation to an algebraic equation greatly simplifying mathematical calculations. In Chapters 9 and 10 we will present specific approaches to the design of feedback controllers for physical systems based on their Laplace transform model. All of this requires a thorough background knowledge of Laplace transforms which we cover in this chapter.

Definition 1 *Laplace Transform*
Let $f(t)$ for $t \geq 0$ denote a time function. Define the Laplace transform $\mathcal{L}\{f\}$ of f as

$$\mathcal{L}\{f(t)\} \triangleq \int_0^\infty e^{-st} f(t) dt, \qquad (2.1)$$

where $s = \sigma + j\omega$ is a complex number.

Notation The symbol "$\triangleq$" means *by definition*. So $\mathcal{L}\{f(t)\}$ is *by definition* the quantity $\int_0^\infty e^{-st} f(t) dt$.

We also use $F(s)$ to denote $\mathcal{L}\{f(t)\}$, i.e., $F(s) \triangleq \int_0^\infty e^{-st} f(t) dt$.

Remark *Region of Convergence*
The integral in (2.1) does not exist for *all* $s = \sigma + j\omega \in \mathbb{C}$. The *region of convergence* is the set of values of s for which the integral exists. This is illustrated in the following examples.

Example 1 *Unit Step Function $u_s(t)$*
The unit step function $u_s(t)$ is defined as

$$u_s(t) \triangleq \begin{cases} 1, & t \geq 0, \\ 0, & t < 0. \end{cases}$$

We compute its Laplace transform according to the definition as

$$\mathcal{L}\{u_s(t)\} = \int_0^\infty e^{-st} u_s(t) dt = \int_0^\infty e^{-st} dt = \left. \frac{e^{-st}}{-s} \right|_{t=0}^\infty$$

$$= \lim_{t\to\infty} \frac{e^{-st}}{-s} - \left. \frac{e^{-st}}{-s} \right|_{t=0}$$

$$= \lim_{t\to\infty} \frac{e^{-st}}{-s} + \frac{1}{s}.$$

An Introduction to System Modeling and Control, First Edition. John Chiasson.
© 2022 John Wiley & Sons, Inc. Published 2022 by John Wiley & Sons, Inc.
Companion website: www.wiley.com/go/chiasson/anintroductiontosystemmodelingandcontrol

2 Laplace Transforms

Now
$$\lim_{t\to\infty} e^{-st} = \lim_{t\to\infty} e^{-(\sigma+j\omega)t} = \lim_{t\to\infty} e^{-\sigma t} e^{-j\omega t}.$$

Recalling Euler's formula
$$e^{j\omega t} = \cos(\omega t) + j\sin(\omega t)$$

this becomes
$$\lim_{t\to\infty} e^{-st} = \lim_{t\to\infty} e^{-\sigma t} e^{-j\omega t} = \lim_{t\to\infty} e^{-\sigma t}\big(\cos(\omega t) - j\sin(\omega t)\big)$$
$$= \begin{cases} 0, & \sigma > 0, \\ \text{does not exist}, & \sigma \leq 0. \end{cases}$$

Thus
$$\mathcal{L}\{u_s(t)\} = \frac{1}{s} \text{ for } \sigma = \text{Re}\{s\} > 0. \tag{2.2}$$

Example 2 $f(t) = e^{2t}$

Consider the function $f(t) = e^{2t}$. Using the definition of a Laplace transform, we have

$$\mathcal{L}\{e^{2t}\} = \int_0^\infty e^{-st} e^{2t} dt = \int_0^\infty e^{-(s-2)t} dt$$
$$= \left.\frac{e^{-(s-2)t}}{-(s-2)}\right|_{t=0}^\infty$$
$$= \lim_{t\to\infty} \frac{e^{-(s-2)t}}{-(s-2)} - \left.\frac{e^{-(s-2)t}}{-(s-2)}\right|_{t=0}$$
$$= \lim_{t\to\infty} \frac{e^{-(s-2)t}}{-(s-2)} + \frac{1}{(s-2)}.$$

As
$$\lim_{t\to\infty} e^{-(s-2)t} = \lim_{t\to\infty} e^{-(\sigma-2+j\omega)t}$$
$$= \lim_{t\to\infty} e^{-(\sigma-2)t} e^{-j\omega t}$$
$$= \lim_{t\to\infty} e^{-(\sigma-2)t}\big(\cos(-\omega t) + j\sin(-\omega t)\big)$$
$$= \begin{cases} 0, & \sigma > 2, \\ \text{does not exist}, & \sigma \leq 2, \end{cases}$$

we have
$$\mathcal{L}\{e^{2t}\} = \frac{1}{s-2} \text{ for } \sigma = \text{Re}\{s\} > 2. \tag{2.3}$$

Example 3 $f(t) = e^{(\sigma_0+j\omega_0)t} = e^{\sigma_0 t}\big(\cos(\omega_0 t) + j\sin(\omega_0 t)\big)$

In this example we consider the Laplace transform of the complex-valued function
$$f(t) = e^{(\sigma_0+j\omega_0)t} = e^{\sigma_0 t}\big(\cos(\omega_0 t) + j\sin(\omega_0 t)\big).$$

We compute
$$\mathcal{L}\{e^{(\sigma_0+j\omega_0)t}\} = \int_0^\infty e^{-st}e^{(\sigma_0+j\omega_0)t}dt$$
$$= \int_0^\infty e^{-[s-(\sigma_0+j\omega_0)]t}dt$$
$$= \left.\frac{e^{-[s-(\sigma_0+j\omega_0)]t}}{-[s-(\sigma_0+j\omega_0)]}\right|_0^\infty$$
$$= \lim_{t\to\infty}\frac{e^{-[s-(\sigma_0+j\omega_0)]t}}{-[s-(\sigma_0+j\omega_0)]} - \left.\frac{e^{-[s-(\sigma_0+j\omega_0)]t}}{-[s-(\sigma_0+j\omega_0)]}\right|_0$$
$$= \lim_{t\to\infty}\frac{e^{-(s-\sigma_0)t+j\omega_0 t}}{-[s-(\sigma_0+j\omega_0)]} + \frac{1}{s-(\sigma_0+j\omega_0)}.$$

Now, as
$$\lim_{t\to\infty} e^{-(s-\sigma_0)t+j\omega_0 t} = \lim_{t\to\infty} e^{-(\sigma-\sigma_0)t}e^{j(\omega_0-\omega)t}$$
$$= \lim_{t\to\infty} e^{-(\sigma-\sigma_0)t}\bigl(\cos((\omega_0-\omega)t) + j\sin((\omega_0-\omega)t)\bigr)$$
$$= \begin{cases} 0, & \sigma > \sigma_0, \\ \text{does not exist}, & \sigma \leq \sigma_0, \end{cases}$$

it follows for $\sigma = \text{Re}\{s\} > \sigma_0 = \text{Re}\{\sigma_0+j\omega_0\}$ that
$$\mathcal{L}\{e^{(\sigma_0+j\omega_0)t}\} = \frac{1}{s-(\sigma_0+j\omega_0)}$$
$$= \frac{1}{s-\sigma_0-j\omega_0}\frac{s-\sigma_0+j\omega_0}{s-\sigma_0+j\omega_0}$$
$$= \frac{s-\sigma_0+j\omega_0}{(s-\sigma_0)^2+\omega_0^2}.$$

Example 4 $f(t) = e^{(\sigma_0+j\omega_0)t}$ *(continued)*
For $\text{Re}\{s\} > \sigma_0$, we just showed that
$$\mathcal{L}\{e^{(\sigma_0+j\omega_0)t}\} = \mathcal{L}\{e^{\sigma_0 t}\cos(\omega_0 t) + je^{\sigma_0 t}\sin(\omega_0 t)\}$$
$$= \frac{s-\sigma_0}{(s-\sigma_0)^2+\omega_0^2} + j\frac{\omega_0}{(s-\sigma_0)^2+\omega_0^2}.$$

This implies that
$$\mathcal{L}\{e^{\sigma_0 t}\cos(\omega_0 t)\} = \frac{s-\sigma_0}{(s-\sigma_0)^2+\omega_0^2} \tag{2.4}$$
$$\mathcal{L}\{e^{\sigma_0 t}\sin(\omega_0 t)\} = \frac{\omega_0}{(s-\sigma_0)^2+\omega_0^2}. \tag{2.5}$$

In particular, for $\sigma_0 = 0$ we have
$$\mathcal{L}\{\cos(\omega_0 t)\} = \frac{s}{s^2+\omega_0^2} \tag{2.6}$$
$$\mathcal{L}\{\sin(\omega_0 t)\} = \frac{\omega_0}{s^2+\omega_0^2}. \tag{2.7}$$

2.1 Laplace Transform Properties

We next recall some properties of the Laplace transform that will simplify many calculations. The first property is differentiation with respect to the Laplace variable s.

Property 1 $\mathcal{L}\{tf(t)\} = -\dfrac{d}{ds}F(s)$

Let
$$\mathcal{L}\{f(t)\} = F(s) \text{ for } \mathrm{Re}\{s\} > \sigma$$

then
$$\mathcal{L}\{tf(t)\} = -\frac{d}{ds}F(s) \text{ for } \mathrm{Re}\{s\} > \sigma. \tag{2.8}$$

Proof.
$$F(s) = \int_0^\infty e^{-st} f(t)\, dt$$

then
$$-\frac{d}{ds}F(s) = -\frac{d}{ds}\int_0^\infty e^{-st} f(t)\, dt = -\int_0^\infty (-t)e^{-st} f(t)\, dt$$
$$= \int_0^\infty t e^{-st} f(t)\, dt.$$
∎

As an example of this property, we show how to obtain the Laplace transform of $t^n/n!$ for arbitrary n.

Example 5 $\mathcal{L}\left\{\dfrac{t^n}{n!}\right\} = \dfrac{1}{s^{n+1}}$

We have already shown in a previous example that
$$\mathcal{L}\{u_s(t)\} = \int_0^\infty e^{-st}\, dt = \frac{1}{s} \text{ for } \mathrm{Re}\{s\} > 0.$$

Then
$$\frac{d}{ds}\int_0^\infty e^{-st}\, dt = \frac{d}{ds}\frac{1}{s}$$

or
$$\int_0^\infty t e^{-st}\, dt = \frac{1}{s^2}.$$

Thus
$$\mathcal{L}\{t\} = \frac{1}{s^2} \text{ for } \mathrm{Re}\{s\} > 0. \tag{2.9}$$

Continuing with
$$\int_0^\infty t e^{-st}\, dt = \frac{1}{s^2},$$

we differentiate both sides with respect to s to obtain
$$\int_0^\infty (-t) t e^{-st}\, dt = -\frac{2}{s^3}$$

or
$$\int_0^\infty \frac{t^2}{2} e^{-st} dt = \frac{1}{s^3} \text{ for Re}\{s\} > 0. \tag{2.10}$$

In a similar fashion, we have for any $n = 0, 1, 2, \ldots$ that
$$\mathcal{L}\left\{\frac{t^n}{n!}\right\} = \frac{1}{s^{n+1}}. \tag{2.11}$$

Example 6 $\mathcal{L}\{t\cos(\omega t)\} = \dfrac{s^2 - \omega^2}{(s^2 + \omega^2)^2}$

We showed in a previous example that
$$\mathcal{L}\{\cos(\omega t)\} = \frac{s}{s^2 + \omega^2} \text{ for Re}\{s\} > 0.$$

Then by the property (2.8) we have
$$\mathcal{L}\{t\cos(\omega t)\} = -\frac{d}{ds}\frac{s}{s^2 + \omega^2} = -\frac{1}{s^2 + \omega^2} + \frac{s(2s)}{(s^2 + \omega^2)^2}$$
$$= -\frac{s^2 + \omega^2}{(s^2 + \omega^2)^2} + \frac{s(2s)}{(s^2 + \omega^2)^2}$$
$$= \frac{s^2 - \omega^2}{(s^2 + \omega^2)^2}. \tag{2.12}$$

Property 2 $\mathcal{L}\{e^{\alpha t} f(t)\} = F(s - \alpha)$

Let
$$\mathcal{L}\{f(t)\} = F(s) \text{ for Re}\{s\} > \sigma$$

then
$$\mathcal{L}\{e^{\alpha t} f(t)\} = F(s - \alpha) \text{ for Re}\{s\} > \sigma + \alpha. \tag{2.13}$$

Proof.
$$\mathcal{L}\{e^{\alpha t} f(t)\} = \int_0^\infty e^{-st} e^{\alpha t} f(t) dt = \int_0^\infty e^{-(s-\alpha)t} f(t) dt$$
$$= F(s - \alpha).$$
∎

Example 7 $f(t) = \cos(\omega t)$

Knowing
$$\mathcal{L}\{\cos(\omega t)\} = \frac{s}{s^2 + \omega^2} \text{ for Re}\{s\} > 0$$

it follows immediately that
$$\mathcal{L}\{e^{\alpha t} \cos(\omega t)\} = \frac{s - \alpha}{(s - \alpha)^2 + \omega^2} \text{ for Re}\{s\} > \alpha. \tag{2.14}$$

2 Laplace Transforms

Property 3 $\mathcal{L}\left\{\dfrac{d}{dt}f(t)\right\} = sF(s) - f(0)$

Let
$$\mathcal{L}\{f(t)\} = F(s) \text{ for } \operatorname{Re}\{s\} > \sigma$$

then
$$\mathcal{L}\left\{\dfrac{d}{dt}f(t)\right\} = sF(s) - f(0) \text{ for } \operatorname{Re}\{s\} > \sigma. \tag{2.15}$$

Proof. By definition of the Laplace transform we have
$$\mathcal{L}\left\{\dfrac{d}{dt}f(t)\right\} \triangleq \int_0^\infty e^{-st}f'(t)dt.$$

We next integrate by parts by letting
$$u = e^{-st}, \ dv = f'(t)dt$$

and
$$du = -se^{-st}dt, \ v = f(t)$$

so that
$$\int_0^\infty \underbrace{e^{-st}f'(t)dt}_{udv} = \underbrace{e^{-st}f(t)|_0^\infty}_{uv} - \int_0^\infty \underbrace{(-s)e^{-st}f(t)dt}_{vdu}$$
$$= \lim_{t\to\infty} e^{-st}f(t) - f(0) + s\int_0^\infty e^{-st}f(t)dt.$$

Now for $\operatorname{Re}\{s\} > \sigma$, the Laplace transform of $f(t)$ exists so that $\lim_{t\to\infty} e^{-st}f(t) = 0$ for $\operatorname{Re}\{s\} > \sigma$.[1] Thus this last equation becomes
$$\int_0^\infty e^{-st}f'(t)dt = sF(s) - f(0) \text{ for } \operatorname{Re}\{s\} > \sigma.$$
∎

Example 8 $f(t) = \cos(\omega t)$

Consider $f(t) = \cos(\omega t)$ and its derivative, that is,
$$f(t) = \cos(\omega t)$$
$$f'(t) = -\omega \sin(\omega t).$$

The Laplace transforms of $f(t), f'(t)$ exist for $\operatorname{Re}\{s\} > 0$. Then using
$$\mathcal{L}\{f'(t)\} = s\mathcal{L}\{f(t)\} - f(0),$$

we have
$$\mathcal{L}\{-\omega\sin(\omega t)\} = s\mathcal{L}\{\cos(\omega t)\} - 1$$
$$= s\frac{s}{s^2+\omega^2} - \frac{s^2+\omega^2}{s^2+\omega^2}$$
$$= -\frac{\omega^2}{s^2+\omega^2}$$

---

[1] To be precise, we will only be concerned with functions that are of the form $t^m e^{\sigma t}\cos(\omega t + \theta)$ where $m \geq 0$ is an integer and σ, ω, θ are constants. For all such functions we have $\lim_{t\to\infty} e^{-st}f(t) = 0$ for $\operatorname{Re}\{s\} > \sigma$.

or, after some rearrangement, we obtain

$$\mathcal{L}\{\sin(\omega t)\} = \frac{\omega}{s^2 + \omega^2} \text{ for Re}\{s\} > 0.$$

Example 9 *Solving a Differential Equation*
Consider the first-order differential equation given by

$$\frac{dx}{dt} + ax = u_s(t),$$

where the input u_s is a *step* input. With

$$X(s) \triangleq \mathcal{L}\{x(t)\}$$

we have

$$\mathcal{L}\{\dot{x}(t)\} = sX(s) - x(0)$$

and

$$\mathcal{L}\{u_s(t)\} = \frac{1}{s}.$$

Taking the Laplace transform of both sides of the differential equation, that is,

$$\mathcal{L}\left\{\frac{dx}{dt} + ax\right\} = \mathcal{L}\{u_s(t)\}$$

we obtain

$$sX(s) - x(0) + aX(s) = \frac{1}{s}.$$

Collecting terms with $X(s)$ to the left side this becomes

$$(s+a)X(s) = x(0) + \frac{1}{s}$$

or finally

$$X(s) = \frac{x(0)}{s+a} + \frac{1}{s(s+a)}.$$

To determine $x(t)$ we compute

$$\begin{aligned}
x(t) &= \mathcal{L}^{-1}\left\{\frac{x(0)}{s+a} + \frac{1}{s(s+a)}\right\} \\
&= \mathcal{L}^{-1}\left\{\frac{x(0)}{s+a} + \frac{1}{a}\frac{1}{s} - \frac{1}{a}\frac{1}{s+a}\right\} \\
&= x(0)e^{-at} + \frac{1}{a}u_s(t) - \frac{1}{a}e^{-at},
\end{aligned}$$

where the second equation followed by doing a partial-fraction expansion, which is our next topic.

2.2 Partial Fraction Expansion

In solving differential equations using the Laplace transform, the final step is often finding the inverse Laplace transform of a rational function[2] in the Laplace variable s. In the previous example, we used the fact that

$$\frac{1}{s(s+a)} = \frac{1}{a}\frac{1}{s} - \frac{1}{a}\frac{1}{s+a}.$$

That is, the rational function $\frac{1}{s(s+a)}$ was decomposed into two simpler functions whose inverse transforms are known. This decomposition process is known as a *partial fraction expansion*. We illustrate how to do this by a series of examples.

Example 10 $F(s) = \dfrac{1}{(s+2)(s+3)}$

Write

$$F(s) = \frac{1}{(s+2)(s+3)} = \frac{A}{s+2} + \frac{B}{s+3}.$$

Then

$$(s+2)F(s) = \frac{1}{s+3} = A + B\frac{s+2}{s+3}$$

so that

$$\lim_{s \to -2}(s+2)F(s) = \underbrace{\lim_{s \to -2}\frac{1}{s+3}}_{1} = \lim_{s \to -2}\left(A + B\frac{s+2}{s+3}\right) = A.$$

Thus

$$A = 1.$$

Similarly,

$$(s+3)F(s) = \frac{1}{s+2} = A\frac{s+3}{s+2} + B$$

so that

$$\lim_{s \to -3}(s+3)F(s) = \underbrace{\lim_{s \to -3}\frac{1}{s+2}}_{-1} = \lim_{s \to -3}\left(A\frac{s+3}{s+2} + B\right) = B$$

or

$$B = -1.$$

We then have

$$F(s) = \frac{1}{(s+2)(s+3)} = \frac{1}{s+2} - \frac{1}{s+3}.$$

By the table of Laplace transforms we know that

$$\mathcal{L}\{e^{-at}\} = \frac{1}{s+a}$$

---

[2] A rational function is simply the ratio of two polynomials.

and thus
$$f(t) = \mathcal{L}^{-1}\left\{\frac{1}{s+2} - \frac{1}{s+3}\right\} = (e^{-2t} - e^{-3t})u_s(t).$$

Example 11 $F(s) = \dfrac{2s+12}{s^2+2s+5}$

We want to find the inverse Laplace transform of
$$F(s) = \frac{2s+12}{s^2+2s+5}.$$

The roots of
$$s^2 + 2s + 5 = 0$$
are
$$s = \frac{-2 \pm \sqrt{2^2 - 4 \times 5}}{2} = \frac{-2 \pm \sqrt{-16}}{2} = -1 \pm j2.$$

We call the roots of the denominator of $F(s)$ the *poles* of $F(s)$. We may now write
$$\begin{aligned}s^2 + 2s + 5 &= \big[s - (-1 + 2j)\big]\big[s - (-1 - 2j)\big] \\ &= (s + 1 - 2j)(s + 1 + 2j) \\ &= (s+1)^2 + 4.\end{aligned}$$

From the Laplace transform table we know
$$\mathcal{L}\{e^{\sigma t}\sin(\omega t)u_s(t)\} = \frac{\omega}{(s-\sigma)^2 + \omega^2}$$
$$\mathcal{L}\{e^{\sigma t}\cos(\omega t)u_s(t)\} = \frac{s-\sigma}{(s-\sigma)^2 + \omega^2}.$$

Identify $\sigma = -1, \omega = 2$ and rewrite $F(s)$ as follows:
$$\begin{aligned}F(s) &= \frac{2s+12}{s^2+2s+5} = \frac{2s+12}{(s+1)^2+4} \\ &= \frac{2(s+1)+10}{(s+1)^2+4} \\ &= 2\frac{s+1}{(s+1)^2+4} + 5\frac{2}{(s+1)^2+4}.\end{aligned}$$

Using the Laplace transform table we have
$$f(t) = 2e^{-t}\cos(2t)u_s(t) + 5e^{-t}\sin(2t)u_s(t).$$

We can check this result using MATLAB:

```
%Compute the Inverse Laplace Transform
syms F2 s t
F2 = (2*s+12)/(s^2+2*s+5)
ilaplace(F2,s,t)
```

MATLAB should return:
$$\exp(-t)(2\cos(2t) + 5\sin(2t)).$$

We will need to manipulate complex numbers so we now give a brief review of such computations.

Digression on Complex Numbers

Let
$$c = a + jb$$
denote a complex number where a and b are real and $j = \sqrt{-1}$. The *complex conjugate* of c is denoted by c^* and defined as
$$c^* \triangleq a - jb.$$

Note that
$$(c^*)^* = (a - jb)^* = a + jb = c.$$

The magnitude of c is defined as
$$|c| = \sqrt{cc^*} = \sqrt{(a+jb)(a-jb)} = \sqrt{a^2 + b^2}$$
and this equals the magnitude of c^* as
$$|c^*| = \sqrt{c^*(c^*)^*} = \sqrt{(a-jb)(a+jb)} = \sqrt{a^2 + b^2}.$$

We can also represent complex numbers in polar coordinate form. We define
$$\angle c \triangleq \tan^{-1}(b, a)$$
as indicated in Figure 2.1. The expression $\tan^{-1}(b, a)$ is expressed in most computer languages as `atan2(b,a)`. This gives the angle of $c = a + bj$ in the correct quadrant. For example, let $c_1 = -1 + j$ then
$$\angle c_1 = \tan^{-1}(1, -1) = \texttt{atan2(1,-1)} = 3\pi/4 = 2.3562.$$

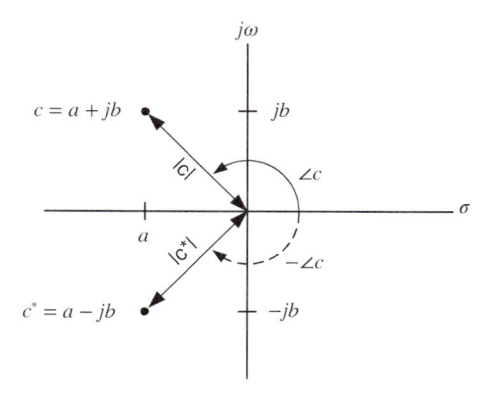

Figure 2.1. $c = a + jb = |c|e^{j\angle c}$ and $c^* = a - jb = |c|e^{-j\angle c}$.

2.2 Partial Fraction Expansion

In contrast, if we consider $c_2 = 1 - j$ then

$$\angle c_2 = \tan^{-1}(-1, 1) = \texttt{atan2(-1,1)} = -\pi/4 = -0.7854.$$

We can now write c in polar coordinate form as

$$c = |c|e^{j\angle c}$$
$$= |c|\cos(\angle c) + j|c|\sin(\angle c).$$

Then we have

$$c^* = \big(|c|\cos(\angle c) + j|c|\sin(\angle c)\big)^* = |c|\cos(\angle c) - j|c|\sin(\angle c)$$
$$= |c|\cos(-\angle c) + j|c|\sin(-\angle c)$$
$$= |c|e^{-j\angle c}.$$

That is,

$$|c^*| = |c|$$
$$\angle c^* = -\angle c.$$

We show that $(c_1 c_2)^* = c_1^* c_2^*$.

$$(c_1 c_2)^* = \big((a_1 + jb_1)(a_2 + jb_2)\big)^* = \big(|c_1|e^{j\angle c_1}|c_2|e^{j\angle c_2}\big)^*$$
$$= \big(|c_1||c_2|e^{j(\angle c_1 + \angle c_2)}\big)^*$$
$$= |c_1||c_2|e^{-j(\angle c_1 + \angle c_2)}$$
$$= |c_1|e^{-j\angle c_1}|c_2|e^{-j\angle c_2}$$
$$= c_1^* c_2^*.$$

Similarly,

$$\left(\frac{c_1}{c_2}\right)^* = \frac{c_1^*}{c_2^*}$$
$$(c_1 + c_2)^* = c_1^* + c_2^*.$$

End of Digression on Complex Numbers

Example 12 $F(s) = \dfrac{2s + 12}{s^2 + 2s + 5}$ *(continued)*

We redo the previous example by doing a partial fraction expansion in terms of the complex conjugate pair of poles of $F(s)$.

$$F(s) = \frac{2s + 12}{[s - (-1 + 2j)][s - (-1 - 2j)]} = \frac{\beta_1}{s - (-1 + 2j)} + \frac{\beta_2}{s - (-1 - 2j)}.$$

Then

$$[s - (-1 + 2j)]F(s) = [s - (-1 + 2j)]\frac{2s + 12}{s^2 + 2s + 5}$$
$$= \frac{2s + 12}{s - (-1 - 2j)}$$

and
$$\left[s-(-1+2j)\right]F(s)=\beta_1+\frac{\beta_2\left[s-(-1+2j)\right]}{s-(-1-2j)}.$$

Therefore
$$\lim_{s\to -1+2j}\left[s-(-1+2j)\right]F(s)=\beta_1$$

and also
$$\lim_{s\to -1+2j}\left[s-(-1+2j)\right]F(s)=\lim_{s\to -1+2j}\frac{2s+12}{s-(-1-2j)}$$
$$=\frac{2(-1+2j)+12}{-1+2j-(-1-2j)}$$
$$=\frac{10+4j}{4j}$$
$$=1-2.5j.$$

Thus $\beta_1=1-2.5j$. Similarly,
$$\beta_2=\lim_{s\to -1-2j}\left[s-(-1-2j)\right]F(s)=\lim_{s\to -1-2j}\frac{2s+12}{s-(-1+2j)}$$
$$=\frac{2(-1-2j)+12}{-1-2j-(-1+2j)}$$
$$=\frac{10-4j}{-4j}$$
$$=1+2.5j.$$

An important point to note here is that
$$\beta_2=\beta_1^*.$$

This will always be the case! Returning back to the partial fraction expansion, we have shown
$$F(s)=\frac{1-2.5j}{s-(-1+2j)}+\frac{1+2.5j}{s-(-1-2j)}.$$

This result can be checked using MATLAB:

```
%Compute the Partial Fraction Expansion
% F2(s) = (2s+12)/(s^2+2s+5)
Fnum = [2 12];
Fden = [1 2 5];
[beta,poles,k] = residue(Fnum,Fden)
```

MATLAB should return:

```
beta =
1.0000 - 2.5000i
1.0000 + 2.5000i
poles =
-1.0000 + 2.0000i
-1.0000 - 2.0000i
k = []
```

2.2 Partial Fraction Expansion

We now put $\beta_1 = 1 - 2.5j$ into polar coordinate form (see Figure 2.2).

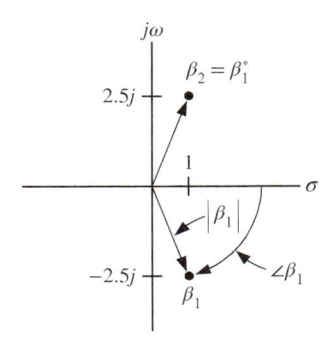

Figure 2.2. Converting $1 - 2j$ to polar coordinate form.

We have

$$\begin{aligned}
\beta_1 = |\beta_1|e^{j\angle\beta_1} &= |1 - 2.5j|e^{j\angle(1-2.5j)} \\
&= \sqrt{1^2 + (2.5)^2}\, e^{j\tan^{-1}(-2.5,1)} \\
&= 2.693 e^{-j1.19} \\
&= 2.693 e^{-j68.2°}.
\end{aligned}$$

We again check our answers using MATLAB:

```
c = 1.0000 - 2.5000i
c_mag = abs(c)
c_angle = angle(c)
c_angle2 = atan2(imag(c),real(c))
c_angle3 = c_angle*180/pi
```

MATLAB should return:

```
c = 1.0000 - 2.5000i
c_mag = 2.6926
c_angle = -1.1903
c_angle2 = -1.1903
c_angle3 = -68.1986
```

Putting $\beta_1, \beta_2 = \beta_1^*$ in polar coordinate form we have

$$\begin{aligned}
f(t) = \mathcal{L}^{-1}\{F(s)\} &= \mathcal{L}^{-1}\left\{\frac{\beta_1}{s-(-1+2j)} + \frac{\beta_1^*}{s-(-1-2j)}\right\} \\
&= \beta_1 e^{(-1+2j)t} + \beta_1^* e^{(-1-2j)t} \\
&= |\beta_1|e^{j\angle\beta_1} e^{-t} e^{2jt} + |\beta_1|e^{-j\angle\beta_1} e^{-t} e^{-2jt} \\
&= |\beta_1|e^{-t}\left(e^{j(2t+\angle\beta_1)} + e^{-j(2t+\angle\beta_1)}\right) \\
&= 2|\beta_1|e^{-t}\cos(2t+\angle\beta_1) \text{ by Euler's formula} \\
&= 2|\beta_1|e^{-t}\cos(2t)\cos(\angle\beta_1) - 2|\beta_1|e^{-t}\sin(2t)\sin(\angle\beta_1).
\end{aligned}$$

In Example 11 we obtained

$$f(t) = 2e^{-t}\cos(2t)u_s(t) + 5e^{-t}\sin(2t)u_s(t).$$

So it better be that

$$2 = +2|\beta_1|\cos(\angle\beta_1) = +2 \times 2.6932\cos(-1.19)$$
$$5 = -2|\beta_1|\sin(\angle\beta_1) = -2 \times 2.6932\sin(-1.19)$$

which does indeed hold (Use MATLAB to check this!).

Example 13 *Multiple Roots*
Let

$$F(s) = \frac{1}{s(s+2)^2}.$$

The partial fraction expansion of $F(s)$ is written as

$$F(s) = \frac{1}{s(s+2)^2} = \frac{A_0}{s} + \frac{A_1}{s+2} + \frac{A_2}{(s+2)^2}.$$

We first multiply through by $s(s+2)^2$ to obtain

$$1 = A_0(s+2)^2 + A_1 s(s+2) + A_2 s$$

or

$$1 = (A_0 + A_1)s^2 + (4A_0 + 2A_1 + A_2)s + 4A_0$$

Equating coefficients of s we have

$$A_0 = 1/4,\ A_1 = -1/4,\ A_2 = -1/2.$$

Thus

$$\frac{1}{s(s+2)^2} = \frac{1/4}{s} - \frac{1/4}{s+2} - \frac{1/2}{(s+2)^2}$$

and finally[3]

$$f(t) = \frac{1}{4}u_s(t) - \frac{1}{4}e^{-2t} - \frac{1}{2}te^{-2t}.$$

Example 14 $F(s) = \dfrac{1}{s(s+2)^2}$

A simpler way to compute the partial fraction expansion of Example 13 is as follows:

$$F(s) = \frac{1}{s(s+2)^2} = \frac{A_0}{s} + \frac{A_1}{s+2} + \frac{A_2}{(s+2)^2}.$$

Multiply through by $(s+2)^2$ to obtain

$$\frac{1}{s} = \frac{A_0}{s}(s+2)^2 + A_1(s+2) + A_2$$

---

[3] Recall that $\mathcal{L}\{t\} = 1/s^2$ and $\mathcal{L}\{e^{\alpha t}f(t)\} = F(s-\alpha)$ so that with $\alpha = -2$ we have $\mathcal{L}\{te^{-2t}\} = 1/(s+2)^2$.

2.2 Partial Fraction Expansion

and then set $s = -2$, which gives $A_2 = -1/2$. Then we have

$$\frac{1}{s(s+2)^2} = \frac{A_0}{s} + \frac{A_1}{s+2} - \frac{1/2}{(s+2)^2}$$

which is rearranged to give

$$\frac{1}{s(s+2)^2} + \frac{(1/2)s}{s(s+2)^2} = \frac{A_0}{s} + \frac{A_1}{s+2}$$

or

$$\frac{1}{2}\frac{1}{s(s+2)} = \frac{A_0}{s} + \frac{A_1}{s+2}.$$

As in Example 13 this results in $A_0 = 1/4, A_1 = -1/4$.

Example 15 $F(s) = \dfrac{1}{s(s^2+s+1)}$

Using the quadratic formula we can solve

$$s^2 + s + 1 = 0$$

to obtain

$$s = \frac{-1 \pm \sqrt{1^2 - 4(1)(1)}}{2} = -\frac{1}{2} \pm j\frac{\sqrt{3}}{2}.$$

Then $F(s)$ can be written as

$$F(s) = \frac{1}{s(s^2+s+1)} = \frac{1}{s\left[s-(-1/2+j\sqrt{3}/2)\right]\left[s-(-1/2-j\sqrt{3}/2)\right]} \quad (2.16)$$

$$= \frac{1}{s\left[s+1/2-j\sqrt{3}/2\right]\left[s+1/2+j\sqrt{3}/2\right]}$$

$$= \frac{1}{s\left[(s+1/2)^2+3/4\right]}. \quad (2.17)$$

We can then do the expansion in terms of the complex conjugate poles using (2.16). However there is an alternative approach using (2.17), which we now show. We write[4]

$$F(s) = \frac{1}{s\left[(s+1/2)^2+3/4\right]} = A_0\frac{1}{s} + A_1\underbrace{\frac{s+1/2}{(s+1/2)^2+3/4}}_{\mathcal{L}\{e^{-(1/2)t}\cos(\sqrt{3}/2\,t)\}} + A_2\underbrace{\frac{\sqrt{3}/2}{(s+1/2)^2+3/4}}_{\mathcal{L}\{e^{-(1/2)t}\sin(\sqrt{3}/2\,t)\}}.$$

---

[4]In partial fraction expansion theory it is often taught to write

$$F(s) = \frac{1}{s(s^2+s+1)} = B_0\frac{1}{s} + \frac{B_1 s + B_2}{(s+1/2)^2+3/4}.$$

This is equivalent to writing

$$F(s) = \frac{1}{s(s^2+s+1)} = A_0\frac{1}{s} + A_1\frac{s+1/2}{(s+1/2)^2+3/4} + A_2\frac{1/2}{(s+1/2)^2+3/4}.$$

by setting $B_1 s + B_2 = A_1(s+1/2) + A_2(\sqrt{3}/2) = A_1 s + A_1/2 + A_2(\sqrt{3}/2)$. However, $\dfrac{s+1/2}{(s+1/2)^2+3/4}$ and $\dfrac{\sqrt{3}/2}{(s+1/2)^2+3/4}$ are in the Laplace transform table making it easier doing the expansion in terms of A_1, A_2.

Multiplying through by $s(s^2 + s + 1)$ to clear fractions, we obtain

$$1 = A_0(s^2 + s + 1) + A_1 s(s + 1/2) + A_2 s\sqrt{3}/2$$

or

$$1 = A_0 + \left(A_0 + \frac{1}{2}A_1 + \frac{1}{2}\sqrt{3}A_2\right)s + (A_0 + A_1)s^2.$$

This results in

$$A_0 = 1$$
$$A_1 = -1$$
$$A_2 = -\frac{A_0 + A_1/2}{\sqrt{3}/2} = -\frac{1}{\sqrt{3}}$$

so that

$$f(t) = u_s(t) - e^{-(1/2)t}\cos(\sqrt{3}/2\, t) - \sqrt{1/3}\, e^{-(1/2)t}\sin(\sqrt{3}/2\, t). \tag{2.18}$$

Example 16 $F(s) = \dfrac{1}{s(s^2 + s + 1)}$ (again)

Let's redo the previous example the "hard" way using (2.16). We have

$$F(s) = \frac{1}{s\left[s - (-1/2 + j\sqrt{3}/2)\right]\left[s - (-1/2 - j\sqrt{3}/2)\right]} = A\frac{1}{s} + \frac{\beta_1}{s - (-1/2 + j\sqrt{3}/2)} + \frac{\beta_1^*}{s - (-1/2 - j\sqrt{3}/2)}.$$

Then

$$A = \lim_{s \to 0} sF(s) = \lim_{s \to 0} \frac{1}{s^2 + s + 1} = 1$$

and

$$\beta_1 = \lim_{s \to -1/2 + j\sqrt{3}/2} \left[s - (-1/2 + j\sqrt{3}/2)\right] F(s)$$

$$= \lim_{s \to -1/2 + j\sqrt{3}/2} \frac{1}{s\left[s - (-1/2 - j\sqrt{3}/2)\right]}$$

$$= \frac{1}{(-1/2 + j\sqrt{3}/2)\left[(-1/2 + j\sqrt{3}/2) - (-1/2 - j\sqrt{3}/2)\right]}$$

$$= \frac{1}{(-1/2 + j\sqrt{3}/2)\, 2j\sqrt{3}/2}$$

$$= \frac{1}{-1/2 + j\sqrt{3}/2}\frac{(-1/2 - j\sqrt{3}/2)}{(-1/2 - j\sqrt{3}/2)}\frac{-j}{\sqrt{3}}$$

$$= \frac{(-1/2 - j\sqrt{3}/2)}{1/4 + 3/4}\frac{-j}{\sqrt{3}}$$

$$= -\frac{1}{2} + j\frac{1}{2\sqrt{3}}.$$

2.2 Partial Fraction Expansion

Of course

$$\beta_1^* = -\frac{1}{2} - j\frac{1}{2\sqrt{3}}.$$

To obtain the same form of the answer given in Example 15 we must convert β_1 to polar coordinate form (Figure 2.3). We have

$$|\beta_1| = \sqrt{1/4 + 1/12} = \sqrt{4/12} = \sqrt{\frac{1}{3}}$$

and

$$\angle \beta_1 = \tan^{-1}\left(\frac{1}{2\sqrt{3}}, -\frac{1}{2}\right) = 90° + \tan^{-1}\left(\frac{\frac{1}{2}}{\frac{1}{2\sqrt{3}}}\right) = 90° + \tan^{-1}(\sqrt{3})$$

$$= 150° \text{ or } 5\pi/6 \text{ radians}.$$

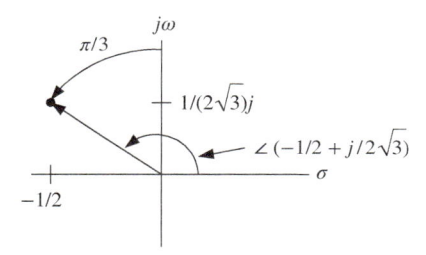

Figure 2.3. $\angle \beta_1 = \pi/2 + \pi/3$.

Recall we defined $\tan^{-1}(b, a)$ to be the same as the computer language command `atan2(b,a)`. Finally

$$f(t) = u_s(t) + \sqrt{\frac{1}{3}}e^{-(1/2)t}e^{j(\sqrt{3}/2t + 5\pi/6)} + \sqrt{\frac{1}{3}}e^{-(1/2)t}e^{-j(\sqrt{3}/2t + 5\pi/6)}$$

$$= u_s(t) + \sqrt{\frac{1}{3}}e^{-(1/2)t}\left(e^{j(\sqrt{3}/2t + 5\pi/6)} + e^{-j(\sqrt{3}/2t + 5\pi/6)}\right)$$

$$= u_s(t) + \sqrt{\frac{1}{3}}e^{-(1/2)t}2\cos(\sqrt{3}/2t + 5\pi/6) \text{ Euler's formula}$$

$$= u_s(t) + \sqrt{\frac{1}{3}}e^{-(1/2)t}2\Big(\cos(\sqrt{3}/2t)\cos(5\pi/6) - \sin(\sqrt{3}/2t)\sin(5\pi/6)\Big)$$

$$= u_s(t) + \sqrt{\frac{1}{3}}e^{-(1/2)t}2\left(\cos(\sqrt{3}/2t)\frac{-\sqrt{3}}{2} - \sin(\sqrt{3}/2t)\frac{1}{2}\right)$$

$$= u_s(t) - e^{-(1/2)t}\cos(\sqrt{3}/2t) - \sqrt{1/3}e^{-(1/2)t}\sin(\sqrt{3}/2t) \qquad (2.19)$$

which is the same as (2.18).

Non Strictly Proper Rational Functions

In all of the examples with partial fractions we had *strictly proper* rational functions. That is, we had

$$F(s) = \frac{b(s)}{a(s)}, \quad \deg\{b(s)\} < \deg\{a(s)\}.$$

Suppose we considered

$$F(s) = \frac{(s+2)(s+3)}{(s+1)(s+4)} = \frac{s^2 + 5s + 6}{s^2 + 5s + 4},$$

where $F(s)$ is proper as $\deg\{b(s)\} \leq \deg\{a(s)\}$, but *not* strictly proper as $\deg\{b(s)\} = \deg\{a(s)\}$. To have the partial fraction expansion method work, we must first divide the numerator by the denominator as follows:

$$F(s) = \frac{s^2 + 5s + 6}{s^2 + 5s + 4}$$
$$= \frac{s^2 + 5s + 4}{s^2 + 5s + 4} + \frac{2}{s^2 + 5s + 4}$$
$$= 1 + \frac{2}{s^2 + 5s + 4}$$
$$= 1 + \frac{2}{(s+1)(s+4)}$$
$$= 1 + \frac{A}{s+1} + \frac{B}{s+4}$$
$$= 1 + \frac{2/3}{s+1} - \frac{2/3}{s+4}.$$

In MATLAB we would write the program

```
Fnum = [1 5 6]
Fden = [1 5 4]
[beta,poles,k] = residue(Fnum,Fden)
```

and it should return

```
beta = -0.6667 0.6667
poles = -4 -1
k = 1
```

Example 17 $F(s) = \dfrac{s^3}{s^2 + 5s + 4}$

As another example let $F(s) = \dfrac{s^3}{s^2 + 5s + 4}$, which is *not* proper. We first use long division to divide $s^2 + 5s + 4$ into s^3 as follows:

$$\begin{array}{r}
s - 5 \\
s^2 + 5s + 4 \; \big| \; \overline{s^3 } \\
s^3 + 5s^2 + 4s \\
\overline{0 - 5s^2 - 4s} \\
-5s^2 - 25s - 20 \\
\overline{21s + 20}
\end{array}$$

Thus
$$F(s) = \frac{s^3}{s^2 + 5s + 4} = s - 5 + \frac{21s + 20}{s^2 + 5s + 4}.$$

We next do a partial fraction expansion of $\frac{21s + 20}{s^2 + 5s + 4}$ to obtain

$$\frac{21s + 20}{s^2 + 5s + 4} = \frac{A}{s + 1} + \frac{B}{s + 4} = -\frac{1/3}{s + 1} + \frac{64/3}{s + 4}.$$

Finally
$$F(s) = \frac{s^3}{s^2 + 5s + 4} = s - 5 - \frac{1/3}{s + 1} + \frac{64/3}{s + 4}.$$

In MATLAB we would write the program

```
Fnum = [1 0 0 0]
Fden = [1 5 4]
[beta,poles,k] = residue(Fnum,Fden)
```

and it should return

```
beta = 21.3333 -0.3333
poles = -4 -1
k = 1 -5
```

Remark In our application of Laplace transforms to physical systems $F(s)$ will always turn out to be strictly proper.

2.3 Poles and Zeros

In the above we computed the inverse Laplace transforms of

$$F_1(s) \triangleq \frac{1}{(s + 2)(s + 3)}$$

and

$$F_2(s) \triangleq \frac{2s + 12}{s^2 + 2s + 5}.$$

Both of these are the ratio of two polynomials (rational functions) in s. Note also that the degree of the numerator polynomial is less than or equal to the degree of the denominator.
  In general, we write
$$F(s) = \frac{b(s)}{a(s)},$$

where $b(s)$ is the numerator polynomial and $a(s)$ is the dominator polynomial. In practice we are dealing with physical systems and, as a consequence, it will turn out that

$$\deg\{b(s)\} < \deg\{a(s)\}.$$

Definition 2 *Poles of $F(s)$*
  The *poles* of $F(s)$ are the roots of $a(s) = 0$.

Definition 3 *Zeros of $F(s)$*

The *zeros* of $F(s)$ are the roots of $b(s) = 0$.

Example 18 $F_1(s) \triangleq \dfrac{1}{(s+2)(s+3)}$

The poles of
$$F_1(s) \triangleq \dfrac{1}{(s+2)(s+3)}$$
are
$$s = -2, s = -3.$$

$F_1(s)$ has no zeros as the numerator can never be zero.

Example 19 $F_2(s) \triangleq \dfrac{2s+12}{s^2+2s+5}$

The poles of
$$F_2(s) \triangleq \dfrac{2s+12}{s^2+2s+5} = \dfrac{2s+12}{\left[s-(-1+2j)\right]\left[s-(-1-2j)\right]}$$
are
$$s = -1+2j, s = -1-2j.$$

$F_2(s)$ has one zero at
$$s = -6.$$

2.4 Poles and Partial Fractions

Consider $F(s)$ to have the form
$$F(s) = \dfrac{b(s)}{a(s)} = \dfrac{(s-z_1)(s-z_2)}{(s-p_1)(s-p_2)(s-p_3)}.$$

With p_1, p_2, p_3 distinct, a partial fraction expansion of $F(s)$ gives
$$F(s) = \dfrac{A_1}{s-p_1} + \dfrac{A_2}{s-p_2} + \dfrac{A_3}{s-p_3}$$
and
$$f(t) = A_1 e^{p_1 t} + A_2 e^{p_2 t} + A_3 e^{p_3 t}.$$

The point here is that we only needed to know the poles of $F(s)$ in order to determine the *form* of the time response. By form of the time response is simply meant the functions $e^{p_1 t}, e^{p_2 t}, e^{p_3 t}$. The zeros play a role in the values of A_1, A_2, A_3 but do *not* have any role in the form of the time response.

Similarly, if
$$F(s) = \dfrac{(s-z_1)(s-z_2)}{(s-p_1)(s-p_2)^2}$$
then
$$F(s) = \dfrac{A_1}{s-p_1} + \dfrac{A_2}{s-p_2} + \dfrac{A_3}{(s-p_2)^2}.$$

2.4 Poles and Partial Fractions

Without evaluating A_1, A_2, A_3 we know that

$$f(t) = A_1 e^{p_1 t} + A_2 e^{p_2 t} + A_3 t e^{p_2 t}.$$

Again, the point here is that we only needed to know the poles of $F(s)$ in order to determine the form of the time response.

Example 20 $F(s) = \dfrac{s+3}{(s+2)(s+6)}$

By the partial fraction expansion method we know that

$$F(s) = \frac{(s+3)}{(s+2)(s+6)} = \frac{A}{s+2} + \frac{B}{s+6}.$$

That is, the expansion is in terms of the poles of $F(s)$. Without even evaluating A, B we know that

$$f(t) = Ae^{-2t} + Be^{-6t}$$

and thus that $f(t)$ dies out as $t \to \infty$.

Example 21 $F(s) = \dfrac{s+3}{(s+2)(s-6)}$

By the partial fraction expansion method we know that

$$F(s) = \frac{s+3}{(s+2)(s-6)} = \frac{A}{s+2} + \frac{B}{s-6}.$$

That is, the expansion is in terms of the poles of $F(s)$. Without even evaluating A, B we know that

$$f(t) = Ae^{-2t} + Be^{6t}$$

and thus that $f(t)$ does *not* die out as $t \to \infty$.

Example 22 $F(s) = \dfrac{2s+12}{s^2+2s+5}$

Figure 2.4 on the next page is a *pole–zero plot* for $F(s)$, that is, it marks the two poles at $-1 \pm 2j$ by an × and the zero at -6 by an ◦. By the partial fraction expansion method we know that

$$F(s) = \frac{2s+12}{[s-(-1+2j)][s-(-1-2j)]} = \frac{\beta_1}{s-(-1+2j)} + \frac{\beta_1^*}{s-(-1-2j)}.$$

Without even evaluating β_1, β_1^* we know that

$$f(t) = \beta_1 e^{-t} e^{2jt} + \beta_1^* e^{-t} e^{-2jt} = 2|\beta_1| e^{-t} \cos(2t + \angle \beta_1)$$

and thus that $f(t)$ dies out as $t \to \infty$.

The poles of $F(s)$ are $-1 \pm 2j$ where the real part of the pole determines the rate of decay as e^{-t} and the imaginary part of the pole determines the oscillation rate as $\cos(2t + \angle \beta_1)$.

Example 23 $F(s) = \dfrac{2s+12}{s^2-2s+5}$

The pole–zero plot for this $F(s)$ is shown in Figure 2.5 on the next page.

2 Laplace Transforms

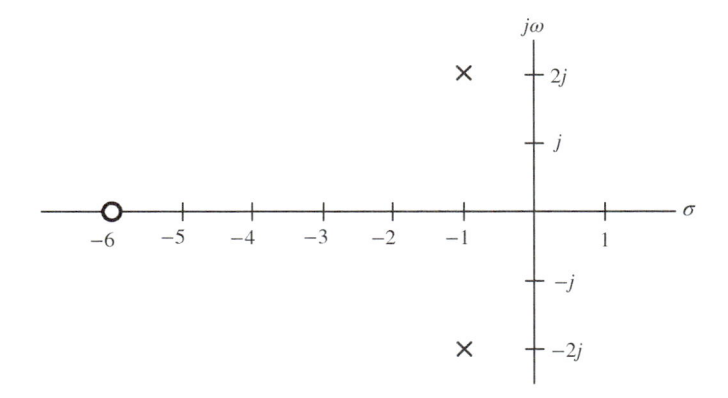

Figure 2.4. Location of the poles and zero of $F(s)$.

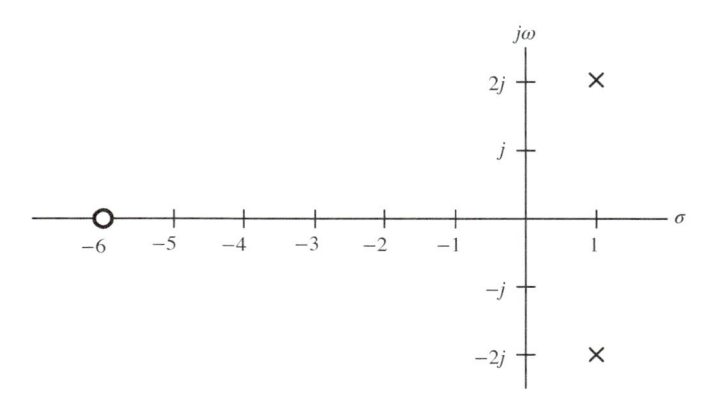

Figure 2.5. Location of the poles and zero of $F(s)$.

By the partial fraction expansion method we know that

$$F(s) = \frac{2s+12}{[s-(1+2j)][s-(1-2j)]} = \frac{\beta_1}{s-(1+2j)} + \frac{\beta_1^*}{s-(1-2j)}.$$

Without even evaluating β_1, β_1^* we know that

$$f(t) = \beta_1 e^t e^{2jt} + \beta_1^* e^t e^{-2jt} = 2|\beta_1| e^t \cos(2t + \angle \beta_1)$$

and thus that $f(t)$ does *not* die out as $t \to \infty$. The real parts of the poles of $F(s)$ are both 1 resulting in $f(t)$ having the factor e^t which grows without bound.

Definition 4 *Open Left Half-Plane*

Let $s = \sigma + j\omega$ so that $\text{Re}\{s\} = \sigma$ and $\text{Im}\{s\} = \omega$. As illustrated in Figure 2.6, the open left half-plane is where

$$\sigma = \text{Re}\{s\} < 0.$$

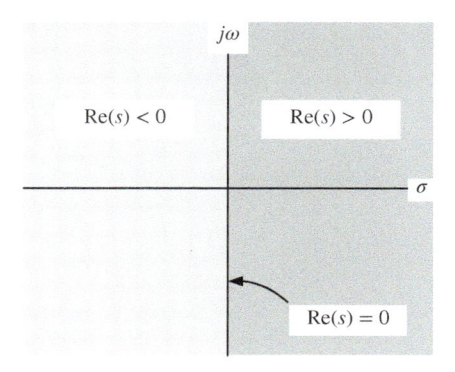

Figure 2.6. The open left half-plane is where Re{s} < 0.

Theorem 1 *Asymptotic Response of $f(t)$*
Let $F(s) = \mathcal{L}\{f(t)\}$ be strictly proper and rational. Then

$$f(t) \to 0 \text{ as } t \to \infty \tag{2.20}$$

if and only if all the poles of $F(s)$ are in the open left half-plane.

Proof. By the above examples this follows from the method of partial fractions in computing the inverse Laplace transform. ∎

Appendix: Exponential Function

One way to define the exponential function e^s is as a power series. We follow this approach here as follows:

$$e^s \triangleq 1 + s + \frac{s^2}{2!} + \frac{s^3}{3!} + \frac{s^4}{4!} + \cdots$$

$$= \sum_{n=0}^{\infty} \frac{s^n}{n!}.$$

As this infinite series converges for all values of $s = \sigma + j\omega \in \mathbb{C}$ (not proven here) this definition is valid for all $s \in \mathbb{C}$. With $s = 0$ it follows from this definition that $e^0 = 1$.

Next note that

$$\frac{d}{ds}e^s = \frac{d}{ds}\left(1 + s + \frac{s^2}{2!} + \frac{s^3}{3!} + \frac{s^4}{4!} + \cdots\right)$$

$$= 0 + 1 + s + \frac{s^2}{2!} + \frac{s^3}{3!} + \cdots$$

$$= 1 + s + \frac{s^2}{2!} + \frac{s^3}{3!} + \cdots$$

$$= \sum_{n=0}^{\infty} \frac{s^n}{n!}$$

$$= e^s.$$

2 Laplace Transforms

Recall from elementary algebra that, for example,

$$2^3 2^7 = 2^{10}.$$

This is a property of the exponents of numbers with the same base (the base is 2 in this example). We now show this property holds for e^s. First we calculate

$$e^{s_1} e^{s_2} = \left(1 + s_1 + \frac{s_1^2}{2!} + \frac{s_1^3}{3!} + \frac{s_1^4}{4!} + \cdots\right)\left(1 + s_2 + \frac{s_2^2}{2!} + \frac{s_2^3}{3!} + \frac{s_2^4}{4!} + \cdots\right)$$

$$= 1 + s_1 + s_2 + \frac{s_1^2}{2!} + s_1 s_2 + \frac{s_2^2}{2!} + \frac{s_1^3}{3!} + \frac{s_1 s_2^2}{2!} + \frac{s_1^2 s_2}{2!} + \frac{s_2^3}{3!} + \cdots$$

$$= 1 + s_1 + s_2 + \frac{1}{2!}(s_1^2 + 2 s_1 s_2 + s_2^2) + \frac{1}{3!}(s_1^3 + 3 s_1 s_2^2 + 3 s_1^2 s_2 + s_2^3) + \cdots.$$

Next we calculate

$$e^{s_1 + s_2} = 1 + s_1 + s_2 + \frac{(s_1 + s_2)^2}{2!} + \frac{(s_1 + s_2)^3}{3!} + \frac{(s_1 + s_2)^4}{4!} + \cdots$$

$$= 1 + s_1 + s_2 + \frac{1}{2!}(s_1^2 + 2 s_1 s_2 + s_2^2) + \frac{1}{3!}(s_1^3 + 3 s_1 s_2^2 + 3 s_1^2 s_2 + s_2^3) + \cdots.$$

By inspection we see that

$$e^{s_1} e^{s_2} = e^{s_1 + s_2}.$$

This property is the reason we call $e^s \triangleq 1 + s + \frac{s^2}{2!} + \frac{s^3}{3!} + \frac{s^4}{4!} + \cdots$ the *exponential* function. In particular, we have

$$e^s e^{-s} = e^{s-s} = e^0 = 1$$

or

$$e^{-s} = \frac{1}{e^s}.$$

Euler's Formula

Let $s = j\omega$ be a purely imaginary number. Then

$$e^{j\omega} = 1 + j\omega + \frac{(j\omega)^2}{2!} + \frac{(j\omega)^3}{3!} + \frac{(j\omega)^4}{4!} + \frac{(j\omega)^5}{5!} + \frac{(j\omega)^6}{6!} \cdots$$

$$= 1 + j\omega - \frac{\omega^2}{2!} - j\frac{\omega^3}{3!} + \frac{\omega^4}{4!} + j\frac{\omega^5}{5!} + \frac{\omega^6}{6!} + \cdots$$

$$= 1 - \frac{\omega^2}{2!} + \frac{\omega^4}{4!} - \frac{\omega^6}{6!} + \cdots + j\left(\omega - \frac{\omega^3}{3!} + \frac{\omega^5}{5!} + \cdots\right)$$

$$= \cos(\omega) + j \sin(\omega)$$

as the power series expansion for $\cos(\omega)$ and $\sin(\omega)$ are

$$\cos(\omega) = \sum_{i=0}^{\infty} (-1)^i \frac{\omega^{2i}}{(2i)!} = 1 - \frac{\omega^2}{2!} + \frac{\omega^4}{4!} - \frac{\omega^6}{6!} + \cdots$$

$$\sin(\omega) = \sum_{i=1}^{\infty} (-1)^{i+1} \frac{\omega^{2i-1}}{(2i+1)!} = \omega - \frac{\omega^3}{3!} + \frac{\omega^5}{5!} + \cdots.$$

The expression
$$e^{j\omega} = \cos(\omega) + j\sin(\omega)$$
is referred to as *Euler's formula*.

Remark With m a non-negative integer, i.e., $m \in \{0, 1, 2, \ldots\}$ and using Euler's formula we may write
$$t^m e^{(\sigma+j\omega)t} = t^m e^{\sigma t}\cos(\omega t) + jt^m e^{\sigma t}\sin(\omega t).$$

It turns out that linear time-invariant differential equations have solutions composed only of terms of the form $At^m e^{\sigma t}\cos(\omega t)$ and $Bt^m e^{\sigma t}\sin(\omega t)$. This is the reason why the exponential function appears so much in linear systems theory!

Real Valued Exponential Function

Let's look at e^s with $s = \sigma$ a *real* number.
$$e^\sigma = 1 + \sigma + \frac{\sigma^2}{2!} + \frac{\sigma^3}{3!} + \frac{\sigma^4}{4!} + \cdots$$
$$= \sum_{n=0}^{\infty} \frac{\sigma^n}{n!}.$$

As shown above for $\sigma = 0$ we have $e^0 = 1$. Also we have
$$e^\sigma > 0 \text{ for } \sigma > 0$$
simply because each term in the power series expansion is positive for $\sigma > 0$. Still with $\sigma > 0$ we have (as previously shown)
$$e^{-\sigma} = \frac{1}{e^\sigma} > 0 \text{ as } e^\sigma > 0 \text{ for } \sigma > 0.$$

Therefore
$$e^\sigma > 0 \text{ for all } -\infty < \sigma < +\infty.$$

Figure 2.7 on the next page is a graph of the exponential function where σ has been replaced by x. From the point of view of the stick man in the figure is the graph of the *inverse* of $y = e^x$.

Natural Logarithm Function

As was just shown, the exponential function that takes $x \in \mathbb{R}$ to $y = e^x$ is always positive. For any $y > 0$ define the natural logarithm function $\ln(y)$ as the *inverse function* of $y = e^x$. A graph of $x = \ln(y)$ is shown in Figure 2.8. As $1 = e^0$ we have $\ln(1) = 0$.

We now show that
$$\frac{d}{dy}\ln(y) = \frac{1}{y} \text{ for } y > 0.$$

To explain, the exponential function takes x to $y = e^x$ so its inverse must take $y = e^x$ to x, that is,
$$\ln(e^x) = x.$$

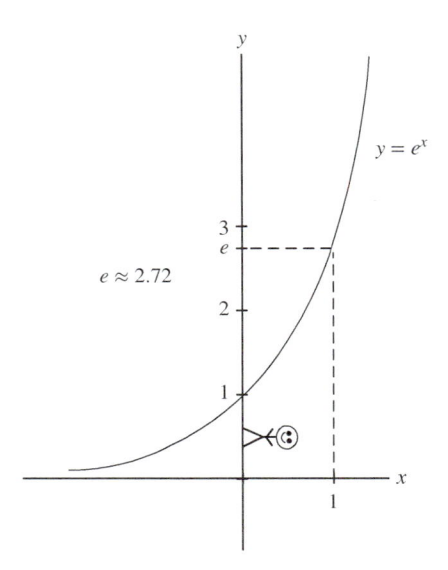

Figure 2.7. Graph of $y = e^x$.

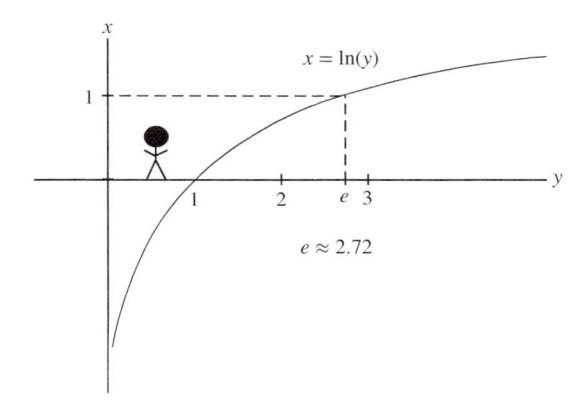

Figure 2.8. Graph of $x = \ln(y)$.

This just says the natural logarithm is the *inverse function* of $y = e^x, x \in \mathbb{R}$. Then differentiating both sides of $x = \ln(e^x)$ with respect to x we obtain

$$1 = \frac{d}{dx}x = \frac{d}{dx}\ln(e^x) = \frac{d}{dy}\ln(y)\bigg|_{y=e^x}\frac{dy}{dx} = \frac{d}{dy}\ln(y)\bigg|_{y=e^x}\frac{d}{dx}e^x = \frac{d}{dy}\ln(y)\bigg|_{y=e^x}e^x$$

or

$$\frac{d}{dy}\ln(y)\bigg|_{y=e^x} = \frac{1}{e^x} \implies \frac{d}{dy}\ln(y) = \frac{1}{y}.$$

Problems

Problem 1 *Euler's Identity*

(a) $e^{(\sigma+j\omega)t} = a(t) + jb(t)$, $a(t) = ?$ $b(t) = ?$

(b) $\dfrac{d}{dt} e^{(\sigma+j\omega)t} = c(t) + jd(t)$, $c(t) = ?$ $d(t) = ?$

(c) $(\sigma + j\omega) e^{(\sigma+j\omega)t} = c(t) + jd(t)$, $c(t) = ?$ $d(t) = ?$

Problem 2 *Trigonometry*

(a) $\dfrac{d}{dt}\cos(\omega t) = ?$, $\dfrac{d}{dt}\sin(\omega t) = ?$, $\cos(\pi) = ?$, $\sin(\pi) = ?$, $\cos(\pi/2) = ?$

(b) $\sin(\pi/2) = ?$, $\cos(\pi/6) = ?$, $\sin(\pi/6) = ?$, $\cos(\pi/3) = ?$, $\sin(\pi/3) = ?$

Problem 3 *Complex Numbers*

(a) $(2 + 3j)^* = ?$

(b) $\dfrac{1}{2 - j2} = a + jb$, $a = ?$ $b = ?$

(c) $(2 - 2j)(1 + 2j) = a + jb$, $a = ?$ $b = ?$

(d) $2 - 2j = re^{j\theta}$, $r = ?$, $\theta = ?$

Problem 4 *Calculus*

$\dfrac{d}{ds} \int_0^\infty e^{-st} x(t)\,dt = ?$

Problem 5 *Inverse Laplace Transform*

Let
$$F(s) = \dfrac{3s^2 + s + 4}{s^3}$$
and compute $f(t) = \mathcal{L}^{-1}\{F(s)\}$. Check your result using MATLAB.

Problem 6 *Inverse Laplace Transform*

Let
$$F(s) = \dfrac{3s^2 + 4s + 24}{(s - 1)(s + 2)(s + 5)}$$
and compute its inverse Laplace transform. Check your result using MATLAB.

Problem 7 *Inverse Laplace Transform*

Let
$$F(s) = \dfrac{4}{s^2(s + 1)}$$
and find its inverse Laplace transform. Check your result using MATLAB.

Problem 8 *Inverse Laplace Transform*
Let
$$F(s) = \frac{3s}{s^2 + 2s + 26}$$
and find its inverse Laplace transform. Check your result using MATLAB.

Problem 9 $\mathcal{L}\left\{\int_0^t x(\tau)d\tau\right\} = \frac{1}{s}X(s)$
Let
$$X(s) = \mathcal{L}\{x(t)\}.$$

Show that
$$\mathcal{L}\left\{\int_0^t x(\tau)d\tau\right\} = \frac{1}{s}X(s).$$

Hint: Let $y(t) = \int_0^t x(\tau)d\tau$ and apply property 3 of the Laplace transform to $y(t)$.

Problem 10 $\mathcal{L}\left\{\int_0^\infty g(t-\tau)u(\tau)d\tau\right\} = G(s)U(s)$
Let
$$U(s) = \mathcal{L}\{u(t)\}, \quad G(s) = \mathcal{L}\{g(t)\}.$$

Show that
$$\mathcal{L}\left\{\int_0^\infty g(t-\tau)u(\tau)d\tau\right\} = G(s)U(s).$$

Hint:
$$\mathcal{L}\left\{\int_0^\infty g(t-\tau)u(\tau)d\tau\right\} = \int_0^\infty e^{-st}\int_0^\infty g(t-\tau)u(\tau)d\tau dt$$
$$= \int_0^\infty \int_0^\infty e^{-s(t-\tau)}g(t-\tau)e^{-s\tau}u(\tau)d\tau dt = \cdots.$$

Problem 11 *Inverse Laplace Transform*
Compute the inverse Laplace transform of
$$F(s) = \frac{s+1}{s^2}.$$

Problem 12 *Inverse Laplace Transform*
Compute the inverse Laplace transform of
$$F(s) = \frac{10}{s(s+1)}.$$

Problem 13 *Inverse Laplace Transform*
Compute the inverse Laplace transform of
$$F(s) = \frac{-10}{s(s-1)}.$$

Problem 14 *Inverse Laplace Transform*
Compute the inverse Laplace transform of
$$F(s) = \frac{s+1}{s(s+2)(s+3)}.$$

Problem 15 *Inverse Laplace Transform*
Compute the inverse Laplace transform of
$$F(s) = -\frac{s+1}{s(s+2)(s-3)}.$$

Problem 16 *Inverse Laplace Transform*
Compute the inverse Laplace transform of
$$F(s) = \frac{s+1}{(s+2)(s+3)}.$$

Problem 17 *Inverse Laplace Transform*
Compute the inverse Laplace transform of
$$F(s) = \frac{2s+12}{s^2-2s+5} = \frac{2s+12}{(s-1)^2+4} = A_1\frac{s-1}{(s-1)^2+4} + A_2\frac{2}{(s-1)^2+4}$$
by first computing A_1, A_2.

Problem 18 *Inverse Laplace Transform*
Compute the inverse Laplace transform of
$$F(s) = \frac{2s+12}{s(s^2-2s+5)} = \frac{A}{s} + \frac{\beta}{s-(1+2j)} + \frac{\beta^*}{s-(1-2j)}$$
by first computing A, β.

Laplace Transform Pairs

f(t)	$\mathcal{L}\{f(t)\}$	Poles	ROC[5]
$u_s(t)$	$\dfrac{1}{s}$	$p=0$	$\text{Re}\{s\} > 0$
$tu_s(t)$	$\dfrac{1}{s^2}$	$p=0,0$	$\text{Re}\{s\} > 0$
$t^n u_s(t)$	$\dfrac{n!}{s^{n+1}}$	$p=0\ (n+1\ \text{times})$	$\text{Re}\{s\} > 0$
$e^{pt}u_s(t)$	$\dfrac{1}{s-p}$	$p=\sigma+j\omega$	$\text{Re}\{s\} > \sigma$

---

[5] Region Of Convergence.

f(t)	$\mathcal{L}\{f(t)\}$	Poles	ROC[5]
$t^n e^{\sigma t} u_s(t)$	$\dfrac{n!}{(s-\sigma)^{n+1}}$	$p = \sigma$ $(n+1$ times$)$	$\text{Re}\{s\} > \sigma$
$\sin(\omega t) u_s(t)$	$\dfrac{\omega}{\underbrace{s^2+\omega^2}_{(s-j\omega)(s+j\omega)}}$	$p = \pm j\omega$	$\text{Re}\{s\} > 0$
$\cos(\omega t) u_s(t)$	$\dfrac{s}{\underbrace{s^2+\omega^2}_{(s-j\omega)(s+j\omega)}}$	$p = \pm j\omega$	$\text{Re}\{s\} > 0$
$e^{\sigma t}\sin(\omega t) u_s(t)$	$\dfrac{\omega}{\underbrace{(s-\sigma)^2+\omega^2}_{[s-(\sigma+j\omega)][s-(\sigma-j\omega)]}}$	$p = \sigma \pm j\omega$	$\text{Re}\{s\} > \sigma$
$e^{\sigma t}\cos(\omega t) u_s(t)$	$\dfrac{s-\sigma}{\underbrace{(s-\sigma)^2+\omega^2}_{[s-(\sigma+j\omega)][s-(\sigma-j\omega)]}}$	$p = \sigma \pm j\omega$	$\text{Re}\{s\} > \sigma$
$t\sin(\omega t) u_s(t)$	$\dfrac{2\omega s}{\underbrace{(s^2+\omega^2)^2}_{(s-j\omega)^2(s+j\omega)^2}}$	$p = \pm j\omega, \pm j\omega$	$\text{Re}\{s\} > 0$
$t\cos(\omega t) u_s(t)$	$\dfrac{s^2-\omega^2}{\underbrace{(s^2+\omega^2)^2}_{(s-j\omega)^2(s+j\omega)^2}}$	$p = \pm j\omega, \pm j\omega$	$\text{Re}\{s\} > 0$

Laplace Transforms and the ROC

Any strictly proper rational function $F(s) = b(s)/a(s)$ corresponds to a unique time function $f(t)$ defined on the semi-infinite time interval $[0, \infty)$. Because of this fact we will usually not be concerned with the region of convergence. That is, we will convert differential equations into algebraic equations using the Laplace transform to work in the s domain for the design of feedback controllers. After the design is completed in the s domain, a Laplace transform table is used to find the corresponding (unique) time function on $[0, \infty)$. However, the region of convergence must be (and is) taken into account in the *Overshoot* and *Undershoot* appendices of Chapter 10.

Laplace Transform Properties

$$\mathcal{L}\left\{\int_0^t f(\tau)d\tau\right\} = \frac{1}{s}F(s)$$

$$\mathcal{L}\left\{\frac{d}{dt}f(t)\right\} = sF(s) - f(0)$$

$$\mathcal{L}\{e^{at}f(t)\} = F(s-a)$$

$$\mathcal{L}\{tf(t)\} = -\frac{d}{ds}F(s)$$

$$\mathcal{L}\left\{\int_0^\infty f_2(t-\tau)f_1(\tau)d\tau\right\} = F_2(s)F_1(s)$$

Trigonometric Table

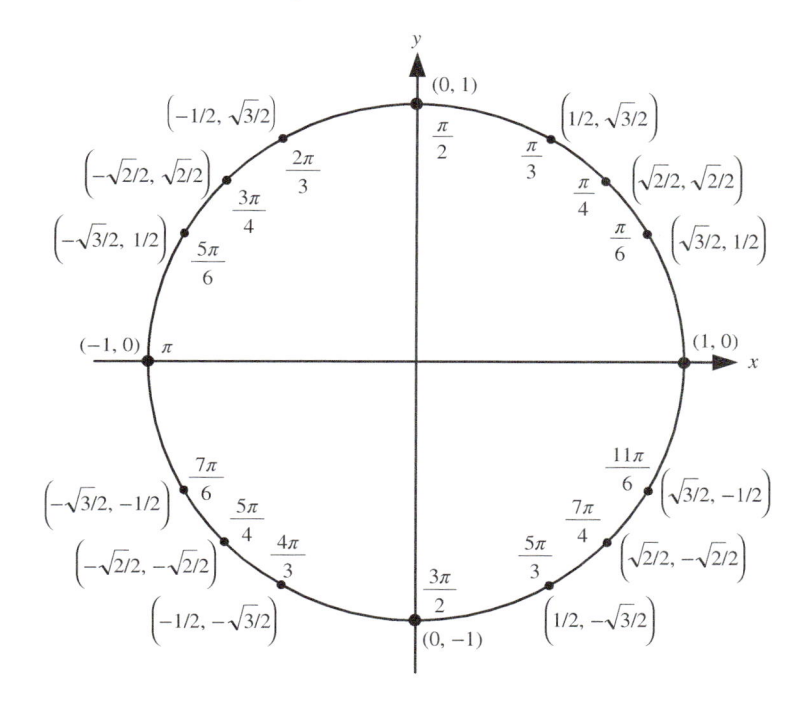

Trigonometric Identities

$$e^{j\theta} = \cos(\theta) + j\sin(\theta)$$

$$a\cos(\theta) + b\sin(\theta) = \sqrt{a^2+b^2}\cos\left(\theta + \tan^{-1}(b/a)\right)$$

$$\sin(2\theta) = 2\sin(\theta)\cos(\theta)$$

$$\cos(2\theta) = \cos^2(\theta) - \sin^2(\theta) = 2\cos^2(\theta) - 1 = 1 - 2\sin^2(\theta)$$

$$\cos^2(\theta) = \frac{1+\cos(2\theta)}{2}$$

$$\sin^2(\theta) = \frac{1-\cos(2\theta)}{2}$$

$$\cos(\theta_1 \pm \theta_2) = \cos(\theta_1)\cos(\theta_2) \mp \sin(\theta_1)\sin(\theta_2)$$

$$\sin(\theta_1 \pm \theta_2) = \sin(\theta_1)\cos(\theta_2) \pm \cos(\theta_1)\sin(\theta_2)$$

$$\cos(\theta_1)\cos(\theta_2) = \frac{1}{2}\cos(\theta_1 + \theta_2) + \frac{1}{2}\cos(\theta_1 - \theta_2)$$

$$\sin(\theta_1)\sin(\theta_2) = \frac{1}{2}\cos(\theta_1 - \theta_2) - \frac{1}{2}\cos(\theta_1 + \theta_2)$$
$$\sin(\theta_1)\cos(\theta_2) = \frac{1}{2}\sin(\theta_1 + \theta_2) + \frac{1}{2}\sin(\theta_1 - \theta_2)$$
$$\cos(\theta_1)\sin(\theta_2) = \frac{1}{2}\sin(\theta_1 + \theta_2) - \frac{1}{2}\sin(\theta_1 - \theta_2).$$

3

Differential Equations and Stability

In later chapters we will derive models of motors, carts, inverted pendulums, etc. using first principles of physics. The model of each of these systems will turn out to be a differential equation. To later deal with these models we use this chapter to see how differential equations are solved using Laplace transforms. We will next look at the stability of these differential equation system models. For example, keeping an inverted pendulum upright is the problem of making sure the closed-loop differential equation describing the inverted pendulum is stable.

3.1 Differential Equations

Consider the differential equation

$$\ddot{x} + \dot{x} + x = u. \qquad (3.1)$$

Recall that if $X(s) = \mathcal{L}\{x(t)\}$ then $\mathcal{L}\{\dot{x}\} = sX(s) - x(0)$. It then follows that

$$\begin{aligned}\mathcal{L}\{\ddot{x}\} &= s\mathcal{L}\{\dot{x}\} - \dot{x}(0) \\ &= s\big(sX(s) - x(0)\big) - \dot{x}(0) \\ &= s^2 X(s) - sx(0) - \dot{x}(0).\end{aligned}$$

We next take the Laplace transform of both sides of (3.1) to obtain

$$\mathcal{L}\{\ddot{x} + \dot{x} + x\} = U(s)$$

or

$$\underbrace{s^2 X(s) - sx(0) - \dot{x}(0)}_{\mathcal{L}\{\ddot{x}\}} + \underbrace{sX(s) - x(0)}_{\mathcal{L}\{\dot{x}\}} + X(s) = \underbrace{U(s)}_{\mathcal{L}\{u\}}.$$

Collecting terms this becomes

$$(s^2 + s + 1)X(s) - sx(0) - \dot{x}(0) - x(0) = U(s)$$

or finally

$$X(s) = \underbrace{\frac{1}{s^2 + s + 1}}_{G(s)} \underbrace{U(s)}_{\text{Input}} + \underbrace{\frac{sx(0) + \dot{x}(0) + x(0)}{s^2 + s + 1}}_{\text{zero input response}}. \qquad (3.2)$$

The zero input response is the inverse Laplace transform of

$$\frac{sx(0) + \dot{x}(0) + x(0)}{s^2 + s + 1}.$$

An Introduction to System Modeling and Control, First Edition. John Chiasson.
© 2022 John Wiley & Sons, Inc. Published 2022 by John Wiley & Sons, Inc.
Companion website: www.wiley.com/go/chiasson/anintroductiontosystemmodelingandcontrol

If the initial conditions are zero, i.e., $\dot{x}(0) = x(0) = 0$, then

$$X(s) = \underbrace{\frac{1}{s^2 + s + 1}}_{G(s)} \underbrace{U(s)}_{\text{Input}}.$$

The *transfer function* is defined to be the ratio

$$G(s) \triangleq \left. \frac{X(s)}{U(s)} \right|_{\dot{x}(0)=x(0)=0} = \frac{1}{s^2 + s + 1}. \tag{3.3}$$

That is, the transfer function is the ratio of $X(s)/U(s)$ with the initial conditions set to zero.

The *order* of a transfer function is the degree of its denominator. The transfer function $G(s)$ given in (3.3) has order 2.

With zero initial conditions let's set the input to be a step function, i.e., $u(t) = u_s(t)$. Then

$$X(s) = G(s)U(s) = \frac{1}{s^2 + s + 1} \frac{1}{s}$$

and, by (2.18) or (2.19) of Chapter 2, we have

$$x(t) = u_s(t) - e^{-(1/2)t} \cos(\sqrt{3}/2 t) u_s(t) - \sqrt{1/3} e^{-(1/2)t} \sin(\sqrt{3}/2 t) u_s(t).$$

Note that $\lim_{t \to \infty} x(t) = 1$.

Sinusoidal Steady-State Response

Let's now consider an example with a sinusoidal input. That is, consider the differential equation

$$\frac{d^2 x}{dt^2} + \frac{dx}{dt} + x = u(t) \tag{3.4}$$

with input

$$u(t) = U_0 \cos(\omega t) \tag{3.5}$$

and initial conditions

$$\dot{x}(0) = x(0) = 0.$$

We have

$$\mathcal{L}\{U_0 \cos(\omega t)\} = \frac{U_0 s}{s^2 + \omega^2}$$

so that

$$\mathcal{L}\left\{ \frac{d^2 x}{dt^2} + \frac{dx}{dt} + x \right\} = \frac{U_0 s}{s^2 + \omega^2}.$$

By the assumption of zero initial conditions this becomes

$$(s^2 + s + 1)X(s) = \frac{U_0 s}{s^2 + \omega^2}$$

or

$$X(s) = \underbrace{\frac{1}{s^2 + s + 1}}_{G(s)} \underbrace{\frac{U_0 s}{s^2 + \omega^2}}_{U(s)}. \tag{3.6}$$

3.1 Differential Equations

In order to do a partial fraction expansion of $X(s)$ we factor its denominator to obtain

$$X(s) = \frac{U_0 s}{\left[s-(-1/2+j\sqrt{3}/2)\right]\left[s-(-1/2-j\sqrt{3}/2)\right](s-j\omega)(s+j\omega)}.$$

A partial fraction expansion results in

$$X(s) = \underbrace{\frac{\beta}{s-(-1/2+j\sqrt{3}/2)} + \frac{\beta^*}{s-(-1/2-j\sqrt{3}/2)}}_{\text{From the poles of } G(s)} + \underbrace{\frac{k}{s-j\omega} + \frac{k^*}{s+j\omega}}_{\text{From the poles of } U(s)}.$$

We next explicitly compute k and k^*, while we won't be interested in the explicit values of β and β^*. We have

$$k = \lim_{s \to j\omega}(s-j\omega)X(s) = \lim_{s \to j\omega}(s-j\omega)G(s)U(s) = \lim_{s \to j\omega}(s-j\omega)G(s)\frac{U_0 s}{(s-j\omega)(s+j\omega)}$$

$$= \lim_{s \to j\omega} G(s)\frac{U_0 s}{s+j\omega}$$

$$= G(j\omega)\frac{U_0 j\omega}{2j\omega}$$

$$= \frac{U_0}{2}G(j\omega).$$

Then

$$k^* = \frac{U_0}{2}G^*(j\omega) = \frac{U_0}{2}G(-j\omega), \tag{3.7}$$

where we used the fact that

$$G^*(j\omega) = \left(\frac{1}{(j\omega)^2 + j\omega + 1}\right)^* = \frac{1}{\left((j\omega)(j\omega)\right)^* + (j\omega)^* + 1} = \frac{1}{(-j\omega)^2 + (-j\omega) + 1}$$

$$= G(-j\omega).$$

We now have

$$X(s) = \frac{\beta}{s-(-1/2+j\sqrt{3}/2)} + \frac{\beta^*}{s-(-1/2-j\sqrt{3}/2)} + \frac{U_0}{2}G(j\omega)\frac{1}{s-j\omega}$$

$$+ \frac{U_0}{2}G^*(j\omega)\frac{1}{s+j\omega} \tag{3.8}$$

and therefore

$$x(t) = \left(\beta e^{(-1/2+j\sqrt{3}/2)t} + \beta^* e^{(-1/2-j\sqrt{3}/2)t}\right)u_s(t) + \left(\frac{U_0}{2}G(j\omega)e^{j\omega t} + \frac{U_0}{2}G^*(j\omega)e^{-j\omega t}\right)u_s(t)$$

$$= \left(|\beta|e^{j\angle\beta}e^{-1/2t}e^{j(\sqrt{3}/2)t} + |\beta|e^{-j\angle\beta}e^{-1/2t}e^{-j(\sqrt{3}/2)t}\right)u_s(t)$$

$$+ \left(\frac{U_0}{2}|G(j\omega)|\,e^{j\angle G(j\omega)}e^{j\omega t} + \frac{U_0}{2}|G(j\omega)|e^{-j\angle G(j\omega)}e^{-j\omega t}\right)u_s(t)$$

$$= |\beta|e^{-1/2t}\left(e^{j(\sqrt{3}/2t+\angle\beta)} + e^{-j(\sqrt{3}/2t+\angle\beta)}\right)u_s(t)$$

$$+ \frac{U_0}{2}|G(j\omega)|\left(e^{j\left(\omega t+\angle G(j\omega)\right)} + e^{-j\left(\omega t+\angle G(j\omega)\right)}\right)u_s(t)$$

$$= 2|\beta|e^{-1/2t}\cos(\sqrt{3}/2t+\angle\beta)u_s(t) + U_0\,|G(j\omega)|\cos\bigl(\omega t+\angle G(j\omega)\bigr)u_s(t). \tag{3.9}$$

Since $e^{-1/2t} \to 0$ as $t \to \infty$ we have

$$x(t) \to x_{ss}(t) \triangleq U_0 |G(j\omega)| \cos(\omega t + \angle G(j\omega)) u_s(t). \quad (3.10)$$

If the poles of the transfer function $G(s)$ have negative real parts and a sinusoidal input $u(t) = U_0 \cos(\omega t)$ is applied to the system, the sinusoidal steady-state output $x_{ss}(t)$ is simply the input with its magnitude multiplied by $|G(j\omega)|$ and its phase offset by $\angle G(j\omega)$. In Chapter 4 the SIMULINK simulation tool is introduced and Problem 10 of that chapter asks you to simulate this example. (Or ask your instructor to run this simulation for you now!)

Example 1 *Unstable System*

Suppose our system is

$$\frac{d^2x}{dt^2} - \frac{dx}{dt} + x = u(t), \quad (3.11)$$

where

$$u(t) = U_0 \cos(\omega t) \quad (3.12)$$

and

$$\dot{x}(0) = x(0) = 0.$$

Then

$$X(s) = \underbrace{\frac{1}{s^2 - s + 1}}_{G(s)} \underbrace{\frac{U_0 s}{s^2 + \omega^2}}_{U(s)}. \quad (3.13)$$

A partial fraction expansion of $X(s)$ gives[1]

$$X(s) = \underbrace{\frac{\beta}{s - (1/2 + j\sqrt{3}/2)} + \frac{\beta^*}{s - (1/2 - j\sqrt{3}/2)}}_{\text{From the poles of } G(s)} + \underbrace{\frac{k}{s - j\omega} + \frac{k^*}{s + j\omega}}_{\text{From the poles of } U(s)}. \quad (3.14)$$

Note that the poles of $G(s)$ are in the right half-plane. The time response is given by

$$x(t) = \left(\beta e^{(1/2 + j\sqrt{3}/2)t} + \beta^* e^{(1/2 - j\sqrt{3}/2)t}\right) u_s(t) + \left(\frac{U_0}{2} G(j\omega) e^{j\omega t} + \frac{U_0}{2} G^*(j\omega) e^{-j\omega t}\right) u_s(t)$$

$$= 2|\beta| e^{1/2t} \cos(\sqrt{3}/2 t + \angle \beta) u_s(t) + U_0 |G(j\omega)| \cos(\omega t + \angle G(j\omega)) u_s(t). \quad (3.15)$$

In this example $x(t)$ does *not* have a sinusoidal steady-state response as $e^{1/2t}$ does not die out with the result that the term $2|\beta| e^{1/2t} \cos(\sqrt{3}/2 t + \angle \beta) u_s(t)$ eventually oscillates between $\pm\infty$! Problem 11 of Chapter 4 asks you to do a SIMULINK simulation of this example. (Or, again, ask your instructor to run this simulation for you now!)

3.2 Phasor Method of Solution

Previously we applied the input $u(t) = U_0 \cos(\omega t) u_s(t)$ to a couple of differential equations and showed that

$$U_0 |G(j\omega)| \cos(\omega t + \angle G(j\omega)) u_s(t)$$

---

[1] Note that $s^2 - s + 1 = [s - (1/2 + j\sqrt{3}/2)][s - (1/2 - j\sqrt{3}/2)]$.

was part of the solution. With $G(j\omega)$ the transfer function of the differential equation, we now show that this expression is always a solution to the differential equation with *special* initial conditions. This is done using phasors.

Example 2 *Phasor Method of Solution*
Let's again consider the differential equation example given by (3.4) and (3.5), but this time using the complex phasor method of solution. To proceed, consider the differential equation

$$\frac{d^2x}{dt^2} + \frac{dx}{dt} + x = u(t) \tag{3.16}$$

with input

$$u(t) = U_0 \cos(\omega t) \text{ for } -\infty < t < \infty.$$

We first solve this differential equation by letting the input be the *complex phasor* function given by

$$\mathbf{u}(t) = U_0 e^{j\omega t} = U_0 \cos(\omega t) + jU_0 \sin(\omega t) \text{ for } -\infty < t < \infty. \tag{3.17}$$

The bold $\mathbf{u}$ is used to signify it is *complex* valued. Our original input $u(t)$ is simply given by

$$u(t) = \operatorname{Re}\{\mathbf{u}(t)\} = \operatorname{Re}\{U_0 e^{j\omega t}\}.$$

We now look for a complex phasor solution of the form

$$\mathbf{x}(t) = \mathbf{A} e^{j\omega t}, \tag{3.18}$$

where $\mathbf{A} = |\mathbf{A}| e^{j\angle \mathbf{A}} \in \mathbb{C}$ is a complex constant.[2] That is, we want to find $\mathbf{A} \in \mathbb{C}$ such that $\mathbf{x}(t) = \mathbf{A} e^{j\omega t}$ is a solution to the differential equation

$$\frac{d^2\mathbf{x}}{dt^2} + \frac{d\mathbf{x}}{dt} + \mathbf{x} = \mathbf{u}(t). \tag{3.19}$$

Substituting (3.17) and (3.18) into this equation gives

$$\frac{d^2}{dt^2} \mathbf{A} e^{j\omega t} + \frac{d}{dt} \mathbf{A} e^{j\omega t} + \mathbf{A} e^{j\omega t} = U_0 e^{j\omega t}.$$

As the derivative of $\mathbf{A} e^{j\omega t}$ is simply $j\omega \mathbf{A} e^{j\omega t}$ this becomes

$$(j\omega)^2 \mathbf{A} e^{j\omega t} + j\omega \mathbf{A} e^{j\omega t} + \mathbf{A} e^{j\omega t} = U_0 e^{j\omega t}.$$

The simplicity of this computation is the reason for the phasor approach! Combining terms we have

$$((j\omega)^2 + j\omega + 1)\mathbf{A} e^{j\omega t} = U_0 e^{j\omega t}$$

so that

$$\mathbf{A} = \underbrace{\frac{1}{(j\omega)^2 + j\omega + 1}}_{G(j\omega)} U_0.$$

That is, the solution to the differential equation (3.19) with the complex function input $\mathbf{u}(t) = U_0 e^{j\omega t}$ is simply

$$\mathbf{x}(t) = G(j\omega) U_0 e^{j\omega t}.$$

---

[2] The terminology "phasor" can refer to either $\mathbf{A} e^{j\omega t}$ or just to $\mathbf{A}$.

Consequently, the solution to (3.16) with input $u(t) = U_0 \cos(\omega t)$ is

$$x(t) = \text{Re}\{G(j\omega)U_0 e^{j\omega t}\} = \text{Re}\{|G(j\omega)| e^{j\angle G(j\omega)} U_0 e^{j\omega t}\}$$
$$= |G(j\omega)|U_0 \cos(\omega t + \angle G(j\omega)). \qquad (3.20)$$

In more detail write

$$\mathbf{x}(t) = \underbrace{G(j\omega)U_0 e^{j\omega t}}_{\mathbf{A}} = |G(j\omega)| e^{j\angle G(j\omega)} U_0 e^{j\omega t} = \underbrace{|G(j\omega)|U_0 \cos(\omega t + \angle G(j\omega))}_{x_R(t)}$$
$$+ j\underbrace{|G(j\omega)|U_0 \sin(\omega t + \angle G(j\omega))}_{x_I(t)}$$

$$\mathbf{u}(t) = U_0 e^{j\omega t} = \underbrace{U_0 \cos(\omega t)}_{u_R(t)} + j\underbrace{U_0 \sin(\omega t)}_{u_I(t)}.$$

We just showed that $\mathbf{x}(t), \mathbf{u}(t)$ satisfy

$$\underbrace{\frac{d^2}{dt^2}\left(x_R(t) + jx_I(t)\right) + \frac{d}{dt}\left(x_R(t) + jx_I(t)\right) + x_R(t) + jx_I(t)}_{G(j\omega)U_0 e^{j\omega t}} = \underbrace{u_R(t) + ju_I(t)}_{U_0 e^{j\omega t}}.$$

Equating real and imaginary parts of both sides we have

$$\frac{d^2}{dt^2}x_R(t) + \frac{d}{dt}x_R(t) + x_R(t) = u_R(t)$$
$$\frac{d^2}{dt^2}x_I(t) + \frac{d}{dt}x_I(t) + x_I(t) = u_I(t).$$

As $u_R(t) = U_0 \cos(\omega t)$ it follows that $x_R(t) = \text{Re}\{\mathbf{x}(t)\} = \text{Re}\{G(j\omega)U_0 e^{j\omega t}\}$ is a solution of (3.16).

Now we want to look at solutions to (3.16), but starting at $t = 0$. For $t > 0$ and input $u(t) = U_0 \cos(\omega t)u_s(t)$ it follows that

$$x(t) = |G(j\omega)|U_0 \cos(\omega t + \angle G(j\omega))u_s(t) \qquad (3.21)$$

satisfies the differential equation (3.16) with the initial conditions simply given by

$$x(0) = \quad |G(j\omega)|U_0 \cos(\angle G(j\omega))$$
$$\dot{x}(0) = -\omega|G(j\omega)|U_0 \sin(\angle G(j\omega)).$$

That is, $x(t)$ given by (3.21) is the *unique* solution to the differential equation (3.16) with these initial conditions at $t = 0$ and input $U_0 \cos(\omega t)u_s(t)$. On the other hand, $x(t)$ given by (3.9) is the unique solution to (3.16) with *zero* initial conditions.[3]

Example 3 *Phasor Method of Solution*

Let's reconsider the differential equation given by (3.11) with input (3.12) using the complex phasor method of solution. That is, consider

$$\frac{d^2 x}{dt^2} - \frac{dx}{dt} + x = u(t) \qquad (3.22)$$

---

[3]Equations (3.16) and (3.4) are the *same* differential equation.

with input
$$u(t) = U_0 \cos(\omega t) u_s(t).$$

In order to solve this equation we first take the input to be the *complex phasor* function given by
$$\mathbf{u}(t) = U_0 e^{j\omega t} = U_0 \cos(\omega t) + jU_0 \sin(\omega t) \quad \text{for} \quad -\infty < t < \infty.$$

With $\mathbf{u}(t) = U_0 e^{j\omega t}$ we look for a solution $\mathbf{x}(t)$ of the differential equation
$$\frac{d^2\mathbf{x}}{dt^2} - \frac{d\mathbf{x}}{dt} + \mathbf{x} = \mathbf{u}(t) \tag{3.23}$$

in the phasor form
$$\mathbf{x}(t) = \mathbf{A} e^{j\omega t}. \tag{3.24}$$

Substituting (3.24) into (3.23) gives
$$\frac{d^2}{dt^2} \mathbf{A} e^{j\omega t} - \frac{d}{dt} \mathbf{A} e^{j\omega t} + \mathbf{A} e^{j\omega t} = U_0 e^{j\omega t}.$$

Computing the derivatives gives
$$(j\omega)^2 \mathbf{A} e^{j\omega t} - j\omega \mathbf{A} e^{j\omega t} + \mathbf{A} e^{j\omega t} = U_0 e^{j\omega t}$$

or
$$\left((j\omega)^2 - j\omega + 1\right) \mathbf{A} e^{j\omega t} = U_0 e^{j\omega t}.$$

The complex phasor $\mathbf{A}$ is given by
$$\mathbf{A} = \underbrace{\frac{1}{(j\omega)^2 - j\omega + 1}}_{G(j\omega)} U_0.$$

Then the solution is simply
$$\mathbf{x}(t) = G(j\omega) U_0 e^{j\omega t}$$

and the solution to (3.22) with input $u(t) = U_0 \cos(\omega t)$ is
$$x(t) = \text{Re}\{G(j\omega) U_0 e^{j\omega t}\} = \text{Re}\left\{|G(j\omega)| e^{j \angle G(j\omega)} U_0 e^{j\omega t}\right\} = |G(j\omega)| U_0 \cos(\omega t + \angle G(j\omega)).$$

Finally, starting at $t = 0$ with input $u(t) = U_0 \cos(\omega t) u_s(t)$, the solution of the differential equation
$$\frac{d^2 x}{dt^2} - \frac{dx}{dt} + x = U_0 \cos(\omega t) u_s(t)$$

is
$$x(t) = |G(j\omega)| U_0 \cos(\omega t + \angle G(j\omega)) u_s(t)$$

with initial conditions
$$x(0) = |G(j\omega)| U_0 \cos(\angle G(j\omega))$$
$$\dot{x}(0) = -\omega |G(j\omega)| U_0 \sin(\angle G(j\omega)).$$

In contrast, (3.15) is the solution to the same differential equation, but with *zero* initial conditions. In (3.15) the part of the response due to the poles of $G(s)$ (i.e. $1/2 \pm j\sqrt{3}/2$)

does *not* die out so the solution obtained from the phasor method is *not* a sinusoidal steady-state solution.

Summary

Consider a general differential equation

$$\frac{d^3x}{dt^3} + a_2\frac{d^2x}{dt^2} + a_1\frac{dx}{dt} + a_0 x = b_2\frac{d^2u}{dt^2} + b_1\frac{du}{dt} + b_0 u$$

with transfer function

$$G(s) = \frac{X(s)}{U(s)} = \frac{b_2 s^2 + b_1 s + b_0}{s^3 + a_2 s^2 + a_1 s + a_0}.$$

With $u(t) = U_0 \cos(\omega t)$, the phasor solution $x_{ph}(t)$ is given by

$$x_{ph}(t) \triangleq |G(j\omega)| U_0 \cos(\omega t + \angle G(j\omega)) \quad \text{for } -\infty < t < \infty.$$

However, starting at time $t = 0$ with $u(t) = U_0 \cos(\omega t) u_s(t)$, it follows that

$$x(t) = |G(j\omega)| U_0 \cos(\omega t + \angle G(j\omega)) u_s(t)$$

is a solution to the above differential equation with the special initial conditions

$$\begin{aligned} x(0) &= |G(j\omega)|U_0 \cos(\angle G(j\omega)) \\ \dot{x}(0) &= -\omega |G(j\omega)|U_0 \sin(\angle G(j\omega)) \\ \ddot{x}(0) &= -\omega^2 |G(j\omega)|U_0 \cos(\angle G(j\omega)). \end{aligned}$$

Further, if the poles of $G(s)$ are in the open left half-plane, then this is *also* the sinusoidal steady-state solution meaning that for any solution $x(t)$ with arbitrary initial conditions we have

$$x(t) \to x_{ss}(t) = |G(j\omega)| U_0 \cos(\omega t + \angle G(j\omega)) u_s(t).$$

Remark If $u(t) = U_0 \cos(\omega t + \theta_0)$ then with $\mathbf{U}_0 \triangleq U_0 e^{j\theta_0}$ we set

$$\mathbf{u}(t) = \mathbf{U}_0 e^{j\omega t} = U_0 e^{j\omega t + \theta_0} = U_0 \cos(\omega t + \theta_0) + jU_0 \sin(\omega t + \theta_0).$$

Similar to the above (see Problem 10) it then follows that

$$\mathbf{x}(t) = \mathbf{A}e^{j\omega t} = G(j\omega)U_0 e^{j\theta_0} e^{j\omega t} = G(j\omega)\mathbf{U}_0 e^{j\omega t}$$

and

$$\begin{aligned} x(t) &= \text{Re}\{G(j\omega)\mathbf{U}_0 e^{j\omega t}\} = \text{Re}\{G(j\omega)U_0 e^{j\theta_0} e^{j\omega t}\} \\ &= \text{Re}\{|G(j\omega)| e^{j\angle G(j\omega)} U_0 e^{j\theta_0} e^{j\omega t}\} \\ &= |G(j\omega)|U_0 \cos(\omega t + \theta_0 + \angle G(j\omega)). \end{aligned}$$

Definition 1 *Frequency Response*

With $G(s)$ the transfer function of a differential equation,

$$G(j\omega) \triangleq G(s)|_{s=j\omega} \quad \text{for } -\infty < \omega < \infty$$

is the *frequency response* function.

3.3 Final Value Theorem 57

For any sinusoidal input $u(t) = U_0 \cos(\omega t + \theta_0)$ the frequency response function $G(j\omega)$ can be used to obtain the solution $x(t) = |G(j\omega)|U_0 \cos(\omega t + \theta_0 + \angle G(j\omega))$ to the corresponding differential equation. The frequency response plays a large role in the analysis of control systems as explained in Chapter 11.

3.3 Final Value Theorem

We now present the final value theorem (FVT) as it will turn out to be a key tool used throughout this text. We do this presentation using a series of examples.

Example 4 $F(s) = \dfrac{10}{s(s+1)}$

With $F(s) = \dfrac{10}{s(s+1)}$ we want to look at the behavior of $f(t)$ as $t \to \infty$. Doing a partial fraction expansion of $F(s)$ gives

$$F(s) = \frac{10}{s(s+1)} = \frac{A}{s} + \frac{B}{s+1}.$$

We know immediately that

$$f(t) = Au_s(t) + Be^{-t}u_s(t)$$

and therefore

$$\lim_{t \to \infty} f(t) = A.$$

By the method of partial fractions we know that we can evaluate A as

$$A = \lim_{s \to 0} sF(s) = \lim_{s \to 0} s\frac{10}{s(s+1)} = 10$$

and thus

$$\lim_{t \to \infty} f(t) = \lim_{s \to 0} sF(s) = 10.$$

Example 5 $F(s) = \dfrac{-10}{s(s-1)}$

With $F(s) = \dfrac{-10}{s(s-1)}$ we want to look at the behavior of $f(t)$ as $t \to \infty$. Doing a partial fraction expansion of $F(s)$ gives

$$F(s) = \frac{-10}{s(s-1)} = \frac{A}{s} + \frac{B}{s-1}.$$

We know immediately that

$$f(t) = Au_s(t) + Be^{t}u_s(t)$$

which shows that $\lim_{t \to \infty} f(t)$ does *not* exist. Again, by the method of partial fractions, we evaluate A as

$$A = \lim_{s \to 0} sF(s) = \lim_{s \to 0} s\frac{-10}{s(s-1)} = 10.$$

58 3 Differential Equations and Stability

In this example $\lim_{s \to 0} sF(s) = A$ is simply the coefficient of the unit step function in the computation of the inverse Laplace transform. However, the time behavior of $f(t)$ as $t \to \infty$ is dominated by the growing exponential term e^t.

Example 6 $F(s) = \dfrac{s+1}{s(s+2)(s+3)}$

We want to look at the behavior of $f(t)$ as $t \to \infty$. A partial fraction expansion of $F(s)$ gives

$$F(s) = \frac{s+1}{s(s+2)(s+3)} = \frac{A}{s} + \frac{B}{s+2} + \frac{C}{s+3}.$$

We know immediately that

$$f(t) = A u_s(t) + B e^{-2t} u_s(t) + C e^{-3t} u_s(t)$$

and therefore

$$\lim_{t \to \infty} f(t) = A.$$

Again, by the method of partial fractions, we evaluate A as

$$A = \lim_{s \to 0} sF(s) = \lim_{s \to 0} s \frac{s+1}{s(s+2)(s+3)} = \frac{1}{6}.$$

Thus

$$\lim_{t \to \infty} f(t) = \lim_{s \to 0} sF(s) = 1/6.$$

The point here is that $\lim_{s \to 0} sF(s)$ is simply the coefficient of the $1/s$ term in the partial fraction expansion of $F(s)$. In this example it is also the final value, i.e., $\lim_{t \to \infty} f(t)$ because the part of the response due to the poles at -2 and -3 die out.

Example 7 $F(s) = -\dfrac{s+1}{s(s+2)(s-3)}$

We want to look at the behavior of $f(t)$ as $t \to \infty$. A partial fraction expansion of $F(s)$ gives

$$F(s) = -\frac{s+1}{s(s+2)(s-3)} = \frac{A}{s} + \frac{B}{s+2} + \frac{C}{s-3}.$$

We know immediately that

$$f(t) = A u_s(t) + B e^{-2t} + C e^{3t}$$

which shows that $\lim_{t \to \infty} f(t)$ does *not* exist. Again, by the method of partial fractions, we know that we can evaluate A as

$$A = \lim_{s \to 0} sF(s) = \lim_{s \to 0} s \frac{-(s+1)}{s(s+2)(s-3)} = \frac{1}{6}.$$

The point here is that $\lim_{s \to 0} sF(s)$ is simply the coefficient of the $1/s$ term in the partial fraction expansion of $F(s)$ (equivalently, the coefficient of the unit step function in the time response). However, the time behavior of $f(t)$ as $t \to \infty$ is dominated by the growing exponential term e^{3t} and so $f(t)$ does not have a final value.

Example 8 $F(s) = \dfrac{2s+12}{s(s^2-2s+5)}$

We look at the behavior of $f(t)$ as $t \to \infty$. By the partial fraction expansion method, we have

$$F(s) = \frac{2s+12}{s\left[s-(1+2j)\right]\left[s-(1-2j)\right]} = \frac{A}{s} + \frac{\beta}{s-(1+2j)} + \frac{\beta^*}{s-(1-2j)}.$$

Immediately we have

$$f(t) = Au_s(t) + \beta e^t e^{2jt} + \beta^* e^t e^{-2jt} = Au_s(t) + 2|\beta|e^t \cos(2t + \angle\beta)$$

showing that $\lim_{t\to\infty} f(t)$ does not exist. However, we still have that

$$A = \lim_{s\to 0} sF(s) = \lim_{s\to 0} s\frac{2s+12}{s(s^2-2s+5)} = \frac{12}{5}.$$

Again, the point here is that $\lim_{s\to 0} sF(s)$ is simply the coefficient of the $1/s$ term in the partial fraction expansion of $F(s)$ However, the time behavior of $f(t)$ as $t \to \infty$ is dominated by the term $e^t \cos(2t + \angle\beta)$, which does not die out.

Example 9 $F(s) = \dfrac{s+1}{(s+2)(s+3)}$

In this example $F(s)$ does not have a pole at $s = 0$. The partial fraction expansion has the form

$$F(s) = \frac{s+1}{(s+2)(s+3)} = \frac{A}{s+2} + \frac{B}{s+3}$$

and immediately we have

$$f(t) = Ae^{-2t}u_s(t) + Be^{-3t}u_s(t).$$

Thus $\lim_{t\to\infty} f(t) = 0$. Also, $\lim_{s\to 0} sF(s) = 0$ as there is no pole at $s = 0$ in the partial fraction expansion.

We can finally state the FVT that will be used throughout the remainder of this book.

Theorem 1 *Final Value Theorem (FVT)*
Let
$$F(s) = \frac{b(s)}{a(s)}$$

be a rational function[4] and strictly proper, i.e.,

$$\deg\{b(s)\} < \deg\{a(s)\}.$$

Let $f(t)$ denote the inverse Laplace transform of $F(s)$.
Then $\lim_{t\to\infty} f(t)$ exists and is given by

$$\lim_{t\to\infty} f(t) = \lim_{s\to 0} sF(s)$$

if and only if all the poles of $sF(s)$ are in the open left half-plane.

---
[4]Rational function simply means it is the ratio of two polynomials.

3 Differential Equations and Stability

Proof. (sketch) Let

$$F(s) = \frac{(s-z_1)(s-z_2)}{s(s-p_1)(s-p_2)} = \frac{A}{s} + \frac{\beta_1}{s-p_1} + \frac{\beta_2}{s-p_2}.$$

Then

$$f(t) = Au_s(t) + \beta_1 e^{p_1 t} + \beta_2 e^{p_2 t}.$$

The poles of $sF(s)$ are p_1 and p_2, and $A = \lim_{s \to 0} sF(s)$. So $e^{p_1 t} \to 0, e^{p_2 t} \to 0$ if and only if the real parts of p_1 and p_2 are *negative* or, equivalently, if and only if p_1, p_2 are in the *open left half-plane*.

Alternatively consider

$$F(s) = \frac{s-z_2}{(s-p_1)(s-p_2)} = \frac{\beta_1}{s-p_1} + \frac{\beta_2}{s-p_2}.$$

Then

$$f(t) = (\beta_1 e^{p_1 t} + \beta_2 e^{p_2 t})u_s(t).$$

The poles of $sF(s)$ are p_1 and p_2, and $\lim_{s \to 0} sF(s) = 0$. So $\lim_{t \to \infty} f(t) = 0 = \lim_{s \to 0} sF(s)$ if and only if $e^{p_1 t} \to 0$ and $e^{p_2 t} \to 0$, which is true if and only if p_1 and p_2 are in the *open left half-plane*. ∎

Remark Another way to state the FVT is to say that $\lim_{t \to \infty} f(t)$ exists if and only if $F(s)$ has *at most* one pole at $s = 0$ and the rest of its poles are in the open left half-plane.

Example 10 $F(s) = \dfrac{10}{s(s+1)}$

In this example

$$sF(s) = \frac{10}{s+1}$$

and the pole of $sF(s)$ is in the open left half-plane. Therefore $\lim_{t \to \infty} f(t) = \lim_{s \to 0} sF(s)$. This is seen directly by seeing that the partial-fraction expansion of $F(s)$ has the form

$$F(s) = \frac{A}{s} + \frac{B}{s+1}$$

with corresponding time function

$$f(t) = Au_s(t) + Be^{-t}u_s(t).$$

Then $\lim_{s \to 0} sF(s) = A$ is the coefficient of the unit step and also is the final value because the second term dies out (its pole is at -1 that is in the open left half-plane).

Example 11 $F(s) = \dfrac{-10}{s(s-1)}$

In this example

$$sF(s) = -\frac{10}{s-1}$$

and the pole of $sF(s)$ is *not* in the open left half-plane. As a result the $\lim_{t\to\infty} f(t)$ does *not* exist. To see this directly note that partial-fraction expansion of $F(s)$ has the form

$$F(s) = \frac{A}{s} + \frac{B}{s-1}$$

with corresponding time function

$$f(t) = Au_s(t) + Be^t u_s(t).$$

Here $\lim_{s\to 0} sF(s) = A$ is simply the coefficient of the unit step. It is not the final value because the second term does not die out (its pole is at 1 which is in the right half-plane).

Example 12 $F(s) = \dfrac{2s+12}{s^2+2s+5}$

In this example

$$sF(s) = s\frac{2s+12}{\big[s-(-1+2j)\big]\big[s-(-1-2j)\big]}$$

showing the poles of $sF(s)$ are in the open left half-plane. By the FVT $\lim_{t\to\infty} f(t) = \lim_{s\to 0} sF(s) = 0$. We can show directly by noting that $F(s)$ has the partial-fraction expansion

$$F(s) = \frac{\beta}{s-(-1+2j)} + \frac{\beta^*}{s-(-1-2j)}$$

with corresponding time function

$$f(t) = (\beta e^{-t}e^{2jt} + \beta^* e^{-t}e^{-2jt})u_s(t) = 2|\beta|e^{-t}\cos(2t + \angle\beta)u_s(t).$$

The two terms due to the complex conjugate pair of poles die out as these poles are in the open left half-plane. The limit $\lim_{s\to 0} sF(s) = 0$ as there is not a $1/s$ term in the partial fraction expansion.

Example 13 $F(s) = \dfrac{2s+12}{s^2-2s+5}$

In this example

$$sF(s) = s\frac{2s+12}{\big[s-(1+2j)\big]\big[s-(1-2j)\big]}$$

and the poles of $sF(s)$ are *not* in the open left half-plane. Therefore $\lim_{t\to\infty} f(t)$ does *not* exist. This is shown directly by seeing that the partial-fraction expansion of $F(s)$ has the form

$$F(s) = \frac{\beta}{s-(1+2j)} + \frac{\beta^*}{s-(1-2j)}$$

with corresponding time function

$$f(t) = (\beta e^t e^{2jt} + \beta^* e^t e^{-2jt})u_s(t) = 2|\beta|e^t\cos(2t + \angle\beta)u_s(t).$$

The limit $\lim_{s\to 0} sF(s) = 0$ as there is not a $1/s$ term in the partial fraction expansion. However, the two terms due to the complex conjugate pair of poles do *not* die out as these poles are in the right half-plane.

Example 14 $F(s) = \dfrac{s+1}{s^2}$

In this example

$$sF(s) = s\dfrac{s+1}{s^2} = \dfrac{s+1}{s}$$

and $sF(s)$ has a pole at $s = 0$, which is *not* in the open left half-plane. Therefore $\lim_{t \to \infty} f(t)$ does *not* exist. We can also see this directly from the partial expansion of $F(s)$, which has the form

$$F(s) = \dfrac{s+1}{s^2} = \dfrac{A}{s} + \dfrac{B}{s^2}$$

with corresponding time function

$$f(t) = Au_s(t) + Btu_s(t).$$

The $\lim_{t \to \infty} tu_s(t) = \infty$ showing $f(t)$ does not have a final value. In this example $\lim_{s \to 0} sF(s) = \infty$ is *not* even the coefficient of the $1/s$ term in the partial fraction expansion.

3.4 Stable Transfer Functions

Consider the general third-order differential equation given by

$$\dddot{y} + a_2\ddot{y} + a_1\dot{y} + a_0 y = b_2\ddot{u} + b_1\dot{u} + b_0 u.$$

With zero initial conditions, i.e., $y(0) = \dot{y}(0) = \ddot{y}(0) = 0, u(0) = \dot{u}(0) = 0$, we have

$$(s^3 + a_2 s^2 + a_1 s + a_0)Y(s) = (b_2 s^2 + b_1 s + b_0)U(s)$$

or

$$G(s) \triangleq \left.\dfrac{Y(s)}{U(s)}\right|_{\text{zero initial conditions}} = \dfrac{b_2 s^2 + b_1 s + b_0}{s^3 + a_2 s^2 + a_1 s + a_0}.$$

This is a typical example of a transfer function considered in this book. That is, for the physical models we consider the transfer function will be *rational* and *strictly proper*. In other words,

$$G(s) = b(s)/a(s),$$

where $a(s)$ and $b(s)$ are polynomials in s (making $G(s)$ rational) and $\deg\{b(s)\} < \deg\{a(s)\}$ (making $G(s)$ strictly proper).

Definition 2 *Stable Transfer Functions*
We say that a strictly proper rational transfer function $G(s) = b(s)/a(s)$ is *stable* if the poles of $G(s)$, i.e., the roots of $a(s) = 0$, are in the open left half-plane.

Example 15 Let

$$G_1(s) = \dfrac{1}{s^2 - s + 1}.$$

$G_1(s)$ is not stable as its poles are $1/2 \pm j\sqrt{3}/2$, which are *not* in the open left half-plane.

Let
$$G_2(s) = \frac{1}{s(s+2)}.$$

$G_2(s)$ is not stable as its poles are $0, -2$ and the pole at 0 is *not* in the open left half-plane.
Let
$$G_3(s) = \frac{1}{s^2+s+1}.$$

$G_3(s)$ is stable as its poles are $-1/2 \pm j\sqrt{3}/2$, which are in the open left half-plane.

Definition 3 *Stable Polynomial*

We say that a polynomial $a(s)$ is *stable* if the roots of $a(s) = 0$ are in the open left half-plane.

Example 16 Let
$$a_1(s) = s^2 - s + 1.$$

$a_1(s)$ is *not* stable as its roots are $1/2 \pm j\sqrt{3}/2$, which are *not* in the open left half-plane.
Let
$$a_2(s) = s(s+2).$$

$a_2(s)$ is *not* stable as its roots are $0, -2$ and the root at 0 is *not* in the open left half-plane.
Let
$$a_3(s) = s^2 + s + 1.$$

$a_3(s)$ is stable as its roots are $-1/2 \pm j\sqrt{3}/2$, which are in the open left half-plane.

Final Value Theorem and Stable Transfer Functions

Consider the differential equation
$$\ddot{x}(t) + 2\dot{x}(t) = u(t)$$

with transfer function
$$G(s) = \frac{X(s)}{U(s)} = \frac{1}{s(s+2)}.$$

This transfer function is not stable as it has a pole at $s = 0$. Suppose $U(s) = 1/s$ so that $X(s)$ is given by
$$X(s) = G(s)U(s) = \frac{1}{s(s+2)}\frac{1}{s} = \frac{1}{s^2(s+2)}.$$

Then $sX(s) = \frac{1}{s(s+2)}$ has a pole at $s = 0$ and the FVT tells us the $\lim_{t \to \infty} x(t)$ does *not* exist.

Now consider the stable transfer function
$$G(s) = \frac{1}{s+2}$$

which corresponds to the differential equation
$$\dot{x}(t) + 2x(t) = u(t).$$

64   3 Differential Equations and Stability

Then with a step input $U(s) = 1/s$ we have

$$X(s) = \frac{1}{s+2}\frac{1}{s}$$

and

$$x(\infty) = \lim_{t \to \infty} x(t) = \lim_{s \to} sX(s) = \lim_{s \to} s\frac{1}{s+2}\frac{1}{s} = \frac{1}{2}.$$

The usual situation is to apply an input to a differential equation and use the FVT on its output (solution).

Summary

Let a system have the differential equation model

$$\ddot{y} + a_1\dot{y} + a_0 y = b_1\dot{u} + b_0 u.$$

Taking the Laplace transform of both sides gives[5]

$$s^2 Y(s) - sy(0) - \dot{y}(0) + a_1(sY(s) - y(0)) + a_0 Y(s) = b_1(sU(s) - u(0)) + b_0 U(s).$$

Rearranging we have

$$Y(s) = \underbrace{\frac{b_1 s + b_0}{s^2 + a_1 s + a_0}}_{G(s)}\underbrace{U(s)}_{\text{Input}} + \underbrace{\frac{sy(0) + \dot{y}(0) + a_1 y(0) - b_1 u(0)}{s^2 + a_1 s + a_0}}_{\text{Same denominator as } G(s)!}.$$

- Let $G(s)$ be stable so the roots of

$$s^2 + a_1 s + a_0 = (s - p_1)(s - p_2) = 0$$

are in the *open left half-plane*. By partial fraction expansion theory we have, for *any* initial conditions, that

$$\mathcal{L}^{-1}\left\{\frac{sy(0) + \dot{y}(0) + a_1 y(0) - b_1 u(0)}{s^2 + a_1 s + a_0}\right\} = \mathcal{L}^{-1}\left\{\frac{sy(0) + \dot{y}(0) + a_1 y(0) - b_1 u(0)}{(s - p_1)(s - p_2)}\right\}$$

$$= Ae^{p_1 t}u_s(t) + Be^{p_2 t}u_s(t) \to 0.$$

- Apply a step input $u_s(t)$.

$$Y(s) = G(s)\frac{1}{s} + \frac{sy(0) + \dot{y}(0) + a_1 y(0) - b_1 u(0)}{s^2 + a_1 s + a_0}$$

$$= \frac{b_1 s + b_0}{s^2 + a_1 s + a_0}\frac{1}{s} + \frac{sy(0) + \dot{y}(0) + a_1 y(0) - b_1 u(0)}{s^2 + a_1 s + a_0}.$$

Then

$$sY(s) = s\frac{b_1 s + b_0}{(s - p_1)(s - p_2)}\frac{1}{s} + s\left(\frac{sy(0) + \dot{y}(0) + a_1 y(0) - b_1 u(0)}{(s - p_1)(s - p_2)}\right)$$

---

[5] $\mathcal{L}\{\ddot{y}\} = s\mathcal{L}\{\dot{y}\} - \dot{y}(0) = s(sY(s) - y(0)) - \dot{y}(0) = s^2 Y(s) - sy(0) - \dot{y}(0).$

is *stable*. By the FVT

$$\lim_{t\to\infty} y(t) = \lim_{s\to 0} sY(s) = \lim_{s\to 0} sG(s)\frac{1}{s} = G(0) = \frac{b_0}{a_0}. \quad (3.25)$$

If $G(s)$ is *not* stable then (3.25) is not valid!

- Apply a sinusoidal input $u(t) = U_0 \cos(\omega t) u_s(t)$.

$$Y(s) = \underbrace{\frac{b_1 s + b_0}{s^2 + a_1 s + a_0}}_{G(s)} \underbrace{U_0 \frac{s}{s^2 + \omega^2}}_{\text{Input}} + \underbrace{\frac{sy(0) + \dot{y}(0) + a_1 y(0) - b_1 u(0)}{s^2 + a_1 s + a_0}}_{\frac{A}{s-p_1} + \frac{B}{s-p_2}}.$$

A partial fraction expansion of the first term (with the input) is

$$\frac{b_1 s + b_0}{s^2 + a_1 s + a_0} U_0 \frac{s}{s^2 + \omega^2} = \frac{b_1 s + b_0}{(s-p_1)(s-p_2)} U_0 \frac{s}{s^2 + \omega^2}$$

$$= \frac{C}{s-p_1} + \frac{D}{s-p_2} + \frac{k}{s-j\omega} + \frac{k^*}{s+j\omega}.$$

The inverse Laplace transform has the form

$$\mathcal{L}^{-1}\left\{\frac{b_1 s + b_0}{s^2 + a_1 s + a_0} U_0 \frac{s}{s^2 + \omega^2}\right\} = Ce^{p_1 t} u_s(t) + De^{p_2 t} u_s(t)$$

$$+ \underbrace{U_0 |G(j\omega)| \cos(\omega t + \angle G(j\omega))}_{\text{Phasor solution}}$$

$$\to U_0 |G(j\omega)| \cos(\omega t + \angle G(j\omega)) u_s(t).$$

That is, as $G(s)$ is stable, we have

$$y(t) \to y_{ph}(t) \triangleq \text{Re}\{G(j\omega) U_0 e^{j\omega t}\} = |G(j\omega)| U_0 \cos(\omega t + \angle G(j\omega)). \quad (3.26)$$

If $G(s)$ is *not* stable then (3.26) is not valid!

3.5 Routh–Hurwitz Stability Test

We now present the Routh–Hurwitz test for stability. Given any polynomial this test allows us to straightforwardly check whether or not its roots are in the open left half-plane.

Recall that a strictly proper transfer function

$$G(s) = \frac{b(s)}{a(s)}, \quad \deg\{b(s)\} < \deg\{a(s)\} \quad (3.27)$$

is *stable* if and only if the roots of

$$a(s) = s^n + a_{n-1} s^{n-1} + \cdots + a_1 s + a_0 = 0 \quad (3.28)$$

are in the open left half-plane. With reference to Figure 3.1 on the next page this is the same as saying $G(s)$ is stable if and only if

$$a(s) \neq 0 \text{ for } \text{Re}\{s\} \geq 0. \quad (3.29)$$

3 Differential Equations and Stability

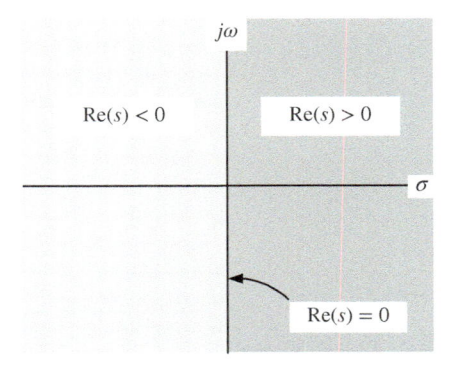

Figure 3.1. $a(s)$ is stable if and only if all of its roots are in $\text{Re}(s) < 0$, that is, in the open left half-plane.

For an arbitrary polynomial of the form

$$a(s) = s^n + a_{n-1}s^{n-1} + \cdots + a_1 s + a_0$$

it is not obvious where its roots are located.

Let's first consider the second-order polynomial

$$a(s) = s^2 + a_1 s + a_0.$$

Suppose $a(s)$ is stable and its roots are

$$p_i = -1 \pm j2.$$

Then

$$a(s) = \bigl(s - (-1 + j2)\bigr)\bigl(s - (-1 - j2)\bigr) = s^2 + 2s + 5.$$

Note that both coefficients of $a(s)$ are positive.

On the other hand, suppose $a(s)$ is not stable and its roots are

$$p_i = 1 \pm j2.$$

Then

$$a(s) = \bigl(s - (1 + j2)\bigr)\bigl(s - (1 - j2)\bigr) = s^2 - 2s + 5.$$

Note that the coefficient a_1 is *negative*.

More generally, let $a(s)$ have the complex conjugate pair of roots

$$p_i = \sigma \pm j\omega.$$

so that

$$a(s) \triangleq \bigl(s - (\sigma + j\omega)\bigr)\bigl(s - (\sigma - j\omega)\bigr) = s^2 - 2\sigma s + \sigma^2 + \omega^2.$$

Then $a(s)$ is stable if and only if $\sigma < 0$, which in this case means both coefficients $a_1 = -2\sigma, a_2 = \sigma^2 + \omega^2$ are positive.

Now any polynomial can be factored down into its real roots and its complex conjugate pairs of roots. For example, suppose $a(s)$ has degree 3 with one real root p_1, and one pair of complex conjugate roots $\sigma_1 \pm j\omega_1$. Then we can write

$$a(s) = (s - p_1)(s^2 - 2\sigma_1 s + \sigma_1^2 + \omega_1^2). \tag{3.30}$$

3.5 Routh–Hurwitz Stability Test

If $a(s)$ is *stable*, then $p_1 < 0$, and $\sigma_1 < 0$. Clearly then $-p_1 > 0$ and $-\sigma_1 > 0$ resulting in all the coefficients in each factor of (3.30) being *positive*. Thus, after this is multiplied out, we have

$$a(s) = (s - p_1)\left(s^2 + 2(-\sigma_1)s + \sigma_1^2 + \omega_1^2\right)$$
$$= s^3 + \underbrace{(2(-\sigma_1) + (-p_1))}_{a_2} s^2 + \underbrace{(\sigma_1^2 + 2(-p_1)(-\sigma_1) + \omega_1^2)}_{a_1} s + \underbrace{(-p_1)(\sigma_1^2 + \omega_1^2)}_{a_0}. \quad (3.31)$$

The coefficients a_i must be positive. We have just shown the following theorem.

Theorem 2 *Necessary Condition for Stability*
Let
$$a(s) = s^n + a_{n-1}s^{n-1} + \cdots + a_1 s + a_0$$
be a polynomial of degree n. A *necessary* condition that $a(s) \neq 0$ for $\text{Re}\{s\} \geq 0$ (all its roots are in the open left half-plane) is that all of its coefficients be positive, that is, $a_i > 0$ for $i = 0, \ldots, n-1$.

Remark 1 *The Condition $a_i > 0$ is NOT Sufficient for Stability*
Consider
$$a(s) = s^3 + s^2 + s + 1$$
which has all of its coefficients positive. However, it factors as
$$a(s) = s^3 + s^2 + s + 1 = (s+1)(s^2 + 1)$$
showing that it has two roots at $\pm j$ and it is therefore not stable.
The polynomial
$$a(s) = s^3 + s^2 + s + 2 = 0$$
has roots $0.177 \pm j1.203$ and -1.353 showing that it is not stable.

Routh–Hurwitz Criterion for Stability

We now present a necessary and sufficient condition for stability known as the Routh–Hurwitz condition. We start with a general fourth order polynomial given by
$$a(s) = s^4 + a_3 s^3 + a_2 s^2 + a_1 s + a_0.$$

Next we form the Routh array defined as

s^4	1	a_2	a_0
s^3	a_3	a_1	0
s^2	$c_1 \triangleq \dfrac{a_3 \cdot a_2 - a_1 \cdot 1}{a_3}$	$c_2 \triangleq \dfrac{a_3 \cdot a_0 - 0 \cdot 1}{a_3}$	0
s	$d_1 \triangleq \dfrac{c_1 \cdot a_1 - c_2 \cdot a_3}{c_1}$	$d_2 \triangleq \dfrac{c_1 \cdot 0 - 0 \cdot a_3}{c_1}$	
s^0	$e_1 \triangleq \dfrac{d_1 \cdot c_2 - d_2 \cdot c_1}{d_1}$		

We can now state (but won't prove!) the important Routh–Hurwitz stability criterion.

68   3 Differential Equations and Stability

Theorem 3 *Routh–Hurwitz Stability Criterion*

Main result: Form the Routh array. Then $a(s)$ has all its roots in the open left half-plane (i.e. $a(s) \neq 0$ for $\text{Re}\{s\} \geq 0$) if and only if all the elements of the first column of the Routh array are *positive*.

Secondary result: If all the elements of the first column of the Routh array are *non zero*, the number of sign changes in the first column equals the number of roots of $a(s)$ in the right half-plane.

Proof. The proof is omitted. A nice explanation of this result is given in [11]. ∎

Remark In practice a control system must be stable to work. If this is not the case we are not typically interested in the number of poles it has in the right half-plane.

Example 17 *Routh–Hurwitz Stability Criterion*
Consider the polynomial
$$a(s) = 4s^5 + 2s^4 + 9s^3 + 4s^2 + 5s + 1.$$

To determine whether or not all of its roots are in the open left half-plane we first form the Routh array.

$$
\begin{array}{c|cccc}
s^5 & 4 & 9 & 5 \\
s^4 & 2 & 4 & 1 \\
s^3 & \dfrac{2 \cdot 9 - 4 \cdot 4}{2} = 1 & \dfrac{2 \cdot 5 - 1 \cdot 4}{2} = 3 & 0 \\
s^2 & \dfrac{1 \cdot 4 - 3 \cdot 2}{1} = -2 & \dfrac{1 \cdot 1 - 0 \cdot 2}{1} = 1 & 0 \\
s & \dfrac{-2 \cdot 3 - 1 \cdot 1}{-2} = 3.5 & \dfrac{-2 \cdot 0 - 0 \cdot 1}{-2} = 0 & \\
s^0 & \dfrac{3.5 \cdot 1 - 0 \cdot (-2)}{3.5} = 1 & &
\end{array}
$$

The first column has a negative element in the s^2 row equal to -2. Thus $a(s)$ is not stable. $a(s)$ has two roots in the right half-plane. This is because there are two sign changes in the first column. The first sign change is from 1 to -2 (going from the s^3 row to the s^2 row) and the second sign change is from -2 to 3.5 (going from the s^2 row to the s row).
The roots of
$$a(s) = 4s^5 + 2s^4 + 9s^3 + 4s^2 + 5s + 1 = 0$$
can be found numerically and are
$$0.072 \pm j1.22, \; -0.212 \pm j0.849, \; -0.22.$$

Example 18 *Routh–Hurwitz Stability Criterion* [12]
Let
$$a(s) = s^3 + \alpha s^2 + s + 1.$$

3.5 Routh–Hurwitz Stability Test

We use the Routh–Hurwitz criterion to determine the values of α for which the roots of $a(s)$ are in the open left half-plane. To do so, we first form the Routh array.

$$
\begin{array}{c|ccc}
s^3 & 1 & & 1 \\
s^2 & \alpha & & 1 \\
s & \dfrac{\alpha \cdot 1 - 1 \cdot 1}{\alpha} = \dfrac{\alpha - 1}{\alpha} & & \dfrac{\alpha \cdot 0 - 0 \cdot 1}{\alpha} = 0 \\
s^0 & \dfrac{\dfrac{\alpha-1}{\alpha} \cdot 1 - 0 \cdot \alpha}{\dfrac{\alpha-1}{\alpha}} = 1 & &
\end{array}
$$

The first column is positive if and only if

$$\alpha > 0$$

and

$$\frac{\alpha - 1}{\alpha} > 0.$$

This reduces to

$$\alpha > 0$$
$$\alpha - 1 > 0$$

or finally

$$\alpha > 1.$$

So for $\alpha > 1$, $a(s)$ is stable, that is, all three of its roots are in the open left half-plane.

For $0 < \alpha < 1$ there are two sign changes in the first column. That is, going from the s^2 row to the s row the sign changes from $+$ to $-$ and then going from the s row to s^0 row the sign changes from $-$ to $+$. Thus for $0 < \alpha < 1$ there are two roots of $a(s)$ in the right half-plane.

Example 19 *Routh–Hurwitz Stability Criterion* [12]

We use the Routh–Hurwitz test to check for which values of K the roots of

$$a(s) \triangleq s^3 + 5s^2 + 2s + K - 8$$

are all in the open left half-plane. First form the Routh array.

$$
\begin{array}{c|ccc}
s^3 & 1 & & 2 \\
s^2 & 5 & & K - 8 \\
s & \dfrac{5 \cdot 2 - (K-8) \cdot 1}{5} = \dfrac{18 - K}{5} & & \dfrac{5 \cdot 0 - 0 \cdot 1}{5} = 0 \\
s^0 & \dfrac{\dfrac{18-K}{5}(K-8) - 0 \cdot 5}{\dfrac{18-K}{5}} = K - 8 & &
\end{array}
$$

The elements of the first column are positive if and only if

$$\frac{18 - K}{5} > 0$$
$$K - 8 > 0$$

or

$$18 > K$$
$$K > 8$$

or

$$8 < K < 18.$$

For $K > 18$ there are two sign changes in the first column so that for these values of K there are two roots of $a(s)$ in the right half-plane.

Example 20 *Routh–Hurwitz Stability Criterion*
Let
$$a(s) = s^3 + 3Ks^2 + (K+2)s + 4.$$

We use the Routh–Hurwitz criterion to determine for which values of K the roots of $a(s)$ are all in the open left half-plane. To do so, we first form the Routh array.

$$\begin{array}{cllll}
s^3 & 1 & K+2 & 0 \\
s^2 & 3K & 4 & 0 \\
s & \dfrac{3K(K+2) - 4}{3K} = \dfrac{3(K+2.528)(K-0.528)}{3K} & \dfrac{3K \cdot 0 - 0 \cdot 1}{3K} = 0 & 0 \\
s^0 & \dfrac{\dfrac{3K(K+2)-4}{3K} \cdot 4 - 0 \cdot 3K}{\dfrac{3K(K+2)-4}{3K}} = 4 &
\end{array}$$

Looking at the s^2 row we must have $K > 0$ for $a(s)$ to be stable. The s row requires

$$\frac{3K(K+2) - 4}{3K} > 0$$

or, as $K > 0$ is already required for stability, this condition reduces to

$$3K(K+2) - 4 > 0.$$

Using the quadratic formula we solve

$$3K(K+2) - 4 = 3K^2 + 6K - 4 = 0$$

to obtain
$$K = \frac{-6 \pm \sqrt{36 - 4(3)(-4)}}{2(3)} = -1 \pm \frac{\sqrt{21}}{3} = -2.528, 0.528.$$

Then
$$3K^2 + 6K - 4 = 3(K + 2.528)(K - 0.528)$$
and
$$3(K + 2.528)(K - 0.528) > 0 \text{ for } K > 0.528 \text{ or } K < -2.528.$$

So the s^2 row requires $K > 0$ while the s row requires either $K > 0.528$ or $K < -2.528$. Thus for both rows to have positive elements in the first column we must have $K > 0.528$ for $a(s)$ to be stable.

For $0 < K < 0.528$, in the first column the s^2 row is positive while the s row is negative. Thus there are two sign changes showing that $a(s)$ has two roots in the right half-plane.

What happens at $K = 0.528$? We reason that $a(s)$ has two roots on the $j\omega$ axis as follows. For K a little less than 0.528, $a(s)$ has two roots in the right half-plane while for K just greater than 0.528 it has no roots in the right half-plane. So at $K = 0.528$ these two of the roots of $a(s)$ are on the $j\omega$ axis.

Example 21 *General Second-Order Polynomial*
Let
$$a(s) = s^2 + a_1 s + a_0.$$

We use the Routh–Hurwitz criterion to determine for which values of a_1 and a_0 that $a(s)$ is stable. The Routh array is

$$\begin{array}{ccc} s^2 & 1 & a_0 \\ s & a_1 & 0 \\ s^0 & \dfrac{a_1 \cdot a_0 - 0 \cdot 1}{a_1} = a_0 & 0 \end{array}$$

The first column is positive if and only if
$$a_1 > 0 \text{ and } a_0 > 0.$$

That is, a second-order polynomial is stable if and only if both of its coefficients are positive.

Example 22 *General Third-Order Polynomial*
Let
$$a(s) = s^3 + a_2 s^2 + a_1 s + a_0.$$

We use the Routh–Hurwitz criterion to determine the values of $a_0, a_1,$ and a_2 for which $a(s)$ is stable. The Routh array is

$$\begin{array}{ccc} s^3 & 1 & a_1 \\ s^2 & a_2 & a_0 \end{array}$$

$$\begin{array}{cc} s & \dfrac{a_2 \cdot a_1 - a_0 \cdot 1}{a_2} & 0 \end{array}$$

$$\begin{array}{cc} s^0 & \dfrac{\dfrac{a_2 \cdot a_1 - a_0 \cdot 1}{a_2} \cdot a_0 - 0 \cdot a_2}{\dfrac{a_2 \cdot a_1 - a_0 \cdot 1}{a_2}} = a_0 \end{array}$$

72   3 Differential Equations and Stability

The first column is positive if and only if

$$a_2 > 0, a_0 > 0, \text{ and } \frac{a_2 \cdot a_1 - a_0}{a_2} > 0.$$

Equivalently, this reduces to

$$a_2 > 0, a_0 > 0, \text{ and } a_2 a_1 - a_0 > 0.$$

Special Case – A Row of the Routh Array Has All Zeros*

Here we consider the special case where the Routh array has all zeros in a row. Of course this means it has a zero in the first column and therefore the polynomial $a(s)$ is not stable. In this case we will show that $a(s)$ must have roots on the $j\omega$ axis. As previously mentioned we are usually only interested if a system is stable or not so this result does not have a lot value. However, in Chapter 12 it is one of several tools for sketching the *root locus* of a closed-loop system.

Example 23 *Row of Zeros* [12]
   Recall Example 18 where we considered

$$a(s) = s^3 + \alpha s^2 + s + 1.$$

Its Routh array is

$$\begin{array}{c|cc} s^3 & 1 & 1 \\ s^2 & \alpha & 1 \\ s & \dfrac{\alpha - 1}{\alpha} & 0 \\ s^0 & 1 & \end{array}$$

and it was shown to be stable for $\alpha > 1$. Further it was also shown to have two right half-plane roots for $0 < \alpha < 1$.

What about $\alpha = 1$? Well for α just less than 1 there are two roots in the open right half-plane while for α just greater than 1 all the roots are in the open left half-plane. We guess that at $\alpha = 1$ two of the roots of $a(s)$ are on the $j\omega$ axis. This is indeed true, but the proof will not be given. We now show how to find the location of the two roots on $j\omega$ axis for $\alpha = 1$. With $\alpha = 1$ the Routh array becomes

$$\begin{array}{c|cc} s^3 & 1 & 1 \\ s^2 & 1 & 1 \\ s & 0 & 0 \\ s^0 & 1 & \end{array}$$

Note that now the s row only has 0s in it. We go to the row above it, which is the s^2 row. Recalling how the Routh array is formed (see page 68), the first element of the s^2 row is 1, which corresponds to s^2 while the second element of the s^2 row is also 1 and corresponds to s^0. We use these two coefficients to form the *auxiliary polynomial* defined by

$$1 \cdot s^2 + 1 \cdot s^0 = s^2 + 1.$$

The roots of this auxiliary polynomial are

$$s = \pm j$$

and they are *also* the locations of the two roots of $a(s)$ that are on the $j\omega$ axis for $\alpha = 1$. Specifically, with $\alpha = 1$ we have

$$a(s) = s^3 + s^2 + s + 1 = (s+1)(s^2+1)$$

explicitly showing that the roots of the auxiliary equation are also roots of $a(s)$.

Example 24 *Row of Zeros* [12]
Recall Example 19 where we considered the stability of

$$a(s) \triangleq s^3 + 5s^2 + 2s + K - 8.$$

Its Routh array is

$$\begin{array}{c|cc} s^3 & 1 & 2 \\ s^2 & 5 & K-8 \\ s & \dfrac{18-K}{5} & 0 \\ s^0 & K-8 & \end{array}$$

which shows $a(s)$ is stable for $8 < K < 18$. For $K > 18$ there are two roots of $a(s)$ in the right half-plane.

What about $K = 18$? Well for K just less than 18 all the roots of $a(s)$ are in the open left half-plane while for K greater than 18 it has two roots in the right half-plane. So for $K = 18$ we expect to have two roots on the $j\omega$ axis. To find the location of these two roots we set $K = 18$ in the Routh array to obtain

$$\begin{array}{c|cc} s^3 & 1 & 2 \\ s^2 & 5 & 10 \\ s & 0 & 0 \\ s^0 & 10 & \end{array}$$

Note that the s row only has 0s in it. We go to the row above it, which is the s^2 row. The first element in the s^2 row is 5 corresponding to s^2 and the second element in the s^2 row is 10 corresponding to s^0. We use these two elements to form the *auxiliary polynomial* defined by

$$5 \cdot s^2 + 10 \cdot s^0 = 5(s^2 + 2).$$

The roots of this auxiliary polynomial are

$$s = \pm j\sqrt{2}$$

and these are *also* the locations of the two roots of $a(s)$, which are on the $j\omega$ axis for $K = 18$. Explicitly, with $K = 18$ we have

$$a(s) = s^3 + 5s^2 + 2s + 10 = (s^2+2)(s+5)$$

showing that the roots of the auxiliary polynomial are also roots of $a(s)$.
What about $K = 8$? The Routh array for $K = 8$ is

$$\begin{array}{c|cc} s^3 & 1 & 2 \\ s^2 & 5 & 0 \\ s & 2 & 0 \\ s^0 & 0 & 0 \end{array}$$

74   3 Differential Equations and Stability

For $K < 8$ there is one root in the open right half-plane while for K just greater than 8 all the roots are in the open left half-plane. So we expect that for $K = 8$ there is one root on the $j\omega$ axis. This is easy to see as

$$a(s) \triangleq s^3 + 5s^2 + 2s + K - 8|_{K=8} = s^3 + 5s^2 + 2s$$

showing that $a(s)$ has a root at $s = 0$.

Example 25 *Row of Zeros*

Recall Example 20 where we considered

$$a(s) = s^3 + 3Ks^2 + (K+2)s + 4.$$

Its Routh array is

s^3	1	$K+2$
s^2	$3K$	4
s	$\dfrac{3(K+2.582)(K-0.528)}{3K}$	0
s^0	4	

Using this array it was shown that $a(s)$ is stable for $K > 0.528$. For $0 < K < 0.528$, it was shown that $a(s)$ has two roots in the right half-plane.

What happens at $K = 0.528$? We reasoned that $a(s)$ has two roots on the $j\omega$ axis. This is because for K a little less than 0.528 the polynomial $a(s)$ has two roots in the right half-plane while for K just greater than 0.528 it has no roots in the right half-plane. We expect that at $K = 0.528$ two of the roots of $a(s)$ are on the $j\omega$ axis. To find these roots we set $K = 0.528$ into the Routh array to obtain

s^3	1	2.528
s^2	3(0.528)	4
s	0	0
s^0	4	

Note that the s row only has 0s in it. We go to the row above it, which is the s^2 row and form the auxiliary polynomial defined by

$$3(0.528)s^2 + 4 = 3(0.528)(s^2 + 2.52).$$

The roots of this auxiliary polynomial are

$$s = \pm j1.6$$

and these are also the locations of the two roots of $a(s)$ that are on the $j\omega$ axis for $K = 0.528$. In fact for $K = 0.528$ we have

$$\begin{aligned}a(s) &= s^3 + 3Ks^2 + (K+2)s + 4|_{K=0.528} \\ &= s^3 + 3(0.528)s^2 + (2.528)s + 4 \\ &= (s + 1.58)(s^2 + 2.52)\end{aligned}$$

showing explicitly that the roots of the auxiliary polynomial are also roots of $a(s)$.

3.5 Routh–Hurwitz Stability Test

Example 26 *Row of Zeros*
Consider the polynomial
$$a(s) = s^3 + 2s^2 + s + 2.$$

The Routh array is then

$$\begin{array}{c|cc} s^3 & 1 & 1 \\ s^2 & 2 & 2 \\ s & 0 & 0 \\ s^0 & & \end{array}$$

The s row is a row of zeros. We form the auxiliary equation
$$2s^2 + 2 = 2(s^2 + 1)$$

which has roots
$$s = \pm j.$$

As the roots of the auxiliary equation are also roots of $a(s)$ it follows that $\pm j$ are the roots of $a(s)$ on the $j\omega$ axis. In fact
$$a(s) = s^3 + 2s^2 + s + 2 = (s+2)(s^2+1).$$

Special Case – Zero in First Column, but the Row Is Not Identically Zero*

Here we consider the situation where there is zero in the first column, but the row is not identically zero. Let's consider this case by doing some examples.

Example 27 *Zero in the First Column*
Consider the polynomial
$$a(s) = s^4 + s^3 + 5s^2 + 5s + 2.$$

We use the Routh–Hurwitz criterion to determine if it is stable or not. If it is not stable, we will also find out how many roots are in the right half-plane. The Routh array is

$$\begin{array}{c|ccc} s^4 & 1 & 5 & 2 \\ s^3 & 1 & 5 & 0 \\ s^2 & 0 & \dfrac{1\cdot 2 - 0\cdot 1}{1} = 2 & 0 \\ s & & & \\ s^0 & & & \end{array}$$

Due to the zero in the s^2 row we cannot complete the Routh array. However, we do know that $a(s)$ is not stable. We now want to find out how many roots are in the right half-plane. To proceed one replaces the 0 in the first column of the s^2 row with a small $\epsilon > 0$ and uses this to complete the array. We have

$$\begin{array}{c|ccc} s^4 & 1 & 5 & 2 \\ s^3 & 1 & 5 & 0 \\ s^2 & \epsilon > 0 & \dfrac{1\cdot 2 - 0\cdot 1}{1} = 2 & 0 \\ s & \dfrac{\epsilon\cdot 5 - 2\cdot 1}{\epsilon} = 5 - \dfrac{2}{\epsilon} & 0 & \\ s^0 & \dfrac{(5 - 2/\epsilon)2 - 0.6}{5 - 2/\epsilon} = 2 & & \end{array}$$

With ϵ small we have $5 - 2/\epsilon < 0$ and therefore there are two sign changes in the first column that tells us that $a(s)$ has two roots in the right half-plane.

Example 28 *Zero in the First Column*
Consider
$$a(s) = s^3 - 3s + 2.$$

We know immediately that it is not stable because the coefficient of the s term is negative and the coefficient of s^2 is zero. We use the Routh–Hurwitz criterion to find out how many roots are in the right half-plane. The Routh array is

$$\begin{array}{c|cc} s^3 & 1 & -3 \\ s^2 & 0 & 2 \\ s & & \\ s^0 & & \end{array}$$

We cannot continue to fill out the Routh array due to the zero in first column of the s^2 row. To proceed we replace the 0 in the first column of the s^2 row with a small $\epsilon > 0$ and then complete the array as follows.

$$\begin{array}{c|ccc} s^3 & 1 & & -3 \\ s^2 & \epsilon & & 2 \\ s & \dfrac{\epsilon \cdot (-3) - 2 \cdot 1}{\epsilon} = -3 - 2/\epsilon & & 0 \\ s^0 & \dfrac{(-3 - 2/\epsilon) \cdot 2 - 0 \cdot \epsilon}{-3 - 2/\epsilon} = 2 & & \end{array}$$

With $\epsilon > 0$ and small there are two sign changes in the first column telling us that $a(s)$ has two roots in the right half-plane. In fact, $a(s) = s^3 - 3s + 2 = (s+2)(s-1)^2$ in agreement with the Routh–Hurwitz test.

Example 29 *Routh–Hurwitz Criterion*
Consider again
$$a(s) = s^3 - 3s + 2.$$

Again we know immediately that it is not stable because the coefficient of the s term is negative and the coefficient of the s^2 is zero. Next we define the new polynomial

$$\bar{a}(s) \triangleq (s+3)a(s) = (s+3)(s^3 - 3s + 2) = s^4 + 3s^3 - 3s^2 - 7s + 6.$$

Of course we still know that $\bar{a}(s)$ is not stable because the s^2 and s terms have negative coefficients. We also know that $\bar{a}(s)$ has the same number of right half-plane zeros as $a(s)$. The Routh array is below.

s^4	1	-3	6
s^3	3	-7	0
s^2	$\dfrac{3\cdot(-3)-(-7)\cdot 1}{3} = -\dfrac{2}{3}$	6	0
s	$\dfrac{-\dfrac{2}{3}\cdot(-7)-6\cdot 3}{-\dfrac{2}{3}} = 20$	0	
s^0	$\dfrac{20\cdot 6 - 0\cdot \left(-\dfrac{2}{3}\right)}{20} = 6$		

There are two sign changes in the first column that tell us that $\bar{a}(s)$, and therefore $a(s)$, have two roots in the right half-plane.

Problems

Problem 1 *Differential Equations*

Given the differential equation

$$\ddot{x} + 2\dot{x} + x = 0$$

with initial conditions $x(0) = 1$, $\dot{x}(0) = 1$ use Laplace transforms to find the solution. Check your result using MATLAB.

Problem 2 *Differential Equations*

Given the differential equation

$$\ddot{x} - 2\dot{x} + x = 0$$

with initial conditions $x(0) = -1$, $\dot{x}(0) = +1$ use Laplace transforms to find the solution. Check your result using MATLAB.

Problem 3 *Laplace Transform and the Final Value Theorem*

Let

$$F(s) = \frac{2}{s(s-1)(s-2)}.$$

(a) Find $f(t) = \mathcal{L}^{-1}\{F(s)\}$ by the method of partial fraction expansion. Check your result using MATLAB.

(b) Can you use the final value theorem to find $\lim_{t\to\infty} f(t)$? If so, do so and explain. If not, explain why not.

Problem 4 Laplace Transform
Let
$$F(s) = \frac{2}{s(s+1)(s+2)}.$$

(a) Find $f(t) = \mathcal{L}^{-1}\{F(s)\}$.

(b) Can you use the final value theorem to find $\lim_{t\to\infty} f(t)$? If so, do so and explain. If not, explain why not.

Problem 5 Differential Equations
Consider the differential equation
$$\frac{d^2y}{dt^2} + 4y = u(t).$$

(a) Compute the transfer function of this differential equation.

(b) Assuming zero initial conditions and with $u(t) = u_s(t)$ a step input, find the solution to this differential equation using the partial fraction expansion method of Laplace transforms. Check your result using MATLAB.

(c) Can you use the final value theorem to find $\lim_{t\to\infty} y(t)$? If so, do so and explain. If not, explain why not.

Problem 6 Laplace Transform
Let
$$F(s) = \frac{1}{(s^2+1)(s+1)}.$$

(a) Find $f(t) = \mathcal{L}^{-1}\{F(s)\}$ using the partial-fraction expansion method. Check your result using MATLAB.

(b) Can you use the final value theorem to find $\lim_{t\to\infty} f(t)$? If so, do so and explain. If not, explain.

Problem 7 Differential Equations
Consider the differential equation
$$\frac{d^2y}{dt^2} + 5\frac{dy}{dt} + 6y = u(t).$$

(a) Compute the transfer function of this differential equation.

(b) Assuming zero initial conditions and with $u(t) = u_s(t)$ a step input, find the solution to this differential equation. Check your result using MATLAB.

(c) Can you use the final value theorem to find $\lim_{t\to\infty} y(t)$? If so, do so. If not, explain why not.

Problem 8 Transfer Functions and Stability
Let a system be modeled by the differential equation
$$\ddot{x} + 2\dot{x} + 5x = \dot{u} + 10u$$

which has transfer function

$$G(s) = \frac{X(s)}{U(s)} = \frac{s+10}{s^2+2s+5}.$$

(a) Is this transfer function stable? Explain and show work.

(b) With $U(s) = 1/s$, $X(s)$ is given by

$$X(s) = G(s)\frac{1}{s} = \frac{s+10}{s^2+2s+5}\frac{1}{s}.$$

Can you use the final value theorem to find $\lim_{t\to\infty} x(t)$? If so, do so and explain. If not, explain why not.

(c) With $U(s) = \frac{s}{s^2+\omega^2}$, $X(s)$ is given by

$$X(s) = G(s)\frac{s}{s^2+\omega^2} = \frac{s+10}{s^2+2s+5}\frac{s}{s^2+\omega^2}.$$

Can you use the final value theorem to find $\lim_{t\to\infty} x(t)$? If so, do so and explain. If not, explain why not.

Problem 9 *Differential Equations*
Consider the differential equation

$$\dot{x} - 2x = u_s(t)$$

with $u_s(t)$ the unit step function and initial condition $x(0) = 1$.

(a) Use Laplace transforms to find the solution.

(b) Consider again the differential equation in part (a), that is,

$$\dot{x} - 2x = u(t).$$

Compute the transfer function $G(s) = X(s)/U(s)$ of this system.

(c) In your answer to part (b) let $u(t) = u_s(t)$ a step input so that

$$X(s) = G(s)\frac{1}{s}.$$

Can you use the final value theorem to find $\lim_{t\to\infty} x(t)$? Explain why or why not.

Problem 10 *Phasor Method*
Consider the general third-order differential equation given by

$$\dddot{y} + a_2\ddot{y} + a_1\dot{y} + a_0 y = b_2\ddot{u} + b_1\dot{u} + b_0 u. \tag{3.32}$$

Let

$$u(t) = U_0\cos(\omega t + \theta_0) = \text{Re}\{U_0 e^{j\theta_0} e^{j\omega t}\} = \text{Re}\{\mathbf{U}_0 e^{j\omega t}\}, \tag{3.33}$$

where

$$\mathbf{U}_0 \triangleq U_0 e^{j\theta_0}.$$

(a) Compute the transfer function $G(s)$ of this system.

(b) With $\mathbf{u}(t) \triangleq \mathbf{U}_0 e^{j\omega t}$ the input to the above differential equation set $\mathbf{y}(t) = \mathbf{A}e^{j\omega t}$ and show that $\mathbf{A} = \mathbf{U}_0 G(j\omega)$ makes $\mathbf{y}(t)$ a solution. Use this to show that

$$y(t) = \mathrm{Re}\{\mathbf{y}(t)\} = \mathrm{Re}\{\mathbf{A}e^{j\omega t}\} = |G(j\omega)| U_0 \cos(\omega t + \theta_0 + \angle G(j\omega))$$

is a solution to (3.32) with input (3.33).

Problem 11 *Phasor Solution vs. Sinusoidal Steady-State Solution*
Consider the differential equation

$$\frac{dx}{dt} - 2x = 2u \qquad (3.34)$$

with input

$$u(t) = U_0 \cos(\omega t) \text{ for } -\infty < t < \infty.$$

To solve this equation, first consider the complex valued input given by $\mathbf{u}(t) = U_0 e^{j\omega t}$. Note that

$$u(t) = \mathrm{Re}\{\mathbf{u}(t)\} = \mathrm{Re}\{U_0 e^{j\omega t}\} = U_0 \cos(\omega t).$$

(a) With $\mathbf{u}(t)$ the input to the above differential equation and $\mathbf{A}$ a constant complex number, show that a solution of the form

$$\mathbf{x}(t) = \mathbf{A}e^{j\omega t}$$

is possible. Give your steps on how you obtained $\mathbf{A}$.

(b) What is the relationship between $\mathbf{A}$ and the transfer function $G(s) = X(s)/U(s)$ of the differential equation?

(c) With $\omega = 2$ compute the phasor solution $x(t) = |G(j2)| U_0 \cos(2t + \angle G(j2))$ to (3.34). With input $u(t) = U_0 \cos(2t) u_s(t)$ it follows that $x(t) = |G(j2)| U_0 \cos(2t + \angle G(j2)) u_s(t)$ is the solution to (3.34) for $t \geq 0$ with a particular value for the initial condition $x(0)$. Find that value.

(d) Is your answer in part (c) the sinusoidal steady-state solution for the input $u(t) = U_0 \cos(2t) u_s(t)$? Explain why or why not. Hint: Is $G(s)$ stable?

Problem 12 *Phasor Solution vs. Sinusoidal Steady-State Solution*
Let a system be defined by the transfer function

$$\frac{C(s)}{R(s)} = G_1(s) = \frac{6\sqrt{2}}{s+3}.$$

(a) Compute the phasor solution with input $r(t) = 10\cos(3t)$ for $-\infty < t < \infty$.

(b) With $c(0) = 0$, use the phasor solution to find the sinusoidal steady-state response of this system with input $r(t) = 10\cos(3t) u_s(t)$? Hint: Is $G_1(s)$ stable?

(c) With $c(0) = 10$, use the phasor solution to find the sinusoidal steady-state response of this system with input $r(t) = 10\cos(3t) u_s(t)$? Hint: See your answer to part (a).

(d) Now let $G_2(s) = \dfrac{6\sqrt{2}}{s-3}$, but still with the input $r(t) = 10\cos(3t)u_s(t)$. With $c(0) = 0$ what is the sinusoidal steady-state response of this system due to this input? Hint: No computation is needed.

Problem 13 *Phasor Method*

Consider the differential equation

$$\frac{d^2x}{dt^2} + a_1 \frac{dx}{dt} + a_0 x = b_1 \frac{du}{dt} + b_0 u$$

with input

$$u(t) = U_0 \cos(\omega t) \quad \text{for } -\infty < t < \infty.$$

Let

$$\mathbf{u}(t) = U_0 e^{j\omega t}$$

so that

$$u(t) = \operatorname{Re}\{\mathbf{u}(t)\} = \operatorname{Re}\{U_0 e^{j\omega t}\} = U_0 \cos(\omega t).$$

(a) With $\mathbf{u}(t)$ the input to the above differential equation, show that a solution of the form

$$\mathbf{x}(t) = \mathbf{A} e^{j\omega t}$$

is possible. Show your steps on how you obtained $\mathbf{A}$.

(b) What is the relationship between $\mathbf{A}$ and the transfer function $G(s) = X(s)/U(s)$ of the differential equation?

(c) Use your answers in parts (a) and (b) to give an expression for the phasor solution when the input is $u(t) = U_0 \cos(\omega t)$.

(d) Suppose you start the system at $t = 0$ with input $u(t) = U_0 \cos(\omega t) u_s(t)$. What condition on $G(s)$ is needed for the phasor solution to be the sinusoidal steady-state solution?

Problem 14 *Sinusoidal Steady-State Response*

Let a system be given by

$$\frac{dy}{dt} + 4y = 8\sqrt{2}\,u$$

so that its transfer function is

$$G_1(s) = \frac{8\sqrt{2}}{s+4}.$$

(a) Is $G_1(s)$ stable?

(b) Let the input be $u(t) \triangleq 10\cos(4t)u_s(t)$ so that

$$U(s) = 10 \frac{s}{s^2 + 4^2}.$$

Compute a solution using the phasor approach. Is this also the sinusoidal steady-state solution to the system? Explain briefly.

(c) Now consider the system
$$\frac{dy}{dt} - 4y = 8\sqrt{2}u$$
with transfer function
$$G_2(s) = \frac{8\sqrt{2}}{s-4}.$$
Is $G_2(s)$ stable?

(d) For the system of part (c) let the input be $u(t) \triangleq 10\cos(4t)u_s(t)$ so that
$$U(s) = 10\frac{s}{s^2 + 4^2}.$$
Compute a solution to the system of part (c) using the phasor approach. Is this also a sinusoidal steady-state solution? Explain briefly.

Problem 15 *Stability*

Suppose
$$\ddot{y} + 2\dot{y} + 2y = 3\dot{u} + 4u$$
so that
$$Y(s) = G(s)U(s),$$
where $G(s)$ is the transfer function.

(a) Compute $G(s)$ from the differential equation.

(b) Let $U(s) = \frac{U_0 s}{s^2 + \omega^2}$, i.e., $u(t) = U_0 \cos(\omega t)u_s(t)$. Is $|G(j\omega)|U_0 \cos(\omega t + \angle G(j\omega))u_s(t)$ a solution to the above differential equation? Explain briefly. No calculations are needed.

(c) Let $U(s) = 1/s$, i.e., $u(t) = u_s(t)$. With $y(t)$ the solution to the above differential equation with arbitrary initial conditions, is it true that
$$\lim_{t \to \infty} y(t) = \lim_{s \to 0} sY(s)?$$
Explain briefly.

(d) Let $U(s) = \frac{U_0 s}{s^2 + \omega^2}$, i.e., $u(t) = U_0 \cos(\omega t)u_s(t)$. With $y(t)$ the corresponding solution and arbitrary initial conditions, is it true that
$$y(t) \to |G(j\omega)|U_0 \cos(\omega t + \angle G(j\omega)) \text{ as } t \to \infty?$$
Explain briefly.

(e) Let $U(s) = 1/s^2$, i.e., $u(t) = tu_s(t)$. With $y(t)$ the solution to the above differential equation with *zero* initial conditions, is it true that
$$\lim_{t \to \infty} y(t) = \lim_{s \to 0} sY(s)?$$
Explain briefly.

Problem 16 Stability
Suppose
$$\ddot{y} - 2\dot{y} + 2y = 3\dot{u} + 4u$$

so that
$$Y(s) = G(s)U(s),$$

where $G(s)$ is the transfer function.

(a) Compute $G(s)$ from the differential equation.

(b) Let $U(s) = \dfrac{U_0 s}{s^2 + \omega^2}$, i.e., $u(t) = U_0 \cos(\omega t) u_s(t)$. Is $|G(j\omega)| U_0 \cos(\omega t + \angle G(j\omega)) u_s(t)$ a solution to the above differential equation? Explain briefly. No calculations are needed.

(c) Let $U(s) = 1/s$, i.e., $u(t) = u_s(t)$. With $y(t)$ the solution to the above differential equation with arbitrary initial conditions, is it true that
$$\lim_{t \to \infty} y(t) = \lim_{s \to 0} sY(s)?$$

Explain briefly.

(d) Let $U(s) = \dfrac{U_0 s}{s^2 + \omega^2}$, i.e., $u(t) = U_0 \cos(\omega t) u_s(t)$. With $y(t)$ the corresponding solution and arbitrary initial conditions, is it true that
$$y(t) \to |G(j\omega)| U_0 \cos(\omega t + \angle G(j\omega)) \quad \text{as } t \to \infty?$$

Explain briefly.

(e) Let $U(s) = 1/s^2$, i.e., $u(t) = tu_s(t)$. With $y(t)$ the solution to the above differential equation with *zero* initial conditions, is it true that
$$\lim_{t \to \infty} y(t) = \lim_{s \to 0} sY(s)?$$

Explain briefly.

Problem 17 Stability
Suppose
$$\dddot{y} + a_2 \ddot{y} + a_1 \dot{y} + a_0 y = b_2 \ddot{u} + b_1 \dot{u} + b_0 u$$

so that
$$Y(s) = G(s)U(s),$$

where $G(s)$ is the transfer function.

(a) Compute $G(s)$ from the differential equation.

(b) Let $U(s) = \dfrac{U_0 s}{s^2 + \omega^2}$, i.e., $u(t) = U_0 \cos(\omega t) u_s(t)$. Is it always true that
$$|G(j\omega)| U_0 \cos(\omega t + \angle G(j\omega)) u_s(t)$$
is a solution to the differential equation? Explain briefly.

84   3 Differential Equations and Stability

(c) Let $U(s) = 1/s$, i.e., $u(t) = u_s(t)$. With $y(t)$ the solution to the differential equation with arbitrary initial conditions is it always true that

$$\lim_{t \to \infty} y(t) = \lim_{s \to 0} sY(s)?$$

Explain briefly.

(d) Suppose $G(s)$ is *stable*, $U(s) = 1/(s^2 + 1)$, i.e., $u(t) = \sin(t)u_s(t)$ and let $y(t)$ be the solution to the above differential equation with *zero* initial conditions. Is it always true that

$$\lim_{t \to \infty} y(t) = \lim_{s \to 0} sY(s)?$$

Explain briefly.

(e) Let $U(s) = U_0 \dfrac{2}{s^2 + 2^2}$, i.e., $u(t) = U_0 \sin(2t)u_s(t)$ and suppose $G(s)$ is given by

$$G(s) = \frac{s^2 + 4}{(s+2)(s+4)(s+5)}.$$

With $Y(s) = G(s)U(s)$ is it true that $\lim_{t \to \infty} y(t) = \lim_{s \to 0} sY(s)$? Explain briefly.

Problem 18 *Stability*

Suppose

$$\ddot{y} + a_1 \dot{y} + a_0 y = b_1 \dot{u} + b_0 u$$

so that

$$Y(s) = G(s)U(s),$$

where $G(s)$ is the transfer function.

(a) Compute $G(s)$ from the differential equation.

(b) Let $U(s) = \dfrac{U_0 s}{s^2 + \omega^2}$, i.e., $u(t) = U_0 \cos(\omega t)u_s(t)$. Is it always true that

$$|G(j\omega)| U_0 \cos(\omega t + \angle G(j\omega))u_s(t)$$

is a solution to the differential equation? Explain briefly.

(c) Let $U(s) = 1/s$, i.e., $u(t) = u_s(t)$. With $y(t)$ the solution to the differential equation with arbitrary initial conditions is it always true that

$$\lim_{t \to \infty} y(t) = \lim_{s \to 0} sY(s)?$$

Explain briefly.

(d) Let $U(s) = \dfrac{U_0 s}{s^2 + \omega^2}$, i.e., $u(t) = U_0 \cos(\omega t)u_s(t)$. As $t \to \infty$ is it always true that

$$y(t) \to |G(j\omega)| U_0 \cos(\omega t + \angle G(j\omega))u_s(t)?$$

Explain briefly.

(e) Let $U(s) = 1/s$, i.e., $u(t) = u_s(t)$. With $y(t)$ the solution to the differential equation with *zero* initial conditions is it always true that

$$\lim_{t \to \infty} y(t) = \lim_{s \to 0} sY(s)?$$

Problem 19 $s^2 + a_1 s + a_0$ *is stable iff* $a_1 > 0$ *and* $a_0 > 0$
Use the quadratic formula to show that the solutions of the polynomial equation

$$s^2 + a_1 s + a_0 = 0$$

are in the open left half-plane *if and only if* $a_1 > 0$ *and* $a_0 > 0$.

Problem 20 *Stability Tests* [12]
If you can determine the stability of the following polynomials without using the Routh–Hurwitz criterion or solving for the roots, do so. Otherwise use the Routh–Hurwitz criterion.

(a) $s^3 + s + 2$

(b) $s^4 + s^2 + 1$

(c) $s^4 - 1$

(d) $-s^2 - 2s - 2$

(e) $-s^3 - 2s^2 - 3s - 1$

(f) $s^3 + 2s^2 + 3s + 1$

(g) $s^3 + 2s^2 + 3s - 1$

Problem 21 *Routh–Hurwitz Test* [12]
For what values of α is the polynomial $s^3 + s^2 + \alpha s + 1$ stable?

Problem 22 *Routh–Hurwitz Test* [12]
For what values of α is the polynomial $s^3 + s^2 + s + \alpha$ stable?

Problem 23 *Routh–Hurwitz Criterion* [12]
Let

$$a(s) = s^3 + (14 - K)s^2 + (6 - K)s + 79 - 18K.$$

Find the range of values of K for which $a(s)$ is stable.

Problem 24 *Final Value Theorem*
Let

$$X(s) = \frac{10}{s\left(s^3 + 5s^2 + (K-6)s + K\right)}.$$

Find the final value of $x(t)$ as a function of K.

Problem 25 *Final Value Theorem*
Let

$$C(s) = \frac{(s+10)(s+2)}{s^3 + 12s^2 + 20s + 20K} \frac{R_0}{s}.$$

Find the final value of $c(t)$ in terms of R_0 and K. Show work.

86   3   Differential Equations and Stability

Problem 26 *Routh–Hurwitz Criterion*
Consider the polynomial
$$a(s) = s^3 + 2s^2 + s + 2.$$

Use the Routh–Hurwitz criterion to show whether or not $a(s)$ is stable.

Problem 27 *Routh–Hurwitz Criterion*
Let
$$a(s) = s^4 + 2s^3 + 5s^2 + 4s + 6.$$

Use the Routh–Hurwitz criterion to show whether or not $a(s)$ is stable.

Problem 28 *Routh–Hurwitz Criterion*
For what values of K does
$$a(s) = s^3 + 5s^2 + (5K + 6)s + 3K = 0$$

have all its roots in the open left half-plane?

Problem 29 *Routh–Hurwitz Criterion*
Let
$$G(s) = \frac{s^2(s + 10)}{s^3 + 10s^2 + Ks + K}.$$

(a) For what values of K is $G(s)$ stable?
(b) With $R(s) = R_0/s$ let
$$E(s) \triangleq G(s)R(s) = \frac{s^2(s + 10)}{s^3 + 10s^2 + Ks + K} \frac{R_0}{s}.$$

For what values of K will $e(t) \to 0$?

Problem 30 *Final Error*
Let
$$G(s) \triangleq \frac{2s(s + 10)}{s^2(s + 6)(s + 10) + 2K(s + 2)}$$

and, with $D(s) = D_0/s$, let
$$E(s) \triangleq G(s)D(s) = \frac{2s(s + 10)}{s^2(s + 6)(s + 10) + 2K(s + 2)} \frac{D_0}{s}.$$

For what values of K will $e(\infty) = \lim_{t \to \infty} e(t) = 0$? Explain and show work.

Problem 31 *Routh–Hurwitz Criterion*
Let
$$G(s) = \frac{K}{s^4 + 3s^3 + 3s^2 + 2s + K}.$$

For what values of K is $G(s)$ stable?

Problem 32 *Row of Zeros in the Routh Array*
Let
$$G(s) = \frac{K}{s^4 + 3s^3 + 3s^2 + 2s + K}.$$

For what values of K does $G(s)$ have poles of the $j\omega$ axis? What are the values of these poles on the $j\omega$ axis?

Problem 33 *Row of Zeros in the Routh Array*
Let
$$a(s) = s^4 + 2s^3 + 5s^2 + 4s + 6.$$

Find any roots of $a(s)$ on the $j\omega$ axis.

Problem 34 *Row of Zeros in the Routh Array*
Consider the polynomial
$$a(s) = s^3 + 2s^2 + s + 2.$$

Find any roots of $a(s)$ on the $j\omega$ axis.

Problem 35 *Row of Zeros in the Routh Array* [12]
Let
$$a(s) = s^3 + (14 - K)s^2 + (6 - K)s + 79 - 18K.$$

Find the values of K for which $a(s)$ has roots on the $j\omega$ axis. Also find the corresponding roots.

4

Mass–Spring–Damper Systems

A key first step in control design is to find a mathematical model (usually as a differential equation) of the physical system. In this chapter we apply Newton's laws of motion to derive differential equation models of masses connected to springs and dampers (dashpots). We also take a first look at simulating differential equations in software.

4.1 Mechanical Work

Let a mass m be moving in one dimension along the x-axis with a force F acting on it as indicated in Figure 4.1.

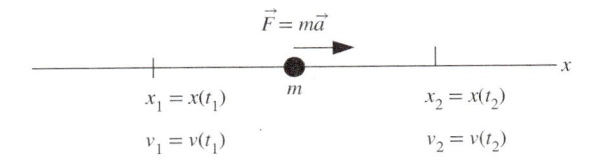

Figure 4.1. Force acting on a mass m.

With $x(t)$ the mass's position at time t, $v = dx/dt$ its velocity, and $a = d^2x/dt^2$ its acceleration, recall that Newton's law says that $F = ma$ or equivalently

$$F = ma = m\frac{dv}{dt} = m\frac{d^2x}{dt^2}. \tag{4.1}$$

Also recall that by definition, the work W done on the mass as the force acts on it while the mass goes from x_1 to x_2 is

$$W \triangleq \int_{x_1}^{x_2} F(x)dx. \tag{4.2}$$

In this integral we are assuming we know the force as a function of the position x.

We next make a change of variables in this integral to make the integrand a function of time t. To do this, let t_1 be the time that mass is at position x_1 so that $x_1 \triangleq x(t_1)$. Correspondingly, the velocity at time t_1 is given by $v_1 \triangleq \frac{dx}{dt}\bigg|_{t_1} = v(t_1)$. Similarly, with t_2 the time the mass is at x_2 we have $x_2 \triangleq x(t_2)$, $v_2 \triangleq \frac{dx}{dt}\bigg|_{t_2} = v(t_2)$.

We now rewrite the work W as follows:

$$W \triangleq \int_{x_1}^{x_2} F\,dx = \int_{t_1}^{t_2} \underbrace{m\frac{dv}{dt}}_{F}\underbrace{vdt}_{dx}.$$

An Introduction to System Modeling and Control, First Edition. John Chiasson.
© 2022 John Wiley & Sons, Inc. Published 2022 by John Wiley & Sons, Inc.
Companion website: www.wiley.com/go/chiasson/anintroductiontosystemmodelingandcontrol

Then

$$W \triangleq \int_{t_1}^{t_2} m\frac{dv}{dt}vdt = \int_{t_1}^{t_2} \frac{d}{dt}\left(\frac{mv^2}{2}\right)dt = \frac{mv^2}{2}\bigg|_{v_1}^{v_2} = \frac{1}{2}mv_2^2 - \frac{1}{2}mv_1^2. \quad (4.3)$$

The kinetic energy (KE) of the mass is defined to be $\frac{1}{2}mv^2$. So we have shown that the work done on m equals the change in its KE. That is,

$$\int_{x_1}^{x_2} Fdx = \frac{1}{2}mv_2^2 - \frac{1}{2}mv_1^2. \quad (4.4)$$

4.2 Modeling Mass–Spring–Damper Systems

Example 1 *Horizontal Mass–Spring–Damper System*
Consider the spring–mass–damper system in Figure 4.2.

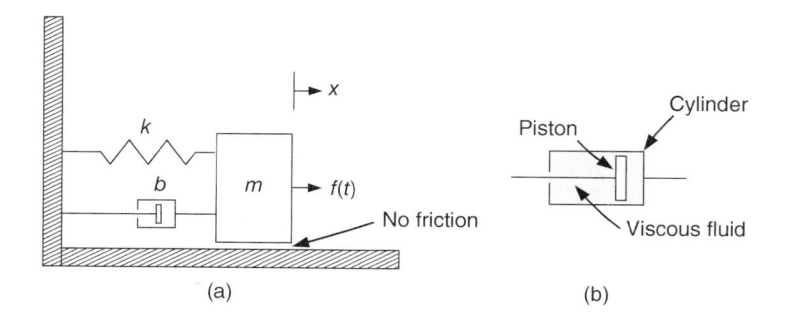

Figure 4.2. (a) Mass–spring–damper system. (b) Damper cross section.

If $x = 0$ then the spring is neither compressed nor stretched. In this case we say the spring is *relaxed*. The force on the mass m due to the spring is

$$F_{mk} = -kx.$$

Note that if $x > 0$ the spring is pulling m to the left while if $x < 0$ the spring is pushing m to the right.

A cross section of the damper (dashpot) pictured in Figure 4.2b shows that it consists of a piston encased in a sealed cylinder which is filled with a viscous fluid. If the piston is to move, it must do so by forcing the fluid around its outer surface from one side to the other. The force of the damper on m is proportional to its velocity $\dot{x}$ and is given by

$$F_{mb} = -b\dot{x}.$$

Note that if $\dot{x} > 0$ then the mass m along with the cylinder is moving to the right with the piston providing a resistive force in the $-x$ direction to oppose this motion. Similarly, if $\dot{x} < 0$ then the mass m along with the cylinder is moving to the left with the piston

4.2 Modeling Mass–Spring–Damper Systems

providing resistive force now in the $+x$ direction to oppose this motion. This damper force whose opposing force is proportional to speed is also referred to as *viscous friction*.

$f(t)$ denotes an external force. The equation of motion is then

$$m\frac{d^2x}{dt^2} = -kx - b\frac{dx}{dt} + f(t)$$

which is rearranged to obtain

$$m\frac{d^2x}{dt^2} + b\frac{dx}{dt} + kx = f(t). \tag{4.5}$$

Taking the Laplace transform with zero initial conditions gives

$$(ms^2 + bs + k)X(s) = F(s)$$

or

$$X(s) = \frac{1}{ms^2 + bs + k}F(s). \tag{4.6}$$

If the initial conditions are not zero, you may show that

$$X(s) = \underbrace{\frac{1}{ms^2 + bs + k}}_{\text{Transfer function}}\underbrace{F(s)}_{\text{Input}} + \underbrace{\frac{mx(0)s + bx(0) + m\dot{x}(0)}{ms^2 + bs + k}}_{\text{Initial condition response}}. \tag{4.7}$$

Example 2 *Vertical Mass–Spring–Damper System*
Figure 4.3 shows a mass-spring-damper system hung from a ceiling.

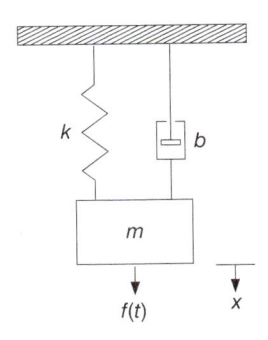

Figure 4.3. Vertical mass–spring–mass–damper system.

The equations of motion are then

$$m\frac{d^2x}{dt^2} = -kx - b\frac{dx}{dt} + mg + f(t)$$

which are rearranged to obtain

$$m\frac{d^2x}{dt^2} + b\frac{dx}{dt} + kx = mg + f(t). \tag{4.8}$$

Taking the Laplace transform with zero initial conditions gives

$$(ms^2 + bs + k)X(s) = F(s) + \frac{mg}{s}$$

or
$$X(s) = \frac{1}{ms^2 + bs + k}\left(F(s) + \frac{mg}{s}\right). \quad (4.9)$$

If the initial conditions are not zero, you may show that

$$X(s) = \underbrace{\frac{1}{ms^2 + bs + k}}_{\text{Transfer function}}\left(F(s) + \frac{mg}{s}\right) + \frac{mx(0)s + bx(0) + m\dot{x}(0)}{ms^2 + bs + k}. \quad (4.10)$$

Equilibrium Conditions

Suppose $f(t) = 0$ so that the equations of motion are now

$$m\frac{d^2 x}{dt^2} + b\frac{dx}{dt} + kx = mg.$$

Equilibrium means the mass m is at rest (not moving) so that $\dot{x} = \ddot{x} \equiv 0$. At equilibrium we have

$$kx_0 = mg$$

or the mass is at

$$x_0 = \frac{mg}{k}. \quad (4.11)$$

This just says the spring has to be stretched so that it can provide an upward force to cancel out the gravitational force.

Equations of Motion About the Equilibrium Point

Let
$$\Delta x = x - x_0$$

and substitute $x = \Delta x + x_0$ into (4.8) to obtain

$$m\frac{d^2}{dt^2}(\Delta x + x_0) + b\frac{d}{dt}(\Delta x + x_0) + k(\Delta x + x_0) = mg + f(t).$$

Rearranging

$$m\frac{d^2}{dt^2}\Delta x + b\frac{d}{dt}\Delta x + k\Delta x + \underbrace{m\frac{d^2}{dt^2}x_0 + b\frac{d}{dt}x_0}_{0} + \underbrace{kx_0}_{mg} = mg + f(t)$$

or finally

$$m\frac{d^2}{dt^2}\Delta x + b\frac{d}{dt}\Delta x + k\Delta x = f(t).$$

So if we write the equations of motion with respect to the equilibrium position, the gravity term is gone. However, we have to remember that the solution Δx to this equation is with respect to the equilibrium position as indicated in Figure 4.4.

4.2 Modeling Mass–Spring–Damper Systems

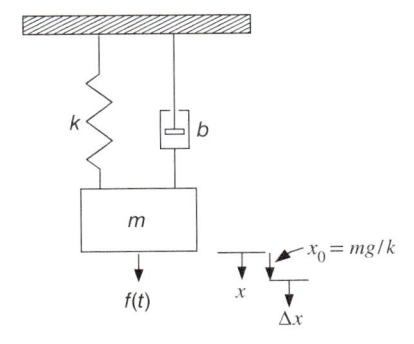

Figure 4.4. Equations of motion about the equilibrium point.

Often books interpret the reference position for x in Figure 4.3 to be with respect to the equilibrium position. Then the equations of motion are simply

$$m\frac{d^2}{dt^2}x + b\frac{d}{dt}x + kx = f(t),$$

but one must remember to interpret x in this way!

Example 3 *Mass–Spring–Damper System with Two Masses*
Figure 4.5 shows two masses M and m connected together by a spring and a damper.

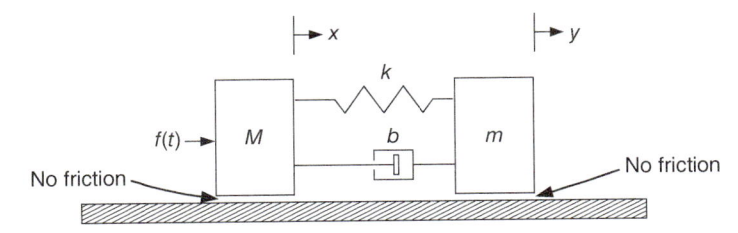

Figure 4.5. Two masses connected by a spring and a damper.

We consider the external force $f(t)$ to be the input and the position y of mass m to be the output.

The references x and y for the positions of M and m, respectively, are interpreted that if $x = y$ the spring is relaxed (neither compressed nor stretched). So the spring forces F_{mk} and F_{Mk} on m and M, respectively, are given by

$$F_{mk} = -k(y - x)$$
$$F_{Mk} = +k(y - x).$$

That is, if $y - x > 0$ then the spring is pulling m in the $-y$ direction and pulling M in the $+x$ direction.

Let's consider the damper (dashpot) forces F_{mb} and F_{Mb} on m and M, respectively. These depend on the relative velocity $\dot{y} - \dot{x}$ between m and M. We have

$$F_{mb} = -b(\dot{y} - \dot{x})$$
$$F_{Mb} = +b(\dot{y} - \dot{x}).$$

For example, if $\dot{y} - \dot{x} > 0$ then the cylinder is moving to the right faster than the piston. Thus the piston is producing a resistive force on the cylinder (and therefore m) to oppose its motion to the right. Under these same conditions, the cylinder is dragging the piston (and therefore M) to the right.

The equations of motion are then

$$m\frac{d^2y}{dt^2} = F_{mk} + F_{mb}$$

$$M\frac{d^2x}{dt^2} = F_{Mk} + F_{Mb} + f(t)$$

or

$$m\frac{d^2y}{dt^2} = -k(y-x) - b(\dot{y}-\dot{x})$$

$$M\frac{d^2x}{dt^2} = k(y-x) + b(\dot{y}-\dot{x}) + f(t).$$

We rearrange this to obtain

$$m\frac{d^2y}{dt^2} + b\dot{y} + ky = kx + b\dot{x}$$

$$M\frac{d^2x}{dt^2} + b\dot{x} + kx = ky + b\dot{y} + f(t).$$

Taking Laplace transforms with zero initial conditions we obtain

$$(ms^2 + bs + k)Y(s) = (bs + k)X(s)$$

$$(Ms^2 + bs + k)X(s) = (bs + k)Y(s) + F(s).$$

As we are taking the input to be $f(t)$ and the output to be $y(t)$, we want to eliminate $X(s)$ from these two equations. Solving for $X(s)$ in the first equation and substituting into the second equation we obtain

$$(Ms^2 + bs + k)\frac{ms^2 + bs + k}{bs + k}Y(s) = (bs + k)Y(s) + F(s).$$

We next solve this for $Y(s)$ as follows:

$$(Ms^2 + bs + k)(ms^2 + bs + k)Y(s) = (bs + k)^2 Y(s) + (bs + k)F(s)$$

or

$$Y(s) = \frac{bs + k}{(Ms^2 + bs + k)(ms^2 + bs + k) - (bs + k)^2} F(s)$$

$$= \underbrace{\frac{bs + k}{Mms^4 + (Mb + bm)s^3 + (Mk + km)s^2}}_{\text{Transfer function}} F(s)$$

Example 4 *Mass–Spring–Damper System with Massless Point*

Figure 4.6 shows a spring–mass–damper system all connected in series. The point A showing the connection point of the spring and piston is referenced by y. Here we consider $f(t)$ as the input and the position y of A as the output.

4.2 Modeling Mass–Spring–Damper Systems

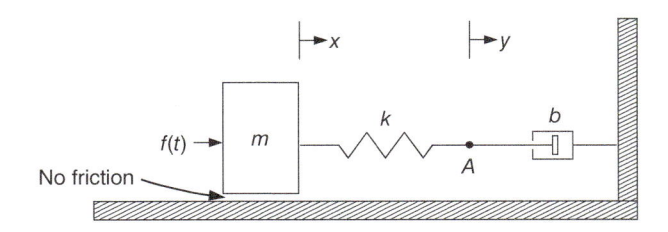

Figure 4.6. Mass–spring–damper system all connected in series.

There is no mass at point A and this is a problem in terms of using Newton's equations. To get around this, we simply consider the point A to have a small mass m_A. After setting up the equations, we will let $m_A \to 0$. We have

$$m\frac{d^2x}{dt^2} = k(y-x) + f(t)$$

$$m_A\frac{d^2y}{dt^2} = -k(y-x) - b\frac{dy}{dt}$$

or

$$m\frac{d^2x}{dt^2} + kx = ky + f(t)$$

$$m_A\frac{d^2y}{dt^2} + b\frac{dy}{dt} + ky = kx.$$

We let $m_A \to 0$ and take the Laplace transforms with zero initial conditions to obtain

$$(ms^2 + k)X(s) = kY(s) + F(s)$$

$$(bs + k)Y(s) = kX(s).$$

As y is given to be the output, we eliminate $X(s)$ as follows:

$$(ms^2 + k)\frac{bs+k}{k}Y(s) = kY(s) + F(s)$$

or

$$(ms^2 + k)(bs + k)Y(s) = k^2 Y(s) + kF(s)$$

or

$$Y(s) = \frac{k}{(ms^2 + k)(bs + k) - k^2}F(s)$$

$$= \underbrace{\frac{k}{bms^3 + kms^2 + bks}}_{\text{Transfer function}}F(s).$$

Example 5 *Position Input and Massless Point*

In Figure 4.7 the input to the spring–mass–damper system is the position y and the output is the position x of the mass m. The right side of the spring k_1 does not show any mass so we will let m_p denote the mass of the piston to which the spring is attached. Then later we will let $m_p \to 0$.

4 Mass–Spring–Damper Systems

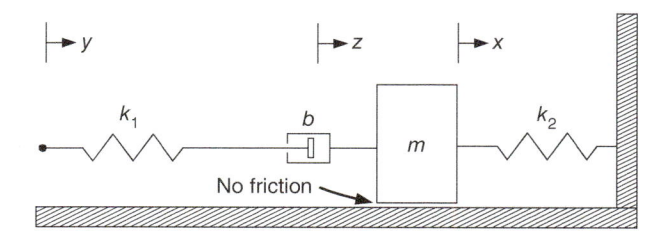

Figure 4.7. Mass–spring–damper system with the position y as input.

We let z denote the position of the piston (mass m_p). If $y = z$ then the spring is relaxed. The relative velocity between the piston and the cylinder of the damper is $\dot{z} - \dot{x}$.

We now compute the equations of motion for m and m_p. We have

$$m\ddot{x} = -k_2 x + b(\dot{z} - \dot{x})$$
$$m_p \ddot{z} = -b(\dot{z} - \dot{x}) - k_1(z - y)$$

or

$$m\ddot{x} + k_2 x + b\dot{x} = b\dot{z}$$
$$m_p \ddot{z} + b\dot{z} + k_1 z = b\dot{x} + k_1 y.$$

Remember that y is the input and x is the output. We now take the Laplace transforms of these equations to obtain

$$(ms^2 + bs + k_2)X(s) = bsZ(s)$$
$$(m_p s^2 + bs + k_1)Z(s) = bsX(s) + k_1 Y(s).$$

Eliminating $Z(s)$ gives

$$(m_p s^2 + bs + k_1)\frac{ms^2 + bs + k_2}{bs}X(s) = bsX(s) + k_1 Y(s)$$

or

$$(m_p s^2 + bs + k_1)(ms^2 + bs + k_2)X(s) = b^2 s^2 X(s) + k_1 bs Y(s).$$

Letting $m_p = 0$ and solving for $X(s)$ we finally obtain

$$X(s) = \frac{k_1 bs}{(bs + k_1)(ms^2 + bs + k_2) - b^2 s^2} Y(s)$$
$$= \underbrace{\frac{k_1 bs}{bms^3 + mk_1 s^2 + (bk_1 + bk_2)s + k_1 k_2}}_{\text{Transfer function}} Y(s).$$

4.3 Simulation

Consider a first-order differential equation system given by

$$\frac{dx}{dt} = -ax + bu$$
$$x(0) = x_0.$$

To implement this equation on a digital computer, it needs to be converted to *discrete time*. Let $x(kT), u(kT)$ denote the values of $x(t), u(t)$ at time kT. We approximate the derivative at time $(k+1)T$ using the backward difference given by

$$\left.\frac{dx}{dt}\right|_{t=(k+1)T} = \frac{x((k+1)T) - x(kT)}{T}.$$

Then the discrete time version of the system is taken to be

$$\frac{x((k+1)T) - x(kT)}{T} = -ax(kT) + bu(kT)$$

$$x(0) = x_0.$$

Rearrange to obtain

$$x((k+1)T) = (1 - aT)x(kT) + Tbu(kT)$$

$$x(0) = x_0.$$

For example, suppose $x_0 = 1$ and $u(t) = \cos(t)$. Then

$$x(T) = (1 - aT)x(0) + Tb\cos(0)$$
$$x(2T) = (1 - aT)x(T) + Tb\cos(T)$$
$$x(3T) = (1 - aT)x(2T) + Tb\cos(2T)$$
$$\vdots = \vdots$$

This recursive discrete-time model is used to implement the original continuous-time system on a digital computer.

With the input u taken to be a unit step input, a SIMULINK diagram for this system is given in Figure 4.8.

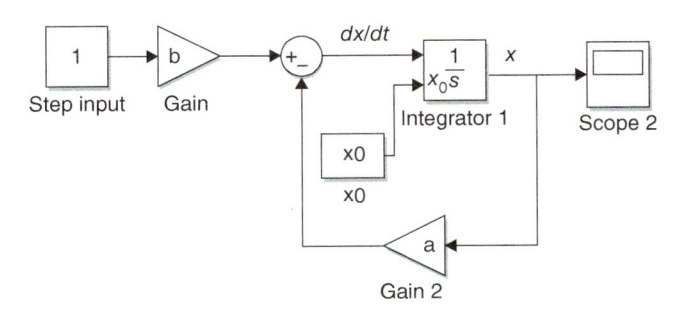

Figure 4.8. SIMULINK block diagram for $\dfrac{dx}{dt} = -ax + bu, x(0) = x_0$.

When the simulation is run, SIMULINK converts the block diagram to a discrete-time form and runs the program. The input u was chosen to be a unit step input where the block above with u in it is a `constant` block from the SIMULINK library with a "1" inside. Even though SIMULINK uses a $1/s$ block (integrator), the simulation is a time

4 Mass–Spring–Damper Systems

domain process! Perhaps the block should have the integral sign $\int$ rather than $1/s$, but the tradition is to use $1/s$.

We can also represent

$$\frac{dx}{dt} = -ax + bu$$

in the Laplace domain (assuming zero initial conditions) as

$$X(s) = \frac{b}{s+a}U(s).$$

The SIMULINK block diagram can then be drawn as shown in Figure 4.9.

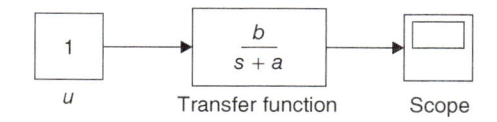

Figure 4.9. SIMULINK block diagram for $X(s) = \dfrac{b}{s+a}U(s)$.

SIMULINK then converts this to

$$x\big((k+1)T\big) = (1 - aT)x(kT) + Tbu(kT)$$

and then generates C code to run on a digital computer. When using a transfer function block, SIMULINK assumes zero initial conditions.

In these two examples we are taking u to be a unit step input so a "1" is put inside the `constant` block from the SIMULINK library so that $u(kT) = 1$ for $k = 0, 1, 2, \ldots$.

Figure 4.10 shows the SIMULINK block diagram for simulating the horizontal spring–mass–damper system of Example 1. Note that the upper left-hand side of this block diagram shows `Simulation` selected. To the right of this is `DEBUG` followed by `MODELING`. Click on `MODELING` to select it and then click on `Model Settings` to open up the `Configuration_Parameters` dialog box shown in Figure 4.11.

In the `Simulation time` block of Figure 4.11 we have set `Start time: 0.0` and `Stop time: 10` (seconds). Note that the `Solver options` are set as `Type: fixed-step`, and `Solver: ode1(Euler)` and `Fixed-step size (fundamental sample time): 0.001` (seconds). The vertical black bar going into the `To File` block of Figure 4.10 is a Mux block and is found in the `Signal Routing` folder of the SIMULINK Library. Double clicking on the `To File` block opens up the dialog box shown in Figure 4.12. The data is stored in a MATLAB data file called `output.mat`. The data in this file is called `outputdata` and consists of an array (matrix) of three rows. The first row is the time, the second row is the position x and the third row is the input force f.

Before running the SIMULINK file, run the following MATLAB program:

```
% Data file
m = 2; b = 4; k = 16; f = 8;
```

4.3 Simulation 99

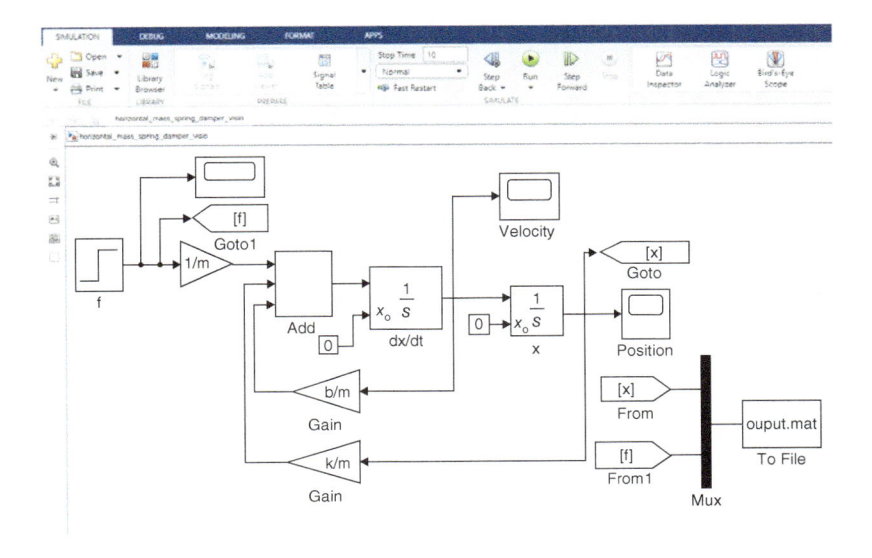

Figure 4.10. Simulation of $m\dfrac{d^2x}{dt^2} = -kx - b\dfrac{dx}{dt} + f(t)$.

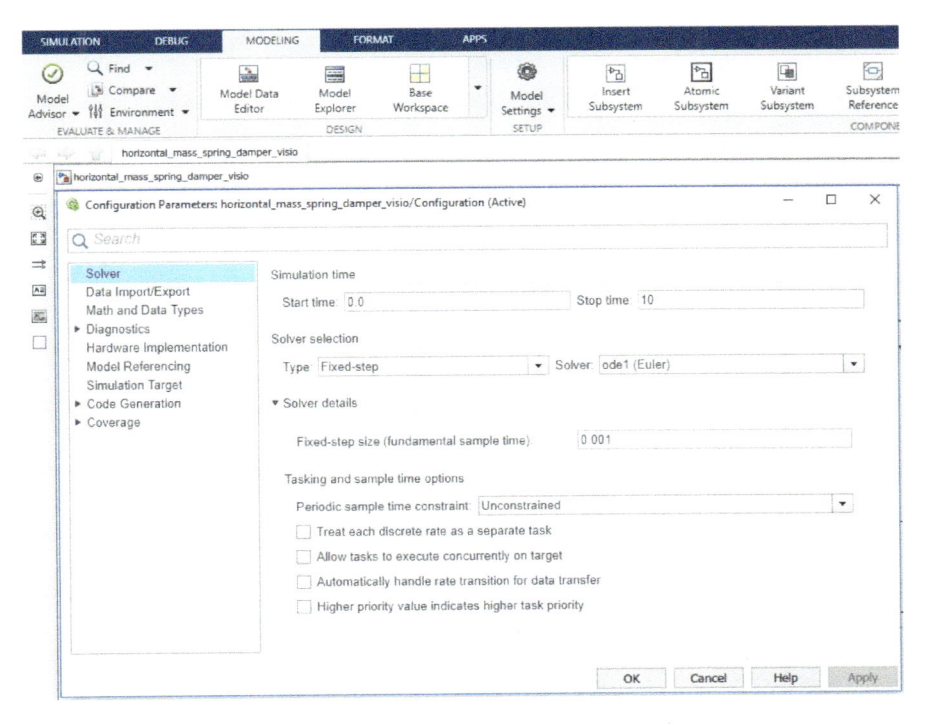

Figure 4.11. Dialog box for the configuration parameters.

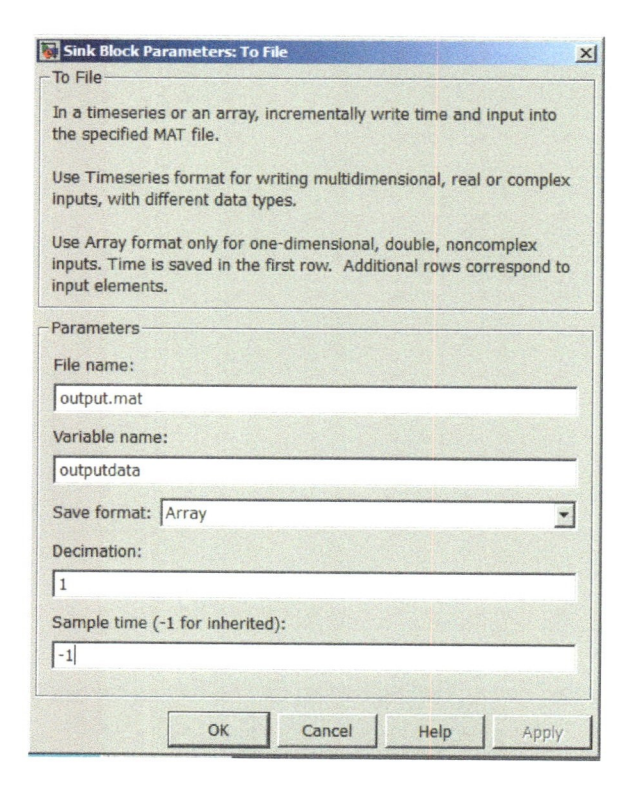

Figure 4.12. Dialog box for the To File block.

Run the SIMULINK program, then go to MATLAB and run the following program to plot the data.

```
% Data Processing file
load output.mat % Brings the outputdata into the workspace
t = outputdata(1,:); x = outputdata(2,:); f_input = outputdata(3,:);
% t is the time, x is the position output and f_input is the force input.
% The following lines plot the position output
p1 = plot(t,x,'b.'); % For more plotting options type "help plot"
set(gca,'FontSize',11)
title('x(t)','FontSize',18)
ylabel('x(t)','FontSize',16)
xlabel ('Time in seconds','FontSize',16)
% Set the line width for plotting
set(p1,'LineWidth',2); set(p1,'MarkerSize',10);
```

Problems

Problem 1 *Vertical Mass–Spring System*

Consider the spring–mass–damper system in Figure 4.13.

(a) Write down the equations of motion of the mass m using Newton's laws of motion.

(b) With $f(t) = 0$, compute the equilibrium point x_0 of the system.

(c) Let $\Delta x \triangleq x - x_0$ and rewrite the equations of motion in terms of Δx.

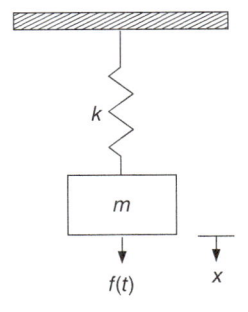

Figure 4.13. Vertical mass–spring system.

Problem 2 *Vertical Mass–Spring–Damper System*

Figure 4.14 shows a spring–mass–damper system with the spring attached to the ceiling and the mass m, and the damper attached to the floor and the mass m. $f(t)$ is an external input force. The position of m is y and, as usual, $y = 0$ means the spring is relaxed.

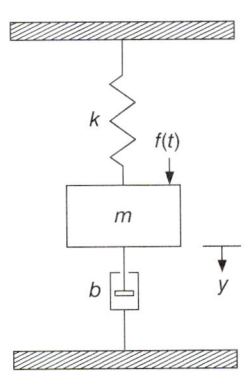

Figure 4.14. Vertical mass–spring–damper system.

(a) Write the equations of motion for m. Compute the transfer function from the input $f(t) + mg$ to the output $y(t)$. Remember that in computing a transfer function the initial conditions are set to zero.

(b) With $f(t) = 0$, find the equilibrium position y_0 of m.

(c) With $\Delta y = y - y_0$ and $f(t) \neq 0$ rewrite the equations of motion in terms of the position Δy relative to the equilibrium point. Compute the transfer function from the input $f(t)$ to the output $\Delta y(t)$. Remember that in computing a transfer function the initial conditions are set to zero.

(d) With $m = 2$ kg, $b = 4$ N/(m/s), $k = 16$ N/m and the input $f(t) = 2u_s(t)$, does $\Delta y(t)$ have a final value? Explain why or why not. If it does have a final value, compute it using the final value theorem.

(e) Make a SIMULINK simulation of this system. Hand in a print out of the SIMULINK diagram, and a plot of $\Delta y(t)$ with the initial conditions set to zero. Does $\Delta y(t)$ settle out to the final value found in part (d)?

4 Mass–Spring–Damper Systems

Problem 3 *Moving Mass–Spring–Damper System*
Figure 4.15 shows a spring–mass–damper system in the back of a truck where the position x of the truck is the input and the position y of the mass m is the output.

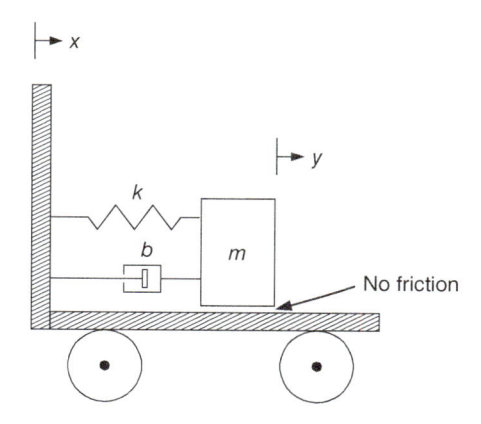

Figure 4.15. Moving mass–spring–damper system.

(a) Find the equations of motion of the mass m.
(b) Compute the transfer function from the input $x(t)$ to the output $y(t)$.
(c) Set $m = 2$ $kg, b = 4$ $N/(m/s), k = 16$ N/m and let the input be

$$x(t) = \begin{cases} 2t, & 0 \leq t \leq 1 \\ 2, & t > 1. \end{cases}$$

Does $y(t)$ have a final value? Explain why or why not. If it does have a final value, compute it using the final value theorem.

(d) Make a SIMULINK simulation of this system. Hand in a print out of the SIMULINK diagram, and a plot of $x(t)$ and $y(t)$ with the initial conditions set to zero. Does $y(t)$ settle out to the final value found in part (c)?

Problem 4 *Vertical Mass–Spring–Damper System*
Consider the mass–spring–damper system of Figure 4.16. Let m_p be the mass of the damper's piston which we will later let go to zero.

(a) Write down the equations of motion for this spring–mass–damper system.
(b) Find the equilibrium point (x_{01}, x_{02}) for this system.
(c) Rewrite the equations of motion in terms of y_1 and y_2 given by

$$y_1 = x_1 - x_{01}$$
$$y_2 = x_2 - x_{02}.$$

(d) Let $m_p = 0$ and further let $y_2(0) = \dot{y}_2(0) = 0$ and $y_1(0) = 0$, but $\dot{y}_1(0)$ be arbitrary. Compute $Y_1(s)$.

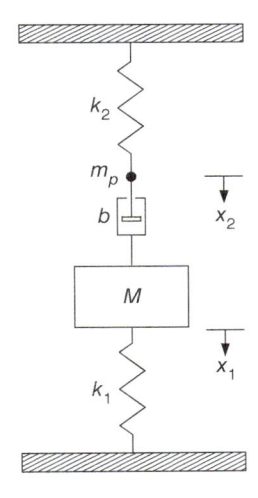

Figure 4.16. Vertical mass–spring–damper system.

(e) Make a SIMULINK simulation of this system. Let $M = 2$ kg, $b = 4$ N/(m/s), $k_1 = 16$ N/m, and $k_2 = 16$ N/m. Hand in a print out of the SIMULINK diagram, and a plot of $y_1(t)$ and $y_2(t)$ with $y_2(0) = \dot{y}_2(0) = 0, y_1(0) = 0$, but $\dot{y}_1(0) = 1$.

Problem 5 *Mass–Spring–Damper Model for a Wheel Assembly* [1]
Figure 4.17 shows a model of a car wheel attached to the body of the car.

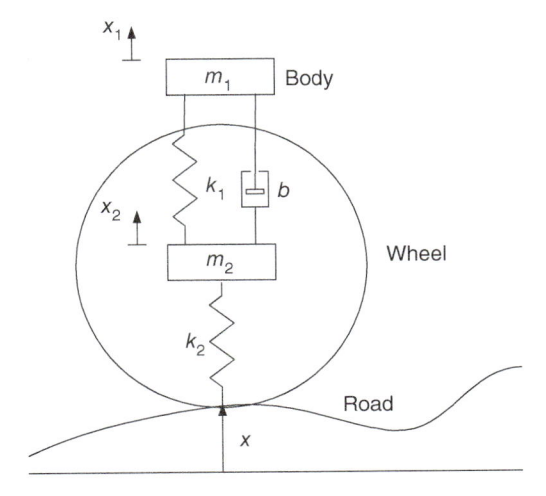

Figure 4.17. Vertical mass–spring–damper modeling a suspension system.

The spring k_2 is used to model the tire's flexibility with the mass m_2 modeling wheel assembly to which the tire is attached. The wheel assembly is attached to the car body using a spring (modeled by k_1) and a shock absorber (modeled by b). Finally m_1 is the mass of that part of the car body suspended by this assembly. The (disturbance) input is x which is the height of the road above some fixed position and the output is x_1 which gives the position of the car body. As usual, if $x = x_2$ then the spring k_2 is relaxed, while if $x_2 = x_1$ then the spring k_1 is relaxed.

(a) Write down the equations of motion.

(b) With the input $x = 0$, find the equilibrium point (x_{01}, x_{02}) for this system.

(c) Let
$$y_1 \triangleq x_1 - x_{01}$$
$$y_2 \triangleq x_2 - x_{02}$$
and find the equations of motion in terms of y_1 and y_2.

(d) Find the transfer function from the input x to the output y_1.

(e) With $m_1 = 100, m_2 = 10, b = 1, k_1 = 16, k_2 = 16, x(t) = 0.1u_s(t)$ and zero initial conditions, make a SIMULINK simulation of this system. Hand in a print out of the SIMULINK diagram and a plot of $y_1(t)$.

Problem 6 *Horizontal Mass–Spring–Damper System*

Consider the spring–mass–damper system in Figure 4.18.

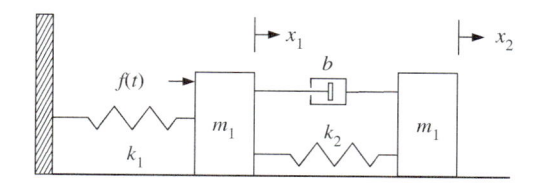

Figure 4.18. Horizontal mass–spring–damper system.

(a) Write down the equations of motion of the two masses m_1, m_2 using Newton's laws of motion.

(b) Compute the transfer function from the input $f(t)$ to the output $x_2(t)$.

(c) With $m_1 = 4, m_2 = 2, b = 4, k_1 = 16, k_2 = 16, f(t) = 8u_s(t)$ and zero initial conditions, make a SIMULINK simulation of this system. Hand in a print out of the SIMULINK diagram and a plot of $x_1(t)$ and $x_2(t)$.

Problem 7 *Vertical Mass–Spring–Damper System*

Consider the spring–mass–damper system in Figure 4.19.

(a) Write down the equations of motion of the two masses m_1, m_2 using Newtons's laws of motion.

(b) With $f(t) = 0$, compute the equilibrium point (x_{01}, x_{02}) of the system.

(c) Rewrite the equations of motion in part (a) in terms of y_1, y_2 given by
$$y_1 = x_1 - x_{01}$$
$$y_2 = x_2 - x_{02}.$$

(d) With $m_1 = 10, m_2 = 2, b = 1, k_1 = 16, k_2 = 16, f(t) = u_s(t)$ and zero initial conditions, make a SIMULINK simulation of this system. Hand in a print out of the SIMULINK diagram and a plot of $y_1(t)$ and $y_2(t)$.

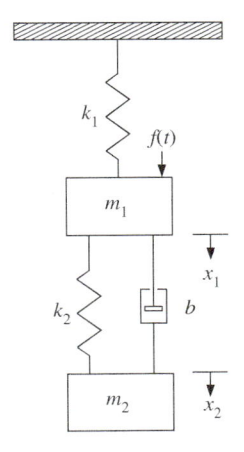

Figure 4.19. Vertical mass–spring–damper system.

Problem 8 *Final Value Theorem*

In Example 5 we showed the transfer function of the system given in Figure 4.7 was given by
$$G(s) = \frac{X(s)}{Y(s)} = \frac{k_1 bs}{bms^3 + mk_1 s^2 + (bk_1 + bk_2)s + k_1 k_2}.$$

Recall that $Y(s)$ is the (input) position of the left side of the spring while $X(s)$ is the (output) position of the mass m. Let $m = 2$ kg, $b = 4$ N/(m/s), $k_1 = k_2 = 16$ N/m.

(a) Use MATLAB to compute the poles of $G(s)$. Is $G(s)$ stable?

(b) Let $Y(s) = y_0/s$ be a step input. Can you use the final value theorem to compute $\lim_{t \to \infty} x(t)$? If so, do so. If not, explain why not.

(c) In Example 5 it was shown that
$$(m_p s^2 + bs + k_1)Z(s) = bsX(s) + k_1 Y(s)$$

or (setting $m_p = 0$)
$$Z(s) = \frac{bs}{bs + k_1} G(s) Y(s) + \frac{k_1}{bs + k_1} Y(s).$$

Can you use the final value theorem to compute $\lim_{t \to \infty} z(t)$? If so, do so. If not, explain why not.

(d) Give a physical interpretation of your answers in parts (b) and (c).

Problem 9 *Simulation of a DC Motor*

The differential equations describing a DC (direct current) motor are

$$L\frac{di}{dt} = -Ri - K_b \omega_R + V_S$$

$$J\frac{d\omega_R}{dt} = K_T i - f\omega_R - \tau_L \quad (4.12)$$

$$\frac{d\theta_R}{dt} = \omega_R.$$

4 Mass–Spring–Damper Systems

The parameter values are $J = 6 \times 10^{-5}$ kg/m^2, $K_T = K_b = 0.07$ Nm/A (V/rad/s), $f = 0.4 \times 10^{-3}$ Nm/rad/s, $R = 2$ Ω, $L = 0.002$ Henrys, $V_{\max} = 40$ V, $I_{\max} = 5$ A. For a step input voltage set $V_S = 10$ V and for a ramp input voltage set $V_S(t) = t$. Set the start time to 0 and the stop time to 0.2 seconds. Use a fixed-step Euler integration with a step size of 0.001 second. The SIMULINK block diagram is shown in Figure 4.20.

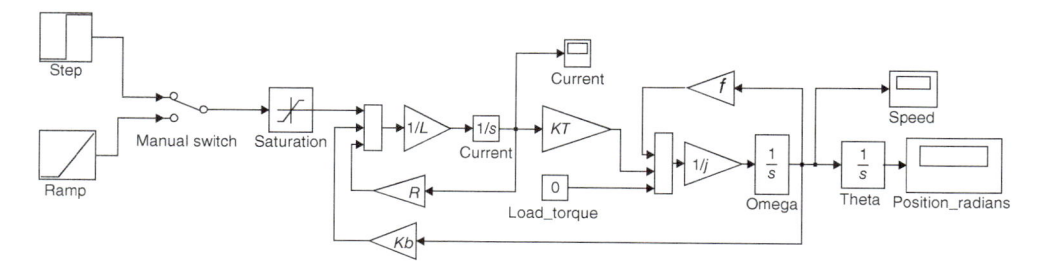

Figure 4.20. SIMULINK simulation of a DC motor.

The `saturation` block is found in the SIMULINK library in the `Discontinuities` folder. If you double click on this block the dialog window shown in Figure 4.21 comes up.

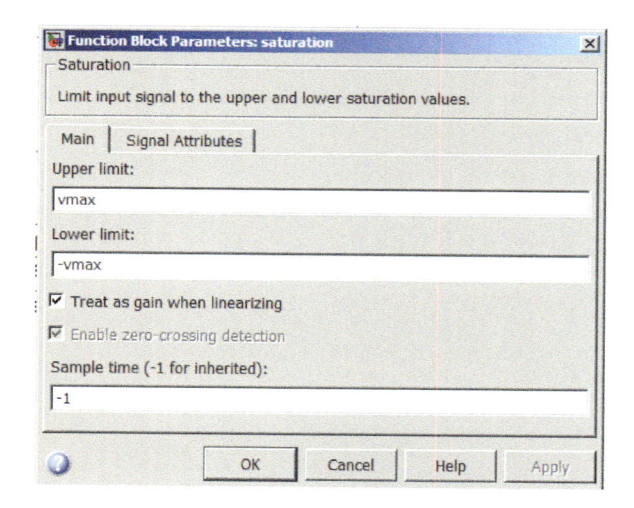

Figure 4.21. Dialog box for the `Saturation` block.

The `manual switch` block is found in the SIMULINK library in the `Signal Routing` folder.

(a) Write a .m MATLAB file named `DC_sim_data.m` to set all the parameter values.

(b) Write a .slx SIMULINK file named `DC_sim.slx` that reproduces the block diagram of Figure 4.20.

(c) Run your simulation with the start time set to 0 and the stop time set to 0.2 seconds. Use a fixed-step Euler integration with a step size of 0.001 seconds.

(d) With the simulation debugged, add a `To File` SIMULINK block (in the `Sinks` folder of the library) to store the input voltage, the current, the speed, and the position.

(e) For the step input case, hand in a printout of your .m file, a screenshot of your SIMULINK block diagram (should look similar to Figure 4.20), and a MATLAB plot of each of the stored variables vs. time.

Remark Notice that the simulation diagram of Figure 4.20 does not use I_{max}. Usually, for protection, an amplifier has a current sensor inside to measure the current. If the magnitude of the current exceeds I_{max} the amplifier is designed to immediately shut down.

Problem 10 *Simulation of a Stable Differential Equation*
In Chapter 3 we found a solution to the differential equation

$$\frac{d^2x}{dt^2} + \frac{dx}{dt} + x = U_0 \cos(\omega t) \tag{4.13}$$

using the phasor method. With

$$X(s) = \underbrace{\frac{1}{s^2 + s + 1}}_{G(s)} U_0 \underbrace{\frac{s}{s^2 + \omega^2}}_{U(s)}$$

the phasor solution is given by $x_{ph}(t) = |G(j\omega)| U_0 \cos(\omega t + \angle G(j\omega))$. This is in fact the unique solution to (4.13) with the initial conditions

$$x(0) = |G(j\omega)| U_0 \cos(\angle G(j\omega))$$
$$\dot{x}(0) = -|G(j\omega)| \omega U_0 \sin(\angle G(j\omega)).$$

As $G(s)$ is stable, $x_{ph}(t)$ is also the steady-state solution, i.e., for *arbitrary* initial conditions $x(t) \to x_{ph}(t)$. Figure 4.22 is a SIMULINK block diagram for this system. To run this simulation you will need to first run the following .m file.

```
U0 = 2; omega = 2;
% G(jw) = 1/(jw)^2+jw+1)
G = 1/((i*omega)^2 + i*omega + 1);
Gmag = abs(G); Gphase = angle(G); Gphase_deg = Gphase*180/pi;
```

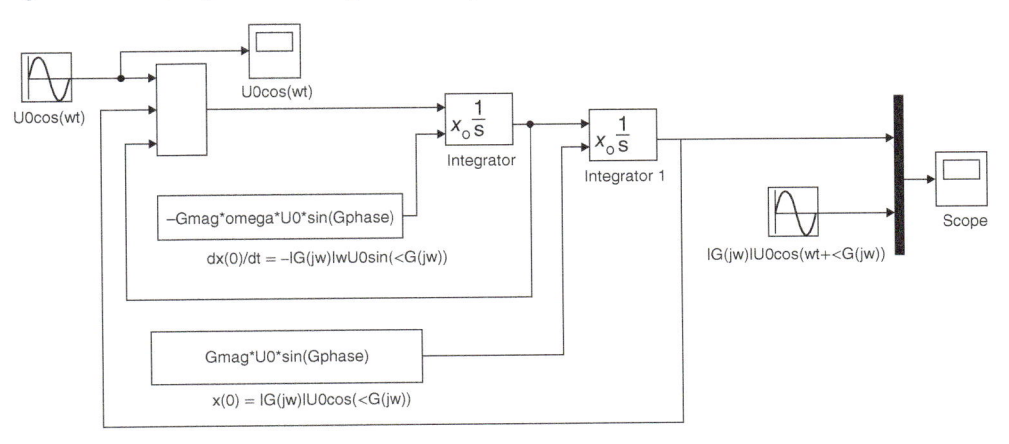

Figure 4.22. SIMULINK block diagram for $\dfrac{d^2x}{dt^2} + \dfrac{dx}{dt} + x = U_0 \cos(\omega t)$.

108   4 Mass–Spring–Damper Systems

To implement the input $U_0\cos(\omega t)$ open the dialog box of the `Sine Wave` block labeled as `U0cos(wt)` and fill it in as shown in Figure 4.23.

Figure 4.23. Dialog block for the `Sine Wave` source block for $U_0\sin(\omega t + \pi/2) = U_0\cos(\omega t)$.

To implement the input phasor solution $|G(j\omega)|\,U_0\cos(\omega t + \angle G(j\omega))$ open the dialog box of the `Sine Wave` block labeled as `|G(jw)|U0cos(wt+<G(jw))` and fill it in as shown in Figure 4.24.

(a) Implement this simulation and run it for 15 seconds. Hand in the plot from the output scope.

(b) Set the initial conditions to zero and run the simulation for 15 seconds. Hand in the plot from the output scope.

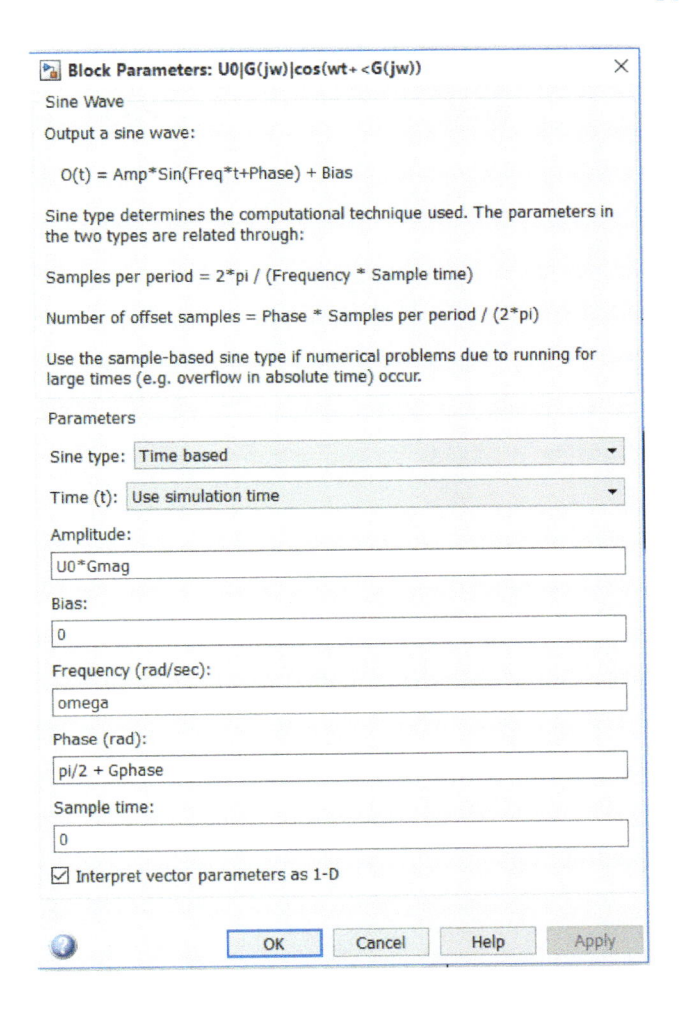

Figure 4.24. Dialog block for Sine Wave source for $|G(j\omega)| U_0 \sin(\omega t + \pi/2 + \angle G(j\omega)) = |G(j\omega)| U_0 \cos(\omega t + \angle G(j\omega))$

Problem 11 *Simulation of an Unstable Differential Equation*

This problem is similar to Problem 10 except with an *unstable* differential equation. In Chapter 3 we found a solution to the differential equation

$$\frac{d^2x}{dt^2} - \frac{dx}{dt} + x = U_0 \cos(\omega t) \tag{4.14}$$

using the phasor method. With

$$X(s) = \underbrace{\frac{1}{s^2 - s + 1}}_{G(s)} U_0 \underbrace{\frac{s}{s^2 + \omega^2}}_{U(s)}$$

the phasor solution is given by $x_{ph}(t) = |G(j\omega)|\, U_0 \cos(\omega t + \angle G(j\omega))$. This is the unique solution to (4.14) with the initial conditions

$$x(0) = |G(j\omega)|\, U_0 \cos(\angle G(j\omega)) \tag{4.15}$$

$$\dot{x}(0) = -|G(j\omega)|\, \omega U_0 \sin(\angle G(j\omega)). \tag{4.16}$$

As $G(s)$ is *unstable*, there is no steady-state solution, i.e., for *arbitrary* initial conditions $x(t) \nrightarrow x_{ph}(t)$. Modify your SIMULINK simulation from Problem 10 to simulate this system. In particular, you will need to run the following .m file.

```
U0 = 2; omega = 2;
% G(jw) = 1/(jw)^2-jw+1)
G = 1/((i*omega)^2 - i*omega + 1);
Gmag = abs(G); Gphase = angle(G); Gphase_deg = Gphase*180/pi;
```

(a) In the Simulation/Model Configuration Parameters dialog box make sure that fixed step size is 0.001. Run it for 10 seconds. Hand in the plot from the output scope. You should see that the simulation starts to diverge from the phasor solution at about five seconds.

(b) In the Simulation/Model Configuration Parameters dialog box change the fixed step size to 0.0001. Run it for 10 seconds. Hand in the plot from the output scope. You should see that the simulation starts to diverge from the phasor solution at about 10 seconds.

Remark $|G(j\omega)|\, U_0 \cos(\omega t + \angle G(j\omega))$ is the solution to (4.14) with the initial conditions (4.15) and (4.16). However, because the numerical integration of the differential equation is not exact, the SIMULINK solution diverges from $|G(j\omega)|\, U_0 \cos(\omega t + \angle G(j\omega))$. As a consequence the output response contains terms corresponding to the unstable poles of the transfer function which grow unbounded. As part (b) shows, even if the step size is very small, the integration is still not exact and the numerically computed output response in SIMULINK of this unstable differential equation again goes unbounded.

5
Rigid Body Rotational Dynamics

5.1 Moment of Inertia

The equations of motion of a rigid body that is constrained to rotate about a fixed axis are reviewed here briefly. Consider the cylinder shown in Figure 5.1.

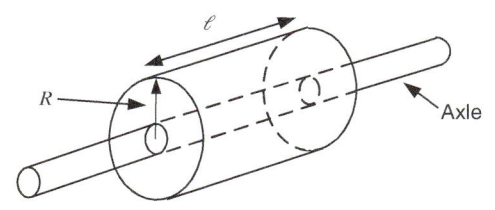

Figure 5.1. Cylinder constrained to rotate about a fixed axis.

The approach here is to obtain the equations of motion of the cylinder by first obtaining an expression for its kinetic energy. To do so, denote the angular speed of the cylinder by ω and the mass density of the material making up the cylinder by ρ. Then consider the cylinder to be made up of a large number n of small pieces of material Δm_i where the ith piece has mass

$$\Delta m_i = \rho r_i \Delta\theta \Delta\ell \Delta r.$$

This is illustrated in Figure 5.2. Each piece of mass Δm_i is rotating at the same angular speed ω so that the *linear* speed of Δm_i is $v_i = r_i \omega$ where r_i is the distance of Δm_i from the axis of rotation. The kinetic energy KE_i of Δm_i is given by

$$KE_i = \frac{1}{2}\Delta m_i v_i^2 = \frac{1}{2}\Delta m_i (r_i \omega)^2.$$

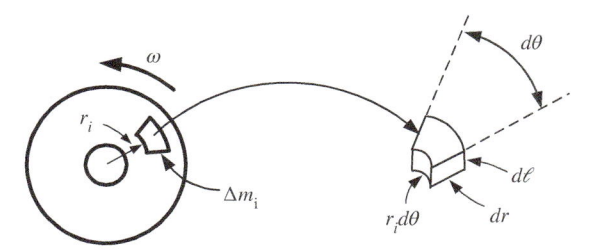

Figure 5.2. Cylinder is considered to be made up of small masses Δm_i.

The total kinetic energy is then

$$KE = \sum_{i=1}^{n}(KE)_i = \sum_{i=1}^{n}\frac{1}{2}\Delta m_i v_i^2 = \sum_{i=1}^{n}\frac{1}{2}\Delta m_i (r_i \omega)^2 = \frac{1}{2}\omega^2 \sum_{i=1}^{n}\Delta m_i r_i^2.$$

An Introduction to System Modeling and Control, First Edition. John Chiasson.
© 2022 John Wiley & Sons, Inc. Published 2022 by John Wiley & Sons, Inc.
Companion website: www.wiley.com/go/chiasson/anintroductiontosystemmodelingandcontrol

112   5 Rigid Body Rotational Dynamics

Dividing the cylinder into finer and finer pieces so that $n \to \infty$ and $\Delta m_i \to 0$, the sum

$$\sum_{i=1}^{n} \Delta m_i r_i^2$$

becomes the integral

$$J = \iiint_{cylinder} r^2 dm.$$

The quantity J is called the *moment of inertia*. Using J the kinetic energy of the cylinder may now be written as

$$KE = \frac{1}{2} J \omega^2.$$

Taking the axle radius to be zero, the moment of inertia of the cylinder (assuming the mass density ρ is constant) is computed to be

$$J = \int_0^R \int_0^\ell \int_0^{2\pi} r^2 \rho r \, d\theta \, d\ell \, dr = \frac{1}{2}(\pi R^2 \ell \rho) R^2 = \frac{1}{2} M R^2,$$

where M is the total mass of the cylinder.

5.2 Newton's Law of Rotational Motion

The kinetic energy is now used to derive a relationship between torque and angular acceleration. Recall from elementary mechanics that the work done on a mass by an external force equals the change in its kinetic energy. In particular, consider an external force $\vec{F}$ acting on the cylinder as shown in Figure 5.3.

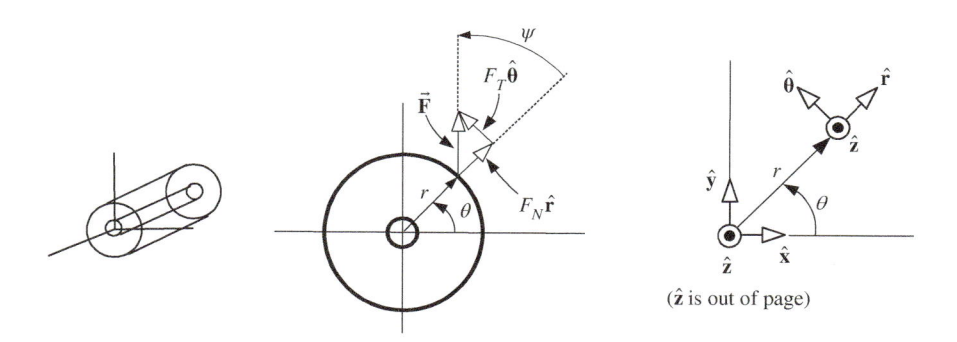

Figure 5.3. Force $\vec{F}$ applied to the cylinder is resolved into a normal and tangential component using polar coordinates.

The cylinder is on an axle and therefore constrained to rotate about the z axis. Figure 5.3 shows the force $\vec{F}$ applied to the cylinder at the position (r, θ) (in polar coordinates) resolved into a tangential component F_T (tangent to the rotational motion) and a normal component F_N. Using polar coordinates, we write this force as $\vec{F} = F_N \hat{r} + F_T \hat{\theta}$ where $\hat{r}$ is a unit vector in the increasing r direction and $\hat{\theta}$ is a unit vector in the increasing θ direction. Similarly, $\hat{x}$, $\hat{y}$, and $\hat{z}$ are unit vectors in the increasing x, y, z directions, respectively. The torque about an axis is defined as the cross product $\vec{\tau} \triangleq \vec{r} \times \vec{F}$ where $\vec{r}$

5.2 Newton's Law of Rotational Motion

is the vector from the axis to the point of application of the force and $\vec{\mathbf{F}}$ is the applied force. We then have

$$\vec{\tau} \triangleq \vec{\mathbf{r}} \times \vec{\mathbf{F}} = r\hat{\mathbf{r}} \times (F_N \hat{\mathbf{r}} + F_T \hat{\boldsymbol{\theta}})$$

$$= rF_N \underbrace{\hat{\mathbf{r}} \times \hat{\mathbf{r}}}_{\vec{0}} + rF_T \underbrace{\hat{\mathbf{r}} \times \hat{\boldsymbol{\theta}}}_{\hat{\mathbf{z}}}$$

$$= rF_T \hat{\mathbf{z}}$$

$$= r|\vec{\mathbf{F}}|\sin(\psi)\hat{\mathbf{z}}, \tag{5.1}$$

where ψ is the angle from $\vec{\mathbf{r}}$ to $\vec{\mathbf{F}}$. Recall from elementary mechanics that the *magnitude* of the cross product $\vec{\mathbf{r}} \times \vec{\mathbf{F}}$ is defined as $|\vec{\mathbf{r}}||\vec{\mathbf{F}}|\sin(\psi) = r|\vec{\mathbf{F}}|\sin(\psi)$ and the *direction* of $\vec{\mathbf{r}} \times \vec{\mathbf{F}}$ is perpendicular to both $\vec{\mathbf{r}}$ and $\vec{\mathbf{F}}$ along the axis of rotation determined by the right-hand Rule.[1] $F_T = |\vec{\mathbf{F}}|\sin(\psi)$ is the *tangential* component and we have just shown the torque is given by

$$\vec{\tau} = \tau \hat{\mathbf{z}} = rF_T \hat{\mathbf{z}}$$

or in scalar form by

$$\tau = rF_T.$$

The motivation for the definition of torque as given by (5.1) is that it is the cause of *rotational* motion (angular acceleration). Specifically, the rotational motion about an axis is caused by the applied tangential force F_T and the further away from the axis of rotation that the tangential force F_T is applied, the easier it is to get rotational motion. That is, the torque (cause of rotational motion) increases if either r or F_T increases, which corresponds to one's experience (e.g. opening doors).

To summarize, $\vec{\tau}$ is a vector pointing along the axis of rotation with its magnitude given by

$$|\vec{\tau}| = |\tau| = |rF_T|.$$

(Recall that the angular velocity vector $\vec{\omega} = \omega \hat{\mathbf{z}}$ also points along the axis of rotation where ω is the angular speed.)

Let the external force $\vec{\mathbf{F}}$ act on the cylinder to move (rotate) it by a displacement $d\vec{\mathbf{s}} \triangleq ds\hat{\boldsymbol{\theta}} = rd\theta\hat{\boldsymbol{\theta}}$. The change in work done on the cylinder by this force is then

$$dW = \vec{\mathbf{F}} \cdot d\vec{\mathbf{s}} = F_T r d\theta = \tau d\theta.$$

Dividing by dt, the power (rate of work) delivered to the cylinder is given by

$$\frac{dW}{dt} = \tau \frac{d\theta}{dt} = \tau\omega.$$

As the rate of work done equals the rate of change of kinetic energy, it follows that

$$\frac{dW}{dt} = \frac{d}{dt}\left(\frac{1}{2}J\omega^2\right) = \tau \frac{d\theta}{dt} \tag{5.2}$$

or

$$J\omega \frac{d\omega}{dt} = \tau\omega.$$

---

[1] Curl the fingers of your right hand in the direction from the first vector $\vec{\mathbf{r}}$ to the second vector $\vec{\mathbf{F}}$. Then your thumb points in the direction of $\vec{\mathbf{r}} \times \vec{\mathbf{F}}$.

5 Rigid Body Rotational Dynamics

This gives the fundamental relationship between torque and angular acceleration:

$$\tau = J\frac{d\omega}{dt}. \tag{5.3}$$

That is, the applied torque equals the moment of inertia times the angular acceleration. This is the basic equation for rigid body rotational dynamics about a fixed axis.

Viscous Rotational Friction

Almost always there are frictional forces, and therefore, frictional torques acting between the axle and the bearings.[2] This is illustrated in Figure 5.4.

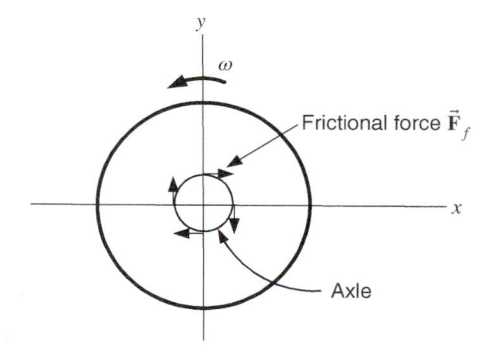

Figure 5.4. Viscous friction torque.

Often the frictional force is proportional to the angular speed and this model of friction is called *viscous friction*, which is expressed mathematically as

$$\vec{\tau} = -f\vec{\omega} = -f\omega\hat{\mathbf{z}}$$

or, in scalar form,

$$\tau = -f\omega,$$

where $f > 0$ is the *coefficient of viscous friction*.

Sign Convention for Torque

Suppose the axis of rotation is along the z axis. The torque is

$$\vec{\tau} \triangleq \vec{r} \times \vec{F} = r|\vec{F}|\sin(\psi)\hat{\mathbf{z}} = \underbrace{rF_T}_{\tau}\hat{\mathbf{z}} = \tau\hat{\mathbf{z}},$$

where ψ is the angle from $\vec{r}$ to $\vec{F}$. The cross product $\vec{r} \times \vec{F}$ has the component $r|\vec{F}|\sin(\psi)$ in the $\hat{\mathbf{z}}$ direction. Notice that direction of $\vec{r} \times \vec{F}$ is perpendicular to both $\vec{r}$ and $\vec{F}$.

In engineering applications, the systems are designed so that the applied force is tangential to the rotational motion, i.e., $\psi = \pi/2$ so that $F_T = |\vec{F}|\sin(\psi) = |\vec{F}|$. Further, in engineering texts the sign convention for torque is indicated by a curved arrow as shown in Figure 5.5.

---

[2] An interesting exception are magnetic bearings where the axel is levitated by magnetic fields so that there is no mechanical contact.

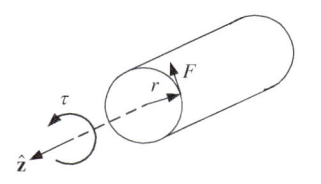

Figure 5.5. Sign convention for torque.

If $\tau = rF_T > 0$ then the torque will cause the cylinder to rotate around the z axis in the direction indicated by the curved arrow. On the other hand, if $\tau = rF_T < 0$ then the torque will cause the cylinder to rotate around the z axis in the direction *opposite* to that indicated by the curved arrow. Physics texts prefer to write $\vec{\tau} \triangleq \tau\hat{\mathbf{z}}$.

Example 1 *Rack and Pinion System*
A rack and pinion system is illustrated in Figure 5.6, which is used to convert rotary motion to linear motion and vice versa. In Figure 5.6 the axis of the pinion (gear wheel) is considered to be fixed in space. A torque τ applied to the shaft causes the pinion to rotate, which moves the rack in the x direction through the contact of their teeth.

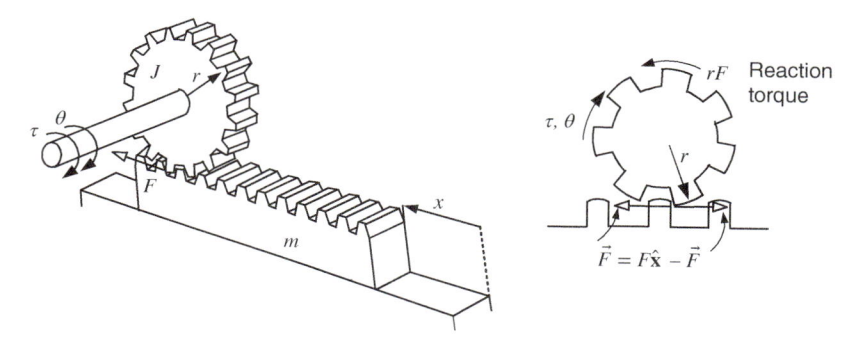

Figure 5.6. Rack and pinion system. F is the force of the pinion tooth on the rack tooth. $-F$ is the reaction force of the rack tooth on the pinion tooth.

We take the input to be the torque τ (produced by a motor) and the output to be the position x. Because the teeth on the pinion wheel and rack are meshed together, there is an algebraic relationship between the angle θ and position x given by $x = r\theta$. m is the mass of the rack and J is the inertia of the pinion gear wheel. Let F be the force of the pinion tooth on the rack tooth in the x direction so that $-F$ is the reaction force of the rack tooth on the pinion tooth. Applying Newton's law to the mass of the rack we have

$$m\ddot{x} = F.$$

Applying Newton's law for rotational motion to the pinion gear we have

$$J\ddot{\theta} = \tau - Fr,$$

where $-Fr$ is the reaction torque on the pinion gear produced by the rack. Multiply the first equation by r and add to the second to obtain

$$J\ddot{\theta} + mr\ddot{x} = \tau.$$

116   5 Rigid Body Rotational Dynamics

In this equation we eliminate θ using the algebraic constraint $\theta = x/r$ and rearrange it to obtain
$$(J + mr^2)\ddot{x} = r\tau.$$

The transfer function is then
$$\frac{X(s)}{\tau(s)} = \frac{r}{J + mr^2}\frac{1}{s^2}.$$

Conservation of Energy

Let's rederive this transfer function of the rack and pinion system using conservation of energy. The kinetic energy of the rack and pinion system is
$$KE = \frac{1}{2}J\dot{\theta}^2 + \frac{1}{2}m\dot{x}^2 = \frac{1}{2}J\dot{x}^2/r^2 + \frac{1}{2}m\dot{x}^2 = \frac{1}{2}(J/r^2 + m)\dot{x}^2.$$

As shown in (5.2), the rate of work done equals the rate of change of kinetic energy. Consequently we have,
$$\tau\frac{d\theta}{dt} = \frac{d(KE)}{dt}$$

or
$$\tau\frac{1}{r}\frac{dx}{dt} = (J/r^2 + m)\dot{x}\ddot{x}$$

or
$$r\tau = (J + mr^2)\ddot{x}.$$

The transfer function from τ to X is then
$$\frac{X(s)}{\tau(s)} = \frac{r}{J + mr^2}\frac{1}{s^2}.$$

Example 2 *Rack and Pinion System Connected to a Spring*

Figure 5.7 shows a rack and pinion system with the rack connected to the wall through a spring and also some viscous friction damping between the rack and the support surface. We consider the torque τ (produced by a motor) to be the input and we take the position x to be the output. r is the radius of the pinion (gear), J_p is the moment of inertia of the

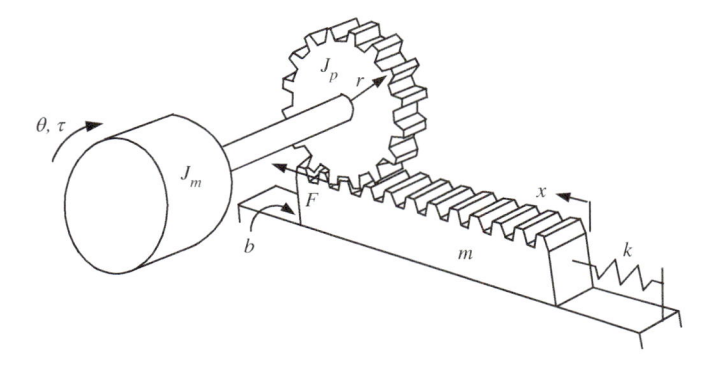

Figure 5.7. Rack and pinion system.

5.2 Newton's Law of Rotational Motion

pinion about its center, J_m is the moment of inertia of the motor, b is the viscous friction coefficient between the rack and the surface it is on, k is the spring constant, and m is the mass of the rack.

The input is the torque τ and the position x is the output. Let's first find the transfer function using Newton's laws. Let F be the force of the pinion wheel tooth on the rack gear tooth. Then the reaction force of the rack gear tooth on the pinion wheel is $-F$ resulting in a reaction torque on the pinion wheel of $-rF$. The equations modeling this system are then

$$(J_m + J_p)\ddot{\theta} = \tau - Fr$$
$$m\ddot{x} = F - b\dot{x} - kx$$

along with the algebraic constraint

$$x = r\theta.$$

F is not known and typically not measurable. Consequently, we eliminate F from these two equations to obtain

$$(J_m + J_p)\ddot{\theta} + rm\ddot{x} = \tau - r(b\dot{x} + kx).$$

Substituting $\theta = x/r$ (and thus $\dot{x} = r\dot{\theta}$ and $\ddot{x} = r\ddot{\theta}$) we obtain

$$(J_m + J_p)\ddot{x}/r + rm\ddot{x} = \tau - rb\dot{x} - rkx.$$

Rearranging this becomes

$$(J_m + J_p + mr^2)\ddot{x} + r^2 b\dot{x} + kr^2 x = \tau r.$$

Taking the Laplace transform with zero initial conditions gives

$$(J_m + J_p + mr^2)s^2 X(s) + r^2 bs X(s) + kr^2 X(s) = r\tau(s).$$

The transfer function is then

$$X(s) = \frac{r}{(J_m + J_p + mr^2)s^2 + r^2 bs + kr^2} \tau(s).$$

Let's now find the equations of motion using conservation of energy. Referring back to (5.3), we see that if the external torque τ rotates the system by $d\theta$, the work dW done on the system is $\tau d\theta$. By conservation of energy, this is equal to the change in the kinetic energy of $J_m + J_p$ and m plus the change in the spring's potential energy plus the heat dissipated due to the viscous friction. In terms of the input power $P = dW/dt$ put into the system by τ, we have $P = dW/dt = \tau d\theta/dt$ so that

$$\underbrace{\tau \frac{d\theta}{dt}}_{\text{Input mech pwr}} = \frac{d}{dt}\underbrace{\left(\frac{1}{2}(J_m + J_p)\dot{\theta}^2 + \frac{1}{2}m\dot{x}^2 + \frac{1}{2}kx^2\right)}_{\text{KE} \qquad\qquad \text{PE}} + \underbrace{(b\dot{x})\dot{x}}_{\text{Pwr dissipated as heat}}$$

$$= (J_m + J_p)\dot{\theta}\ddot{\theta} + m\dot{x}\ddot{x} + kx\dot{x} + b\dot{x}^2.$$

118   5 Rigid Body Rotational Dynamics

Using the algebraic constraint $x = r\theta$ this becomes

$$\tau \frac{d(x/r)}{dt} = (J_m + J_p)(\dot{x}/r)(\ddot{x}/r) + m\dot{x}\ddot{x} + kx\dot{x} + b\dot{x}^2.$$

Multiplying through by r^2 and canceling the common factor $\dot{x}$ we have

$$\tau r = (J_m + J_p + mr^2)\ddot{x} + r^2 b\dot{x} + kr^2 x$$

which will result in the same transfer function as before.

Example 3 *Satellite with Solar Panels*

Following [3] and [13] let's consider a simple model of a satellite, which has solar panels attached to it to provide electric power as shown in Figure 5.8a. The panels are flexible in order to make them light as possible for launching into orbit. The two solar panels are attached to the same shaft going through the satellite main body and a motor produces a torque τ to turn the panels to line them up with sun. However, as the panels are not rigid, the torque causes the panels to oscillate about their axis of rotation. θ is the angle of the motor shaft with respect to the satellite body while θ_p is the angle of the end of the solar panel with respect to the satellite body. We model this system using a torsional spring and rotational damper system as shown in Figure 5.8b.

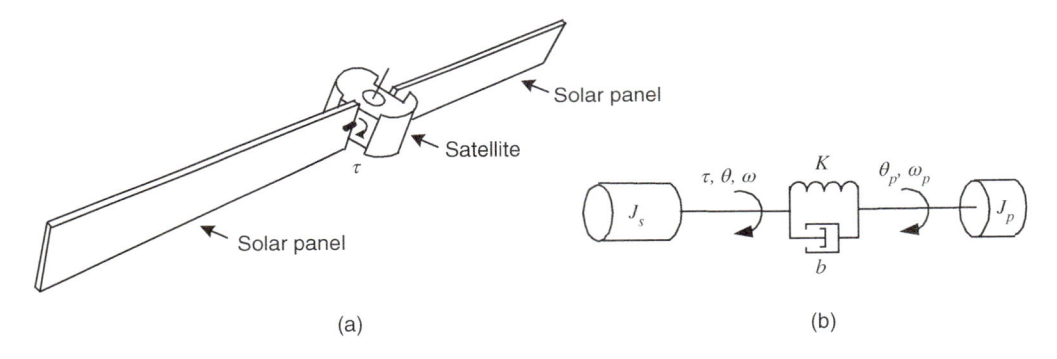

Figure 5.8. (a) Satellite with solar panels for power. (b) Lumped parameter model.

The torsional spring constant is K, the rotational damper constant is b (typically very small), the moment of inertia of the satellite main body is J_s, and J_p is the moment of inertia of both solar panels. The solar panels are not rigid so they really do not have a moment of inertia. However J_p is taken to be an "equivalent" moment of inertia with the expectation that this lumped parameter model is a "good" approximation for the dynamics of the flexible solar panels.[3] The torsional spring constant K is used to model the torque produced when the shaft of the solar panels is twisted by the motor's torque. This torque on the motor shaft is $-K(\theta - \theta_p)$ while the (reaction) torque on the solar panel shaft is $K(\theta - \theta_p)$. When the motor turns (twists) the solar panel shaft, the panel oscillates (twists back and forth) with smaller and smaller amplitude until it finally stops. The rotational damper is used to model this internal damping mechanism. The energy loss due to this damping is dissipated as heat. This damping torque on the motor shaft

---

[3] The values of K, b, J_p that (approximately) model the mechanical dynamics of the flexible panels can be found experimentally using measured data.

5.2 Newton's Law of Rotational Motion

is $-b\left(\frac{d\theta}{dt} - \frac{d\theta_p}{dt}\right)$ while the (reaction) torque on the solar panel shaft is $b\left(\frac{d\theta}{dt} - \frac{d\theta_p}{dt}\right)$.
Putting this altogether the dynamic equations of this satellite system are

$$J_s \frac{d^2\theta}{dt^2} = -K(\theta - \theta_p) - b\left(\frac{d\theta}{dt} - \frac{d\theta_p}{dt}\right) + \tau \tag{5.4}$$

$$J_p \frac{d^2\theta_p}{dt^2} = K(\theta - \theta_p) + b\left(\frac{d\theta}{dt} - \frac{d\theta_p}{dt}\right). \tag{5.5}$$

Computing the Laplace transforms (with zero initial conditions) and rearranging gives

$$s^2 J_s \theta(s) + bs\theta(s) + K\theta(s) = bs\theta_p(s) + K\theta_p(s) + \tau(s)$$

$$s^2 J_p \theta_p(s) + bs\theta_p(s) + K\theta_p(s) = bs\theta(s) + K\theta(s).$$

Some more rearrangement gives

$$\theta(s) = \frac{bs + K}{s^2 J_s + bs + K}\theta_p(s) + \frac{1}{s^2 J_s + bs + K}\tau(s) \tag{5.6}$$

$$\theta_p(s) = \frac{bs + K}{s^2 J_p + bs + K}\theta(s). \tag{5.7}$$

Eliminating $\theta_p(s)$ we obtain

$$\theta(s) = \frac{bs + K}{s^2 J_s + bs + K}\frac{bs + K}{s^2 J_p + bs + K}\theta(s) + \frac{1}{s^2 J_s + bs + K}\tau(s).$$

Solving for $\theta(s)$ we have

$$\theta(s) = \frac{s^2 J_p + bs + K}{s^2\left(J_p J_s s^2 + b(J_p + J_s)s + K(J_p + J_s)\right)}\tau(s).$$

Then $\theta_p(s)$ is simply found as

$$\theta_p(s) = \frac{bs + K}{s^2 J_p + bs + K}\theta(s) = \frac{bs + K}{s^2\left(J_p J_s s^2 + b(J_p + J_s)s + K(J_p + J_s)\right)}\tau(s).$$

Summarizing we have

$$\theta(s) = \frac{s^2 J_p + bs + K}{s^2\left(J_p J_s s^2 + b(J_p + J_s)s + K(J_p + J_s)\right)}\tau(s) \tag{5.8}$$

$$\theta_p(s) = \frac{bs + K}{s^2\left(J_p J_s s^2 + b(J_p + J_s)s + K(J_p + J_s)\right)}\tau(s). \tag{5.9}$$

Suppose the angle θ of the motor shaft is measured. Then the sensor for θ is located on the same rigid body as the actuator (motor) and we refer to this as the *collocated* sensor and actuator case. On the other hand, suppose the solar panel angle θ_p is measured. Then this angle sensor is located at the end of one of the solar panels. As the actuator and sensor are not on the same rigid body we refer to this as the *noncollocated* sensor and actuator case.

120   5 Rigid Body Rotational Dynamics

5.3 Gears

This presentation is adapted from that given in [1]. Using the elementary rigid body dynamics developed previously, the model of the two gear system illustrated in Figure 5.9 is now developed.

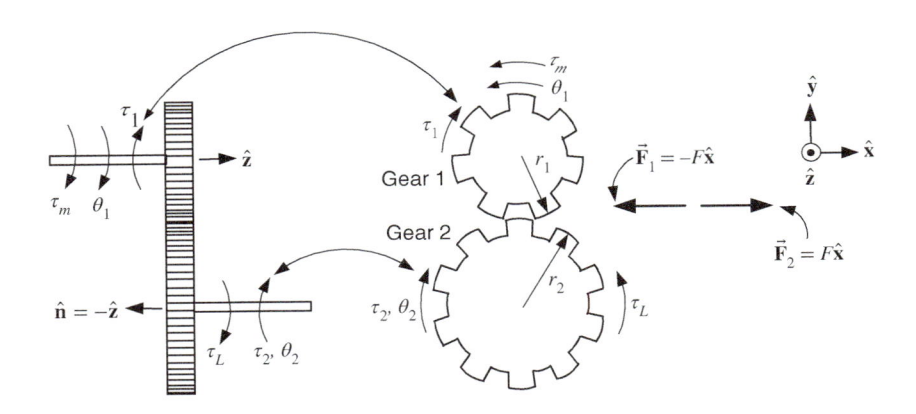

Figure 5.9. Two gear system. Source: Courtesy of Sharon Katz.

In Figure 5.9,

τ_1 is the torque exerted on gear 1 by gear 2.

$\vec{F}_1$ is the force exerted on gear 1 by gear 2.

τ_2 is the torque exerted on gear 2 by gear 1.

$\vec{F}_2$ is the force exerted on gear 2 by gear 1.

θ_1 is the angle rotated by gear 1.

θ_2 is the angle rotated by gear 2.

n_1 is the number of teeth on gear 1.

n_2 is the number of teeth on gear 2.

r_1 is the radius of gear 1.

r_2 is the radius of gear 2.

In Figure 5.9 we have written $\vec{F}_1 = -F\hat{x}$ and $\vec{F}_2 \triangleq F\hat{x}$ since by Newton's third law $\vec{F}_1$ and $\vec{F}_2$ are equal in magnitude, but opposite in direction. With $\vec{F}_1 \triangleq F(-\hat{x})$ we see that $F > 0$ means the force on gear 1 is in the $-\hat{x}$ direction as shown in Figure 5.9. Also, let $\vec{r}_1 \triangleq r_1(-\hat{y})$ so that $\vec{\tau}_1 = \vec{r}_1 \times \vec{F}_1 = r_1 F(-\hat{y}) \times (-\hat{x}) = r_1 F(-\hat{z}) = r_1 F \hat{n}$. That is, $\vec{\tau}_1 = \tau_1 \hat{n}$ where $\tau_1 = r_1 F$ and $\hat{n} \triangleq -\hat{z}$ is a unit vector. Similarly, with $\vec{F}_2 \triangleq F\hat{x}$ we see that $F > 0$ means the force on gear 2 is in the $\hat{x}$ direction. Writing $\vec{r}_2 \triangleq r_2 \hat{y}$ it follows that $\vec{\tau}_2 = \vec{r}_2 \times \vec{F}_2 = r_2 F(-\hat{z}) = -\tau_2 \hat{z} = \tau_2 \hat{n}$ with $\tau_2 = r_2 F$.

Algebraic Relationships Between Two Gears

There are three important algebraic relationships between the gears.

(1) The gears have different radii, but the teeth on each gear are the same size in order that they mesh together properly. Consequently, the number of teeth on the surface of each gear is proportional to the radius of each gear. For example, if $r_2 = 2r_1$, then $n_2 = 2n_1$. In general,
$$\frac{r_2}{r_1} = \frac{n_2}{n_1}.$$

(2) As $\tau_1 = r_1 F$ and $\tau_2 = r_2 F$ it then follows that
$$\frac{\tau_2}{\tau_1} = \frac{r_2}{r_1}.$$

(3) As the teeth on each gear are meshed together at the point of contact, the distance traveled along the circumference of each gear is the same. In other words, $\theta_1 r_1 = \theta_2 r_2$ or
$$\frac{\theta_2}{\theta_1} = \frac{r_1}{r_2}.$$

The first two algebraic relationships can be summarized as
$$\frac{\tau_2}{\tau_1} = \frac{r_2}{r_1} = \frac{n_2}{n_1},$$
and these ratios are easily remembered by thinking of gear 2 as larger in radius than gear 1. Then the number of teeth on gear 2 must also be larger (because its circumference is larger) and the torque on gear 2 is also larger (because its radius is larger).

The last algebraic relationship is summarized as
$$\frac{r_1}{r_2} = \frac{\theta_2}{\theta_1} = \frac{\omega_2}{\omega_1},$$
but it is more easily remembered by writing $\theta_1 r_1 = \theta_2 r_2$, which just states the distance traveled along the surface of each gear is the same as they are meshed together.

Dynamic Relationships Between Two Gears

Consider the two gear system shown in Figure 5.10 on the next page. The motor torque τ_m acts on gear 1 and the torque τ_L is a load torque acting on gear 2.

In Figure 5.10, the following notation is used.

J_1 is the moment of inertia of the motor shaft.

J_2 is the moment of inertia of the output shaft.

f_1 is the viscous friction coefficient of the motor shaft.

f_2 is the viscous friction coefficient of the output shaft.

θ_1 is the angle rotated by gear 1.

122   5   Rigid Body Rotational Dynamics

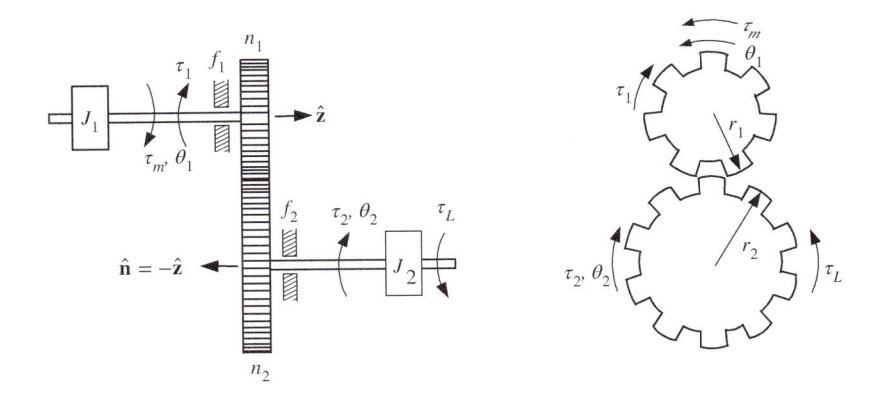

Figure 5.10. Dynamic equations for a two gear system. Source: Courtesy of Sharon Katz.

θ_2 is the angle rotated by gear 2.

ω_1 is the angular speed of gear 1.

ω_2 is the angular speed of gear 2.

τ_1 is the torque exerted on gear 1 by gear 2.

τ_2 is the torque exerted on gear 2 by gear 1.

The sign conventions for the torques $\tau_m, \tau_1, \tau_2, \tau_L$ are indicated in Figure 5.10. In particular, if $\tau_m > 0, \tau_1 > 0$ then they oppose each other and similarly, if $\tau_2 > 0, \tau_L > 0$ then these two torques oppose each other. A load torque is illustrated in Figure 5.11 in which the load torque on gear 2 is $\tau_L = r_2 mg$ with r_2 the radius of the pickup reel (gear 2).

We now put everything together to write down the differential equations that characterize the dynamic behavior of the gears. Recall that the fundamental equation of rigid body dynamics is given by

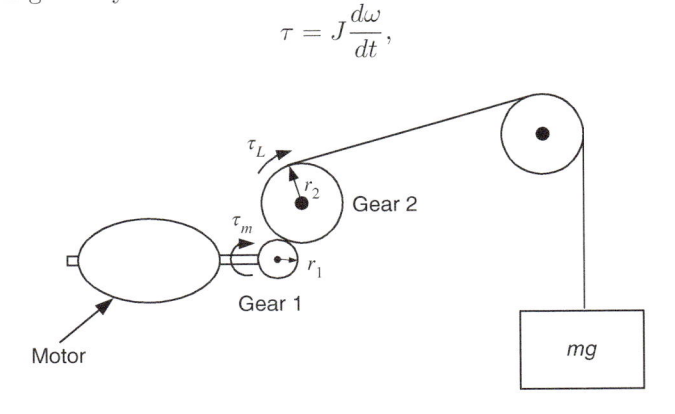

$$\tau = J \frac{d\omega}{dt},$$

Figure 5.11. Illustration of load torque. Source: Courtesy of Sharon Katz.

where τ is the *total* torque on the rigid body, J is the moment of inertia of the rigid body and $d\omega/dt$ is its angular acceleration about the fixed axis of rotation. The equations of motion for the two gears are given by

$$\begin{aligned}\tau_m - \tau_1 - f_1\omega_1 &= J_1\frac{d\omega_1}{dt} \\ \tau_2 - \tau_L - f_2\omega_2 &= J_2\frac{d\omega_2}{dt}.\end{aligned} \quad (5.10)$$

Typically, the input (motor) torque τ_m is known, and the output position θ_2 and speed ω_2 are measured. Consequently, the variables τ_1, τ_2, ω_1 need to be eliminated, which is done as follows:

$$\begin{aligned}\tau_2 = \frac{n_2}{n_1}\tau_1 &= \frac{n_2}{n_1}\left(\tau_m - f_1\omega_1 - J_1\frac{d\omega_1}{dt}\right) = \frac{n_2}{n_1}\left(\tau_m - f_1\left(\frac{n_2}{n_1}\omega_2\right) - J_1\frac{d}{dt}\left(\frac{n_2}{n_1}\omega_2\right)\right) \\ &= \frac{n_2}{n_1}\tau_m - \left(\frac{n_2}{n_1}\right)^2 f_1\omega_2 - \left(\frac{n_2}{n_1}\right)^2 J_1\frac{d\omega_2}{dt}. \quad (5.11)\end{aligned}$$

Substituting this expression for τ_2 into the second equation of (5.10) results in

$$\frac{n_2}{n_1}\tau_m - \left(\frac{n_2}{n_1}\right)^2 f_1\omega_2 - \left(\frac{n_2}{n_1}\right)^2 J_1\frac{d\omega_2}{dt} - \tau_L - f_2\omega_2 = J_2\frac{d\omega_2}{dt}.$$

Rearranging, the desired result is

$$\frac{n_2}{n_1}\tau_m = \underbrace{\left(J_2 + (n_2/n_1)^2 J_1\right)}_{J}\frac{d\omega_2}{dt} + \underbrace{\left(f_2 + (n_2/n_1)^2 f_1\right)}_{f}\omega_2 + \tau_L. \quad (5.12)$$

Let $n = n_2/n_1$ denote the *gear ratio*, $J \triangleq J_2 + n^2 J_1$ denote the total moment of inertia reflected to the output shaft, and $f \triangleq f_2 + n^2 f_1$ denote the total viscous friction coefficient reflected to the output shaft. Equation (5.12) can now be written succinctly as

$$n\tau_m = J\frac{d\omega_2}{dt} + f\omega_2 + \tau_L. \quad (5.13)$$

The net effect of the gears is to increase the motor torque from τ_m on the motor shaft to $n\tau_m$ on the output shaft, to add the quantity $n^2 J_1$ to the moment of inertia of the output shaft and to add $n^2 f_1$ to the viscous friction coefficient of the output shaft.

Remark Everything could have been referred to the motor shaft instead of the output (load) shaft. To do so, first substitute $\omega_2 = (n_1/n_2)\omega_1$ into (5.12) to obtain

$$\frac{n_2}{n_1}\tau_m = \left(J_2 + \left(\frac{n_2}{n_1}\right)^2 J_1\right)\frac{d}{dt}\left(\frac{n_1}{n_2}\omega_1\right) + \left(f_2 + \left(\frac{n_2}{n_1}\right)^2 f_1\right)\left(\frac{n_1}{n_2}\omega_1\right) + \tau_L.$$

Multiplying both sides by n_1/n_2 results in

$$\tau_m = \left(J_2 + \left(\frac{n_2}{n_1}\right)^2 J_1\right)\left(\frac{n_1}{n_2}\right)^2 \frac{d\omega_1}{dt} + \left(f_2 + \left(\frac{n_2}{n_1}\right)^2 f_1\right)\left(\frac{n_2}{n_1}\right)^2 \omega_1 + \frac{n_1}{n_2}\tau_L$$

or finally

$$\tau_m = \left(\left(\frac{n_1}{n_2}\right)^2 J_2 + J_1\right)\frac{d\omega_1}{dt} + \left(\left(\frac{n_1}{n_2}\right)^2 f_2 + f_1\right)\omega_1 + \frac{n_1}{n_2}\tau_L.$$

In this formulation, the load torque on the input shaft is *reduced* by n_1/n_2 from that on the output shaft, and $(n_1/n_2)^2 J_2$ has been added to the inertia of the motor shaft and $(n_1/n_2)^2 f_2$ has been added to the viscous friction coefficient of the motor shaft.

Example 4 *Rolling Mill (Adapted from [3])*

Figure 5.12 illustrates a rolling mill where, e.g., aluminum comes into the rollers at the thickness T and exits with thickness x. The motor torque τ_m exerted on gear 1 results in the torque τ_2 being exerted on gear 2, which in turn produces the force F on the rack and top roller. The force F reduces the aluminum sheet to the thickness x. The control problem is measure the output thickness x and use this value to choose the motor torque τ_m so that $x(t) \to x_d$ where x_d is the desired thickness. We now derive a mathematical model of this system. The upward (reaction) force by the rolled sheet on the top roller is

$$F_L = k(T - x),$$

where k is a constant. The gear ratio is $n = n_2/n_1 = r_2/r_1$ and the viscous friction forces on the gear shafts are taken to be zero, i.e., $f_1 = f_2 = 0$. Let f_{rack} be the coefficient of viscous friction between the rack and the structure (not shown) holding it so that $-f_{rack}\frac{d}{dt}(T-x)$ is the viscous friction force on the rack.

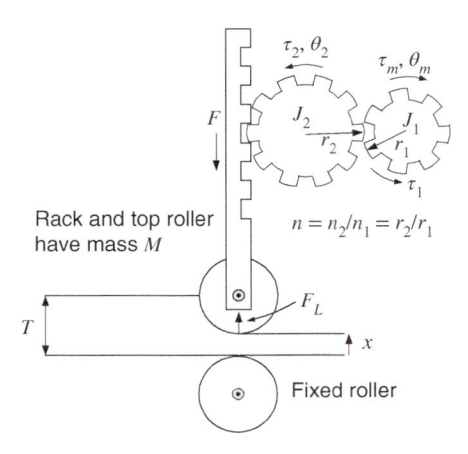

Figure 5.12. Rolling mill.

With $n_2/n_1 = r_2/r_1$ and $f_1 = f_2 = 0$ we use the gear equation (5.12) to write

$$\frac{r_2}{r_1}\tau_m = \left(J_2 + (r_2/r_1)^2 J_1\right)\frac{d\omega_2}{dt} + \tau_L, \tag{5.14}$$

where $\tau_L = r_2 F$. As $T - x = r_2\theta_2 = r_1\theta_m$ the angular velocity ω_2 and acceleration $d\omega_2/dt$ may be written as

$$\omega_2 = \frac{1}{r_2}\frac{d}{dt}(T-x), \quad \frac{d\omega_2}{dt} = \frac{1}{r_2}\frac{d^2(T-x)}{dt^2}.$$

Substituting this expression for $d\omega_2/dt$ into (5.14) along with $\tau_L = r_2 F$ gives

$$\frac{r_2}{r_1}\tau_m = \left(J_2 + (r_2/r_1)^2 J_1\right)\frac{1}{r_2}\frac{d^2(T-x)}{dt^2} + r_2 F.$$

Using Newton's equation applied to the rack and top roller, which have a combined mass M, we have

$$M\frac{d^2(T-x)}{dt^2} = F - f_{rack}\frac{d}{dt}(T-x) - F_L = F - f_{rack}\frac{d}{dt}(T-x) - k(T-x).$$

Eliminating F from these last two equations we finally have a differential equation model for the thickness x given by

$$\underbrace{\left(J_2 + (r_2/r_1)^2 J_1 + r_2^2 M\right)}_{J_{eq}}\frac{d^2(T-x)}{dt^2} + r_2^2 f_{rack}\frac{d}{dt}(T-x) + r_2^2 k(T-x) = \frac{r_2^2}{r_1}\tau_m.$$

With $y \triangleq T - x$, $a_1 \triangleq r_2^2 f_{rack}/J_{eq}$, $a_0 \triangleq r_2^2 k/J_{eq}$, $b_0 \triangleq r_2^2/(r_1 J_{eq})$, and $u \triangleq \tau_m$, the model is simply expressed by

$$\ddot{y} + a_1 \dot{y} + a_0 y = b_0 u.$$

The transfer function is

$$G(s) = \frac{Y(s)}{U(s)} = \frac{b_0}{s^2 + a_1 s + a_0}.$$

As $a_1 > 0, a_0 > 0$ this transfer function is stable. This rather complex mechanical device is described by a second-order transfer function. This can then be used to design a feedback controller that regulates the thickness of the aluminum sheet. Such controllers will be designed in Chapters 9 and 10.

Tension

Figure 5.13 shows a block of mass m tied to a ceiling through a rope or cable. The tension T_1 next to the block of mass m is the upward force on the block by the cable while

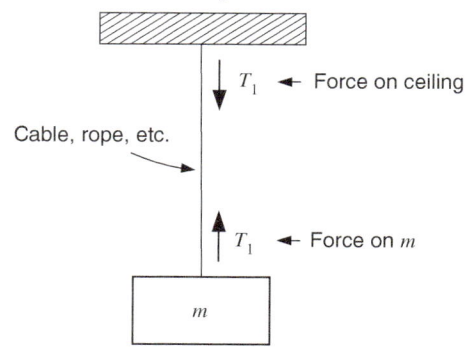

Figure 5.13. T_1 is the tension in the cable or rope.

126   5   Rigid Body Rotational Dynamics

the tension T_1 next to the ceiling is the downward force on the ceiling by the cable. As the mass m is stationary, the tension must equal mg to cancel out gravity. Think of tension in a rope, cable, etc. as a spring that can only be stretched (not compressed), but with a spring constant that is essentially infinite ($k = \infty$). So when the two ends of a cable are pulled apart, the cable does not stretch any distance, but produces restoring forces.

Example 5 *Single Pulley*
Figure 5.14 shows two masses m_1 and m_2 connected about a pulley by a rope of length ℓ.

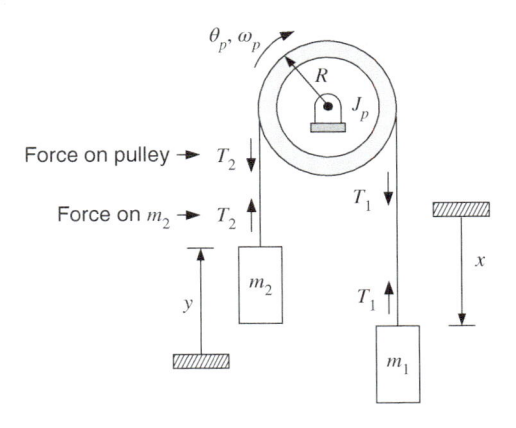

Figure 5.14. Massless pulley.

With the masses on either side the rope is under tension so it imperceptibly stretched providing a restoring force. On the right side of the pulley $T_1 > 0$ means there is an upward force on m_1 and a downward force T_1 on the right side of the pulley. Similarly, on the left side, $T_2 > 0$ is an upward force on m_2 and there is also a downward force T_2 on the left side of the pulley.

The rope of length ℓ is such that when $x = 0$ we have $y = 0$ as well. This implies that $y = x$. Also, $\theta_p = 0$ when $x = y = 0$. The rope does not slip as the pulley rotates so the algebraic constraint is
$$y = x = R\theta_p.$$
The equations of motion are
$$J_p \frac{d\omega_p}{dt} = RT_1 - RT_2$$
$$m_1 \frac{d^2x}{dt^2} = m_1 g - T_1$$
$$m_2 \frac{d^2y}{dt^2} = -m_2 g + T_2.$$

We will often take a pulley to be massless and we do so here, that is, we set $J_p = 0$. Then immediately we have $T_1 = T_2$ and using $y = x$ the equations of motion reduce to
$$m_1 \frac{d^2x}{dt^2} = m_1 g - T_1$$
$$m_2 \frac{d^2x}{dt^2} = -m_2 g + T_1.$$

Adding these two equations to cancel T_1 we finally obtain

$$\frac{d^2x}{dt^2} = \frac{m_1 - m_2}{m_1 + m_2} g.$$

Problem 6 shows why pulleys are so useful.

5.4 Rolling Cylinder

We now develop the equations of motion for a cylinder rolling on a surface without slipping.

Combined Translational and Rotational Motion

Figure 5.15a shows a cylinder of mass m and inertia J moving to the right with translational velocity v, but no rotational velocity about its main axis, i.e., $\omega = 0$. Its kinetic energy is $\frac{1}{2}mv^2$. In Figure 5.15b the same cylinder is rotating about its main axis, but with no translational velocity, i.e., $v = 0$. Its kinetic energy is then $\frac{1}{2}J\omega^2$.

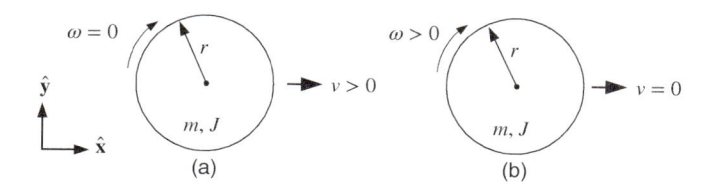

Figure 5.15. (a) A cylinder moving at velocity v with no angular velocity about its axis. (b) A cylinder rotating at angular velocity ω about its axis, but no translational motion.

Figure 5.16 shows this same cylinder with its center of mass (which is on the axis of rotation) moving to the right at velocity v_{cm} while it also rotating about its main axis with angular velocity ω.

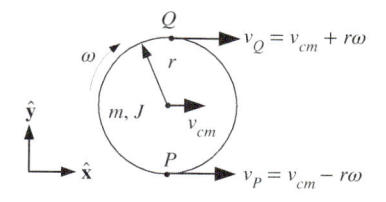

Figure 5.16. Translational and rotational motion.

The point Q at the top of the cylinder is moving at velocity $r\omega \hat{\mathbf{x}}$ with respect to the axis of rotation, but the axis of rotation is moving at velocity $v_{cm}\hat{\mathbf{x}}$ to the right. So the total velocity of the point Q is

$$v_Q \hat{\mathbf{x}} = v_{cm} \hat{\mathbf{x}} + r\omega \hat{\mathbf{x}}.$$

128   5 Rigid Body Rotational Dynamics

The point P of the bottom of the cylinder is moving at velocity $-r\omega\hat{\mathbf{x}}$ with respect to the axis of rotation, but the axis of rotation (center of mass) is moving at velocity $v_{cm}\hat{\mathbf{x}}$ to the right. So the total velocity of the point P is

$$v_P\hat{\mathbf{x}} = v_{cm}\hat{\mathbf{x}} - r\omega\hat{\mathbf{x}}.$$

The total kinetic energy is

$$\frac{1}{2}mv_{cm}^2 + \frac{1}{2}J\omega^2.$$

Rolling on a Flat Surface Without Slipping

We now insert the no slip condition. Consider the cylinder in Figure 5.17, which has mass m and moment of inertia J about its axis of rotation ($-\hat{\mathbf{z}}$ direction in Figure 5.17).

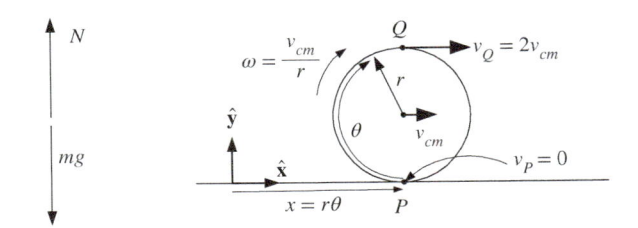

Figure 5.17. Cylinder rolling on a flat surface with no slip.

The center of mass of the cylinder is on the axis of rotation in the midpoint of the cylinder. To say that the cylinder rolls without slipping means that

$$x = r\theta,$$

where θ is the angle the cylinder has rotated and x is the distance the cylinder has rotated along the surface. Differentiating both sides of this with respect to the time t gives

$$\frac{dx}{dt} = r\omega.$$

The surface provides an upward normal force N to cancel the downward force mg of gravity.

As Figure 5.17 indicates, the velocity v_{cm} of the center of mass of the cylinder is given by

$$v_{cm} = \frac{dx}{dt}.$$

This is simply because the cylinder's center of mass is directly above the point of contact P of the cylinder with the level surface and this point of contact is moving to the right at velocity dx/dt.

The cylinder rotates about its center of mass with angular rate

$$\omega = \frac{d\theta}{dt} = \frac{d(x/r)}{dt} = \frac{\dot{x}}{r} = \frac{v_{cm}}{r}.$$

Using the no slip condition we have

$$v_Q = v_{cm} + r\omega = 2v_{cm}$$
$$v_P = v_{cm} - r\omega = 0.$$

Note that the point of contact P of the cylinder's surface with the flat surface has zero speed. As there are no forces or torques on the cylinder its kinetic energy is given by

$$KE = \frac{1}{2}mv_{cm}^2 + \frac{1}{2}J\omega^2 = \frac{1}{2}m\dot{x}^2 + \frac{1}{2}\frac{J}{r^2}\dot{x}^2 = \frac{1}{2}\left(m + \frac{J}{r^2}\right)\dot{x}^2.$$

Cylinder Rolling Down an Inclined Plane

Figure 5.18 shows a cylinder rolling down an inclined plane. The cylinder has mass m and moment of inertia J. The x direction is taken to be positive going up the inclined plane and the y direction is perpendicular to the inclined plane. There is a component of gravity $mg\sin(\phi)$ that is pushing the cylinder in the $-x$ direction down the inclined plane. Due to friction there is a force F_f produced on the cylinder at the point of contact with the inclined plane (and by Newton's third law a force $-F_f$ produced by the cylinder on the inclined plane). It is this force F_f that produces the torque to turn the cylinder.

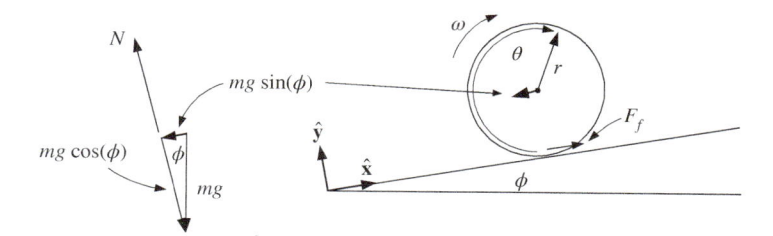

Figure 5.18. Cylinder rolling down an incline under the influence of gravity.

As illustrated in Figure 5.19 we can think of the force F_f as that in a rack and pinion system. Here the rack is the surface of the incline, the pinion is the cylinder, and their meshed teeth model the surfaces interacting.

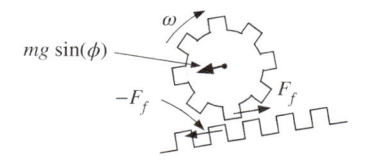

Figure 5.19. The interaction of the surfaces of the incline and cylinder modeled as a rack and pinion system.

We continue to assume the no slip condition so the point of contact between the cylinder and inclined plane has *zero* velocity. Thus the friction there is *static* not viscous and so there is no energy loss. (Think of *viscous* friction as two bodies rubbing/slipping against

130   5 Rigid Body Rotational Dynamics

each other.) As the cylinder rolls without slipping we have

$$x = r\theta$$
$$v_{cm} = \frac{dx}{dt} = r\omega.$$

In this case $\omega < 0$ as it is rolling down the inclined plane with x decreasing.

Static and Kinetic Friction

The friction F_f in Figure 5.18 is *static* friction. We have already discussed viscous friction in Chapter 4, which is the friction between two *lubricated* surfaces sliding against each other. For example, between the piston and cylinder of a damper is a fluid (oil/lubricant).[4] However, when two *dry* (non-lubricated) surfaces are sliding against each other, the friction is modeled as *kinetic* friction as we now describe. Consider Figure 5.20 that shows a box of mass m on an incline.

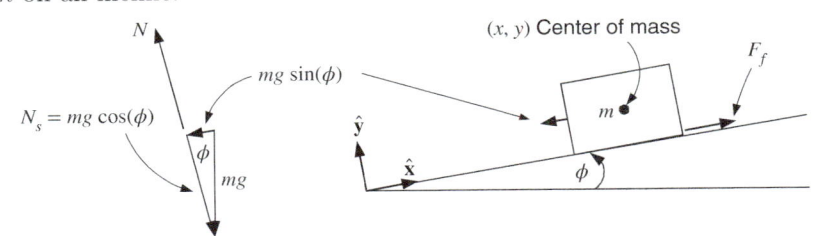

Figure 5.20. Static friction.

From experience we know that if the angle ϕ is too small the box will not slide down the incline. This is explained by saying there is a static friction force F_f between the bottom surface of the box and the surface of the incline. This friction cancels out the gravitational force $mg\sin(\phi)$. The static friction F_f is limited to $0 \leq F_f \leq F_{f\max}$ where

$$F_{f\max} = \mu_s N_s = \mu_s mg\cos(\phi).$$

$F_{f\max}$ is the maximum static friction between the surfaces, $N_s = mg\cos(\phi)$ is the normal (to the incline) force *on* the box and μ_s is an empirically[5] determined constant called the *coefficient of static friction*. As long as the gravitational force $mg\sin(\phi)$ is less than $F_{f\max} = \mu_s mg\cos(\phi)$ the box will sit on the incline and not move. As the angle ϕ increases the static friction will also continue to increase to cancel out the gravitational force $mg\sin(\phi)$.

However, if ϕ is increased enough so that the gravitational force $mg\sin(\phi)$ is now *greater* than the maximum possible static friction $F_{f\max} = \mu_s mg\cos(\phi)$, the box will slide down the incline. As the box slides down, it is opposed by *kinetic* friction between the two *dry* (not lubricated) surfaces. The kinetic friction F_{kf} is modeled as

$$F_{kf} = \mu_k mg\cos(\phi)$$

where μ_k is also an empirically determined constant. It turns out that $\mu_k \ll \mu_s$ so the kinetic friction is much less than the static friction. That is, after the box starts to slide, the friction against it decreases dramatically. As already mentioned, this kinetic friction model is for two *dry* (not lubricated) surfaces sliding against each other.

---
[4] Axels supported by ball bearings are lubricated and so are modeled using viscous friction.
[5] By "empirically" we simply mean it is an experimentally determined constant.

Equations of Motion Using Newton's Laws

Newton's three laws of motion are valid with respect to non-accelerating coordinate systems. The rotational law $\tau = J d\omega/dt$ was derived from the three laws so it also valid in non-accelerating coordinate systems. However, it also turns out that $\tau = J d\omega/dt$ still holds in an *accelerating* coordinate system as long as the axis of rotation is through the *center of mass* of the rigid body [14].

Let's go back to the rolling cylinder in Figure 5.18. The cylinder is not moving in the y direction as there is a normal force $N = mg\cos(\phi)$ in the $+\hat{\mathbf{y}}$ direction that cancels out the gravity component $mg\cos(\phi)$ in the $-\hat{\mathbf{y}}$ direction. There is a static friction force F_f at the point of contact of the cylinder with the inclined plane.[6] Let (x, y) be the coordinates of the center of mass of the cylinder. The equations of motion of the cylinder are

$$m\frac{d^2y}{dt^2} = -mg\cos(\phi) + N = 0 \tag{5.15}$$

$$m\frac{d^2x}{dt^2} = -mg\sin(\phi) + F_f \tag{5.16}$$

$$J\frac{d^2\theta}{dt^2} = -rF_f \tag{5.17}$$

$$x = r\theta. \tag{5.18}$$

Though the axis of rotation of the cylinder is accelerating, it goes through the center of mass of the cylinder so (5.17) is valid. To eliminate the force F_f we multiply the second equation by r and add it to the third equation to obtain

$$rm\frac{d^2x}{dt^2} + J\frac{d^2\theta}{dt^2} = -rmg\sin(\phi).$$

Using the no slip condition $\theta = x/r$ to eliminate θ results in

$$rm\frac{d^2x}{dt^2} + \frac{J}{r}\frac{d^2x}{dt^2} = -rmg\sin(\phi)$$

or

$$\frac{d^2x}{dt^2} = -\underbrace{\frac{mr^2}{mr^2 + J}g\sin(\phi)}_{\text{Constant}}.$$

Summary of Using Newton's Equations of Motion for a Rigid Body

For translational motion of a rigid body Newton's law $F = ma$ is used in an inertial coordinate system[7] to obtain the motion of the center of mass of the rigid body. F is the sum of *all* forces acting on the rigid body. For example, in equations (5.15) and (5.16) the three external forces acting on the cylinder are mg, F_f, and N with (x, y) the coordinates of its center of mass.

For rotational motion of a rigid body we use $\tau = Jd\omega/dt$ where τ is the sum of all torques (with respect to its axis of rotation) acting on the rigid body whose moment of

---

[6] Remember that at the point of contact of the cylinder with the inclined plane, the bottom of the cylinder has zero speed. Thus this is static not viscous friction. As long as the cylinder rolls without slipping, there will be no viscous friction.

[7] A coordinate system that is *not* accelerating.

5 Rigid Body Rotational Dynamics

inertia is J. This equation can be used even if axis of rotation is accelerating as long as the axis of rotation goes through the center of mass of the rigid body [14]. For example, in (5.17) we are taking the axis of rotation to be the axis of the cylinder, which is accelerating. With respect to this axis the only torque is $-rF_f$ as the gravitational force mg acts through the center of mass of the cylinder and so its moment arm is zero.

Equations of Motion Derived from Conservation of Energy

The kinetic energy of the cylinder is given by

$$KE = \frac{1}{2}m\dot{x}^2 + \frac{1}{2}J\omega^2 = \frac{1}{2}m\dot{x}^2 + \frac{1}{2}\frac{J}{r^2}\dot{x}^2 = \frac{1}{2}\left(m + \frac{J}{r^2}\right)\dot{x}^2. \tag{5.19}$$

Referring to Figure 5.21, we see that the axis of the cylinder (which the center of mass lies on) is at the position $x\hat{\mathbf{x}} + y\hat{\mathbf{y}}$. With $d\cos(\phi) = r$ it follows that the axis of the cylinder is at a height $(x - d\sin(\phi))\sin(\phi) + d$ above the horizontal (see Problem 18). Thus the gravitational potential energy of the cylinder is

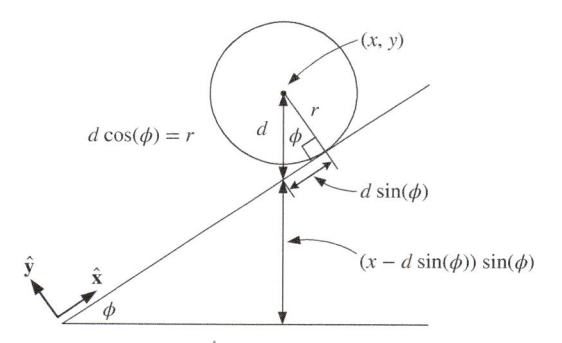

$$PE = mg(x\sin(\phi) - d\sin^2(\phi) + d). \tag{5.20}$$

Figure 5.21. Height of the cylinder axis above the horizontal.

The total energy of the cylinder is then

$$KE + PE = \frac{1}{2}\left(m + \frac{J}{r^2}\right)\dot{x}^2 + mg\big(x\sin(\phi) - d\sin^2(\phi) + d\big).$$

As the total energy is constant we have

$$\frac{d}{dt}(KE + PE) = \left(m + \frac{J}{r^2}\right)\dot{x}\ddot{x} + mg\dot{x}\sin(\phi) = 0.$$

Canceling out $\dot{x}$ and rearranging this becomes

$$\frac{d^2x}{dt^2} = -\frac{mr^2}{mr^2 + J}g\sin(\phi). \tag{5.21}$$

Remark Figure 5.22 shows a block of mass m sliding down an inclined plane with a *frictionless* surface.

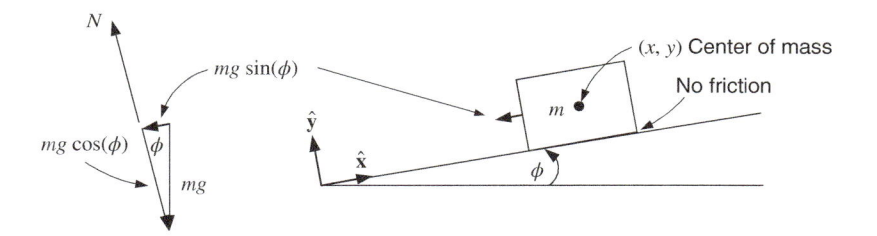

Figure 5.22. Block of mass m sliding down a frictionless inclined plane.

Again with (x, y) denoting the coordinates of the center of mass of the block, the equations of motion are

$$m\frac{d^2y}{dt^2} = -mg\cos(\phi) + N = 0$$

$$m\frac{d^2x}{dt^2} = -mg\sin(\phi)$$

or simply

$$\frac{d^2x}{dt^2} = -g\sin(\phi). \tag{5.22}$$

Comparing (5.21) with (5.22) we see that the rolling cylinder accelerates down the inclined plane slower than the box of the same mass m in Figure 5.22. This is because in the case of the rolling cylinder, some of gravitational potential energy goes into rotational motion (kinetic energy) as well as translational motion (kinetic energy) rather than all of it going into translational motion as in the case of the box in Figure 5.22.

Motorized Cylinder Going Up an Inclined Plane

Suppose the cylinder has a motor inside to produce a torque τ_m. Figures 5.23 and 5.24 on the next page are the same as Figures 5.18 and 5.19, respectively, except that the motor torque τ_m has been added.

As above (x, y) are the coordinates of the center of mass of the cylinder. The equations of motion for the motorized cylinder are then

$$m\frac{d^2y}{dt^2} = -mg\cos(\phi) + N = 0$$

$$m\frac{d^2x}{dt^2} = -mg\sin(\phi) + F_f$$

$$J\frac{d^2\theta}{dt^2} = \tau_m - rF_f$$

$$x = r\theta.$$

Multiplying the second equation by r and adding to the third equation to eliminate F_f we obtain

$$rm\frac{d^2x}{dt^2} + J\frac{d^2\theta}{dt^2} = \tau_m - rmg\sin(\phi).$$

134   5 Rigid Body Rotational Dynamics

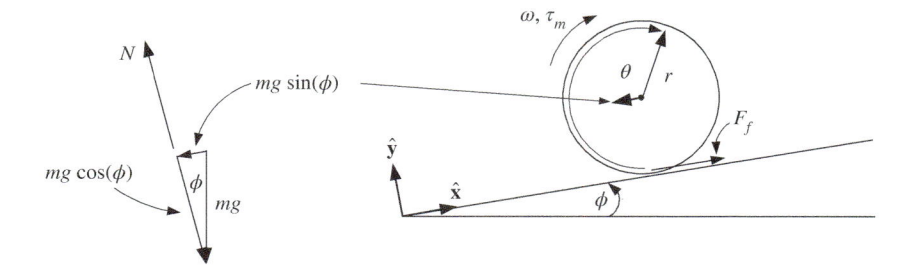

Figure 5.23. Cylinder going up an incline using a motor.

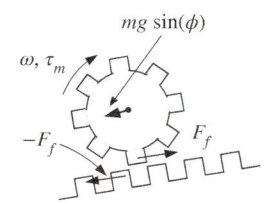

Figure 5.24. The interaction of the surfaces of the incline and cylinder modeled as a rack and pinion system.

Using the no slip condition $x = r\theta$ to eliminate θ we have

$$rm\frac{d^2x}{dt^2} + \frac{J}{r}\frac{d^2x}{dt^2} = \tau_m - rmg\sin(\phi)$$

or finally

$$\frac{d^2x}{dt^2} = \frac{r}{mr^2 + J}\tau_m - \frac{mr^2}{mr^2 + J}g\sin(\phi).$$

Equations of Motion Derived from Conservation of Energy

Using the expressions for the kinetic and potential energies in (5.19) and (5.20), respectively, the mechanical energy of the cylinder in Figure 5.23 is given by

$$KE + PE = \frac{1}{2}\left(m + \frac{J}{r^2}\right)\dot{x}^2 + mg\big(x\sin(\phi) - d\sin^2(\phi) + d\big).$$

The rate of change of the cylinder's energy is equal to the mechanical power $\tau_m\omega$ put into the cylinder by the motor, i.e.,

$$\frac{d}{dt}(KE + PE) = \tau_m\omega.$$

Thus we have

$$\frac{d}{dt}(KE + PE) = \left(m + \frac{J}{r^2}\right)\dot{x}\ddot{x} + mg\dot{x}\sin(\phi) = \tau_m\omega = \tau_m\frac{\dot{x}}{r}.$$

Canceling out $\dot{x}$ from both sides and rearranging this becomes

$$\frac{d^2x}{dt^2} = \frac{\tau_m}{r\left(m + \frac{J}{r^2}\right)} - \frac{m}{m + \frac{J}{r^2}} g \sin(\phi) = \frac{r}{mr^2 + J}\tau_m - \frac{mr^2}{mr^2 + J} g \sin(\phi).$$

Problems

Problem 1 *Moments of Inertia*

(a) *Moment of Inertia of a Pendulum Rod About Its Endpoint*
A cylindrical rod of uniform mass density ρ has length ℓ and a circular cross section of area a. It is allowed to rotate about a pivot as shown in Figure 5.25.

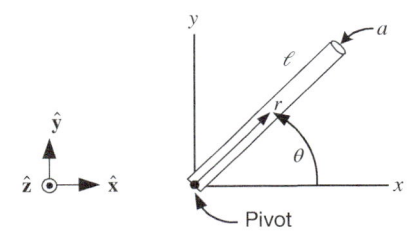

Figure 5.25. Uniform density rod rotating on a pivot at one end.

Show that its moment of inertia about the pivot axis is $m\dfrac{\ell^2}{3}$ where $m = \rho a \ell$ is the mass of the pendulum rod. Compute the torque on the rod due to gravity about the pivot and use this to give the equations of motion of the rod about the pivot.

(b) *Moment of Inertia of a Rod About Its Center of Mass*
A cylindrical rod of uniform mass density ρ has length 2ℓ and a circular cross section of area a. It is allowed to rotate about a pivot at the origin through its center of mass as shown in Figure 5.26.

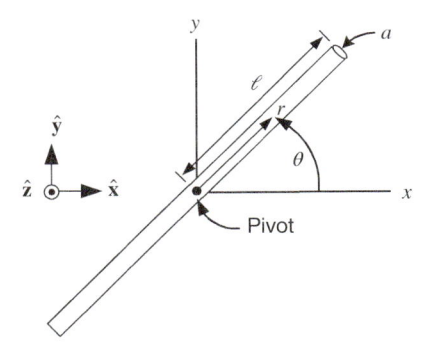

Figure 5.26. Rod of length 2ℓ rotating about its center of mass.

Show that the moment of inertia of the rod about this axis of rotation is $\dfrac{m\ell^2}{3}$. Compute the torque on the rod due to gravity about the pivot and use this to give the equations of motion of the rod about the pivot.

136   5 Rigid Body Rotational Dynamics

Problem 2 *Transfer Function*

In the mechanical system shown in Figure 5.27 the cable wraps around the disk without slipping. The input is the force $f(t)$ and the output is the displacement x of m. y locates the left side of spring k_2. The spring constants k_2 and k_3 are for *linear* springs while the spring constant k_1 is for a *torsional* spring. The springs are relaxed if $\theta = 0, y = x = 0$. There is a viscous friction force between the mass m and the floor that has coefficient of friction b_2. There is a *torsional* friction force between the cable disk and its guide with a coefficient of friction b_1.

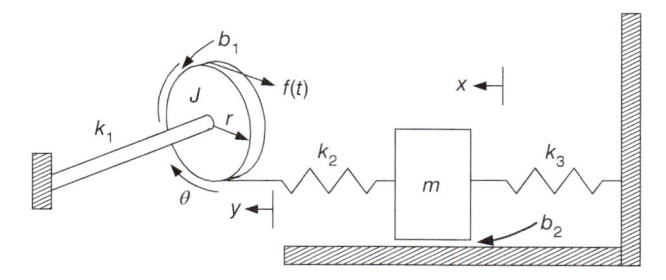

Figure 5.27. Torsional and linear mass–spring–damper system.

(a) Write down the equations of motion for m and J. Hints: Use $f(t)$ to find the input torque. What is the relationship between y and θ?

(b) With $F(s) = \mathcal{L}\{f(t)\}$ and $X(s) = \mathcal{L}\{x(t)\}$ compute the transfer function

$$T(s) \triangleq \frac{X(s)}{F(s)}.$$

Problem 3 *Wind up Cable*

In Figure 5.28 a shaft with a steel drum of total inertia J is used to wind up a cable in order to raise a mass m. The back end of the shaft is connected to a wall through a rotary damper with damping coefficient b. The input to the shaft is the torque τ while the weight mg produces a disturbance (input) torque on the shaft as well. The output is taken to be the angular position of the shaft θ, which has the same sign convention as ω and τ. Ignore the mass of the cable, and assume there is no slip between the cable and the steel drum so that $v = r\dot\theta$ and $\dot v = r\ddot\theta$,

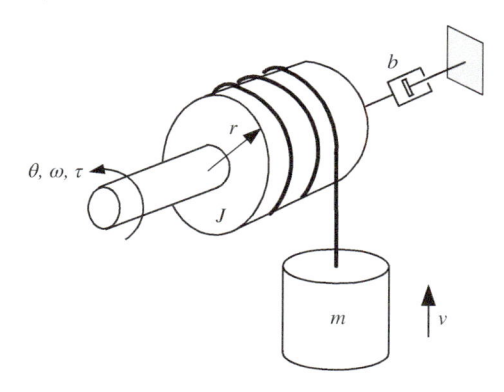

Figure 5.28. Winding up a cable with a load on it.

(a) Give the equation of motion of the shaft. The tension in the cable is $mg + m\dot{v}$

(b) The output is the position θ and the inputs are the torque τ and the (disturbance) gravitational force mg. With zero initial conditions find an expression for the Laplace transform $\theta(s)$ in terms of the Laplace transform of these two inputs.

Problem 4 *Gear Equation*

The gear equation (5.13) can be derived in the Laplace domain starting with the set of Eqs. (5.10). To do so, let the initial conditions be zero and compute the Laplace transform of each of the two equations in (5.10). Eliminate $\omega_1(s)$ and $\tau_1(s)$ in the first equation by substituting $\omega_1(s) = (n_2/n_1)\omega_2(s)$ and $\tau_1(s) = (n_1/n_2)\tau_2(s)$, respectively. Then eliminate $\tau_2(s)$ to have a single (algebraic) equation for $\omega_2(s)$ in terms of $\tau_m(s)$ and $\tau_L(s)$. To simplify your final expression let $J = J_2 + \left(\dfrac{n_2}{n_1}\right)^2 J_1$ and $f = f_2 + \left(\dfrac{n_2}{n_1}\right)^2 f_1$.

Problem 5 *Modeling a Gear System*

Figure 5.29 shows a gear system for an elevator car. The input shaft has a moment of inertia J_1 and a torque (from a motor) τ_m applied to it. The output shaft consisting of gear 2 and a pulley has a moment of inertia J_2. As the pulley is rotated the elevator car of mass M can be raised or lowered (assume the cable does not slip on the pulley). The counter weight has mass m, the pulley has radius r_p, and the position of the elevator car is denoted by z. As there is no slip between the pulley and the cable we may write $z = r_p \theta_2$. At $z = 0$ the elevator is on the ground floor and the counterweight is a distance L above the ground. The position z_c of the counterweight above the ground may thus be written as $z_c = L - z$. The algebraic gear relationships are

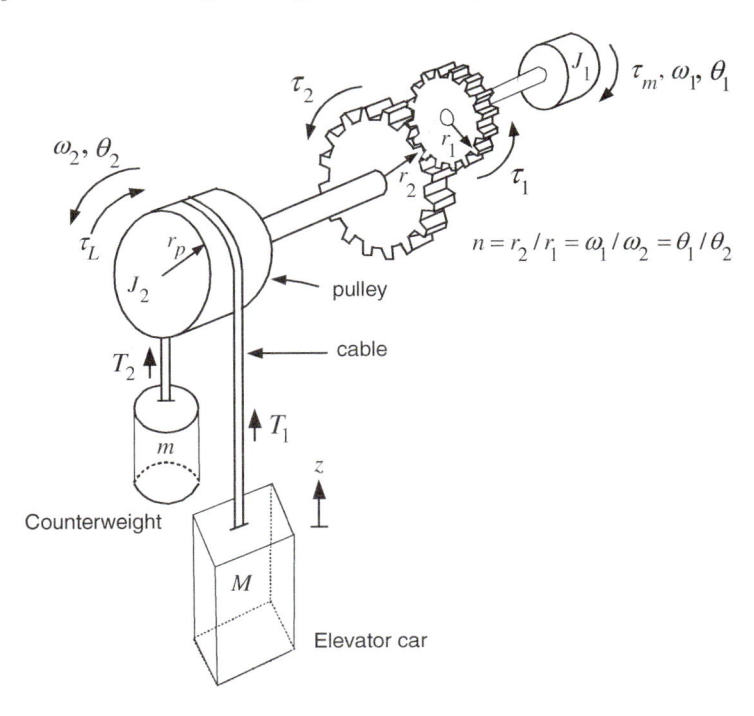

Figure 5.29. Gear system for an elevator. Source: Adapted from Palm [15]. System Dynamics, Second Edition, McGraw-Hill, 2010.

138   5 Rigid Body Rotational Dynamics

$$\frac{\omega_1}{\omega_2} = \frac{\theta_1}{\theta_2} = n = \frac{n_2}{n_1} = \frac{r_2}{r_1} = \frac{T_2}{T_1}.$$

(a) Write down Newton's equation of rotational motion for the input (motor) shaft.

(b) The load torque is $\tau_L = r_p(T_1 - T_2)$ where T_1, T_2 are the tensions in the cable as indicated in Figure 5.29. Using Newton's equations of motion

$$M\frac{d^2 z}{dt^2} = T_1 - Mg$$

$$m\frac{d^2(L-z)}{dt^2} = T_2 - mg$$

it follows that

$$\tau_L = r_p(T_1 - T_2) = r_p(M-m)\frac{d^2 z}{dt^2} + r_p(Mg - mg) = r_p^2(M-m)\frac{d^2\theta_2}{dt^2} + r_p(M-m)g.$$

Use this expression for the load torque to write down the equation of rotational motion for the output shaft angle θ_2

(c) Using your answers in parts (a) and (b), give a single differential equation for θ_2 with input τ_m. That is, eliminate $\tau_1, \tau_2, \omega_1, \theta_1$ from your equations.

(d) Using your answer in part (c) write down the differential equation for the elevator car position z with the motor torque τ_m as input.

Problem 6 *Why Pulleys?*

Figure 5.30 shows a two pulley system used to lift a mass m. We can apply a downward force F to the left pulley in order to raise the right pulley along with the mass m. The setup of this system is such that if $y = 0$ then $x = x_0$ and $\theta_{p2} = \theta_{p1} = 0$. Further $y(t) = R\theta_{p1}(t) = R\theta_{p2}(t)$ and the length of the cable is $\ell = y(t) + 2R\pi + 2x(t) + x_p$, which is constant.

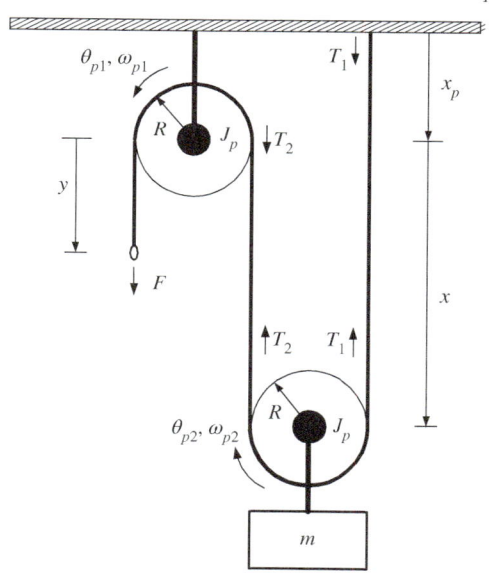

Figure 5.30. Two pulley system.

(a) Set up the equations of motion for the mass m and the two pulleys with moment of inertia J_p in terms of T_1, T_2, F, and x. That is, $\theta_{p_2}, \theta_{p_1}$, and y should be eliminated from your set of equations.

(b) With $J_p = 0$ simplify your answer to part (a) by eliminating T_1 and T_2.

(c) What force F is needed to keep $m\dfrac{d^2x}{dt^2} = 0$? If you only had a single pulley (as in Figure 5.14 of the text), how much force F is needed to maintain $m\dfrac{d^2x}{dt^2} = 0$? The point here is that if you want to lift a weight of mg, the two pulley system only requires a force of $F = mg/2$.

(d) Suppose the right pulley starts at $x(0) = x_0$ so the left pulley is at $y(0) = 0$. The force F is then used to lift the mass m up to $x = 0$ at time t_f, i.e., $x(t_f) = 0$. Show that $y(t_f) = 2x_0$. The point here is that if you lift a weight mg a distance x_0 then the force F must pull the cable down a distance $2x_0$.

Problem 7 *Tension* [16]

Figure 5.31 shows a solid cylinder of uniform density of radius R and mass m. The moment of inertia of such a cylinder is $J = \dfrac{1}{2}mR^2$. Two cords are wrapped around the cylinder near each of the ends with the top of the cords attached to the ceiling. By symmetry the tension in each cord is the same and is denoted as T_1. Note that the tension T_1 is the force exerted on the cylinder by the cord. There is also the downward gravitational force mg on the cylinder. z locates the axis of rotation of the cylinder from the ceiling.

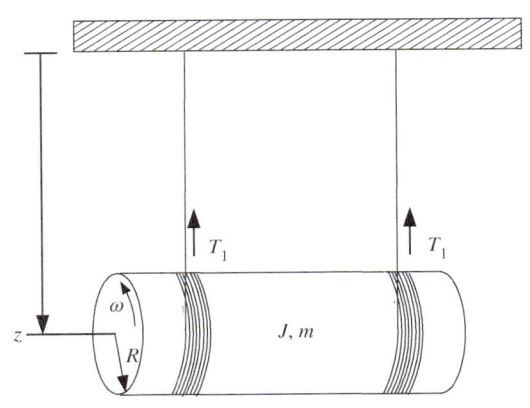

Figure 5.31. Cable wire unwinding due to gravity.

(a) Write down the equations of motion for the cylinder. (Hint: There should be two equations: $md^2z/dt^2 = \cdots$ and $Jd^2\theta/dt^2 = \cdots$.)

Remark Newton's equations for motion are correct if used with respect to an inertial (non-accelerating) coordinate system. However, in this problem we are using $\tau = Jd^2\theta/dt^2$ with respect to the axis of the cylinder (contains the center of mass), which is accelerating downward. It turns out that $\tau = Jd^2\theta/dt^2$ is *still valid* if the rigid body is accelerating as long as it is computed about the center of mass of the rigid body.

(b) As the cylinder falls under the force of gravity assume that the cords unwrap with no slip, that is, $z = R\theta$. Use this to eliminate θ in your answer to part (a) and give the equation of motion for z. Your answer should have T_1 eliminated as well.

(c) What is the linear acceleration of the cylinder downward? (Answer: $2g/3$.)

(d) What is the value of T_1? (Answer: $mg/6$.)

Problem 8 *Rotational Mass–Spring–Damper System*

The left side of Figure 5.32 shows a rotational fluid system used to damp out angular oscillations of a pulley wheel due to a crankshaft that is not completely rigid.

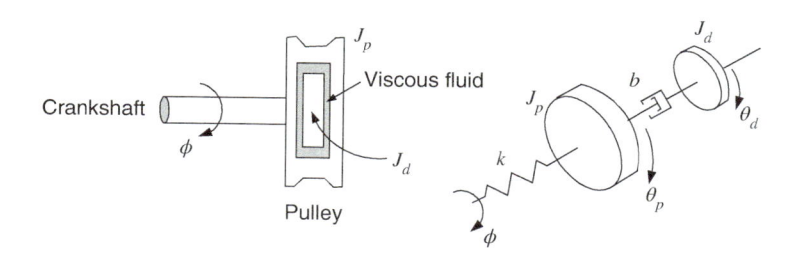

Figure 5.32. Rotational mass–spring–damper System. Source: Adapted from Palm [15]. System Dynamics, Second Edition, McGraw-Hill, 2010.

The crankshaft angle ϕ is the input and angular position θ_p of the pulley wheel is the output. The cylinder inside the pulley wheel has inertia J_d and is surrounded by a fluid that has the effect of damping out vibrations caused by the crankshaft. The rotational mass–spring–damper system shown on the right side of Figure 5.32 is used to model the system. The spring constant k is for the *torsional* spring, which produces a torque when twisted.

(a) Write down the differential equation model of this system.

(b) Compute the transfer function from ϕ to θ_p, i.e., $G(s) = \theta_p(s)/\phi(s)$.

Problem 9 *Simulation of the Satellite System*

In the text a differential equation model of the satellite with solar panels was given as

$$J_s \frac{d^2\theta}{dt^2} = -K(\theta - \theta_p) - b\left(\frac{d\theta}{dt} - \frac{d\theta_p}{dt}\right) + \tau$$

$$J_p \frac{d^2\theta_p}{dt^2} = K(\theta - \theta_p) + b\left(\frac{d\theta}{dt} - \frac{d\theta_p}{dt}\right).$$

Using the parameters from [13] given by $J_s = 5$ kg-m^2, $J_p = 1$ kg-m^2, $K = 0.15$ N-m/rad, $b = 0.05$ Nm/rad/s make a SIMULINK simulation of this system with $\tau = u_s(t) - u_t(t-1)$. Give plots of $\theta_p(t), \theta(t)$ for 50 seconds.

Problem 10 *Simulation of the Satellite System*

In the text it was shown that a transfer function model of the satellite system is given by

$$\theta(s) = \frac{s^2 J_p + bs + K}{s^2\left(J_p J_s s^2 + b(J_p + J_s)s + K(J_p + J_s)\right)}\tau(s)$$

$$\theta_p(s) = \frac{bs + K}{s^2 J_p + bs + K}\theta(s).$$

Using the parameters from [13] given by $J_s = 5$ kg-m^2, $J_p = 1$ kg-m^2, $K = 0.15$ N-m/rad, $b = 0.05$ Nm/rad/s make a SIMULINK simulation of this system with $\tau = u_s(t) - u_t(t-1)$. Give plots of $\theta_p(t), \theta(t)$ for 50 seconds.

Problem 11 *Simulation of the Satellite System*

In the text it was shown that a transfer function model of the satellite system is given by

$$\theta_p(s) = \frac{bs + K}{s^2\left(J_p J_s s^2 + b(J_p + J_s)s + K(J_p + J_s)\right)}\tau(s)$$

$$\theta(s) = \frac{bs + K}{s^2 J_s + bs + K}\theta_p(s) + \frac{1}{s^2 J_s + bs + K}\tau(s).$$

Using the parameters from [13] given by $J_s = 5$ kg-m^2, $J_p = 1$ kg-m^2, $K = 0.15$ N-m/rad, $b = 0.05$ Nm/rad/s make a SIMULINK simulation of this system with $\tau = u_s(t) - u_t(t-1)$. Give plots of $\theta_p(t), \theta(t)$ for 50 seconds.

Problem 12 *Suspended Beam* (Adapted from [17])

As shown in Figure 5.33 a beam is attached to a wall through a pivot and is held up by a spring k_1 attached to it and the ceiling. The beam has a moment of inertia J about the pivot. The applied input force $f(t)$ is always perpendicular to the beam. The deflection angle of the beam is denoted by θ with $\theta = 0$ corresponding to the beam being horizontal. The deflection angle is assumed to be small so that $\sin(\theta) \approx \theta$. The spring k_1 is attached at a distance ℓ_1 from the pivot and the spring k_2 is attached at a distance ℓ_2 from the pivot. With $\theta = 0$ and $w = 0$ both springs are relaxed (neither compressed nor stretched).

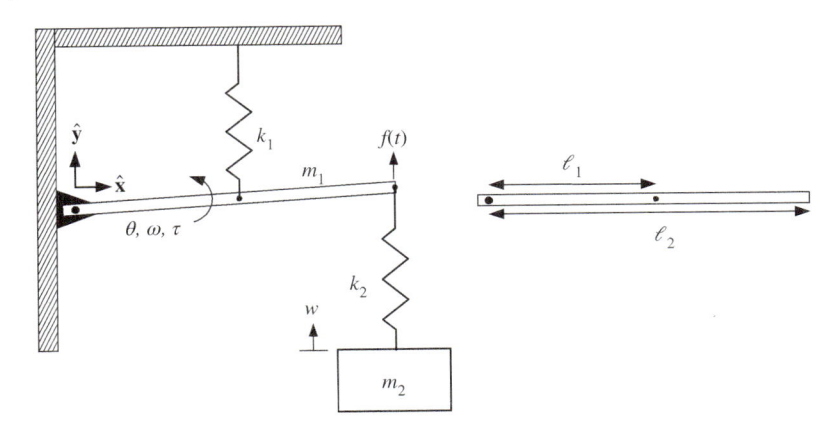

Figure 5.33. Beam attached to a wall and suspended from the ceiling.

142   5 Rigid Body Rotational Dynamics

(a) Show that the equations of motion of the beam and box are

$$J\ddot{\theta} = \underbrace{-k_1(\ell_1 \sin(\theta))\,\ell_1}_{\tau_{k_1}} \underbrace{- m_1 g \cos(\theta) \ell_2/2}_{\tau_{m_1}} + \underbrace{k_2(w - \ell_2 \sin(\theta))\,\ell_2}_{\tau_{k_2}} + \underbrace{\ell_2 f(t)}_{\tau_f}$$

$$m_2 \frac{d^2 w}{dt^2} = -m_2 g - k_2(w - \ell_2 \sin(\theta)).$$

(b) As θ is assumed to be small, $\sin(\theta) \approx \theta$ and $\cos(\theta) \approx 1$. Rewrite the equations of motion using these approximations.

(c) With the external force $f(t)$ set to 0 the equilibrium point of the system is defined by

$$\theta = \theta_0 \text{ (constant)}, \dot{\theta} = 0$$
$$w = w_0 \text{ (constant)}, \dot{w} = 0.$$

Setting $\theta = \theta_0, w = w_0$ and $\dot{\theta} = \dot{w} = 0$, compute the equilibrium values of θ_0, w_0. Your solution should show that both $\theta_0 < 0$ and $w_0 < 0$. Can you give a physical reason why this must hold?

(d) Let

$$\Delta \theta \triangleq \theta - \theta_0$$
$$\Delta w \triangleq w - w_0,$$

where $\Delta \theta, \Delta w$ are the angle of the beam and position of the box *relative* to their equilibrium positions. Note that

$$\frac{d^2 \Delta \theta}{dt^2} = \frac{d^2 \theta}{dt^2}, \frac{d\Delta \theta}{dt} = \frac{d\theta}{dt} \text{ and } \frac{d^2 \Delta w}{dt^2} = \frac{d^2 w}{dt^2}, \frac{d\Delta w}{dt} = \frac{dw}{dt}.$$

By substituting $\theta = \Delta\theta + \theta_0$ and $w = \Delta w + w_0$ into your answer in part (b), rewrite the equations of motion in terms of $\Delta\theta$ and Δw.

(e) Compute the transfer function from the external force $F(s)$ to $\Delta W(s)$.

Problem 13 *Rotational Motion*

Figure 5.34 shows a cylinder and box each on their own incline and connected by a rope passing over a pulley. The pulley has radius R_p and is massless so $m_p = J_p = 0$. The cylinder of radius R has mass m_{cyl} and moment of inertia J about its axis of rotation. Assume the center of mass of the cylinder is also on the axis of rotation and that the cylinder rolls without slipping. There is a friction force F_f exerted on the cylinder by the incline. If $F_f > 0$ then the friction force is producing a torque on the cylinder in the clockwise direction. Because the pulley is massless, the tension force T by the rope on the cylinder axis is the same as the tension force by the rope on the box. The constraint $x + y = d$ must hold because we assume the rope stretches negligibly under tension. Notice that the no slip condition requires $dy/dt = -Rd\theta/dt$.

Write down the equations of motion for the box of mass m_b. On the incline plane with the box the friction between it and the box is taken to be zero. Your answer should be in terms of x with T, F_f, and θ eliminated from the final equation.

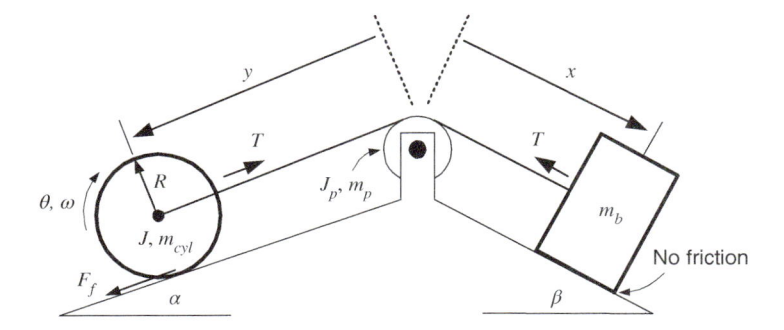

Figure 5.34. Cylinder pulled up an incline by a sliding mass. Source: Adapted from Palm [15]. System Dynamics, Second Edition, McGraw-Hill, 2010.

Problem 14 *Rigid Body Dynamics*

In Figure 5.35 the cylinder has mass m_{cyl}, moment of inertia J_{cyl}, radius R, and rolls without slip so $y = -R\theta$. The box of mass m is located at $x = z + \ell$ where ℓ is constant. The total length of the pulley cord is $2z + y$ and is constant, i.e., $2z + y = d$ with d constant. Then
$$2x + y = d + 2\ell.$$

Pulley 1 has radius R_{p1} and is massless so $J_{p1} = 0$. Similarly, pulley 2 has radius R_{p2} and is also massless so $J_{p2} = 0$ as well.

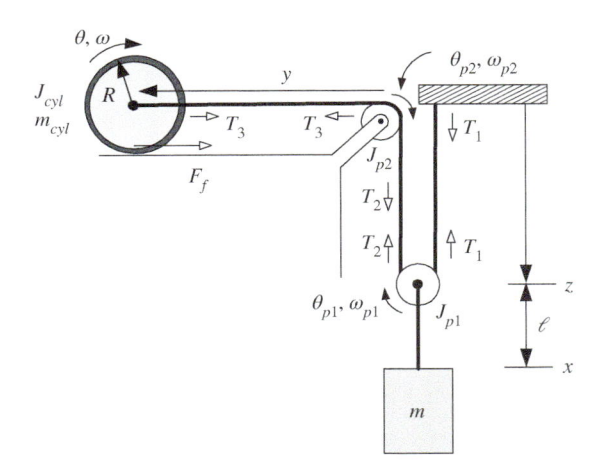

Figure 5.35. Mass, cylinder, and pulley system. Source: Adapted from Palm [15]. System Dynamics, Second Edition, McGraw-Hill, 2010.

(a) Using Newton's law of rotational motion explain why $T_1 = T_2 = T_3$.

(b) Using $T_1 = T_2 = T_3$, show that the equations of motion for the cylinder and box system are
$$m\frac{d^2x}{dt^2} = -2T_1 + mg, \quad J_{cyl}\frac{d^2\theta}{dt^2} = -RF_f, \quad m_{cyl}\frac{d^2y}{dt^2} = -T_1 - F_f.$$

144   5   Rigid Body Rotational Dynamics

(c) Eliminate the tension forces, the static friction force, y, and θ from your equations in part (b) to obtain a single equation for the mass m of the form $\dfrac{d^2x}{dt^2} = \cdots$.

Problem 15 *Rigid Body Dynamics via Conservation of Energy*

In Figure 5.36 the cylinder has mass m_{cyl}, moment of inertia J_{cyl}, radius R and rolls without slip so $y = -R\theta$. The box of mass m is located at $x = z + \ell$ where ℓ is

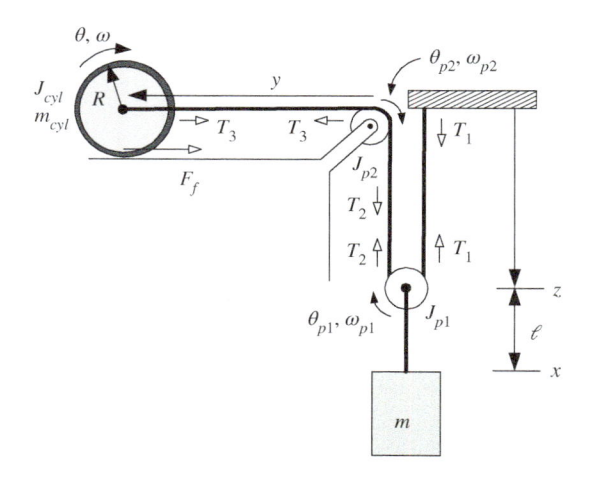

Figure 5.36. Mass, cylinder, and pulley system.

constant. The total length of the pulley cord is $2z + y$ and is constant, i.e., $2z + y = d$ with d constant. Then
$$2x + y = d + 2\ell.$$

Pulley 1 has radius R_{p1} and is massless so $J_{p1} = 0$. Similarly, pulley 2 has radius R_{p2} and is also massless so $J_{p2} = 0$ as well.

(a) Using Newton's law of rotational motion explain why $T_1 = T_2 = T_3$.

(b) By conservation of energy the kinetic energy of the cylinder (rotational and translational) plus the kinetic and potential energy of the mass m is constant. Use this to derive the equation of motion of the mass m. You final answer should have $\omega, \dot{y}$ eliminated.

Problem 16 *Box on an Incline with Kinetic Friction*

Figure 5.37 shows a box sliding on an incline. The coefficient of kinetic friction between the dry surfaces of the box and incline is μ_k.

The two pulleys are massless so $T_1 = T_2 = T_3$ and the pulleys rotate without slipping against the rope. With ℓ the length of the rope we have the constraints
$$x + 2z = \ell$$
$$y + z = \text{constant}.$$

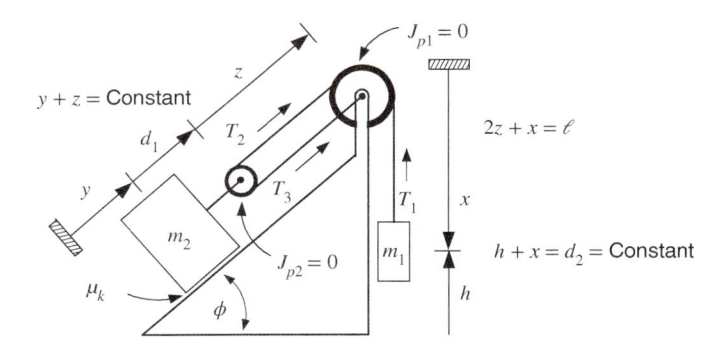

Figure 5.37. Box sliding on an incline being pulled up by $m_1 g$.

(a) Give the equations of motion for m_1 and m_2.

(b) Eliminate the tension and y from the equations of motion and give a single equation in the form
$$\frac{d^2 x}{dt^2} = \cdots$$

(c) Derive your answer in part (b) using conservation of energy. Note that the center of mass of m_2 is $y \sin(\phi)$ above the horizontal and the center of mass of m_1 is $h = d_2 - x$ above the horizontal.

Problem 17 *Static and Kinetic Friction* [16]

Figure 5.38 shows a cylinder of mass m, moment of inertia J, and radius R rolling *without slip* on a level surface at constant angular velocity ω.

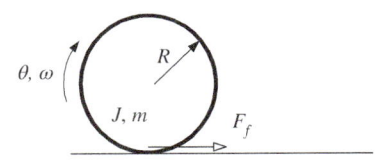

Figure 5.38. Cylinder rolling on a flat surface.

(a) What is the velocity v_{cm} of the center of mass (axis of rotation) of the cylinder?

(b) What is the value of the static friction F_f?

Now consider Figure 5.39. Figure 5.39a shows a cylinder rotating at a constant angular velocity ω_0, but zero translational velocity ($v_{cm} = 0$). Figure 5.39b shows the same cylinder just after it has been let down onto a dry flat surface. The cylinder continues to rotate, but *with* slip. The kinetic friction force $F_{kf} = \mu_k m g$ between the two dry surfaces slows down the angular velocity of the cylinder until it rotates *without* slip.

(c) The kinetic friction force is constant so the torque on the cylinder is constant. That is, we have
$$J \frac{d\omega}{dt} = -R \mu_k m g$$

146   5 Rigid Body Rotational Dynamics

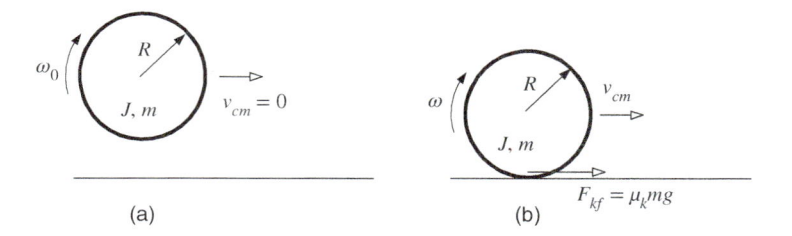

Figure 5.39. Kinetic friction and a slipping cylinder.

with initial condition $\omega(t) = \omega_0$. Further

$$m\frac{dv_{cm}}{dt} = \mu_k mg$$

with initial condition $v_{cm}(0) = 0$. Solve these equations for the time t_f that the cylinder stops slipping, that is, the time $v_{cm}(t_f) = R\omega(t_f)$. Simplify your expression by using the fact that $J = \frac{1}{2}mR^2$ for a cylinder.

(d) Use your answer in part (c) to compute $v_{cm}(t_f)$ and $\omega(t_f)$.

Problem 18 *Cylinder Rolling Down an Incline*
Figure 5.40 shows a cylinder of radius r rolling down an incline. With $d = r/\cos(\phi)$ show that the center of mass is a distance $(x - d\sin(\phi))\sin(\phi) + d$ above the horizontal. Hint: The location of the axis of the cylinder is at $x\hat{x} + y\hat{y}$. The vertical line through the center of the cylinder intersects the incline at $x - d\sin(\phi)$, that is, at the point $(x - d\sin(\phi))\hat{x} + 0\hat{y}$.

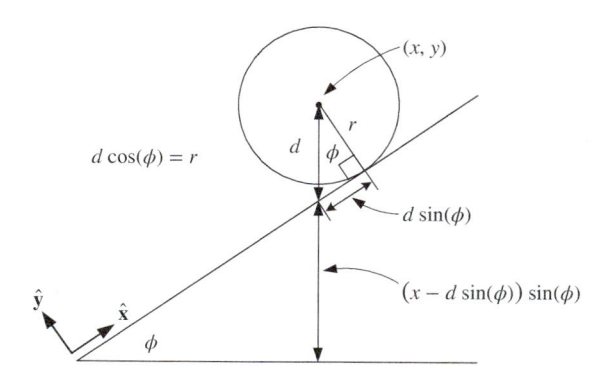

Figure 5.40. Calculation of the vertical height of the axis of rotation.

Problem 19 *Vehicle Equations Via Conservation of Energy*
Consider a vehicle rolling down an incline as shown in Figure 5.41. The body of the vehicle has mass m. The front set of wheels has total mass m_F, total moment of inertia J_F, and radius r_F. The back set of wheels has total mass m_B, total moment of inertia J_B, and radius r_B. The vehicle rolls down the incline without slipping. Let (x, y) be the coordinates of the center of mass of the vehicle body. The center of mass of the front wheels is on its axis and similarly the center of mass of the back wheels is on its axis. The incline is at

an angle ϕ. Figure 5.41 shows that the back wheels are at $x_B = x + \ell_B$ along the incline where x is the location of the center of mass of the cart along the x axis. Similarly, the front wheels are at $x_F = x - \ell_F$ along the incline.

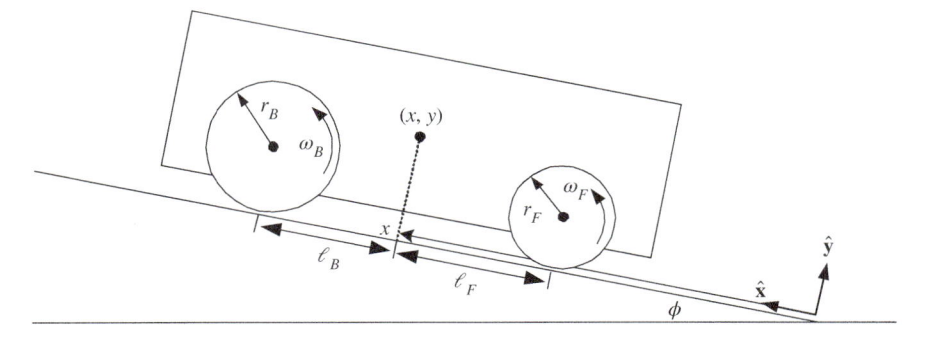

Figure 5.41. Vehicle rolling down an incline.

(a) Write down the total kinetic energy of the two wheels and the vehicle body.

(b) Write down the total potential energy of the two wheels and the vehicle body. Hint: Look at Figure 5.42 and note the following:

(i) The height of the center of mass of the vehicle body is (see Problem 18)
$$(x - d_m \sin(\phi)) \sin(\phi) + d_m = x \sin(\phi) + \underbrace{d_m \cos^2(\phi)}_{d_0}.$$

(ii) The height of the center of mass of the back wheels is
$$(x + \ell_B - d_B \sin(\phi)) \sin(\phi) + d_B = x \sin(\phi) + \underbrace{\ell_B \sin(\phi) + d_B \cos^2(\phi)}_{d_1}.$$

(iii) The height of the center of mass of the front wheels is
$$(x - \ell_F - d_F \sin(\phi)) \sin(\phi) + d_F = x \sin(\phi) \underbrace{-\ell_F \sin(\phi) + d_F \cos^2(\phi)}_{d_2}.$$

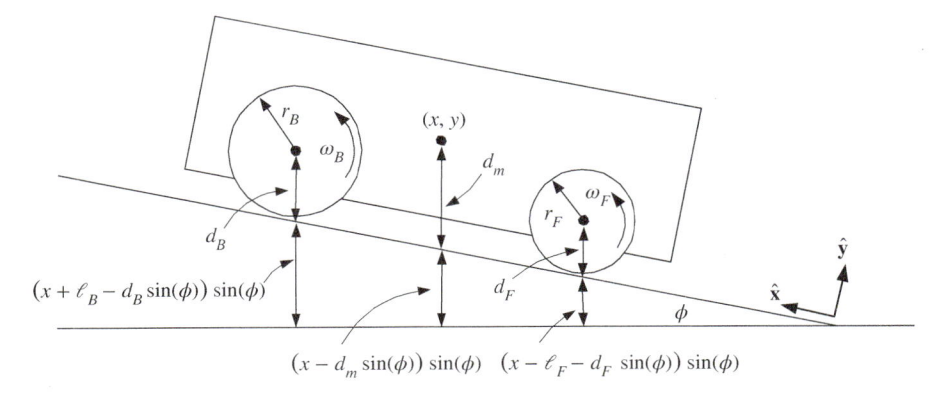

Figure 5.42. Vertical heights of the vehicle's axels and center of mass.

148   5 Rigid Body Rotational Dynamics

(c) Using the no slip condition write down the relationships between ω_B, ω_F, and $\dot{x}$.

(d) Write down the total energy KE with ω_F and ω_B eliminated from the expression.

(e) Using the fact that the total energy is constant, that is, $\frac{d}{dt}(KE + PE) = 0$ to show that the equation of motion is

$$\ddot{x} = -\frac{m_B + m_F + m}{J_B/r_B^2 + m_B + J_F/r_F^2 + m_F + m} g \sin(\phi).$$

Problem 20 *Equations of Motion of a Vehicle*
Redo Problem 19 using Newton's equation for rotational motion for the front and back wheels and Newton's equation for translational motion for the front wheels, the back wheels, and the cart. This means you will start by writing down five differential equations. Figure 5.43 shows the geometric relationships between the vehicle center of mass and the wheels on the incline. The x coordinate of the center of mass of the back wheel is

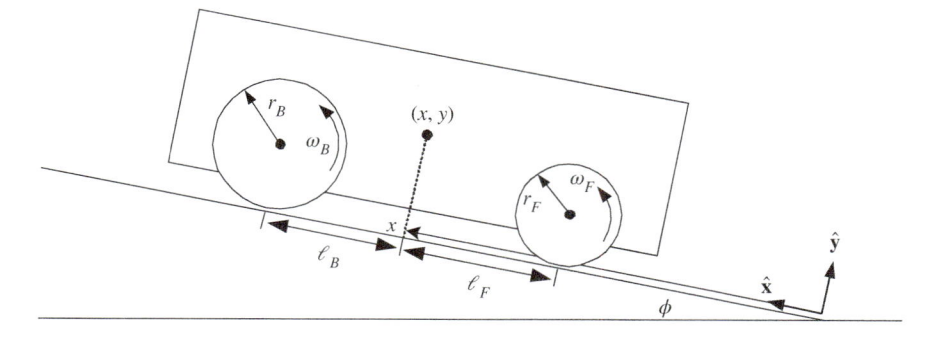

Figure 5.43. Vehicle rolling down an incline.

$x_B = x + \ell_B$ and the x coordinate of the center of mass of the front wheel is $x_F = x - \ell_F$. You must eliminate ω_B, ω_F, F_F, and F_B from your final answer. The no slip condition tells us that

$$x_B = x + \ell_B = r_B \theta_B + \ell_B \Rightarrow \frac{dx_B}{dt} = r_B \omega_B$$

$$x_F = x - \ell_F = r_F \theta_F - \ell_F \Rightarrow \frac{dx_F}{dt} = r_F \omega_F.$$

Does your answer look right? Specifically, is $\ddot{x}$ positive or negative? Are ω_B, ω_F positive or negative?

6
The Physics of the DC Motor

The principles of operation of a direct current (DC) motor are presented based on fundamental concepts from electricity and magnetism contained in any basic physics course. In order to do this we review the concepts of magnetic fields, magnetic force, Faraday's law, and induced electromotive forces (emfs) as they apply to modeling the DC motor. All of the physics concepts referred to in this chapter are contained in the book *Physics* by Halliday and Resnick [18].

6.1 Magnetic Force

Motors work on the basic principle that magnetic fields produce forces on wires carrying a current. In fact, this experimental phenomenon is what is used to define the magnetic field. If one places a current carrying wire between the poles of a magnet as in Figure 6.1, a force is exerted on the wire.

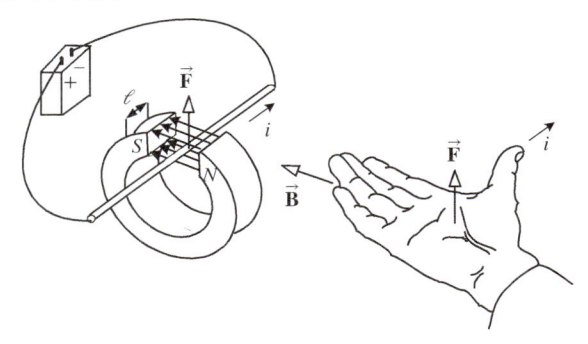

Figure 6.1. Magnetic force law. Source: Adapted from Haber-Schaim et al. [19]. PSSC Physics, 7th Edition, Kendall/Hunt, Dubuque, IA, 1991.

Experimentally, the magnitude of this force is found to be proportional to both the amount of current i in the wire and to the length ℓ of the wire that is between the poles of the magnet. That is, F_{magnetic} is proportional to ℓi. The direction of the magnetic field $\vec{B}$ at any point is defined to be the direction that a small compass needle would point at that location. This direction is indicated by arrows in between the north and south poles in Figure 6.1. With the direction of $\vec{B}$ perpendicular to the wire, the strength (magnitude) of the *magnetic induction field* $\vec{B}$ is defined to be

$$B = |\vec{B}| \triangleq \frac{F_{\text{magnetic}}}{\ell i},$$

An Introduction to System Modeling and Control, First Edition. John Chiasson.
© 2022 John Wiley & Sons, Inc. Published 2022 by John Wiley & Sons, Inc.
Companion website: www.wiley.com/go/chiasson/anintroductiontosystemmodelingandcontrol

where F_{magnetic} is the magnetic force, i is the current, and ℓ is the length of wire perpendicular to the magnetic field carrying the current. That is, B is the proportionality constant so that $F_{\text{magnetic}} = i\ell B$. As illustrated in Figure 6.1, the direction of the force can be determined using the right-hand rule. Specifically, using your right hand, point your fingers in the direction of the magnetic field and point your thumb in the direction of the current. Then the direction of the force is out of your palm.

Further experiments show that if the wire is parallel to the $\vec{\mathbf{B}}$ field rather than perpendicular as in Figure 6.1, then no force is exerted on the wire. If the wire is at some angle θ with respect to $\vec{\mathbf{B}}$ as in Figure 6.2, then the force is proportional to the *component* of $\vec{\mathbf{B}}$ perpendicular to the wire; that is, it is proportional to $B_\perp = B\sin(\theta)$. This is summarized in the *magnetic force law*: Let $\vec{\ell}$ denote a vector whose magnitude is the length ℓ of the wire in the magnetic field and whose direction is defined as the positive direction of current in the bar; then the magnetic force on the bar of length ℓ carrying the current i is given by

$$\vec{\mathbf{F}}_{\text{magnetic}} = i\vec{\ell} \times \vec{\mathbf{B}}.$$

In scalar terms we have $F_{\text{magnetic}} = i\ell B \sin(\theta) = i\ell B_\perp$. Again, $B_\perp \triangleq B\sin(\theta)$ is the component of $\vec{\mathbf{B}}$ perpendicular to the wire.[1]

(a)   (b)

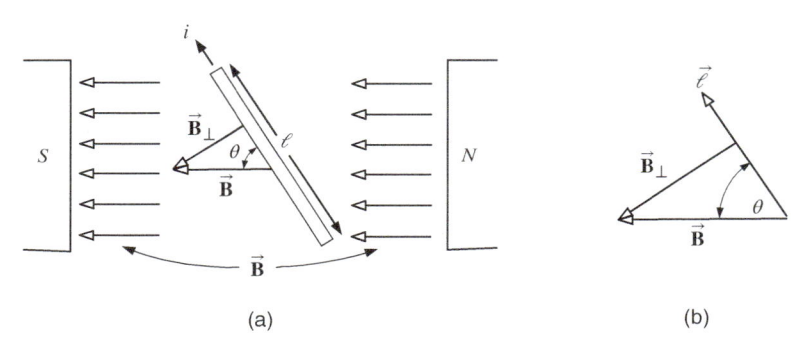

Figure 6.2. Only the component $B_\perp$ of the magnetic field that is perpendicular to the wire produces a force on the current.

Example 1 *A Linear DC Machine* [20]

Consider the simple linear DC machine in Figure 6.3 where a sliding bar rests on a simple circuit consisting of two rails. An external magnetic field is going through the loop of the circuit into the page indicated by the $\otimes$ in the plane of the loop. Closing the switch results in a current flowing around the circuit and the external magnetic field produces a force on the bar, which is free to move. The force on the bar is now computed.

The magnetic field is constant and points into the page (indicated by $\otimes$) so that written in vector notation, $\vec{\mathbf{B}} = -B\hat{\mathbf{z}}$ with $B > 0$. By the right hand rule, the magnetic force on the sliding bar points to the right. Explicitly, with $\vec{\ell} = -\ell\hat{\mathbf{y}}$, the force is given by

$$\vec{\mathbf{F}}_{\text{magnetic}} = i\vec{\ell} \times \vec{\mathbf{B}} = i(-\ell\hat{\mathbf{y}}) \times (-B\hat{\mathbf{z}}) = i\ell B\hat{\mathbf{x}}.$$

To find the equations of motion for the bar, let f be the coefficient of viscous (sliding) friction of the bar so that the friction force is given by $F_f = -f dx/dt$. Then, with m_ℓ

---

[1] Motors are designed so that the conductors are perpendicular to the external magnetic field.

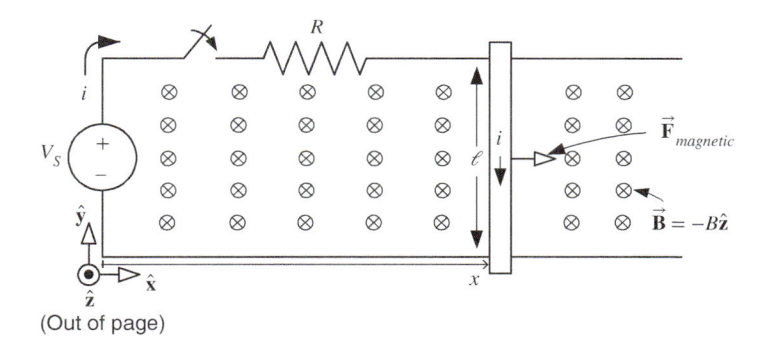

Figure 6.3. A linear DC motor.

denoting the mass of the bar, Newton's law gives

$$i\ell B - f dx/dt = m_\ell d^2 x/dt^2.$$

Just after closing the switch at $t = 0$, but before the bar starts to move, the current is $i(0^+) = V_S(0^+)/R$. However, it turns out that as the bar moves the current does *not* stay at this value, but instead decreases due to electromagnetic induction. This will be explained later.

6.2 Single-Loop Motor

As a first step to modeling a DC motor, a simplistic single-loop motor is considered. It is first shown how torque is produced and then how the current in the single loop can be reversed (commutated) every half turn to keep the torque constant.

Torque Production

Consider the magnetic system in Figure 6.4 on the next page, where a cylindrical core is cut out of a block of a permanent magnet and replaced with a soft iron core. The term "soft" iron refers to the fact that material is easily magnetized (a permanent magnet is referred to as "hard" iron). An important property of soft magnetic materials is that the magnetic field at the surface of such materials tends to be normal (perpendicular) to the surface. Consequently, the cylindrical shape of the surfaces of the soft iron core and the stator permanent magnet has the effect of making the field in the air gap *radially* directed; furthermore, it is reasonably constant (uniform) in magnitude.

A mathematical description of the magnetic field in the air gap due to the permanent magnet is simply

$$\vec{B} = \begin{cases} +B\hat{r} & \text{for } 0 < \theta < \pi \\ -B\hat{r} & \text{for } \pi < \theta < 2\pi, \end{cases}$$

where $B > 0$ is the magnitude or strength of the magnetic field and θ is an arbitrary location in the air gap.

Figure 6.5 on the next page shows a rotor loop wound around the iron core of Figure 6.4. The length of the rotor is ℓ_1 and its diameter is ℓ_2. The torque on this rotor loop is now calculated by considering the magnetic forces on sides a and a' of the loop. On the other two sides of the loop, that is, the front and back sides, the magnetic field has negligible

152   6 The Physics of the DC Motor

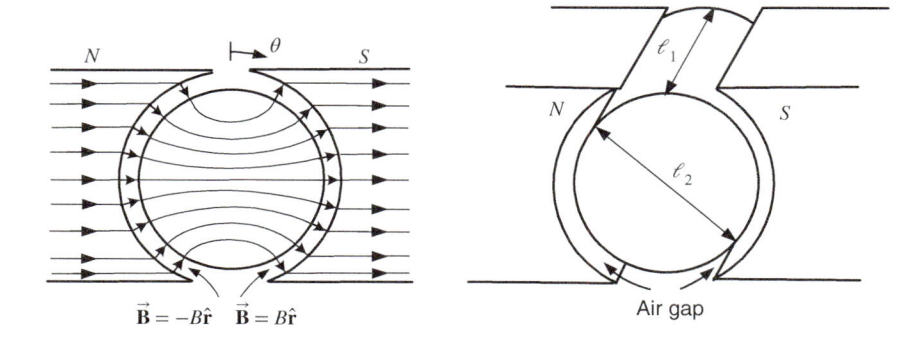

Figure 6.4. Soft iron cylindrical core placed inside a hollowed out permanent magnet to produce a radial magnetic field in the air gap.

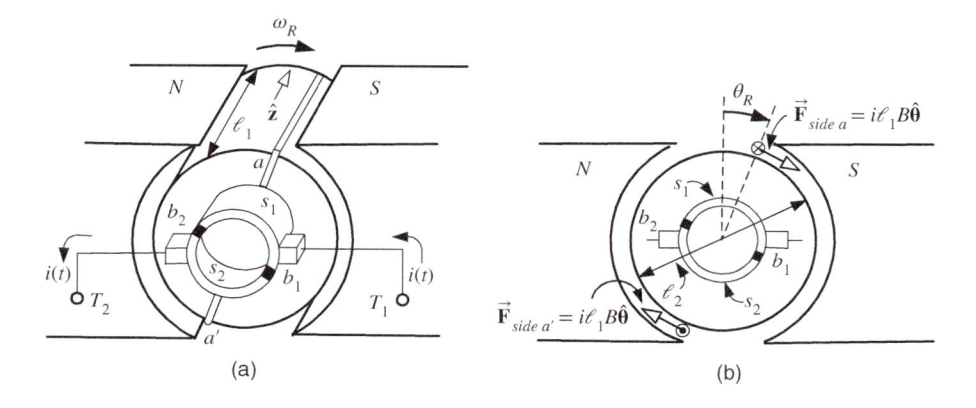

Figure 6.5. A single-loop motor. Source: Adapted from Matsch and Morgan [21]. Electromagnetic and Electromechanical Machines, 3rd edition, John Wiley & Sons, New York, 1986.

strength so that no significant force is produced on these sides. As illustrated in Figure 6.5b, the rotor angular position is taken to be the angle θ_R from the vertical to side a of the rotor loop.

Figure 6.6 shows the cylindrical coordinate system used in Figure 6.5. Here $\hat{\mathbf{r}}, \hat{\boldsymbol{\theta}}, \hat{\mathbf{z}}$ denote unit cylindrical coordinate vectors. The unit vector $\hat{\mathbf{z}}$ points along the rotor axis into the page in Figure 6.5b, $\hat{\boldsymbol{\theta}}$ is in the direction of increasing θ, and $\hat{\mathbf{r}}$ is in the direction of increasing r.

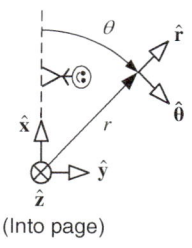

Figure 6.6. Cylindrical coordinate system used in Figure 6.5.

Referring back to Figure 6.5, for $i > 0$, the current in side a of the loop is going into the page (denoted by $\otimes$) and then comes out of the page (denoted by $\odot$) on side a'. Thus, on side a, $\vec{\ell} = \ell_1 \hat{\mathbf{z}}$ (as $\vec{\ell}$ points in the direction of positive current flow) and the magnetic force $\vec{\mathbf{F}}_{\text{side } a}$ on side a is then

$$\vec{\mathbf{F}}_{\text{side } a} = i\vec{\ell} \times \vec{\mathbf{B}} = i(\ell_1 \hat{\mathbf{z}}) \times (B\hat{\mathbf{r}}) = i\ell_1 B \hat{\boldsymbol{\theta}}$$

which is tangential to the motion as shown in Figure 6.5b. The resulting torque is

$$\vec{\tau}_{\text{side } a} = (\ell_2/2)\hat{\mathbf{r}} \times \vec{\mathbf{F}}_{\text{side } a} = (\ell_2/2)i\ell_1 B \hat{\mathbf{r}} \times \hat{\boldsymbol{\theta}} = (\ell_2/2)i\ell_1 B \hat{\mathbf{z}}.$$

Similarly, the magnetic force on side a' of the rotor loop is

$$\vec{\mathbf{F}}_{\text{side } a'} = i\vec{\ell} \times \vec{\mathbf{B}} = i(-\ell_1 \hat{\mathbf{z}}) \times (-B\hat{\mathbf{r}}) = i\ell_1 B \hat{\boldsymbol{\theta}}$$

so that the corresponding torque is

$$\vec{\tau}_{\text{side } a'} = (\ell_2/2)\hat{\mathbf{r}} \times \vec{\mathbf{F}}_{\text{side } a'} = (\ell_2/2)i\ell_1 B \hat{\mathbf{r}} \times \hat{\boldsymbol{\theta}} = (\ell_2/2)i\ell_1 B \hat{\mathbf{z}}.$$

The total torque on the rotor loop is then

$$\vec{\tau}_m = \vec{\tau}_{\text{side } a} + \vec{\tau}_{\text{side } a'} = 2(\ell_2/2)i\ell_1 B \hat{\mathbf{z}} = \ell_1 \ell_2 B i \hat{\mathbf{z}}.$$

The torque points along the z axis, which is the axis of rotation. In scalar form,

$$\tau_m = K_T i,$$

where $K_T \triangleq \ell_1 \ell_2 B$. The force is proportional to the strength B of magnetic field $\vec{\mathbf{B}}$ in the air gap due to the permanent magnet.

Wound Field DC Motor

In order to increase the strength of the magnetic field in the air gap, the permanent magnet can be replaced with a soft iron material with wire wound around its periphery as shown in Figure 6.7a. This winding is referred to as the *field winding*, and the current it carries is called the *field current*. In normal operation, the field current is held constant. The strength of the magnetic field in the air gap is then proportional to the field current i_f at lower current levels (i.e., $B = K_f i_f$) and then saturates as the current increases. This may be written as $B = f(i_f)$ where $f(\cdot)$ is a saturation curve satisfying $f(0) = 0, f'(0) = K_f$ as shown in Figure 6.7b.

Commutation of the Single-Loop Motor

In the derivation of the torque expression $\tau_m = K_T i$ we assumed that the current in the side of the rotor loop[2] under the south pole face is into the page and the current in the side of the loop under the north pole face is out of the page as in Figure 6.8a. In order to make this assumption valid, the direction of the current in the loop must be changed each time the rotor loop passes through the vertical. The process of changing the direction of

---

[2] The rotor loop is also referred to as the *armature* winding and the current in it as the *armature* current.

154   6 The Physics of the DC Motor

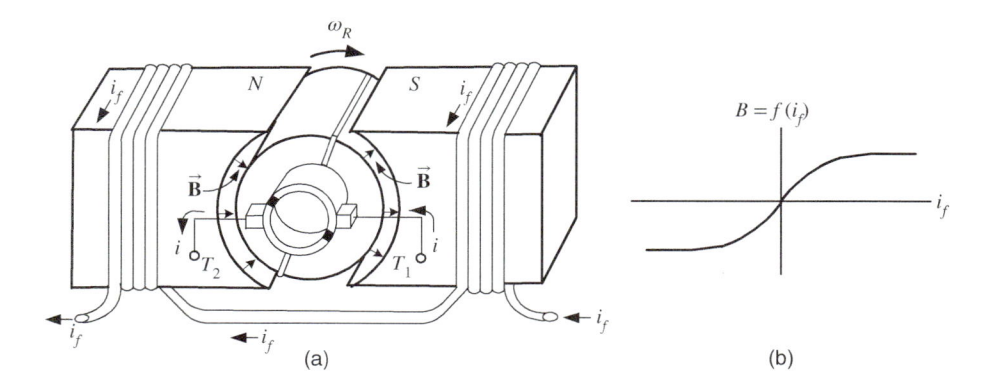

Figure 6.7. (a) DC motor with a field winding. (b) Radial magnetic field strength in the air gap.

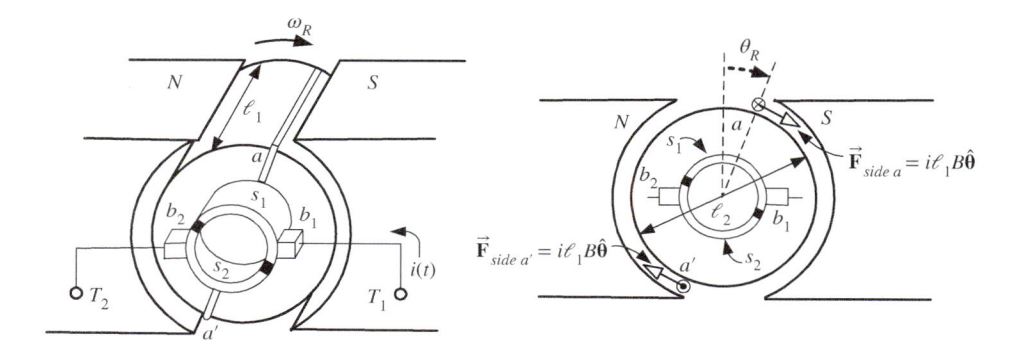

Figure 6.8a. $0 < \theta_R < \pi$. Source: Adapted from Matsch and Morgan [21]. Electromagnetic and Electromechanical Machines, 3rd edition, John Wiley & Sons, New York, 1986.

the current is referred to as *commutation* and is done at $\theta_R = 0$ and $\theta_R = \pi$ through the use of the slip rings s_1, s_2 and the brushes b_1, b_2 shown in Figure 6.8. The slip rings are rigidly attached to the loop and thus rotate with it. The brushes are fixed in space with the slip rings making a sliding electrical contact with the brushes as the loop rotates.

To see how the commutation of the current is accomplished using the brushes and slip rings, consider the sequence of Figures 6.8a–d. As shown in Figure 6.8a, the current goes through brush b_1 into the slip ring s_1. From there, it travels down (into the page $\otimes$) side a of the loop, comes back up side a' (out of the page $\odot$) into the slip ring s_2, and, finally, comes out the brush b_2. Note that side a of the loop is under the south pole face while side a' is under the north pole face. Figure 6.8b shows the rotor loop just before commutation where the same comments as in Figure 6.8a apply.

Figure 6.8c shows that when $\theta_R = \pi$, the slip rings at the ends of the loop are shorted together by the brushes forcing the current in the loop to drop to zero.

Subsequently, as shown in Figure 6.8d, with $\pi < \theta_R < 2\pi$, the current is now going through brush b_1 into slip ring s_2. From there, the current travels down (into the page $\otimes$) side a' of the loop and comes back up (out of the page $\odot$) side a. In other words, the current has *reversed* its direction in the loop from that in Figures 6.8a and 6.8b. This is precisely what is desired, as side a is now under the north pole face and side a' is under the south pole face. As a result of the brushes and slip rings, the current direction in the loop is reversed every half-turn.

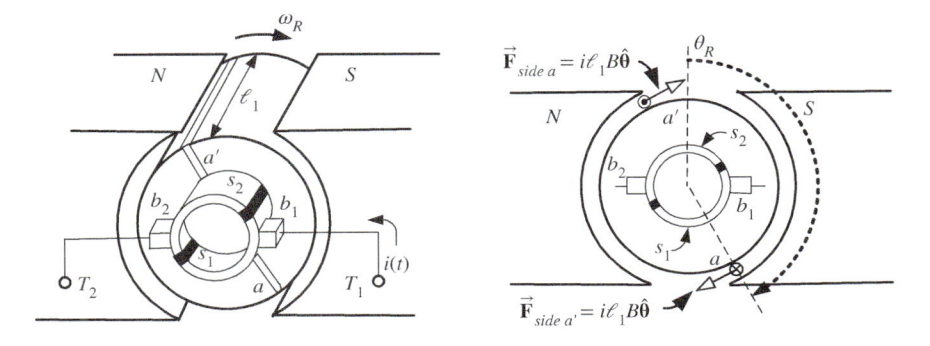

Figure 6.8b. Rotor loop just prior to commutation where $0 < \theta_R < \pi$.

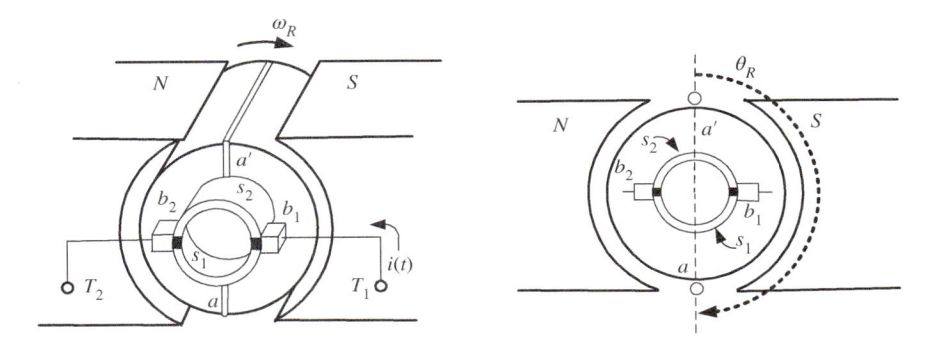

Figure 6.8c. The ends of the rotor loop are shorted when $\theta_R = \pi$.

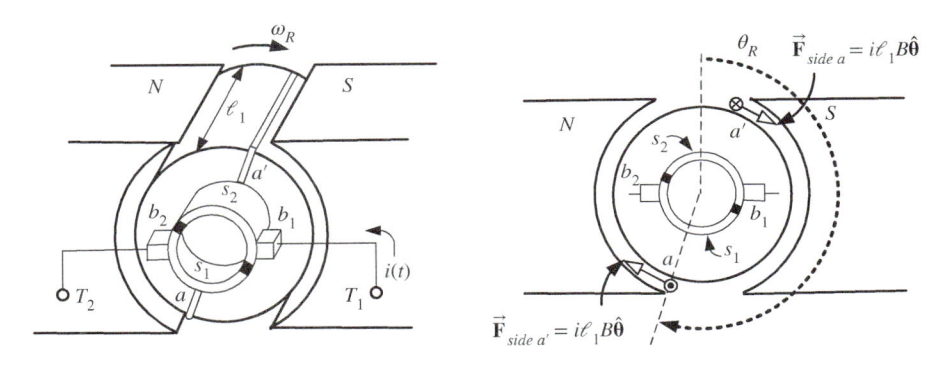

Figure 6.8d. Rotor loop just after commutation where $\pi < \theta_R < 2\pi$.

6.3 Faraday's Law

Figure 6.9 on the next page shows a magnet moving upwards into a wire loop producing a changing magnetic flux in the loop.

Recall that a changing flux within a loop produces an induced $emf\,\xi$ in the loop according to Faraday's law.[3] That is,

$$\xi = -d\phi/dt,$$

---

[3] ξ is the Greek letter "xi" and is pronounced "ksee."

156   6 The Physics of the DC Motor

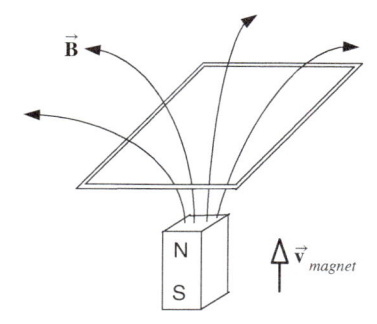

Figure 6.9. A magnet moving upwards produces a changing flux in the loop that in turn results in an induced emf and current in the loop.

where

$$\phi = \int_S \vec{B} \cdot d\vec{S}$$

is the flux in the loop and S is any surface with the loop as its boundary. Faraday's law is now reviewed in some detail.

The Surface Element Vector $d\vec{S}$

The surface element $d\vec{S}$ is a vector whose magnitude is a differential (small) element of area dS and whose direction is normal (perpendicular) to the surface element. As there are two possibilities for the normal to the surface, one must choose the normal in a consistent manner. Furthermore, depending on the particular normal chosen, a convention is used to characterize the positive and negative directions of travel around the surface boundary. To describe this, consider Figure 6.10a, which shows a small surface element with the normal direction taken to be up in the positive z direction. In this case, with $\hat{\mathbf{n}} = \hat{\mathbf{z}}$, $dS = dxdy$, the surface element vector is defined by

$$d\vec{S} \triangleq dxdy\hat{\mathbf{z}}.$$

The corresponding direction of travel around the surface boundary is indicated by the curved arrow in the figure.

In Figure 6.10b a surface element with the normal direction taken to be down in the negative z direction is shown. In this case $\hat{\mathbf{n}} = -\hat{\mathbf{z}}, dS = dxdy$ so that the surface element

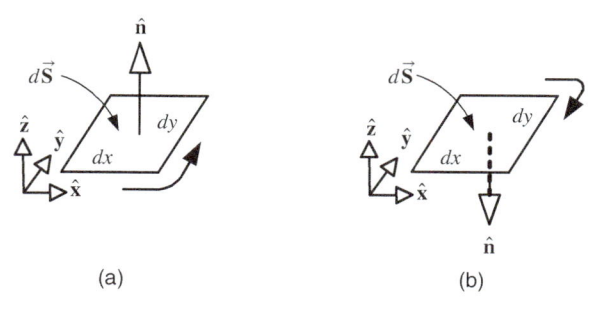

Figure 6.10. (a) Positive direction of travel around a surface element with the normal up. (b) Positive direction of travel around a surface element with the normal down.

vector is defined as
$$d\vec{S} = -dxdy\hat{z}.$$

The direction of positive travel around the surface element is indicated by the curved arrow in Figure 6.10b and is opposite to that of Figure 6.10a.

As illustrated in Figure 6.10, the vector differential surface element $d\vec{S}$ is defined to be a vector whose magnitude is the area of the differential surface element and whose direction is normal to the surface. One may choose either normal, and the corresponding direction of positive travel around the surface is then determined.

Two surface elements may be connected together as in Figure 6.11 and travel around the total surface is defined as shown. Note that along the common boundary of the two joined surface elements, the directions of travel "cancel" out each other, resulting in a net travel path around both surface elements. The normals for the surface elements must both be up or both be down; that is, the normal must be continuous as one goes from one surface element to the next.

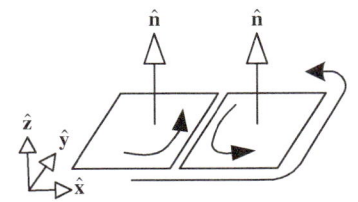

Figure 6.11. Positive direction of travel around two joined surface elements.

Figure 6.12 shows a surface in the x–y plane with a rectangular boundary made up of small surface elements $d\vec{S} \triangleq dxdy\hat{n} = dxdy\hat{z}$. The $\odot$ in the middle of each surface element just indicates the normal is $\hat{n} = \hat{z}$ (out of the page). Only those surface elements that have sides on the rectangular boundary do not get their direction of travel "cancelled" by a neighboring surface element. Consequently, the net direction of travel about the rectangular boundary of this surface is *counterclockwise*, which is taken to be the positive direction of travel around the surface.

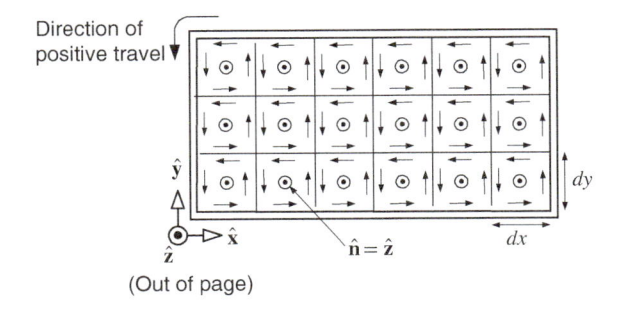

Figure 6.12. Positive direction of travel around a surface boundary.

Interpreting the Sign of ξ

The interpretation of positive and negative values of the induced emf ξ can now be given. Faraday's law says that the induced emf (voltage) in a loop is given by
$$\xi = -d\phi/dt,$$

158   6   The Physics of the DC Motor

where
$$\phi = \int_S \vec{B} \cdot d\vec{S}.$$

If $\xi > 0$, the induced emf will force current in the positive direction of travel around the surface while if $\xi < 0$, the induced emf will force current in the opposite direction. Faraday's law is now illustrated by some examples. Specifically, it is used to compute the induced emf in the linear DC machine, the induced emf in the single-loop machine, and the self-induced voltage in the single-loop machine.

Back Emf in a Linear DC Machine

Figure 6.13 shows the linear DC machine where the back emf it generates is now computed. The magnetic field is constant and points into the page, that is, $\vec{B} = -B\hat{z}$, where $B > 0$. The magnetic force on the bar is $\vec{F}_{magnetic} = i\ell B\hat{x}$. To compute the induced voltage in the loop of the circuit, let $\hat{n} = \hat{z}$ be the normal to the surface so that $d\vec{S} = dxdy\hat{z}$, where $dS = dxdy$. Then

$$\phi = \int_S \vec{B} \cdot d\vec{S} = \int_0^\ell \int_0^x (-B\hat{z}) \cdot (dxdy\hat{z}) = \int_0^\ell \int_0^x -Bdxdy = -B\ell x.$$

The induced (back) emf is therefore given by

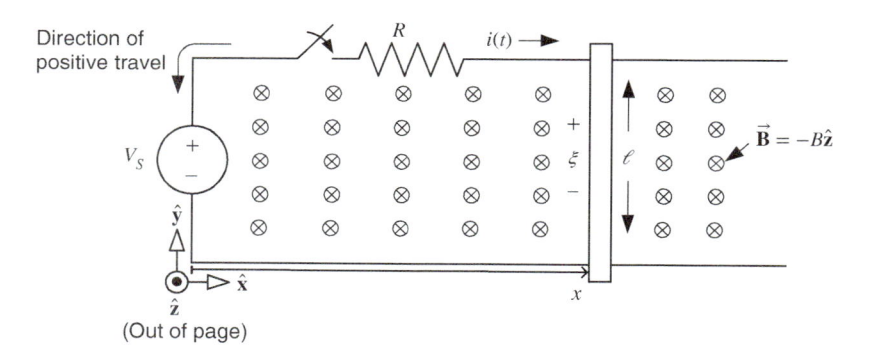

$$\xi = -d\phi/dt = -d(-B\ell x)/dt = B\ell v.$$

Figure 6.13. With $d\vec{S} = dxdy\hat{z}$, the direction of positive travel is in the counterclockwise direction.

In the flux computation, the normal for the surface was taken to be in $+\hat{z}$ direction. By putting together the differential flux surfaces $d\vec{S}$ in a fashion similar to Figure 6.11, the positive direction of travel around the surface is counterclockwise around the loop as indicated in Figure 6.13. Here the sign conventions for source voltage V_S and the back emf ξ are opposite so that, as the back emf $\xi = B\ell v > 0$, it is *opposing* the applied source voltage V_S.

Remark $\phi = -B\ell x$ is the flux in the circuit due to the *external* magnetic field $\vec{B} = -B\hat{z}$. There is also a flux $\psi = Li$ due to the current i in the circuit. For this example, the inductance is small and one just sets $L = 0$.

Electromechanical Energy Conversion

As the back emf $\xi = B\ell v$ opposes the current i, electrical power is being absorbed by this back emf. Specifically, the electrical power absorbed by the back emf is $i\xi = iB\ell v$ while the mechanical power produced is $F_{\text{magnetic}} v = i\ell B v$. That is, the electrical power absorbed by the back emf reappears as mechanical power, as it must by conservation of energy. Another way to view this is to note that $V_S i$ is the electrical power delivered by the source and, as $V_S - B\ell v = Ri$, one may write

$$V_S i = Ri^2 + i(B\ell v) = Ri^2 + F_{\text{magnetic}} v.$$

In words, the power $V_S i$ from the source has the amount Ri^2 dissipated as heat in the resistance R and the amount $F_{\text{magnetic}} v$ converted into mechanical power.

Equations of Motion for the Linear DC Machine

The equations of motion for the bar in the linear DC machine are now derived. With the inductance L of the circuit loop taken to be zero, m_ℓ the mass of the bar, f the coefficient of viscous friction, it follows that

$$V_S - B\ell v = Ri$$

$$m_\ell \frac{dv}{dt} = i\ell B - fv.$$

Eliminating the current i, one obtains

$$m_\ell \frac{d^2 x}{dt^2} = \ell B(V_S - B\ell v)/R - fv = -\left(\frac{B^2 \ell^2}{R} + f\right)\frac{dx}{dt} + \frac{\ell B}{R} V_S$$

or

$$m_\ell \frac{d^2 x}{dt^2} + \left(\frac{B^2 \ell^2}{R} + f\right)\frac{dx}{dt} = \frac{\ell B}{R} V_S. \qquad (6.1)$$

This is the equation of motion for the bar with V_S as the control input and the position x at the measured output.

Back Emf in the Single-Loop Motor

The back emf induced in the single loop motor by the external magnetic field of the permanent magnet is now computed. To do so, consider the flux surface for the rotor loop shown in Figure 6.14 on the next page. The surface is a half-cylinder of radius $\ell_2/2$ and length ℓ_1 with the rotor loop as its boundary. The cylindrical surface is in the air gap, where the magnetic field is known to be radially directed and constant in magnitude, that is,

$$\vec{B} = \begin{cases} +B\hat{\mathbf{r}} & \text{for } 0 < \theta < \pi \\ -B\hat{\mathbf{r}} & \text{for } \pi < \theta < 2\pi. \end{cases} \qquad (6.2)$$

On the cylindrical part of the surface, the surface element is chosen as

$$d\vec{S} = (\ell_2/2) d\theta dz \hat{\mathbf{r}}$$

which is directed outward from the axis of the cylinder as illustrated in Figure 6.15 on the next page. The corresponding direction of positive travel is also indicated in Figure 6.15.

160   6 The Physics of the DC Motor

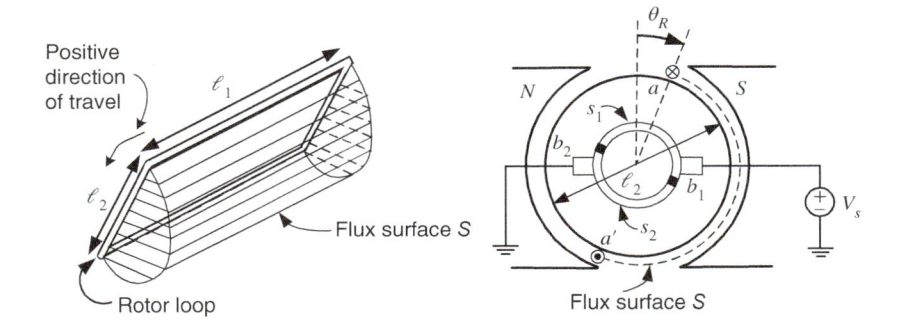

Figure 6.14. Flux surface for the single loop motor.

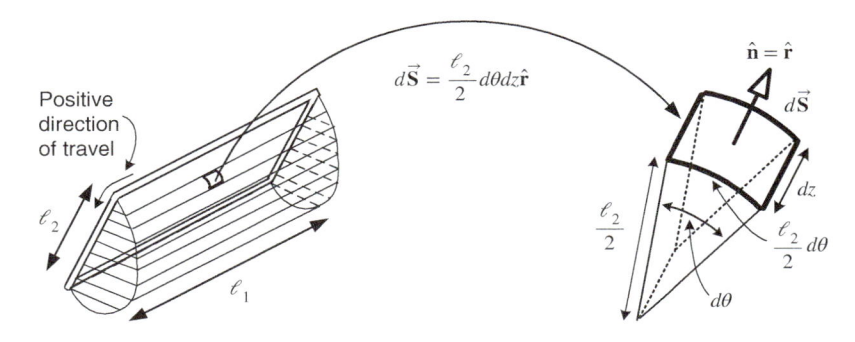

Figure 6.15. Surface element vector for the flux surface of Figure 6.14.

On the two ends (half-disks) of the cylindrical surface, the $\vec{B}$ field is quite weak making the flux through these two half-disks negligible.

Neglecting the flux through the two ends of the surface, the flux $\phi(\theta_R)$ for $0 < \theta_R < \pi$ is given by

$$\begin{aligned}
\phi(\theta_R) &= \int_S \vec{B} \cdot d\vec{S} \\
&= \int_0^{\ell_1} \int_{\theta=\theta_R}^{\theta=\pi} (B\hat{\mathbf{r}}) \cdot \left(\frac{\ell_2}{2} d\theta dz \hat{\mathbf{r}}\right) + \int_0^{\ell_1} \int_{\theta=\pi}^{\theta=\pi+\theta_R} (-B\hat{\mathbf{r}}) \cdot \left(\frac{\ell_2}{2} d\theta dz \hat{\mathbf{r}}\right) \\
&= \int_0^{\ell_1} \int_{\theta=\theta_R}^{\theta=\pi} B\frac{\ell_2}{2} d\theta dz + \int_0^{\ell_1} \int_{\theta=\pi}^{\theta=\pi+\theta_R} -B\frac{\ell_2}{2} d\theta dz \\
&= \frac{\ell_1 \ell_2 B}{2}(\pi - \theta_R) - \frac{\ell_1 \ell_2 B}{2}\theta_R \\
&= -\ell_1 \ell_2 B(\theta_R - \pi/2).
\end{aligned} \qquad (6.3)$$

This derivation is based on the fact that the $\vec{B}$ field is directed radially outward over the length $(\ell_2/2)(\pi - \theta_R)$ and radially inward over the length $(\ell_2/2)\theta_R$ (see Figure 6.14). In Problem 8, the reader is asked to show that

$$\phi(\theta_R) = -\ell_1 \ell_2 B(\theta_R - \pi/2 - \pi) \quad \text{for } \pi < \theta_R < 2\pi. \qquad (6.4)$$

A plot of the flux versus the rotor angle θ_R is given in Figure 6.16.

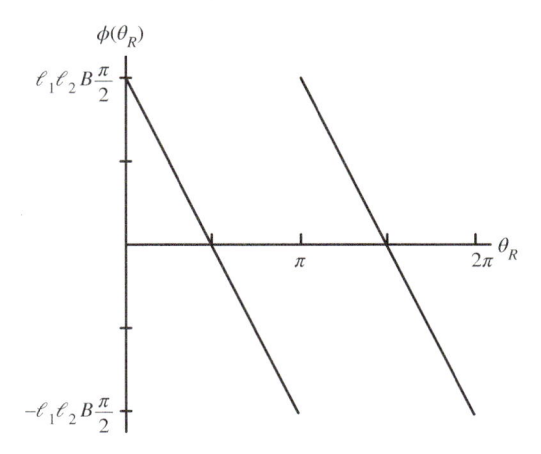

Figure 6.16. The rotor flux $\phi(\theta_R)$ due to the external magnetic field vs. θ_R.

By (6.3) and (6.4), the induced emf in the rotor loop is calculated as

$$\xi = -\frac{d\phi}{dt} = (\ell_1\ell_2 B)\frac{d\theta_R}{dt} = K_b \omega_R,$$

where $K_b \triangleq \ell_1\ell_2 B$ is called the back emf constant.

The total emf in the rotor loop due to the voltage source V_S and external magnetic field is $V_S - K_b\omega_R$. How does one know to subtract ξ from the applied voltage V_S? As shown in Figure 6.15, the positive direction of travel around the loop is in *opposition* to V_S, so that if $\xi > 0$, it is opposing the applied voltage V_S. The standard terminology is to call $\xi \triangleq K_b \omega_R$ the *back emf* of the motor.

Self-Induced Emf in the Single-Loop Motor

The computation of the flux in the rotor loop produced by its own (armature) *current* is now done. To do so, consider the flux surface shown in Figure 6.17.

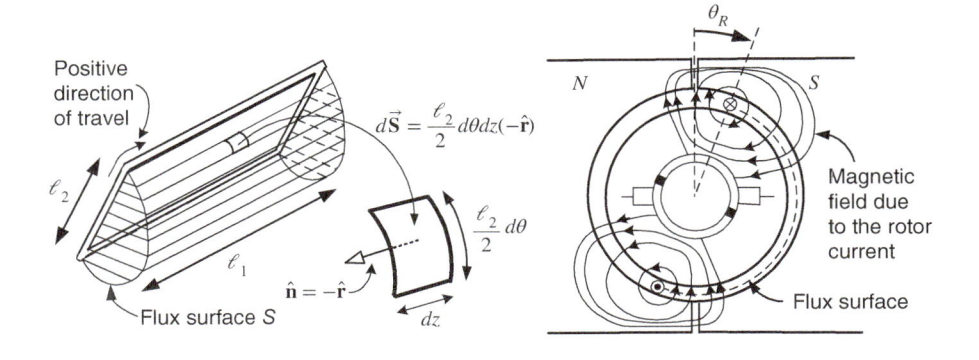

Figure 6.17. Computation of the inductance of the rotor loop. The surface element vector is $d\vec{S} = -r_R d\theta dz \hat{\mathbf{r}}$ with a resulting positive direction of travel as indicated by the curved arrow. This direction coincides with the direction of positive current, that is, $i > 0$.

6 The Physics of the DC Motor

Let $r_R = \ell_2/2$ denote the radius of the rotor and note that the magnetic field on the flux surface due to the *armature current* has the form

$$\vec{B}(r_R, \theta - \theta_R, i) = iK(r_R, \theta - \theta_R)(-\hat{\mathbf{r}}),$$

where

$$K(r_R, \theta - \theta_R) > 0 \quad \text{for } 0 \leq \theta - \theta_R \leq \pi$$
$$K(r_R, \theta - \theta_R) < 0 \quad \text{for } \pi \leq \theta - \theta_R \leq 2\pi.$$

The exact expression for $K(r_R, \theta - \theta_R)$ is not easy to compute, but it is not needed for the analysis here. Rather, the point is that with $i > 0$, the magnetic field $\vec{B}(r_R, \theta - \theta_R, i)$ due to the current in the rotor loop is radially in on the flux surface shown in Figure 6.17 for $\theta_R \leq \theta \leq \theta_R + \pi$. For convenience, the surface element is chosen to be $d\vec{S} = r_R d\theta dz(-\hat{\mathbf{r}})$ so that positive direction of travel around the surface coincides with the positive direction of the current i in the loop.

The flux ψ in the rotor loop is then computed as[4]

$$\psi(i) = \int_S \vec{B} \cdot d\vec{S} = \int_0^{\ell_1} \int_{\theta_R}^{\theta_R + \pi} iK(r_R, \theta - \theta_R)(-\hat{\mathbf{r}}) \cdot (-r_R d\theta dz \hat{\mathbf{r}})$$

$$= i \int_0^{\ell_1} \int_{\theta_R}^{\theta_R + \pi} K(r_R, \theta - \theta_R) r_R d\theta dz$$

$$= Li,$$

where

$$L \triangleq \int_0^{\ell_1} \int_{\theta_R}^{\theta_R + \pi} K(r_R, \theta - \theta_R) r_R d\theta dz > 0. \tag{6.5}$$

This last equation just says the flux in the loop (due to the current in the loop) is proportional to the current i in the loop. The proportionality constant L is the called the *inductance* of the loop. If $-d\psi/dt = -Ldi/dt > 0$, then the induced emf will force current into the page $\otimes$ on side a and out of the page $\odot$ of side a' in Figure 6.17. That is, this induced emf has the same sign convention as the armature current i and the source voltage V_S.

With the rotor locked at some angle θ_R so that the external magnetic field cannot induce an emf in the rotor loop, the equation describing the current i in the rotor loop is given by Kirchhoff's voltage law

$$V_S - Ri - L\frac{di}{dt} = 0$$

or

$$V_S = Ri + L\frac{di}{dt}.$$

Here R is the resistance of the loop and V_S is the source voltage applied to the loop. The loop and its equivalent circuit are shown in Figure 6.18.

---

[4] The notation ψ is used to distinguish this flux from the flux ϕ in the loop due to the *external* permanent magnet. However, the total flux using an inward normal would be $\psi - \phi$ as the *outward* normal was used to compute ϕ in Section 6.3.

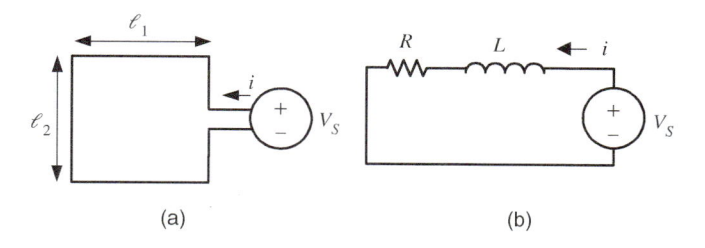

Figure 6.18. (a) Rotor loop. (b) Equivalent circuit.

6.4 Dynamic Equations of the DC Motor

The complete set of equations for a DC motor can now be given. The total voltage in the loop due to the voltage source V_S, the external permanent magnet, and the changing current i in the rotor loop is

$$V_S - K_b \omega_R - L\frac{di}{dt}.$$

This voltage goes into building up the current in the loop against the loop's resistance, that is,

$$V_S - K_b \omega_R - L\frac{di}{dt} = Ri$$

or

$$L\frac{di}{dt} = -Ri - K_b \omega_R + V_S.$$

This relationship is often illustrated by the equivalent circuit given in Figure 6.19. Recall that the torque τ_m on the loop due to the external magnetic field acting on the current in the loop is

$$\tau_m = K_T i,$$

where $K_T \triangleq \ell_1 \ell_2 B$ is called the torque constant. By connecting a shaft and gears to one end of the rotor, this motor torque can be used to do work (lift weight, etc.). Let $-f\omega_R$ model the friction torque (due to the brushes, bearings, etc.) where f is the coefficient of viscous friction and let τ_L be the load torque (e.g., due to a weight being lifted).

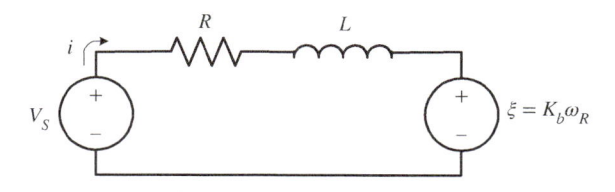

Figure 6.19. Equivalent circuit of the armature electrical dynamics.

Then, by Newton's law,

$$\tau_m - \tau_L - f\omega_R = J\frac{d\omega_R}{dt},$$

where J is the moment of inertia of the rotor. The system of equations characterizing the DC motor is then

$$L\frac{di}{dt} = -Ri - K_b\omega_R + V_S$$
$$J\frac{d\omega_R}{dt} = K_T i - f\omega_R - \tau_L \qquad (6.6)$$
$$\frac{d\theta_R}{dt} = \omega_R.$$

A picture of a DC motor servo system and its associated schematic is shown in Figure 6.20. In the schematic, R is the resistance of the rotor loop, L is the inductance of the rotor loop, $\xi = K_b\omega_R$ is the back emf, $\tau_m = K_T i$ is the motor torque, J is the rotor moment of inertia, and f is the coefficient of viscous friction. The positive directions for τ_m, θ_R, and τ_L are indicated by the curved arrows. The fact that the curved arrow for τ_L is opposite to that of τ_m just means that if the load torque is positive then it opposes a positive motor torque τ_m.

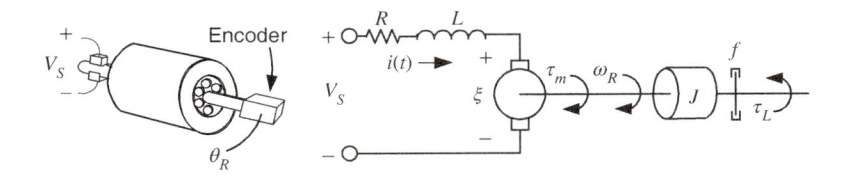

Figure 6.20. DC motor drawing and schematic.

Electromechanical Energy Conversion

The mechanical power produced by the DC motor is $\tau_m\omega_R = K_T i\omega_R = i\ell_1\ell_2 B\omega_R$ while the electrical power absorbed by the back emf is $i\xi = iK_b\omega_R = i\ell_1\ell_2 B\omega_R$. We showed above by direct calculation that $K_T = K_b = \ell_1\ell_2 B$ and this simply shows that conservation of energy holds. That is, the electrical power $iK_b\omega_R$ absorbed by the back emf equals (is converted to) the mechanical power $\tau_m\omega_R$ produced. Another way to view this energy conversion is to write the electrical equation as

$$V_S = Ri + L\frac{di}{dt} + \xi.$$

The power out of the voltage source $V_S(t)$ is given by

$$V_S(t)i(t) = Ri^2 + Li\frac{di}{dt} + iK_b\omega_R$$
$$= Ri^2 + \frac{d}{dt}\left(\frac{1}{2}Li^2\right) + K_T i\omega_R$$
$$= Ri^2 + \frac{d}{dt}\left(\frac{1}{2}Li^2\right) + \tau_m\omega_R.$$

Thus the power $V_S(t)i(t)$ delivered by the source goes into heat loss in the resistance R, into stored magnetic energy in the inductance L of the loop, and the amount $i\xi$ goes into the mechanical energy $\tau_m\omega_R$.

Remark *Voltage and Current Limits*

The amount of voltage V_S that may be applied to the input terminals T_1, T_2 of the motor is limited by capabilities of the amplifier supplying the voltage, that is, $|V_S| \leq V_{\max}$. With $V_c(t)$ be the voltage commanded to the amplifier, the actual voltage V_S out of the amplifier to the motor is limited by $V_{\max}$ as illustrated in Figure 6.21.

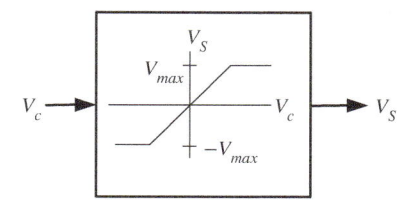

Figure 6.21. Saturation model of an amplifier.

In addition, there is a limit to the amount of current the rotor loop can handle before overheating or causing problems with commutation as previously mentioned. Typically there are two current limits (ratings), the *continuous* current limit $I_{\max\_cont}$ and the *peak* current limit $I_{\max\_peak}$. The continuous current limit $I_{\max\_cont}$ is the amount of current the motor can handle if left in use indefinitely. That is, the amount of heat dissipated in the rotor windings due to ohmic losses is equal to the amount of heat taken away by thermal conduction through the brushes and thermal convection with the air so as to be in a thermal equilibrium. The peak current limit $I_{\max\_peak}$ is the amount of current the motor can handle for short periods of time (typically only a few seconds).

6.5 Optical Encoder Model

A common position sensor used in industry is the optical encoder, which is illustrated in Figure 6.22 on the next page [22, 23]. An optical encoder consists of a set of windows spaced equally around a circular disk with a light source shining through the window when it is aligned with the source. A detector puts out a high voltage when there is light and a low voltage otherwise. For the setup of Figure 6.22a, there are 12 windows (lines or slots) so that for every complete revolution of the circular disk (i.e., of the motor), there will be 12 pulses. Using digital electronic circuitry one can detect a pulse going high or low so that in one revolution with 12 pulses there will be a total of 24 times that a pulse went either high or low. Note that each time a pulse goes high or low the motor has rotated $2\pi/24$ radians or $360°/24 = 15°$. By simply counting the number N of rising and falling edges of the pulse, one can obtain the position of the rotor to within $15°$.

In order to detect the direction of rotation, two light detectors are used as shown in Figure 6.22b. In detail, the length of the windows is the same as the length of the distance between windows. The two light detectors are placed a distance apart equal to 1/2 of a window length. One period of the voltage waveform coming out of the light detector corresponds to the distance from the beginning of one window to the next and this cycle of the voltage waveform is considered to be $360°$ as illustrated in Figure 6.22b. Consequently, the two light detectors are considered to be $90°$ apart or, equivalently, they are said to be in *quadrature*.

Figure 6.23a on the next page shows the voltage waveforms out of the two light detectors when the rotor is turning *clockwise*. The vertical dashed lines in Figure 6.23 refer to a time

6 The Physics of the DC Motor

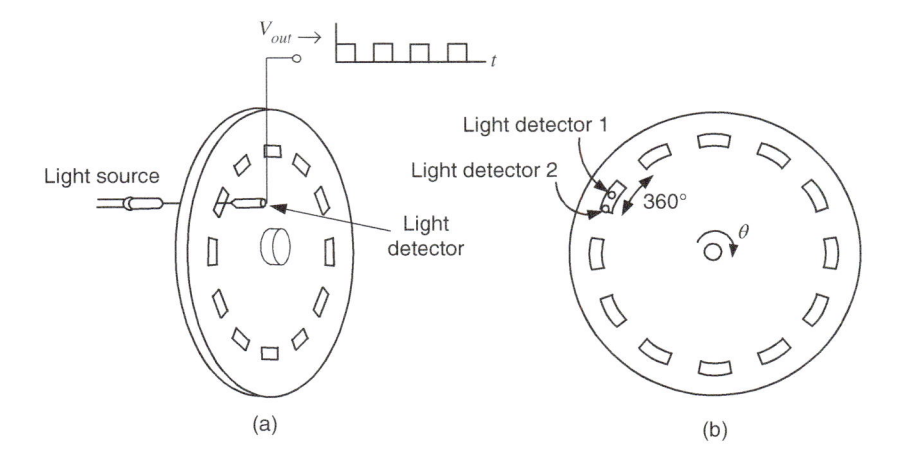

Figure 6.22. Schematic diagram of an optical encoder. Source: Adapted from deSilva [23]. Mechatronics: An Integrative Approach, CRC Press, Boca Raton, FL, 2004.

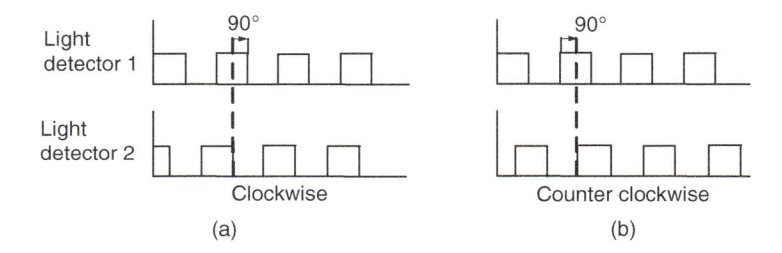

Figure 6.23. (a) Voltage waveforms for clockwise rotation. (b) Voltage waveforms for counterclockwise rotation.

with the encoder disk at the position shown in Figure 6.22b. As shown in Figure 6.23a, for clockwise rotation the voltage of light detector 1 is 90° behind that of light detector 2, that is, the voltage from light detector 2 goes to zero first and then a quarter of a cycle later, the voltage from light detector 1 goes down.

In contrast, Figure 6.23b shows the voltage waveforms out of the two light detectors when the rotor is turning *counterclockwise*. In this case the voltage waveform of light detector 2 is 90° behind that of light detector 1.

The encoder has electronic circuits to detect the relative phase of the two light detector voltage signals and uses that information to determine whether the rising or falling edges of a pulse should increase the count (clockwise motion) or decrease the count (counterclockwise motion).

If an optical encoder has N_w windows (lines/slots), then there are $2N_w$ rising and falling edges per revolution giving a resolution of $2\pi/(2N_w)$ radians. If one counts the voltage pulses from *both* light detectors, there will be $4N_w$ (equally spaced) rising and falling edges per revolution so that a resolution of $2\pi/(4N_w)$ radians is achieved. For example, with $N_w = 500$, the resolution of the encoder is $2\pi/2000$ rad or $360°/2000 = 0.18°$.

Encoder Model

Let N_{enc} denote the number of counts per revolution put out by the encoder. Then the position of the shaft is given by

$$\theta_m(t) = \frac{2\pi}{N_{enc}} N(t) \text{ radians}$$

which is a conversion of the integer counts N to angular position.

Backward Difference Estimation of Speed

The optical encoder gives the position measurement, but not the speed of the cart. However, one can use this measurement to deduce the speed. The most straightforward way is to compute the *backward difference* of the position and divide by the sample period, i.e.,

$$\omega_{bd}(kT) \triangleq \frac{2\pi}{N_{enc}} \left(\frac{N(kT) - N(kT-T)}{T} \right),$$

where T is the time between samples and $N(kT)$ is the optical encoder count at time kT.

The error in estimating the speed by differentiation of the position measurements can be found as follows: At any discrete time kT, $N(kT)$ is in error by at most one encoder count. In particular, $N(kT)$ can be only too small by at most one encoder count. $N(kT)$ is never too large because of the way the encoder works as shown in Figure 6.24.

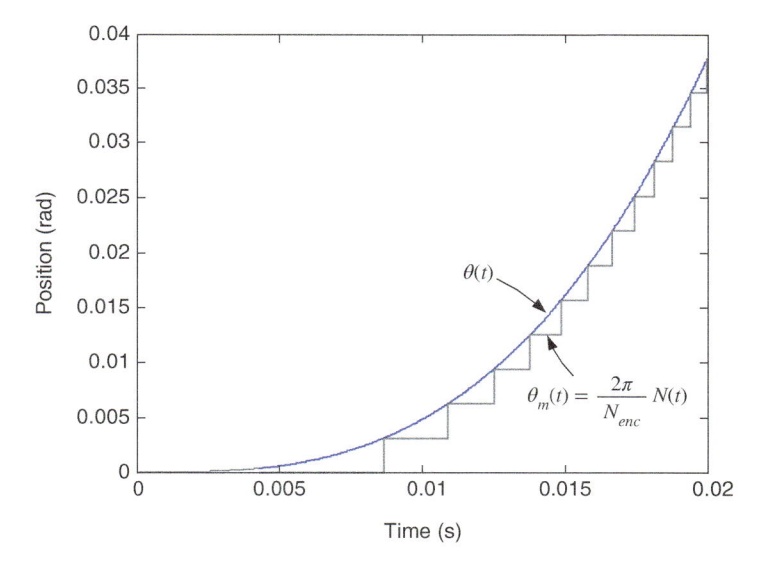

Figure 6.24. Plot of $\theta(t)$ and the encoder output $(2\pi/N_{enc})N(t)$.

Thus, with $\theta(kT)$ the true position in radians, we have

$$\theta(kT) = \frac{2\pi}{N_{enc}} N(kT) + \frac{2\pi}{N_{enc}} e(kT),$$

where $e(kT)$ represents the *positive fractional* count that the encoder cannot sense, that is, $0 \leq e(kT) < 1$ for all k. The speed may then be written as

$$\omega(kT) = \left(\frac{\theta(kT) - \theta(kT-T)}{T}\right)$$
$$= \frac{2\pi}{N_{enc}}\left(\frac{N(kT) - N(kT-T)}{T}\right) + \frac{2\pi}{N_{enc}}\left(\frac{e(kT) - e(kT-T)}{T}\right).$$

As $0 \leq e(kT) \leq 1$ and $0 \leq e(kT-T) \leq 1$, it follows that

$$|e(kT) - e(kT-T)| \leq 1.$$

It is now straightforward to compute a bound on the error in estimating the speed. As the speed estimate is given by

$$\omega_{bd}(kT) = \frac{2\pi}{N_{enc}}\left(\frac{N(kT) - N(kT-T)}{T}\right)$$

and the difference $e(kT) - e(kT-T)$ is bounded by ± 1, it follows that

$$|\text{Error in } \omega_{bd}(kT)| = \frac{2\pi}{N_{enc}}\left|\frac{e(kT) - e(kT-T)}{T}\right| \leq \frac{2\pi/N_{enc}}{T}.$$

As the sampling rate increases (T gets smaller), the error gets larger. On the other hand, as the sampling rate decreases this error becomes smaller, but the backward difference approximation

$$\omega(kT) \approx \frac{2\pi}{N_{enc}}\left(\frac{N(kT) - N(kT-T)}{T}\right)$$

becomes less valid. The error in computing $\omega(kT)$ is due to the encoder resolution $2\pi/N_{enc}$ and the accuracy of the finite difference in approximating the derivative. One way to decrease this error would be to use an encoder with a higher resolution. Such encoders are typically more expensive and cannot operate at higher speeds (as the speed increases, a large number of pulses are coming in so fast that the pulse detection circuitry cannot keep up).

6.6 Tachometer for a DC Machine*

A tachometer is a device for measuring the speed of a DC motor by putting out a voltage proportional to the motor's speed. A tachometer for the simple linear DC machine is considered first.

Tachometer for the Linear DC Machine

Figure 6.25 shows a tachometer added to the linear DC machine. The magnetic field in the DC motor is $\vec{\mathbf{B}}_1 = -B_1\hat{\mathbf{z}}$ with $B_1 > 0$ while in the tachometer it is $\vec{\mathbf{B}}_2 = -B_2\hat{\mathbf{z}}$ with $B_2 > 0$. The two bars are rigidly connected together by the insulating material. The motor force (the magnetic force on the upper bar) is $F_m = i\ell_1 B_1$, and the induced (back) emf in the motor is $\xi = V_b = B_1\ell_1 v$, where v is the speed of the motor (bar).

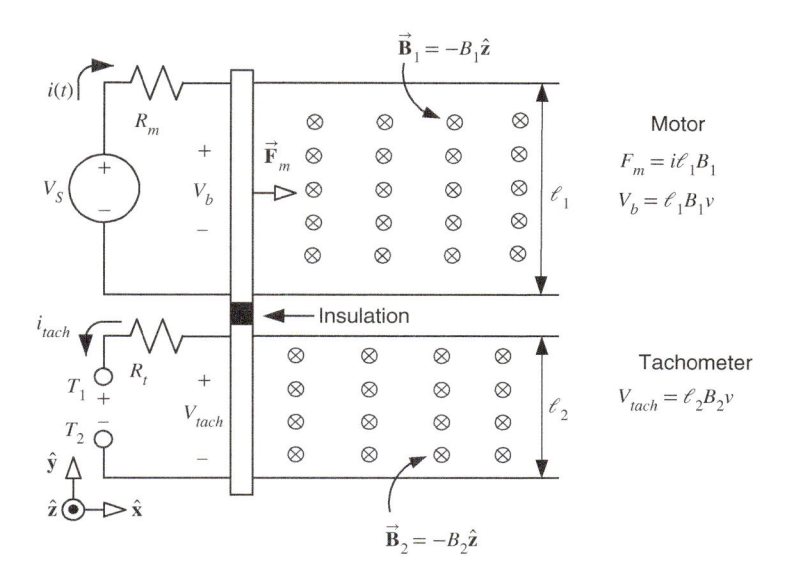

Figure 6.25. DC tachometer (generator).

The induced (back) emf in the tachometer is given by $\xi = V_{tach} = B_2\ell_2 v$ so that by measuring the voltage between the terminals T_1 and T_2, the speed v of the motor can be computed. Note that the tachometer and motor have the same physical structure. In fact, the tachometer is nothing more than a generator putting out a voltage proportional to the speed.

Tachometer for the Single-Loop DC Motor

A tachometer for the single loop DC motor is constructed by attaching another loop to the shaft and rotating it an external magnetic field to act as a DC generator. That is, the changing flux in the tachometer loop produces (generates) an induced emf according to Faraday's law and this emf is proportional to the shaft's speed. To see this, consider Figure 6.26, where a motor loop is driven by a voltage V_S and, attached to the same shaft, is a second loop called a tachometer.

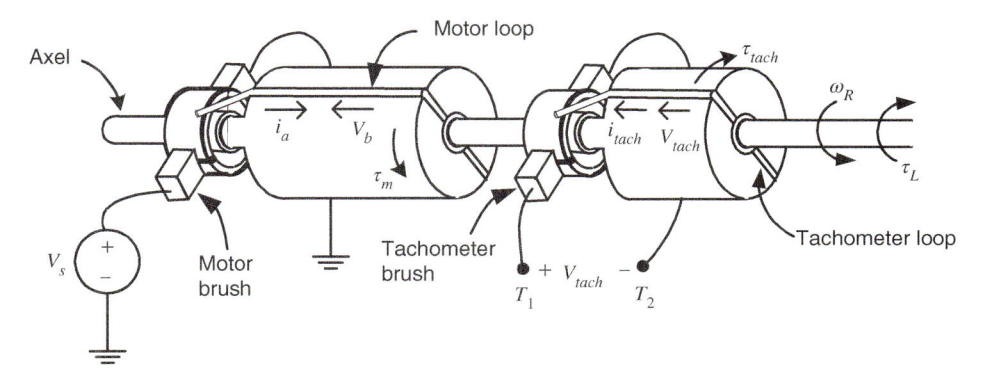

Figure 6.26. Single loop motor and tachometer. Source: Courtesy of Sharon Katz.

170 6 The Physics of the DC Motor

Both loops rotate in an external radial magnetic field, which is not shown in Figure 6.26, but is shown for the tachometer loop in Figure 6.27. It is important to point out that no voltage is applied to the terminals T_1 and T_2 of the tachometer as was the case for the motor. Instead, the voltage V_{tach} between the terminals T_1 and T_2 of the tachometer is measured. Specifically, in the same way the back emf was computed for the DC motor, one can calculate the flux in the loop of the tachometer due to the external magnetic field. This computation is (see Figure 6.27)

$$\phi = \int_S \vec{B} \cdot d\vec{S} = \int_0^{\ell_1} \int_{\theta_R}^{\pi} (B\hat{\mathbf{r}}) \cdot \left(\frac{\ell_2}{2} d\theta dz \hat{\mathbf{r}}\right) + \int_0^{\ell_1} \int_{\pi}^{\pi+\theta_R} (-B\hat{\mathbf{r}}) \cdot \left(\frac{\ell_2}{2} d\theta dz \hat{\mathbf{r}}\right)$$

$$= \int_0^{\ell_1} \int_{\theta_R}^{\pi} B\frac{\ell_2}{2} d\theta dz + \int_0^{\ell_1} \int_{\pi}^{\pi+\theta_R} -B\frac{\ell_2}{2} d\theta dz$$

$$= (\ell_1\ell_2 B/2)(\pi - \theta_R) - (\ell_1\ell_2 B/2)\theta_R$$

$$= -\ell_1\ell_2 B\theta_R + (\ell_1\ell_2 B/2)\pi.$$

The induced emf is then

$$V_{\text{tach}} = -d\phi/dt = (\ell_1\ell_2 B)d\theta_R/dt = K_{b\_\text{tach}}\omega_R,$$

where $K_{b\_\text{tach}} = \ell_1\ell_2 B$ is a constant depending on the dimensions of the tachometer rotor and the strength of the external magnetic field of the tachometer. This shows that the voltage between the terminals T_1 and T_2 is proportional to the angular speed and therefore can be used to measure the speed.

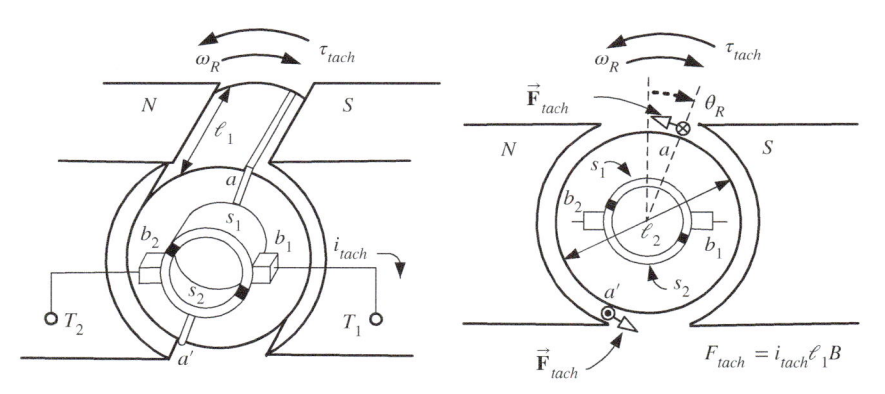

Figure 6.27. Cutaway view of the DC tachometer. Source: Adapted from Matsch and Morgan [21]. Electromagnetic and Electromechanical Machines, 3rd edition, John Wiley & Sons, New York, 1986.

6.7 The Multiloop DC Motor*

The single loop motor of Figure 6.5 was used to illustrate the basic physics of the DC motor. However, it is not a practical motor. This is rectified with the placement of rotor loops around the complete periphery of the rotor so as to produce much more torque. However, these additional rotor loops require a more involved way to commutate their current as will be explained shortly.

6.7 The Multiloop DC Motor*

Increased Torque Production

Figure 6.28 shows the addition of several loops to the motor with each loop similar in form to the loop in Figure 6.5. There are now eight slots in the rotor with two loops placed in each pair of slots (180° apart) for a total of eight loops.

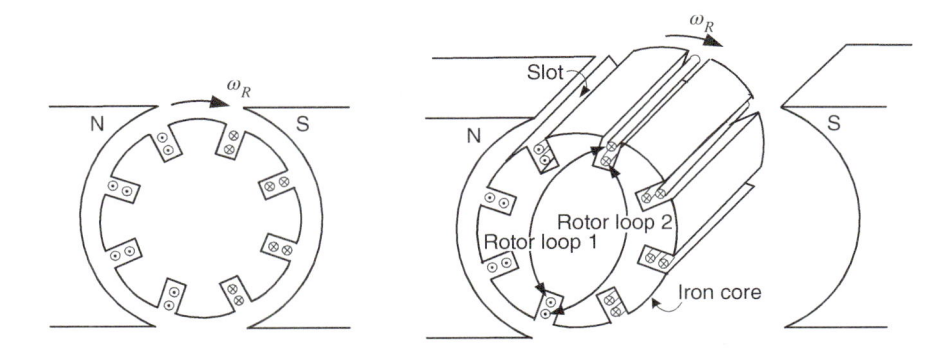

Figure 6.28. A multiloop armature for a DC motor.

The torque on the rotor is now $\tau_m = n\ell_1\ell_2 Bi$, where $n = 8$ is the number of rotor loops and B is the strength of the radial magnetic field in the air gap produced by the external magnetic field. Of course, some method must be found to ensure the current in each loop is reversed every half-turn so that (for positive torque) all the loop sides under the south pole face will have their current going into the page $\otimes$ and all the loop sides under the north pole face will have their current coming out of the page $\odot$. This process is referred to as *commutation* and is considered next.

Commutation of the Armature Current

As seen in Figure 6.28, as a rotor loop rotates clockwise past the vertical position, the current in the top side of the loop must change direction from coming out of the page to going into the page. That is, each rotor loop must have the current in it reversed every half-turn. This is done using a commutator that is illustrated in Figure 6.29 for the rotor shown in Figure 6.28. The commutator for this rotor consists of eight copper segments (labeled a–h in Figure 6.31a), which are separated by insulating material. By connecting each of the ends of the rotor loops of Figure 6.28 to the appropriate copper

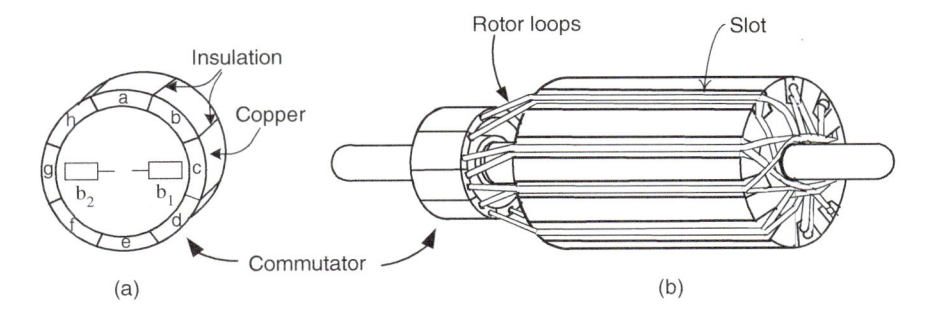

Figure 6.29. Commutator for the rotor in Figure 6.28.

172   6   The Physics of the DC Motor

segments of the commutator, the current will be reversed every half-turn as it rotates past the vertical. To explain all of this, consider Figure 6.31a, which shows explicitly how the ends of the rotor loops are connected to the segments of the commutator. The eight rotor loops of Figure 6.28 are labeled as $1-1', \ldots, 8-8'$ in Figure 6.31a with the ends of each such loop electrically connected (soldered) to a particular pair of commutator segments. For example, the ends of loop $1-1'$ are connected to commutator segments a and b, respectively. The commutator and rotor loops all rotate together rigidly while the two brushes (labeled b_1 and b_2) remain stationary. The brushes are typically made of a carbon material and are mechanically pressed against the commutator surface, making electrical contact.[5] That is, as the commutator rotates, the particular segment that is rubbing against the brush makes electrical contact. Figure 6.30 is a photograph of the rotor of an actual DC motor with a tachometer.

Figure 6.30. Photo of the rotor of a DC motor (left) and its tachometer (right). Note that the slots for the windings of the DC motor are skewed. Source: Photo courtesy of Professor J. D. Birdwell of the University of Tennessee.

As previously explained, to obtain positive torque, it must be that whenever a side of a loop is under a south pole face, the current must be into the page ($\otimes$) and the other side of the loop, which is under the north pole face, must have its current out of the page ($\odot$). When the loop side rotates from being under one pole face to the other pole face, the current in that loop must be reversed (commutated). The mechanism of how this connection between the armature loops, the commutator, and the brushes can reverse the current in each rotor loop every half turn is now explained.

With reference to Figure 6.31a, the armature current i enters brush b_1 and into commutator segment c. By symmetry, half of this armature current (i.e., $i/2$) goes through loop $3-3'$ into commutator segment d, then through loop $4-4'$ into commutator segment e, then through loop $5-5'$ into commutator segment f, then through loop $6-6'$ into commutator segment g, and, finally, out through brush b_2. This path (circuit) of the current is denoted in **bold**. Similarly, there is a parallel path for the other half of the current armature current. Specifically, the other half of the armature current $i/2$ goes through loop $2'-2$ into commutator segment b, then through loop $1'-1$ into commutator segment a, then through loop $8'-8$ into commutator segment h, then into loop $7'-7$ into commutator segment g, and finally, out through brush b_2. This path (circuit) is denoted without bold. So, for the rotor in the position shown in Figure 6.31a, there are two parallel circuits from b_1 to b_2 each made up of four loops connected in series and each circuit carries half of the armature current. The sides of the loops under the south pole face have their current into the page while the other side of these loops (which are under the north pole face) have their current out of the page so that positive torque is produced. The sides of the loops in Figure 6.31a are 45° apart.

---

[5] The figure shows a gap between the brushes and the commutator, but this was done for illustration and there is no gap in reality. Also, for illustrative purposes, the brushes are shown inside the commutator when in fact they are normally pressed against the commutator from the outside.

6.7 The Multiloop DC Motor*

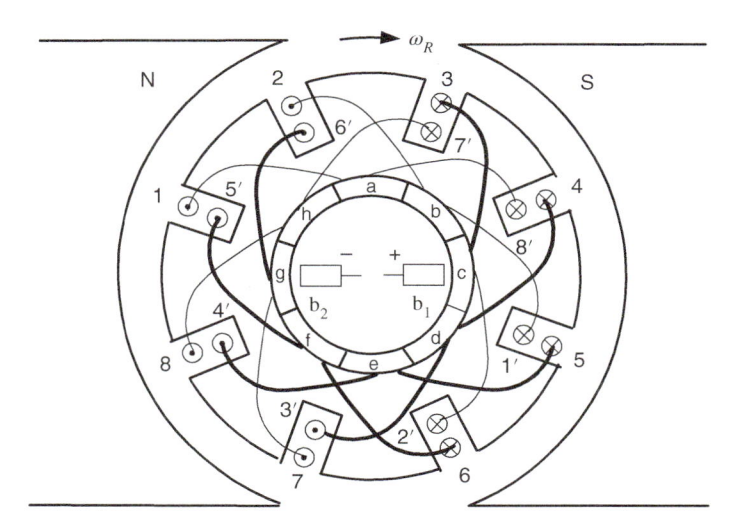

Figure 6.31a. (a). Rotor loops and commutator for four sets of rotor loops. Brushes remained fixed in space, that is, they do not rotate. Source: Adapted from Chapman [20]. Electric Machinery Fundamentals, McGraw-Hill, New York, 1985.

Figure 6.31b shows the rotor turned $45°/2$ with respect to Figure 6.31a. In this case, brush b_1 shorts the two commutator segments b and c together while the brush b_2 shorts together the two commutator segments f and g.

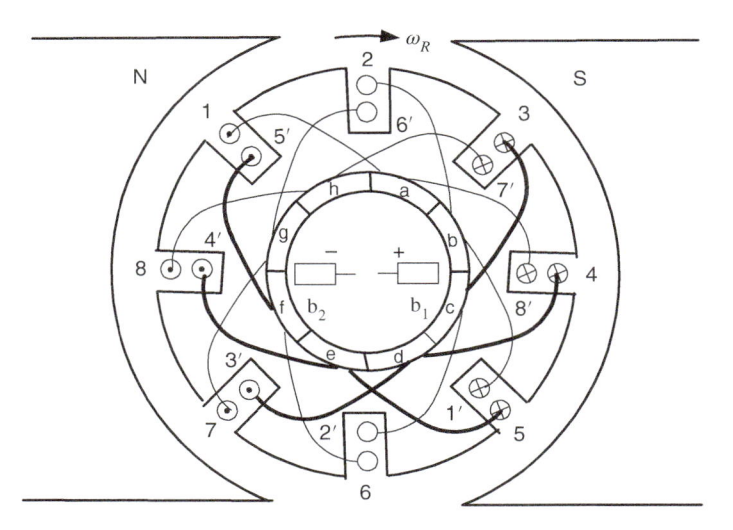

Figure 6.31b. Rotor turned $45°/2$ with respect to Figure 6.31. Source: Adapted from Chapman [20]. Electric Machinery Fundamentals, McGraw-Hill, New York, 1985.

The ends of loop 2–2′ are connected to commutator segments b and c (which are now shorted together) so that the current in this loop is now zero. Similarly, the ends of loop 6–6′ are connected to commutator segments f and g and the current in this loop is also zero. For the remaining loops, $i/2$ goes through loop 3–3′ into commutator segment d, then through loop 4–4′ into commutator segment e, then into loop 5–5′ into commutator segment f, and finally, out brush b_2. These loops are denoted in **bold** in the figure. Similarly, $i/2$ goes through loop 1–1' into commutator segment a, then through loop 8'–8 into commutator segment h, then into loop 7'–7 into commutator segment g, and finally out brush b_2.

The motor continues to rotate and consider it now after it has moved additional $45°/2$ so that it has the position shown in Figure 6.31c. Now the current enters brush b_1 and into commutator segment b. By symmetry, half the current $i/2$ goes through loop 2–2' into commutator segment c, then through loop 3–3' into commutator segment d, then through loop 4–4' into commutator segment e, then through loop 5–5' into commutator segment f, and finally out through brush b_2. This path (circuit) of the current is denoted in **bold**. Similarly, the other half of the current goes through loop 1'–1 into commutator segment a then through loop 8'–8 into commutator segment h then through loop 7'–7 into commutator segment g then into loop 6'–6 into commutator segment f and finally, out through brush b_2. This path (circuit) is denoted without using bold.

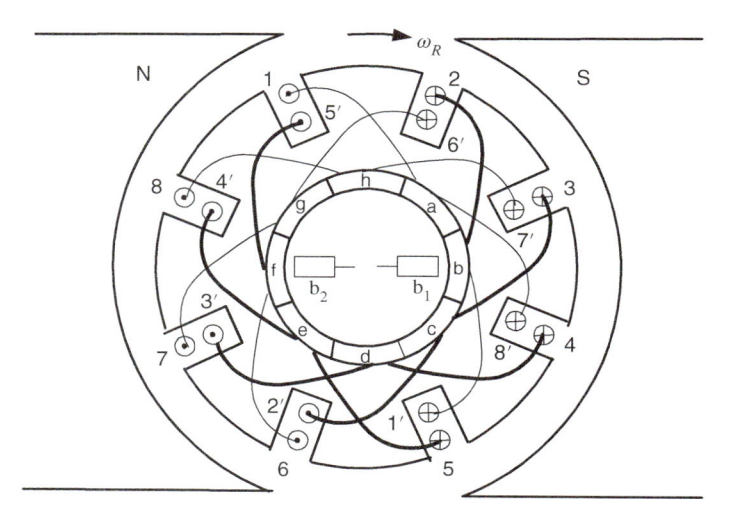

Figure 6.31c. Rotor turned 45° with respect to Figure 6.31a. Source: Adapted from Chapman [20]. Electric Machinery Fundamentals, McGraw-Hill, New York, 1985.

As the sequence of Figures 6.31a–c shows, the current in loops 2–2' and 6–6' were reversed as these two loops rotated past the vertical position. In summary, there are two parallel paths, each consisting of four loops, and when any loop goes to the vertical position, the current in that loop is reversed. In this way, all sides of the loop under the south pole have their current going into the page and all sides under the north pole have their current coming out of the page for positive torque production.

Remark The scheme for current commutation presented here is from [20]. However, there are other schemes and the reader is referred to [20, 21, 24, 25] for an introduction to them.

Brushless DC Motors

The commutator assembly of DC machines requires periodic cleaning and the brushes themselves wear down and must be periodically replaced. Permanent magnet synchronous machines do not have such drawbacks. With the advent of modern power electronics and powerful digital signal processors permanent magnet synchronous machines have been made into precision motion control actuators. Their manufacturers provide internal control loops so that to the end user the equations describing this actuator have the same *form* as the system of equations (6.6) for the DC motor [7]. For this reason they are called brushless DC motors. For quite some time now they have replaced the DC motor as the industry standard for motion control actuation.

Problems

Problem 1 *Faraday's Law*

Consider Figure 6.32 where a magnet is *moving up* into a square planar loop of copper wire.

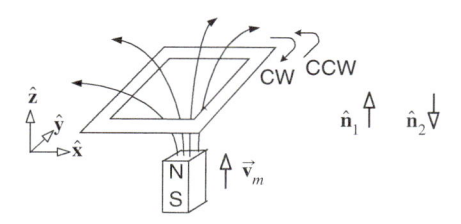

Figure 6.32. Induced emf in a loop due to a moving magnet.

(a) Using the normal $\hat{\mathbf{n}}_1$, roughly sketch the loop's flux $\phi_1(t)$ as a function of t while the magnet is below the copper loop. Is the flux $\phi_1(t)$ in the loop produced by the magnet increasing or decreasing?

(b) Using the normal $\hat{\mathbf{n}}_1$, what is the direction of positive travel around the surface whose boundary is the loop (clockwise or counterclockwise)?

(c) What is the direction of the induced current in Figure 6.32 (clockwise or counterclockwise)? Let $\psi_1(i)$ be the flux in the loop due to the induced current. Is $\psi_1(i)$ positive or negative while the magnet is below the loop? Is $\psi_1(i)$ increasing or decreasing while the magnet is below the loop, but moving up?

(d) Using the normal $\hat{\mathbf{n}}_2$, roughly sketch the loop's flux $\phi_2(t)$ as a function of t while the magnet is below the copper loop. Is the flux in the loop produced by the magnet increasing or decreasing?

(e) Using the normal $\hat{\mathbf{n}}_2$, what is the direction of positive travel around the surface whose boundary is the loop (clockwise or counterclockwise)?

(f) What is the direction of the induced current in Figure 6.32 (clockwise or counterclockwise)? Let $\psi_2(i)$ be the flux in the loop due to the induced current. Is $\psi_2(i)$ positive or negative while the magnet is below the loop? Is $\psi_2(i)$ increasing or decreasing while the magnet is below the loop, but moving up?

Problem 2 *Faraday's Law*

Consider Figure 6.33 on the next page where a magnet is below a planar loop of copper wire and *moving down* away from the loop.

176   6   The Physics of the DC Motor

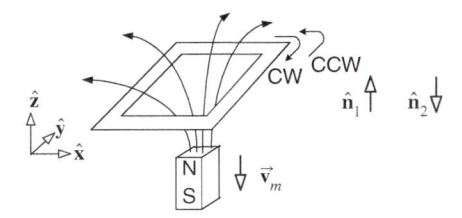

Figure 6.33. Induced emf in a loop due to a moving magnet.

(a) Using the normal $\hat{\mathbf{n}}_1$, roughly sketch the loop's flux $\phi_1(t)$ as a function of t. Is the flux in the loop produced by the magnet increasing or decreasing?

(b) Using the normal $\hat{\mathbf{n}}_1$, what is the direction of positive travel around the surface whose boundary is the loop (clockwise or counterclockwise)?

(c) What is the direction of the induced current in Figure 6.33 (clockwise or counterclockwise)? Let $\psi_1(i)$ be the flux in the loop due to the induced current. Is $\psi_1(i)$ positive or negative while the magnet is below the loop? Is $\psi_1(i)$ increasing or decreasing while the magnet is moving down below the loop?

(d) Using the normal $\hat{\mathbf{n}}_2$, roughly sketch the loop's flux $\phi_2(t)$ as a function of t. Is the flux in the loop produced by the magnet increasing or decreasing?

(e) Using the normal $\hat{\mathbf{n}}_2$, what is the direction of positive travel around the surface whose boundary is the loop (clockwise or counterclockwise)?

(f) What is the direction of the induced current in Figure 6.33 (clockwise or counterclockwise)? Let $\psi_2(i)$ be the flux in the loop due to the induced current. Is $\psi_2(i)$ positive or negative while the magnet is below the loop? Is $\psi_2(i)$ increasing or decreasing while the magnet is moving down below the loop?

Problem 3   *The Linear DC Motor*

Consider the simple linear DC motor of Figure 6.34 where $B > 0$. Take the normal to the surface enclosed by the loop to be $\hat{\mathbf{n}} = -\hat{\mathbf{z}}$.

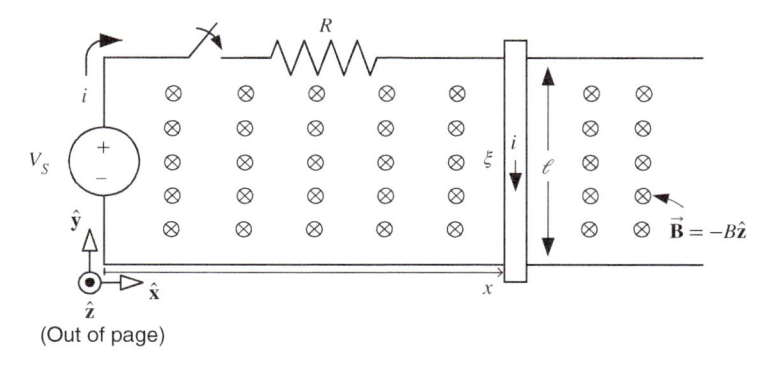

Figure 6.34. A linear DC motor.

(a) What is the magnetic force $\vec{\mathbf{F}}_{\text{magnetic}}$ on the bar?

(b) What is the flux through the surface?

(c) What is the direction of positive travel around this flux surface? (CW or CCW)

(d) What is the induced emf ξ in the loop in terms of B, ℓ and the speed v of the bar?

(e) Do V_S and ξ have the same sign convention? Is ξ positive or negative?

(f) Let R denote the resistance of the circuit, m_ℓ denote the mass of the bar, and f be the viscous friction coefficient between the sliding bar and the rails. Assume the inductance of the circuit loop is zero. Write down the equation for the current i in the machine and the differential equation for the position x of the bar. Similar to (6.1), eliminate the current to obtain the equation of motion of the bar with input V_S.

Problem 4 *The Linear DC Motor*

Consider the simple linear motor in Figure 6.35 where the magnetic field $\vec{B} = B\hat{z}$ ($B > 0$) is up out of the page. Closing the switch causes a current to flow in the wire loop.

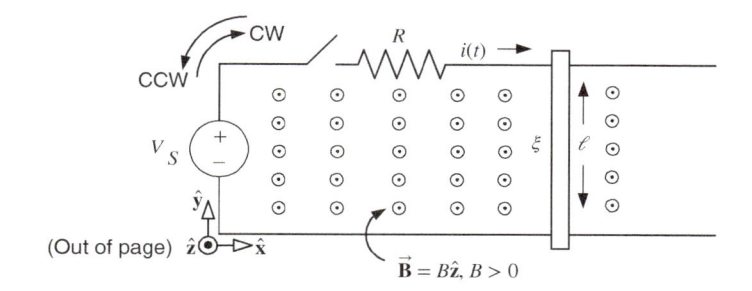

Figure 6.35. Linear DC machine with $\vec{B} = B\hat{z}$, $B > 0$.

(a) What is the magnetic force $\vec{F}_{magnetic}$ on the sliding bar in terms of B, i, and ℓ? Give both the magnitude and direction of $\vec{F}_{magnetic}$.

(b) Take the normal to the surface enclosed by the loop to be $\hat{n} = \hat{z}$. What is the flux through the surface?

(c) What is the induced emf ξ in the loop in terms of B, ℓ, and the speed v of the bar?

(d) What is the sign convention for the induced emf ξ drop around the loop? (That is, if $\xi > 0$, would it act to push current in the clockwise or counterclockwise direction?)

(e) Do V_S and ξ have the same sign convention? Draw + and − signs above and below ξ to indicate the sign convention for ξ.

Problem 5 *Torque in a DC Motor*

Figure 6.36 shows a single loop DC motor.

(a) Using the magnetic force law, compute the force $\vec{F}_{side\ a}$ on side a produced by the radial magnetic field acting on the current in side a.

(b) Give the definition of the torque $\vec{\tau}_{side\ a}$ on side a due to $\vec{F}_{side\ a}$ and compute it.

(c) Use the fact that $\vec{\tau}_{side\ a'} = \vec{\tau}_{side\ a}$ to compute the total torque on the rotor and give the expression for the torque constant K_T of the motor.

178   6   The Physics of the DC Motor

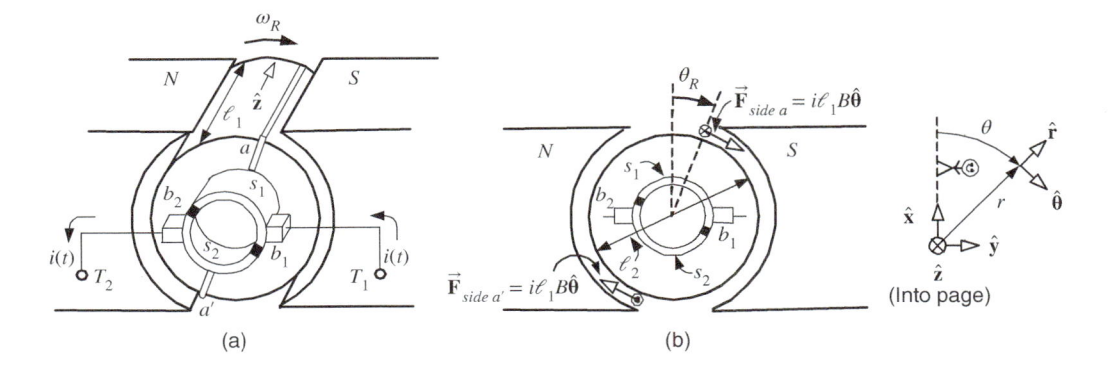

Figure 6.36. Computing the torque produced by a DC motor.

Problem 6 *Back Emf in the Single-Loop Motor*
Consider the single loop motor with the flux surface as indicated in Figure 6.37. A voltage source connected to the brushes is forcing current down side a ($\otimes$) and up side a' ($\odot$).

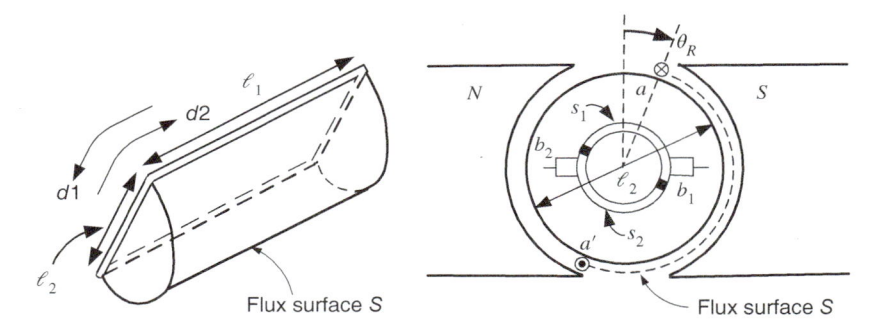

Figure 6.37. Computing the flux with $\hat{\mathbf{n}} = -\hat{\mathbf{r}}$.

(a) With the motor at the angular position θ_R shown, that is, with $0 < \theta_R < \pi$, and using the *inward* normal ($\hat{\mathbf{n}} = -\hat{\mathbf{r}}$), compute the flux through the surface in terms of the magnitude B of the radial magnetic field in the air gap, the axial length ℓ_1 of the motor, the diameter ℓ_2 of the motor and the angle θ_R of the rotor.

(b) What is the positive direction of travel around the flux surface $\mathcal{S}$ (d1 or d2)?

(c) What is the emf ξ induced in the rotor loop? What is the sign convention for the induced emf ξ drop around the loop? (That is, if $\xi > 0$, would it act to push current in d1 direction or the d2 direction?) Do V_S and ξ have the same sign convention? Explain why ξ is now negative. Draw an equivalent circuit of the form in Figure 6.19 for the rotor loop.

Problem 7 *Gauss's Law and Conservation of Flux*
The flux surface in Figure 6.14 was chosen as the half-cylindrical surface with two half disks at either end because the B field is known on the cylindrical surface being given by (6.2) and can be taken to be zero on the two half disks. If the flux surface had been taken to be a flat planar surface with the rectangular loop as its boundary, then it would not

be clear how to compute the flux on this surface as the B field is not known there. Show, using Gauss's law $\phi = \oint_S \vec{B} \cdot d\vec{S} \equiv 0$, that both surfaces give the same flux. In general, by Gauss's law, one can compute the flux using any surface as long as its boundary is the loop.

Problem 8 *Flux in the Single-Loop DC Motor*
Figure 6.38 shows the rotor loop with $\pi < \theta_R < 2\pi$.

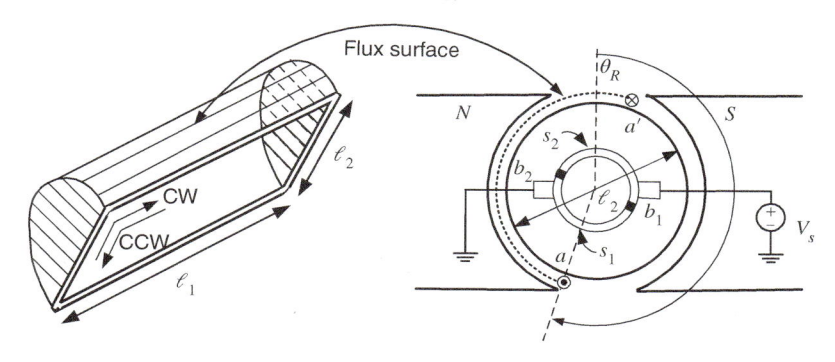

Figure 6.38. Rotor loop where $\pi < \theta_R < 2\pi$.

(a) Using the flux surface shown in Figure 6.39 with $d\vec{S} = (\ell_2/2)d\theta dz \hat{r}$, show that

$$\phi(\theta_R) = (\theta_R - \pi/2 - \pi)\ell_1\ell_2 B \text{ for } \pi < \theta_R < 2\pi.$$

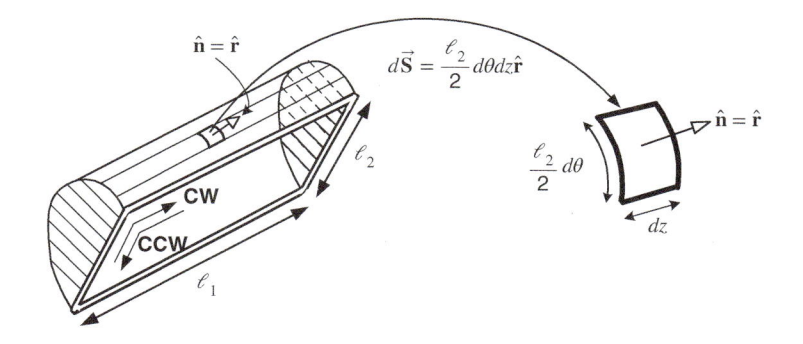

Figure 6.39. Flux surface with the normal radially out.

Plot $\phi(\theta_R)$ for $0 < \theta_R < 2\pi$ (note that $\phi(\theta_R)$ for $0 < \theta_R < \pi$ is computed in the text). Compute the back emf ξ and give its sign convention, that is, if $\xi = -d\phi(\theta_R)/dt > 0$, will it force current in the CW or the CCW direction? Do ξ and V_S have the same sign convention? (Yes! Explain!) Draw an equivalent circuit diagram of the form of Figure 6.19 to illustrate the sign convention.

(b) Using the flux surface shown in Figure 6.40 with $d\vec{S} = (\ell_2/2)d\theta dz(-\hat{r})$, show that

$$\phi(\theta_R) = -(\theta_R - \pi/2 - \pi)\ell_1\ell_2 B \text{ for } \pi < \theta_R < 2\pi.$$

A plot of this is given in Figure 6.16. Compute the back emf and give its sign convention, that is, if $-d\phi(\theta_R)/dt > 0$ will it force current in the CW or the CCW direction?

180   6 The Physics of the DC Motor

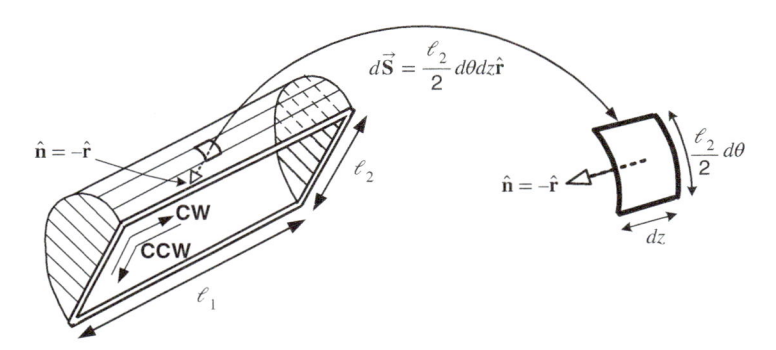

Figure 6.40. Flux surface with the normal radially in.

Do ξ and V_S have the same sign convention? (No, they are opposite! Explain!) Draw an equivalent circuit diagram of the form of Figure 6.19 to illustrate the sign convention.

(c) Note that in part (b) the normal to the flux surface is taken to be radially in while with $0 < \theta_R < \pi$ it was taken to be radially out (see Figure 6.15). Explain why reversing the normal of the flux surface each half turn in this way results in an equivalent circuit of the form shown in Figure 6.19, which is valid for all rotor angular positions.
Hint: Note that the + side of V_S is now electrically connected to side a' of the loop through brush b_1, while when $0 < \theta_R < \pi$ in Figure 6.15, the + side of V_S was electrically connected to side a of the loop through brush b_1. That is, the sign convention for V_S in the loop changes every half-turn. Thus it is also necessary to change the sign convention for the flux and therefore for ξ every half-turn in order that the sign conventions for V_S and ξ have the same relationship to each other for all θ_R.

(d) Show by taking $d\vec{S} = (\ell_2/2)d\theta dz\hat{r}$ for $0 < \theta_R < \pi$ and $d\vec{S} = (\ell_2/2)d\theta dz(-\hat{r})$ for $\pi < \theta_R < 2\pi$, that $\phi(\theta_R) = -(\theta_R \bmod \pi - \pi/2)\ell_1\ell_2 B$ for all θ_R and that $\xi = -d\phi(\theta_R)/dt$ has the sign convention given in Figure 6.19 for all θ_R.

Problem 9 *Simulation of the DC Motor*
Figure 6.41 is a SIMULINK block diagram for the DC motor. Let $V_{\max} = 40$ V, $I_{\max} = 5$ A, $K_b = K_T = 0.07$ V/rad/s (= N-m/A), $J = 6 \times 10^{-5}$ kg-m^2, $R = 2$ Ω, $L = 2$ mH, $\tau_L = 0$ and $f = 0.0004$ N-m/rad/s. Implement this simulation with a step input of $V_S(t) = 10$ V. Give plots of (a) $\theta(t)$, (b) $\omega(t)$, (c) $i(t)$, and (d) $V_S(t)$.

Problem 10 *Simulation of a DC Motor with an Optical Encoder*
Figure 6.42 shows a SIMULINK block diagram for an open-loop DC motor with an encoder model. Let $K_b = K_T = 0.07$ V/rad/s (= N-m/A), $J = 6 \times 10^{-5}$ kg-m^2, $R = 2$ Ω, $L = 2$ mH, $\tau_L = 0$ and $f = 0.0004$ N-m/rad/s. Further, let $N_{enc} = 4 \times 1024 = 4096$ be the number of pulses (rising and fall edges) out of the optical encoder per revolution. The angle π is denoted in MATLAB as "pi". The gray Gain block of $N_{enc}/(2\pi)$ converts the angle into counts. The floor function (denoted by $floor(x) = \lfloor x \rfloor$) in the gray Rounding Function block converts $N_{enc}/(2\pi)$ into the largest *integer* less than $N_{enc}/(2\pi)$.

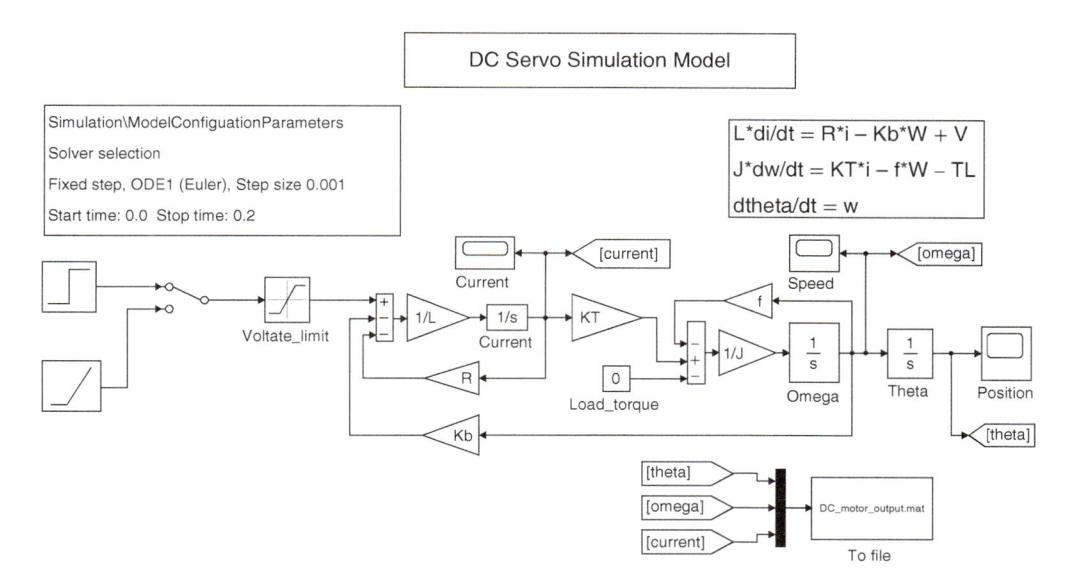

Figure 6.41. DC motor SIMULINK block diagram.

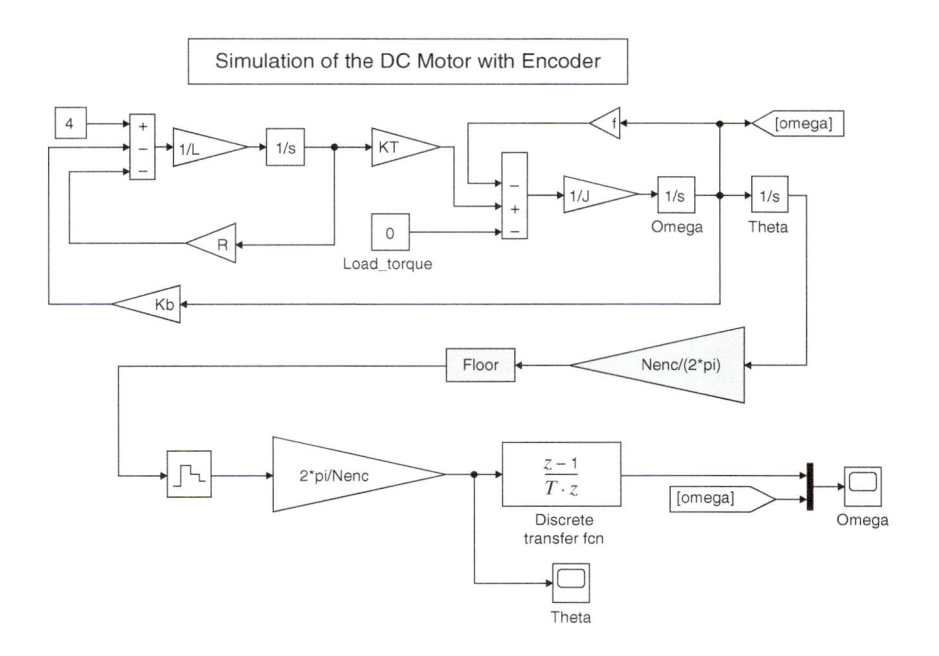

Figure 6.42. SIMULINK block diagram for a DC motor and optical encoder.

182   6   The Physics of the DC Motor

Figure 6.43 is the dialog box for the `Rounding Function` box.

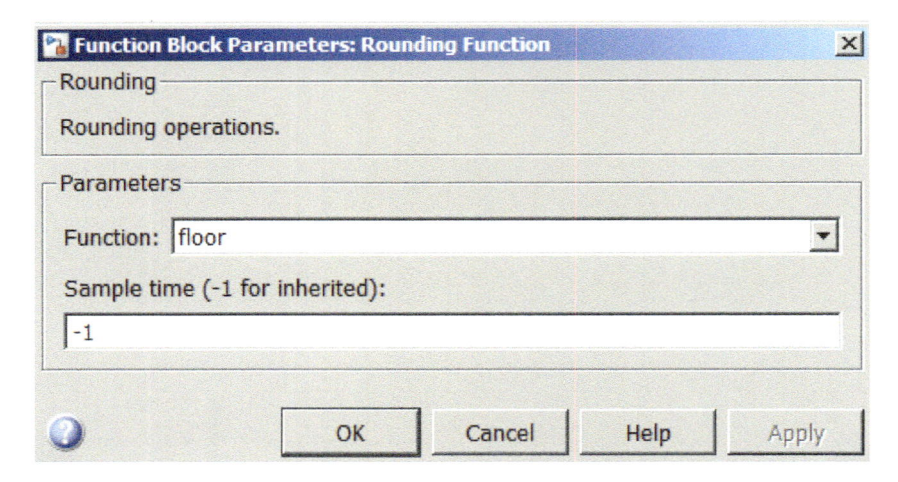

Figure 6.43. Dialog box for the `Rounding Function` block.

Figure 6.44 is the dialog box for the `Discrete Transfer Function` block.

Figure 6.44. Dialog box for the `Discrete Transfer Fcn` block.

Figure 6.45 is the dialog box for the `Zero-Order Hold` block.

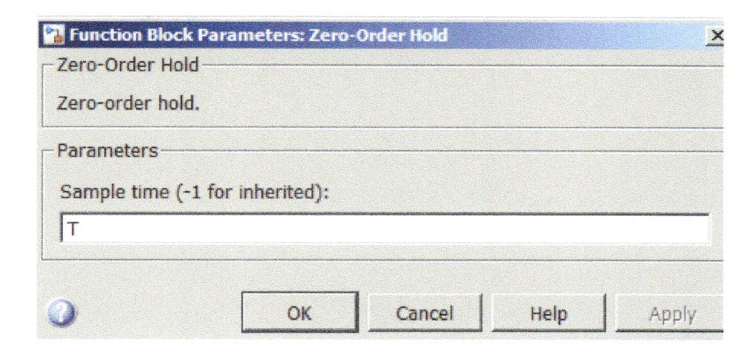

Figure 6.45. Dialog box for the `Zero Order Hold` block.

(a) Implement the simulation of Figure 6.42.

(b) Show in the simulation that the backward difference error is bounded by $\frac{2\pi/N_{enc}}{T}$.

Problem 11 *DC Motor with a Field Winding and Gear System*

Figure 6.46 is a schematic diagram for a DC motor with a field winding (see Figure 6.7 in the text) and a gear system attached. The schematic shows the symbol of a battery attached to a winding producing the field current i_f. The reason for this is to indicate that the voltage applied to the field winding is *constant*. As a result the current in the field winding $i_f(t) = I_f$ is *constant*. Then the back emf "constant" is given by

$$K_b = K_f I_f.$$

What must the torque constant equal? Write down the differential equations that model this system. Your final answer must have τ_1, τ_2, θ_m, and ω_m eliminated.

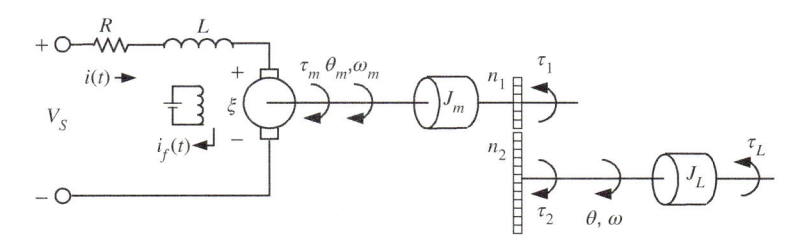

Figure 6.46. DC motor with a field winding and gear system.

Problem 12 *DC Motor and Rolling Mill*

In Example 4 of Chapter 5 the mechanical model of the simple rolling mill shown in Figure 6.47 was developed. The thickness of the incoming material is denoted as T

6 The Physics of the DC Motor

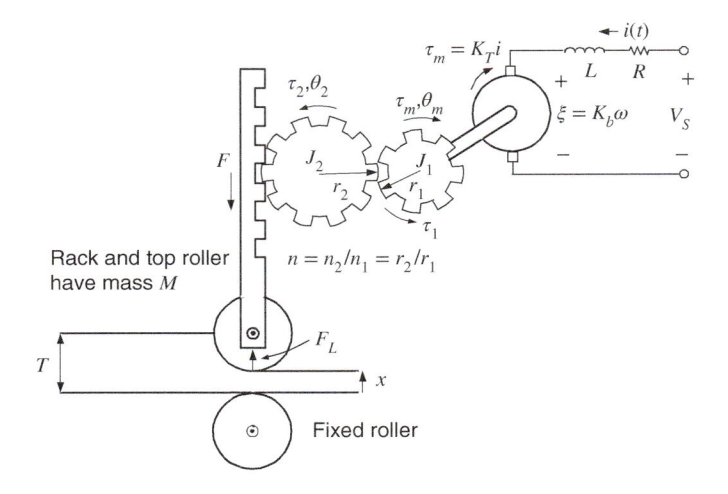

Figure 6.47. Rolling mill powered by a DC motor.

and the thickness of the material leaving the rollers is denoted as x making $T - x$ is the distance the rack moves. The mechanical model of the roller system was derived to be

$$J_{eq}\frac{d^2(T-x)}{dt^2} + r_2^2 f_{rack}\frac{d}{dt}(T-x) + r_2^2 k(T-x) = \frac{r_2^2}{r_1}\tau_m,$$

where $J_{eq} \triangleq J_2 + (r_2/r_1)^2 J_1 + r_2^2 M$. Here f_{rack} is the coefficient of viscous friction between the rack and the structure (not shown) holding it so that $-f_{rack}\frac{d}{dt}(T-x)$ is the corresponding friction force on the rack.

(a) Add the equation for the motor current to this model. Your final answer should be two coupled differential equations with the motor voltage V_S as input, and the output $T - x$ and the motor current i as the variables. That is, it should be of the form

$$J_{eq}\frac{d^2(T-x)}{dt^2} = \cdots$$

$$L\frac{di}{dt} = \cdots$$

(b) Set the motor inductance $L = 0$ and compute the transfer function from V_S to $y = T - x$. Is this transfer function stable?

7

Block Diagrams

The reason for using Laplace transforms is to convert differential equations into algebraic equations. The point of this chapter is to convert these algebraic equations into a block diagram representation, which will be the starting point for the design of feedback controllers for these systems as shown in Chapters 9 and 10.

7.1 Block Diagram for a DC Motor

A DC motor servo (positioning) system typically consists of a DC motor, an amplifier and sensors for position and current. The interest here is to understand how to model this system in the s domain for control purposes. Recall the dynamic equations of the DC motor given as

$$L\frac{di}{dt} = -Ri(t) - K_b\omega(t) + v_a(t)$$
$$J\frac{d\omega}{dt} = -f\omega(t) + K_T i(t) - \tau_L(t) \quad (7.1)$$
$$\frac{d\theta}{dt} = \omega(t).$$

Let's take the Laplace transform of the DC machine equations (7.1) with zero initial conditions to obtain

$$sLI(s) = -RI(s) - K_b\omega(s) + V_a(s)$$
$$sJ\omega(s) = -f\omega(s) + K_T I(s) - \tau_L(s) \quad (7.2)$$
$$s\theta(s) = \omega(s).$$

After some rearrangement we have

$$(sL + R)I(s) = V_a(s) - K_b\omega(s)$$
$$(sJ + f)\omega(s) = K_T I(s) - \tau_L(s) \quad (7.3)$$
$$s\theta(s) = \omega(s).$$

A block diagram of these equations is given in Figure 7.1 on the next page and is simply a graphical way to illustrate the relationships between the Laplace transform variables.

An Introduction to System Modeling and Control, First Edition. John Chiasson.
© 2022 John Wiley & Sons, Inc. Published 2022 by John Wiley & Sons, Inc.
Companion website: www.wiley.com/go/chiasson/anintroductiontosystemmodelingandcontrol

186   7 Block Diagrams

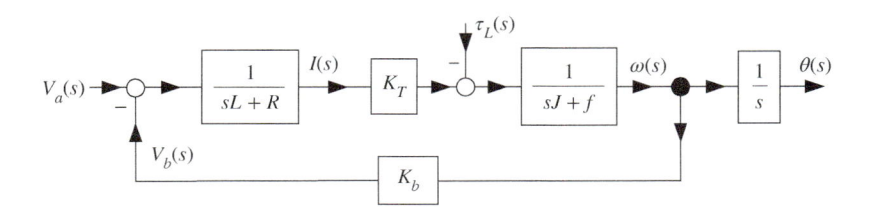

Figure 7.1. Block diagram of a DC motor.

That is, the block diagram represents

$$I(s) = \frac{1}{sL+R}\left(V_a(s) - K_b\omega(s)\right)$$
$$\omega(s) = \frac{1}{sJ+f}\left(K_T I(s) - \tau_L(s)\right) \qquad (7.4)$$
$$\theta(s) = \frac{1}{s}\omega(s).$$

In Figure 7.1 an unfilled circle "∘" denotes a summing junction while a filled circle "•" is a node indicating the same variable is being sent to different blocks.

For a DC motor it often turns out (but not always, see Example 6) that the inductance L is negligible so we set $L = 0$ and the block diagram of Figure 7.1 becomes that of Figure 7.2.

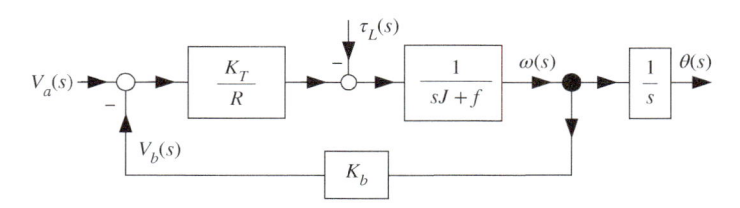

Figure 7.2. Block diagram of the DC motor with $L = 0$.

We next move the load torque (disturbance input) $\tau_L(s)$ to the left side of the K_T/R block to obtain the *equivalent* block diagram of Figure 7.3.

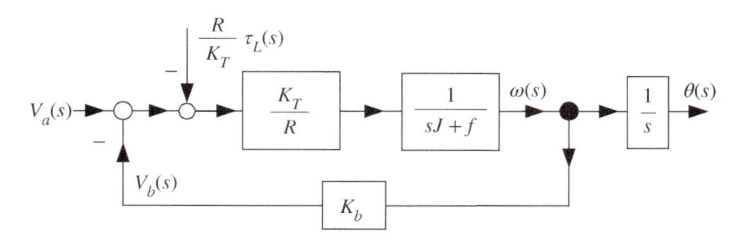

Figure 7.3. Equivalent block diagram of the DC motor with the disturbance torque moved toward the input.

Finally, we can just put the disturbance $R\tau_L(s)/K_T$ into the same summing junction as $V_a(s)$ to end up with the equivalent block diagram of Figure 7.4.

7.2 Block Diagram Reduction

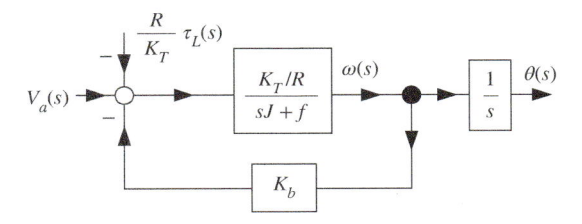

Figure 7.4. Equivalent block diagram of the DC motor with the disturbance torque moved to the input.

Remark Note that $\dfrac{R\tau_L}{K_T}$ has units $\dfrac{(\text{Ohms})(\text{Nm})}{\text{Nm/amperes}} = (\text{Ohms})(\text{amperes}) = \text{Volts}$. This makes sense as $-R\tau_L/K_T$ is going into the same summing junction as the input voltage V_a. The interpretation here is that if a voltage equal to $-R\tau_L/K_T$ is put into the motor then it has the same effect on the outputs ω and θ as the actual load torque τ_L.

7.2 Block Diagram Reduction

In many control systems, the block diagram can be reduced to combinations of the basic block diagram shown in Figure 7.5.

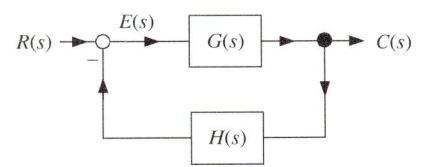

Figure 7.5. Standard block diagram.

For example, the block diagram of Figure 7.4 is of this form if we make the identification

$$G(s) = \frac{K_T/R}{sJ + f}$$
$$H(s) = K_b$$
$$R(s) = V_a(s) - \frac{R}{K_T}\tau_L(s) \tag{7.5}$$
$$C(s) = \omega(s).$$

If we want the transfer function from $R(s)$ to $C(s)$, we compute

$$E(s) = R(s) - H(s)C(s)$$
$$C(s) = G(s)E(s)$$
$$= G(s)R(s) - G(s)H(s)C(s).$$

Rearrange the last equation to obtain

$$\bigl(1 + G(s)H(s)\bigr)C(s) = G(s)R(s)$$

188   7 Block Diagrams

or
$$C(s) = \frac{G(s)}{1 + G(s)H(s)} R(s). \qquad (7.6)$$

We will see that (7.6) is a common expression when dealing with control systems.

We next compute the transfer function from the input $R(s)$ to the error $E(s)$. Proceeding, we have
$$E(s) = R(s) - H(s)C(s)$$
$$= R(s) - H(s)G(s)E(s),$$

which upon rearrangement becomes
$$(1 + H(s)G(s)) E(s) = R(s).$$

Finally, the error $E(s)$ is given by
$$E(s) = \frac{1}{1 + H(s)G(s)} R(s). \qquad (7.7)$$

We will also see that (7.7) is used over and over again to analyze control systems.

Example 1 *Transfer Function of the DC Motor*

Let's look at the block diagram of the DC motor given in Figure 7.4 using the identification given in (7.5). We then have

$$\omega(s) = \frac{\frac{K_T/R}{sJ+f}}{1 + K_b \frac{K_T/R}{sJ+f}} \left(V_a(s) - \frac{R}{K_T}\tau_L(s) \right) = \frac{K_T/R}{sJ + f + K_b K_T/R} \left(V_a(s) - \frac{R}{K_T}\tau_L(s) \right)$$

$$= \frac{\frac{K_T}{RJ}}{s + \frac{f + K_b K_T/R}{J}} \left(V_a(s) - \frac{R}{K_T}\tau_L(s) \right).$$

Then setting $a \triangleq \frac{f + K_b K_T/R}{J}$, $b \triangleq \frac{K_T}{RJ}$, and $K_L \triangleq \frac{R}{K_T}$, the block diagram of the DC motor reduces to that of Figure 7.6.

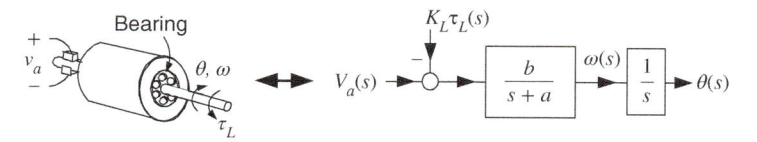

Figure 7.6. Block diagram of a DC motor.

Interestingly, we will show how a simple experiment can be done to estimate the values of a and b so we don't need to know all the individual parameter values of the motor, i.e., L, R, K_T, J, and f.

The control problem is to measure the position $\theta(t)$ and use this to figure out the voltage $v_a(t)$ to apply to the motor at each time t to get the rotor to turn to the desired angle θ_d. It must do this even with a load torque τ_L acting on it. This will be done in the following chapters.

7.2 Block Diagram Reduction

Figure 7.7 is a block diagram model of a simple control system using a DC motor with voltage as input and angular position as output. A sensor (optical encoder) is attached to the motor shaft to provide the position measurement. Inside the dashed box of Figure 7.7 is the DC motor model from Figure 7.6. The dashed box encloses the physical model of the motor including any load torque on it.

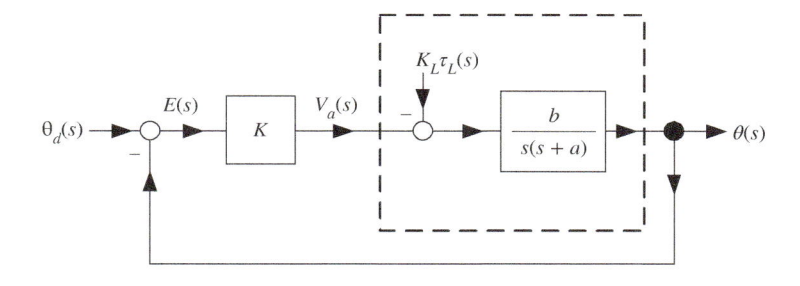

Figure 7.7. Block diagram of simple proportional feedback for a DC motor.

From a control perspective, all we can do is input voltage to the motor and then measure the angular position of the rotor shaft. In the diagram, θ_d is the desired position. For many applications, we will choose θ_d to be a step input given by

$$\theta_d(t) = \theta_0 u_s(t)$$

or

$$\theta_d(s) = \frac{\theta_0}{s}.$$

At the initial time $t = 0$, we take $\theta(0) = 0$ and we want the motor shaft to rotate to the angular position θ_0. The *error* signal is given by

$$e(t) = \theta_0 - \theta(t)$$

or, in the Laplace domain, it is

$$E(s) = \frac{\theta_0}{s} - \theta(s).$$

The controller structure shown in Figure 7.7 shows the voltage applied to the motor is simply

$$v_a(t) = K(\theta_0 - \theta(t))$$

or

$$V_a(s) = K\left(\frac{\theta_0}{s} - \theta(s)\right).$$

That is, the voltage applied to the motor is proportional (through the gain K) to the error between the desired rotor position θ_0 and it actual (measured) position $\theta(t)$. For this reason, this control structure is called *proportional* feedback. We will study this in great detail later.

A general control system can often be put in the form of Figure 7.8 on the next page.

7 Block Diagrams

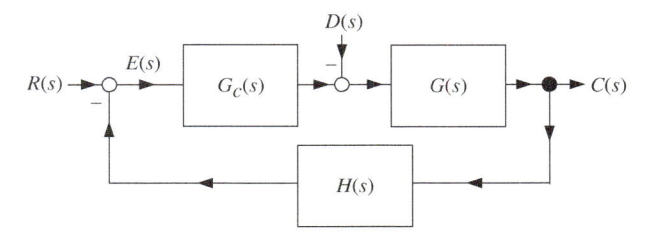

Figure 7.8. Typical block diagram of a control system.

In particular, the control system for the DC motor given in Figure 7.7 has this form with the identification

$$G(s) = \frac{b}{s(s+a)} \quad \text{(open loop transfer function)}$$
$$H(s) = 1$$
$$G_c(s) = K \quad \text{(controller transfer function)}$$
$$D(s) = K_L \tau_L(s) \quad \text{(disturbance input)}$$
$$R(s) = \frac{\theta_0}{s} \quad \text{(reference input)}.$$

Let's use block diagram reduction to compute the output and error transfer functions. To do this, we simply rearrange the block diagram of Figure 7.8 to obtain the equivalent block diagram of Figure 7.9.

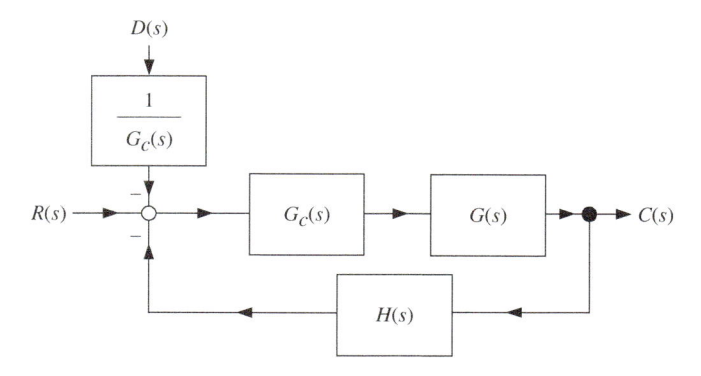

Figure 7.9. Equivalent block diagram of Figure 7.8. $E(s)$ is no longer available on this equivalent block diagram.

Then, just as we derived the expression given in (7.6), we have

$$C(s) = \frac{G_c(s)G(s)}{1 + H(s)G_c(s)G(s)} \left(R(s) - \frac{1}{G_c(s)} D(s) \right)$$
$$= \frac{G_c(s)G(s)}{1 + H(s)G_c(s)G(s)} R(s) - \frac{G(s)}{1 + H(s)G_c(s)G(s)} D(s). \tag{7.8}$$

7.2 Block Diagram Reduction

Example 2 *Closed-Loop Transfer Function*

Let's use expression (7.8) to compute the output $\theta(s)$ of the simple proportional control system of Figure 7.7. We have

$$\theta(s) = \frac{K\dfrac{b}{s(s+a)}}{1+K\dfrac{b}{s(s+a)}}\frac{\theta_0}{s} - \frac{\dfrac{b}{s(s+a)}}{1+K\dfrac{b}{s(s+a)}}\frac{K_L\tau_{L0}}{s}$$

$$= \frac{Kb}{s^2+as+Kb}\frac{\theta_0}{s} - \frac{b}{s^2+as+Kb}\frac{K_L\tau_{L0}}{s}.$$

For the DC motor the parameters a, b are positive, that is, $a > 0$ and $b > 0$. So, with $K > 0$ it follows that

$$s\theta(s) = \frac{Kb}{s^2+as+Kb}\theta_0 - \frac{b}{s^2+as+Kb}K_L\tau_{L0}$$

is stable. By the final value theorem

$$\lim_{t\to\infty}\theta(t) = \lim_{s\to 0} s\theta(s) = \theta_0 - \frac{K_L\tau_{L0}}{K}.$$

If the load torque is zero, i.e., $\tau_{L0} = 0$, then the angular position of the motor goes to the desired value θ_0. On the other hand, if $\tau_{L0} \neq 0$, it does not. We will see in later chapters how to deal with non-zero load torques. For now the point was simply to show how to obtain the closed-loop transfer function from the block diagram.

We will also be making extensive use of the error $E(s)$ when we consider the design of control systems. With reference to Figure 7.8 and using the expression (7.8), the error signal $E(s)$ is computed as[1]

$$E(s) = R(s) - H(s)C(s) = R(s) - \left(\frac{H(s)G_c(s)G(s)}{1+H(s)G_c(s)G(s)}R(s) - \frac{H(s)G(s)}{1+H(s)G_c(s)G(s)}D(s)\right)$$

$$= \frac{1}{1+H(s)G_c(s)G(s)}R(s) + \frac{H(s)G(s)}{1+H(s)G_c(s)G(s)}D(s). \qquad (7.9)$$

Also, we will mostly use *unity feedback*, which just means that $H(s) = 1$. In this case the last expression reduces to

$$E(s) = \frac{1}{1+G_c(s)G(s)}R(s) + \frac{G(s)}{1+G_c(s)G(s)}D(s).$$

We will use this last expression over and over again in Chapters 9 and 10 to design a feedback controller $G_c(s)$. For the remainder of the chapter we do more examples computing the closed-loop transfer function from a block diagram.

Example 3 *Block Diagram Reduction*

Consider the block diagram of Figure 7.10 on the next page. Moving $H_1(s)$ out of the feedback loop gives the block diagram of Figure 7.11 on the next page. Figure 7.11 is straightforwardly reduced to be the equivalent block diagram of Figure 7.12 on the next page.

---

[1] Note in Figure 7.9 the signal going into $G_c(s)$ is no longer $E(s)$. Thus we *cannot* use Eq. (7.7) applied to Figure 7.9 to compute $E(s)$.

7 Block Diagrams

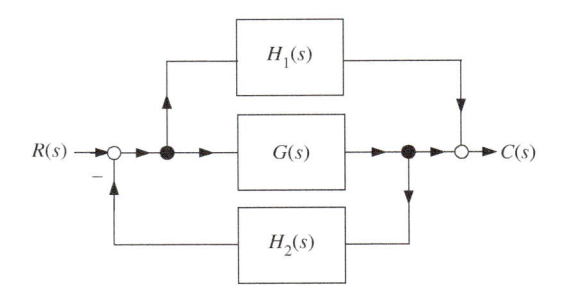

Figure 7.10. Block diagram reduction example.

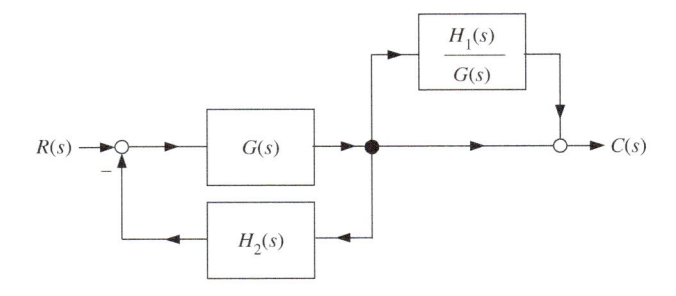

Figure 7.11. Equivalent block diagram to that of Figure 7.10.

Figure 7.12. Reduction of the block diagram of Figure 7.11

Finally, Figure 7.12 reduces to the block diagram of Figure 7.13.

Figure 7.13. $\dfrac{C(s)}{R(s)} = \dfrac{G(s) + H_1(s)}{1 + G(s)H_2(s)}.$

Example 4 *Block Diagram Reduction*

Consider the block diagram of Figure 7.14. We first remove the feedback loop containing H_3 out of the other two feedback loops to obtain the block diagram of Figure 7.15.

We next simplify the two bottom feedback loops in Figure 7.15 to obtain the block diagram of Figure 7.16. Some simple rearrangement of Figure 7.16 gives the block diagram of Figure 7.17.

7.2 Block Diagram Reduction

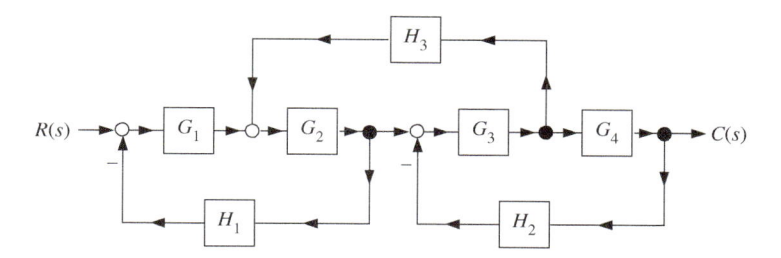

Figure 7.14. Block diagram reduction example.

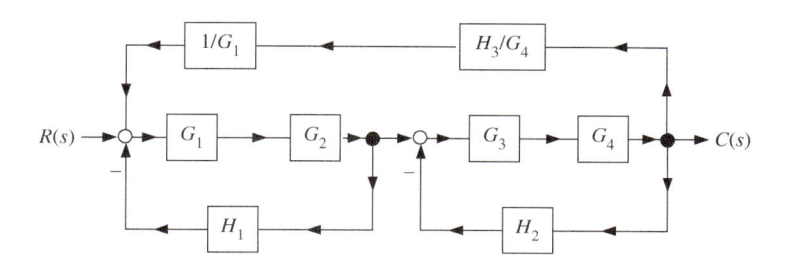

Figure 7.15. Block diagram equivalent to Figure 7.14.

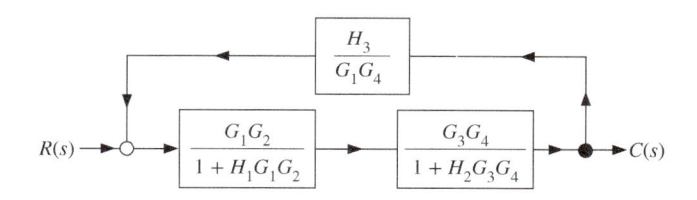

Figure 7.16. Block diagram equivalent to Figure 7.15.

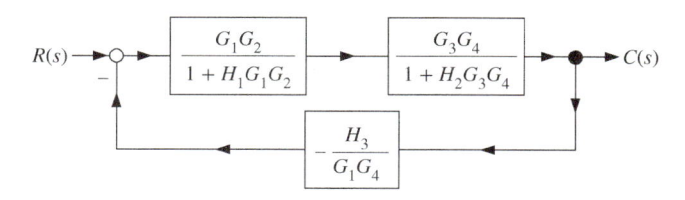

Figure 7.17. Block diagram equivalent to Figure 7.16.

From the block diagram of Figure 7.17 we obtain the output response $C(s)$ as

$$C(s) = \frac{\dfrac{G_1 G_2}{1 + H_1 G_1 G_2} \dfrac{G_3 G_4}{1 + H_2 G_3 G_4}}{1 - \dfrac{H_3}{G_1 G_2} \dfrac{G_1 G_2}{1 + H_1 G_1 G_2} \dfrac{G_3 G_4}{1 + H_2 G_3 G_4}} R(s).$$

194   7   Block Diagrams

Example 5 *Block Diagram Reduction*
Consider the block diagram of Figure 7.18.

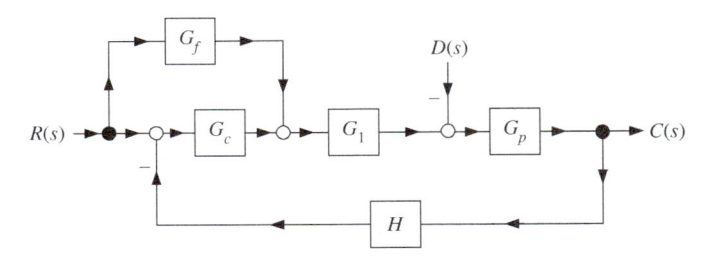

Figure 7.18. Block diagram reduction.

We first rearrange this block diagram so that there are two inputs with $R(s)$ as shown in Figure 7.19.

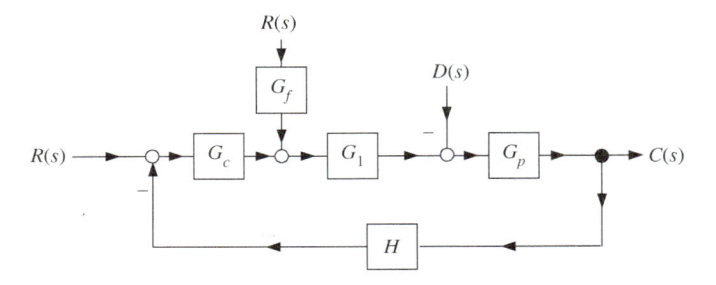

Figure 7.19. Block diagram equivalent to Figure 7.18

Then we move the disturbance input $D(s)$ and the second $R(s)$ input to the first summing junction to obtain the block diagram of Figure 7.20.

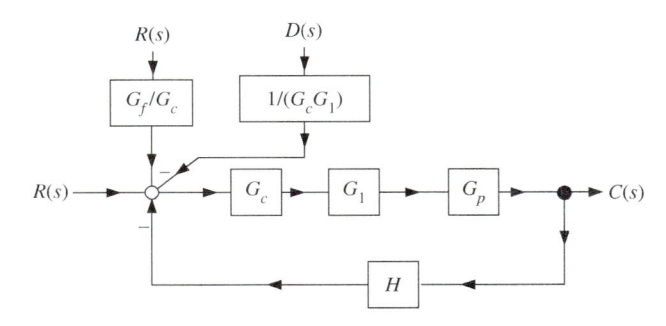

Figure 7.20. Block diagram equivalent to Figure 7.18.

From this last block diagram we obtain the output response $C(s)$ as

$$C(s) = \frac{G_c G_1 G_p}{1 + H G_c G_1 G_p} \left(R(s) + \frac{G_f}{G_c} R(s) - \frac{1}{G_c G_1} D(s) \right)$$

$$= \frac{G_1 G_p (G_c + G_f)}{1 + H G_c G_1 G_p} R(s) - \frac{G_p}{1 + H G_c G_1 G_p} D(s).$$

7.2 Block Diagram Reduction

Example 6 *Current Command Amplifier for a DC Motor*

The input to a DC motor is the voltage v_a. However, the torque equation is $J d\omega/dt = -f\omega + K_T i - \tau_L$, where the motor torque $K_T i$ is proportional to the current. Thus, it would be convenient if the current was the input since we could then specify the motor's torque by specifying the current. To get around the fact that the voltage is the input, an inner current control loop is added that allows direct current command. Specifically, as shown in Figure 7.21, the current is sensed and fed back (typically using analog electronics inside the amplifier) through a proportional controller. Such an amplifier is referred to as a *current command* amplifier. The voltage is forced by this controller to go to whatever value necessary to obtain the desired current. To analyze this setup we go back to Eqs. (7.4) of the DC motor, which are repeated here for convenience.

$$I(s) = \frac{-K_b \omega(s) + V_a(s)}{sL + R}$$

$$\omega(s) = \frac{K_T I(s) - \tau_L(s)}{sJ + f}$$

$$\theta(s) = \frac{1}{s}\omega(s).$$

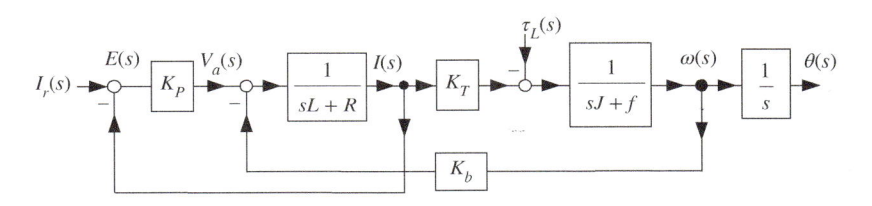

Figure 7.21. DC motor with an inner current control loop.

In Figure 7.21 $I_r(s)$ is the reference (desired) current, $I(s)$ is the measured current in the motor, and $K_P > 0$ is a proportional gain. We want to show that the gain K_p can be chosen to force $i_r(t) \to i(t)$. To do this we must compute the transfer function $G(s) = \omega(s)/I_r(s)$, which is done by rearranging the block diagram of Figure 7.21 to be the equivalent block diagram of Figure 7.22.

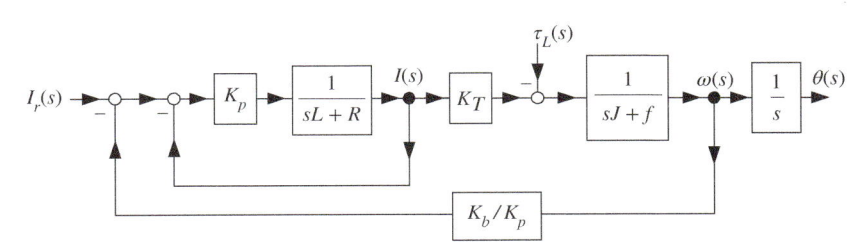

Figure 7.22. Equivalent block diagram.

Computing the transfer function of the inner loop, the block diagram of Figure 7.22 reduces to that of Figure 7.23 on the next page.

The load torque $\tau_L(s)$ is then moved to the same node as $I_r(s)$ to obtain equivalent block diagram of Figure 7.24 on the next page.

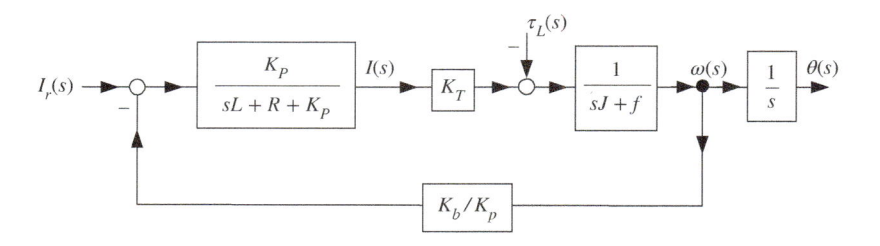

Figure 7.23. Simplified block diagram.

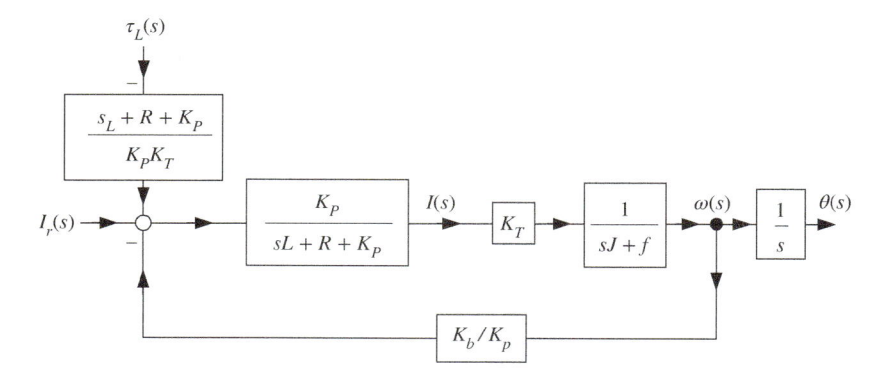

Figure 7.24. Reduced block diagram.

With $\omega(s)$ the output, the block diagram of Figure 7.24 immediately gives

$$\omega(s) = \frac{\dfrac{K_p}{sL+R+K_p}\dfrac{K_T}{sJ+f}}{1+\dfrac{K_b}{K_p}\dfrac{K_p}{sL+R+K_p}\dfrac{K_T}{sJ+f}}\left(I_r(s) - \dfrac{1}{K_pK_T}\tau_L(s)\right).$$

We now use *high-gain* feedback meaning we let $K_p \to \infty$. The expression for $\omega(s)$ then reduces to

$$\omega(s) = \frac{K_T}{sJ+f}\left(I_r(s) - \frac{1}{K_T}\tau_L(s)\right).$$

The corresponding block diagram is shown in Figure 7.25.

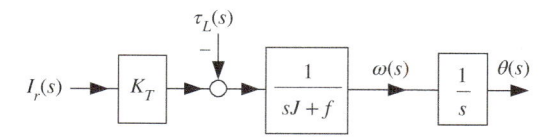

Figure 7.25. Current command model of a DC motor.

To summarize, if the gain K_P can be made large enough, the actual current $i(t)$ is forced to track $i_r(t)$ quite fast so that i_r is essentially equal to i. However, note that one cannot make the gain K_P arbitrarily large. This is easily seen by noting that the voltage commanded into the amplifier is

$$v_a(t) = K_P\left(i_r(t) - i(t)\right).$$

For large gains K_P this could result in $v_a(t)$ being greater than $V_{\max}$ causing the amplifier to saturate. That said, typically a current controller can be designed to force $i(t) \to i_r(t)$ fast enough that $i_r(t)$ can be considered equal to $i(t)$. The following reduced-order system is then be used to design a speed or position controller for the DC motor.

$$\frac{d\omega}{dt} = (K_T/J)i_r(t) - (f/J)\omega(t) - \tau_L/J$$
$$\frac{d\theta}{dt} = \omega.$$

Problems

Problem 1 *Block Diagram Reduction*
Use the block diagram reduction method to compute the transfer function $C(s)/R(s)$ for the system of Figure 7.26. Show work!

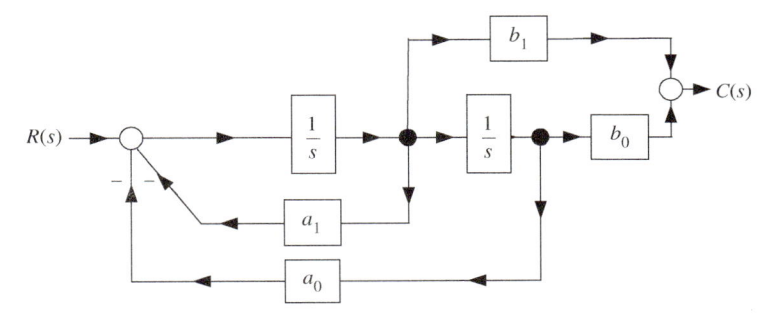

Figure 7.26. Block diagram reduction.

Problem 2 *Block Diagram Reduction*
Compute the transfer function $C(s)/R(s)$ of the system given in Figure 7.27 using block diagram reduction.

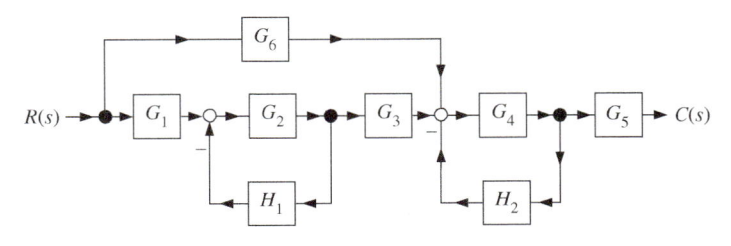

Figure 7.27. Block diagram reduction.

Problem 3 *Block Diagram Reduction*
A block diagram of a control system is shown in Figure 7.28 on the next page.

7 Block Diagrams

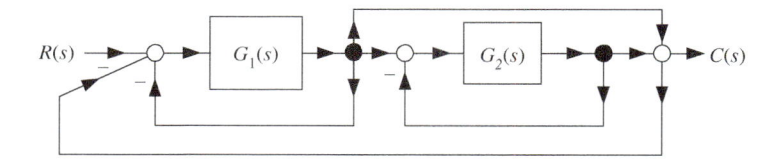

Figure 7.28. Block diagram reduction.

Compute the transfer function $C(s)/R(s)$ using block diagram reduction. Hint: Note this block diagram is equivalent to the block diagram of Figure 7.29.

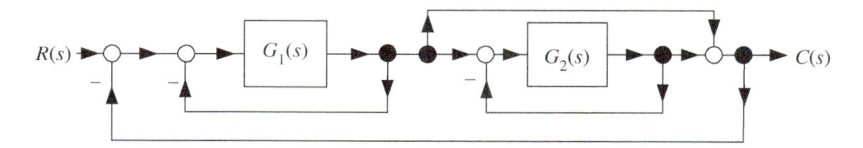

Figure 7.29. Block diagram equivalent to Figure 7.28.

Problem 4 *Modeling and Block Diagram Reduction*

Figure 7.30 is a schematic diagram for a servo motor with moment of inertia J_m connected by a flexible shaft to another rigid body whose moment of inertia is J_l. The flexible shaft is modeled by with a torsional spring whose constant is K. There is viscous friction on the motor shaft (viscous friction coefficient B_m) and viscous friction on the output shaft (viscous friction coefficient B_l). The motor torque is $K_T i_a$ and the back emf is $v_b = K_b \omega_m$. τ_l is the torque on the output shaft produced by the spring and is given by $\tau_l = K(\theta_m - \theta_l)$.

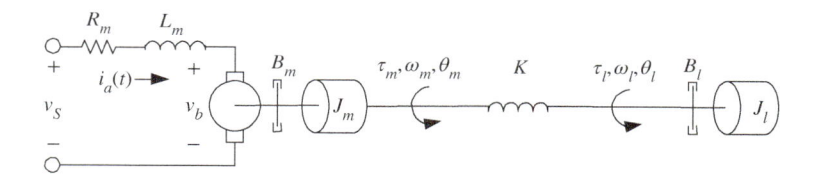

Figure 7.30. Motor whose output shaft is flexible.

(a) Write down the differential equations that describe this model.

(b) With zero initial conditions take the Laplace transform of the differential equations in part (a) and show they are written in the Laplace domain as

$$(sL_m + R_m)I_a(s) = -K_b s\theta_m(s) + V_S(s)$$
$$(J_m s^2 + B_m s + K)\theta_m(s) = K_T I_a(s) + K\theta_l(s) \quad (7.10)$$
$$(J_l s^2 + B_l s + K)\theta_l(s) = K\theta_m(s).$$

(c) A block diagram for the equations in part (b) is given in Figure 7.31. Redraw the block diagram with an equivalent block diagram so that the two loops do not intersect. Hint: Move the line connected at $\theta_m(s)$ to the right of $\theta_l(s)$.

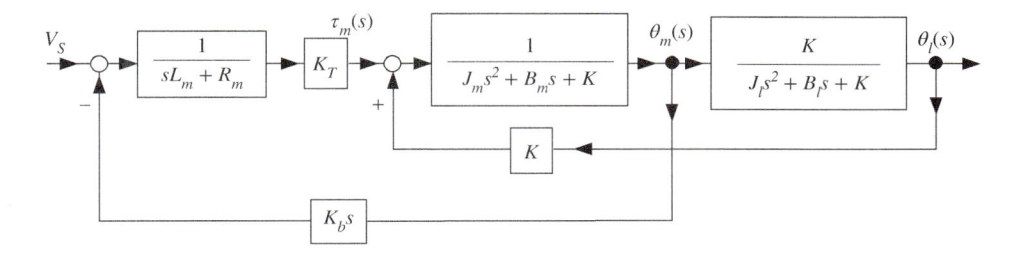

Figure 7.31. Block diagram of a motor whose output shaft is flexible.

(d) Let
$$G_1(s) = \frac{K_T}{sL_m + R_m}, \quad G_2(s) = \frac{1}{J_m s^2 + B_m s + K}, \quad G_3(s) = \frac{K}{J_l s^2 + B_l s + K}.$$

Compute the transfer function from $V_S(s)$ to $\theta_l(s)$ in terms of $G_1(s), G_2(s), G_3(s), K_b s$, and K. Hint: First compute $G_4(s) = \dfrac{\theta_l(s)}{\tau_m(s)}$.

Problem 5 *Block Diagram Reduction*

Compute the transfer function $C(s)/R(s)$ of the system given in Figure 7.32 using block diagram reduction.

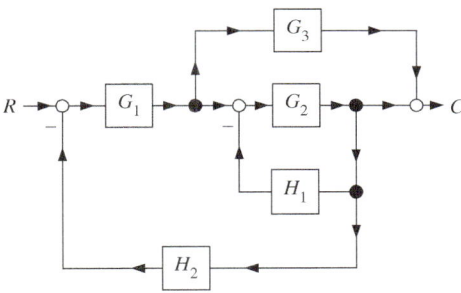

Figure 7.32. Block diagram reduction.

Problem 6 *High-Gain Proportional Plus Integral Current Control*

Consider a proportional plus integral (PI) current controller given by $K_P + K_I/s = K_P(s + \alpha)/s$ where $K_I = \alpha K_P$. This is illustrated in the block diagram of Figure 7.33.

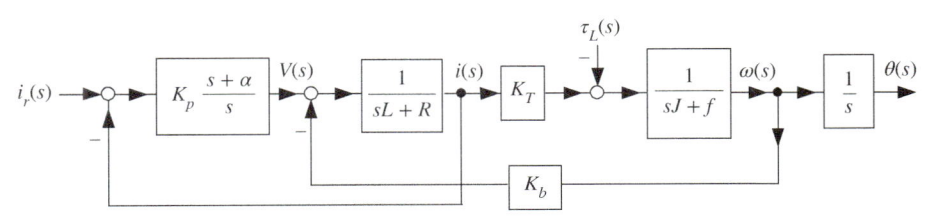

Figure 7.33. PI current controller.

Let $\tau_L = 0$ in parts (a)–(c).

(a) Using block diagram reduction compute $G(s) = \omega(s)/i_r(s)$.

200   7 Block Diagrams

(b) Show that $G(s) = \omega(s)/i_r(s) \to K_T/(sJ+f)$ as $K_P \to \infty$.

(c) Compute the transfer function $i(s)/i_r(s)$ and show that $i(s)/i_r(s) \to 1$ as $K_p \to \infty$.

Problem 7 *Simulation Diagram*

Use block diagram reduction on the simulation diagram of Figure 7.34 to show that

$$Y(s) = \frac{b_3 s^3 + b_2 s^2 + b_1 s + b_0}{s^3 + a_2 s^2 + a_1 s + a_0} R(s).$$

We use the terminology *simulation* diagram when each block is either an integrator or a constant.

Hint: Move the blocks with the $b_j, j = 0, 1, 2, 3$ to the left side of the diagram out of the feedback loops.

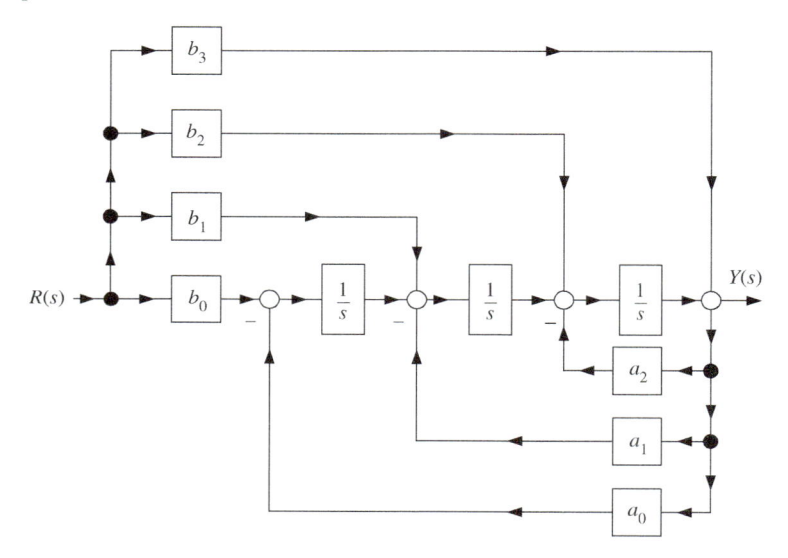

Figure 7.34. Simulation diagram for a third-order transfer function.

Problem 8 *Block Diagram Reduction for a DC Motor*

Consider the block diagram shown in Figure 7.35. Use block diagram reduction to find the output $C(s)$ in terms of $R(s)$ and $D(s)$. Hint: First step is to move $D(s)$ to the node for $V_a(s)$.

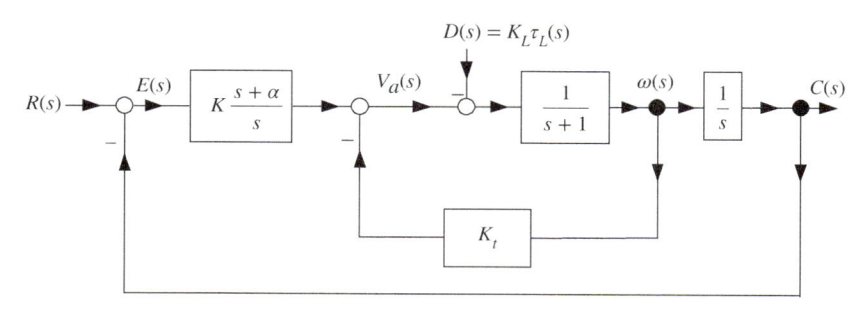

Figure 7.35. Block diagram for a DC motor control system.

Problem 9 *Simulation*

Compare a simulation of the DC motor using the equations given in (7.1) with a simulation using the block diagram of Figure 7.6. A SIMULINK block diagram to do this is shown in Figure 7.36.

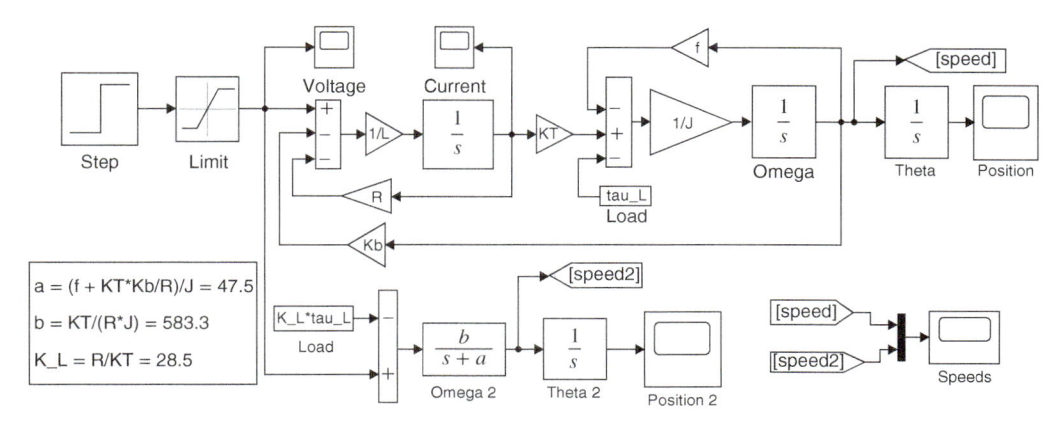

Figure 7.36. SIMULINK block diagram for the DC motor.

Let $V_{\max} = 40$ V, $I_{\max} = 5$ A, $K_b = K_T = 0.07$ V/rad/s ($=$ N-m/A), $J = 6 \times 10^{-5}$ kg-m^2, $R = 2$ Ω, $L = 2$ mH, and $f = 0.0004$ N-m/rad/s. Set $K_L = R/K_T$ and $\tau_L = 0$. Put a step input voltage of $V_S(t) = 10$ V into the motor and plot out the two speeds $\omega(t)$ from the simulation together on the same plot. Recall that in the block diagram model of Figure 7.6 the inductance was set equal to zero. Are the plots close? (Answer: Yes, if you did the simulation correctly!) Hand in your .m file, which should set all the values of the motor parameters and it should compute the values of a, b, K_L. Also hand in a screenshot of your SIMULINK block diagram.

8

System Responses

8.1 First-Order System Response

Recall the block diagram of a DC motor shown in Figure 8.1.

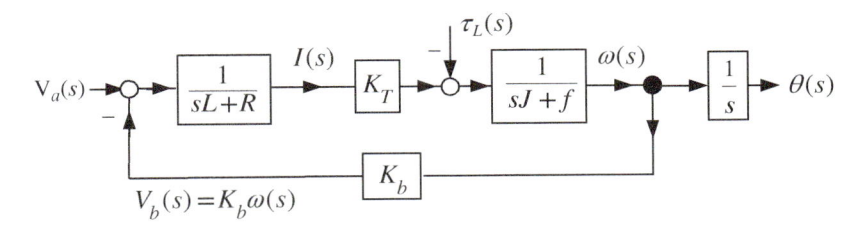

Figure 8.1. Block diagram of a DC motor.

Setting $L = 0$, $a \triangleq \dfrac{f + K_b K_T/R}{J}$, $b \triangleq \dfrac{K_T}{RJ}$, $K_L \triangleq \dfrac{R}{K_T}$, and, along with a block diagram reduction, this simplifies to the block diagram of Figure 8.2.

Figure 8.2. Simplified block diagram of a DC motor.

Our interest now is the response of the motor's angular speed $\omega(t)$ to a step input voltage. Removing the integrator for θ and defining

$$T_m \triangleq 1/a$$
$$K_m \triangleq b/a$$

we obtain a first-order system whose block diagram is given in Figure 8.3.

Figure 8.3. Time constant form for a first-order transfer function.

204   8 System Responses

The open-loop transfer function from input voltage to speed is

$$G(s) = \frac{K_m}{T_m s + 1}.$$

Notice that the denominator's leading coefficient is T_m rather than 1. This is called the *time constant form* of the transfer function where T_m is the *time constant*.

For the present set $\tau_L = 0$ and so $\omega(s)$ is given by

$$\omega(s) = \frac{K_m}{T_m s + 1} V_a(s). \tag{8.1}$$

Let the input voltage be the step input $V_a(s) = V_0/s$. The speed response $\omega(s)$ is

$$\omega(s) = \frac{K_m}{T_m s + 1} \frac{V_0}{s} = V_0 K_m \left(\frac{1}{s} - \frac{1}{s + 1/T_m} \right), \tag{8.2}$$

where the last expression follows from a partial-fraction expansion. Computing the inverse Laplace transform, $\omega(t)$ is given by

$$\omega(t) = V_0 K_m (1 - e^{-t/T_m}) u_s(t). \tag{8.3}$$

Let's plot $\omega(t)$ by making a table of its values.

t	$\dfrac{\omega(t)}{V_0 K_m} = 1 - e^{-t/T_m}$
0	$1 - e^{-0} = 0$
T_m	$1 - e^{-1} = 0.632$
$2T_m$	$1 - e^{-2} = 0.86$
$3T_m$	$1 - e^{-3} = 0.95$
$4T_m$	$1 - e^{-4} = 0.98$

These values are then used to make the graph of Figure 8.4.

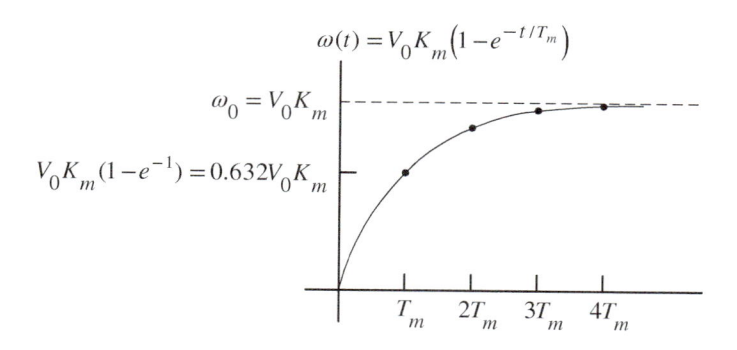

Figure 8.4. Step response of a first-order system.

Note that the *time constant* T_m has units of time. It is a measure of how fast the system responds to an input. The smaller the value of T_m the faster the system responds. For example, if ω_0 is the desired final angular speed, we would choose

$$V_0 = \frac{\omega_0}{K_m}.$$

Then the step input voltage of $V_a(s) = \frac{\omega_0}{K_m}\frac{1}{s}$ gives a speed response of

$$\omega(t) = \omega_0(1 - e^{-t/T_m})u_s(t). \tag{8.4}$$

For $t > 4T_m$, $\omega(t)$ is within 2% of ω_0. That is, $0.98\omega_0 \le \omega_0(1 - e^{-t/T_m}) \le \omega_0$. Hence the smaller the value of T_m, the faster $\omega(t)$ is within 2% of its final value ω_0.

Identification

Suppose the motor has a sensor that can be used to measure the angular speed (e.g., an encoder or a tachometer). Then apply a constant voltage V_0 to the motor and record its speed to obtain a curve similar to that of Figure 8.4. The measured speed curve gives ω_0 and the parameter K_m is given by

$$K_m = \omega_0/V_0. \tag{8.5}$$

The speed curve is also used to determine the time t_m such that $\omega(t_m) = (1 - e^{-1})\omega_0 = 0.632\omega_0$. This is the time constant, i.e., by (8.4) we have

$$T_m = t_m. \tag{8.6}$$

From the experimentally determined values of T_m and K_m, the motor's transfer function is $G(s) = \frac{K_m}{T_m s + 1}$.

8.2 Second-Order System Response

Let's now consider the response of the angular position θ of the motor using the setup of Figure 8.5. As indicated in Figure 8.5, a position sensor (encoder) is attached to the motor and used to bring the value of rotor angle into the computer. With $\theta_d = \theta_0 u_s(t)$ the desired

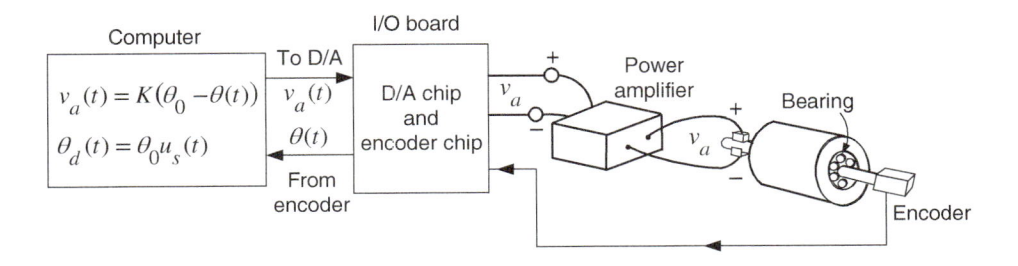

Figure 8.5. Position control of a DC motor.

8 System Responses

rotor angle, the voltage

$$v_a(t) = K\left(\theta_0(t) - \theta(t)\right)$$

is commanded to the amplifier using a digital to analog (D/A) converter.

To study the response $\theta(t)$ we need only use the block diagram model of Figure 8.5, which is given in Figure 8.6.

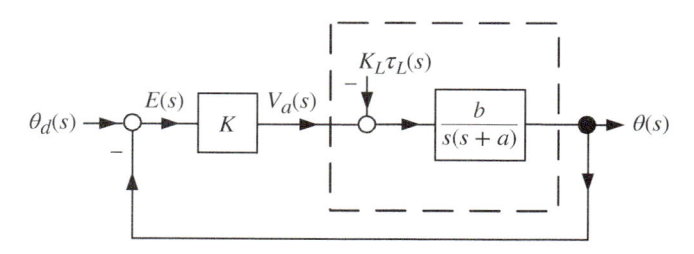

Figure 8.6. Proportional feedback position control of a DC motor.

Before analyzing this simple feedback control system, let's switch to a more general notation by setting $R(s) = \theta_d(s)$ and $C(s) = \theta(s)$. Also, for now, set $\tau_L = 0$. The transfer function from the input reference position $R(s)$ to the output position $C(s)$ is calculated as

$$C(s) = \frac{\dfrac{bK}{s(s+a)}}{1 + \dfrac{bK}{s(s+a)}} R(s) = \frac{bK}{s^2 + as + bK} R(s).$$

We refer to

$$\frac{C(s)}{R(s)} = \frac{bK}{s^2 + as + bK} \qquad (8.7)$$

as the *closed-loop* transfer function.

For this DC motor we have $a > 0, b > 0$ and we choose $K > 0$. With $R(s) = R_0/s$, $\omega_n^2 \triangleq bK$, and $2\zeta\omega_n \triangleq a$ (so $\zeta \triangleq \dfrac{a}{2\omega_n}$), the transfer function (8.7) can be rewritten in the standard form

$$C(s) = \frac{\omega_n^2}{s^2 + 2\zeta\omega_n s + \omega_n^2} \frac{R_0}{s}. \qquad (8.8)$$

$\zeta > 0$ is referred to as the *damping ratio* and $\omega_n > 0$ is referred to as the *natural frequency*. The quantity

$$sC(s) = \frac{\omega_n^2}{s^2 + 2\zeta\omega_n s + \omega_n^2} R_0$$

is stable so that by the final value theorem

$$c(\infty) \triangleq \lim_{t \to \infty} c(t) = \lim_{s \to 0} sC(s) = \lim_{s \to 0} s \frac{\omega_n^2}{s^2 + 2\zeta\omega_n s + \omega_n^2} \frac{R_0}{s} = R_0.$$

That is, $c(t) \to R_0$ as $t \to \infty$.

Transient Response and Closed-Loop Poles

We just showed that $c(t)$ goes to the desired final value of R_0. To find the complete solution $c(t)$ we will do a partial fraction expansion of $C(s)$. Our system is

$$C(s) = \underbrace{\frac{\omega_n^2}{s^2 + 2\zeta\omega_n s + \omega_n^2}}_{\text{Transfer function}} \underbrace{\frac{R_0}{s}}_{\text{Input}} \tag{8.9}$$

and, to do the partial fraction expansion, we must first find the poles p_1, p_2 of the transfer function. These are the roots of

$$s^2 + 2\zeta\omega_n s + \omega_n^2 = 0.$$

By the quadratic formula we have

$$p_i = \frac{-2\zeta\omega_n \pm \sqrt{(2\zeta\omega_n)^2 - 4\omega_n^2}}{2} = -\zeta\omega_n \pm \sqrt{(\zeta^2 - 1)\omega_n^2}$$

$$= -\zeta\omega_n \pm \omega_n\sqrt{\zeta^2 - 1}. \tag{8.10}$$

We are interested in the case where $\zeta > 0$ and $\omega_n > 0$ so that these poles are in the open left-half plane. Including $\zeta = 0$, we consider the following:

$$\begin{aligned} 0 < \zeta < 1 \quad & p_1, p_2 = -\zeta\omega_n \pm j\omega_n\sqrt{1 - \zeta^2} \\ \zeta = 1 \quad & p_1, p_2 = -\omega_n \\ \zeta > 1 \quad & p_1 = -\zeta\omega_n + \omega_n\sqrt{\zeta^2 - 1},\ p_2 = -\zeta\omega_n - \omega_n\sqrt{\zeta^2 - 1} \\ \zeta = 0 \quad & p_1, p_2 = \pm j\omega_n. \end{aligned} \tag{8.11}$$

With $\zeta > 0$ both poles are in the open left half-plane and the closed-loop transfer function

$$G(s) \triangleq \frac{\omega_n^2}{s^2 + 2\zeta\omega_n s + \omega_n^2}$$

is stable. If $\zeta = 0$, then the poles are on the $j\omega$ axis at $\pm j\omega_n$.

Figure 8.7 on the next page shows the pole locations for p_1, p_2 for the case $0 < \zeta < 1$. In the figure $\omega_d \triangleq \omega_n\sqrt{1 - \zeta^2}$ is called the *damped frequency*. Notice that for $0 \leq \zeta \leq 1$ we have

$$\begin{aligned} |p_i|^2 &= \left|-\zeta\omega_n \pm j\omega_n\sqrt{1 - \zeta^2}\right|^2 \\ &= (-\zeta\omega_n)^2 + \left(\pm\omega_n\sqrt{1 - \zeta^2}\right)^2 \\ &= \zeta^2\omega_n^2 + \omega_n^2(1 - \zeta^2) \\ &= \omega_n^2. \end{aligned} \tag{8.12}$$

That is, for $0 \leq \zeta \leq 1$ the poles are on a semicircle centered at the origin with a radius of ω_n. This is illustrated in Figure 8.8 on the next page. In more detail, Figure 8.8 shows

8 System Responses

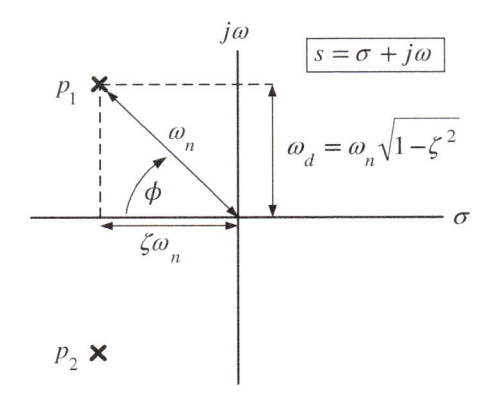

Figure 8.7. The closed-loop poles p_1, p_2 for $0 < \zeta < 1$.

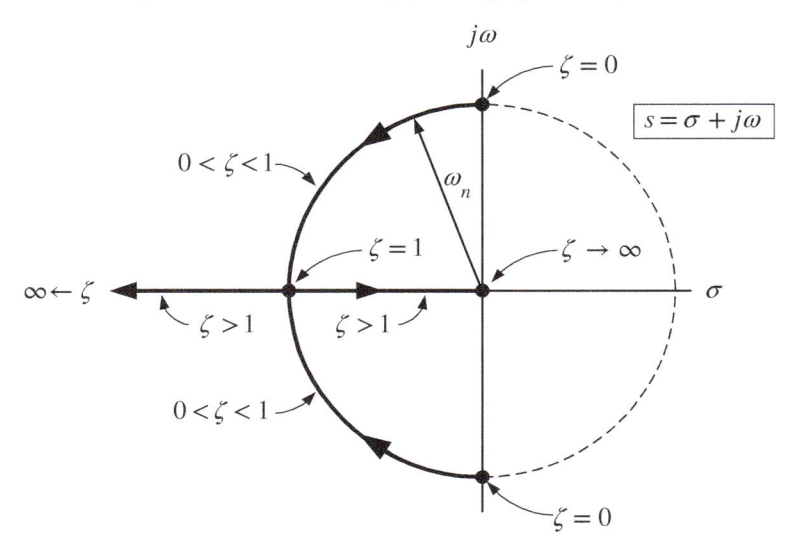

Figure 8.8. The locus of the closed-loop poles p_1, p_2 for $0 \leq \zeta < \infty$.

that for $\zeta = 0$ the two poles are at $\pm j\omega_n$. As ζ increases from 0 to 1 the two poles trace out a semicircle in the left half-plane. At $\zeta = 1$ the two poles are on top of each other at $-\omega_n$. As ζ increases greater than 1 one of the poles heads toward the origin while the other one goes off to $-\infty$.

Let's now proceed with the partial fraction expansion for the case where $0 < \zeta < 1$. In this case we have a complex-conjugate pair of poles given by

$$p_1, p_2 = -\zeta\omega_n \pm j\omega_n\sqrt{1-\zeta^2}.$$

Then

$$C(s) = \frac{\omega_n^2}{s^2 + 2\zeta\omega_n s + \omega_n^2} \frac{R_0}{s}$$

$$= \frac{\omega_n^2}{\left(s - (-\zeta\omega_n + j\omega_n\sqrt{1-\zeta^2})\right)\left(s - (-\zeta\omega_n - j\omega_n\sqrt{1-\zeta^2})\right)} \frac{R_0}{s}$$

$$= \frac{R_0}{s} + \frac{\beta}{s - (-\zeta\omega_n + j\omega_n\sqrt{1-\zeta^2})} + \frac{\beta^*}{s - (-\zeta\omega_n - j\omega_n\sqrt{1-\zeta^2})}. \quad (8.13)$$

8.2 Second-Order System Response

In the time domain this becomes

$$c(t) = R_0 u_s(t) + \underbrace{\beta e^{-\zeta\omega_n t} e^{+j\omega_n\sqrt{1-\zeta^2}t} + \beta^* e^{-\zeta\omega_n t} e^{-j\omega_n\sqrt{1-\zeta^2}t}}_{\text{Transient}} \qquad (8.14)$$

$$= R_0 u_s(t) + \underbrace{2|\beta| e^{-\zeta\omega_n t} \cos\left((\omega_n\sqrt{1-\zeta^2}t) + \angle\beta\right)}_{\text{Transient}}.$$

Note that the transient part of the response dies out exponentially according to the *real part* $-\zeta\omega_n$ of the poles of the transfer function. We want the transient part to die out fast so the $c(t) \to R_0$ fast. That is, we want both poles to have their real parts as far in the left half-plane as possible. From Figure 8.8 this indicates we should take $\zeta = 1$ which results in both poles being at $-\omega_n$. If we take $\zeta > 1$ then one of the poles heads out toward $-\infty$, but the other one heads toward the origin.

It is now a rather messy affair to compute the expressions for β and β^* in (8.13) and (8.14). So we take a different approach to the partial fraction expansion by rewriting $C(s)$ as

$$C(s) = \frac{\omega_n^2}{s^2 + 2\zeta\omega_n s + \omega_n^2} \frac{R_0}{s} = \frac{R_0}{s} + \frac{A_1(s + \zeta\omega_n) + A_2\omega_n\sqrt{1-\zeta^2}}{(s + \zeta\omega_n)^2 + \omega_n^2(1-\zeta^2)}.$$

Multiplying through by $s(s^2 + 2\zeta\omega_n s + \omega_n^2)$ we have

$$\omega_n^2 R_0 = (s^2 + 2\zeta\omega_n s + \omega_n^2) R_0 + s\left(A_1(s + \zeta\omega_n) + A_2\omega_n\sqrt{1-\zeta^2}\right).$$

Equating terms in s requires A_1 and A_2 satisfy

$$s^2 : 0 = (R_0 + A_1)s^2 \Rightarrow A_1 = -R_0$$

$$s^1 : 0 = \left(2\zeta\omega_n R_0 + A_1\zeta\omega_n + A_2\omega_n\sqrt{1-\zeta^2}\right)s \Rightarrow A_2 = -R_0\zeta/\sqrt{1-\zeta^2}$$

$$s^0 : \omega_n^2 R_0 = \omega_n^2 R_0.$$

Then

$$C(s) = \frac{R_0}{s} - R_0 \frac{s + \zeta\omega_n}{(s + \zeta\omega_n)^2 + \omega_n^2(1-\zeta^2)} - \frac{R_0\zeta}{\sqrt{1-\zeta^2}} \frac{\omega_n\sqrt{1-\zeta^2}}{(s + \zeta\omega_n)^2 + \omega_n^2(1-\zeta^2)}. \qquad (8.15)$$

Identifying $\sigma = -\zeta\omega_n$, and $\omega^2 = \omega_n^2(1-\zeta^2)$ the relevant Laplace transforms are

$$e^{\sigma t}\sin(\omega t)u_s(t) \leftrightarrow \frac{\omega}{(s-\sigma)^2 + \omega^2}$$

$$e^{\sigma t}\cos(\omega t)u_s(t) \leftrightarrow \frac{s-\sigma}{(s-\sigma)^2 + \omega^2}.$$

For $t \geq 0$ we then have

$$c(t) = R_0 u - R_0 e^{-\zeta\omega_n t}\cos\left(\omega_n\sqrt{1-\zeta^2}t\right) - R_0\frac{\zeta}{\sqrt{1-\zeta^2}}e^{-\zeta\omega_n t}\sin\left(\omega_n\sqrt{1-\zeta^2}t\right) \qquad (8.16)$$

$$= R_0 - R_0\frac{e^{-\zeta\omega_n t}}{\sqrt{1-\zeta^2}}\left(\sqrt{1-\zeta^2}\cos\left(\omega_n\sqrt{1-\zeta^2}t\right) + \zeta\sin\left(\omega_n\sqrt{1-\zeta^2}t\right)\right)$$

$$= R_0 - R_0\frac{e^{-\zeta\omega_n t}}{\sqrt{1-\zeta^2}}\left(\sin(\phi)\cos\left(\omega_n\sqrt{1-\zeta^2}t\right) + \cos(\phi)\sin\left(\omega_n\sqrt{1-\zeta^2}t\right)\right)$$

$$= R_0 - R_0\frac{e^{-\zeta\omega_n t}}{\sqrt{1-\zeta^2}}\sin\left(\omega_n\sqrt{1-\zeta^2}t + \phi\right), \qquad (8.17)$$

where
$$\sin(\phi) \triangleq \sqrt{1-\zeta^2}, \ \cos(\phi) \triangleq \zeta \qquad (8.18)$$
so that
$$\phi \triangleq \tan^{-1}\left(\frac{\sqrt{1-\zeta^2}}{\zeta}\right). \qquad (8.19)$$

This is the angle ϕ indicated in Figure 8.7. The transient part of (8.17) decays as $e^{-\zeta \omega_n t}$ where $-\zeta \omega_n$ is the real part of the closed-loop poles and it oscillates with frequency $\omega_d = \omega_n \sqrt{1-\zeta^2}$, which is the imaginary part of the closed-loop poles. Recall that ω_d is called the damped frequency. Figure 8.9 shows a plot of the unit step response $c(t)$ for values of ζ between 0.1 and 2.

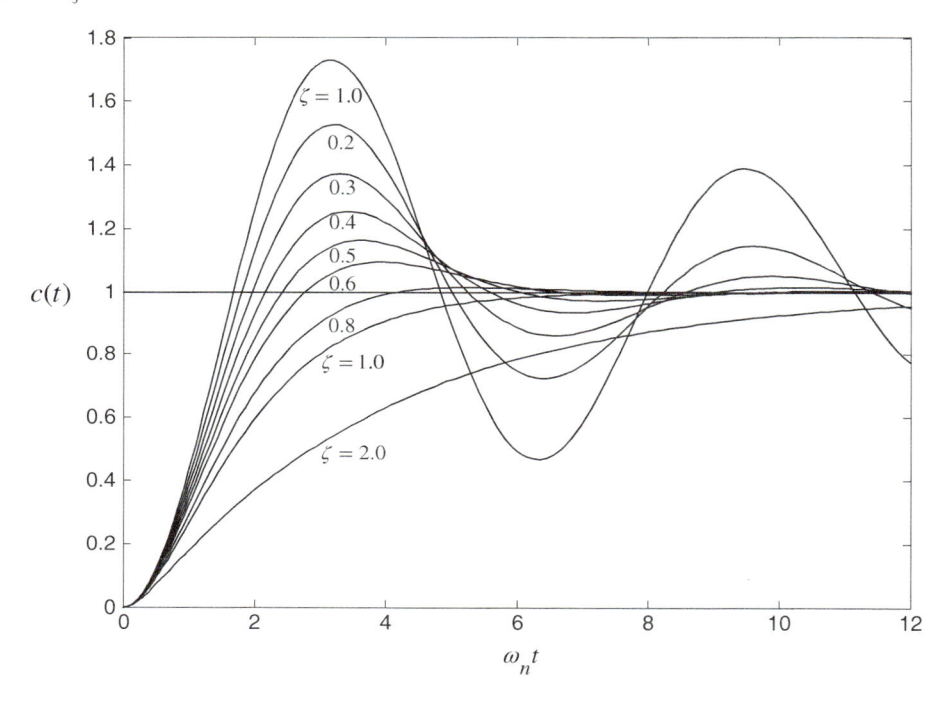

Figure 8.9. The unit step response $c(t)$ for $0 < \zeta \leq 2$.

For $\zeta \geq 1$ the response is not oscillatory and just steadily increases up to the final value of 1 (see Problems 1 and 2). On the other hand, for $0 < \zeta < 1$ the response is oscillatory and becoming more so as ζ gets closer to 0.

Peak Time t_p and Percent Overshoot M_p

For $0 < \zeta < 1$ the output response is oscillatory with frequency $\omega_d = \omega_n \sqrt{1-\zeta^2}$. Figure 8.10 shows the response of such a second-order system to a unit step input. The peak time t_p is the time when the response $c(t)$ obtains its peak value. Using (8.16) we now derive expressions for the peak time and the corresponding peak value $c(t_p)$. We will later show how these expressions can be used to estimate the parameters a and b of the transfer function model $G(s) = \dfrac{b}{s(s+a)}$ from experimental data.

8.2 Second-Order System Response

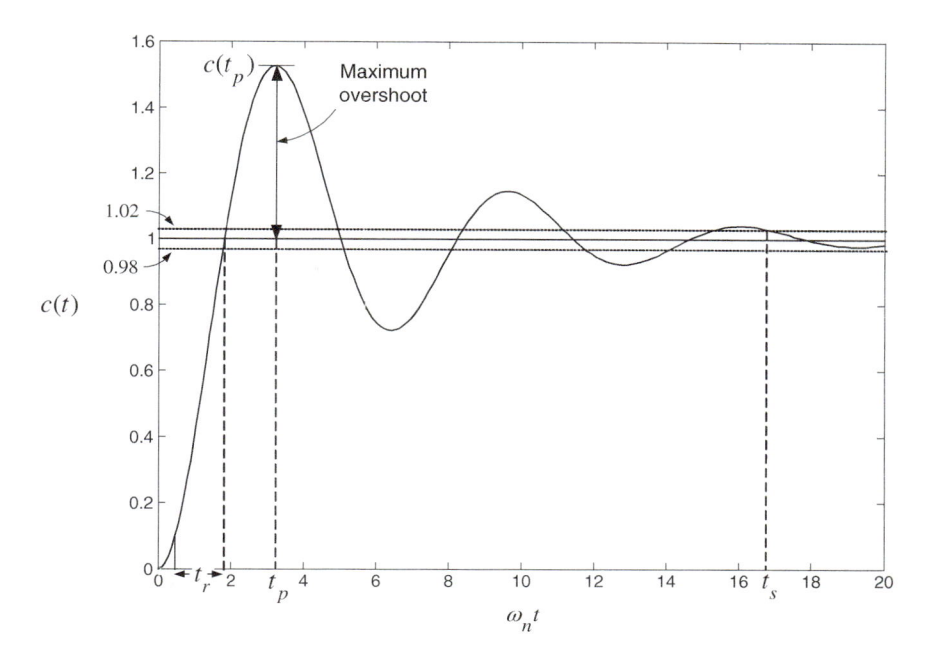

Figure 8.10. Rise time t_r, peak time t_p and settling time t_s.

With $0 < \zeta < 1$ the output response is oscillatory and the peak overshoot occurs at a time t_p when $\dot{c}(t) = 0$. With $\sigma \triangleq -\zeta\omega_n$ and $\omega_d \triangleq \omega_n\sqrt{1-\zeta^2}$, we rewrite (8.16) as

$$c(t) = R_0 - R_0 e^{\sigma t} \cos(\omega_d t) + R_0 \frac{\sigma}{\omega_d} e^{\sigma t} \sin(\omega_d t). \tag{8.20}$$

Differentiating we have

$$\frac{d}{dt}c(t) = -R_0 \sigma e^{\sigma t}\cos(\omega_d t) + R_0 \omega_d e^{\sigma t}\sin(\omega_d t) + R_0 \frac{\sigma^2}{\omega_d} e^{\sigma t}\sin(\omega_d t) + R_0 \sigma e^{\sigma t}\cos(\omega_d t)$$

$$= R_0 e^{\sigma t}\left(\frac{\sigma^2}{\omega_d} + \omega_d\right)\sin(\omega_d t). \tag{8.21}$$

The solutions to $dc/dt = 0$ are the solutions to

$$\sin(\omega_d t_p) = 0.$$

The smallest non zero solution is

$$t_p = \frac{\pi}{\omega_d} = \frac{\pi}{\omega_n\sqrt{1-\zeta^2}}. \tag{8.22}$$

The value of $c(t)$ at t_p is then

$$c(t_p) = R_0 - R_0 e^{\sigma t_p}\cos(\omega_d t_p) + R_0 \frac{\sigma}{\omega_d} e^{\sigma t_p}\sin(\omega_d t_p)$$

$$= R_0 - R_0 e^{\sigma t_p}\cos(\pi)$$

$$= R_0 + R_0 e^{-\pi\zeta/\sqrt{1-\zeta^2}}.$$

Note that $c(\infty) = R_0$. The *fractional overshoot* M_p is defined by

$$M_p \triangleq \frac{c(t_p) - c(\infty)}{c(\infty)} = e^{-\pi\zeta/\sqrt{1-\zeta^2}}, \qquad (8.23)$$

where we used the fact that $c(\infty) = R_0$. The *percent overshoot* is simply $M_p \times 100$. However, we will often refer to M_p as the percent overshoot and the reader should understand that M_p is given by (8.23).

If we plot the step response and then measure t_p and M_p, then ζ and ω_n can be found. To explain, take the natural logarithm of both sides of (8.23) to obtain

$$\ln(M_p) = -\pi\zeta/\sqrt{1-\zeta^2}.$$

After some algebra we obtain

$$\zeta = \sqrt{\frac{\ln^2(M_p)}{\pi^2 + \ln^2(M_p)}}. \qquad (8.24)$$

Next, using (8.22) for the peak time, ω_n is given by

$$\omega_n = \frac{\pi}{t_p \sqrt{1-\zeta^2}}. \qquad (8.25)$$

With these values of ζ and ω_n we can compute the values of a and b. To do, recall

$$C(s) = \frac{Kb}{s^2 + as + Kb} \frac{R_0}{s} = \frac{\omega_n^2}{s^2 + 2\zeta\omega_n s + \omega_n^2} \frac{R_0}{s}$$

so that

$$Kb = \omega_n^2$$
$$a = 2\zeta\omega_n.$$

Finally, as we set the value of the gain K, we use its value to then compute b.

Settling Time t_s

By (8.17) the second-order step response for $0 < \zeta < 1$ is

$$c(t) = R_0 - R_0 \frac{e^{-\zeta\omega_n t}}{\sqrt{1-\zeta^2}} \sin\left(\omega_n \sqrt{1-\zeta^2}\, t + \phi\right). \qquad (8.26)$$

The settling time is usually defined as the time when $c(t)$ is within 2% of its final value and stays within 2% of its final value. See Figure 8.11.

To compute the time that $c(t)$ intersects one of the lines $(1 \pm 0.02)R_0$ and stays within these lines is difficult. One way to simplify the calculation is to compute the time that the *envelope* of $c(t)$ intersects one of the lines $(1 \pm 0.02)R_0$. The envelope of $c(t)$ is given by

$$R_0 \pm R_0 \frac{e^{-\zeta\omega_n t}}{\sqrt{1-\zeta^2}} \qquad (8.27)$$

and is indicated along with a step response in Figure 8.11.

8.2 Second-Order System Response

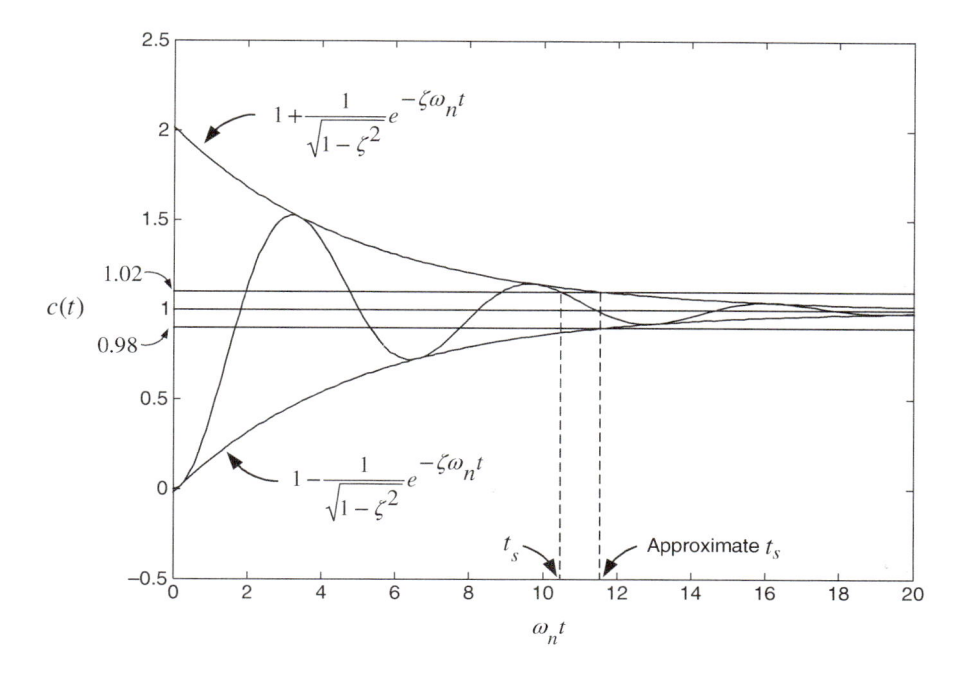

Figure 8.11. Using the envelope of the step response to compute an approximate settling time.

Let's now use the envelope to obtain an *upper bound* t_{sb} on the settling time. For the 2% criterion we set

$$R_0 + R_0 \frac{e^{-\zeta\omega_n t_{sb}}}{\sqrt{1-\zeta^2}} = 1.02 R_0$$

to determine t_{sb}. We solve this as follows.

$$\frac{e^{-\zeta\omega_n t_{sb}}}{\sqrt{1-\zeta^2}} = 0.02$$

$$e^{-\zeta\omega_n t_{sb}} = 0.02\sqrt{1-\zeta^2}$$

$$-\zeta\omega_n t_{sb} = \ln\left(0.02\sqrt{1-\zeta^2}\right).$$

Finally, the upper bound t_{sb} for the settling time is given by

$$t_s \leq t_{sb} = -\frac{\ln\left(0.02\sqrt{1-\zeta^2}\right)}{\zeta\omega_n}. \tag{8.28}$$

However, note that the upper bound t_{sb} goes to ∞ as either $\zeta \to 1$ or as $\zeta \to 0$. Consequently, it is not a useful approximation method for these two limiting cases. This is a problem because for $0 < \zeta < 1$, we are interested in the case when ζ is close to 1 in order to keep the overshoot small. To deal with this, let's consider another approach for obtaining an upper bound on the settling time. As we are interested in the case when ζ is close to 1 we simply do the settling time calculation for $\zeta = 1$. We have

$$C(s) = \frac{\omega_n^2}{s^2 + 2\zeta\omega_n s + \omega_n^2}\frac{R_0}{s} = \frac{\omega_n^2}{(s+\omega_n)^2}\frac{R_0}{s} = \frac{1}{s}R_0 - \frac{R_0}{s+\omega_n} - \omega_n\frac{R_0}{(s+\omega_n)^2}.$$

Then $c(t)$ is given by
$$c(t) = R_0 - R_0(e^{-t\omega_n} + t\omega_n e^{-t\omega_n}).$$

When $t = t_s = \left.\dfrac{4}{\zeta\omega_n}\right|_{\zeta=1} = \dfrac{4}{\omega_n}$ we have

$$c(t_s) = e^{-t_s\omega_n} + t_s\omega_n e^{-t_s\omega_n} = e^{-4} + 4e^{-4} = 5e^{-4} = 0.092 \approx 0.1.$$

The output response $c(t)$ is within 10% of the final value R_0 at this settling time rather than 2%. However, if we use $t_s \triangleq \left.\dfrac{5}{\zeta\omega_n}\right|_{\zeta=1}$ then $c(t_s) = 0.04$ or 4% of the final value while $t_s \triangleq \left.\dfrac{6}{\zeta\omega_n}\right|_{\zeta=1}$ gives $c(t_s) = 0.017$ making it within 2% of the final value. We define the *time constant* of the complex-conjugate pair of closed-loop poles to be

$$\left|\dfrac{1}{\sigma}\right| = \left|\dfrac{1}{-\zeta\omega_n}\right| = \dfrac{1}{\zeta\omega_n}. \tag{8.29}$$

With ζ close to 1, taking $t_s \triangleq \dfrac{4}{\zeta\omega_n}, \dfrac{5}{\zeta\omega_n},$ or $\dfrac{6}{\zeta\omega_n}$ results in $c(t)$ being within 10%, 4%, or 2% respectively, of its final value.

Important Remark The further in the left half-plane the closed-loop poles are located, the faster the transients die out and thus the faster the system reaches its final value.

Rise Time t_r

As indicated in Figure 8.10, the rise time is defined as the amount of time for the response $c(t)$ to go from $0.1R_0$ to the *first* time it reaches R_0.[1] As a crude approximation, one sees for those step responses in Figure 8.9 with $\zeta < 0.5$, the step response $c(t)$ reaches $0.1R_0$ at $\omega_n t_1 \approx 0.2$ and reaches R_0 for the first time at $\omega_n t_2 \approx 2$. Then the rise time is given by

$$t_r = t_2 - t_1 \approx \dfrac{2 - 0.2}{\omega_n} = \dfrac{1.8}{\omega_n}. \tag{8.30}$$

Remark Although this gives us a formula to plug numbers into, it is not very useful because we typically want ζ to be close to 1 (so that the overshoot is not too large) while (8.30) is based on $\zeta < 0.5$.

Summary of M_p, t_p, t_s

With
$$C(s) = \dfrac{\omega_n^2}{s^2 + 2\zeta\omega_n s + \omega_n^2} \dfrac{R_0}{s}$$

the percent overshoot, the peak time, and the settling time of $c(t)$ are as follows.

$$M_p \triangleq \dfrac{c(t_p) - c(\infty)}{c(\infty)} = e^{-\pi\zeta/\sqrt{1-\zeta^2}} \quad \text{where} \quad t_p = \dfrac{\pi}{\omega_n\sqrt{1-\zeta^2}}. \tag{8.31}$$

---

[1] Remember that R_0 is the value of the step input and also equals the final value of $c(t)$.

$$t_s = \frac{4}{\zeta\omega_n}, \frac{5}{\zeta\omega_n}, \frac{6}{\zeta\omega_n} \quad \text{for } 10\%, 4\%, 2\% \text{ of the final value with } \zeta \text{ close to 1.} \quad (8.32)$$

$$t_r = \frac{1.8}{\omega_n} \quad \text{for } \zeta < 0.5. \quad (8.33)$$

The expressions for t_s assume ζ is close to 1. In general, we will be concerned about overshoot, settling time, and rise time. However, Eqs. (8.31)–(8.33) are only for second-order systems without zeros making their usefulness quite limited.

Choosing the Gain of a Proportional Controller

Let's go back to our proportional control system for the angular position of a DC motor with no load torque. This is shown in Figure 8.12. The closed-loop transfer function is given by

$$\frac{\theta(s)}{\theta_d(s)} = \frac{K\frac{b}{s(s+a)}}{1 + K\frac{b}{s(s+a)}} = \frac{Kb}{s^2 + as + Kb}.$$

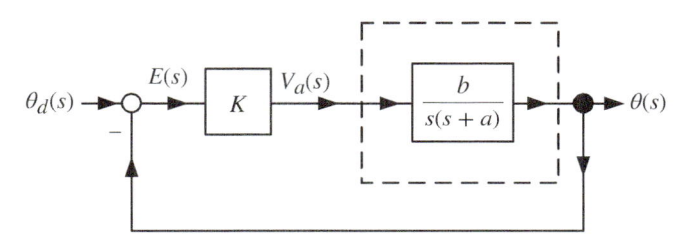

Figure 8.12. Simple proportional feedback control.

Suppose $b = 1$ and $a = 4$ so that for a step input of

$$\theta_d(s) = \frac{\theta_0}{s}$$

we have

$$\theta(s) = \frac{K}{s^2 + 4s + K}\frac{\theta_0}{s}. \quad (8.34)$$

With $K > 0$ the roots of $s^2 + 4s + K$ are in the open left half-plane so that

$$s\theta(s) = \frac{K}{s^2 + 4s + K}\theta_0$$

is stable. By the final value theorem we have

$$\theta(\infty) = \lim_{t \to \infty} \theta(t) = \lim_{s \to 0} s\theta(s) = \theta_0.$$

However, the behavior of the $\theta(t)$ as it goes to θ_0 is determined by the location of the closed-loop poles in the left half-plane. Let's look at these pole locations as the control gain K is varied from 0 to ∞. We set

$$s^2 + 4s + K = 0$$

and by the quadratic formula we have

$$s = \frac{-4 \pm \sqrt{16-4K}}{2} = -2 \pm \sqrt{4-K}.$$

The closed-loop poles are conveniently written as

$$p_i = \begin{cases} -2 \pm \sqrt{4-K}, & 0 \leq K \leq 4 \\ -2 \pm j\sqrt{K-4}, & 4 < K. \end{cases}$$

We can then easily construct the following table of values for the closed-loop poles.

K	Closed-Loop Poles
0	$s = 0, -4$
2	$s = -2 \pm \sqrt{2} = -3.414, -0.596$
4	$s = -2, -2$
8	$s = -2 \pm 2j$
13	$s = -2 \pm 3j$

From this table we obtain the sketch of the closed-loop poles shown in Figure 8.13, which is referred to as a *root locus*. See Problem 10 on how to use MATLAB to generate the plot of Figure 8.13. (The root locus is studied in detail in Chapter 12).

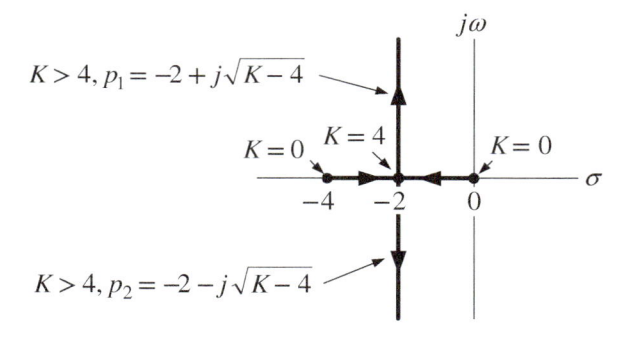

Figure 8.13. Sketch of closed-loop poles as K is varied from 0 to ∞.

Some Observations of the Root Locus

- Both poles are in the open left half-plane for $K > 0$.
- For $K > 4$ we have a pair of complex-conjugate closed-loop poles and thus the response will be oscillatory.
- We typically want the poles as far in the left half-plane as possible which means we would want to have $K \geq 4$.

8.3 Second-Order Systems with Zeros 217

- The value of K must be small enough so the response is not too oscillatory and to avoid saturating the amplifier.

Finally, suppose we want to choose K so that the damping ratio ζ is equal to 0.8. That is, from (8.34) we want to have

$$\theta(s) = \frac{K}{s^2 + 4s + K}\frac{\theta_0}{s} = \frac{\omega_n^2}{s^2 + 2\zeta\omega_n s + \omega_n^2}\frac{\theta_0}{s}$$

with K chosen so that $\zeta = 0.8$. Equating coefficients gives

$$4 = 2\zeta\omega_n = 2(0.8)\omega_n \text{ and } K = \omega_n^2.$$

Solving for K gives

$$K = \left(\frac{4}{2(0.8)}\right)^2 = 6.25.$$

We can also obtain this from the geometry of Figure 8.14 as

$$\tan(\phi) = \left.\frac{\omega_n\sqrt{1-\zeta^2}}{\zeta\omega_n}\right|_{\zeta=0.8} = 0.75 = \frac{\sqrt{K-4}}{2}$$

$$\implies K = (3/2)^2 + 4 = 6.25.$$

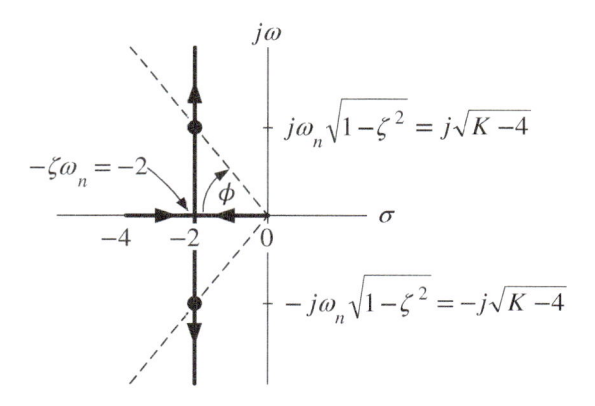

Figure 8.14. Choose K so that $\zeta = 0.8$.

8.3 Second-Order Systems with Zeros

Instead of position control, let's now consider *speed* control of a DC motor as illustrated in Figure 8.15 on the next page. Here ω_0 is the desired final speed and the angular speed $\omega(t)$ is found by (numerically) differentiating the position measurement obtained from the encoder. The voltage commanded to the amplifier is

$$v_a(t) = K\left(\omega_0 - \omega(t)\right) + Kz\int_0^t \left(\omega_0 - \omega(\tau)\right)d\tau.$$

8 System Responses

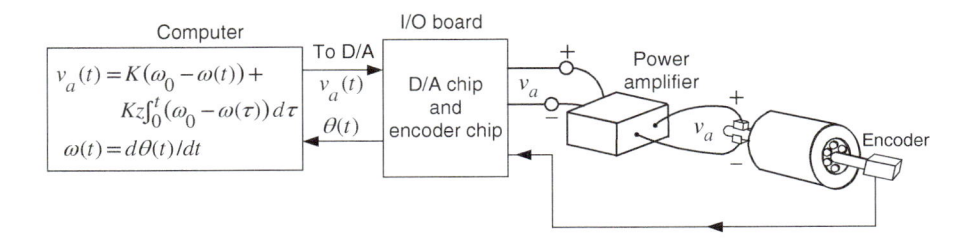

Figure 8.15. Speed control system for a DC motor.

In the Laplace domain the error $E(s)$ is

$$E(s) = \frac{\omega_0}{s} - \omega(s)$$

and voltage $V_a(s)$ is

$$V_a(s) = KE(s) + Kz\frac{1}{s}E(s) = K\frac{s+z}{s}E(s).$$

This setup is called a *proportional plus integral* (PI) controller. The block diagram model of this system is given in Figure 8.16.

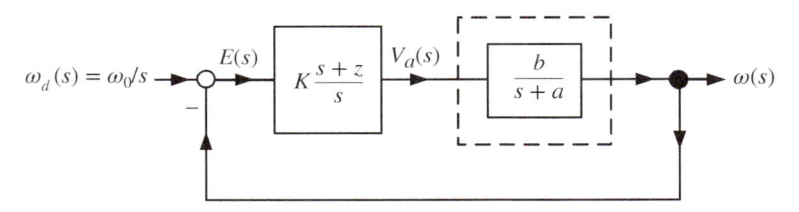

Figure 8.16. Block diagram of a PI speed control system for a DC motor.

The speed $\omega(s)$ is given by

$$\omega(s) = \frac{\frac{K(s+z)}{s}\frac{b}{s+a}}{1 + \frac{K(s+z)}{s}\frac{b}{s+a}}\frac{\omega_0}{s} = \frac{Kb(s+z)}{s^2 + as + Kb(s+z)}\frac{\omega_0}{s}$$

$$= \underbrace{\frac{Kb(s+z)}{s^2 + (a+Kb)s + Kbz}}_{\text{Closed-loop transfer function}}\underbrace{\frac{\omega_0}{s}}_{\text{Input}}. \quad (8.35)$$

The motor parameters are positive so $a > 0$ and $b > 0$. Taking $K > 0$ and $z > 0$ the roots of

$$s^2 + (a+Kb)s + Kbz = 0$$

are in the *open left half-plane*. Thus

$$s\omega(s) = \frac{Kb(s+z)}{s^2 + (a+Kb)s + Kbz}\omega_0$$

8.3 Second-Order Systems with Zeros

is stable and by the final value theorem we have

$$w(\infty) = \lim_{s \to 0} sw(s) = w_0.$$

This is the reason for choosing $G_c(s) = K\dfrac{s+z}{s}$. In Problem 12 you are asked to show that if $G_c(s) = K$, then $w(\infty) \neq w_0$!

Now set $C(s) \triangleq w(s)$ and define

$$w_n^2 \triangleq Kbz, \quad 2\zeta w_n \triangleq a + Kb, \quad \alpha \triangleq \dfrac{z}{\zeta w_n}.$$

Note that $Kb = w_n^2/z = w_n/(\alpha\zeta)$. We may now rewrite (8.35) as

$$C(s) = \underbrace{\frac{w_n}{\alpha\zeta} \frac{s + \alpha\zeta w_n}{s^2 + 2\zeta w_n s + w_n^2}}_{\text{Closed-loop transfer function}} \frac{w_0}{s}. \tag{8.36}$$

We would like to know the effect of the zero at $s = -z = -\alpha\zeta w_n$ on the transient response. Proceeding, the closed-loop transfer function

$$G_{cl}(s) \triangleq \frac{w_n}{\alpha\zeta} \frac{s + \alpha\zeta w_n}{s^2 + 2\zeta w_n s + w_n^2} \tag{8.37}$$

is second-order with a zero at $-\alpha\zeta w_n$. With $0 < \zeta < 1$ a plot of the zero and the two complex-conjugate poles of $G_{cl}(s)$ is given in Figure 8.17.

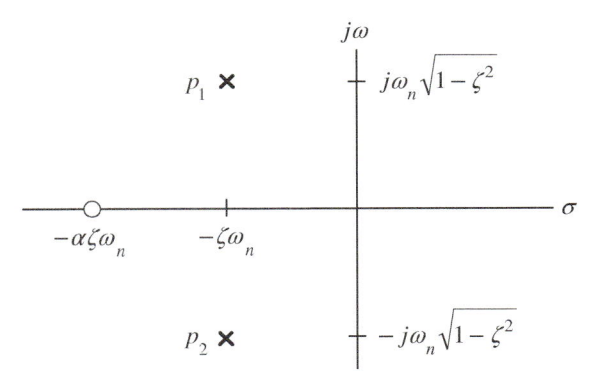

Figure 8.17. Pole–zero plot of $G_{cl}(s) = \dfrac{w_n}{\alpha\zeta} \dfrac{s + \alpha\zeta w_n}{s^2 + 2\zeta w_n s + w_n^2}$ with $\alpha > 0$.

With $w_0 = 1, \zeta = 0.6$ and $w_n = 1$, Figure 8.18 on the next page is a plot of the unit step response $c(t)$ for $\alpha = 1, 2, 4$, and 100. For $0 < \alpha \ll 100$ we see that the step response has much more overshoot compared with a second-order system *without* a zero. To see why there is more overshoot, we first decompose $C(s)$ as

$$C(s) = \frac{w_n^2}{s^2 + 2\zeta w_n s + w_n^2} \frac{1}{s} + \frac{1}{\alpha\zeta w_n} s \frac{w_n^2}{s^2 + 2\zeta w_n s + w_n^2} \frac{1}{s}. \tag{8.38}$$

Note that as $\alpha \to \infty$, this expression for $C(s)$ reduces to that of a second-order system without a zero considered previously. In particular, Figure 8.18 shows the plot for $\alpha = 100$ is essentially the response of a second-order system without a zero.

220   8 System Responses

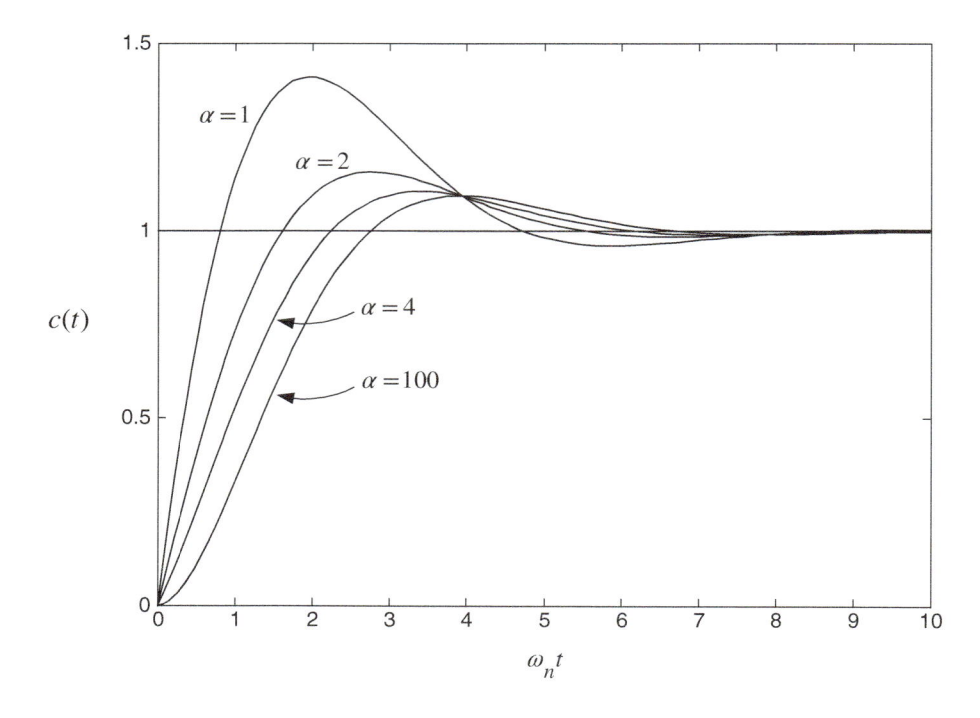

Figure 8.18. Unit step responses for $\alpha = 1, 2, 4, 100$ ($\zeta = 0.6, \omega_n = 1$).

Next let
$$C_1(s) \triangleq \frac{\omega_n^2}{s^2 + 2\zeta\omega_n s + \omega_n^2}\frac{1}{s}$$

so that
$$C(s) = C_1(s) + \frac{1}{\alpha\zeta\omega_n}sC_1(s).$$

In the time domain $c(t)$ may be written as

$$c(t) = c_1(t) + \frac{1}{\alpha\zeta\omega_n}\frac{dc_1(t)}{dt}. \tag{8.39}$$

Figure 8.19 is a plot of $c_1(t)$, its scaled derivative $\frac{1}{\alpha\zeta\omega_n}\frac{dc_1(t)}{dt}$, and their sum $c(t)$. We see that the derivative term $\frac{1}{\alpha\zeta\omega_n}\frac{dc_1(t)}{dt}$ is the source of the increased overshoot which results in larger overshoots as $\alpha > 0$ gets smaller. Here $\alpha > 0$ so that the zero $-\alpha\zeta\omega_n$ is in the open left half-plane (see Figure 8.17). A closed-loop system with *all* of its zeros in the open left half-plane is referred to as a *minimum phase* system.

8.3 Second-Order Systems with Zeros   221

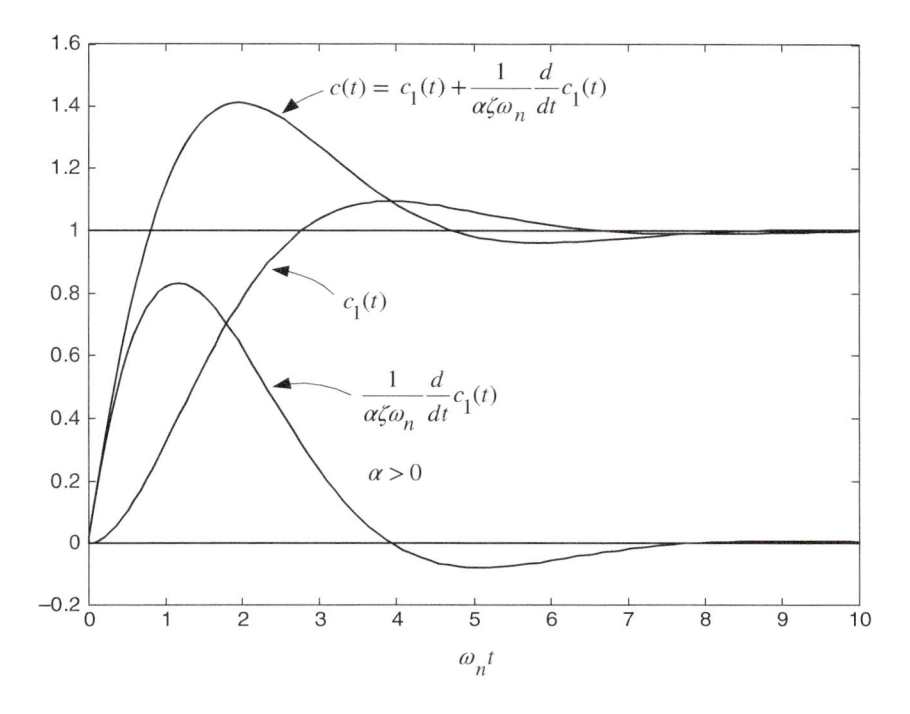

Figure 8.19. Unit step response of a second-order system with a zero in the open left half-plane.

Right Half-Plane Zero

If $\alpha < 0$ then the zero $-\alpha\zeta\omega_n$ is in the open right half-plane. A closed-loop system with *any* of its zeros in the open right half-plane is referred to as a *non-minimum phase* system. A pole–zero plot for this case is shown in Figure 8.20. Recall that zeros have nothing to do with stability, only the poles do. This system is still stable and $c(t) \to 1$ as $t \to \infty$.

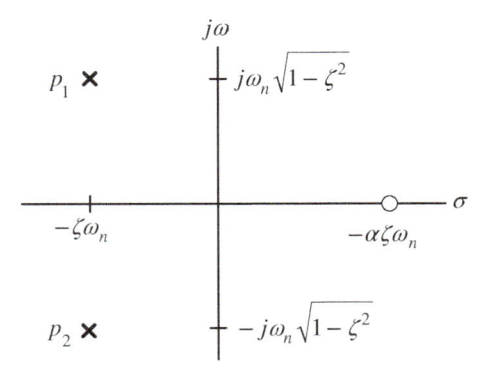

Figure 8.20. Pole–zero plot for $G(s) = \dfrac{\omega_n}{\alpha\zeta} \dfrac{s + \alpha\zeta\omega_n}{s^2 + 2\zeta\omega_n s + \omega_n^2}$ with $\alpha < 0$.

222   8   System Responses

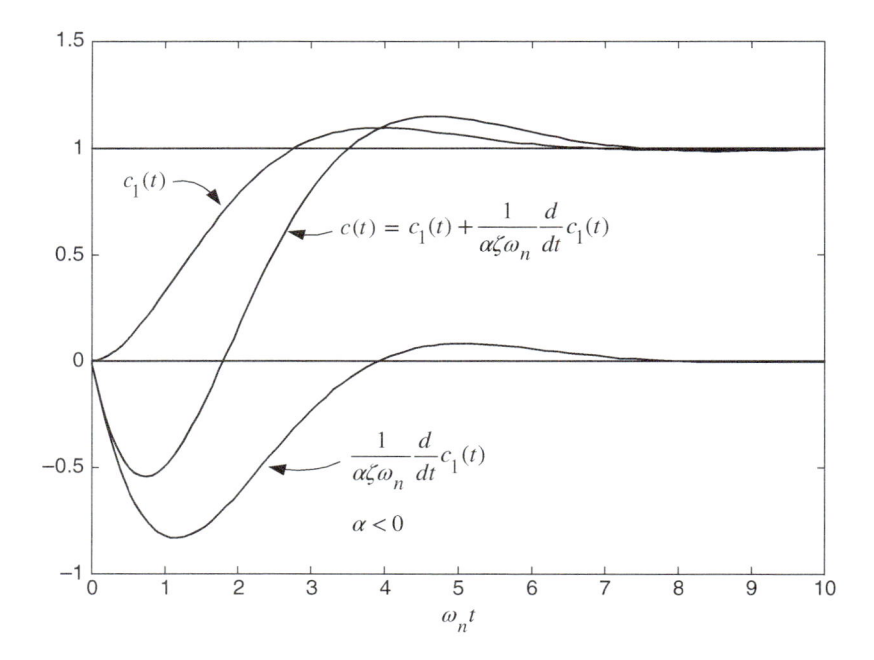

Figure 8.21. Unit step response of a second-order system with a zero in the open right half-plane ($\alpha < 0$).

Figure 8.21 shows the step response where we see that $c(t)$ *first* goes negative before coming back and going to its final value of 1. That is, the response goes in the opposite direction of that desired, but then comes back and settles out to the desired value. This type of response is referred to as *undershoot*.

Remark Any stable closed-loop transfer function with an odd number of *real right half-plane zeros* will have a step response with undershoot [26].

8.4   Third-Order Systems

Finally, let's consider a third-order system which has one real pole and a pair of complex-conjugate poles. Specifically, with $0 < \zeta < 1$ and $\alpha > 0$, consider the third-order closed-loop transfer function given by

$$G(s) = \frac{\alpha\zeta\omega_n}{s + \alpha\zeta\omega_n} \frac{\omega_n^2}{s^2 + 2\zeta\omega_n s + \omega_n^2}. \tag{8.40}$$

The poles of this system are then (see Figure 8.22)

$$p_1 = -\zeta\omega_n + j\omega_n\sqrt{1 - \zeta^2}, \ p_2 = -\zeta\omega_n - j\omega_n\sqrt{1 - \zeta^2}, \ p_3 = -\alpha\zeta\omega_n.$$

Let's look at its unit step response

$$C(s) = G(s)\frac{1}{s} = \frac{\alpha\zeta\omega_n}{s + \alpha\zeta\omega_n} \frac{\omega_n^2}{s^2 + 2\zeta\omega_n s + \omega_n^2} \frac{1}{s}.$$

8.4 Third-Order Systems

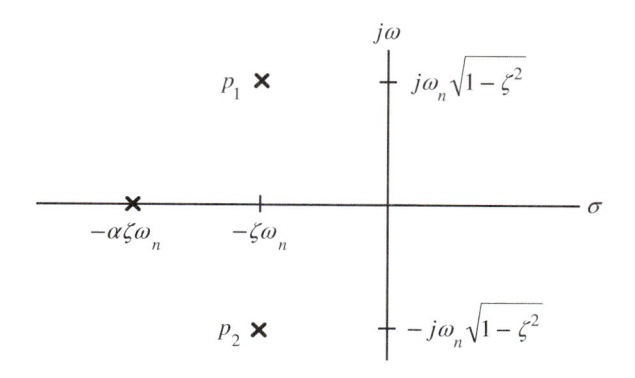

Figure 8.22. Pole–zero plot for $G(s) = \dfrac{\alpha \zeta \omega_n}{s + \alpha \zeta \omega_n} \dfrac{\omega_n^2}{s^2 + 2\zeta \omega_n s + \omega_n^2}$.

As $\zeta > 0, \omega_n > 0$, and $\alpha > 0$, it follows that

$$sC(s) = \frac{\alpha \zeta \omega_n}{s + \alpha \zeta \omega_n} \frac{\omega_n^2}{s^2 + 2\zeta \omega_n s + \omega_n^2}$$

is stable. By the final value theorem we have $c(\infty) = \lim_{t \to \infty} c(t) = \lim_{s \to 0} sC(s) = 1$.

Figure 8.23 shows the step responses for various values of the real pole location. Next we rewrite $C(s)$ as

$$C(s) = \frac{\alpha \zeta \omega_n}{s + \alpha \zeta \omega_n} \frac{\omega_n^2}{s^2 + 2\zeta \omega_n s + \omega_n^2} \frac{1}{s} = \frac{1}{\frac{s}{\alpha \zeta \omega_n} + 1} \frac{\omega_n^2}{s^2 + 2\zeta \omega_n s + \omega_n^2} \frac{1}{s}. \tag{8.41}$$

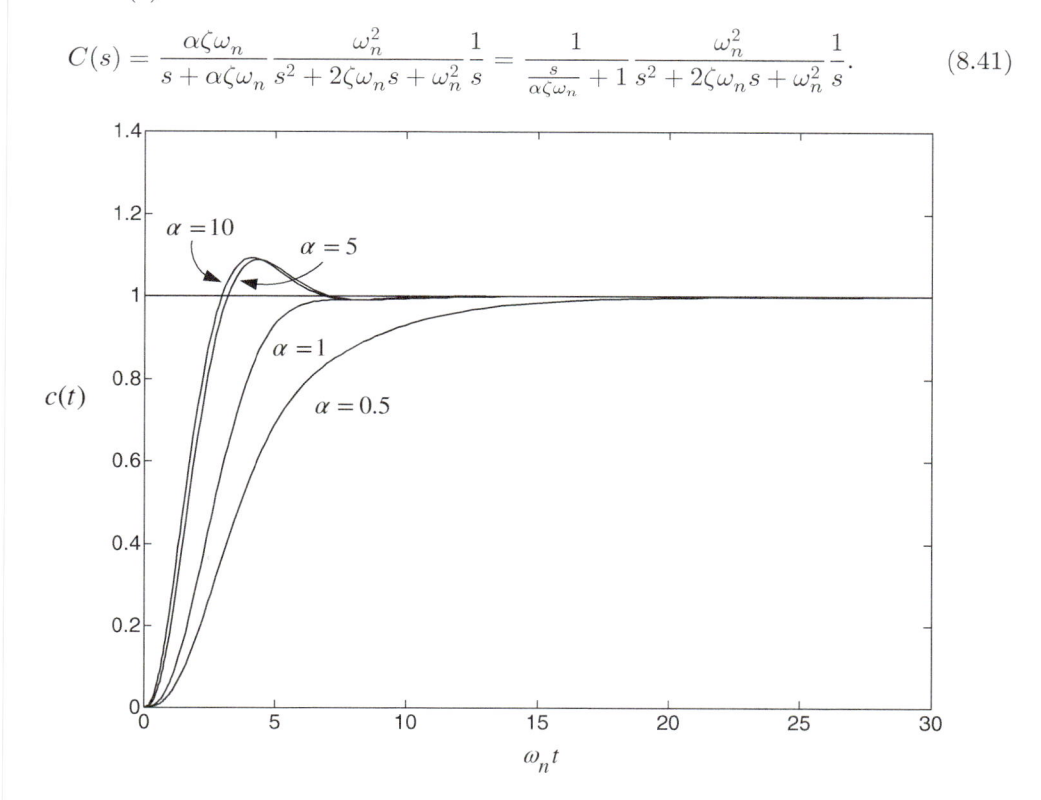

Figure 8.23. Unit step responses of a third-order system which has one real pole and a pair of complex conjugate poles for $\alpha = 0.5, 1, 5, 10$ and $\zeta = 0.6, \omega_n = 1$.

224   8 System Responses

As $\alpha \to \infty$ this expression for $C(s)$ reduces to the second-order system without a zero studied previously. For example, in Figure 8.23, it turns out that the plot for $\alpha = 10$ essentially corresponds to that of just a second-order system. Notice that as α decreases, the pole at $-\alpha\zeta\omega_n$ moves closer to the $j\omega$ axis and the response becomes more sluggish in that it takes longer to settle out to its final value.

Appendix: Root Locus Matlab File

```
% RootLocus for G(s) = K*b/(s*(s+a))
close all; clear; clc
a = 4; b = 1;
% Open loop transfer function G(s) = b/(s*(s+a)) = b/(s ^2+a*s)
den = [1 a 0]; num = [b]; tf_openloop = tf(num,den);
% Plot the CLPs for K going from 0 to 100 in steps of 1.
K = [0:1:100];
rlocus(tf_openloop,K)
% The input to the rlocus command is the OPEN LOOP transfer fn.
% It plots the poles of the CLOSED LOOP transfer fn.
% G_cl(s) = KG(s)/(1 + KG(s)) = Kb/(s^2 + as + Kb)
% Make the linewidth thicker, the marker size and font size bigger.
h = findobj(gca, 'Type', 'line');
set(h, 'LineWidth', 4); set(h, 'MarkerSize', 15); set(gca,'FontSize',20)
% Set range of x-axis [-5,0] and y-axis [-10,10]
v = [-5 0.5 -10 10]; axis(v);
title('G(s)= b/(s^2+as)','FontSize',20)
xlabel('Re(s)','FontSize',20); ylabel ('Im(s)','FontSize',20)
K = 6.25
% The next command gives the CLPs for this value of K
p = rlocus(num,den,K)
```

Problems

Problem 1 *Step Response for $\zeta = 1$*

(a) With $\zeta = 1$, compute the inverse Laplace transform of

$$C(s) = \frac{\omega_n^2}{s^2 + 2\zeta\omega_n s + \omega_n^2}\frac{R_0}{s} = \frac{\omega_n^2}{(s+\omega_n)^2}\frac{R_0}{s}.$$

(b) Show that $dc/dt > 0$ for all $t > 0$.

Problem 2 *Step Response for $\zeta > 1$*

(a) With $\zeta > 1$, show that the inverse Laplace transform of

$$C(s) = \frac{\omega_n^2}{s^2 + 2\zeta\omega_n s + \omega_n^2}\frac{R_0}{s}$$

$$= \frac{\omega_n^2}{\left(s - (-\zeta\omega_n + \omega_n\sqrt{\zeta^2 - 1})\right)\left(s - (-\zeta\omega_n - \omega_n\sqrt{\zeta^2 - 1})\right)}\frac{R_0}{s}$$

is
$$c(t) = R_0 u_s(t) + \frac{\omega_n^2}{p_1 - p_2}\frac{R_0}{p_1}e^{p_1 t} + \frac{\omega_n^2}{p_2 - p_1}\frac{R_0}{p_2}e^{p_2 t},$$

where
$$p_1 = -\zeta\omega_n + \omega_n\sqrt{\zeta^2 - 1}$$
$$p_2 = -\zeta\omega_n - \omega_n\sqrt{\zeta^2 - 1}.$$

(b) Show that $dc/dt > 0$ for all $t > 0$.

Problem 3 *Step Response for $0 < \zeta < 1$*
With $0 < \zeta < 1$, recall the partial fraction expansion given in (8.13), i.e.,

$$C(s) = \frac{\omega_n^2}{s^2 + 2\zeta\omega_n s + \omega_n^2}\frac{R_0}{s}$$
$$= \frac{R_0}{s} + \frac{\beta}{s - \left(-\zeta\omega_n + j\omega_n\sqrt{1-\zeta^2}\right)} + \frac{\beta^*}{s - \left(-\zeta\omega_n - j\omega_n\sqrt{1-\zeta^2}\right)}.$$

Compute β and β^* and then simplify to show that

$$c(t) = R_0 u_s(t) + \beta e^{-\zeta\omega_n t}e^{+j\omega_n\sqrt{1-\zeta^2}t}u_s(t) + \beta^* e^{-\zeta\omega_n t}e^{-j\omega_n\sqrt{1-\zeta^2}t}u_s(t)$$

becomes
$$c(t) = R_0 u_s(t) - R_0\frac{e^{-\zeta\omega_n t}}{\sqrt{1-\zeta^2}}\sin\left(\omega_n\sqrt{1-\zeta^2}t + \phi\right)u_s(t)$$

with
$$\phi = \tan^{-1}\left(\sqrt{1-\zeta^2}/\zeta\right).$$

Problem 4 *Second-Order Under Damped Systems*
Let
$$G(s) = \frac{\omega_n^2}{s^2 + 2\zeta\omega_n s + \omega_n^2}$$

and
$$C(s) = \frac{\omega_n^2}{s^2 + 2\zeta\omega_n s + \omega_n^2}\frac{1}{s}.$$

It is given that $\omega_n > 0$. Give the answer along with a brief explanation to the following:

(a) For what value(s) of ζ is $G(s)$ stable?

(b) For what value(s) of ζ does $c(t)$ have a peak overshoot?

(c) For what value(s) of ζ does $c(t)$ *not* have a peak overshoot?

8 System Responses

(d) For what value(s) of ζ are both poles furthest in the left half-plane?

(e) For what value(s) of ζ are the poles at $\pm j\omega_n$?

Problem 5 *S-Plane Regions*

The standard second-order transfer function is given by

$$G(s) = \frac{\omega_n^2}{s^2 + 2\zeta\omega_n s + \omega_n^2}.$$

(a) Sketch the region in the *s-plane* where $0.6 \leq \zeta \leq 0.8$ and $\omega_n \geq 0.5$.

For parts (b) and (c) assume that a unit step input is applied to the above transfer function to produce a unit step output response.

(b) What are the values of ζ and ω_n in the region specified in part (a) that result in the *minimum* fractional overshoot M_p in the output response? What is the value of the minimum M_p?

(c) What are the values of ζ and ω_n in the region specified in part (a) that have the *longest* settling time t_s for the unit step output response? What is the value of this settling time?

Problem 6 *Model Parameters from the Transient Response*

The step input $r(t) = R_0 u_s(t)$ ($R(s) = R_0/s$) is applied to a system whose block diagram model is given in Figure 8.24. The corresponding step response measurement is shown in Figure 8.25. In this step response, the peak time is $t_p = \frac{\pi}{1.6} \approx 1.96$ with peak value $c(t_p) = 2.2$. The open-loop transfer function $G(s)$ is

$$G(s) = \frac{b}{s(s+a)}$$

where $a > 0$ and $b > 0$ to be determined. The value of $K = 2$ was used to obtain the step response shown in Figure 8.25.

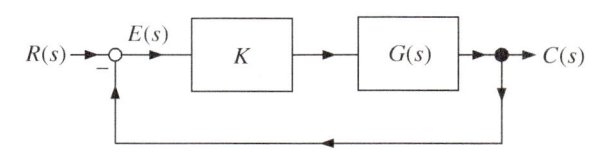

Figure 8.24. Step response of a second-order system.

(a) What are the values of R_0 and M_p?

(b) Compute the output $C(s)$ in terms of the closed-loop transfer function and the reference input $R(s) = R_0/s$.

(c) What are the corresponding values of ζ and ω_n?

(d) What are the values of a and b? (Answer: $a = 2.4, b = 2$.)

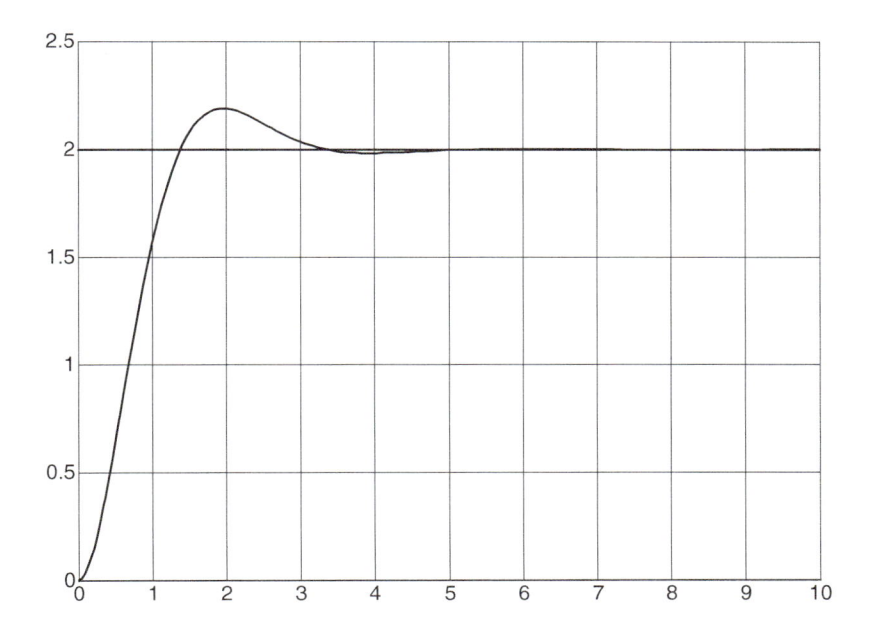

Figure 8.25. $c(t_p) = 2.2$ and $t_p = \pi/1.6$.

Problem 7 *Identification of the Motor Model from the Speed Response*

Problem 9 of Chapter 7 compared two different ways to simulate a DC motor. In this problem we use the transfer function model $G(s) = \dfrac{b}{s(s+a)}$ where $a \triangleq \dfrac{f + K_b K_T/R}{J}$, $b \triangleq \dfrac{K_T}{RJ}$. As in Problem 9 of Chapter 7, set $V_{\max} = 40$ V, $I_{\max} = 5$ A, $K_b = K_T = 0.07$ V/rad/s (N-m/A), $J = 6 \times 10^{-5}$ kg-m^2, $R = 2$ Ω, and $f = 0.0004$ N-m/rad/s. Figure 8.26 is a SIMULINK block diagram to simulate how the direct identification of the parameters a and b in the transfer function model $G(s) = \dfrac{b}{s(s+a)}$ can be carried out. The idea is to put a step input voltage and measure the angular position response of the motor using an optical encoder. The angular position is then differentiated to compute the angular speed ω. From knowing the input and measuring/calculating the output speed, the parameters a and b of the transfer function model can be computed as shown in the first section of this chapter. (See Chapter 6 and especially Problem 10 of that chapter for details on modeling an optical encoder.)

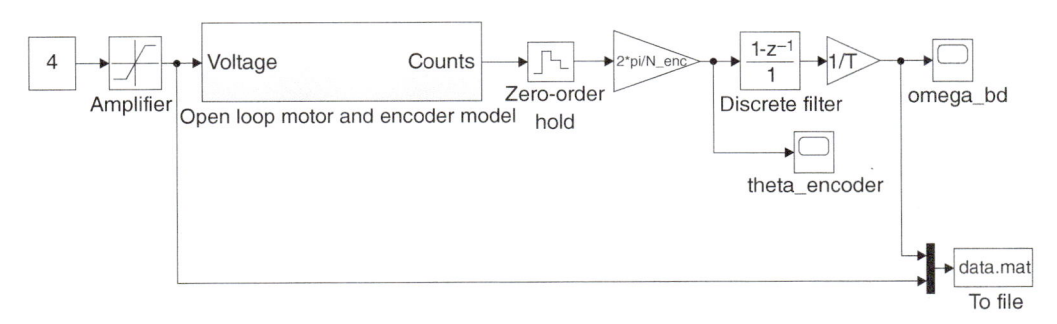

Figure 8.26. SIMULINK simulation using the output of an optical encoder to calculate speed.

228   8 System Responses

(a) Build the SIMULINK simulation illustrated in Figures 8.26 and 8.27. The inside of the *subsystem* block entitled Open Loop Motor & Encoder Model of Figure 8.26 is the SIMULINK block diagram given in Figure 8.27.

To start the build of SIMULINK system, first draw the four blocks (Transfer Fcn, Integrator, Gain, and Rounding Function) of Figure 8.27 *without* the (oval-shaped) input port and the (oval-shaped) output port. Then select these four blocks, right-click and select Create Subsystem from Selection. This action will automatically create the input and output ports. Then do the rest of the diagram as shown in Figure 8.26. The code needed to run this SIMULINK simulation is

```
clear;clc;close all
N_enc = 4*1024;resolution = 360/N_enc;Vmax = 40;Imax = 5;
% step size of the simulation
T = 0.001;
% motor Parameters
R = 2;KT = 0.07;Kb = KT;L = 0.002;f = 0.0004;J = 6e-5;
speed_error = 2*pi/N_enc/T;
a = (f + Kb*KT/R)/J;b = KT/(R*J);
```

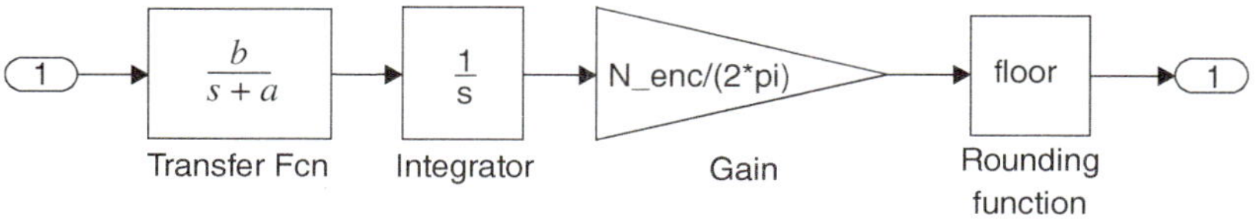

$$\boxed{1} \longrightarrow \boxed{\frac{b}{s+a}} \longrightarrow \boxed{\frac{1}{s}} \longrightarrow \boxed{\text{N\_enc}/(2\text{*pi})} \longrightarrow \boxed{\text{floor}} \longrightarrow \boxed{1}$$

Transfer Fcn   Integrator   Gain   Rounding function

Figure 8.27. SIMULINK setup for a DC motor and optical encoder.

(b) Put a step input voltage $V(t) = 4$ V and run the simulation for 0.2 seconds. Then open up the omega_bd scope and find the final speed ω_0 of the motor and the time t_m that $\omega(t_m) = (1 - e^{-1})\omega_0 = 0.632\omega_0$. Use Eqs. (8.5) and (8.6) to identify K_m, T_m in the transfer function model $G(s) = \dfrac{K_m}{s(T_m s + 1)}$. Then $a = 1/T_m$ and $b = K_m/T_m$ in the transfer function model $G(s) = \dfrac{b}{s(s+a)}$.

(c) Let the parameter values computed in part (b) be denoted as a_{est}, b_{est} where "est" is short for "estimated". Compare these with the values used in the simulation of part (a) by computing the normalized errors $e_a = \dfrac{a - a_{est}}{a}$ and $e_b = \dfrac{b - b_{est}}{b}$.

Problem 8 *Proportional Feedback*

Consider the proportional feedback system of Figure 8.12 with $b = 1$ and $a = 4$. It was shown that a gain of $K = 6.25$ resulted in damping ratio of $\zeta = 0.8$.

(a) What is the value of ω_n for $K = 6.25$?

(b) Simulate this system and plot $c(t)$ for an input of $r(t) = 2u_s(t)$. From the plot, determine the value of M_p and t_p. From these measured values, compute ζ and ω_n. Is this value of ζ equal to 0.8? Does the value of ω_n agree with the value computed in part (a)? (Answer: Yes and yes, but show your work!)

Problem 9 *Proportional Feedback*

Consider a satellite tracking antenna whose equation of motion is given by

$$J\frac{d^2\theta}{dt^2} + f\frac{d\theta}{dt} = T_c,$$

where θ is the angle of elevation the antenna is pointing with respect to the ground, J is the moment of inertia of the antenna, f is the viscous friction of the bearings supporting the antenna, and T_c the torque produced by the motor that rotates the antenna. The open-loop transfer function from $T_c(s)$ to $\theta(s)$ is

$$\theta(s) = \frac{1}{s(Js+f)} T_c(s).$$

Feedback of the elevation angle is needed to keep the antenna pointing in the correct direction. Consider the proportional feedback given by

$$T_c(s) = K(\theta_r(s) - \theta(s)),$$

where the θ_r is the reference (desired) angle of elevation the antenna is to point at. The block diagram for this feedback system is shown in Figure 8.28

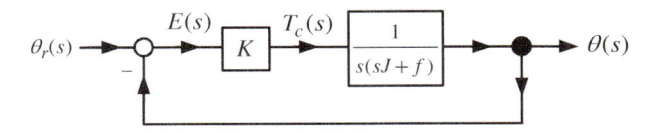

Figure 8.28. Block diagram of an antenna pointing system.

(a) Give the closed-loop transfer function from $\theta_r(s)$ to $\theta(s)$.

(b) With $f = 2$ and $J = 1$, sketch in the s-plane the possible root locations as K varies from 0 to infinity.

(c) Let

$$\theta_r(t) = \theta_0 u_s(t)$$

be a step input reference. For what value(s) of K will the final value of θ equal θ_0? Explain briefly.

(d) For what value(s) of K will the response $\theta(t)$ oscillate before reaching its final value?

(e) For what value of K will the damping ratio ζ equal 0.6 ?

Problem 10 *Root Locus Plot*

Run the MATLAB file Chapter8_RootLocus.m given in the appendix of this chapter to make the plot of Figure 8.13.

Problem 11 *Effects of Right Half-Plane Zeros*

Consider the closed-loop system given in Figure 8.29.

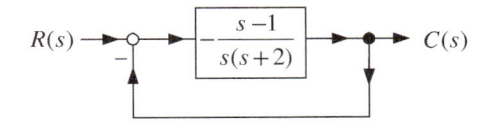

Figure 8.29. System with a right half-plane zero.

(a) Compute the closed-loop transfer function $C(s)/R(s)$ and show that it has the form of (8.36) (page 219) with $\alpha = -2$, $\zeta = 1/2$, and $\omega_n = 1$.

230   8 System Responses

(b) Let $R(s) = 1/s$ and show that $C(s)$ can be written in the form of (8.38).

(c) Figure 8.30 is a plot of $c_1(t) = \mathcal{L}^{-1}\left\{\dfrac{1}{s^2+s+1}\dfrac{1}{s}\right\}$. On this plot, sketch both $c_2(t) \triangleq \mathcal{L}^{-1}\left\{-s\left(\dfrac{1}{s^2+s+1}\dfrac{1}{s}\right)\right\}$ and the complete response $c(t) = c_1(t) + c_2(t)$.

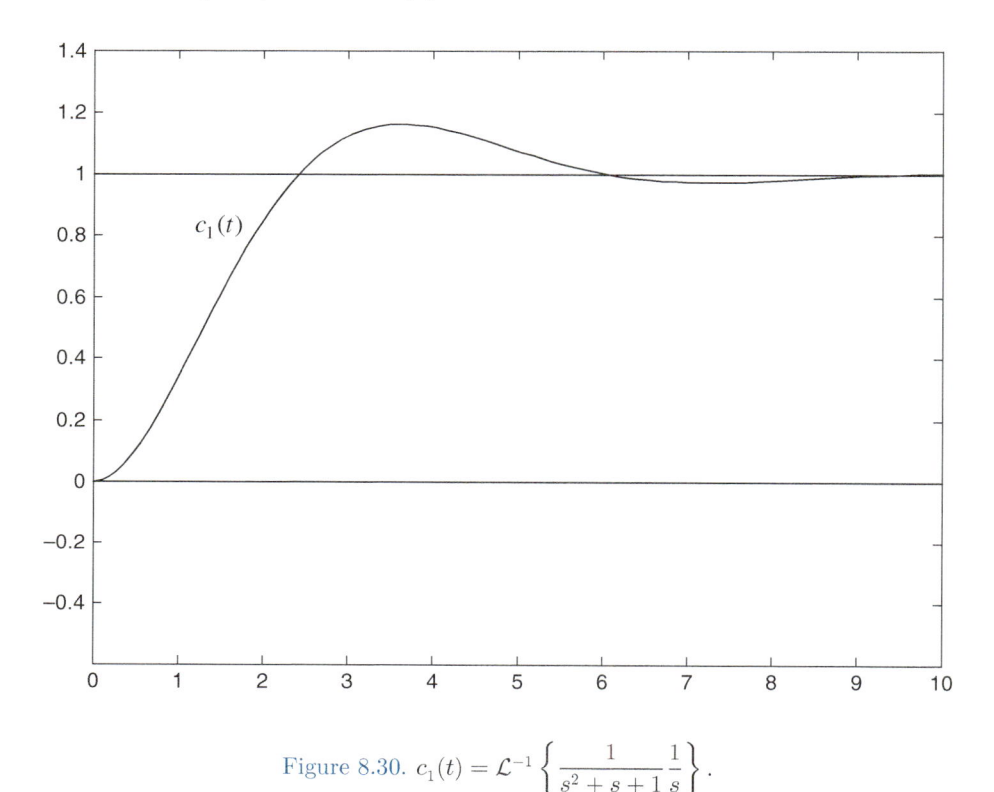

Figure 8.30. $c_1(t) = \mathcal{L}^{-1}\left\{\dfrac{1}{s^2+s+1}\dfrac{1}{s}\right\}$.

Problem 12 *Proportional Speed Control of a DC Motor*

Consider the speed control of a DC motor using just a proportional controller as indicated in Figure 8.31.

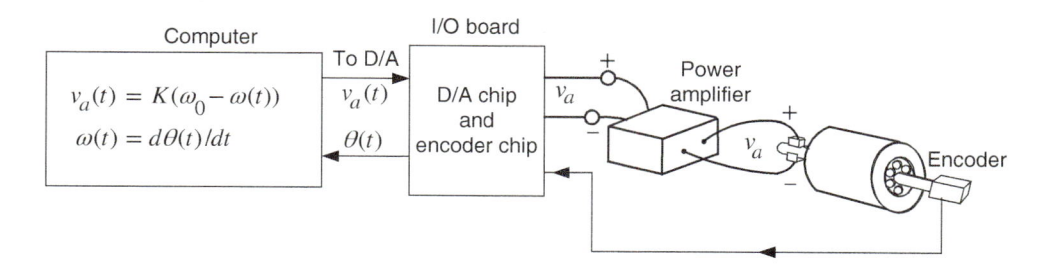

Figure 8.31. Proportional speed control of a DC motor.

The corresponding block diagram is given in Figure 8.32.

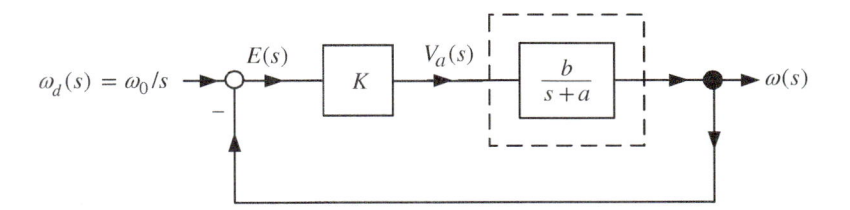

Figure 8.32. Block diagram for proportional speed controller.

(a) Compute $w(s)$.

(b) With $a > 0$ and $b > 0$ show that $sw(s)$ is stable for $K > 0$.

(c) For $K > 0$ show that $w(\infty) \neq w_0$!

(d) With $w_0 = 0.5, \tau_L = 0, a = 11$, and $b = 2.5$ simulate this system in SIMULINK. For $K = 2$ and $K = 50$ plot $w(t)$ and $w_d(t) = w_0 u_s(t)$ on the same graph. For each value of K how close is $w(\infty)$ to w_0? For each value of K what is the value of the closed-loop pole? Suppose $V_{\max} = 5.0$ V, i.e., the input voltage is limited to $-5 \leq v_a(t) \leq 5$. What is the largest value of K one can choose before the saturation limit is reached?

Problem 13 *Disturbance Rejection Using a PI Speed Controller*
Consider again the proportional plus integral (PI) speed controller for the DC motor shown in Figure 8.33. However, now the load torque τ_L is not zero.

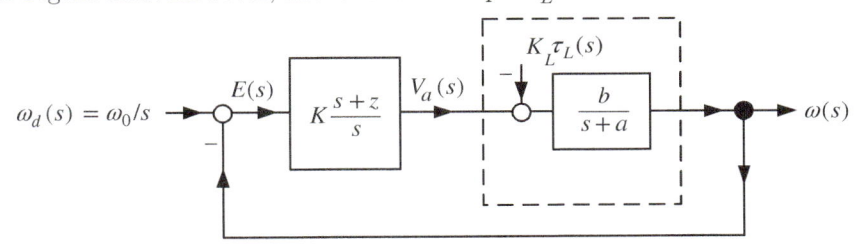

Figure 8.33. Block diagram for PI speed control of a DC motor.

(a) With $w_d(s) = w_0/s$ and $\tau_L(s) = \tau_{L0}/s$, use block diagram reduction to compute $w(s)$.

(b) With $a > 0$ and $b > 0$, show that $sw(s)$ is stable for $K > 0$ and $z > 0$.

(c) Show that $w(\infty) = w_0$ for any value of τ_{L0}. That is, the PI controller eliminates the effect of the load torque on the final speed!

(d) Set $w_0 = 0.5, a = 11, b = 2.5, K_L = 91, \tau_{L0} = 0.014, K = 3.6, z = 22$ and simulate this system. Plot $w(t)$ and $w_d(t) = w_0 u_s(t)$ on the same graph. Does $w(\infty) = w_0$? What are the values of the closed-loop poles? You should find that $w(s) = \dfrac{9(s+22)}{s^2 + 20s + 198} \dfrac{w_0}{s}$ and this fits the form of (8.36) with $w_n = \sqrt{198} = 14.7, \zeta = 0.71$, and $\alpha = 2.2$. Thus significant overshoot is expected.

(e) Redo part (d) with $z = 11$. You should find that
$$w(s) = \frac{9(s+11)}{s^2 + 20s + 99} \frac{w_0}{s} = \frac{9(s+11)}{(s+11)(s+9)} \frac{w_0}{s} = \frac{9}{s+9} \frac{w_0}{s}.$$
Thus no overshoot is expected.

Problem 14 Right Half-Plane and Left Half-Plane Zeros

In this problem we again consider *speed* control of the DC motor. The transfer function is $\frac{\omega(s)}{V_a(s)} = \frac{b}{s+a}$ where $a \triangleq \frac{f+K_bK_T/R}{J}$, $b \triangleq \frac{K_T}{RJ}$ with $K_b = K_T = 0.07$ V/rad/s ($=$ N-m/A), $J = 6 \times 10^{-5}$ kg-m^2, $R = 2$ Ω, $f = 0.0004$ N-m/rad/s, $K_L = R/K_T$, $V_{\max} = 40$ V, and $I_{\max} = 5$ A. Figure 8.34 is a SIMULINK block diagram of the closed-loop system. With these parameter values, it turns out that $a = 47.5$, $b = 583.3$, and $K_L = 28.6$. Take $\tau_L = 0$.

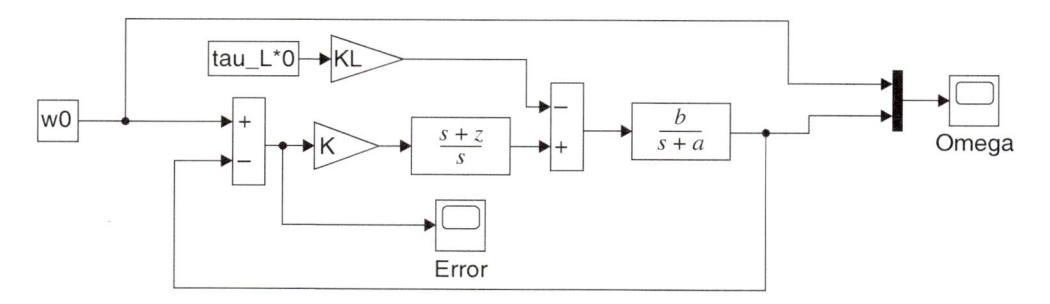

Figure 8.34. Values K and z can result in left-half or right-half plane zeros.

The closed-loop transfer function is then

$$\omega(s) = \frac{K \frac{s+z}{s} \frac{b}{s+a}}{1 + K \frac{s+z}{s} \frac{b}{s+a}} \frac{\omega_0}{s} = \underbrace{\frac{Kb(s+z)}{s^2 + (Kb+a)s + Kbz}}_{G_{CL}(s)} \frac{\omega_0}{s}.$$

The point of this problem is observe the effect that the location of the zero at $-z$ has on the step response $\omega(t)$. Set K and z so that

$$s^2 + (Kb+a)s + Kbz = (s+r_1)(s+r_2) = s^2 + (r_1+r_2)s + r_1r_2.$$

That is,

$$K = \frac{r_1 + r_2 - a}{b}$$

$$z = \frac{r_1 r_2}{Kb} = \frac{r_1 r_2}{r_1 + r_2 - a}.$$

(a) Simulate this system for $r_1 = r_2 = 20$ so the poles of $G_{CL}(s)$ are both -20. You should find that $K = -0.0129$ and $z = -53.3$. Note the undershoot.

(b) Simulate this system for $r_1 = r_2 = 50$ so the poles of $G_{CL}(s)$ are both -50. You should find that $K = 0.09$ and $z = 47.62$. Is there overshoot?

(c) Simulate this system for $r_1 = r_2 = 100$ so the poles of $G_{CL}(s)$ are both -100. You should find out that $K = 0.261$ and $z = 65.6$. Is there overshoot?

9

Tracking and Disturbance Rejection

9.1 Servomechanism

Before developing tracking and disturbance rejection feedback controllers, we first review how a physical system described by a differential equation model is abstracted to a block diagram. In this and following chapters, feedback controllers will be developed using such block diagram models so it is important to keep in mind where they come from and what they represent. Specifically, consider a *servomechanism* (servo system) used for positioning applications such as robot arms and machine tools. The physical hardware consists of a servomotor (DC motor in our case), power amplifier, and a position sensor (encoder) as shown in Figure 9.1. A schematic diagram of a complete servomechanism is given in Figure 9.2 including a simple proportional feedback controller.

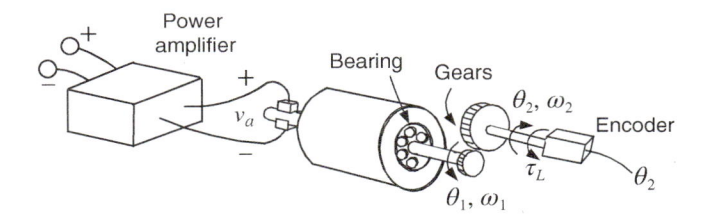

Figure 9.1. DC motor, power amplifier, gears. and encoder of a servo system.

Note that the encoder (position sensor) is on the output shaft rather than the motor shaft. This simple controller does the following: The position of the output shaft θ ($\triangleq \theta_2$) is obtained from the encoder[1] and is subtracted from the desired (reference) position θ_{ref}. This difference $e(t) = \theta_{ref} - \theta$ is called the error. The error is then multiplied by a gain and output as the commanded voltage to the amplifier using a Digital to Analog (D/A) converter. The whole point of control theory is to give a procedure for computing this output voltage based on the error signal. The procedure just described is called *proportional* control.

---

[1] An encoder puts out a pulse every time the rotor rotates a certain distance. For example, an encoder with $N_{enc} = 2000$ pulses/rev puts out a pulse everytime the motor shaft turns $2\pi/2000$ rad. These pulses are counted by the encoder and this (integer) value is sent to the computer controller. Multiplying this integer number of pulses by $K_0 = 2\pi/2000$ gives the position of the motor in radians. See Chapter 6.

An Introduction to System Modeling and Control, First Edition. John Chiasson.
© 2022 John Wiley & Sons, Inc. Published 2022 by John Wiley & Sons, Inc.
Companion website: www.wiley.com/go/chiasson/anintroductiontosystemmodelingandcontrol

234   9 Tracking and Disturbance Rejection

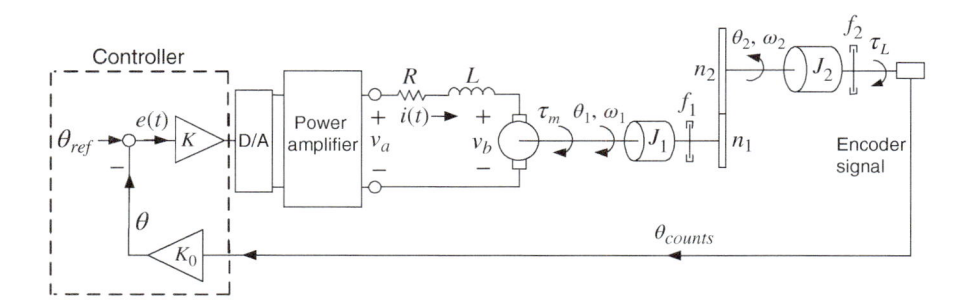

Figure 9.2. Schematic diagram of a servomechanism.

To determine the value of K to be used in this controller, we must first develop a model of the system that captures its dynamic behavior. Using the schematic of the servomechanism given in Figure 9.2, a model is now developed. In the schematic, we have

K_T is the motor torque constant ($\tau_m = K_T i$).

$K_b = K_T$ is the back-emf constant ($v_b = K_b \omega_m$).

K is the amplifier gain.

$K_0 = 2\pi/N_{enc}$ where N_{enc} is the number of encoder pulses per revolution ($\theta = K_0 \theta_{counts}$).

J_1 is the moment of inertia of the motor shaft.

J_2 is the moment of inertia of the output shaft.

f_1 is the viscous-friction coefficient of the motor shaft.

f_2 is the viscous-friction coefficient of the output shaft.

n_1 is the number of gear teeth on the motor shaft.

n_2 is the number of gear teeth on output shaft.

$n = n_2/n_1$ is the gear ratio.

$J = J_1 + n^2 J_2$ is total inertia reflected to motor shaft.

$f = f_1 + n^2 f_2$ is the total viscous friction coefficient reflected to the motor shaft.

To proceed with the mathematical model of this system, recall from Chapter 5 the equations of motion for the two gears given by

$$\tau_m - \tau_1 - f_1 \omega_1 = J_1 \frac{d\omega_1}{dt}$$
$$\tau_2 - \tau_L - f_2 \omega_2 = J_2 \frac{d\omega_2}{dt}. \tag{9.1}$$

τ_1 is the torque exerted on gear 1 by gear 2, while τ_2 is the torque exerted on gear 2 by gear 1. The torques between the two gears are related by

$$\frac{\tau_2}{\tau_1} = \frac{r_2}{r_1} = \frac{n_2}{n_1}. \tag{9.2}$$

Further, as $r_1\theta_1 = r_2\theta_2$ and thus $r_1\omega_1 = r_2\omega_2$, we also have

$$\frac{\theta_2}{\theta_1} = \frac{\omega_2}{\omega_1} = \frac{r_1}{r_2} = \frac{n_1}{n_2}. \tag{9.3}$$

In particular, the angular velocity of the motor shaft ω_1 is related to the output angular velocity ω_2 by

$$\omega_1 = \frac{n_2}{n_1}\omega_2. \tag{9.4}$$

Referring everything to the input (motor) shaft, we have

$$\begin{aligned}
\tau_1 &= \frac{n_1}{n_2}\tau_2 \\
&= \frac{n_1}{n_2}\left(\tau_L + f_2\omega_2 + J_2\frac{d\omega_2}{dt}\right) \\
&= \frac{n_1}{n_2}\left(\tau_L + f_2\left(\frac{n_1}{n_2}\omega_1\right) + J_2\frac{d\left(\frac{n_1}{n_2}\omega_1\right)}{dt}\right) \\
&= \frac{n_1}{n_2}\tau_L + \left(\frac{n_1}{n_2}\right)^2 f_2\omega_1 + \left(\frac{n_1}{n_2}\right)^2 J_2\frac{d\omega_1}{dt}.
\end{aligned} \tag{9.5}$$

Substituting this expression for τ_1 into the first equation of (9.1) results in

$$\tau_m = \underbrace{\left(\left(\frac{n_1}{n_2}\right)^2 J_2 + J_1\right)}_{J}\frac{d\omega_1}{dt} + \underbrace{\left(\left(\frac{n_1}{n_2}\right)^2 f_2 + f_1\right)}_{f}\omega_1 + \frac{n_1}{n_2}\tau_L. \tag{9.6}$$

With $J \triangleq J_1 + (n_1/n_2)^2 J_2$ the total inertia reflected to the input shaft and $f \triangleq f_1 + (n_1/n_2)^2 f_2$, the total viscous friction coefficient reflected to the input shaft, this can be written compactly as

$$\tau_m = J\frac{d\omega_m}{dt} + f\omega_m + \frac{n_1}{n_2}\tau_L, \tag{9.7}$$

where $\omega_m \triangleq \omega_1, \theta_m \triangleq \theta_1$ are now used. The motor electrical equations are given by

$$L\frac{di(t)}{dt} = -Ri(t) - v_b(t) + v_a(t)$$

$$v_b(t) = K_b\omega_m(t)$$

$$\tau_m(t) = K_T i(t).$$

This servo system is then described by

$$L\frac{di(t)}{dt} = -Ri(t) - K_b\omega_m(t) + v_a(t)$$

$$J\frac{d\omega_m(t)}{dt} = K_T i(t) - f\omega_m(t) - \frac{n_1}{n_2}\tau_L(t)$$

9 Tracking and Disturbance Rejection

$$\frac{d\theta_m(t)}{dt} = \omega_m(t)$$

$$\theta(t) \triangleq \theta_2(t) = \frac{n_1}{n_2}\theta_m(t)$$

$$v_a(t) = K\left(\theta_{ref}(t) - \theta(t)\right).$$

It is easier to work with these equations in the s domain. Taking the Laplace transform of these equations and using the more generic notation $r \triangleq \theta_{ref}, c \triangleq \theta$ gives

$$(sL + R)I(s) = V_a(s) - K_b\omega_m(s)$$

$$(Js + f)\omega_m(s) = K_T I(s) - \frac{n_1}{n_2}\tau_L(s)$$

$$s\theta_m(s) = \omega_m(s) \quad (9.8)$$

$$C(s) \triangleq \theta_2(s) = \frac{n_1}{n_2}\theta_m(s)$$

$$V_a(s) = K\left(R(s) - C(s)\right).$$

The block diagram shown in Figure 9.3 is simply a pictorial way to represent the algebraic relationships given in (9.8).

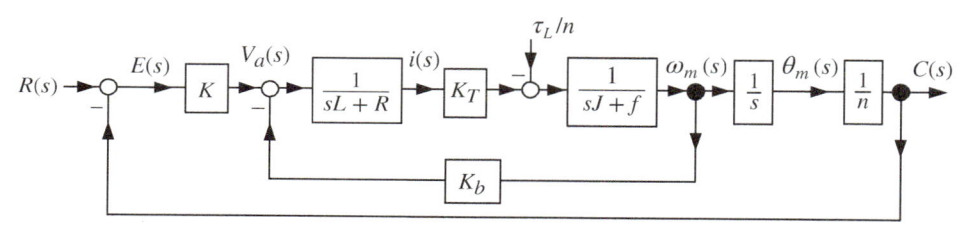

Figure 9.3. Block diagram for the servo system.

One way to simplify this block diagram is to consider the armature inductance to be negligible by setting $L = 0$ and moving the load τ_L/n to the same summing junction as V_a. This results in the equivalent block diagram given in Figure 9.4.

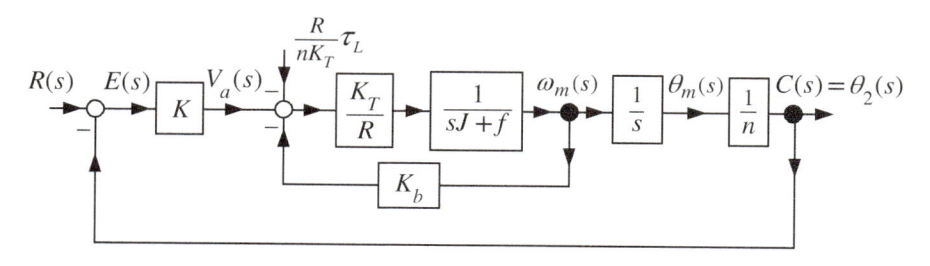

Figure 9.4. Equivalent block diagram of the servo system with $L = 0$.

The transfer function of the inner loop, i.e., from $V_a(s) - \frac{R}{nK_T}\tau_L(s)$ to $\omega_m(s)$, is

$$\frac{\frac{K_T/R}{sJ + f}}{1 + K_b \frac{K_T/R}{sJ + f}} = \frac{b'}{s + a},$$

where
$$b' \triangleq \frac{K_T}{RJ}, \quad a \triangleq \frac{f + K_b K_T/R}{J}.$$

We then have the equivalent block diagram shown in Figure 9.5.

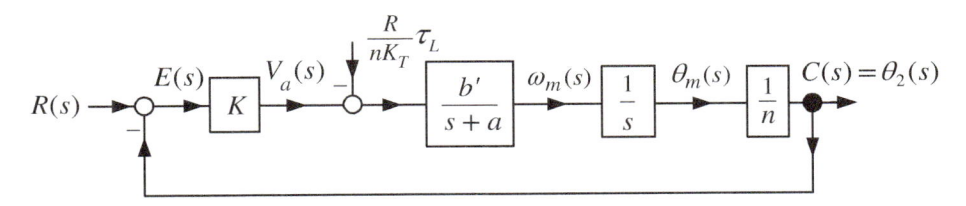

Figure 9.5. Block diagram equivalent to Figure 9.4.

Finally, with
$$b \triangleq \frac{b'}{n}, \quad K_L \triangleq \frac{R}{nK_T}$$

the system block diagram simplifies to that of Figure 9.6.

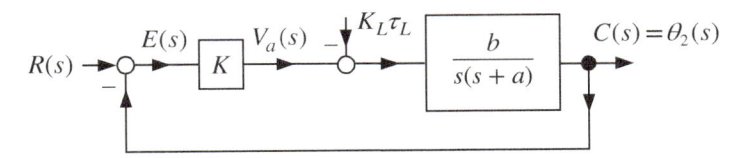

Figure 9.6. A simplified block diagram of the servo system.

It is important to point out that a rather complicated control system has been reduced to this simple block diagram by making the approximation $L = 0$ and some algebraic (block diagram) manipulations. In particular, the system composed of the amplifier, motor, and gears in Figure 9.7a is modeled by the simple block diagram in Figure 9.7b.

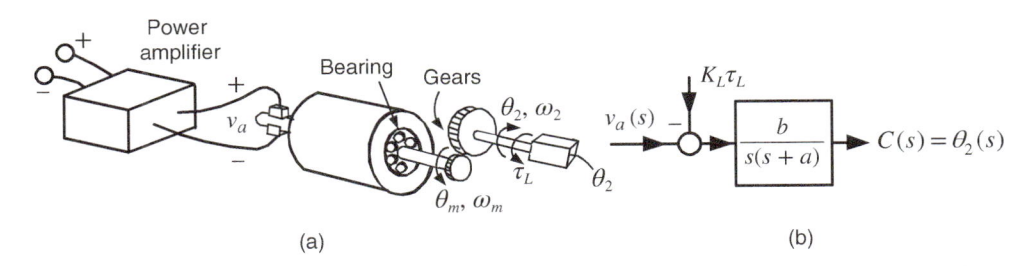

Figure 9.7. (a) DC motor servo system. (b) Block diagram representation.

However, though this system is mathematically equivalent (with $L = 0$) from input V_a to output C of the original system, there is not a one-to-one equivalence inside the block diagram. For example, consider the disturbance $D(s) \triangleq K_L \tau_L(s)$. The constant $K_L = \frac{R}{nK_T}$ has the units $\frac{\text{Ohms}}{\text{Newton-meter/Amps}} = \frac{\text{Ohms-Amps}}{\text{Nt-m}} = \frac{\text{Volts}}{\text{Newton-meter}}$ so that $K_L \tau_L$ has the units of volts! This is consistent with the previous drawing, since at the same summing junction the input voltage $V_a(s) = K(R(s) - C(s))$ is added to the quantity

9 Tracking and Disturbance Rejection

$-K_L \tau_L(s)$. This is not to say that the load is a voltage! Instead, it says that if a voltage given by $-K_L \tau_L$ is applied to the DC motor then the effect on the output position $c(t)$ is the same as the actual load torque.

Finally, let
$$G_m(s) = \frac{b}{s(s+a)}, \quad D(s) = K_L \tau_L(s)$$
so that the block diagram of the servo system is abstracted to that of Figure 9.8.

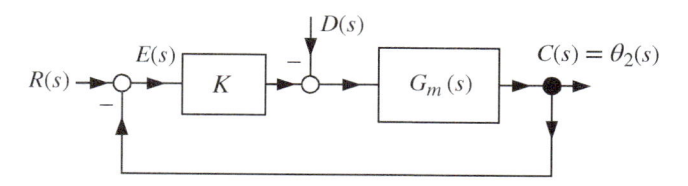

Figure 9.8. Block diagram of servo system in standard form.

The controller gain K is typically replaced by a more general controller specified by a transfer function $G_c(s) = b_c(s)/a_c(s)$ where the subscript "c" denotes controller. In fact, the whole point of automatic control theory is to provide a methodology to choose $G_c(s)$ so that the output $c(t)$ goes to any specified angle despite a load torque acting on the system. The block diagram of Figure 9.9 is a standard form for the analysis of tracking and disturbance rejection of control systems. For example, note that the block diagram for the servo system of Figure 9.8 is in this form with $G_c(s) = K$.

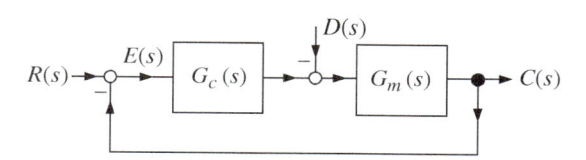

Figure 9.9. Control system in standard block diagram form.

The transfer functions from $R(s)$ and $D(s)$ to $C(s)$ and $E(s)$ in Figure 9.9 are needed to design feedback controllers. To compute them, the block diagram of Figure 9.9 is redrawn as shown in Figure 9.10.

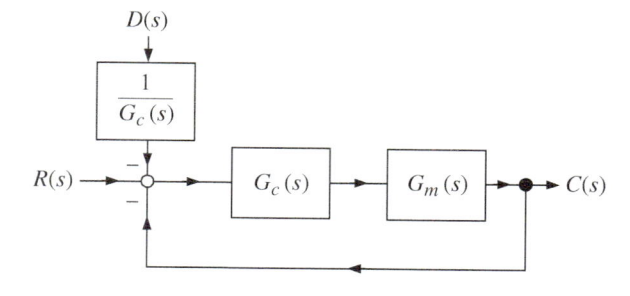

Figure 9.10. Equivalent block diagram to Figure 9.9.

By inspection of Figure 9.10 we have

$$C(s) = \frac{G_c(s)G_m(s)}{1+G_c(s)G_m(s)}\left(R(s) - \frac{1}{G_c(s)}D(s)\right)$$

$$= \frac{G_c(s)G_m(s)}{1+G_c(s)G_m(s)}R(s) - \frac{G_m(s)}{1+G_c(s)G_m(s)}D(s). \tag{9.9}$$

The error $E(s)$ is given by

$$E(s) = R(s) - C(s)$$

$$= \frac{1}{1+G_c(s)G_m(s)}R(s) + \frac{G_m(s)}{1+G_c(s)G_m(s)}D(s). \tag{9.10}$$

Fundamentally, a controller $G_c(s)$ is designed to achieve two objectives for the closed-loop system:

(1) *Tracking*: If the reference is set as $r(t) = \pi/6$ so $R(s) = (\pi/6)/s$, then it is required that $c(t) \to \pi/6$ as $t \to \infty$, i.e., the output must track the input.

(2) *Disturbance rejection*: With the reference set as $r(t) = \pi/6$, then $c(t) \to \pi/6$ no matter what load is on the motor. For example, if a robot arm is to move $\pi/6$, it must do so no matter how much weight it is carrying.

9.2 Control of a DC Servo Motor

The ideas of tracking and disturbance rejection are now illustrated through a series of examples. Specifically, consider a DC servo system with open-loop model $G_m(s) = \dfrac{1}{s(s+1)}$. Various controllers are considered for both tracking and disturbance rejection.

Tracking

A simple proportional controller given by $G_c(s) = K$ is considered first.

Example 1 *Tracking a Step Input*

Consider the unity feedback control system given in the block diagram of Figure 9.11 with $G_m(s) = \dfrac{1}{s(s+1)}$, a constant gain controller K, and $D(s) = 0$.

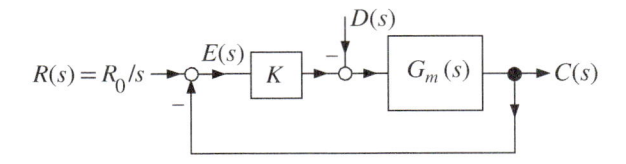

Figure 9.11. Tracking a step input with $D(s) = 0$.

The output and error transfer functions are, respectively,

$$C(s) = \frac{KG_m(s)}{1+KG_m(s)}R(s)$$

$$E(s) = \frac{1}{1+KG_m(s)}R(s).$$

Suppose the objective is to track a step input. With $r(t) = R_0 u_s(t)$ so that $R(s) = R/s$, the objective is to have the error $e(t) \to 0$ as $t \to \infty$ as this implies that $c(t) \to R_0$. Evaluating $E(s)$ we obtain

$$E(s) = \frac{1}{1+K\frac{1}{s(s+1)}}\frac{R_0}{s} = \frac{s(s+1)}{s(s+1)+K}\frac{R_0}{s} = \frac{s+1}{s^2+s+K}R_0.$$

Note that the poles of the forward open-loop transfer function $KG_m(s) = K\frac{1}{s(s+1)}$ reappear in the numerator of $E(s)$ after clearing the fractions in the denominator. In particular the "s" in $K\frac{1}{s(s+1)}$ canceled the "s" in $R(s)$. For $K > 0$, $s^2 + s + K$ is a stable polynomial and thus it follows $sE(s) = \frac{s(s+1)}{s^2+s+K}R_0$ is stable. By the final value theorem we have

$$e(\infty) = \lim_{s \to 0} sE(s) = \lim_{s \to 0} s\frac{s(s+1)}{s(s+1)+K}\frac{R_0}{s} = 0.$$

This controller works in terms of our objective as $c(t) \to R_0$!

Where does stability come in? Let $K = 1$ so that

$$s^2 + s + 1 = \left[s - \left(-\frac{1}{2} + \frac{j\sqrt{3}}{2}\right)\right]\left[s - \left(-\frac{1}{2} - \frac{j\sqrt{3}}{2}\right)\right]$$

$$= (s - p_1)(s - p_2)$$

with $p_1 = -1/2 + j\sqrt{3}/2, p_2 = p_1^* = -1/2 - j\sqrt{3}/2$. Then a partial fraction expansion of $E(s)$ gives

$$E(s) = \frac{s(s+1)}{s(s+1)+K}\frac{R_0}{s} = \frac{s+1}{(s-p_1)(s-p_2)}R_0$$

$$= \frac{\beta}{s-p_1} + \frac{\beta^*}{s-p_2},$$

where $\beta = \beta_1 + j\beta_2 = |\beta|e^{j\angle\beta}$ and $|\beta|^2 = \beta_1^2 + \beta_2^2, \angle\beta = \tan^{-1}(\beta_2/\beta_1)$.

In the time domain we have

$$e(t) = \beta e^{p_1 t} + \beta^* e^{p_2 t}$$

$$= |\beta|e^{j\angle\beta}e^{-(1/2)t+j(\sqrt{3}/2)t} + |\beta|e^{-j\angle\beta}e^{-(1/2)t-j(\sqrt{3}/2)t}$$

$$= 2|\beta|e^{-(1/2)t}\cos\left((\sqrt{3}/2)t + \angle\beta\right) \to 0 \text{ as } t \to \infty.$$

Recall that the poles of the closed-loop transfer function determine the form of the transient response. As the closed-loop poles p_1, p_2 are in the *open left half-plane*, the transients die out. The real part of the closed-loop poles ($\text{Re}\{p_1\} = \text{Re}\{p_2\} = -1/2$) determine how fast the transient $e(t)$ dies out and, as a consequence, one typically chooses the gain K such that the poles of the closed-loop system are as far as possible in the left half-plane.

Example 2 *Tracking a Ramp Input*

Consider the same system as in the previous example, but now let the reference input $r(t)$ be a ramp function in Figure 9.12. Specifically, let $r(t) = \omega_0 t$ where ω_0 is a constant $(R(s) = \omega_0/s^2)$ with $D(s) = 0$.

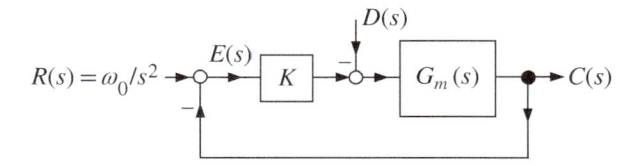

Figure 9.12. Tracking a ramp input with $D(s) = 0$.

The error $E(s)$ is then

$$E(s) = \frac{1}{1+K\frac{1}{s(s+1)}}\frac{\omega_0}{s^2} = \frac{s(s+1)}{s(s+1)+K}\frac{\omega_0}{s^2} = \frac{s+1}{s^2+s+K}\frac{\omega_0}{s}.$$

Note again that the pole of $KG_m(s)$ at $s = 0$ canceled one of the poles at $s = 0$ of $R(s)$. For $K > 0$ it follows that $s^2 + s + K$ is stable and thus $sE(s) = \frac{s+1}{s^2+s+K}\omega_0$ is stable. By the final value theorem we have

$$e(\infty) = \lim_{s \to 0} sE(s) = \lim_{s \to 0} \frac{s+1}{s(s+1)+K}\omega_0 = \frac{\omega_0}{K}.$$

Consequently, asymptotic tracking is not achieved as $e(t)$ does not go to 0 as $t \to \infty$. However, as illustrated in Figure 9.13, the motor does follow the input with a finite error, which can be made small by taking K large.

Where does stability come in? Of course we needed stability be able to use the final value theorem. In more detail, let $K = 1$ so that

$$E(s) = \frac{s(s+1)}{s(s+1)+K}\frac{\omega_0}{s^2}$$

$$= \frac{s+1}{s(s-p_1)(s-p_2)}\omega_0, \; p_i = \left(-1 \pm j\sqrt{3}\right)/2$$

$$= \frac{A}{s} + \frac{\beta}{s-p_1} + \frac{\beta^*}{s-p_2}$$

$$= \frac{\omega_0}{s} + \frac{\beta}{s-p_1} + \frac{\beta^*}{s-p_2} \text{ as } \omega_0 = \lim_{s \to 0} sE(s).$$

As the poles are in the open left half-plane $(\text{Re}(p_1) = \text{Re}(p_2) < 0)$, it follows that

$$e(t) = \omega_0 u_s(t) + \beta e^{p_1 t} + \beta^* e^{p_2 t} \to \omega_0.$$

Again, with the closed-loop poles in the left-hand plane, the transients die out.

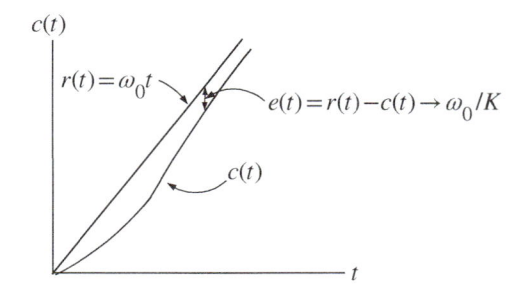

Figure 9.13. Error with a ramp input and using a proportional controller.

The simple proportional controller was not adequate to track a ramp input $r(t) = \omega_0 t$ with zero final error. The answer to asymptotically achieving zero error for this system lies in considering a different controller!

Example 3 *Integral Controller*
Consider the system of Figure 9.14 with the controller $G_c(s) = K/s$, i.e., an integrator.

Figure 9.14. An integral controller.

Then

$$E(s) = \frac{1}{1 + \dfrac{K}{s}\dfrac{1}{s(s+1)}}\frac{\omega_0}{s^2} = \frac{s^2(s+1)}{s^2(s+1) + K}\frac{\omega_0}{s^2} = \frac{s+1}{s^3 + s^2 + K}\omega_0,$$

where we note that the denominator of $\dfrac{K}{s}\dfrac{1}{s(s+1)}$ reappeared in the numerator of $E(s)$ after clearing fractions. As a consequence, the factor s^2 in the numerator of $E(s)$ then cancels the s^2 in denominator of the input $R(s)$. If $s^3 + s^2 + K$ was stable the final value theorem would give

$$e(\infty) = \lim_{s \to 0} sE(s) = \lim_{s \to 0} s\frac{s+1}{s^3 + s^2 + K}\omega_0 = 0.$$

The difficulty here is that $s^3 + s^2 + K$ is *not* stable for any value of K. To see this, just note that $s^3 + s^2 + K = s^3 + s^2 + 0s + K$ and recall that a necessary condition for stability is that all the coefficients be positive. To emphasize the stability aspect, let $K = 1$ for which $s^3 + s^2 + 1$ has roots $-1.47, 0.23 \pm j0.79$. The error response is found by doing a partial fraction expansion of $E(s)$ as

$$E(s) = \frac{s+1}{s^3 + s^2 + 1}\omega_0 = \frac{s+1}{(s+1.47)(s - (0.23 + j0.79))(s - (0.23 - j0.79))}\omega_0$$

$$= \frac{A}{s+1.47} + \frac{\beta}{s - (0.23 + j0.79)} + \frac{\beta^*}{s - (0.23 - j0.79)}.$$

9.2 Control of a DC Servo Motor

The time response is then

$$e(t) = Ae^{-1.47t} + \beta e^{0.23t}e^{j0.79t} + \beta^* e^{0.23t}e^{-j0.79t}$$
$$= Ae^{-1.47t} + 2|\beta|e^{0.23t}\cos(0.79t + \angle\beta).$$

The error $e(t)$ does not go to zero due to the complex-conjugate pair of unstable closed-loop poles at $0.23 \pm j0.79$. Remember, the fact that $\lim_{s\to 0} sE(s) = 0$ says nothing about the final value unless $sE(s)$ is stable!

As we have shown by example, the difficult problem is not in getting $\lim_{s\to 0} sE(s) = 0$, but rather it is in making the closed-loop system stable. As shown earlier, a constant gain (proportional) controller will give closed-loop stability, but not $\lim_{s\to 0} sE(s) = 0$ with a ramp input. On the other hand, an integral controller gives $\lim_{s\to 0} sE(s) = 0$, but not closed-loop stability. Let's combine the two and see if that will work. Specifically, let

$$G_c(s) = K\frac{s+\alpha}{s} = K + \frac{\alpha K}{s}.$$

This is called a proportional (K) plus integral $(\alpha K/s)$ controller or PI controller.

Example 4 *PI Controller*
Again, consider the same system as in the previous examples, except we use a PI controller as illustrated in Figure 9.15.

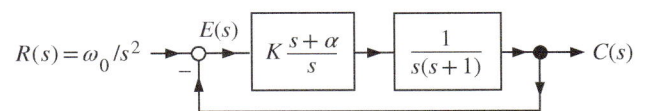

Figure 9.15. A proportional plus integral (PI) controller.

From the block diagram, the error $E(s)$ is seen to be given by

$$E(s) = \frac{1}{1 + K\frac{s+\alpha}{s}\frac{1}{s(s+1)}}R(s) = \frac{s^2(s+1)}{s^2(s+1) + K(s+\alpha)}\frac{\omega_0}{s^2}$$

$$= \frac{s+1}{s^3 + s^2 + Ks + \alpha K}\omega_0.$$

The denominator of the forward open-loop system $K\dfrac{s+\alpha}{s}\dfrac{1}{s(s+1)}$ reappears in the numerator of $E(s)$. Consequently, the factor of "s^2" in the numerator of $E(s)$ cancels the "s^2" in the denominator of $R(s)$. Then, *if* $s^3 + s^2 + Ks + \alpha K$ is stable, the final value theorem gives

$$e(\infty) = \lim_{s\to 0} sE(s) = \lim_{s\to 0} s\frac{s+1}{s^3+s^2+Ks+\alpha K}\omega_0 = 0.$$

To check stability of $s^3 + s^2 + Ks + \alpha K$ we use the Routh–Hurwitz test. The Routh table is

$$\begin{array}{c|cc} s^3 & 1 & K \\ s^2 & 1 & \alpha K \\ s & \dfrac{K-\alpha K}{1} & 0 \\ s^0 & \alpha K & \end{array}$$

9 Tracking and Disturbance Rejection

The first column is positive if and only if $\alpha K > 0$ and $K - \alpha K = K(1-\alpha) > 0$ or

$$K > 0 \text{ and } 0 < \alpha < 1.$$

For example, with $K = 1$ and $\alpha = 1/2$, the denominator of $E(s)$ is

$$s^3 + s^2 + s + 1/2 = (s+0.65)\left(s - [-0.176 + j0.861]\right)\left(s - [-0.176 - j0.861]\right).$$

Then, with $r = -0.65$ and $p_i = -0.176 \pm j0.861$, the partial fraction expansion of $E(s)$ gives

$$E(s) = \frac{s+1}{(s-r)(s-p_1)(s-p_2)}\omega_0 = \frac{A}{s-r} + \frac{\beta}{s-p_1} + \frac{\beta^*}{s-p_2}.$$

The corresponding time response is then

$$e(t) = Ae^{-0.65t} + \beta e^{-0.176t}e^{j0.861t} + \beta^* e^{-0.176t}e^{-j0.861t}$$
$$= Ae^{-0.65t} + 2|\beta|e^{-0.176t}\cos(0.861t + \angle\beta) \to 0 \text{ as } t \to \infty.$$

In general, adding integrators to the controller tends to destabilize (make unstable) the closed-loop system while adding proportional (constant gain) control tends to stabilize the closed-loop system.

Disturbance Rejection

Consider now the problem of getting the motor (robot arm) to move a specified number of degrees with an external load (weight) acting on it.

Example 5 *Constant Load Torque with a Proportional Controller*
Consider the control system in the block diagram of Figure 9.16. Let

$$G_m(s) = \frac{1}{s(s+1)}$$

and let the disturbance be the constant load torque (see Figure 9.17).

$$\tau_L(t) = \tau_{L0}u_s(t).$$

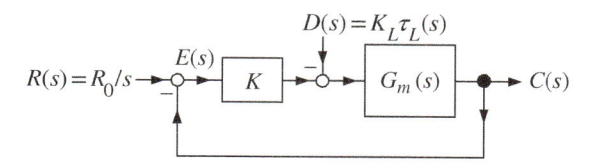

Set $D(s) = K_L\tau_L(s) = K_L\tau_{L0}/s$.

Figure 9.16. A proportional controller with a load acting on the system.

It was shown in (9.10) that $E(s) = E_R(s) + E_D(s)$ where

$$E_R(s) = \frac{1}{1 + KG_m(s)}R(s)$$

$$E_D(s) = \frac{G_m(s)}{1 + KG_m(s)}D(s).$$

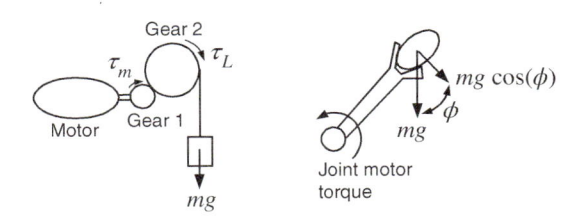

Figure 9.17. Examples of torque loads on DC motors.

For $K > 0$ and $R(s) = R_0/s$, Example 1 showed that $e_R(t) = \mathcal{L}^{-1}\{E_R(s)\} \to 0$. The interest here is disturbance rejection, that is, whether or not $e_D(t) = \mathcal{L}^{-1}\{E_D(s)\} \to 0$ as $t \to \infty$. $E_D(s)$ is explicitly given by

$$E_D(s) = \frac{G_m(s)}{1 + KG_m(s)} D(s) = \frac{\frac{1}{s(s+1)}}{1 + K\frac{1}{s(s+1)}} K_L \tau_L(s) = \frac{K_L}{s^2 + s + K} \frac{\tau_{L0}}{s}.$$

$E_D(s)$ is the error in the position response due to the load torque. The load torque is $\tau_L(s) = \tau_{L0}/s$ and, as $s^2 + s + K$ is stable for $K > 0$, the final position error due to this load torque is

$$e_D(\infty) = \lim_{s \to 0} sE_D(s) = \lim_{s \to 0} s \frac{K_L}{s^2 + s + K} \frac{\tau_{L0}}{s} = \frac{K_L \tau_{L0}}{K}.$$

The error

$$r(t) - c(t) = e(t) = e_R(t) + e_D(t) \to e_R(\infty) + e_D(\infty)$$
$$= 0 + K_L \tau_{L0}/K.$$

Rearranging, we have

$$c(\infty) = r(\infty) - e(\infty) = R_0 - K_L \tau_{L0}/K.$$

The conclusion here is that the final output position depends on the value of the load torque τ_{L0}. This is usually not acceptable as typically the load torque is unknown and it is important to precisely position the motor regardless of the load.[2]

We now show that a PI controller can achieve zero final error for this system even with an unknown constant load torque acting on it.

Example 6 *Constant Load Torque with a PI Controller*

Let $G_c(s) = K_p + K_I/s = K\frac{s + \alpha}{s}$ where $K_p = K, K_I = \alpha K$. This is illustrated in the block diagram of Figure 9.18.

---

[2]Note that if the gain K can be made large, then the disturbance term $K_L \tau_{L0}/K$ could be made negligible.

246 9 Tracking and Disturbance Rejection

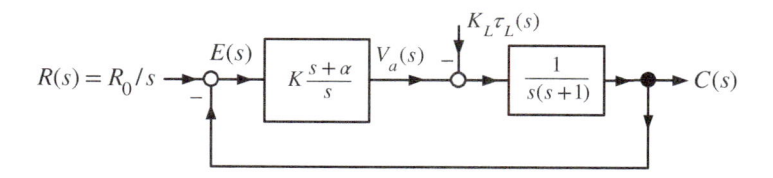

Figure 9.18. PI controller for disturbance rejection.

The error $E_D(s)$ is given by

$$E_D(s) = \frac{G_m(s)}{1 + G_c(s)G_m(s)}D(s) = \frac{\frac{1}{s(s+1)}}{1 + K\frac{s+\alpha}{s}\frac{1}{s(s+1)}}\frac{K_L \tau_{L0}}{s}$$

$$= \frac{s}{s^2(s+1) + K(s+\alpha)}\frac{K_L \tau_{L0}}{s}$$

$$= \frac{1}{s^3 + s^2 + Ks + \alpha K}K_L \tau_{L0}.$$

Note that, in contrast to the tracking case, only the denominator of $G_c(s)$, i.e., "s", reappears in the numerator after clearing the fractions in $E_D(s)$. This "s" cancels the $1/s$ in $\tau_L(s)$. Now, by the Routh test, $s^3 + s^2 + Ks + \alpha K$ is stable for $K > 0$ and $0 < \alpha < 1$, and so $sE_D(s)$ is also stable. The final value theorem then gives

$$e_D(\infty) = \lim_{s \to 0} sE_D(s) = \lim_{s \to 0} s\frac{K_L}{s^3 + s^2 + Ks + \alpha K}\tau_{L0} = 0.$$

Consequently, the load torque τ_{L0} has no effect on the final position.

Interpretation of the PI Controller

We just analyzed the system given in the block diagram of Figure 9.19. Specifically, with $R(s) = R_0/s$, $\tau_L(s) = \tau_{L0}/s$, K and α can be chosen so that the closed-loop system is stable resulting in $e(\infty) = \lim_{s \to 0} sE(s) = 0$.

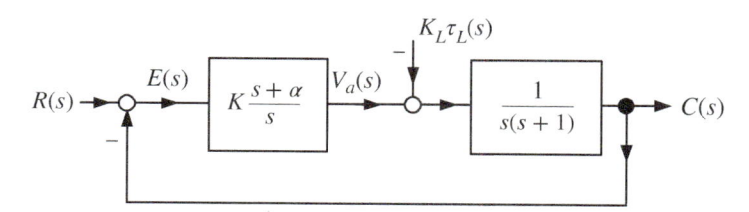

Figure 9.19. $v_a(t) \to K_L \tau_{L0}$ to cancel out the effect of the load torque.

$V_a(s)$ is the voltage applied to the motor and is given by

$$V_a(s) = K\frac{s+\alpha}{s}E(s) = KE(s) + \frac{\alpha K}{s}E(s). \tag{9.11}$$

9.2 Control of a DC Servo Motor

In the time domain we have

$$v_a(t) = Ke(t) + \alpha K \int_0^t e(\tau)d\tau.$$

To compute $v_a(t)$ we need to know the error $e(t)$. With $R(s) = R_0/s$ and $\tau_L(s) = \tau_{L0}/s$ we have

$$E(s) = \underbrace{\frac{s^2(s+1)}{s^3 + s^2 + Ks + \alpha K}\frac{R_0}{s}}_{E_R(s)} + \underbrace{\frac{K_L s}{s^3 + s^2 + Ks + \alpha K}\frac{\tau_{L0}}{s}}_{E_D(s)} \qquad (9.12)$$

and we note that $E(s)$ is stable as we are taking $K > 0$ and $0 < \alpha < 1$. Then $sV_a(s) = sKE(s) + \alpha KE(s)$ is also stable and by the final value theorem we have

$$\begin{aligned}v_a(\infty) &= \lim_{t \to \infty} v_a(t) = \lim_{s \to 0} sV_a(s) \\ &= \lim_{s \to 0} \left(sKE(s) + \alpha KE(s)\right) \\ &= \alpha KE(0) \\ &= K_L \tau_{L0}\end{aligned}$$

which is due only to the output of the integrator. As $v_a(\infty) = K_L \tau_{L0}$ we see that the output of the integrator goes to exactly that voltage needed to cancel the load torque disturbance (see Figure 9.20)!

Figure 9.20. Shaded area is $\int_0^\infty e(t)dt$ and $v_a(\infty) = \alpha K \int_0^\infty e(t)dt = K_L \tau_0$.

Summary of the PI Controller for a DC Servo

It has been shown that the controller $G_c(s) = K(s+\alpha)/s$ with $K > 0$ and $0 < \alpha < 1$ will force the motor with transfer function $G_m(s) = \dfrac{1}{s(s+1)}$ to track inputs of the form $r(t) = R_0 + \omega_0 t$ with zero final error in spite of any constant load disturbances. The control system designer then wants to choose K and α so that the closed-loop poles are far in the left half-plane in order have the transients die out quickly. Remember, the closed-loop poles determine the transient response! Recall from (9.9) and (9.10) that

$$C(s) = \frac{G_c(s)G_m(s)}{1 + G_c(s)G_m(s)} R(s) - \frac{G_m(s)}{1 + G_c(s)G_m(s)} K_L \tau_L(s)$$

$$E(s) = \underbrace{\frac{1}{1 + G_c(s)G_m(s)} R(s)}_{E_R(s)} + \underbrace{\frac{G_m(s)}{1 + G_c(s)G_m(s)} K_L \tau_L(s)}_{E_D(s)}.$$

Evaluating the error $E(s)$ gives

$$E(s) = \frac{s^2(s+1)}{s^3 + s^2 + Ks + \alpha K}\left(\frac{R_0}{s} + \frac{\omega_0}{s^2}\right) + \frac{K_L s}{s^3 + s^2 + Ks + \alpha K}\frac{\tau_{L0}}{s}.$$

The gains K and α are chosen so that the poles of the closed-loop transfer function

$$s^3 + s^2 + Ks + \alpha K = (s-r)(s-p_1)(s-p_2)$$

satisfy $r < 0$, $\text{Re}(p_1) = \text{Re}(p_2) < 0$. The error $E(s)$ is then

$$E(s) = \frac{(s+1)(R_0 s + \omega_0)}{(s-r)(s-p_1)(s-p_2)} + \frac{1}{(s-r)(s-p_1)(s-p_2)}K_L \tau_{L0}$$

or, in the time domain, we have

$$e(t) = Ae^{rt} + Be^{p_1 t} + B^* e^{p_2 t} + (Ce^{rt} + De^{p_1 t} + D^* e^{p_2 t})K_L \tau_{L0}.$$

As $r < 0$, and $\text{Re}(p_1) = \text{Re}(p_2) < 0$ it follows that

$$e(t) \to 0 \text{ as } t \to \infty.$$

Proportional plus Integral plus Derivative Control

A proportional plus integral plus derivative (PID) controller is defined by

$$G_c(s) = K_P + \frac{K_I}{s} + K_D s = \frac{K_D s^2 + K_P s + K_I}{s}$$

or, in the time domain,

$$v_a(t) = K_P e(t) + K_I \int_0^t e(\tau)d\tau + K_D \frac{de}{dt}.$$

This is illustrated in the block diagram of Figure 9.21.

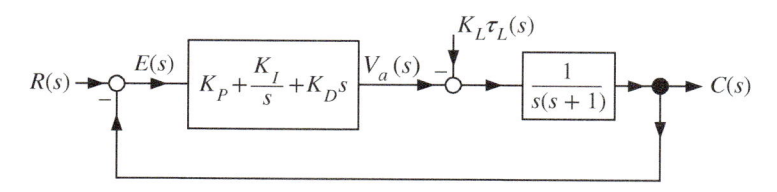

Figure 9.21. A proportional plus integral plus derivative controller.

The advantages of proportional plus integral controllers $K_P + K_I/s$ have already been discussed. The derivative control term $K_D s$ is used to force the transients to die out faster. That is, as will be shown in the following, the term $K_D s$ allows the control designer to put the closed-loop poles further in the left half-plane. However, before this is done, a practical issue concerning the derivative controller $K_D s$ is discussed.

Practical Problem with Derivative Controllers

Differentiation of a signal with noise amplifies the noise! Most signals contain high-frequency low-amplitude noise. For example, a DC power supply takes a 60 Hz AC

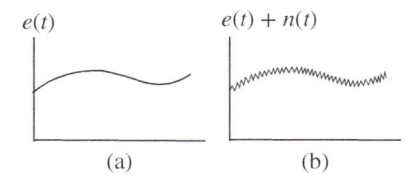

Figure 9.22. (a) An error signal. (b) An error signal with high-frequency low-amplitude noise.

signal and rectifies it into a DC voltage. Due to this process (full-wave rectification) there is a small amplitude 120 Hz signal (i.e., noise) on the DC output (see Figure 9.22).

If one measures the error signal $e(t)$ in the DC servomotor system, the signal $e(t) + n(t)$ is actually obtained where $n(t)$ is a noise term. Numerically differentiating this signal gives the approximation

$$\dot{e}(t) + \dot{n}(t) \approx \frac{e(t) - e(t - \Delta t) + n(t) - n(t - \Delta t)}{\Delta t}.$$

As the difference $n(t) - n(t - \Delta t)$ may be of the same order (or higher) of magnitude as $e(t) - e(t - \Delta t)$, it follows that $\dot{n}(t)$ can be of the same order of magnitude as $\dot{e}(t)$. For example, let this noise[3] be given by $n(t) = 0.01\sin(2\pi(120t)) \approx 0.01\sin(754t)$. Thus, even though $n(t)$ is quite small, $\dot{n}(t) = 7.54\cos(754t)$ is significant. In Chapter 6 optical encoders for measuring angular position were discussed and modeled. With Δt the time between samples of $\theta(t)$, it was shown the angular speed $\omega(t) = \dot{\theta}(t)$ can be calculated using $\dot{\theta}(t) = (\theta(t) - \theta(t - \Delta t))/\Delta t$, but that this estimate does contain high frequency noise.

Practical Implementation of Derivative Feedback

One possibility to deal with the fact that differentiation amplifies high frequency noise is to replace $K_D s$ with $K_D s/(\tau s + 1)$ as shown in Figure 9.23. With $\tau > 0$ (typically small) we have

$$\frac{K_D s}{\tau s + 1} \approx \begin{cases} K_D s, & |s| \ll 1/\tau \\ K_D/\tau, & |s| \gg 1/\tau. \end{cases}$$

At low frequencies $K_D s/(\tau s + 1)$ acts as a differentiator, while at high frequencies it acts as a proportional gain. This PID controller is illustrated in Figure 9.23.

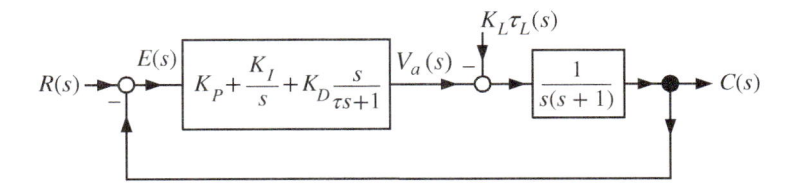

Figure 9.23. Differentiation followed by low pass filtering.

---

[3]This could be a model for the noise on the DC bus of an amplifier when it uses a full-wave rectifier to convert the 120 Hz AC outlet to DC.

250   9   Tracking and Disturbance Rejection

The closed-loop transfer function (taking $\tau = 0$) is then

$$\frac{C(s)}{R(s)} = \frac{\dfrac{K_D s^2 + K_P s + K_I}{s}\dfrac{1}{s(s+1)}}{1 + \dfrac{K_D s^2 + K_P s + K_I}{s}\dfrac{1}{s(s+1)}} = \frac{K_D s^2 + K_P s + K_I}{s^3 + (1+K_D)s^2 + K_P s + K_I}. \qquad (9.13)$$

Note that closed-loop transfer now has two zeros due to the zeros of the PID controller. Recall from Chapter 8 that zeros in the open left half-plane contribute to overshoot in the step response.

With the reference $R(s)$ a step input, the PID control setup of Figure 9.23 results in differentiating this step function which is not differentiable at $t = 0$. This is avoided by using the PI-D control architecture shown in Figure 9.24. The PI-D notation is used to indicate the PI controller is in the forward path while the D controller is in the feedback path.

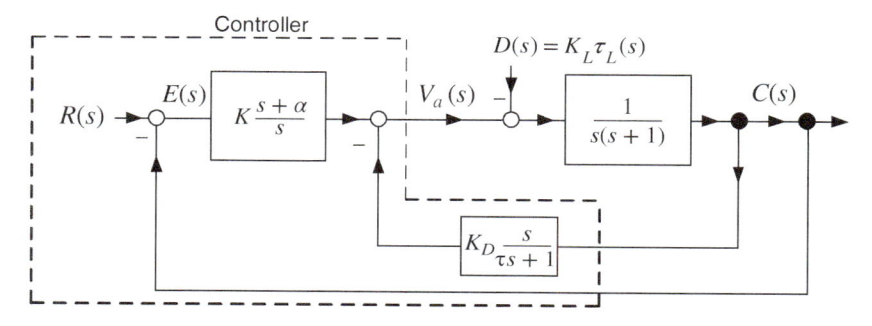

Figure 9.24. PI-D implementation of derivative feedback.

To analyze this setup we first do a simple block diagram manipulation to obtain the equivalent system of Figure 9.25.

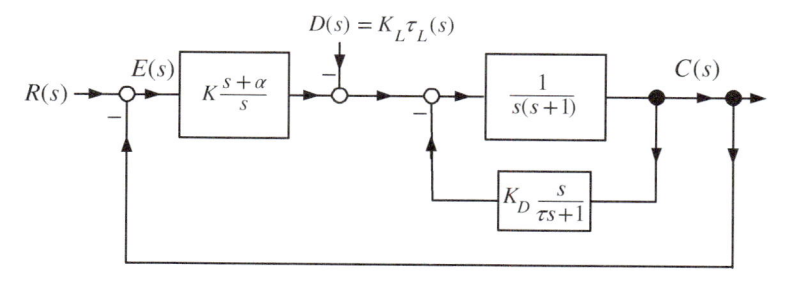

Figure 9.25. Equivalent block diagram of Figure 9.24.

In order to set the values of K, α, and K_D, we take $\tau = 0$ as it is small. Of course in the implementation $K_D s/(\tau s + 1)$ is used. As a result the block diagram of Figure 9.25 reduces to that of Figure 9.26.

Taking $D(s) = 0$ the closed-loop transfer function is

$$\frac{C(s)}{R(s)} = \frac{\dfrac{K(s+\alpha)}{s}\dfrac{1}{s(s+1+K_D)}}{1 + \dfrac{K(s+\alpha)}{s}\dfrac{1}{s(s+1+K_D)}} = \frac{K(s+\alpha)}{s^3 + (1+K_D)s^2 + Ks + \alpha K}.$$

9.2 Control of a DC Servo Motor

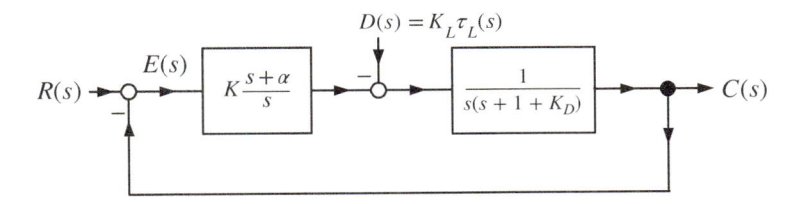

Figure 9.26. Equivalent block diagram of Figure 9.25 with $\tau = 0$.

Note that using PI-D control architecture there is now only one zero in the closed-loop transfer function in contrast to the two zeros in (9.13) using the PID architecture of Figure 9.23.

Using the PI-D controller $E_R(s)$ and $E_D(s)$ are, respectively, given by

$$E_R(s) = \frac{1}{1 + K\frac{s+\alpha}{s}\frac{1}{s+1+K_D}\frac{1}{s}} R(s) = \frac{s^2(s+1+K_D)}{s^3 + (1+K_D)s^2 + Ks + \alpha K} R(s)$$

and

$$E_D(s) = \frac{\frac{1}{s+1+K_D}\frac{1}{s}}{1 + K\frac{s+\alpha}{s}\frac{1}{s+1+K_D}\frac{1}{s}} D(s) = \frac{s}{s^3 + (1+K_D)s^2 + Ks + \alpha K} D(s).$$

As the parameters K_D, K, and α are chosen by the control engineer, the coefficients of $s^3 + (1+K_D)s^2 + Ks + \alpha K$ may be chosen arbitrarily. In other words, the location of the closed-loop poles can be placed anywhere in the left half-plane by the control system designer! This is the reason for using derivative feedback in a DC servomotor controller.

In summary, the PI-D controller for a DC servomotor allows one to achieve tracking of step and ramp inputs, rejection of constant load torque disturbances, and the capability to place the closed-loop poles at any desired location. Typically, one desires the closed-loop poles to be as far in the left half-plane as possible so that the transients die out quickly. However, this usually means the control gains must be quite large. To explain, let's put in some actual numbers in the equations. The error $E_R(s)$ is given by

$$E_R(s) = \frac{s^2(s+1+K_D)}{s^3 + (1+K_D)s^2 + Ks + \alpha K} R(s).$$

Suppose it is desired to put the closed-loop poles at $-10, -10+j10$, and $-10-j10$. Set

$$s^3 + (1+K_D)s^2 + Ks + \alpha K = (s+10)\left(s - (-10+j10)\right)\left(s - (-10-j10)\right)$$
$$= (s+10)(s^2 + 20s + 200)$$
$$= s^3 + 30s^2 + 400s + 2000.$$

This means we must choose $1 + K_D = 30$ or $K_D = 29, K = 400$, and $\alpha K = 2000$ or $\alpha = 5$ as shown in Figure 9.27. Let the system start from rest with $\theta(0) = 0$ and $\omega(0) = 0$. Apply the step reference input $r(t) = R_0 u_s(t)$, at $t = 0$ we have $c(0^+) = \theta(0^+) = 0, \omega(0^+) = 0$ and

$$e(0^+) = r(0^+) - c(0^+) = R_0.$$

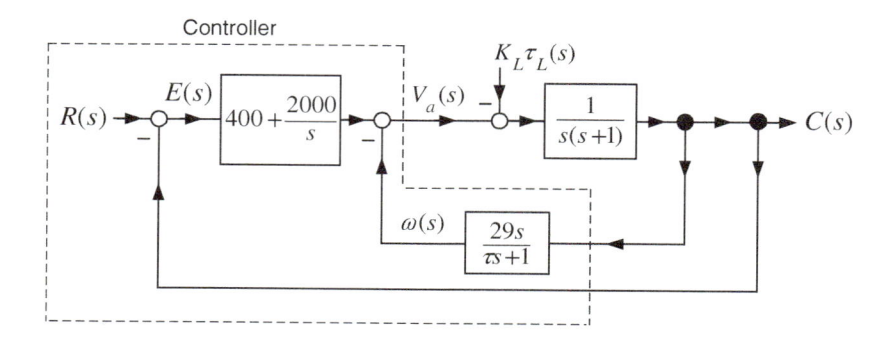

Figure 9.27. Example of a PI-D controller design.

Then the voltage at $t = 0$ applied to the motor is given by

$$v_a(0) = 400e(0) + 2000\int_0^0 e(t)dt - 29\omega(0) = 400e(0) = 400R_0.$$

This shows that just after the reference input $r(t) = R_0 u_s(t)$ is applied, the output of the amplifier (the input to the motor) is required to be $400R_0$. In particular, if $R_0 = 0.1$ rad ($5.7°$), the amplifier is required to put out 40 V, which is about the saturation limit of a small DC servo amplifier. The point here is that, in practice, one cannot place the closed-loop poles arbitrarily, but is limited by physical constraints such as amplifier saturation. These physical constraints were not included in the model on which the design was based.

9.3 Theory of Tracking and Disturbance Rejection

The examples of the previous section give the background to present a general approach to tracking and disturbance rejection of step inputs. The following definitions are needed to present the general approach to tracking and disturbance rejection.

Definition 1 *Type Number*
Let

$$G(s) = \frac{b(s)}{s^j \bar{a}(s)}, \qquad (9.14)$$

where $\bar{a}(0) \neq 0$. Then $G(s)$ is said to be a type j system which simply means it has j poles at the origin.

Example 7 *Transfer Functions and Their Type Numbers*
The following table gives some transfer functions and their type numbers.

$$G(s) = \frac{1}{s^2} \qquad \text{Type 2} \quad \bar{a}(s) = 1$$

$$G(s) = \frac{1}{s(s+1)^2} \qquad \text{Type 1} \quad \bar{a}(s) = (s+1)^2$$

9.3 Theory of Tracking and Disturbance Rejection

$$G(s) = \frac{1}{s+2} \qquad \text{Type 0} \quad \bar{a}(s) = s+2$$

$$G(s) = \frac{s+1}{s(s^2+2s+2)} \qquad \text{Type 1} \quad \bar{a}(s) = s^2+2s+2.$$

Definition 2 *Type Number of Inputs*

If the reference input $R(s)$ to a system is given by

$$R(s) = \frac{R_0}{s^j},$$

then $R(s)$ is said to be a type j reference input.

Similarly, if the disturbance input $D(s)$ to a system is given by

$$D(s) = \frac{D_0}{s^j},$$

then $D(s)$ is said to be a type j disturbance input.

Theorem 1 *Tracking with Zero Steady-State Error*

Consider now the general tracking problem as setup in the block diagram of Figure 9.28.

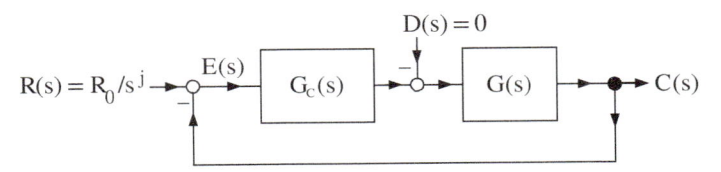

Figure 9.28. Block diagram for tracking a type j input.

Here the transfer function of the plant (physical system)

$$G(s) = b(s)/a(s), \quad \deg\{b(s)\} < \deg\{a(s)\}$$

is assumed to be strictly proper and the transfer function of the controller

$$G_c(s) = b_c(s)/a_c(s), \quad \deg\{b_c(s)\} \le \deg\{a_c(s)\}$$

is assumed to be proper. The tracking error $E_R(s)$ is given by

$$E_R(s) = \frac{1}{1+G_c(s)G(s)}R(s) = \frac{1}{1+\dfrac{b_c(s)}{a_c(s)}\dfrac{b(s)}{a(s)}}R(s) = \frac{a_c(s)a(s)}{a_c(s)a(s)+b_c(s)b(s)}R(s).$$

For some positive integer j let the reference input be

$$R(s) = \frac{R_0}{s^j} \quad \text{or} \quad r(t) = R_0\frac{t^{j-1}}{(j-1)!}.$$

Then the error $e_R(t)$ goes to 0, i.e.,

$$\lim_{t\to\infty} e_R(t) = \lim_{t\to\infty}(r(t)-c(t)) = 0,$$

if

(1) The closed-loop system is stable, i.e.,

$$a_c(s)a(s) + b_c(s)b(s)$$

has all of its roots in the open left half-plane.

(2) The type number of the forward open-loop transfer function $G_c(s)G(s)$ is j or greater.

Remark The input is $R(s) = R_0/s^j$, and condition (2) says that the forward open-loop transfer function $G_c(s)G(s)$ must also contain at least a factor of $1/s^j$.

Proof. We are given that conditions (1) and (2) hold. To show that $e(t) \to 0$ as $t \to \infty$, we compute the error $E(s)$ given by

$$E_R(s) = \frac{1}{1 + G_c(s)G(s)} R(s) = \frac{1}{1 + \frac{b_c(s)}{a_c(s)} \frac{b(s)}{a(s)}} \frac{R_0}{s^j} = \frac{a_c(s)a(s)}{a_c(s)a(s) + b_c(s)b(s)} \frac{R_0}{s^j}.$$

By condition (2) $G_c(s)G(s)$ is at least type j, so that $a_c(s)a(s) = s^j \bar{a}(s)$. The factor s^j cancels the denominator of $R(s)$ resulting in

$$E_R(s) = \frac{\bar{a}(s)}{a_c(s)a(s) + b_c(s)b(s)} R_0.$$

By condition (1) the closed-loop system is stable. That is, the roots of the polynomial

$$a_c(s)a(s) + b_c(s)b(s) = (s - p_1) \cdots (s - p_n)$$

satisfy $\operatorname{Re}(p_i) < 0$ for $i = 1, \ldots, n$. A partial fraction expansion of $E(s)$ then gives[4]

$$E_R(s) = \frac{\bar{a}(s)}{(s - p_1) \cdots (s - p_n)} R_0 = \frac{A_1}{s - p_1} + \frac{A_2}{s - p_2} + \cdots + \frac{A_n}{s - p_n}.$$

In the time domain this becomes

$$e_R(t) = A_1 e^{p_1 t} + \cdots + A_n e^{p_n t}.$$

Again, as $\operatorname{Re}(p_i) < 0$ for $i = 1, \ldots, n$ it follows that $e_R(t) \to 0$. ∎

Theorem 2 *Disturbance Rejection with Zero Steady-State Error*

Consider now the disturbance rejection problem. The block diagram of Figure 9.29 illustrates the setup.

The transfer function $G(s)$ is assumed to be strictly proper, i.e.,

$$G(s) = b(s)/a(s), \quad \deg\{b(s)\} < \deg\{a(s)\}.$$

---

[4] This is assuming the poles are distinct. If they are not, the proof is easily modified with the same final result.

9.3 Theory of Tracking and Disturbance Rejection

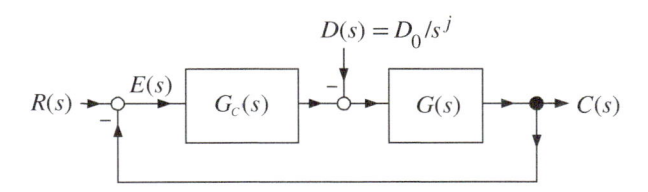

Figure 9.29. Block diagram for rejecting a type j disturbance.

The controller transfer function is assumed to be proper, i.e.,

$$G_c(s) = b_c(s)/a_c(s), \ \deg\{b_c(s)\} \leq \deg\{a_c(s)\}.$$

The error $E(s)$ is then

$$E(s) = \underbrace{\frac{1}{1 + G_c(s)G(s)} R(s)}_{E_R(s)} + \underbrace{\frac{G(s)}{1 + G_c(s)G(s)} D(s)}_{E_D(s)}.$$

The error $E_D(s)$ due to the disturbance is

$$E_D(s) = \frac{G(s)}{1 + G_c(s)G(s)} D(s) = \frac{\frac{b(s)}{a(s)}}{1 + \frac{b_c(s)}{a_c(s)} \frac{b(s)}{a(s)}} D(s) = \frac{a_c(s)b(s)}{a_c(s)a(s) + b_c(s)b(s)} D(s). \quad (9.15)$$

For some positive integer j let the disturbance be given by

$$D(s) = D_0/s^j \ \text{ so } \ d(t) = D_0 \frac{t^{j-1}}{(j-1)!}.$$

Then the error $e_D(t) \triangleq \mathcal{L}^{-1}\{E_D(s)\}$ goes to 0, i.e.,

$$\lim_{t \to \infty} e_D(t) = 0,$$

if

(1) The closed-loop system is stable, i.e.,

$$a_c(s)a(s) + b_c(s)b(s)$$

has all of its roots in the open left half-plane.

(2) The type number of the controller transfer function $G_c(s)$ is j or greater.

Remark In contrast to the tracking problem where $G_c(s)G(s)$ must be type j, the controller $G_c(s)$ *by itself* must be type j.

Proof. We are given that conditions (1) and (2) above hold. The error $E_D(s)$ is given by

$$E_D(s) = \frac{a_c(s)b(s)}{a_c(s)a(s) + b_c(s)b(s)} \frac{D_0}{s^j}.$$

By condition (2) of this theorem $a_c(s) = s^j \bar{a}_c(s)$ and therefore

$$E_D(s) = \frac{\bar{a}_c(s)b(s)}{a_c(s)a(s) + b_c(s)b(s)} D_0 = \frac{\bar{a}_c(s)b(s)}{(s-p_1)\cdots(s-p_n)} D_0$$

$$= \frac{A_1}{s-p_1} + \frac{A_2}{s-p_2} + \cdots + \frac{A_n}{s-p_n}.$$

By condition (1) the closed-loop system is stable so $\text{Re}\{p_i\} < 0$ for $i = 1, \ldots, n$. Consequently, as $t \to \infty$

$$e_D(t) = A_1 e^{p_1 t} + \cdots + A_n e^{p_n t} \to 0.$$

∎

9.4 Internal Model Principle

The proofs of the tracking and disturbance rejection theorems suggest how one can achieve the same results with more general types of reference and disturbance signals. This is illustrated in the following example.

Example 8 *Rejecting a Sinusoidal Disturbance*

The control system of Figure 9.30 has open-loop transfer function $G(s) = \frac{1}{s(s+1)}$. A controller is to be designed that tracks $R(s) = R_0/s$ and rejects the sinusoidal disturbance $D(s) = \frac{D_0}{s^2+1}$, i.e., $d(t) = D_0 \sin(t)$. Consider the controller $G_c(s) = K\frac{s+\alpha}{s^2+1}$, which was chosen so that its poles contained the poles of $D(s)$.

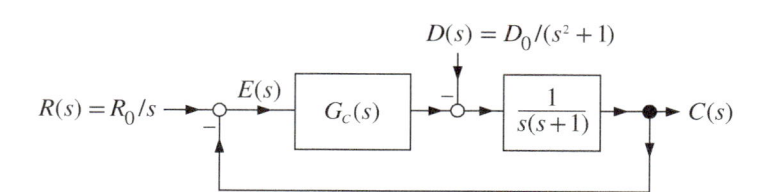

Figure 9.30. Asymptotically rejecting a sinusoidal disturbance.

With this choice for $G_c(s)$ the error $E_D(s)$ due to the disturbance is

$$E_D(s) = \frac{G(s)}{1 + G_c(s)G(s)} D(s) = \frac{\frac{1}{s(s+1)}}{1 + K\frac{s+\alpha}{s^2+1}\frac{1}{s(s+1)}} \frac{D_0}{s^2+1}$$

$$= \frac{s^2+1}{s^4 + s^3 + s^2 + (K+1)s + \alpha K} \frac{D_0}{s^2+1}$$

$$= \frac{1}{s^4 + s^3 + s^2 + (K+1)s + \alpha K} D_0.$$

The factor $s^2 + 1$ in the denominator of $G_c(s)$ reappears in the numerator of $E_D(s)$ to cancel the factor $s^2 + 1$ in the denominator of $D(s)$. Then $e_D(t) \to 0$ if $s^4 + s^3 + s^2 + (K+$

$1)s + \alpha K$ is stable. To check this we form its Routh table.

s^4	1		1	αK
s^3	1		$K+1$	0
s^2	$\dfrac{1-(K+1)}{1} = -K$		αK	0
s	$\dfrac{-K(K+1)-\alpha K}{-K} = \dfrac{K(K+1+\alpha)}{K}$		0	
s^0	αK			

Stability requires $-K > 0, K+1+\alpha > 0$ and $\alpha K > 0$ or

$$K < 0, -(K+\alpha) < 1, \alpha < 0.$$

For example, choose $\alpha = -0.25$ and $K = -0.6$ so that $s^4 + s^3 + s^2 + 0.4s + 0.15$ has roots $-0.17 \pm j0.63, -0.33 \pm j0.49$. The error response is

$$e_D(t) = 4.7e^{-0.17t}\sin(0.63t + 0.37) - 4.1e^{-0.33t}\sin(0.49t + 0.4) \to 0.$$

However, the transients die out slowly. We also have

$$\begin{aligned}
E_R(s) &= \frac{1}{1+G_c(s)G(s)}R(s) \\
&= \frac{1}{1+K\dfrac{s+\alpha}{s^2+1}\dfrac{1}{s(s+1)}}\frac{R_0}{s} \\
&= \frac{s(s+1)(s^2+1)}{s^4+s^3+s^2+(K+1)s+\alpha K}\frac{R_0}{s} \\
&= \frac{(s+1)(s^2+1)}{s^4+s^3+s^2+(K+1)s+\alpha K}R_0.
\end{aligned}$$

$E_R(s)$ has the same denominator as $E_D(s)$, which is stable for $\alpha = -0.25, K = -0.6$, and so $e_R(t) \to 0$ as well.

Remark $G_c(s) = -0.6\dfrac{s-0.25}{s^2+1}$ isn't a very good controller as the transient response dies out slowly. The reader should work Problem 11, which previews Chapter 10 by showing how a controller can be designed to place the closed-loop poles at *any* location in the open left half-plane.

This example and the previous examples can be summarized as the *Internal Model Principle* [27–29]. This just says that in order to track a given reference signal, the forward open-loop transfer function $G_c(s)G(s)$ must contain the same *unstable* poles as the reference signal and the closed-loop system must be stable. Similarly, in order to achieve asymptotic rejection of a disturbance, the forward open-loop controller $G_c(s)$ must (by

258   9   Tracking and Disturbance Rejection

itself) contain the same unstable poles as the disturbance signal along with the closed-loop system being stable.

An approach to feedback design for phase-locked loops from the point of view of the internal model principle is nicely presented in [30].

9.5 Design Example: PI-D Control of Aircraft Pitch

In [31] the transfer function from the elevator angle $\delta(s)$ in radians to the pitch angle $\theta(s)$ in radians of a small aircraft is given to be (see Figure 9.31)

$$\frac{\theta(s)}{\delta(s)} = G(s) = \frac{1.51s + 0.1774}{s^3 + 0.739s^2 + 0.921s}.$$

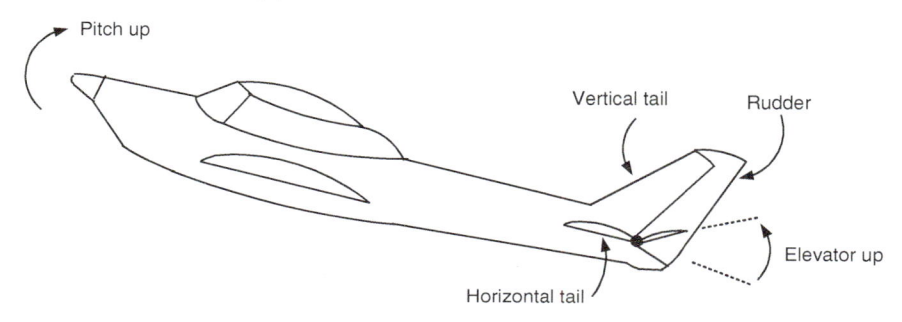

Figure 9.31. Using the elevators to pitch the aircraft up.

With a step input of 0.2 rad, the design specifications are (i) overshoot less than 10%, (ii) rise time less than two seconds, (iii) settling time less than 10 seconds, (iv) final error less than 2%. We take the maximum elevator deflection to be 25° (0.436 rad), that is, $-25° \leq \delta \leq 25°$.

This is a type one system so the final error should be zero as long as the closed-loop system is stable. However, an integrator is desired to reject disturbances. Air turbulence on the aircraft body and wings causes its pitch to change. As shown in Figure 9.32 this disturbance is modeled as an equivalent (unknown) deflection of the elevator angle. We consider a unity-feedback PI-D controller as given in Figure 9.32. Let $R(s) = R_0/s$ with $R_0 = 0.2$ rad or $(0.2)(180/\pi) = 11.5$ degrees, and take $D(s) = 0$.

How does one choose the feedback gains? We want the closed-loop poles far in the left half-plane so that the transients die out quickly, but we want to make sure the controller

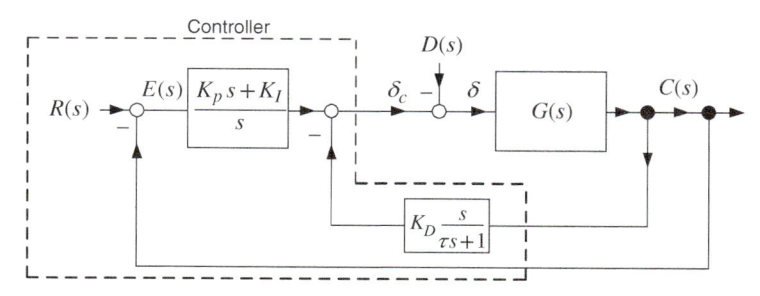

Figure 9.32. PI-D unity feedback control system.

9.5 Design Example: PI-D Control of Aircraft Pitch

gains are not so big as to saturate the actuator (elevator angle). There can be significant overshoot due to complex conjugate closed-loop poles or due to the zeros of the closed-loop system. To start, notice that commanded elevator angle δ_c (output of the controller) is

$$\delta_c(t) = K_p e(t) + K_I \int_0^t e(t')dt' + K_D \frac{dc(t)}{dt},$$

where we set $\tau = 0$ to simplify the presentation. At $t = 0$ this becomes

$$\delta_c(0) = K_p e(0) + K_I \int_0^0 e(t')dt' + K_D \dot{c}(0) = K_p(R_0 - c(0)) = K_p R_0.$$

Now, as $-0.436 \leq \delta \leq 0.436$ and $R_0 = 0.2$ rad it follows that $K_p \leq 0.436/0.2 = 2.18$ to avoid saturation of the actuator. We next compute the closed-loop transfer function $G_{CL}(s)$ (again with $\tau = 0$ to simplify the calculations). First set

$$G_2(s) \triangleq \frac{\dfrac{1.51s + 0.1774}{s^3 + 0.739s^2 + 0.921s}}{1 + K_D s \dfrac{1.51s + 0.1774}{s^3 + 0.739s^2 + 0.921s}}$$

$$= \frac{1.51s + 0.1774}{s^3 + (1.51K_D + 0.739)s^2 + (0.1774K_D + 0.921)s}.$$

The closed-loop transfer function is then

$$G_{CL}(s) = \frac{\left(\dfrac{K_p s + K_I}{s}\right) \dfrac{1.51s + 0.1774}{s^3 + (1.51K_D + 0.739)s^2 + (0.1774K_D + 0.921)s}}{1 + \left(\dfrac{K_p s + K_I}{s}\right) \dfrac{1.51s + 0.1774}{s^3 + (1.51K_D + 0.739)s^2 + (0.1774K_D + 0.921)s}}$$

$$= \frac{(K_p s + K_I)(1.51s + 0.1774)}{s\left(s^3 + (1.51K_D + 0.739)s^2 + (0.1774K_D + 0.921)s\right) + (K_p s + K_I)(1.51s + 0.1774)}$$

$$= \frac{(K_p s + K_I)(1.51s + 0.1774)}{s^4 + (1.51K_D + 0.739)s^3 + (1.51K_p + 0.1774K_D + 0.921)s^2 + (0.1774K_p + 1.51K_I)s + 0.1774K_I}.$$

The zeros of $G_{CL}(s)$ consists of the open-loop zero of $G(s)$ and the zero of the PI-D controller. The closed-loop poles are the roots of

$$a_{CL}(s) = s^4 + (1.51K_D + 0.739)s^3 + (1.51K_p + 0.1774K_D + 0.921)s^2$$
$$+ (0.1774K_p + 1.51K_I)s + 0.1774K_I.$$

Note that we could use K_I to set the value of the last coefficient, then use K_p to then set the value of the coefficient of s term, and finally set the coefficient of the s^2 term using K_D. However, we then cannot set the value of the s^3 term. More generally, we can arbitrarily set only three of the four coefficients of $a_{CL}(s)$. A fundamental problem in feedback control is to figure out how to choose the gains. In this application we want to find the values

260   9 Tracking and Disturbance Rejection

of the gains that achieve the above specifications.[5] One procedure to choose the gains is given in [32] as follows.

(1) Set $K_I = K_D = 0$ and adjust the value of K_p "until the closed-loop response oscillates."
    This assumes the closed-loop system can be made stable for small values of $K_p > 0$.

(2) With this value of K_p adjust the value of K_I so that the error goes to zero.
    Typically, the output will become even more oscillatory as K_I is increased and may even go unstable.

(3) With these values of K_p and K_D adjust the value of K_D to damp out the oscillatory response and reduce overshoot.

There is no assurance that this heuristic procedure will result in a satisfactory response.

Set the proportional gain as $K_p = 2.18$, which is the largest value it can have without saturating the actuator with a step reference of 0.2 rad. With K_p fixed at this value, K_I is varied from 0.2 to 2. Choosing $K_I = 1$ gives the response shown in Figure 9.33.

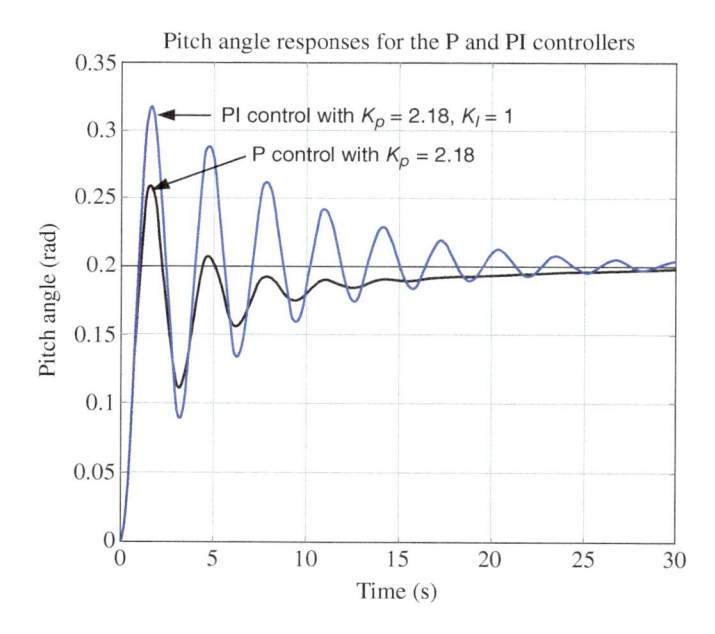

Figure 9.33. Output responses in radians for the P and PI controllers.

Finally the derivative gain K_D is varied from 1 to 2 (with $\tau = 0.05$) and the value $K_D = 1.5$ is chosen. The output response is shown in Figure 9.34.

The response shown in Figure 9.34 does *not* quite meet specifications. The fractional overshoot is $(0.223 - 0.2)/0.2 = 0.115$ or 11.5% (not the specified 10%) and at 10 seconds the output is within $(0.207 - 0.2)/0.2 = 0.035$ or 3.5% of the final value (not the 2%

---

[5]Presumably these specifications are considered necessary by pilots of the aircraft. When such specifications are first presented it is not obvious that a controller can be found to meet them.

9.5 Design Example: PI-D Control of Aircraft Pitch

Pitch angle responses for the PI-D controller

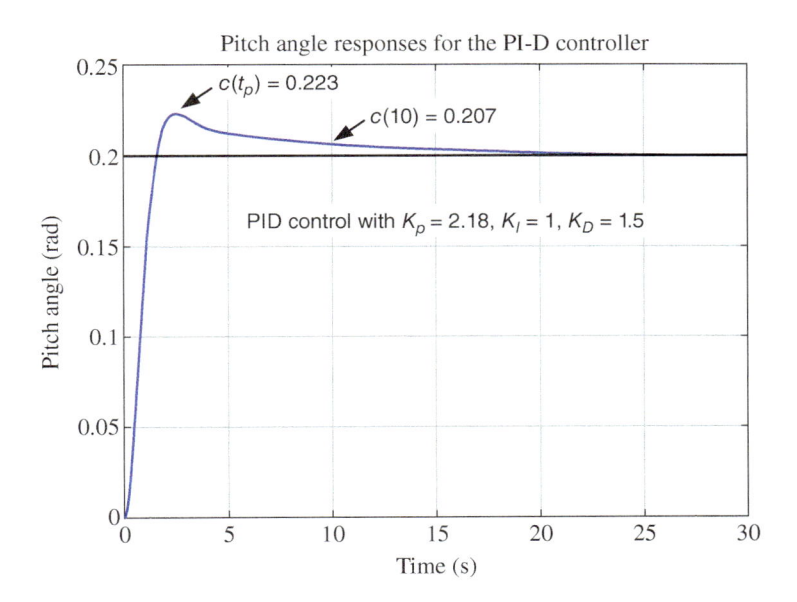

$c(t_p) = 0.223$
$c(10) = 0.207$
PID control with $K_p = 2.18$, $K_I = 1$, $K_D = 1.5$

Figure 9.34. Output response in radians of the PI-D controller with no actuator constraint.

Elevator angle δ_c

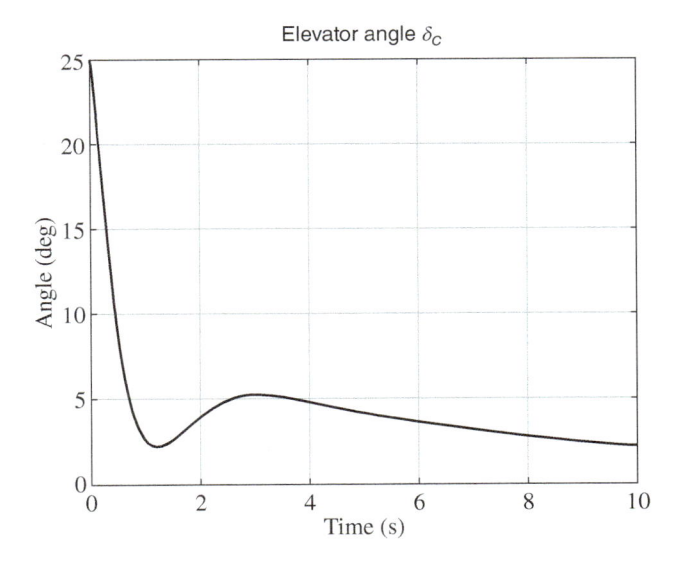

Figure 9.35. Elevator command δ_c in degrees.

specified). However, the rise time is about 1.6 seconds, which is within the two seconds specification. The corresponding command to the elevator is given in Figure 9.35, which does not saturate.

The main difficulty in trying to achieve the specification is the limitation on the proportional gain in order to prevent saturation of the elevator angle. It turns out one could meet the specifications by just increasing the gains and allowing the elevator angle to saturate.

9 Tracking and Disturbance Rejection

However, as soon as an actuator is in saturation, its output is stuck at the maximum value until it gets out of saturation. For example, if the pilot pitched up and then decided to immediately pitch down instead, the plane would not react to the new command until the actuator left saturation.

Rather than saturate the actuator, another approach is to simply have the reference ramp up to 0.2 rad. Set the reference input as

$$r(t) = \begin{cases} \dfrac{0.2}{1.5}t & \text{for } 0 \leq t \leq 1.5 \\ 0.2 & \text{for } t > 1.5. \end{cases}$$

With this reference and, after many trials, the gains were set as $K_P = 12, K_I = 0.25, K_D = 2$. Note that using a ramp as reference input allowed the proportional gain to be much larger. The pitch angle response $c(t)$ and reference $r(t)$ are as shown in Figure 9.36. The corresponding elevator angle for this response is given in Figure 9.37. The closed-loop transfer function is

$$G_{CL}(s) = \frac{(K_p s + K_I)(1.51s + 0.1774)}{s^4 + s^3(1.51K_D + 0.739) + s^2(1.51K_p + 0.1774K_D + 0.921) + s(0.1774K_p + 1.51K_I) + 0.1774K_I}.$$

Note that $G_{CL}(0) = 1$ so that

$$\lim_{t \to \infty} c(t) = \lim_{s \to 0} sC(s) = \lim_{s \to 0} sG_{CL}(s)\frac{0.2}{s} = 0.2.$$

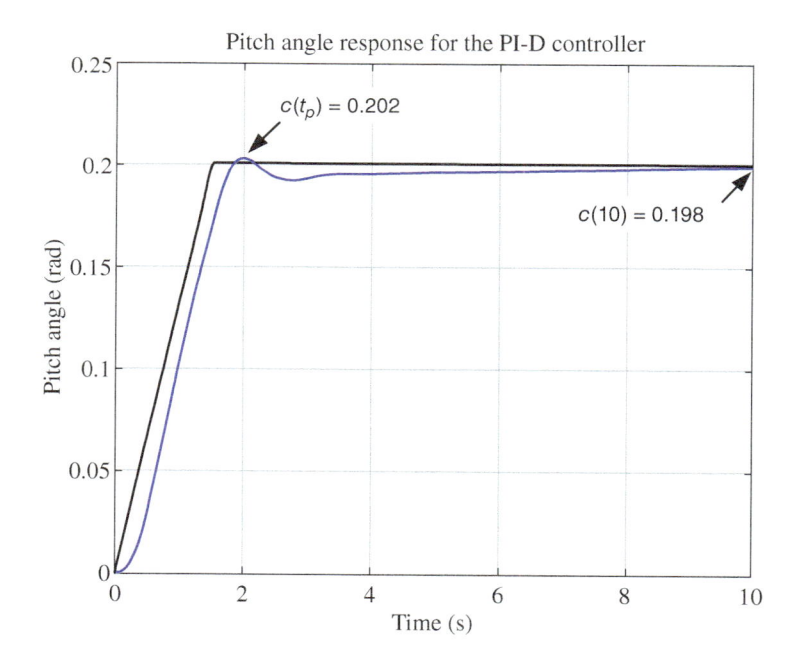

Figure 9.36. Pitch angle and ramp reference.

9.5 Design Example: PI-D Control of Aircraft Pitch

Elevator angle command δ_c

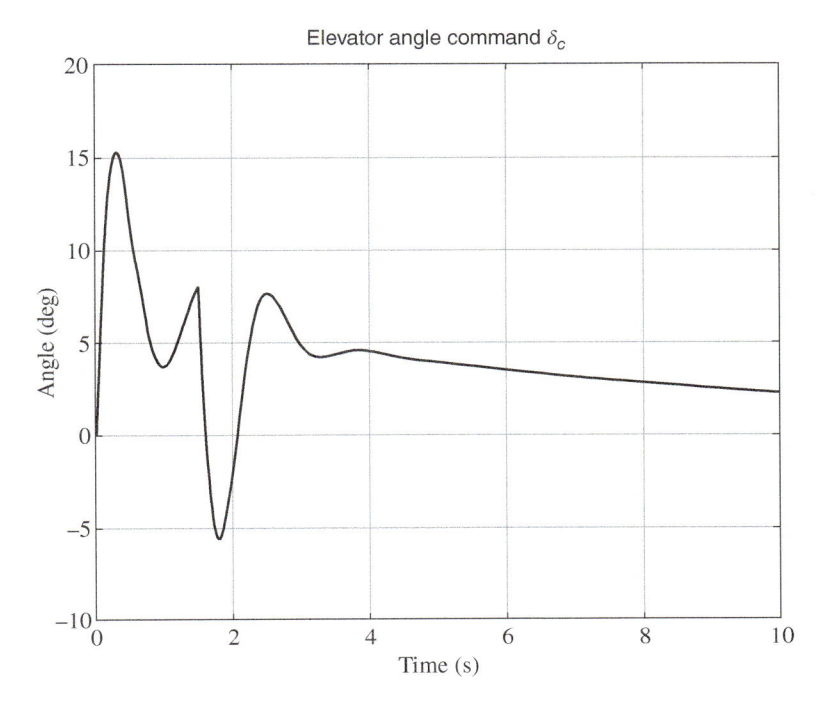

Figure 9.37. Elevator angle vs. time when using the ramp reference input.

Of course, as the open-loop system is type 1, we already knew step reference input would be tracked. Substituting the values of the gains gives

$$G_{CL}(s) = \frac{(12s + 0.25)(1.51s + 0.1774)}{s^4 + 3.759s^3 + 19.396s^2 + 2.5063s + 0.044\,35}$$

$$= \frac{18.12(s + 0.0208)(s + 0.1175)}{(s^2 + 3.627s + 18.92)(s + 0.111)(s + 0.0211)}.$$

An important observation of $G_{CL}(s)$ is that its two zeros at $-0.0208, -0.1175$ essentially cancel its two poles at -0.0211 and -0.111, respectively.[6] This means that in the partial fraction expansion of $C(s) = G_{CL}(s)\frac{0.2}{s}$ the coefficients of the $\frac{1}{s+0.111}$ and $\frac{1}{s+0.0211}$ terms will be small (why?). In fact, the inverse Laplace transform is

$$c(t) = \mathcal{L}^{-1}\left\{G_{CL}(s)\frac{0.2}{s}\right\}$$

$$= 0.2u_s(t) + 0.0029e^{-0.0211t} - 0.0115e^{-0.111t}$$

$$\quad -0.191e^{-1.81t}\left(\cos(3.95t) + 0.4603\sin(3.95t)\right),$$

where the coefficient 0.0029 of $e^{-0.0211t}$ and the coefficient -0.0115 of $e^{-0.111t}$ are more than 20 times smaller in magnitude than the coefficient -0.191 in front of $e^{-1.81t}$. Thus, though

---

[6]This cancellation is between stable poles and zeros. This situation is not uncommon and can be best understood from root locus theory. See Chapter 12.

$e^{-0.0211t}$ and $e^{-0.111t}$ die out slowly, their effect on the transient response are relatively insignificant.

Parametric Uncertainty (Robustness)

Up to this point we have emphasized that feedback control can provide tracking and disturbance rejection. However, an extremely important property of feedback control is that it will often work quite well even if the model parameters used in the design are uncertain (as they almost always are). For example, suppose the truth model[7] of the transfer function is

$$G_{truth}(s) = \frac{\theta(s)}{\delta(s)} = \frac{s + 0.12}{s^3 + 0.4s^2 + s}. \quad (9.16)$$

Using the same gains found using the design model $G(s)$, Figure 9.38 shows the pitch angle response using the both the truth model $G_{truth}(s)$ and the original model $G(s)$. Figure 9.39 shows the corresponding elevator command angles for the two models.

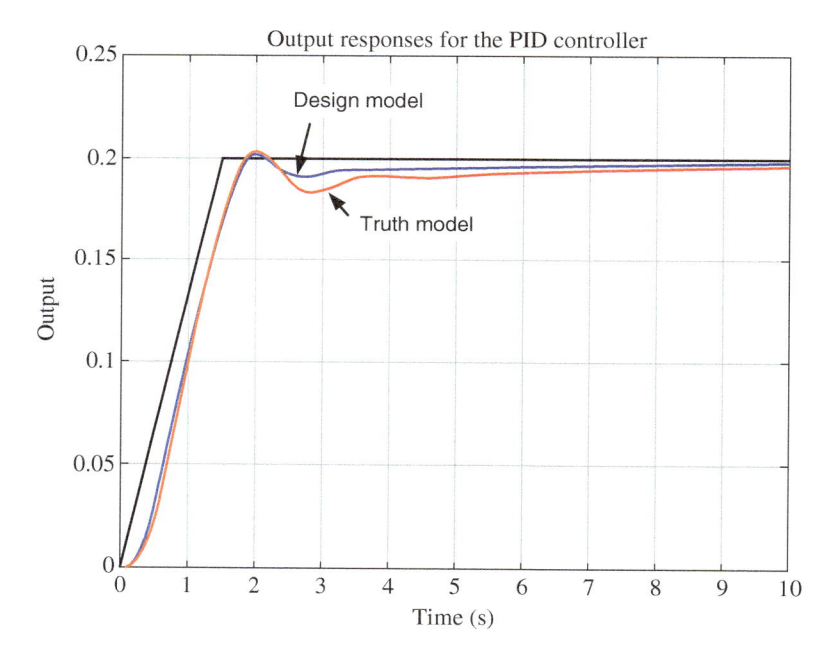

Figure 9.38. Pitch response using the design model $G(s)$ and the truth model $G_{truth}(s)$.

The two responses are quite close even though the parameters (numerator and denominator coefficients of $G(s)$ and G_{truth}) are quite different. This is a fundamental advantage of using closed-loop feedback! That is, even if the controller is designed using a model whose parameter values are somewhat off their actual values, the response of the actual system can still be quite good. Problem 17 asks you to show that the specifications can be met for a *step* input (without saturating the actuator) if an I-PD control architecture is used.

---

[7] By "truth" model we simply mean we are taking this to be the exact model of $\theta(s)/\delta(s)$. However, the "exact" model is never known.

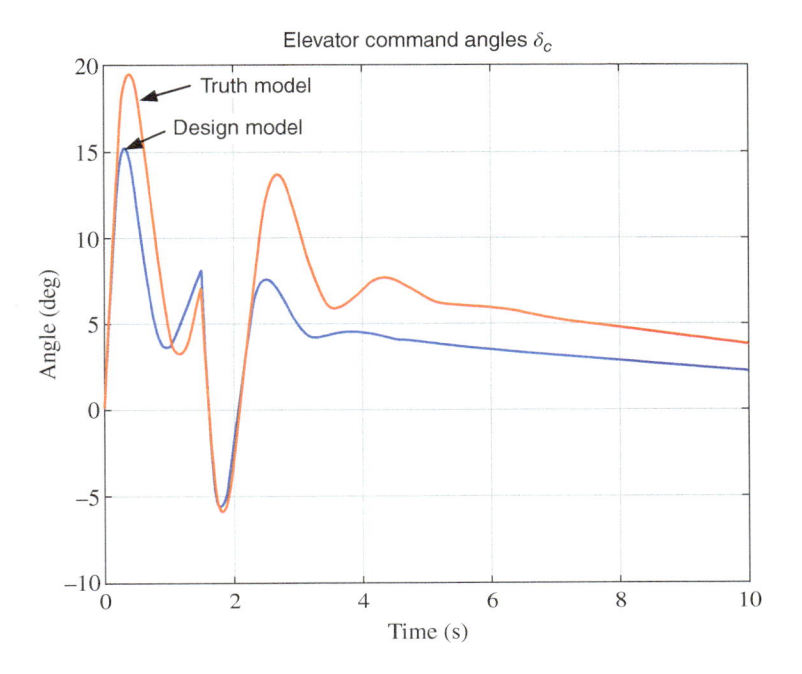

Figure 9.39. Elevator commands δ_c using the design model $G(s)$ and the truth model $G_{truth}(s)$.

9.6 Model Uncertainty and Feedback*

We have seen how feedback can provide a way to force a system to asymptotically track a step or ramp input and reject a constant disturbance, that is, to do tracking and disturbance rejection. We now take a look at another reason why feedback is such an important tool in the control of practical systems; it significantly reduces the effect of uncertainty in the model on the performance of the system. To explain we again look at the DC motor of Figure 9.40 whose block diagram model is given in Figure 9.41.[8]

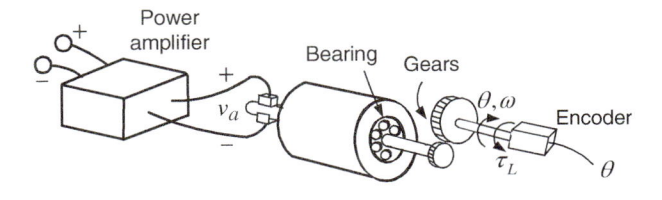

Figure 9.40. DC motor servo system.

Taking the speed as output, setting $L = 0$, and defining $K_m = \dfrac{K_T}{Rf + K_b K_T}$, $T_m = \dfrac{RJ}{Rf + K_b K_T}$, $K_L = \dfrac{R}{K_T}$, the block diagram of Figure 9.41 reduces to that of Figure 9.42.

---

[8] This presentation is adapted from [33].

266   9 Tracking and Disturbance Rejection

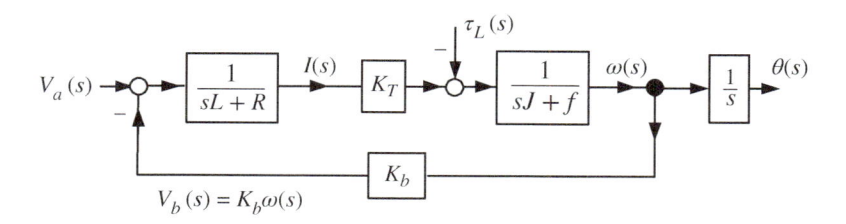

Figure 9.41. Block diagram of the DC motor servo system. Source: Adapted from G. F. Franklin, J. D. Powell, and A. Emami-Naeini, Feedback Control of Dynamic Systems, Addison Wesley, Reading, MA, 1986.

Figure 9.42. Simplified block diagram of a DC motor servo system.

Open-Loop Control

Let's first look at open-loop speed control. With $V_a(s) = V_0/s$ and $\tau_L(s) = \tau_{L0}/s$, $\omega(s)$ is given by

$$\omega(s) = \underbrace{\frac{K_m}{T_m s + 1}}_{G(s)} \frac{V_0}{s} - \frac{K_m}{T_m s + 1} \frac{K_L \tau_{L0}}{s}.$$

As $T_m > 0$ it follows that $s\omega(s)$ is stable and therefore

$$\omega(\infty) = \lim_{s \to 0} s\omega(s) = K_m V_0 - K_m K_L \tau_{L0}.$$

The first problem we see is that the final speed depends on the load torque τ_{L0}. For now let's take $\tau_{L0} = 0$. Then, with ω_0 the desired final speed, set

$$V_0 = \frac{\omega_0}{K_m}$$

so that $\omega(\infty) = \omega_0$. We see that an accurate value of K_m is required for this open-loop control to work.

Closed-Loop Control

Figure 9.43 is a block diagram of a simple proportional feedback control system for speed. Note that this requires adding a sensor (tachometer or optical encoder) to the motor in order to measure/calculate speed.

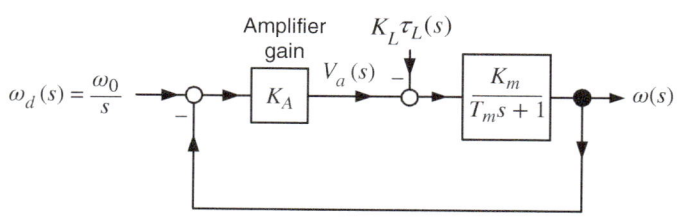

Figure 9.43. Simple proportional feedback control of speed.

9.6 Model Uncertainty and Feedback*

With $\omega_d(s) = \omega_0/s$ and $\tau_L(s) = \tau_{L0}/s$ we have

$$\omega(s) = \frac{\frac{K_A K_m}{T_m s + 1}}{1 + \frac{K_A K_m}{T_m s + 1}} \frac{\omega_0}{s} - \frac{\frac{K_m}{T_m s + 1}}{1 + \frac{K_A K_m}{T_m s + 1}} \frac{K_L \tau_{L0}}{s}$$

$$= \frac{K_A K_m}{T_m s + 1 + K_A K_m} \frac{\omega_0}{s} - \frac{K_m}{T_m s + 1 + K_A K_m} \frac{K_L \tau_{L0}}{s}.$$

As $T_m > 0$, $K_m > 0$, and taking $K_A > 0$ it follows that $s\omega(s)$ is stable and therefore

$$\omega(\infty) = \lim_{s \to 0} s\omega(s) = \frac{K_A K_m}{1 + K_A K_m} \omega_0 - \frac{K_m}{1 + K_A K_m} K_L \tau_{L0}.$$

Taking K_A large enough so that $1 + K_A K_m \gg 1$ and $K_L \tau_{L0}/K_A \ll 1$ we have

$$\omega(\infty) = \frac{K_A K_m}{1 + K_A K_m} \omega_0 - \frac{K_m}{1 + K_A K_m} K_L \tau_{L0} \approx \omega_0 - \frac{1}{K_A} K_L \tau_{L0} \approx \omega_0.$$

Thus we have approximate tracking even though there is uncertainty in the value of K_m and a load torque acting on the system.

Speeding Up the Response

To continue comparing open-loop with closed-loop control, let's look at speeding up response of the output. Figure 9.44 shows an open-loop control system where the open-loop controller is $G_c(s) = \frac{T_m s + 1}{T_d s + 1}$. With ω_0 the desired final speed and $\tau_L = 0$ set $R(s) = \frac{\omega_0}{K_m} \frac{1}{s}$. Then

$$\omega(s) = \frac{T_m s + 1}{T_d s + 1} \frac{K_m}{T_m s + 1} \frac{\omega_0}{K_m} \frac{1}{s} = \frac{K_m}{T_d s + 1} \frac{\omega_0}{K_m} \frac{1}{s}.$$

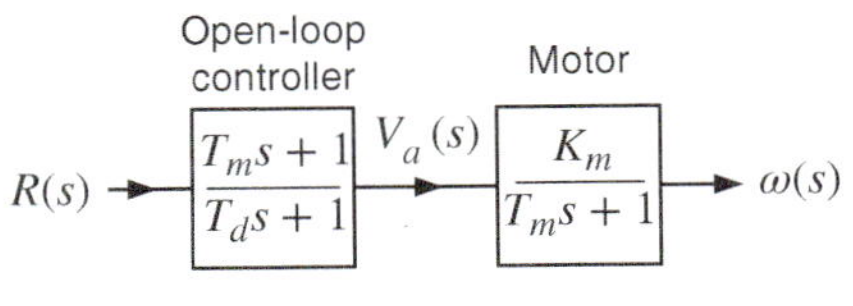

Figure 9.44. Open-loop controller to speed up the response.

As $T_m > 0$ this was a stable pole–zero cancellation. In the time domain we have

$$\omega(t) = (1 - e^{-t/T_d})\omega_0.$$

Without the open-loop controller $G_c(s) = \frac{T_m s + 1}{T_d s + 1}$, but the same reference input $R(s) = \frac{\omega_0}{K_m} \frac{1}{s}$, the output response is

$$\omega(t) = (1 - e^{-t/T_m})\omega_0.$$

268   9 Tracking and Disturbance Rejection

To be specific set $T_d = T_m/10$ so that

$$\omega(t) = (1 - e^{-t/T_d})\omega_0 = (1 - e^{-10t/T_m})\omega_0.$$

Then $\omega(t) \to \omega_0$ ten times faster using the open-loop controller $G_c(s) = \dfrac{T_m s + 1}{T_d s + 1}$. However, an accurate value of T_m is needed to obtain this speed up in the response in addition to an accurate value of K_m. Most importantly, if there is a load torque the open-loop controller cannot reduce its effect on the output response.

On the other hand, let's go back to the closed-loop control system of Figure 9.43 (with $\tau_L = 0$ and $R(s) = \omega_0/s$) where

$$\omega(s) = \frac{\dfrac{K_A K_m}{T_m s + 1}}{1 + \dfrac{K_A K_m}{T_m s + 1}} \frac{\omega_0}{s} = \frac{K_A K_m}{T_m s + 1 + K_A K_m} \frac{\omega_0}{s} = \frac{\dfrac{K_A K_m}{1 + K_A K_m}}{\underbrace{\dfrac{T_m}{1 + K_A K_m}}_{T_d} s + 1} \frac{\omega_0}{s}.$$

Taking $K_A \gg 1/K_m$ it follows that

$$\frac{K_A K_m}{1 + K_A K_m} \approx 1,\ T_d \triangleq \frac{T_m}{1 + K_A K_m} \approx \frac{T_m}{K_A K_m} \ll T_m.$$

Thus $\omega(t) = (1 - e^{-t/T_d})\omega_0 \to \omega_0$ much faster due to the feedback. With this closed-loop feedback we do not need accurate values of T_m or K_m; we need only take the feedback gain K_A to be large. This is often referred to as *high-gain* feedback.

Sensitivity Reduction via Feedback

Let's continue to look at the speed control of the DC motor with output

$$\omega(s) = \underbrace{\frac{K_m}{T_m s + 1}}_{G(s)} \frac{V_0}{s}.$$

Suppose the actual value of K_m is $K_m + \Delta K_m$ making ΔK_m the error in our estimate of K_m. Then

$$\Delta G(s) = \frac{K_m + \Delta K_m}{T_m s + 1} - \frac{K_m}{T_m s + 1} = \frac{\Delta K_m}{T_m s + 1}.$$

With the open-loop controller $G_c(s) = \dfrac{T_m s + 1}{T_d s + 1}$ the presumed response is $\omega(s) = G(s)G_c(s)R(s)$ (see Figure 9.45).

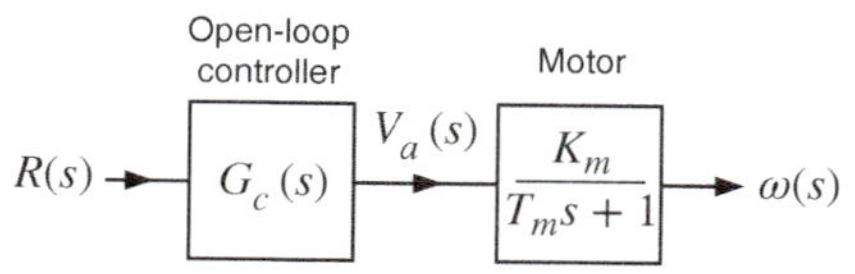

Figure 9.45. Open-loop speed control.

9.6 Model Uncertainty and Feedback*

However, as illustrated in Figure 9.46, it is actually

$$\omega(s) + \Delta\omega(s) = \big(G(s) + \Delta G(s)\big)G_c(s)R(s) = G(s)G_c(s)R(s) + \Delta G(s)G_c(s)R(s).$$

Figure 9.46. Open-loop speed control with uncertainty in K_m.

The error is then
$$\Delta\omega(s) = \Delta G(s)G_c(s)R(s)$$

making the fractional error

$$\frac{\Delta\omega(s)}{\omega(s)} = \frac{\Delta G(s)G_c(s)R(s)}{G(s)G_c(s)R(s)} = \frac{\Delta G(s)}{G(s)} = \frac{\dfrac{\Delta K_m}{T_m s+1}}{\dfrac{K_m}{T_m s+1}} = \frac{\Delta K_m}{K_m}.$$

The fractional change in the (Laplace transform of the) output response is the same as the fractional change in K_m. In particular consider the final values of the speed due to the step input $R(s) = R_0/s$. We have

$$\frac{\Delta\omega(\infty)}{\omega(\infty)} = \frac{\lim_{s\to 0} s\Delta\omega(s)}{\lim_{s\to 0} s\omega(s)} = \frac{\lim_{s\to 0} s\Delta G(s)G_c(s)R_0/s}{\lim_{s\to 0} sG(s)G_c(s)R_0/s} = \frac{\Delta G(0)G_c(0)R_0}{G(0)G_c(0)R_0} = \frac{\Delta G(0)}{G(0)} = \frac{\Delta K_m}{K_m}.$$

The fractional (percentage) error in the final speed is the same as the fractional (percentage) error in K_m.

Let's repeat this calculation, but using a proportional feedback controller. The feedback structure is shown in Figure 9.47. With $G_c(s) = K_A$ and $G(s) = \dfrac{K_m}{T_m s+1}$, $\omega(s)$ is given by

$$\omega(s) = \frac{G_c(s)G(s)}{1+G_c(s)G(s)}R(s) = \frac{K_A \dfrac{K_m}{T_m s+1}}{1+K_A \dfrac{K_m}{T_m s+1}}R(s) = \frac{K_A K_m}{T_m s+1+K_A K_m}R(s).$$

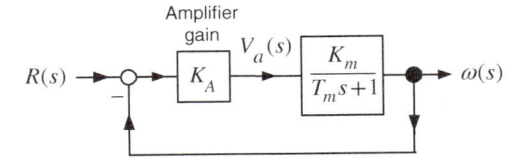

Figure 9.47. Proportional feedback speed control.

Again consider an error in the value of K_m given by ΔK_m so that $\Delta G(s) = \dfrac{\Delta K_m}{T_m s+1}$. Let's first compute the change in the closed-loop transfer function due to a change $\Delta G(s)$

in the open-loop system in a more general setting. To proceed the closed-loop transfer function is

$$G_{CL}(s) \triangleq \frac{G_c(s)G(s)}{1+G_c(s)G(s)}$$

and

$$\omega(s) = G_{CL}(s)R(s).$$

The change in $G_{CL}(s)$ due to a change in $G(s)$ is

$$\Delta G_{CL}(s) = \frac{\Delta G_{CL}(s)}{\Delta G(s)}\Delta G(s) \approx \left(\frac{d}{dG}G_{CL}(s)\right)\Delta G(s) = \frac{G_c(s)}{\left(1+G_c(s)G(s)\right)^2}\Delta G(s),$$

where we computed

$$\frac{d}{dG}\left(\frac{G_c(s)G(s)}{1+G_c(s)G(s)}\right) = \frac{G_c(s)}{1+G_c(s)G(s)} - \frac{G_c(s)G(s)}{\left(1+G_c(s)G(s)\right)^2}G(s) = \frac{G_c(s)}{\left(1+G_c(s)G(s)\right)^2}.$$

Then we can write

$$\omega(s) + \Delta\omega(s) = \left(G_{CL}(s) + \Delta G_{CL}(s)\right)R(s)$$

so that

$$\Delta\omega(s) = \Delta G_{CL}(s)R(s) \approx \frac{G_c(s)\Delta G(s)}{\left(1+G_c(s)G(s)\right)^2}R(s)$$

$$= \underbrace{\frac{G_c(s)G(s)}{1+G_c(s)G(s)}R(s)}_{\omega(s)}\frac{1}{1+G_c(s)G(s)}\frac{\Delta G(s)}{G(s)}.$$

Rearranging this last equation gives

$$\frac{\Delta\omega(s)}{\omega(s)} = \frac{1}{1+G_c(s)G(s)}\frac{\Delta G(s)}{G(s)}. \tag{9.17}$$

For the proportional speed control feedback system of Figure 9.47 this reduces to

$$\frac{\Delta\omega(s)}{\omega(s)} = \frac{1}{1+K_A\dfrac{K_m}{T_m s+1}}\dfrac{\dfrac{\Delta K_m}{T_m s+1}}{\dfrac{K_m}{T_m s+1}} = \frac{1}{1+K_A\dfrac{K_m}{T_m s+1}}\frac{\Delta K_m}{K_m}.$$

With a step reference input $R(s) = R_0/s$ and $1 + K_A K_m \gg 1$ we have

$$\frac{\Delta\omega(\infty)}{\omega(\infty)} = \frac{\lim_{s\to 0} s\Delta\omega(s)}{\lim_{s\to 0} s\omega(s)} = \frac{1}{1+K_A K_m}\frac{\Delta K_m}{K_m} \ll \frac{\Delta K_m}{K_m}.$$

The important observation of

$$\frac{1}{1+K_A K_m}\frac{\Delta K_m}{K_m} \ll \frac{\Delta K_m}{K_m}$$

is that the left side is the percentage change in the final speed for the closed-loop system while the right side is the percentage change in the motor speed for the open-loop system.

Conclusion

The effect of parameter uncertainty ΔK_m on the final speed response $\omega(\infty)$ is much less for the closed-loop system compared with the open-loop system.

Remark The quantity $S(s) \triangleq \dfrac{1}{1+G_c(s)G(s)}$ in Eq. (9.17) is called the *sensitivity* function. We have much more to say about this function in Chapter 17.

Example 9 *Operational Amplifiers*

Consider the simplified model of an operational amplifier (Op Amp) feedback control system given in Figure 9.48.

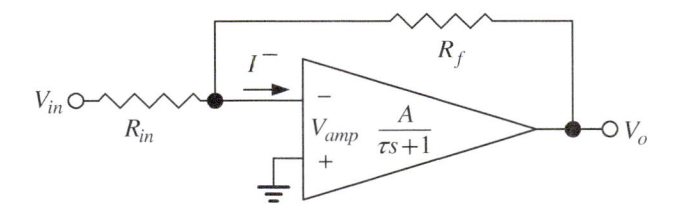

Figure 9.48. Operational amplifier feedback system.

The amplifier gain is written as

$$\frac{V_o(s)}{V_{amp}(s)} = \frac{A}{\tau s + 1},$$

where A is positive and very large (10^5 or higher) and τ is very small ($1/10^5$ or smaller). We first compute the transfer function $V_o(s)/V_{in}(s)$. Let V_{amp} be the voltage drop from the + to the − terminals of the operational amplifier where we note that V_{amp} is very small ($\approx 10^{-5}$ Volts). Proceeding, the current I^- into the − terminal of the Op Amp is

$$I^- = \frac{V_{in} - (-V_{amp})}{R_{in}} + \frac{V_o - (-V_{amp})}{R_f}. \tag{9.18}$$

However, the input impedance of an Op Amp is very large ($10^6\ \Omega$ or larger) so $I^- \approx 0$. Using $V_o = \dfrac{A}{\tau s + 1} V_{amp}$ and setting $I^- = 0$, Eq. (9.18) becomes

$$\frac{V_{in} + \dfrac{V_o}{A/(\tau s + 1)}}{R_{in}} + \frac{V_o + \dfrac{V_o}{A/(\tau s + 1)}}{R_f} = 0.$$

Upon solving for V_o we obtain

$$V_o = -\frac{\dfrac{A}{\tau s + 1}\dfrac{R_f}{R_{in}+R_f}}{1 + \dfrac{A}{\tau s + 1}\dfrac{R_{in}}{R_{in}+R_f}} V_{in} = -\frac{\dfrac{A}{\tau s + 1}\dfrac{R_f}{R_{in}+R_f}}{1 + \underbrace{\left(\dfrac{R_{in}}{R_f}\right)}_{\beta}\underbrace{\left(\dfrac{A}{\tau s + 1}\dfrac{R_f}{R_{in}+R_f}\right)}_{G(s)}} V_{in}. \tag{9.19}$$

272   9 Tracking and Disturbance Rejection

As A is very large we have for $|s| < 1/\tau$

$$V_o = -\frac{\dfrac{A}{\tau s+1}\dfrac{R_f}{R_{in}+R_f}}{1+\dfrac{A}{\tau s+1}\dfrac{R_{in}}{R_{in}+R_f}}V_{in} \approx -\frac{\dfrac{A}{\tau s+1}\dfrac{R_f}{R_{in}+R_f}}{\dfrac{A}{\tau s+1}\dfrac{R_{in}}{R_{in}+R_f}}V_{in} = -\frac{R_f}{R_{in}}V_{in}.$$

On the other hand, for $|s| \gg 1/\tau$ it follows that $V_0 \to 0$. Summarizing we have

$$V_o \approx \begin{cases} -\dfrac{R_f}{R_{in}}V_{in}, & |s| < 1/\tau \\ 0, & |s| \gg 1/\tau. \end{cases}$$

With $\beta \triangleq R_{in}/R_f$, Eq. (9.19) can be represented in block diagram form as shown in Figure 9.49.

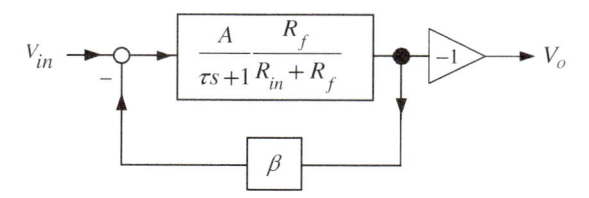

Figure 9.49. Block diagram representation of feedback for an Op Amp.

Using the block diagram we get back the transfer function V_o/V_{in} as follows.

$$\frac{V_o}{V_{in}} = -\frac{\dfrac{A}{\tau s+1}\dfrac{R_f}{R_{in}+R_f}}{1+\beta\left(\dfrac{A}{\tau s+1}\dfrac{R_f}{R_{in}+R_f}\right)} = -\frac{A\dfrac{R_f}{R_{in}+R_f}}{\tau s+1+\beta A\dfrac{R_f}{R_{in}+R_f}}.$$

This is stable as $\beta A > 0$. Further, as A is very large, for $|s| < 1/\tau$ this reduces to

$$\frac{V_o}{V_{in}} = -\frac{A\dfrac{R_f}{R_{in}+R_f}}{\tau s+1+\beta A\dfrac{R_f}{R_{in}+R_f}} \approx -\frac{1}{\beta} = -\frac{R_f}{R_{in}}.$$

With the values of the resistors R_f and R_{in} known, we have a constant gain amplifier over the range $|s| < 1/\tau$ despite not accurately knowing the values of A and τ. For example, with $R_f/R_{in} = 10$ we have $V_0 = -10V_{in}$ though A and τ vary considerably with temperature. In summary, we have given up having a large gain ($A \sim 10^5$) that varies with temperature to having a lower gain ($1/\beta = 10$) that is reliably constant. What we have just discussed is known as the *negative feedback amplifier* invented by Harold S. Black in 1927 for use on vacuum tube amplifiers. This was a critical invention for the Bell Telephone company at the time for them to be able to transmit telephone calls over long distances without distortion. It remains a critical invention now for use with operational amplifiers.

Problems

Problem 1 *Tracking*

The open-loop transfer is a double integrator given by $G(s) = 1/s^2$ set in the unity feedback control system of Figure 9.50.

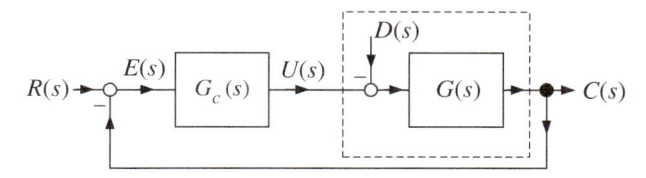

Figure 9.50. Block diagram for a double integrator control system.

With $D(s) = 0$ the error is given by

$$E(s) = \frac{1}{1 + G_c(s)G(s)} R(s).$$

Consider the controller with transfer function

$$G_c(s) = K \frac{s+1}{s+10}.$$

(a) For what values of K is the closed-loop transfer function $E(s)/R(s)$ stable?

(b) Let the reference input be

$$R(s) = \frac{R_0}{s}.$$

For what values of K does

$$\lim_{t \to \infty} e(t) = 0?$$

Problem 2 *Tracking and Disturbance Rejection*

Consider the control system of Figure 9.51 where $G(s) = \dfrac{1}{s-1}$.

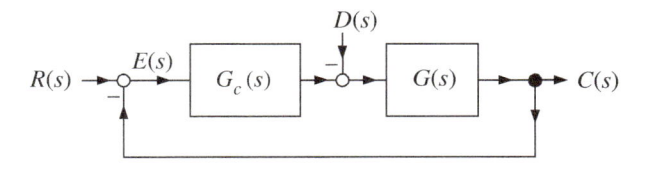

Figure 9.51. Tracking and disturbance rejection of an unstable system.

(a) With $R(s) = R_0/s$, $D(s) = 0$, and $G_c(s) = K/s$ compute $E(s)$. For what values K does $e(t) \to 0$? Show work to explain!

(b) With $R(s) = R_0/s$, $D(s) = 0$, and $G_c(s) = K$ compute $E(s)$. What is value of the final error $e(\infty) = \lim_{t \to \infty} e(t)$ in terms of K? Show work to explain!

(c) With $R(s) = 0$, $D(s) = D_0/s$, and $G_c(s) = K$ compute $E(s)$. What is value of the final error $e(\infty) = \lim_{t \to \infty} e(t)$ in terms of K? Show work to explain!

9 Tracking and Disturbance Rejection

Problem 3 *Tracking and Disturbance Rejection*

Consider the control system of Figure 9.52 where the open-loop system is

$$G(s) = \frac{b}{s+a}, \quad a > 0, b > 0.$$

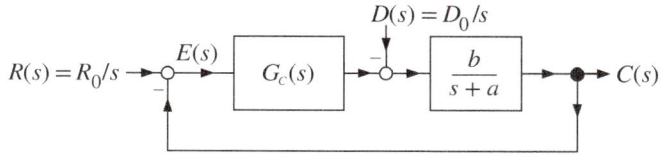

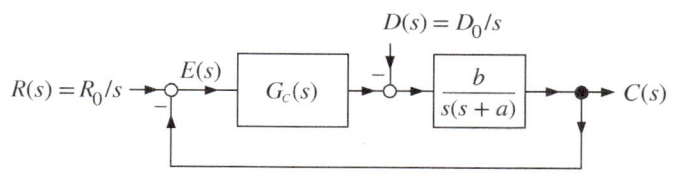

Figure 9.52. Pole-placement and disturbance rejection.

(a) With $G_c(s) = \dfrac{K}{s}$ compute $E(s)$. For what values K does $e(t) \to 0$? Show work to explain!

(b) With $G_c(s) = K\dfrac{s+\alpha}{s}$ compute $E(s)$. For what values K and α are the closed-loop poles at $-r_1, -r_2$ where $r_1 > 0, r_2 > 0$? Note that

$$(s+r_1)(s+r_2) = s^2 + (r_1+r_2)s + r_1 r_2.$$

(c) Using the controller designed in part (b), what is the final error $e(\infty) = \lim_{t \to \infty} e(t)$ if the disturbance is a ramp given by $D(s) = D_0/s^2$?

Problem 4 *Tracking and Disturbance Rejection*

Consider the control system in Figure 9.53 where the open-loop system is

$$G(s) = \frac{b}{s(s+a)}, \quad a > 0, b > 0.$$

Figure 9.53. Pole-placement and disturbance rejection.

(a) Let $G_c(s) = \dfrac{K}{s}$ and compute $E_D(s)$. For what values of K does the error due the step disturbance go to zero? Show work to explain.

(b) Let $G_c(s) = \dfrac{b_1 s + b_0}{s + a_0}$ and compute $E(s) = E_R(s) + E_D(s)$. For what values of b_1, b_0, and a_0 are the closed-loop poles at $-r_1, -r_2, -r_3$ where $r_1 > 0, r_2 > 0, r_3 > 0$? Note that

$$(s+r_1)(s+r_2)(s+r_3) = s^3 + (r_1+r_2+r_3)s^2 + (r_1 r_2 + r_1 r_3 + r_2 r_3)s + r_1 r_2 r_3.$$

(c) Using the controller designed in part (b), compute the final error $e(\infty)$ with $R(s) = R_0/s$ and $D(s) = D_0/s$.

Problem 5 *Tracking and Disturbance Rejection*
With $a > 0$ and $b > 0$, consider the control system in Figure 9.54.

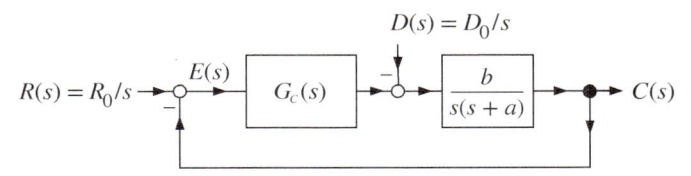

Figure 9.54. Pole-placement and disturbance rejection.

(a) Let $G_c(s) = K\dfrac{s+\alpha}{s}$ and compute $E_D(s)$. For what values of K and α does $e_D(t) \to 0$?

(b) For what values of K and α does $e_R(t) \to 0$ with $R(s) = R_0/s$?

(c) Now suppose $r(t)$ is a ramp input, i.e., $R(s) = R_0/s^2$. For what values of K and α does $e_R(t) \to 0$?

(d) Now suppose $r(t)$ is a step input, i.e., $R(s) = R_0/s$, but with disturbance given by $D(s) = D_0/s^2$? Compute $e(\infty) = \lim_{t\to\infty} e(t)$ as a function of K and α.

Problem 6 *Tracking and Disturbance Rejection*
In the feedback control system of Figure 9.55 let the controller be $G_c(s) = K\dfrac{s+1/2}{s}$.

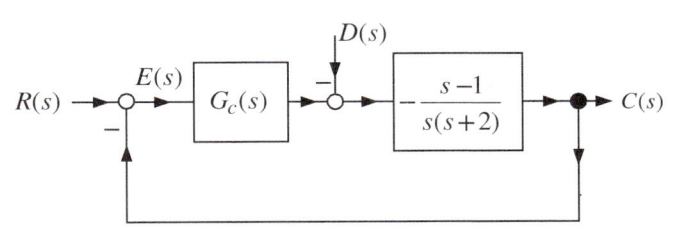

Figure 9.55. Controller design for a non-minimum phase system.

(a) Let $R(s) = R_0/s$ and $D(s) = 0$. For what values of K will $e(t) \to 0$?

(b) Let $R(s) = R_0/s$ and $D(s) = D_0/s$. For what values of K will $e(t) \to 0$? Just give the answer and a brief explanation. No new calculations are needed!

(c) Let $R(s) = R_0/s^2$ and $D(s) = D_0/s$. For what values of K will $e(t) \to 0$? Just give the answer and a brief explanation. No new calculations are needed!

Problem 7 *Disturbance Rejection*
Figure 9.56 shows the setup for a DC motor servo control system.

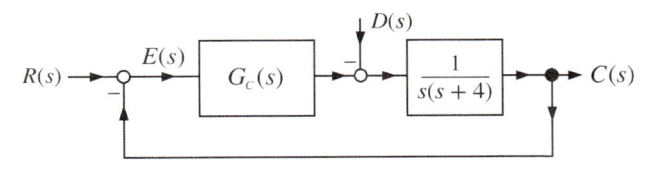

Figure 9.56. Disturbance rejection for a DC motor.

In parts (a)–(d), consider just the error due to the disturbance $D(s)$.

(a) Let $G_c(s) = K$ and $D(s) = D_0/s$. Compute $e_D(\infty) = \lim_{t \to \infty} e_D(t)$.

(b) Let $G_c(s) = K\dfrac{s+2}{s}$ and $D(s) = D_0/s$. Compute $e_D(\infty) = \lim_{t \to \infty} e_D(t)$.

(c) Let $G_c(s) = K\dfrac{s+2}{s}$ and $D(s) = D_0/s^2$. Compute $e_D(\infty) = \lim_{t \to \infty} e_D(t)$.

(d) Let $G_c(s) = K\dfrac{s+2}{s^2}$ and $D(s) = D_0/s^2$. Compute $e_D(\infty) = \lim_{t \to \infty} e_D(t)$.

Problem 8 *Tracking and Disturbance Rejection for a Satellite*

The satellite is at just the right distance from the earth such that it revolves around the earth at the same rate as the earth rotates about its own axis. Consequently, the satellite is always over the same position of the earth. It is required that the antenna always point at the same position on the earth so that signals are relayed between the United States and Europe correctly (see Figure 9.57).

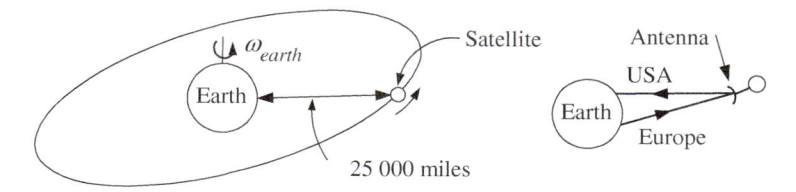

Figure 9.57. Geosynchronous satellite. $\omega_{earth} = \omega_{satellite} = 1$ rev/day.

A mathematical model of the satellite is given by

$$J\frac{d^2\theta}{dt^2} = \tau,$$

where J is the moment of inertia of the satellite, $\tau = 2rF$ is the torque due to the jets, and θ is the angular position of the satellite's antenna. The torque produced by the jets is the control input (see Figure 9.58). In the s domain, the model is

$$\frac{\theta(s)}{\tau(s)} = G_p(s) = \frac{1}{J}\frac{1}{s^2}.$$

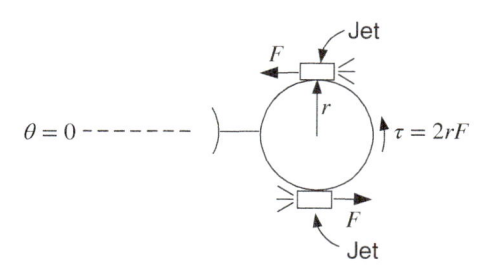

Figure 9.58. Satellite control jets.

A block diagram for this control system is shown in Figure 9.59 where $D(s)$ represents a disturbance torque such as radiation pressure.

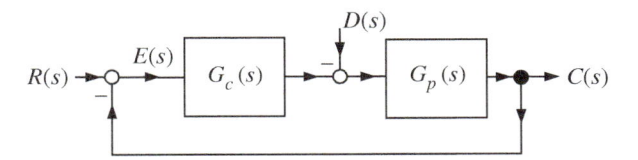

Figure 9.59. Block diagram for the satellite pointing control system.

(a) $C(s) = G_1(s)R(s) + G_2(s)D(s)$. Find G_1, G_2 in terms of G_c, G_p.
(b) $E(s) = T_1(s)R(s) + T_2(s)D(s)$. Find T_1, T_2 in terms of G_c, G_p.

Tracking: Assume $D(s) = 0$ in parts (c)–(i).

(c) Let $R(s) = 1/s^2$, i.e., $r(t) = t$. What type number must $G_c(s)$ have to track this input?
(d) Will the controller $G_c(s) = K$ work to track $R(s) = 1/s^2$ in part (c)? Explain.
(e) Will the controller $G_c(s) = K(s+a)/s$ work to track $R(s) = 1/s^2$ in part (c)? Explain.

Consider a gyroscope to stabilize the closed-loop system. The gyroscope is an instrument that measures angular speed. The new control structure is shown in Figure 9.60. Assume $D(s) = 0$ and $K_g > 0$.

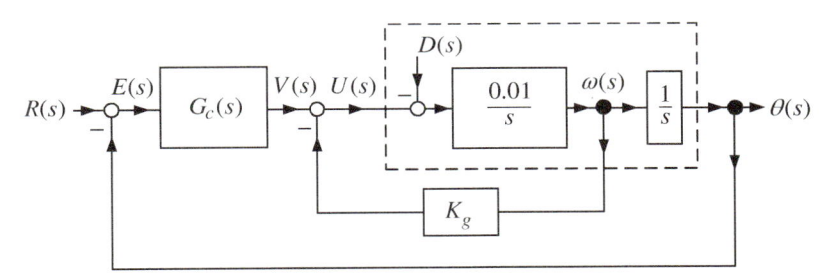

Figure 9.60. Satellite pointing control system with a gyroscope added.

(f) Show that $\omega(s) = \dfrac{0.01}{s + 0.01K_g}(V(s) - D(s))$ and, as a consequence, the block diagram of Figure 9.60 reduces to that of Figure 9.61.

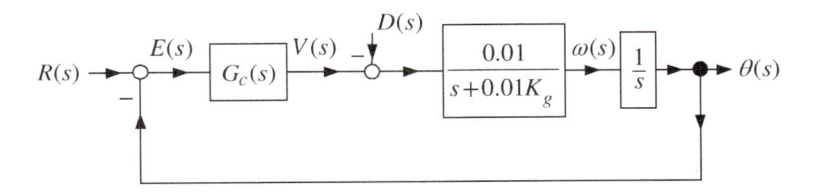

Figure 9.61. Block diagram reduction of the satellite control system with gyro.

(g) What type number must $G_c(s)$ have to track $r(t) = tu_s(t)$, i.e., $R(s) = 1/s^2$?
(h) Will $G_c(s) = K/s$ work to track $R(s) = 1/s^2$ in part (g)? Explain.

278 9 Tracking and Disturbance Rejection

(i) Will $G_c(s) = K(s+a)/s$ work to track $R(s) = 1/s^2$ in part (g)? For what values of K, K_g, and α?

Disturbance rejection: For parts (j)–(k) use the block diagram of Figure 9.61 with $R(s) = 0$.

(j) What type number must $G_c(s)$ have to reject $D(s) = 1/s$?

(k) Will $G_c(s) = K/s$ work to reject $D(s) = 1/s$ in part (j)? Explain.

(l) Will $G_c(s) = K(s+a)/s$ work to reject $D(s) = 1/s$ in part (j)? Explain. No need to use the Routh–Hurwitz test as you can use the controller parameters to place the three closed-loop poles at $-r_1, -r_2, -r_3$!

Problem 9 *Internal Model Principle*
Consider the control system given in Figure 9.62.

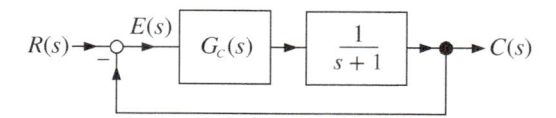

Figure 9.62. Tracking of *stable* reference signals.

(a) Let $R(s) = 1/(s+2)$, i.e., $r(t) = e^{-2t}$. Will $G_c(s) = K > 0$ work to asymptotically track $R(s)$? Explain.

(b) Let $R(s) = \dfrac{1}{(s^2 + 2s + 2)(s+3)(s+6)}$. Will $G_c(s) = K > 0$ work to track $R(s)$? Explain.

(c) What conclusion can you reach about tracking stable reference signals that asymptotically go to zero? Give a brief explanation of your conclusion.

Problem 10 *Rejecting a Sinusoidal Disturbance*
Consider the feedback system of Figure 9.63 with a step reference input $r(t) = R_0 u_s(t)$ and a sinusoidal disturbance $d(t) = D_0 \sin(t)$.

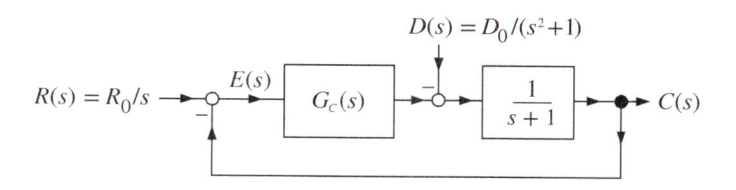

Figure 9.63. Rejection of a sinusoidal disturbance.

(a) Consider the controller $G_c(s) = K\dfrac{s+\alpha}{s^2+1}$ which was chosen to have a denominator that was the *same* as that of the disturbance. For what values of K and α will the final disturbance error $e_D(\infty)$ be zero?

(b) Keeping $D(s) = \dfrac{D_0}{s^2+1}$ and using the same controller designed in part (a), let $R(s) = R_0/s$. For what values of K and α is $e(\infty)$ finite? Compute $e(\infty)$ in terms of R_0, K, and α.

Problem 11 *Rejecting a Sinusoidal Disturbance*

Let's reconsider Example 8 whose block diagram is repeated in Figure 9.64.

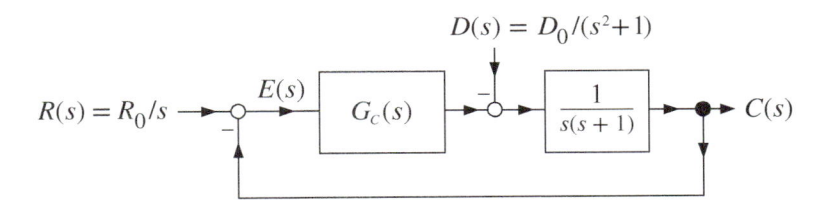

Figure 9.64. Asymptotically rejecting a sinusoidal disturbance.

We give a "sneak preview" of the next chapter by letting the controller have the form

$$G_c(s) = \frac{b_c(s)}{a_c(s)} = \underbrace{\frac{b_3 s^3 + b_2 s^2 + b_1 s + b_0}{s + a_0}}_{\overline{G}_c(s)} \cdot \frac{1}{s^2+1}.$$

Note that $G_c(s)$ is proper, but not strictly proper. The factor $\dfrac{1}{s^2+1}$ is required by the internal model principle in order to asymptotically reject the disturbance. As explained in Chapter 10, the factor $\overline{G}_c(s)$ will allow us place the closed-loop poles in any desired location.

(a) Compute $E_D(s)$.

(b) Use your answer in part (a) to show that the controller parameters b_3, b_2, b_1, b_0, a_0 can be chosen so that the denominator of $E_D(s)$ is $s^5 + f_4 s^4 + f_3 s^3 + f_2 s^2 + f_1 s + f_0$ where coefficients $f_4, \ldots, f_0$ can be arbitrarily specified.

Problem 12 *Tracking and Disturbance Rejection*

Consider the missile in Figure 9.65 where the control problem is to keep the angle θ zero so the missile axis is aligned with its velocity vector $\vec{v}$. The control input is the thrust angle δ.

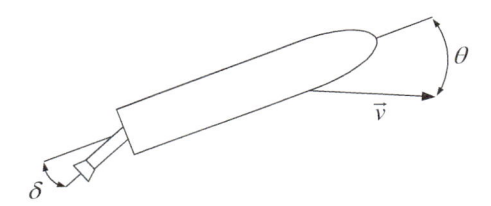

Figure 9.65. Missile attitude control.

280 9 Tracking and Disturbance Rejection

The block diagram for this missile control system is given in Figure 9.66.

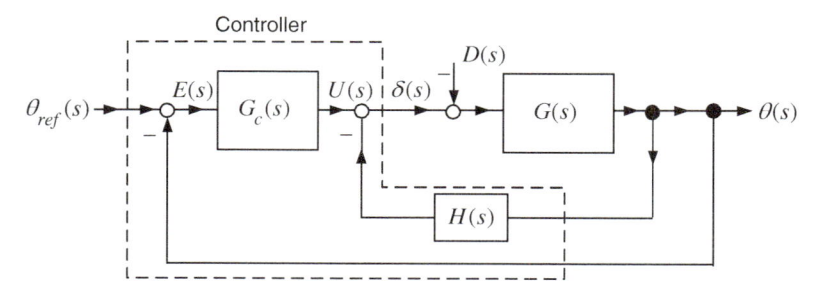

Figure 9.66. Missile control system.

The transfer function between the thrust angle δ and θ is

$$G(s) \triangleq \frac{\theta(s)}{\delta(s)} = \frac{1}{s^2 - 1}.$$

The gyro feedback is given by $H(s) = K_t s$. The problem is to keep the missile pointing in the same direction as its velocity, that is, keep $\theta = 0$ by varying the thrust angle δ. Due to aerodynamic forces on the missile, the velocity and attitude get misaligned. We can model these aerodynamic forces as the disturbance $D(s)$ as shown in the block diagram. In all that follows, let $\theta_{ref}(s) = 0$. The objective is design a controller that eliminates (or at least reduces the effect of) the disturbance $D(s)$ on maintaining $\theta = 0$.

(a) Show that

$$E_D(s) = \frac{G(s)}{1 + (G_c(s) + H(s))\,G(s)} D(s).$$

(b) Let $G_c(s) = K$, $K_t = 0$, and $D(s) = D_0/s$. Compute $e_D(\infty)$. Explain and show work.

(c) Let $G_c(s) = K$, $K_t = 1$, and $D(s) = D_0/s$. Compute $e_D(\infty)$. Explain and show work.

(d) Let $G_c(s) = K(s+\alpha)/s$, $K_t > 0$, and $D(s) = D_0/s$. Can $e_D(\infty)$ be made to go to zero? Explain and show work.

Problem 13 *Tracking and Disturbance Rejection*

Consider the control system given in Figure 9.67.

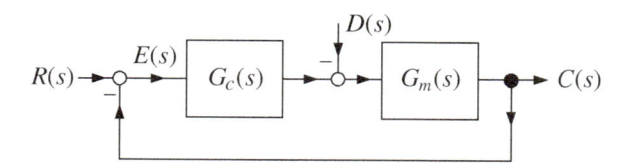

Figure 9.67. Unity feedback control system.

(a) The error can be written in the form $E(s) = T_1(s)R(s) + T_2(s)D(s)$. Find $T_1(s)$ and $T_2(s)$ in terms of $G_c(s)$ and $G_m(s)$.

Let $G_m(s) = \dfrac{2}{s(s+6)}$ in the remaining parts (b)–(g).

(b) With $D(s) = 0$, what is the minimum type number $G_c(s)$ must have to track $r(t) = R_0 u_s(t)$?

(c) With $D(s) = 0$, $R(s) = R_0/s$, will the controller $G_c(s) = K/s$ work to make $e(\infty) = 0$? Explain why or why not. Show work.

(d) With $D(s) = 0$, $R(s) = R_0/s^2$ so that $r(t) = R_0 t u_s(t)$, will the controller $G_c(s) = K\dfrac{s+2}{s+10}\dfrac{1}{s}$ work to make $e(\infty) = 0$? If so, for what values of K? Explain and show work.

(e) Let $R(s) = 0$, $D(s) = D_0/s$ so that $d(t) = D_0 u_s(t)$. What is the minimum type number that $G_c(s)$ must have to reject this disturbance?

(f) Let $R(s) = 0$ and $D(s) = D_0/s$. Will the controller $G_c(s) = K\dfrac{s+2}{s+10}\dfrac{1}{s}$ work to make $e(\infty) = 0$? Explain and show work.

(g) Let $R(s) = 0$ and $D(s) = D_0/s^2$. Using the controller $G_c(s) = K\dfrac{s+2}{s+10}\dfrac{1}{s}$, compute $e(\infty)$. Explain and show work.

Problem 14 *A Lead Controller Approximates a PD Controller*
A lead controller $G_c(s)$ is given by

$$G_c(s) = K_D \frac{s+\alpha}{s/20+1} = 20 K_D \frac{s+\alpha}{s+20} \quad \text{with} \quad \alpha < 20.$$

For $|s| \ll 20$, i.e., $|s/20| \ll 1$, we have $G_c(s) \approx K_D(s+\alpha) = K_D s + \alpha K_D$ showing that it is approximately a proportional plus derivative (PD) controller. For $|s/20| \gg 1$, $G_c(s) = K_D(s+\alpha)/(s/20+1) \approx 20 K_D$. In other words, at low frequencies, $G_c(s)$ acts as a proportional plus derivative controller while at high frequencies it is a proportional controller. This is in contrast to a differentiator $K_D s$ whose gain goes to ∞ as $|s| \to \infty$. Consider the control system given in Figure 9.68.

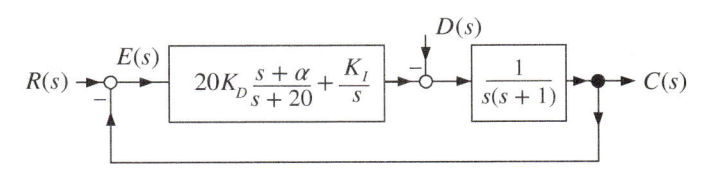

Figure 9.68. Control system with both a lead and an integral controller.

As indicated in Figure 9.69, for the purposes of design use the approximation

$$G_c(s) = K_D \frac{s+\alpha}{s/20+1} + \frac{K_I}{s} \approx K_D s + \alpha K_D + \frac{K_I}{s}.$$

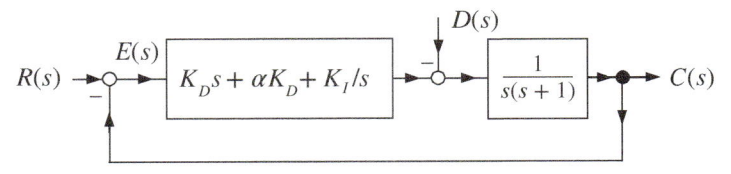

Figure 9.69. Lead controller replaced by a PD controller to determine the gain values.

9 Tracking and Disturbance Rejection

(a) For the controller $K_D s + \alpha K_D + \frac{K_I}{s}$ compute K_D, α, and K_I that place the closed-loop poles at $-3, -1 \pm j$.

(b) For the values of K_D, α, and K_I found in (a), use MATLAB to find the closed-loop poles of the actual system, that is, with the controller $K_D \frac{s+\alpha}{s/20+1} + \frac{K_I}{s}$.

In the following use $G_c(s) = K_D \frac{s+\alpha}{s/20+1} + \frac{K_I}{s}$.

(c) Set $D(s) = 0$ and $R(s) = R_0/s^2$. Is $e(\infty) = 0$? Explain.

(d) Set $D(s) = 0$ and $R(s) = R_0/s^3$ ($r(t) = R_0 t^2/2$). Compute $e(\infty)$.

(e) Set $R(s) = 0$ and $D(s) = D_0/s$. Is $e_D(\infty) = 0$? Explain.

(f) Simulate this system with $r(t) = 2t$ and $d(t) = (1/2)u_s(t)$. Plot the error $e(t)$ and $v_a(t)$ on separate graphs.

Problem 15 *Internal Model Principle*
Consider the control system in Figure 9.70.

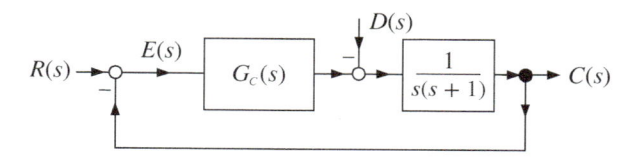

Figure 9.70. Internal model principle.

(a) With $R(s) = \frac{1}{s^2+1}$ and $D(s) = \frac{D_0}{s}$ can the controller $G_c(s) = K\frac{s+\alpha}{s^2+1}$ be used to have $e(t) \to 0$? Explain your reasoning.

(b) With $R(s) = \frac{2}{s} + \frac{1}{s^2+1}$ and $D(s) = 0$ can the controller $G_c(s) = K\frac{s+\alpha}{s^2+1}$ be used to have $e(t) \to 0$? Explain your reasoning.

(c) With the input $R(s) = \frac{1}{s^2+1}$, consider the controller

$$G_c(s) = \frac{b_c(s)}{a_c(s)} \triangleq \frac{(f_3 - f_4 + 1)s^3 + (f_2 - f_4 + f_6)s^2 + (f_1 - f_4 + f_5)s + f_0}{s^3 + (f_6 - 1)s^2 + (f_5 - f_6)s + f_4 - f_5} \frac{1}{s^2+1}.$$

Show that the error $E(s)$ is given by

$$E(s) = \frac{(s^3 + (f_6-1)s^2 + (f_5 - f_6)s + f_4 - f_5) s(s+1)}{s^7 + f_6 s^6 + f_5 s^5 + f_4 s^4 + f_3 s^3 + f_2 s^2 + f_1 s + f_0} R_0.$$

Using the f_i the closed-loop poles can be chosen as desired. The procedure to obtain this controller $G_c(s)$ will be given in Chapter 10.

Problems 283

Problem 16 *Internal Model Principle*

Consider the DC motor control system of Figure 9.71 with open-loop transfer function $G(s) = \dfrac{1}{s(s+1)}$. Let $r(t) = \sin(t)$ or $R(s) = \dfrac{1}{s^2+1}$. Using the controller $G_c(s) = K\dfrac{s+\alpha}{s^2+1}$ it was shown in Example 8 that this reference input could be asymptotically tracked. However, the performance was poor as the transient response died out slowly.

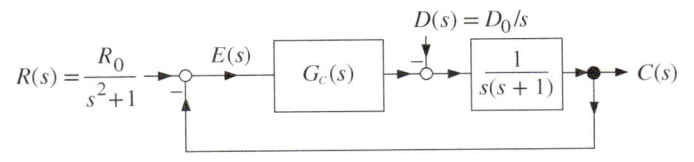

Figure 9.71. Tracking a sinusoidal signal.

Let's add a tachometer to the shaft of the DC motor to provide a speed measurement and see if it can be used to obtain better performance. The output of the tachometer is a voltage V_t proportional to the motor's speed, i.e., $V_{tach} = K_t \omega$ where K_t is an adjustable gain. See Figure 9.72.

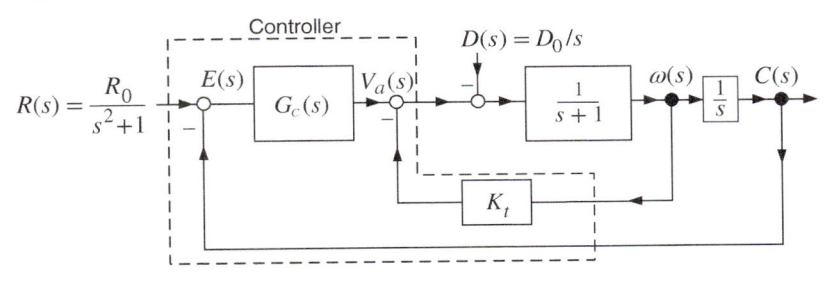

Figure 9.72. Using a tachometer to speed up the system response.

(a) Along with the tachometer, let $G_c(s) = K_p + K\dfrac{s+\alpha}{s^2+1}$. Show that using the gains $K, \alpha, K_t,$ and K_p the closed-loop poles may be placed at any desired location.

(b) Repeat (a) with $G_c(s) = K_p + \dfrac{K_I}{s} + K\dfrac{s+\alpha}{s^2+1}$. What is the advantage of including an integrator in $G_c(s)$?

Problem 17 *Pitch Control with a I-PD Controller*

As in the text, again consider the pitch control of the aircraft with transfer function

$$\frac{\theta(s)}{\delta(s)} = G(s) = \frac{1.51s + 0.1774}{s^3 + 0.739s^2 + 0.921s}.$$

The design specifications are still (i) overshoot less than 10%, (ii) rise time less than two seconds, (iii) settling time less than 10 seconds, (iv) final error less than 2%. We take the maximum elevator deflection to be $25°$ (0.436 rad), that is, $-25° \leq \delta \leq 25°$. Here the controller is using the I-PD architecture shown in Figure 9.73. This results in the elevator command being given by (taking $\tau = 0$)

$$\delta_c(t) = K_I \int_0^t e(t')dt' + K_p c(t) + K_D \frac{dc(t)}{dt}.$$

284   9 Tracking and Disturbance Rejection

In particular, at $t=0$ this becomes $\delta_c(0) = K_I \int_0^0 e(t')dt' + K_p c(0) + K_D \dot{c}(0) = 0$. As a consequence, K_p and K_D may be taken to be much larger without saturating the actuator compared with their corresponding values in the PI-D controller. Further, the closed-loop system will have only the single zero of the open-loop transfer function $G(s)$.

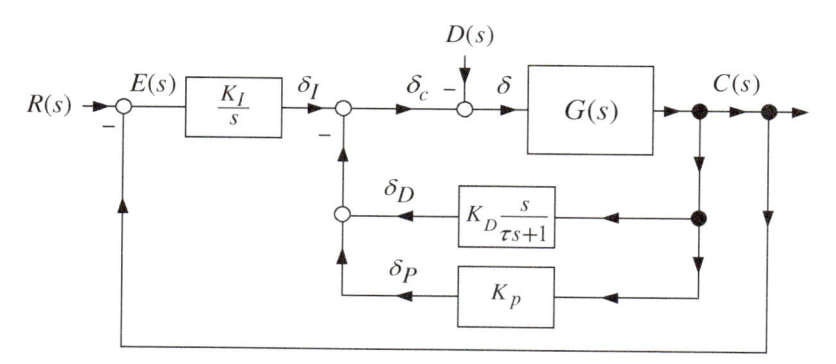

Figure 9.73. I-PD unity feedback control system.

Let $R(s) = R_0/s$ with $R_0 = 0.2$ rad (11.5 degrees) and take $D(s) = 0$.

(a) Compute the inner loop transfer function $G_2(s) = C(s)/\delta_I(s)$.

(b) Compute the overall closed-loop transfer function $G_{CL}(s) = \dfrac{G_2(s)G(s)}{1+G_2(s)G(s)}$.

(c) Are the closed-loop poles the same as those obtained using the PI-D and PID controllers?

(d) Make and run a simulation of this system. What are the values of K_p, K_D, and K_I you chose? What was the rise time (time to $0.9R_0$), settling time (2% criterion), and overshoot of the response with $D(s) = 0$?

(e) Compute $G_{CL}(s)$ with the values of gains you chose in part (d). Give it in pole–zero form. Are there any stable pole–zero cancellations?

10

Pole Placement, 2 DOF Controllers, and Internal Stability

Given a transfer function model $G(s)$ we first show how to design a controller $G_c(s)$ so that the closed-loop poles can be placed at *any* desired location! It turns out to be a quite straightforward procedure. However, the zeros of $G_c(s)$ along with those of $G(s)$ often result in significant overshoot in the closed-loop step response. Consequently, we also show how this overshoot can often be eliminated by passing the step reference input through a filter $G_f(s)$ (transfer function). This use of two transfer functions $G_c(s)$ and $G_f(s)$ is referred to as a *two degree of freedom* (2 DOF) controller. Finally we present the notion of *internal stability*, which requires there be no *unstable* pole–zero cancellations between the controller transfer function $G_c(s)$ and the model transfer function $G(s)$.

10.1 Output Pole Placement

In using the internal model principle to track a reference signal $R(s)$ or reject a disturbance $D(s)$, the fundamental difficulty was to make the closed-loop system stable as it is straightforward to make $G_c(s)G(s)$ contain the poles of $R(s)$, and $G_c(s)$ contain the poles of $D(s)$. Given a physical system described by a strictly proper transfer function, we now show how to design a controller that guarantees closed-loop stability while achieving our tracking and disturbance rejection objectives.

Example 1 *Pole Placement for a Second-Order Control System*
Consider the control system of Figure 10.1 with the open-loop transfer function given by $G(s) = \dfrac{b}{s(s+a)}$.

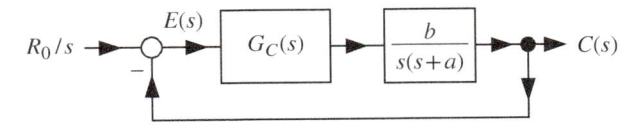

Figure 10.1. Pole placement for a second-order system.

In this example we just want to be able to track a step input. Because $G(s)$ is type 1 this is accomplished by simply designing $G_c(s)$ so that the closed-loop system is stable. The pole placement controller is given by

$$G_c(s) = \frac{b_c(s)}{a_c(s)} = \frac{b_1 s + b_0}{s + a_0}.$$

An Introduction to System Modeling and Control, First Edition. John Chiasson.
© 2022 John Wiley & Sons, Inc. Published 2022 by John Wiley & Sons, Inc.
Companion website: www.wiley.com/go/chiasson/anintroductiontosystemmodelingandcontrol

10 Pole Placement, 2 DOF Controllers, and Internal Stability

To explain, the denominator of $G_c(s)$ is chosen to be a *monic*[1] polynomial of degree 1 less than the denominator polynomial of $G(s) = \dfrac{b}{s(s+a)}$. The numerator of $G_c(s)$ is then chosen to have the same degree as its denominator.[2] Choosing $G_c(s)$ to have this form we now show that b_1, b_0, and a_0 can be found to put the closed-loop poles at any arbitrarily chosen location. We have

$$E(s) = \frac{1}{1 + \dfrac{b_1 s + b_0}{s + a_0} \dfrac{b}{s(s+a)}} \frac{R_0}{s}$$

$$= \frac{(s+a_0)s(s+a)}{s(s+a)(s+a_0) + (b_1 s + b_0)b} \frac{R_0}{s}$$

$$= \frac{(s+a_0)(s+a)}{s^3 + (a+a_0)s^2 + (aa_0 + bb_1)s + bb_0} R_0.$$

With

$$s^3 + f_2 s^2 + f_1 s + f_0$$

the desired denominator of $E_R(s)$ we must set

$$s^3 + (a+a_0)s^2 + (aa_0 + bb_1)s + bb_0 = s^3 + f_2 s^2 + f_1 s + f_0.$$

Equating coefficients of the powers of s gives

$$bb_0 = f_0$$
$$aa_0 + bb_1 = f_1$$
$$a + a_0 = f_2$$

or

$$b_0 = f_0/b$$
$$a_0 = f_2 - a$$
$$b_1 = \frac{f_1 - aa_0}{b} = \frac{f_1 - af_2 + a^2}{b}.$$

If we want the three closed-loop poles at $-r_1, -r_2, -r_2$ we simply set

$$s^3 + f_2 s^2 + f_1 s + f_0 = (s+r_1)(s+r_2)(s+r_3)$$
$$= s^3 + \underbrace{(r_1 + r_2 + r_3)}_{f_2}s^2 + \underbrace{(r_1 r_2 + r_1 r_3 + r_2 r_3)}_{f_1}s + \underbrace{r_1 r_2 r_3}_{f_0}.$$

Going from the second to the third line we did a multinomial expansion of third order. See the Appendix on *Multinomial Expansions* at the end of this chapter for multinomial expansions of orders 2 through 7. This control system is illustrated in Figure 10.2.

---

[1] A monic polynomial is one whose leading coefficient is 1.
[2] So $G_c(s)$ is proper, but not strictly proper.

10.1 Output Pole Placement

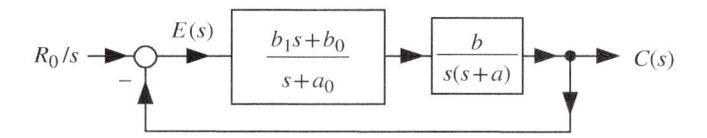

Figure 10.2. Pole placement for a second-order system.

Example 2 *Disturbance Rejection for a Second-Order Control System*

Let's consider the previous example where we also want to reject a constant disturbance as well. The control system is shown in Figure 10.3.

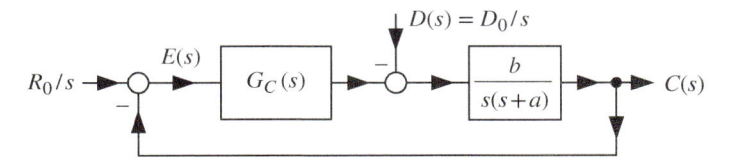

Figure 10.3. Disturbance rejection for a second-order system.

In order to arbitrarily place the closed-loop poles, $G_c(s)$ is chosen to have the form

$$G_c(s) = \frac{b_c(s)}{a_c(s)} = \underbrace{\frac{b_2 s^2 + b_1 s + b_0}{s + a_0}}_{\bar{G}_c(s)} \frac{1}{s}.$$

To explain, $G_c(s)$ is written as the product of the two factors $\bar{G}_c(s)$ and $1/s$. It must have the factor $1/s$ to reject the step disturbance according to the internal model principle. The factor $\bar{G}_c(s)$ is determined as follows. We look at $G(s) = \dfrac{b}{s(s+a)}$ and note the denominator of $G(s)$ has degree 2. The denominator of $\bar{G}_c(s)$ is then set to be a polynomial of degree 1 *less* than the denominator of $G(s)$, which in this case is $s + a_0$. The denominator of $G_c(s)$ is now $s(s + a_0)$, which is degree 2. The numerator of $G_c(s) = \bar{G}_c(s)\dfrac{1}{s}$ is then set to be a polynomial of this same degree, which in this example is $b_2 s^2 + b_1 s + b_0$. Note that $G_c(s)$ is proper, but not strictly proper.

Let's compute $E_D(s)$. We have

$$E_D(s) = \frac{G(s)}{1 + G_c(s)G(s)} \frac{D_0}{s} = \frac{\dfrac{b}{s(s+a)}}{1 + \dfrac{b_2 s^2 + b_1 s + b_0}{s + a_0}\dfrac{1}{s}\dfrac{b}{s(s+a)}} \frac{D_0}{s}$$

$$= \frac{s(s + a_0)b}{s^2(s + a_0)(s + a) + bb_2 s^2 + bb_1 s + bb_0} \frac{D_0}{s}$$

$$= \frac{(s + a_0)b}{s^4 + (a + a_0)s^3 + (aa_0 + bb_2)s^2 + bb_1 s + bb_0} D_0.$$

With

$$s^4 + f_3 s^3 + f_2 s^2 + f_1 s + f_0$$

the desired denominator of $E_D(s)$ set

$$s^4 + (a + a_0)s^3 + (aa_0 + bb_2)s^2 + bb_1 s + bb_0 = s^4 + f_3 s^3 + f_2 s^2 + f_1 s + f_0.$$

288   10 Pole Placement, 2 DOF Controllers, and Internal Stability

This requires setting the coefficients of $G_c(s)$ as

$$b_0 = f_0/b$$
$$b_1 = f_1/b$$
$$a_0 = f_3 - a$$
$$b_2 = (f_2 - aa_0)/b = (f_2 - af_3 + a^2)/b.$$

Example 3 *Tracking a Sinusoid*

Consider the control system shown in Figure 10.4. It is required that this system asymptotically track the reference signal $r(t) = R_0 \sin(t)$, i.e., $R(s) = R_0/(s^2+1)$.

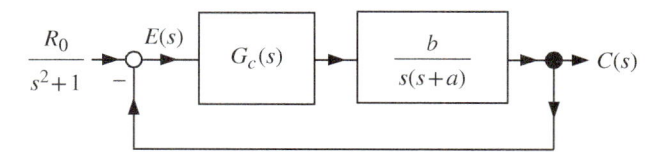

Figure 10.4. Asymptotic tracking of a sinusoidal input.

In order to arbitrarily place the closed-loop poles, $G_c(s)$ is chosen to have the form

$$G_c(s) = \frac{b_c(s)}{a_c(s)} = \underbrace{\frac{b_3 s^3 + b_2 s^2 + b_1 s + b_0}{s + a_0}}_{\bar{G}_c(s)} \frac{1}{s^2+1}.$$

To explain, $G_c(s)$ must have the factor $\frac{1}{s^2+1}$ to track the sinusoidal reference input according to the internal model principle. The denominator of $\bar{G}_c(s)$ is chosen to have degree 1 less than the denominator of $G(s)$. Finally, the numerator of $G_c(s)$ is set to have the same degree as its denominator. Note that $G_c(s)$ is proper. Proceeding, the error is given by

$$E(s) = \frac{1}{1 + G_c(s)G(s)} R(s)$$

$$= \frac{1}{1 + \frac{b_3 s^3 + b_2 s^2 + b_1 s + b_0}{s + a_0} \frac{1}{s^2+1} \frac{b}{s(s+a)}} \frac{R_0}{s^2+1}$$

$$= \frac{(s+a_0)s(s+a)(s^2+1)}{(s+a_0)s(s+a)(s^2+1) + bb_3 s^3 + bb_2 s^2 + bb_1 s + bb_0} \frac{R_0}{s^2+1}$$

$$= \frac{(s+a_0)s(s+a)}{s^5 + (a+a_0)s^4 + (aa_0 + bb_3 + 1)s^3 + (a + a_0 + bb_2)s^2 + (aa_0 + bb_1)s + bb_0} R_0.$$

With

$$s^5 + f_4 s^4 + f_3 s^3 + f_2 s^2 + f_1 s + f_0$$

the desired denominator of $E_R(s)$ set

$$s^5 + (a + a_0)s^4 + (aa_0 + bb_3 + 1)s^3 + (a + a_0 + bb_2)s^2 + (aa_0 + bb_1)s + bb_0$$
$$= s^5 + f_4 s^4 + f_3 s^3 + f_2 s^2 + f_1 s + f_0.$$

This requires setting the coefficients of $G_c(s)$ as

$$b_0 = \frac{f_0}{b}$$
$$a_0 = f_4 - a$$
$$b_1 = \frac{f_1 - aa_0}{b} = \frac{f_1 - af_4 + a^2}{b}$$
$$b_2 = \frac{f_2 - a - a_0}{b} = \frac{f_2 - f_4}{b}$$
$$b_3 = \frac{f_3 - aa_0 - 1}{b} = \frac{f_3 - af_4 + a^2 - 1}{b}.$$

Example 4 *Pole Placement*

Consider the control system given in Figure 10.5 where

$$G(s) = \frac{s-2}{(s-1)(s-3)}.$$

It is required to track step inputs and reject step disturbances.

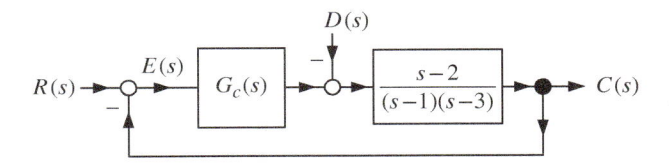

Figure 10.5. An unstable system with a right half-plane zero.

To arbitrarily place the closed-loop poles, $G_c(s)$ is chosen to have the form

$$G_c(s) = \underbrace{\frac{b_2 s^2 + b_1 s + b_0}{s + a_0}}_{\bar{G}_c(s)} \frac{1}{s}.$$

$G_c(s)$ must have the factor $1/s$ to track the step input and reject the step disturbance according to the internal model principle. The denominator of $\bar{G}_c(s)$ is chosen to have degree 1 less than the degree of $G(s)$. Finally, the numerator of $G_c(s)$ is set to have the same degree as its denominator. Then

$$E_R(s) = \frac{1}{1 + G_c(s)G(s)} R(s)$$

$$= \frac{1}{1 + \frac{b_2 s^2 + b_1 s + b_0}{s + a_0} \frac{1}{s} \frac{s-2}{(s-1)(s-3)}} \frac{R_0}{s}$$

$$= \frac{s(s+a_0)(s-1)(s-3)}{s(s+a_0)(s-1)(s-3) + (b_2 s^2 + b_1 s + b_0)(s-2)} \frac{R_0}{s}$$

$$= \frac{(s+a_0)(s-1)(s-3)}{s^4 + (b_2 + a_0 - 4)s^3 + (b_1 - 2b_2 - 4a_0 + 3)s^2 + (3a_0 + b_0 - 2b_1)s - 2b_0} R_0.$$

The equation that must then be solved is

$$s^4 + (b_2 + a_0 - 4)s^3 + (b_1 - 2b_2 - 4a_0 + 3)s^2 + (3a_0 + b_0 - 2b_1)s - 2b_0$$
$$= s^4 + f_3 s^3 + f_2 s^2 + f_1 s + f_0.$$

In matrix form this is written as

$$\begin{bmatrix} f_3 \\ f_2 \\ f_1 \\ f_0 \end{bmatrix} = \begin{bmatrix} 1 & 0 & 0 & 1 \\ -2 & 1 & 0 & -4 \\ 0 & -2 & 1 & 3 \\ 0 & 0 & -2 & 0 \end{bmatrix} \begin{bmatrix} b_2 \\ b_1 \\ b_0 \\ a_0 \end{bmatrix} + \begin{bmatrix} -4 \\ 3 \\ 0 \\ 0 \end{bmatrix}.$$

Solving for the coefficients of $G_c(s)$ we obtain

$$\begin{bmatrix} b_2 \\ b_1 \\ b_0 \\ a_0 \end{bmatrix} = \begin{bmatrix} 1 & 0 & 0 & 1 \\ -2 & 1 & 0 & -4 \\ 0 & -2 & 1 & 3 \\ 0 & 0 & -2 & 0 \end{bmatrix}^{-1} \left(\begin{bmatrix} f_3 \\ f_2 \\ f_1 \\ f_0 \end{bmatrix} - \begin{bmatrix} -4 \\ 3 \\ 0 \\ 0 \end{bmatrix} \right)$$

$$= \begin{bmatrix} 5 & 2 & 1 & 1/2 \\ -6 & -3 & -2 & -1 \\ 0 & 0 & 0 & -1/2 \\ -4 & -2 & -1 & -1/2 \end{bmatrix} \left(\begin{bmatrix} f_3 \\ f_2 \\ f_1 \\ f_0 \end{bmatrix} - \begin{bmatrix} -4 \\ 3 \\ 0 \\ 0 \end{bmatrix} \right).$$

This simplifies to

$$\begin{bmatrix} b_2 \\ b_1 \\ b_0 \\ a_0 \end{bmatrix} = \begin{bmatrix} f_0/2 + f_1 + 2f_2 + 5f_3 + 14 \\ -f_0 - 2f_1 - 3f_2 - 6f_3 - 15 \\ -f_0/2 \\ -f_0/2 - f_1 - 2f_2 - 4f_3 - 10 \end{bmatrix}. \quad (10.1)$$

To set the closed-loop poles at $-r_1, -r_2, -r_3, -r_4$ we simply set[3]

$$s^4 + f_3 s^3 + f_2 s^2 + f_1 s + f_0 = (s+r_1)(s+r_2)(s+r_3)(s+r_4)$$
$$= s^4 + \underbrace{(r_1 + r_2 + r_3 + r_4)}_{f_3} s^3$$
$$+ \underbrace{(r_1 r_2 + r_1 r_3 + r_1 r_4 + r_2 r_3 + r_2 r_4 + r_3 r_4)}_{f_2} s^2$$
$$+ \underbrace{(r_1 r_2 r_3 + r_1 r_2 r_4 + r_1 r_3 r_4 + r_2 r_3 r_4)}_{f_1} s + \underbrace{r_1 r_2 r_3 r_4}_{f_0}.$$

Using (10.1) with the four closed-loop poles all placed at -1, the controller is

$$G_c(s) = \frac{50.5 s^2 - 66 s - 0.5}{s - 42.5} \frac{1}{s} = \frac{50.5(s - 1.3145)(s + 0.0075)}{s - 42.5} \frac{1}{s}. \quad (10.2)$$

Figure 10.6 shows the unit step response of the closed-loop system. The overshoot is unacceptably high and will be addressed in Section 10.2. See Figure 10.24 and Problem 9.

---

[3] See the Appendix at the end of this chapter on *Multinomial Expansions*.

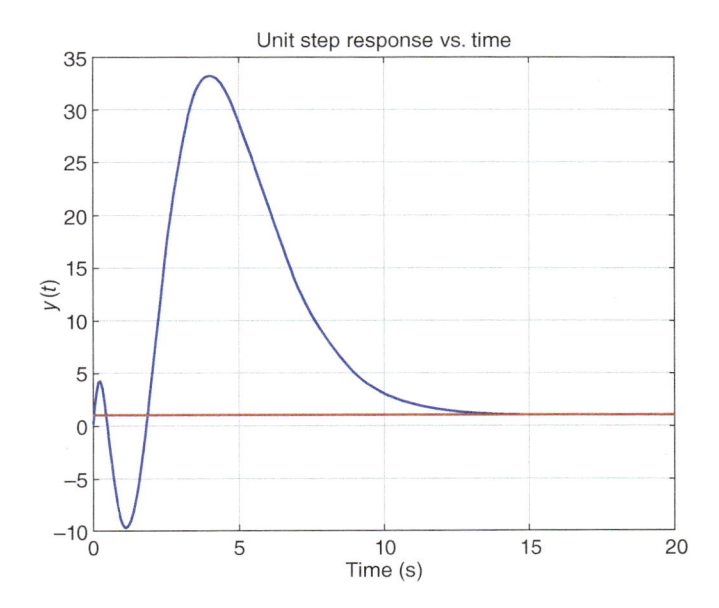

Figure 10.6. Step response with all poles at -1.

Disturbance Model

Our approach to disturbance rejection depends on being able to model the disturbance as going into the system at the same location as the control input. Though this may seem quite restrictive, it is often possible to do this as is now shown. Recall the equations of the DC motor given by

$$L\frac{di}{dt} = -Ri(t) - K_b\omega(t) + u(t)$$

$$J\frac{d\omega}{dt} = K_T i(t) - f\omega(t) - \tau_L(t)$$

$$\frac{d\theta}{dt} = \omega(t).$$

Here the physical input is the voltage $u(t)$ in the first equation while the disturbance (load torque) is in the second equation. In this model the disturbance does not enter the physical system at the same place as the input $u(t)$. We now show how to determine an equivalent disturbance that enters in the same place as the input voltage. Taking the Laplace transform of these equations with zero initial conditions we obtain

$$sLI(s) = -RI(s) - K_b\omega(s) + V_a(s)$$

$$sJ\omega(s) = K_T I(s) - f\omega(s) - \tau_L(s)$$

$$s\theta(s) = \omega(s).$$

After some rearrangement this set of equations can be written as

$$I(s) = \frac{1}{sL + R}\left(V_a(s) - K_b\omega(s)\right)$$

10 Pole Placement, 2 DOF Controllers, and Internal Stability

$$\omega(s) = \frac{1}{sJ+f}\left(K_T I(s) - \tau_L(s)\right)$$

$$\theta(s) = \frac{1}{s}\omega(s).$$

The corresponding block diagram is given in Figure 10.7.

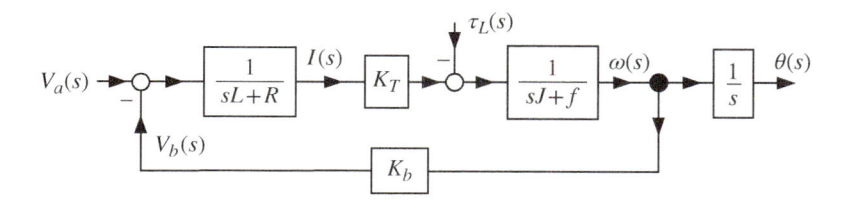

Figure 10.7. Block diagram of a DC motor.

A simple block diagram manipulation results in Figure 10.8.

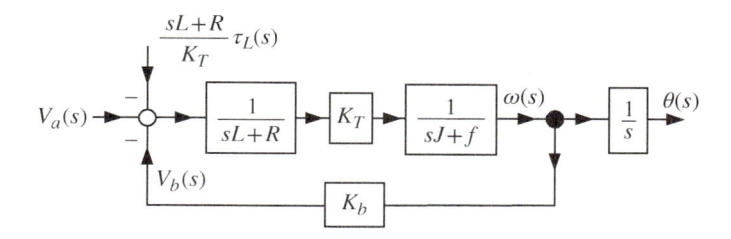

Figure 10.8. Block diagram with the disturbance entering via the input.

From Figure 10.8 the equations of the DC motor can be written in the equivalent form

$$L\frac{di}{dt} = -Ri(t) - K_b\omega(t) + v_a(t) - \left(\frac{L}{K_T}\dot{\tau}_L(t) + \frac{R}{K_T}\tau_L(t)\right)$$

$$J\frac{d\omega}{dt} = K_T i(t) - f\omega(t)$$

$$\frac{d\theta}{dt} = \omega(t).$$

The equivalent disturbance at the voltage input is now $d(t) = \frac{L}{K_T}\dot{\tau}_L(t) + \frac{R}{K_T}\tau_L(t)$. As $\tau_L(t) = \tau_{L0}u_s(t)$ making $\dot{\tau}_L(t) = 0$ for $t > 0$ we take $d(t) = \frac{R}{K_T}\tau_L(t)$. The equations of the DC motor with an equivalent voltage disturbance are given by

$$L\frac{di}{dt} = -Ri(t) - K_b\omega(t) + v_a(t) - \frac{R}{K_T}\tau_{L0}$$

$$J\frac{d\omega}{dt} = K_T i(t) - f\omega(t)$$

$$\frac{d\theta}{dt} = \omega(t).$$

10.1 Output Pole Placement

Remark A constant load torque τ_{L0} is equivalent to a constant current disturbance of τ_{L0}/K_T, which in turn is equivalent to a constant voltage disturbance of $R\tau_{L0}/K_T$!

Effect of the Initial Conditions on the Control Design

We have always taken the initial conditions to be zero to design a feedback controller. Let's see what happens if they are not zero. Let a system be described by

$$\ddot{y} - 2\dot{y} + 2y = \dot{u} - u.$$

The Laplace transform of this equation is then

$$s^2 Y(s) - sy(0) - \dot{y}(0) - 2sY(s) + 2y(0) + 2Y(s) = sU(s) - u(0) - U(s)$$

or

$$Y(s) = \underbrace{\frac{s-1}{s^2 - 2s + 2}}_{G(s)} U(s) + \frac{sy(0) + \dot{y}(0) - 2y(0) - u(0)}{s^2 - 2s + 2}.$$

A key observation here is that the denominator of the initial condition term is the *same* as the denominator of the open-loop transfer function. A block diagram representation is given in Figure 10.9.

Figure 10.9. Including the initial conditions in the block diagram of a system.

Consider a unity feedback controller for this system as shown in Figure 10.10.

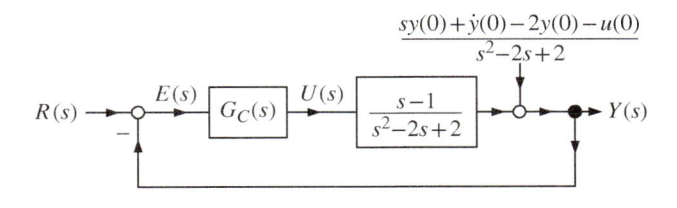

Figure 10.10. Feedback control system with the initial conditions included.

Let's design $G_c(s)$ so that the closed-loop transfer function $Y(s)/R(s)$ is stable. The controller

$$G_c(s) = \frac{-2s + 13}{s + 7}$$

places the closed-loop poles at $-1, -1, -1$. With nonzero initial conditions we then have

$$Y(s) = G(s)U(s) + \frac{sy(0) + \dot{y}(0) - 2y(0) - u(0)}{s^2 - 2s + 2}$$

$$= G(s)G_c(s)\left(R(s) - Y(s)\right) + \frac{sy(0) + \dot{y}(0) - 2y(0) - u(0)}{s^2 - 2s + 2}.$$

10 Pole Placement, 2 DOF Controllers, and Internal Stability

Solving for $Y(s)$ we have

$$Y(s) = \underbrace{\frac{G(s)G_c(s)}{1+G(s)G_c(s)}}_{G_{CL}(s)} R(s) + \underbrace{\frac{1}{1+G(s)G_c(s)} \frac{sy(0)+\dot{y}(0)-2y(0)-u(0)}{s^2-2s+2}}_{Y_{IC}(s)}.$$

The initial condition response $Y_{IC}(s)$ is then

$$\begin{aligned} Y_{IC}(s) &\triangleq \frac{1}{1+\dfrac{s-1}{s^2-2s+2}\dfrac{-2s+13}{s+7}} \frac{sy(0)+\dot{y}(0)-2y(0)-u(0)}{s^2-2s+2} \\ &= \frac{(s^2-2s+2)(s+7)}{s^3+3s^2+3s+1} \frac{sy(0)+\dot{y}(0)-2y(0)-u(0)}{s^2-2s+2} \\ &= \frac{(s+7)\bigl(sy(0)+\dot{y}(0)-2y(0)-u(0)\bigr)}{(s+1)(s+1)(s+1)}. \end{aligned}$$

Note the cancellation of the denominator of the initial condition term by the denominator of the open-loop transfer function. We have

$$y_{IC}(t) = \mathcal{L}^{-1}\{Y_{IC}(s)\} = \mathcal{L}^{-1}\left\{\frac{(s+7)\bigl(sy(0)+\dot{y}(0)-2y(0)-u(0)\bigr)}{(s+1)(s+1)(s+1)}\right\} \to 0.$$

Further

$$\begin{aligned} Y(s)/R(s) = G_{CL}(s) &= \frac{G(s)G_c(s)}{1+G(s)G_c(s)} = \frac{\dfrac{s-1}{s^2-2s+2}\dfrac{-2s+13}{s+7}}{1+\dfrac{s-1}{s^2-2s+2}\dfrac{-2s+13}{s+7}} \\ &= \frac{(s-1)(-2s+13)}{(s+1)(s+1)(s+1)}. \end{aligned}$$

Conclusion

The closed-loop initial condition response $Y_{IC}(s)$ will always have the *same* denominator as closed-loop transfer function $Y(s)/R(s) = G_{CL}(s)$. So, as long as the closed-loop system is stable, the initial condition response asymptotically goes to zero.

Figure 10.11 is an equivalent block diagram to that of Figure 10.10. In this block diagram it looks like the initial condition term $\dfrac{sy(0)+\dot{y}(0)-2y(0)-u(0)}{s-1}$ is now a "disturbance". Does $G_c(s)$ need to contain a pole at $s = 1$ to reject this "disturbance"? The answer is no!

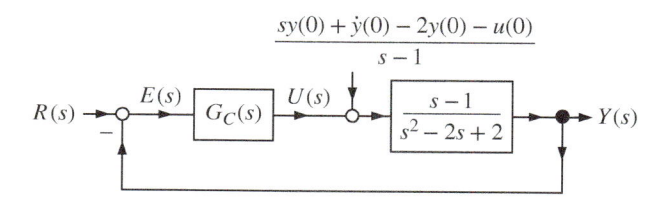

Figure 10.11. Block diagram equivalent to Figure 10.10.

10.1 Output Pole Placement

In this example the initial condition "disturbance" $\dfrac{sy(0)+\dot{y}(0)-u(0)}{s-1}$ has a pole at $s=1$. However, $G(s)$ has a zero at $s=1$ so that the closed-loop transfer function $G_{CL}(s)$ also has a zero at $s=1$, i.e.,

$$Y_{IC}(s) = \dfrac{\dfrac{s-1}{s^2-2s+2}}{1+\dfrac{-2s+13}{s+7}\dfrac{s-1}{s^2-2s+2}}\dfrac{sy(0)+\dot{y}(0)-2y(0)-u(0)}{s-1}$$

$$= \underbrace{\dfrac{(s+7)(s-1)}{s^3+3s^2+3s+1}}_{G_{CL}(s)} \dfrac{sy(0)+\dot{y}(0)-2y(0)-u(0)}{s-1}.$$

This cancellation of the factor $s-1$ is simply because we divided $sy(0)+\dot{y}(0)-2y(0)-u(0)$ by $s-1$ so that Figure 10.11 would be equivalent to Figure 10.10. It may perhaps be helpful to keep with the block diagram of Figure 10.10 for representing initial conditions in order to avoid confusion with disturbance rejection.

Example 5 *Inverted Pendulum*

Figure 10.12 shows an inverted pendulum on a cart. The pendulum rod of length ℓ is free to rotate about the cart pivot. The control objective is to apply an input force u to the cart to keep the pendulum angle at $\theta = 0$.

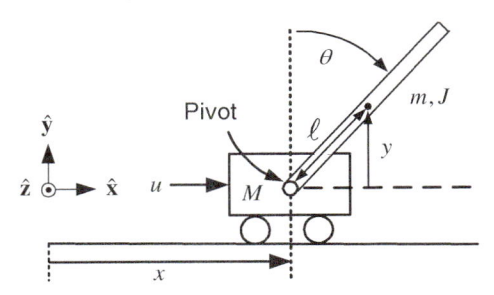

Figure 10.12. Inverted pendulum on a cart. The center of mass of the pendulum rod is at $(x+\ell\sin(\theta), y)$.

In Chapter 13 we derive a mathematical model of the inverted pendulum. To summarize that model, M is the mass of the cart, m is the mass of the pendulum rod of length 2ℓ, $J = m\ell^2/2$ is the moment of inertia of the pendulum rod about its center of mass at $(x+\ell\sin(\theta), y)$ and u is the input force. The position x of the cart and the angle θ of the rod are both measured. With θ small the center of mass of the rod is well approximated to be at $(x+\ell\theta, y)$. With an abuse of notation we also let y denote the system output as defined by

$$y \triangleq x + \left(\ell + \dfrac{J}{m\ell}\right)\theta.$$

As shown in Chapter 13 the transfer function from $U(s)$ to $Y(s)$ is

$$Y(s) = X(s) + \left(\ell + \dfrac{J}{m\ell}\right)\theta(s) = \underbrace{-\dfrac{\kappa mg\ell}{s^2(s^2-\alpha^2)}}_{G_Y(s)}U(s) + \dfrac{p_Y(s)}{s^2(s^2-\alpha^2)},$$

10 Pole Placement, 2 DOF Controllers, and Internal Stability

where

$$p_Y(s) = ((J/(m\ell) + \ell)s^2 - \kappa g(m\ell)^2)(s\theta(0) + \dot{\theta}(0)) + (s^2 - \alpha^2)(sx(0) + \dot{x}(0)),$$

$$\alpha^2 = \frac{mg\ell(M+m)}{Mm\ell^2 + J(M+m)}, \quad \kappa = \frac{1}{Mm\ell^2 + J(M+m)}.$$

Consider the unity feedback control system for the inverted pendulum shown in Figure 10.13.

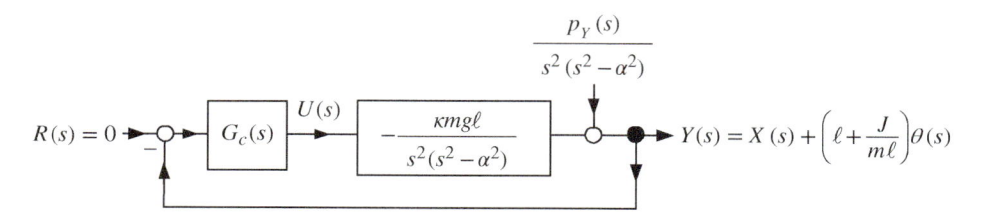

Figure 10.13. Closed-loop controller for the inverted pendulum.

With no reference or disturbance input, the controller need only stabilize the closed-loop system. We design a minimum order controller that allows arbitrary pole placement to stabilize the closed-loop system. This requires choosing $G_c(s)$ to have the form

$$G_c(s) = \frac{b_c(s)}{a_c(s)} = \frac{b_3 s^3 + b_2 s^2 + b_1 s + b_0}{s^3 + a_2 s^2 + a_1 s + a_0}.$$

With $U(s) = -G_c(s)Y(s)$ the output $Y(s)$ satisfies

$$Y(s) = -G_Y(s)G_c(s)Y(s) + \frac{p_Y(s)}{s^2(s^2 - \alpha^2)}.$$

Solving this for $Y(s)$ we obtain

$$Y(s) = \frac{1}{1 + G_Y(s)G_c(s)} \frac{p_Y(s)}{s^2(s^2 - \alpha^2)}$$

$$= \frac{1}{1 + \dfrac{b_3 s^3 + b_2 s^2 + b_1 s + b_0}{s^3 + a_2 s^2 + a_1 s + a_0} \dfrac{-\kappa mg\ell}{s^2(s^2 - \alpha^2)}} \frac{p_Y(s)}{s^2(s^2 - \alpha^2)}$$

$$= \frac{(s^3 + a_2 s^2 + a_1 s + a_0)p_Y(s)}{(s^3 + a_2 s^2 + a_1 s + a_0)s^2(s^2 - \alpha^2) + (-\kappa mg\ell)(b_3 s^3 + b_2 s^2 + b_1 s + b_0)}.$$

Collecting terms in s the denominator of $Y(s)$ is rewritten as

$$s^7 + a_2 s^6 + (a_1 - \alpha^2)s^5 + (a_0 - \alpha^2 a_2)s^4 + (-a_1 \alpha^2 - \kappa mg\ell b_3)s^3$$
$$+ (-a_0 \alpha^2 - \kappa mg\ell b_2)s^2 - \kappa mg\ell b_1 s - \kappa mg\ell b_0.$$

With $s^7 + f_6 s^6 + f_5 s^5 + f_4 s^4 + f_3 s^3 + f_2 s^2 + f_1 s + f_0$ the desired denominator of $Y(s)$ we must set

$$-\kappa mg\ell b_0 = f_0$$

10.1 Output Pole Placement

$$-\kappa mg\ell b_1 = f_1$$
$$-a_0\alpha^2 - \kappa mg\ell b_2 = f_2$$
$$-a_1\alpha^2 - \kappa mg\ell b_3 = f_3$$
$$a_0 - a_2\alpha^2 = f_4$$
$$a_1 - \alpha^2 = f_5$$
$$a_2 = f_6.$$

In matrix form this becomes

$$\begin{bmatrix} -\kappa mg\ell & 0 & 0 & 0 & 0 & 0 & 0 \\ 0 & -\kappa mg\ell & 0 & 0 & 0 & 0 & 0 \\ 0 & 0 & -\kappa mg\ell & 0 & -\alpha^2 & 0 & 0 \\ 0 & 0 & 0 & -\kappa mg\ell & 0 & -\alpha^2 & 0 \\ 0 & 0 & 0 & 0 & 1 & 0 & -\alpha^2 \\ 0 & 0 & 0 & 0 & 0 & 1 & 0 \\ 0 & 0 & 0 & 0 & 0 & 0 & 1 \end{bmatrix} \begin{bmatrix} b_0 \\ b_1 \\ b_2 \\ b_3 \\ a_0 \\ a_1 \\ a_2 \end{bmatrix} = \begin{bmatrix} f_0 \\ f_1 \\ f_2 \\ f_3 \\ f_4 \\ f_5 + \alpha^2 \\ f_6 \end{bmatrix}.$$

Inverting the matrix gives

$$\begin{bmatrix} b_0 \\ b_1 \\ b_2 \\ b_3 \\ a_0 \\ a_1 \\ a_2 \end{bmatrix} = \frac{1}{\kappa mg\ell} \begin{bmatrix} -1 & 0 & 0 & 0 & 0 & 0 & 0 \\ 0 & -1 & 0 & 0 & 0 & 0 & 0 \\ 0 & 0 & -1 & 0 & -\alpha^2 & 0 & -\alpha^4 \\ 0 & 0 & 0 & -1 & 0 & -\alpha^2 & 0 \\ 0 & 0 & 0 & 0 & \kappa mg\ell & 0 & \alpha^2\kappa mg\ell \\ 0 & 0 & 0 & 0 & 0 & \kappa mg\ell & 0 \\ 0 & 0 & 0 & 0 & 0 & 0 & \kappa mg\ell \end{bmatrix} \begin{bmatrix} f_0 \\ f_1 \\ f_2 \\ f_3 \\ f_4 \\ f_5 + \alpha^2 \\ f_6 \end{bmatrix}.$$

The parameter values for the QUANSER [34] inverted pendulum results in $\kappa mg\ell = 36.5705, \alpha^2 = 29.256$ so

$$G_Y(s) = \frac{-36.5705}{s^2(s^2 - 29.256)} = \frac{-36.5705}{s^2(s + 5.4089)(s - 5.4089)}.$$

Choosing the seven closed-loop poles to be at -5 the controller is given by

$$G_c(s) = -\frac{1041.6s^3 + 6113.7s^2 + 2990.8s + 2136.3}{s^3 + 35s^2 + 554.3s + 5399}$$
$$= -\frac{1041.6(s + 5.4088)(s^2 + 0.4308s + 0.3792)}{(s + 20.833)(s^2 + 14.1672s + 259.1533)}.$$

See Problem 22 where you are asked to simulate this control system.

Warning

This controller results in a closed-loop system having what are called small *stability margins* (see Section 11.9). This means that if the parameter values of $G_Y(s)$ are off by a little bit from their actual values, this controller, which was designed based on $G_Y(s)$, may not stabilize the closed-loop system. Further, this closed-loop system also has a high

298   10 Pole Placement, 2 DOF Controllers, and Internal Stability

sensitivity which means small disturbances may cause the pendulum angle to swing so far from $\theta = 0$ that the linear model $G_Y(s)$ is no longer a valid approximation of the nonlinear pendulum model. Consequently this controller $G_c(s)$ based on $G_Y(s)$ may not be able to return the pendulum to the upright position. Don't despair! Chapters 13 and 15 show how to design a statespace controller for the inverted pendulum which has good stability margins and low sensitivity as explained in Chapter 17.

Remarks on Pole Placement In all of the examples the controller was proper, but not strictly proper. Often it is desirable to have a strictly proper controller (see Section 11.6). This is easy to do as shown in Problems 13 and 14. The general statement and proof of the pole placement algorithm is given in the Appendix *Output Pole Placement* of this chapter.

10.2 Two Degrees of Freedom Controllers

We have seen that a proportional plus integral (PI) controller typically results in a left half-plane zero in the closed-loop transfer function and in Chapter 8 we saw that such a zero indicates there will be overshoot in the step response. More generally, $G_c(s)G(s)$ having zeros implies significant overshoot in the closed-loop step response. In this section we show how this overshoot can often be eliminated by passing the step reference input through a filter $G_f(s)$ (transfer function). This use of two transfer functions $G_c(s)$ and $G_f(s)$ is referred to as a 2 DOF controller. To show how this is done, consider the PI-D servo motor control system given in Figure 10.14. You may think of the measurement of ω as being obtained from a tachometer (see Figures 6.26 and 6.30) or from the backward difference computation of an optical encoder signal (see Section 6.5).

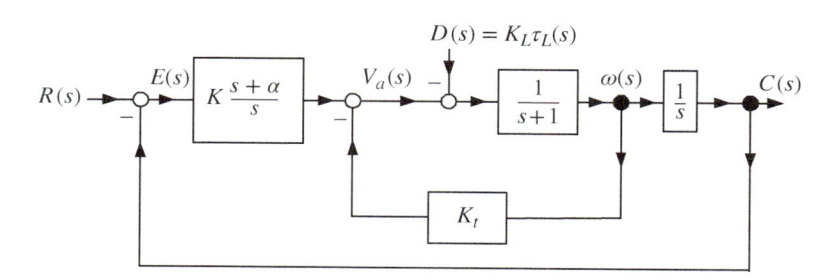

Figure 10.14. PI-D controller.

A simple block diagram manipulation gives the equivalent system block diagram of Figure 10.15.

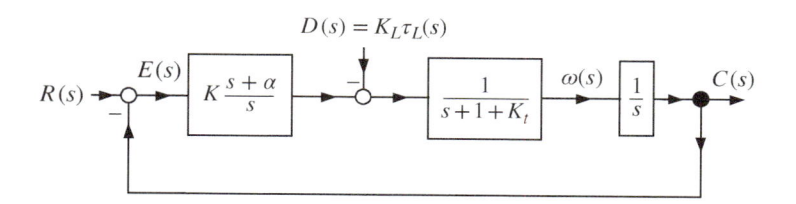

Figure 10.15. Equivalent block diagram of Figure 10.14.

10.2 Two Degrees of Freedom Controllers

First let $D(s) = K_L \tau_L(s) = 0$. We have

$$E_R(s) = \frac{1}{1 + K\dfrac{s+\alpha}{s}\dfrac{1}{s+1+K_t}\dfrac{1}{s}} R(s) = \frac{s^2(s+1+K_t)}{s^3 + (1+K_t)s^2 + Ks + \alpha K} R(s)$$

As the parameters K_t, K, and α are chosen by the control engineer, the coefficients of $s^3 + (1+K_t)s^2 + Ks + \alpha K$ may be chosen arbitrarily. In other words, using the controller gains K_t, K, and α we can place the closed-loop poles at any desired location. Let the desired location of the closed-loop poles be $-r_1, -r_2, -r_3$, which requires choosing these controller gains so that

$$s^3 + (1+K_t)s^2 + Ks + \alpha K = (s+r_1)(s+r_2)(s+r_3)$$
$$= s^3 + (r_1+r_2+r_3)s^2 + (r_1 r_2 + r_1 r_3 + r_2 r_3)s + r_1 r_2 r_3.$$

That is, we set

$$K_t = r_1 + r_2 + r_3 - 1$$
$$K = r_1 r_2 + r_1 r_3 + r_2 r_3 \qquad (10.3)$$
$$\alpha = \frac{r_1 r_2 r_3}{K} = \frac{r_1 r_2 r_3}{r_1 r_2 + r_1 r_3 + r_2 r_3}.$$

Let's now take a look at the output $C(s)$ due to a step input. In particular we are interested in whether or not there is overshoot in the step response. With $D(s) = 0$ we have

$$C(s) = \frac{K\dfrac{s+\alpha}{s}\dfrac{1}{s+1+K_t}\dfrac{1}{s}}{1 + K\dfrac{s+\alpha}{s}\dfrac{1}{s+1+K_t}\dfrac{1}{s}} \frac{R_0}{s} = \frac{K(s+\alpha)}{s^3 + (1+K_t)s^2 + Ks + \alpha K} \frac{R_0}{s}$$

$$= \frac{K(s+\alpha)}{(s+r_1)(s+r_2)(s+r_3)} \frac{R_0}{s}.$$

Using the fact that $sC(s)$ is stable the final value theorem gives

$$c(\infty) = \lim_{s \to 0} sC(s) = \lim_{s \to 0} s \frac{K(s+\alpha)}{s^3 + (1+K_t)s^2 + Ks + \alpha K} \frac{R_0}{s} = R_0.$$

We are able to put the closed-loop poles at $-r_1, -r_2, -r_3$ and achieve asymptotic tracking. However, as seen in Figure 10.16, $c(t)$ will exhibit *overshoot*.

As we now explain, this control system will always have overshoot in a step response.

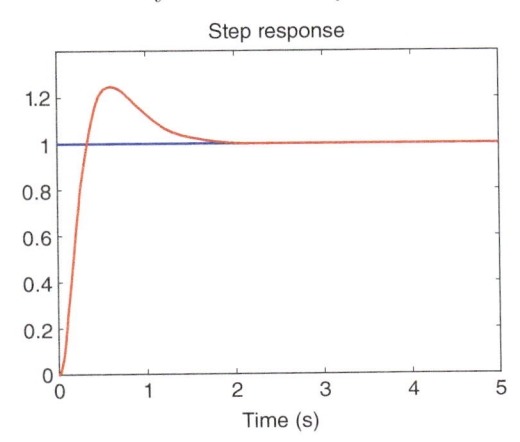

Figure 10.16. Step response of the system of Figure 10.14 with the three closed-loop poles all at -5.

Stable Type 2 Systems Have Overshoot

The step response of a type 2 closed-loop stable system will have overshoot. With

$$L(s) \triangleq K \frac{s+\alpha}{s} \frac{1}{s+1+K_t} \frac{1}{s},$$

Figure 10.15 may be redrawn as shown in Figure 10.17.

$L(s)$ is type 2 and the closed-loop transfer function given by $\frac{L(s)}{1+L(s)}$ is *stable* by choosing the controller gains α, K, and K_t as previously shown. By Theorem 2 in the *Overshoot* Appendix of this chapter, the step response of this system must have overshoot.

Figure 10.17. Stable type 2 system with a step input.

Stable Systems with Real Poles and No Zeros Do Not Have Overshoot

What kind of closed-loop systems do *not* have overshoot in their step responses? Theorem 4 of the *Overshoot* Appendix shows that a step response will not overshoot if the closed-loop transfer function is (i) stable, (ii) has *real* poles, and (iii) *no* zeros. In our example the closed-loop system is

$$C(s) = \underbrace{\frac{K(s+\alpha)}{(s+r_1)(s+r_2)(s+r_3)}}_{\text{Closed-loop transfer function}} \frac{R_0}{s}.$$

We choose the $r_i, i = 1, 2, 3$ to be real and positive, but we need to get rid of the zero at $-\alpha$ where $\alpha = \frac{r_1 r_2 r_3}{r_1 r_2 + r_1 r_3 + r_2 r_3}$. As $\alpha > 0$ the transfer function

$$\frac{\alpha}{s+\alpha}$$

is *stable* and we use it as an *reference input filter*. Specifically, consider the 2 DOF control system shown in Figure 10.18.

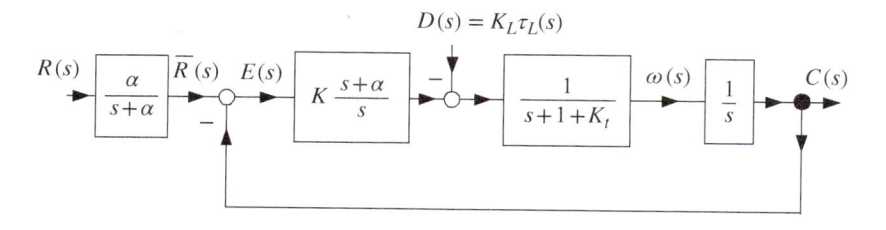

Figure 10.18. Two degree of freedom controller to eliminate overshoot.

10.2 Two Degrees of Freedom Controllers

With $D(s) = K_L \tau_L(s) = 0$, we then have

$$C(s) = \frac{K\dfrac{s+\alpha}{s}\dfrac{1}{s+1+K_t}\dfrac{1}{s}}{1+K\dfrac{s+\alpha}{s}\dfrac{1}{s+1+K_t}\dfrac{1}{s}}\dfrac{\alpha}{s+\alpha}\dfrac{R_0}{s}$$

$$= \frac{K(s+\alpha)}{s^3 + (1+K_t)s^2 + Ks + \alpha K}\dfrac{\alpha}{s+\alpha}\dfrac{R_0}{s}$$

$$= \frac{\alpha K}{(s+r_1)(s+r_2)(s+r_3)}\dfrac{R_0}{s}. \tag{10.4}$$

(Note that this final expression for $C(s)$ involved a *stable* pole–zero cancellation.) The step response of the control system of Figure 10.18 is shown in Figure 10.19. This has no overshoot as guaranteed by Theorem 4 in the *Overshoot* Appendix.

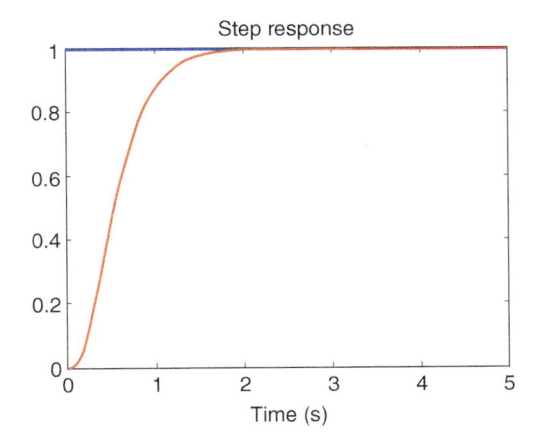

Figure 10.19. Step response of the system of Figure 10.18 with the closed-loop poles at -5.

We still have asymptotic tracking because, as $sC(s)$ is stable, it follows by the final value theorem that

$$c(\infty) = \lim_{s \to 0} sC(s) = \lim_{s \to 0} \underbrace{\frac{K(s+\alpha)}{s^3 + (1+K_t)s^2 + Ks + \alpha K}}_{\to 1} \underbrace{\frac{\alpha}{s+\alpha}}_{\to 1} R_0 = R_0.$$

Another way to see this asymptotic tracking is to define

$$\bar{R}(s) \triangleq \frac{\alpha}{s+\alpha}\frac{R_0}{s} = \frac{R_0}{s} - \frac{R_0}{s+\alpha}$$

so that $\bar{r}(t) = R_0 u_s(t) - R_0 e^{-\alpha t} u_s(t)$ is the input to feedback loop and $\bar{r}(t) \to r(t) = R_0 u_s(t)$. Also, as the controller $G_c(s)$ is type 1, this controller also rejects constant disturbances! This method works as long as the poles of the closed-loop transfer function are *real* and the zeros are in the *open* left half-plane.

10 Pole Placement, 2 DOF Controllers, and Internal Stability

Will this 2 DOF controller still track a ramp input? Let $R(s) = \omega_0/s^2$ be the ramp input so that

$$\bar{R}(s) = \frac{\alpha}{s+\alpha}\frac{\omega_0}{s^2} = \frac{\omega_0}{s^2} - \frac{\omega_0/\alpha}{s} + \frac{\omega_0}{\alpha}\frac{1}{s+\alpha}$$

$$= \underbrace{\mathcal{L}\left\{\omega_0 t - \frac{\omega_0}{\alpha}u_s(t) + \frac{\omega_0}{\alpha}e^{-\alpha t}\right\}}_{\bar{r}(t)}.$$

$\bar{r}(t) = \omega_0 t - \frac{\omega_0}{\alpha}u_s(t) + \frac{\omega_0}{\alpha}e^{-\alpha t}$ is the input to the feedback loop and, as $L(s)$ is type 2, the output $c(t)$ will asymptotically track $\omega_0 t - \frac{\omega_0}{\alpha}u_s(t)$. However, if the reference input is $R(s) \triangleq \omega_0\frac{1}{s^2} + \frac{\omega_0}{\alpha}\frac{1}{s}$ then the output will asymptotically track $\omega_0 t$ (why?).

Stable Systems with Real Poles and a Single Right Half-Plane (RHP) Zero Do Not Have Overshoot

Let's now consider an example with a right half-plane zero.

Example 6 *One Right Half-Plane Zero*

Consider the feedback control system of Figure 10.20 where the open-loop system transfer function is

$$G(s) = -\frac{s-1}{s(s+2)}.$$

To arbitrarily place the closed-loop poles, $G_c(s)$ is chosen to have form

$$G_c(s) = \frac{b_1 s + b_0}{s + a_0}.$$

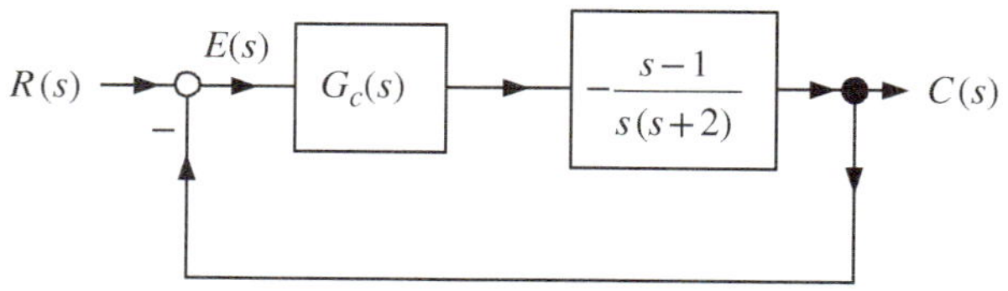

Figure 10.20. Open-loop system with a right half-plane zero.

$G(s)$ is type 1 and, with a stable closed-loop, step inputs will be tracked. Set $R(s) = R_0/s$ and calculate

$$E(s) = \frac{1}{1 + G(s)G_c(s)}R(s)$$

$$= \frac{1}{1 - \frac{s-1}{s(s+2)}\frac{b_1 s + b_0}{s + a_0}}\frac{R_0}{s}$$

$$= \frac{s(s+2)(s+a_0)}{s(s+2)(s+a_0) - (s-1)(b_1 s + b_0)}\frac{R_0}{s}$$

$$= \frac{s(s+2)(s+a_0)}{s^3 + (a_0 - b_1 + 2)s^2 + (2a_0 - b_0 + b_1)s + b_0}\frac{R_0}{s}$$

10.2 Two Degrees of Freedom Controllers

$$= \frac{(s+2)(s+a_0)}{s^3 + (a_0 - b_1 + 2)s^2 + (2a_0 - b_0 + b_1)s + b_0} R_0.$$

Let the desired closed-loop poles be at $-r_1, -r_2, -r_3$, which requires

$$\begin{aligned}s^3 + (a_0 - b_1 + 2)s^2 + (2a_0 - b_0 + b_1)s + b_0 &= (s + r_1)(s + r_2)(s + r_3) \\ &= s^3 + (r_1 + r_2 + r_3)s^2 + (r_1 r_2 + r_1 r_3 + r_2 r_3)s \\ &\quad + r_1 r_2 r_3.\end{aligned}$$

Solving the set of linear equations

$$a_0 - b_1 + 2 = r_1 + r_2 + r_3$$
$$2a_0 + b_1 - b_0 = r_1 r_2 + r_1 r_3 + r_2 r_3$$
$$b_0 = r_1 r_2 r_3,$$

the coefficients of the controller $G_c(s)$ are

$$b_1 = \frac{-2(r_1 + r_2 + r_3) + r_1 r_2 + r_1 r_3 + r_2 r_3 + r_1 r_2 r_3 + 4}{3}$$

$$a_0 = \frac{r_1 + r_2 + r_3 + r_1 r_2 + r_1 r_3 + r_2 r_3 + r_1 r_2 r_3 - 2}{3}$$

$$b_0 = r_1 r_2 r_3.$$

The closed-loop response $C(s)$ is then

$$C(s) = \frac{G(s) G_c(s)}{1 + G(s) G_c(s)} R(s) = \frac{\dfrac{s-1}{s(s+2)} \dfrac{b_1 s + b_0}{s + a_0}}{1 - \dfrac{s-1}{s(s+2)} \dfrac{b_1 s + b_0}{s + a_0}} \frac{R_0}{s} = \frac{-(s-1) b_1 (s + b_0/b_1)}{(s + r_1)(s + r_2)(s + r_3)} \frac{R_0}{s}.$$

Choose the location of the closed-loop poles so that the zero at $-b_0/b_1$ is in the open left half-plane.[4] Then the transfer function

$$\frac{b_0/b_1}{s + b_0/b_1}$$

is stable, and used as a reference input filter as shown in Figure 10.21.

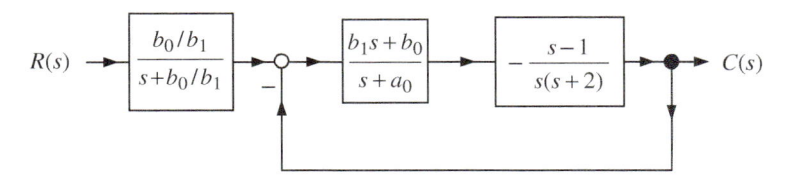

Figure 10.21. Two degree of freedom controller to eliminate overshoot.

---

[4] For example if $r_1 = r_2 = r_3 = r$ then $b_1(r) = (-6r + 3r^2 + r^3 + 4)/3$ and some calculations show that $b_1(r) > 0$ for $r > 0$. Thus for $r > 0$ we have $b_0/b_1 > 0$ and the zero $-b_0/b_1$ is in the open left half-plane.

The closed-loop response $C(s)$ is now

$$C(s) = \frac{G(s)G_c(s)}{1+G(s)G_c(s)} \frac{b_0/b_1}{s+b_0/b_1} R(s)$$

$$= \frac{-\dfrac{s-1}{s(s+2)} \dfrac{b_1 s+b_0}{s+a_0}}{1 - \dfrac{s-1}{s(s+2)} \dfrac{b_1 s+b_0}{s+a_0}} \frac{b_0/b_1}{s+b_0/b_1} \frac{R_0}{s}$$

$$= \frac{-(s-1)b_1(s+b_0/b_1)}{s^3+(a_0-b_1+2)s^2+(2a_0+b_1-b_0)s+b_0} \frac{b_0/b_1}{s+b_0/b_1} \frac{R_0}{s}$$

$$= \underbrace{-\frac{(s-1)b_0}{(s+r_1)(s+r_2)(s+r_3)}}_{G_{CL}(s)} \frac{R_0}{s}.$$

As $G_{CL}(s)$ is stable with all *real* poles and only one right half-plane zero, Theorem 5 in the *Overshoot* Appendix of this chapter guarantees there will be no overshoot. However, the system will have undershoot. Recall from Chapter 8 (see page 222) that any stable closed-loop transfer function with an odd number of *real right half-plane zeros* will have a step response with undershoot [26].

Example 7 *Two Right Half-Plane Zeros*
Recall Example 4 where we considered the control system shown in Figure 10.22.

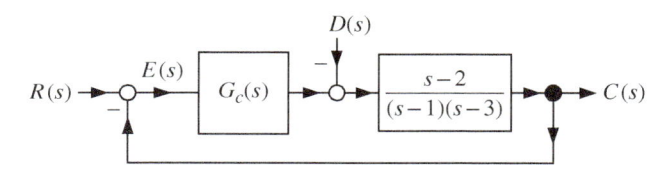

Figure 10.22. Closed-loop system with two right half-plane zeros.

We found that a controller of the form

$$\frac{b_2 s^2+b_1 s+b_0}{s+a_0}\frac{1}{s}$$

allows us to place the closed-loop poles arbitrarily in the open left half-plane while being able to track step inputs and reject constant disturbances. However, as shown in Figure 10.6, the overshoot in the step response was unacceptable. We now show how a reference input filter can be used to eliminate overshoot. With $D(s) = 0$, and choosing the gains as in (10.1) of Example 4, we have

$$C(s) = \frac{G_c(s)G(s)}{1+G_c(s)G(s)} R(s)$$

$$= \frac{(b_2 s^2+b_1 s+b_0)(s-2)}{s^4+(a_0+b_2-4)s^3+(b_1-4a_0-2b_2+3)s^2+(3a_0+b_0-2b_1)s-2b_0} \frac{R_0}{s}$$

$$= \frac{(b_2 s^2+b_1 s+b_0)(s-2)}{s^4+f_3 s^3+f_2 s^2+f_1 s+f_0} \frac{R_0}{s}.$$

10.2 Two Degrees of Freedom Controllers

Note that the zeros of the closed-loop transfer function consists of the zeros of the controller $G_c(s)$ and those of the open-loop model $G(s)$. As usual let

$$f_3 = r_1 + r_2 + r_3 + r_4$$
$$f_2 = r_1r_2 + r_1r_3 + r_1r_4 + r_2r_3 + r_2r_4 + r_3r_4$$
$$f_1 = r_1r_2r_3 + r_1r_2r_4 + r_1r_3r_4 + r_2r_3r_4$$
$$f_0 = r_1r_2r_3r_4.$$

Setting $r_1 = r_2 = r_3 = r_4 = r$ the closed-loop response $C(s)$ is now

$$C(s) = \frac{(b_2s^2 + b_1s + b_0)(s-2)}{(s+r)^4} R_0.$$

As shown in (10.2), with the four closed-loop poles all set at -1, the controller is

$$G_c(s) = \frac{50.5s^2 - 66s - 0.5}{s - 42.5}\frac{1}{s} = \frac{50.5(s - 1.3145)(s + 0.0075)}{s - 42.5}\frac{1}{s}.$$

Thus the closed-loop transfer function has two right half-plane zeros at $z_1 = 2$ and $z_2 = 1.3145$ and one left half-plane zero at $z_3 = -0.0075$. To develop a reference filter for the two *right half-plane zeros* z_1, z_2, we form the polynomial

$$(s-z_1)(s-z_2) = (s-1.3145)(s-2) = s^2 - 3.3145s + \underbrace{2.629}_{\omega_0^2}. \tag{10.5}$$

Then the filter

$$\frac{\omega_0^2}{(s+\omega_0)^2}$$

will prevent overshoot from the two right half-plane zeros at 2 and 1.3145 (see Theorem 6 in the *Overshoot* Appendix). Finally the filter

$$\frac{\alpha}{s+\alpha} = \frac{0.0075}{s+0.0075}$$

will prevent overshoot due to the left half-plane zero at -0.0075. The complete block diagram for this control system is given in Figure 10.23.

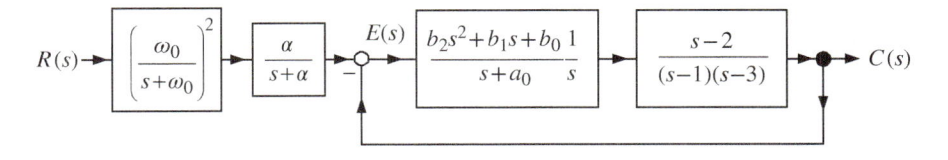

Figure 10.23. Reference filter for a system with 2 right half-plane zeros.

Figure 10.24 on the next page is the unit step response of this 2 DOF controller with the closed-loop poles at -1. Problem 9 asks you to simulate the step response of this system with and without the reference input filter.

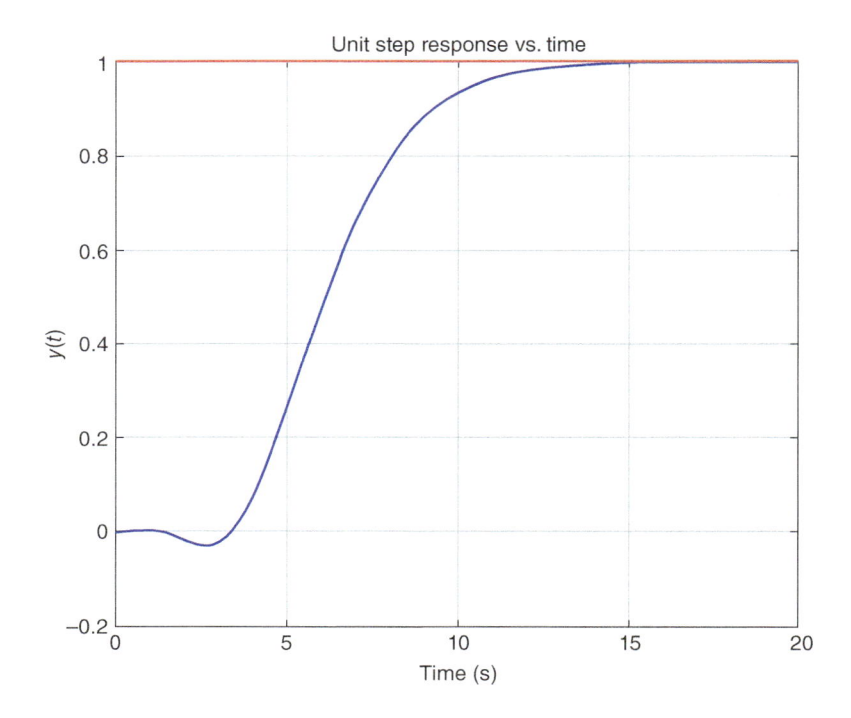

Figure 10.24. Step response of the 2 DOF control system of Figure 10.23.

Important Remarks

(1) Given any open-loop model of a physical system, there are fundamental limitations that a feedback controller cannot overcome. These limitations especially are revealed as a robustness issue when trying to control a system which has poles in the *open right half-plane* [35, 36]. This means that a controller designed based on the model $G(s) = \dfrac{s-2}{(s-1)(s-3)}$ may result in the closed-loop system being *unstable* if the actual system is only a little different. For example, Problem 9 part (c) asks you to use the same controller $G_c(s)$ as earlier, but set $G(s) = \dfrac{s-2}{(s-0.9)(s-3)}$. You should see that the closed-loop system is *unstable*. This will be discussed in more detail later after studying Nyquist theory. (See Problem 30 of Chapter 11.)

(2) This "toy" example also has a zero in the open RHP which is close to the two poles there. As explained in Chapter 17 this is a strong indication there is *no* controller that can achieve a robust closed-loop system.

Eliminating Overshoot with a 2 DOF Controller

With reference to Figure 10.25, which is a block diagram of a 2 DOF controller, let's look more generally how a reference filter $G_f(s)$ is used to eliminate overshoot [5, 6, 37, 38].

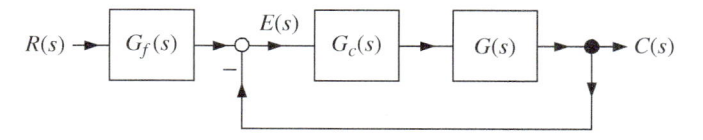

Figure 10.25. Two DOF controller.

10.2 Two Degrees of Freedom Controllers

Let $G_c(s)G(s)$ be at least type 1 and let

$$G_{CL}(s) = \frac{G_c(s)G(s)}{1+G_c(s)G(s)} = \frac{n(s)}{d(s)},$$

where $d(s)$ has *real* roots which are all in the *open left half-plane*.

(a) $G_{CL}(s)$ has no right half-plane zeros.

Let $n(s)$ have all of its roots in the open left half-plane. Set

$$G_f(s) = \frac{n(0)}{n(s)}.$$

Then the inverse Laplace transform of

$$C(s) = G_f(s)G_{CL}(s)\frac{R_0}{s}$$

will *not* have overshoot and $c(\infty) = R_0$. See Theorem 4 in the *Overshoot* Appendix for the proof.

Note that the roots of $n(s)$ may be real or complex; they need only be in the open left half-plane.

(b) $G_{CL}(s)$ has one right half-plane zero.

Let $n(s) = \bar{n}(s)(s-z)$ where the roots of $\bar{n}(s)$ are all in the open left half-plane and $z > 0$. That is, $G_{CL}(s)$ has one *right half-plane* zero. Set

$$G_f(s) = \frac{\bar{n}(0)}{\bar{n}(s)}.$$

Then the inverse Laplace transform of

$$C(s) = G_f(s)G_{CL}(s)\frac{R_0}{s}$$

will *not* have overshoot and $c(\infty) = R_0$. See Theorem 5 in the *Overshoot* Appendix for the proof. However, this step response will have undershoot. Also, the further one places the closed-loop poles in the left half-plane (to decrease the settling time) the larger the magnitude of the undershoot. See Theorem 10 in the *Undershoot* Appendix for the proof.

(c) $G_{CL}(s)$ has two right half-plane complex conjugate zeros.

Let $n(s) = \bar{n}(s)(s^2 - 2\zeta\omega_0 s + \omega_0^2)$ with $\zeta > 0, \omega_0 > 0$ and with the roots of $\bar{n}(s)$ in the open left half-plane. Then $G_{CL}(s)$ has a complex pair of zeros in the open *right* half-plane at $\zeta\omega_0 \pm j\omega_0\sqrt{1-\zeta^2}$. Set the reference input filter as

$$G_f(s) = \frac{\bar{n}(0)}{\bar{n}(s)}\left(\frac{\omega_0}{s+\omega_0}\right)^2.$$

Then the inverse Laplace transform of

$$C(s) = G_f(s)G_{CL}(s)\frac{R_0}{s}$$

will *not* have overshoot and $c(\infty) = R_0$. See Theorem 6 in the *Overshoot* Appendix for the proof.

(d) $G_{CL}(s)$ has two right half-plane real zeros.

Let $n(s) = \bar{n}(s)(s-z_1)(s-z_2)$ where the roots of $\bar{n}(s)$ are all in the open left half-plane and $z_1 > 0, z_2 > 0$. That is, $G_{CL}(s)$ has two real zeros in the open *right half-plane*. With

$$(s-z_1)(s-z_2) = s^2 - (z_1+z_2)s + z_1z_2$$

and $\omega_0 \triangleq \sqrt{z_1 z_2}$ set the reference input filter as

$$G_f(s) = \frac{\bar{n}(0)}{\bar{n}(s)}\left(\frac{\omega_0}{s+\omega_0}\right)^2.$$

Then the inverse Laplace transform of

$$C(s) = G_f(s)G_{CL}(s)\frac{R_0}{s}$$

will *not* have overshoot and $c(\infty) = R_0$. See Theorem 6 in the *Overshoot* Appendix for the proof.

Remark Actually part (d) can be taken into account in part (c) by simply letting $\zeta \geq 1$. That is, set $2\zeta\omega_0 = z_1 + z_2$ and $\omega_0 \triangleq \sqrt{z_1 z_2}$ so that

$$\zeta = \frac{z_1+z_2}{2\sqrt{z_1 z_2}} \geq 1 \text{ for } z_1 > 0, z_2 > 0.$$

(e) $G(0) \neq 0 \implies$ Overshoot can always be eliminated [37].

Let $G(s)$ be the open-loop transfer function model of the system with $G(0) \neq 0$. Then there exists a 2 DOF controller such that the step response does not have overshoot and $c(\infty) = R_0$.

The proof of this is given in [37]. Also see page 163 of [5]. The proof of this result is constructive, but it turns out that in general one cannot arbitrarily place the closed-loop poles as in cases (a)–(d).

Remark If $G(s)$ has all its zeros in the open left half-plane and the reference input $r(t)$ is smooth enough (it can be continuously differentiated enough times), then there is a 2 DOF controller that can track the reference input with zero error with the closed-loop poles arbitrarily placed in the open left half-plane. This is shown in [39] (see pages 274–278). However, in general, such a design is not robust to changes in the controller parameters [38].

10.3 Internal Stability

Figure 10.26 shows the standard unity feedback control system.

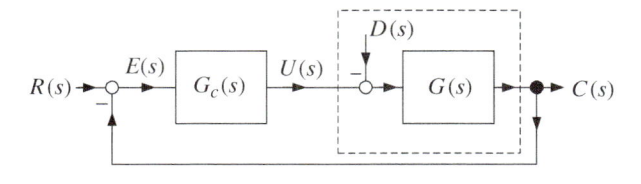

Figure 10.26. Unity feedback control system.

There are two external inputs to this system, the reference input $R(s)$, and the disturbance input $D(s)$. We are interested in their effect on $E(s), U(s)$, and $C(s)$, which is a total of six transfer functions. These six transfer functions are given in Eqs. (10.6)–(10.8). This is the first time we have considered the transfer functions from the external inputs $R(s)$ and $D(s)$ to the input $U(s)$ of the physical system. These transfer functions must also be stable as we don't want the input to the physical system to go unbounded. In fact we require all six of these transfer functions to be stable.

$$E(s) = \frac{1}{1 + G_c(s)G(s)} R(s) + \frac{G(s)}{1 + G_c(s)G(s)} D(s) \tag{10.6}$$

$$C(s) = \frac{G_c(s)G(s)}{1 + G_c(s)G(s)} R(s) - \frac{G(s)}{1 + G_c(s)G(s)} D(s) \tag{10.7}$$

$$U(s) = \frac{G_c(s)}{1 + G_c(s)G(s)} R(s) + \frac{G_c(s)G(s)}{1 + G_c(s)G(s)} D(s). \tag{10.8}$$

Remark There are six transfer functions in (10.6)–(10.8) with two of them equal and two of them differing by a minus sign. So there are in effect only four different transfer functions and these four transfer functions are referred to as the "gang of four" [40].

Definition 1 *Internal Stability*

The system of Figure 10.26 is said to have *internal stability* (or be *internally stable*) if the six transfer functions of (10.6)–(10.8) are stable.

Let the system model transfer function and the controller transfer function be, respectively,

$$G(s) = \frac{b(s)}{a(s)}, \quad G_c(s) = \frac{b_c(s)}{a_c(s)}. \tag{10.9}$$

$G(s)$ is assumed to be strictly proper, i.e. deg$\{b(s)\}$ < deg$\{a(s)\}$ and $G_c(s)$ is assumed to be proper, i.e., deg$\{b_c(s)\}$ ≤ deg$\{a_c(s)\}$. Let's now substitute the expressions (10.9) into (10.6)–(10.8) and clear the fractions to obtain

$$E(s) = \frac{a_c(s)a(s)}{a_c(s)a(s) + b_c(s)b(s)} R(s) + \frac{a_c(s)b(s)}{a_c(s)a(s) + b_c(s)b(s)} D(s)$$

$$C(s) = \frac{b_c(s)b(s)}{a_c(s)a(s) + b_c(s)b(s)} R(s) - \frac{a_c(s)b(s)}{a_c(s)a(s) + b_c(s)b(s)} D(s)$$

$$U(s) = \frac{a(s)b_c(s)}{a_c(s)a(s) + b_c(s)b(s)} R(s) + \frac{b_c(s)b(s)}{a_c(s)a(s) + b_c(s)b(s)} D(s).$$

The polynomial

$$a_c(s)a(s) + b_c(s)b(s)$$

is called the *characteristic polynomial* of the closed-loop system.[5] This shows that all of the transfer functions of (10.6)–(10.8) are stable if all the roots of the characteristic polynomial are in the *open left half-plane*.

Observation Note that $\dfrac{C(s)}{R(s)} = \dfrac{b_c(s)b(s)}{a_c(s)a(s) + b_c(s)b(s)}$, which shows that the zeros of the closed-loop are simply the zeros of the open-loop system $G(s)$ and the zeros of the controller $G_c(s)$. Feedback does not change the locations of the zeros.

Though up to now we have not stated it explicitly, we have been assuming that both $\{a(s), b(s)\}$ and $\{a_c(s), b_c(s)\}$ are coprime. We now define what we mean by *coprime*.

Definition 2 *Coprime Polynomials*
Two polynomials $\{a(s), b(s)\}$ are *coprime* if there is no s_0 such that $a(s_0) = b(s_0) = 0$.

Example 8 *Coprime Polynomials*
Let
$$a(s) = (s+1)(s-1)$$
$$b(s) = s - 1.$$

Then the polynomials $a(s)$ and $b(s)$ are *not* coprime as
$$a(1) = b(1) = 0.$$

Note they both have the factor $s - 1$.

Example 9 *Coprime Polynomials*
Let
$$a(s) = (s+1)(s^2 + 1)$$
$$b(s) = s - 1.$$

Then the polynomials $a(s)$ and $b(s)$ are coprime as $b(s) = 0$ only if $s = 1$ and $a(1) \neq 0$.

Note that $a(s)$ and $b(s)$ are coprime if and only if they do *not* have a common factor. When we write
$$G(s) = \frac{b(s)}{a(s)}$$

and say that $\{a(s), b(s)\}$ are coprime, this is the same as saying that $b(s)$ and $a(s)$ have no factors in common. Similarly for $G_c(s) = \dfrac{b_c(s)}{a_c(s)}$.

Definition 3 *Coprime Transfer Functions*
We say that the transfer function
$$G(s) = \frac{b(s)}{a(s)}$$

is a *coprime* transfer function if $a(s)$ and $b(s)$ are coprime.

---

[5] The word "characteristic" comes from working with control systems in the statespace (see Chapter 15) rather than with transfer functions. A characteristic value (eigenvalue) is simply a root of the characteristic equation.

10.3 Internal Stability

Remark As mentioned previously we are going to always assume that the transfer functions $G(s)$ and $G_c(s)$ are each coprime transfer functions, but we will use this terminology at times just for emphasis. This is a very sensible assumption. For example, suppose the transfer function of a physical system is

$$G(s) = \frac{s+3}{s(s+1)(s+2)}.$$

Then there is no sense in writing this as

$$G(s) = \frac{(s+3)(s+4)}{s(s+1)(s+2)(s+4)}$$

and the same holds for a controller transfer function.

Unstable Pole–Zero Cancellation Inside the Loop (Bad)

Recall that we are assured all transfer functions are stable if all the roots of

$$a_c(s)a(s) + b_c(s)b(s) = 0$$

are in the open left half-plane. However, we have to be careful that when computing the closed-loop transfer functions we actually have the full characteristic polynomial in the denominator. We can sometimes be fooled when there is a pole–zero cancellation between the controller transfer function $G_c(s)$ and the model transfer function $G(s)$. We explain this issue in the following examples.

Example 10 *Unstable Pole–Zero Cancellation*

With $R(s) = R_0/s$ and $D(s) = D_0/s$ consider the control system of Figure 10.27 whose open-loop transfer function is

$$G(s) = \frac{b(s)}{a(s)} = \frac{1}{s-1}.$$

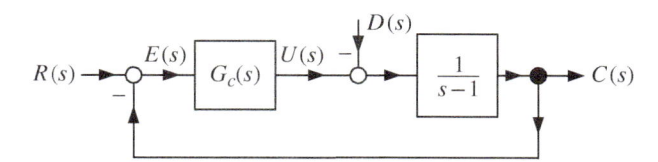

Figure 10.27. Unstable pole–zero cancellation.

The error $E(s)$ is

$$E(s) = \frac{1}{1 + G_c(s)G(s)} R(s) + \frac{G(s)}{1 + G_c(s)G(s)} D(s).$$

Suppose we choose the controller to be

$$G_c(s) = \frac{b_c(s)}{a_c(s)} = \frac{s-1}{s(s+1)}.$$

10 Pole Placement, 2 DOF Controllers, and Internal Stability

The error $E(s)$ becomes

$$E(s) = \frac{1}{1 + \dfrac{s-1}{s(s+1)}\dfrac{1}{s-1}} R(s) + \frac{\dfrac{1}{s-1}}{1 + \dfrac{s-1}{s(s+1)}\dfrac{1}{s-1}} D(s)$$

$$= \frac{s(s+1)(s-1)}{s(s+1)(s-1)+(s-1)} R(s) + \frac{s(s+1)}{s(s+1)(s-1)+(s-1)} D(s)$$

$$= \frac{s(s+1)(s-1)}{(s^2+s+1)(s-1)} R(s) + \frac{s(s+1)}{(s^2+s+1)(s-1)} D(s)$$

$$= \underbrace{\frac{s(s+1)}{s^2+s+1}}_{E_R(s)} R(s) + \underbrace{\frac{s(s+1)}{s^2+s+1}\frac{1}{s-1}}_{E_D(s)} D(s).$$

First note that the closed-loop characteristic polynomial

$$a_c(s)a(s) + b_c(s)b(s) = (s^2+s+1)(s-1)$$

is *not* stable. However, in computing $E_R(s)$ we calculated

$$E_R(s) = \frac{s(s+1)(s-1)}{(s^2+s+1)(s-1)}\frac{R_0}{s} = \frac{s+1}{s^2+s+1} R_0.$$

This calculation used an *unstable pole–zero cancellation* to get rid of the pole at $s = 1$. This is also seen in

$$G_c(s)G(s) = \frac{s-1}{s(s+1)}\frac{1}{s-1} = \frac{1}{s(s+1)},$$

where the unstable zero in $G_c(s)$ cancels the unstable pole in $G(s)$.

On the other hand, in computing $E_D(s)$ we calculated

$$E_D(s) = \frac{s(s+1)}{(s^2+s+1)(s-1)}\frac{D_0}{s} = \frac{s+1}{(s^2+s+1)(s-1)} D_0.$$

There is no unstable pole–zero cancellation and thus $E_D(s)$ is unstable resulting in $|e_D(t)| \to \infty$ due to the unstable pole of $E_D(s)$ at $s = 1$. Recall in Definition 1 that internal stability requires all the transfer functions of (10.6)–(10.8) to be stable. In this example, $E_R(s)/R(s)$ is stable (due to unstable pole–zero cancellation), but $E_D(s)/D(s)$ is not stable.

Example 11 *Unstable Pole–Zero Cancellation*

Let's reconsider the previous example with $D(s) = 0$. Even with the disturbance zero we show that this pole–zero cancellation will not work. To explain, the model of the physical system is

$$G(s) = \frac{1}{s-1}.$$

However, no model is perfect and, in particular, we do not know the pole is exactly at $s = 1$. Our model is really

$$G(s) \triangleq \frac{1}{s-p},$$

10.3 Internal Stability

where $p \approx 1$. That is, the model used for the system says that $p = 1$ while we only know that $p \approx 1$, i.e., it is close to, but not exactly equal to 1. In the previous example we chose the controller as
$$G_c(s) = \frac{s-1}{s(s+1)}.$$

We then have
$$E_R(s) = \frac{1}{1 + \dfrac{s-1}{s(s+1)}\dfrac{1}{s-p}} R(s) = \frac{s(s+1)(s-p)}{s^3 + (1-p)s^2 + (1-p)s - 1}\frac{R_0}{s}.$$

The polynomial
$$s^3 + (1-p)s^2 + (1-p)s - 1$$
is unstable for all p as the coefficient of the s^0 term is negative (Routh–Hurwitz criterion). Further
$$s^3 + (1-p)s^2 + (1-p)s - 1\big|_{p=1} = (s^2 + s + 1)(s-1)$$
so that the numerator term $s - p|_{p=1} = s - 1$ cancels the $s - 1$ in the denominator only if p is *exactly* equal to 1. This will not happen in practice!

Example 12 *Unstable Pole–Zero Cancellation*

Consider the closed-loop control system of Figure 10.28 in which the open-loop system has transfer function
$$G(s) = \frac{b(s)}{a(s)} = \frac{s-1}{s(s+1)}.$$

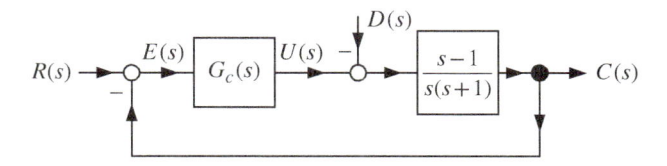

Figure 10.28. Unstable pole–zero cancellation.

The error $E(s)$ is given by
$$E(s) = \frac{1}{1 + G_c(s)G(s)} R(s) + \frac{G(s)}{1 + G_c(s)G(s)} D(s).$$

Suppose we make the (unrealistic) assumption that this model of the system is exact. Then, if one chooses
$$G_c(s) = \frac{K}{s-1},$$
it follows that
$$E(s) = \frac{1}{1 + \dfrac{K}{s-1}\dfrac{s-1}{s(s+1)}} R(s) + \frac{\dfrac{s-1}{s(s+1)}}{1 + \dfrac{K}{s-1}\dfrac{s-1}{s(s+1)}} D(s)$$

$$= \frac{s(s+1)(s-1)}{(s^2+s+K)(s-1)}\frac{R_0}{s} + \frac{(s-1)^2}{(s^2+s+K)(s-1)}\frac{D_0}{s}$$

$$= \frac{s+1}{s^2+s+K}R_0 + \frac{s-1}{s^2+s+K}\frac{D_0}{s}.$$

With $K > 0$ and this "perfect" unstable pole–zero cancellation between the numerator of $G(s)$ and the denominator of $G_c(s)$ we have $sE(s)$ stable so that by the final value theorem

$$e(\infty) = -\frac{D_0}{K}.$$

We might think that if we choose the gain K large enough, then our final error will be insignificant and this controller will be acceptable. However, this is not so! To explain, consider

$$U(s) = \frac{G_c(s)}{1+G_c(s)G(s)}R(s) + \frac{G_c(s)G(s)}{1+G_c(s)G(s)}D(s)$$

$$= \frac{\frac{K}{s-1}}{1+\frac{K}{s-1}\frac{s-1}{s(s+1)}}R(s) + \frac{\frac{K}{s-1}\frac{s-1}{s(s+1)}}{1+\frac{K}{s-1}\frac{s-1}{s(s+1)}}D(s)$$

$$= \frac{s(s+1)K}{(s^2+s+K)(s-1)}\frac{R_0}{s} + \frac{K}{s^2+s+K}\frac{D_0}{s}.$$

This shows that a step reference input $R(s) = R_0/s$ will cause the actual input $u(t)$ to the physical system to be unbounded. Again, even with perfect (impossible) pole–zero cancellation, the controller $G_c(s) = \frac{K}{s-1}$ is not viable.

Example 13 *Unstable Pole–Zero Cancellation*

Again consider the previous example where

$$G(s) = \frac{b(s)}{a(s)} = \frac{s-1}{s(s+1)}.$$

In reality our model is

$$G(s) = \frac{b(s)}{a(s)} = \frac{s-z}{s(s+1)},$$

where z is known to be close to 1, but it will not be exactly 1. With $K = 1$ we have

$$E_R(s) = \frac{1}{1+\frac{1}{s-1}\frac{s-z}{s(s+1)}}R(s) = \frac{s(s+1)(s-1)}{s^3-z}R_0$$

which is clearly unstable. Only if $z = 1$ does $s^3 - z$ equal $(s-1)(s^2+s+1)$ and the unstable pole and zero cancel.

To reiterate, by definition the system of Figure 10.26 is *internally* stable if and only if the six transfer functions of (10.6)–(10.8) are stable. Even if there is an "exact" *unstable* pole–zero cancellation in $G_c(s)G(s)$, at least one of the six transfer functions will be unstable. (See the Appendix to this chapter on *Unstable Pole–Zero Cancellations*.) On the other hand, if there are *no* unstable pole–zero cancellations and one of the six transfer functions is stable, then all of the others will be stable as well.

10.3 Internal Stability

Unstable Pole–Zero Cancellation Outside the Loop (Good)

We showed earlier that unstable pole–zero cancellation between the model transfer function $G(s)$ and the controller transfer function $G_c(s)$ will never result in a stable control system. However, we have been doing unstable pole–zero cancellations between the *closed-loop* transfer function $G_{CL}(s)$ and $R(s)$ & $D(s)$, i.e., outside the closed-loop. Let's look into this in some more detail. Consider the control system depicted in Figure 10.29.

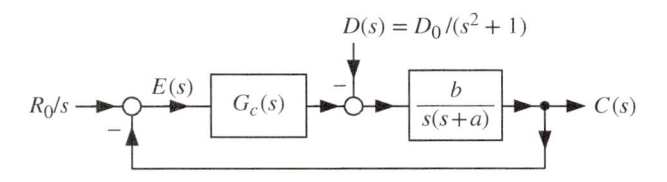

Figure 10.29. Cancellation between the closed-loop transfer function and the disturbance.

The internal model principal instructs us to choose a controller of the form

$$G_c(s) = \frac{b_3 s^3 + b_2 s^2 + b_1 s + b_0}{s + a_0} \frac{1}{s^2 + 1}.$$

Then

$$E_D(s) = \frac{G(s)}{1 + G_c(s)G(s)} D(s)$$

$$= \frac{\dfrac{b}{s(s+a)}}{1 + \dfrac{b_3 s^3 + b_2 s^2 + b_1 s + b_0}{s + a_0} \dfrac{1}{s^2 + 1} \dfrac{b}{s(s+a)}} \frac{D_0}{s^2 + 1}$$

$$= \frac{(s + a_0)(s^2 + 1)b}{\underbrace{s^5 + (a + a_0)s^4 + (aa_0 + bb_3 + 1)s^3 + (a + a_0 + bb_2)s^2 + (aa_0 + bb_1)s + bb_0}_{G_{CL}(s)}} \frac{D_0}{s^2 + 1}$$

$$= \frac{(s + a_0)b}{s^5 + f_4 s^4 + f_3 s^3 + f_2 s^2 + f_1 s + f_0} D_0.$$

Based on the internal model principle we forced $G_c(s)$ to have the factor $s^2 + 1$ in its denominator so that it ends up in the numerator of the *closed-loop* transfer function $G_{CL}(s)$ to cancel the $s^2 + 1$ in the denominator of $D(s)$. Though this is an unstable pole–zero cancellation, it goes on *outside* the closed-loop. What happens if there is not an exact cancellation? For example, suppose the disturbance is really

$$D(s) = \frac{D_0}{s^2 + 1 + \epsilon}$$

so that

$$E_D(s) = \frac{(s + a_0)(s^2 + 1)b}{s^5 + f_4 s^4 + f_3 s^3 + f_2 s^2 + f_1 s + f_0} \frac{D_0}{s^2 + 1 + \epsilon}$$

$$= \frac{(s+a_0)(s^2+1+\epsilon-\epsilon)b}{s^5+f_4s^4+f_3s^3+f_2s^2+f_1s+f_0}\frac{D_0}{s^2+1+\epsilon}$$

$$= \frac{(s+a_0)b}{s^5+f_4s^4+f_3s^3+f_2s^2+f_1s+f_0}D_0$$

$$- \underbrace{\frac{(s+a_0)b}{s^5+f_4s^4+f_3s^3+f_2s^2+f_1s+f_0}}_{H(s)}\frac{\epsilon D_0}{s^2+1+\epsilon}.$$

As $H(s)$ is stable the steady-state response is

$$e_D(t) \to -\epsilon D_0 \left|H(j\sqrt{1+\epsilon})\right| \sin\left((\sqrt{1+\epsilon})t + \angle H(j\sqrt{1+\epsilon})\right).$$

The error $e_D(t)$ does not go to zero, but it does remain *bounded* and small (assuming ϵ is small). The closed-loop characteristic polynomial

$$s^5 + (a+a_0)s^4 + (aa_0+bb_3+1)s^3 + (a+a_0+bb_2)s^2 + (aa_0+bb_1)s + bb_0$$
$$= s^5 + f_4s^4 + f_3s^3 + f_2s^2 + f_1s + f_0$$

remained the same so that the six closed-loop system transfer functions are stable. Summarizing, we are making the following points.

- No physical system can handle unbounded reference or disturbance inputs; thus we consider them to be bounded in applications.

- Bounded disturbances $D(s)$ have simple poles on the $j\omega$ axis, e.g., $D(s) = D_0/s$, $D(s) = D_0/(s^2+\omega^2)$. So an inexact cancellation between zeros of a stable closed-loop transfer function and the poles of the disturbance still results in a bounded error signal as shown.

- Bounded reference inputs also have simple poles on the $j\omega$ axis such as $R(s) = R_0/s, R(s) = R_0/(s^2+\omega^2)$. As in the case of bounded disturbances an inexact cancellation between the zeros of the closed-loop transfer function and the reference input poles results in the error signal remaining bounded. We have also considered ramp reference inputs that are unbounded, i.e., $R(s) = \omega_0/s^2$ or equivalently $r(t) = tu_s(t)$. However, in any application, the ramp will only be applied for a finite time. For example, a typical reference input with a ramp component is

$$r(t) = \begin{cases} \omega_0 t, & 0 \le t \le \theta_0/\omega_0, \\ \theta_0, & \theta_0/\omega_0 < t. \end{cases}$$

This is a bounded reference signal.

10.4 Design Example: 2 DOF Control of Aircraft Pitch

As in Chapter 9 consider control of the pitch angle of a small aircraft (Figure 10.30) whose transfer function is as follows [31]:

$$\frac{\theta(s)}{\delta(s)} = G(s) = \frac{1.51s + 0.1774}{s^3 + 0.739s^2 + 0.921s}.$$

10.4 Design Example: 2 DOF Control of Aircraft Pitch

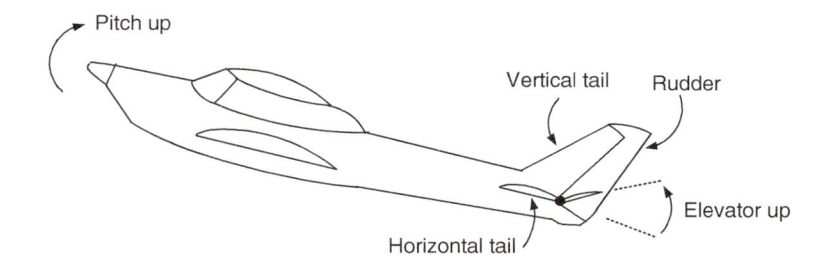

Figure 10.30. Using the elevators to pitch the aircraft up.

With a step input of 0.2 rad (11.5 degrees) the design specifications are (i) overshoot less than 10%, (ii) rise time less than 2 seconds, (iii) settling time less than 10 seconds, and (iv) final error less than 2%. We take the maximum elevator deflection to be 25° (0.436 rad), that is, $-25° \leq \delta \leq 25°$. The elevator angle δ is in radians as is the pitch angle $c(t) = \theta(t)$. Any disturbances to the pitch angle due to wind gusts, etc. on the aircraft are modeled as an equivalent disturbance input to the elevator.

Figure 10.31 shows the 2 DOF control architecture that will be used. $G_c(s)$ will be used to place the closed-loop poles and $G_f(s)$ is used to eliminate the effect of left half-plane zeros on the pitch response.

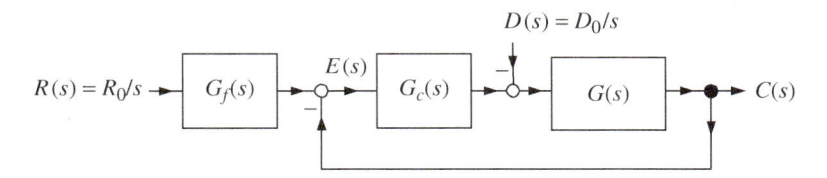

Figure 10.31. 2 DOF controller using pole placement.

The minimum order controller to achieve arbitrary pole placement has the form

$$G_c(s) = \frac{b_3 s^3 + b_2 s^2 + b_1 s + b_0}{s^2 + a_1 s + a_0} \frac{1}{s}.$$

Then

$$\frac{C(s)}{R(s)} = \frac{\dfrac{b_3 s^3 + b_2 s^2 + b_1 s + b_0}{s^2 + a_1 s + a_0} \dfrac{1}{s} \dfrac{1.51s + 0.1774}{s^3 + 0.739 s^2 + 0.921 s}}{1 + \dfrac{b_3 s^3 + b_2 s^2 + b_1 s + b_0}{s^2 + a_1 s + a_0} \dfrac{1}{s} \dfrac{1.51s + 0.1774}{s^3 + 0.739 s^2 + 0.921 s}}$$

$$= \frac{(b_2 s^2 + b_1 s + b_0)(1.51s + 0.1774)}{s(s^2 + a_1 s + a_0)(s^3 + 0.739 s^2 + 0.921 s) + (b_3 s^3 + b_2 s^2 + b_1 s + b_0)(1.51s + 0.1774)}$$

$$= \frac{(b_2 s^2 + b_1 s + b_0)(1.51s + 0.1774)}{a_{CL}(s)},$$

where

$$a_{CL}(s) = s^6 + (a_1 + 0.739)s^5 + (a_0 + 0.739 a_1 + 1.51 b_3 + 0.921)s^4$$

$$+ (0.739a_0 + 0.921a_1 + 1.51b_2 + 0.1774b_3)s^3$$
$$+ (0.921a_0 + 1.51b_1 + 0.1774b_2)s^2 + (1.51b_0 + 0.1774b_1)s + 0.1774b_0.$$

Set the desired closed-loop characteristic polynomial as

$$a_{CL}(s) = s^6 + f_5 s^5 + f_4 s^4 + f_3 s^3 + f_2 s^2 + f_1 s + f_0,$$

where the $f_i, i = 0, 1, \ldots, 5$ are still to be determined. The parameters of the controller are chosen to satisfy

$$\begin{bmatrix} f_5 \\ f_4 \\ f_3 \\ f_2 \\ f_1 \\ f_0 \end{bmatrix} = \begin{bmatrix} 0 & 0 & 0 & 0 & 1 & 0 \\ 1.51 & 0 & 0 & 0 & 0.739 & 1 \\ 0.1774 & 1.51 & 0 & 0 & 0.921 & 0.739 \\ 0 & 0.1774 & 1.51 & 0 & 0 & 0.921 \\ 0 & 0 & 0.1774 & 1.51 & 0 & 0 \\ 0 & 0 & 0 & 0.1774 & 0 & 0 \end{bmatrix} \begin{bmatrix} b_3 \\ b_2 \\ b_1 \\ b_0 \\ a_1 \\ a_0 \end{bmatrix} + \begin{bmatrix} 0.739 \\ 0.921 \\ 0 \\ 0 \\ 0 \\ 0 \end{bmatrix}$$

and thus

$$\begin{bmatrix} b_3 \\ b_2 \\ b_1 \\ b_0 \\ a_1 \\ a_0 \end{bmatrix} = \begin{bmatrix} 0 & 0 & 0 & 0 & 1 & 0 \\ 1.51 & 0 & 0 & 0 & 0.739 & 1 \\ 0.1774 & 1.51 & 0 & 0 & 0.921 & 0.739 \\ 0 & 0.1774 & 1.51 & 0 & 0 & 0.921 \\ 0 & 0 & 0.1774 & 1.51 & 0 & 0 \\ 0 & 0 & 0 & 0.1774 & 0 & 0 \end{bmatrix}^{-1}$$
$$\times \left(\begin{bmatrix} f_5 \\ f_4 \\ f_3 \\ f_2 \\ f_1 \\ f_0 \end{bmatrix} - \begin{bmatrix} 0.739 \\ 0.921 \\ 0 \\ 0 \\ 0 \\ 0 \end{bmatrix} \right). \qquad (10.10)$$

Important Comments on Pole Placement

It turns out that when using the pole placement algorithm the choice of location for the closed-loop poles is crucial for good performance. Though one can arbitrarily place the closed-loop poles, the zeros of the controller *cannot* be chosen. If possible we don't want the controller to have any zeros in the right half-plane as this will result in undershoot. The basic objective is to put the closed-loop poles far enough in the left half-plane to obtain a fast response (without saturating the actuator) while still having the zeros of the controller in the open left half-plane so their effect on the response can be eliminated by a reference input filter. For these reasons it is often not any easy task to choose the location of the closed-loop poles.

10.4 Design Example: 2 DOF Control of Aircraft Pitch

Let's proceed with choosing the closed-loop poles. We put one of the closed-loop poles at $-0.1774/1.51 = -0.1175$ to cancel the zero of $G(s)$ at this location. Then the closed-loop characteristic polynomial has the form

$$a_{CL}(s) = (s + 0.1175)(s + r)(s^2 + 2\zeta_1\omega_{n1}s + \omega_{n1}^2)(s^2 + 2\zeta_2\omega_{n2}s + \omega_{n2}^2).$$

Note that choosing $\zeta = 1$ results in two poles at $-\omega_n$, i.e., identical and real. The MATLAB code to convert this expression to the form

$$a_{CL}(s) = s^6 + f_5 s^5 + f_4 s^4 + f_3 s^3 + f_2 s^2 + f_1 s + f_0$$

is given in the Appendix entitled *Multinomial Expansions*. The "tuning" process now consists of varying $r, \zeta_1, \omega_{n1}, \zeta_2, \omega_{n2}$. After a choice of these values one must first check if the zeros of the controller $G_c(s)$, i.e., the roots of $b_3 s^3 + b_2 s^2 + b_1 s + b_0 = 0$, are in the open left-half plane. If not, then a new set of values for $r, \zeta_1, \omega_{n1}, \zeta_2, \omega_{n2}$ are chosen. If the zeros of $G_c(s)$ are in the open left-half plane then next check if the specifications are met and that the actuator does not saturate. Effort was given to having all the closed-loop poles real and negative, but this kept resulting in zeros of $G_c(s)$ in the open *right* half-plane. This forced the consideration of having complex conjugate pairs of poles. After quite a bit of trial and error the closed-loop characteristic polynomial was chosen to be

$$a_{CL}(s) = (s + 0.1175)(s + 1.6)(s^2 + 5s + 25)(s^2 + 5s + 25).$$

The controller is then

$$G_c(s) = \frac{54.34s^3 + 238.4s^2 + 678.8s + 662.3}{s^2 + 10.98s + 1.28} \frac{1}{s}$$

$$= \frac{54.34(s + 1.49)(s - [-1.45 + j2.47])(s - [-1.45 - j2.47])}{(s + 0.1175)(s + 10.86)} \frac{1}{s}.$$

Note that the zeros of $G_c(s)$ are all in the open *left-half* plane. As already mentioned we used the controller $G_c(s)$ to place one of the closed-loop poles at the zero -0.1175 of $G(s)$. This results in $G_c(s)$ having a pole at this zero as expected by Problem 16.

The reference input filter is then chosen as

$$G_f(s) = \frac{b_0}{b_3 s^3 + b_2 s^2 + b_1 s + b_0} = \frac{662.3}{54.34s^3 + 238.4s^2 + 678.8s + 662.3}$$

to cancel the zeros of $G_c(s)$. Using this 2 DOF controller the pitch angle response $\theta(t)$ and the corresponding elevator command $\delta_c(t)$ are as shown in Figure 10.32 on the next page. The rise time is about 1.5 seconds, the overshoot is zero and the 2% settling time is 3.1 seconds. As done in Chapter 9 (see (9.16) on page 264), let's replace the model $G(s) = \frac{1.51s + 0.1774}{s^3 + 0.739s^2 + 0.921s}$ by a "truth" model $G_{truth}(s) = \frac{s + 0.12}{s^3 + 0.4s^2 + s}$. Figure 10.33 on the next page shows the controller still achieves the given specifications with a rise time of 1.53 seconds, zero overshoot, and a 2% settling time of 3.15 seconds.

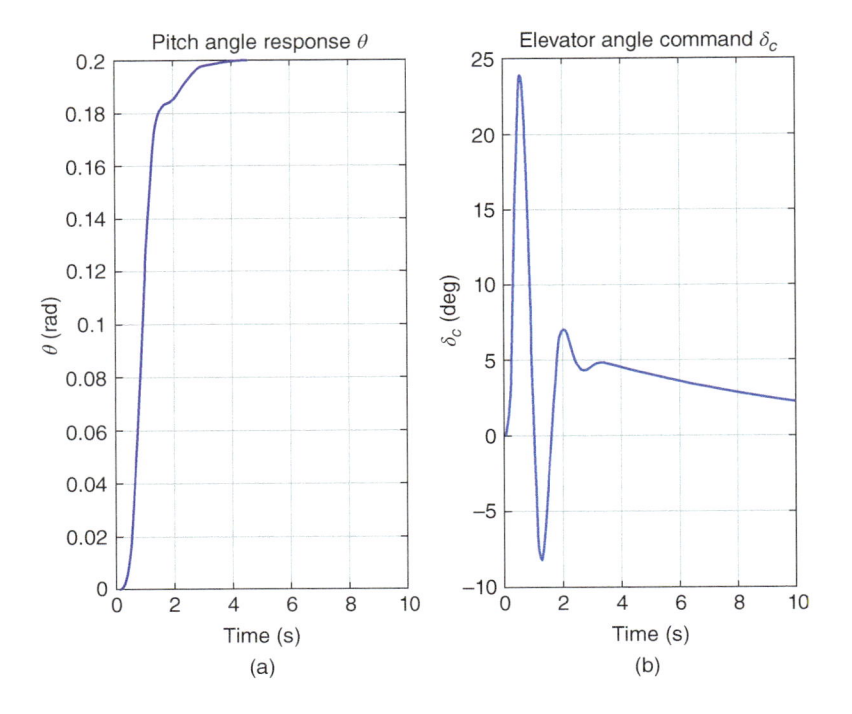

320   10   Pole Placement, 2 DOF Controllers, and Internal Stability

Figure 10.32. (a) Pitch angle $\theta(t)$. (b) Elevator command $\delta_c(t)$.

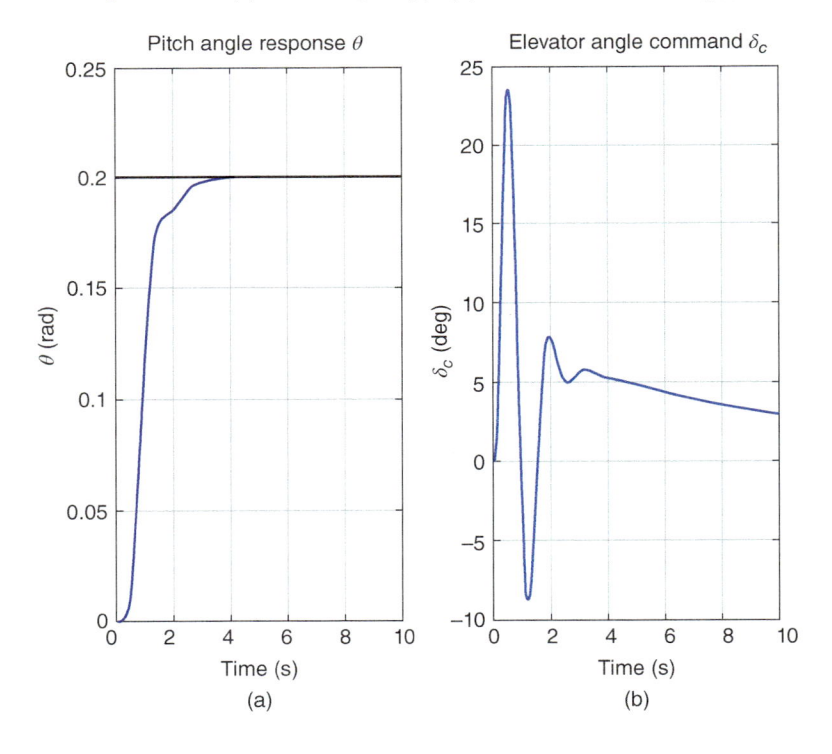

Figure 10.33. Truth model. (a) Pitch angle $\theta(t)$. (b) Elevator command $\delta_c(t)$.

10.5 Design Example: Satellite with Solar Panels (Collocated Case)

In Chapter 5 a simple mass–spring–damper model of a satellite with solar panels was developed (see Figure 10.34 on page 321). The differential equations characterizing this lumped parameter model are

$$J_s \frac{d^2\theta}{dt^2} = -K(\theta - \theta_p) - b\left(\frac{d\theta}{dt} - \frac{d\theta_p}{dt}\right) + \tau$$

$$J_p \frac{d^2\theta_p}{dt^2} = K(\theta - \theta_p) + b\left(\frac{d\theta}{dt} - \frac{d\theta_p}{dt}\right).$$

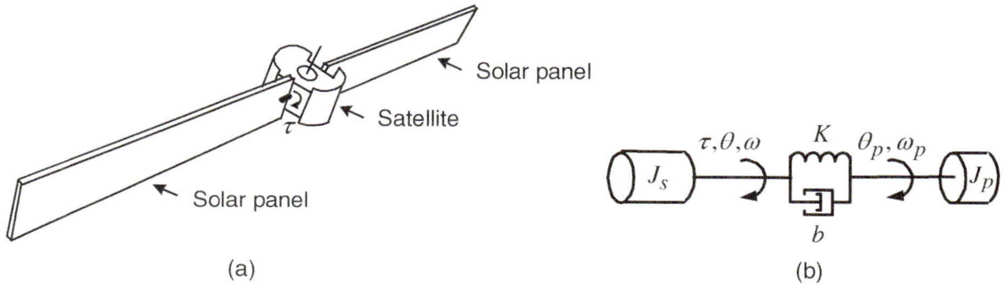

Figure 10.34. (a) Satellite with solar panels for power. (b) Lumped parameter model.

Taking Laplace transforms this reduces to

$$\theta(s) = \frac{bs + K}{s^2 J_s + bs + K}\theta_p(s) + \frac{1}{s^2 J_s + bs + K}\tau(s) \tag{10.11}$$

$$\theta_p(s) = \frac{bs + K}{s^2 J_p + bs + K}\theta(s). \tag{10.12}$$

For the collocated case the sensor is located on the actuator, that is, the motor shaft angle θ is measured. Solving (10.11) and (10.12) for $\theta(s)$ gives

$$G(s) = \frac{\theta(s)}{\tau(s)} = \frac{s^2 J_p + bs + K}{s^2 \left(J_p J_s s^2 + b(J_p + J_s)s + K(J_p + J_s)\right)}$$

$$= \frac{(1/J_s)s^2 + \left(b/(J_s J_p)\right)s + K/(J_s J_p)}{s^2 \left(s^2 + b(1/J_p + 1/J_s)s + K(1/J_p + 1/J_s)\right)}$$

$$= \frac{\beta_2 s^2 + \beta_1 s + \beta_0}{s^2(s^2 + \alpha_1 s + \alpha_0)}$$

with the obvious definitions for $\beta_2, \beta_1, \beta_0, \alpha_1, \alpha_2$. We take $J_s = 5$ kg-m^2, $J_p = 1$ kg-m^2, $K = 0.15$ N-m/rad, $b = 0.05$ Nm/rad/s with $|\tau| \leq 5$ as in [13].

A 2 DOF control system is to be used whose block diagram is shown in Figure 10.35. Using (10.12) the extra block is included in Figure 10.35 in order to obtain θ_p from θ. Though only θ is being fed back, the goal is to design $G_c(s)$ and $G_f(s)$ to obtain a "good" response for $\theta_p(t)$. It turns out that a good response is obtained by having θ turn slowly in order to "gently" turn θ_p to the desired value.

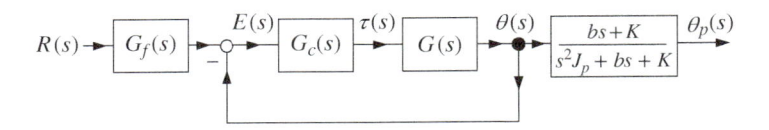

10 Pole Placement, 2 DOF Controllers, and Internal Stability

$$R(s) \rightarrow \boxed{G_f(s)} \xrightarrow{E(s)} \bigcirc \rightarrow \boxed{G_c(s)} \xrightarrow{\tau(s)} \boxed{G(s)} \xrightarrow{\theta(s)} \boxed{\frac{bs+K}{s^2 J_p + bs + K}} \xrightarrow{\theta_p(s)}$$

Figure 10.35. Two DOF controller for the collocated case.

The minimum order controller that achieves arbitrary pole placement and rejects constant disturbances has the form

$$G_c(s) = \frac{b_4 s^4 + b_3 s^3 + b_2 s^2 + b_1 s + b_0}{s^3 + a_2 s^2 + a_1 s + a_0} \frac{1}{s}.$$

We choose two (of the seven) closed-loop poles to be placed at the two zeros of $G(s)$, i.e. the roots of $s^2 + (b/J_p)s + K/J_p$. This is done to prevent overshoot from these open-loop zeros. This cancellation is done inside the loop as this is less sensitive to the uncertainty in the zeros of $G(s)$. (See Section 9.6 on *Model Uncertainty and Feedback*.) The remaining six closed-loop poles are chosen to obtain a "good" response for $\theta_p(t)$ meaning a "fast" rise time with "small" overshoot, and a "short" settling time. The quotes around "good", "fast", "small", and "short" mean that we don't have a precise specification for them because one doesn't know *a priori* how fast, small, or short we can make them. The difference $\theta_p(t) - \theta(t)$ is the amount the solar panel shaft is twisted and it is also of utmost importance that this be kept small so the shaft does not break.

With the input denoted as $R(s)$ the error $E(s) = R(s) - \theta(s)$ may be written as

$$E(s) = \frac{1}{1 + G_c(s)G(s)} R(s)$$

$$= \frac{1}{1 + \frac{b_4 s^4 + b_3 s^3 + b_2 s^2 + b_1 s + b_0}{s^3 + a_2 s^2 + a_1 s + a_0} \frac{1}{s} \frac{\beta_2 s^2 + \beta_1 s + \beta_0}{s^2(s^2 + \alpha_1 s + \alpha_0)}} R(s)$$

$$= \frac{(s^3 + a_2 s^2 + a_1 s + a_0) s^2 (s^2 + \alpha_1 s + \alpha_0)}{(s^3 + a_2 s^2 + a_1 s + a_0) s^3 (s^2 + \alpha_1 s + \alpha_0) + (b_4 s^4 + b_3 s^3 + b_2 s^2 + b_1 s + b_0)(\beta_2 s^2 + \beta_1 s + \beta_0)} R(s).$$

Expanding the denominator of $E(s)$ in powers of s gives

$$a_{CL}(s) = s^8 + (\alpha_1 + a_2)s^7 + (\alpha_0 + a_1 + \alpha_1 a_2 + \beta_2 b_4)s^6$$
$$+ (a_0 + \alpha_1 a_1 + \alpha_0 a_2 + \beta_1 b_4 + \beta_2 b_3)s^5$$
$$+ (\alpha_0 a_1 + \alpha_1 a_0 + \beta_0 b_4 + \beta_1 b_3 + \beta_2 b_2)s^4 + (\alpha_0 a_0 + \beta_0 b_3 + \beta_1 b_2 + \beta_2 b_1)s^3$$
$$+ (\beta_0 b_2 + \beta_1 b_1 + \beta_2 b_0)s^2 + (\beta_0 b_1 + \beta_1 b_0)s + \beta_0 b_0.$$

With the desired closed-loop polynomial chosen to be $s^8 + f_7 s^7 + f_6 s^6 + f_5 s^5 + f_4 s^4 + f_3 s^3 + f_2 s^2 + f_1 s + f_0$, the controller coefficients $b_4, b_3, b_2, b_1, b_0, a_2, a_1, a_0$ must satisfy

$$\begin{bmatrix} f_7 \\ f_6 \\ f_5 \\ f_4 \\ f_3 \\ f_2 \\ f_1 \\ f_0 \end{bmatrix} = \begin{bmatrix} 0 & 0 & 0 & 0 & 0 & 1 & 0 & 0 \\ \beta_2 & 0 & 0 & 0 & 0 & \alpha_1 & 1 & 0 \\ \beta_1 & \beta_2 & 0 & 0 & 0 & \alpha_0 & \alpha_1 & 1 \\ \beta_0 & \beta_1 & \beta_2 & 0 & 0 & 0 & \alpha_0 & \alpha_1 \\ 0 & \beta_0 & \beta_1 & \beta_2 & 0 & 0 & 0 & \alpha_0 \\ 0 & 0 & \beta_0 & \beta_1 & \beta_2 & 0 & 0 & 0 \\ 0 & 0 & 0 & \beta_0 & \beta_1 & 0 & 0 & 0 \\ 0 & 0 & 0 & 0 & \beta_0 & 0 & 0 & 0 \end{bmatrix} \begin{bmatrix} b_4 \\ b_3 \\ b_2 \\ b_1 \\ b_0 \\ a_2 \\ a_1 \\ a_0 \end{bmatrix} + \begin{bmatrix} \alpha_1 \\ \alpha_0 \\ 0 \\ 0 \\ 0 \\ 0 \\ 0 \\ 0 \end{bmatrix}.$$

10.5 Design Example: Satellite with Solar Panels (Collocated Case)

We choose $a_{CL}(s) = s^8 + f_7 s^7 + f_6 s^6 + f_5 s^5 + f_4 s^4 + f_3 s^3 + f_2 s^2 + f_1 s + f_0$ by setting it equal to

$$a_{CL}(s) = (s^2 + (b/J_p)s + K/J_p)(s^2 + 2\zeta_2 \omega_{n2} s + \omega_{n2}^2)(s^2 + 2\zeta_3 \omega_{n3} s + \omega_{n3}^2)$$
$$\times (s^2 + 2\zeta_4 \omega_{n4} s + \omega_{n4}^2).$$

Substituting in the given parameter values the zeros of $s^2 + (b/J_p)s + K/J_p = s^2 + 0.05s + 0.15$ are $-0.025 \pm j0.387$. There are still have six more closed-loop poles to place. However, before doing this, note that Figure 10.35 shows $\theta(s)$ is the input to the transfer function $\dfrac{bs + K}{s^2 J_p + bs + K}$ which has output $\theta_p(s)$. This suggests a smooth non-oscillatory response for $\theta(t)$ might result in a response for $\theta_p(t)$ that is not very oscillatory. Thus, to keep $\theta(t)$ from oscillating, we choose the reference input $r(t)$ to rise up to 1 in 50 seconds and then hold constant at 1 as shown in Figure 10.36. The "tuning" process now consists of varying $\zeta_2, \omega_{n2}, \zeta_3, \omega_{n3}, \zeta_4, \omega_{n4}$. As in the pitch control design, after making a choice of these values, one must check if the zeros of the controller $G_c(s)$ (i.e., the roots of $b_4 s^4 + b_3 s^3 + b_2 s^2 + b_1 s + b_0 = 0$) are in the open left-half plane. If not, then a new set of values for $\zeta_2, \omega_{n2}, \zeta_3, \omega_{n3}, \zeta_4, \omega_{n4}$ are considered. If the zeros of $G_c(s)$ are in the open left-half plane, we then check if the specifications are met, and that the actuator does not saturate. After some iterations, it was found that $\zeta_2 = 1, \omega_{n2} = 0.5, \zeta_3 = 1, \omega_{n3} = 0.5, \zeta_4 = 1, \omega_{n4} = 0.5$ (i.e., placing six closed-loop poles all at -0.5) gave the response for θ and θ_p shown in Figure 10.36. The resulting zeros of the controller $G_c(s)$ are $-0.143 \pm j0.087, -0.147 \pm j0.377$. The ramp for the reference is critical in having $\theta_p(t)$ be a low amplitude oscillation. Note that difference $\theta_p(t) - \theta(t)$ is kept small. With the given values for the model parameters the open-loop transfer function is

$$G(s) = \frac{\beta_2 s^2 + \beta_1 s + \beta_0}{s^2(s^2 + \alpha_1 s + \alpha_0)} = \frac{0.2(s^2 + 0.05s + 0.15)}{s^2(s^2 + 0.06s + 0.18)}.$$

With the closed-loop poles chosen as just discussed the controller transfer function is

$$G_c(s) = \frac{16.97 s^4 + 9.854 s^3 + 4.6875 s^2 + 0.9375 s + 0.0781}{s^4 + 2.99 s^3 + 0.297 s^2 + 0.441 s}$$
$$= \frac{16.97(s^2 + 0.286\ 78s + 0.028\ 087)(s^2 + 0.2939 s + 0.163\ 85)}{s(s + 2.94)(s^2 + 0.05 s + 0.15)}.$$

Note that poles of $G_c(s)$ include the two zeros of $G(s)$ which, by Problem 16, is to be expected. For the *noncollocated* case see Problem 24.

Observation The forward transfer function $G_c(s)G(s)$ is type 3, yet $\theta(t)$ is not tracking the type 2 reference input $r(t)$ in Figure 10.36. Why? Would it track $r(t)$ if you set $G_f(s) = 1$? (Hint: See page 301.)

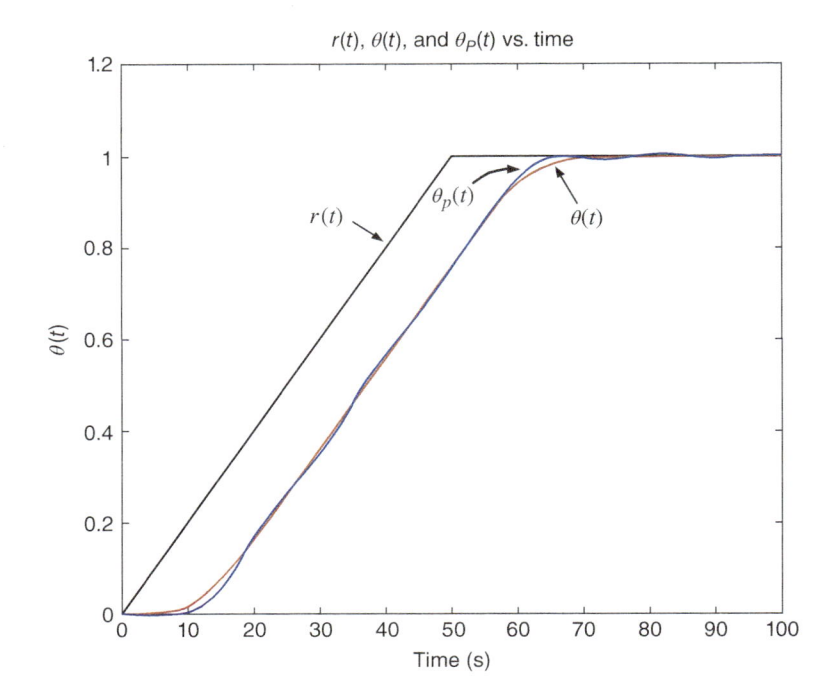

Figure 10.36. The responses $\theta(t)$ and $\theta_p(t)$ due to the reference input $r(t)$.

Appendix: Output Pole Placement

All of the signals $R(s), D(s)$ and the transfer functions $G(s)$ that have been considered up to this point have had Laplace transforms that are the ratio of two polynomials in s. That is, they are of the form $R(s) = n_R(s)/d_R(s), D(s) = n_D(s)/d_D(s), G(s) = n(s)/d(s)$ where $n_R(s), d_R(s), n_D(s), d_D(s), n(s), d(s)$ are all *polynomial* functions in s. We refer to $R(s)$ and $D(s)$ as *rational* signals and $G(s)$ as a *rational* transfer function. We will continue to assume we are working with rational signals and transfer functions. We also assume that $G(s)$ is strictly proper (true of models of physical systems), and the same for $R(s)$ and $D(s)$ (the author cannot think of a physical situation where this is not true). Recall that we have previously considered inputs of the form

$$R(s) = \frac{1}{s}, \ R(s) = \frac{1}{s^2}, \ R(s) = \frac{1}{s^2+1}$$

and disturbances $D(s)$ of the same form. Of course we could also have signals such as

$$D(s) = \frac{\alpha s + \beta}{s^2 + \omega_0^2}.$$

We also assume that $d_R(s), d_D(s)$ are *monic*, that is, their leading (highest degree) coefficient is 1. For example, if $R(s) = \dfrac{1}{2s^2+8}$ we rewrite it as $R(s) = \dfrac{1/2}{s^2+4}$ and let $d_R(s) = s^2 + 4$.

Let $d_{RD}(s)$ be the monic polynomial that we must include in the denominator of $G_c(s)$ to achieve our tracking and disturbance objectives. For example, suppose

$G(s) = \dfrac{1}{s(s+1)}, R(s) = \dfrac{R_0}{s^2+1}$ and $D(s) = \dfrac{D_0}{s}$. Then by the internal model principle the controller $G_c(s)$ must contain $s(s^2+1)$ in its denominator to track $R(s)$ and reject $D(s)$. Thus $d_{RD}(s) = s(s^2+1)$.

Consider the general single-input single-output (SISO) control system of Figure 10.37.

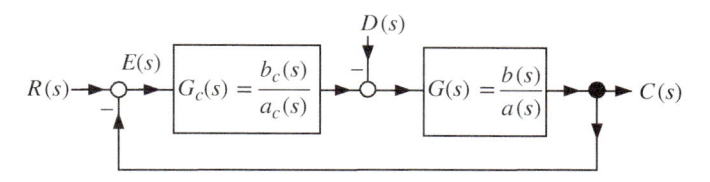

Figure 10.37. General tracking and disturbance rejection problem.

We are looking at *asymptotic tracking*, that is, we want $e(t) \to 0$ as $t \to \infty$. We have the following theorem (see [5, 6, 17]).

Theorem 1 *Tracking and Disturbance Rejection with Pole Placement*

Let the transfer function model of the open-loop system be

$$G(s) = \frac{b(s)}{a(s)} = \frac{b_{n-1}s^{n-1} + \cdots + b_0}{s^n + a_{n-1}s^{n-1} + \cdots + a_0}, \qquad (10.13)$$

where $b(s)$ and $a(s)$ are coprime. Let $d_{RD}(s)$ be the monic polynomial that the controller must have in its denominator to achieve the tracking and disturbance rejection objectives. With

$$k \triangleq \deg\{d_{RD}(s)\},$$

let the controller have the form

$$G_c(s) = \frac{b_c(s)}{a_c(s)} = \underbrace{\frac{\bar{b}_c(s)}{\bar{a}_c(s)}}_{\bar{G}_c(s)} \frac{1}{d_{RD}(s)} = \frac{\bar{b}_{n+k-1}s^{n+k-1} + \cdots + \bar{b}_0}{s^{n-1} + \bar{a}_{n-2}s^{n-2} + \cdots + \bar{a}_1 s + \bar{a}_0} \frac{1}{d_{RD}(s)}. \qquad (10.14)$$

Note that we have set $\bar{b}_c(s) \triangleq b_c(s)$. We also require that $d_{RD}(s)$ and $b(s)$ be coprime so that there is no unstable pole-zero cancellation between the controller and model transfer functions. As a consequence the polynomials

$$b(s) = b_{n-1}s^{n-1} + \cdots + b_0$$

and

$$d_{RD}(s)a(s) = d_{RD}(s)(s^n + a_{n-1}s^{n-1} + \cdots + a_0)$$

are coprime (have no factors in common).

Then one can choose the controller parameters

$$\bar{a}_0, \bar{a}_1, \ldots, \bar{a}_{n-2}, \bar{b}_0, \bar{b}_1, \ldots, \bar{b}_{n+k-1}$$

of $\bar{G}_c(s)$ in such a way that the closed-loop poles can be arbitrarily assigned.

Remark Note that $\bar{a}_c(s)$ is chosen to have degree $n-1$, which is 1 less than the degree of the denominator of $G(s)$. Thus the degree of the denominator of $G_c(s)$, i.e. the degree of $d_{RD}(s)\bar{a}_c(s)$ is $n-1+k$. The numerator of $G_c(s)$, i.e. $\bar{b}_c(s) = b_c(s)$ is chosen to have the same degree as its denominator, which is $n-1+k$.

Proof. For purposes of exposition we do the proof assuming $d_{RD}(s) = s$ (so that $k=1$) and let
$$G(s) = \frac{b_2 s^2 + b_1 s + b_0}{s^3 + a_2 s^2 + a_1 s + a_0}.$$

Then the controller is chosen to have the form
$$G_c(s) = \bar{G}_c(s)\frac{1}{s} = \frac{\bar{b}_3 s^3 + \bar{b}_2 s^2 + \bar{b}_1 s + \bar{b}_0}{s^2 + \bar{a}_1 s + \bar{a}_0}\frac{1}{s}.$$

The error is then
$$E(s) = \frac{1}{1 + \dfrac{\bar{b}_3 s^3 + \bar{b}_2 s^2 + \bar{b}_1 s + \bar{b}_0}{s^2 + \bar{a}_1 s + \bar{a}_0}\dfrac{1}{s}\dfrac{b_2 s^2 + b_1 s + b_0}{s^3 + a_2 s^2 + a_1 s + a_0}} R(s)$$

$$= \frac{1}{1 + \dfrac{\bar{b}_3 s^3 + \bar{b}_2 s^2 + \bar{b}_1 s + \bar{b}_0}{s^2 + \bar{a}_1 s + \bar{a}_0}\dfrac{b_2 s^2 + b_1 s + b_0}{\underbrace{s^4 + a_2 s^3 + a_1 s^2 + a_0 s}_{d_{RD}(s)a(s)}}} R(s)$$

$$= \frac{(s^2 + \bar{a}_1 s + \bar{a}_0)(s^4 + a_2 s^3 + a_1 s^2 + a_0 s)}{(s^2 + \bar{a}_1 s + \bar{a}_0)(s^4 + a_2 s^3 + a_1 s^2 + a_0 s) + (\bar{b}_3 s^3 + \bar{b}_2 s^2 + \bar{b}_1 s + \bar{b}_0)(b_2 s^2 + b_1 s + b_0)} R(s).$$

To assign the closed-loop poles, we must have
$$\underbrace{(s^2 + \bar{a}_1 s + \bar{a}_0)(s^4 + a_2 s^3 + a_1 s^2 + a_0 s)}_{d_{RD}(s)a(s)} + (\bar{b}_3 s^3 + \bar{b}_2 s^2 + \bar{b}_1 s + \bar{b}_0)\underbrace{(b_2 s^2 + b_1 s + b_0)}_{b(s)}$$
$$= s^6 + f_5 s^5 + f_4 s^4 + f_3 s^3 + f_2 s^2 + f_1 s + f_0 \tag{10.15}$$

for arbitrarily chosen f_i. Equation (10.15) is referred to as the *Diophantine* equation. Equating powers of s this last equation is equivalent to the matrix equation

$$\begin{matrix} 1 \\ s \\ s^2 \\ s^3 \\ s^4 \\ s^5 \\ s^6 \end{matrix} \underbrace{\begin{bmatrix} 0 & 0 & 0 & b_0 & 0 & 0 & 0 \\ a_0 & 0 & 0 & b_1 & b_0 & 0 & 0 \\ a_1 & a_0 & 0 & b_2 & b_1 & b_0 & 0 \\ a_2 & a_1 & a_0 & 0 & b_2 & b_1 & b_0 \\ 1 & a_2 & a_1 & 0 & 0 & b_2 & b_1 \\ 0 & 1 & a_2 & 0 & 0 & 0 & b_2 \\ 0 & 0 & 1 & 0 & 0 & 0 & 0 \end{bmatrix}}_{S} \begin{bmatrix} \bar{a}_0 \\ \bar{a}_1 \\ 1 \\ \bar{b}_0 \\ \bar{b}_1 \\ \bar{b}_2 \\ \bar{b}_3 \end{bmatrix} = \begin{bmatrix} f_0 \\ f_1 \\ f_2 \\ f_3 \\ f_4 \\ f_5 \\ 1 \end{bmatrix}. \tag{10.16}$$

The matrix S is referred to as the *Sylvester resultant* matrix. If S is invertible then we can solve (10.16) for coefficients of the controller and achieve pole placement.

Appendix: Output Pole Placement

We now show that the matrix S is invertible if and only if the two polynomials $d_{RD}(s)a(s) = s(s^3 + a_2s^2 + a_1s + a_0)$ and $b(s) = b_3s^3 + b_2s^2 + b_1s + b_0$ are coprime.

(*If*) Suppose $s(s^3 + a_2s^2 + a_1s + a_0)$ and $b_3s^3 + b_2s^2 + b_1s + b_0$ are coprime. We show that S is invertible. Consider the minor modification of (10.15) obtained by replacing $s^2 + \bar{a}_1s + \bar{a}_0$ with $\bar{a}_2s^2 + \bar{a}_1s + \bar{a}_0$. Then we may write

$$(\bar{a}_2s^2 + \bar{a}_1s + \bar{a}_0)\underbrace{(s^4 + a_2s^3 + a_1s^2 + a_0s)}_{d_{RD}(s)a(s)} + (\bar{b}_3s^3 + \bar{b}_2s^2 + \bar{b}_1s + \bar{b}_0)\underbrace{(b_2s^2 + b_1s + b_0)}_{b(s)}$$
$$= f_6s^6 + f_5s^5 + f_4s^4 + f_3s^3 + f_2s^2 + f_1s + f_0. \qquad (10.17)$$

Equation (10.16) becomes

$$\begin{matrix} 1 \\ s \\ s^2 \\ s^3 \\ s^4 \\ s^5 \\ s^6 \end{matrix} \underbrace{\begin{bmatrix} 0 & 0 & 0 & b_0 & 0 & 0 & 0 \\ a_0 & 0 & 0 & b_1 & b_0 & 0 & 0 \\ a_1 & a_0 & 0 & b_2 & b_1 & b_0 & 0 \\ a_2 & a_1 & a_0 & 0 & b_2 & b_1 & b_0 \\ 1 & a_2 & a_1 & 0 & 0 & b_2 & b_1 \\ 0 & 1 & a_2 & 0 & 0 & 0 & b_2 \\ 0 & 0 & 1 & 0 & 0 & 0 & 0 \end{bmatrix}}_{S} \begin{bmatrix} \bar{a}_0 \\ \bar{a}_1 \\ \bar{a}_2 \\ \bar{b}_0 \\ \bar{b}_1 \\ \bar{b}_2 \\ \bar{b}_3 \end{bmatrix} = \begin{bmatrix} f_0 \\ f_1 \\ f_2 \\ f_3 \\ f_4 \\ f_5 \\ f_6 \end{bmatrix}. \qquad (10.18)$$

Note that this is the same Sylvester resultant matrix S as in (10.16). The proof is by contradiction. Suppose S is not invertible so that its columns are linear dependent. Then there is a *nonzero* vector

$$\begin{bmatrix} \bar{a}_0 & \bar{a}_1 & \bar{a}_2 & \bar{b}_0 & \bar{b}_1 & \bar{b}_2 & \bar{b}_3 \end{bmatrix}^T$$

such that

$$\begin{bmatrix} 0 & 0 & 0 & b_0 & 0 & 0 & 0 \\ a_0 & 0 & 0 & b_1 & b_0 & 0 & 0 \\ a_1 & a_0 & 0 & b_2 & b_1 & b_0 & 0 \\ a_2 & a_1 & a_0 & 0 & b_2 & b_1 & b_0 \\ 1 & a_2 & a_1 & 0 & 0 & b_2 & b_1 \\ 0 & 1 & a_2 & 0 & 0 & 0 & b_2 \\ 0 & 0 & 1 & 0 & 0 & 0 & 0 \end{bmatrix} \begin{bmatrix} \bar{a}_0 \\ \bar{a}_1 \\ \bar{a}_2 \\ \bar{b}_0 \\ \bar{b}_1 \\ \bar{b}_2 \\ \bar{b}_3 \end{bmatrix} = \begin{bmatrix} 0 \\ 0 \\ 0 \\ 0 \\ 0 \\ 0 \\ 0 \end{bmatrix}.$$

For these particular values of $\bar{a}_0, \bar{a}_1, \bar{a}_2, \bar{b}_0, \bar{b}_1, \bar{b}_2, \bar{b}_3$ (which are not all zero) we have

$$(\bar{a}_2s^2 + \bar{a}_1s + \bar{a}_0)(s^4 + a_2s^3 + a_1s^2 + a_0s) + (\bar{b}_3s^3 + \bar{b}_2s^2 + \bar{b}_1s + \bar{b}_0)(b_2s^2 + b_1s + b_0) = 0.$$

It follows immediately that $\bar{a}_2 = 0$ so this reduces to

$$(\bar{a}_1s + \bar{a}_0)(s^4 + a_2s^3 + a_1s^2 + a_0s) + (\bar{b}_3s^3 + \bar{b}_2s^2 + \bar{b}_1s + \bar{b}_0)(b_2s^2 + b_1s + b_0) = 0$$

or

$$(\bar{a}_1s + \bar{a}_0)\underbrace{(s^4 + a_2s^3 + a_1s^2 + a_0s)}_{d_{RD}(s)a(s) \text{ has degree } n+k} = -\underbrace{(\bar{b}_3s^3 + \bar{b}_2s^2 + \bar{b}_1s + \bar{b}_0)}_{\bar{b}_c(s) \text{ has degree } n-1+k}(b_2s^2 + b_1s + b_0). \qquad (10.19)$$

We have already pointed out that $\bar{a}_2 = 0$. If $\bar{a}_1 = \bar{a}_0 = \bar{b}_3 = \bar{b}_2 = \bar{b}_1 = \bar{b}_0 = 0$ then this last equation obviously holds. However, the assumption that S is singular means that we could (and did) choose them such that they are not all zero.

Now the roots of $s^4 + a_2 s^3 + a_1 s^2 + a_0 s$ (in general $d_{RD}(s)a(s)$ with degree $n+k$) must divide the right side of (10.19). However $\bar{b}_3 s^3 + \bar{b}_2 s^2 + \bar{b}_1 s + \bar{b}_0$ (in general $\bar{b}_c(s)$ with degree $n+k-1$) has only three factors (in general $n-1+k$ factors) while $s^4 + a_2 s^3 + a_1 s^2 + a_0 s$ has four factors (in general $n+k$ factors). Thus the four factors of $s^4 + a_2 s^3 + a_1 s^2 + a_0 s$ cannot all divide out the three factors of $\bar{b}_3 s^3 + \bar{b}_2 s^2 + \bar{b}_1 s + \bar{b}_0$. So at least one of them must divide $b_2 s^2 + b_1 s + b_0$ which contradicts the assumption that $d_{RD}(s)a(s) = s^4 + a_2 s^3 + a_1 s^2 + a_0 s$ and $b(s) = b_2 s^2 + b_1 s + b_0$ are coprime. With S singular we have a contradiction. Thus S must be nonsingular.

(*Only If*) Let S be invertible and we show that $d_{RD}(s)a(s) = s(s^3 + a_2 s^2 + a_1 s + a_0)$ and $b(s) = b_3 s^3 + b_2 s^2 + b_1 s + b_0$ are coprime. The proof is by contradiction. Suppose they are not coprime so there is an s_0 such that $s_0(s_0^3 + a_2 s_0^2 + a_1 s_0 + a_0) = 0$ and $b_3 s_0^3 + b_2 s_0^2 + b_1 s_0 + b_0 = 0$. Setting $s = s_0$, the left side of (10.15) will be zero no matter what we choose for $s^2 + \bar{a}_1 s + \bar{a}_0$ and $\bar{b}_3 s^3 + \bar{b}_2 s^2 + \bar{b}_1 s + \bar{b}_0$! To obtain a contradiction, let s_1 be any number that is *not* a common zero of these same two polynomials (in particular, $s_1 \neq s_0$). Choose the coefficients f_i such that $s^6 + f_5 s^5 + f_4 s^4 + f_3 s^3 + f_2 s^2 + f_1 s + f_0 = (s - s_1)^6$. The fact that S is invertible means the right side of

$$\begin{bmatrix} \bar{a}_0 \\ \bar{a}_1 \\ 1 \\ \bar{b}_0 \\ \bar{b}_1 \\ \bar{b}_2 \\ \bar{b}_3 \end{bmatrix} = \begin{bmatrix} 0 & 0 & 0 & b_0 & 0 & 0 & 0 \\ a_0 & 0 & 0 & b_1 & b_0 & 0 & 0 \\ a_1 & a_0 & 0 & b_2 & b_1 & b_0 & 0 \\ a_2 & a_1 & a_0 & 0 & b_2 & b_1 & b_0 \\ 1 & a_2 & a_1 & 0 & 0 & b_2 & b_1 \\ 0 & 1 & a_2 & 0 & 0 & 0 & b_2 \\ 0 & 0 & 1 & 0 & 0 & 0 & 0 \end{bmatrix}^{-1} \begin{bmatrix} f_0 \\ f_1 \\ f_2 \\ f_3 \\ f_4 \\ f_5 \\ 1 \end{bmatrix}$$

exists and thus we can find the coefficients of the polynomials $s^2 + \bar{a}_1 s + \bar{a}_0$ and $\bar{b}_3 s^3 + \bar{b}_2 s^2 + \bar{b}_1 s + \bar{b}_0$ such that (10.15) holds. However, the right side of (10.15) is now $(s - s_1)^6$ which is not zero for $s = s_0$ while the left side is zero for $s = s_0$. This contradiction shows that if S is invertible then $s(s^3 + a_2 s^2 + a_1 s + a_0)$ and $b_3 s^3 + b_2 s^2 + b_1 s + b_0$ must be coprime. ∎

Appendix: Multinomial Expansions

See the `Chapter10_students` folder in the MATLAB/SIMULINK files for the .m files for the multinomial expansions of orders 3 through 7. The first two are

$$(s + r_1)(s + r_2) = s^2 + (r_1 + r_2)s + r_1 r_2$$
$$(s + r_1)(s + r_2)(s + r_3) = s^3 + (r_1 + r_2 + r_3)s^2 + (r_1 r_2 + r_1 r_3 + r_2 r_3)s + r_1 r_2 r_3.$$

Next consider the coefficients for a 4th-order expansion. That is, compute the coefficients $f_i, i = 0, \ldots, 3$ for

$$(s + r_1)(s + r_2)(s + r_3)(s + r_4) = s^4 + f_3 s^3 + f_2 s^2 + f_1 s + f_0$$

or
$$(s^2 + 2\zeta_1\omega_{n1}s + \omega_{n1}^2)(s^2 + 2\zeta_2\omega_{n2}s + \omega_{n2}^2) = s^4 + f_3 s^3 + f_2 s^2 + f_1 s + f_0.$$

The MATLAB code to do this is as follows.

```
% Fourth-Order Multinomial Expansion
close all; clc; clear
s = sym('s');
r1 = 2; r2 = 2; r3 = 2; r4 = 2;
zeta1 = 0.5; wn1 = 5; zeta2 = 0.6; wn2 = 4;
a_cl = (s+r1)*(s+r2)*(s+r3)*(s+r4);
% a_cl = (s^2+2*zeta1*wn1*s+wn1^2)*(s^2+2*zeta2*wn2*s+wn2^2);
a_cl = expand(a_cl); pretty(a_cl)
ff = sym2poly(a_cl);
f3 = ff(2); f2 = ff(3); f1 = ff(4); f0 = ff(5);
f = [f3; f2; f1; f0];
```

Appendix: Overshoot

Theorem 2 *Overshoot in Type 2 Systems* [5, 6]

Consider the block diagram of Figure 10.38 where $L(s) = G_c(s)G(s)$ is a coprime transfer function of type 2, that is,
$$L(s) = \frac{b_L(s)}{a_L(s)} = \frac{b_L(s)}{s^2 \bar{a}_L(s)}$$

with $\bar{a}_L(0) \neq 0$. (Note that $b_L(0) \neq 0$ as $b_L(s)$ and $a_L(s)$ are coprime.) Further, suppose the closed-loop system is *stable*.

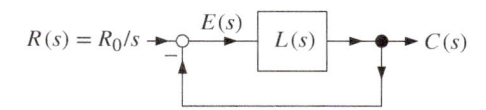

Figure 10.38. Stable type 2 system with a step input.

Then the step response of this system will always have overshoot. That is, the step response
$$c(t) = \mathcal{L}^{-1}\left\{\frac{L(s)}{1+L(s)} \frac{R_0}{s}\right\}$$

will have overshoot no matter where the closed-loop poles are situated in the open left half-plane.

Proof.
$$E(s) = \frac{1}{1+L(s)} \frac{R_0}{s} = \frac{s^2 \bar{a}_L(s)}{s^2 \bar{a}_L(s) + b_L(s)} \frac{R_0}{s} = \frac{s \bar{a}_L(s)}{s^2 \bar{a}_L(s) + b_L(s)} R_0.$$

We are given that the closed-loop system is stable, which means that the roots $p_1, \ldots, p_n$ of $s^2 \bar{a}_L(s) + b_L(s) = 0$ satisfy $\text{Re}\{p_i\} < 0$ for $i = 1, \ldots, n$. Assuming these roots are distinct we then have

$$E(s) = \frac{s\bar{a}_L(s)}{s^2\bar{a}_L(s) + b_L(s)} R_0 = \frac{s\bar{a}_L(s)}{(s-p_1)\cdots(s-p_n)} R_0 = \sum_{i=1}^{n} \frac{A_i}{s-p_i}.$$

Computing the inverse Laplace transform gives $e(t) = \sum_{i=1}^{n} A_i e^{p_i t}$ as

$$\frac{1}{s-p_i} = \int_0^\infty e^{-st} e^{p_i t} dt \quad \text{for} \quad \text{Re}\{s\} > \text{Re}\{p_i\}.$$

As $E(s)$ is stable it follows that $\alpha \triangleq \max\{\text{Re}\{p_i\}\} < 0$ (see Figure 10.39). Thus for $\text{Re}\{s\} > \alpha$ the integral

$$E(s) = \int_0^\infty e^{-st} e(t) dt$$

exists. In particular, the integral exists for $s = 0$ and so

$$\int_0^\infty e(t) dt = E(s)|_{s=0} = E(0) = \frac{0 \cdot \bar{a}_L(0)}{0^2 \cdot \bar{a}_L(0) + b_L(0)} R_0 = 0.$$

Now $e(t) = c(t) - r(t)$ with $e(0) = 0 - R_0 < 0$ if $R_0 > 0$. Thus the only way the integral $\int_0^\infty e(t) dt$ can be zero is that $e(t) = c(t) - R_0$ must eventually become positive which means that $c(t)$ must eventually be greater than R_0 and therefore there is overshoot. A similar argument applies for $R_0 < 0$.

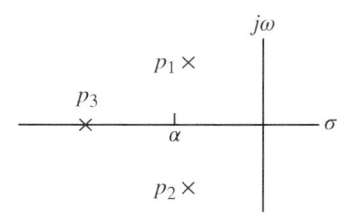

Figure 10.39. As $E(s)$ is stable it follows that $\alpha \triangleq \max\{\text{Re}\{p_i\}\} < 0$. Thus for $\text{Re}\{s\} > \alpha$ the integral $E(s) = \int_0^\infty e^{-st} e(t) dt$ exists, which includes $s = 0$. ∎

Theorem 3 *Cascade of Stable Systems without Overshoot [5]*
Let $G_1(s)$ and $G_2(s)$ be two *stable* transfer functions. Suppose

$$G_1(0) > 0 \text{ with } g_1(t) \triangleq \mathcal{L}^{-1}\{G_1(s)\} > 0 \text{ for all } t$$

and

$$G_2(0) > 0 \text{ where } c_2(t) \triangleq \mathcal{L}^{-1}\{G_2(s)/s\} \text{ has no overshoot.}$$

Then the unit step response of the system $G_1(s)G_2(s)$ has no overshoot, that is, $c(t)$ given by

$$c(t) \triangleq \mathcal{L}^{-1}\left\{G_2(s)G_1(s)\frac{1}{s}\right\}$$

has no overshoot.

Appendix: Overshoot   331

Proof. Now $c_2(0) = 0$ and, as $G_2(s)$ is stable, it follows by the final value theorem that $c_2(\infty) = \lim_{s \to 0} sG_2(s)\frac{1}{s} = G_2(0)$. So saying that $c_2(t)$ has no overshoot is the same as writing
$$c_2(t) \leq G_2(0) \text{ for all } t.$$

As $G_2(s)G_1(s)$ is stable we have
$$c(\infty) = \lim_{s \to 0} sG_2(s)G_1(s)\frac{1}{s} = G_2(0)G_1(0) > 0$$

with $c(0) = 0$. We want to show that
$$c(t) \leq G_2(0)G_1(0) \text{ for all } t.$$

To do this we write
$$c(t) = \int_0^t g_1(t - \tau)c_2(\tau)d\tau.$$

As $c_2(t) \leq G_2(0)$ and $g_1(t - \tau) > 0$ for all $0 \leq \tau \leq t$ it follows that
$$g_1(t - \tau)c_2(t) \leq g_1(t - \tau)G_2(0).$$

Then
$$c(t) = \int_0^t g_1(t - \tau)c_2(\tau)d\tau \leq \int_0^t g_1(t - \tau)d\tau G_2(0)$$
$$= \int_0^t g_1(\tau)d\tau G_2(0)$$
$$\leq \int_0^\infty g_1(\tau)d\tau G_2(0) \text{ as } g_1(\tau) > 0 \text{ for all } \tau$$
$$= G_1(0)G_2(0)$$
$$= c(\infty).$$

Therefore $c(t)$ does not have overshoot. ∎

Theorem 4 *Stable System with Real Poles and No Zeros [5]*
Suppose the closed-loop transfer function $G_{CL}(s)$ is stable with *real* poles and *no* zeros. Then the step response has no overshoot.

Proof. We are given that the closed-loop transfer function is of the form
$$G_{CL}(s) = G_{CL}(0)\frac{-p_1}{s - p_1}\frac{-p_2}{s - p_2} \cdots \frac{-p_n}{s - p_n},$$

where for $i = 1, \ldots, n$ the $p_i < 0$ (real and negative). The step response is the inverse Laplace transform of
$$C(s) = G_{CL}(0)\frac{-p_1}{s - p_1}\frac{-p_2}{s - p_2} \cdots \frac{-p_n}{s - p_n}\frac{1}{s}.$$

Let
$$G_1(s) = \frac{-p_1}{s - p_1}, \ G_2(s) = \frac{-p_2}{s - p_2}.$$

Then
$$g_1(t) = \mathcal{L}^{-1}\{G_1(s)\} = -p_1 e^{p_1 t} > 0 \text{ for all } t \geq 0$$

and, as p_2 is real and negative, $c_2(t)$ given by
$$c_2(t) \triangleq \mathcal{L}^{-1}\left\{G_2(s)\frac{1}{s}\right\} = 1 - e^{p_1 t}$$

does not overshoot. Thus, by the previous theorem (Theorem 3), $c_{12}(t)$ given by
$$c_{12}(t) \triangleq \mathcal{L}^{-1}\left\{G_1(s)G_2(s)\frac{1}{s}\right\}$$

does not overshoot. We then continue by applying Theorem 3 to $G_3(s)$ ($g_3(t) = -p_3 e^{p_3 t}$) and $G_{12}(s) \triangleq G_1(s)G_2(s)$ to conclude that
$$c_{123}(t) \triangleq \mathcal{L}^{-1}\left\{G_1(s)G_2(s)G_3(s)\frac{1}{s}\right\}$$

has no overshoot. Continuing in this manner it follows that
$$c(t) \triangleq \mathcal{L}^{-1}\left\{G_{CL}(0)G_1(s)G_2(s)\cdots G_n(s)\frac{1}{s}\right\} = G_{CL}(0)\mathcal{L}^{-1}\left\{\frac{-p_1}{s-p_1}\frac{-p_2}{s-p_2}\cdots\frac{-p_n}{s-p_n}\frac{1}{s}\right\}$$

has no overshoot. ∎

Theorem 5 *Stable System with Real Poles and One RHP Zero [5]*

Suppose the closed-loop transfer function $G_{CL}(s)$ is stable with *real* poles and has one zero in the *open right half-plane*.

Then the step response has no overshoot.

Remark Though this step response will not have overshoot, it will have *undershoot* as it has a single right half-plane zero (see page 222 of Chapter 8).

Proof. We are given that the closed-loop transfer function is of the form
$$G_{CL}(s) = G_{CL}(0)\frac{s-z}{-z}\frac{-p_1}{s-p_1}\frac{-p_2}{s-p_2}\cdots\frac{-p_n}{s-p_n},$$

where for $i = 1, \ldots, n$ the $p_i < 0$ (real and negative) and $z > 0$. The step response is the inverse Laplace transform of
$$C(s) = G_{CL}(0)\frac{s-z}{-z}\frac{-p_1}{s-p_1}\frac{-p_2}{s-p_2}\cdots\frac{-p_n}{s-p_n}\frac{1}{s}.$$

With
$$G_1(s) = \frac{s-z}{-z}\frac{-p_1}{s-p_1}, \ G_2(s) = \frac{-p_2}{s-p_2}$$

Appendix: Overshoot   333

we see that
$$g_2(t) = \mathcal{L}^{-1}\{G_2(s)\} = -p_2 e^{p_2 t} > 0 \text{ for all } t \geq 0$$

and
$$c_{1z}(t) \triangleq \mathcal{L}^{-1}\left\{\frac{s-z}{-z}\frac{-p_1}{s-p_1}\frac{1}{s}\right\} = \mathcal{L}^{-1}\left\{\frac{1}{s} - \frac{z-p_1}{z}\frac{1}{s-p_1}\right\} = 1 - \frac{z-p_1}{z}e^{p_1 t}.$$

As $\dfrac{z-p_1}{z} > 0$ it follows that $c_{1z}(t)$ increases monotonically from $1 - \dfrac{z-p_1}{z}$ to 1 as t goes from 0 to ∞. Thus $c_{1z}(t)$ has no overshoot. Again using Theorem 3 it follows that

$$c_{12z}(t) \triangleq \mathcal{L}^{-1}\left\{G_1(s)G_2(s)\frac{1}{s}\right\} = \mathcal{L}^{-1}\left\{\frac{s-z}{-z}\frac{-p_1}{s-p_1}\frac{-p_2}{s-p_2}\frac{1}{s}\right\}$$

has no overshoot. Continuing in this manner we see that

$$c(t) \triangleq G_{CL}(0)\mathcal{L}^{-1}\left\{\frac{s-z}{-z}\frac{-p_1}{s-p_1}\frac{-p_2}{s-p_2}\cdots\frac{-p_n}{s-p_n}\frac{1}{s}\right\}$$

has no overshoot.   ■

Theorem 6 *Stable System with Real Poles and Two RHP Zeros [5]*

Suppose the closed-loop transfer function $G_{CL}(s)$ is stable with *real* poles and two *right half-plane* zeros. That is,

$$G_{CL}(s) = G_{CL}(0)\frac{s^2 - 2\zeta\omega_0 s + \omega_0^2}{\omega_0^2}\frac{-p_1}{s-p_1}\frac{-p_2}{s-p_2}\cdots\frac{-p_n}{s-p_n}$$

where $\omega_0 > 0, \zeta > 0$. With the addition of the reference input filter

$$\frac{\omega_0^2}{(s+\omega_0)^2},$$

the unit step response $\mathcal{L}^{-1}\left\{G_{CL}(s)\dfrac{\omega_0^2}{(s+\omega_0)^2}\dfrac{1}{s}\right\}$ has no overshoot.

Remark If $0 < \zeta < 1$ then the two right half-plane zeros are a complex conjugate pair. If $\zeta \geq 1$ then the two right half-plane zeros are real.

Proof. We want to show that the unit step response of

$$G_{CL}(s) = G_{CL}(0)\frac{\omega_0^2}{(s+\omega_0)^2}\frac{s^2 - 2\zeta\omega_0 s + \omega_0^2}{\omega_0^2}\frac{-p_1}{s-p_1}\frac{-p_2}{s-p_2}\cdots\frac{-p_n}{s-p_n}$$

has no overshoot. To prove this we first show that the unit step response

$$\mathcal{L}^{-1}\left\{\frac{s^2 - 2\zeta\omega_0 s + \omega_0^2}{(s+\omega_0)^2}\frac{1}{s}\right\} = 1 - 2\omega_0 t e^{-t\omega_0}(1+\zeta)$$

has no overshoot. Any extrema point of this step response is a root of

$$\frac{d}{dt}\left(1 - 2\omega_0 t e^{-t\omega_0}(1+\zeta)\right) = 0.$$

The only solution is at $t = 1/\omega_0$ which corresponds to the minimum value of this step response. It has a maximum value of 1 at both $t = 0$ and $t = \infty$. Thus it does not have overshoot. With

$$G_1(s) = \frac{s^2 - 2\zeta\omega_0 s + \omega_0^2}{(s+\omega_0)^2}\frac{1}{s}, G_2(s) = \frac{-p_2}{s - p_2},$$

Theorem 3 again applies to show that

$$\mathcal{L}^{-1}\left\{G_1(s)G_2(s)\frac{1}{s}\right\}$$

has no overshoot. Continuing in this manner it follows that

$$c(t) \triangleq G_{CL}(0)\mathcal{L}^{-1}\left\{\frac{s^2 - 2\zeta\omega_0 s + \omega_0^2}{(s+\omega_0)^2}\frac{1}{s}\frac{-p_1}{s-p_1}\frac{-p_2}{s-p_2}\cdots\frac{-p_n}{s-p_n}\right\}$$

has no overshoot. ∎

Theorem 7 *Overshoot Due to Left Half-Plane Zeros [6]*

Let $G(s)$ be the transfer function of a physical system. Suppose the controller $G_c(s)$ makes $\frac{G_c(s)G(s)}{1 + G_c(s)G(s)}$ stable with the closed-loop poles p_i satisfying $\text{Re}\{p_i\} < -\alpha$ for all $i = 1, 2, \ldots, n$ and some $\alpha > 0$. Further, let z be a zero of $G(s)$ (i.e., $G(z) = 0$) with $-\alpha < z < 0$. In words, z is a zero of $G(s)$ in the open left-half plane, but is to the right of all the closed-loop poles. Then

$$c(t) = \mathcal{L}^{-1}\{C(s)\} = \mathcal{L}^{-1}\left\{\frac{G_c(s)G(s)}{1 + G_c(s)G(s)}\frac{1}{s}\right\}$$

will have overshoot.

Proof. We have

$$E(s) = R(s) - C(s) = \frac{1}{1 + G_c(s)G(s)}\frac{1}{s}$$

and for $\text{Re}\{s\} > -\alpha$

$$\int_0^\infty e(t)e^{-st}dt = \frac{1}{1 + G_c(s)G(s)}\frac{1}{s}.$$

In particular, as $z > -\alpha$ and $G(z) = 0$, it follows that

$$\int_0^\infty e(t)e^{-zt}dt = \frac{1}{1 + G_c(z)G(z)}\frac{1}{z} = \frac{1}{z} < 0 \text{ as } z < 0.$$

Then

$$\int_0^\infty \left(r(t) - c(t)\right)e^{-zt}dt = \int_0^\infty e(t)e^{-zt}dt < 0.$$

As $r(t) = u_s(t)$ and $c(0) = 0$ it follows that $e(0) = 1$. So $e(t)$ starts out positive, but its integral over all time is negative so $e(t)$ itself must go negative. Finally $e(t)$ negative means $c(t) > r(t) = 1$ and thus there is overshoot. ∎

Appendix: Unstable Pole-Zero Cancellation

Theorem 8 *Unstable Pole-Zero Cancellation*
Let the model transfer function and the controller transfer function be, respectively,

$$G(s) = \frac{b(s)}{a(s)}, \quad G_c(s) = \frac{b_c(s)}{a_c(s)}.$$

$G(s)$ is assumed to be strictly proper and $G_c(s)$ is assumed to be proper.

In the unity feedback control system of Figure 10.40 suppose there is an unstable pole-zero cancellation between $G(s)$ and $G_c(s)$. Then, even with *exact* pole-zero cancellation, one (or more) of the following six transfer functions will be unstable:

$$E(s) = \frac{a_c(s)a(s)}{a_c(s)a(s) + b_c(s)b(s)} R(s) + \frac{a_c(s)b(s)}{a_c(s)a(s) + b_c(s)b(s)} D(s)$$

$$C(s) = \frac{b_c(s)b(s)}{a_c(s)a(s) + b_c(s)b(s)} R(s) - \frac{a_c(s)b(s)}{a_c(s)a(s) + b_c(s)b(s)} D(s)$$

$$U(s) = \frac{a(s)b_c(s)}{a_c(s)a(s) + b_c(s)b(s)} R(s) + \frac{b_c(s)b(s)}{a_c(s)a(s) + b_c(s)b(s)} D(s).$$

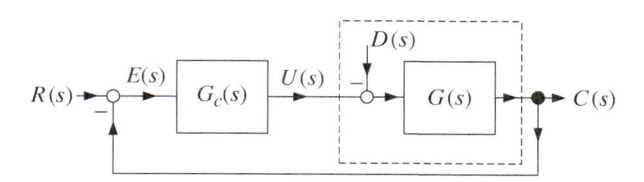

Figure 10.40. Standard unity feedback control system.

Proof. Suppose

$$G(s) = \frac{b(s)}{a(s)} = \frac{b(s)}{(s-1)\bar{a}(s)}, G_c(s) = \frac{b_c(s)}{a_c(s)} = \frac{(s-1)\bar{b}_c(s)}{a_c(s)}.$$

so that in the expression $1 + G(s)G_c(s)$ the unstable pole of $G(s)$ is canceled by an unstable zero of $G_c(s)$. The six transfer functions may then written as

$$E(s) = \frac{a_c(s)(s-1)\bar{a}(s)}{a_c(s)(s-1)\bar{a}(s) + (s-1)\bar{b}_c(s)b(s)} R(s) + \frac{a_c(s)b(s)}{a_c(s)(s-1)\bar{a}(s) + (s-1)\bar{b}_c(s)b(s)} D(s)$$

$$C(s) = \frac{(s-1)\bar{b}_c(s)b(s)}{a_c(s)(s-1)\bar{a}(s) + (s-1)\bar{b}_c(s)b(s)} R(s) - \frac{a_c(s)b(s)}{a_c(s)(s-1)\bar{a}(s) + (s-1)\bar{b}_c(s)b(s)} D(s)$$

$$U(s) = \frac{(s-1)\bar{a}(s)(s-1)\bar{b}_c(s)}{a_c(s)(s-1)\bar{a}(s) + (s-1)\bar{b}_c(s)b(s)} R(s) + \frac{(s-1)\bar{b}_c(s)b(s)}{a_c(s)(s-1)\bar{a}(s) + (s-1)\bar{b}_c(s)b(s)} D(s).$$

After the "exact" cancellations we have

$$E(s) = \frac{a_c(s)\bar{a}(s)}{a_c(s)\bar{a}(s) + \bar{b}_c(s)b(s)} R(s) + \frac{a_c(s)b(s)}{(s-1)\left(a_c(s)\bar{a}(s) + \bar{b}_c(s)b(s)\right)} D(s)$$

$$C(s) = \frac{\bar{b}_c(s)b(s)}{a_c(s)\bar{a}(s) + \bar{b}_c(s)b(s)} R(s) - \frac{a_c(s)b(s)}{(s-1)\left(a_c(s)\bar{a}(s) + \bar{b}_c(s)b(s)\right)} D(s)$$

$$U(s) = \frac{\bar{a}(s)(s-1)\bar{b}_c(s)}{a_c(s)\bar{a}(s) + \bar{b}_c(s)b(s)} R(s) + \frac{\bar{b}_c(s)b(s)}{a_c(s)\bar{a}(s) + \bar{b}_c(s)b(s)} D(s).$$

Even with exact cancellation, both $E_D(s)/D(s)$ and $C_D(s)/D(s)$ have a pole at $s = 1$ and are therefore unstable. Similarly, if

$$G(s) = \frac{b(s)}{a(s)} = \frac{(s-1)\bar{b}(s)}{a(s)}, G_c(s) = \frac{b_c(s)}{a_c(s)} = \frac{b_c(s)}{(s-1)\bar{a}_c(s)}$$

we have

$$U(s) = \underbrace{\frac{a(s)b_c(s)}{a_c(s)a(s) + b_c(s)b(s)}}_{U_R(s)} R(s) + \underbrace{\frac{b_c(s)b(s)}{a_c(s)a(s) + b_c(s)b(s)}}_{U_D(s)} D(s)$$

$$= \frac{a(s)b_c(s)}{(s-1)\bar{a}_c(s)a(s) + b_c(s)(s-1)\bar{b}(s)} R(s) + \frac{b_c(s)(s-1)\bar{b}(s)}{(s-1)\bar{a}_c(s)a(s) + b_c(s)(s-1)\bar{b}(s)} D(s)$$

$$= \frac{a(s)b_c(s)}{(s-1)\left(\bar{a}_c(s)a(s) + b_c(s)\bar{b}(s)\right)} R(s) + \frac{b_c(s)\bar{b}(s)}{\bar{a}_c(s)a(s) + b_c(s)\bar{b}(s)} D(s).$$

Thus $U_R(s)/R(s)$ has a pole at $s = 1$ and is therefore unstable. ∎

Appendix: Undershoot

It was remarked in Chapter 8 (page 222) that the step response of a stable system will exhibit undershoot if it has an odd number of right half-plane zeros. For example, suppose $c(t) \to 1$ then by undershoot it was meant that this response initially (immediately) goes negative at time 0 before going positive to ultimately converge to 1. See Vidyasagar [26] for a precise statement and proof of this result. However, here we will consider undershoot to simply mean that the unit step response of a type 1 (or higher) system went negative at *any* time before converging to its final value of 1.

Theorem 9 *Undershoot in Systems with Right Half-Plane Zeros [5, 6]*
Consider the block diagram of Figure 10.41 where $L(s) = G_c(s)G(s)$ is type 1 or higher. With the closed-loop system *stable* let z_0 denote a zero of $L(s)$ in the open right half-plane, i.e., $L(z_0) = G_c(z_0)G(z_0) = 0$ with $\text{Re}\{z_0\} > 0$. Then the unit step response will have undershoot.

Appendix: Undershoot 337

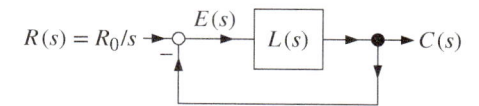

Figure 10.41. Step response of a stable system with a right half-plane zero.

Proof. With a unit step input we have $C(s) = \dfrac{G_c(s)G(s)}{1+G_c(s)G(s)}\dfrac{1}{s}$ with $C(z_0) = 0$. As the closed-loop system is stable there is an $\alpha > 0$ such that all the closed-loop poles p_i satisfy $\text{Re}\{p_i\} < -\alpha < 0$. Consequently

$$C(s) = \int_0^\infty c(t)e^{-st}dt \text{ for } \text{Re}\{s\} > -\alpha.$$

In particular

$$\int_0^\infty c(t)e^{-z_0 t}dt = C(z_0) = \frac{G_c(z_0)G(z_0)}{1+G_c(z_0)G(z_0)}\frac{1}{z_0} = 0.$$

The initial value is $c(0) = 0$ and, as $L(s)$ is at least type 1, we also have $c(t) \to 1$ as $t \to \infty$. This shows $c(t)$ starts at 0 and ends up positive. So there must be some length of time that $c(t)$ is negative in order to satisfy $\int_0^\infty c(t)e^{-z_0 t}dt = 0$. That is, the step response $c(t)$ will have *undershoot* no matter where the closed-loop poles are placed in the open left half-plane. ∎

In the step response of a stable closed-loop system with a single right half-plane zero, there is a tradeoff between the magnitude of the undershoot $|y(t_{\min})|$ and the settling time t_s, where $y(t_{\min})$ and t_s are as shown in Figure 10.42.

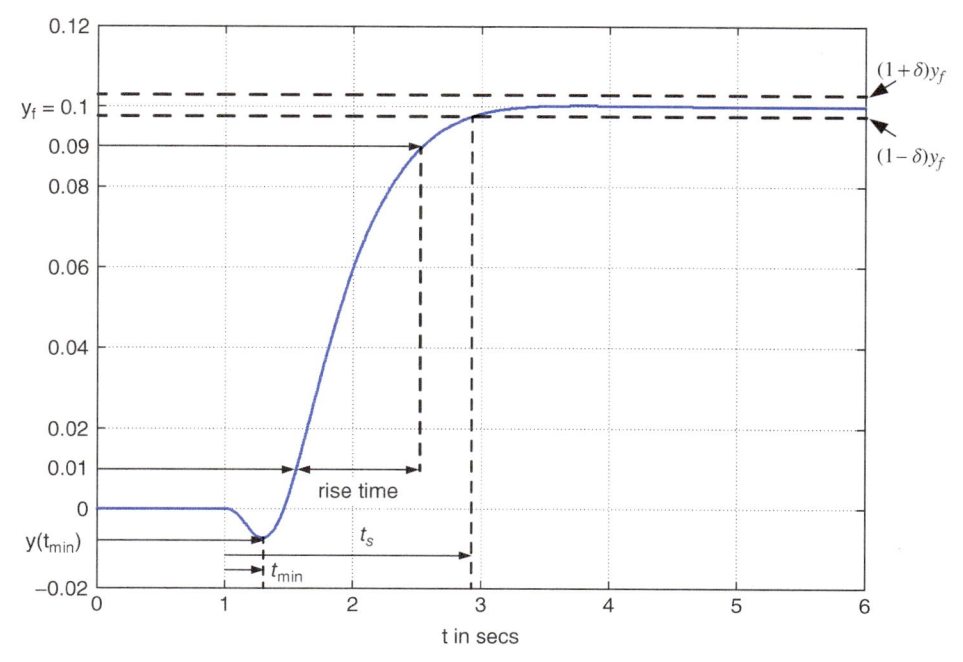

Figure 10.42. Relation between $y(t_{\min})$ and the settling time t_s.

Let δ denote the error tolerance which is often taken to be 0.02 or 2%. Further, let y_f denote the final value of $y(t)$, i.e., $\lim_{t \to \infty} y(t) = y_f$. Recall that the *settling time* is the smallest time t_s such that for $t > t_s$ we have

$$\left| \frac{y(t) - y_f}{y_f} \right| < \delta.$$

We can rewrite this condition as

$$(1 - \delta)y_f < y(t) < (1 + \delta)y_f \quad \text{for} \quad t > t_s.$$

We have the following theorem connecting $y(t_{\min})$ and t_s.

Theorem 10 *Magnitude of Undershoot vs. Settling Time [5, 6]*
Let

$$G(s) = \frac{\bar{n}(s)(s - z)}{d(s)},$$

where the roots of $d(s)$ are in the open left half-plane and $z > 0$. Let δ be the error tolerance for the settling time. Then

$$\frac{|y(t_{\min})|}{y_f} \geq \frac{1 - \delta}{e^{zt_s} - 1}.$$

Note that as $z > 0$ we have $e^{zt_s} \geq 1$ for $t_s > 0$ and $e^{zt_s} \to 1$ as $t_s \to 0$. Thus putting the closed-loop poles further in the left half-plane to make t_s small results in $\dfrac{|y(t_{\min})|}{y_f} \to \infty$!

Proof. With a unit step input the output is given by

$$Y(s) = G(s)\frac{1}{s} = \frac{\bar{n}(s)(s - z)}{d(s)}\frac{1}{s} = \int_0^\infty y(t)e^{-st}dt.$$

At $s = z$ we have

$$\int_0^\infty y(t)e^{-zt}dt = \left.\frac{\bar{n}(s)(s - z)}{d(s)}\frac{1}{s}\right|_{s=z} = 0.$$

Then

$$\int_0^\infty y(t)e^{-zt}dt = \int_0^{t_s} y(t)e^{-zt}dt + \int_{t_s}^\infty y(t)e^{-zt}dt = 0.$$

Further

$$\int_0^{t_s} y(t)e^{-zt}dt \geq \int_0^{t_s} y(t_{\min})e^{-zt}dt = y(t_{\min})\int_0^{t_s} e^{-zt}dt = y(t_{\min})\frac{1 - e^{-zt_s}}{z}$$

$$\int_{t_s}^\infty y(t)e^{-zt}dt \geq \int_{t_s}^\infty (1 - \delta)y_f e^{-zt}dt = (1 - \delta)\frac{e^{-zt_s}}{z}y_f.$$

Thus

$$0 = \int_0^{t_s} y(t)e^{-zt}dt + \int_{t_s}^\infty y(t)e^{-zt}dt \geq y(t_{\min})\frac{1 - e^{-zt_s}}{z} + (1 - \delta)\frac{e^{-zt_s}}{z}y_f$$

or
$$-y(t_{\min})\frac{1-e^{-zt_s}}{z} \geq (1-\delta)\frac{e^{-zt_s}}{z}y_f$$

or
$$\frac{-y(t_{\min})}{y_f} \geq \frac{1-\delta}{e^{zt_s}-1}.$$

As there is undershoot we have $y(t_{\min}) < 0$ so then $|y(t_{\min})| = -y(t_{\min})$ and finally

$$\frac{|y(t_{\min})|}{y_f} \geq \frac{1-\delta}{e^{zt_s}-1}.$$

∎

Problems

Problem 1 *Minimum Order Controller*

A system with open-loop transfer function $G(s)$ is in a unity feedback control architecture as shown in Figure 10.43. In parts (a)–(d), give the form of the *minimum* order controller $G_c(s)$ that allows arbitrary pole placement while achieving asymptotic tracking of the reference signal and asymptotic rejection of the disturbance input.

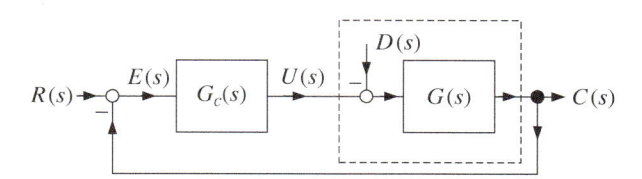

Figure 10.43. Standard unity feedback control system.

(a) $G(s) = \dfrac{s+2}{s(s+4)}, R(s) = R_0/s, D(s) = D_0/s.$

(b) $G(s) = \dfrac{s+2}{s(s+4)}, R(s) = R_0/s, D(s) = 0.$

(c) $G(s) = \dfrac{s+2}{s(s+4)}, R(s) = R_0/s, D(s) = D_0/(s^2+\omega^2).$

(d) $G(s) = \dfrac{s+2}{s(s+4)}, R(s) = R_0/(s^2+\omega^2), D(s) = D_0/s.$

Problem 2 *Minimum Order Controller*

A system with open-loop transfer function $G(s)$ is in a unity feedback control architecture as shown in Figure 10.44. For parts (a)–(d), give the form of the *minimum* order controller $G_c(s)$ that allows arbitrary pole placement while achieving asymptotic tracking of the reference signal and asymptotic rejection of the disturbance input.

(a) $G(s) = \dfrac{1}{s(s^2-2s+2)}, R(s) = R_0/s, D(s) = D_0/s.$

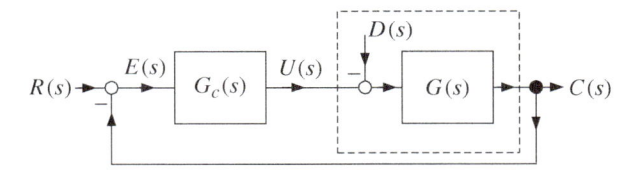

10 Pole Placement, 2 DOF Controllers, and Internal Stability

Figure 10.44. Standard unity feedback control system.

(b) $G(s) = \dfrac{1}{s(s^2 - 2s + 2)}, R(s) = R_0/s, D(s) = 0.$

(c) $G(s) = \dfrac{1}{s(s^2 - 2s + 2)}, R(s) = R_0/s, D(s) = D_0/(s^2 + \omega^2).$

(d) $G(s) = \dfrac{1}{s(s^2 - 2s + 2)}, R(s) = R_0/(s^2 + \omega^2), D(s) = D_0/s.$

Problem 3 *Pole Placement*

Consider the feedback system in Figure 10.45 where $G_c(s) = \dfrac{b_1 s + b_0}{s + a_0}, G(s) = \dfrac{1}{s^2 - 1}, R(s) = \dfrac{R_0}{s}.$

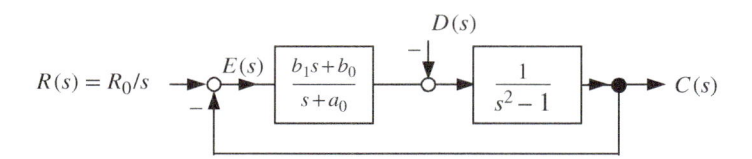

Figure 10.45. Pole placement for an unstable system.

(a) With $D(s) = 0$, compute $E(s)$.

(b) Find the values of $b_1, b_0,$ and a_0 such that the closed-loop poles are at $-1, -1, -1$. Note that

$$(s + r_1)(s + r_2)(s + r_3) = s^3 + (r_1 + r_2 + r_3)s^2 + (r_1 r_2 + r_1 r_3 + r_2 r_3)s + r_1 r_2 r_3$$

(c) With $D(s) = 0$ and using the controller designed in part (b), calculate $sE(s)$. Is $sE(s)$ stable? Explain briefly. If $sE(s)$ is stable what is the value of $e(\infty)$?

Problem 4 *Pole Placement for a First-Order Control System*

Consider the block diagram of Figure 10.46 that shows a control system for the system

$$G(s) = \dfrac{b}{s + a}.$$

This could model the transfer function of a DC motor from voltage input to *speed* output.

(a) With $r_1 > 0, r_2 > 0$ let the desired closed-loop characteristic polynomial be

$$s^2 + f_1 s + f_0 = (s + r_1)(s + r_2) = s^2 + (r_1 + r_2)s + r_1 r_2.$$

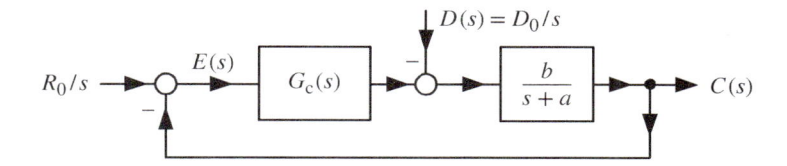

Figure 10.46. Tracking and disturbance rejection with pole placement.

Design the minimum order controller $G_c(s)$ that rejects a step disturbance and places the poles at $-r_1, -r_2$.

(b) Compute the closed-loop transfer function $C_R(s)/R(s)$. Show that it has a zero at $z = -b_0/b_1 = -f_0/(f_1 - a) = -r_1 r_2/(r_1 + r_2 - a)$. Thus for $r_1 + r_2 > a$ the zero is in the open LHP while for $r_1 + r_2 < a$ the zero is in the open RHP.

(c) Simulate this control system with $R_0 = 1, D_0 = 0, a = 10, b = 2$. Do the simulation for two cases: (i) The zero is in the open left half-plane. (ii) The zero is in the open right half-plane. Which of these two has undershoot in the step response? Note: In practice undershoot is not desirable.

Problem 5 *Stable Type 2 Systems and Overshoot*
Let a control system be as given in Figure 10.47.

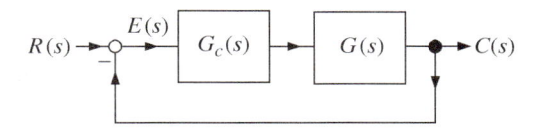

Figure 10.47. Unity feedback control system.

Then
$$C(s) = \underbrace{\frac{G_c(s)G(s)}{1 + G_c(s)G(s)}}_{G_{CL}(s)} R(s).$$

(a) Let
$$G(s) = \frac{1}{(s+3)} \frac{1}{s}, \quad G_c(s) = \frac{3(s + 1/3)}{s}.$$

Compute the closed-loop transfer function $G_{CL}(s)$. Will a step response have overshoot?

(b) Figure 10.48 shows a reference filter added to the control system. Can you design a reference filter $G_f(s)$ so that a step response does not have overshoot? If so, do so.

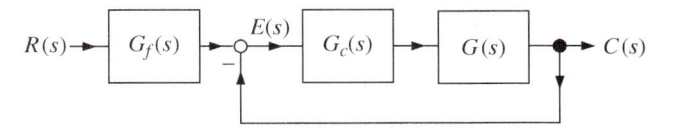

Figure 10.48. Reference input filter for a 2 DOF controller.

10 Pole Placement, 2 DOF Controllers, and Internal Stability

Problem 6 *Rejecting a Sinusoidal Disturbance*
Consider the control system shown in Figure 10.49.

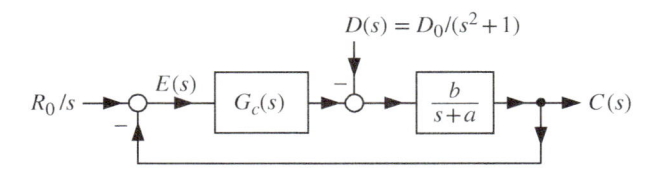

Figure 10.49. Control system with a sinusoidal disturbance.

(a) Give the form of the minimum order controller $G_c(s)$ that will result in tracking the step input, rejecting the sinusoidal disturbance while allowing one to arbitrarily place the closed-loop poles.

(b) Compute $E_D(s)$. Your final answer must be the ratio of two polynomials.

(c) Compute the values of the controller parameters so that the closed-loop characteristic polynomial is $s^4 + f_3 s^3 + f_2 s^2 + f_1 s + f_0$. Explicitly solve for the controller parameters in terms of a, b, and the $f_i, i = 0, \ldots, 3$.

(d) With $a = 10, b = 2$ place all of the closed-loop poles at -3. Simulate the closed-loop system.

Problem 7 *Eliminating Overshoot Using a 2 DOF Controller*
Consider the control system shown in Figure 10.50 with

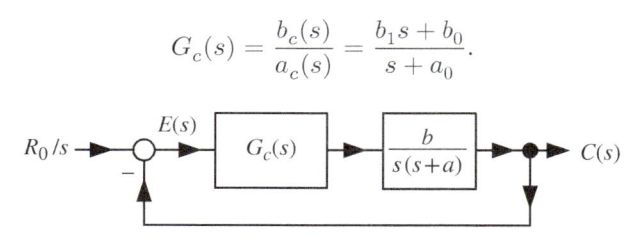

$$G_c(s) = \frac{b_c(s)}{a_c(s)} = \frac{b_1 s + b_0}{s + a_0}.$$

Figure 10.50. Elimination of overshoot with a 2 DOF controller.

(a) Compute the closed-loop transfer function from $R(s)$ to $C(s)$.

(b) Show how to specify the controller parameters a_0, b_1, and b_0 so that the closed-loop characteristic polynomial is

$$s^3 + f_2 s^2 + f_1 s + f_0.$$

(c) Show how to choose the f_0, f_1, f_2 such that the closed-loop poles are at $-r_1, -r_2, -r_3$ with $r_1 > 0, r_2 > 0, r_3 > 0$. With this choice, add a reference input filter (2 DOF controller) to ensure that the step response has no overshoot. Draw a block diagram that includes your reference input filter. Is your designed reference input filter stable for the values of r_1, r_2, r_3 you chose?

(d) Simulate your 2 DOF controller design with $R_0 = 0.1, b = 2$, and $a = 10$.

Problem 8 *Control Design*

Consider the control system shown in Figure 10.51.

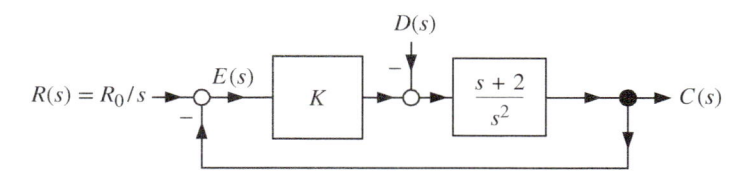

Figure 10.51. Elimination of overshoot.

(a) With $R(s) = R_0/s$ and $D(s) = 0$, for what values of K will $e(t) \to 0$?

(b) For what values of K does the closed-loop system have stable *real* poles?

(c) With $D(s) = D_0/s$ and $R(s) = R_0/s$ compute $C(s)$.

(d) Recall that the step response of any *stable* type 2 system will have overshoot. With $D(s) = 0$ can you add a reference input filter to the control system of Figure 10.51 to ensure there will be no overshoot? If so, draw a block diagram of your new control system and briefly explain why it works.

(e) In part (c) you computed the response $C(s)$ due to a step reference input $R(s) = R_0/s$ and a step disturbance $D(s) = D_0/s$. Let $C_D(s)$ denote the output response due to just the disturbance, i.e., set the reference input $R(s)$ to zero. Can you add a filter to the disturbance input so that the response $C_D(s)$ to a step disturbance has no overshoot? If so, draw a block diagram to show how. If not, explain briefly why not.

Problem 9 *Eliminating Overshoot Using a 2 DOF Controller*

Consider the control system shown in Figure 10.52 which was considered in Example 4 on page 289. Let $D(s) = 0$ in this problem.

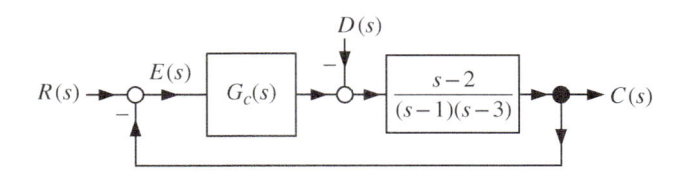

Figure 10.52. Elimination of overshoot for a system with a RHP zero.

(a) Simulate the step response of this system with $G_c(s)$ chosen as in Example 4 that places all the closed-loop poles at -1. Plot $r(t)$ and $c(t)$ on the same graph.

(b) Add to the simulation the reference input filter given in Example 7 on page 304. Simulate the step response as in part (a). Plot $r(t)$ and $c(t)$ on the same graph.

(c) Using the same $G_c(s)$ designed in part (a) and the reference input filter from part (b), let $G(s) = \dfrac{s-2}{(s-0.9)(s-3)}$. Simulate the step response. You should see that it is unstable! The point here is that the controller was designed based on the model being $G(s) = \dfrac{s-2}{(s-1)(s-3)}$, but if $G(s) = \dfrac{s-2}{(s-0.9)(s-3)}$ is closer to representing

10 Pole Placement, 2 DOF Controllers, and Internal Stability

the actual open-loop system, the closed-loop system will be unstable. We say this control system is *not* robust because a small change in the open-loop system model can result in the closed-loop system being unstable. Whenever the open-loop system has a pole in the open right half-plane, it should be a "red flag" to the control designer that robustness will be a problem. This will be explained later after studying Nyquist theory.

Problem 10 *Tracking and Disturbance Rejection with Pole Placement*
Consider the control system given in Figure 10.53 with $a > 0$ and $b > 0$.

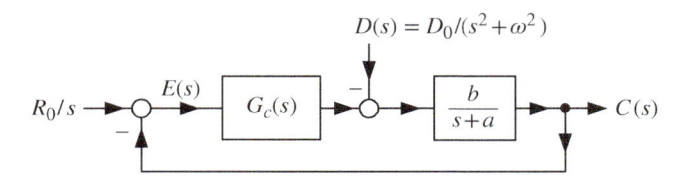

Figure 10.53. Tracking a step input while rejecting a sinusoidal disturbance.

(a) Design a controller of minimum order that achieves tracking and disturbance rejection of $R(s)$ and $D(s)$, respectively, while allowing arbitrary pole placement.

(b) Simulate your design with $a = 10, b = 2, \omega = 1, R_0 = 1$, and $D_0 = 0.1$. Plot $c(t)$ and $r(t)$ on the same graph.

(c) Set the disturbance $D(s)$ to 0 and using the controller designed in part (a), add an input reference filter that results in no overshoot (nor undershoot) in the step response. Did your design result in any zeros being in the open right half-plane? If so, try changing the location of your closed-loop poles. (Hint: You should find putting all the closed-loop poles at -10 results in all the zeros being in the open left half-plane. However, putting all the poles at -1 results in all the zeros being in the open right half-plane!) Simulate your design and plot $c(t)$ and $r(t)$ on the same graph.

Problem 11 *Tracking and Disturbance Rejection with Pole Placement*
Consider the control system given in Figure 10.54.

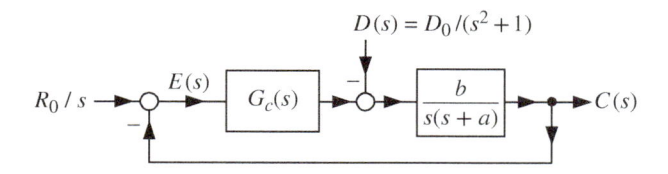

Figure 10.54. Tracking a step input while rejecting a sinusoidal disturbance.

(a) Design a controller of minimum order which achieves tracking and disturbance rejection of $R(s)$ and $D(s)$, respectively, while allowing arbitrary pole placement.

(b) Simulate this control with $a = 10, b = 2, R_0 = 1$, and $D_0 = 0.1$. Plot $c(t)$ and $r(t)$ on the same graph.

(c) Add a reference input filter that results in no overshoot in the step response with the disturbance set to zero. Plot $c(t)$ and $r(t)$ on the same graph.

Remark Rejecting sinusoidal disturbances is a practical problem in motion control systems. See, e.g., Aerotech Inc. [41, 42]. This approach based on the internal model principle is referred to as *harmonic cancellation* [43, 44].

Problem 12 *Pole Placement for a Double Integrator System*
Consider the control system in Figure 10.55 where
$$G(s) = \frac{1}{s^2}.$$

The objective is to track step inputs and reject step disturbances.

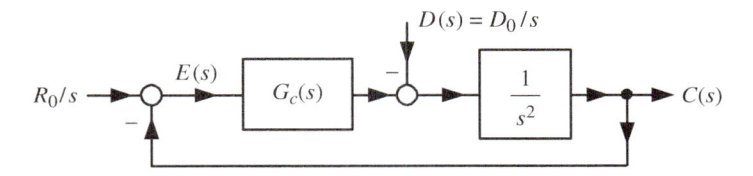

Figure 10.55. Pole placement for a double integrator system.

(a) Give the form of the minimum order controller that will track step inputs, reject step disturbances while allowing one to arbitrarily assign the closed-loop poles.

(b) Compute $E_R(s)$. You must reduce $E_R(s)$ to be the ratio of two polynomials.

(c) Choose the parameters of your controller so that the closed-loop characteristic polynomial is $s^4 + f_3 s^3 + f_2 s^2 + f_1 s + f_0$.

(d) Choose the four closed-loop poles to all be at -1 so $r_1 = r_2 = r_3 = r_4 = 1$. Note that

$$(s+r_1)(s+r_2)(s+r_3)(s+r_4) = s^4 + (r_1 + r_2 + r_3 + r_4)s^3$$
$$+ (r_1 r_2 + r_1 r_3 + r_1 r_4 + r_2 r_3 + r_2 r_4 + r_3 r_4)s^2$$
$$+ (r_1 r_2 r_3 + r_1 r_2 r_4 + r_1 r_3 r_4 + r_2 r_3 r_4)s + r_1 r_2 r_3 r_4.$$

Use this expression to find the numerical values of the parameters of the controller you designed in parts (a)–(c), and explicitly give $G_c(s)$.

(e) With $R(s) = R_0/s$ and $D(s) = D_0/s$, compute $C(s)$.

(f) With $D_0 = 0$ so $D(s) = 0$ can you design a reference input filter to ensure that the step response $C_R(s)$ due to $R(s) = R_0/s$ has no overshoot? If so, do so by drawing a block diagram of your complete control system and explain briefly why it works. That is, what conditions ensure there is no overshoot and does your design fulfill those conditions? On the other hand, if it is not possible to prevent overshoot with a reference input filter explain why not.

346   10 Pole Placement, 2 DOF Controllers, and Internal Stability

Problem 13 *Strictly Proper $G_c(s)$*

Figure 10.56 is a unity feedback control system for the open-loop system $G(s) = \dfrac{b}{s(s+a)}$.

Figure 10.56. Pole placement with a strictly proper $G_c(s)$.

One often desires a strictly proper controller so that $|G(s)G_c(s)|$ dies out fast for large $|s|$ to avoid any effect of model uncertainty at high frequencies. To do this consider the controller
$$G_c(s) = \frac{b_1 s + b_0}{s^2 + a_1 s + a_0}.$$

To explain, a step input is to be tracked and $G(s)$ has the factor $1/s$ so the controller is not required to have this factor. The denominator of $G_c(s)$ is chosen to be a monic polynomial of the *same* degree as the denominator polynomial of $G(s) = \dfrac{b}{s(s+a)}$. The numerator of $G_c(s)$ is then chosen to have degree one less than its denominator to make it strictly proper.

(a) Find the values of the controller parameters a_0, a_1, b_1, b_0 so that the closed-loop characteristic polynomial is given by
$$s^4 + f_3 s^3 + f_2 s^2 + f_1 s + f_0.$$

(b) Using the controller of part (a), simulate this control system with $a = 2, b = 1.5, R_0 = 1$ and the closed-loop poles at -2.

Problem 14 *Strictly Proper $G_c(s)$*

Figure 10.57 shows a unity feedback control system for $G(s) = \dfrac{b}{s(s+a)}$. Let's design a strictly proper pole placement controller that also rejects constant disturbances. To do this, let the controller have the form
$$G_c(s) = \underbrace{\frac{b_2 s^2 + b_1 s + b_0}{s^2 + a_1 s + a_0}}_{\overline{G}_c(s)} \frac{1}{s}.$$

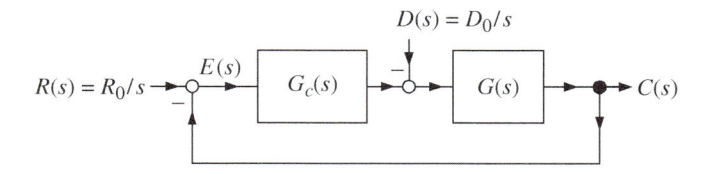

Figure 10.57. Disturbance rejection with a strictly proper $G_c(s)$.

To explain, $G_c(s)$ is written as the product of the two factors $\bar{G}_c(s)$ and $1/s$. It must have the factor $1/s$ to reject the step disturbance according to the internal model principle. The factor $\bar{G}_c(s)$ is determined as follows. We look at $G(s) = \dfrac{b}{s(s+a)}$ and note the denominator of $G(s)$ has degree two. The denominator of $\bar{G}_c(s)$ is then set to be a polynomial of the *same* degree as the denominator of $G(s)$. The numerator of $G_c(s) = \bar{G}_c(s)\dfrac{1}{s}$ is then set to be a polynomial of one degree less than its denominator so that it is strictly proper.

(a) Find the controller parameters a_1, a_0, b_2, b_1, b_0 so that the closed-loop characteristic polynomial is
$$s^5 + f_4 s^4 + f_3 s^3 + f_2 s^2 + f_1 s + f_0.$$

(b) Using the controller of part (a), simulate this control system with $a = 2, b = 1.5, R_0 = 1, D_0 = 1$, and the closed-loop poles at -5.

Problem 15 *Stable Pole-Zero Cancellation*

Consider the closed-loop control system of Figure 10.58 in which the open-loop system has transfer function
$$G(s) = \dfrac{b(s)}{a(s)} = \dfrac{s+2}{s(s+1)}.$$

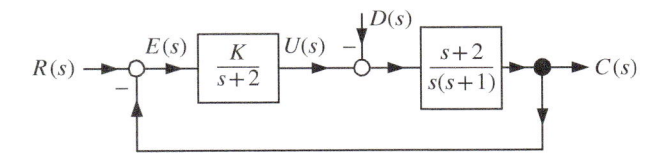

Figure 10.58. Stable pole–zero cancellation.

Let
$$G_c(s) = \dfrac{K}{s+2}.$$

(a) Show that the six transfer functions that characterize internal stability are stable for $K > 0$. Note that you need only show that the four transfer functions $E(s)/R(s), E(s)/D(s), C(s)/R(s), U(s)/R(s)$ are stable.

(b) Now let the open-loop system have the transfer function $G(s) = \dfrac{s+z}{s(s+1)}$ where $z > 0$ is known to be close to 2, but not necessarily equal to 2. With $G_c(s) = \dfrac{K}{s+2}$ show that all six transfer functions characterizing internal stability are stable for $K > 0$ and $0 < z \le 3$. This means that though we took $z = 2$ it can vary as $0 < z \le 3$ with the closed-loop system still being stable.

Problem 16 *Stable Pole-Zero Cancellation*

Consider the closed-loop control system of Figure 10.59 in which the open-loop system has transfer function
$$G(s) = \dfrac{b(s)}{a(s)} = \dfrac{s+2}{s(s+1)(s+3)}.$$

348   10   Pole Placement, 2 DOF Controllers, and Internal Stability

$$R(s) = R_0/s \longrightarrow \bigotimes \xrightarrow{E(s)} \boxed{G_c(s)} \longrightarrow \boxed{\frac{s+2}{s(s+1)(s+3)}} \longrightarrow C(s)$$

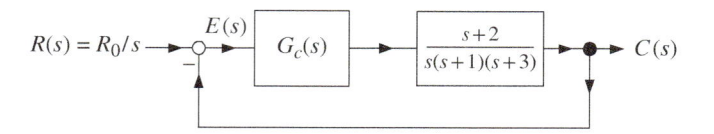

Figure 10.59. Stable pole-zero cancellation.

(a) Design a minimum order controller that places the closed-loop poles at $-2, -4, -8, -6,$ and -10. Note that one of the closed-loop poles was chosen to cancel the open-loop zero at -2.

(b) If you did part (a) correctly you should have found out that

$$G_c(s) = \frac{b_c(s)}{a_c(s)} = \frac{185s^2 + 1160s + 1920}{(s+2)(s+24)}.$$

The point here is that the denominator of $G_c(s)$ must contain the factor $s+2$ in order to cancel the open-loop zero at -2. Show why this must be so. Hint: We want

$$C(s) = \frac{\frac{b_c(s)}{a_c(s)} \frac{b(s)}{a(s)}}{1 + \frac{b_c(s)}{a_c(s)} \frac{b(s)}{a(s)}} R(s) = \frac{b_c(s)b(s)}{a_c(s)a(s) + b_c(s)b(s)} R(s)$$

$$= \frac{(185s^2 + 1160s + 1920)(s+2)}{(s+2)(s+4)(s+6)(s+8)(s+10)} R(s).$$

This requires finding $a_c(s), b_c(s)$ such that

$$a_c(s)a(s) + b_c(s)b(s) = (s+2)(s+4)(s+6)(s+8)(s+10).$$

Explain why it must be that $a_c(-2) = 0$.

Problem 17 *Unbounded References and Disturbances*

It has been pointed out that no control system can track unbounded references or disturbances. Let's look at a control system of Figure 10.60.

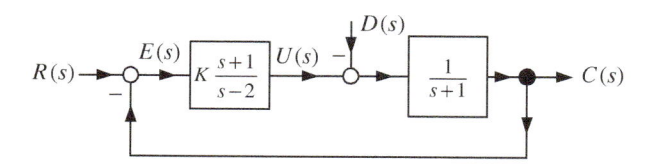

$$R(s) \longrightarrow \bigotimes \xrightarrow{E(s)} \boxed{K\frac{s+1}{s-2}} \xrightarrow{U(s)} \bigotimes \xleftarrow{D(s)} \boxed{\frac{1}{s+1}} \longrightarrow C(s)$$

Figure 10.60. Unbounded references and disturbances.

Let $D(s) = 0$ and $R(s) = \dfrac{R_0}{s-2}$ so that $r(t) = R_0 e^{2t}$ is an *unbounded* reference input. Let the controller be

$$G_c(s) = K(s+2)/(s-2).$$

(a) Compute $E_R(s)$ and show, assuming perfect cancellation between the pole of $R(s)$ and the zero of the closed-loop transfer function, that K can be chosen to achieve $e_R(t) \to 0$.

(b) Compute $U(s)$ and show $u(t)$ goes unbounded.

(c) Show that with $G_c(s) = K(s+2)/(s-2)$ and $K > 1$ this closed-loop system is *internally* stable.

Remark The control system is not able to track $R_0 e^{2t}$ because $u(t)$ goes unbounded. That is, the problem with not being able to track $R_0 e^{2t}$ is that it is *unbounded* as the closed-loop system is internally stable.

Problem 18 *Non-Minimum Phase System*

Consider the control system of Figure 10.61 with the open-loop system from [45] given by

$$G(s) = \frac{s-1}{s^2 - s - 2}.$$

Figure 10.61. Pole placement for an unstable non-minimum phase system.

In this problem we look at just stabilizing the closed-loop system. Tracking a step input and rejecting a step disturbance are considered in the next problem.

(a) Design a first order controller so that the closed-loop transfer function has a characteristic polynomial given by

$$s^3 + f_2 s^2 + f_1 s + f_0.$$

(b) With $r > 0$ put all the closed-loop poles at $-r$. Compute $e_R(\infty)$ and $e_D(\infty)$. Simulate this control system with $R_0 = 1$, $D_0 = 0.2$ and the closed-loop poles at -5.

(c) Now consider a 2 DOF controller as indicated in Figure 10.62. With $D(s) = 0$ can you design a reference input filter so that there is no overshoot in a step response? If so, do so.

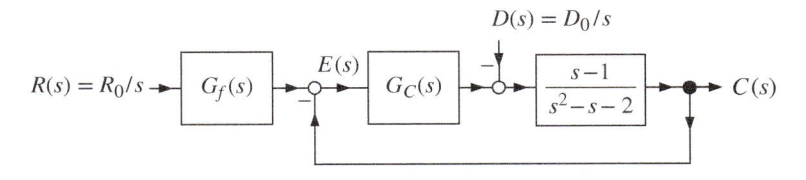

Figure 10.62. Reference input filter to eliminate overshoot.

(d) With $D(s) = 0$ and using the control system of part (c), will there be undershoot? Explain why or why not.

(e) Simulate the control system of part (c) with $R_0 = 1, D_0 = 0$ and the closed-loop poles at -5.

Problem 19 *Non-Minimum Phase System*

Consider the control system of Figure 10.63 with the open-loop transfer function from [45] given by
$$G(s) = \frac{s-1}{s^2 - s - 2}.$$

Figure 10.63. Tracking and disturbance rejection of an unstable non-minimum phase system.

(a) Design a controller that tracks step inputs, rejects constant disturbances, and with the closed-loop characteristic polynomial given by
$$s^4 + f_3 s^3 + f_2 s^2 + f_1 s + f_0.$$

(b) With $r > 0$ put all the closed-loop poles at $-r$. Compute $e_R(\infty)$ and $e_D(\infty)$. Simulate this control system with $R_0 = 1$ and $D_0 = 0.1$. Plot $r(t)$ and $c(t)$ on the same plot.

(c) With $D(s) = 0$ and using the controller designed in parts (a) and (b), can you design a reference input filter so that there is no overshoot in a step response. If so, do so. If not, explain why not (see Figure 10.64).

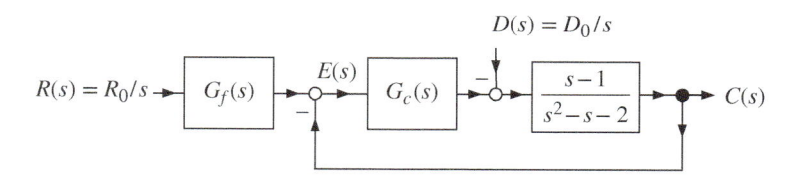

Figure 10.64. Reference input filter to eliminate overshoot.

(d) With $D(s) = 0$ and using the control system of part (b), will the step response have undershoot? Explain why or why not.

(e) Simulate the control system of part (c) with $R_0 = 1, D_0 = 0$ and the closed-loop poles at -5. Plot $r(t)$ and $c(t)$ on the same plot.

Problem 20 *Noncollocated Control* [46]

Two masses m_1 and m_2 are connected by a spring and damper as shown in Figure 10.65. The mass m_1 has a force (control) input $u(t)$. The objective is to control the position x_2 of the second mass.

In this setup the actuator (control input u) is on the mass m_1 while we are assuming there is a sensor to measure the position of m_2, i.e., the output is $y(t) = x_2(t)$. This case

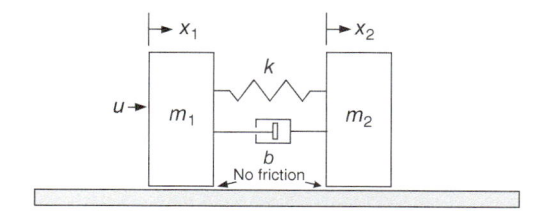

Figure 10.65. Control of a mass–spring–damper system.

where the actuator and sensor are *not* located on the same rigid body is referred to as *noncollocated* control.

(a) Show that the transfer function is

$$G(s) = \frac{Y(s)}{U(s)} = \frac{bs + k}{s^2\left(m_1 m_2 s^2 + (m_1 + m_2)bs + (m_1 + m_2)k\right)}.$$

(b) Let $m_1 = 1, m_2 = 1, k = 2, b = 0.1$. Give the minimum order unity feedback controller $G_c(s) = b_c(s)/a_c(s)$ that will track a step input $R(s) = R_0/s$ and allow arbitrary placement of the closed-loop poles.

(c) Using the minimum order controller find the error $E(s)$.

(d) With the desired closed-loop polynomial of the form $a_{CL}(s) = s^7 + f_6 s^6 + \cdots + f_2 s^2 + f_1 s + f_0$, write down the set of linear equations for the coefficients f_i in terms of the coefficients of your minimum order controller and the coefficients of $G(s)$.

(e) Rewrite your answer from part (d) in matrix form.

(f) In a MATLAB program specify the coefficients $f_i, i = 1, \ldots, 6$ so that the seven closed-loop poles are at $-r_1, -r_2, -r_3, -r_4, -r_5, -r_7$. Then add to this program the code to solve the matrix equation of part (e) for the coefficients of $b_c(s)$ and $a_c(s)$.

(g) Make a SIMULINK simulation of the complete system. The plant model $G(s)$ and the controller $G_c(s)$ should be in separate SIMULINK blocks.
If all the closed-loop poles are set to -1 then the zeros of the $G_c(s)$ turn out to be $-0.145, -0.484, 0.498$.

Problem 21 *Collocated Control*

Two masses m_1 and m_2 are connected by a spring and damper as shown in Figure 10.66 on the next page. The mass m_1 has a force (control) input $u(t)$. The objective is to control the position x_1 of the first mass.

In this setup the actuator (control input u) is on the mass m_1 and we are assuming there is a sensor to measure the position of m_1, i.e., the output is $y(t) = x_1(t)$. This situation where the actuator and sensor are located on the same rigid body is referred to as *collocated* control.

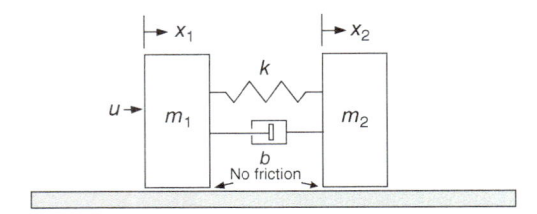

352   10 Pole Placement, 2 DOF Controllers, and Internal Stability

Figure 10.66. Control of a mass–spring–damper system.

(a) Show that the transfer function is

$$G(s) = \frac{Y(s)}{U(s)} = \frac{m_2 s^2 + bs + k}{s^2\left(m_1 m_2 s^2 + (m_1 + m_2)bs + (m_1 + m_2)k\right)}.$$

(b) Let $m_1 = 1$, $m_2 = 1$, $k = 2$, $b = 0.1$. What is the minimum order unity feedback controller $G_c(s) = b_c(s)/a_c(s)$ that will track a step input $R(s) = R_0/s$ and allow arbitrary placement of the closed-loop poles?

(c) Using the minimum order controller find the error $E(s)$.

(d) With the desired closed-loop polynomial of the form $a_{CL}(s) = s^7 + f_6 s^6 + \cdots + f_2 s^2 + f_1 s + f_0$ write down the set of linear equations for the coefficients f_i in terms of the coefficients of your minimum order controller and the coefficients of $G(s)$.

(e) Rewrite your answer from part (d) in matrix form.

(f) In a MATLAB program specify the coefficients $f_i, i = 1, \ldots, 6$ so that the closed-loop poles are at $-r_1, -r_2, -r_3, -r_4, -r_5, -r_7$. Then add to this program the code to solve the matrix equation of part (e) for the coefficients for $b_c(s), a_c(s)$.

(g) Do a SIMULINK simulation of the complete system. The plant model $G(s)$ and the controller $G_c(s)$ should be in separate SIMULINK blocks.
If all the closed-loop poles are set to -1 then the zeros of the $G_c(s)$ turn out to be $-0.0047 \pm j0.46, -0.145$.
If all the closed-loop poles are set to be -2 then the zeros of the $G_c(s)$ turn out to be $267.95, -1.38, -0.366$.
If all the closed-loop poles are set to be -3 then the zeros of the $G_c(s)$ turn out to be $0.508 \pm j2.98, -0.418$.

Problem 22 *Inverted Pendulum*

Simulate the inverted pendulum feedback control system of Example 5. Take $\alpha^2 = 29.256$ and $\beta_0 = \kappa mg\ell = 36.5705$. These values correspond to the parameters of the inverted pendulum of QUANSER [34]. Figure 10.67 is a SIMULINK block diagram to carry out the simulation.

SIMULINK assumes zero initial conditions when using the transfer function blocks. In order to excite the system away from the zero initial conditions, an impulsive force[6] is applied to the horizontal motion of the cart as shown in Figure 10.67. When placing the closed-loop poles make sure the maximum commanded horizontal force is no more than 10 N in absolute value, i.e., $|u(t)| \leq 10$. Hint: Try placing all the closed-loop poles at -5.

---

[6] A force that acts for only a very short time, in this case, from $t = 0.2$ to $t = 0.3$ seconds.

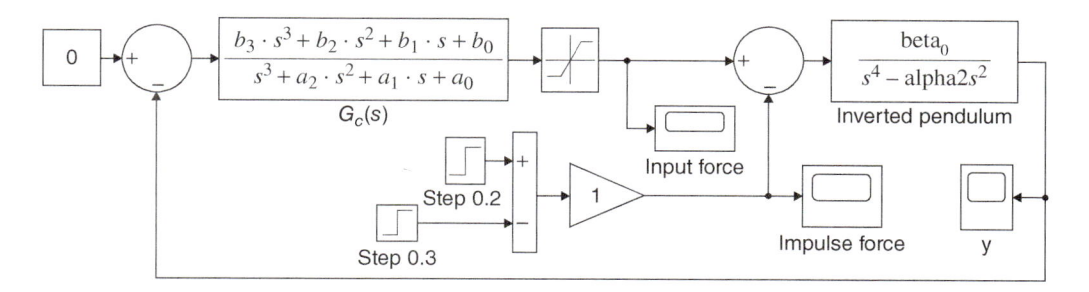

Figure 10.67. SIMULINK block diagram for the inverted pendulum.

Problem 23 *Control of Aircraft Pitch*

In Section 10.4 a controller for the pitch of a small aircraft was designed based on the transfer function model given by [31]

$$\frac{\theta(s)}{\delta(s)} = G(s) = \frac{1.51s + 0.1774}{s^3 + 0.739s^2 + 0.921s}.$$

In this problem the controller is not required to reject step disturbances. $G(s)$ is a type one system so the final error with a step reference input will be zero as long as the closed-loop system is stable.

The design specifications are (i) overshoot less than 10%, (ii) rise time less than two seconds, (iii) settling time less than 10 seconds, (iv) final error less than 2%.

As in [31] take the reference input to be 0.2 rad (11.5 degrees). The elevator angle (actuator) is limited to ± 0.436 rad or $\pm 25°$.

(a) Using the unity feedback control structure given in Figure 10.68, design a minimum order controller $G_c(s)$ that places the closed-loop poles at $-r_1, \ldots, -r_5$ where $r_i > 0$.

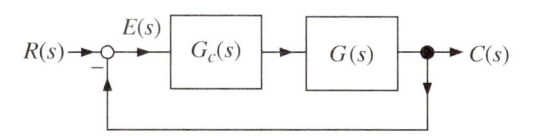

Figure 10.68. Unity feedback control system.

(b) Let $r_i = 5$ for $i = 1, \ldots, 5$ so the closed-loop poles are all at -5. What are the values of the three zeros?

(c) For the design in parts (a) and (b) can you find a reference input filter so there is no overshoot in the step response? If so, do so.

(d) Simulate your design. Does it meet the specifications? If not, how close is it to meeting the specifications? See Figure 10.42 for a definition of rise time when undershoot is present. See Figure 4.6 on page 134 of [5] for a more general definition of rise time.

Problem 24 *Satellite with Solar Panels (Noncollocated Case)*

Consider the control of the solar panels of a satellite as in the text, but for the *noncollocated* case. That is, the sensor is located at the end of a solar panel to measure θ_p. It was shown in the text that the Laplace transform variables $\theta(s), \theta_p(s), \tau(s)$ satisfy

$$\theta(s) = \frac{bs + K}{s^2 J_s + bs + K}\theta_p(s) + \frac{1}{s^2 J_s + bs + K}\tau(s)$$

$$\theta_p(s) = \frac{bs + K}{s^2 J_p + bs + K}\theta(s).$$

The transfer function from $\tau(s)$ to $\theta_p(s)$ is then

$$G_p(s) = \frac{\theta_p(s)}{\tau(s)} = \frac{(b/(J_s J_p))\, s + K/(J_s J_p)}{s^2\left((s^2 + b(1/J_p + 1/J_s)s + K(1/J_p + 1/J_s)\right)} = \frac{\beta_1 s + \beta_0}{s^2(s^2 + \alpha_1 s + \alpha_0)}$$

with the obvious definitions for $\beta_2, \beta_1, \beta_0, \alpha_1, \alpha_2$. Again set $J_s = 5$ kg-m^2, $J_p = 1$ kg-m^2, $K = 0.15$ N-m/rad, $b = 0.05$ Nm/rad/s as in [13]. The zero of $G(s)$ is the root of $bs + K = 0$ which is $-K/b = -0.15/0.05 = -3$.

The objective of this problem is to design a 2 DOF controller for the control of the satellite solar panels. That is, to determine $G_f(s), G_c(s)$ in the block diagram of Figure 10.69 to obtain a satisfactory step input response for θ_p.

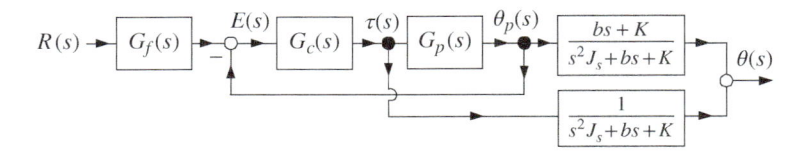

Figure 10.69. Two DOF controller for the non-collocated case.

Use the simplified block diagram of Figure 10.70 to design $G_c(s)$.

Figure 10.70. Two DOF controller for the non-collocated case.

(a) Give the minimum order controller $G_c(s)$ that allows arbitrary placement of the closed-loop poles and will provide asymptotic rejection of constant disturbances.

(b) With $G_p(s) = \dfrac{\beta_1 s + \beta_0}{s^2(s^2 + \alpha_1 s + \alpha_0)}$, the expression for $G_c(s)$ from part (a) and with $G_f(s) = 1$ for now, compute $E(s)/R(s)$ written as the ratio of two polynomials.

(c) Let the desired closed-loop polynomial taken to be $s^8 + f_7 s^7 + f_6 s^6 + f_5 s^5 + f_4 s^4 + f_3 s^3 + f_2 s^2 + f_1 s + f_0$ and define

$$f \triangleq \begin{bmatrix} f_7 & f_6 & f_5 & f_4 & f_3 & f_2 & f_1 & f_0 \end{bmatrix}^T \in \mathbb{R}^8.$$

Let $c \in \mathbb{R}^8$ be the controller coefficients written as

$$c \triangleq \begin{bmatrix} b_4 & b_3 & b_2 & b_1 & b_0 & a_2 & a_1 & a_0 \end{bmatrix}^T \in \mathbb{R}^8.$$

Find the matrix $A \in \mathbb{R}^{8\times 8}$ and vector $d \in \mathbb{R}^8$ such that
$$f = Ac + d.$$

(d) Choose locations for the closed-loop poles and do a SIMULINK simulation using the same reference input $r(t)$ as shown in Figure 10.36 for the collocated case. You should choose the closed-loop poles such that the zeros of the controller $G_c(s)$, i.e., the roots of $b_4 s^4 + b_3 s^3 + b_2 s^2 + b_1 s + b_0 = 0$, are all in the open left-half plane. Choose the reference input filter to be

$$G_f(s) = \frac{b_0}{b_4 s^4 + b_3 s^3 + b_2 s^2 + b_1 s + b_0}.$$

Can you choose the closed-loop poles to all be real? How far in the open left-half plane can you place the poles without saturating the torque τ? Show $r(t), \theta(t), \theta_p(t)$ together on a single plot. On a separate plot show the difference $\theta_p(t) - \theta(t)$.

Problem 25 *DC Motor with a Flexible Shaft* [3]

In [3] a DC motor with a flexible shaft is modeled by a transfer function from voltage input to the angular position of the flexible shaft given by

$$G(s) = \frac{1}{s(s+1)} \frac{2500}{(s^2+s+2500)} = \frac{2500}{s^4 + 2s^3 + 2501 s^2 + 2500 s}.$$

In this problem use a 2 DOF control system as illustrated in Figure 10.71.

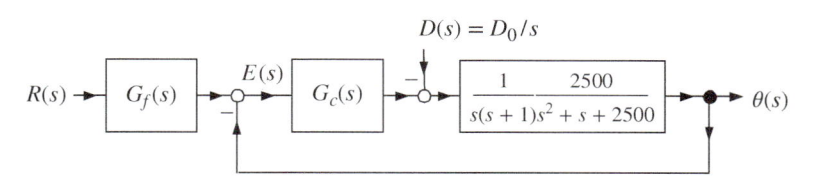

Figure 10.71. Two DOF control system for a DC motor with a flexible shaft.

(a) Give the form of the minimum order that rejects constant disturbances and allows for arbitrary pole placement.

(b) With $D(s) = 0$ and $G_f(s) = 1$, compute the transfer function $E(s)/R(s)$.

(c) Let the desired closed-loop polynomial taken to be $s^8 + f_7 s^7 + f_6 s^6 + f_5 s^5 + f_4 s^4 + f_3 s^3 + f_2 s^2 + f_1 s + f_0$ and define

$$f \triangleq \begin{bmatrix} f_7 & f_6 & f_5 & f_4 & f_3 & f_2 & f_1 & f_0 \end{bmatrix}^T \in \mathbb{R}^8.$$

Let $c \in \mathbb{R}^8$ be the controller coefficients written as

$$c \triangleq \begin{bmatrix} b_4 & b_3 & b_2 & b_1 & b_0 & a_2 & a_1 & a_0 \end{bmatrix}^T \in \mathbb{R}^8.$$

Find the matrix $A \in \mathbb{R}^{8\times 8}$ and vector $d \in \mathbb{R}^8$ such that

$$f = Ac + d.$$

(d) Choose locations for the closed-loop poles so the settling time (2% criterion) is one second or less with no more than 10% overshoot. This can take some time as one must try choosing real poles, complex conjugate poles, or some mixture of them with the objective that the zeros of the controller are in the open left-half plane as well as meeting the specifications. So try choosing four of them at -10 and two sets of complex conjugate poles with $\zeta = 0.05$ and $\omega_n = 50$. That is, set

$$a_{CL}(s) = s^8 + f_7 s^7 + f_6 s^6 + f_5 s^5 + f_4 s^4 + f_3 s^3 + f_2 s^2 + f_1 s + f_0$$
$$= (s+r_1)(s+r_2)(s+r_3)(s+r_4)(s^2 + 2\zeta_3 \omega_{n3} s + \omega_{n3}^2)(s^2 + 2\zeta_4 \omega_{n4} s + \omega_{n4}^2)$$

with $r_1 = r_2 = r_3 = r_4 = 10, \zeta_3 = 0.05, \omega_{n3} = 50, \zeta_4 = 0.05, \omega_{n4} = 50$. You will need to modify code in the Appendix of this chapter for *Multinomial Expansions* to have MATLAB automatically solve for the $f_i, i = 1, \ldots, 7$. Check that the zeros of $G_c(s)$ are in the open left-half plane.

(e) Make a SIMULINK simulation using the reference input $r(t) = u_s(t)$. You should choose the closed-loop poles such that the zeros of the controller $G_c(s)$, i.e., the roots of

$$b_4 s^4 + b_3 s^3 + b_2 s^2 + b_1 s + b_0 = 0,$$

are all in the open left-half plane. Choose the reference input filter to be

$$G_f(s) = \frac{b_0}{b_4 s^4 + b_3 s^3 + b_2 s^2 + b_1 s + b_0}.$$

Plot $r(t)$ and $\theta(t)$ on the same graph.

Problem 26 *DC Motor with a Flexible Shaft* [3]

In [3] a DC motor with a flexible shaft is modeled by the transfer function from voltage input to the angular position of the flexible shaft given by

$$G(s) = \frac{1}{s(s+1)} \frac{2500}{(s^2 + s + 2500)} = \frac{2500}{s^4 + 2s^3 + 2501s^2 + 2500s}.$$

Further the controller chosen in [3] was a cascade of a lead, a lag, and a notch controller given by

$$G_c(s) = 91 \frac{s+2}{s+13} \frac{s+0.05}{s+0.01} \frac{s^2 + 0.8s + 3600}{s^2 + 120s + 3600}. \qquad (10.20)$$

This controller still works well if the lag controller $\frac{s+0.05}{s+0.01}$ is replaced by the PI controller $\frac{s+0.05}{s}$. Doing this the controller is now

$$G_c(s) = 91 \frac{s+2}{s+13} \frac{s+0.05}{s} \frac{s^2 + 0.8s + 3600}{s^2 + 120s + 3600}$$
$$= \frac{91s^4 + 259.35s^3 + 3.2776 \times 10^5 s^2 + 6.7159 \times 10^5 s + 32760}{s^4 + 133s^3 + 5160s^2 + 46800s}. \qquad (10.21)$$

Consider the control system of Figure 10.72.

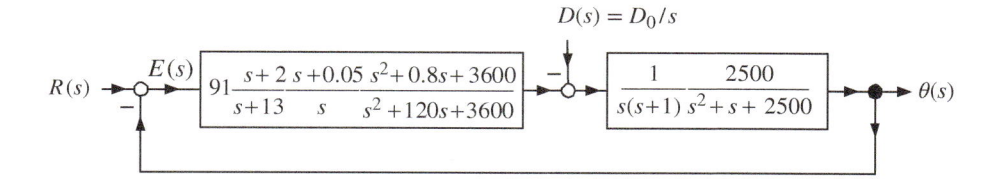

Figure 10.72. Cascade of a lead, a PI, and a notch controller.

This results in the closed-loop characteristic polynomial being

$$a_{CL}(s) = s^8 + 135s^7 + 7927s^6 + 392253s^5 + 13558760s^4 + 130595175s^3 \\ + 936395850s^2 + 1\,678968200s + 81900000,$$

with the closed-loop poles at

$$-61.632 \pm j9.0903, -0.65275 \pm j49.941, -4.0148 \pm j7.4573, -2.3505, -0.050174 \quad (10.22)$$

(a) The minimum order pole placement controller that achieves rejection of constant disturbances is

$$G_c(s) = \frac{b_4 s^4 + b_3 s^3 + b_2 s^2 + b_1 s + b_0}{s^3 + a_2 s^2 + a_1 s + a_0} \frac{1}{s}.$$

Using the control setup of Figure 10.73 compute the controller coefficients that place the closed-loop poles as given in (10.22). It will turn out that $G_c(s)$ is given by (10.21).

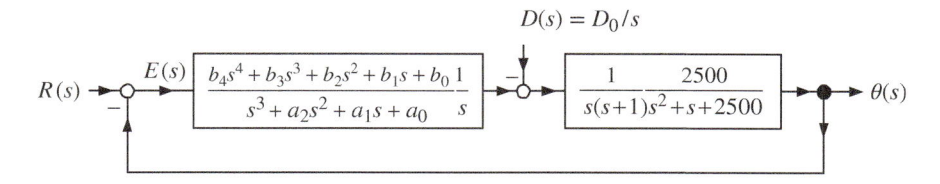

Figure 10.73. Pole placement controller.

(b) Simulate the system of part (a) with a unit step reference input.

Remark The controller $G_c(s)$ for a single-input single-output system in unity feedback form is uniquely specified by the location of the closed-loop poles. As another example, compute the eight closed-loop poles using the controller given in (10.20). Then the pole placement algorithm with $G_c(s)$ of the form (not minimum order)

$$G_c(s) = \frac{b_4 s^4 + b_3 s^3 + b_2 s^2 + b_1 s + b_0}{s^4 + a_3 s^3 + a_2 s^2 + a_1 s + a_0},$$

and designed to have these same eight closed-loop poles would result in the controller being given by (10.20).

Problem 27 *Control of Pitch Angle*

The transfer function model of a small aircraft from elevator angle δ in degrees to pitch angle θ in degrees is given in [3] to be

$$\frac{\theta(s)}{\delta(s)} = G(s) = \frac{160(s+2.5)(s+0.7)}{(s^2+5s+40)(s^2+0.03s+0.06)} = 160\frac{s^2+3.2s+1.75}{s^4+5.03s^3+40.21s^2+1.5s+2.4}.$$

For a 5 degree step input reference the performance specification is that the rise time is one second or less and no more than 10% overshoot. In [3] (pages 260–267) their design resulted in the feedback controller (and no reference input filter)

$$G_{PID}(s) \triangleq 1.5\frac{s+3}{s+20}\left(1+\frac{0.15}{s}\right) = \frac{1.5}{20}\frac{s+3}{s/20+1}\left(1+\frac{0.15}{s}\right)$$

$$\approx 0.075s + \frac{0.034}{s} + 0.236 \text{ for } |s| \ll 20.$$

That is, it is an approximate PID controller. In the following you are to design a 2 DOF controller using pole placement as indicated in Figure 10.74.

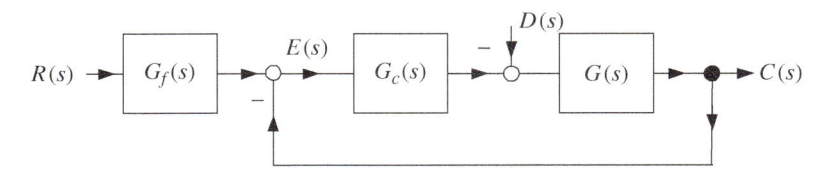

Figure 10.74. Two DOF controller.

(a) Design a minimum order controller $G_c(s)$ that asymptotically rejects constant disturbances and allows arbitrary placement of the closed-loop poles.

(b) Let the desired closed-loop characteristic polynomial have the form

$$f(s) = s^8 + f_7 s^7 + f_6 s^6 + f_5 s^5 + f_4 s^4 + f_3 s^3 + f_2 s^2 + f_1 s + f_0.$$

Determine the coefficients of your controller in terms of $f_i, i = 0, \ldots, 7$.

(c) As in the pitch control design in the text, the choice of location of the closed-loop poles is critical to meeting the performance specifications. An approach is to let $a_{CL}(s)$ have the form

$$a_{CL}(s) = (s^2 + 2\zeta_1\omega_{n1}s + \omega_{n1}^2)(s^2 + 2\zeta_2\omega_{n2}s + \omega_{n2}^2)(s^2 + 3.2s + 1.75)(s+r_1)(s+r_2).$$

Note that two of the closed-loop poles are specified to cancel (stable pole/zero cancellations) the two zeros of the open-loop pitch model in the hope of eliminating their effect on the step response. As in the pitch control example of Section 10.4, there are an additional two pairs of complex conjugate poles which need to be placed "far enough" in the left half-plane. This requires a few iterations to get it right. Try setting $\zeta_1 = \zeta_2 = 0.5$, $\omega_{n1} = 10, \omega_{n2} = 20$ so that

$$(s^2 + 2\zeta_1\omega_{n1}s + \omega_{n1}^2)(s^2 + 2\zeta_2\omega_{n2}s + \omega_{n2}^2) = (s^2 + 20s + 400)(s^2 + 10s + 100).$$

Note the real part of these two pairs of complex poles are -10 and -20. Then put the two real poles at $-5, -6$ (or replace them with another complex conjugate pair of poles). Finally

$$\begin{aligned} a_{CL}(s) &= (s^2 + 2\zeta_1\omega_{n1}s + \omega_{n1}^2)(s^2 + 2\zeta_2\omega_{n2}s + \omega_{n2}^2)(s^2 + 3.2s + 1.75)(s + r_1)(s + r_2) \\ &= (s^2 + 20s + 400)(s^2 + 10s + 100)(s^2 + 3.2s + 1.75)(s + 5)(s + 6) \\ &= s^8 + 44.2s^7 + 1193s^6 + 18064s^5 + 1.7558 \times 10^5 s^4 + 1.0520 \times 10^6 s^3 \\ &\quad + 3.4063 \times 10^6 s^2 + 4.925 \times 10^6 s + 2.1 \times 10^6. \end{aligned}$$

With $a_{CL}(s)$ chosen this way compute the zeros of $G_c(s)$, that is, the roots of $b_4 s^4 + b_3 s^3 + b_2 s^2 + b_1 s + b_0 = 0$, and verify they are in the open left-half plane. Use this fact to specify an input reference filter $G_f(s)$ to cancel the zeros of $G_c(s)$. With this 2 DOF design the closed-loop transfer function will not have zeros, but the step response could still have overshoot because the closed-loop poles are not all real.

With $R(s) = 5/s$ (5 degrees of pitch angle), simulate your designed 2 DOF control system. Show the reference input and output on the same plot in degrees. On a different plot show the elevator angle in degrees.

Remark For the actual issue addressed in [3] the 2 DOF controller designed in this problem is not applicable. In [3] the application requires that the final controller have the form $G_c(s) = \bar{G}_c(s)\dfrac{s+\alpha}{s}$. That is, a PI controller can be factored out of $G_c(s)$ and this is not the case with the 2 DOF controller as it has no real zeros. In the 2 DOF design the elevator command is $\delta(s) = G_c(s)E(s)$. However, in [3] there are two elevators with δ_e set by the pilot and δ_t (trim) set by the autopilot. These two elevator commands are given by $\delta_e(s) = 1.5\dfrac{s+3}{s+20}E(s)$ and $\delta_t(s) = 1.5\dfrac{s+3}{s+20}\dfrac{0.15}{s}E(s)$ so the total elevator command is $\delta(s) = \delta_e(s) + \delta_t(s) = 1.5\dfrac{s+3}{s+20}\dfrac{s+0.15}{s}E(s)$.

11

Frequency Response Methods

11.1 Bode Diagrams

In Chapter 3 transfer functions of differential equations were introduced. For example, consider the third-order differential equation given by

$$\dddot{y} + a_2\ddot{y} + a_1\dot{y} + a_0 y = b_2\ddot{u} + b_1\dot{u} + b_0 u.$$

Taking the Laplace transform with zero initial conditions, i.e., $y(0) = \dot{y}(0) = \ddot{y}(0) = 0$, $u(0) = \dot{u}(0) = 0$, we have

$$(s^3 + a_2 s^2 + a_1 s + a_0)Y(s) = (b_2 s^2 + b_1 s + b_0)U(s)$$

resulting in the transfer function

$$G(s) \triangleq \left.\frac{Y(s)}{U(s)}\right|_{\text{zero initial conditions}} = \frac{b_2 s^2 + b_1 s + b_0}{s^3 + a_2 s^2 + a_1 s + a_0}.$$

With the input given by $u(t) = U_0 \cos(\omega t)$ for $-\infty < t < \infty$, it was also shown in Chapter 3 that

$$y(t) = |G(j\omega)|\, U_0 \cos(\omega t + \angle G(j\omega))$$

is a solution to the differential equation.[1]

The frequency response $G(j\omega)$ shows how the differential equation processes any sinusoidal input $U_0 \cos(\omega t)$. It has been found that considerable insight into a system (differential equation) can be found by plotting its frequency response. Specifically, let $G(j\omega)$ be written in polar coordinate form as

$$G(j\omega) = |G(j\omega)|\, e^{j\angle G(j\omega)}. \tag{11.1}$$

To plot $G(j\omega)$ we do it in the form of the *Bode diagram* (Bode plot) due to H.W. Bode [47]. The Bode diagram is a plot of

$$20 \log_{10}|G(j\omega)| \quad \text{vs.} \quad \log_{10}(\omega) \tag{11.2}$$

and

$$\angle G(j\omega) \quad \text{vs.} \quad \log_{10}(\omega). \tag{11.3}$$

We will call the units of $20 \log_{10}|G(j\omega)|$ decibels or dB (named after Alexander Graham Bell) and the units of $\angle G(j\omega)$ will be degrees.

---

[1] If $G(s)$ is stable then it is also the steady-state solution for any set of initial conditions.

An Introduction to System Modeling and Control, First Edition. John Chiasson.
© 2022 John Wiley & Sons, Inc. Published 2022 by John Wiley & Sons, Inc.
Companion website: www.wiley.com/go/chiasson/anintroductiontosystemmodelingandcontrol

11 Frequency Response Methods

Let's consider a specific example of a first-order system with a stable pole given by

$$G(j\omega) = \frac{1}{\tau j\omega + 1}, \quad |G(j\omega)| = \left|\frac{1}{\tau j\omega + 1}\right| = \frac{1}{\sqrt{(\tau\omega)^2 + 1}} \quad \text{with } \tau > 0. \tag{11.4}$$

Notice that we have written $G(j\omega)$ in *time constant* form in contrast to $G(j\omega) = \dfrac{1/\tau}{j\omega + 1/\tau}$, which is pole–zero form. It turns out that the time constant form is more convenient for drawing Bode diagrams. The pole $p = 1/\tau$ is referred to as the *break point* or *corner frequency*.

Bode Magnitude Plot

A table of values of $20\log_{10}|G(j\omega)|$ vs. $\log_{10}(\omega)$ are given below. Note that the frequencies in the table are chosen to be powers of 10 multiples of the break point $1/\tau$.

ω	$\log_{10}(\omega)$	$20\log_{10}|G(j\omega)|$ in dB
0	$-\infty$	$20\log_{10}|1/1| = 0$
$0.1/\tau$	$-1 + \log_{10}(1/\tau)$	$20\log_{10}\left|\dfrac{1}{0.1j+1}\right| \approx 0$
$1/\tau$	$\log_{10}(1/\tau)$	$20\log_{10}\left|\dfrac{1}{j+1}\right| = 20\log_{10}|1/\sqrt{2}| = -3$
$10/\tau$	$1 + \log_{10}(1/\tau)$	$20\log_{10}\left|\dfrac{1}{10j+1}\right| = 20\log_{10}|1/\sqrt{101}| \approx -20$
$100/\tau$	$2 + \log_{10}(1/\tau)$	$20\log_{10}\left|\dfrac{1}{100j+1}\right| \approx 20\log_{10}|1/100| = -40$
$1000/\tau$	$3 + \log_{10}(1/\tau)$	$20\log_{10}\left|\dfrac{1}{1000j+1}\right| \approx 20\log_{10}|1/1000| = -60$
∞	∞	$20\log_{10}\left|\dfrac{1}{\infty j+1}\right| = 20\log_{10}|0| = -\infty$

Semilog Graphs

With $G(j\omega)$ given in (11.4) and using the previous table,[2] a plot $20\log_{10}|G(j\omega)|$ vs. $\log_{10}\omega$ is as shown in Figure 11.1.

This plot is a *semilog* graph, which simply means the distance along the abscissa (horizontal axis) is $\log_{10}(\omega)$ while the ordinate (vertical axis) is a linear scale. However, though the horizontal distances are $\log_{10}(\omega)$, the label is the value of ω as shown in Figure 11.1. For example the point labeled $\omega = 10^0 = 1$ has $\log(10^0) = 0$ so it is the zero point of the abscissa. The point labeled $\omega = 5$ has $\log_{10}(5) = 0.7$ so it is drawn a distance of 0.7 from

---
[2] Actually, using MATLAB!

the point labeled 10^0. Similarly, the point labeled 10^1 has $\log(10^1) = 1$ so it is drawn a distance of 1 from the point labeled 10^0.

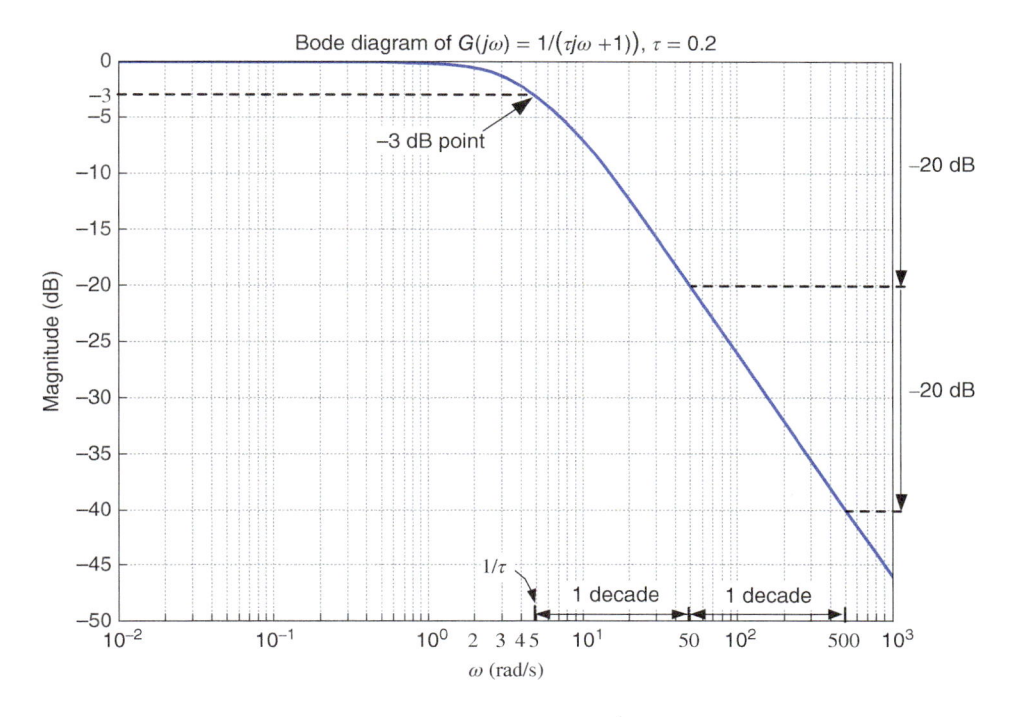

Figure 11.1. Bode magnitude plot of $G(s) = \dfrac{1}{j\omega/5 + 1}$ where $\tau = 1/5 = 0.2$.

Asymptotic Magnitude Plot

To better understand the Bode magnitude plot of Figure 11.1, we note that

$$20\log_{10}\left|\dfrac{1}{\tau j\omega + 1}\right| \approx \begin{cases} 0, & \omega < 1/\tau \\ -3, & \omega = 1/\tau \\ 20\log_{10}\left(\dfrac{1}{\omega\tau}\right) = -20\log_{10}\left(\dfrac{\omega}{1/\tau}\right), & \omega > 1/\tau. \end{cases} \quad (11.5)$$

The *asymptotic* magnitude plot is given by (11.5) for the two regions $\omega < 1/\tau$ and $\omega > 1/\tau$ and is drawn in Figure 11.2.

Every factor of 10 is referred to as *decade*. For example, $10^2(1/\tau)$ would be two decades *above* $1/\tau$ while $10^{-1}(1/\tau)$ would be one decade *below* $1/\tau$. With $k \geq 1$ so that $\omega = 10^k(1/\tau)$ is k decades above $1/\tau$, we have

$$-20\log_{10}\left(\dfrac{\omega}{1/\tau}\right) = -20\log_{10}\left(\dfrac{10^k(1/\tau)}{1/\tau}\right) = -20k \text{ dB}$$

and

$$\log_{10}\omega = \log_{10}\left(10^k\dfrac{1}{\tau}\right) = k + \log_{10}\left(\dfrac{1}{\tau}\right).$$

364   11   Frequency Response Methods

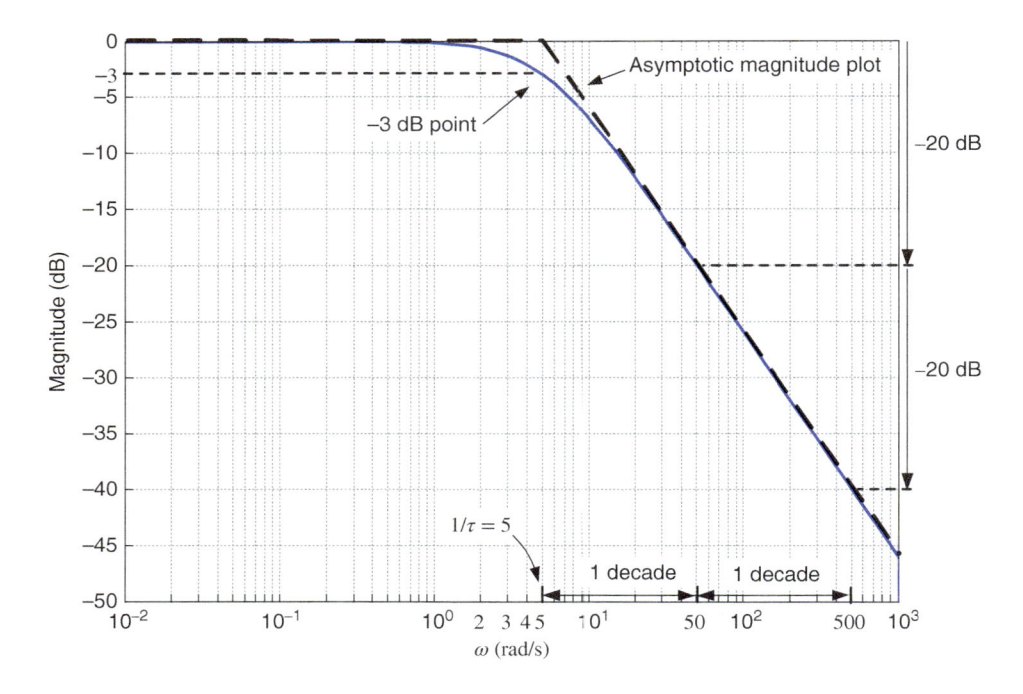

Figure 11.2. Asymptotic magnitude plot of $G(j\omega) = \dfrac{1}{\tau j\omega + 1}$ where $\tau = 1/5 = 0.2$.

This says that for every decade above $1/\tau$ the magnitude goes down by 20 dB. Again, for emphasis, Figure 11.2 shows that the graph goes down -20 dB for every decade (factor of 10) past the break point $1/\tau$.

Bode Phase Plot

The Bode phase plot is $\angle G(j\omega)$ vs. $\log_{10}\omega$. We first make the following table.

ω	$\log_{10}(\omega)$	$\angle G(j\omega) = \angle \dfrac{1}{\tau j\omega + 1} = -\tan^{-1}\left(\dfrac{\omega}{1/\tau}\right)$
0	$-\infty$	$-\tan^{-1}(0) = 0°$
$(0.1)(1/\tau)$	$-1 + \log_{10}(1/\tau)$	$-\tan^{-1}(0.1) = -5.7°$
$1/\tau$	$\log_{10}(1/\tau)$	$-\tan^{-1}(1) = -45°$
$10(1/\tau)$	$1 + \log_{10}(1/\tau)$	$-\tan^{-1}(10) = -84.3°$
$100(1/\tau)$	$2 + \log_{10}(1/\tau)$	$-\tan^{-1}(100) \approx -90°$

The phase plot is shown in Figure 11.3.

11.1 Bode Diagrams

Bode diagram of $G(j\omega) = 1/(\tau j\omega + 1)$, $\tau = 0.2$

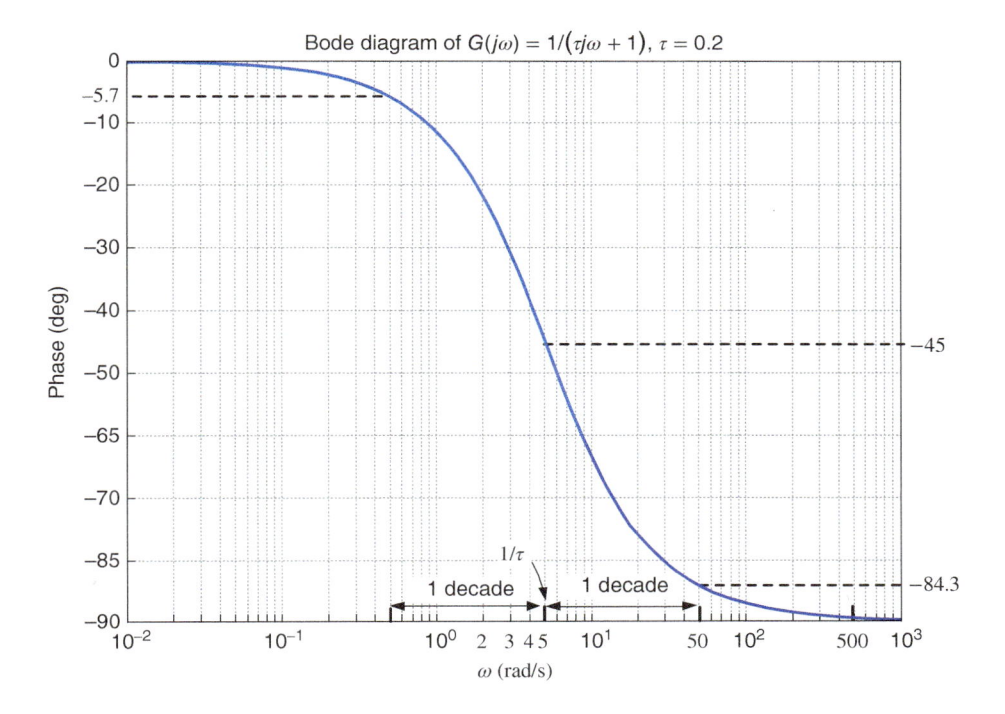

Figure 11.3. Bode phase diagram of $G(j\omega) = \dfrac{1}{\tau j\omega + 1}$ where $1/\tau = 5$.

Simple Examples

Bode Diagram of a First-Order Unstable Pole

Let's now consider the Bode diagram of $G(j\omega) = \dfrac{1}{-\tau j\omega + 1}$. In particular, let $\tau = 0.2$ so that $1/\tau = 5$ and $G(j\omega) = \dfrac{1}{-0.2j\omega + 1}$. In this case

$$|G(j\omega)| = \left|\dfrac{1}{-\tau j\omega + 1}\right| = \dfrac{1}{\sqrt{(\tau\omega)^2 + 1}}$$

and so the magnitude plot is the same as Figure 11.2. For the angle plot we have

$$\angle G(j\omega) = \angle \dfrac{1}{-\tau j\omega + 1} = -\tan^{-1}\left(-\dfrac{\omega}{1/\tau}\right) = \tan^{-1}\left(\dfrac{\omega}{1/\tau}\right)$$

which is the *negative* of angle plot of Figure 11.3. Thus the Bode diagram of $G(j\omega) = \dfrac{1}{-\tau j\omega + 1}$ is as shown in Figure 11.4.

366   11 Frequency Response Methods

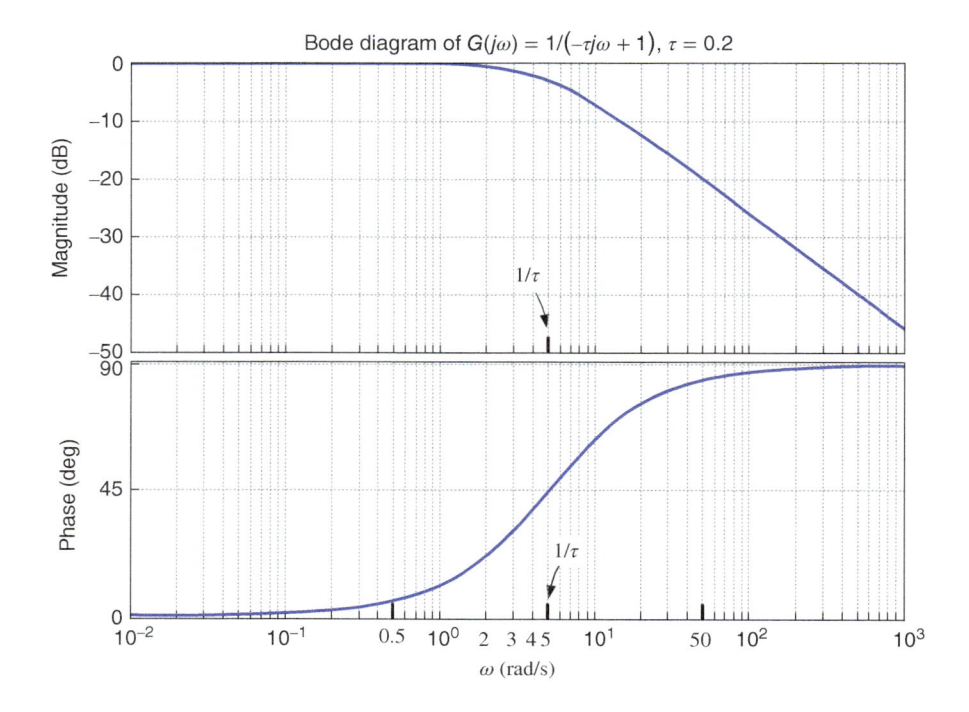

Figure 11.4. Magnitude and phase plots of $G(j\omega) = \dfrac{1}{-\tau j\omega + 1}$ with $\tau = 1/5 = 0.2$.

Bode Diagram of a First-Order Zero

Let's now consider the Bode diagram of

$$G(j\omega) = \tau j\omega + 1.$$

In this case

$$20\log_{10}\big|G(j\omega)\big| = 20\log_{10}|\tau j\omega + 1| = 20\log_{10}\sqrt{(\tau\omega)^2 + 1} = -20\log_{10}\dfrac{1}{\sqrt{(\tau\omega)^2 + 1}}.$$

With $\tau = 0.2$ this is the *negative* of the magnitude plot of Figure 11.2. For the angle plot we have

$$\angle G(j\omega) = \angle(\tau j\omega + 1) = \tan^{-1}\left(\dfrac{\omega}{1/\tau}\right),$$

which is the *negative* of angle plot of Figure 11.3. Thus the Bode diagram of $G(j\omega) = \tau j\omega + 1$ is as shown in Figure 11.5.

With $\tau = 0.2$, Problem 1 asks the reader to sketch the Bode diagram of

$$G(j\omega) = -\tau j\omega + 1.$$

11.1 Bode Diagrams

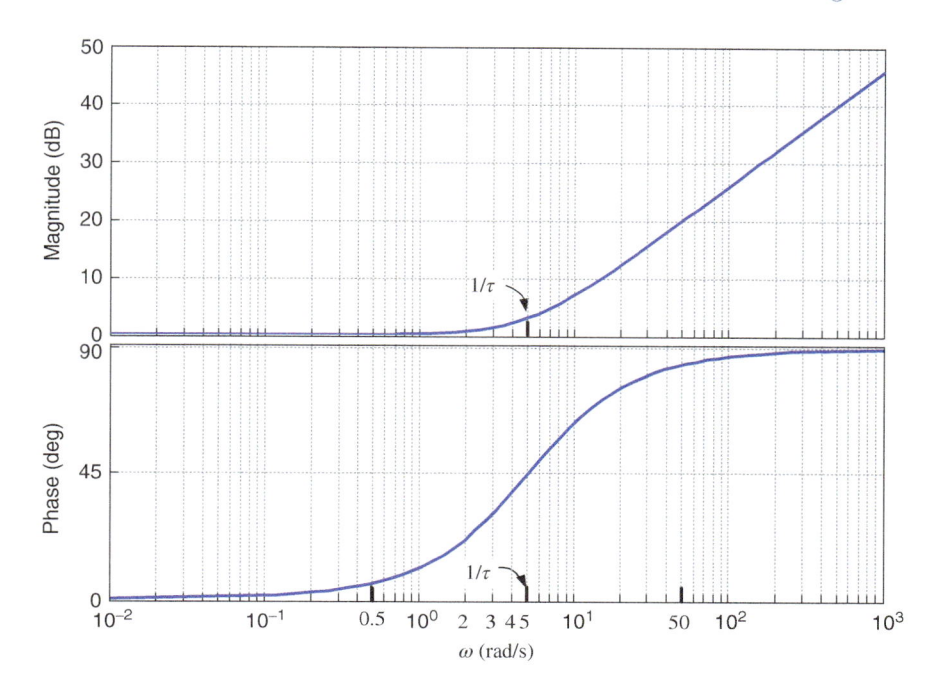

Figure 11.5. Bode magnitude and phase plots of $G(j\omega) = \tau j\omega + 1$ with $\tau = 1/5 = 0.2$.

Bode Diagram of a Pole at the Origin

Let's now consider $G(j\omega) = \dfrac{1}{j\omega}$. We have

$$20 \log_{10} |G(j\omega)| = 20 \log_{10} \left|\frac{1}{j\omega}\right| = -20 \log_{10} \omega$$

and

$$\angle G(j\omega) = -90°.$$

An easy way to do the magnitude plot is to start at $\omega = 10^0 = 1$ where $20 \log_{10} |G(j1)| = -20 \log_{10} |1| = 0$ and go to $\omega = 10$ where $20 \log_{10} |G(j\omega)| = -20 \log_{10} |10| = -20$. Then just draw a straight line through these two points. See Figure 11.6.

More Bode Diagram Examples

Example 1 $G(j\omega) = 1/j\omega(j\omega + 5)$

Let's now consider the transfer function,

$$G(j\omega) = \frac{1}{j\omega(j\omega + 5)} = \frac{1}{5}\frac{1}{j\omega(0.2j\omega + 1)}.$$

The first step to remember in graphing a Bode diagram is to put any real poles in time constant form, that is, we rewrite $\dfrac{1}{j\omega + 5}$ as $\dfrac{1}{5}\dfrac{1}{0.2j\omega + 1}$. Then, by a property of logarithms,

368   11   Frequency Response Methods

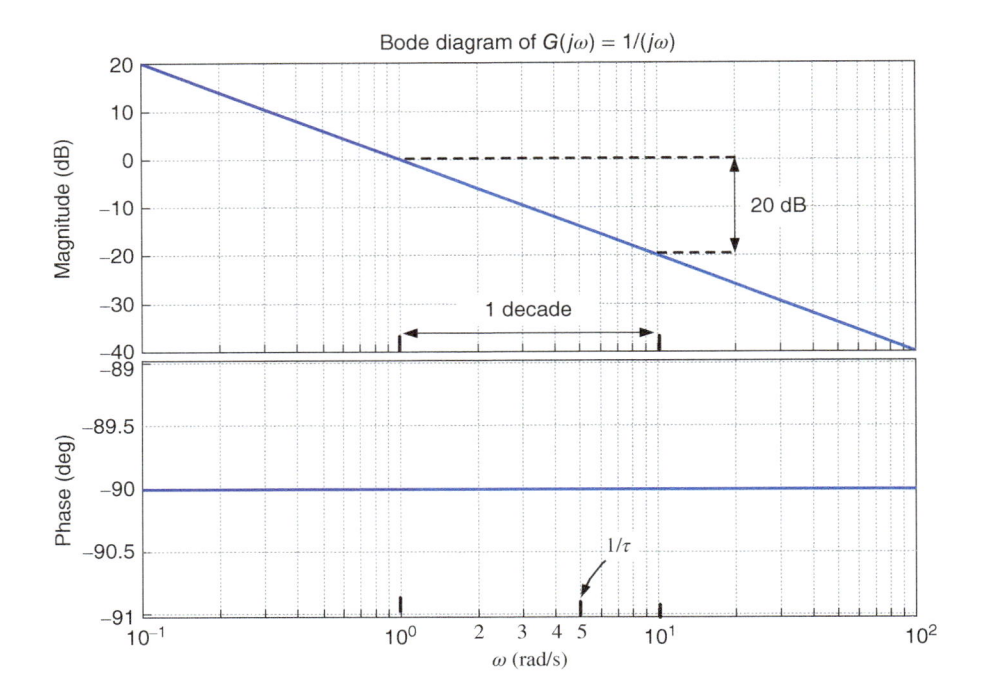

Figure 11.6. Bode diagram of $G(j\omega) = \dfrac{1}{j\omega}$.

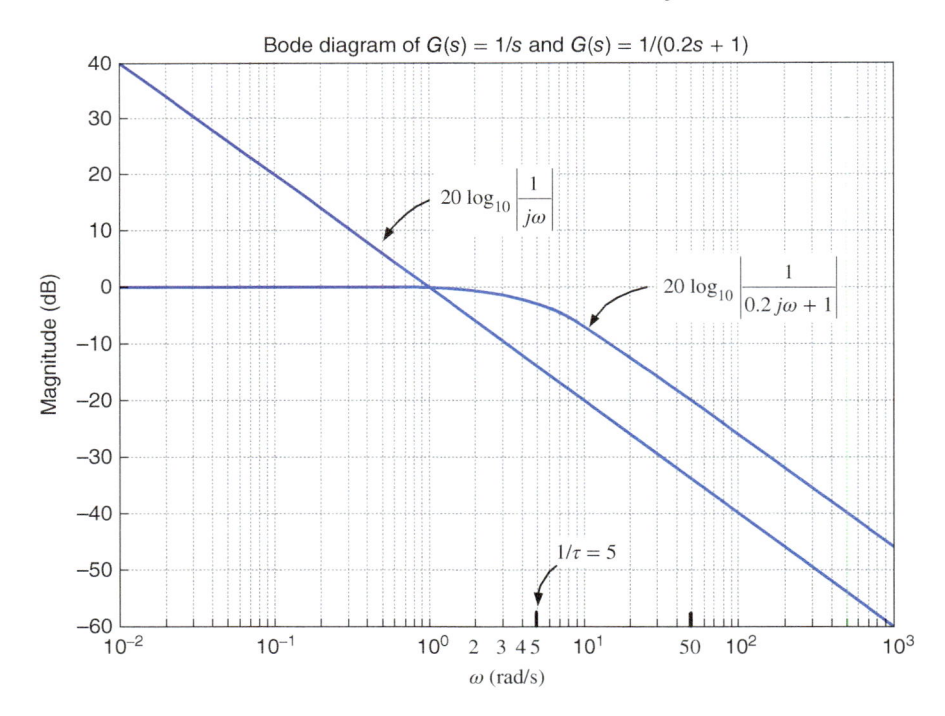

Figure 11.7. Bode plot of $20\log_{10}\left|\dfrac{1}{j\omega}\right|$ and $20\log_{10}\left|\dfrac{1}{0.2j\omega+1}\right|$.

11.1 Bode Diagrams

we have

$$20 \log_{10}|G(j\omega)| = 20 \log_{10}\left|\frac{1}{5}\frac{1}{j\omega(0.2j\omega+1)}\right|$$

$$= 20 \log_{10}\left|\frac{1}{5}\right| + 20 \log_{10}\left|\frac{1}{j\omega}\right| + 20 \log_{10}\left|\frac{1}{0.2j\omega+1}\right|.$$

Figure 11.7 shows the magnitude plot of $20 \log_{10}\left|\frac{1}{j\omega}\right|$ and $20 \log_{10}\left|\frac{1}{0.2j\omega+1}\right|$.

As shown in Figure 11.8 the next step is to graph

$$20 \log_{10}\left|\frac{1}{j\omega}\right| + 20 \log_{10}\left|\frac{1}{0.2j\omega+1}\right|.$$

One simply does the plot of $20 \log_{10}\left|\frac{1}{j\omega}\right|$ until the break point at $1/\tau = 5$ because $20 \log_{10}\left|\frac{1}{0.2j\omega+1}\right| \approx 0$ for $\omega < 1/\tau = 5$. (The asymptotic plot of $20 \log_{10}\left|\frac{1}{0.2j\omega+1}\right|$ is zero for $\omega < 1/\tau = 5$.) After the break point the graph now goes down at -40 dB/decade because both $20 \log_{10}\left|\frac{1}{j\omega}\right|$ and $20 \log_{10}\left|\frac{1}{0.2j\omega+1}\right|$ are going down at -20 dB/decade and we are adding them.

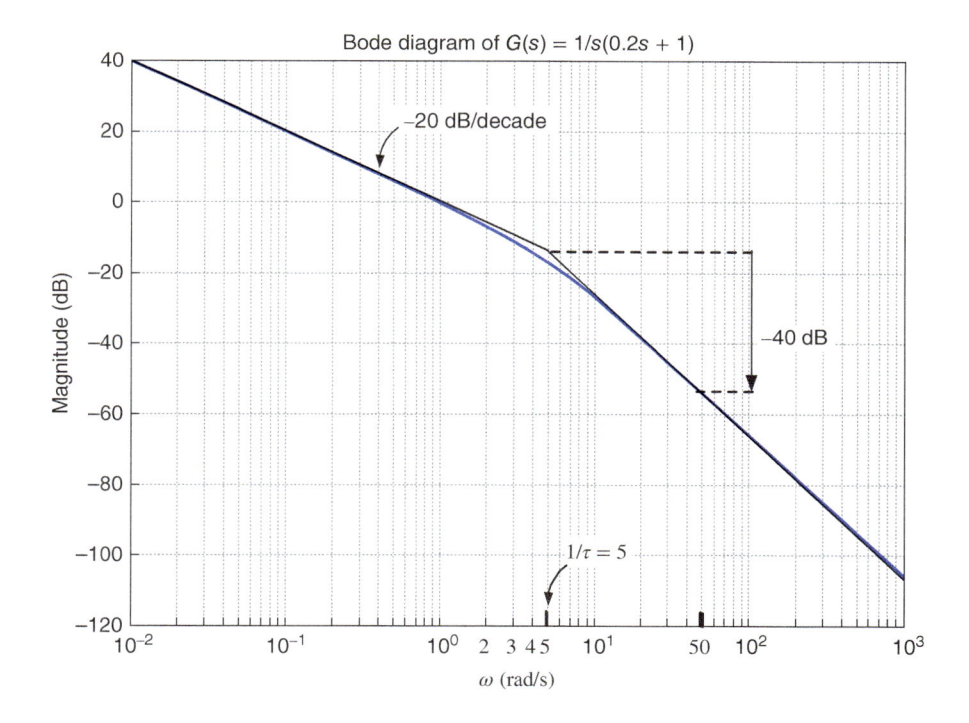

Figure 11.8. Bode plot of $20 \log_{10}\left|\frac{1}{j\omega}\right| + 20 \log_{10}\left|\frac{1}{0.2j\omega+1}\right|$.

370   11 Frequency Response Methods

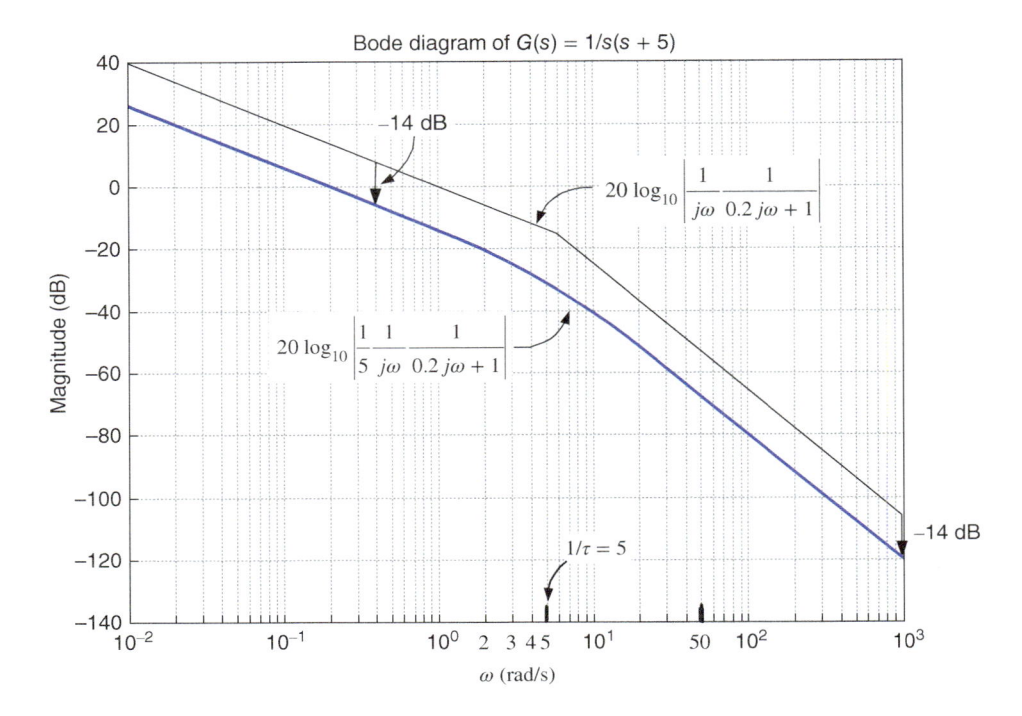

Figure 11.9. Bode magnitude plot $20\log_{10}\left|\dfrac{1}{j\omega(j\omega+5)}\right| = 20\log_{10}\left|\dfrac{1}{5}\dfrac{1}{j\omega(0.2j\omega+1)}\right|$.

To finish the plot we simply add the constant term $20\log_{10}|1/5| = -14$ dB to the graph of Figure 11.8 to obtain the Bode diagram of Figure 11.9. In Figure 11.9 we also show the plot of Figure 11.8 to indicate that Figure 11.9 is obtained by lowering the plot in Figure 11.8 by 14 dB.

To do the Bode phase plot of $\dfrac{1}{j\omega(j\omega+5)}$, note that

$$\angle\dfrac{1}{j\omega(j\omega+5)} = \angle\dfrac{1}{5}\dfrac{1}{j\omega(0.2j\omega+1)} = \angle\dfrac{1}{5} + \angle\dfrac{1}{j\omega} + \angle\dfrac{1}{0.2j\omega+1}$$

$$= 0° - 90° + \angle\dfrac{1}{0.2j\omega+1}.$$

Thus we need only shift the Bode phase plot of Figure 11.3 down by $-90°$ to obtain the Bode phase plot of $\angle\dfrac{1}{j\omega(j\omega+5)}$. This is shown in Figure 11.10. In particular, at the break point $1/\tau = 5$, we have

$$\left.\angle\dfrac{1}{j\omega(j\omega+5)}\right|_{\omega=1/\tau} = -90° - 45° = -135°.$$

[Bode phase diagram of $G(s) = 1/s(s+5)$]

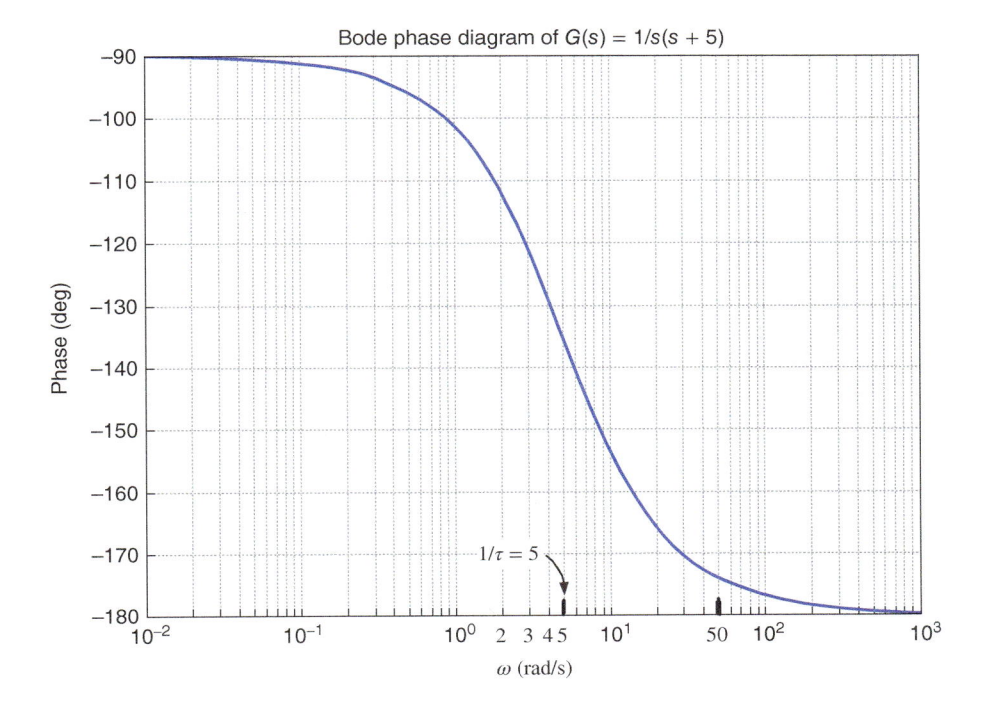

Figure 11.10. $\angle \dfrac{1}{j\omega(j\omega + 5)}$ vs. $\log_{10}\omega$.

Example 2 $G_c(j\omega) = \dfrac{j\omega + 0.1}{j\omega + 0.01}$

Let's now do the Bode diagram of the lag compensator[3] given by $G_c(s) = \dfrac{s + 0.1}{s + 0.01}$.
We have

$$20\log_{10}|G_c(j\omega)| = 20\log_{10}\left|\dfrac{0.1}{0.01}\dfrac{j\omega/0.1 + 1}{j\omega/0.01 + 1}\right|$$

$$= \underbrace{20\log_{10}|10|}_{20 \text{ dB}} + 20\log_{10}|j\omega/0.1 + 1| + 20\log_{10}\left|\dfrac{1}{j\omega/0.01 + 1}\right|.$$

The asymptotic approximations to each term are

$$20\log_{10}|10| = 20 \text{ dB}$$

$$20\log_{10}\left|\dfrac{1}{j\omega/0.01 + 1}\right| = \begin{cases} 0, & \omega < 0.01 \\ -20\log_{10}\left(\dfrac{\omega}{0.01}\right), & \omega > 0.01 \end{cases}$$

$$20\log_{10}|j\omega/0.1 + 1| = \begin{cases} 0, & \omega < 0.1 \\ 20\log_{10}\left(\dfrac{\omega}{0.1}\right), & \omega > 0.1 \end{cases}$$

---

[3] Compensator is just another word for controller.

11 Frequency Response Methods

[Bode diagram of $G(j\omega) = (j\omega + 0.1)/(j\omega + 0.01)$ showing magnitude in dB vs ω (rad/s), with asymptotic magnitude plot and actual plot of $20\log_{10}\left|10\dfrac{j\omega/0.1 + 1}{j\omega/0.01 + 1}\right|$]

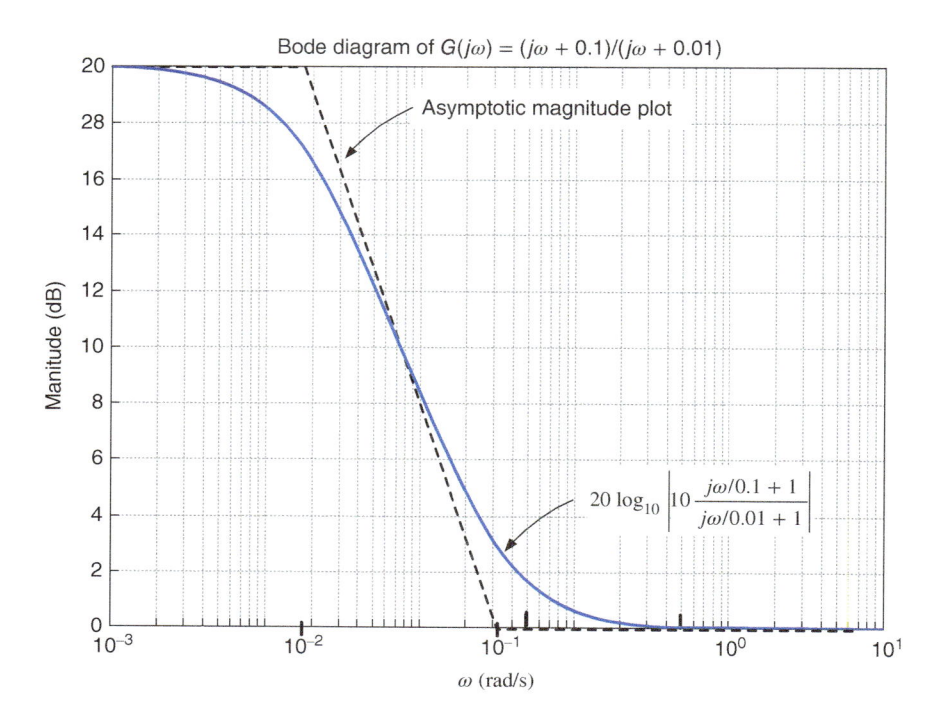

Figure 11.11. $20\log_{10}\left|10\dfrac{j\omega/0.1 + 1}{j\omega/0.01 + 1}\right|$ vs. $\log_{10}\omega$.

so that the asymptotic magnitude plot is given by

$$20\log_{10}|G_c(j\omega)| = \begin{cases} 20\log_{10}|10|, & \omega < 0.01 \\ 20\log_{10}|10| - 20\log_{10}\left(\dfrac{\omega}{0.01}\right), & 0.01 < \omega < 0.1 \\ \underbrace{20\log_{10}|10| - 20\log_{10}\left(\dfrac{\omega}{0.01}\right) + 20\log_{10}\left(\dfrac{\omega}{0.1}\right)}_{0 \text{ dB}}, & 0.1 < \omega. \end{cases}$$

The asymptotic magnitude plot along with the actual magnitude plot are shown in Figure 11.11.

We now consider the phase plot, that is,

$$\angle G_c(j\omega) = \angle(j\omega/0.1 + 1) + \angle\dfrac{1}{j\omega/0.01 + 1} \quad \text{vs.} \log_{10}(\omega).$$

This is given in Figure 11.12. Because the pole at $s = -0.01$ and the zero at $s = 0.1$ are separated by a decade (factor of 10), the angle plot is straightforward to plot by using the following table.

11.1 Bode Diagrams 373

ω	$\angle(j\omega/0.1+1)$	$\angle\dfrac{1}{j\omega/0.01+1}$	$\angle(j\omega/0.1+1)+\angle\dfrac{1}{j\omega/0.01+1}$
0.001	≈ 0	$\approx -5.7°$	$\approx -5.7°$
0.01	$\approx 5.7°$	$\approx -45°$	$\approx -39.3°$
0.1	$\approx 45°$	$\approx -90°$	$\approx -45°$
1	$\approx 90°$	$\approx -90°$	$\approx 0°$

Though the table did not capture this point, Figure 11.12 shows that the phase is at a minimum at approximately $\omega = 0.03$.

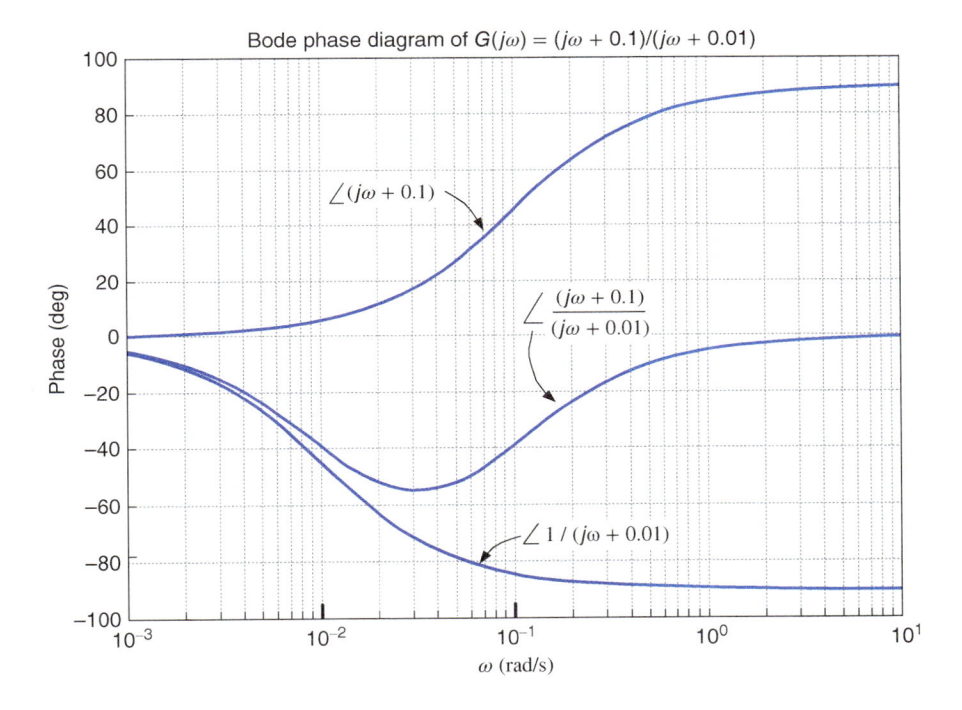

Figure 11.12. $\angle G_c(j\omega) = \angle(j\omega/0.1+1) + \angle\dfrac{1}{j\omega/0.01+1}$ vs. ω.

Example 3 $G_c(j\omega) = \dfrac{j\omega+2}{j\omega+4}$

Let's now do the Bode diagram of a lead compensator given by

$$G_c(s) = \dfrac{s+2}{s+4}.$$

11 Frequency Response Methods

We have
$$20\log_{10}|G_c(j\omega)| = 20\log_{10}\left|\frac{2}{4}\frac{j\omega/2+1}{j\omega/4+1}\right|$$
$$= \underbrace{20\log_{10}\left|\frac{1}{2}\right|}_{-6\text{ dB}} + 20\log_{10}|j\omega/2+1| + 20\log_{10}\left|\frac{1}{j\omega/4+1}\right|.$$

The asymptotic approximations to each term are
$$20\log_{10}|1/2| = -6\text{ dB}$$
$$20\log_{10}|j\omega/2+1| = \begin{cases} 0, & \omega < 2 \\ 20\log_{10}\left(\frac{\omega}{2}\right), & \omega > 2 \end{cases}$$
$$20\log_{10}\left|\frac{1}{j\omega/4+1}\right| = \begin{cases} 0, & \omega < 4 \\ -20\log_{10}\left(\frac{\omega}{4}\right), & \omega > 4. \end{cases}$$

The asymptotic magnitude plot is then
$$20\log_{10}|G_c(j\omega)| = \begin{cases} 20\log_{10}|1/2|, & \omega < 2 \\ 20\log_{10}|1/2| + 20\log_{10}\left(\frac{\omega}{2}\right), & 2 < \omega < 4 \\ \underbrace{20\log_{10}|1/2| + 20\log_{10}\left(\frac{\omega}{2}\right) - 20\log_{10}\left(\frac{\omega}{4}\right)}_{0}, & 4 < \omega \end{cases}$$

and is drawn in Figure 11.13.

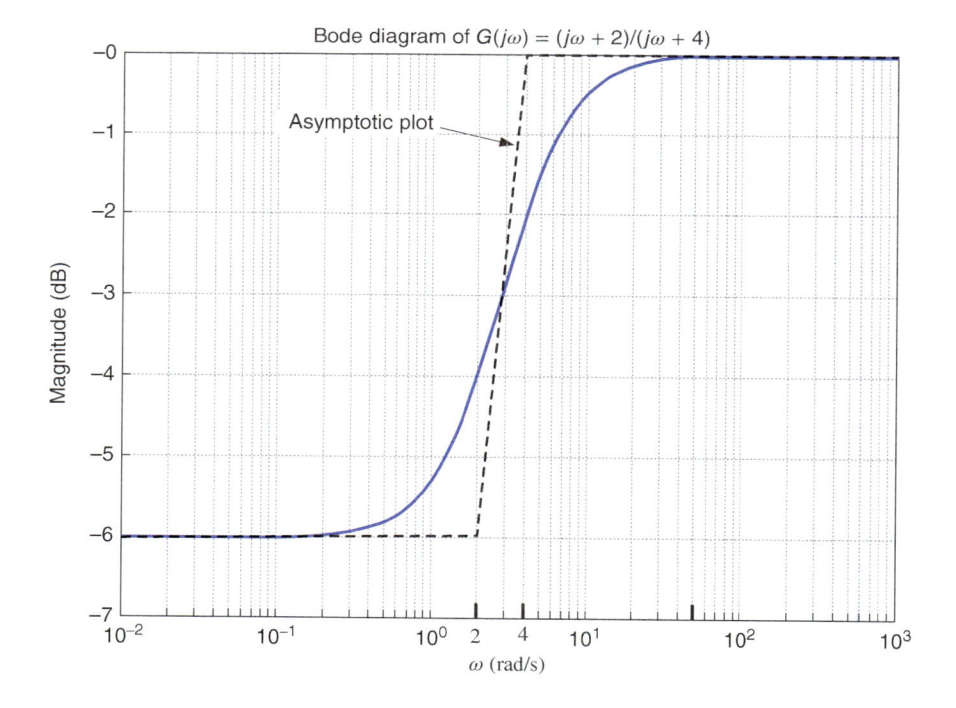

Figure 11.13. $20\log_{10}\left|\dfrac{j\omega+2}{j\omega+4}\right|$ vs. $\log_{10}\omega$.

11.1 Bode Diagrams

Bode phase diagram of $G(j\omega) = (j\omega + 2)/(j\omega + 4)$

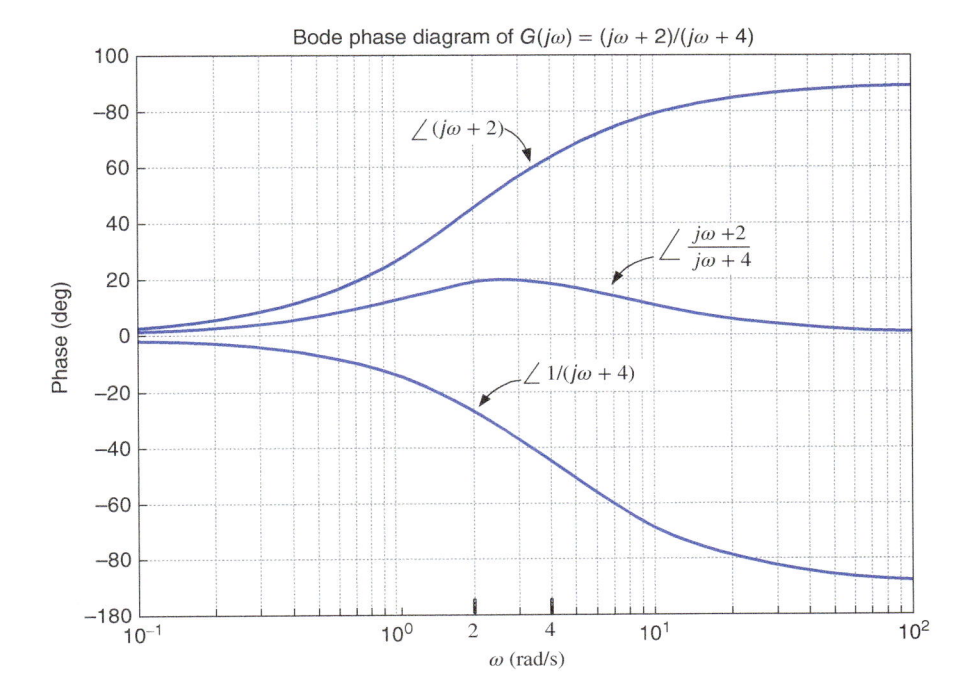

Figure 11.14. $\angle \dfrac{j\omega + 2}{j\omega + 4}$ vs. $\log_{10}\omega$.

Figure 11.14 is a plot of $\angle \dfrac{j\omega + 2}{j\omega + 4} = \angle(j\omega/2 + 1) + \angle \dfrac{1}{j\omega/4 + 1}$ vs. $\log \omega$. The plot also includes $\angle(j\omega/2 + 1) = \angle(j\omega + 2)$ and $\angle \dfrac{1}{j\omega/4 + 1} = \angle \dfrac{1}{j\omega + 4}$. These phase plots are sketched using the following table.

ω	$\angle(j\omega/2 + 1)$	$\angle(j\omega/4 + 1)$	$\angle(j\omega/2 + 1) + \angle \dfrac{1}{j\omega/4 + 1}$
0.2	5.7°	0	5.7°
2	45°	$\tan^{-1}(1/2) = 26.6°$	18.4°
4	$\tan^{-1}(2) = 63.2°$	45°	18.2
20	84.3°	$\tan^{-1}(5) = 78.7°$	5.6°
40	90°	84.3°	5.7°
∞	90°	90°	0°

The maximum phase occurs for ω between 2 and 4.

Example 4 $G(j\omega) = \dfrac{(j\omega + 40)(j\omega/0.4 + 1)}{(j\omega/400 + 1)(j\omega + 4)^3}$

Let's consider the transfer function

$$G(j\omega) = \dfrac{(j\omega + 40)(j\omega/0.4 + 1)}{(j\omega/400 + 1)(j\omega + 4)^3}.$$

11 Frequency Response Methods

The first step to remember is to put it into time constant form which is

$$G(j\omega) = \frac{40}{4^3} \frac{(j\omega/40 + 1)(j\omega/0.4 + 1)}{(j\omega/400 + 1)(j\omega/4 + 1)^3}.$$

We plot

$$20\log_{10}|G(j\omega)| = \underbrace{20\log_{10}|40/64|}_{-4.1 \text{ dB}} + 20\log_{10}|j\omega/40 + 1| + 20\log_{10}|j\omega/0.4 + 1|$$

$$+ 20\log_{10}|1/(j\omega/400 + 1)| + 60\log_{10}|1/(j\omega/4 + 1)|,$$

which is shown on the top half of Figure 11.15. Let's outline how we did the asymptotic magnitude plot.

- Starting at $\omega = 0.01$ only the constant term contributes so the magnitude is $20\log_{10}|40/64| = -4.1$ dB.
- At the first (zero) breakpoint at $\omega = 0.4$ we go up at the rate of 20 dB/decade until the second breakpoint at $\omega = 4$. We are then at $-4.1 + (20 \text{ dB/decade}) \times (1 \text{ decade}) \approx 16$ dB.
- Due to the triple (pole) breakpoint at $\omega = 4$, we now go down at -40 dB/decade. (We were going up at 20 dB/decade and the triple breakpoint has us go down at -60 dB/decade so the sum is going down at -40 dB/decade.) At $\omega = 40$ we are at 16 dB $+(-40 \text{ dB/decade}) \times (1 \text{ decade}) = -24$ dB.
- Then due to the (zero) breakpoint at $\omega = 40$ we go down at -20 dB/decade until $\omega = 400$. The magnitude is then -24 dB $+(-20 \text{ dB/decade}) \times (1 \text{ decade}) = -44$ dB.
- The final (pole) breakpoint at $\omega = 400$ has the magnitude go down an additional -20 dB/decade so that at $\omega = 4000$ the magnitude is at -44 dB $+(-40 \text{ dB/decade}) \times (1 \text{ decade}) = -84$ dB.

To see that the phase plot makes sense we make the following table.

ω	$\angle G(j\omega)$
0.04	0
0.4	$\approx \angle(j\omega/0.4 + 1)\vert_{\omega=0.4} = 45°$
4	$\approx \angle(j\omega/0.4 + 1)\vert_{\omega=4} - 3 \times \angle(j\omega/4 + 1)\vert_{\omega=4}$
	$= 90° - 3 \times 45° = -45°$
40	$\approx \angle(j\omega/0.4 + 1)\vert_{\omega=40} - 3 \times \angle(j\omega/4 + 1)\vert_{\omega=40} + \angle(j\omega/40 + 1)\vert_{\omega=40}$
	$= 90° - 3 \times 90° + 45° = -135°$
400	$\approx \angle(j\omega/0.4 + 1)\vert_{\omega=400} - 3 \times \angle(j\omega/4 + 1)\vert_{\omega=400} + \angle(j\omega/40 + 1)\vert_{\omega=400}$
	$-\angle(j\omega/400 + 1)\vert_{\omega=400} = 90° - 3 \times 90° + 90° - 45° = -135°$
4000	$\approx \angle(j\omega/0.4 + 1)\vert_{\omega=4000} - 3 \times \angle(j\omega/4 + 1)\vert_{\omega=4000} + \angle(j\omega/40 + 1)\vert_{\omega=4000}$
	$-\angle(j\omega/400 + 1)\vert_{\omega=4000} = 90° - 3 \times 90° + 90° - 90° = -180°$

11.1 Bode Diagrams 377

Bode diagram of G(s) = (s + 40)(s/0.4 + 1)/(s/400 + 1)(s + 4)³

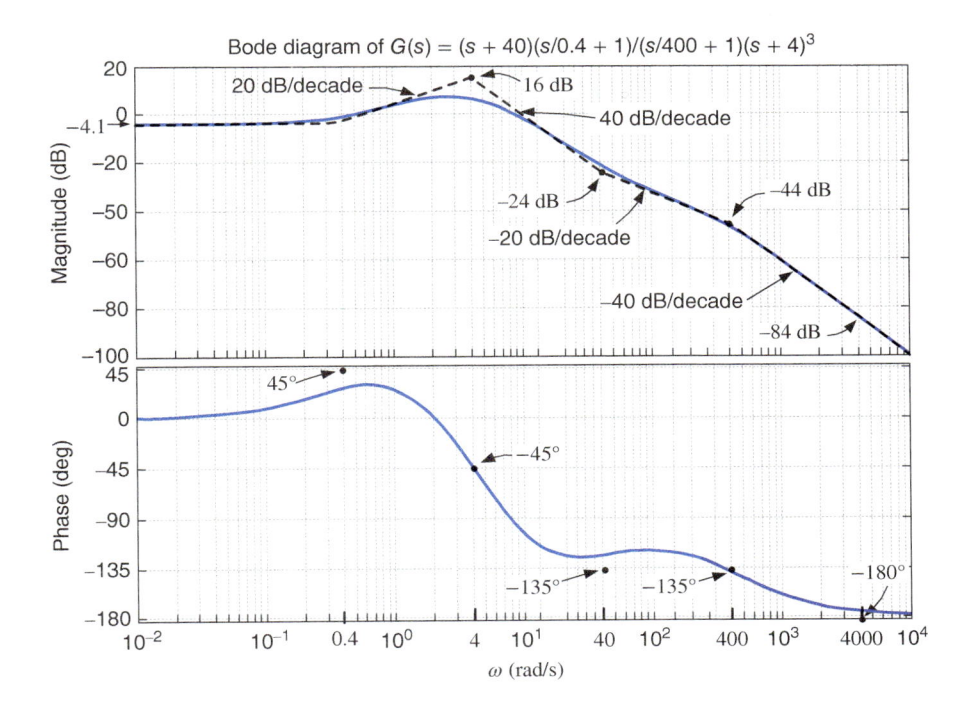

Figure 11.15. Bode diagram of $G(j\omega) = \dfrac{(j\omega + 40)(j\omega/0.4 + 1)}{(j\omega/400 + 1)(j\omega + 4)^3}$.

In the phase plot (the bottom plot of Figure 11.15) it is seen at $\omega = 0.4$ that the angle is not close to the approximate angle of $45°$. This is because at $\omega = 0.4$ we have $\angle(j\omega/0.4 + 1)|_{\omega=0.4} = 45°$ and we took $\angle(j\omega/4 + 1)|_{\omega=0.4} \approx 0$. However, $\angle(j\omega/4 + 1)|_{\omega=0.4} = 5.7°$ so that

$$\angle G(j\omega)|_{\omega=0.4} = \angle(j\omega/0.4 + 1)|_{\omega=0.4} - 3\angle(j\omega/4 + 1)|_{\omega=0.4}$$
$$= 45° - 3 \times 5.7°$$
$$= 27.9°.$$

Similarly, at $\omega = 40$, in the expression

$$\angle G(j\omega)|_{\omega=40} = \angle(j\omega/0.4 + 1)|_{\omega=40} - 3\angle(j\omega/4 + 1)|_{\omega=40} + \angle(j\omega/40 + 1)|_{\omega=40}$$

we used $\angle(j\omega/4 + 1)|_{\omega=40} = 90°$, but actually it is $\angle(j\omega/4 + 1)|_{\omega=40} = 84.3°$. Further, we took $\angle(j\omega/400 + 1)|_{\omega=40} = 0$, but it is actually $\angle(j\omega/400 + 1)|_{\omega=40} = 5.7°$. Thus the phase at $\omega = 40$ is

$$\angle G(j\omega)|_{\omega=40} = \angle(j\omega/0.4 + 1)|_{\omega=40} - 3\angle(j\omega/4 + 1)|_{\omega=40} + \angle(j\omega/40 + 1)|_{\omega=40}$$
$$-\angle(j\omega/400 + 1)|_{\omega=40}$$
$$= 90° - 3(84.3°) + 45° - 5.7° = -123.6°.$$

Bode Diagram with Complex Poles

Let's now look at the Bode diagram of

$$G(s) = \dfrac{\omega_n^2}{s^2 + 2\zeta\omega_n s + \omega_n^2},$$

378   11 Frequency Response Methods

where $0 < \zeta < 1$ so the poles form a complex conjugate pair. Then

$$G(j\omega) = \frac{1}{(j\omega/\omega_n)^2 + 2\zeta j\omega/\omega_n + 1} = \frac{1}{1 - (\omega/\omega_n)^2 + j2\zeta\omega/\omega_n}.$$

We have directly

$$G(j0) = 1, \ G(j\omega_n) = \frac{1}{2\zeta j}$$

and, as

$$G(j\omega) \approx \frac{1}{(j\omega/\omega_n)^2} \quad \text{for } \omega \gg \omega_n,$$

the asymptotic magnitude plot is given by

$$20\log_{10}\left|\frac{1}{(j\omega/\omega_n)^2}\right| = \begin{cases} 0, & \omega < \omega_n \\ 20\log_{10}\left|\frac{1}{(j\omega/\omega_n)^2}\right| = -40\log_{10}|\omega/\omega_n|, & \omega > \omega_n. \end{cases}$$

The Bode magnitude plot is given in Figure 11.16. We still consider ω_n as a (double) breakpoint and note that the magnitude plot decreases at -40 dB/decade after the breakpoint.

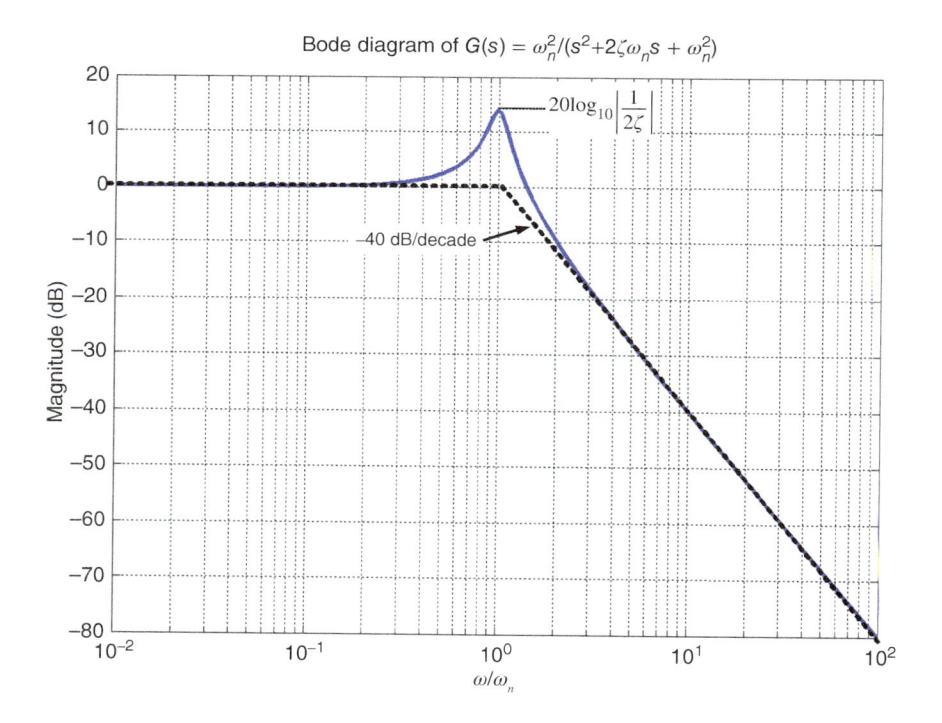

Figure 11.16. $20\log\left|\frac{1}{(j\omega/\omega_n)^2 + 2\zeta j\omega/\omega_n + 1}\right|$ vs. $\log\omega$ for $\omega_n = 1, \zeta = 0.1$.

Peak and Resonant Values

The *resonant frequency* ω_r is the frequency at which the Bode magnitude plot is a maximum. Setting

$$\frac{d}{d\omega}\left|\frac{1}{(j\omega/\omega_n)^2 + 2\zeta j\omega/\omega_n + 1}\right| = 0$$

and solving for ω gives the resonant frequency at

$$\omega_r \triangleq \omega_n\sqrt{1 - 2\zeta^2} \text{ for } 0 < \zeta < 1/\sqrt{2} = 0.707.$$

The corresponding peak value of the magnitude plot is then found to be

$$|G(j\omega_r)| = \frac{1}{2\zeta\sqrt{1-\zeta^2}}.$$

For $\zeta \ll 1$ we have $\omega_r \approx \omega_n$ and $|G(j\omega_r)| \approx \dfrac{1}{2\zeta}$. Figure 11.17a shows the Bode magnitude plots for $\zeta = 0.1, 0.2, 0.3, 0.5, 0.7, 1.0$.

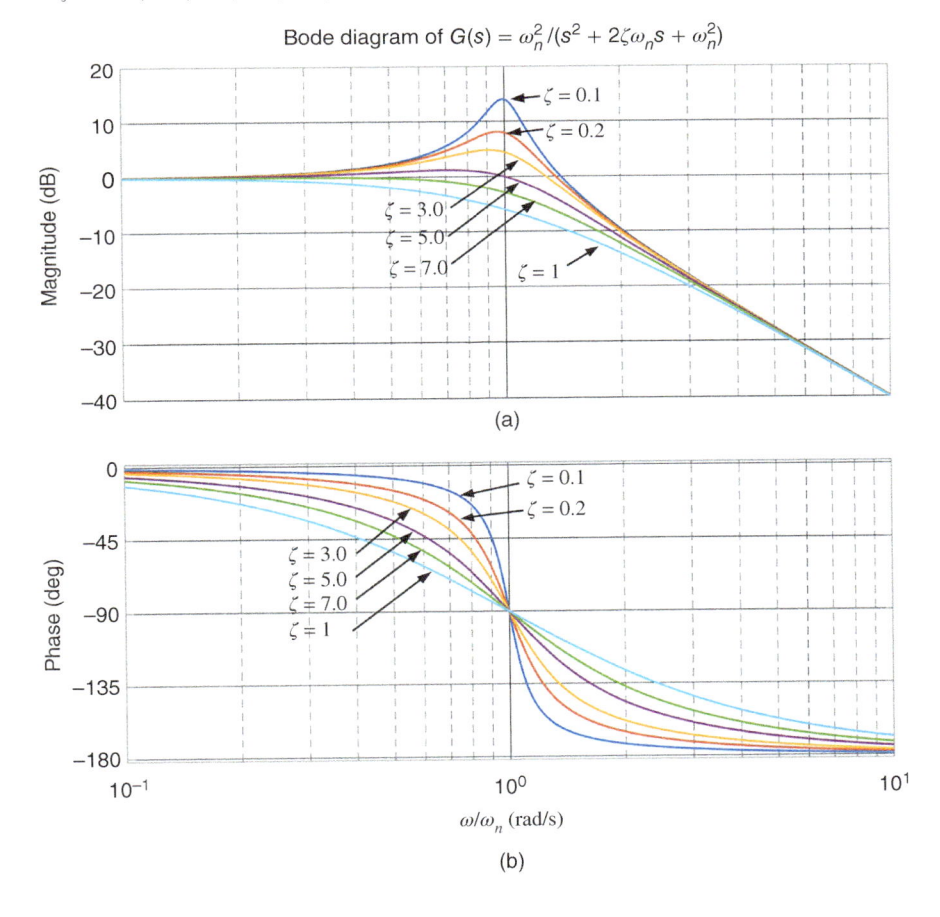

Figure 11.17. (a) $20\log_{10}\left|\dfrac{1}{(j\omega/\omega_n)^2 + 2\zeta j\omega/\omega_n + 1}\right|$ vs. $\log_{10}\omega/\omega_n$. (b) $\angle\dfrac{1}{(j\omega/\omega_n)^2 + 2\zeta j\omega/\omega_n + 1}$ vs. $\log\omega/\omega_n$.

Phase Plot

The phase diagram is a plot of

$$\angle G(j\omega) = \angle \frac{1}{(j\omega/\omega_n)^2 + 2\zeta j\omega/\omega_n + 1} = -\tan^{-1}\left(\frac{2\zeta(\omega/\omega_n)}{1-(\omega/\omega_n)^2}\right) \quad \text{vs. } \log_{10}\omega/\omega_n.$$

Figure 11.18 indicates how $\angle G(j\omega)$ is computed.

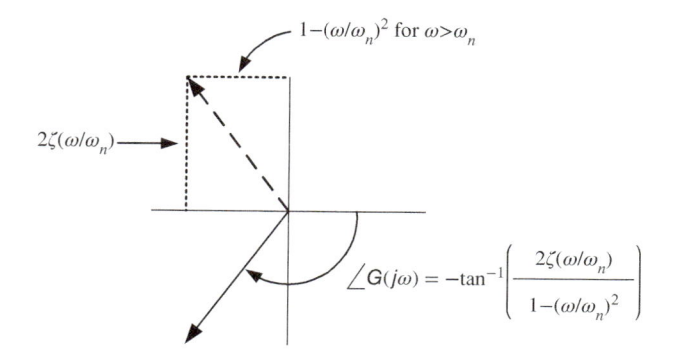

Figure 11.18. $\angle G(j\omega) = -\tan^{-1}\left(\frac{2\zeta(\omega/\omega_n)}{1-(\omega/\omega_n)^2}\right)$.

We next make a table of values.

ω	$\log_{10}\omega$	$\angle G(j\omega) = -\tan^{-1}\left(\frac{2\zeta(\omega/\omega_n)}{1-(\omega/\omega_n)^2}\right)$
0	$-\infty$	0
$0.1\omega_n$	$-1 + \log_{10}\omega_n$	$-\tan^{-1}\left(\frac{2\zeta/10}{1-1/100}\right) \approx -\tan^{-1}(\zeta/5)$
ω_n	$\log_{10}\omega_n$	$-\tan^{-1}\left(\frac{2\zeta}{0}\right) \approx -90°$
$10\omega_n$	$1 + \log_{10}\omega_n$	$-\tan^{-1}\left(\frac{20\zeta}{1-100}\right) \approx -180° + \tan^{-1}(\zeta/5)$
∞	∞	$-180°$

The phase plot with $\omega_n = 1$ and $\zeta = 0.1$ is shown in Figure 11.19. Figure 11.17b shows the Bode phase plots for $\zeta = 0.1, 0.2, 0.3, 0.5, 0.7, 1.0$.

Bode phase diagram of $G(s) = \omega_n^2/(s^2 + 2\zeta\omega_n s + \omega_n^2)$

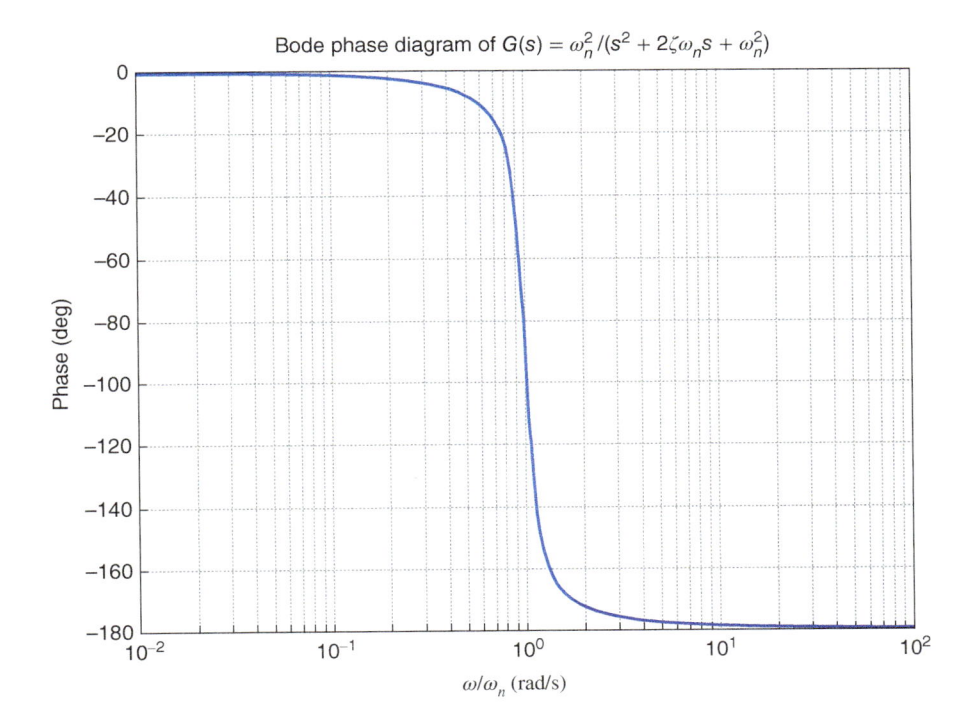

Figure 11.19. $\angle G(j\omega) = -\tan^{-1}\left(\dfrac{2\zeta(\omega/\omega_n)}{1 - (\omega/\omega_n)^2}\right)$ vs. $\log_{10}\omega/\omega_n$ with $\zeta = 0.1$.

Example 5 $G(j\omega) = \dfrac{j\omega/3 + 1}{(j\omega)^2/2 + j\omega/2 + 1}$

We consider the Bode diagram of

$$G(j\omega) = \dfrac{j\omega/3 + 1}{(j\omega)^2/2 + j\omega/2 + 1}.$$

There is a breakpoint at 3 due to the zero. We have to match up

$$\dfrac{1}{(j\omega)^2/2 + j\omega/2 + 1} = \dfrac{1}{(j\omega/\omega_n)^2 + 2\zeta j\omega/\omega_n + 1}$$

which requires $\omega_n^2 = 2$, $2\zeta/\omega_n = 1/2$ or $\omega_n = \sqrt{2}$, $\zeta = 1/(2\sqrt{2}) = 0.35$.

As shown in Figure 11.20, with $\omega > \sqrt{2}$ (double pole breakpoint), the magnitude decreases at -40 dB/decade until the (zero) breakpoint is reached and then it decreases at only -20 dB/decade. The phase diagram of $\dfrac{j\omega/3 + 1}{(j\omega)^2/2 + j\omega/2 + 1}$ is shown in Figure 11.21.

382   11   Frequency Response Methods

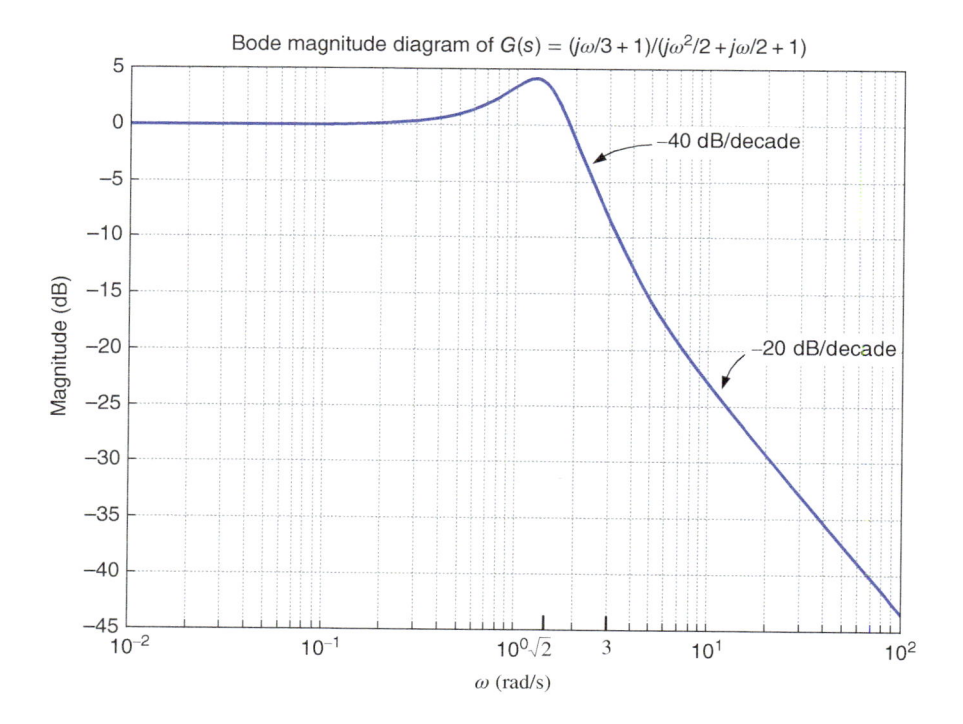

Figure 11.20. $20\log_{10}\left|\dfrac{j\omega/3+1}{(j\omega)^2/2+j\omega/2+1}\right|$ vs. $\log_{10}\omega$.

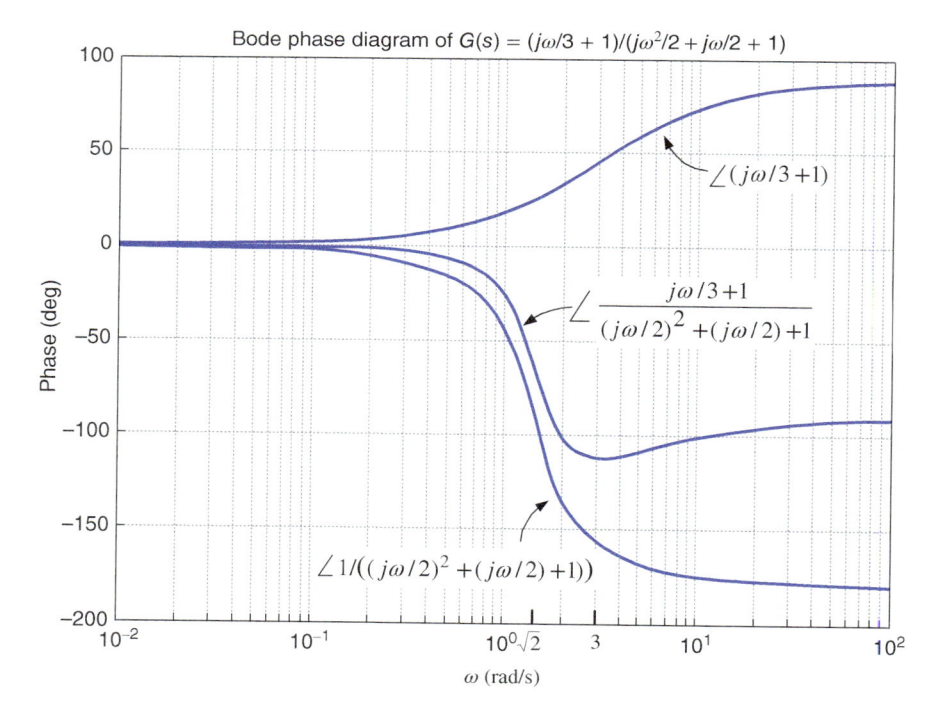

Figure 11.21. $\angle\dfrac{j\omega/3+1}{(j\omega)^2/2+j\omega/2+1}$ vs. $\log_{10}\omega$.

11.2 Nyquist Theory

Nyquist theory is concerned with stability. It allows us to check closed-loop stability of a system by inspection of the open-loop Bode diagram. H. Nyquist [48] developed this fundamentally important test for stability. What makes Nyquist theory so important is that it provides a measure of the *relative* stability of the closed-loop system. By this is meant that it can be used to find out if the controller will keep the closed-loop system stable despite uncertainty in the open-loop model $G(s)$ that was used to design the controller.

To develop the Nyquist stability test, we first need to understand the *principle of the argument* for complex rational functions. We do this by looking at a series of examples.

Principle of the Argument

Let
$$G(s) = s + 1 = s - (-1) = |s+1|e^{j\angle(s+1)}$$

and consider the curve $\mathcal{C}$ shown on the left side of Figure 11.22. The right side of the same figure is a plot of $G(s)$ as "s" goes around $\mathcal{C}$. In more detail, let "s" travel around $\mathcal{C}$ once in the *clockwise* direction so it successively goes through s_0, s_1, s_2, s_3 and back to s_0 with $\angle(s_i+1) = 0, -\pi/2, -\pi, -3\pi/2, -2\pi$, respectively. The image $G(s)$ goes around the *origin* once in the *clockwise* direction where successively $\angle G(s_i) = 0, -\pi/2, -\pi, -3\pi/2, -2\pi$.

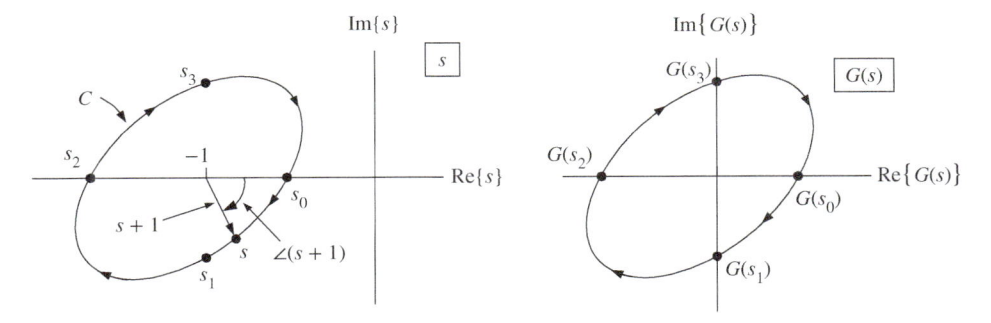

Figure 11.22. $G(s) = s + 1 = s - (-1) = |s+1|e^{j\angle(s+1)}$.

As a second example, let
$$G(s) = \frac{1}{s+1} = \frac{1}{|s+1|}e^{-j\angle(s+1)}$$

and consider the curve $\mathcal{C}$ shown on the left side of Figure 11.23. The right side of this figure is a plot of $G(s)$ as "s" goes around $\mathcal{C}$. Specifically, let "s" travel around $\mathcal{C}$ once in the *clockwise* direction so it successively goes through s_0, s_1, s_2, s_3 and back to s_0 with $\angle(s_i+1) = \angle(s_i+1) = 0, -\pi/2, -\pi, -3\pi/2, -2\pi$, respectively. The image $G(s)$ goes around the *origin* once in the *counterclockwise* direction where successively $\angle G(s_i) = 0, \pi/2, \pi, 3\pi/2, 2\pi$.

In this third example we again let
$$G(s) = s + 1 = s - (-1) = |s+1|e^{j\angle(s+1)},$$

but consider the closed curve $\mathcal{C}$ shown in Figure 11.24.

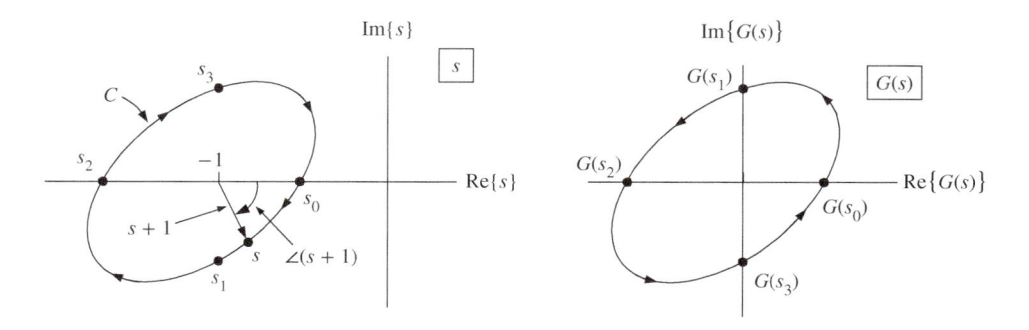

Figure 11.23. $G(s) = \dfrac{1}{s+1} = \dfrac{1}{|s+1|}e^{-j\angle(s+1)}$.

(a) (b)

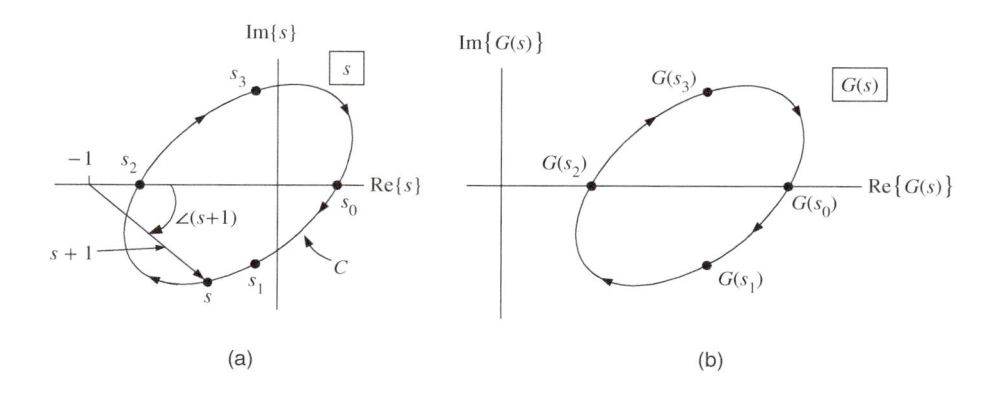

Figure 11.24. $G(s) = |s+1|e^{j\angle(s+1)}$.

Figure 11.24 is similar to that of Figure 11.22 except $s = -1$ is no longer *inside* the curve. The right side of the same figure is a plot of $G(s)$ as "s" goes around $\mathcal{C}$. As "s" travels around $\mathcal{C}$ once in the *clockwise* direction, it successively goes through s_0, s_1, s_2, s_3 and back to s_0 with $\angle(s_i + 1) = 0, -\pi/4, 0, \pi/4, 0$ respectively. The image $G(s)$ does *not* go around the *origin* as the angle $\angle(s+1)$ does not change by 2π as "s" goes around $\mathcal{C}$.

As a more general example consider Figure 11.25 and the transfer function

$$G(s) = \frac{(s-z_1)(s-z_2)}{(s-p_1)(s-p_2)} = \frac{|s-z_1|e^{j\angle(s-z_1)}|s-z_2|e^{j\angle(s-z_2)}}{|s-p_1|e^{j\angle(s-p_1)}|s-p_2|e^{j\angle(s-p_2)}}$$

$$= \frac{|s-z_1||s-z_2|}{|s-p_1||s-p_2|} e^{j\angle(s-z_1)} e^{j\angle(s-z_2)} e^{-j\angle(s-p_1)} e^{-j\angle(s-p_2)}.$$

Note that z_2 is outside the curve, while z_1, p_1, p_2 are all inside the closed contour.

The following table gives the change in angle for each pole and zero of $G(s)$ as "s" goes around the contour.

11.2 Nyquist Theory

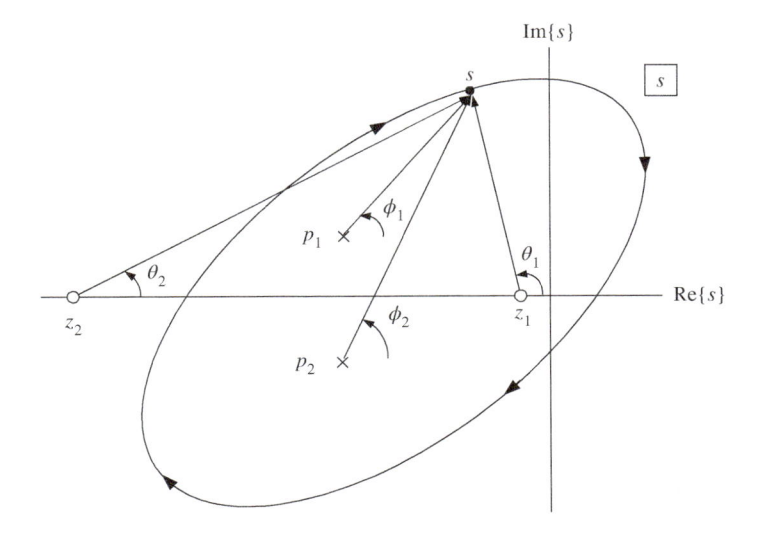

Figure 11.25. $G(s) = \dfrac{|s-z_1||s-z_2|}{|s-p_1||s-p_2|} e^{j\angle(s-z_1)} e^{j\angle(s-z_2)} e^{-j\angle(s-p_1)} e^{-j\angle(s-p_2)}$.

Pole/zero	Change in angle	Causes $G(s)$ to go around the origin
$\angle \dfrac{1}{s-p_1} = -\phi_1$	$+2\pi$	Once in the CCW direction
$\angle \dfrac{1}{s-p_2} = -\phi_2$	$+2\pi$	Once in the CCW direction
$\angle(s-z_1) = \theta_1$	-2π	Once in the CW direction
$\angle(s-z_2) = \theta_2$	0	Zero

From the table we see that as "s" goes around the closed curve once in the *clockwise* direction the image $G(s)$ goes around the origin once in the *counterclockwise* direction.

Theorem 1 *Principle of the Argument*
Let $G(s)$ be a rational function of s (ratio of two polynomials in s). For example

$$G(s) = \frac{(s-z_1)(s-z_2)}{(s-p_1)(s-p_2)}.$$

Let $\mathcal{C}$ be a simple closed curve[4] in the complex plane. Let s go around the curve $\mathcal{C}$ once in the clockwise direction. Then the number of times N the image $G(s)$ of the curve goes around the origin in the clockwise direction is

$$N = Z - P,$$

where Z is the number of zeros *inside* the closed curve $\mathcal{C}$ and P is the number of poles *inside* the closed curve $\mathcal{C}$.

Proof. The general proof is a straightforward generalization of that given for the example in Figure 11.24. ∎

---
[4] The terminology "simple" means that the curve $\mathcal{C}$ does not cross itself, e.g., it is not a figure 8.

Example 6 *Principle of the Argument*

Let
$$G(s) = \frac{1}{(s+1)(s+3)}$$

and choose the closed curve $\mathcal{C}$ to enclose the right half-plane as shown in Figure 11.26. For $\omega \geq 0$ we have

$$G(j\omega) = \frac{1}{(j\omega+1)(j\omega+3)} = \frac{(1-j\omega)(3-j\omega)}{(1+\omega^2)(9+\omega^2)} = \frac{3-\omega^2-j4\omega}{(1+\omega^2)(9+\omega^2)}.$$

We calculate

$$G(j0) = 1/3$$

$$G(j\sqrt{3}) = \frac{-j4\sqrt{3}}{(4)(12)} = -j0.144$$

$$G(j\omega) \approx -\frac{1}{\omega^2} \text{ for } \omega \text{ large.}$$

On the semicircle part of the curve $\mathcal{C}$ we have $s = Re^{j\theta}$ with $-\pi/2 \leq \theta \leq \pi/2$ and we may write

$$G(Re^{j\theta}) = \frac{1}{(Re^{j\theta}+1)(Re^{j\theta}+3)} \approx \frac{1}{R^2}e^{-j2\theta} \text{ as } R \text{ is large.}$$

Let's now make a table of values of $G(Re^{j\theta})$ for $-\pi/2 \leq \theta \leq \pi/2$.

θ	$G(Re^{j\theta}) \approx e^{-j2\theta}/R^2$
$\pi/2$	$e^{-j\pi}/R^2 = -1/R^2$
$\pi/4$	$e^{-j\pi/2}/R^2 = -j/R^2$
0	$e^{j0}/R^2 = 1/R^2$
$-\pi/4$	$e^{j\pi/2}/R^2 = j/R^2$
$-\pi/2$	$e^{j\pi}/R^2 = -1/R^2$

Finally, for $\omega \leq 0$, the image $G(j\omega)$ is simply the complex conjugate of the image for $\omega \geq 0$.

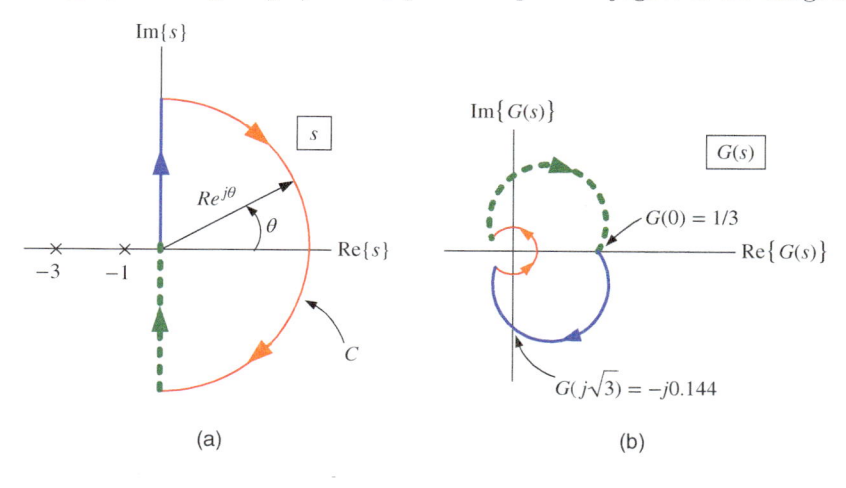

Figure 11.26. $G(s) = \dfrac{1}{(s+1)(s+3)}$.

11.2 Nyquist Theory

We now apply the principle of the argument to this example. There are no poles or zeros of $G(s)$ *inside* the closed curve $\mathcal{C}$ so $P = Z = 0$. Further, by inspection of the right side of Figure 11.26 we see the image of $G(s)$ on this curve does *not* go around the origin so $N = 0$. Thus we have directly verified that $N = Z - P$.

Example 7 *Principle of the Argument*
We again consider the transfer

$$G(s) = \frac{1}{(s+1)(s+3)},$$

but we now choose the closed curve $\mathcal{C}$ to enclose the left half-plane as shown in Figure 11.27.

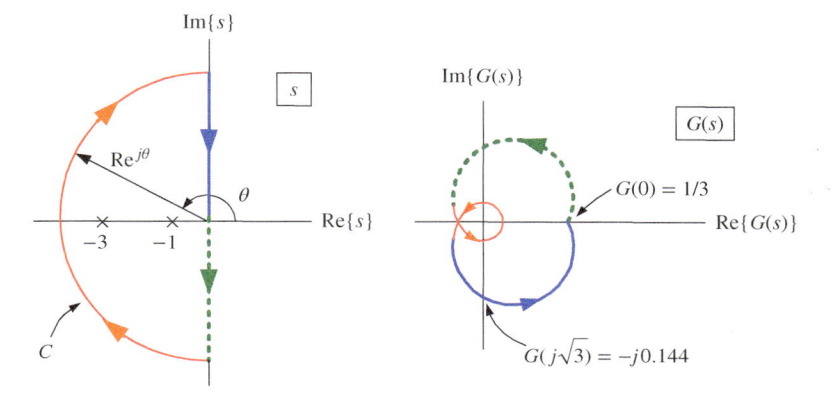

Figure 11.27. Nyquist polar plot of $\frac{1}{(s+1)(s+3)}$.

For $-\infty < \omega < \infty$ the image $G(j\omega)$ is the same as Example 6 except with the direction arrows reversed as we are going in the opposite direction along the $j\omega$ axis compared with the previous example. On the semicircle part of the curve $\mathcal{C}$, we have $s = Re^{j\theta}$ with $-\pi/2 \le \theta \le -3\pi/2$, and we may write

$$G(Re^{j\theta}) = \frac{1}{(Re^{j\theta}+1)(Re^{j\theta}+3)} \approx \frac{1}{R^2}e^{-j2\theta} \text{ as } R \text{ is large.}$$

We make a table of values of $G(Re^{j\theta})$ for $-\pi/2 \le \theta \le -3\pi/2$.

θ	$G(Re^{j\theta}) \approx e^{-j2\theta}/R^2$
$-\pi/2$	$e^{j\pi}/R^2 = -1/R^2$
$-3\pi/4$	$e^{j3\pi/2}/R^2 = -j/R^2$
$-\pi$	$e^{j2\pi}/R^2 = 1/R^2$
$-5\pi/4$	$e^{j5\pi/2}/R^2 = j/R^2$
$-3\pi/2$	$e^{j3\pi}/R^2 = -1/R^2$

We now apply the principle of the argument to this example. There are two poles of $G(s)$ *inside* the closed curve $\mathcal{C}$ and no zeros so $P = 2, Z = 0$. Further, by inspection of the right side of Figure 11.26, we see the image of $G(s)$ along this curve *does* go around the origin twice in the *counterclockwise* direction so $N = -2$. Thus $N = Z - P$ as both sides equal -2.

Example 8 *Principle of the Argument*
Let
$$G(s) = \frac{10(s+1)}{s(s-10)} = -\frac{s+1}{s(-s/10+1)}.$$

This transfer function has a pole at $s = 10$ and $s = 0$. As shown in Figure 11.28, the closed curve $\mathcal{C}$ is taken to enclose the right half-plane. To avoid the pole at $s = 0$ we take the path to be $s = re^{j\theta}$ with $-\pi/2 \leq \theta \leq \pi/2$ and then we let $r \to 0$. To enclose the right half-plane let $s = Re^{j\theta}$ with $-\pi/2 \leq \theta \leq \pi/2$ and $R \to \infty$.

For $s = j\omega$ we have

$$G(j\omega) = -\frac{j\omega+1}{j\omega(-j\omega/10+1)} = -\frac{(j\omega+1)(-j\omega)(j\omega/10+1)}{\omega^2(\omega^2/100+1)}$$

$$= \frac{-\omega^2(1+1/10) + j\omega(1-\omega^2/10)}{\omega^2(\omega^2/100+1)}$$

$$\to -(1+1/10) + \frac{j}{\omega} \text{ as } \omega \to 0.$$

Therefore
$$G(j\sqrt{10}) = -\frac{10(1+1/10)}{10(10/100+1)} = -1$$

$$G(j\omega) \approx -(1+1/10) + j/\omega \text{ for } 0 < \omega \ll 1$$

$$G(j\omega) \approx -j10/\omega \text{ for } \omega \gg 1.$$

For $s = re^{j\theta}$ we have
$$G(re^{j\theta}) = \frac{10(re^{j\theta}+1)}{re^{j\theta}(re^{j\theta}-10)} \approx \frac{10}{re^{j\theta}(-10)} = -e^{-j\theta}/r.$$

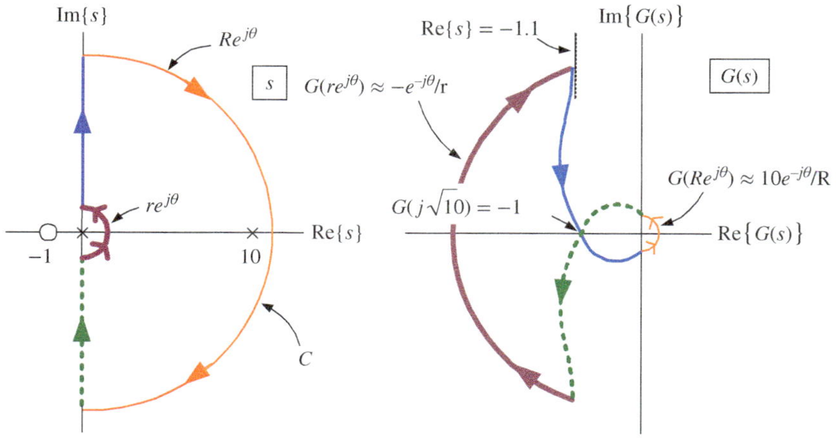

Figure 11.28. Nyquist polar plot of $G(s) = \dfrac{10(s+1)}{s(s-10)}$. The plot of $G(j\omega)$ (blue) is asymptotic to the vertical line $\text{Re}\{s\} = -(1/10+1)$ as $\omega \to 0$.

Making a table of values of $G(re^{j\theta})$ for $-\pi/2 \leq \theta \leq \pi/2$ gives

θ	$G(re^{j\theta}) \approx -e^{-j\theta}/r$
$-\pi/2$	$-e^{j\pi/2}/r = -j/r$
$-\pi/4$	$-e^{j\pi/4}/r$
0	$-1/r$
$+\pi/4$	$-e^{-j\pi/4}/r$
$+\pi/2$	$-e^{-j\pi/2}/r = j/r$

Finally, for $s = Re^{j\theta}$ with $-\pi/2 \leq \theta \leq \pi/2$, we have

$$G(Re^{j\theta}) = \frac{10(Re^{j\theta}+1)}{Re^{j\theta}(Re^{j\theta}-10)} \approx \frac{10Re^{j\theta}}{Re^{j\theta}Re^{j\theta}} = 10e^{-j\theta}/R.$$

A table of values is then

θ	$G(Re^{j\theta}) \approx 10e^{-j\theta}/R$
$+\pi/2$	$-j10/R$
0	$10/R$
$-\pi/2$	$j10/R$

Using these results we get the image $G(s)$ as shown on the right side of Figure 11.28.

We now apply the principle of the argument to this example. There is one pole of $G(s)$ *inside* the closed curve $\mathcal{C}$ and no zeros of $G(s)$ inside $\mathcal{C}$. So $P=1, Z=0$. Further, by inspection of the right side of Figure 11.28 we see the image of $G(s)$ on this curve *does* go around the origin *once* in the *counterclockwise* direction so $N=-1$. Thus $N = Z - P$ as both sides equal -1.

Nyquist Polar Plots

Definition 1 *Nyquist Contour*
The Nyquist contour is a simple closed-curve that encloses the right half-plane and has its direction of travel oriented to be in the clockwise direction.

Example 9 *Nyquist Contour*
Figure 11.29 shows two examples of Nyquist contours. In both examples we let $R \to \infty$. The example on the right side shows the case of an open-loop pole on the $j\omega$ axis at the origin, which is bypassed using a semicircular detour to the right for which we let $r \to 0$.

390   11 Frequency Response Methods

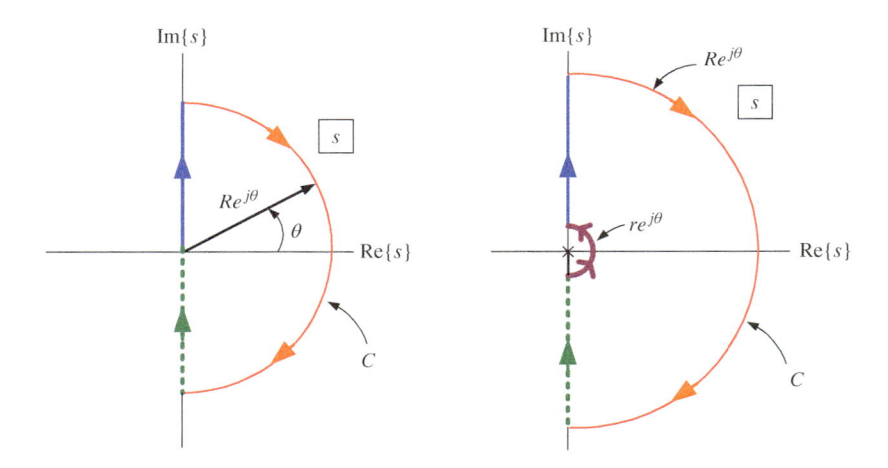

Figure 11.29. Examples of Nyquist contours.

Definition 2 *Nyquist Polar Plot*
The Nyquist polar plot is a plot of $G(s)$ as s goes around the Nyquist contour in a clockwise fashion.

Example 10 *Nyquist Polar Plot*
Let $G(s) = \dfrac{1}{(s+1)(s+3)}$ as in Example 6, which has the polar plot shown on the right side of Figure 11.30. This is the same as Figure 11.26 except we have set $R = \infty$ so that the image of $s = Re^{j\theta}$ is mapped to the origin, that is, $G(Re^{j\theta})|_{R=\infty} = 0$.

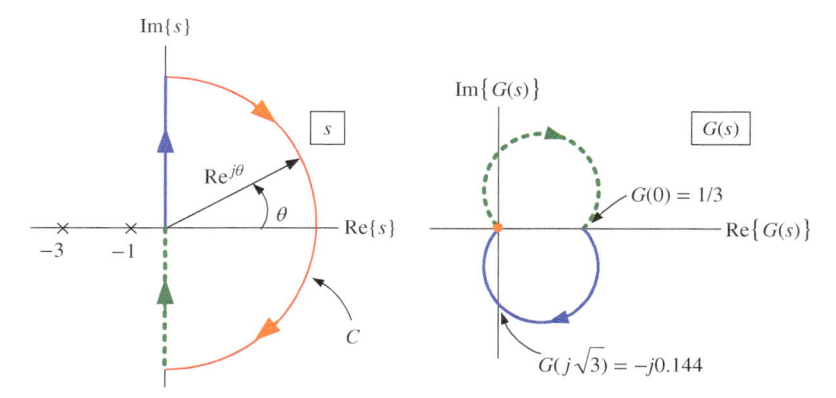

Figure 11.30. Nyquist polar plot of $G(s) = \dfrac{1}{(s+1)(s+3)}$.

A key observation of the Nyquist polar plot is that the part due to the $j\omega$ axis can be drawn from the Bode diagram of $G(j\omega)$, which is given in Figure 11.31.

See the Appendix to this chapter entitled *Bode and Nyquist Plots in Matlab*, which gives the MATLAB code to graph polar plots. The MATLAB command `nyquist` only does the part of the polar plot corresponding to $G(j\omega)$ for $-\infty < \omega < \infty$. (If $G(s)$ has a pole at $s = 0$ the `nyquist` command will not draw the part of the polar plot corresponding to $G(re^{j\theta})$ with $r \to 0$.)

11.2 Nyquist Theory 391

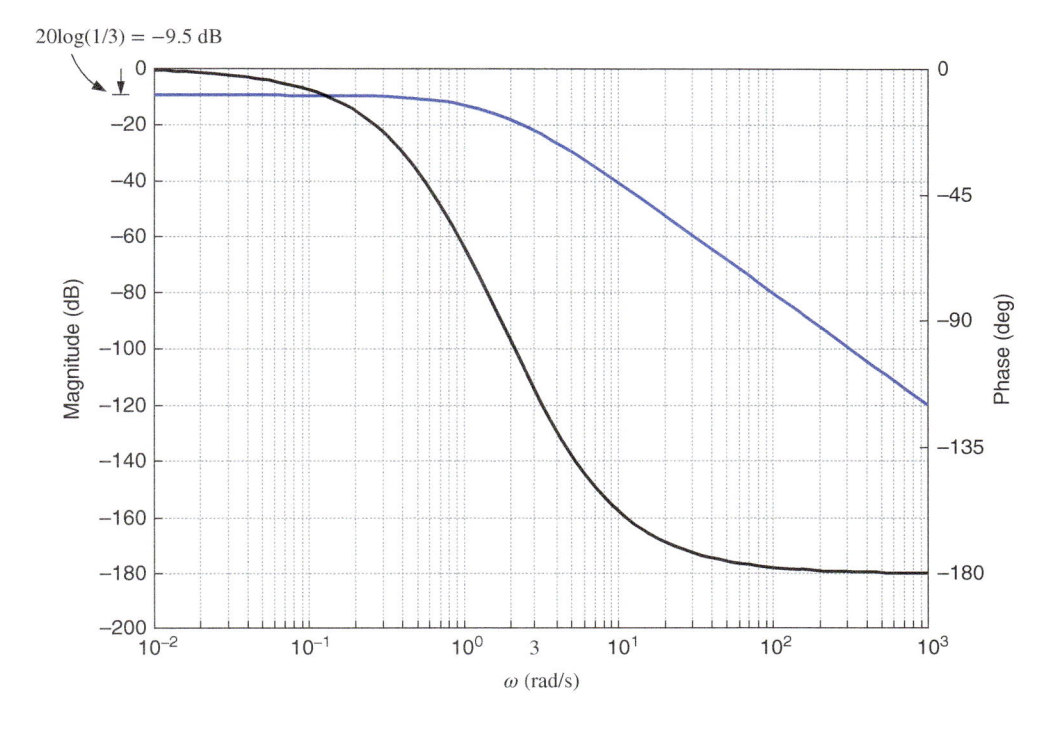

Figure 11.31. Bode diagram of $G(s) = \dfrac{1}{(s+1)(s+3)}$.

Example 11 *Nyquist Polar Plot*

Let $G(s) = \dfrac{10(s+1)}{s(s-10)} = -\dfrac{s+1}{s(-s/10+1)}$ as in Example 8.

Figure 11.32 is the same as Figure 11.28 except we have set $R = \infty$ as the image of $s = Re^{j\theta}$ is mapped to the origin, that is, $G(Re^{j\theta})|_{R=\infty} = 0$. As pointed out in Example 10, the part of the Nyquist polar plot due to the $j\omega$ axis can be sketched using the Bode diagram of $G(s)$, which is given in Figure 11.33. As explained in the appendix use the MATLAB command `nyquist` to draw the polar plot and compare it with the sketch in Figure 11.32.

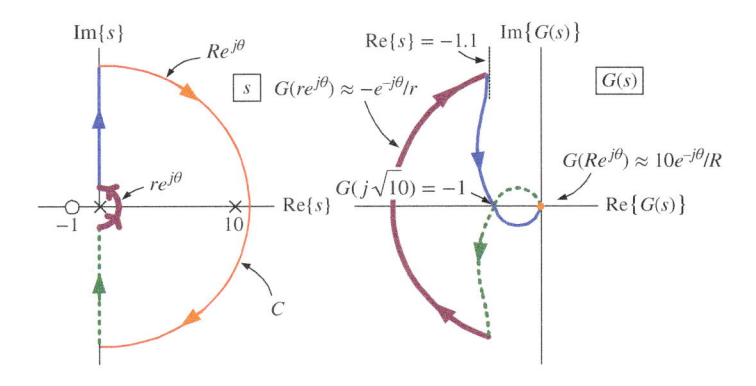

Figure 11.32. Nyquist polar plot of $G(s) = \dfrac{10(s+1)}{s(s-10)}$.

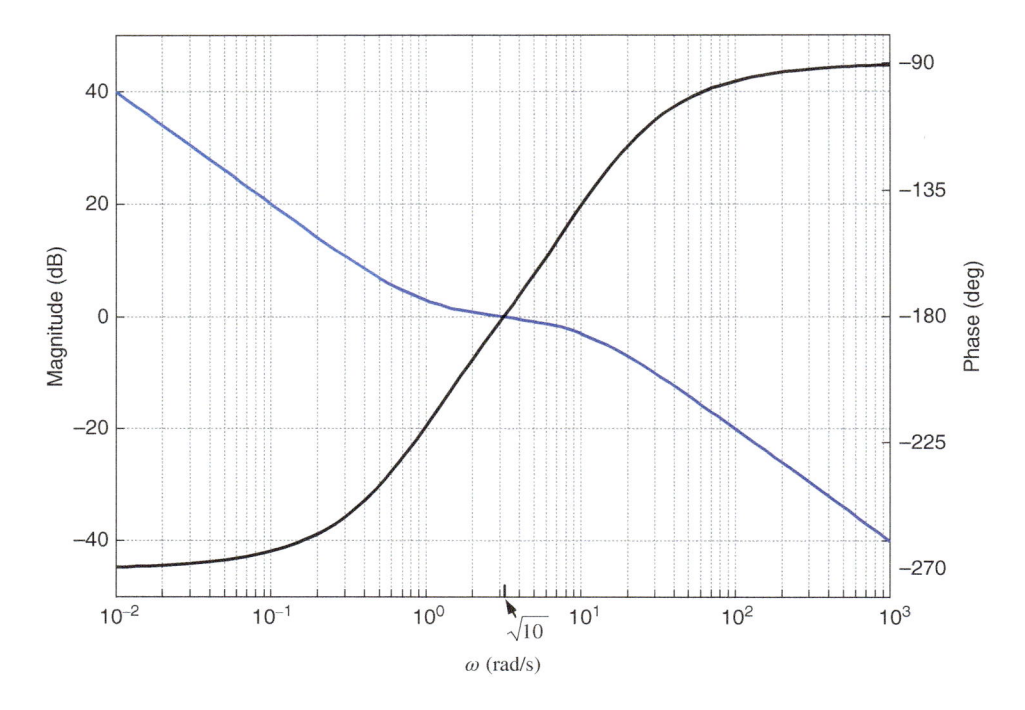

Figure 11.33. Bode diagram of $G(s) = \dfrac{10(s+1)}{s(s-10)}$, $G(j\sqrt{10}) = -1$.

Nyquist Test for Stability

Let the open-loop transfer function be $G(s) = \dfrac{1}{(s+1)(s+3)}$ and let's consider a simple proportional control in a unity feedback system as shown in Figure 11.34.

The closed-loop transfer function is

$$\frac{C(s)}{R(s)} = \frac{KG(s)}{1+KG(s)} = \frac{\dfrac{K}{(s+1)(s+3)}}{1+\dfrac{K}{(s+1)(s+3)}} = \frac{K}{(s+1)(s+3)+K}.$$

We see that the values of s for which

$$1+KG(s) = 1 + \frac{K}{(s+1)(s+3)} = \frac{(s+1)(s+3)+K}{(s+1)(s+3)} = 0$$

are the *closed-loop poles*. Thus the closed-loop system is *stable* if and only if

$$1+KG(s) \neq 0 \text{ for } \operatorname{Re}\{s\} \geq 0$$

Figure 11.34. Proportional feedback control system.

11.2 Nyquist Theory

or, equivalently,

$$\frac{1}{K} + G(s) \neq 0 \text{ for } \operatorname{Re}\{s\} \geq 0.$$

The Nyquist stability test is carried out by applying the principle of the argument to

$$\frac{1}{K} + G(s).$$

Specifically, let the curve $\mathcal{C}$ enclose the right half-plane as shown on the left side of Figure 11.35. The corresponding polar plot is shown on the right side of Figure 11.35. We have let $R \to \infty$ in order to enclose the complete right half-plane inside $\mathcal{C}$ and therefore $G(Re^{j\theta})|_{R\to\infty}$ is mapped to the origin on the right side of Figure 11.35.

With $K > 0$, we next shift the polar plot to the right by $1/K$ to obtain the plot of Figure 11.36.

By inspection we see that the poles of $1/K + G(s)$ are just the poles of $G(s)$ and they are not inside the curve $\mathcal{C}$. Further, with $K > 0$, we see that the image $1/K + G(s)$ does not go around the origin so $N = 0$. By the principle of the argument we have

$$N = Z - P.$$

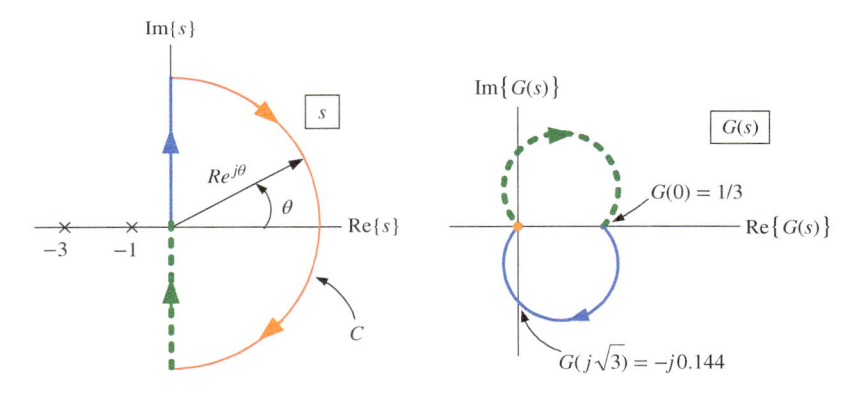

Figure 11.35. Nyquist plot of $G(s) = \dfrac{1}{(s+1)(s+3)}$.

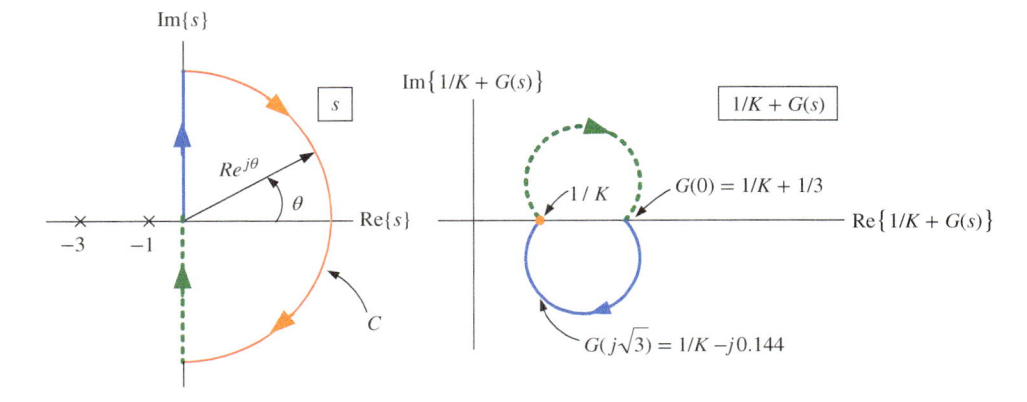

Figure 11.36. Nyquist plot of $1 + KG(s)$ for $K > 0$.

394   11   Frequency Response Methods

Consequently
$$Z = N + P = 0 + 0 = 0.$$

So $1/K + G(s)$ has no zeros in the right half-plane and therefore the closed-loop system is stable.

There is a slightly easier way to do the test as indicated in Figure 11.37. We simply observe that $1/K + G(s)$ goes around the origin if and only if $G(s)$ goes around $-1/K$. Thus, as shown in Figure 11.37, we draw the Nyquist plot of $G(s)$ and then mark the $-1/K$ point on the plot. We see that with $K > 0$ the polar plot of $G(s)$ does not go around $-1/K$ and thus $N = 0$.

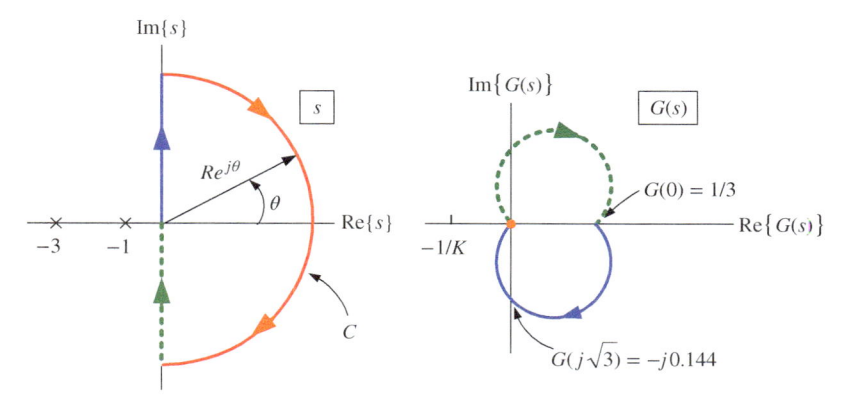

Figure 11.37. Nyquist plot of $1/K + G(s)$ for $K > 0$.

Example 12 *Nyquist Stability Test [3]*

Let
$$G(s) = \frac{10(s+1)}{s(s-10)} = -\frac{s+1}{s(-s/10+1)}.$$

As always we take the curve $\mathcal{C}$ to enclose the right half-plane as shown on the left side of Figure 11.38. The image of $G(s)$ as $\mathcal{C}$ is traversed once is shown on the right side of Figure 11.38. Figure 11.38 is the same as Figure 11.28 except we have set $R = \infty$ so $G(Re^{j\theta}) = 0$. The Nyquist stability test is carried out by applying the principle of the argument to
$$\frac{1}{K} + G(s).$$

The polar plot is shown on the right side of Figure 11.38. $G(s)$ and therefore $1/K + G(s)$ has one pole inside $\mathcal{C}$ at $s = 10$. Therefore $P = 1$. The number of times $1/K + G(s)$ goes around the origin is the same as the number of times $G(s)$ goes around $-1/K$. We break up the Nyquist test into two cases:

Case (1): As shown on the right side of Figure 11.38, consider $-\infty < -1/K < -1$ ($0 < K < 1$). Then $N = 1$ as $G(s)$ goes around $-1/K$ once in the clockwise direction.

The principle of the argument tells us that
$$N = Z - P$$

so
$$Z = N + P = 1 + 1 = 2$$

and therefore $1/K + G(s) = 0$ has two zeros in the right half-plane. The closed-loop system is unstable for $0 < K < 1$.

11.2 Nyquist Theory

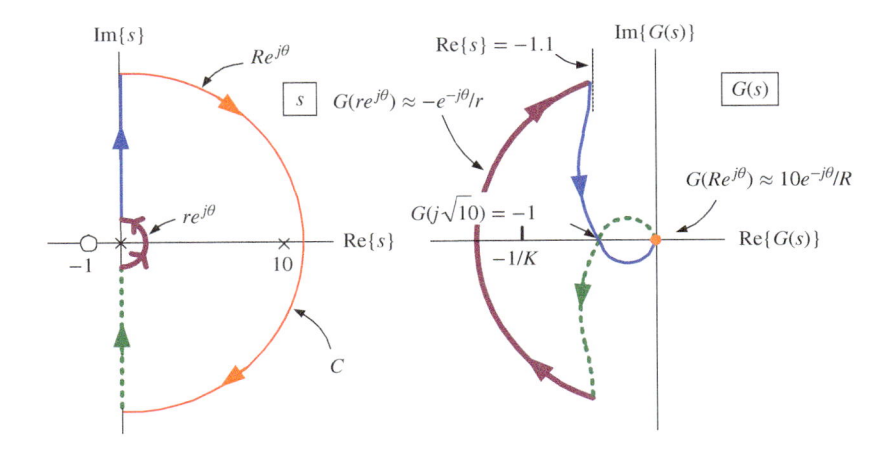

Figure 11.38. Nyquist plot of $1/K + G(s)$ for $0 < K < 1$ or $-\infty < -1/K < -1$.

Case (2): Now consider $-1 < -1/K < 0$ ($K > 1$) as shown in Figure 11.39. Then $N = -1$ as $G(s)$ goes around $-1/K$ once in the counterclockwise direction. By the principle of the argument $N = Z - P$ so

$$Z = N + P = -1 + 1 = 0.$$

Therefore $1/K + G(s) = 0$ has no zeros in the right half-plane. The closed-loop system is stable for $K > 1$.

Check: Apply the Routh–Hurwitz test to $s(s-10) + 10K(s+1) = s^2 + 10(K-1)s + 10K = 0$.

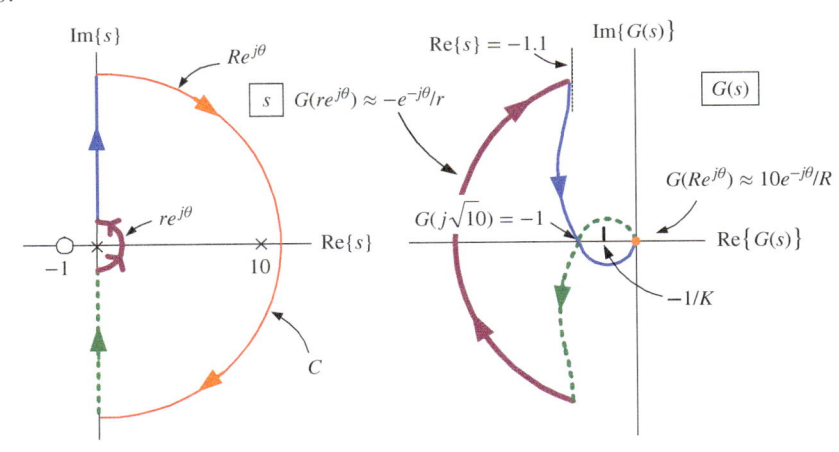

Figure 11.39. Nyquist plot of $1/K + G(s)$ for $K > 1$ or $-1 < -1/K < 0$.

Remark The Nyquist stability test seems (because it is) much more complicated compared with the Routh–Hurwitz test. However, as explained shortly, the Nyquist test based on the polar plot will give us a measure of the relative stability of the closed-loop system.

Example 13 *Nyquist Stability Test*

In this example we take the open-loop system to be $G(s) = \dfrac{1}{(s+1)(s^2 + \sqrt{2}s + 1)}$ where we note that $s^2 + \sqrt{2}s + 1 = \left[s - (-1/\sqrt{2} + j/\sqrt{2})\right]\left[s - (-1/\sqrt{2} - j/\sqrt{2})\right]$. We want to know for what values of $K > 0$ are the roots of

$$1 + KG(s) = 0$$

396   11   Frequency Response Methods

all in the open left half-plane. Equivalently, for what values of $K > 0$ are the zeros of $\frac{1}{K} + G(s) = 0$ *not* in the right half-plane.

To proceed consider the closed curve $\mathcal{C}$ shown on the left side of Figure 11.40, which encloses the right half-plane. The right side of Figure 11.40 shows the image $G(s)$ as s goes around the curve $\mathcal{C}$. As $R \to \infty$ we see that $G(Re^{j\theta}) \to 0$. With $s = j\omega$ we use the Bode diagram given in Figure 11.41 to sketch $G(j\omega)$ for $\omega \geq 0$. In particular, the Bode diagram shows that $\angle G(j1.55) = -180°$, $20\log_{10}|G(j1.55)| = 20\log_{10}|-1/4.8| = -13.6$ dB or $G(j1.55) = -1/4.8$.

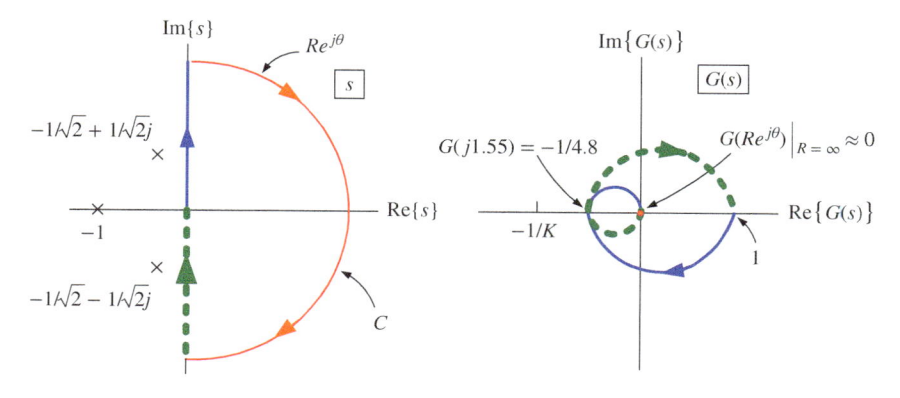

Figure 11.40. Nyquist plot of $G(s) = \dfrac{1}{(s+1)(s^2 + \sqrt{2}s + 1)}$.

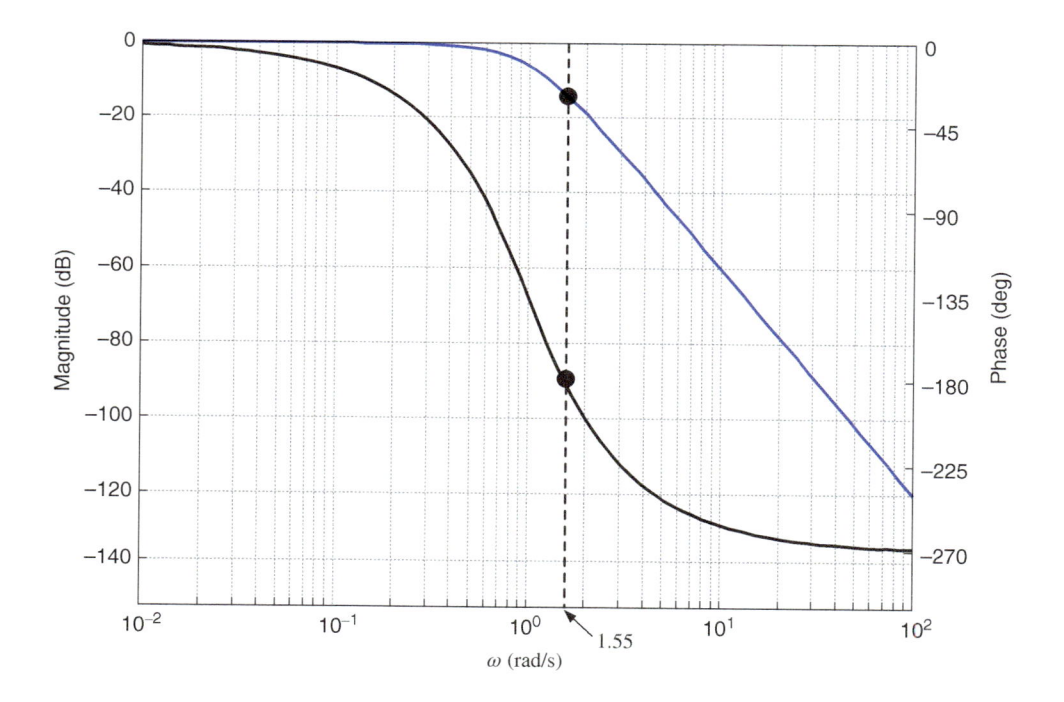

Figure 11.41. Bode diagram of $G(s) = \dfrac{1}{(s+1)(s^2 + \sqrt{2}s + 1)}$.

11.2 Nyquist Theory

Case (1): Consider $-1/K < -1/4.8$ ($0 < K < 4.8$). From the right side of Figure 11.40, we see that $N = 0$ and, as $P = 0$, we have

$$Z = N + P = 0.$$

Thus $1/K + G(s)$ has no zeros in the right half-plane for $0 < K < 4.8$ and the closed-loop system is stable.

Case (2): Next consider $-1/4.8 < -1/K$ ($K > 4.8$) as indicated in Figure 11.42. From the right side of Figure 11.42 we see that $N = 2$ and, as $P = 0$, we have

$$Z = N + P = 2 + 0 = 2.$$

Thus $1/K + G(s)$ has two zeros in the right half-plane for $K > 4.8$ and thus the closed-loop system is unstable.

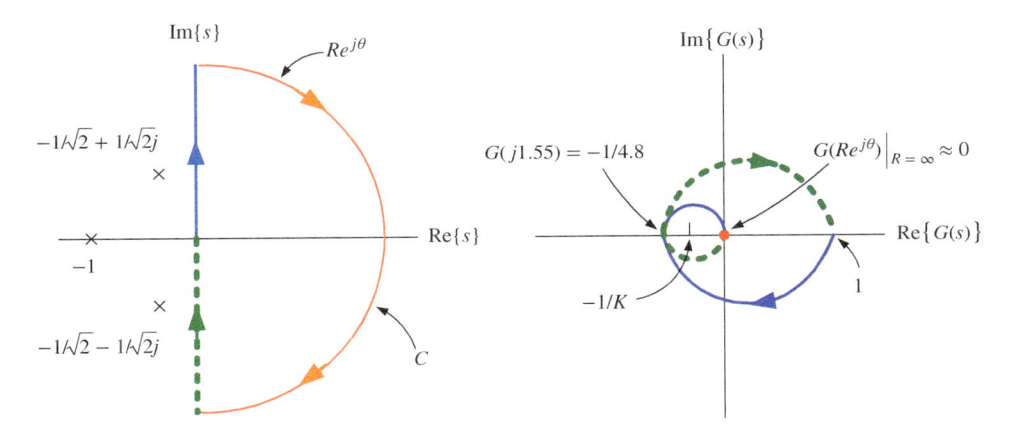

Figure 11.42. Nyquist plot of $G(s) = \dfrac{1}{(s+1)(s^2 + \sqrt{2}s + 1)}$.

Check: Apply the Routh–Hurwitz test to $(s+1)(s^2 + \sqrt{2}s + 1) + K = 0$.

Example 14 *Nyquist Stability Test*

In this example we take the open-loop system to be

$$G(s) = \frac{1}{s(s+1)^2}.$$

We want to know for which values of K does $1 + KG(s)$ have all of its zeros in the open left-half plane. Equivalently, we want to know for which values of K does $1/K + G(s)$ have *no* zeros in the right half-plane. The left side of Figure 11.43 shows the closed curve $\mathcal{C}$ which contains the right half-plane as we let $r \to 0$ and $R \to \infty$. As $R \to \infty$ we see that $G(Re^{j\theta}) \to 0$. With ω small we have

$$G(j\omega) = \frac{1}{j\omega(j\omega+1)^2} = \frac{(-j\omega+1)^2}{j\omega(j\omega+1)^2(-j\omega+1)^2} = \frac{-2\omega - j(1-\omega^2)}{\omega(\omega^2+1)^2} \approx -2 - \frac{j}{\omega}.$$

11 Frequency Response Methods

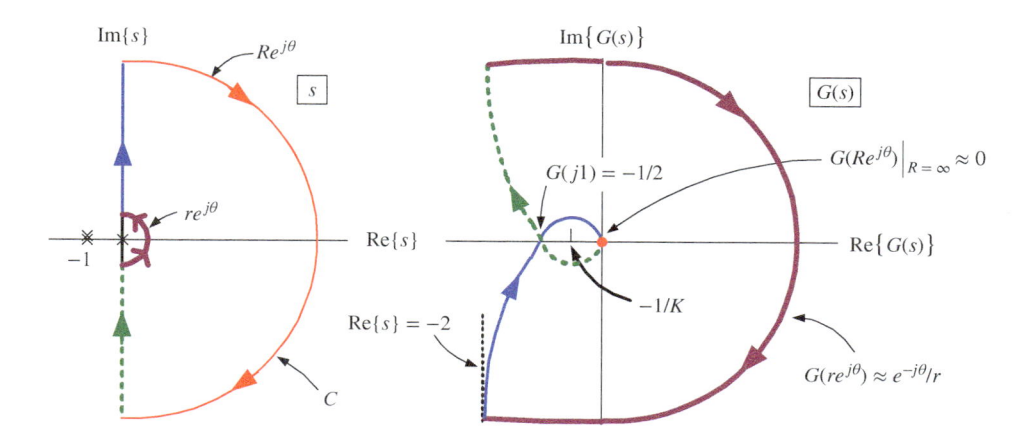

Figure 11.43. Nyquist plot of $G(s) = \dfrac{1}{s(s+1)^2}$.

The sketch of $G(j\omega)$ for $\omega > 0$ given on the right side of Figure 11.43 is done using the Bode diagram of Figure 11.44. Specifically, the Bode diagram shows that $G(j\omega)$ starts out with $\angle G(j\omega) \approx -90°$ and $|G(j\omega)| \to \infty$. At $\omega = 1$ we have $G(j1) = -1/2$ or $\angle G(j\omega) = -180°, 20\log_{10}|G(j1)| = 20\log_{10}(1/2) = -6$ dB. As $\omega \to \infty$ we have $\angle G(j\omega) \to -270°$ and $|G(j\omega)| \to 0$. To complete the plot of $G(s)$ we write

$$G(re^{j\theta}) = \frac{1}{re^{j\theta}(re^{j\theta}+1)^2} \approx \frac{1}{re^{j\theta}} = \frac{1}{r}e^{-j\theta}.$$

Making a table of values of $G(re^{j\theta})$ for $-\pi/2 \leq \theta \leq \pi/2$ gives

θ	$G(re^{j\theta}) \approx e^{-j\theta}/r$
$-\pi/2$	$e^{j\pi/2}/r = j/r$
$-\pi/4$	$e^{j\pi/4}/r$
0	$1/r$
$+\pi/4$	$e^{-j\pi/4}/r$
$+\pi/2$	$e^{-j\pi/2}/r = -j/r$

Putting this all altogether we obtain the Nyquist plot on the right side of Figure 11.43.

Case (1) Consider $-1/2 < -1/K < 0$ ($K > 2$) as shown in Figure 11.43. Then we see that $N = 2, P = 0$ so that

$$Z = N + P = 2$$

which tells us that $1/K + G(s)$ has two zeros in the right half-plane for $K > 2$.

Case (2) Next consider $-1/K < -1/2$ ($0 < K < 2$) as shown in Figure 11.45. Then we see that $N = 0, P = 0$ so that

$$Z = N + P = 0$$

which tells us that $1/K + G(s)$ has no zeros in the right half-plane for $0 < K < 2$. The closed-loop system is stable for $0 < K < 2$.

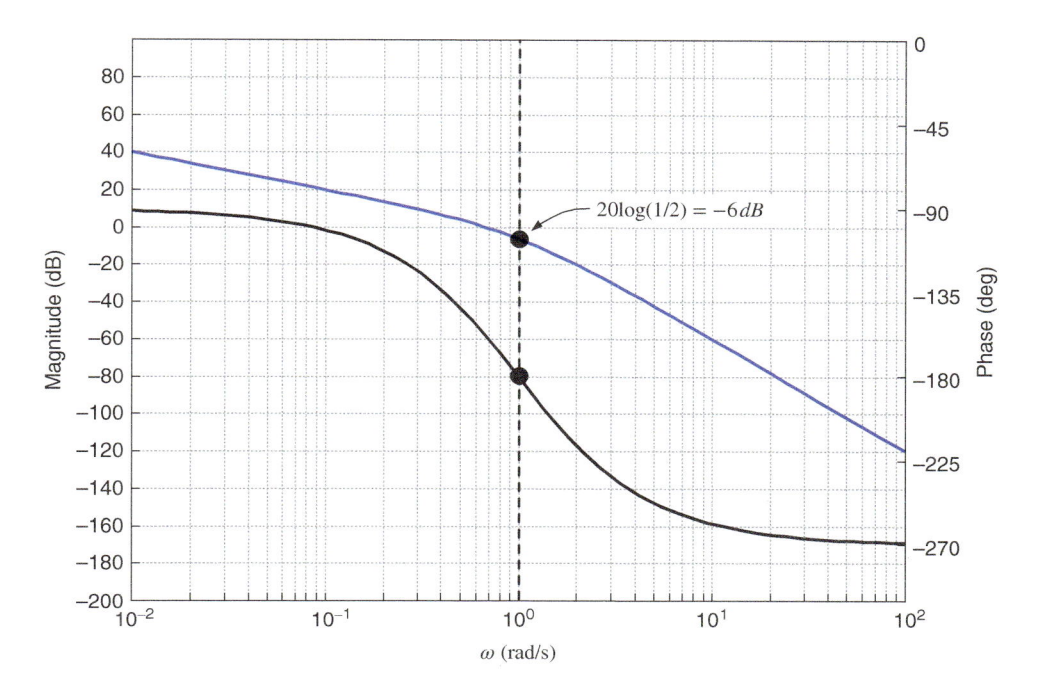

Figure 11.44. Bode diagram of $G(s) = \dfrac{1}{s(s+1)^2}$.

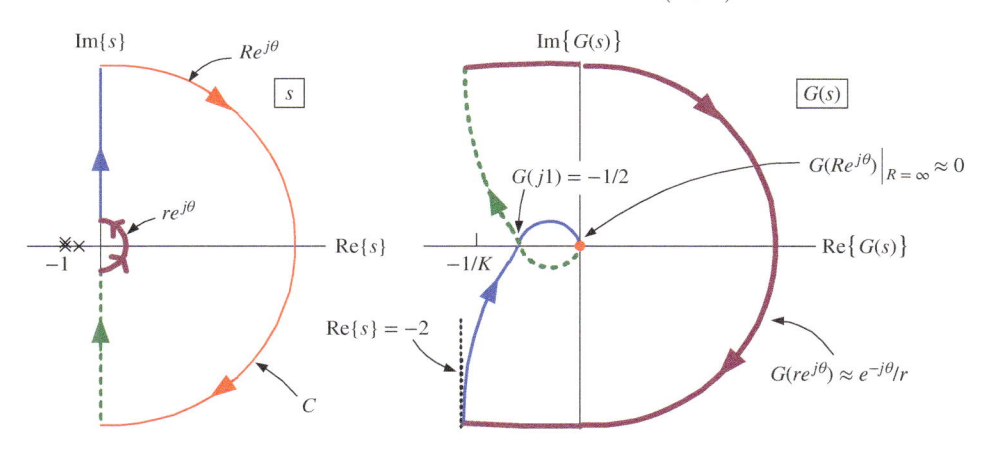

Figure 11.45. Nyquist plot of $G(s) = \dfrac{1}{s(s+1)^2}$.

Example 15 *Nyquist Stability Test [3]*

Let a system have an open-loop transfer function given by

$$G(s) = \frac{100(s/10 + 1)}{s(s-1)(s/100 + 1)} = -\frac{100(s/10 + 1)}{s(-s + 1)(s/100 + 1)}.$$

We want to determine the values of K for which the zeros of

$$1 + KG(s) = 0$$

are in the open left half-plane. Equivalently, we want to ensure that

$$1/K + G(s) = 0$$

has no zeros in the right half-plane. As usual, we choose a closed curve in the s plane that encloses the right half-plane as shown on the left side of Figure 11.46. We of course let $r \to 0$ and $R \to \infty$.

For $G(j\omega)$ we have

$$\begin{aligned} G(j\omega) &= -\frac{100(j\omega/10 + 1)}{j\omega(-j\omega + 1)(j\omega/100 + 1)} \\ &= -\frac{100(j\omega/10 + 1)(j\omega + 1)(-j\omega/100 + 1)}{j\omega(-j\omega + 1)(j\omega/100 + 1)(j\omega + 1)(-j\omega/100 + 1)} \\ &= \frac{-\left(109\omega + \frac{1}{10}\omega^3\right) + j\left(-\frac{89}{10}\omega^2 + 100\right)}{\frac{1}{10000}\omega^5 + \frac{10001}{10000}\omega^3 + \omega} \\ &\to -109 + j\frac{100}{\omega} \text{ as } \omega \to 0. \end{aligned}$$

We use the Bode diagram in Figure 11.47 to sketch $G(j\omega)$ for $\omega > 0$. For ω small we have $\angle G(j\omega) \approx 90°$ ($-90°$ due to the pole at $s = 0$ and $+180°$ due to the constant -1) and $|G(j\omega)| = \infty$. As ω increases, the magnitude $|G(j\omega)|$ decreases while the angle $\angle G(j\omega)$ increases to $180°$ at $\omega = 3.35$ with $|G(j3.35)| = 9$. As shown in Figure 11.46, as $\omega \to \infty$, the angle $\angle G(j\omega)$ ends up at $180°$ while $|G(j\omega)| \to 0$.

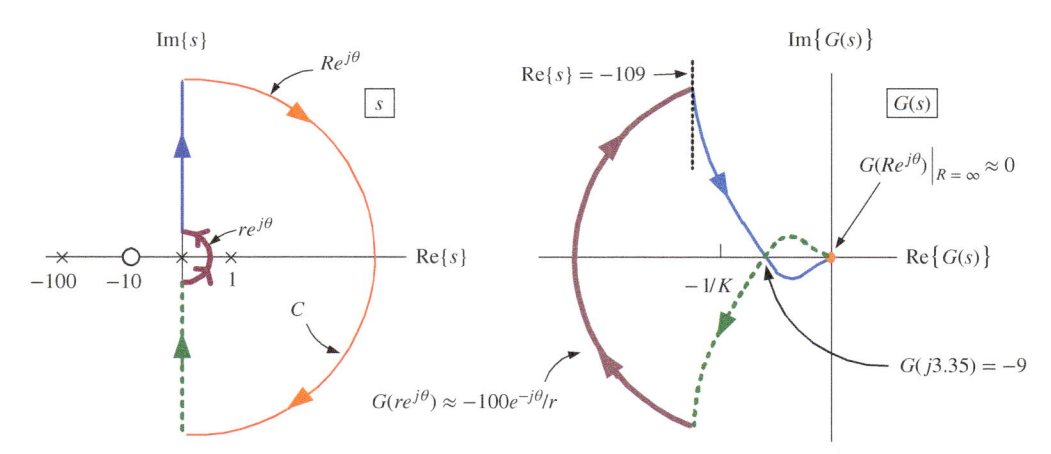

Figure 11.46. Nyquist plot of $G(s) = \dfrac{100(s/10 + 1)}{s(s-1)(s/100 + 1)}$.

For $s = Re^{j\theta}$ we see that $G(Re^{j\theta}) = 0$ for $R = \infty$. To go around the pole at the origin we set $s = re^{j\theta}$. Then

$$G(re^{j\theta}) = \frac{100(re^{j\theta}/10 + 1)}{re^{j\theta}(re^{j\theta} - 1)(re^{j\theta}/100 + 1)} \approx -\frac{100}{re^{j\theta}} = -\frac{100}{r}e^{-j\theta}.$$

11.2 Nyquist Theory

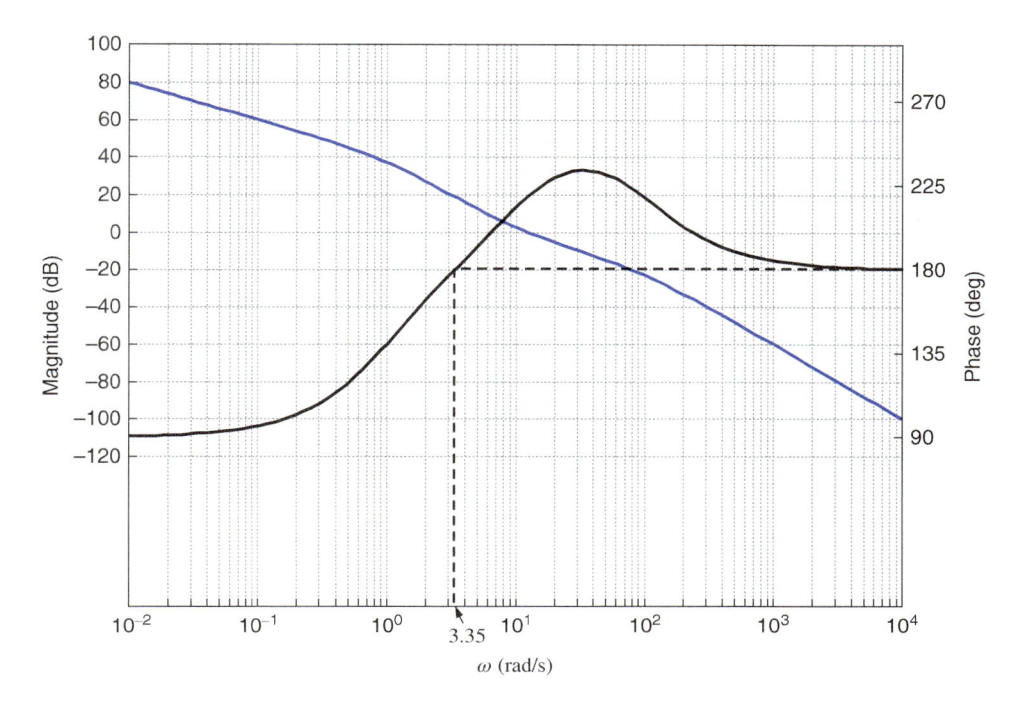

Figure 11.47. Bode plot of $G(s) = \dfrac{100(s/10+1)}{s(s-1)(s/100+1)}$, $G(j3.352) = -9$.

To plot $G(re^{j\theta}) \approx -100e^{-j\theta}/r$ for $-\pi/2 \leq \theta \leq \pi/2$ we make the following table of its values.

θ	$G(re^{j\theta}) \approx -100e^{-j\theta}/r$
$-\pi/2$	$-100e^{j\pi/2}/r = -j100/r$
$-\pi/4$	$-100e^{j\pi/4}/r$
0	$-100/r$
$\pi/4$	$-100e^{-j\pi/4}/r$
$\pi/2$	$-100e^{-j\pi/2}/r = j100/r$

This table is then used to complete the Nyquist plot as shown on the right side of Figure 11.46. We now consider two cases to check for closed-loop stability.

Case (1): Consider $-1/K < -9$ ($0 < K < 1/9$). Then $P=1$ (as there is an open-loop pole at 1) and $N=1$. Thus
$$Z = N + P = 2$$
showing that $1/K + G(s)$ has two zeros in the right half-plane. The closed-loop system is unstable for $0 < K < 1/9$.

Case (2): Next consider $-9 < -1/K < 0$ ($K > 1/9$). Then $P=1$ (as there is an open-loop pole at 1) and $N=-1$. Thus
$$Z = N + P = -1 + 1 = 0$$

showing that $1/K + G(s)$ has *no* zeros in the right half-plane. The closed-loop system is stable for $K > 1/9$.

11.3 Relative Stability: Gain and Phase Margins

We first observe the following equivalent ways of looking at the Nyquist encirclement criterion. Specifically, for $K > 0$, the following statements are equivalent:

(1) $G(j\omega), -\infty < \omega < \infty$ goes around the $-1/K + j0$ point N times.[5]
(2) $1/K + G(j\omega), -\infty < \omega < \infty$ goes around the origin N times.
(3) $1 + KG(j\omega), -\infty < \omega < \infty$ goes around the origin N times.
(4) $KG(j\omega), -\infty < \omega < \infty$ goes around the $-1 + j0$ point N times.

Remember that we are thinking of $G(s)$ as the open-loop transfer function in a feedback system as given in Figure 11.48.

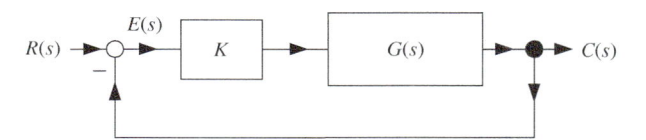

Figure 11.48. Unity feedback controller.

Statement (3) was the original formulation we used to describe the Nyquist stability test. This requires the most effort in terms of redrawing $1 + KG(j\omega)$ each time K changes. Then we observed statement (3) was equivalent to statement (2), which means we draw $G(j\omega), -\infty < \omega < \infty$ and then shift this plot by $1/K$. Finally we did the Nyquist stability tests using statement (1) because we just draw $G(j\omega), -\infty < \omega < \infty$ once and then shift around $-1/K + j0$. Statement (4) provides the most straightforward way to explain the notions of *gain and phase margins*, i.e., of relative stability. To do so, consider an open-loop system given by

$$G(s) = \frac{2}{(s+1)(s^2 + \sqrt{2}s + 1)},$$

which is the same as Example 13 except for the factor of 2 in the numerator. Figure 11.49 shows the Nyquist plot of $K \frac{2}{(s+1)(s^2 + \sqrt{2}s + 1)}$. In this example $P = 0$. As the right side of Figure 11.49 shows, the Nyquist plot does not encircle the $-1 + j0$ point for $0 < K < 2.4$ so $1 + KG(s) = 0$ does not have any zeros in the right half-plane. In words, the closed-loop system is stable for $0 < K < 2.4$. As the gain K can be increased up to 2.4 before the system becomes unstable, we say the *gain margin* is 2.4. Typically the gain margin is given in decibels which is $20 \log_{10} 2.4 = 7.6$ dB.

The right side of Figure 11.49 shows a dashed circle of radius 1 centered at the origin. A line from the origin to the point where this dashed circle intersects $G(j\omega)$ is also shown.

---

[5] $-1/K + j0$ is the same as $-1/K$. We write the term $j0$ to emphasize we are in the complex plane.

11.3 Relative Stability: Gain and Phase Margins

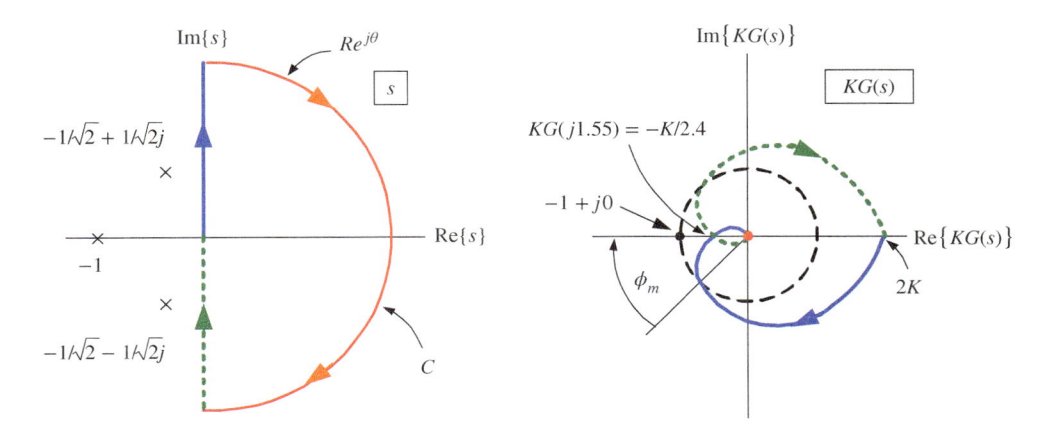

Figure 11.49. $KG(s) = K \dfrac{2}{(s+1)(s^2 + \sqrt{2}s + 1)}$, $G(j1.55) = -1/2.4$, $|G(j1)| = 1$.

The angle ϕ_m between this line and the negative real axis is called the *phase margin*. It is the amount the Nyquist plot can be rotated before it encloses the $-1 + j0$ point. We can also write

$$\phi_m = \angle G(j\omega_g) + 180°,$$

where ω_g is the value of ω for which $|G(j\omega_g)| = 1$. Note that in Figure 11.49 that $\angle G(j\omega_g) < 0$ which is the typical case.

We have the following definition.

Definition 3 *Gain Crossover Frequency ω_g*
A *gain crossover* frequency is a frequency ω_g at which

$$|G(j\omega_g)| = 1 \iff 20 \log |G(j\omega_g)| = 0.$$

Example 16 *Gain Crossover Frequency ω_g and the Phase Margin*

With reference to the Bode diagram of Figure 11.50 on the next page, we see that at the gain crossover frequency $\omega_g = 1$, $20 \log_{10} |G(j1)| = 0$ dB and $\angle G(j1) = -135°$ so that phase margin is

$$\phi_m = \angle G(j\omega_g) + 180° = -135° + 180° = 45°$$

We can also obtain the gain margin from the open-loop Bode diagram. We first make the definition of the phase crossover frequency.

Definition 4 *Phase Crossover Frequency ω_ϕ*
A *phase crossover* frequency is a frequency ω_ϕ at which $\angle G(j\omega_\phi) = -180°$.

Example 17 *Gain and Phase Margins*

With reference to the Bode diagram of Figure 11.50 we see that the phase crossover frequency is $\omega_\phi = 1.55$ and $20 \log_{10} |G(j1.55)| = 20 \log_{10} |-1/2.4| = -7.6$ dB. So the gain margin is 7.6 dB.

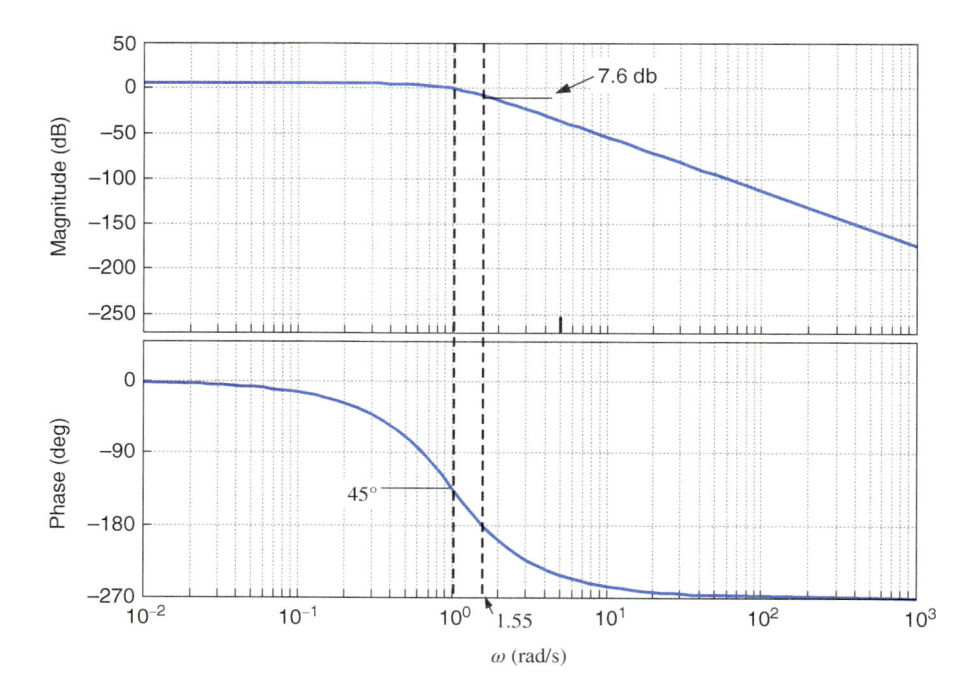

Figure 11.50. $G(s) = \dfrac{2}{(s+1)(s^2 + \sqrt{2}s + 1)}$. $\angle G(j1.55) = 180°$ and $20 \log_{10}|G(j1)| = 0$.

Example 18 *Gain and Phase Margins*

Let's consider again Example 14 where the open-loop transfer function is $G(s) = \dfrac{1}{s(s+1)^2}$. Figure 11.51 is the Nyquist plot of $KG(s)$ (not just $G(s)$ as in Example 14).

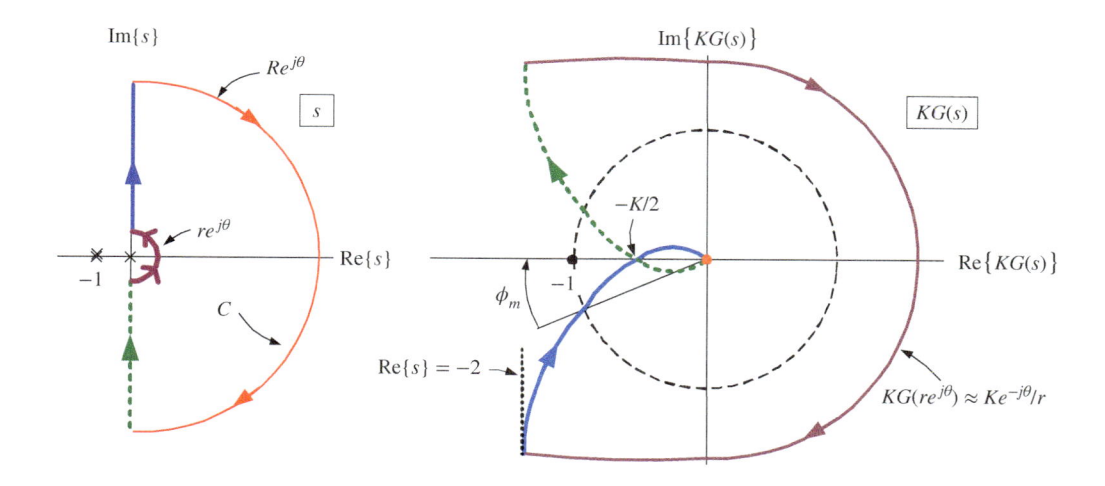

Figure 11.51. Nyquist plot of $K\dfrac{1}{s(s+1)^2}$, $G(j1) = -1/2$, $G(j0.68) = 1e^{-j159°}$.

11.3 Relative Stability: Gain and Phase Margins

In this example $P = 0$ and for $0 < K < 2$ it follows from the right side of Figure 11.51 that $N = 0$ so the closed-loop system is stable as $Z = N + P = 0$. For $K > 2$ we see that the Nyquist plot encircles the $-1 + j0$ point. Thus the gain margin is 2 or $20 \log_{10} 2 = 6$ dB. The phase margin is

$$\phi_m = \angle G(j\omega_g) + 180 = -159° + 180° = 21°.$$

The Bode diagram in Figure 11.52 shows the crossover frequency is $\omega_\phi = 1$. We may write

$$-20 \log_{10} |G(j\omega_\phi)| = -20 \log_{10}(1/2) = 20 \log_{10}(2) = 6 \text{ dB}.$$

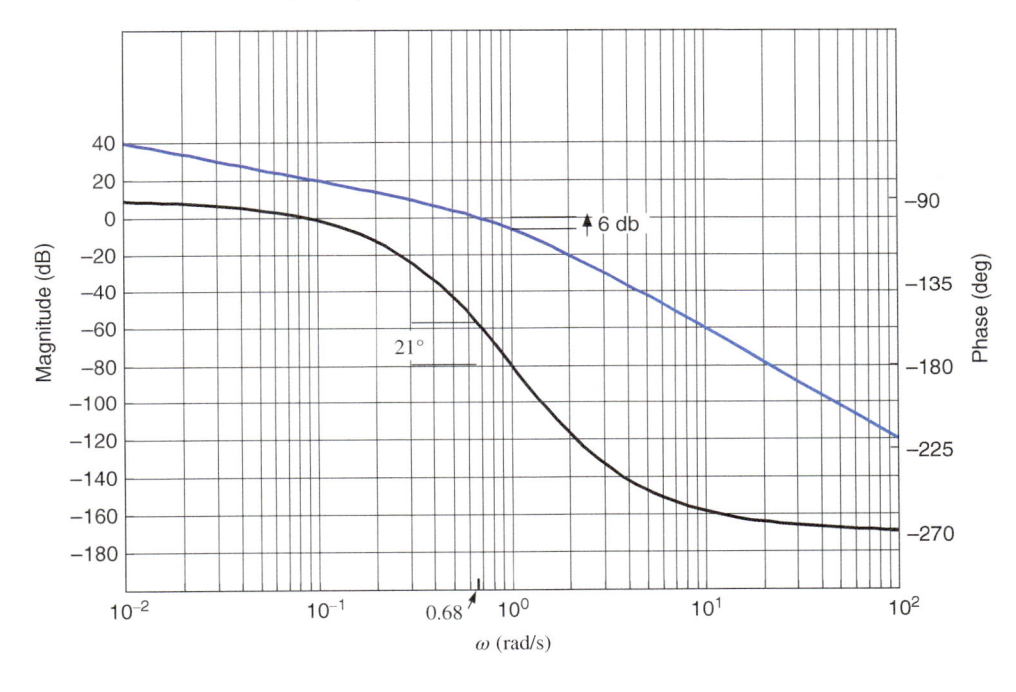

Figure 11.52. Bode diagram of $G(s) = \dfrac{1}{s(s+1)^2}$.

Summarizing, the important information at the crossover frequencies is

$\omega_g = 0.68$	$20 \log_{10}|G(j\omega_g)| = 0$ dB	$\angle G(j\omega_g) = -159°$
$\omega_\phi = 1$	$20 \log_{10}|G(j\omega_\phi)| = 20 \log_{10}(1/2) = -6$ dB	$\angle G(j\omega_\phi) = -180°$

Example 19 *Gain and Phase Margins*

Now we consider Example 15 where $G(s) = \dfrac{100(s/10 + 1)}{s(s-1)(s/100 + 1)}$. In Figure 11.53 we give the Nyquist plot of $KG(s)$ (not $G(s)$ as in Example 15). The *difference* in

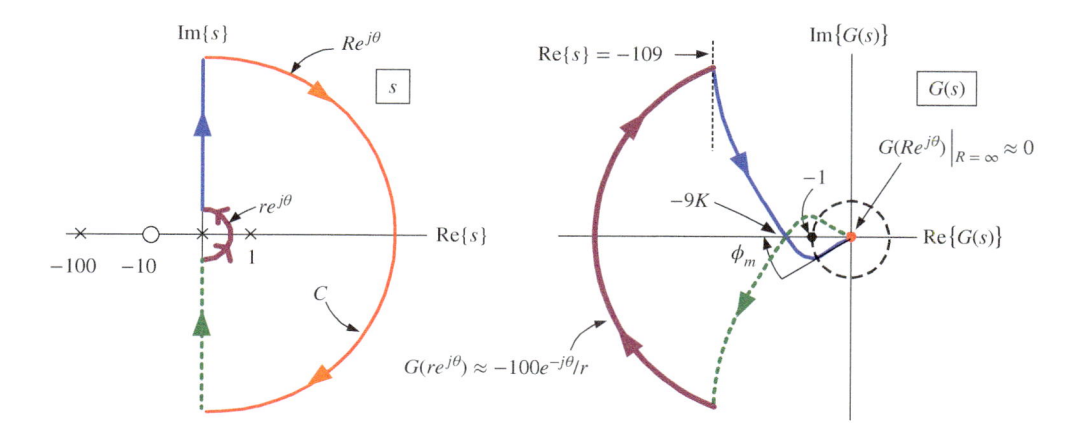

Figure 11.53. $K\dfrac{100(s/10+1)}{s(s-1)(s/100+1)}$, $G(j3.35) = -9$, $G(j12.6) = 1e^{-j140°}$.

this example compared with the previous example is that $G(re^{j\theta}), -\pi/2 \leq \theta \leq \pi/2$ now encloses the left half-plane as shown on the right side of Figure 11.53. The gain crossover and phase crossover frequencies along with $G(j\omega_g) = G(j\omega_g)|e^{j\angle G(j\omega_g)}$ and $G(j\omega_\phi) = G(j\omega_\phi)|e^{j\angle G(j\omega_\phi)}$ are given by

$\omega_g = 12.6$	$20\log_{10}|G(j\omega_g)| = 0$ dB	$\angle G(j\omega_g) = -140°$
$\omega_\phi = 3.35$	$20\log_{10}|G(j\omega_\phi)| = 20\log_{10}(|-9|) = 19.1$ dB	$\angle G(j\omega_\phi) = -180°$

As $P = 1$ stability requires the Nyquist plot encircle $-1 + j0$ once in the counterclockwise direction, that is, we must have $N = -1$. Therefore $K > 1/9$ is required for closed-loop stability.

The gain margin in this case must be interpreted as the amount K can *reduced* before the closed-loop system becomes unstable. This amount is $20\log_{10}(1/9) = -19.1$ dB. (Compare with the previous examples in which K can be *increased* by no more than the gain margin to maintain closed-loop stability.) This gain margin is indicated on the Bode diagram given in Figure 11.54, but note that it is shown with an arrow going down to emphasize that 19 dB is how much than gain can be reduced before the closed-loop system becomes unstable.

At the gain crossover frequency $\angle G(j\omega_g) = -140°$ so that the Nyquist polar plot can be rotated by $40°$ before the Nyquist plot *no longer* encircles $-1 + j0$ and thus the closed-loop system would no longer be stable. The phase margin is $40°$ and is indicated on the Bode diagram given in Figure 11.54.

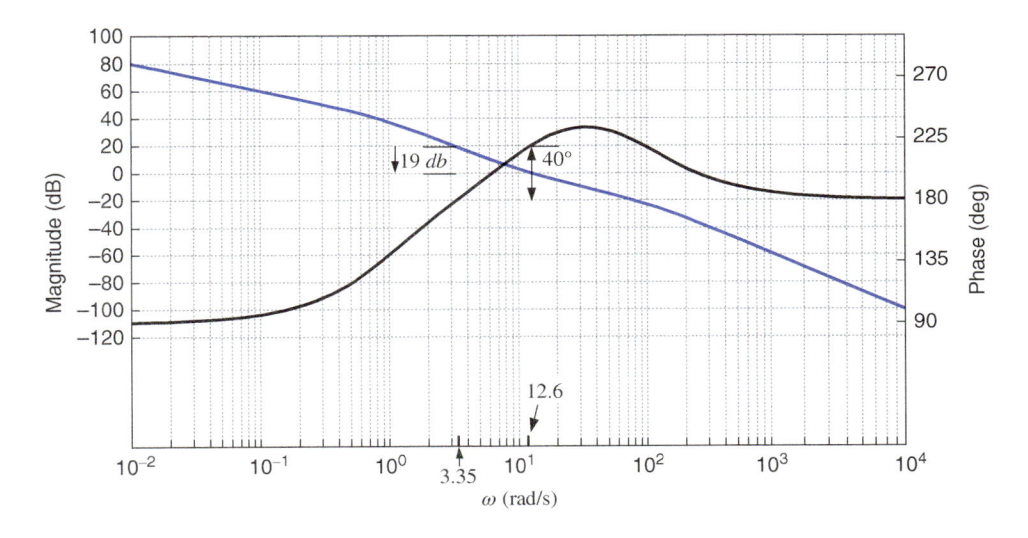

Figure 11.54. Bode diagram of $\dfrac{100(s/10+1)}{s(s-1)(s/100+1)}$.

Relative Stability

The gain and phase margins give a measure of relative stability. That is, the larger the gain margin the more we can increase the gain before the closed-loop system becomes unstable. Similarly, the larger the phase margin the more the Nyquist plot can be rotated before the closed-loop system becomes unstable. As a simple example, let the nominal model of the physical system be

$$G_o(s) = \frac{s-2}{(s-1)(s-3)}.$$

But, unknown to the control designer, the transfer function

$$G(s) = \frac{s-2}{(s-0.9)(s-3)}$$

is the "true" model of the system. (Note that $G_o(s)$ and $G(s)$ have the same number of right half-plane poles.) To characterize this difference we write

$$G(s) = G_o(s)\frac{G(s)}{G_o(s)} = G_o(s)\Delta(s) \text{ where } \Delta(s) \triangleq \frac{G(s)}{G_o(s)} = \frac{s-1}{s-0.9}.$$

The controller is designed based on $G_o(s)$, but the actual system is $G_o(s)\Delta(s)$. Figure 11.55 is the Bode diagram for $\Delta(s)$.

The Nyquist plot is done with $G_c(j\omega)G_o(j\omega)$. The quantity $\Delta(j\omega) = |\Delta(j\omega)|e^{j\angle\Delta(j\omega)}$ both rotates and scales the original Nyquist plot $G_c(j\omega)G_o(j\omega)$. If the phase and gain

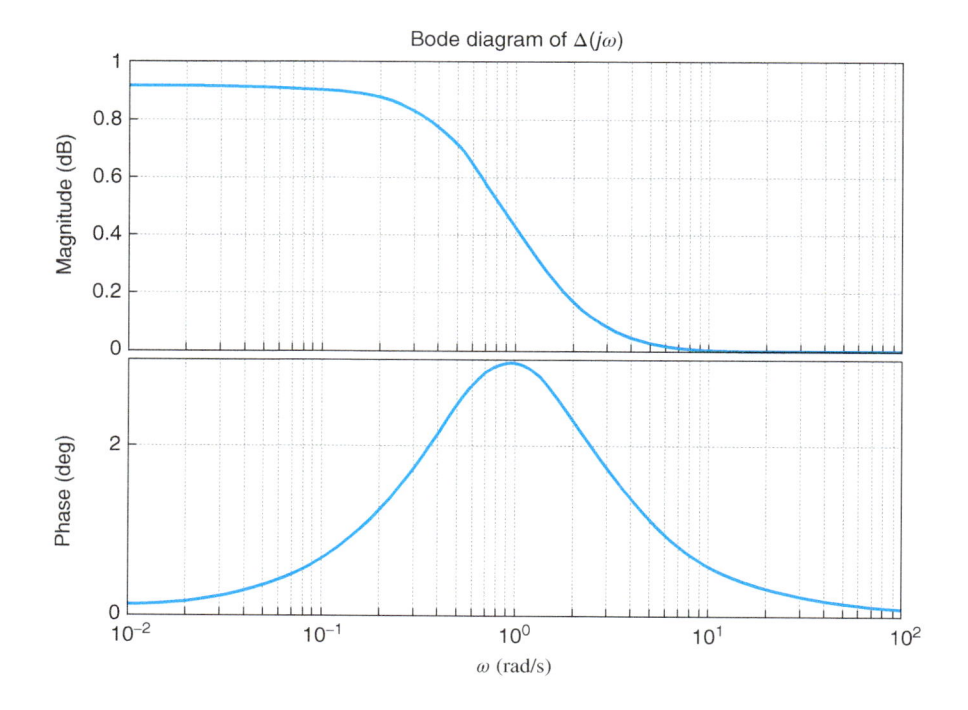

Figure 11.55. Bode diagram of $\Delta(j\omega)$.

margins are too small, it can happen that the closed-loop system is unstable due to the (unknown) uncertainty $\Delta(j\omega)$ of the model. It is desired that the gain and phase margins be "large enough" to ensure the closed-loop system is stable despite uncertainty in the model $G(s)$ of the physical system. How large should these margins be? Well it is impossible to know! If a specification is given that the system have a phase margin of at least 35°, it just means that the specifier thinks that this will ensure closed-loop stability despite uncertainty in the model.

Problem 30 considers a Nyquist analysis for this system using the controller designed in Chapter 10 (see page 289 and page 304) that placed all the closed-loop poles at -1. There it is seen that this controller results in very small stability margins. In fact, if this controller is used for the system model $G(s) = G_o(s)\Delta(s) = \dfrac{s-2}{(s-0.9)(s-3)}$ the resulting closed-loop system is unstable! The more model uncertainty a controller can deal with (by keeping the closed-loop system stable), the more *robust* we say it is. As long as $G(s)$ has the same number of right half-plane poles as $G_o(s)$ and $G_c(j\omega)G(j\omega)$ encircles $-1 + j0$ the same number of times as $G_c(j\omega)G_o(j\omega)$, the closed-loop system will be stable.

As an alternative to gain and phase margins, let Figure 11.56 be the Nyquist plot of a system that is closed-loop stable. To take into account uncertainty in the model $G_o(j\omega)$, it is important to make sure $G_c(j\omega)G_o(j\omega)$ isn't too close to $-1 + j0$. The vector (complex number) from $-1 + j0$ to the Nyquist plot is $G_c(j\omega)G_o(j\omega) - (-1 + j0) = 1 + G_c(j\omega)G_o(j\omega)$. With the definition

$$\beta \triangleq \min_{\omega \geq 0} \left|1 + G_c(j\omega)G_o(j\omega)\right|$$

the criteria is to find $G_c(j\omega)$ so that β is large enough. The size of β (just like the amount of gain and phase margin) that ensures closed-loop stability in spite of model uncertainty is not known. However, for each controller choice one can compute β as a relative measure of how well the controller will deal with model uncertainty.

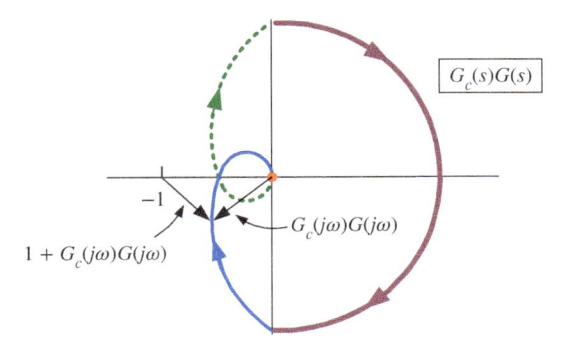

Figure 11.56. The vector (complex number) from $-1 + j0$ to $G_c(j\omega)G(j\omega)$ is $1 + G_c(j\omega)G(j\omega)$.

The area of H_∞ (H-infinity) control was developed to give quantitative conditions to ensure the closed-loop system is stable despite model uncertainty. The designer develops a model of the uncertainties expected for the application at hand and H_∞ control theory gives testable conditions to determine if the closed-loop system remains stable despite these uncertainties. Elementary introductions to this theory are given in [5, 49].

11.4 Closed-Loop Bandwidth

With $G(s)$ the transfer function of a physical system, let $G_c(s)$ be the transfer function of a controller that stabilizes the closed-loop system with the closed-loop transfer function $T(s)$ given by

$$T(s) = \frac{G_c(s)G(s)}{1 + G_c(s)G(s)}.$$

In Section 11.5 we discuss designing the controller based on the open-loop Bode diagram of $G(j\omega)$. That is, we are looking at the design problem based just on the frequency response $G(j\omega), -\infty < \omega < \infty$. To do so we need to introduce the *Fourier transform*. Let the closed-loop system be third-order and write

$$T(s) = \frac{b_2 s^2 + b_1 s + b_0}{s^3 + a_2 s^2 + a_1 s + a_0}.$$

With $r(t)$ the reference input and $y(t)$ the output, the closed-loop system is represented in the time domain by the differential equation

$$\frac{d^3}{dt^2}y(t) + a_2\frac{d^2}{dt}y(t) + a_1\frac{d}{dt}y(t) + a_0 y(t) = b_2\frac{d^2}{dt}r(t) + a_1\frac{d}{dt}r(t) + a_0 r(t). \qquad (11.6)$$

The Fourier transform of $y(t)$ and $r(t)$ are, respectively, defined by

$$Y(j\omega) \triangleq \int_{-\infty}^{+\infty} y(t)e^{-j\omega t}dt$$

$$R(j\omega) \triangleq \int_{-\infty}^{+\infty} r(t)e^{-j\omega t}dt.$$

Though the definition of the Fourier transform is with $-\infty < t < \infty$, we continue to take all signals to be zero for $t < 0$, i.e., $r(t) = 0$ for $t < 0$ and $y(t) = 0$ for $t < 0$. With this convention it seen that

$$Y(j\omega) = Y(s)|_{s=j\omega} \quad \text{and} \quad R(j\omega) = R(s)|_{s=j\omega}.$$

The signals $y(t)$ and $r(t)$ can be recovered using the *inverse Fourier transform* given by (see [50, 51])

$$y(t) = \frac{1}{2\pi} \int_{-\infty}^{+\infty} Y(j\omega)e^{j\omega t} d\omega$$

$$r(t) = \frac{1}{2\pi} \int_{-\infty}^{+\infty} R(j\omega)e^{j\omega t} d\omega.$$

The interpretation here is that $y(t)$ is made up of the sum (integral) of the pure sinusoids $Y(j\omega)e^{j\omega t} d\omega$ from $\omega = -\infty$ to $\omega = +\infty$. $Y(j\omega)d\omega$ is the *frequency content* of $y(t)$ between ω and $\omega + d\omega$. A similar interpretation holds between $r(t)$ and $R(j\omega)d\omega$. Differentiating both sides of these inverse Fourier transforms with respect to t gives

$$\frac{d}{dt}y(t) = \frac{1}{2\pi} \int_{-\infty}^{+\infty} j\omega Y(j\omega)e^{j\omega t} d\omega$$

$$\frac{d}{dt}r(t) = \frac{1}{2\pi} \int_{-\infty}^{+\infty} j\omega R(j\omega)e^{j\omega t} d\omega.$$

That is, the Fourier transform of dy/dt is $j\omega Y(j\omega)$ and the Fourier transform of dr/dt is $j\omega R(j\omega)$. As a consequence, taking the Fourier transform of both sides of (11.6) gives

$$(j\omega)^3 Y(j\omega) + a_2(j\omega)^2 Y(j\omega) + a_1(j\omega)Y(j\omega) + a_0 Y(j\omega) = b_2(j\omega)^2 R(j\omega) + a_1(j\omega)^2 R(j\omega)$$
$$+ a_0 R(j\omega)$$

or

$$T(j\omega) = \frac{Y(j\omega)}{R(j\omega)} = \frac{b_2(j\omega)^2 + b_1(j\omega) + b_0}{(j\omega)^3 + a_2(j\omega)^2 + a_1(j\omega) + a_0}.$$

This is simply the transfer function $T(s)$ evaluated at $s = j\omega$. The Fourier transform is now used to interpret how the closed-loop system (differential equation) processes the reference input $r(t)$ to produce the output $y(t)$. Specifically, the output $y(t)$ may be written as

$$y(t) = \frac{1}{2\pi}\int_{-\infty}^{+\infty} Y(j\omega)e^{j\omega t} d\omega = \frac{1}{2\pi}\int_{-\infty}^{+\infty} T(j\omega)R(j\omega)e^{j\omega t} d\omega.$$

$R(j\omega)d\omega$ is the *frequency content* of $r(t)$ between ω and $\omega + d\omega$. This expression shows that the *frequency content* of $y(t)$ in this same interval is simply $T(j\omega)R(j\omega)d\omega$. That is $y(t)$ is the sum (integral) of the pure sinusoids $T(j\omega)R(j\omega)e^{j\omega t} d\omega$ from $\omega = -\infty$ to $\omega = +\infty$.

Let the goal be to have $y(t)$ track $r(t)$. To do this we now show that a controller $G_c(j\omega)$ must be found so that the closed-loop transfer function

$$T(j\omega) = \frac{G_c(j\omega)G(j\omega)}{1 + G_c(j\omega)G(j\omega)}$$

satisfies

$$T(j\omega) \approx \begin{cases} 1, & |\omega| \leq \omega_B \\ 0, & |\omega| > \omega_B. \end{cases} \quad (11.7)$$

We refer to ω_B as the *bandwidth* of the closed-loop system. See Figure 11.57. To see how tracking is achieved, suppose $R(j\omega) \approx 0$ for $|\omega| > \omega_B$. Then we have

$$y(t) = \frac{1}{2\pi}\int_{-\infty}^{+\infty} T(j\omega)R(j\omega)e^{j\omega t} d\omega$$

$$\approx \frac{1}{2\pi}\int_{-\omega_B}^{+\omega_B} T(j\omega)R(j\omega)e^{j\omega t} d\omega \text{ as } R(j\omega) \approx 0 \text{ for } |\omega| > \omega_B$$

11.4 Closed-Loop Bandwidth

$$\approx \frac{1}{2\pi} \int_{-\omega_B}^{+\omega_B} R(j\omega) e^{j\omega t} d\omega \text{ as } T(j\omega) \approx 1 \text{ for } |\omega| \leq \omega_B$$

$$\approx \frac{1}{2\pi} \int_{-\infty}^{+\infty} R(j\omega) e^{j\omega t} d\omega \text{ as } R(j\omega) \approx 0 \text{ for } |\omega| > \omega_B$$

$$= r(t).$$

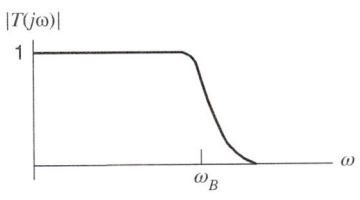

Figure 11.57. $|T(j\omega)|$ vs. ω.

That is, if the controller can be designed so that $T(j\omega) \approx 1$ ($|T(j\omega)| \approx 1, \angle T(j\omega) \approx 0$) for $|\omega| \leq \omega_B$ with $R(j\omega) \approx 0$ outside this frequency interval, then we should have good tracking. With $T(s)$ stable and $r(t) = R_0 \cos(\omega t)$, it was shown in Chapter 3 that

$$y(t) \to y_{ss}(t) = |T(j\omega)| R_0 \cos\left(\omega t + \angle T(j\omega)\right) \approx R_0 \cos(\omega t) \text{ for } |\omega| \leq \omega_B.$$

However, a typical input reference is a step input. If $r(t) = u_s(t)$ is a unit step, then its Fourier transform does not exist as

$$\int_{-\infty}^{\infty} u_s(t) e^{-j\omega t} dt = \int_{0}^{\infty} u_s(t) e^{-j\omega t} dt = \left.\frac{e^{-j\omega t}}{-j\omega}\right|_0^{\infty} = \underbrace{\lim_{t \to \infty} \frac{e^{-j\omega t}}{-j\omega}}_{\text{Does not converge}} + \frac{1}{j\omega}.$$

To get around this let's replace $r(t) = u_s(t)$ with $r(t) = e^{-\alpha t} u_s(t)$ and think of α as very small so that $e^{-\alpha t} \approx 1$ for the time interval of interest. The Fourier transform of $e^{-\alpha t} u_s(t)$ is

$$\int_{-\infty}^{\infty} e^{-\alpha t} u_s(t) e^{-j\omega t} dt = \frac{1}{\alpha + j\omega}.$$

If the controller is designed so that closed-loop bandwidth is $\omega_B = 10$, then for $|\omega| > \omega_B$ it follows that $|R(j\omega)| = \left|\frac{1}{\alpha + j\omega}\right| \leq \frac{1}{10} \ll \frac{1}{\alpha}$ where $\frac{1}{\alpha}$ is the peak value of $|R(j\omega)|$ (remember that α is very small). In words, if the designed controller results in a closed-loop bandwidth of 10, then we expect good tracking of step functions.

Remark Using Cauchy's residue theorem it can be shown that $\frac{1}{2\pi} \int_{-\infty}^{+\infty} \frac{1}{\alpha + j\omega} e^{j\omega t} d\omega = e^{-\alpha t} u_s(t)$ [52].

Definition 5 *Closed-Loop Bandwidth*

Let a stable closed-loop system have a magnitude plot of the type shown in Figure 11.57. That is, $T(j0) = 1$, $|T(j\omega)| \approx 1$ for $|\omega| \leq \omega_B$, and $|T(j\omega)| \to 0$ rapidly for $|\omega| > \omega_B$. Define the closed-loop bandwidth ω_B as the frequency that satisfies

$$|T(j\omega_B)| = |T(0)|/\sqrt{2} = 1/\sqrt{2}$$

or, equivalently,

$$20 \log_{10}|T(j\omega_B)| = 20 \log_{10}(1/\sqrt{2}) = -3 \text{ dB}.$$

11 Frequency Response Methods

Example 20 *Closed-Loop Bandwidth*
Take a look at Figure 11.2 which is the Bode diagram of ($\tau = 0.2$)

$$T(j\omega) = \frac{1}{\tau j\omega + 1} = \frac{1/\tau}{j\omega + 1/\tau}.$$

Let's consider this to be the Bode diagram of a closed-loop system. At the pole (break point) $p = 1/\tau = 5$ we see that magnitude plot is down -3 dB (a factor of $1/\sqrt{2}$) from its low frequency value of 0 dB (magnitude 1). So the bandwidth of this system is $\omega_B = 5$. If $\tau = 0.1$ so $1/\tau = 10$ then the bandwidth would be $\omega_B = 10$. The further the pole $p = -1/\tau$ resides in the left-half plane, the larger the bandwidth.

Figure 11.3 shows that the frequency must be below $0.1\omega_B$ for $\angle T(j\omega) \approx 0$. That is, $T(j\omega) \approx 1$ is only reasonable for $|\omega| < 0.1\omega_B$.

Example 21 *Closed-Loop Bandwidth*
With $\omega_n = 1$ and $\zeta = 0.1$ Figure 11.16 is the Bode diagram of

$$T(s) = \frac{\omega_n^2}{s^2 + 2\zeta\omega_n s + \omega_n^2}.$$

The closed-loop poles are at $p_i = -\zeta\omega_n \pm j\omega_n\sqrt{1-\zeta^2}$. The magnitude peak in Figure 11.16 is quite large because the damping coefficient $\zeta = 0.1$ is small. In placing the closed-loop poles one would normally not let ζ be any smaller than about 0.6. At the (double) break point ω_n we have $|G(j\omega_n)| = \frac{1}{2\zeta}$. If ζ is close to 1, the magnitude plot is down by $1/2$ (-6 dB), while if $\zeta = 0.6$ the magnitude plot is down by $\frac{1}{1.2}$ (-1.6 dB). Rather than compute the 3 dB bandwidth for each value of ζ we just take the bandwidth to be $\omega_B = \omega_n$. Then $|T(j\omega)| \approx 1$ for $|\omega| \leq \omega_B = \omega_n$ and $|T(j\omega)| \to 0$ rapidly as $|\omega|$ increases past $\omega_B = \omega_n$. The larger ω_n the further the poles are in the left half-plane and the larger the bandwidth. For $\zeta \geq 0.6$, the phase plot Figure 11.17 shows that the frequency must be below $0.1\omega_B = 0.1\omega_n$ for $\angle T(j\omega) \approx 0$. That is, $T(j\omega) \approx 1$ is only reasonable for $|\omega| < 0.1\omega_B$.

The Bode diagrams of Figures 11.2 and 11.16 used in Examples 20 and 21, respectively, have the desired shapes for a *closed-loop* system as given by (11.7). Let's now look at the properties the open-loop transfer function $G_c(s)G(s)$ should have so the closed-loop transfer function $T(s) = \frac{G_c(s)G(s)}{1 + G_c(s)G(s)}$ has the desired shape. Consider the general unity feedback control system of Figure 11.58.

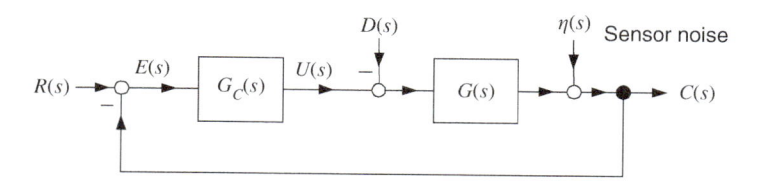

Figure 11.58. Closed-loop control system.

We have included the additive sensor noise η in the block diagram. Such noise is typically bounded and of high-frequency. The output $C(s)$ is

$$C(s) = \underbrace{\frac{G_c(s)G(s)}{1+G_c(s)G(s)}}_{T(s)} R(s) - \frac{G(s)}{1+G_c(s)G(s)} D(s) + \underbrace{\frac{1}{1+G_c(s)G(s)}}_{S(s)} \eta(s).$$

As $G(s)$ is strictly proper and $G_c(s)$ is (at least) proper, we have

$$\left|G_c(j\omega)G(j\omega)\right| \to 0 \text{ as } |\omega| \to \infty.$$

Another way to look at the bandwidth ω_B is as the frequency such that $\left|G_c(j\omega)G(j\omega)\right| \gg 1$ for $|\omega| \leq \omega_B$. If this holds it follows that

$$T(j\omega) = \frac{G_c(j\omega)G(j\omega)}{1+G_c(j\omega)G(j\omega)} \approx \frac{G_c(j\omega)G(j\omega)}{G_c(j\omega)G(j\omega)} = 1 \text{ for } |\omega| \leq \omega_B.$$

In particular, if the controller is designed to have an integrator for disturbance rejection then, as $\omega \to 0$, we have $\left|G_c(j\omega)G(j\omega)\right| \to \infty$ and $T(j\omega) \to 1$.

For $|\omega| > \omega_B$ we want $\left|G_c(j\omega)G(j\omega)\right| \to 0$ fast as ω increases so that

$$T(j\omega) = \frac{G_c(j\omega)G(j\omega)}{1+G_c(j\omega)G(j\omega)} \approx 0 \text{ for } |\omega| > \omega_B.$$

Recall $G(j\omega)$ represents a physical system so it is strictly proper. In Chapter 10, using the pole placement algorithm, we took $G_c(j\omega)$ to be just proper. One may want to specify that $G_c(j\omega)$ be strictly proper so that $\left|G_c(j\omega)G(j\omega)\right|$ goes to zero faster for $|\omega| > \omega_B$. As shown in Problems 13 and 14 of Chapter 10, it is straightforward to modify the pole placement method so that $G_c(j\omega)$ is strictly proper.

We want the bandwidth ω_B to contain most of the frequency content of $R(j\omega)$, that is, $R(j\omega) \approx 0$ for $|\omega| > \omega_B$. If this holds it follows that

$$C_R(j\omega) \triangleq \frac{G_c(j\omega)G(j\omega)}{1+G_c(j\omega)G(j\omega)} R(j\omega) \approx R(j\omega).$$

In words, the closed-loop system will do a good job tracking a reference input $R(j\omega)$ as long as the frequency content of $R(j\omega)$ is within the bandwidth ω_B of $T(j\omega)$.

Disturbance rejection follows if $\left|G_c(j\omega)\right| \gg 1$ for $|\omega| < \omega_B$ as

$$C_D(j\omega) \triangleq -\frac{G(j\omega)}{1+G_c(j\omega)G(j\omega)} D(s) \approx -\frac{1}{G_c(j\omega)} D(j\omega) \approx 0 \text{ for } |\omega| < \omega_B.$$

Further, $|G(j\omega)| \to 0$ as $|\omega| \to \infty$ and so $\left|G_c(j\omega)G(j\omega)\right| \to 0$ as well. Consequently $\left|C_D(j\omega)\right| \to 0$ as $|\omega| \to \infty$.

Conclusion

Good tracking and disturbance rejection is achieved if a stabilizing controller $G_c(j\omega)$ is designed with $\left|G_c(j\omega)\right| \gg 1$ for $|\omega| < \omega_B$ and $\left|G_c(j\omega)\right| \to 0$ rapidly for $|\omega| > \omega_B$.

The effect of the measurement noise on the output is

$$C_\eta(j\omega) \triangleq \frac{1}{1+G_c(j\omega)G(j\omega)} \eta(j\omega) \approx \begin{cases} 0 & |\omega| \leq \omega_B \\ \eta(j\omega) & |\omega| > \omega_B. \end{cases}$$

414   11   Frequency Response Methods

This shows the low-frequency noise on the output is attenuated by feedback while the high frequency noise is essentially unchanged. However, even with the feedback, the low-frequency sensor noise is a problem for the actuator. To explain, the output of the controller $U_\eta(s)$ (actuator command) due to just the measurement noise of the sensor is

$$U_\eta(s) \triangleq -\frac{G_c(j\omega)}{1+G_c(j\omega)G(j\omega)}\eta(j\omega) \approx \begin{cases} -\dfrac{1}{G(j\omega)}\eta(j\omega) & |\omega| \le \omega_B \\ -G_c(j\omega)\eta(j\omega) \to 0 & |\omega| > \omega_B. \end{cases}$$

As $G(s)$ is strictly proper, $1/G(s)$ is non proper resulting in the low-frequency noise being differentiated.[6] As explained in Chapter 9 this amplifies the noise (see page 249). Consequently it is very important to have sensors with essentially no noise for $|\omega| \le \omega_B$. In other words, the closed-loop bandwidth must be chosen so that it does not contain any significant frequency content of $\eta(j\omega)$.

11.5   Lead and Lag Compensation

Consider the standard unity feedback controller configuration given in Figure 11.59.

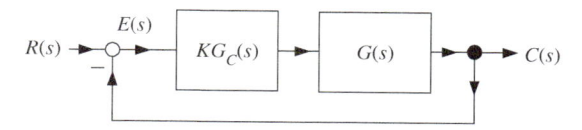

Figure 11.59. Standard unity feedback control configuration.

Let's first consider the case where $G_c(s) = 1$ so it is simply a proportional feedback control problem. Then $E(s) = \dfrac{1}{1+KG(s)}R(s)$. To be more concrete, let

$$G(s) = \frac{1}{(s+1)(s+2)(s+4)}, \quad R(s) = \frac{1}{s}.$$

If the closed-loop system is stable, then

$$e(\infty) = \lim_{s \to 0} s \frac{1}{1+K\dfrac{1}{(s+1)(s+2)(s+4)}} \frac{1}{s} = \lim_{s \to 0} \frac{1}{1+K\dfrac{1}{(s+1)(s+2)(s+4)}} = \frac{1}{1+K/8}.$$

The larger the gain the smaller the steady-state error as long as the closed-loop system is stable. However, increasing the open-loop gain K usually *decreases* the phase margin. If the gain is increased enough the closed-loop system will go unstable. This is illustrated in Figure 11.60 where $K > 1$ showing the phase margin decreases for $KG(j\omega)$ compared with $G(j\omega)$.

---

[6]For example with $G(s) = \dfrac{b}{s(s+a)}$ then $-\eta(s)/G(s) = -\dfrac{1}{b}(s^2+as)\eta(s)$ so the sensor noise $\eta(s)$ is differentiated twice.

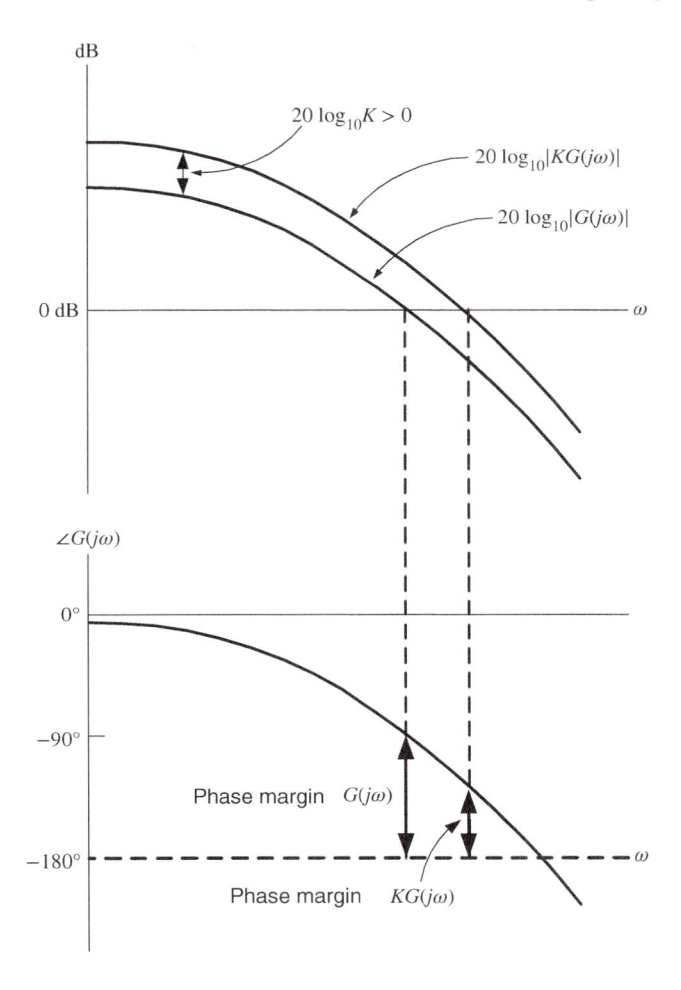

Figure 11.60. How increasing the gain can decrease the phase margin.

Lag Compensation

Consider a compensator given by

$$G_c(s) = \frac{0.01}{0.1} \frac{s + 0.1}{s + 0.01} = \frac{s/0.1 + 1}{s/0.01 + 1}.$$

We then have

$$G_c(0) = 1 \text{ and } G_c(j\omega) \to \frac{0.01}{0.1} = \frac{1}{10} \text{ as } \omega \to \infty.$$

Then

$$KG_c(0) = K \text{ and } KG_c(j\omega) \to \frac{K}{10} \text{ as } \omega \to \infty.$$

The Bode diagram of $G_c(j\omega)$ is shown in Figure 11.61 on the next page where $\omega_p = 0.01$ and $\omega_z = 0.1$. A lag compensator (controller) has both of its break points at low frequencies with the pole first (i.e., $\omega_p < \omega_z$) and both are well below the gain crossover frequency ω_g ($|G(j\omega_g)| = 1$), i.e., $\omega_p < \omega_z \ll \omega_g$. Figure 11.62 shows Bode diagram of $KG_c(j\omega)G(j\omega)$

11 Frequency Response Methods

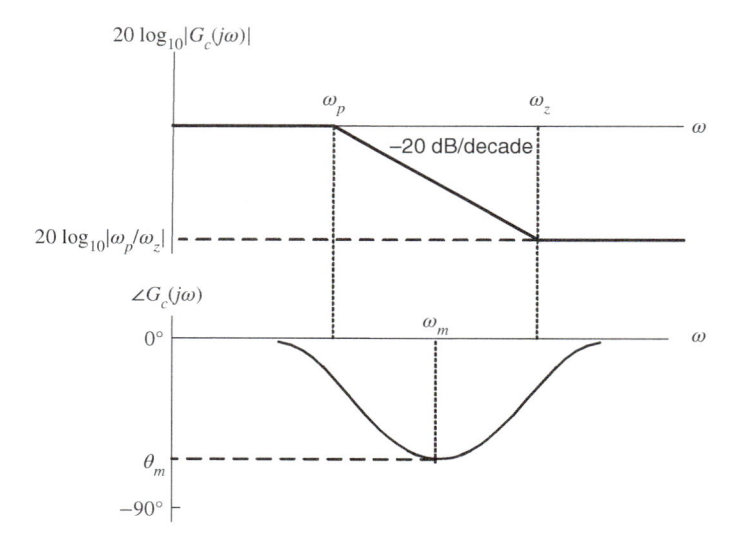

Figure 11.61. Bode diagram of a lag compensator $G_c(s) = \dfrac{s/\omega_z + 1}{s/\omega_p + 1}$, $\omega_p < \omega_z$.

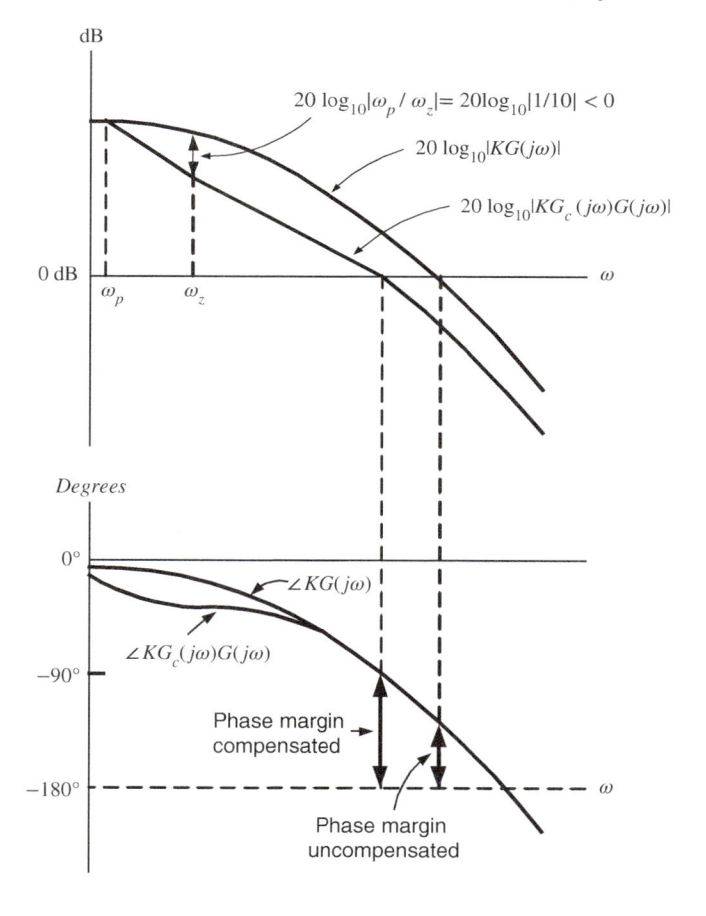

Figure 11.62. $KG_c(j\omega)G(j\omega)$ and $KG(j\omega)$ for lag compensation.

along with the Bode diagram of $KG(j\omega)$. The factor $KG_c(j\omega)$ has gain K and phase 0 at low frequency while it has gain $K(\omega_p/\omega_z) = K/10$ and phase 0 for frequencies a decade above ω_z. Consequently, as Figure 11.62 shows, the gain crossover frequency of $KG_c(j\omega)G(j\omega)$ is reduced compared to that of $KG(j\omega)$ resulting in an increase in the phase margin.

Lead Compensation

A lead compensator $KG_c(s)$ is of the form

$$KG_c(s) = K\frac{s/\omega_z + 1}{s/\omega_p + 1} = K\frac{\omega_p}{\omega_z}\frac{s+\omega_z}{s+\omega_p},$$

where

$$\omega_z < \omega_p.$$

At low frequencies where $|s| \ll \omega_p$ we have

$$KG_c(s) = K\frac{s/\omega_z + 1}{s/\omega_p + 1} = K\frac{1}{\omega_z}\frac{s+\omega_z}{s/\omega_p+1} \approx K\frac{1}{\omega_z}(s+\omega_z) = K\underbrace{\frac{1}{\omega_z}s}_{K_D} + \underbrace{K}_{K_P}.$$

This is approximately a proportional plus derivative or PD controller. At high frequencies we have

$$KG_c(s) = K\frac{s/\omega_z + 1}{s/\omega_p + 1} \approx K\frac{\omega_p}{\omega_z}$$

which becomes a proportional controller and thus we avoid the amplification of high-frequency noise (due to a derivative controller). In summary, the lead compensator can be considered an approximation to a PD compensator without the noise amplification problem at high frequencies. The Bode diagram of $G_c(j\omega) = \dfrac{s/\omega_z + 1}{s/\omega_p + 1}$ is shown in Figure 11.63 where (in contrast to the lag compensator) we have $\omega_z < \omega_p$.

It turns out that the maximum value of $\angle G(j\omega)$, denoted by θ_m, occurs at $\omega_m \triangleq \sqrt{\omega_z \omega_p}$ or $\log_{10}\omega_m = \dfrac{1}{2}(\log_{10}\omega_z + \log_{10}\omega_p)$. Note that

$$KG_c(0) = K$$
$$KG_c(j\omega) \to \frac{\omega_p}{\omega_z}K > K \text{ as } \omega \to \infty.$$

The key observation of a lead compensator (in contrast to the lag compensator) is to have the gain crossover frequency ω_g of $KG(j\omega)$ ($|G(j\omega_g)| = 1$) to be above ω_z and below ω_p, that is,

$$\omega_z < \omega_g < \omega_p.$$

Figure 11.64 shows the lead compensated Bode diagram of $KG_c(j\omega)G(j\omega)$ along with the uncompensated Bode diagram of $KG(j\omega)$. As Figure 11.64 shows, $\angle G_c(j\omega)$ adds positive phase around the (original) gain crossover frequency ω_g, but is also adds gain $|G(j\omega)|$ so that the (new) gain crossover frequency ω_{g2} of $KG_c(j\omega)G(j\omega)$ moves to the right just past ω_p. However, enough phase is added that the phase margin of $KG_c(j\omega)G(j\omega)$ is greater than the phase margin of $KG(j\omega)$.

418   11   Frequency Response Methods

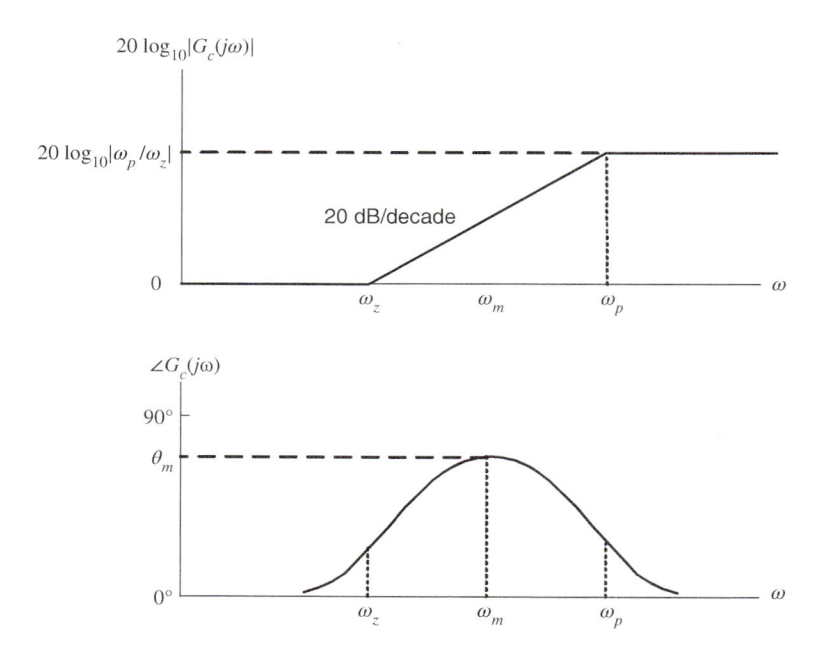

Figure 11.63. Lead compensator $G_c(s) = \dfrac{s/\omega_z + 1}{s/\omega_p + 1}$, $\omega_z < \omega_p$.

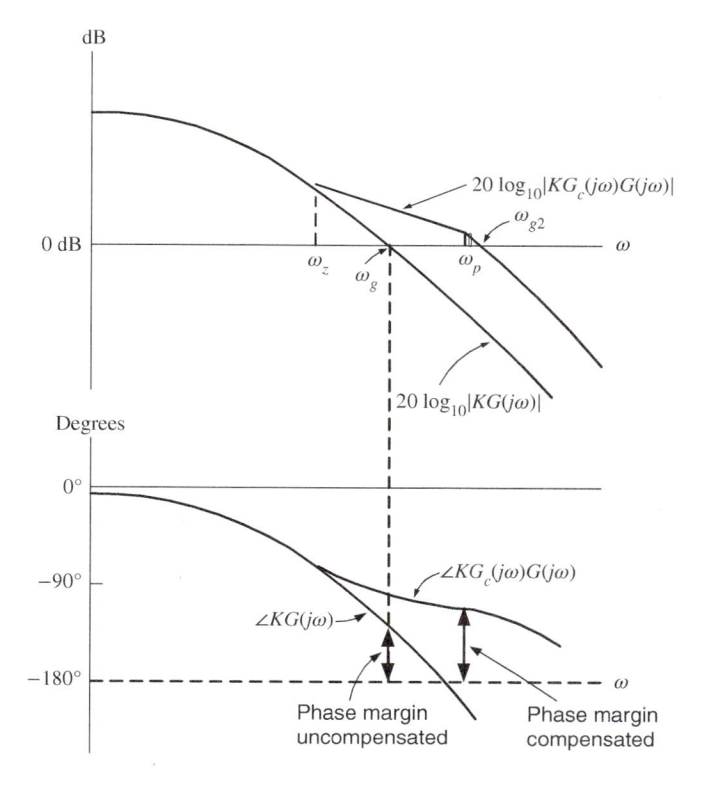

Figure 11.64. $KG_c(j\omega)G(j\omega)$ and $KG(j\omega)$ for lead compensation.

11.6 Double Integrator Control via Lead-Lag Compensation

Let the open-loop system be a double integrator, i.e.,

$$G(s) = \frac{1}{s^2}.$$

Consider the unity feedback control structure of Figure 11.65.

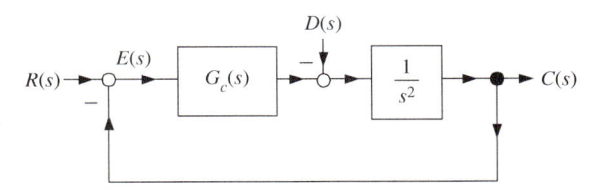

Figure 11.65. Unity feedback control structure.

The Bode diagram of $G(j\omega) = \dfrac{1}{(j\omega)^2}$ is given in Figure 11.66.

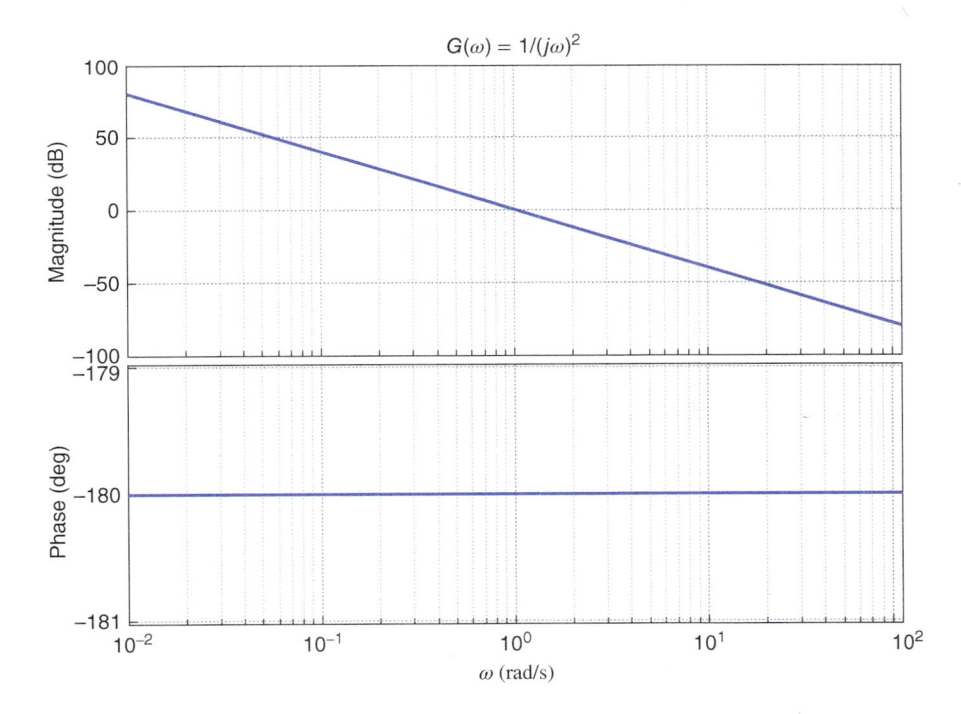

Figure 11.66. Bode diagram of $G(j\omega) = \dfrac{1}{(j\omega)^2}$.

The open-loop system is unstable and we need to make the closed-loop system stable. Consider a lead controller $G_c(s) = K \dfrac{s + z_c}{s + p_c}$ to do this. The gain crossover frequency is

420   11   Frequency Response Methods

$\omega_{cg} = 1$ rad/s so we choose the zero to be $z_c = 0.1$ so that it is a decade below the gain crossover frequency. We next choose the pole to be $p_c = 10$ so that it is a decade above the gain crossover frequency. This way $\dfrac{s+0.1}{s+10}$ contributes about 90° (actually only 78.6°) at $\omega = \omega_{cg} = 1$ and has corresponding gain of $\left|\dfrac{j\omega+0.1}{j\omega+10}\right|_{\omega=\omega_{cg}=1} = 0.1$. We choose $K = 10$ so that the gain of the lead compensator is 1 thus keeping the gain crossover frequency at $\omega = 1$. That is, choose the controller to be

$$G_c(s) = 10\frac{s+0.1}{s+10}.$$

The Bode diagram of $G_c(j\omega)G(j\omega)$ is given Figure 11.67. It shows that at $\omega = \omega_{cg} = 1$ we have $G_c(j\omega)G(j\omega) = 1e^{-j101.4°}$ so the phase margin is $180° - 101.4° = 78.6°$.

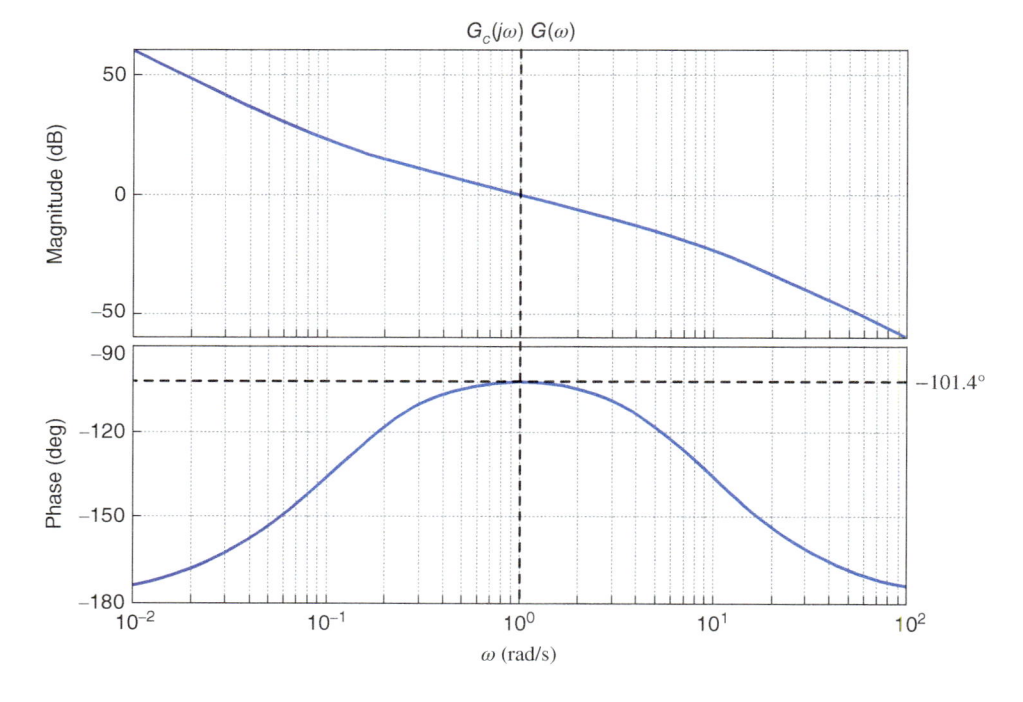

Figure 11.67. Bode diagram for $G_c(j\omega)G(j\omega) = 10\dfrac{j\omega+0.1}{j\omega+10}\dfrac{1}{(j\omega)^2}$.

Let's now sketch the Nyquist plot. The Nyquist contour for $G_c(s)G(s) = 10\dfrac{s+0.1}{s+10}\dfrac{1}{s^2}$ is given in Figure 11.68.

On the part of the Nyquist contour where $s = re^{j\theta}$ with r small, we have

$$G_c(re^{j\theta})G(re^{j\theta}) = 10\frac{re^{j\theta}+0.1}{re^{j\theta}+10}\frac{1}{r^2e^{j2\theta}} \approx \frac{0.1}{r^2}e^{-j2\theta},$$

11.6 Double Integrator Control via Lead-Lag Compensation

which we use to compute the following table of values.

θ	$G(re^{j\theta})G_c(re^{j\theta}) \approx 0.1e^{-j2\theta}/r^2$
$-\pi/2$	$0.1e^{j180°}/r^2$
$-\pi/4$	$0.1e^{j90°}/r^2$
0	$0.1/r^2$
$+\pi/4$	$0.1e^{-j90°}/r^2$
$+\pi/2$	$0.1e^{-j180°}/r^2$

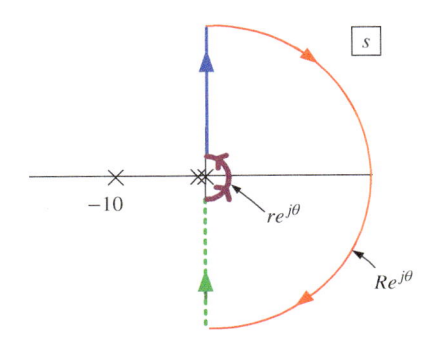

Figure 11.68. Nyquist contour for $G_c(s)G(s) = 10\dfrac{s+0.1}{s+10}\dfrac{1}{s^2}$.

From the table and the Bode diagram we obtain the sketch of the polar plot shown in Figure 11.69.[7]

From the Nyquist contour of Figure 11.68 it is seen that $P = 0$. The polar plot of Figure 11.69 shows $N = 0$ so we have $Z = N + P = 0$. For all $K > 0$ the polar plot of $KG_c(s)G(s)$ has the same form as in Figure 11.69 and thus the gain margin is infinite. The Nyquist plot is not drawn to scale, but it turns out the phase margin is 78.6°.

Let's look at the step response. With a unit step input we have

$$C(s) = \underbrace{\frac{G_c(s)G(s)}{1+G_c(s)G(s)}}_{G_{CL}(s)} R(s) = \frac{10\dfrac{s+0.1}{s+10}\dfrac{1}{s^2}}{1+10\dfrac{s+0.1}{s+10}\dfrac{1}{s^2}}\frac{1}{s} = \frac{10(s+0.1)}{s^3+10s^2+10s+1}\frac{1}{s}.$$

---

[7] Here's a little more detail on the Nyquist plot as $\omega \to 0$. We have

$$G_c(j\omega)G(j\omega) = 10\frac{j\omega+0.1}{j\omega+10}\frac{1}{(j\omega)^2} = -10\frac{(j\omega+0.1)(-j\omega+10)}{(j\omega+10)(-j\omega+10)}\frac{1}{\omega^2} = -10\frac{\omega^2+1+j9.9\omega}{\omega^2+100}\frac{1}{\omega^2}$$

$$\approx -\frac{1}{10}\frac{1}{\omega^2} - j0.99\frac{1}{\omega} \text{ for } \omega \text{ small.}$$

This calculation shows (see Figure 11.69) the imaginary part of $G_c(j\omega)G(j\omega)$ is negative for all $\omega > 0$. At $\omega = 0$, the plot $G_c(j\omega)G(j\omega)$ for $\omega > 0$ and the plot $G_c(j\omega)G(j\omega)$ for $\omega < 0$ connect to $G_c(re^{j\theta})G(re^{j\theta})|_{\theta=\pi/2, r=0}$ and $G_c(re^{j\theta})G(re^{j\theta})|_{\theta=-\pi/2, r=0}$, respectively.

11 Frequency Response Methods

$G_c(re^{j\theta})G(re^{j\theta}) \approx \dfrac{0.1}{r^2} e^{-j2\theta}$

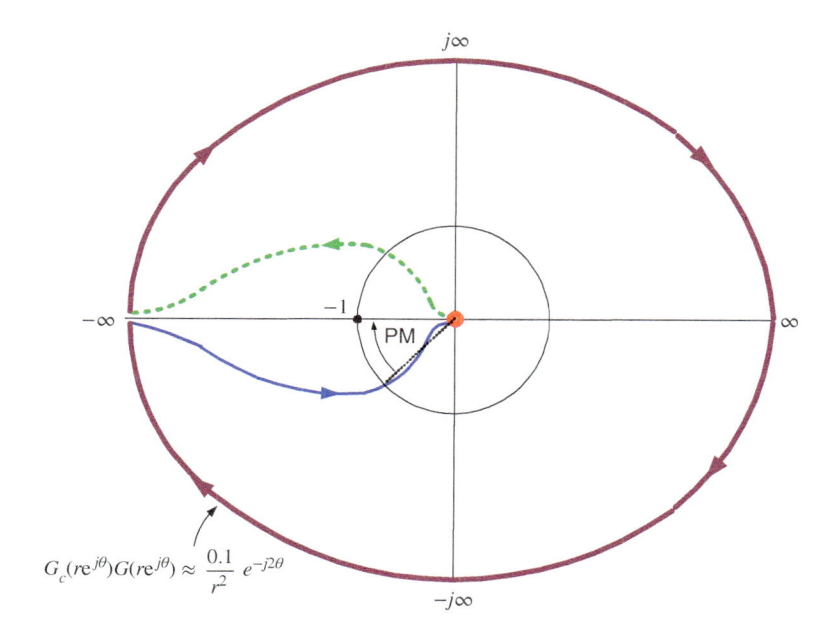

Figure 11.69. Polar plot for $G_c(s)G(s) = 10\dfrac{s+0.1}{s+10}\dfrac{1}{s^2}$.

The closed-loop poles are $-0.11, -1,$ and -8.9. The partial fraction expansion of $C(s)$ has the form

$$C(s) = \frac{R_0}{s} + \frac{A_1}{s+1} + \frac{A_2}{s+8.9} + \frac{A_3}{s+0.11}.$$

With a closed-loop pole at -0.11 it is a concern that the part of the transient response $A_3 e^{-0.11t}$ due to it dies out slowly. However there is a zero at -0.1, which approximately cancels the pole at -0.11 resulting in A_3 being small. Specifically,

$$A_3 = \lim_{s \to -0.11}(s+0.11)\frac{10(s+0.1)}{(s+1)(s+8.9)(s+0.11)}\frac{1}{s} = \frac{10(s+0.1)}{(s+1)(s+8.9)}\frac{1}{s}\bigg|_{s=-0.11} = 0.116.$$

The complete response is

$$c(t) = u_s(t) + 0.116 e^{-0.11t} - 1.3 e^{-1.0t} + 0.14 e^{-8.9t}.$$

The lead controller is $G_c(s) = K\dfrac{s+0.1}{s+10}$ with $K = 10$. The root locus method given in Chapter 11 shows that the larger the value K the closer one of the closed-loop poles is to the zero of $G_c(s)$. Thus this approximate pole/zero cancellation is not surprising after studying the root locus method.

The error $E_D(s)$ due to a step disturbance $D(s) = D_0/s$ is

$$E_D(s) = \frac{\dfrac{1}{s^2}}{1 + 10\dfrac{s+0.1}{s+10}\dfrac{1}{s^2}}\frac{D_0}{s} = \frac{1}{s^2 + 10\dfrac{s+0.1}{s+10}}\frac{D_0}{s}$$

11.6 Double Integrator Control via Lead-Lag Compensation

giving a final error of

$$e_D(\infty) = \lim_{s \to 0} sE_D(s) = \lim_{s \to 0} s \frac{1}{s^2 + 10\frac{s+0.1}{s+10}} \frac{D_0}{s} = 10D_0.$$

Quite big! Let's now include a lag compensator $\frac{s+0.1}{s+0.01}$ for its disturbance attenuation capability. This lag compensator was chosen because

$$\left.\frac{s+0.1}{s+0.01}\right|_{s=0} = 10$$

and

$$\frac{j\omega + 0.1}{j\omega + 0.01} \approx 1 \quad \text{for} \quad j\omega \gg 0.1.$$

That is, it provides a gain of 10 at low frequencies, but does not change the gain crossover point or phase margin. The (lead-lag) controller is now

$$G_c(j\omega) = \frac{s+0.1}{s+0.01} 10\frac{s+0.1}{s+10}.$$

With a step disturbance we see that

$$E_D(s) = \frac{\frac{1}{s^2}}{1 + \frac{s+0.1}{s+0.01} 10\frac{s+0.1}{s+10}\frac{1}{s^2}} \frac{D_0}{s} = \frac{1}{s^2 + \frac{s+0.1}{s+0.01} 10\frac{s+0.1}{s+10}} \frac{D_0}{s}$$

so

$$e_D(\infty) = \lim_{s \to 0} s \frac{1}{s^2 + \frac{s+0.1}{s+0.01} 10\frac{s+0.1}{s+10}} \frac{D_0}{s} = D_0.$$

The final disturbance error has been reduced by a factor of 10. The closed-loop response due a unit step input is now

$$C(s) = \frac{\frac{s+0.1}{s+0.01} 10\frac{s+0.1}{s+10}\frac{1}{s^2}}{1 + \frac{s+0.1}{s+0.01} 10\frac{s+0.1}{s+10}\frac{1}{s^2}} \frac{1}{s} = \frac{10s^2 + 2s + 0.1}{s^4 + 10.01s^3 + 10.1s^2 + 2s + 0.1}\frac{1}{s}$$

$$= \frac{10(s+0.1)^2}{(s+0.08)(s+0.16)(s+0.87)(s+8.9)}\frac{1}{s}.$$

Note that the two closed-loop poles -0.08 and -0.16 are close to the two zeros at -0.1. Let's look at the new gain and phase margins. The Bode diagram for $G_c(s)G(s) = \frac{s+0.1}{s+0.01} 10\frac{s+0.1}{s+10}\frac{1}{s^2}$ is given in Figure 11.70 on the next page.

This Bode diagram shows the phase margin is $73.5°$. Note that $\left|\frac{j\omega + 0.1}{j\omega + 0.01}\right|_{\omega=\omega_{cg}=1} = 1.005$. However, $\angle\left.\frac{j\omega + 0.1}{j\omega + 0.01}\right|_{\omega=\omega_{cg}=1} = -5.1°$, which accounts for the $5°$ decrease in phase

11 Frequency Response Methods

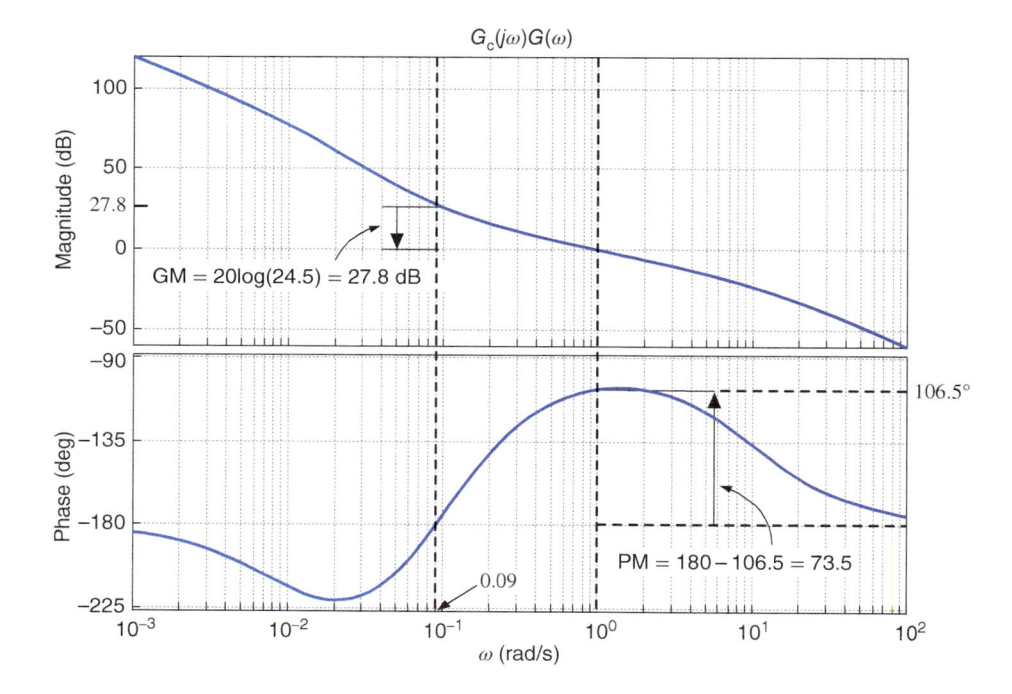

Figure 11.70. Bode diagram for $G_c(s)G(s) = \dfrac{s+0.1}{s+0.01} 10 \dfrac{s+0.1}{s+10} \dfrac{1}{s^2}$.

margin compared with Figure 11.67. To understand the gain margin let's take a look at the Nyquist plot. The Nyquist contour is shown in Figure 11.71.

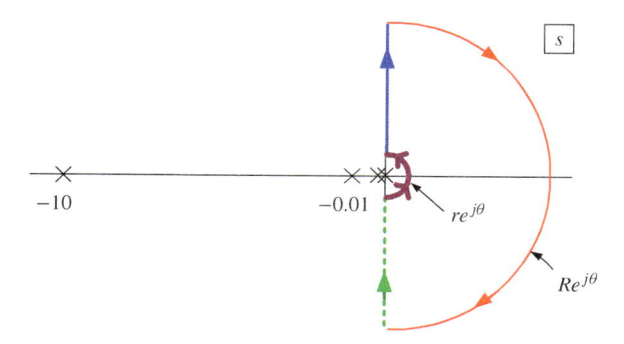

Figure 11.71. Nyquist contour for $G_c(s)G(s) = \dfrac{s+0.1}{s+0.01} 10 \dfrac{s+0.1}{s+10} \dfrac{1}{s^2}$.

In this case

$$G(re^{j\theta})G_c(re^{j\theta}) = \frac{re^{j\theta}+0.1}{re^{j\theta}+0.01} 10 \frac{re^{j\theta}+0.1}{re^{j\theta}+10} \frac{1}{r^2 e^{j2\theta}} \approx \frac{1}{r^2} e^{-j2\theta}$$

11.6 Double Integrator Control via Lead-Lag Compensation

which has the same shape as in Figure 11.69. Using this and the Bode diagram of Figure 11.70, the polar plot is as sketched in Figure 11.72.[8] Figure 11.71 shows that $P = 0$. Figure 11.72 shows that the polar plot encircles $-1 + j0$ once in the counterclockwise direction and once in the clockwise direction so $N = 0$. We then have $Z = N + P = 0$ and thus the closed-loop system is stable. If we multiply $G_c(s)G(s)$ by any factor less than $1/24.5 = 0.0408$, the polar plot no longer encircles the $-1 + j0$ in the counterclockwise direction, only in the clockwise direction resulting in $N = 1$. In this case $Z = N + P = 1$ making the closed-loop system unstable. Thus the gain margin is 0.0408 or $20\log(0.0408) = -27.8$ dB. The negative sign indicates that the gain plot can be *decreased* by this amount before the closed-loop system is unstable. However, it is usual to indicate this gain margin as the positive number 27.8 dB and use an arrow as in Figure 11.70 to indicate that if the gain plot is lowered by this amount (or more) the closed-loop system becomes unstable. The addition of the lag compensator reduces the error due to a constant disturbance by a factor of 10, but the gain margin is now 27.8 dB instead of being infinite as it was without the lag compensator.

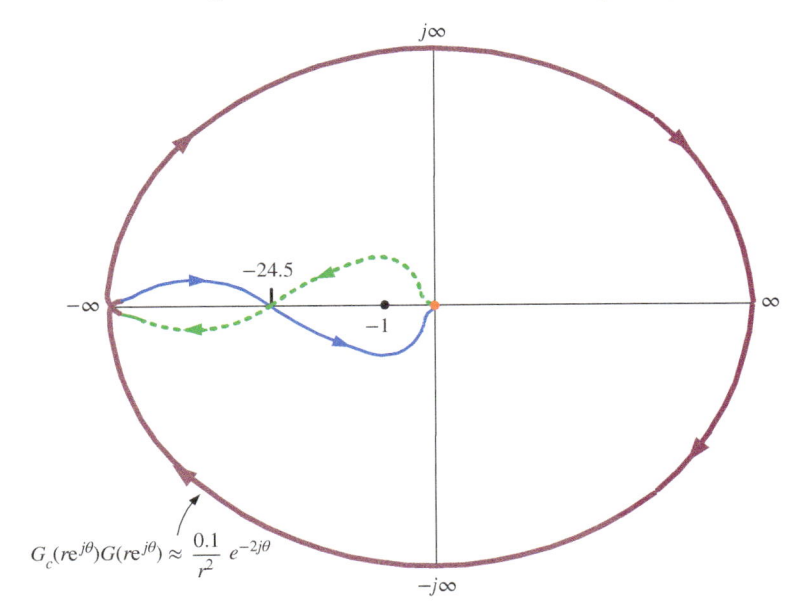

Figure 11.72. Polar plot of $G_c(s)G(s) = \dfrac{s+0.1}{s+0.01} 10 \dfrac{s+0.1}{s+10} \dfrac{1}{s^2}$.

---

[8] A bit more detail on the Nyquist plot as $\omega \to 0$. We have

$$G_c(j\omega)G(j\omega) = \frac{j\omega+0.1}{j\omega+0.01} 10 \frac{j\omega+0.1}{j\omega+10} \frac{1}{(j\omega)^2} = -\frac{(j\omega+0.1)(-j\omega+0.01)(j\omega+0.1)(-j\omega+10)}{(j\omega+0.01)(-j\omega+0.01)(j\omega+10)(-j\omega+10)} \frac{1}{\omega^2}$$

$$= -\frac{\omega^4 + 1.892\omega^2 + 0.001 + j(9.81\omega^3 - 0.0801\omega)}{\omega^4 + 100\omega^2 + 0.01} \frac{1}{\omega^2}$$

$$\approx -\frac{0.1}{\omega^2} + j\frac{8.01}{\omega} \text{ for } \omega \text{ small.}$$

This calculation shows (see Figure 11.72) the imaginary part of $G_c(j\omega)G(j\omega)$ for $\omega > 0$ is positive as $\omega \to 0$. From the Nyquist contour shown in Figure 11.71 this connects to $G_c(re^{j\theta})G(re^{j\theta})|_{\substack{\theta=\pi/2 \\ r=0}}$.

Why are we concerned about gain and phase margins? The model above is $G(s) = 1/s^2$, but with the sensor it might be k/s^2 where k is nominally 1. That is, the model is accurate and the sensor is calibrated accurately to have $k = 1$, but perhaps over time the electronic components degrade so k is no longer exactly 1. The gain margin tells us that as long as $0.0408 \leq k \leq \infty$ the closed-loop system remains stable.

11.7 Inverted Pendulum with Output $Y(s) = X(s) + \left(\ell + \dfrac{J}{m\ell}\right)\theta(s)$

In Example 5 of Chapter 10 a feedback controller using pole placement was designed for the inverted pendulum using the output

$$y(t) = x(t) + \left(\ell + \frac{J}{m\ell}\right)\theta(t).$$

The transfer function from $U(s)$ to $Y(s)$ is (see Chapter 13)

$$Y(s) = X(s) + \left(\ell + \frac{J}{m\ell}\right)\theta(s) = \underbrace{-\frac{\kappa m g \ell}{s^2(s^2 - \alpha^2)}}_{G_Y(s)} U(s) + \frac{p_Y(s)}{s^2(s^2 - \alpha^2)},$$

where

$$p_Y(s) \triangleq \left((\ell + J/(m\ell))\, s^2 - \kappa g(m\ell)^2\right)\left(s\theta(0) + \dot{\theta}(0)\right) + (s^2 - \alpha^2)\left(sx(0) + \dot{x}(0)\right)$$

$$\alpha^2 = \frac{mg\ell(M+m)}{Mm\ell^2 + J(M+m)},\ \kappa = \frac{1}{Mm\ell^2 + J(M+m)}.$$

We considered a unity feedback control system for the inverted pendulum as shown in Figure 11.73. Using the QUANSER parameter values, the open-loop transfer function of the inverted pendulum is ($\kappa mg\ell = 36.57, \alpha^2 = 29.26$)

$$G_Y(s) = \frac{-36.57}{s^2(s^2 - 29.26)} = \frac{-36.57}{s^2(s+5.409)(s-5.409)}.$$

In order to arbitrarily place the closed-loop poles with a minimum order controller, $G_c(s)$ was chosen to be of the form

$$G_c(s) = \frac{b_c(s)}{a_c(s)} = \frac{b_3 s^3 + b_2 s^2 + b_1 s + b_0}{s^3 + a_2 s^2 + a_1 s + a_0}.$$

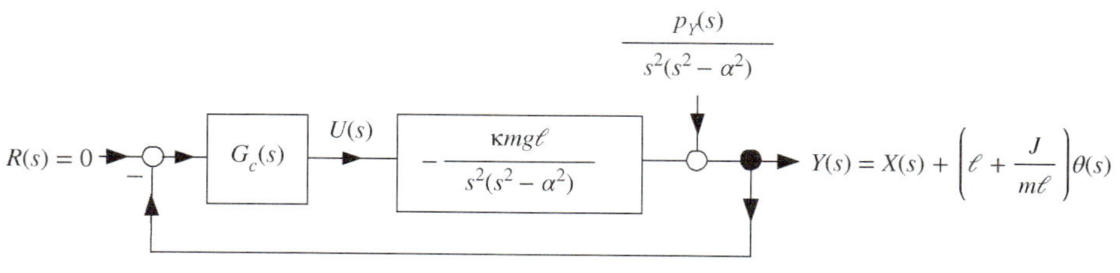

Figure 11.73. Closed-loop controller for the inverted pendulum.

With the closed-loop poles all placed at -5, it was shown in Example 5 of Chapter 10 to result in the controller

$$G_c(s) = -\frac{1041.6s^3 + 6113.7s^2 + 2990.8s + 2136.3}{s^3 + 35s^2 + 554.3s + 5399}$$

$$= -1041.6\frac{(s+5.409)(s^2+0.46s+0.3792)}{(s+20.835)(s^2+14.17s+259.13)}.$$

We now look at what Nyquist theory can tell us about this controller design. Figure 11.74a is the Nyquist contour and Figure 11.74b is the corresponding Nyquist plot of $G_c(s)G_Y(s)$. From the Nyquist plot it is seen that the closed-loop transfer function

$$\frac{KG_c(s)G_Y(s)}{1+KG_c(s)G_Y(s)}$$

is stable for $-1.135 < -\frac{1}{K} < -0.76$ or $0.881 = \frac{1}{1.135} < K < \frac{1}{0.76} = 1.282$. The phase margin turns out to be about $7°$. These small gain and phase margins indicate it would be difficult (probably unlikely) that this controller would work in practice, i.e., if the model parameters are a bit off then the closed-loop system will likely not be stable using this controller. We will say much more about this in Chapter 17, and show that a statespace controller (Chapter 15) for the inverted pendulum turns out to be quite robust!

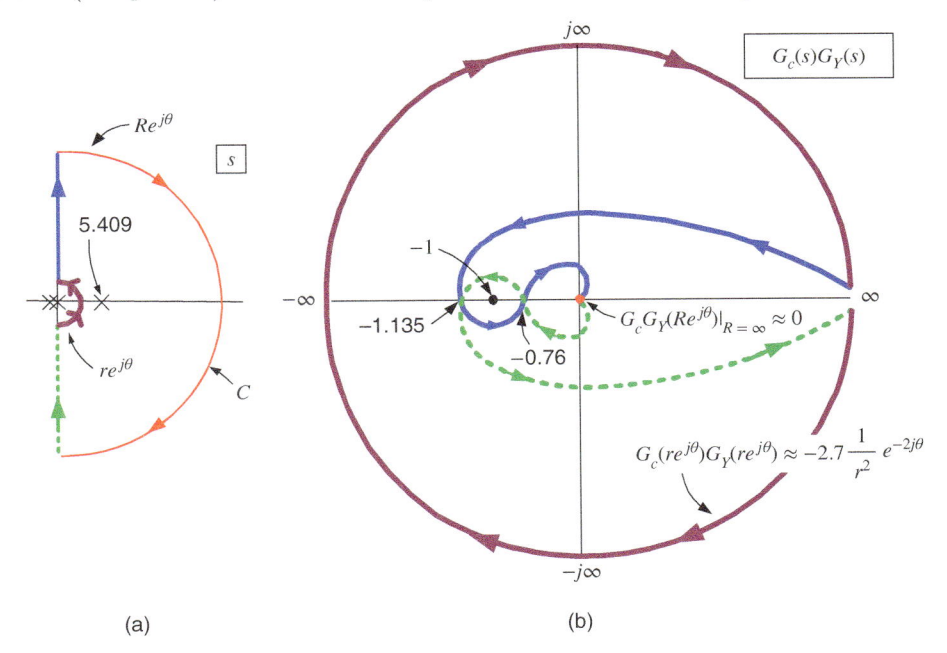

Figure 11.74. (a) Nyquist contour. (b) Nyquist plot.

Appendix: Bode and Nyquist Plots in Matlab

```
% Example 1 Magnitude and Phase on a single plot
close all; clear; clc
% Open-loop transfer function G(s) = (s+0.1)/(s+0.01)
```

```
den = [1 0.01]; num = [1 0.1]; sys = tf(num,den);
[mag phase wout] = bode(sys,{0.0001 100});
h = plotyy(wout,20*log10(squeeze(mag)),wout,squeeze(phase),'semilogx','semilogx');
grid on; set(h(1),'FontSize',20);set(h(2),'FontSize',20)
set(get(h(1),'Children'),'LineWidth',5)
%set(get(h(1),'Children'),'LineStyle',Dashed)
set(get(h(2),'Children'),'LineWidth',5)
set(h(1),'YTick',[-10, 0,5,10,15,20]); set(h(2),'YTick',[-90, -67.5, -45,-22.5 0])
set(h(1),'Ycolor',[0 0 1])
% Set color of right side to same color as left side
h1_color = get(h(1),'YColor'); set(h(2),'YColor',h1_color)
set(get(h(2),'Children'),'Color',h1_color)
title('Bode Diagram of G(j\omega)= (1+j0.1\omega) / (1+j0.01\omega)','FontSize',20)
xlabel('\omega rad/sec','FontSize',20)
set(get(h(1),'Ylabel'),'String','Magnitude in dB','FontSize',20)
set(get(h(2),'Ylabel'),'String','Phase in degrees','FontSize',20)

% Example 2 Separate Magnitude and Phase Plot
close all; clear; clc
% Gc(s) = (s+0.1)/(s+0.01)
den = [1 0.01]; num = [1 0.1];tf_openloop = tf(num,den);
h = bodeplot(tf_openloop); grid on
ax = findobj(gcf,'type','axes'); set(ax,'LineWidth',0.5);
ax = findall(gcf,'Type','axes'); set(ax,'GridColor','black')
options = getoptions(h);
options.MagVisible = 'on'; options.PhaseVisible = 'on';
options.Title.String = 'Bode Diagram of G(j\omega)= (1+j0.1\omega) / (1+j0.01\omega)';
options.XLabel.String = 'omega'; options.YLabel.String = {'Magnitude', 'Phase'};
options.Xlim ={[0.0001,100]};
options.TickLabel.FontSize = 25;
options.Title.FontSize = 25; options.XLabel.FontSize = 25; options.YLabel.FontSize = 25;
setoptions(h,options);
% workarounds for line width
lin_width = findobj(gcf,'type','line');
set(lin_width,'linewidth',4);

%Example 1 Nyquist
close all; clear; clc
% Gc(s) = 1/(s+1)(s+3)
den1 = [1 4 3]; num1 = [1];tf1 = tf(num1,den1);w = {0.01,100};
nyquist(tf1,w)

% Example 2 Nyquist
figure
% Gc(s) = 10*(s+1)/s(s-10)
den2 = [1 -10 0]; num2 = [10 10];tf2 = tf(num2,den2);w = {0.1,1000};
nyquist(tf2,w);
```

Problems

Problem 1 *Bode Diagram of a First-Order Unstable Zero*

Draw the Bode diagram (magnitude and phase) of $G(j\omega) = -\tau j\omega + 1$ with $\tau = -0.2$. You should be able to do this quickly based on the examples given in the text.

In Problems 2–7 find the transfer function $G(s)$ corresponding to the given Bode plot. Using MATLAB, replot the Bode diagram from the expression you found for $G(s)$. In each of these Bode plots the breakpoints for the poles and zeros are integers, and are separated by at least a decade. This means, e.g., there cannot be a Bode diagram of $G(s) = (s+2)/(s+6)$ as the breakpoints at 2 and 6 are not separated by a decade.

Problem 2 *Bode Plot*
In Figure 11.75 $G(s)$ has a single real pole.

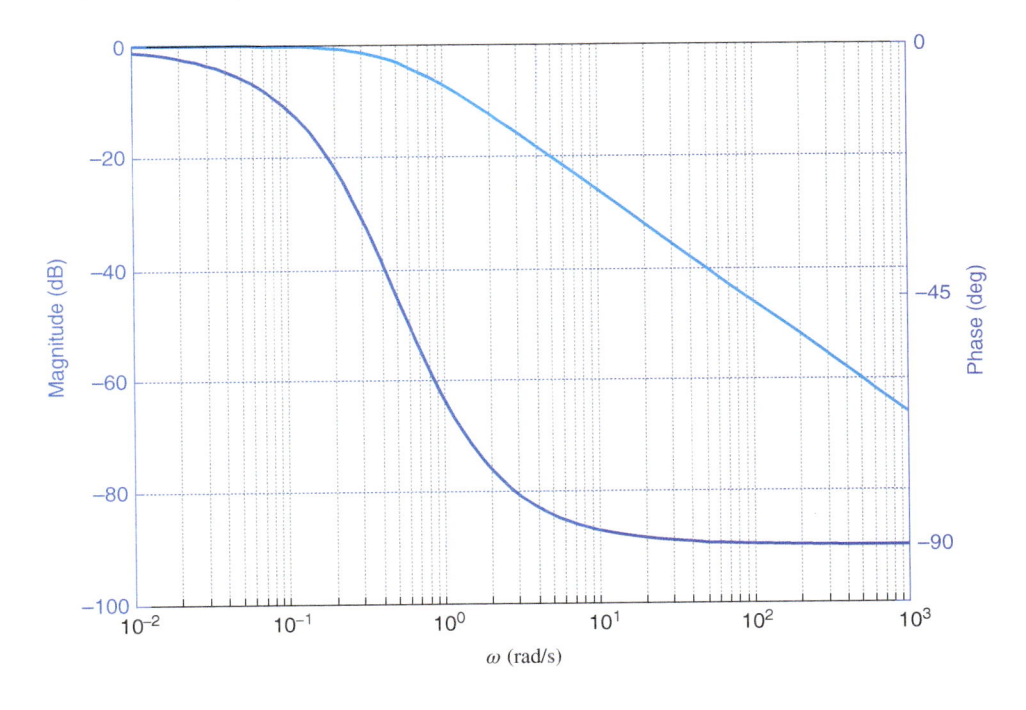

Figure 11.75. $G(s) = ?$

Problem 3 *Bode Plot*
$G(s)$ has two real poles. Use the Bode diagram of Figure 11.76 on the next page to determine $G(s)$.

Problem 4 *Bode Plot*
$G(s)$ has two real poles. Use the Bode diagram of Figure 11.77 on the next page to determine $G(s)$.

Problem 5 *Bode Plot*
$G(s)$ has a complex conjugate pair of poles with a breakpoint at 10. Use the Bode diagram of Figure 11.78 on page 431 to determine $G(s)$. From the Bode diagram $20\log_{10}|G(j10)| = 1.94$.

Problem 6 *Bode Plot*
$G(s)$ has a real zero and a complex conjugate pair of poles whose breakpoint is at 10. Use the Bode diagram of Figure 11.79 on page 431 to determine $G(s)$.

430   11 Frequency Response Methods

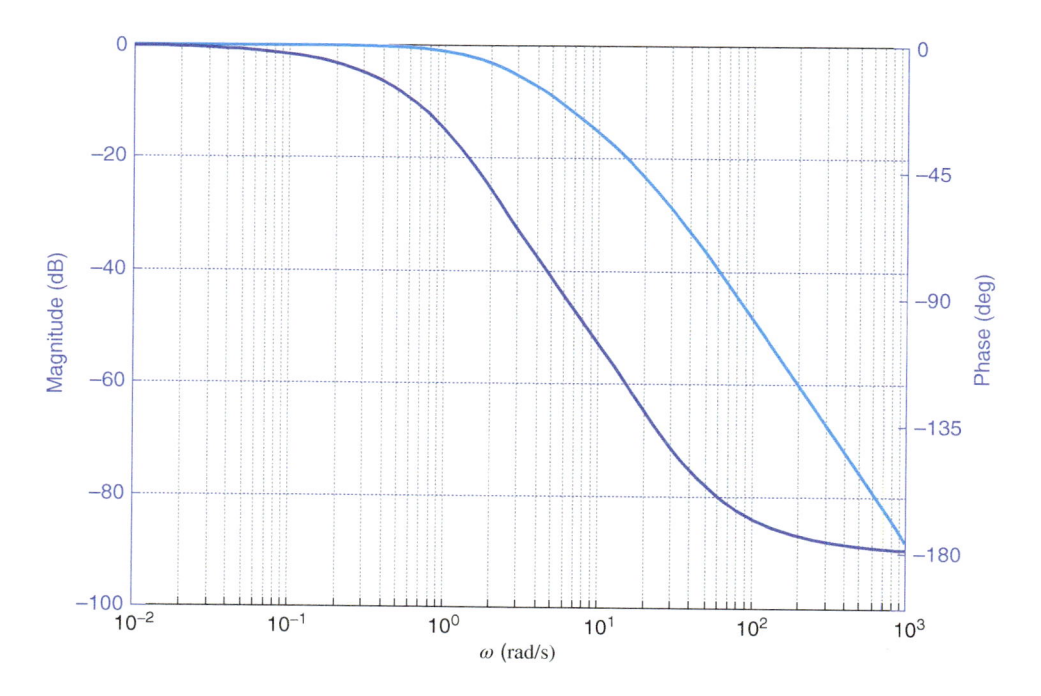

Figure 11.76. $G(s) = ?$

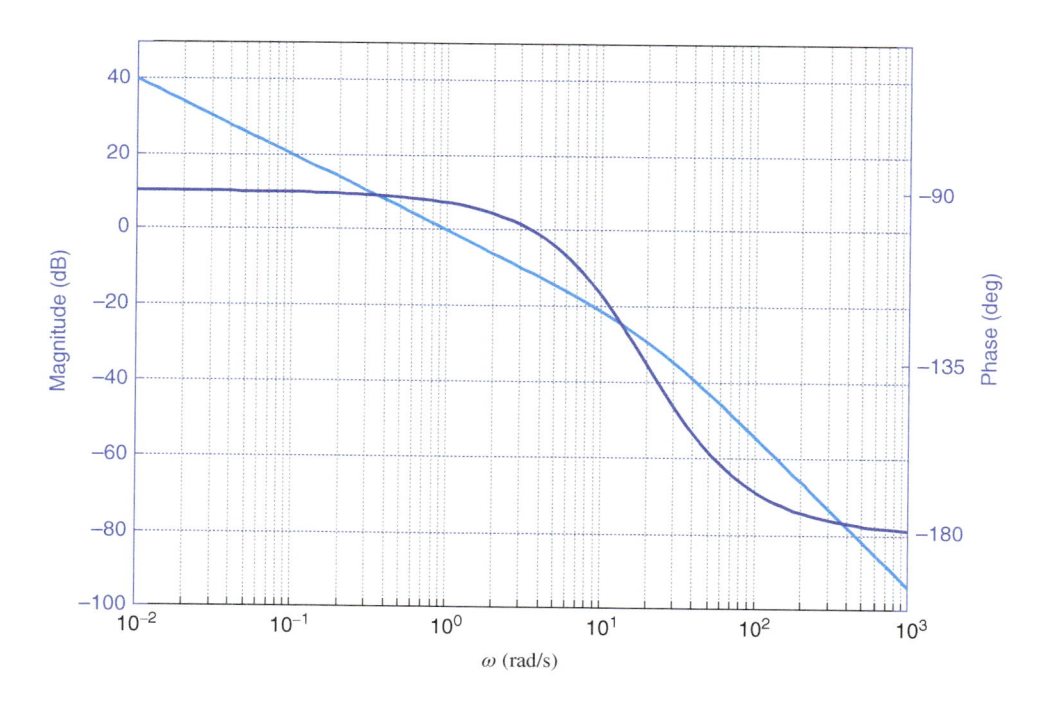

Figure 11.77. $G(s) = ?$

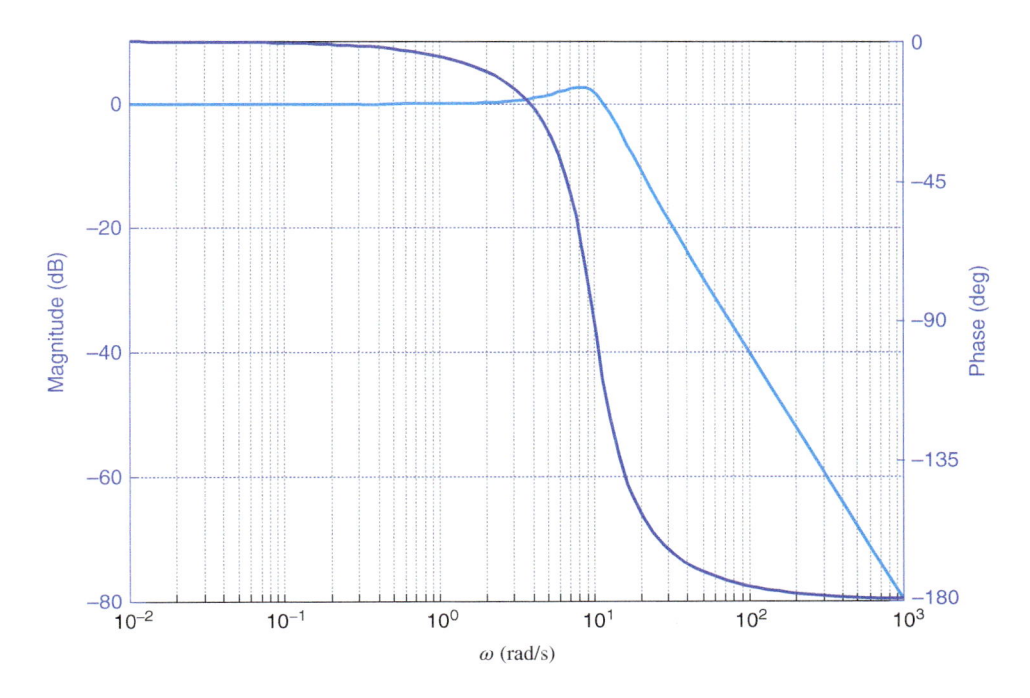

Figure 11.78. $G(s) = ?$

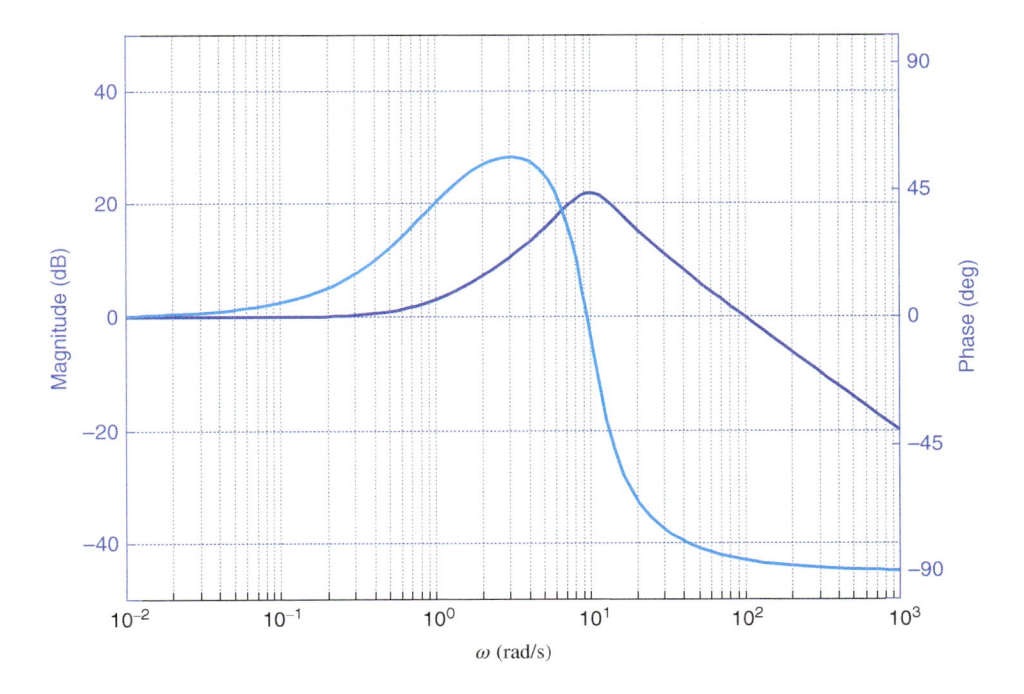

Figure 11.79. $G(s) = ?$

432   11 Frequency Response Methods

Problem 7 *Bode Plot*

$G(s)$ has a real zero and two real poles. Use the Bode diagram of Figure 11.80 to determine $G(s)$.

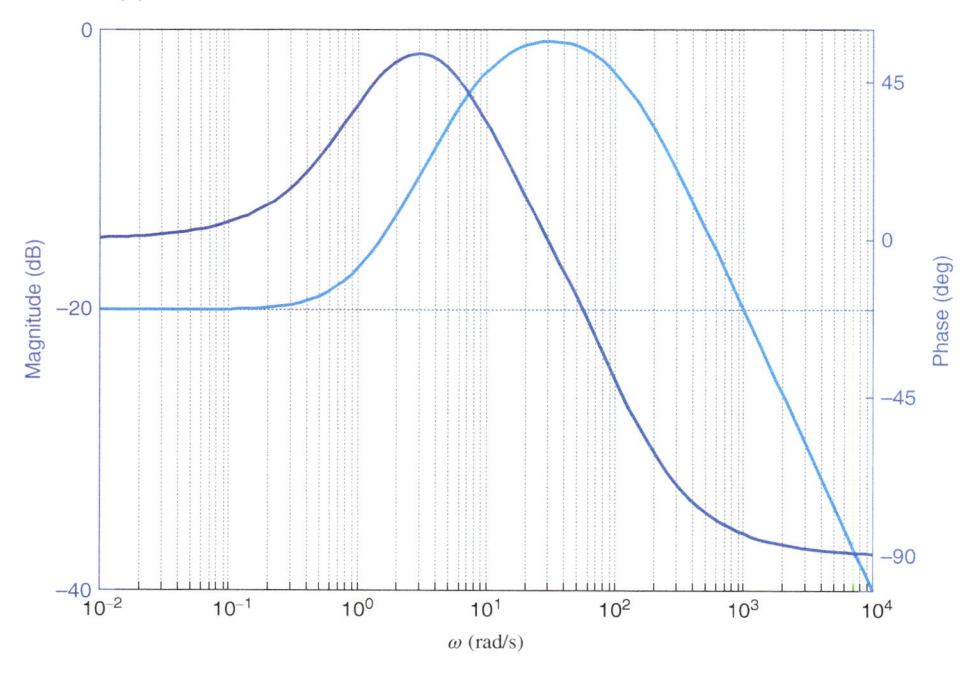

Figure 11.80. $G(s) = ?$

Problem 8 *Bode Plot*

Consider a transfer function $G(s)$ whose asymptotic Bode magnitude plot is shown in Figure 11.81.

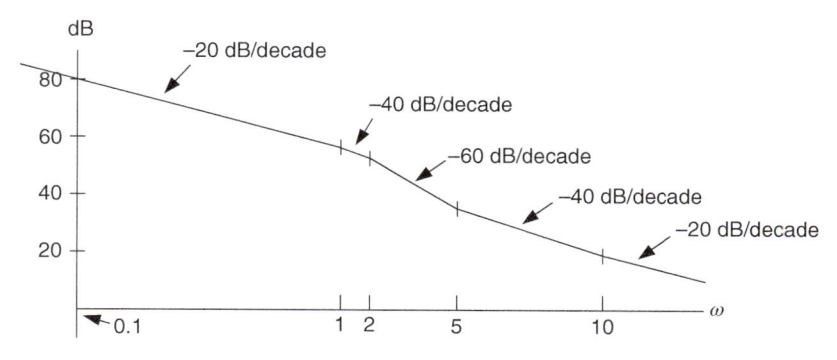

Figure 11.81. Asymptotic Bode magnitude plot.

(a) Let $G_1(s)$ have the asymptotic Bode magnitude plot given in Figure 11.81. Suppose all the poles and zeros of $G_1(s)$ are in the open *left* half-plane or on the $j\omega$ axis. Find the transfer function $G_1(s)$.

(b) Let $G_2(s)$ have the asymptotic Bode magnitude plot given in Figure 11.81. Suppose all the poles and zeros of $G_1(s)$ are in the open *right* half-plane or on the $j\omega$ axis. Find the transfer function $G_2(s)$.

Problem 9 *Principle of the Argument*

Let $G(s) = 1/(s-1)$ and consider the Nyquist contour of Figure 11.82. Following the procedure of Example 6, draw the corresponding Nyquist plot and use it to verify the principle of the argument.

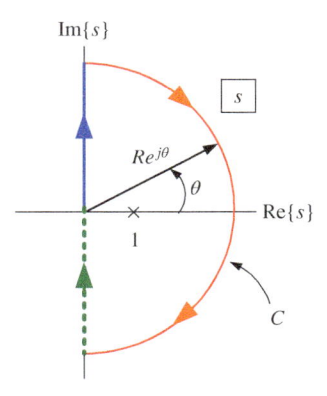

Figure 11.82. Nyquist contour.

Problem 10 *Principle of the Argument*

Let $G(s) = 1/(s-1)$ and consider the Nyquist contour of Figure 11.83. Following the procedure of Example 7, draw the corresponding Nyquist plot and use it to verify the principle of the argument.

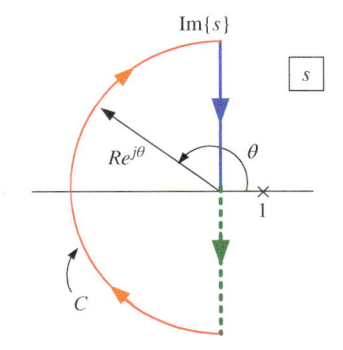

Figure 11.83. Nyquist contour.

Problem 11 *Routh–Hurwitz Test of* $1 + K\dfrac{10(s+1)}{s(s-10)}$

Check the answer of Example 12 using the Routh–Hurwitz test.

Problem 12 *Polar Plot of* $G(s) = \dfrac{1}{(s+1)(s^2 + \sqrt{2}s + 1)}$

(a) The *phase crossover* frequency is the value of frequency ω_ϕ such that $\angle G(j\omega_\phi) = 180°$. For the transfer function in Example 13 show that $\omega_\phi = 1.55$ and $G(j1.55) = -1/4.8$ by computing $G(j\omega)$ and setting the imaginary part to 0.

(b) Show that $G(j1) = \dfrac{1}{2}e^{-j3\pi/4}$.

434   11 Frequency Response Methods

Problem 13 *Routh–Hurwitz Test of* $1 + K\dfrac{1}{(s+1)(s^2+\sqrt{2}s+1)}$

Check Example 13 using the Routh–Hurwitz test.

Problem 14 *Polar Plot of* $\dfrac{1}{s(s+1)^2}$

(a) The *phase crossover* frequency is the value of frequency ω_ϕ such that $\angle G(j\omega_\phi) = 180°$. For the transfer function of Example 14 show that $\omega_\phi = 1$ and $G(j\omega_\phi) = -1/2$.

(b) The *gain crossover* frequency is the value of ω_g such that $|G(j\omega_g)| = 1$. For the transfer function of Example 14 show that $\omega_g = 0.68$ and $\angle G(j\omega_g) = -159°$.

Problem 15 *Routh–Hurwitz Test of* $1 + K\dfrac{1}{s(s+1)^2}$

Check the results of Example 14 using the Routh–Hurwitz test.

Problem 16 *Polar Plot of* $G(s) = \dfrac{100(s/10 + 1)}{s(s-1)(s/100+1)}$

(a) The *phase crossover* frequency is the value of frequency ω_ϕ such that $\angle G(j\omega_\phi) = -180°$. For the transfer function of Example 15 show that $\omega_\phi = 3.35$ and $G(j\omega_\phi) = -9$.

(b) The *gain crossover* frequency is the value of ω_g such that $|G(j\omega_g)| = 1$. For the transfer function of Example 15 compute ω_g and $\angle G(j\omega_g)$.

Problem 17 *Routh–Hurwitz Test for* $1 + KG(s) = \dfrac{100(s/10+1)}{s(s-1)(s/100+1)}$

Check the results of Example 15 using the Routh–Hurwitz test.

Problem 18 *Nyquist Stability Test*
Consider the simple proportional feedback system in Figure 11.84.

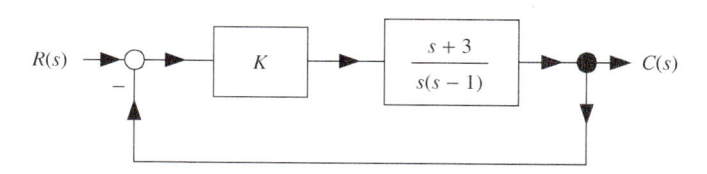

Figure 11.84. Proportional feedback control system.

The Bode diagram for $G(s)$ is given in Figure 11.85.

(a) Draw the Nyquist contour for $G(s)$ in the s-plane.

(b) Draw the Nyquist plot for $G(s)$ in the $G(s)$-plane.

(c) Do a stability analysis using Nyquist theory to find the values of K for which the closed-loop transfer function has zero poles in the open right half-plane, one pole in the open right half-plane, and two poles in the open right half-plane. Hint: Plot $G(j\omega)$ and then check how many times it goes around the $-1/K$ point for $-\infty < K < \infty$.

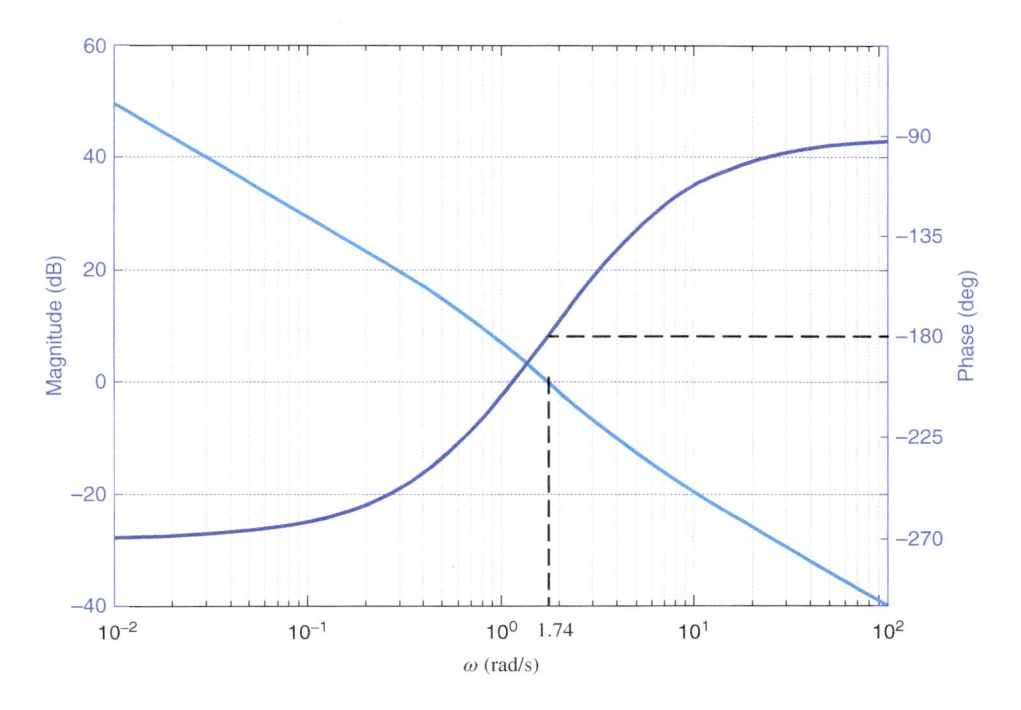

Figure 11.85. $G(j1.74) = 1e^{-j180°}$.

Problem 19 *Nyquist Stability Test for a Non-Minimum Phase System*
Consider the simple proportional feedback system in Figure 11.86.

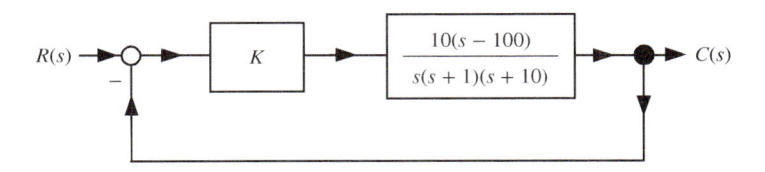

Figure 11.86. Proportional feedback control system.

The Bode diagram for $G(s)$ is given in Figure 11.87.

(a) Draw the Nyquist contour for $G(s)$ in the s-plane.

(b) Draw the Nyquist plot for $G(s) = \dfrac{10(s-100)}{s(s+1)(s+10)}$ in the $G(s)$-plane.

(c) Do a stability analysis using Nyquist theory to find the values of K for which the closed-loop transfer function has zero poles in the open right half-plane, one pole in the open right half-plane, and two poles in the open right half-plane. Hint: Plot $G(j\omega)$ and then check how many times it goes around the $-1/K$ point for $-\infty < K < \infty$.

436   11   Frequency Response Methods

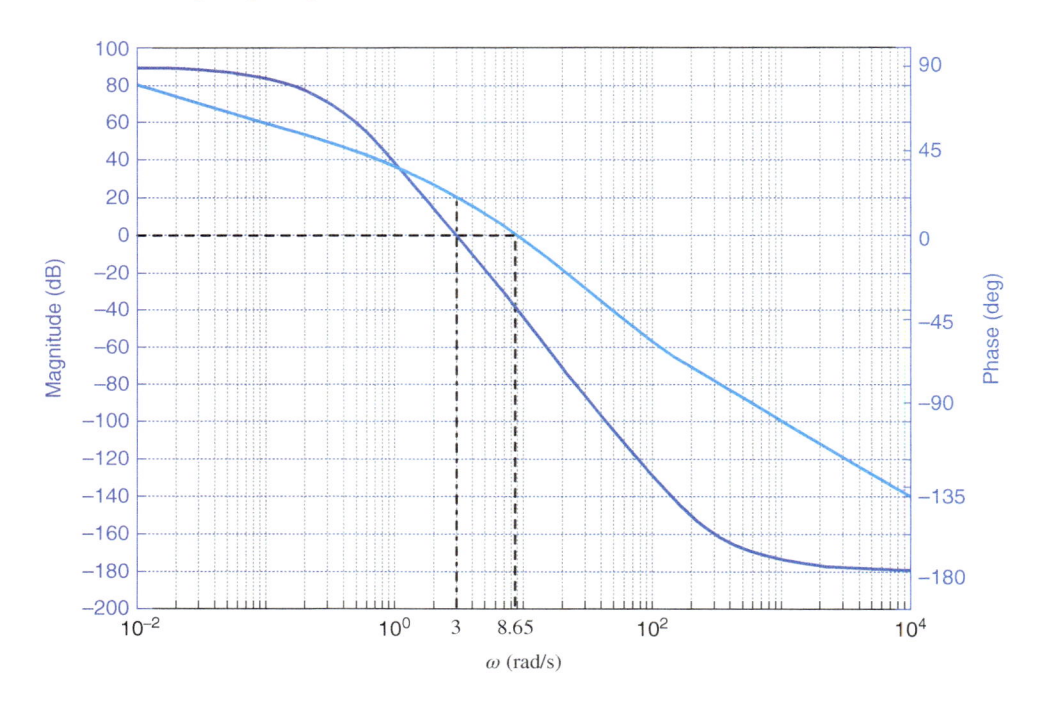

Figure 11.87. $G(j3) = 10e^{j0°}$ and $G(j8.65) = 1e^{-j39°}$.

Problem 20 *Nyquist Stability Test*

Consider the simple proportional feedback system in Figure 11.88. The Bode diagram for $G(s) = \dfrac{1}{s(s+2)^2}$ is given in Figure 11.89.

Figure 11.88. Simple proportional feedback control system.

(a) Draw the Nyquist contour for $G(s)$ in the s-plane.

(b) Draw the Nyquist plot for $G(s) = \dfrac{1}{s(s+2)^2}$ in the $G(s)$-plane.

(c) Do a stability analysis using Nyquist theory to find the values of K for which the closed-loop transfer function has zero poles in the open right half-plane, one pole in the open right half-plane, and two poles in the open right half-plane. Hint: Plot $G(j\omega)$ and then check how many times it goes around the $-1/K$ point for $-\infty < K < \infty$.

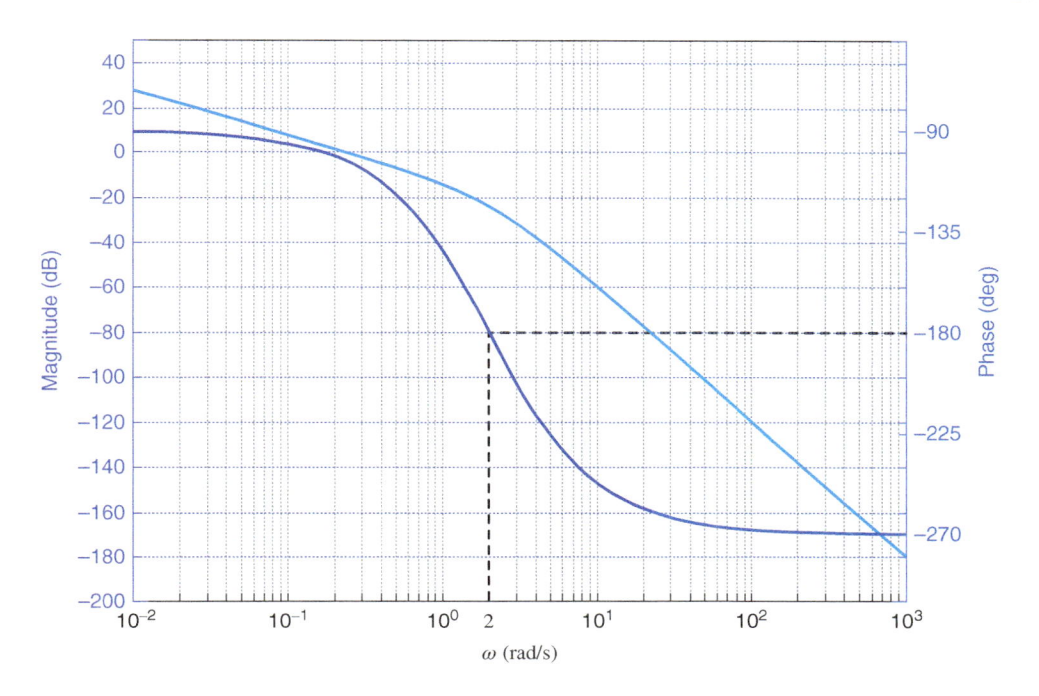

Figure 11.89. $G(j2) = \frac{1}{16}e^{-j180°}$.

Problem 21 *Nyquist Stability Test for a Non-Minimum Phase System*

Let
$$G(s) = \frac{1}{2}\frac{s-1}{s(s+1)}.$$

(a) Make a Bode diagram for $G(s)$. Use this Bode diagram to help draw the Nyquist polar plot for this transfer function.

(b) Do a stability analysis using Nyquist theory to find the values of K for which the closed-loop transfer function has zero poles in the open right half-plane, one pole in the open right half-plane, and two poles in the open right half-plane. Hint: Plot $G(j\omega)$ and then check how many times it goes around the $-1/K$ point for $-\infty < K < \infty$.

(c) Check your answer to part (b) using the Routh–Hurwitz test.

(d) If you did part (c) correctly you found that the system is not closed-loop stable for $K > 0$. In order to give the system a standard gain and phase margin interpretation we simply put a minus sign in the open-loop system block and proceed with $K > 0$. See Figure 11.90.

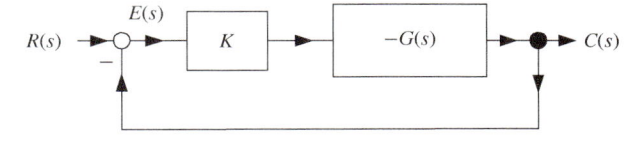

Figure 11.90. Gain and phase margin.

Draw the Bode diagram for $-G(s)$ and use it to sketch the Nyquist plot of $-G(s)$. Then find the gain and phase margins of the closed-loop system.

Problem 22 *Gain and Phase Margin*

Figure 11.91 is the Bode diagram of some open-loop system.

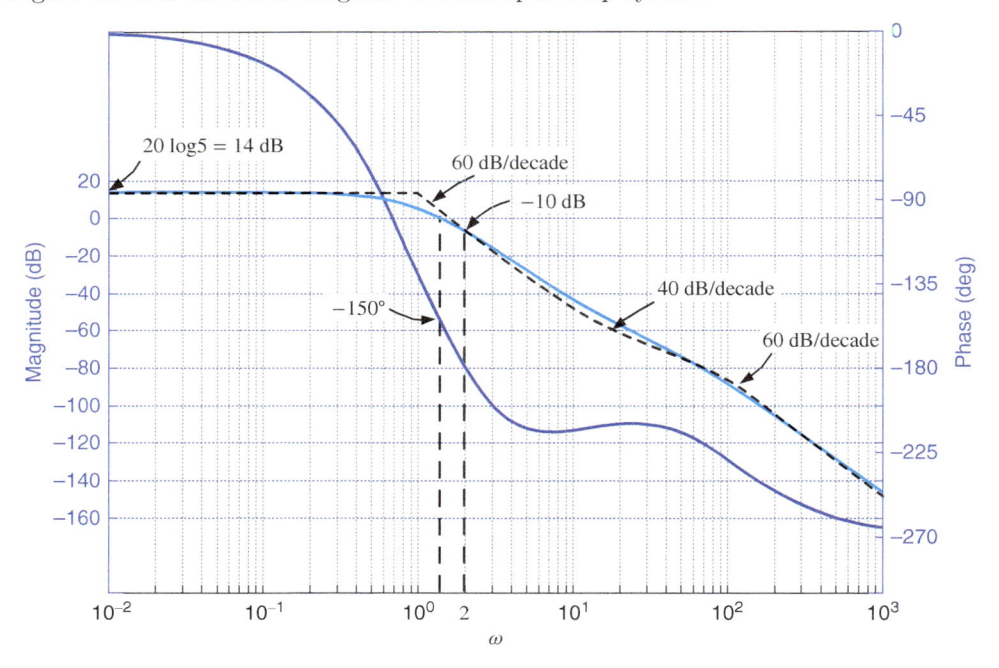

Figure 11.91. Bode diagram of an open-loop system.

(a) Find the transfer function of this system.

(b) Let $G(s)$ be in a unity feedback configuration as shown in Figure 11.92.

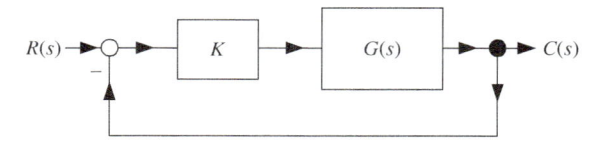

Figure 11.92. Unity feedback control configuration.

With $K = 1$ compute the closed-loop gain and phase margins.

Problem 23 *Gain and Phase Margin*

Consider a system similar to Example 12 given by

$$G(s) = \frac{20(s+1)}{s(s-10)} = -\frac{2(s+1)}{s(-s/10+1)}.$$

(a) Draw the Nyquist contour and Nyquist polar plot for $G(s)$.

(b) Fill in the following table.

$\omega_g = ?$	$20\log_{10}|G(j\omega_g)| = 0$ dB	$\angle G(j\omega_g) = ?$
$\omega_\phi = ?$	$20\log_{10}|G(j\omega_\phi)| = ?$ dB	$\angle G(j\omega_\phi) = -180°$

(c) What are the gain and phase margins? Mark the gain and phase margins on your Bode diagram.

Problem 24 *Gain and Phase Margin*

Let $G(s) = \dfrac{1}{s(s+1)}$

(a) Make a Bode diagram for $G(s)$. Use the Bode diagram to draw the Nyquist polar plot for this transfer function. Fill in the following table.

$\omega_g = ?$	$20\log_{10}|G(j\omega_g)| = 0$ dB	$\angle G(j\omega_g) = ?$
$\omega_\phi = ?$	$20\log_{10}|G(j\omega_\phi)| = ?$ dB	$\angle G(j\omega_\phi) = -180°$

(b) What are the gain and phase margins for this system. Mark them on your Bode diagram.

Problem 25 MATLAB *command* nyquist

Use the nyquist command from MATLAB to draw the polar plot of Example 13 (see the Appendix to this chapter). Compare this plot with the sketch in Figure 11.40.

Problem 26 MATLAB *command* nyquist

Use the nyquist command from MATLAB to draw the polar plot of Example 14 (see the Appendix to this chapter). Compare this plot with the sketch in Figure 11.43.

Problem 27 MATLAB *Command* nyquist

Use the nyquist command from MATLAB to draw the polar plot of Example 15 (see the Appendix to this chapter). Compare this plot with the sketch in Figure 11.46.

Problem 28 *Gain and Phase Margin for a Double Integrator Using Pole Placement*

In Chapter 10 a pole placement controller was designed for the double integrator system shown in Figure 11.93.

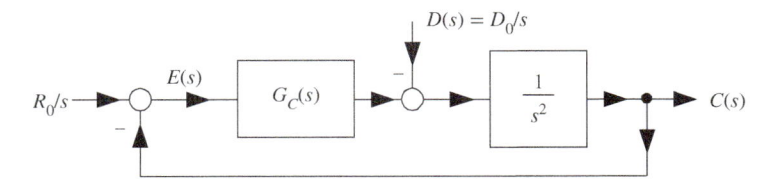

Figure 11.93. Pole placement for a double integrator system.

There it was shown that the controller

$$\frac{6s^2 + 4s + 11}{s+4}\frac{1}{s}$$

placed all four closed-loop poles at -1.

(a) Plot the Bode diagram for $G_c(s)G(s) = \dfrac{6s^2 + 4s + 11}{s+4}\dfrac{1}{s}\dfrac{1}{s^2}$ and use this to compute the gain and phase margin.

(b) Sketch the Nyquist contour for $G_c(s)G(s) = \dfrac{6s^2 + 4s + 11}{s+4}\dfrac{1}{s}\dfrac{1}{s^2}$.

(c) Show that $G(re^{j\theta})G_c(re^{j\theta}) \approx 0.25e^{-j3\theta}/r^3$ for r small and use this to fill out the following table.

θ	$G(re^{j\theta})G_c(re^{j\theta}) \approx 0.25e^{-j3\theta}/r^3$
$-\pi/2$	
$-\pi/3$	
$-\pi/4$	
0	
$+\pi/4$	
$+\pi/3$	
$+\pi/2$	

(d) Use your answers for parts (a),(b), and (c) to sketch the polar plot for $G_c(s)G(s) = \dfrac{6s^2 + 4s + 11}{s+4}\dfrac{1}{s}\dfrac{1}{s^2}$.

Problem 29 *Bode Plots and Nyquist Theory*

Figure 11.94 shows the Bode magnitude plot for some system.

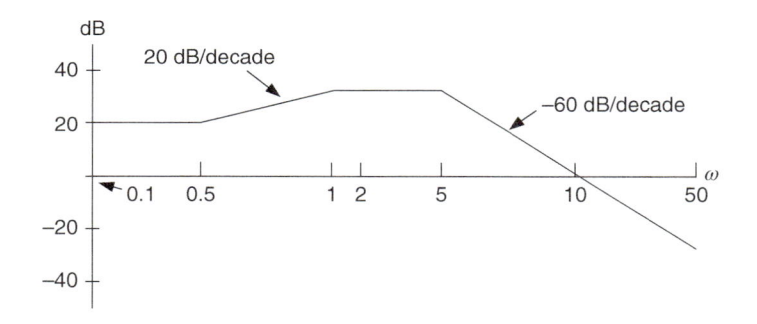

Figure 11.94. $G(s) = ?$

(a) Find the transfer function $G(s)$. Assume that *all* the poles and zeros are in the open left-half plane.

(b) It turns out that $G(j9.05) = 2.25e^{-j180°}$ and $G(j12.6) = 1e^{-j202.8°}$. Sketch the Nyquist plot.

(c) Is the closed-loop system stable? Explain why or why not using Nyquist theory.

Problem 30 *System with Two Right Half-Plane Poles and a Right Half-Plane Zero.*

In Example 7 of Chapter 10 the control system of Figure 11.95 was considered.

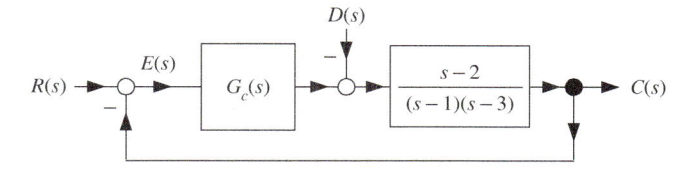

Figure 11.95. System with right half-plane poles and zeros.

The open-loop system model is $G(s) = \dfrac{s-2}{(s-1)(s-3)}$ and the controller

$$G_c(s) = \frac{50.5s^2 - 66s - 0.5}{s - 42.5}\frac{1}{s} = \frac{50.5(s - 1.3145)(s + 0.0075)}{s - 42.5}\frac{1}{s}$$

places all four closed-loop poles at -1. It was pointed out that this controller was not robust, that is, a small change in $G(s)$ results in the closed-loop system not being stable (see page 306). Nyquist theory is now used to look at this in more detail.

The Bode diagram for $G_c(s)G(s)$ is given in Figure 11.96. The Nyquist contour and the corresponding Nyquist plot for $G_c(s)G(s)$ are given in Figure 11.97 on the next page. On the three loops of the Nyquist plot encircling -1.

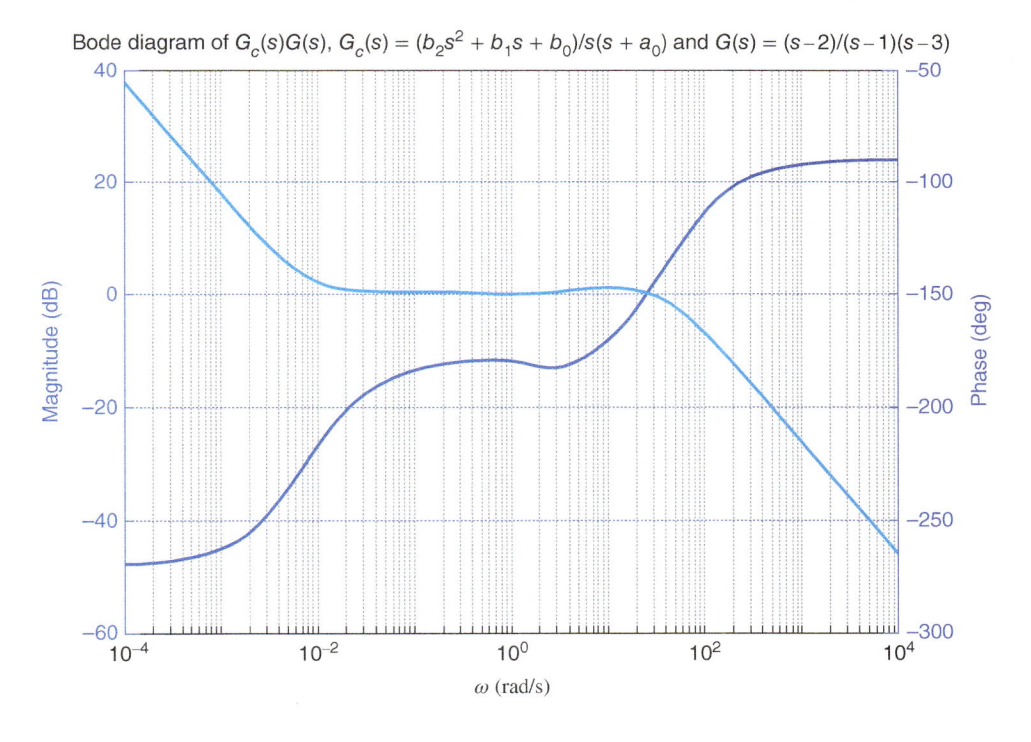

Figure 11.96. Bode diagram for $G_c(s)G(s) = \dfrac{50.5s^2 - 66s - 0.5}{s - 42.5}\dfrac{1}{s}\dfrac{s-2}{(s-1)(s-3)}$.

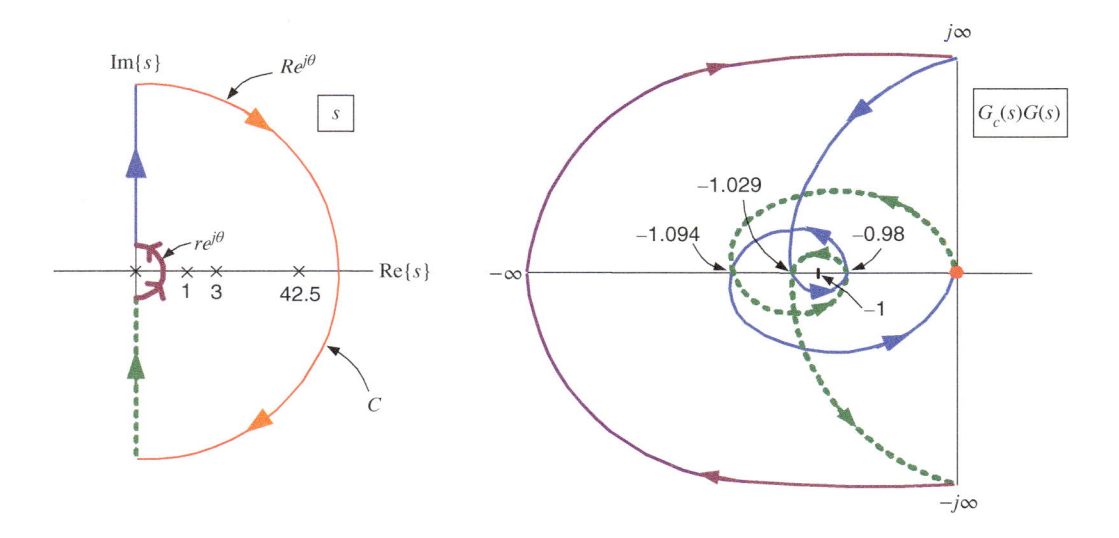

Figure 11.97. Nyquist contour and Nyquist plot for $G_c(s)G(s)$.

There are frequencies where $|G(j\omega)|$ is quite close to 1 and $\angle G(j\omega)$ is quite close to $-180°$. (The Nyquist plot is not drawn to scale!) This of course corresponds to the part of Bode plot where the magnitude is about 0 dB and the phase is about $-180°$.

(a) Using Nyquist theory find the values of K such that the closed-loop system $\dfrac{KG_c(s)G(s)}{1+KG_c(s)G(s)}$ is stable.

(b) Find the intervals of K for which the closed-loop system is unstable. For each interval give the number of closed-loop poles in the right half-plane.

Problem 31 *Gain and Phase Margin of the Satellite with Solar Panels (Collocated Case)*

In Chapter 10 a satellite with solar panels with the sensor and actuator collocated had the transfer function model

$$G(s) = \frac{\theta(s)}{\tau(s)} = \frac{0.2(s^2+0.05s+0.15)}{s^2(s^2+0.06s+0.18)}.$$

A feedback controller was designed whose transfer function was

$$G_c(s) = \frac{16.97s^4+9.854s^3+4.6875s^2+0.9375s+0.0781}{s^4+2.99s^3+0.297s^2+0.441s}$$

$$= \frac{16.97(s^2+0.28678s+0.028087)(s^2+0.2939s+0.16385)}{s(s+2.94)(s^2+0.05s+0.15)},$$

which placed the eight closed-loop poles at $-0.025\pm j0.387, -0.5, -0.5, -0.5, -0.5, -0.5, -0.5$.

(a) The Nyquist contour for $G_c(s)G(s)$ is given in Figure 11.98.

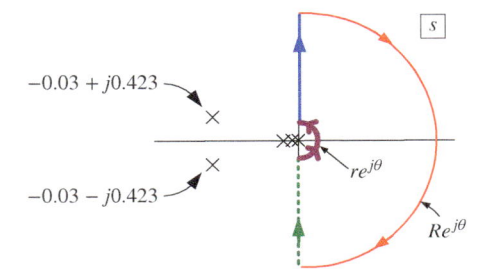

Figure 11.98. Nyquist contour for $G_c(s)G(s)$.

The corresponding polar plot is given in Figure 11.99.

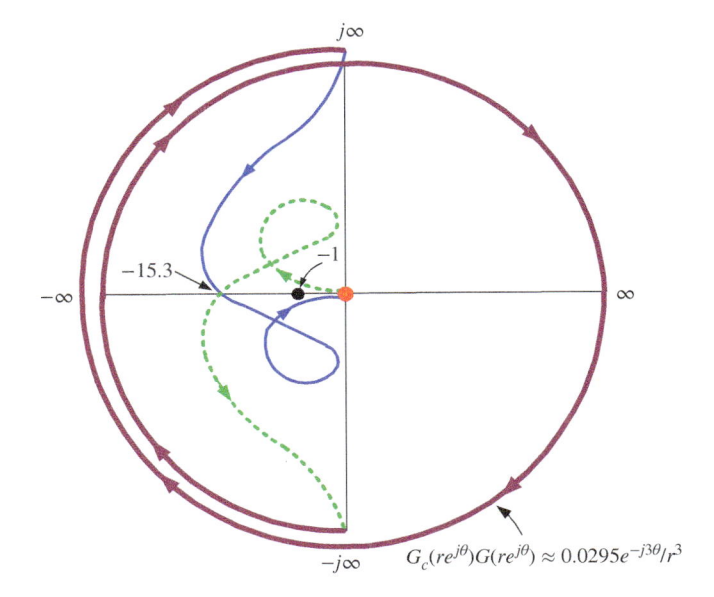

Figure 11.99. Polar plot of $G_c(s)G(s)$.

Use Nyquist theory to determine if the closed-loop system is stable or not. That is, determine P and N to compute $Z = N + P$.

(b) Use MATLAB to plot the Bode diagram of $G_c(j\omega)G(j\omega)$ for $-\infty < \omega < \infty$. Using the Bode diagram and the polar plot, compute the gain and phase margins of the system and indicate them on the Bode diagram. The `margin` command in MATLAB will also be helpful.

Problem 32 *Gain and Phase Margins of a 2 DOF Controller for the DC Motor with a Flexible Shaft*

In Problem 25 of Chapter 10 a 2 DOF controller for a DC motor with a flexible shaft was considered. This control system is given again in Figure 11.100.

444   11   Frequency Response Methods

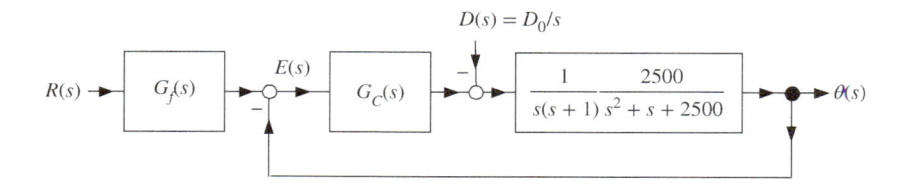

Figure 11.100. Two DOF control system for a DC motor with a flexible shaft.

The transfer function model from voltage input to the angular position of the flexible shaft is (see [3]) (Figure 11.100)

$$G(s) = \frac{1}{s(s+1)} \frac{2500}{(s^2+s+2500)} = \frac{2500}{s^4 + 2s^3 + 2501s^2 + 2500s}.$$

The controller turned out to be

$$G_c(s) = \frac{563.352s^4 + 4013.3616s^3 + 1453504s^2 + 10100000s + 25000000}{s^4 + 48s^3 + 3428s^2 + 106596s}$$

with the zeros of the controller at $-06.3408 \pm j50.615, -3.4986 \pm j2.2543$. The Nyquist contour for $G_c(s)G(s)$ is given in Figure 11.101 and the Nyquist plot is shown in Figure 11.102.

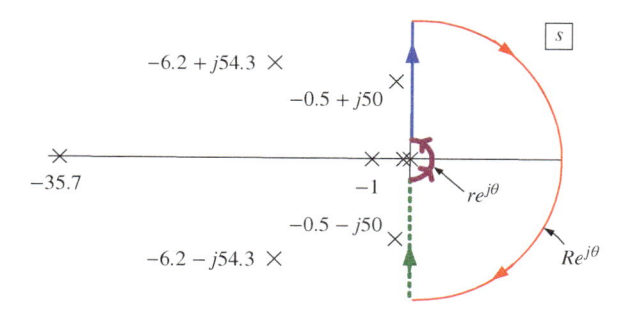

Figure 11.101. Nyquist contour for $G_c(s)G(s)$.

(a) Using the Nyquist contour and polar plot, determine N and P to compute Z to show the closed-loop system is stable.

(b) Graph the Bode diagram of $G_c(s)G(s)$ and use it along with the polar plot to determine the gain and phase margins.

Problem 33 *Gain and Phase Margins for a PID Notch Controller of a DC Motor with a Flexible Shaft*

In Problem 26 of Chapter 10 a PID controller with a notch filter for a DC motor with a flexible shaft was considered. The transfer function from input to the angular position of the flexible shaft is (see [3])

$$G(s) = \frac{1}{s(s+1)} \frac{2500}{(s^2+s+2500)} = \frac{2500}{s^4 + 2s^3 + 2501s^2 + 2500s}.$$

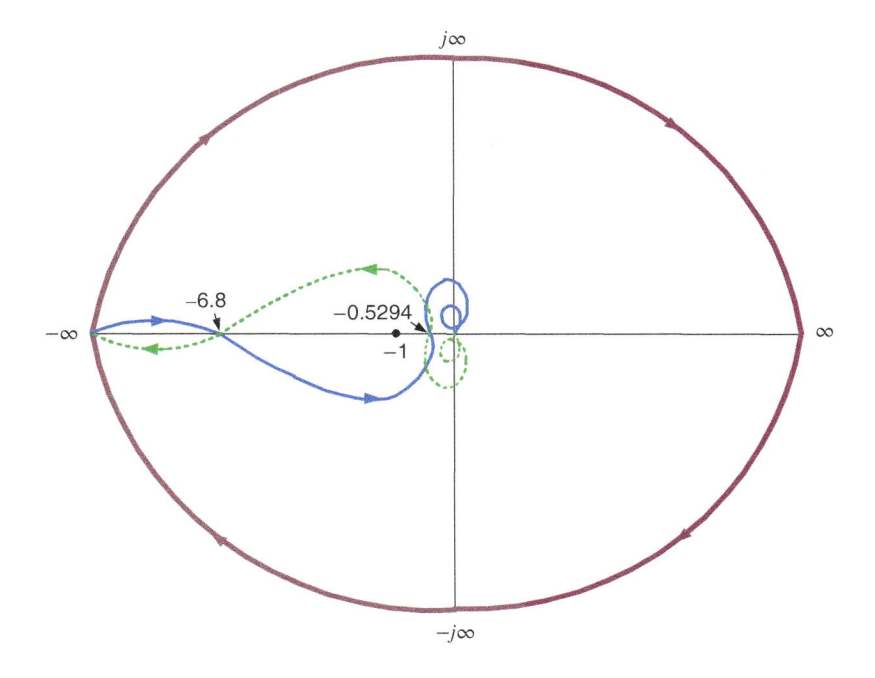

Figure 11.102. Polar plot for $G_c(s)G(s)$.

The controller is

$$G_c(s) = 91\frac{s+2}{s+13}\frac{s+0.05}{s}\frac{s^2+0.8s+3600}{s^2+120s+3600}$$

and is a modification of the controller given in [3] (see Problem 26 of Chapter 10). Figure 11.103 is a block diagram for this control system.

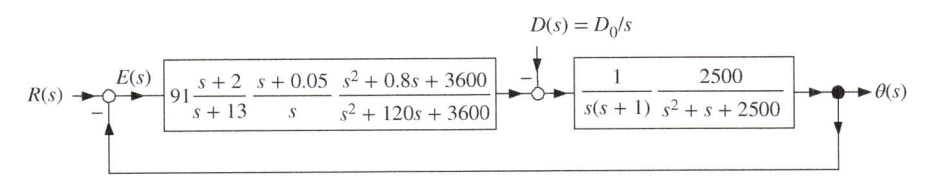

Figure 11.103. PID controller with the notch filter $\frac{s^2+0.8s+3600}{s^2+120s+3600}$.

(a) Graph the Bode diagram of $G_c(s)G(s)$ and determine the gain and phase margins.

(b) Graph the Bode diagram of $G_{notch}(s) = \frac{s^2+0.8s+3600}{s^2+120s+3600}$ for $1 \leq \omega \leq 10^4$ to see the "notch" in the amplitude plot at $\omega = 60$ rad/s.

12
Root Locus

With $m < n$ let an open-loop transfer function $G(s)$ be given by

$$G(s) = \frac{b(s)}{a(s)} = \frac{b_m s^m + b_{m-1} s^{m-1} + \cdots + b_1 s + b_0}{s^n + a_{n-1} s^{n-1} + \cdots + a_1 s + a_0}. \tag{12.1}$$

Consider a proportional feedback controller as given in Figure 12.1.

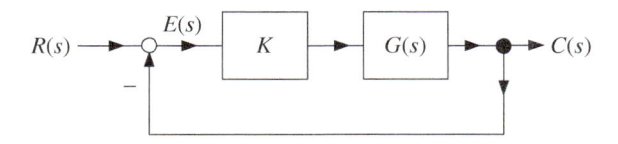

Figure 12.1. System block diagram.

The closed-loop transfer function is

$$\frac{C(s)}{R(s)} = \frac{KG(s)}{1+KG(s)} = \frac{K\frac{b(s)}{a(s)}}{1+K\frac{b(s)}{a(s)}} = \frac{Kb(s)}{a(s)+Kb(s)}. \tag{12.2}$$

In this simple proportional feedback control system we are interested in the movement of the closed-loop poles as the value of K is varied. Specifically, as K varies from 0 to ∞, we want to determine the location of the roots of

$$a(s) + Kb(s) = 0.$$

If one desires the root locations for $K < 0$, simply replace $G(s)$ by $-G(s)$ and still take $K \geq 0$.

We refer to the polynomial $a(s) + Kb(s)$ as the *closed-loop characteristic* polynomial and its roots are the closed-loop poles. The *root locus* is a plot of the roots of $a(s) + Kb(s) = 0$ as K is varied from 0 to ∞. Let's do a simple example by "brute force," that is, by directly solving for the roots.

Example 1 *Root Locus for* $\dfrac{1}{s(s+4)}$

Let

$$G(s) = \frac{1}{s(s+4)}$$

An Introduction to System Modeling and Control, First Edition. John Chiasson.
© 2022 John Wiley & Sons, Inc. Published 2022 by John Wiley & Sons, Inc.
Companion website: www.wiley.com/go/chiasson/anintroductiontosystemmodelingandcontrol

448   12 Root Locus

so that the closed-loop transfer function is then

$$\frac{C(s)}{R(s)} = \frac{KG(s)}{1+KG(s)} = \frac{K\frac{1}{s(s+4)}}{1+K\frac{1}{s(s+4)}} = \frac{K}{s^2+4s+K}.$$

We want to compute the roots of

$$s^2 + 4s + K = 0$$

for $K \geq 0$. Using the quadratic formula these roots are

$$s = \frac{-4 \pm \sqrt{16-4K}}{2} = -2 \pm \sqrt{4-K}.$$

The closed-loop poles as function of K are

$$p_1, p_2 = \begin{cases} -2+\sqrt{4-K}, -2-\sqrt{4-K}, & 0 \leq K \leq 4 \\ -2 \pm j\sqrt{K-4}, & 4 \leq K. \end{cases}$$

A table of values for $K = 0, 2, 4, 8$ is

K	p_1, p_2
0	$0, -4$
2	$-2+\sqrt{2}, -2-\sqrt{2}$
4	$-2, -2$
8	$-2 \pm j2$

Using this table we sketch the *root locus* shown in Figure 12.2.

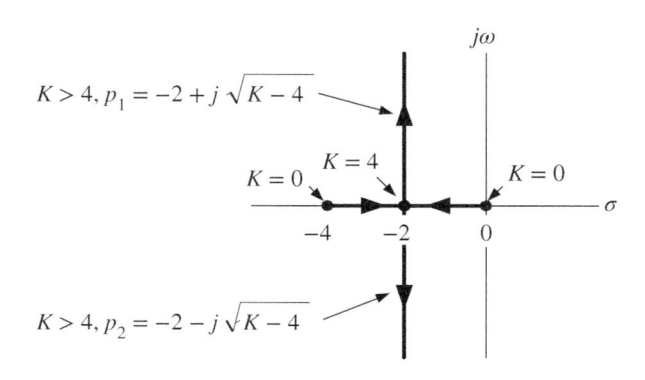

Figure 12.2. A plot of the roots of $s^2 + 4s + K$ for $K \geq 0$.

12.1 Angle Condition and Root Locus Rules

We would like to have a procedure to sketch the root locus *without* having to solve for the roots. Of course a computer can be used to numerically find the roots and draw the root locus. However, by learning to sketch it we will gain insight into the information it contains. This procedure is based on the *angle condition* due to W. Evans [53]. To explain, let's look back at (12.2) and consider the *return difference* defined by

$$1 + KG(s) = 1 + K\frac{b(s)}{a(s)} = \frac{a(s) + Kb(s)}{a(s)}.$$

In regards to the return difference we make the following straightforward observations:

(1) The *zeros* of the return difference are the roots of

$$a(s) + Kb(s) = 0$$

and are the *closed-loop* poles.

(2) The *poles* of the return difference are the roots of

$$a(s) = 0$$

and are the *open-loop* poles.

The root locus is a plot of the *zeros* of $1 + KG(s) = 0$ for $K \geq 0$. If s_0 satisfies

$$1 + KG(s_0) = 0,$$

then, equivalently, it satisfies

$$G(s_0) = -\frac{1}{K}.$$

In general, s_0 and $G(s_0)$ are complex numbers. For s_0 to be *on* the root locus we must have $G(s_0)$ equal to a *negative real* number. That is, in polar coordinates, s_0 must satisfy

$$|G(s_0)| e^{j\angle G(s_0)} = -\frac{1}{K} = \frac{1}{K} e^{j\pi}.$$

That is, if s_0 is on the root locus, then $G(s_0)$ satisfies the angle condition given by

$$\angle G(s_0) = \pi(2\ell + 1), \quad \ell = 0, \pm 1, \pm 2, \ldots$$

and the corresponding value of K is simply given by

$$K = \frac{1}{|G(s_0)|}.$$

450   12 Root Locus

The key observation here is we need only know the *open-loop* transfer function $G(s)$ to determine whether or not s_0 is a *closed-loop* pole.

Let's redo Example 1 using this observation.

Example 2 *Angle Condition for* $\dfrac{1}{s(s+4)}$

We have
$$G(s) = \frac{1}{s(s+4)} = \frac{1}{|s|e^{j\angle s}|s+4|e^{j\angle(s+4)}} = \frac{1}{|s||s+4|}e^{-j\angle s}e^{-j\angle(s+4)}$$

so that
$$\angle G(s) = -\angle s - \angle(s+4).$$

Figure 12.3a shows a point "s" on the root locus. We are now thinking of s and $s+4$ as vectors where s is a vector from the origin to s while $s+4 = s-(-4)$ is a vector from -4 to s. Figure 12.3b shows the angles $\angle s$ and $\angle(s+4)$.

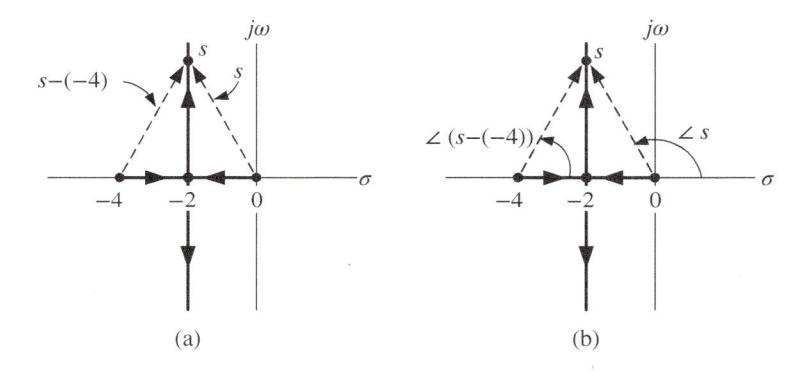

Figure 12.3. (a) The vectors "s" and "$s-(-4)$". (b) The angles $\angle s$ and $\angle\,(s-(-4))$.

From Figure 12.3b we see that
$$\angle s + \angle(s+4) = \pi$$

so that
$$\angle G(s) = -\angle s - \angle(s+4) = -\pi.$$

For example, choosing
$$s = -2 + 2j$$

which is *on* the root locus we have
$$G(-2+2j) = \frac{1}{(-2+j2)(-2+2j+4)} = \frac{1}{(-2+2j)(2+2j)} = -\frac{1}{8}.$$

The angle is
$$\angle G(-2+2j) = \angle(-1/8) = \pi.$$

On the other hand, choosing
$$s = -4 + 2j$$

12.1 Angle Condition and Root Locus Rules

which is *not* on the root locus we have

$$G(-4+2j) = \frac{1}{(-4+j2)(-4+2j+4)} = \frac{1}{(-4+2j)(2j)} = -\frac{1}{4+8j},$$

where $\angle G(-4+2j) = \angle -\dfrac{1}{4+8j} = 117°$. So

$$\angle G(-4+2j) \neq \pi(2\ell+1), \quad \ell = 0, \pm 1, \pm 2, \ldots.$$

Now let $s = -1$ so that $\angle s = \pi, \angle(s-(-1)) = 0$ and thus $\angle G(s) = -\angle s - \angle(s+4) = -\pi$. Thus $s = -1$ is *on* the root locus.

On the other hand, if $s = -5$ then $\angle s = \pi, \angle(s-(-5)) = \pi$ and thus $\angle G(s) = -\angle s - \angle(s-(-5)) = -2\pi$ which is *not* on the root locus.

Let's summarize what we now know from this example. The return difference is given by

$$1 + KG(s) = 1 + K\frac{b(s)}{a(s)} = \frac{a(s) + Kb(s)}{a(s)}$$

and its *zeros* are the closed-loop poles.

(1) The root locus starts on the open-loop poles. This is simply because for $K = 0$ the closed-loop poles are the roots of

$$a(s) + Kb(s) = a(s) = s(s+4) = 0.$$

(2) The number of branches of the root locus is 2, which is simply the degree of the closed-loop characteristic polynomial $a(s) + Kb(s) = s^2 + 4s + K$.

(3) The root locus is symmetric with respect to the real axis simply because the complex roots of

$$a(s) + Kb(s) = s^2 + 4s + K = 0$$

come in complex conjugate pairs.

(4) For any s_0 on the root locus,

$$\angle G(s_0) = \angle \frac{1}{s_0(s_0+4)} = \pi(2\ell+1), \quad \ell = 0, \pm 1, \pm 2, \ldots.$$

(5) A *breakaway point* occurs at $s = -2$ for $K = 4$, which is where the closed-loop characteristic polynomial $s^2 + 4s + K = 0$ has a *double* root. That is,

$$1 + K\frac{1}{s(s+4)}\bigg|_{K=4} = \frac{s^2 + 4s + K}{s(s+4)}\bigg|_{K=4} = \frac{(s+2)^2}{s(s+4)}.$$

This is simply because when the root locus breaks away from the real axis, the two roots must come together so they can leave the real axis as a complex conjugate pair. Note that at this breakaway point

12 Root Locus

$$\frac{d}{ds}\left(1 + 4\frac{1}{s(s+4)}\right)\bigg|_{s=-2} = \frac{d}{ds}\left(\frac{(s+2)^2}{s(s+4)}\right)\bigg|_{s=-2}$$

$$= -\left(\frac{1}{s(s+4)}\right)^2 \left(\frac{d}{ds}(s^2+4s)\right)(s+2)^2\bigg|_{s=-2}$$

$$+ \left(\frac{1}{s(s+4)}\frac{d}{ds}(s+2)^2\right)\bigg|_{s=-2}$$

$$= 0.$$

Let's consider another (rather long) example to illustrate how the angle condition is used to sketch a root locus.

Example 3 *Root Locus for* $\dfrac{1}{s(s+2)(s+4)}$

Consider the closed-loop system of Figure 12.4.

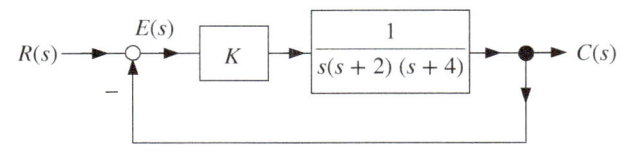

Figure 12.4. System block diagram.

The open-loop transfer function is

$$G(s) = \frac{1}{s(s+2)(s+4)} = \frac{1}{|s||s+2||s+4|}e^{-j(\angle s + \angle(s+2) + \angle(s+4))}.$$

The closed-loop poles are the *zeros* of

$$1 + KG(s) = 1 + K\frac{1}{s(s+2)(s+4)} = \frac{s(s+2)(s+4) + K}{s(s+2)(s+4)}$$

or, equivalently,

$$G(s) = -\frac{1}{K} \implies \angle G(s) = \pi(2\ell+1), \quad \ell = 0, \pm 1, \pm 2, \ldots.$$

Let's sketch the root locus.

(1) $K = 0$. Then
$$s(s+2)(s+4) + K = 0$$
shows that $s = 0, -2, -4$. The root locus starts on the *open-loop* poles.

(2) $K = \infty$. From
$$G(s) = \frac{1}{s(s+2)(s+4)} = -\frac{1}{K}$$
we see that as $K \to \infty$ results in the three roots going to infinity, i.e., $|s| \to \infty$. Notice that as $|s| \to \infty$
$$|G(s)| \approx \frac{1}{|s|^3} \to 0.$$

We say that $G(s)$ has three zeros at infinity.

12.1 Angle Condition and Root Locus Rules

(3) *Real axis root locus*

(i) Consider s real with $-2 < s < 0$ (see Figure 12.5). Then $\angle s = \pi, \angle(s+2) = 0$, $\angle(s+4) = 0$ and therefore

$$\angle G(s) = -\angle s - \angle(s+2) - \angle(s+4) = -\pi$$

so these values of s are *on* the root locus. This is indicated by the thick line between -2 and 0 in Figure 12.5.

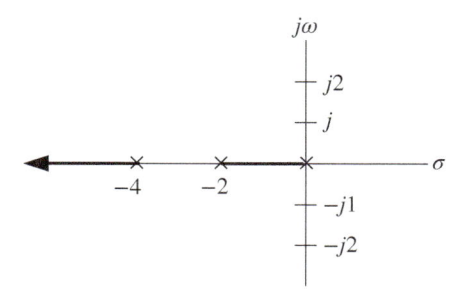

Figure 12.5. Real axis root locus.

(ii) Consider s real with $-4 < s < -2$ so that $\angle s = \pi, \angle(s+2) = \pi, \angle(s+4) = 0$ and therefore

$$\angle G(s) = -\angle s - \angle(s+2) - \angle(s+4) = -2\pi$$

so these values of s are *not* on the root locus.

(iii) Consider s real with $s < -4$ so that $\angle s = \pi, \angle(s+2) = \pi, \angle(s+4) = \pi$ and therefore

$$\angle G(s) = -\angle s - \angle(s+2) - \angle(s+4) = -3\pi$$

so these values of s are *on* the root locus. This is indicated by the thick line from -4 to $-\infty$ in Figure 12.5.

(iv) Finally, consider s real with $s > 0$ so that $\angle s = 0, \angle(s+2) = 0, \angle(s+4) = 0$ and therefore

$$\angle G(s) = -\angle s - \angle(s+2) - \angle(s+4) = 0$$

so these values of s are *not* on the root locus.

(4) *Breakaway point*

A breakaway point occurs at a value of K for which

$$1 + K \frac{1}{s(s+2)(s+4)} = \frac{s(s+2)(s+4) + K}{s(s+2)(s+4)}$$

has *multiple* zeros. That is, s_b is a breakaway point for a value of K_b such that

$$1 + K_b \frac{1}{s(s+2)(s+4)} = \frac{s(s+2)(s+4) + K_b}{s(s+2)(s+4)} = \frac{(s-s_b)^2(s-s_1)}{s(s+2)(s+4)}.$$

This is simply because two branches breaking away from the real axis must first come together (double root) and then leave the real axis as a complex conjugate pair. We see immediately that

$$\frac{d}{ds}\left(\frac{(s-s_b)^2(s-s_1)}{s(s+2)(s+4)}\right)\bigg|_{s=s_b} = 0.$$

So at a double root s_b we have

$$\frac{d}{ds}\left(1 + K_b\frac{1}{s(s+2)(s+4)}\right)\bigg|_{s=s_b} = K_b\frac{d}{ds}\underbrace{\left(\frac{1}{s(s+2)(s+4)}\right)}_{G(s)}\bigg|_{s=s_b} = 0.$$

At a breakaway point we must have $1 + K_b G(s_b) = 0$ and $\dfrac{dG(s)}{ds}\bigg|_{s_b} = 0$. In this example the candidate breakaway points are the solutions to

$$\frac{d}{ds}G(s) = -\left(\frac{1}{s(s+2)(s+4)}\right)^2 \frac{d}{ds}\left(s(s+2)(s+4)\right)$$

$$= -\left(\frac{1}{s(s+2)(s+4)}\right)^2 \frac{d}{ds}\left(s^3 + 6s^2 + 8s\right)$$

$$= 0.$$

Solving

$$3s^2 + 12s + 8 = 0$$

results in the candidate breakaway points

$$s = \frac{-12 \pm \sqrt{(12)^2 - 4(3)(8)}}{6} = -2 \pm 2/\sqrt{3} = -0.845, -3.146.$$

As shown in Figure 12.6, only $s = -0.845$ is on the real axis root locus and is therefore the only breakaway point. That is, only $s = -0.845$ also satisfies $1 + KG(s) = 0$ for some $K \geq 0$.

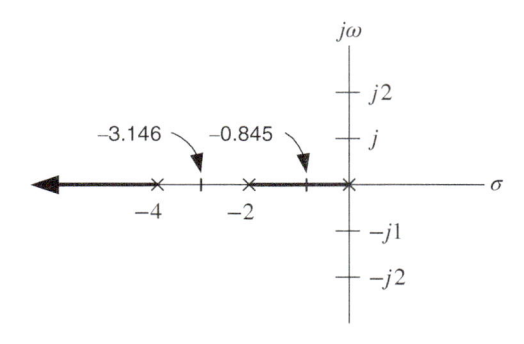

Figure 12.6. $dG/ds = 0$ at $-2 \pm 2/\sqrt{3} = -0.845, -3.146$.

(5) *jω axis intercepts*

We now use the Routh–Hurwitz test to check if any of the branches of the root locus intercepts (crosses) the $j\omega$ axis. That is, are there any values of $K \geq 0$ such that the

zeros of
$$1 + K\frac{1}{s(s+2)(s+4)} = \frac{s(s+2)(s+4) + K}{s(s+2)(s+4)} = 0$$
are on the $j\omega$ axis? To find out we apply the Routh–Hurwitz test to
$$s(s+2)(s+4) + K = s^3 + 6s^2 + 8s + K.$$

The Routh table is

s^3	1	8
s^2	6	K
s	$\frac{48-K}{6}$	0
1	K	

which is stable for $0 < K < 48$ and has two right half-plane poles for $K > 48$. Thus at $K = 48$ there are two poles on the $j\omega$ axis. It turns out that to find these poles we first set $K = 48$ in the Routh table noting that the s row is all zeros. Just above this row is the s^2 row and we use it to form the *auxiliary* equation
$$6s^2 + 48 = 0.$$

The roots of this equation are $s = \pm j2\sqrt{2} = \pm j2.83$ and are the poles on the $j\omega$ axis.[1] These are indicated on Figure 12.7.

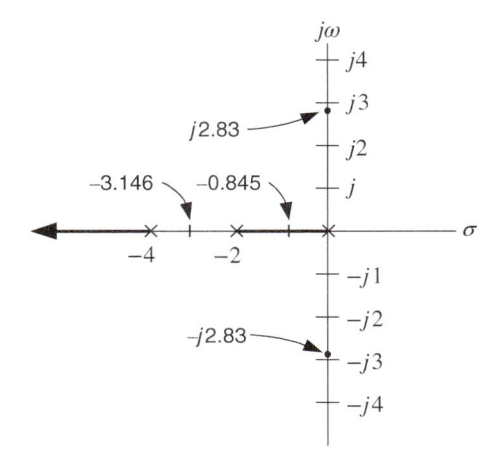

Figure 12.7. $j\omega$ axis intercepts at $s = \pm j2\sqrt{2} = \pm j2.83$.

At this point we can give a sketch of the root locus, which is shown in Figure 12.8. Actually this plot was done using the MATLAB code given on the next page.

---

[1] See [1, 2], or [5] for a proof.

456   12   Root Locus

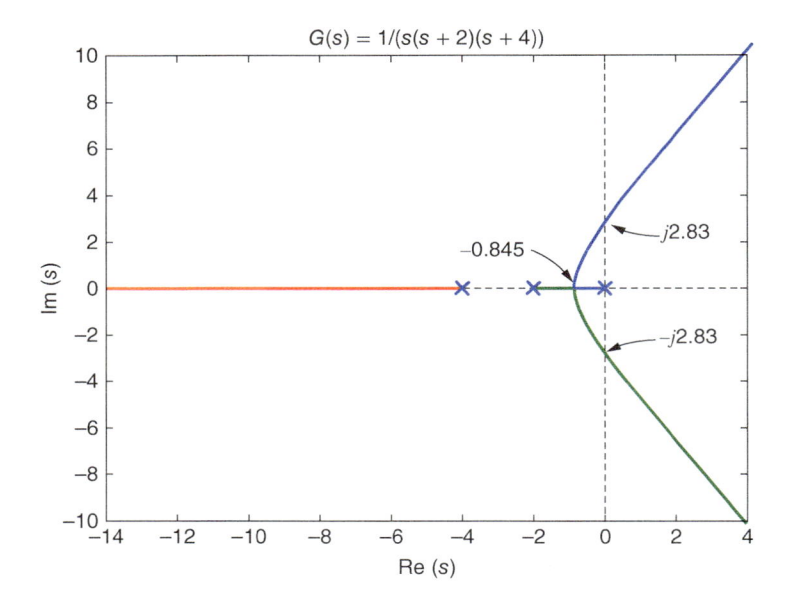

Figure 12.8. Root locus plot of $\dfrac{1}{s(s+2)(s+4)}$.

```
% Root Locus of G(s) = 1/(s(s+2)(s+4)) = 1/(s^3+6s^2+8s)
den = [1 6 8 0];
num = [1];
rlocus(tf(num,den))
% Pretty up the plot by making the linewidth thicker,
% the marker size and font size bigger.
h = findobj(gca, 'Type', 'line');
set(h, 'LineWidth', 6)
set(h, 'MarkerSize', 20)
set(gca,'FontSize',24)
% Set range of x-axis [-10,0] and y-axis [-2,2]
v = [-10 10 -10 10];
axis(v)
axis square
% Title the plot and label the axes
str_G = 'G(s) = 1/(s(s+2)(s+4))';
title(str_G,'FontSize',20)
xlabel('Re(s)','FontSize',20)
ylabel ('Im(s)','FontSize',20)
```

This example illustrated a sequence of rules to draw a root locus. The root locus rules found so far are:

(1) The root locus starts on the open-loop poles.

(2) The root locus ends on the open-loop zeros.

(3) The root locus is symmetric with respect to the real axis.
(4) Any s on the root locus satisfies the angle condition.
(5) A breakaway point from the real axis is a solution to $dG/ds = 0$.
(6) The $j\omega$ axis intercepts may be found using the Routh–Hurwitz test.

12.2 Asymptotes and Their Real Axis Intersection

We next consider the root locus *asymptotes* and their *intersection on the real axis*.
With $m < n$ let the open-loop transfer function be written as

$$G(s) = \frac{b(s)}{a(s)} = \frac{b_m s^m + b_{m-1} s^{m-1} + \cdots + b_1 s + b_0}{s^n + a_{n-1} s^{n-1} + \cdots + a_1 s + a_0}$$

$$= \frac{b_m \left(s^m + \frac{b_{m-1}}{b_m} s^{m-1} + \cdots + \frac{b_1}{b_m} s + \frac{b_0}{b_m} \right)}{s^n + a_{n-1} s^{n-1} + \cdots + a_1 s + a_0}.$$

For large values of the magnitude of s, i.e., for $|s|$ large, the branches of the root locus are asymptotic to straight lines called asymptotes whose angles with respect to the real axis are given by

$$\theta_\ell = \frac{2\ell + 1}{n - m} \pi, \quad \ell = 0, 1, 2, \ldots, n - m - 1.$$

Further these asymptotes intersect on the real axis at (see [1, 2], or [5] for a proof)

$$\sigma_1 = -\frac{a_{n-1} - b_{m-1}/b_m}{n - m}.$$

In the special case where $m = 0$, that is,

$$G(s) = \frac{b_0}{s^n + a_{n-1} s^{n-1} + \cdots + a_1 s + a_0}$$

we take $b_{-1} = 0$ so that

$$\sigma_1 = -\frac{a_{n-1} - b_{-1}/b_0}{n - m} = -\frac{a_{n-1} - 0}{n - m}.$$

This rule is best explained by an example.

Example 4 *Asymptotes for* $G(s) = \dfrac{1}{s(s+2)(s+4)}$

Let

$$G(s) = \frac{1}{s(s+2)(s+4)} = \frac{1}{s^3 + 6s^2 + 8s}$$

be put in the closed-loop system of Figure 12.9.

12 Root Locus

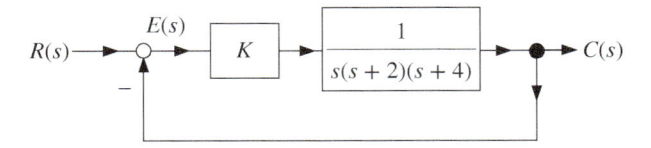

Figure 12.9. System block diagram.

The *angles of the asymptotes* with respect to the real axis are ($n = 3, m = 0$)

$$\theta_\ell = \frac{(2\ell + 1)\pi}{3 - 0}, \quad \ell = 0, 1, 2$$

$$= \frac{\pi}{3}, \frac{3\pi}{3}, \frac{5\pi}{3}$$

$$= \frac{\pi}{3}, \pi, -\frac{\pi}{3}$$

and these asymptotes intersect the real axis at

$$\sigma_1 = -\frac{a_{n-1} - b_{m-1}/b_m}{n - m}, \quad n = 3, \ m = 0$$

$$= -\frac{6 - 0}{3 - 0} \quad \text{as } a_2 = 6$$

$$= -2.$$

Figure 12.10 shows the two asymptotes at $\pi/3$ and $-\pi/3$ added to Figure 12.7. From the geometry we see that these asymptotes intersect the $j\omega$ axis at $\pm j2\tan(\pi/3) = \pm 2\sqrt{3} = \pm j3.46$. The third asymptote has an angle of π and just follows the real axis root locus branch at π.

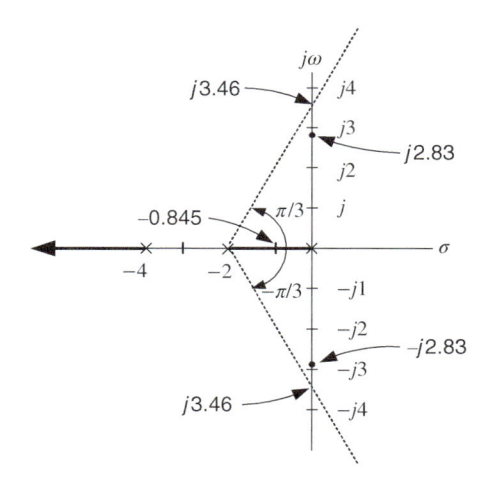

Figure 12.10. Asymptotes and their intercept on the real axis.

We can also add these two asymptotes to Figure 12.8 to obtain Figure 12.11.

12.2 Asymptotes and Their Real Axis Intersection

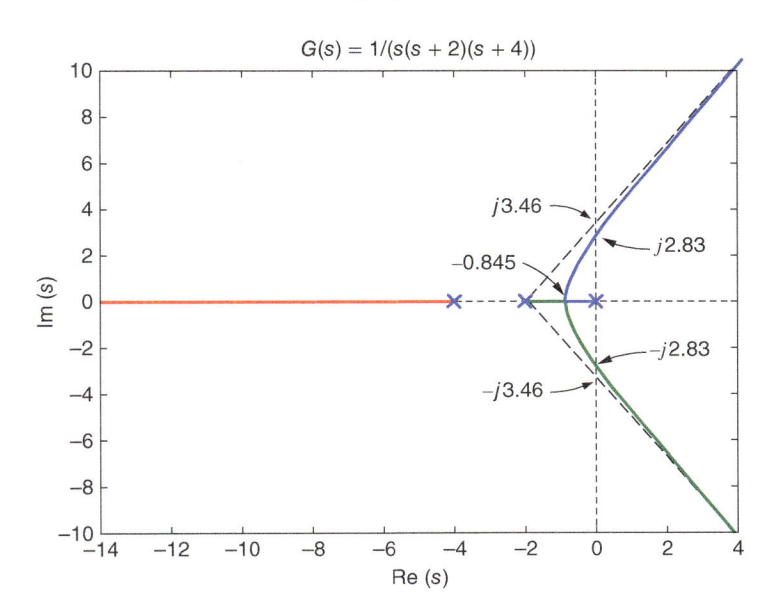

Figure 12.11. Root locus plot with the three asymptotes showing their intersection at -2.

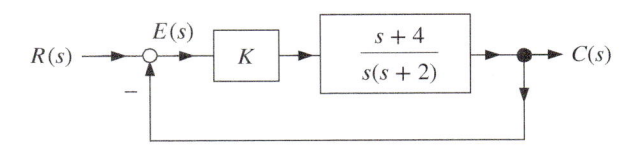

Figure 12.12. System block diagram.

Example 5 *Root Locus of* $G(s) = \dfrac{s+4}{s(s+2)}$

Consider the closed-loop system given in Figure 12.12.
The open-loop transfer function is given by

$$G(s) = \frac{s+4}{s(s+2)} = \frac{|s+4|}{|s||s+2|}e^{j(\angle(s+4)-\angle s-\angle(s+2))}.$$

The closed-loop poles are the *zeros* of

$$1 + KG(s) = 1 + K\frac{s+4}{s(s+2)} = \frac{s(s+2) + K(s+4)}{s(s+2)}.$$

In terms of the angle condition the closed-loop poles satisfy

$$G(s) = -\frac{1}{K}$$
$$\implies \angle G(s) = \pi(2\ell + 1), \quad \ell = 0, \pm 1, \pm 2, \ldots.$$
$$\implies \angle(s+4) - \angle s - \angle(s+2) = \pi(2\ell + 1), \quad \ell = 0, \pm 1, \pm 2, \ldots.$$

Let's sketch the root locus.

(1) $K = 0$. Then
$$s(s+2) + K(s+4) = s(s+2) = 0$$
shows that $s = 0, -2$. The root locus starts on the *open-loop* poles.

(2) $K = \infty$. From
$$G(s) = \frac{s+4}{s(s+2)} = -\frac{1}{K}$$
we see that as $K \to \infty$ we must have the open-loop transfer function $G(s) \to 0$. This requires $s \to -4$ or $|s| \to \infty$. Notice that as $|s| \to \infty$
$$|G(s)| \approx \frac{1}{|s|} \to 0.$$
Thus the open-loop transfer function $G(s)$ has one zero at -4 and one zero at $|s| = \infty$. The root locus ends on the open-loop zeros.

(3) *Real axis root locus*
   (i) Consider s real with $-2 < s < 0$ (see Figure 12.13). Then $\angle s = \pi$, $\angle(s+2) = 0$, $\angle(s+4) = 0$ and therefore
   $$\angle G(s) = \angle(s+4) - \angle s - \angle(s+2) = -\pi.$$
   These values of s are *on* the root locus. This is indicated by the thick line between -2 and 0 in Figure 12.5.

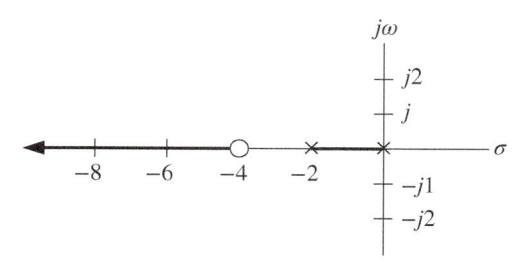

Figure 12.13. Real axis root locus.

   (ii) Consider s real with $-4 < s < -2$ so that $\angle s = \pi$, $\angle(s+2) = \pi$, $\angle(s+4) = 0$ and therefore
   $$\angle G(s) = \angle(s+4) - \angle s - \angle(s+2) = 0.$$
   These values of s are *not* on the root locus.

   (iii) Consider s real with $s < -4$ so that $\angle s = \pi$, $\angle(s+2) = \pi$, $\angle(s+4) = \pi$ and therefore
   $$\angle G(s) = \angle(s+4) - \angle s - \angle(s+2) = -\pi.$$
   These values of s are *on* the root locus. This is indicated by the thick line from -4 to $-\infty$ in Figure 12.5.

   (iv) Finally, consider s real with $s > 0$ so that $\angle s = 0$, $\angle(s+2) = 0$, $\angle(s+4) = 0$ and therefore
   $$\angle G(s) = -\angle s - \angle(s+2) - \angle(s+4) = 0.$$
   These values of s are *not* on the root locus.

12.2 Asymptotes and Their Real Axis Intersection

(4) Breakaway point

A breakaway point occurs at a value of K for which

$$1 + K\frac{s+4}{s(s+2)} = \frac{s(s+2) + K(s+4)}{s(s+2)(s+4)}$$

has *multiple* zeros. As shown in the previous two examples, this implies a breakaway point satisfies

$$\frac{d}{ds}\underbrace{\left(\frac{s+4}{s(s+2)}\right)}_{G(s)} = 0.$$

As

$$\frac{d}{ds}G(s) = \frac{1}{s(s+2)} - \frac{s+4}{(s(s+2))^2}\frac{d}{ds}(s^2+2s) = \frac{s(s+2)-(s+4)(2s+2)}{(s(s+2))^2}$$

$$= -\frac{s^2+8s+8}{(s(s+2))^2}$$

any breakaway point must be a solution to

$$s^2 + 8s + 8 = 0.$$

The candidate breakaway points are then

$$s = \frac{-8 \pm \sqrt{64-32}}{2} = -4 \pm 2\sqrt{2} = -6.828, -1.172.$$

As shown in Figure 12.14 both roots of $dG/ds = 0$ are on the real axis root locus and therefore both are breakaway points.

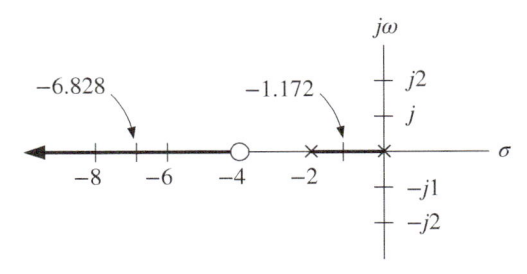

Figure 12.14. $dG/ds = 0$ at $-2 \pm 2\sqrt{2} = -6.828, -1.172$.

(5) $j\omega$ axis intercepts

We now use the Routh–Hurwitz test to check if any of the branches of the root locus intercepts the $j\omega$ axis. That is, are there any values of $K \geq 0$ such that the zeros of

$$1 + K\frac{s+4}{s(s+2)} = \frac{s(s+2) + K(s+4)}{s(s+2)} = 0$$

are on the $j\omega$ axis? To find out we apply the Routh–Hurwitz test to

$$s(s+2) + K(s+4) = s^2 + (K+2)s + 4K.$$

Note this is a second-order polynomial so it is stable for $K > 0$. As we are only interested in the root locus for $K \geq 0$ it follows that the root locus cannot intercept the $j\omega$ axis.

12 Root Locus

(6) *Asymptotes*
As $n = 2, m = 1$, we have

$$\theta_\ell = \frac{\pi(2\ell + 1)}{n - m}, \quad \ell = 0, 1, \ldots, n - m - 1$$

$$= \frac{\pi(2\ell + 1)}{2 - 1}, \quad \ell = 0$$

$$= \pi.$$

Thus there is only one asymptote which makes an angle $\theta_0 = \pi$ with respect to the real axis. The intersection with the real axis is

$$\sigma_1 = -\frac{a_{n-1} - b_{m-1}/b_m}{n - m}, \quad n = 3, \ m = 1$$

$$= -\frac{2 - 4/1}{2 - 1} \text{ as } a_2 = 2$$

$$= 2.$$

In this example the asymptote tells us nothing new because we already knew from the real axis root locus that the root locus went out along the negative real axis.

(7) *Sketch*
To sketch the root locus set $K = 0$ to see that the two branches start at the open-loop poles $0, -2$, respectively. Increasing K we then have the two branches move toward each other on the real-axis root locus to the breakaway point at $-4 + 2\sqrt{2}$. Next the two branches break off the real axis and must go toward the other breakaway point at $-4 - 2\sqrt{2}$. After coming together at the second breakaway point, one of the branches heads toward the open-loop zero at -4 while the other heads toward the open-loop zero at $|s| = \infty$. However, using the root locus rules we cannot ascertain the shape of the branches off of the real axis. We drew it as a circle in Figure 12.15, but this is not obvious.

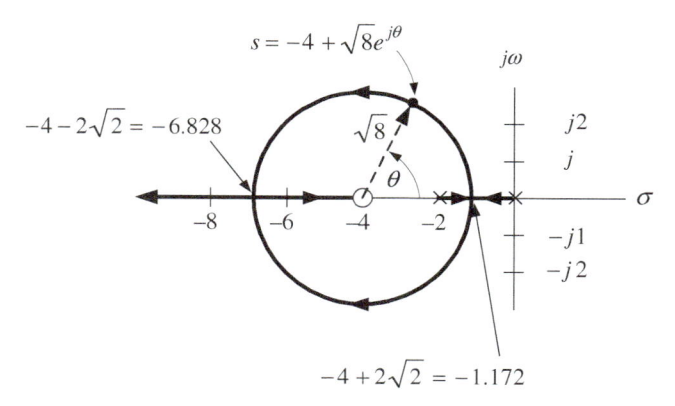

Figure 12.15. Root locus plot of $\dfrac{s + 4}{s(s + 2)}$.

Using the angle condition we can prove that the root locus that is off the real axis is a circle! To do so note that the circle in Figure 12.15 may be written as

$$s = -4 + \sqrt{8}e^{j\theta}, \ 0 \leq \theta \leq 2\pi.$$

12.3 Angles of Departure

To show that these points are on the root locus we compute

$$G(s)\big|_{s=-4+\sqrt{8}e^{j\theta}} = \frac{s+4}{s(s+2)}\bigg|_{s=-4+\sqrt{8}e^{j\theta}} = \frac{\sqrt{8}e^{j\theta}}{(-4+\sqrt{8}e^{j\theta})(-2+\sqrt{8}e^{j\theta})}$$

$$= \frac{\sqrt{8}e^{j\theta}}{8 - 6\sqrt{8}e^{j\theta} + 8e^{2j\theta}}$$

$$= \frac{\sqrt{8}}{8e^{-j\theta} - 6\sqrt{8} + 8e^{j\theta}}$$

$$= \frac{\sqrt{8}}{16\cos(\theta) - 6\sqrt{8}}$$

$$= -\frac{\sqrt{8}}{\underbrace{6\sqrt{8}}_{16.97} - 16\cos(\theta)}.$$

That is,

$$G(-4+\sqrt{8}e^{j\theta}) = -\frac{\sqrt{8}}{\underbrace{16.97 - 16\cos(\theta)}_{>0 \text{ for all } \theta}}$$

so that $\angle G(-4+\sqrt{8}e^{j\theta}) = \pi$ and thus the points

$$s = -4 + \sqrt{8}e^{j\theta},\ 0 \le \theta \le 2\pi$$

are *on* the root locus.

12.3 Angles of Departure

We introduce the final root locus rule known as the angles of departure. We do this through the following example.

Example 6 *Root Locus of* $G(s) = \dfrac{1}{s(s^2+2s+2)}$

Consider the closed-loop system of Figure 12.16.

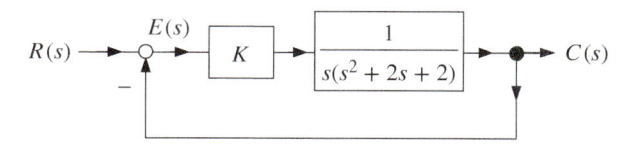

Figure 12.16. Closed-loop system block diagram.

The open-loop transfer function is

$$G(s) = \frac{1}{s(s^2+2s+2)} = \frac{|s+4|}{|s||s-(-1+j)||s-(-1-j)|} e^{-j\left(\angle s + \angle(s-(-1+j)) + \angle s-(-1-j)\right)}.$$

464   12   Root Locus

The closed-loop poles are the *zeros* of

$$1 + KG(s) = 1 + K\frac{1}{s(s^2+2s+2)} = \frac{s^3+2s^2+2s+K}{s(s^2+2s+2)}.$$

In terms of the angle condition the closed-loop poles must satisfy

$$G(s) = -\frac{1}{K}$$
$$\implies \angle G(s) = \pi(2\ell+1), \quad \ell = 0, \pm 1, \pm 2, \ldots$$

or

$$-\angle s - \angle\bigl[(s-(-1+j)\bigr] - \angle\bigl[s-(-1-j)\bigr] = \pi(2\ell+1), \quad \ell = 0, \pm 1, \pm 2, \ldots.$$

Let's sketch the root locus.

(1) $K = 0$. Then

$$s^3 + 2s^2 + 2s + K = s(s^2+2s+2) = s\bigl[s-(-1+j)\bigr]\bigl[s-(-1-j)\bigr] = 0$$

shows that $s = 0, -1 \pm j$. The root locus starts on the *open-loop* poles.

(2) $K = \infty$. From

$$G(s) = \frac{1}{s(s^2+2s+2)} = -\frac{1}{K}$$

we see that as $K \to \infty$ we must have the open-loop transfer function $G(s) \to 0$. This requires $|s| \to \infty$. Notice that as $|s| \to \infty$

$$|G(s)| \approx \frac{1}{|s|^3} \to 0.$$

Thus the open-loop transfer function $G(s)$ has three zeros at $|s| = \infty$. The root locus ends on the open-loop zeros.

(3) *Real axis root locus*

(i) Consider s real with $-\infty < s < 0$ as indicated in Figure 12.17. Then $\angle s = \pi$, $\angle\bigl[(s-(-1+j)\bigr] = -\angle\bigl[(s-(-1-j)\bigr]$ and therefore

$$\angle G(s) = -\angle s \underbrace{-\angle\bigl[(s-(-1+j)\bigr] - \angle\bigl[s-(-1-j)\bigr]}_{0} = -\angle s = -\pi.$$

These values of s are *on* the root locus.

(ii) Consider s real with $s > 0$ so that $\angle s = 0$, $\angle\bigl[(s-(-1+j)\bigr] = -\angle\bigl[(s-(-1-j)\bigr]$ and therefore

$$\angle G(s) = -\angle s \underbrace{-\angle\bigl[(s-(-1+j)\bigr] - \angle\bigl[s-(-1-j)\bigr]}_{0} = -\angle s = 0.$$

These values of s are *not* on the root locus.

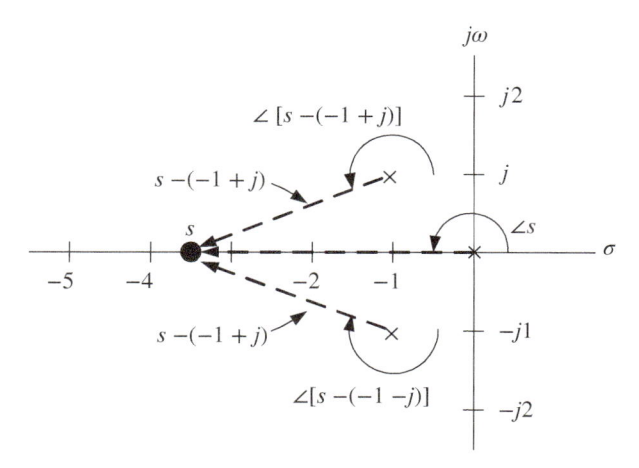

12.3 Angles of Departure 465

Figure 12.17. $\angle[s-(-1+j)]+\angle[s-(-1-j)] = 0$ on the real axis.

Figure 12.18 shows the real axis root locus.

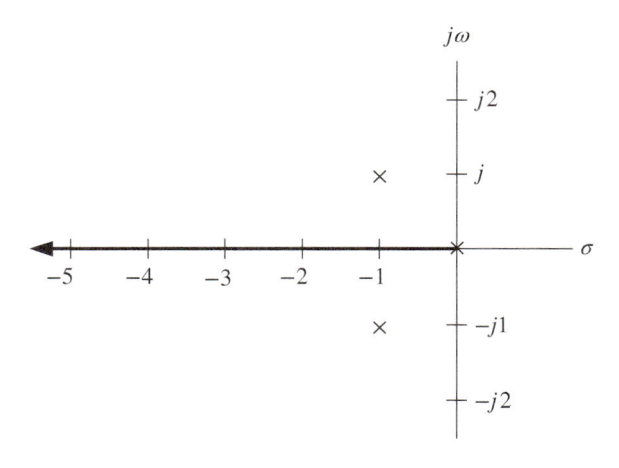

Figure 12.18. Real axis root loci.

(4) *Breakaway point*

A breakaway point occurs at a value of K for which

$$1 + K\frac{1}{s(s^2+2s+2)} = \frac{s^3+2s^2+2s+K}{s(s^2+2s+2)}$$

has *multiple* zeros. As previously explained, a breakaway point must satisfy

$$\frac{d}{ds}\left(1+K\frac{1}{s(s^2+2s+2)}\right) = K\frac{d}{ds}\underbrace{\left(\frac{1}{s(s^2+2s+2)}\right)}_{G(s)} = 0.$$

We have
$$\frac{d}{ds}G(s) = -\frac{1}{[s(s^2+2s+2)]^2}\frac{d}{ds}(s^3+2s^2+2s)$$
$$= -\frac{1}{[s(s^2+2s+2)]^2}(3s^2+4s+2)$$
$$= 0$$

which requires solving
$$3s^2 + 4s + 2 = 0.$$

We obtain
$$s = \frac{-4 \pm \sqrt{16-24}}{6} = -\frac{2}{3} \pm j\frac{\sqrt{2}}{3}$$

which are not on the real axis. Thus there are no breakaway points.

(5) *$j\omega$ axis intercepts*

We now use the Routh–Hurwitz test to check if any of the branches of the root locus intercepts (crosses) the $j\omega$ axis. That is, are there any values of $K \geq 0$ such that the zeros of
$$1 + K\frac{1}{s(s^2+2s+2)} = \frac{s^3+2s^2+2s+K}{s(s^2+2s+2)}$$

are on the $j\omega$ axis? To find out we apply the Routh–Hurwitz test to
$$s^3 + 2s^2 + 2s + K = 0.$$

The Routh table is

s^3	1	2
s^2	2	K
s	$\frac{4-K}{2}$	0
1	K	

which shows that closed-loop system is stable for $0 < K < 4$ and has two right half-plane roots for $K > 4$. Thus for $K = 4$ there are two roots on the $j\omega$ axis. The auxiliary equation for these two roots is constructed using the s^2 row of the Routh table. We have
$$2s^2 + K = 2s^2 + 4 = 0$$

which gives
$$s = \pm j\sqrt{2}.$$

These intercept points are indicated in Figure 12.19.

(6) *Asymptotes*

The angles of the asymptotes with respect to the real axis are given by ($n = 3, m = 0$)
$$\theta_\ell = \frac{\pi(2\ell+1)}{n-m}, \quad \ell = 0, 1, \ldots, n-m-1$$
$$= \frac{\pi(2\ell+1)}{3-0}, \quad \ell = 0, 1, 2.$$

12.3 Angles of Departure

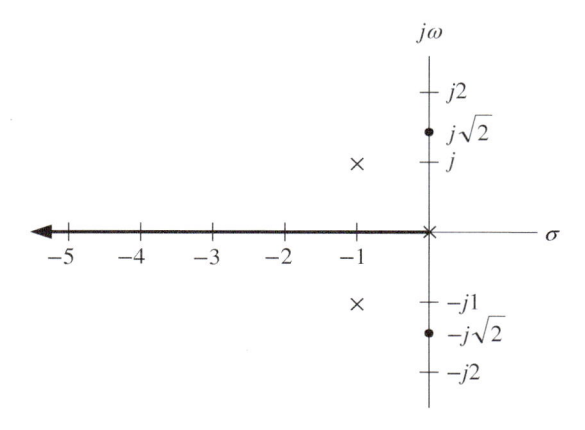

Figure 12.19. $j\omega$ axis intercept points.

The angles of the three asymptotes are

$$\theta_0 = \pi/3, \theta_1 = \pi, \theta_2 = 5\pi/3 \, (-\pi/3).$$

Their intersection with the real axis is at

$$\sigma_1 = -\frac{a_{n-1} - b_{m-1}/b_m}{n - m}, \quad n = 3, \, m = 0$$

$$= -\frac{2 - 0}{3 - 0} \text{ as } a_2 = 2, b_0 = 1$$

$$= -2/3.$$

Figure 12.20 shows the two asymptotes at $\pi/3$ and $-\pi/3$. From the geometry we see that these two asymptotes intersect the $j\omega$ axis at $\pm j(2/3)\tan(\pi/3) = \pm j1.155$. The third asymptote has an angle of π and just follows the branch at π found from the real axis root loci.

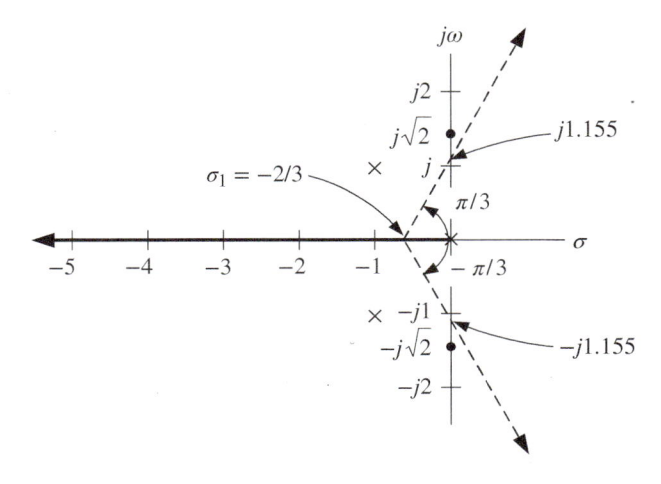

Figure 12.20. Asymptotes and their real axis intercept.

12 Root Locus

(7) *Angles of departure*

The fact that the open-loop transfer function has a complex conjugate pair of poles leads to the idea of computing the angles of departure from these poles using the angle condition. As the name suggests, we are going to determine the angle that a root locus branch leaves an open-loop complex pole. For $K = 0$ the roots are at $s = 0, s = -1 + j$, and $s = -1 - j$. As K is increased slightly, the closed-loop poles (root locus branches) will leave these open-loop poles. We want to compute the angles that the two branches starting on the complex-conjugate pair of poles leave them. Consider K just greater than zero so that the root locus branch that starts at $-1 + j$ is slightly away from this point. This is indicated by the point "s" in Figure 12.21. The key observation here is that s is taken to be very close $-1 + j$ (think of s being much closer to $-1 + j$ than as drawn in Figure 12.21). Then we have

$$\angle s \approx \pi/2 + \pi/4 = 3\pi/4$$
$$\angle[s - (-1 - j)] \approx \pi/2.$$

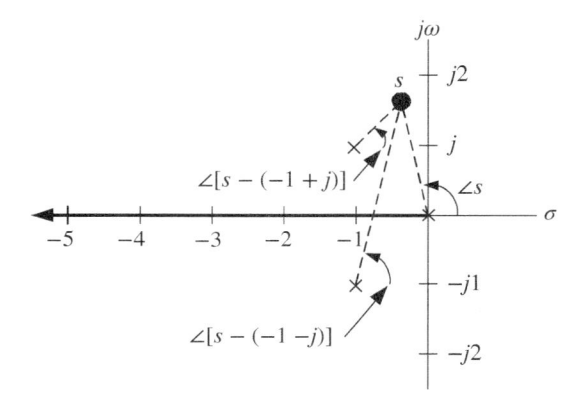

Figure 12.21. Angle of departure from the open-loop pole at $-1 + j$.

The angle condition is

$$-\angle s - \angle[(s - (-1 + j)] - \angle[s - (-1 - j)] = \pi(2\ell + 1), \quad \ell = 0, \pm 1, \pm 2, \ldots.$$

As we are considering K just greater than zero, the point "s" on the root locus is close to $-1 + j$. The angle condition becomes

$$-\pi/2 - \angle[(s - (-1 + j)] - 3\pi/4 = \pi(2\ell + 1), \quad \ell = 0, \pm 1, \pm 2, \ldots.$$

Solving for $\angle[(s - (-1 + j)]$ results in

$$\angle[(s - (-1 + j)] = -5\pi/4 - \pi(2\ell + 1), \ell = 0, \pm 1, \pm 2, \ldots$$

We can take ℓ to be any integer, so let's just take it to be 0. Then

$$\angle[(s - (-1 + j)] = -5\pi/4 - \pi = -\pi/4 - 2\pi$$

or $\angle[(s - (-1 + j)] = -\pi/4$. Since the root locus is symmetric with respect to the real axis, we have $\angle[(s - (-1 - j)] = \pi/4$. Figure 12.22 indicates the angles of departure on the full root locus.

12.3 Angles of Departure

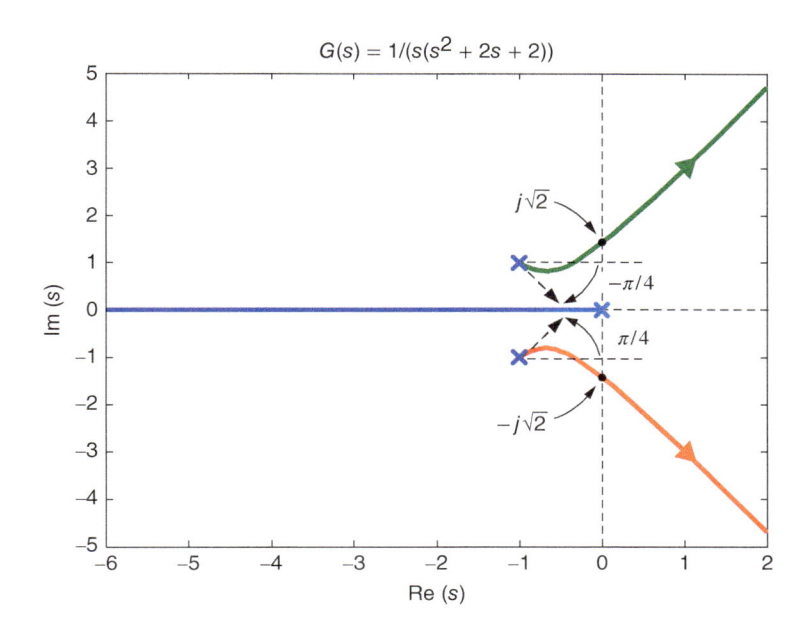

Figure 12.22. The angles of departure from a complex conjugate pair of open-loop poles.

(8) *Sketch*

The root locus starts on the open-loop poles. As there are no breakaway points, the root locus branch that started at $s = 0$ just continues out along the negative real axis to $-\infty$. This is just the real axis root locus. The root locus branch that starts at $-1 + j$ leaves that point at an angle of $-\pi/4$. However, as K increases it must intercept the $j\omega$ axis at $j\sqrt{2}$ and then as $K \to \infty$ it must approach its asymptote. Similar comments apply to the branch that starts at $-1 - j$. Of course Figure 12.22 was drawn using a computer program (MATLAB), but knowing the root locus rules allows us to not be surprised by the computer plot!

Example 7 *Root Locus of* $G(s) = \dfrac{s + 2}{s^2 + 2s + 2}$

Consider the closed-loop system of Figure 12.23.

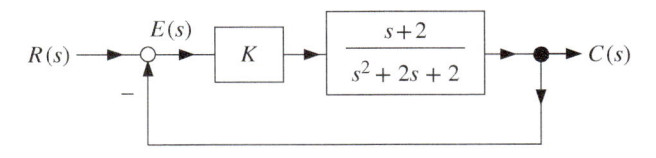

Figure 12.23. System block diagram.

The open-loop transfer function is given by

$$G(s) = \frac{s + 2}{s^2 + 2s + 2}$$

$$= \frac{|s + 2|}{|s| \, |s - (-1 + j)| \, |s - (-1 - j)|} e^{j\left(\angle(s+2) - \angle[s - (-1+j)] - \angle[s - (-1-j)]\right)}.$$

12 Root Locus

The closed-loop poles are the *zeros* of

$$1 + KG(s) = 1 + K\frac{s+2}{s^2+2s+2} = \frac{s^2+(K+2)s+2K}{s^2+2s+2}$$

or, equivalently,

$$G(s) = -\frac{1}{K} \implies \angle G(s) = \pi(2\ell+1), \quad \ell = 0, \pm 1, \pm 2, \ldots.$$

The angle condition is then

$$\angle(s+2) - \angle[s-(-1+j)] - \angle[s-(-1-j)] = \pi(2\ell+1), \quad \ell = 0, \pm 1, \pm 2, \ldots.$$

Let's sketch the root locus.

(1) $K=0$. Then the roots of $s^2+2s+2+K(s+2)$ with $K=0$ are $s=-1\pm j$. The root locus starts on the *open-loop* poles.

(2) $K=\infty$. From

$$G(s) = \frac{s+2}{s^2+2s+2} = -\frac{1}{K}$$

we see that as $K \to \infty$ we must have the open-loop transfer function $G(s) \to 0$. This requires either $s \to -2$ or $|s| \to \infty$. Thus the open-loop transfer function $G(s)$ has one zero at $s=-2$ and another zero at $|s|=\infty$. The root locus ends on the open-loop zeros.

(3) *Real axis root locus*

(i) Consider s real with $-\infty < s < -2$. Then $\angle(s+2) = \pi, \angle[(s-(-1+j)] = -\angle[(s-(-1-j)]$ as indicated in Figure 12.24. Then

$$\angle G(s) = \angle(s+2)\underbrace{-\angle[(s-(-1+j)] - \angle[s-(-1-j)]}_{0} = \pi.$$

These values of s are *on* the root locus.

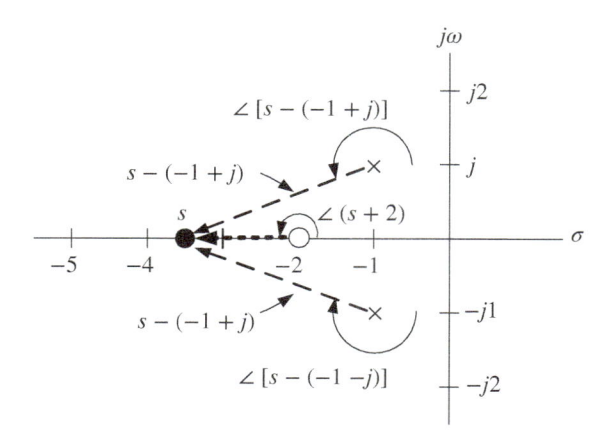

Figure 12.24. $\angle[s-(-1+j)] + \angle[s-(-1-j)] = 0$ on the real axis.

(ii) Consider s real with $s > -2$ so that $\angle(s+2) = 0, \angle[(s-(-1+j)] = -\angle[(s-(-1-j))]$ and therefore

$$\angle G(s) = -\angle(s+2) \underbrace{-\angle[(s-(-1+j)] - \angle[s-(-1-j)]}_{0} = 0.$$

These values of s are *not* on the root locus.
The real axis root locus is given in Figure 12.25.

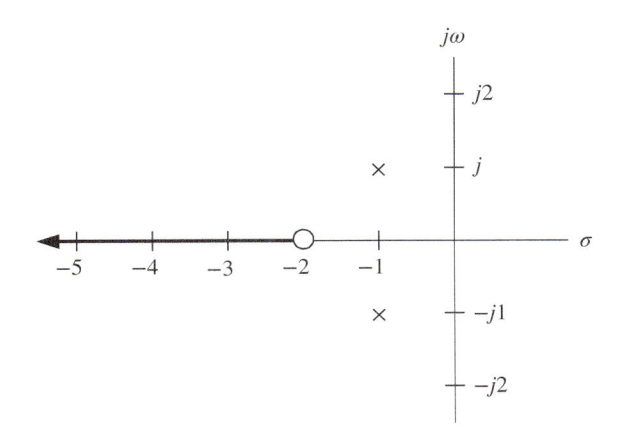

Figure 12.25. Real axis root locus.

(4) *Breakaway point*
A breakaway point occurs at a value of K for which

$$1 + K\frac{s+2}{s^2+2s+2} = \frac{s^2 + (K+2)s + 2K}{s^2+2s+2}$$

has *multiple* zeros. This implies

$$\frac{d}{ds}\left(1 + K\frac{s+2}{s^2+2s+2}\right) = K\frac{d}{ds}\underbrace{\left(\frac{s+2}{s^2+2s+2}\right)}_{G(s)} = 0.$$

We have

$$\frac{d}{ds}G(s) = -\frac{s+2}{(s^2+2s+2)^2}\frac{d}{ds}(s^2+2s+2) + \frac{1}{s^2+2s+2}\frac{d}{ds}(s+2)$$

$$= -\frac{(s+2)(2s+2) + s^2 + 2s + 2}{(s^2+2s+2)^2}$$

$$= -\frac{3s^2 + 8s + 6}{(s^2+2s+2)^2}$$

which requires solving

$$3s^2 + 8s + 6 = 0.$$

We obtain
$$s = \frac{-4 \pm \sqrt{16-8}}{6} = -2 \pm \sqrt{2} = -3.1414, -0.586,$$
where only -3.1414 is on the real axis root locus and thus is the only breakaway point.

(5) *$j\omega$ axis intercepts*

We now use the Routh–Hurwitz test to check if any of the branches of the root locus intercepts (crosses) the $j\omega$ axis. That is, are there any values of $K \geq 0$ such that the zeros of
$$1 + K\frac{s+2}{s^2 + 2s + 2} = \frac{s^2 + (K+2)s + 2K + 2}{s^2 + 2s + 2}$$
are on the $j\omega$ axis? Note that the roots of
$$s^2 + (K+2)s + 2K + 2 = 0$$
are in the open left half-plane for all $K \geq 0$ so there are no $j\omega$ axis intercepts.

(6) *Asymptotes*

The angles of the asymptotes with respect to the real axis are given by
$$\theta_\ell = \frac{\pi(2\ell+1)}{n-m}, \quad \ell = 0, 1, \ldots, n-m-1$$
$$= \frac{\pi(2\ell+1)}{2-1}, \quad \ell = 0$$
$$= \pi.$$

There is a single asymptote whose angle with respect to the real axis is $\theta_0 = \pi$. Its intercept on the real axis is
$$\sigma_1 = -\frac{a_{n-1} - b_{m-1}/b_m}{n-m}, \quad n=2, m=1$$
$$= -\frac{2-2}{2-1} \quad \text{as } a_2 = 2, b_1 = 1, b_0 = 2$$
$$= 0.$$

There is only one asymptote and it coincides with the real-axis root locus.

(7) *Angles of departure*

For $K = 0$ the roots are at $s = 0, s = -1+j$, and $s = -1-j$. As K is increased slightly, the closed-loop poles will leave these roots and we compute the angle that the two branches leave the two complex conjugate poles. Consider K just greater than zero so that the root locus branch that starts at $-1+j$ is slightly away from this point which is indicated by the point "s" in Figure 12.26. The key observation here is that s is taken to be very close $-1+j$ (think of s being much closer to $-1+j$ than as drawn in Figure 12.26). As a result we have
$$\angle(s+2) \approx \pi/4$$
$$\angle[s-(-1-j)] \approx \pi/2.$$

12.3 Angles of Departure

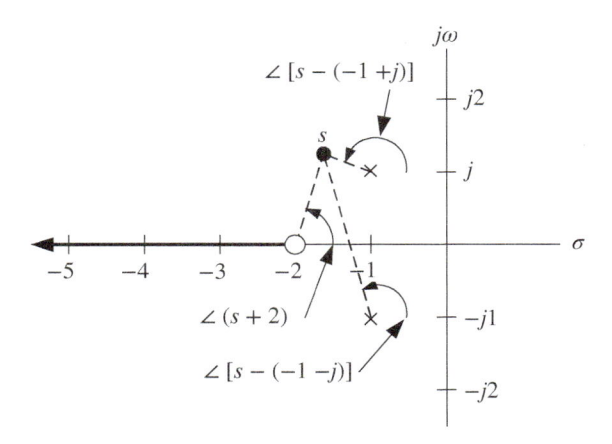

Figure 12.26. Angle of departure from the open-loop pole at $-1+j$.

The angle condition is

$$\angle(s+2) - \angle[s-(-1+j)] - \angle[s-(-1-j)] = \pi(2\ell+1), \quad \ell = 0, \pm 1, \pm 2, \ldots.$$

For K just greater than zero we can write the angle condition approximately as

$$\pi/4 - \angle[(s-(-1+j)] - \pi/2 = \pi(2\ell+1), \quad \ell = 0, \pm 1, \pm 2, \ldots.$$

Solving for $\angle[(s-(-1+j)]$ gives

$$\angle[s-(-1+j)] = -\pi/4 - \pi(2\ell+1), \quad \ell = 0, \pm 1, \pm 2, \ldots.$$

We can take ℓ to be any integer, so let's just take it to be 0. Then

$$\angle[(s-(-1+j)] = -5\pi/4$$

or $\angle[s-(-1+j)] = 3\pi/4$. Since the root locus is symmetric with respect to the real axis, it follows that $\angle[(s-(-1-j)] = -3\pi/4$. Figure 12.27 indicates the angles of departure on the full root locus.

(8) *Sketch*

The root locus starts on the open-loop poles at $-1 \pm j$ leaving them at the calculated angles of departure. We don't really know the shape off of the real axis, which we have drawn as circular in Figure 12.27 on the next page. However, the branches must be symmetric with respect to the real axis and they must come together at the breakaway (or break-in) point at $-2 - \sqrt{2} = -3.414$. After that, one of the branches must head out on the real axis to $-\infty$ and the other one must go to the open-loop zero at -3.

Using the angle condition we show the part of the root locus that is off the real axis is circular! Specifically, we show that the points given by

$$s = -2 + \sqrt{2}e^{j\theta}, \quad 3\pi/4 < \theta < 7\pi/4$$

are on the root locus.

474   12   Root Locus

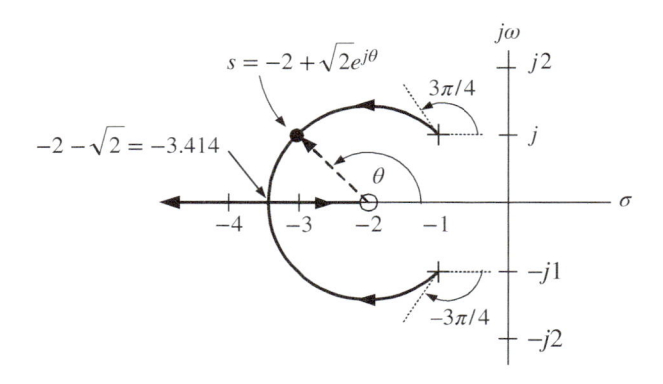

Figure 12.27. Root locus plot of $\dfrac{s+2}{s^2+2s+2}$.

We have

$$G(s)|_{s=-2+\sqrt{2}e^{j\theta}} = \left.\frac{s+2}{s^2+2s+2}\right|_{s=-2+\sqrt{2}e^{j\theta}} = \frac{\sqrt{2}e^{j\theta}}{\left(-2+\sqrt{2}e^{j\theta}\right)^2 + 2\left(-2+\sqrt{2}e^{j\theta}\right) + 2}$$

$$= \frac{\sqrt{2}e^{j\theta}}{4 - 4\sqrt{2}e^{j\theta} + 2e^{2j\theta} - 4 + 2\sqrt{2}e^{j\theta} + 2}$$

$$= \frac{\sqrt{2}}{-2\sqrt{2} + 2e^{j\theta} + 2e^{-j\theta}}$$

$$= \frac{\sqrt{2}}{4\cos(\theta) - 2\sqrt{2}}$$

$$= -\frac{\sqrt{2}/4}{1/\sqrt{2} - \cos(\theta)}.$$

Now $1/\sqrt{2} - \cos(\theta) > 0$ for $3\pi/4 < \theta < 7\pi/4$ so that we can summarize this calculation as

$$G(-2+\sqrt{2}e^{j\theta}) = -\underbrace{\frac{\sqrt{2}/4}{1/\sqrt{2} - \cos(\theta)}}_{>0 \text{ for } 3\pi/4 < \theta < 7\pi/4}.$$

That is, for $3\pi/4 < \theta < 7\pi/4$, we have

$$\angle G(-2+\sqrt{2}e^{j\theta}) = \pi$$

showing that the points

$$s = -2 + \sqrt{2}e^{j\theta}, \; 3\pi/4 < \theta < 7\pi/4$$

are *on* the root locus.

Example 8 *Root Locus of* $G(s) = \dfrac{s+3}{s(s+5)(s+6)(s^2+2s+2)}$

We have the closed-loop system of Figure 12.28.

12.3 Angles of Departure

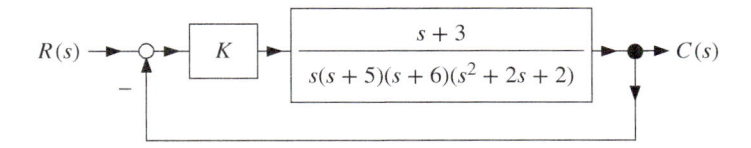

Figure 12.28. Proportional control of $G(s) = \dfrac{s+3}{s(s+5)(s+6)(s^2+2s+2)}$.

The open-loop transfer function is given by

$$G(s) = \frac{s+3}{s(s+5)(s+6)(s^2+2s+2)}$$

$$= \frac{|s+3|}{|s||s+5||s+6||s-(-1+j)||s-(-1-j)|}$$

$$\times e^{j\left(\angle(s+3)-\angle s-\angle(s+5)-\angle(s+6)-\angle[s-(-1+j)]-\angle[s-(-1-j)]\right)}.$$

The closed-loop poles are the *zeros* of

$$1 + KG(s) = 1 + K\frac{s+3}{s(s+5)(s+6)(s^2+2s+2)}$$

$$= \frac{s^5 + 13s^4 + 54s^3 + 82s^2 + 60s + K(s+3)}{s(s+5)(s+6)(s^2+2s+2)}.$$

Equivalently, the closed-loop poles must satisfy

$$G(s) = -\frac{1}{K}.$$

This immediately gives the angle condition

$$\angle G(s) = \pi(2\ell + 1), \quad \ell = 0, \pm 1, \pm 2, \ldots.$$

or

$$\angle(s+3) - \angle s - \angle(s+5) - \angle(s+6) - \angle[s-(-1+j)] - \angle[s-(-1-j)]$$
$$= \pi(2\ell + 1), \quad \ell = 0, \pm 1, \pm 2, \ldots.$$

(1) $K = 0$. With $K = 0$ we have

$$s(s+5)(s+6)(s^2+2s+2) + K(s+3) = s(s+5)(s+6)\big[s-(-1+j)\big]\big[s-(-1-j)\big]$$

with roots $s = 0, -5, -6, -1 \pm j$. The root locus starts on the *open-loop* poles.

(2) $K = \infty$. From

$$G(s) = \frac{s+3}{s(s+5)(s+6)(s^2+2s+2)} = -\frac{1}{K},$$

we see that as $K \to \infty$ the open-loop transfer function $G(s) \to 0$. This requires either $s \to -3$ or $|s| \to \infty$. Notice that as $|s| \to \infty$ we have

$$|G(s)| \approx \frac{1}{|s|^4} \to 0.$$

Thus the open-loop transfer function $G(s)$ has one zero at $s = -3$ and four zeros at $|s| = \infty$. The root locus ends on the open-loop zeros.

(3) *Real axis root locus*

(i) Consider s real with $-3 < s < 0$ (see Figure 12.29). Then $\angle s = \pi, \angle(s+3) = 0$, $\angle(s+5) = \angle(s+6) = 0, \angle[(s-(-1+j)] = -\angle[(s-(-1-j)]$ and therefore

$$\angle G(s) = \angle(s+3) - \angle s - \angle(s+5) - \angle(s+6) - \angle[s-(-1+j)]$$
$$- \angle[s-(-1-j)]$$
$$= -\pi.$$

These values of s are *on* the root locus.

(ii) Consider s real with $-5 < s < -3$ so that $\angle s = \pi, \angle(s+3) = \pi, \angle(s+5) = \angle(s+6) = 0, \angle[(s-(-1+j)] = -\angle[s-(-1-j)]$ and therefore

$$\angle G(s) = \angle(s+3) - \angle s - \angle(s+5) - \angle(s+6) - \angle[s-(-1+j)]$$
$$- \angle[s-(-1-j)]$$
$$= 0.$$

These values of s are *not* on the root locus.

(iii) Consider s real with $-6 < s < -5$ so that $\angle s = \pi, \angle(s+3) = \pi, \angle(s+5) = \pi, \angle(s+6) = 0, \angle[(s-(-1+j)] = -\angle[(s-(-1-j)]$ and therefore

$$\angle G(s) = -\pi.$$

These values of s are *on* the root locus.

(iv) Consider s real with $s < -6$ so that $\angle s = \pi, \angle(s+3) = \pi, \angle(s+5) = \pi, \angle(s+6) = \pi, \angle[(s-(-1+j)] = -\angle[(s-(-1-j)]$ and therefore

$$\angle G(s) = -2\pi.$$

These values of s are *not* on the root locus.

(v) Finally, with s real and $s > 0$ we see that

$$\angle G(s) = 0$$

and therefore these values are *not* on the root locus.
The real axis root locus is given in Figure 12.29.

12.3 Angles of Departure

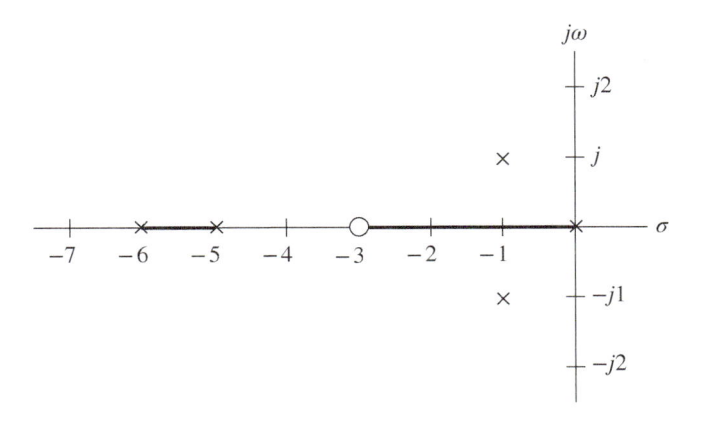

Figure 12.29. Real axis root locus.

(4) Breakaway point

A breakaway point occurs at a value of K for which

$$1 + K\frac{s+3}{s(s+5)(s+6)(s^2+2s+2)} = \frac{s^5 + 13s^4 + 54s^3 + 82s^2 + 60s + K(s+3)}{s(s+5)(s+6)(s^2+2s+2)}$$

has *multiple* zeros. As a consequence any breakaway point must satisfy

$$\frac{d}{ds}\underbrace{\left(\frac{s+3}{s^5 + 13s^4 + 54s^3 + 82s^2 + 60s}\right)}_{G(s)} = 0.$$

After some calculations this reduces to solving

$$g(s) \triangleq s^5 + 13.5s^4 + 66s^2 + 142s^2 + 123s + 45 = 0.$$

The roots of this are $s = -5.53, -0.656 \pm j0.468, -3.33 \pm j1.2$ so that the only breakaway point is -5.53 as indicated on Figure 12.30.

We could have guessed this answer without having to solve $g(s) = 0$. To do so recall that breakaway points can only occur on the real-axis root locus. For s on the real axis root locus and between -5 and -6 we make the table of values of $g(s)$ given by

s	$g(s)$
-5	42.50
-5.5	3.75
-6	-117.0

Note that $g(s)$ changes sign going from -5 to -6 and thus must have a root in that interval.

12 Root Locus

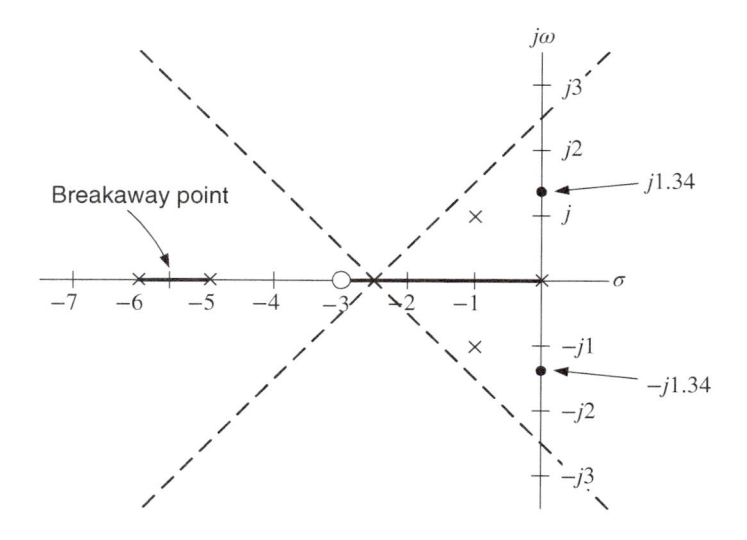

Figure 12.30. Breakaway point, $j\omega$ axis intercepts, and asymptotes.

Similarly, for s on the real axis root locus and between 0 and -3 we make the table of values of $g(s)$ given by

s	$g(s)$
0	45.0
-1	10.5
-2	23.0
-3	22.5

In this case there is no change in sign of $g(s)$ and thus we would guess that $g(s)$ does not have a root in this interval and thus the root locus does not have a breakaway point between 0 and -3.

(5) *$j\omega$ axis intercepts*

We now use the Routh–Hurwitz test to check if any of the branches of the root locus intercepts (crosses) the $j\omega$ axis. That is, are there any values of $K \geq 0$ such that the roots of

$$s^5 + 13s^4 + 54s^3 + 82s^2 + 60s + K(s+3) = 0$$

are on the $j\omega$ axis? We form the Routh table:

s^5	1	54	$60 + K$
s^4	13	82	$3K$
s^3	47.7	$60 + 0.77K$	0
s^2	$0.212(309 - K)$	$3K$	0
s	$\dfrac{3940 - 105K - 0.163K^2}{0.202(309 - K)}$	0	
1	$3K$		

12.3 Angles of Departure

For stability the s^2 row requires $K < 309$. As $3940 - 105K - 0.163K^2 = -0.163(K - 35)(K + 679.7)$ the s row requires $K < 35$ while the last row requires $K > 0$. Thus for

$$0 < K < 35$$

the system is stable. For both $35 < K < 309$ and $K > 309$ there are two sign changes in the first column and thus two right half-plane poles. Set $K = 35$ in the s^2 row of the Routh table to obtain the auxiliary equation

$$0.212(309 - 35)s^2 + 3(35) = 0.$$

This has roots

$$s = \pm j\sqrt{1.8} = \pm j1.34.$$

These intercepts are indicated on Figure 12.30.

(6) *Asymptotes*
The open-loop transfer function is

$$G(s) = \frac{s+3}{s^5 + 13s^4 + 54s^3 + 82s^2 + 60s}.$$

The angles of the asymptotes with respect to the real axis are given by ($n = 5, m = 1$)

$$\theta_\ell = \frac{\pi(2\ell + 1)}{n - m}, \quad \ell = 0, 1, \ldots, n - m - 1$$

$$= \frac{\pi(2\ell + 1)}{5 - 1}, \quad \ell = 0, 1, \ldots, 3$$

or

$$\theta_0 = \pi/4, \theta_1 = 3\pi/4, \theta_2 = 5\pi/4, \theta_3 = 7\pi/4.$$

Their intersection on the real axis is

$$\sigma_1 = -\frac{a_{n-1} - b_{m-1}/b_m}{n - m}, \quad n = 2, m = 1$$

$$= -\frac{13 - 3}{5 - 1} \text{ as } a_4 = 13, b_1 = 1, b_0 = 3$$

$$= -2.5.$$

These asymptotes are drawn on Figure 12.30.

(7) *Angles of departure*
We now compute the angles of departure from the complex conjugate pair of open-loop poles at $-1 \pm j$. We consider the branch that starts at $-1 + j$ with K just greater than 0 so that the root is still close to this point (except think of the point s as much closer to $-1 + j$ than as drawn in Figure 12.31). We then may write (approximately)

$$\underbrace{\angle(s+3)}_{\tan^{-1}(1/2)} - \underbrace{\angle s}_{3\pi/4} - \underbrace{\angle(s+5)}_{\tan^{-1}(1/4)} - \underbrace{\angle(s+6)}_{\tan^{-1}(1/5)} - \angle[s - (-1+j)] - \underbrace{\angle[s - (-1-j)]}_{\pi/2}$$

$$= \pi(2\ell + 1), \quad \ell = 0, \pm 1, \pm 2, \ldots.$$

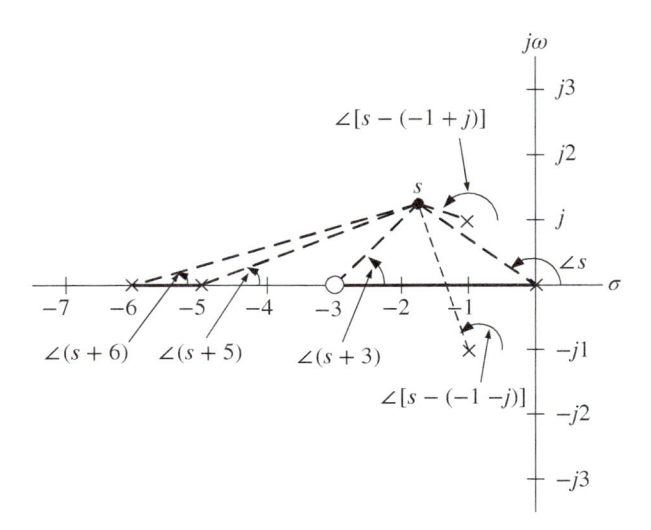

Figure 12.31. Angle of departure from $-1+j$.

Rearranging we have

$$\angle[s-(-1+j)] = \tan^{-1}(1/2) - 3\pi/4 - \tan^{-1}(1/4) - \tan^{-1}(1/5) - \pi/2 - \pi(2\ell+1),$$

$$\ell = 0, \pm 1, \pm 2, \ldots.$$

$$= -3.9057 - \pi \text{ (taking } \ell = 0\text{)}$$

$$= -7.0473 \text{ radians } (-43.8°).$$

(8) *Sketch*

Figure 12.30 is used to sketch the root locus shown in Figure 12.32.

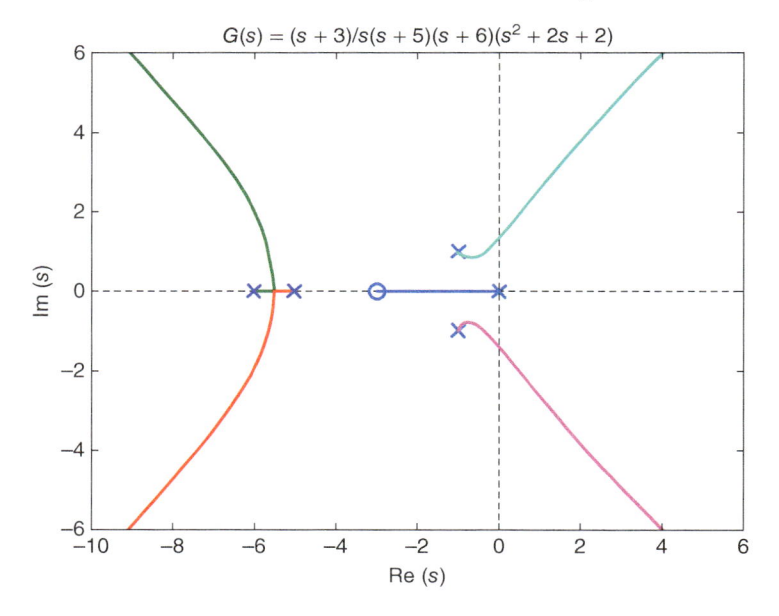

Figure 12.32. Root locus plot of $\dfrac{s+3}{s(s+5)(s+6)(s^2+2s+2)}$.

The two branches that start at open-loop poles at -5 and -6, respectively, come together to the breakaway point at -5.53. These two branches then asymptotically go to the two asymptotes whose angles are $3\pi/4$ and $5\pi/4$, respectively. The branch that starts on the open-loop pole at 0 just goes to the open-loop zero at -3. Finally, the two branches that start at the complex-conjugate pair of poles at $-1+j$ and $-1-j$, respectively, depart from them at $-43.8°$ and $43.8°$, respectively. They then cross the $j\omega$ axis at $j1.34$ and $-j1.34$, respectively, and go toward the two asymptotes at $\pi/4$ and $-\pi/4$, respectively.

Figure 12.32 was drawn using a computer program (MATLAB), but again knowing the root locus rules allows us to not be surprised by the computer plot!

12.4 Effect of Open-Loop Poles on the Root Locus

Figure 12.33 shows the root locus for different open-loop systems using proportional control. Each system has a different set of open-loop poles and no zeros. These four examples are:

(1) $G_c(s)G(s) = K\dfrac{1}{s(s+2)}$

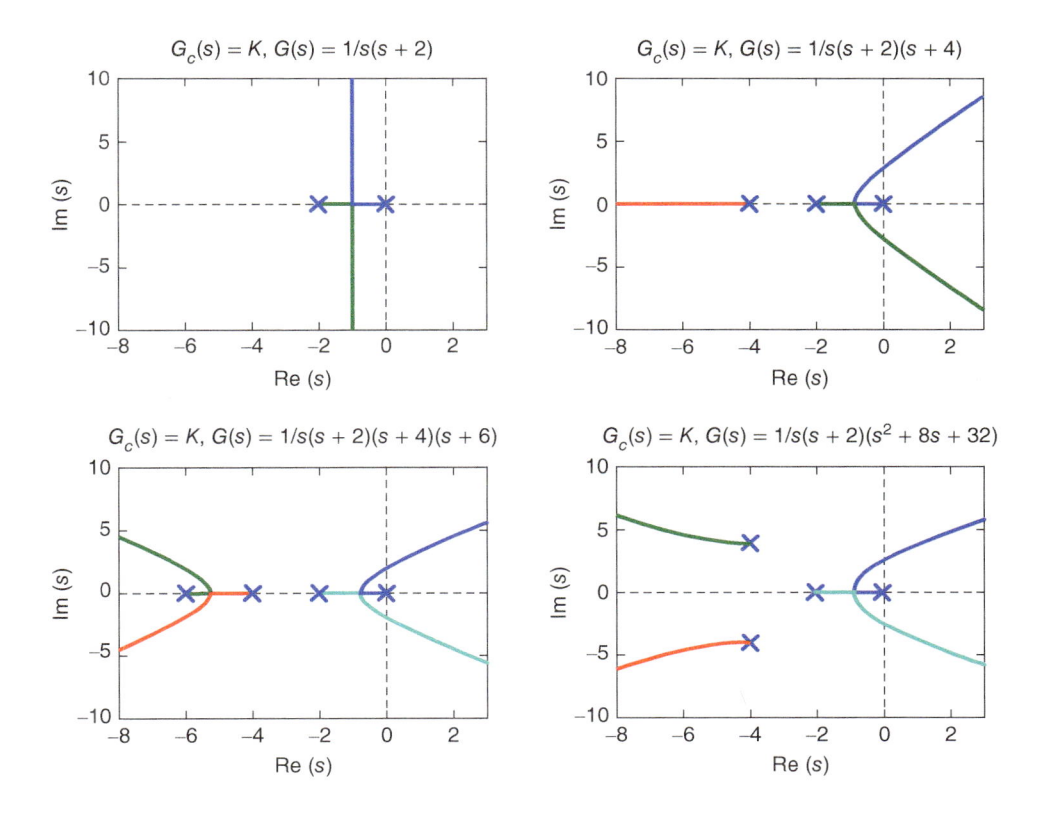

(2) $G_c(s)G(s) = K\dfrac{1}{s(s+2)(s+4)}$

Figure 12.33. Effect of open-loop poles on the root locus.

(3) $G_c(s)G(s) = K\dfrac{1}{s(s+2)(s+4)(s+6)}$

(4) $G_c(s)G(s) = K\dfrac{1}{s(s+2)(s^2+8s+32)}$

12.5 Effect of Open-Loop Zeros on the Root Locus

Figure 12.34 shows the root locus for each of four different controllers for the open-loop system $G(s) = \dfrac{1}{s(s+2)}$. These four examples are:

(1) $G_c(s)G(s) = K\dfrac{1}{s(s+2)}$ Proportional control.

(2) $G_c(s)G(s) = K(s+4)\dfrac{1}{s(s+2)}$ Proportional plus derivative control.

(3) $G_c(s)G(s) = K\dfrac{s+4}{s+6}\dfrac{1}{s(s+2)}$ Lead controller.

(4) $G_c(s)G(s) = K\dfrac{s+1}{s}\dfrac{1}{s(s+2)}$ Proportional plus integral controller.

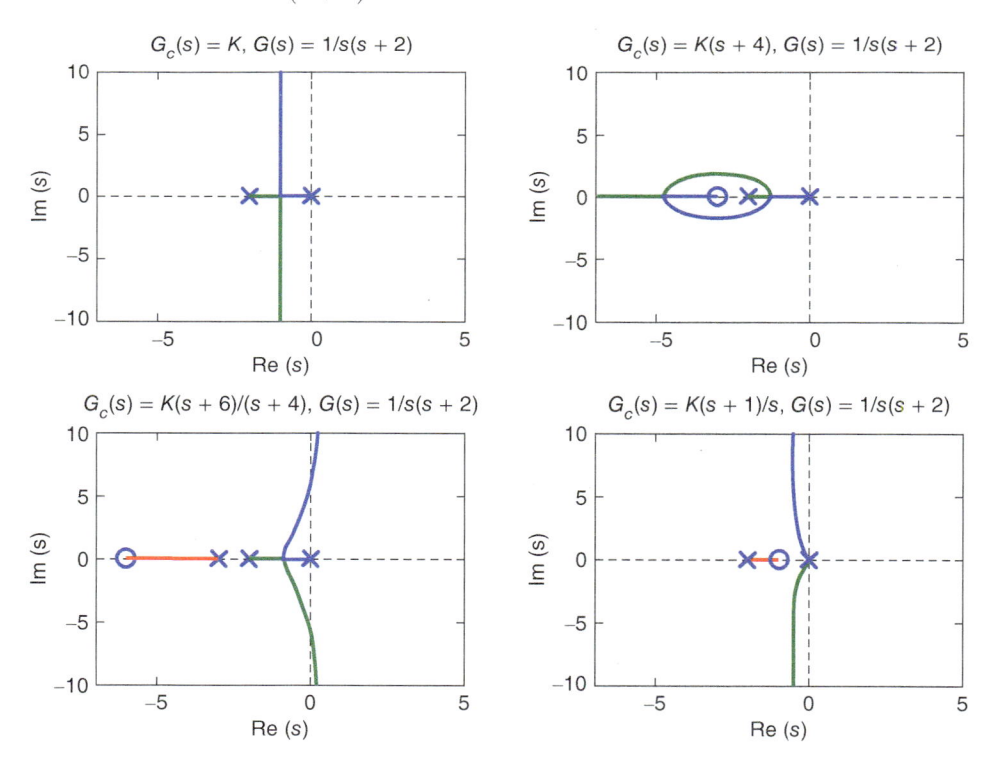

Figure 12.34. Effect of open-loop zeros on the root locus.

12.6 Breakaway Points and the Root Locus

Figure 12.35 shows the root locus of four open-loop systems of the form $G(s) = \dfrac{1}{s(s+a)}$ each using the same controller $G_c(s) = K\dfrac{s+1}{s}$. The value of the open-loop pole at $-a$ has a big effect on the location of the breakaway point [2]. These four examples are:

(1) $G_c(s)G(s) = K\dfrac{s+1}{s}\dfrac{1}{s(s+10)}$

(2) $G_c(s)G(s) = K\dfrac{s+1}{s}\dfrac{1}{s(s+9)}$

(3) $G_c(s)G(s) = K\dfrac{s+1}{s}\dfrac{1}{s(s+8)}$

(4) $G_c(s)G(s) = K\dfrac{s+1}{s}\dfrac{1}{s(s+3)}$

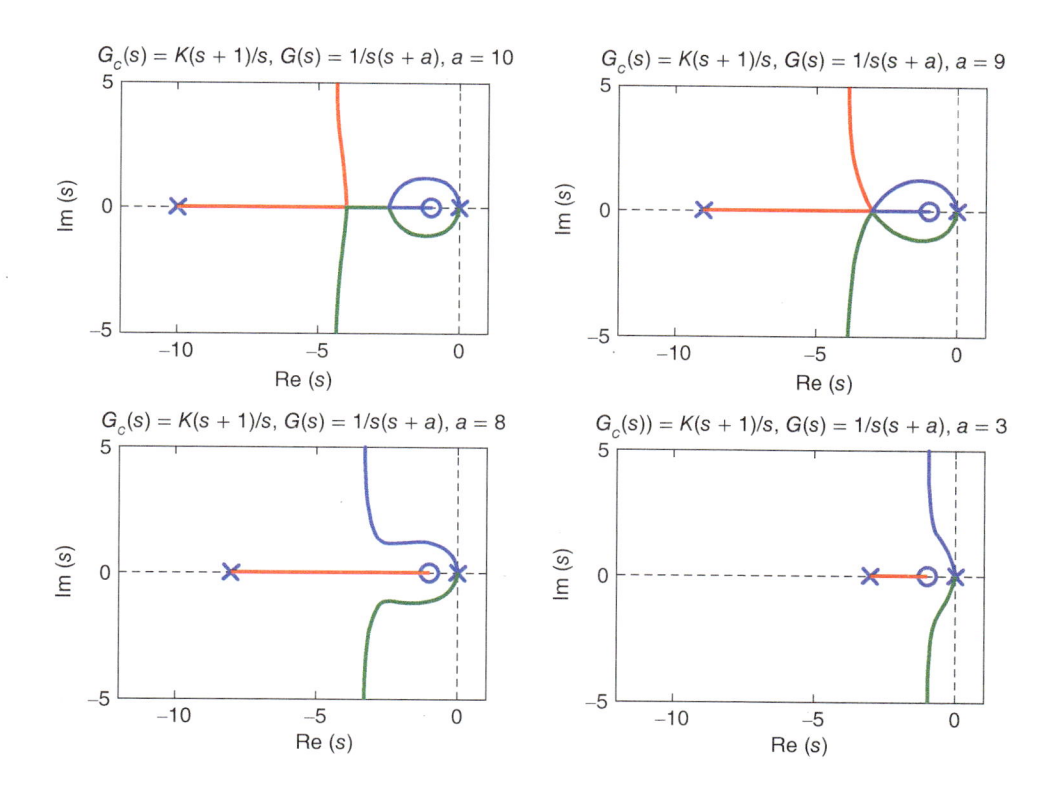

Figure 12.35. Effect of the breakaway point on the root locus. Source: Adapted from Kuo [2]. Automatic Control Systems, Prentice-Hall, Englewood Cliffs, NJ, 1987.

12.7 Design Example: Satellite with Solar Panels (Noncollocated)

In Chapter 5 a model of a satellite with solar panels was developed and given by (see Figure 12.36)

$$\theta(s) = \frac{bs + K}{J_s s^2 + bs + K}\theta_p(s) + \frac{1}{J_s s^2 + bs + K}\tau(s) \tag{12.3}$$

$$\theta_p(s) = \frac{bs + K}{J_p s^2 + bs + K}\theta(s). \tag{12.4}$$

θ is the angle of the actuator (motor) with respect to the satellite, θ_p is angle of the tip of the solar panels with respect to the satellite, and τ is the motor torque used to turn the panels. We now consider the root locus approach to design a controller for this system. We take the sensor to be located at the end of a solar panel, that is, the angle of the solar panel angle θ_p is measured (noncollocated case). Solving (12.3) and (12.4) for $\theta_p(s)$ gives the transfer function from $\tau(s)$ to $\theta_p(s)$ as

$$G_p(s) = \frac{\theta_p(s)}{\tau(s)} = \frac{(b/(J_s J_p))\, s + K/(J_s J_p)}{s^2\left(s^2 + b(1/J_p + 1/J_s)s + K(1/J_p + 1/J_s)\right)}.$$

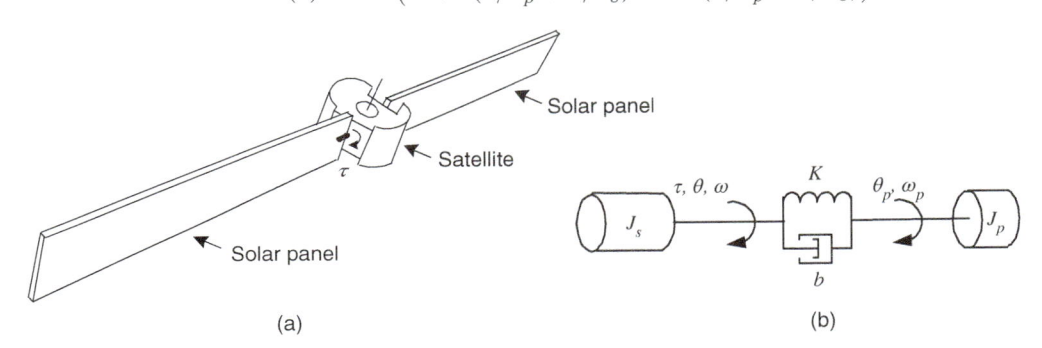

Figure 12.36. (a) Satellite with solar panels for power. (b) Lumped parameter model.

Again with $J_s = 5$ kg-m^2, $J_p = 1$ kg-m^2, $K = 0.15$ N-m/rad, $b = 0.05$ Nm/rad/s, $|\tau| \leq 5$ as in [13] it follows that

$$G_p(s) = \frac{0.01(s+3)}{s^2 + 0.06s + 0.18}\frac{1}{s^2}.$$

A block diagram for the closed-loop system is given in Figure 12.37.

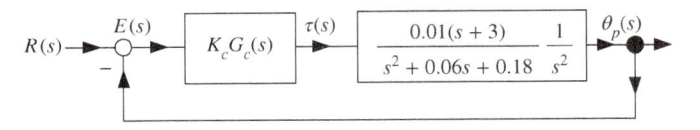

Figure 12.37. Controller for $G_p(s)$.

12.7 Design Example: Satellite with Solar Panels (Noncollocated)

As a first step consider a *notch* filter $G_{notch}(s)$ by

$$G_{notch}(s) \triangleq \frac{s^2 + 0.06s + 0.18}{(s+3)^2}.$$

We then have

$$G_{notch}(s)G_p(s) \triangleq \frac{s^2 + 0.06s + 0.18}{(s+3)^2} \frac{0.01(s+3)}{s^2 + 0.06s + 0.18} \frac{1}{s^2} = \frac{0.01(s+3)}{(s+3)^2}\frac{1}{s^2} = \frac{0.01}{(s+3)s^2}.$$

The notch filter was chosen to cancel the two lightly-damped poles of $G_p(s)$ at $-0.03 \pm j0.42$ (roots of $s^2 + 0.06s + 0.18 = 0$) and replace them by two real poles at -3. The location -3 was chosen simply to cancel the zero of $G_p(s)$. The transfer function model $G_p(s)$ is not known exactly so these stable pole–zero cancellations are not exact. In Chapter 11 it was seen that a lead compensator of the form $G_{lead}(s) = \frac{s+0.1}{s+1}$ can be used to stabilize a double integrator $1/s^2$ so this motivates the use of using a lead compensator to stabilize $G_{notch}(s)G_p(s) = \frac{0.01}{(s+3)s^2}$. The proposed controller is then

$$G_c(s) = \frac{s+0.1}{s+1} \frac{s^2 + 0.06s + 0.18}{(s+3)^2}.$$

The next step would be to do a root locus of $K_c G_c(s) G_p(s)$ to pick a value for K_c. However, we first must deal with the fact that there is not exact cancellation. In doing so we will see the important insight the root locus gives to determining the notch filter. To proceed we modify the notch filter to be

$$G_{notch}(s) = \frac{s^2 + g(0.06)s + g(0.18)}{(s+3.3)^2}.$$

The nominal value of g is 1 and we have now put the poles of $G_{notch}(s)$ at -3.3 so that they are little further in the left half-plane than the zero of $G_p(s)$ at -3. We consider $g = 0.8$ and $g = 1.2$ to see the effect of g on the location of the closed-loop poles. With $g = 0.8$ the notch filter is then

$$G_{notch}(s) = \frac{s^2 + 0.048s + 0.144}{(s+3.3)^2}.$$

A root locus of $K_c G_c(s) G_p(s) = K_c G_{lead}(s) G_{notch}(s) G_p(s)$ for $0 \leq K_c \leq 1500$ is given in Figure 12.38 on the next page.

Figure 12.39 on the next page is a zoomed in version of Figure 12.38 around the $j\omega$ axis. Look at the part of Figure 12.39 showing the two closed-loop poles which start at $-0.03 \pm j0.42$ (open-loop poles of $G_p(s)$) and migrate to the zeros of the notch filter at $-0.024 + j0.38$ as K_c increases from 0. These two poles *stay* in the left-half plane as K_c increases. Thus, even with uncertainty in the poles of $G_p(s)$, there is confidence that the closed-loop poles will not cross over to the right half-plane as K_c is varied.

486   12 Root Locus

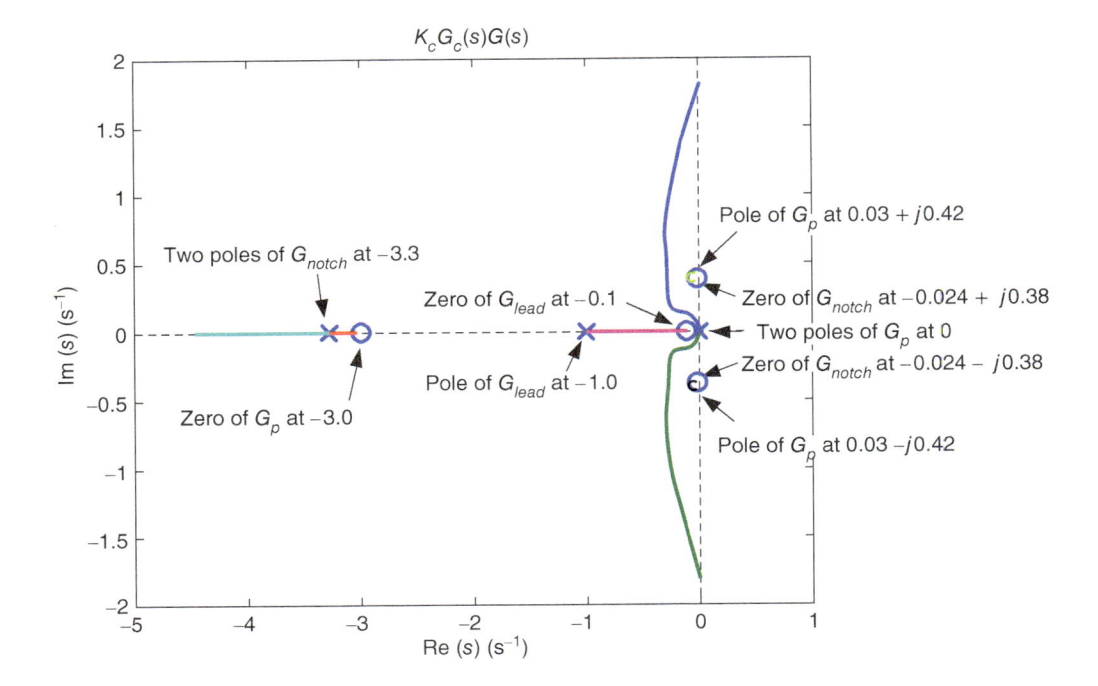

Figure 12.38. Root locus of $K_c G_{lead}(s) G_{notch}(s) G_p(s)$.

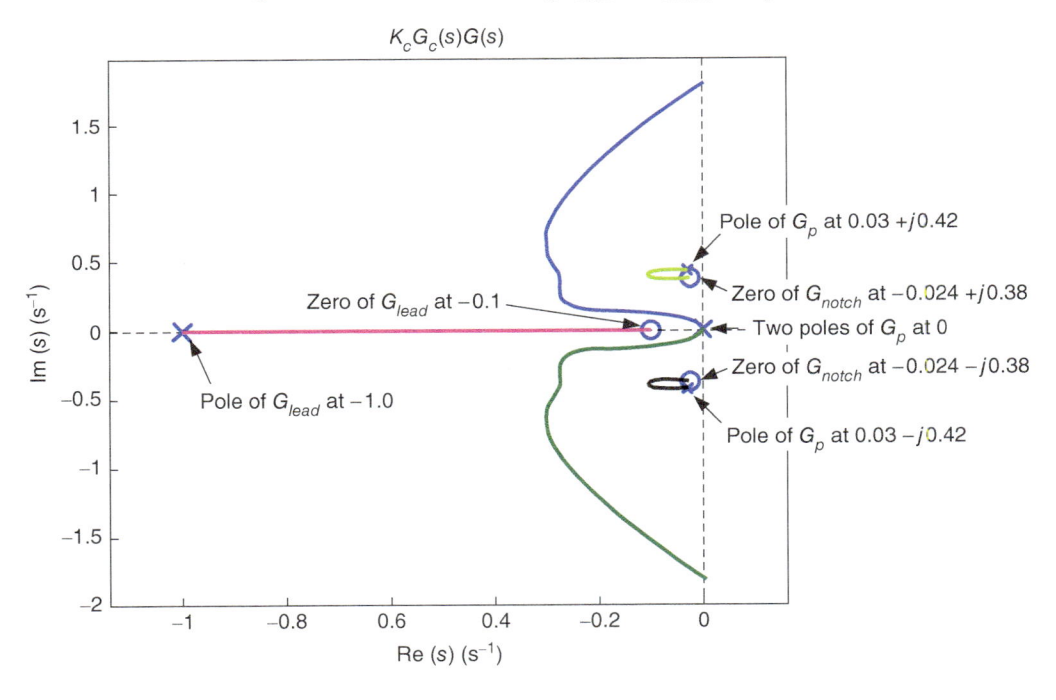

Figure 12.39. Zoomed in root locus of $K_c G_{lead}(s) G_{notch}(s) G_p(s)$ with
$$G_{notch}(s) = \frac{s^2 + (0.8)(0.06)s + (0.8)(0.18)}{(s+3.3)^2}.$$

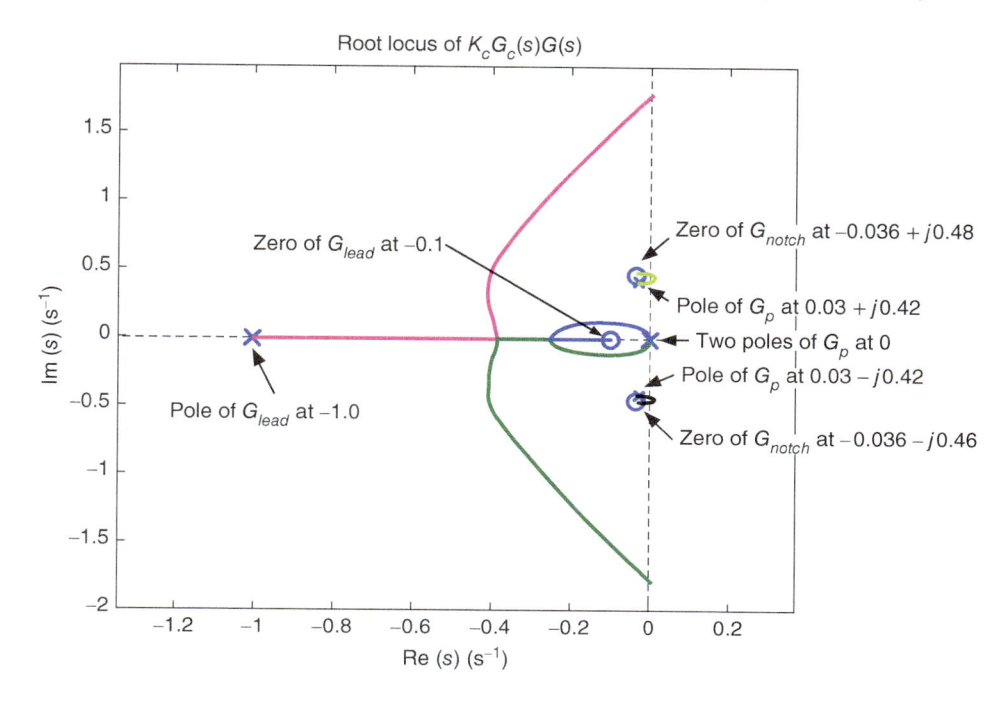

12.7 Design Example: Satellite with Solar Panels (Noncollocated)

[Root locus plot showing: Zero of G_{lead} at -0.1; Pole of G_{lead} at -1.0; Zero of G_{notch} at $-0.036 + j0.48$; Pole of G_p at $0.03 + j0.42$; Two poles of G_p at 0; Pole of G_p at $0.03 - j0.42$; Zero of G_{notch} at $-0.036 - j0.46$.]

Figure 12.40. Root locus of $K_c G_{lead}(s) G_{notch}(s) G_p(s)$ with
$$G_{notch}(s) = \frac{s^2 + (1.2)(0.06)s + (1.2)(0.18)}{(s+3.3)^2}.$$

With $K_c = 50$ ($g = 0.8$) the closed–loop poles are at

$$3.46, -3.19, -0.82, -0.044 \pm j0.43, -0.054 \pm j0.1.$$

Let's now redo the root locus with $g = 1.2$. In this case

$$G_{notch}(s) = \frac{s^2 + (1.2)(0.06)s + (1.2)(0.18)}{(s+3.3)^2} = \frac{s^2 + 0.072s + 0.216}{(s+3.3)^2},$$

where the zeros of G_{notch} are now $-0.036 \pm j0.46$. The root locus (zoomed in around the $j\omega$ axis) is shown in Figure 12.40. The point here is to look at the two closed-loop poles which start at $-0.03 \pm j0.42$ (open-loop poles of $G_p(s)$) and migrate to the zeros of the notch filter at $-0.036 + j0.46$. As K_c increases from 0 these two poles cross into the right half-plane before looping back to the zeros of the notch filter! Thus, with uncertainty in the (open-loop) poles of $G_p(s)$, we cannot be confident that the closed-loop poles will be in the left half-plane for our chosen value of K_c. For this reason $g = 0.8$ is chosen.

With $g = 0.8$, $K_c = 50$ the simulated responses of $\theta_p(t)$ and $\theta(t)$ along with the reference input $r(t)$ are shown in Figure 12.41. Note that the difference $|\theta_p(t) - \theta(t)|$ is small which is important because the larger $|\theta_p(t) - \theta(t)|$ the more mechanical stress within solar panel shaft and we don't want the shaft to break! It turns out that $|\tau(t)| < 0.1$, which is well within the torque limit of 5.

12 Root Locus

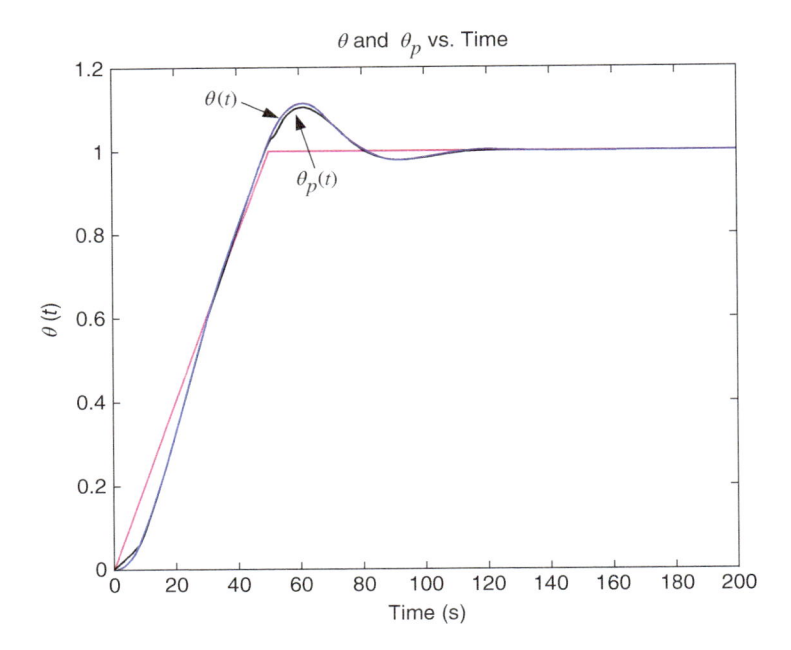

Figure 12.41. Simulation of the responses $\theta_p(t)$ and $\theta(t)$ along with the reference input $r(t)$.

Problems

Problem 1 $G_c(s)G(s) = \dfrac{1}{s(s+1)(s+10)}$

Consider the closed-loop system of Figure 12.42.

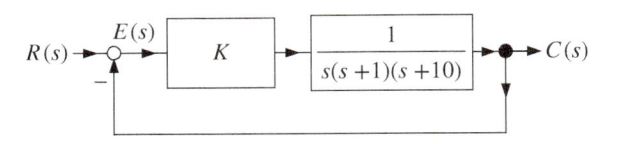

Figure 12.42. Root locus of $1 + K\dfrac{1}{s(s+1)(s+10)}$.

Sketch the root locus of the closed-loop poles for $K \geq 0$ by using the root locus rules as listed.

(1) Closed-loop poles for $K = 0$
(2) Closed-loop poles for $K = \infty$
(3) Real axis root locus
(4) Breakaway point
(5) $j\omega$ axis intercepts
(6) Asymptotes
(7) Angles of departure
(8) Sketch the root locus

Problem 2 $G_c(s)G(s) = \dfrac{1}{s(s^2+2s+10)}$

Consider the closed-loop system of Figure 12.43.

Sketch the root locus of the closed-loop poles for $K \geq 0$ by using the root locus rules as listed.

(1) Closed-loop poles for $K = 0$
(2) Closed-loop poles for $K = \infty$

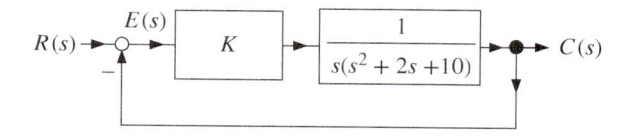

Figure 12.43. Root locus of $1 + K\dfrac{1}{s(s^2+2s+10)}$.

(3) Real axis root locus
(4) Breakaway points
(5) $j\omega$ axis intercepts
(6) Asymptotes
(7) Angles of departure
(8) Sketch the root locus

Problem 3 $G_c(s)G(s) = \dfrac{1}{s^2(s+10)}$

Consider the closed-loop system of Figure 12.44.

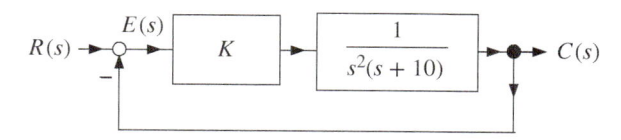

Figure 12.44. Root locus of $1 + K\dfrac{1}{s^2(s+10)}$.

Sketch the root locus of the closed-loop poles for $K \geq 0$ by using the root locus rules as listed.

(1) Closed-loop poles for $K = 0$
(2) Closed-loop poles for $K = \infty$
(3) Real axis root locus
(4) Breakaway points
(5) $j\omega$ axis intercepts
(6) Asymptotes
(7) Angles of departure
(8) Sketch the root locus

Problem 4 $G_c(s) = \dfrac{s+4}{s+20}, G(s) = \dfrac{2}{s^2-4}$

$G(s) = \dfrac{2}{s^2-4}$ has the transfer function form for a magnetically levitated steel ball. Consider the closed-loop control system of Figure 12.45.

12 Root Locus

$$R(s) \rightarrow \underset{-}{\otimes} \xrightarrow{E(s)} \boxed{K \frac{s+4}{s+20}} \rightarrow \boxed{\frac{2}{s^2 - 4}} \rightarrow C(s)$$

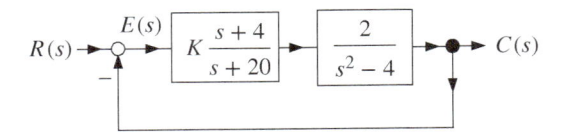

Figure 12.45. Control system for a magnetically levitated steel ball.

Sketch the root locus of the closed-loop poles for $K \geq 0$ by using the root locus rules as listed.
(1) Closed-loop poles for $K = 0$
(2) Closed-loop poles for $K = \infty$
(3) Real axis root locus
(4) Breakaway points
(5) $j\omega$ axis intercepts
(6) Asymptotes
(7) Angles of departure
(8) Sketch the root locus

Problem 5 $G_c(s) = \dfrac{s+4}{s+20}\dfrac{1}{s}, G(s) = \dfrac{2}{s^2 - 4}$

$G(s) = \dfrac{2}{s^2 - 4}$ has the transfer function form for a magnetically levitated steel ball. Consider the closed-loop system of Figure 12.46.

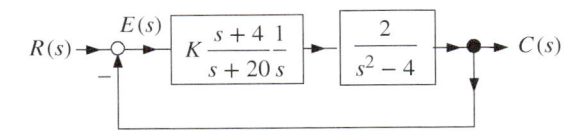

$$R(s) \rightarrow \underset{-}{\otimes} \xrightarrow{E(s)} \boxed{K \frac{s+4}{s+20}\frac{1}{s}} \rightarrow \boxed{\frac{2}{s^2 - 4}} \rightarrow C(s)$$

Figure 12.46. Control system for a magnetically levitated steel ball.

An integrator has been included in the controller in the hope of being able to track step inputs.

(a) Use the Routh–Hurwitz test that there is no value of K that will result in the closed-loop system being stable.

(b) As illustrated in Figure 12.47, consider the controller

$$G_c(s) = \frac{b_2 s^2 + b_1 s + b_0}{s + a_0} \frac{1}{s}.$$

$$R(s) \rightarrow \underset{-}{\otimes} \xrightarrow{E(s)} \boxed{\frac{b_2 s^2 + b_1 s + b_0}{s + a_0}\frac{1}{s}} \rightarrow \boxed{\frac{2}{s^2 - 4}} \rightarrow C(s)$$

Figure 12.47. Pole placement controller for a magnetic levitation system.

Show that the controller parameters b_2, b_1, b_0, a_0 can be chosen such that closed-loop transfer function has its denominator equal to $s^4 + f_3 s^3 + f_2 s^2 + f_1 s + f_0$.

Problem 6 $G_c(s) = \dfrac{s+1/2}{s}, G(s) = \dfrac{1}{s(s+1)}$

Consider the control system of Figure 12.48.

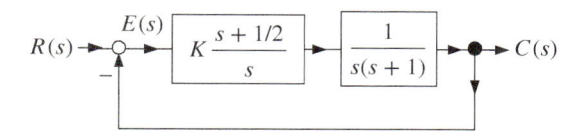

Figure 12.48. Root locus for a PI controller.

Sketch the root locus of the closed-loop poles for $K \geq 0$ by using the root locus rules as listed.
(1) Closed-loop poles for $K = 0$
(2) Closed-loop poles for $K = \infty$
(3) Real axis root locus
(4) Breakaway points
(5) $j\omega$ axis intercepts
(6) Asymptotes
(7) Angles of departure
(8) Sketch the root locus

Problem 7 $G_c(s) = \dfrac{s+0.1}{s+0.01}, G(s) = \dfrac{1}{s(s+1)}$

$G_c(s) = K\dfrac{s+0.1}{s+0.01}$ is a lag controller. It is used to provide high gain at low frequency. Consider the control system of Figure 12.49.

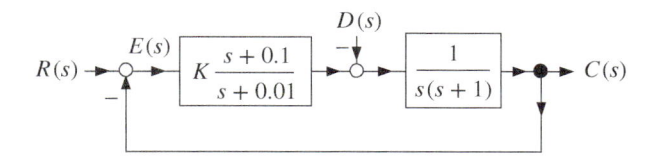

Figure 12.49. Root locus with a lag controller.

To see its use, we compute

$$E_D(s) = \dfrac{\dfrac{1}{s(s+1)}}{1 + K\dfrac{s+0.1}{s+0.01}\dfrac{1}{s(s+1)}} \dfrac{D_0}{s}.$$

If the closed-loop system is stable for some K, then

$$e_D(\infty) = \lim_{s \to 0} sE_D(s) = \lim_{s \to 0} \frac{\frac{1}{s(s+1)}}{1 + K\frac{s+0.1}{s+0.01}\frac{1}{s(s+1)}} D_0$$

$$= \lim_{s \to 0} \frac{\frac{1}{s+1}}{s + K\frac{s+0.1}{s+0.01}\frac{1}{s+1}} D_0 = \frac{1}{10K} D_0.$$

If we only had a simple gain K for the controller, then

$$e_D(\infty) = \lim_{s \to 0} sE_D(s) = \lim_{s \to 0} \frac{\frac{1}{s(s+1)}}{1 + K\frac{1}{s(s+1)}} D_0 \quad \text{(stable for } K > 0\text{)}$$

$$= \lim_{s \to 0} \frac{\frac{1}{s+1}}{s + K\frac{1}{s+1}} D_0$$

$$= \frac{1}{K} D_0.$$

So, for the same value of the gain K, the lag controller decreases the final error by a factor of 10.

Using the lag controller sketch the root locus of the closed-loop poles for $K \geq 0$ by using the root locus rules as listed.

(1) Closed-loop poles for $K = 0$
(2) Closed-loop poles for $K = \infty$
(3) Real axis root locus
(4) Breakaway points
(5) $j\omega$ axis intercepts
(6) Asymptotes
(7) Angles of departure
(8) Sketch the root locus

Problem 8 $G_c(s) = \dfrac{3s+1}{s+3}, G(s) = \dfrac{1}{s^2}$ [46]

Consider the closed-loop system of Figure 12.50.

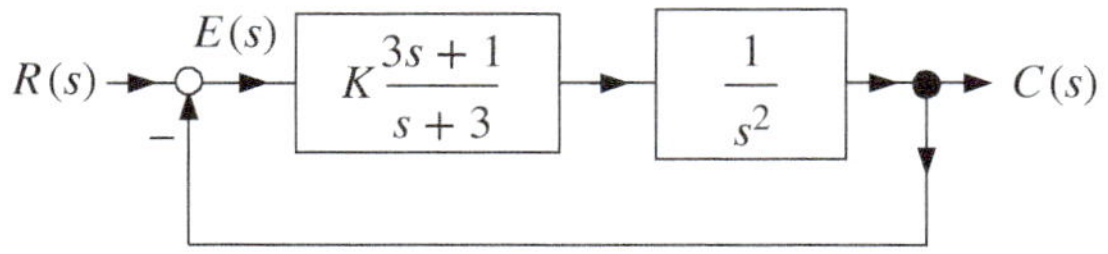

$R(s) \to \bigotimes \to \boxed{K\dfrac{3s+1}{s+3}} \to \boxed{\dfrac{1}{s^2}} \to C(s)$

$E(s)$

Figure 12.50. Root locus for $1 + K\dfrac{3s+1}{s+3}\dfrac{1}{s^2}$.

Note that $G_c(s) = \dfrac{3s+1}{s+3} = 3\dfrac{s+1/3}{s+3} = K\dfrac{s+z}{s+p}$ with $K = 3, z = 1/3, p = 3$. A controller of the form $K\dfrac{s+z}{s+p}$ with $p \gg z > 0$ is called a *lead controller*. Note that

$$K\dfrac{s+z}{s+p} = \dfrac{K}{p}\dfrac{s+z}{s/p+1} \approx \dfrac{K}{p}(s+z) \quad \text{for } |s| \ll p$$

showing that a lead controller is an approximate PD controller.

Sketch the root locus of the closed-loop poles for $K \geq 0$ by using the root locus rules as listed.

(1) Closed-loop poles for $K = 0$
(2) Closed-loop poles for $K = \infty$
(3) Real axis root locus
(4) Breakaway points
(5) $j\omega$ axis intercepts
(6) Asymptotes
(7) Angles of departure
(8) Sketch the root locus

Problem 9 $G_c(s) = \dfrac{s+1}{s-7}, G(s) = \dfrac{s-1}{(s-2)(s+1)}$ [46]

Consider the closed-loop system of Figure 12.51.

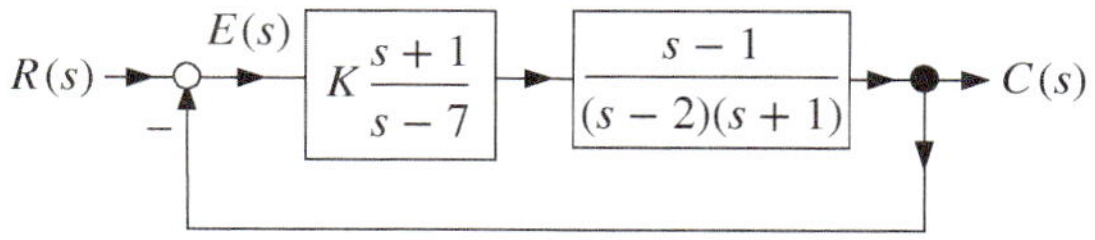

Figure 12.51. Root locus for $1 + K\dfrac{s+1}{s-7}\dfrac{s-1}{(s-2)(s+1)}$.

Sketch the root locus of the closed-loop poles for $K \geq 0$ by using the root locus rules as listed.

(1) Closed-loop poles for $K = 0$
(2) Closed-loop poles for $K = \infty$
(3) Real axis root locus
(4) Breakaway points
(5) $j\omega$ axis intercepts
(6) Asymptotes
(7) Angles of departure
(8) Sketch the root locus
(9) In [46] the design chooses $K = 12$. Where are the closed-loop poles for this value of K?

Problem 10 $G_c(s) = K\dfrac{s+2}{s+10}, G(s) = -\dfrac{s-2}{(s-1)(s+2)}$ [46]

Consider the closed-loop system of Figure 12.52.

12 Root Locus

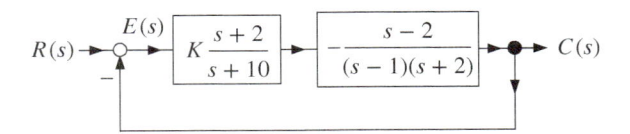

Figure 12.52. Root locus for $1 + K\dfrac{s+2}{s+10}\left(-\dfrac{s-2}{(s-1)(s+2)}\right)$.

In this control system there is a *stable* pole-zero cancellation between $G_c(s)$ and $G(s)$ so that $G_c(s)G(s)$ reduces to

$$G_c(s)G(s) = K\frac{s+2}{s+10}\left(-\frac{s-2}{(s-1)(s+2)}\right) = K\frac{-1}{s+10}\frac{s-2}{s-1}.$$

Sketch the root locus of the closed-loop poles for $K \geq 0$ by using the root locus rules as listed.

(1) Closed-loop poles for $K = 0$
(2) Closed-loop poles for $K = \infty$
(3) Real axis root locus
(4) Breakaway points
(5) $j\omega$ axis intercepts
(6) Asymptotes
(7) Angles of departure
(8) Sketch the root locus
(9) In [46] the design chooses $K = 6$. Where are the closed-loop poles for this value of K?

Problem 11 Root Locus of $G_c(s)G(s) = \dfrac{4(s+0.2)}{s}\dfrac{1}{(s-1)(s+2)}$

Consider the closed-loop system of Figure 12.53.

Figure 12.53. Root locus for $1 + K\dfrac{4(s+0.2)}{s}\dfrac{1}{(s-1)(s+2)}$.

Sketch the root locus of the closed-loop poles for $K \geq 0$ by using the root locus rules as listed.

(1) Closed-loop poles for $K = 0$
(2) Closed-loop poles for $K = \infty$
(3) Real axis root locus
(4) Breakaway points
(5) $j\omega$ axis intercepts
(6) Asymptotes
(7) Angles of departure
(8) Sketch the root locus

Problem 12 $G_c(s) = \dfrac{s + 1/2}{s}, G(s) = -\dfrac{s - 1}{s(s + 2)}$

Consider the closed-loop control system of Figure 12.54.

Figure 12.54. Root locus of $K\dfrac{s+1/2}{s}\left(-\dfrac{s-1}{s(s+2)}\right)$.

Sketch the root locus of the closed-loop poles for $K \geq 0$ by using the root locus rules as listed.
(1) Closed-loop poles for $K = 0$
(2) Closed-loop poles for $K = \infty$
(3) Real axis root locus
(4) Breakaway points
(5) $j\omega$ axis intercepts
(6) Asymptotes
(7) Angles of departure
(8) Sketch the root locus

Problem 13 $G_c(s) = \dfrac{b_1 s + b_0}{s + a_0}, G(s) = \dfrac{2}{s^2 - 4}$

In Problem 5 we considered the control system of Figure 12.55.

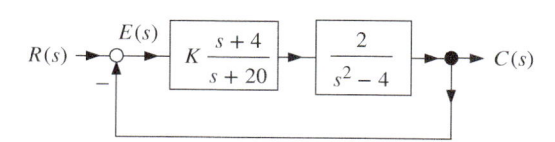

Figure 12.55. Control system for a magnetically levitated steel ball.

It was noted that a controller of the form $K\dfrac{s+z}{s+p}$ ($p \gg z$) is referred to as a lead controller. Consider the controller $G_c(s) = \dfrac{b_1 s + b_0}{s + a_0}$ shown in Figure 12.56, where we make the identification $b_1 = K, b_0 = Kz, a_0 = p$.

Figure 12.56. Pole placement controller for the magnetic levitation system.

Show that the controller gains b_1, b_0, a_0 can be chosen so the closed-loop poles of $C(s)/R(s)$ are at $-r_1, -r_2, -r_3$.

12 Root Locus

Problem 14 *Notch Filter*

In the design example for the satellite with solar panels (see page 484) the notch filter $G_{notch}(s) = \dfrac{s^2 + 0.06s + 0.18}{s^2 + 6s + 9}$ is used. Graph its Bode diagram for $0.01 \leq \omega \leq 100$ to see the "notch" in the amplitude plot at $\omega = 0.42$ rad/s.

13

Inverted Pendulum, Magnetic Levitation, and Cart on a Track

13.1 Inverted Pendulum

Figure 13.1 shows a rod on a cart that is free to rotate about a pivot. The objective is to keep the rod upright ($\theta = 0$) by pushing/pulling on the cart using the input force $u(t)$. We will refer to this as the inverted pendulum control problem. The first thing to do is to derive the mathematical model of this system.

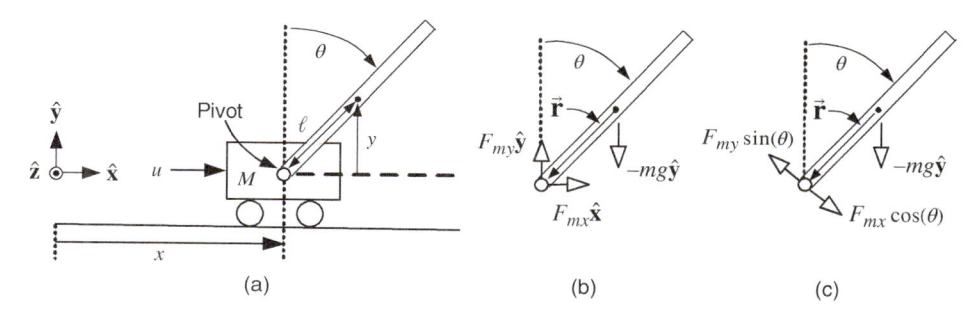

Figure 13.1. (a) Inverted pendulum. (b) Free body diagram. (c) $F_{my}\sin(\theta)$ and $F_{mx}\cos(\theta)$.

Mathematical Model of the Inverted Pendulum

Let $u(t)$ denote an external force on the cart in the horizontal direction. The cart has mass M and the angle between the vertical and the pendulum rod of length 2ℓ is denoted by θ. The center of mass of the rod is a distance ℓ from the pivot. The rod is free to rotate about the pivot and it is assumed that there is a sensor to measure the angle θ. With m denoting the mass of the rod, the force $\boldsymbol{F}_m$ exerted by the pivot on the pendulum rod is written as

$$\boldsymbol{F}_m = F_{mx}\hat{\mathbf{x}} + F_{my}\hat{\mathbf{y}}.$$

There is (of course) the reaction force $-\boldsymbol{F}_m = -F_{mx}\hat{\mathbf{x}} - F_{my}\hat{\mathbf{y}}$ that the pendulum rod exerts on the cart of mass M through the pivot. The rod has a uniform density ρ, uniform cross section a, and a length 2ℓ so its mass is given by $m = \rho a(2\ell)$. The quantity $x + \ell\sin(\theta)$ is the horizontal position of the center of mass of the rod and $\ell\cos(\theta)$ is the vertical position of the center of mass of the rod. The translational motion of the center of mass of any rigid body is determined by the sum of the external forces acting on it. As $(x + \ell\sin(\theta), \ell\cos(\theta))$ locates the center of mass of the pendulum rod, Newton's equations

---

An Introduction to System Modeling and Control, First Edition. John Chiasson.
© 2022 John Wiley & Sons, Inc. Published 2022 by John Wiley & Sons, Inc.
Companion website: www.wiley.com/go/chiasson/anintroductiontosystemmodelingandcontrol

for the rod's translational motion are

$$m\frac{d^2}{dt^2}\left(x + \ell\sin(\theta)\right) = F_{mx}$$

$$m\frac{d^2}{dt^2}\left(\ell\cos(\theta)\right) = F_{my} - mg.$$

We take the axis of rotation of the rod to be in the $-\hat{\mathbf{z}}$ direction through its center of mass.[1] The moment of inertia of the rod about this axis of rotation is

$$J = \int_{rod} r^2 dm = \int_{-\ell}^{\ell} r^2 \underbrace{(\rho a)dr}_{dm} = \rho a \left.\frac{r^3}{3}\right|_{-\ell}^{\ell} = 2\rho a \frac{\ell^3}{3} = \underbrace{\rho a(2\ell)}_{m}\frac{\ell^2}{3} = \frac{m\ell^2}{3}.$$

Newton's equation for rotational motion tell us that the sum of the torques acting on a rigid body equals its moment of inertia times its angular acceleration, that is,

$$\tau = J\frac{d^2\theta}{dt^2}.$$

This equation holds in an inertial reference frame (a reference frame that is not accelerating) *or* in an accelerating frame if the computations are done with respect to an axis of rotation that goes through the *center of mass* of the rigid body. As we have chosen the axis of rotation to be through the center of mass of the rod, this equation is valid here. This was necessary to do as the pendulum rod will be accelerating.

The gravitational force $-mg\hat{\mathbf{y}}$ does *not* produce any *torque* on the pendulum rod as it acts through its center of mass making its moment arm zero (see Figure 13.1c). The vector $\tilde{\vec{r}} = -\left(\ell\sin(\theta)\hat{\mathbf{x}} + \ell\cos(\theta)\hat{\mathbf{y}}\right)$ ($\|\tilde{\vec{r}}\| = \ell$) is a vector from the center of mass of the rod to the point of contact of the force $\vec{F}_\ell = F_{mx}\hat{\mathbf{x}} + F_{my}\hat{\mathbf{y}}$ acting on the rod. The torque about the rod's center of mass due to this force is then

$$\tilde{\vec{\tau}} = \tilde{\vec{r}} \times \vec{F}_\ell = -\left(\ell\sin(\theta)\hat{\mathbf{x}} + \ell\cos(\theta)\hat{\mathbf{y}}\right) \times \left(F_{mx}\hat{\mathbf{x}} + F_{my}\hat{\mathbf{y}}\right)$$

$$= -\ell F_{my}\sin(\theta)\underbrace{\hat{\mathbf{x}}\times\hat{\mathbf{y}}}_{\hat{\mathbf{z}}} - \ell F_{mx}\cos(\theta)\underbrace{\hat{\mathbf{y}}\times\hat{\mathbf{x}}}_{-\hat{\mathbf{z}}}$$

$$= \left(\ell F_{my}\sin(\theta) - \ell F_{mx}\cos(\theta)\right)(-\hat{\mathbf{z}}).$$

As indicated in Figure 13.1c, the component of the force $F_{my}\hat{\mathbf{y}}$ perpendicular to the rod is $F_{my}\sin(\theta)$ producing the torque $\ell F_{my}\sin(\theta)$. This will turn the rod in the positive θ direction if $F_{my} > 0$. Similarly, the component of the force $F_{mx}\hat{\mathbf{x}}$ perpendicular to the rod is $F_{mx}\cos(\theta)$ producing the torque $\ell F_{mx}\cos(\theta)$ and will turn the rod in the $-\theta$ direction if $F_{mx} > 0$. Newton's equation of rotational motion about the rod's center of mass is then

$$J\frac{d^2\theta}{dt^2} = \ell F_{my}\sin(\theta) - \ell F_{mx}\cos(\theta).$$

The equation of motion for the cart is simply

$$M\frac{d^2x}{dt^2} = u(t) - F_{mx}.$$

---

[1] If your right thumb is pointing in the $-\hat{\mathbf{z}}$ direction, then your fingers are pointing in the positive θ direction.

13.1 Inverted Pendulum

The equations of motion for the inverted pendulum system are

$$m\frac{d^2x}{dt^2} - m\ell \sin(\theta)\left(\frac{d\theta}{dt}\right)^2 + m\ell \cos(\theta)\frac{d^2\theta}{dt^2} = F_{mx} \tag{13.1}$$

$$-m\ell \cos(\theta)\left(\frac{d\theta}{dt}\right)^2 - m\ell \sin(\theta)\frac{d^2\theta}{dt^2} = F_{my} - mg \tag{13.2}$$

$$\ell F_{my}\sin(\theta) - \ell F_{mx}\cos(\theta) = J\frac{d^2\theta}{dt^2} \tag{13.3}$$

$$u(t) - F_{mx} = M\frac{d^2x}{dt^2}. \tag{13.4}$$

To eliminate the (unknown) reaction forces, we substitute the expressions for F_{mx} and F_{my} from (13.1) and (13.2) into (13.3) and (13.4) to obtain

$$mg\ell \sin(\theta) - m\ell \cos(\theta)\frac{d^2x}{dt^2} = (J + m\ell^2)\frac{d^2\theta}{dt^2} \tag{13.5}$$

$$(M+m)\frac{d^2x}{dt^2} + m\ell \cos(\theta)\frac{d^2\theta}{dt^2} - m\ell\omega^2 \sin(\theta) = u(t). \tag{13.6}$$

These equations are nonlinear due to the $\sin(\theta)$, $\omega^2 \sin(\theta)$, etc. so we cannot use the Laplace transform yet. Before we make a linear approximation of these equations, let's first rewrite them in a more standard form. Equations (13.5) and (13.6) can be written in matrix form as

$$\begin{bmatrix} J + m\ell^2 & m\ell \cos(\theta) \\ m\ell \cos(\theta) & M+m \end{bmatrix} \begin{bmatrix} d^2\theta/dt^2 \\ d^2x/dt^2 \end{bmatrix} = \begin{bmatrix} mg\ell \sin(\theta) \\ m\ell\omega^2 \sin(\theta) + u(t) \end{bmatrix}.$$

Digression *Inverse of a 2 × 2 matrix*

$$\text{If } M = \begin{bmatrix} a & b \\ c & d \end{bmatrix} \text{ then } M^{-1} = \frac{1}{\underbrace{ad - bc}_{\det M}} \begin{bmatrix} d & -b \\ -c & a \end{bmatrix}.$$

Proof: Simply compute MM^{-1} to find out it equals $I_{2\times 2}$.

End of Digression

Solving for the second-order derivatives results in

$$\begin{bmatrix} d^2\theta/dt^2 \\ d^2x/dt^2 \end{bmatrix} = \begin{bmatrix} J + m\ell^2 & m\ell \cos(\theta) \\ m\ell \cos(\theta) & M+m \end{bmatrix}^{-1} \begin{bmatrix} mg\ell \sin(\theta) \\ m\ell\omega^2 \sin(\theta) + u(t) \end{bmatrix}$$

$$= \frac{1}{JM + Jm + Mm\ell^2 + m^2\ell^2\sin^2(\theta)} \begin{bmatrix} M+m & -m\ell \cos\theta \\ -m\ell \cos\theta & J + m\ell^2 \end{bmatrix}$$

$$\times \begin{bmatrix} mg\ell \sin(\theta) \\ m\ell\omega^2 \sin(\theta) + u \end{bmatrix}$$

$$= \frac{1}{JM + Jm + Mm\ell^2 + m^2\ell^2\sin^2(\theta)}$$

$$\times \left(\begin{bmatrix} gm\ell(M+m)\sin\theta - m^2\ell^2\omega^2\cos\theta\sin\theta \\ m\ell(J+m\ell^2)\omega^2\sin\theta - gm^2\ell^2\cos\theta\sin\theta \end{bmatrix} + \begin{bmatrix} -m\ell\cos\theta \\ J+m\ell^2 \end{bmatrix} u \right). \quad (13.7)$$

With $v \triangleq dx/dt, \omega = \dot{\theta} = d\theta/dt$ we may rewrite (13.7) as

$$\frac{dx}{dt} = v$$

$$\frac{dv}{dt} = \frac{m\ell(J+m\ell^2)\omega^2\sin\theta - gm^2\ell^2\cos\theta\sin\theta}{JM+Jm+Mm\ell^2+m^2\ell^2\sin^2(\theta)} + \frac{J+m\ell^2}{JM+Jm+Mm\ell^2+m^2\ell^2\sin^2(\theta)} u$$

$$\frac{d\theta}{dt} = \omega \quad (13.8)$$

$$\frac{d\omega}{dt} = \frac{gm\ell(M+m)\sin\theta - m^2\ell^2\omega^2\cos\theta\sin\theta}{JM+Jm+Mm\ell^2+m^2\ell^2\sin^2(\theta)} - \frac{m\ell\cos\theta}{JM+Jm+Mm\ell^2+m^2\ell^2\sin^2(\theta)} u.$$

Remark This is called a *statespace* model and refers to the fact that on the left side of each equation is only a first-order derivative, while on the right side of each equation there are no derivatives. The *state* z is simply the vector

$$z \triangleq \begin{bmatrix} x \\ v \\ \theta \\ \omega \end{bmatrix},$$

where x, v, θ, and ω are called the *state variables*.

Linear Approximate Model

The reason for modeling the inverted pendulum is to use it to design a feedback controller that keeps the rod upright. However, to design such a controller based on the nonlinear differential equation model is not available. To get around this a *linear* statespace approximate model of the inverted pendulum is found, which is valid for $|\theta|, |\omega|$ small. A controller can then be designed that stabilizes this linear model. As long as this controller keeps (θ, ω) close to $(0,0)$ (where the linear model is valid), it will also work for the nonlinear model. To develop the linear model we use the following approximations for θ, ω small.

$$\sin(\theta) \approx \theta$$
$$\cos(\theta) \approx 1$$
$$\omega^2 \sin(\theta) \approx \omega^2 \theta \approx 0$$
$$\cos\theta \sin\theta \approx 1 \cdot \theta = \theta \quad (13.9)$$
$$\sin^2(\theta) \approx \theta^2 \approx 0$$
$$\omega^2 \cos\theta \sin\theta \approx \omega^2 \cdot 1 \cdot \theta \approx 0.$$

The idea here is the following. Suppose $\omega = 0.01$, which is small, then $\omega^2 = 0.0001$ is very small. So we keep a term with just ω, but not the really small term ω^2. Using these approximations the model in (13.8) becomes

$$\frac{dx}{dt} = v$$

$$\frac{dv}{dt} = -\frac{gm^2\ell^2}{Mm\ell^2 + J(M+m)}\theta + \frac{J+m\ell^2}{Mm\ell^2 + J(M+m)}u$$

$$\frac{d\theta}{dt} = \omega \qquad (13.10)$$

$$\frac{d\omega}{dt} = \frac{mg\ell(M+m)}{Mm\ell^2 + J(M+m)}\theta - \frac{m\ell}{Mm\ell^2 + J(M+m)}u.$$

In matrix form we have

$$\underbrace{\begin{bmatrix} dx/dt \\ dv/dt \\ d\theta/dt \\ d\omega/dt \end{bmatrix}}_{dz/dt} = \underbrace{\begin{bmatrix} 0 & 1 & 0 & 0 \\ 0 & 0 & -\frac{gm^2\ell^2}{Mm\ell^2 + J(M+m)} & 0 \\ 0 & 0 & 0 & 1 \\ 0 & 0 & \frac{mg\ell(M+m)}{Mm\ell^2 + J(M+m)} & 0 \end{bmatrix}}_{A} \underbrace{\begin{bmatrix} x \\ v \\ \theta \\ \omega \end{bmatrix}}_{z} + \underbrace{\begin{bmatrix} 0 \\ \frac{J+m\ell^2}{Mm\ell^2 + J(M+m)} \\ 0 \\ -\frac{m\ell}{Mm\ell^2 + J(M+m)} \end{bmatrix}}_{b} u.$$

(13.11)

This is a linear statespace (approximate) model of the inverted pendulum. The adjective *linear* refers to fact that the right side of

$$\frac{dz}{dt} = Az + bu$$

is linear in z and u. That is, Az is a linear in z and bu is linear in u. The control problem is to use the measured angle $\theta(t)$ and the measured cart position $x(t)$ to determine the input $u(t)$ that keeps (θ, ω) close to $(0,0)$, i.e., keeps the pendulum rod upright.

Transfer Function Model

Using the last equation of (13.11) with $\frac{d\omega}{dt} = \frac{d^2\theta}{dt^2}$, we have

$$\frac{d^2\theta}{dt^2} = \frac{mg\ell(M+m)}{Mm\ell^2 + J(M+m)}\theta - \frac{m\ell}{Mm\ell^2 + J(M+m)}u. \qquad (13.12)$$

Computing the Laplace transform gives[2]

$$s^2\theta(s) - s\theta(0) - \dot\theta(0) = \underbrace{\frac{mg\ell(M+m)}{Mm\ell^2 + J(M+m)}}_{\alpha^2}\theta(s) - \underbrace{\frac{m\ell}{Mm\ell^2 + J(M+m)}}_{\kappa m\ell}U(s).$$

With the definitions

$$\alpha^2 \triangleq \frac{mg\ell(M+m)}{Mm\ell^2 + J(M+m)}, \quad \kappa \triangleq \frac{1}{Mm\ell^2 + J(M+m)}$$

---

[2]$\mathcal{L}\{\ddot\theta(t)\} = s\mathcal{L}\{\dot\theta(t)\} - \dot\theta(0) = s\left(s\mathcal{L}\{\theta(t)\} - \theta(0)\right) - \dot\theta(0) = s^2\theta(s) - s\theta(0) - \dot\theta(0).$

13 Inverted Pendulum, Magnetic Levitation, and Cart on a Track

we can write $\theta(s)$ as

$$\theta(s) = \underbrace{-\frac{\kappa m\ell}{s^2 - \alpha^2}}_{G_\theta(s)} U(s) + \frac{s\theta(0) + \dot{\theta}(0)}{s^2 - \alpha^2}. \tag{13.13}$$

The open-loop poles of the transfer function $G_\theta(s) = \theta(s)/U(s)$ are

$$\alpha = \sqrt{\frac{mg\ell(M+m)}{Mm\ell^2 + J(M+m)}} \quad \text{and} \quad -\alpha = -\sqrt{\frac{mg\ell(M+m)}{Mm\ell^2 + J(M+m)}}$$

showing it is unstable. Similarly, using the second equation of (13.11) with $\frac{dv}{dt} = \frac{d^2x}{dt^2}$, we have

$$\frac{d^2x}{dt^2} = -\frac{mg(m\ell^2)}{Mm\ell^2 + J(M+m)}\theta + \frac{J + m\ell^2}{Mm\ell^2 + J(M+m)}u.$$

Computing the Laplace transform gives

$$s^2 X(s) - sx(0) - \dot{x}(0) = -\frac{gm^2\ell^2}{Mm\ell^2 + J(M+m)}\theta(s) + \frac{J + m\ell^2}{Mm\ell^2 + J(M+m)}U(s)$$

or

$$X(s) = -\frac{gm^2\ell^2}{Mm\ell^2 + J(M+m)}\frac{1}{s^2}\theta(s) + \frac{J + m\ell^2}{Mm\ell^2 + J(M+m)}\frac{1}{s^2}U(s) + \frac{sx(0) + \dot{x}(0)}{s^2}.$$

Substitution of the expression for $\theta(s)$ in (13.13) and a lot of algebra results in (see Problem 3)

$$X(s) = \underbrace{\frac{\kappa(J + m\ell^2)s^2 - \kappa mg\ell}{s^2(s^2 - \alpha^2)}}_{G_X(s)} U(s) + \frac{p\left(s, \theta(0), \dot{\theta}(0), x(0), \dot{x}(0)\right)}{s^2(s^2 - \alpha^2)}, \tag{13.14}$$

where

$$p\left(s, \theta(0), \dot{\theta}(0), x(0), \dot{x}(0)\right) = x(0)s^3 + \dot{x}(0)s^2 - \left(\kappa mgm\ell^2\theta(0) + \alpha^2 x(0)\right)s$$
$$- \left(\kappa mgm\ell^2\dot{\theta}(0) + \dot{x}(0)\alpha^2\right).$$

Note that $G_X(s)$ has poles at $0, 0, \alpha, -\alpha$ and zeros at $\pm\sqrt{\frac{mg\ell}{J + m\ell^2}} = \pm\sqrt{\frac{g}{\ell}}\sqrt{\frac{1}{J/(m\ell^2) + 1}}$.

In Chapters 9 and 10 we only considered controller designs for single-input single-output (SISO) systems. To obtain the SISO model of the inverted pendulum considered in Example 5 of Chapter 10, take the output to be [49]

$$y \triangleq x + \left(\ell + \frac{J}{m\ell}\right)\theta.$$

After some algebraic manipulation one obtains (see Problem 3)

$$Y(s) = X(s) + \left(\ell + \frac{J}{m\ell}\right)\theta(s)$$

$$= -\underbrace{\frac{\kappa m g \ell}{s^2(s^2 - \alpha^2)}}_{G_Y(s)} U(s) + \frac{p_Y(s, x(0), \dot{x}(0), \theta(0), \dot{\theta}(0))}{s^2(s^2 - \alpha^2)}, \qquad (13.15)$$

where

$$p_Y(s) = \left(x(0) + \theta(0)\left(\ell + \frac{J}{m\ell}\right)\right)s^3 + \left(\dot{x}(0) + \dot{\theta}(0)\left(\ell + \frac{J}{m\ell}\right)\right)s^2$$
$$+ \left(-g\kappa m^2 \ell^2 \theta(0) - x(0)\alpha^2\right)s - g\kappa m^2 \ell^2 \dot{\theta}(0) + \dot{x}(0)\alpha^2.$$

Note that $G_Y(s)$ does not have any zeros. This turns out to be the reason for this particular choice of output as explained in Chapter 17.

Inverted Pendulum Control Using Nested Feedback Loops

To obtain a workable controller we go back to the transfer function (13.14) and factor it as

$$\frac{X(s)}{U(s)} = \frac{\kappa(J + m\ell^2)s^2 - \kappa m g \ell}{s^2(s^2 - \alpha^2)} = \underbrace{\frac{-\kappa m \ell}{s^2 - \alpha^2}}_{\theta(s)/U(s)} \underbrace{\frac{-1}{m\ell} \frac{(J + m\ell^2)s^2 - m g \ell}{s^2}}_{X(s)/\theta(s)}.$$

This suggests the nested loop feedback structure shown in Figure 13.2. (See page 119 of [5] or Problem 5.39 of [3].)

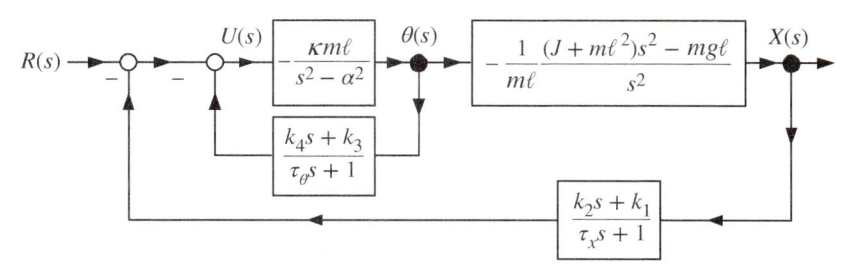

Figure 13.2. Control structure to feedback θ and x separately.

Note that θ and x are each fed back to the input through a proportional plus derivative (PD) controller with low pass filtering (see Chapter 9). The time constants τ_θ and τ_x are taken to be small so we can write

$$U(s) = -\frac{k_1 X(s) + k_2 s X(s)}{\tau_x s + 1} - \frac{k_3 \theta(s) + k_4 s \theta(s)}{\tau_\theta s + 1} + R(s)$$
$$\approx -k_1 X(s) - k_2 s X(s) - k_3 \theta(s) - k_4 s \theta(s) + R(s).$$

In the time domain this is

$$u(t) \approx -k_1 x(t) - k_2 \dot{x}(t) - k_3 \theta(t) - k_4 \dot{\theta}(t) + r(t).$$

This nested loop controller for the inverted pendulum turns out to be a *state feedback controller*. By this is meant that the feedback part of $u(t)$ given by $-k_1 x(t) - k_2 \dot{x}(t) - k_3 \theta(t) - k_4 \dot{\theta}(t)$ is a linear combination of the state variables of the system (13.11).[3] To simplify the analysis we set $\tau_\theta = 0$ and $\tau_x = 0$. The inner loop of Figure 13.2 reduces to

$$\frac{-\dfrac{\kappa m \ell}{s^2 - \alpha^2}}{1 - \dfrac{\kappa m \ell}{s^2 - \alpha^2}(k_4 s + k_3)} = -\frac{\kappa m \ell}{s^2 - \kappa m \ell k_4 s - (\alpha^2 + \kappa m \ell k_3)}$$

resulting in the equivalent block diagram of Figure 13.3.

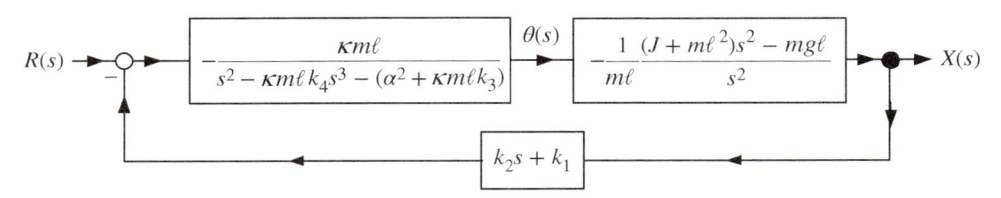

Figure 13.3. Block diagram equivalent to Figure 13.2.

The closed-loop transfer function $X(s)/R(s)$ is given by

$$\frac{X(s)}{R(s)} = \frac{\dfrac{\kappa m \ell}{s^2 - m\kappa \ell k_4 s - (\alpha^2 + m\kappa \ell k_3)} \dfrac{1}{m\ell} \dfrac{(J + m\ell^2)s^2 - mg\ell}{s^2}}{1 + \dfrac{\kappa m \ell}{s^2 - m\kappa \ell k_4 s - (\alpha^2 + m\kappa \ell k_3)} \dfrac{1}{m\ell} \dfrac{(J + m\ell^2)s^2 - mg\ell}{s^2}(k_2 s + k_1)}$$

$$= \frac{\kappa (J + m\ell^2) s^2 - \kappa m g \ell}{s^4 - \kappa m \ell k_4 s^3 - (\alpha^2 + \kappa m \ell k_3) s^2 + \kappa((J + m\ell^2)s^2 - mg\ell)(k_2 s + k_1)}$$

$$= \frac{\kappa (J + m\ell^2) s^2 - \kappa m g \ell}{s^4 + (\kappa k_2(m\ell^2 + J) - m\kappa \ell k_4) s^3 + (\kappa k_1(m\ell^2 + J) - \alpha^2 - m\kappa \ell k_3) s^2}. \quad (13.16)$$
$$\phantom{=\frac{\kappa (J + m\ell^2) s^2 - \kappa m g \ell}{}} -\kappa g m \ell k_2 s - \kappa g m \ell k_1$$

With $s^4 + f_3 s^3 + f_2 s^2 + f_1 s + f_0$ the desired closed-loop characteristic polynomial set

$$f_3 = \kappa k_2 (m\ell^2 + J) - \kappa m \ell k_4$$

---

[3] In Chapter 15 it is shown how to design a state feedback controller based directly on the statespace model (13.11).

13.1 Inverted Pendulum

$$f_2 = \kappa k_1(m\ell^2 + J) - \alpha^2 - \kappa m\ell k_3$$
$$f_1 = -g\kappa m\ell k_2$$
$$f_0 = -g\kappa m\ell k_1$$

or

$$\begin{bmatrix} f_3 \\ f_2 \\ f_1 \\ f_0 \end{bmatrix} = \begin{bmatrix} 0 & \kappa(m\ell^2 + J) & 0 & -\kappa m\ell \\ \kappa(m\ell^2 + J) & 0 & -\kappa m\ell & 0 \\ 0 & -\kappa m g\ell & 0 & 0 \\ -\kappa m g\ell & 0 & 0 & 0 \end{bmatrix} \begin{bmatrix} k_1 \\ k_2 \\ k_3 \\ k_4 \end{bmatrix} + \begin{bmatrix} 0 \\ -\alpha^2 \\ 0 \\ 0 \end{bmatrix}.$$

Then

$$\begin{bmatrix} k_1 \\ k_2 \\ k_3 \\ k_4 \end{bmatrix} = \begin{bmatrix} 0 & \kappa(m\ell^2 + J) & 0 & -\kappa m\ell \\ \kappa(m\ell^2 + J) & 0 & -\kappa m\ell & 0 \\ 0 & -\kappa m g\ell & 0 & 0 \\ -\kappa m g\ell & 0 & 0 & 0 \end{bmatrix}^{-1} \begin{bmatrix} f_3 \\ f_2 + \alpha^2 \\ f_1 \\ f_0 \end{bmatrix}. \qquad (13.17)$$

As done in Chapter 10, to have closed-loop poles at $-r_1, -r_2, -r_3, -r_4$ $(r_i > 0)$, we set the closed-loop characteristic polynomial to be

$$s^4 + f_3 s^3 + f_2 s^2 + f_1 s + f_0 = (s+r_1)(s+r_2)(s+r_3)(s+r_4)$$

$$= s^4 + \underbrace{(r_1 + r_2 + r_3 + r_4)}_{f_3} s^3$$
$$+ \underbrace{(r_1 r_2 + r_1 r_3 + r_1 r_4 + r_2 r_3 + r_2 r_4 + r_3 r_4)}_{f_2} s^2$$
$$+ \underbrace{(r_1 r_2 r_3 + r_1 r_2 r_4 + r_1 r_3 r_4 + r_2 r_3 r_4)}_{f_1} s + \underbrace{r_1 r_2 r_3 r_4}_{f_0}.$$

Referring to Figure 13.2 or 13.3 it is seen that the forward path of this control system is type 2. However this is not in unity feedback form due to $\dfrac{k_2 s + k_1}{\tau_x s + 1} \approx k_2 s + k_1$ in the feedback path. To account for this the reference input filter $G_f(s) = \dfrac{k_2 s + k_1}{\tau_x s + 1}$ is added to the control system as shown in Figure 13.4.

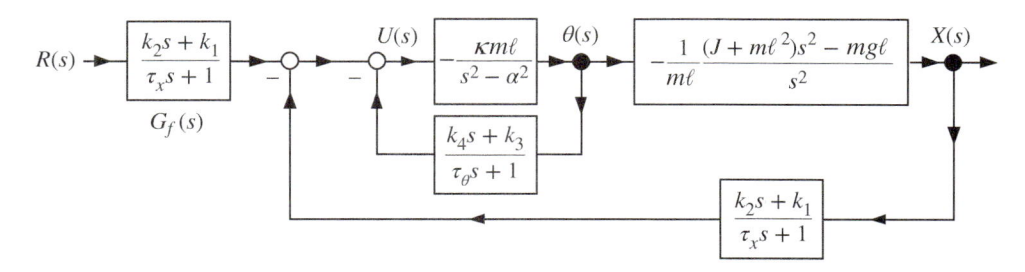

Figure 13.4. Reference input filter added for output tracking of steps and ramps.

A simple rearrangement of the block diagram of Figure 13.4 gives the equivalent block diagram shown in Figure 13.5.

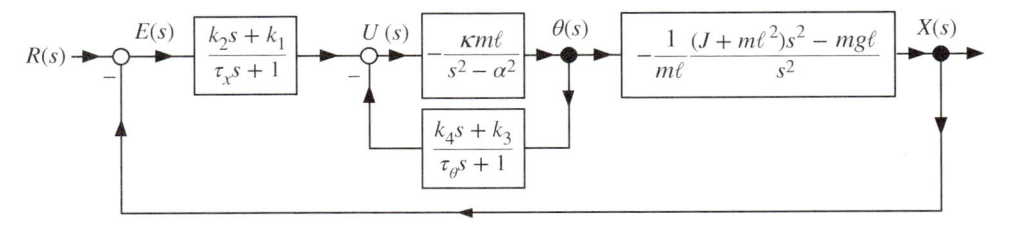

Figure 13.5. Block diagram used to compute $E(s)$.

With $\tau_x = 0, \tau_\theta = 0$ and $R(s) = R_0/s$, the error $E(s)$ is given by

$$E(s) = \cfrac{1}{1 + (k_2 s + k_1)\cfrac{\kappa m\ell}{s^2 - m\kappa\ell k_4 s - (\alpha^2 + m\kappa\ell k_3)}\cfrac{1}{m\ell}\cfrac{(J+m\ell^2)s^2 - mg\ell}{s^2}} \frac{R_0}{s}$$

$$= \frac{(s^2 - m\kappa\ell k_4 s - \alpha^2 - m\kappa\ell k_3)s^2}{s^4 + \bigl(\kappa k_2(m\ell^2 + J) - m\kappa\ell k_4\bigr)s^3 + \bigl(\kappa k_1(m\ell^2 + J) - \alpha^2 - m\kappa\ell k_3\bigr)s^2 - \kappa gm\ell k_2 s - \kappa gm\ell k_1} \frac{R_0}{s}$$

$$= \frac{(s^2 - m\kappa\ell k_4 s - \alpha^2 - m\kappa\ell k_3)s}{(s + r_1)(s + r_2)(s + r_3)(s + r_4)} R_0.$$

We see that $sE(s)$ is stable so that by the final value theorem we have $e(t) \to 0$. Similarly, if $R(s) = R_0/s^2$ we have $e(t) \to 0$. That is, the cart will track step or ramp inputs while keeping the pendulum rod upright.

Remark Problem 4 asks you to simulate this control system using the linear statespace model of the inverted pendulum while Problems 5 and 6 have you do this using the nonlinear statespace model. Problem 13 shows you how to modify the control structure of Figure 13.4 so that the pendulum cart tracks the sinusoidal reference $r(t) = X_0 \sin(\omega_0 t + \phi_0)$. As explained in Chapter 17, this state feedback controller using nested loops results in a control system for the inverted pendulum that is robust. This means it will keep the pendulum rod upright in spite of inaccurate model parameters or small disturbances acting on the cart. This is in contrast to the unity feedback controller of Example 5 of Chapter 10 which was shown in Section 11.9 to have small stability margins.

13.2 Linearization of Nonlinear Models

In the model of the inverted pendulum we were able to use our knowledge of trigonometric functions to come up with a linear model of the inverted pendulum about $(\theta, \dot\theta) = (0,0)$. The reason linear models are so important is because it is known how to design stabilizing feedback controllers for them. In this section we present a systematic way to obtain a linear approximation of a nonlinear model.

Suppose we are given the nonlinear system

$$\frac{dx}{dt} = \sin(x).$$

13.2 Linearization of Nonlinear Models

An equilibrium point (operating point) for this differential equation is a constant value x_{eq} for which $\sin(x_{eq}) = 0$. An equilibrium point is a solution to the differential equation as

$$\underbrace{\frac{dx_{eq}}{dt}}_{0} = \underbrace{\sin(x_{eq})}_{0}.$$

The equilibrium points (operating points) x_{eq} of this differential equation are

$$x_{eq} = 0, \pm\pi, \pm 2\pi, \ldots$$

The general definition is now given.

Definition 1 *Equilibrium Point*
Consider the differential equation

$$\frac{dx}{dt} = f(x).$$

An *equilibrium (operating) point* is any value x_{eq} such that $f(x_{eq}) = 0$.

Remark If x_{eq} is an equilibrium point of the differential equation, then $x(t) = x_{eq}$ is a solution with the initial condition $x(0) = x_{eq}$.

Example 1 *Equilibrium Point*
Let

$$\frac{dx}{dt} = f(x) = -x^3 + 1.$$

Then $x_{eq} = 1$ is the only equilibrium point.

Example 2 *Equilibrium Point*
Let

$$\frac{dx}{dt} = f(x) = -e^{-x} + 1.$$

Then $x_{eq} = 0$ is the only equilibrium point.

Example 3 *Equilibrium Point for a System with an Input*
Let

$$\frac{dx}{dt} = f(x, u) = -x^3 + u,$$

where now we have an input to the system. Suppose it is desired to have the system operate at $x = 2$, that is, the desired equilibrium point is $x_{eq} = 2$. Then we must choose u so that

$$f(2, u) = -(2)^3 + u = 0.$$

Solving gives

$$u_{eq} = 8.$$

$x_{eq} = 2$ is an equilibrium point with the constant input $u_{eq} = 8$. In other words

$$x(t) = x_{eq} = 2$$

is a solution to
$$\frac{dx}{dt} = -x^3 + u$$
with
$$u(t) = u_{eq} = 8, \quad x(0) = 2.$$

Let's next look at how to construct a linear approximate model about an equilibrium point. Consider again the differential equation
$$\frac{dx}{dt} = -x^3 + 1$$
whose equilibrium point is $x_{eq} = 1$. With $f(x) = -x^3 + 1$ a Taylor series expansion of $f(x)$ about $x_{eq} = 1$ is

$$f(x) = f(x_{eq}) + f'(x_{eq})(x - x_{eq}) + f''(x_{eq})\frac{(x - x_{eq})^2}{2!} + \cdots$$
$$= \underbrace{-x_{eq}^3 + 1}_{0} + (-3x_{eq}^2)(x - x_{eq}) + (-6x_{eq})\frac{(x - x_{eq})^2}{2!} + \cdots$$

The first term $f(x_{eq})$ is zero because x_{eq} is an equilibrium point. We want an approximate model for x close to x_{eq}. If $x - x_{eq}$ is small, then $(x - x_{eq})^2, (x - x_{eq})^3$, etc. are very small. We refer to the terms containing $(x - x_{eq})^n$ for $n \geq 2$ as the *higher-order* terms. Setting the higher-order terms to zero we have
$$f(x) \approx f'(x_{eq})(x - x_{eq}) = (-3x_{eq}^2)(x - x_{eq}).$$
Then
$$\frac{d}{dt}(x - x_{eq}) = f(x) - f(x_{eq}) \approx f'(x_{eq})(x - x_{eq}) = (-3x_{eq}^2)(x - x_{eq}).$$
To make this look simpler set
$$z \triangleq x - x_{eq}$$
so that the linear approximate model becomes
$$\frac{dz}{dt} = -3z.$$

Example 4 *Linearization of a Nonlinear System*
Consider the system
$$\frac{dx}{dt} = -x^3 + 2u^2$$
and suppose it is desired to have the system operate at $x_{eq} = 2$. Then we require $u(t) = u_{eq} = 2$ so that $x_{eq} = 2$ is an equilibrium point. With
$$f(x, u) = -x^3 + 2u^2$$

13.2 Linearization of Nonlinear Models 509

a Taylor series expansion of $f(x,u)$ about (x_{eq}, u_{eq}) is

$$f(x,u) = \underbrace{f(x_{eq}, u_{eq})}_{0} + \frac{\partial f(x_{eq}, u_{eq})}{\partial x}(x - x_{eq}) + \frac{\partial f(x_{eq}, u_{eq})}{\partial u}(u - u_{eq})$$

$$+ \text{ higher-order terms}$$

Neglecting the higher-order terms we have

$$\frac{d}{dt}(x - x_{eq}) = f(x,u) - f(x_{eq}, u_{eq})$$

$$\approx \frac{\partial f(x_{eq}, u_{eq})}{\partial x}(x - x_{eq}) + \frac{\partial f(x_{eq}, u_{eq})}{\partial u}(u - u_{eq})$$

$$= (-3x_{eq}^2)(x - x_{eq}) + (4u_{eq})(u - u_{eq})$$

$$= -12(x - x_{eq}) + 8(u - u_{eq}).$$

Set

$$z \triangleq x - x_{eq}, \quad w = u - u_{eq}$$

so that the linear approximate model is then given by

$$\frac{dz}{dt} = -12z + 8w.$$

Example 5 *Linearization of a Nonlinear System*
Consider the system

$$\frac{dx}{dt} = -x^2 u^2 + 1$$

and suppose it is desired to have the system operate at $x_{eq} = 2$. This requires $u(t) = u_{eq} = 1/2$. With

$$f(x,u) = -x^2 u^2 + 1$$

a Taylor series expansion of $f(x,u)$ about $(x_{eq}, u_{eq}) = (2, 1/2)$ is

$$f(x,u) = \underbrace{f(x_{eq}, u_{eq})}_{0} + \frac{\partial f(x_{eq}, u_{eq})}{\partial x}(x - x_{eq}) + \frac{\partial f(x_{eq}, u_{eq})}{\partial u}(u - u_{eq})$$

$$+ \underbrace{\frac{1}{2!}\frac{\partial^2 f(x_{eq}, u_{eq})}{\partial x^2}(x - x_{eq})^2 + \frac{\partial^2 f(x_{eq}, u_{eq})}{\partial x \partial u}(x - x_{eq})(u - u_{eq}) + \cdots}_{\text{Higher-order terms}}$$

Neglecting the higher-order terms results in

$$\frac{d}{dt}(x - x_{eq}) = f(x,u) - f(x_{eq}, u_{eq})$$

$$\approx \frac{\partial f(x_{eq}, u_{eq})}{\partial x}(x - x_{eq}) + \frac{\partial f(x_{eq}, u_{eq})}{\partial u}(u - u_{eq})$$

$$= (-2x_{eq} u_{eq}^2)(x - x_{eq}) + (-2x_{eq}^2 u_{eq})(u - u_{eq})$$

$$= -(x - x_{eq}) - 4(u - u_{eq}).$$

Set
$$z \triangleq x - x_{eq}, w = u - u_{eq}$$
so that the linear approximate model is then written as
$$\frac{dz}{dt} = -z - 4w.$$

13.3 Magnetic Levitation

Here we consider the problem of levitating a steel ball under an electromagnet. A schematic diagram is shown in Figure 13.6. The electromagnet consists of a wire (coil) wrapped around a cylindrical core of soft iron. A voltage u applied to the coil results in a current which produces a magnetic field that attracts the steel ball upwards against the gravitational force pulling down on it. A photograph of such a magnetic levitation system is shown in Figure 1.21.

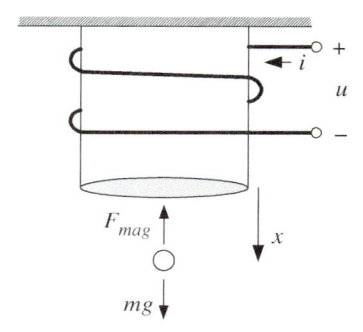

Figure 13.6. Schematic diagram of a magnetically levitated steel ball.

To develop a differential equation model of this system, we first model the flux linkage in the coil. The flux linkage is the sum of all the fluxes in each of the coils wound around the iron core. It is given by
$$\lambda(x, i) \triangleq L(x)i.$$

To explain this expression for $\lambda(x, i)$, consider the center of mass of the steel ball fixed at some position x. The coil is just like an inductor so we expect the flux linkage λ to be proportional to the current i based on elementary physics. The dependence of λ on the ball position x is not as easy to motivate. It is explained in Section 13.3.1 that a reasonable model for the inductance is given by
$$L(x) = L_0 + \frac{rL_1}{x}, \qquad (13.18)$$
where r is the radius of the steel ball. Note that x locates the *center of mass* of the steel ball with respect to the bottom of the electromagnet and increases in the downward direction. Further x can be no smaller than the radius of the steel ball, i.e., $x \geq r$. It is typically true that $L_1 \ll L_0$ and will be assumed here. By Faraday's law and Ohm's law we have
$$u - \frac{d}{dt}\underbrace{(L(x)i)}_{\lambda} - Ri = 0.$$

13.3 Magnetic Levitation

With the self-inductance of the coil given by (13.18), a conservation of energy argument (see below) shows that the magnetic force has the form

$$F_{mag} = -C\frac{i^2}{x^2}, \qquad (13.19)$$

where C is a constant parameter. Conversely, if the force F_{mag} has the form (13.19), then the self-inductance of the coil must have the form (13.18). The expression (13.19) for the magnetic force says that if you double the coil current then the force will increase by a factor of 4. On the other hand, if you double the distance of the steel ball from the bottom of the electromagnet, then the force goes down by a factor of 4.

Putting this all together the model we consider for the suspension of a steel ball beneath the electromagnet is given by

$$\frac{d}{dt}(L(x)i) + Ri = u \qquad (13.20)$$

$$\frac{dx}{dt} = v \qquad (13.21)$$

$$m\frac{dv}{dt} = mg - C\frac{i^2}{x^2}. \qquad (13.22)$$

Conservation of Energy

We now show that if (13.20) holds with $L(x)$ given by (13.18) then conservation of energy implies (13.22) must hold with $C = rL_1/2$. To proceed, we multiply (13.20) by the current i to obtain the electrical power into the electromagnet given by

$$iu = i\frac{d}{dt}(L(x)i) + Ri^2 = L(x)i\frac{di}{dt} + i^2\frac{\partial L(x)}{\partial x}v + Ri^2. \qquad (13.23)$$

The quantity

$$\frac{1}{2}L(x)i^2$$

is the energy stored in the magnetic field of the coil produced by the current in the coil. The rate of change of this magnetic energy is

$$\frac{d}{dt}\left(\frac{1}{2}L(x)i^2\right) = L(x)i\frac{di}{dt} + \frac{1}{2}i^2\frac{\partial L(x)}{\partial x}v. \qquad (13.24)$$

We next solve for $L(x)i\frac{di}{dt}$ in (13.24) and substitute this expression into (13.23) to obtain

$$iu = \frac{d}{dt}\left(\frac{1}{2}L(x)i^2\right) - \frac{1}{2}i^2\frac{\partial L(x)}{\partial x}v + i^2\frac{\partial L(x)}{\partial x}v + Ri^2$$

$$= \frac{d}{dt}\left(\frac{1}{2}L(x)i^2\right) + Ri^2 + \frac{1}{2}i^2\frac{\partial L(x)}{\partial x}v. \qquad (13.25)$$

Equation (13.25) shows that the electrical power iu supplied by voltage source goes into changing the magnetic field energy $\frac{1}{2}L(x)i^2$ of the coil and steel ball, dissipation as heat

Ri^2, and a third term $\frac{1}{2}i^2\frac{\partial L(x)}{\partial x}v$. By conservation of energy this third term must represent the mechanical power supplied to the steel ball. To model the mechanical energy of the steel ball we take its gravitational potential energy $V(x)$ to be zero at $x = 0$ so that it may be written as

$$V(x) = -mgx.$$

Note that $V(x)$ is decreasing as x increases, i.e., as it should as the ball goes downward. The total mechanical energy of the steel ball is then

$$\frac{1}{2}mv^2 + V(x).$$

The rate of change of this mechanical energy is the mechanical power and therefore

$$\frac{d}{dt}\left(\frac{1}{2}mv^2 + V(x)\right) = \frac{1}{2}i^2\frac{\partial L(x)}{\partial x}v$$

or

$$mv\frac{dv}{dt} - mgv = \frac{1}{2}i^2\frac{\partial}{\partial x}\left(L_0 + \frac{rL_1}{x}\right)v = -\frac{1}{2}i^2\frac{rL_1}{x^2}v.$$

Canceling the velocity v and rearranging we have

$$m\frac{dv}{dt} = mg - \frac{rL_1}{2}\frac{i^2}{x^2} \tag{13.26}$$

which verifies (13.22) with $C = rL_1/2$. Summarizing, conservation of energy requires the magnetic force be given by (13.19) with $C = rL_1/2$ if the flux linkage is given by $\lambda(x, i) = (L_0 + rL_1/x)i$. We can turn this argument around by doing some experiments in which we measure the force as a function of (x, i). We would find out experimentally that the force has the form given by (13.19) for some constant C. Then, as a consequence of conservation of energy, $L(x) = (L_0 + rL_1/x)i$ with $L_1 = 2C/r$ (see Problem 8).

Statespace Model

Our model is now

$$L(x)\frac{di}{dt} + i\frac{\partial L(x)}{\partial x}v + Ri = u \tag{13.27}$$

$$\frac{dx}{dt} = v \tag{13.28}$$

$$m\frac{dv}{dt} = mg - \frac{rL_1}{2}\frac{i^2}{x^2}. \tag{13.29}$$

Upon some rearrangement this becomes

$$\frac{di}{dt} = -i\frac{1}{L(x)}\left(-\frac{rL_1}{x^2}\right)v - \frac{R}{L(x)}i + \frac{1}{L(x)}u \tag{13.30}$$

$$\frac{dx}{dt} = v \tag{13.31}$$

$$\frac{dv}{dt} = g - \frac{rL_1}{2m}\frac{i^2}{x^2}, \tag{13.32}$$

where $L(x) = L_0 + rL_1/x$ and we substituted $\partial L(x)/\partial x = -rL_1/x^2$. The system of Eqs. (13.30), (13.31), and (13.32) is a nonlinear differential equation model of the magnetic levitation system.

Remark The model (13.30), (13.31), and (13.32) is called a *statespace* model. As in the case of the inverted pendulum, this refers to the fact that only first-order derivatives are on left side of these equations, while the right side of these equations has no derivatives. The *state* is simply
$$\begin{bmatrix} i \\ x \\ v \end{bmatrix}$$
and i, x, and v are called the *state variables*.

We can simplify the equations a little by recalling our assumption that $L_1 \ll L_0$ and $x \geq r$ so that
$$L(x) = L_0 + rL_1/x \leq L_0 + L_1 \approx L_0.$$
Substituting L_0 for $L(x)$ in (13.30) we obtain
$$\frac{di}{dt} = \frac{rL_1}{L_0}\frac{i}{x^2}v - \frac{R}{L_0}i + \frac{1}{L_0}u \tag{13.33}$$
$$\frac{dx}{dt} = v \tag{13.34}$$
$$\frac{dv}{dt} = g - \frac{rL_1}{2m}\frac{i^2}{x^2}. \tag{13.35}$$

The expressions $\frac{i}{x^2}v$ in (13.33) and $\frac{i^2}{x^2}$ in (13.34) are nonlinear and thus this is still a nonlinear model.

Current Command Input Model

Before developing a linear model we can first simplify the model by using a current command amplifier. This is simply an amplifier that allows one to command a current to it rather than a voltage. Such an amplifier is indicated in Figure 13.7.

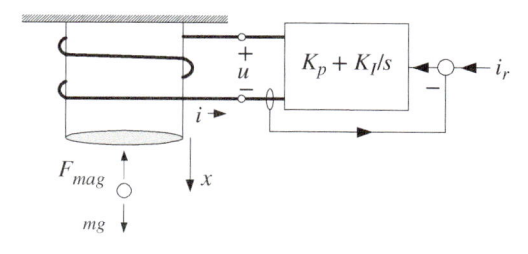

Figure 13.7. Current command amplifier for the magnetic levitation system.

That is, a PI controller forces $i(t) \to i_r(t)$ so fast (compared with the motion of the steel ball) that the current $i(t)$ can be considered as the input. With $i(t)$ the input the model

514 **13 Inverted Pendulum, Magnetic Levitation, and Cart on a Track**

reduces to
$$\frac{dx}{dt} = v \tag{13.36}$$

$$\frac{dv}{dt} = g - \frac{rL_1}{2m}\frac{i^2}{x^2}. \tag{13.37}$$

Linearization About an Equilibrium Point

An equilibrium point for the system is a *constant* solution to the system Eqs. (13.36) and (13.37). We choose an equilibrium point to be where the steel ball is at a set distance x_{eq} below the electromagnet and not moving, that is, we have $x = x_{eq}$ and $v = v_{eq} = 0$. Then (13.36) and (13.37) are simply

$$\frac{dx_{eq}}{dt} = v_{eq}$$

$$\frac{dv_{eq}}{dt} = g - \frac{rL_1}{2m}\frac{i_{eq}^2}{x_{eq}^2}.$$

The first equation holds because x_{eq} is a constant and $v_{eq} = 0$. In the second equation the left side is zero because $v_{eq} = 0$ is constant. For the right side to be zero we must have

$$g = \frac{rL_1}{2m}\frac{i_{eq}^2}{x_{eq}^2} \quad \text{or} \quad i_{eq} = x_{eq}\sqrt{\frac{2mg}{rL_1}}. \tag{13.38}$$

The model (13.36) and (13.37) is nonlinear due to the expression $\frac{i^2}{x^2}$ in (13.37). To find a linear approximate model a Taylor series expansion of

$$f(x,i) = g - \frac{rL_1}{2m}\frac{i^2}{x^2}$$

is done about the equilibrium point (x_{eq}, v_{eq}) with input i_{eq}. First note that

$$f(x_{eq}, i_{eq}) = g - \frac{rL_1}{2m}\frac{i_{eq}^2}{x_{eq}^2} = 0$$

as i_{eq} was chosen to make this true. The Taylor series expansion of $f(x,i)$ about (x_{eq}, i_{eq}) is then (dropping higher-order terms)

$$f(x,i) \approx f(x_{eq}, i_{eq}) + \left.\frac{\partial f(x,i)}{\partial x}\right|_{(x_{eq}, i_{eq})}(x - x_{eq}) + \left.\frac{\partial f(x,i)}{\partial i}\right|_{(x_{eq}, i_{eq})}(i - i_{eq})$$

$$= \underbrace{g - \frac{rL_1}{2m}\frac{i_{eq}^2}{x_{eq}^2}}_{0} + \frac{rL_1}{m}\frac{i_{eq}^2}{x_{eq}^3}(x - x_{eq}) - \frac{rL_1}{m}\frac{i_{eq}}{x_{eq}^2}(i - i_{eq})$$

$$= \frac{rL_1}{m}\frac{i_{eq}^2}{x_{eq}^3}(x - x_{eq}) - \frac{rL_1}{m}\frac{i_{eq}}{x_{eq}^2}(i - i_{eq})$$

$$= \frac{2g}{x_{eq}}(x - x_{eq}) - \frac{2g}{i_{eq}}(i - i_{eq}), \tag{13.39}$$

where in the third line we used (13.38) to obtain

$$\frac{2g}{x_{eq}} = \frac{rL_1}{m}\frac{i_{eq}^2}{x_{eq}^3}, \quad \frac{2g}{i_{eq}} = \frac{rL_1}{m}\frac{i_{eq}}{x_{eq}^2}. \tag{13.40}$$

The model (13.36) and (13.37) becomes

$$\frac{dx}{dt} = v \tag{13.41}$$

$$\frac{dv}{dt} = \frac{2g}{x_{eq}}(x - x_{eq}) - \frac{2g}{i_{eq}}(i - i_{eq}). \tag{13.42}$$

As $v_{eq} = 0, dx_{eq}/dt = dv_{eq}/dt = 0$ we can rewrite this as

$$\frac{d}{dt}(x - x_{eq}) = v - v_{eq} \tag{13.43}$$

$$\frac{d}{dt}(v - v_{eq}) = \frac{2g}{x_{eq}}(x - x_{eq}) - \frac{2g}{i_{eq}}(i - i_{eq}). \tag{13.44}$$

With $\Delta x \triangleq x - x_{eq}, \Delta v \triangleq v - v_{eq}$, and $\Delta i \triangleq i - i_{eq}$ we have

$$\frac{d}{dt}\Delta x = \Delta v \tag{13.45}$$

$$\frac{d}{dt}\Delta v = \frac{2g}{x_{eq}}\Delta x - \frac{2g}{i_{eq}}\Delta i \tag{13.46}$$

or

$$\frac{d}{dt}\begin{bmatrix}\Delta x \\ \Delta v\end{bmatrix} = \underbrace{\begin{bmatrix}0 & 1 \\ \frac{2g}{x_{eq}} & 0\end{bmatrix}}_{A}\begin{bmatrix}\Delta x \\ \Delta v\end{bmatrix} + \underbrace{\begin{bmatrix}0 \\ -\frac{2g}{i_{eq}}\end{bmatrix}}_{b}\Delta i. \tag{13.47}$$

Remark The model (13.47) is called a *linear* statespace model. The adjective linear refers to fact that it is written in the form

$$\frac{d}{dt}\begin{bmatrix}\Delta x \\ \Delta v\end{bmatrix} = A\begin{bmatrix}\Delta x \\ \Delta v\end{bmatrix} + b\Delta i,$$

where the right side is a linear matrix equation. Here the *state* is

$$\begin{bmatrix}\Delta x \\ \Delta v\end{bmatrix}$$

with Δx and Δv the *state variables* and $\Delta i = i - i_{eq}$ the input.

Transfer Function Model

Going back to (13.45) and (13.46), rewrite the model as a single equation given by

$$\frac{d^2}{dt^2}\Delta x = \frac{2g}{x_{eq}}\Delta x - \frac{2g}{i_{eq}}\Delta i.$$

The Laplace transform is

$$s^2 \Delta X(s) - s\Delta x(0) - \Delta \dot{x}(0) = \frac{2g}{x_{eq}} \Delta X(s) - \frac{2g}{i_{eq}} \Delta I(s)$$

or

$$\Delta X(s) = \underbrace{-\frac{2g/i_{eq}}{s^2 - 2g/x_{eq}} \Delta I(s)}_{\text{Transfer function}} + \frac{s\Delta x(0) + \Delta \dot{x}(0)}{s^2 - 2g/x_{eq}}.$$

This expression leads to the block diagram of Figure 13.8. As always, the denominator of the transfer function is the *same as* the denominator of the initial condition term. The transfer function has a pole at $\sqrt{2g/x_{eq}}$ and is therefore unstable.

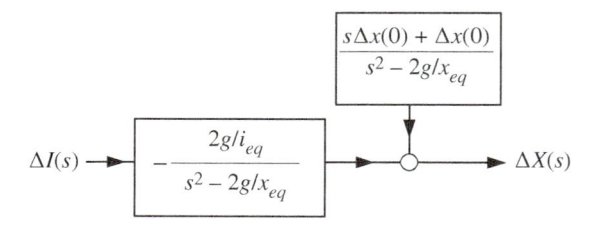

Figure 13.8. Block diagram of the linear model of the magnetic levitation system.

13.4 Cart on a Track System

We now develop a model for the QUANSER [34] cart on a track system illustrated in Figure 13.9. The smaller wheel is powered by a DC motor. The larger wheel is connected to an encoder that gives the position of the cart.

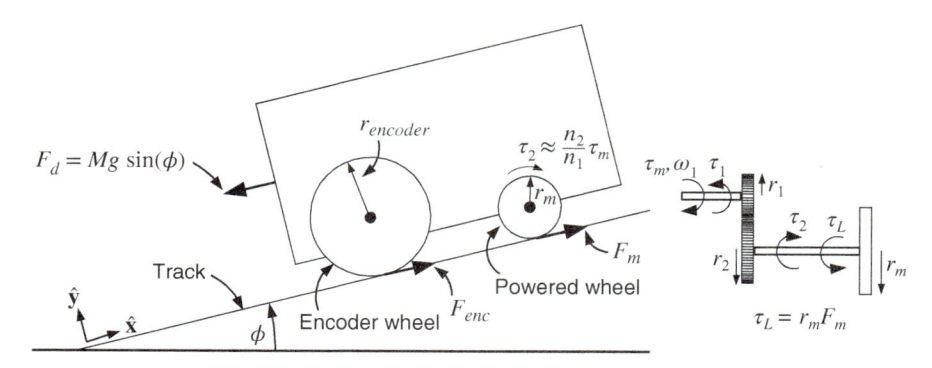

Figure 13.9. Cart on a track with one end of the track raised an angle ϕ.

The following notation is used.

K_T is the motor torque constant ($\tau = K_T i$).

$K_b = K_T$ is the motor back-emf constant ($v_b = K_b \omega_1$).

R is the motor armature resistance.

13.4 Cart on a Track System 517

L is the motor armature inductance (negligible).

M is the mass of the cart and wheels.

r_m is the radius of the powered (motor) wheel.

r_1 is the radius of the gear on the motor shaft.

r_2 is the radius of the gear on the wheel shaft.

J_1 is the moment of inertia of the motor shaft.

J_2 is the moment of inertia of the output shaft and powered wheels.

r_{enc} is the radius of the encoder wheel.

J_{enc} is the moment of inertia of the encoder shaft and wheels.

n_1 is the number of gear teeth on the motor shaft.

n_2 is the number of gear teeth on the output shaft.

$N = n_2/n_1 = r_2/r_1$ is the gear ratio.

$F_d = Mg\sin(\phi)$ is the (disturbance) force of gravity along the direction of the track.

Mechanical Equations

To model the mechanical part of the system, both the rotational motion of the motor, gears, and wheels as well as the translational motion of cart must be taken into account.

Denote by $F_m\hat{\mathbf{x}}$ the force exerted on the powered wheels by the track so that (Newton's third law) $-F_m\hat{\mathbf{x}}$ is the force exerted on the track by the powered wheels. Similarly, let $F_{enc}\hat{\mathbf{x}}$ be the force exerted on the encoder wheels by the track so that $-F_{enc}\hat{\mathbf{x}}$ is the force exerted on the track by the encoder wheels. The motor shaft with gear 1 has moment of inertia J_1, while the wheel shaft with gear 2 has moment of inertia J_2. The equations of motion for the motor and wheel shafts are, respectively,

$$J_1 \frac{d\omega_1}{dt} = \tau_m - \tau_1 \tag{13.48}$$

$$J_2 \frac{d\omega_2}{dt} = \tau_2 - r_m F_m, \tag{13.49}$$

where τ_1 is the torque exerted on gear 1 by gear 2 and τ_2 is the (reaction) torque exerted on gear 2 by gear 1. The quantity $r_m F_m$ is the torque exerted on the powered wheel by the track. Using the gear relationships $\omega_1 = \frac{n_2}{n_1}\omega_2$ and $\tau_2 = \frac{n_2}{n_1}\tau_1$, Eqs. (13.48) and (13.49) reduce to

$$\underbrace{\left(J_1\left(\frac{n_2}{n_1}\right)^2 + J_2\right)}_{J}\frac{d\omega_2}{dt} = \frac{n_2}{n_1}\tau_m - F_m r_m \tag{13.50}$$

or

$$J \frac{d\omega_2}{dt} = \frac{n_2}{n_1}\tau_m - F_m r_m. \tag{13.51}$$

518   13 Inverted Pendulum, Magnetic Levitation, and Cart on a Track

The encoder wheels have a moment of inertia J_{enc}, radius r_{enc}, and angular velocity ω_{enc}. The equation of motion for the encoder wheel is

$$J_{enc}\frac{d\omega_{enc}}{dt} = -F_{enc}r_{enc}, \tag{13.52}$$

where the quantity $r_{enc}F_{enc}$ is the torque exerted on the encoder wheels by the track. However, it turns out that J_{enc} is negligible so we take it to be zero. The equation for the encoder wheel given by (13.52) then simply goes away as $J_{enc} = 0$ results in $F_{enc} = 0$.

The linear velocity along the circumference of the encoder wheels is $r_{enc}\omega_{enc}$ and is the same as the linear velocity along the circumference of the powered wheels $r_m\omega_2$ as they both make contact with the inclined plane without slipping. This common linear velocity is simply the cart velocity $v = dx/dt$. Putting this together we have

$$r_{enc}\omega_{enc} = r_m\omega_2 = v. \tag{13.53}$$

With $F_d = Mg\sin(\phi)$ the force of gravity on the cart the equation of motion for the cart body is

$$M\frac{dv}{dt} = F_m - F_d. \tag{13.54}$$

We put Eqs. (13.51) and (13.54) together to describe the mechanical motion as

$$J\frac{d\omega_2}{dt} = \frac{n_2}{n_1}\tau_m - F_m r_m \tag{13.55}$$

$$M\frac{dv}{dt} = F_m - F_d. \tag{13.56}$$

Using the no slip condition (13.53) for the powered wheel, i.e., $\omega_2 = v/r_m$, and eliminating F_m from (13.55) and (13.56), we obtain a single equation for the velocity v of the cart given by

$$\left(\frac{J}{r_m^2} + M\right)\frac{dv}{dt} = \frac{1}{r_m}\frac{n_2}{n_1}\tau_m - F_d.$$

Electrical Equations

The motor electrical equations are given by (see Figure 13.10)

$$L\frac{di}{dt} = -Ri(t) - K_b\omega_1(t) + v_a(t)$$

with the motor torque given by $\tau_m = K_T i(t)$.

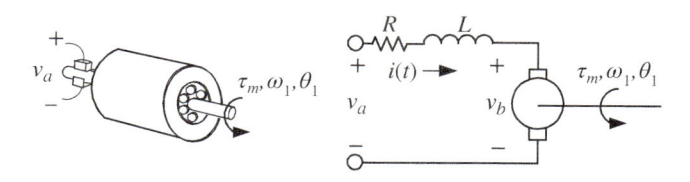

Figure 13.10. DC motor schematic. $\tau_m = K_T i$ and $v_b = K_b\omega_1 = K_b\dfrac{n_2}{n_1}\dfrac{v}{r_m}$.

With no slip between the powered wheel and the track, we use the gear relationships to obtain

$$\omega_1 = \frac{n_2}{n_1}\omega_2 = \frac{n_2}{n_1}\frac{v}{r_m}.$$

The electrical equation may now be rewritten as

$$L\frac{di}{dt} = -Ri(t) - \underbrace{\frac{K_b}{r_m}\frac{n_2}{n_1}v(t)}_{v_b(t)} + v_a(t).$$

Equations of Motion and Block Diagram

Collecting all of the equations together, the cart on the track system is described by[4]

$$L\frac{di}{dt} = -Ri(t) - \underbrace{\frac{K_b}{r_m}\frac{n_2}{n_1}v(t)}_{v_b} + v_a(t)$$

$$\left(\frac{J}{r_m^2} + M\right)\frac{dv}{dt} = \left(\frac{n_2}{n_1}\frac{1}{r_m}\right)K_T i(t) - F_d \tag{13.57}$$

$$\frac{dx}{dt} = v.$$

It is easier to work with these equations in the s domain. With zero initial conditions we take the Laplace transform of the system (13.57) to obtain

$$I(s) = \frac{V_a(s) - V_b(s)}{sL + R}$$

$$V_b(s) = \underbrace{K_b\frac{n_2}{n_1}\frac{1}{r_m}}_{K_b'}v(s)$$

$$s\underbrace{\left(\frac{J}{r_m^2} + M\right)}_{M'}v(s) = \underbrace{\frac{n_2}{n_1}\frac{1}{r_m}K_T}_{K_T'}I(s) - \frac{F_d}{s}.$$

Simplifying the notation by setting

$$K_T' \triangleq K_T\frac{n_2}{n_1}\frac{1}{r_m}, \quad K_b' \triangleq K_b\frac{n_2}{n_1}\frac{1}{r_m}, \quad M' \triangleq \frac{J}{r_m^2} + M,$$

these equations are equivalently represented by the block diagram of Figure 13.11 on the next page.

Setting $L = 0$ (negligible) and moving the disturbance force $F_d/s = Mg\sin(\phi)/s$ to the input summing junction, the block diagram of Figure 13.11 reduces to that of Figure 13.12 on the next page where

$$K_D \triangleq \frac{R}{K_T'} = \frac{R}{K_T\dfrac{n_2}{n_1}\dfrac{1}{r_m}}. \tag{13.58}$$

---

[4]Remember that $v(t) = dx/dt$ is the cart *velocity*, while $v_a(t)$ and $v_b(t)$ are the input and back-emf *voltages*, respectively.

520   13 Inverted Pendulum, Magnetic Levitation, and Cart on a Track

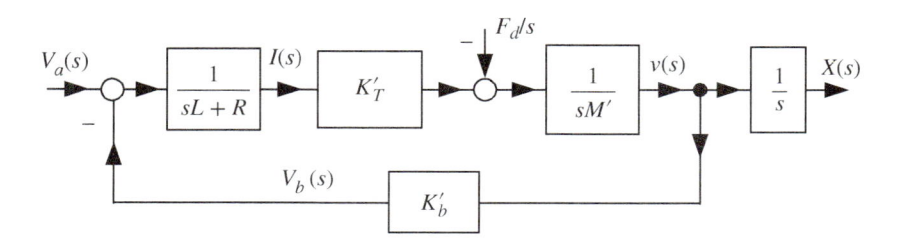

Figure 13.11. Block diagram of the cart on a track.

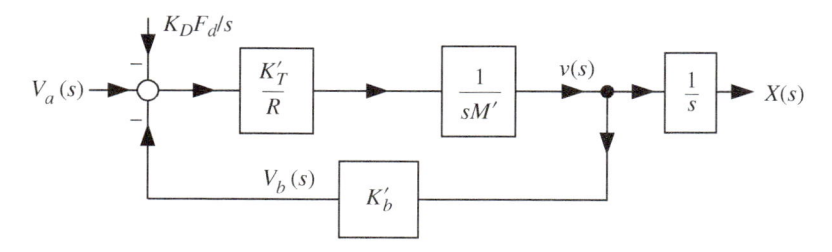

Figure 13.12. Equivalent block diagram for the cart and track system.

We can simplify this block diagram further by defining $G(s)$ as

$$G(s) \triangleq \frac{\frac{K'_T}{R}\frac{1}{sM'}\frac{1}{s}}{1+K'_b\frac{K'_T}{R}\frac{1}{sM'}} = \frac{K'_T}{sRM'+K'_bK'_T}\frac{1}{s} = \frac{\frac{K'_T}{RM'}}{s+\frac{K'_bK'_T}{RM'}}\frac{1}{s} = \frac{b}{s(s+a)},$$

where

$$a \triangleq \frac{K'_bK'_T}{RM'} = \frac{K_bK_T\left(\frac{n_2}{n_1}\frac{1}{r_m}\right)^2}{RJ/r_m^2+RM}, \quad b \triangleq \frac{K'_T}{RM'} = \frac{K_T\frac{n_2}{n_1}\frac{1}{r_m}}{RJ/r_m^2+RM}. \quad (13.59)$$

Using the parameters $a, b,$ and K_D we may now write $X(s)$ as

$$X(s) = \frac{b}{s(s+a)}V_a(s) - \frac{b}{s(s+a)}K_D\frac{F_d}{s}. \quad (13.60)$$

$K_D F_d$ is an input voltage disturbance that has the same effect on the output as the actual load force disturbance F_d. Note that

$$K_D F_d = \frac{R}{K_T\frac{n_2}{n_1}\frac{1}{r_m}}F_d = \frac{n_2}{n_1}\frac{r_m F_D}{K_T}R$$

has units of $(\text{dimensionless}) \times \frac{\text{torque}}{\text{torque/Amp}} \times \text{Ohm} = \text{Ohm} \times \text{Amp} = \text{Volts}$. The block diagram Figure 13.12 reduces to that given in Figure 13.13.

As $bK_D F_d = \frac{K'_T}{RM'}\frac{R}{K'_T}Mg\sin(\phi) = \frac{Mg\sin(\phi)}{M'} = \frac{Mg\sin(\phi)}{J/r_m^2+M}$ we can write (13.60) in the time domain as

$$\frac{d^2x(t)}{dt^2} = -ax(t) + bv_a(t) - \frac{Mg\sin(\phi)}{J/r_m^2+M}u_s(t).$$

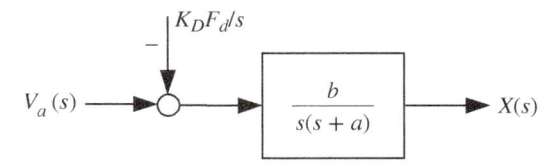

Figure 13.13. Equivalent transfer function model of the cart and track system.

The equivalent disturbance $bK_D F_d = \dfrac{Mg\sin(\phi)}{J/r_m^2 + M}$ is simply the linear acceleration of the cart due to gravity.

Problems

Problem 1 *Pendulum Down*

In the text it was shown that $z_{eq} = \begin{bmatrix} x_{eq} & 0 & 0 & 0 \end{bmatrix}^T$ is an equilibrium point of the inverted pendulum for any value of x_{eq}, that is, of the system of Eq. (13.8).

(a) Show that $z_{eq} = \begin{bmatrix} x_{eq} & v_{eq} & \theta_{eq} & \omega_{eq} \end{bmatrix}^T = \begin{bmatrix} x_{eq} & 0 & \pi & 0 \end{bmatrix}^T$ with $u_{eq} = 0$ is also an equilibrium point of the system of Eq. (13.8) for any value of x_{eq}.

(b) Write $\theta = \delta + \pi$ so $\omega = d\theta/dt = d\delta/dt$. Compute the linear approximate model around $z_{eq} = \begin{bmatrix} x_{eq} & 0 & \pi & 0 \end{bmatrix}^T$ in terms of the state variables x, v, δ, and ω.

Problem 2 $\tau = J\dfrac{d^2\theta}{dt^2}$ *and Reference Systems*

Newton's law for rotational motion of a rigid body about a single axis is $\tau = J\dfrac{d^2\theta}{dt^2}$. It has been pointed in this chapter (as well as Chapter 5) that this equation is valid if the torque is computed in an inertial (non-accelerating) coordinate system or if the torque is computed about the center of mass of the rigid body (even if the center of mass is accelerating). Parts (a), (b), and (c) refer to the inverted pendulum on the cart system.

(a) Suppose the cart is held stationary so that the pivot axis is stationary and therefore in an inertial reference frame. With respect to the *pivot axis* show that the total torque on the pendulum rod is $\tau = mg\ell\sin(\theta)$ where m is the mass of the rod.

(b) With a the cross sectional area of the rod, ρ the mass density of the rod, and 2ℓ the length of the rod, show that its mass is given by $m = 2\ell a\rho$ and its moment of inertia about the pivot axis is $J_{pivot} = \dfrac{4}{3}m\ell^2$.

(c) By parts (a) and (b) the rotational motion of the pendulum rod on the stationary cart is described by

$$J_{pivot}\dfrac{d^2\theta}{dt^2} = mg\ell\sin(\theta). \tag{13.61}$$

Let the cart no longer be held stationary so it can move under the influence of the applied force $u(t)$. Suppose (incorrectly!) that we still compute the motion of the pendulum angle about the pivot axis with the cart being accelerated by the force $u(t)$ (cart is no longer an inertial reference system). What equation would you get for the motion of the pendulum angle? Answer: Same as Eq. (13.61). Explain why this cannot be correct.

Problem 3 *Pendulum Transfer Functions*

Verify the expressions for $X(s)$ and $Y(s) = X(s) + \left(\ell + \dfrac{J}{m\ell}\right)\theta(s)$ given on page 503 in the text.

Problem 4 *Inverted Pendulum Control via Nested Feedback Loops – Linear Model*

A SIMULINK block diagram for the nested loop control system is shown in Figure 13.14. Use the linear statespace model shown in Figure 13.14 for the open-loop inverted pendulum. This model is valid for θ and $\omega = \dot{\theta}$ small. A SIMULINK block diagram to simulate this linear model is shown in Figure 13.15.

$$\begin{bmatrix} dx/dt \\ dv/dt \\ d\theta/dt \\ d\omega/dt \end{bmatrix} = \begin{bmatrix} 0 & 1 & 0 & 0 \\ 0 & 0 & -\dfrac{gm^2\ell^2}{Mm\ell^2 + J(M+m)} & 0 \\ 0 & 0 & 0 & 1 \\ 0 & 0 & \dfrac{mg\ell(M+m)}{Mm\ell^2 + J(M+m)} & 0 \end{bmatrix} \begin{bmatrix} x \\ v \\ \theta \\ \omega \end{bmatrix} + \begin{bmatrix} 0 \\ \dfrac{J+m\ell^2}{Mm\ell^2 + J(M+m)} \\ 0 \\ -\dfrac{m\ell}{Mm\ell^2 + J(M+m)} \end{bmatrix} u.$$

The parameter values for the QUANSER [34] inverted pendulum system are given by $M = 0.57$ kg (cart mass), $m = 0.23$ kg (pendulum rod mass), $\ell = 0.6412/2 = 0.3206$ m (pendulum rod half-length), $g = 9.81$ m/s^2 (acceleration due to gravity), and $J = m\ell^2/3 = 7.88 \times 10^{-3}$ kg-m^2 (moment of inertia of the pendulum rod about its center of mass). Use the Euler integration algorithm with a step size of $T = 0.001$ second. You will need to write a .m file to set these parameter values, to set the time constants $\tau_x = 1/30$ and $\tau_\theta = 1/30$, and to compute the gains k_1, k_2, k_3, k_4 according to (13.17). Run your simulation for both a step and a ramp input. Plot $x(t)$ and $\theta(t)$.

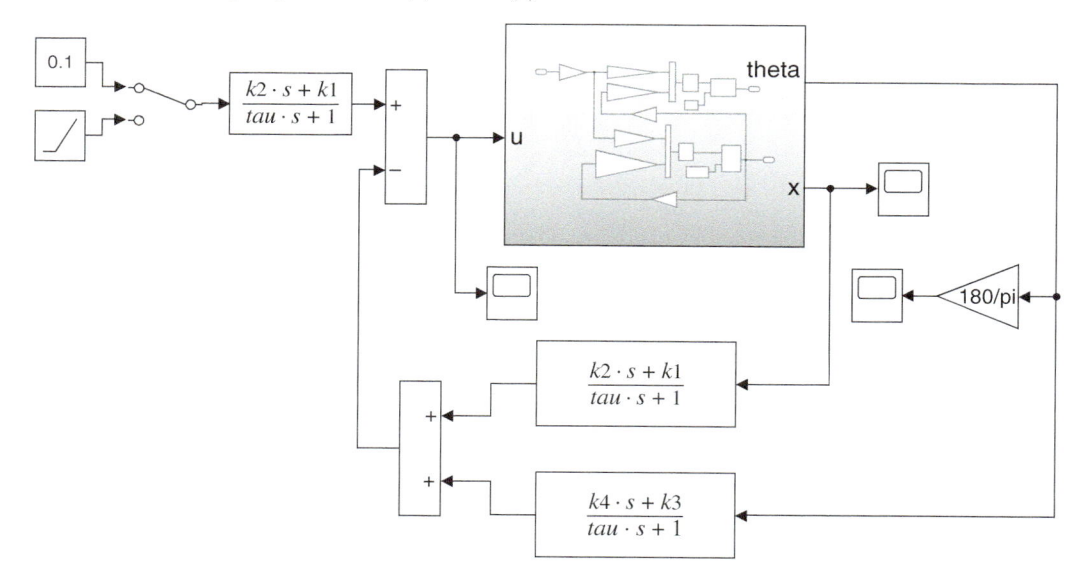

Figure 13.14. Control of the inverted pendulum using nested loops.

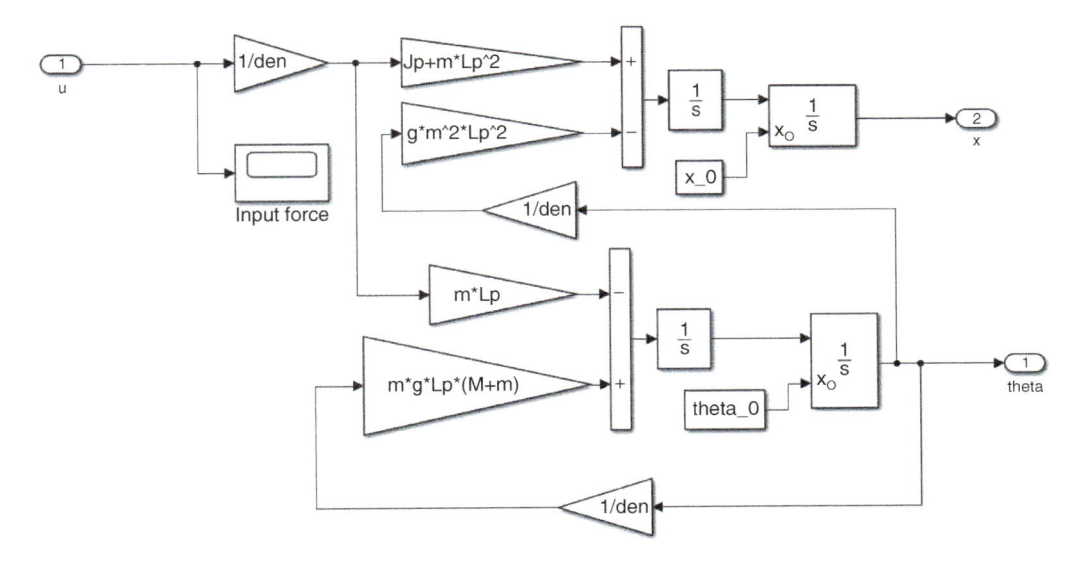

Figure 13.15. Inside the `Linear Statespace Model` of Figure 13.14.

Problem 5 *Inverted Pendulum Control via Nested Feedback Loops – Nonlinear Model*

Redo Problem 4 replacing the linear statespace model with the *nonlinear* statespace model given in (13.8). Figure 13.16 is a SIMULINK block diagram for nested loop control structure, while Figure 13.17 on page 524 is a SIMULINK block diagram for the nonlinear inverted pendulum model. Simulate both a step and a ramp input.

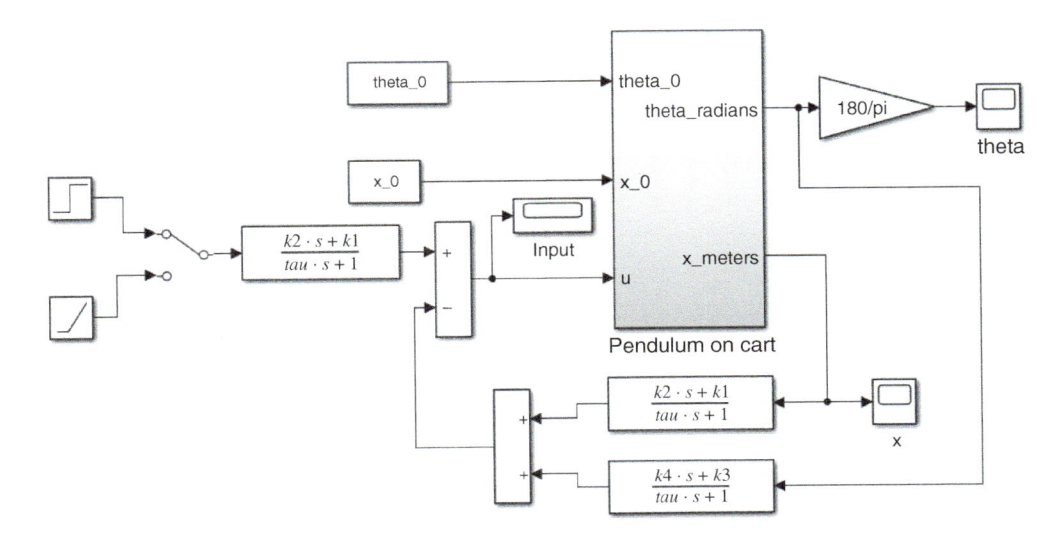

Figure 13.16. Nested loop control system with nonlinear statespace model of the inverted pendulum.

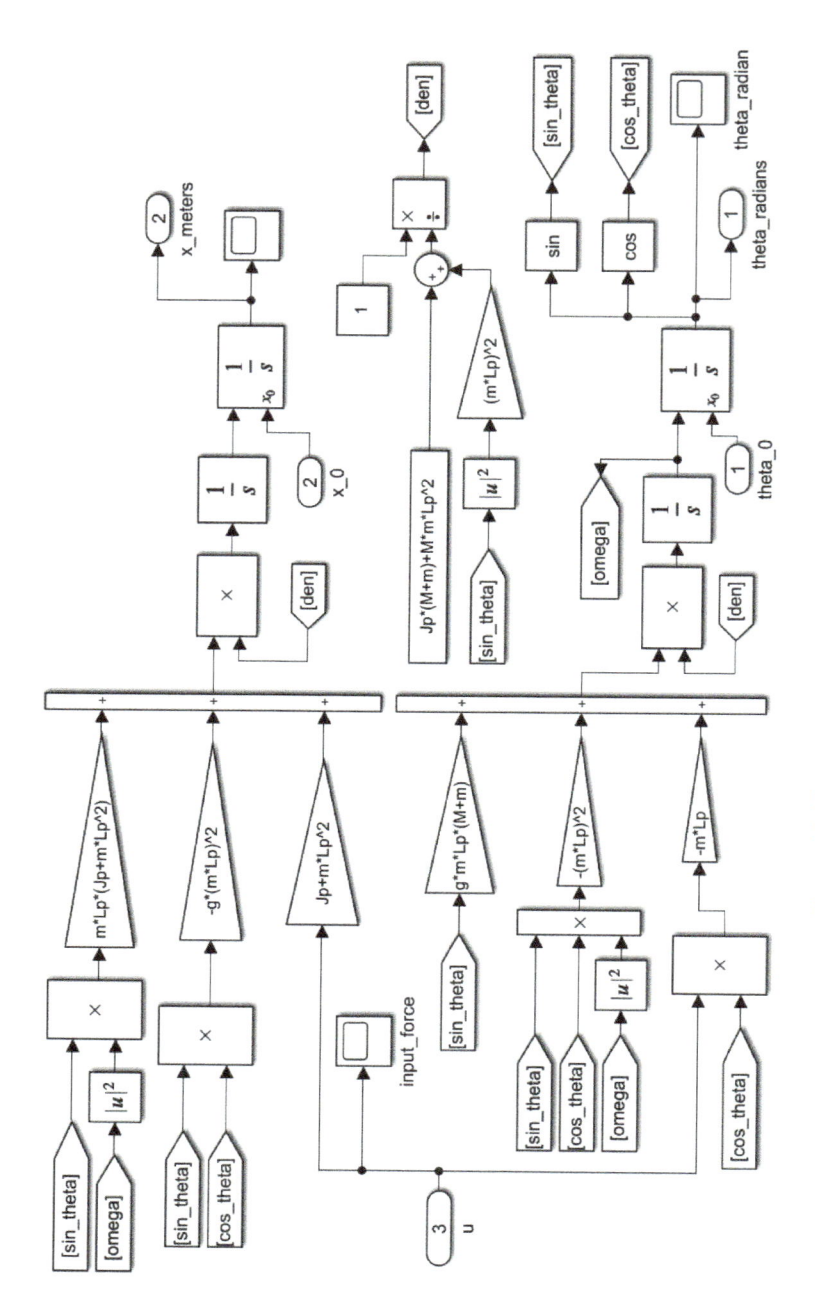

524   13 Inverted Pendulum, Magnetic Levitation, and Cart on a Track

Figure 13.17. Inside the Pendulum on Cart block.

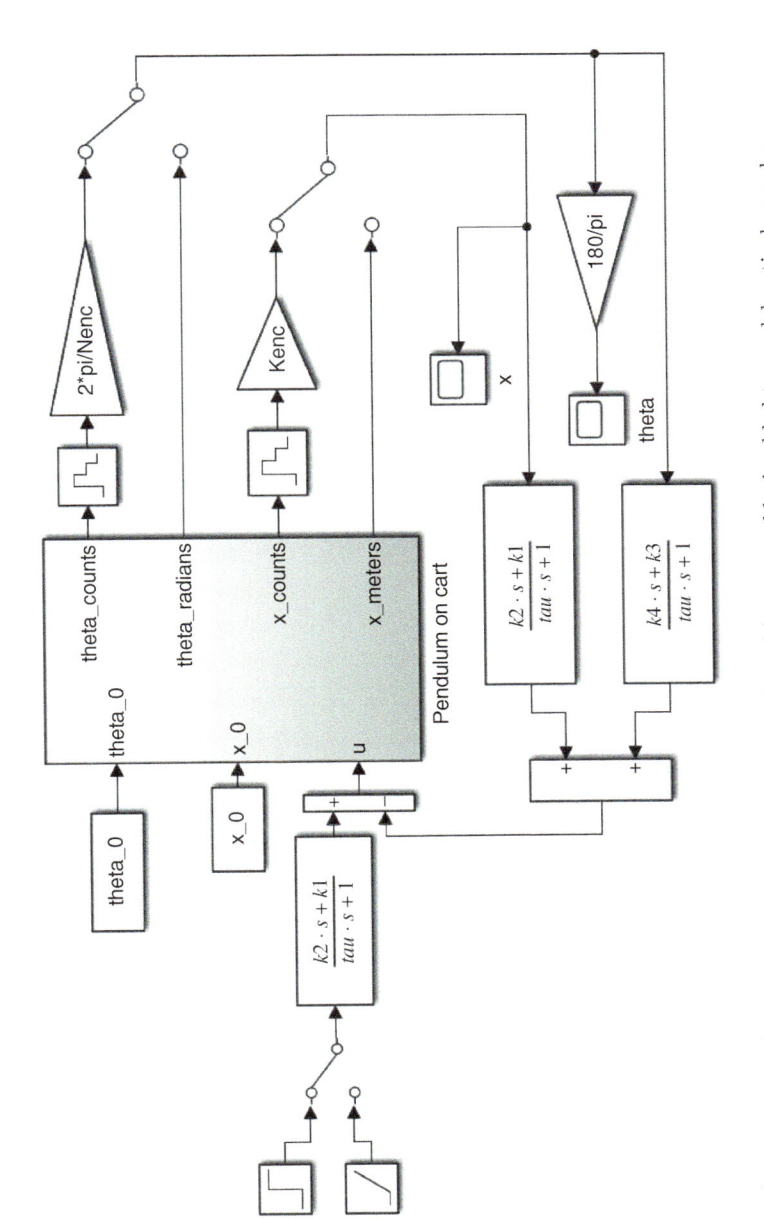

Figure 13.18. A zero_order_hold bock followed by a gain block added to model optical encoders.

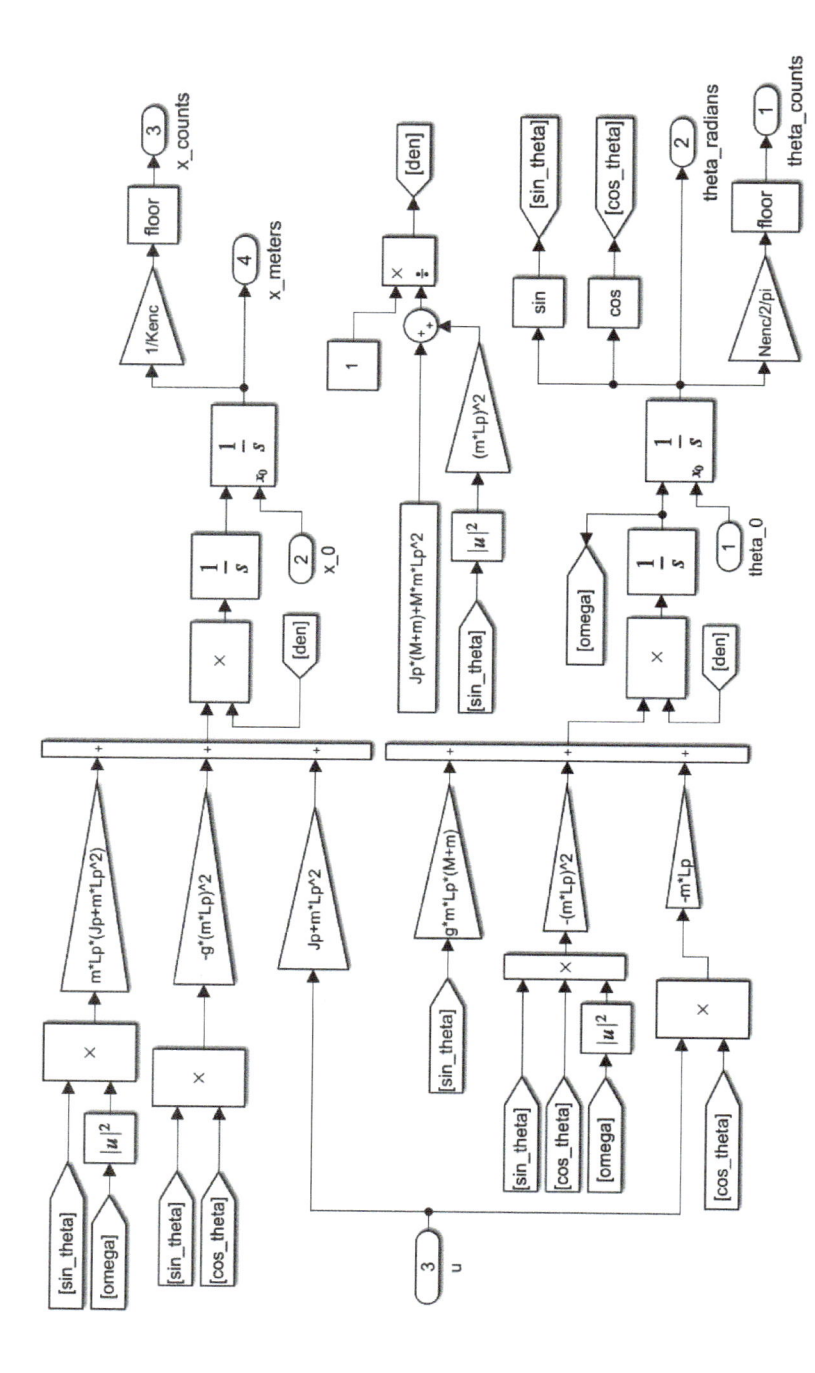

526   13 Inverted Pendulum, Magnetic Levitation, and Cart on a Track

Figure 13.19. A gain block followed by a floor block added to model optical encoders.

Problem 6 *Inverted Pendulum Control via Nested Feedback Loops – Nonlinear Model with Encoders*

The inverted pendulum system built by QUANSER [34] uses optical encoders to measure the pendulum rod angle and the cart position. In this problem you are to add models of these optical encoders to the simulation of Problem 5. (See Chapter 6 for details on the operation of optical encoders.) With N_x the number of counts from the encoder wheel, r_{enc} the radius of the encoder wheel, and N_{enc} the number of pulses/rev produced by the encoder, the cart position x is given by $x = \dfrac{2\pi}{N_{enc}} r_{enc} N_x = K_{enc} N_x$. Similarly, with N_θ the number of counts out of the pendulum rod encoder, the angular position of the pendulum is given by $\theta = \dfrac{2\pi}{N_{enc}} N_\theta$. The QUANSER encoders have $N_{enc} = 4096$ pulses/rev and their encoder wheel has a radius of $r_{enc} = 0.0148$ m.

Figure 13.18 on page 525 is a SIMULINK block diagram of the nested loop controller for the inverted pendulum. The upper right side of Figure 13.18 shows a zero order hold block followed by a gain block of $2\pi/K_{enc}$ to convert the position counts to meters. Similarly the lower right side of Figure 13.18 shows a zero order hold block followed by a gain block of $2\pi/N_{enc}$ to convert the rod angle counts to radians.

Figure 13.19 on page 526 is a SIMULINK block diagram of the nonlinear differential equation model of the inverted pendulum. The upper right side of Figure 13.19 shows a gain block $1/K_{enc}$ followed by a floor block to model the counts (pulses) from the cart position encoder. The lower right side of Figure 13.19 shows a gain block of $N_{enc}/(2\pi)$ followed by a floor block to model the counts (pulses) from the pendulum rod angle encoder.

Run the simulation for both a step and a ramp input.

Problem 7 *Inverted Pendulum Controller with Three Nested Loops*

Figure 13.20 shows an integrator added to the forward path of the nested loop controller to provide the capability to reject disturbances in the input. For example, raising one end of the track would be such a disturbance.

(a) Show how to compute k_0, k_1, k_2, k_3, k_4 so that the closed-loop characteristic polynomial is $s^5 + f_4 s^4 + f_3 s^3 + f_2 s^2 + f_1 s + f_0$ where the f_i can be specified arbitrarily. Hint: Start with the transfer function given in (13.16), which is the transfer function $X(s)/Z_0(s)$ in Figure 13.20.

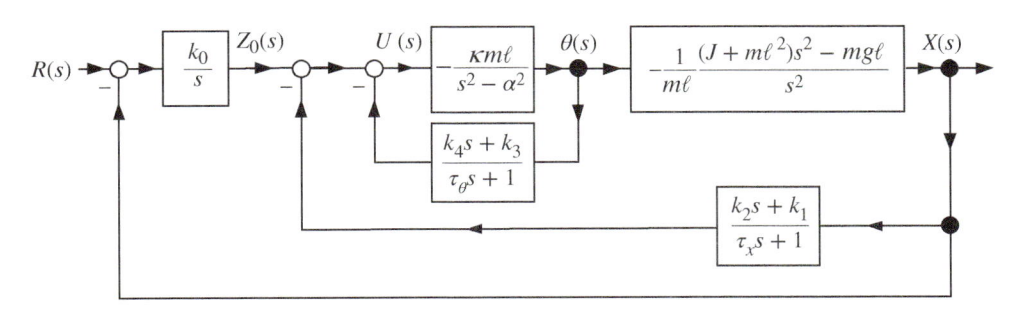

Figure 13.20. Controller for the inverted pendulum with three nested loops.

(b) Modify the simulation from Problem 5 or Problem 6 to implement this controller.

13 Inverted Pendulum, Magnetic Levitation, and Cart on a Track

Problem 8 *Flux Linkage for the Magnetic Levitation System*

Suppose empirically one could measure the magnetic force of the magnetic levitation system. For example, consider the procedure in [10, 54]. Put the ball on a (non-magnetic) pedestal with the ball's center of mass a distance x below the magnet. Then increase the current until the ball just lifts off the pedestal. As the mass m is known and the gravitational constant g is known, mg equals the electromagnet force for that (measured) position and that (measured) current. Carrying this out for several positions and currents it is found that $mg = Ci^2/x^2$ is a good fit to the data for some constant C. With this empirically found value of C conservation of energy requires

$$\frac{1}{2}i^2\frac{\partial L(x)}{\partial x} = -C\frac{i^2}{x^2}.$$

For what value of L_1 does $L(x) = L_0 + rL_1/x$ satisfy this equation.

Problem 9 *Linearization About an Equilibrium Point*

In this problem the third-order model of the magnetic levitation system represented by (13.33)–(13.35) is to be linearized. That is, you are asked to find a linear approximate model for it.

(a) With the steel ball required to be at the fixed location x_{eq} below the electromagnet, show that the equilibrium point and equilibrium input voltage are given by

$$\begin{bmatrix} i_{eq} \\ x_{eq} \\ v_{eq} \end{bmatrix} = \begin{bmatrix} x_{eq}\sqrt{\frac{2mg}{rL_1}} \\ x_{eq} \\ 0 \end{bmatrix}, \quad u_{eq} = Ri_{eq}.$$

(b) The equation for the current $i(t)$ is given by (13.33) and repeated here.

$$\frac{di}{dt} = g(i, x, v, u) = \frac{rL_1}{L_0}\frac{i}{x^2}v - \frac{R}{L_0}i + \frac{1}{L_0}u.$$

Compute the Taylor series expansion of $g(i, x, v, u)$ about the equilibrium point and drop higher-order terms.

(c) The equation for the acceleration of the steel ball is given by (13.35) and repeated here.

$$\frac{dv}{dt} = f(x, i) = g - \frac{rL_1}{2m}\frac{i^2}{x^2}.$$

Compute the Taylor series expansion of $f(i, x)$ about the equilibrium point and drop the higher-order terms.

(d) Show that the approximate linear model about the equilibrium point is then

$$\frac{d}{dt}\begin{bmatrix} i - i_{eq} \\ x - x_{eq} \\ v - v_{eq} \end{bmatrix} = \underbrace{\begin{bmatrix} -\frac{R}{L_0} & 0 & \frac{rL_1}{L_0}\frac{i_{eq}}{x_{eq}^2} \\ 0 & 0 & 1 \\ -\frac{2g}{i_{eq}} & \frac{2g}{x_{eq}} & 0 \end{bmatrix}}_{A}\begin{bmatrix} i - i_{eq} \\ x - x_{eq} \\ v - v_{eq} \end{bmatrix} + \underbrace{\begin{bmatrix} \frac{1}{L_0} \\ 0 \\ 0 \end{bmatrix}}_{b}(u - u_{eq}). \quad (13.62)$$

Remark The model (13.62) is called a *linear* statespace model. The adjective linear refers to fact that it is written in the form

$$\frac{d}{dt}\begin{bmatrix} i - i_{eq} \\ x - x_{eq} \\ v - v_{eq} \end{bmatrix} = A \begin{bmatrix} i - i_{eq} \\ x - x_{eq} \\ v - v_{eq} \end{bmatrix} + b(u - u_{eq}),$$

where the right side is a linear matrix equation. The *state* is $\begin{bmatrix} i - i_{eq} & x - x_{eq} & v - v_{eq} \end{bmatrix}^T$ with $i - i_{eq}, x - x_{eq}$, and $v - v_{eq}$ the corresponding *state variables* and $u - u_{eq}$ the input.

Problem 10 *Simulation of the Nonlinear Magnetic Levitation System Model*

This problem asks you to simulate a control system for the magnetic levitation system.

Figure 13.23 is a SIMULINK block diagram of the overall control system consisting of two nested loops. The inner loop is a current command controller given by $V_S(t) = K_p (i_{ref}(t) - i(t)) + K_I \int_0^t (i_{ref}(\tau) - i(\tau))d\tau$ where $i_{ref}(t) = \Delta i(t) + i_{eq}$. As shown in Figure 13.21, with $\Delta x = x(t) - x_{eq}$ and $\Delta i = i(t) - i_{eq}$, the transfer function from $\Delta I(s)$ to $\Delta X(s)$ is given by $\frac{\Delta X(s)}{\Delta I(s)} = -\frac{b}{s^2 - a^2}$ where $a^2 = 2g/x_{eq}, b = 2g/i_{eq}$. Figure 13.24 shows the inside of the Maglev Model block in Figure 13.23. This is a SIMULINK block diagram of Eqs. (13.30)–(13.32) given in the text. The parameter values from QUANSER [34] are $L_0 = 0.4125$ H, $R = 11$ Ω, $r = 1.27 \times 10^{-2}$ m (12.7 mm), $m = 0.068$ kg, $g = 9.81$ m/s^2, $x_{eq} = r + 0.006$ m, $v_{eq} = 0$, $i_{eq} = x_{eq}\sqrt{2mg/(rL_1)} = 0.845$ A, and $V_{max} = 24$ V. Set $rL_1 = 6.5308 \times 10^{-4}$ Nm^2/A^2.[5]

(a) With the open loop transfer function $G(s) = \frac{\Delta X(s)}{\Delta I(s)} = -\frac{b}{s^2 - a^2}$ use the controller $G_c(s) = \frac{\Delta I(s)}{\Delta E(s)} = \frac{b_1 s + b_0}{s + a_0}$ to place the three closed-loop poles at -100.

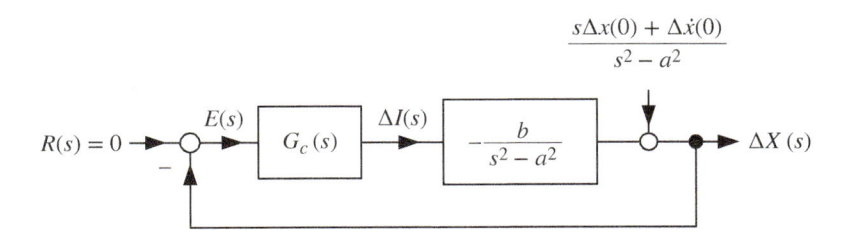

Figure 13.21. Output feedback controller for the magnetic levitation system.

---

[5]QUANSER denotes the force constant as K_m instead of $rL_1/2$ with the value $K_m = 6.5308 \times 10^{-5}$ Nm^2/A^2. However, in the magnetic force expression QUANSER considers the position x of the steel ball to be the distance from the bottom of the magnet to the *top* of the steel ball rather than to its center of mass. Instead of setting $rL_1/2$ equal to this value of K_m, we set $rL_1/2 = 5K_m$ (or $rL_1 = 10K_m$) to compensate for this difference in the reference point for x.

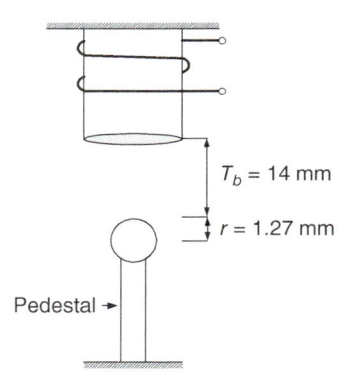

530   13 Inverted Pendulum, Magnetic Levitation, and Cart on a Track

Figure 13.22. Steel ball resting on a pedestal to start.

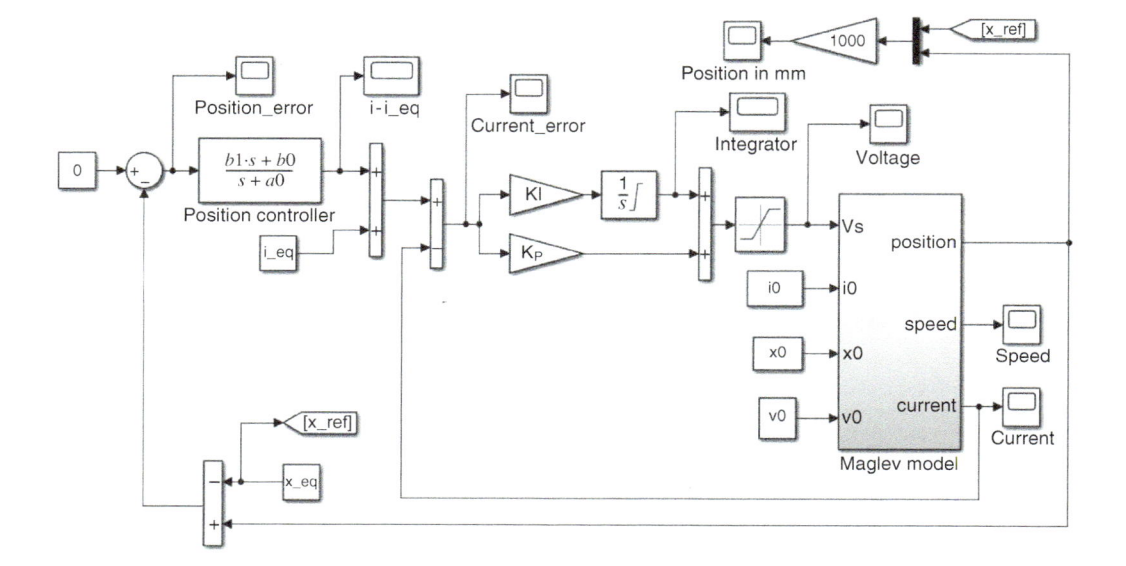

Figure 13.23. SIMULINK block diagram for the closed-loop controller.

(b) With reference to Figure 13.23, set the PI current controller gains as $K_p = 1000$, $K_I = 9000$, and the saturation limits on the integrator and input voltage to be $\pm V_{\max}$. As shown in Figure 13.22, the QUANSER steel ball is set on a pedestal to start. Sitting on the pedestal the center of mass of the steel ball is $0.014 + r$ meters below the bottom of the magnet. Set the initial conditions as $x(0) = 0.014 + r$, $v(0) = 0$, and $i(0) = x(0)\sqrt{2mg/(rL_1)}$. Simulate the complete control system of Figure 13.23 with the controller $G_c(s)$ designed in part (a). Use the Euler integration algorithm with a step size of $T = 0.001$ second. Plot $x(t)$ in millimeters and $i(t)$ in amperes.

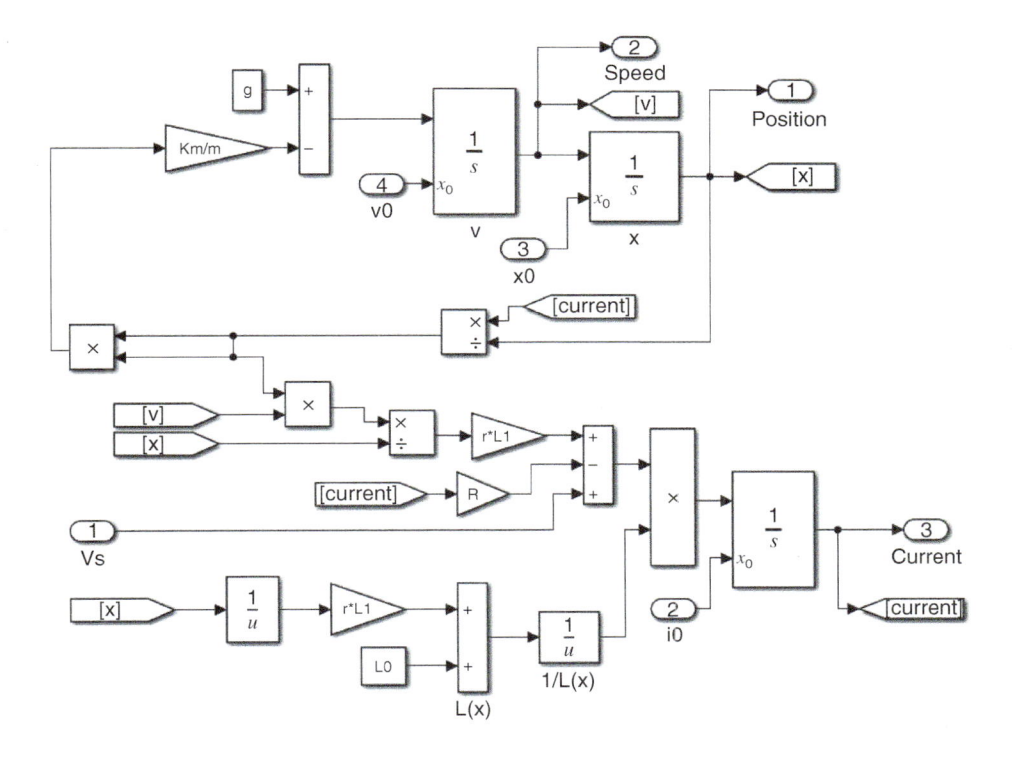

Figure 13.24. SIMULINK model for the open-loop magnetic levitation system.

Problem 11 *Control of the Third-Order Magnetic Levitation System*

Reconsider Problem 10 by designing the controller based on the third-order linear states-pace model of Problem 9 part (c) given by

$$\frac{d}{dt}\begin{bmatrix} i - i_{eq} \\ x - x_{eq} \\ v - v_{eq} \end{bmatrix} = \underbrace{\begin{bmatrix} -\frac{R}{L_{eq}} & 0 & \frac{rL_1 \, i_{eq}}{L_{eq} \, x_{eq}^2} \\ 0 & 0 & 1 \\ -\frac{2g}{i_{eq}} & \frac{2g}{x_{eq}} & 0 \end{bmatrix}}_{A} \begin{bmatrix} i - i_{eq} \\ x - x_{eq} \\ v - v_{eq} \end{bmatrix} + \underbrace{\begin{bmatrix} \frac{1}{L_0} \\ 0 \\ 0 \end{bmatrix}}_{b} (u - u_{eq})$$

$$\Delta x = x - x_{eq} = \underbrace{\begin{bmatrix} 0 & 1 & 0 \end{bmatrix}}_{c} \begin{bmatrix} i - i_{eq} \\ x - x_{eq} \\ v - v_{eq} \end{bmatrix}.$$

The overall block diagram of the control system is shown in Figure 13.25 on the next page.

(a) With $\Delta u = u - u_{eq}, \Delta i = i - i_{eq}, \Delta v = v - v_{eq}$, and $\Delta x = x - x_{eq}$, show that the transfer function from $\Delta U(s)$ to $\Delta X(s)$ is $G(s) = \dfrac{\Delta X(s)}{\Delta U(s)} = \dfrac{\beta_0}{s^3 + \alpha_2 s^2 + \alpha_1 s + \alpha_0}$ where $\beta_0 = -\dfrac{2g}{i_{eq}}\dfrac{1}{L_0}, \alpha_0 = -\dfrac{R}{L_0}\dfrac{2g}{x_{eq}}, \alpha_1 = \dfrac{2g}{i_{eq}}\dfrac{rL_1 \, i_{eq}}{L_0 \, x_{eq}^2} - \dfrac{2g}{x_{eq}}, \alpha_2 = \dfrac{R}{L_0}$.

13 Inverted Pendulum, Magnetic Levitation, and Cart on a Track

(b) Design an output pole placement controller to stabilize the closed-loop system. Where do you place the three closed-loop poles?

(c) Simulate this control system with the same parameter values given in Problem 10. Set $x_{eq} = r + 0.006$ m, $i_{eq} = x_{eq}\sqrt{\frac{2mg}{rL_1}}$, and $v_{eq} = 0$. As in Problem 10 set $x(0) = 0.014 + r$ meters, and $v(0) = 0$. Also set $i(0) = x(0)\sqrt{\frac{2mg}{rL_1}}$. Plot $x(t)$ in millimeters and $i(t)$ in amperes. The inside of the SIMULINK block Maglev Model in Figure 13.25 is given in Figure 13.21.

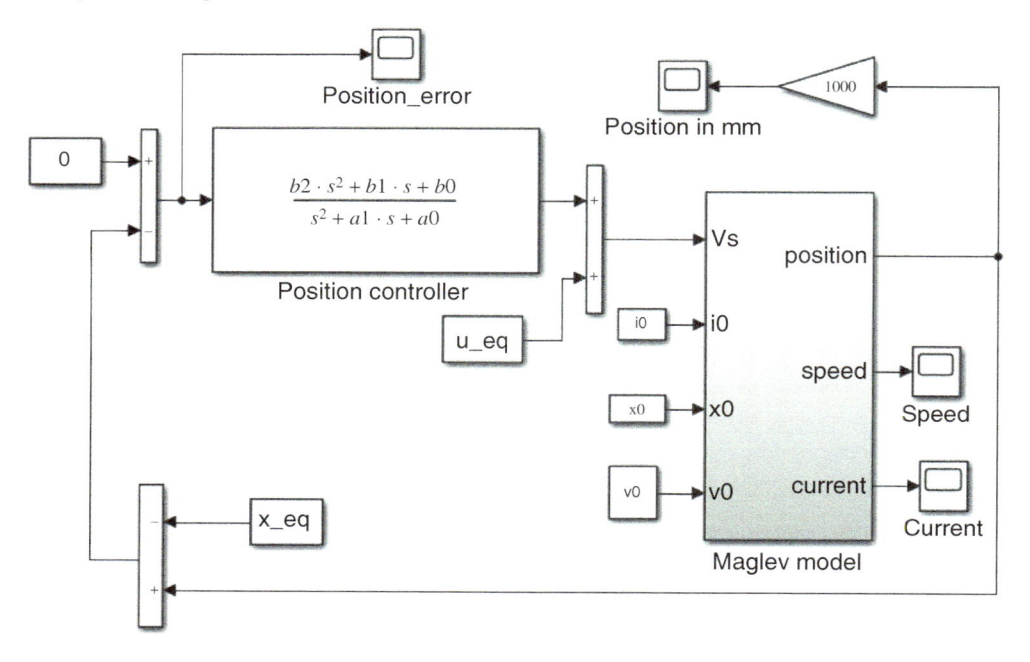

Figure 13.25. Controller for the third-order magnetic levitation model.

(d) Figure 13.26 shows a SIMULINK block diagram using the transfer function model for the control system simulation. Since the initial conditions are set to zero in the

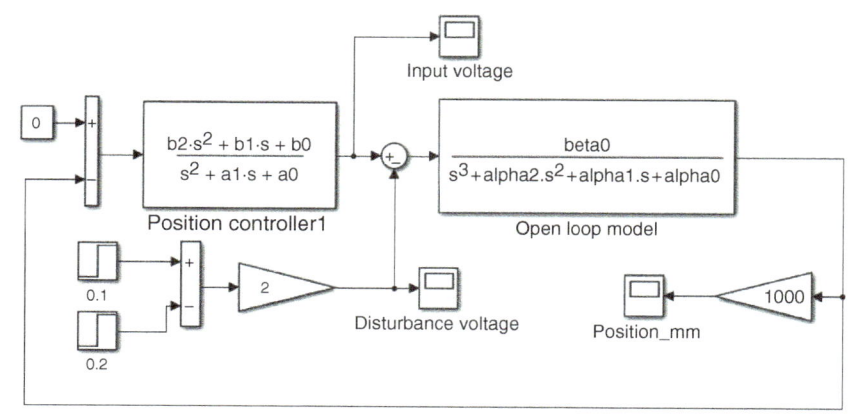

Figure 13.26. Simulation of the transfer function control system model.

transfer function model, a 2 V input voltage disturbance was applied from $t = 0.1$ to $t = 0.2$ seconds to move the steel ball from its equilibrium position. Do the simulation. Plot $\Delta x(t)$ in millimeters and $\Delta u(t)$ in volts.

Problem 12 *Simulation of the Cart on the Track*

This problem asks you to simulate the cart on the track control system shown in Figure 13.27.

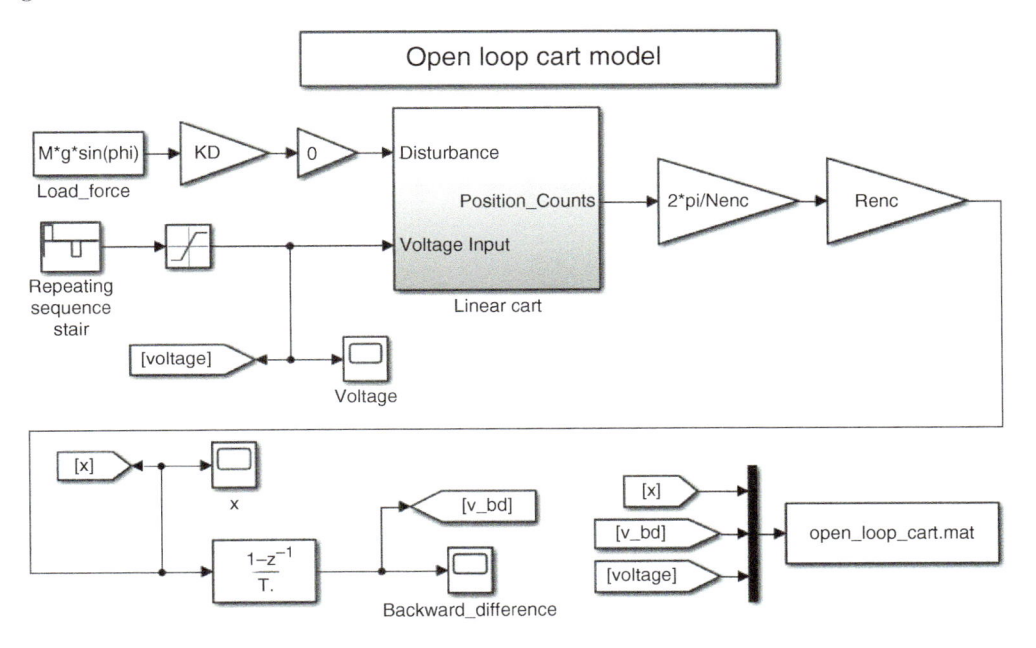

Figure 13.27. SIMULINK block diagram for the open loop cart model.

The inside of the Linear Cart block of Figure 13.27 is shown in Figure 13.28. The parameter values from QUANSER [34] are $K_T = 7.67 \times 10^{-3}$ Nm/A, $R = 2.6$ Ω, $L = 0.18$ mH, $M = 0.57$ kg, $r_m = 6.35 \times 10^{-3}$ m, $J_1 = J_2 = J_{enc} = 0$, $f_1 = f_2 = f_{enc} = 0$, $r_g = n_2/n_1 = 3.71$, $N_{enc} = 4 \times 1024 = 4096$ counts/rev, $R_{enc} = r_{encoder} = 0.01483$ m, $g = 9.8$ m/s^2, $K_{enc} = (2\pi/N_{enc})r_{enc} = 2.275 \times 10^{-5}$ m/count, $V_{\max} = 5$ V, and $T = 0.001$ second.

Use (13.59) to set the values of a and b in the cart transfer function $\dfrac{b}{s(s+a)}$. Set K_D according to (13.58). See Chapter 6, Section 6.5 as well as Problem 10 of that chapter for how to simulate an optical encoder and the backward difference calculation.

(a) Implement the cart on the track simulation as shown in Figures 13.27 and 13.28. In the dialog block for the Repeating Sequence Stair (Figure 13.27) set the input *voltage* to the cart motor as given in Figure 13.29.

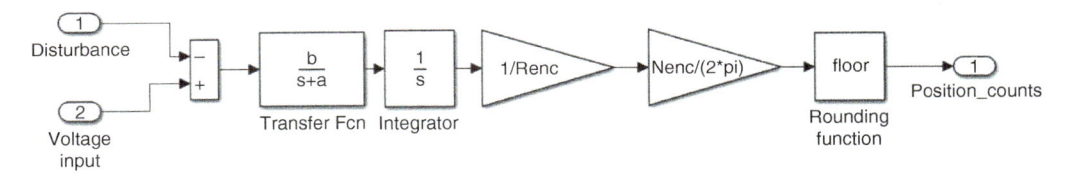

Figure 13.28. Inside the Linear Cart block of Figure 13.27.

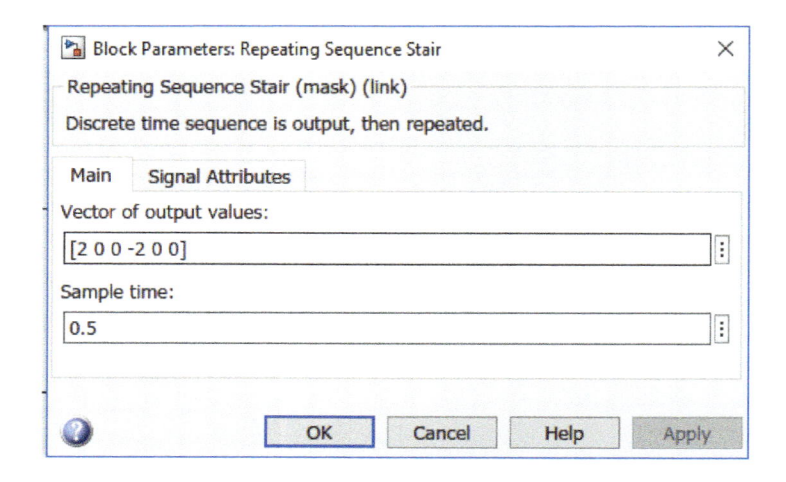

Figure 13.29. Voltage input to the cart motor.

(b) Plot the speed calculated by the backward difference computation and, from the plot, show that the error is bounded by $\dfrac{2\pi r_{enc}}{N_{enc}}\dfrac{1}{T} = \dfrac{K_{enc}}{T}$.

(c) Design a minimum order controller $G_c(s)$ for the cart on the track system that rejects constant disturbances and allows for arbitrary pole placement. Then add this controller to the simulation of part (a). To set the reference input for the *position* of the cart use the Repeating Sequence Stair block and set its dialog block as given in Figure 13.30. This reference input commands the cart to move forward 0.1 m, move backward 0.1 m, and then repeat.

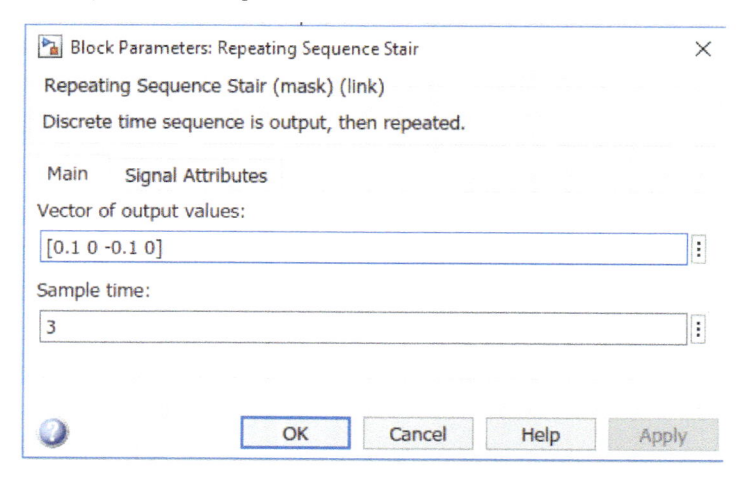

Figure 13.30. Reference input for the cart position.

Problem 13 *Inverted Pendulum Tracking a Sinusoidal Reference Input*

Figure 13.31 shows the controller transfer function $\dfrac{k_{01}s + k_{00}}{s^2 + \omega_0^2}$ added to the forward path to be able track the sinusoidal reference signal $r(t) = X_0 \sin(\omega_0 t + \phi_0)$ as required by the internal model principle. Note that $\mathcal{L}\{X_0 \sin(\omega_0 t + \phi_0)\} = X_0 \dfrac{\sin(\phi_0)s + \cos(\phi_0)\omega_0}{s^2 + \omega_0^2}$.

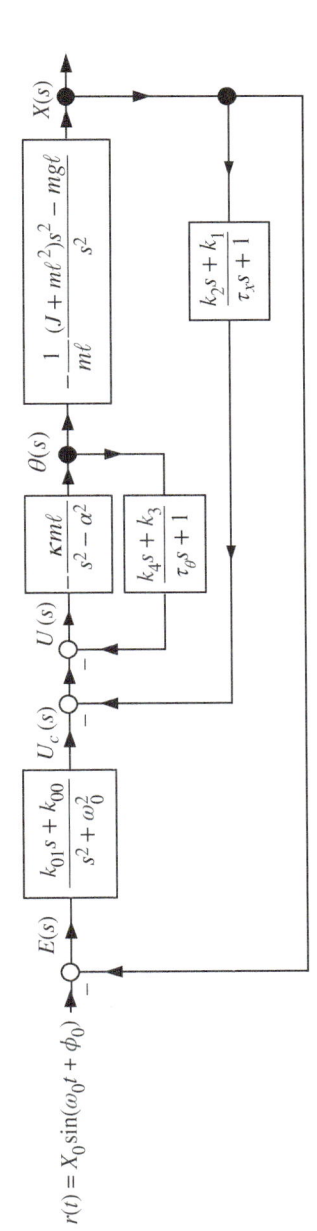

Figure 13.31. Sinusoidal tracking of cart position by the inverted pendulum.

(a) Compute $E(s)/R(s)$ and show that the gains $k_{00}, k_{01}, k_1, k_2, k_3, k_4$ can be used to place the closed-loop poles at any desired location in the open left-half plane. Hint: Note that the transfer function $X(s)/U_c(s)$ has already been computed in the text. Use it to compute $E(s)/R(s)$.

(b) With $r(t) = 0.05\sin(2\pi t)$ simulate this control system using the QUANSER parameter values.

14
State Variables

14.1 Statespace Form

Consider the mass–spring–damper system in Figure 14.1.

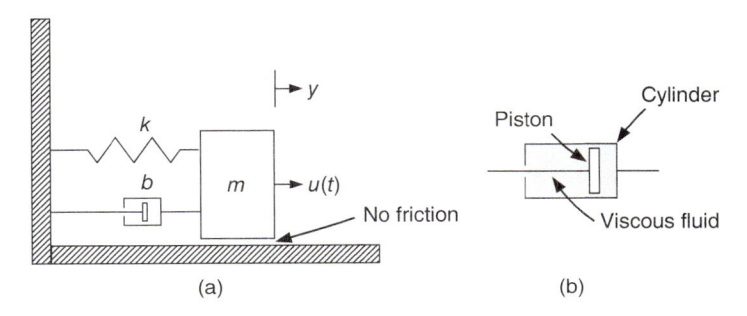

Figure 14.1. (a) Spring–mass–damper system. (b) Damper cross section.

If $y = 0$ then the spring is neither compressed nor stretched, which is to say the spring is *relaxed*. The force on the mass m due to the spring is

$$F_k = -ky.$$

Note that if $y > 0$ the spring is pulling m to the left while if $y < 0$ the spring is pushing m to the right. The damper (dashpot) is illustrated in more detail on Figure 14.1b. The force of the damper on m is proportional to its velocity $\dot{y}$ and is given by

$$F_b = -b\dot{y}.$$

Note that if $\dot{y} > 0$ then the mass m along with the cylinder is moving to the right with the piston providing a resistive force in the $-y$ direction to oppose this motion. Similarly, if $\dot{y} < 0$ then the mass m along with the cylinder is moving to the left with the piston providing resistive force now in the $+y$ direction to oppose this motion. $u(t)$ denotes an external force. The equations of motion are given by

$$m\frac{d^2y}{dt^2} = -ky - b\frac{dy}{dt} + u(t).$$

Let

$$x_1 = y$$
$$x_2 = \dot{y}$$

An Introduction to System Modeling and Control, First Edition. John Chiasson.
© 2022 John Wiley & Sons, Inc. Published 2022 by John Wiley & Sons, Inc.
Companion website: www.wiley.com/go/chiasson/anintroductiontosystemmodelingandcontrol

so that we may write

$$\frac{dx_1}{dt} = x_2$$

$$\frac{dx_2}{dt} = -\frac{k}{m}x_1 - \frac{b}{m}x_2 + \frac{1}{m}u.$$

This is called the *statespace* form. By statespace form we mean that the left-side of the equation has only first-order derivatives and the right-side has no derivatives. The output equation is

$$y = x_1.$$

Using matrix notation, these statespace equations may be written as

$$\frac{d}{dt}\begin{bmatrix} x_1 \\ x_2 \end{bmatrix} = \underbrace{\begin{bmatrix} 0 & 1 \\ -k/m & -b/m \end{bmatrix}}_{A} \underbrace{\begin{bmatrix} x_1 \\ x_2 \end{bmatrix}}_{x} + \underbrace{\begin{bmatrix} 0 \\ 1/m \end{bmatrix}}_{b} u$$

$$y = \underbrace{\begin{bmatrix} 1 & 0 \end{bmatrix}}_{c} \begin{bmatrix} x_1 \\ x_2 \end{bmatrix}.$$

More compactly we write

$$\frac{dx}{dt} = Ax + bu$$

$$y = cx.$$

The quantities x_1 and x_2 are called the *state variables* and

$$x \triangleq \begin{bmatrix} x_1 \\ x_2 \end{bmatrix}$$

is called the *state* of the system.

Example 1 *DC Motor*

Figure 14.2 shows a DC motor with its schematic diagram.

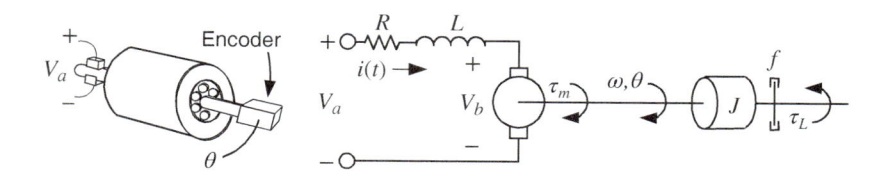

Figure 14.2. DC motor and its schematic diagram.

The equations of motion are

$$V_a = Ri + L\frac{di}{dt} + V_b$$

$$V_b = K_b\omega$$

$$J\frac{d\omega}{dt} = \tau_m - f\omega - \tau_L$$

$$\tau_m = K_T i$$
$$\frac{d\theta}{dt} = \omega$$

or

$$L\frac{di}{dt} = -Ri - K_b\omega + V_a$$
$$J\frac{d\omega}{dt} = K_T i - f\omega - \tau_L$$
$$\frac{d\theta}{dt} = \omega.$$

Rewritten in statespace form we have

$$\frac{di}{dt} = -\frac{R}{L}i - \frac{K_b}{L}\omega + \frac{1}{L}V_a$$
$$\frac{d\omega}{dt} = \frac{K_T}{J}i - \frac{f}{J}\omega - \frac{1}{J}\tau_L$$
$$\frac{d\theta}{dt} = \omega.$$

Using matrix notation this becomes

$$\frac{d}{dt}\begin{bmatrix} i \\ \omega \\ \theta \end{bmatrix} = \underbrace{\begin{bmatrix} -R/L & -K_b/L & 0 \\ K_T/J & -f/J & 0 \\ 0 & 1 & 0 \end{bmatrix}}_{A}\underbrace{\begin{bmatrix} i \\ \omega \\ \theta \end{bmatrix}}_{x} + \underbrace{\begin{bmatrix} 1/L \\ 0 \\ 0 \end{bmatrix}}_{b}V_a + \underbrace{\begin{bmatrix} 0 \\ -1/J \\ 0 \end{bmatrix}}_{p}\tau_L.$$

We consider $u \triangleq V_a$ as the control input and τ_L as the disturbance input. More compactly we have

$$\frac{d}{dt}x = Ax + bu + p\tau_L.$$

The quantities i, ω, θ are called the *state variables* and

$$x \triangleq \begin{bmatrix} i \\ \omega \\ \theta \end{bmatrix}$$

is the *state* of the system.

14.2 Transfer Function to Statespace

Consider the transfer function

$$G(s) = \frac{Y(s)}{U(s)} = \frac{1}{s^2 + 3s + 2}. \qquad (14.1)$$

Clearing fractions in (14.1) we have

$$(s^2 + 3s + 2)Y(s) = U(s).$$

540   14   State Variables

In the time domain this is
$$\frac{d^2y}{dt^2} + 3\frac{dy}{dt} + 2y = u(t).$$

This is referred to as an *input–output* representation since the differential equation contains only the input u and the output y. To put this model into statespace form we let the state variables be
$$x_1 = y$$
$$x_2 = \dot{y}$$

so that
$$\frac{dx_1}{dt} = x_2 \tag{14.2}$$
$$\frac{dx_2}{dt} = -2x_1 - 3x_2 + u \tag{14.3}$$
$$y = x_1. \tag{14.4}$$

This statespace representation can also be represented by the following *simulation* block diagram of Figure 14.3. The adjective "simulation" refers to being able to immediately write down the statespace Eqs. (14.2), (14.3) by inspection from the block diagram.

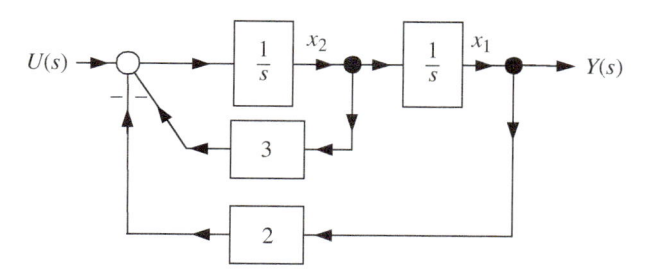

Figure 14.3. Simulation block diagram for $Y(s) = \dfrac{1}{s^2 + 2s + 2}U(s)$.

The outputs of the two integrators are the state variables x_1 and x_2. Observe that the block diagram of Figure 14.3 consists of two nested loops. A straightforward block diagram reduction shows that its transfer function $Y(s)/U(s)$ is simply (14.1). In matrix notation we have
$$\frac{d}{dt}\begin{bmatrix} x_1 \\ x_2 \end{bmatrix} = \underbrace{\begin{bmatrix} 0 & 1 \\ -2 & -3 \end{bmatrix}}_{A} \underbrace{\begin{bmatrix} x_1 \\ x_2 \end{bmatrix}}_{x} + \underbrace{\begin{bmatrix} 0 \\ 1 \end{bmatrix}}_{b} u$$
$$y = \underbrace{\begin{bmatrix} 1 & 0 \end{bmatrix}}_{c} \begin{bmatrix} x_1 \\ x_2 \end{bmatrix}.$$

However, this approach to finding a statespace representation only works if the transfer function has no zeros. To see this let
$$G(s) = \frac{Y(s)}{U(s)} = \frac{s+1}{s^2 + 3s + 2}.$$

14.2 Transfer Function to Statespace

We may write
$$(s^2 + 3s + 2)Y(s) = (s+1)U(s)$$

which in the time domain is
$$\frac{d^2y}{dt^2} + 3\frac{dy}{dt} + 2y = u(t) + \frac{du}{dt}$$

with $y(0) = \dot{y}(0) = 0$. Proceeding as before, we let
$$x_1 = y$$
$$x_2 = \dot{y}$$

which gives
$$\frac{dx_1}{dt} = x_2$$
$$\frac{dx_2}{dt} = -2x_1 - 3x_2 + u + \frac{du}{dt}$$
$$y = x_1$$

or, in matrix form we have
$$\frac{d}{dt}\begin{bmatrix} x_1 \\ x_2 \end{bmatrix} = \begin{bmatrix} 0 & 1 \\ -2 & -3 \end{bmatrix}\begin{bmatrix} x_1 \\ x_2 \end{bmatrix} + \begin{bmatrix} 0 \\ 1 \end{bmatrix}u + \begin{bmatrix} 0 \\ 1 \end{bmatrix}\frac{du}{dt}$$
$$y = \begin{bmatrix} 1 & 0 \end{bmatrix}\begin{bmatrix} x_1 \\ x_2 \end{bmatrix}.$$

This is not a statespace representation due to the derivative du/dt on the right-hand side.

Control Canonical Form

We now give a method that converts any strictly proper transfer function into statespace form.[1] To proceed, let
$$G(s) = \frac{Y(s)}{U(s)} = \frac{b_2 s^2 + b_1 s + b_0}{s^3 + a_2 s^2 + a_1 s + a_0} \quad (14.5)$$

and consider the transfer function
$$\frac{X_1(s)}{U(s)} \triangleq \frac{1}{s^3 + a_2 s^2 + a_1 s + a_0}. \quad (14.6)$$

Using transfer function (14.6) we can draw the *simulation* diagram shown in Figure 14.4. A simulation diagram is a special case of a block diagram in which only integrators $1/s$ or constants are allowed in each block. Furthermore, the integrator blocks are cascaded in series as shown in the block diagram. Observe that the simulation block diagram of Figure 14.4 consists of three nested loops. Using block diagram reduction the reader should verify that $X_1(s)/U(s) = \dfrac{1}{s^3 + a_2 s^2 + a_1 s + a_0}$.

---

[1] See Problem 14.12 for the case of a proper, but not strictly proper transfer function.

542   14   State Variables

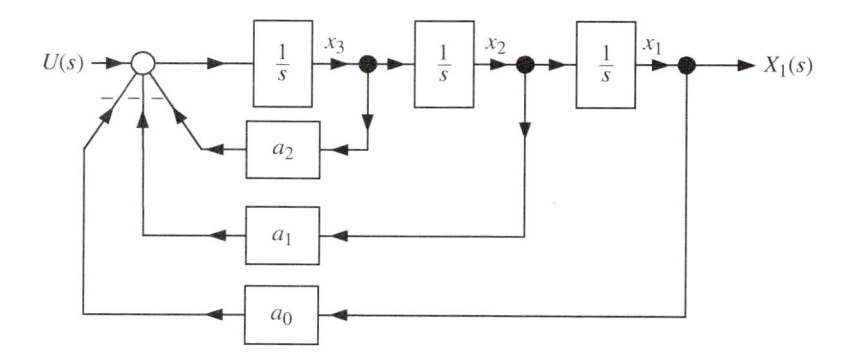

Figure 14.4. Simulation diagram for $\dfrac{X_1(s)}{U(s)} \triangleq \dfrac{1}{s^3 + a_2 s^2 + a_1 s + a_0}$.

As Figure 14.4 indicates, we are taking the state variables to be the outputs of the integrators so that $X_2(s) = sX_1(s)$ and $X_3(s) = sX_2(s) = s^2 X_1(s)$. We see from (14.5) that

$$Y(s) = (b_2 s^2 + b_1 s + b_0) X_1(s) = b_2 s^2 X_1(s) + b_1 s X_1(s) + b_0 X_1(s)$$
$$= b_2 X_3(s) + b_1 X_2(s) + b_0 X_1(s)$$

which leads to the simulation diagram for $G(s)$ given in Figure 14.5.

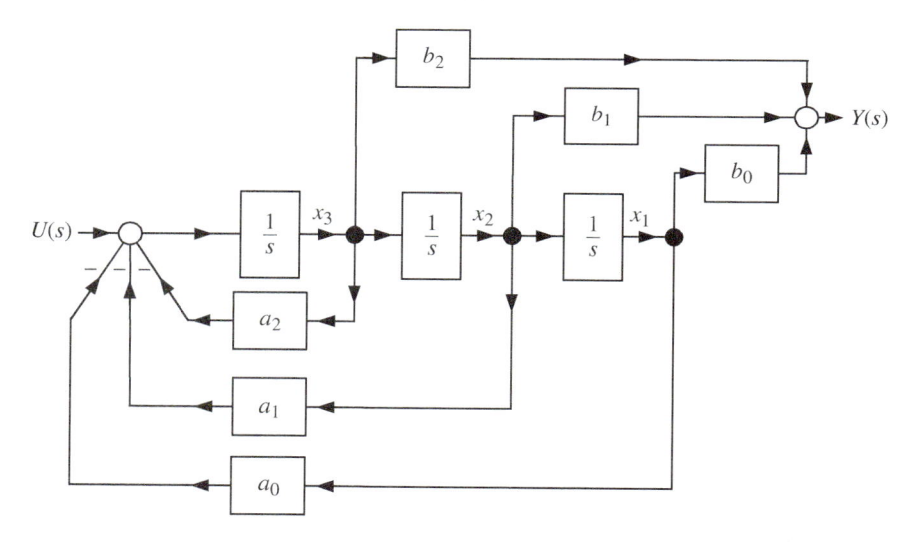

Figure 14.5. Simulation diagram $G(s) = \dfrac{b_2 s^2 + b_1 s + b_0}{s^3 + a_2 s^2 + a_1 s + a_0}$.

From the simulation diagram we have

$$\frac{dx_1}{dt} = x_2$$
$$\frac{dx_2}{dt} = x_3$$

14.2 Transfer Function to Statespace

$$\frac{dx_3}{dt} = -a_0 x_1 - a_1 x_2 - a_2 x_3 + u$$

$$y = b_0 x_1 + b_1 x_2 + b_2 x_3$$

or

$$\frac{d}{dt}\begin{bmatrix} x_1 \\ x_2 \\ x_3 \end{bmatrix} = \underbrace{\begin{bmatrix} 0 & 1 & 0 \\ 0 & 0 & 1 \\ -a_0 & -a_1 & -a_2 \end{bmatrix}}_{A} \underbrace{\begin{bmatrix} x_1 \\ x_2 \\ x_3 \end{bmatrix}}_{x} + \underbrace{\begin{bmatrix} 0 \\ 0 \\ 1 \end{bmatrix}}_{b} u \quad (14.7)$$

$$y = \underbrace{\begin{bmatrix} b_0 & b_1 & b_2 \end{bmatrix}}_{c} \begin{bmatrix} x_1 \\ x_2 \\ x_3 \end{bmatrix}. \quad (14.8)$$

We call this statespace representation a *realization* of $G(s)$, that is, starting from a transfer function $G(s)$ and ending up with a statespace model is a realization of $G(s)$. This special form of the A and b matrices is referred to as the *control canonical* form. As will be shown later, this form is very convenient for state feedback control.

Example 2 *Transfer Function Realization*

Let
$$\frac{Y(s)}{U(s)} = \frac{10s + 8}{s^2 + 5s + 6}$$

and
$$\frac{X_1(s)}{U(s)} = \frac{1}{s^2 + 5s + 6}.$$

Then $Y(s) = 10s X_1(s) + 8 X_1(s)$.

We first draw the simulation diagram for $X_1(s)/U(s)$ and then use the outputs of the two integrators to form $Y(s)$. The simulation diagram is as shown in Figure 14.6.

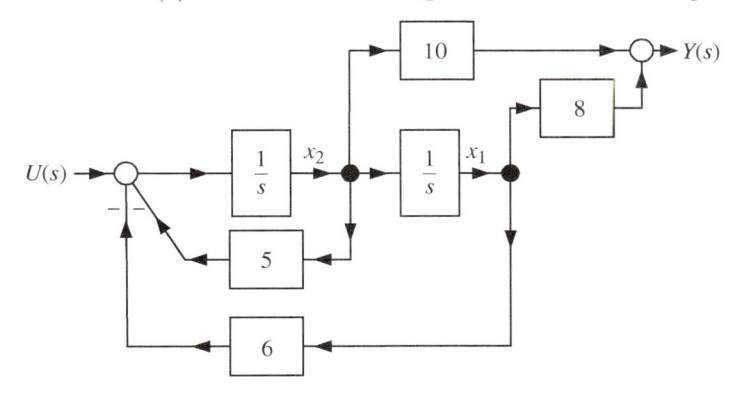

Figure 14.6. Simulation diagram for $G(s) = \dfrac{10s + 8}{s^2 + 5s + 6}$.

The equations of the system is statespace form are

$$\frac{dx_1}{dt} = x_2$$

$$\frac{dx_2}{dt} = -6x_1 - 5x_2 + u$$

$$y = 8x_1 + 10x_2$$

or, in matrix notation, they may be written as

$$\frac{d}{dt}\begin{bmatrix} x_1 \\ x_2 \end{bmatrix} = \underbrace{\begin{bmatrix} 0 & 1 \\ -6 & -5 \end{bmatrix}}_{A} \underbrace{\begin{bmatrix} x_1 \\ x_2 \end{bmatrix}}_{x} + \underbrace{\begin{bmatrix} 0 \\ 1 \end{bmatrix}}_{b} u$$

$$y = \underbrace{\begin{bmatrix} 8 & 10 \end{bmatrix}}_{c} \begin{bmatrix} x_1 \\ x_2 \end{bmatrix}.$$

Example 3 *Discrete-Time Implementation of a Transfer Function*
The derivative of the state variables in Example 2 may be approximated by

$$\begin{bmatrix} \dfrac{dx_1(t)}{dt} \\ \dfrac{dx_2(t)}{dt} \end{bmatrix}_{t=kT} \approx \frac{1}{T}\begin{bmatrix} x_1((k+1)T) - x_1(kT) \\ x_2((k+1)T) - x_2(kT) \end{bmatrix}. \tag{14.9}$$

This is referred to as the Euler approximation for the derivatives. Using this approximation a discrete statespace realization for the system of Example 2 is

$$\frac{1}{T}\left(x\left((k+1)T\right) - x(kT)\right) = Ax(kT) + bu(kT)$$

or

$$x\left((k+1)T\right) = \left(I_{2\times 2} + TA\right) x(kT) + Tbu(kT).$$

More explicitly we have

$$\begin{bmatrix} x_1((k+1)T) \\ x_2((k+1)T) \end{bmatrix} = \left(\begin{bmatrix} 1 & 0 \\ 0 & 1 \end{bmatrix} + T\begin{bmatrix} 0 & 1 \\ -6 & -5 \end{bmatrix}\right)\begin{bmatrix} x_1(kT) \\ x_2(kT) \end{bmatrix} + T\begin{bmatrix} 0 \\ 1 \end{bmatrix} u(kT). \tag{14.10}$$

For example, in SIMULINK when a `transfer function` block with $\dfrac{10s+8}{s^2+5s+6}$ inside of it is used, it is converted to a recursive equation like (14.10) and the simulation runs this recursion.

We can check that the correctness of the simulation block diagram of Figure 14.6 using block diagram reduction. Taking the constant "10" outside the feedback loops gives the equivalent block diagram of Figure 14.7.

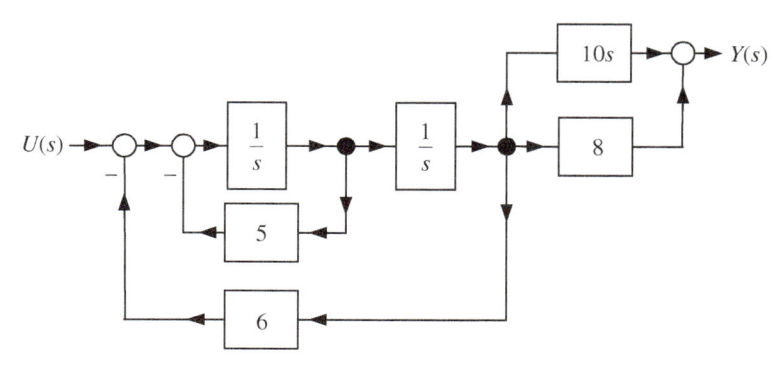

Figure 14.7. First step of a block diagram reduction of Figure 14.6.

14.2 Transfer Function to Statespace

Simplifying the inner feedback loop results in Figure 14.8.

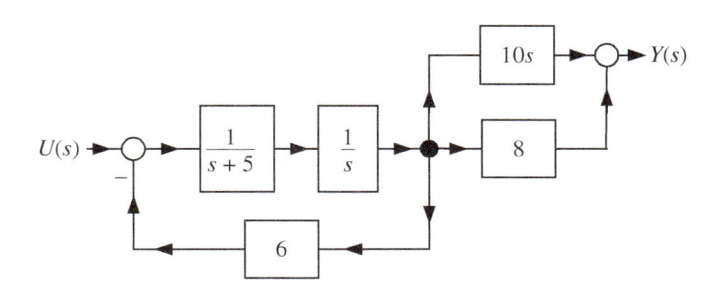

Figure 14.8. Second step of a block diagram reduction of Figure 14.6.

Simplifying the feedback loop in Figure 14.8, and combining the remaining blocks gives the block diagram of Figure 14.9.

Figure 14.9. Block diagram reduction of Figure 14.6.

This shows
$$\frac{Y(s)}{U(s)} = \frac{10s + 8}{s^2 + 5s + 6}.$$

We can also check the statespace realization directly. Before doing this we digress to recall the computation of the inverse of a 2 × 2 matrix.

Digression – *Computation of the Inverse of a 2 × 2 Matrix*

Let
$$A = \begin{bmatrix} a & b \\ c & d \end{bmatrix}$$

and assume its determinant $\det A \neq 0$, that is,

$$\det A = \det \begin{bmatrix} a & b \\ c & d \end{bmatrix} = ad - bc \neq 0.$$

Then A has an inverse given by

$$A^{-1} = \frac{1}{\det A} \begin{bmatrix} d & -b \\ -c & a \end{bmatrix} = \frac{1}{ad - bc} \begin{bmatrix} d & -b \\ -c & a \end{bmatrix}.$$

To check this we simply compute

$$A^{-1}A = \frac{1}{ad - bc} \begin{bmatrix} d & -b \\ -c & a \end{bmatrix} \begin{bmatrix} a & b \\ c & d \end{bmatrix} = \frac{1}{ad - bc} \begin{bmatrix} ad - bc & db - bd \\ -ac + ac & -bc + ad \end{bmatrix}$$

$$= \frac{1}{ad-bc}\begin{bmatrix} ad-bc & 0 \\ 0 & ad-bc \end{bmatrix}$$

$$= \begin{bmatrix} 1 & 0 \\ 0 & 1 \end{bmatrix}.$$

End of Digression

We now compute the transfer function of the statespace model

$$\frac{d}{dt}\begin{bmatrix} x_1 \\ x_2 \end{bmatrix} = \begin{bmatrix} 0 & 1 \\ -6 & -5 \end{bmatrix}\begin{bmatrix} x_1 \\ x_2 \end{bmatrix} + \begin{bmatrix} 0 \\ 1 \end{bmatrix}u$$

$$y = \begin{bmatrix} 8 & 10 \end{bmatrix}\begin{bmatrix} x_1 \\ x_2 \end{bmatrix}.$$

Taking the Laplace transform of both sides gives

$$\begin{bmatrix} sX_1(s) - x_1(0) \\ sX_2(s) - x_2(0) \end{bmatrix} = \begin{bmatrix} 0 & 1 \\ -6 & -5 \end{bmatrix}\begin{bmatrix} X_1(s) \\ X_2(s) \end{bmatrix} + \begin{bmatrix} 0 \\ 1 \end{bmatrix}U(s)$$

$$Y(s) = \begin{bmatrix} 8 & 10 \end{bmatrix}\begin{bmatrix} X_1(s) \\ X_2(s) \end{bmatrix}.$$

We are computing the transfer function $Y(s)/U(s)$ so we set the initial conditions to zero, i.e., $x_1(0) = x_2(0) = 0$. Rearranging gives

$$\left(s\underbrace{\begin{bmatrix} 1 & 0 \\ 0 & 1 \end{bmatrix}}_{I} - \underbrace{\begin{bmatrix} 0 & 1 \\ -6 & -5 \end{bmatrix}}_{A}\right)\begin{bmatrix} X_1(s) \\ X_2(s) \end{bmatrix} = \underbrace{\begin{bmatrix} 0 \\ 1 \end{bmatrix}}_{b}U(s).$$

Solving for $X_1(s)$ and $X_2(s)$ we obtain

$$\begin{bmatrix} X_1(s) \\ X_2(s) \end{bmatrix} = \underbrace{\left(s\begin{bmatrix} 1 & 0 \\ 0 & 1 \end{bmatrix} - \begin{bmatrix} 0 & 1 \\ -6 & -5 \end{bmatrix}\right)^{-1}}_{(sI-A)^{-1}}\underbrace{\begin{bmatrix} 0 \\ 1 \end{bmatrix}}_{b}U(s) = \begin{bmatrix} s & -1 \\ 6 & s+5 \end{bmatrix}^{-1}\begin{bmatrix} 0 \\ 1 \end{bmatrix}U(s)$$

$$= \frac{1}{s(s+5) - (-1)(6)}\begin{bmatrix} s+5 & 1 \\ -6 & s \end{bmatrix}\begin{bmatrix} 0 \\ 1 \end{bmatrix}U(s)$$

$$= \frac{1}{s^2+5s+6}\begin{bmatrix} s+5 & 1 \\ -6 & s \end{bmatrix}\begin{bmatrix} 0 \\ 1 \end{bmatrix}U(s)$$

$$= \frac{1}{s^2+5s+6}\begin{bmatrix} 1 \\ s \end{bmatrix}U(s).$$

Finally,

$$Y(s) = \begin{bmatrix} 8 & 10 \end{bmatrix}\begin{bmatrix} X_1(s) \\ X_2(s) \end{bmatrix} = \begin{bmatrix} 8 & 10 \end{bmatrix}\frac{1}{s^2+5s+6}\begin{bmatrix} 1 \\ s \end{bmatrix}U(s) = \frac{8+10s}{s^2+5s+6}U(s).$$

We next consider a transfer function that is proper, but not strictly proper.

14.2 Transfer Function to Statespace

Example 4 *Transfer Function Realization*

Let
$$G(s) = \frac{s^2 + 10s + 5}{s^2 + 5s + 6}$$

which is proper, but not strictly proper. We rewrite this as

$$G(s) = \frac{s^2 + 5s + 6 - (5s + 6) + 10s + 5}{s^2 + 5s + 6} = \frac{s^2 + 5s + 6 + 5s - 1}{s^2 + 5s + 6} = 1 + \underbrace{\frac{5s - 1}{s^2 + 5s + 6}}_{G_{sp}}.$$

We let
$$G_{sp}(s) \triangleq \frac{5s - 1}{s^2 + 5s + 6}$$

denote the strictly proper part of $G(s)$. We then do a realization (find a statespace model) of $G_{sp}(s)$. We let

$$\frac{X_1(s)}{U(s)} \triangleq \frac{1}{s^2 + 5s + 6}$$

and draw a simulation diagram for $X_1(s)/U(s)$ (see Figure 14.10). Then we use the outputs $X_2(s), X_1(s)$ of the two integrators to make the simulation diagram for $Y_{sp}(s)/U(s)$. Finally, as $Y(s) = Y_{sp}(s) + U(s)$ we simply add a feed forward signal line from the input $U(s)$ to sum it with $Y_{sp}(s)$ to obtain a simulation diagram for $Y(s)/U(s)$ as shown in Figure 14.10.

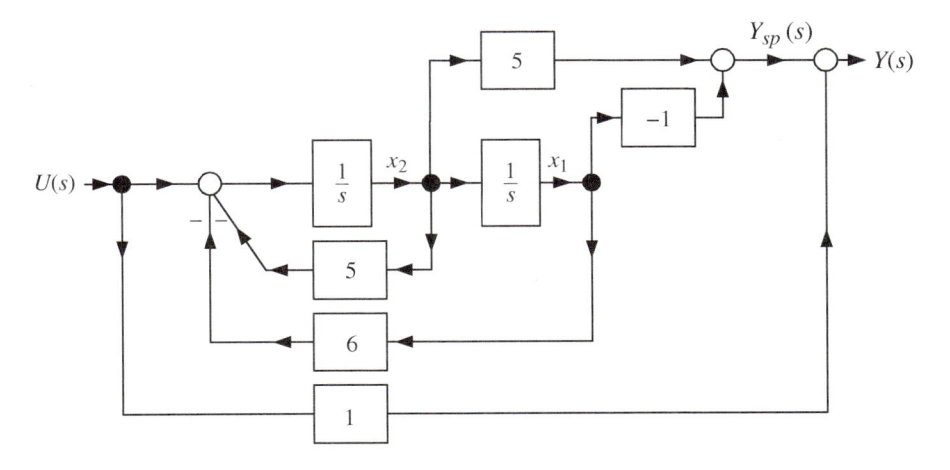

Figure 14.10. Simulation diagram of $G(s) = \dfrac{s^2 + 10s + 5}{s^2 + 5s + 6}$.

The statespace realization is

$$\frac{dx_1}{dt} = x_2$$

$$\frac{dx_2}{dt} = -6x_1 - 5x_2 + u$$

$$y = -x_1 + 5x_2 + u$$

or

$$\frac{d}{dt}\begin{bmatrix} x_1 \\ x_2 \end{bmatrix} = \begin{bmatrix} 0 & 1 \\ -6 & -5 \end{bmatrix}\begin{bmatrix} x_1 \\ x_2 \end{bmatrix} + \begin{bmatrix} 0 \\ 1 \end{bmatrix} u$$

$$y = \begin{bmatrix} -1 & 5 \end{bmatrix}\begin{bmatrix} x_1 \\ x_2 \end{bmatrix} + u.$$

We can check this as follows:

$$\begin{bmatrix} sX_1(s) - \underbrace{x_1(0)}_{0} \\ sX_2(s) - \underbrace{x_2(0)}_{0} \end{bmatrix} = \begin{bmatrix} 0 & 1 \\ -6 & -5 \end{bmatrix}\begin{bmatrix} X_1(s) \\ X_2(s) \end{bmatrix} + \begin{bmatrix} 0 \\ 1 \end{bmatrix} U(s)$$

$$Y(s) = \begin{bmatrix} -1 & 5 \end{bmatrix}\begin{bmatrix} X_1(s) \\ X_2(s) \end{bmatrix} + U(s)$$

or

$$\begin{bmatrix} X_1(s) \\ X_2(s) \end{bmatrix} = \left(s\begin{bmatrix} 1 & 0 \\ 0 & 1 \end{bmatrix} - \begin{bmatrix} 0 & 1 \\ -6 & -5 \end{bmatrix}\right)^{-1}\begin{bmatrix} 0 \\ 1 \end{bmatrix} U(s) = \frac{1}{s^2 + 5s + 6}\begin{bmatrix} 1 \\ s \end{bmatrix} U(s)$$

with

$$Y(s) = \begin{bmatrix} -1 & 5 \end{bmatrix}\begin{bmatrix} X_1(s) \\ X_2(s) \end{bmatrix} + U(s).$$

Finally we have

$$Y(s) = \begin{bmatrix} -1 & 5 \end{bmatrix}\begin{bmatrix} X_1(s) \\ X_2(s) \end{bmatrix} + U(s) = \begin{bmatrix} -1 & 5 \end{bmatrix}\frac{1}{s^2 + 5s + 6}\begin{bmatrix} 1 \\ s \end{bmatrix} U(s) + U(s)$$

$$= \frac{5s - 1}{s^2 + 5s + 6} U(s) + U(s)$$

$$= \frac{s^2 + 10s + 5}{s^2 + 5s + 6} U(s).$$

Example 5 *Implementing a Feedback Controller*
Consider the feedback system in Figure 14.11.

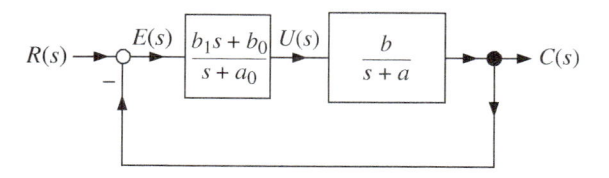

Figure 14.11. Unity feedback control system.

Suppose we want to implement the controller $G_c(s) = \frac{b_1 s + b_0}{s + a_0}$ on a microprocessor. The first thing we do is obtain a realization (statespace model) of $G_c(s)$. We have

$$G_c(s) = \frac{b_1 s + b_0}{s + a_0} = \frac{b_1(s + a_0) - b_1 a_0 + b_0}{s + a_0} = b_1 + \frac{b_0 - b_1 a_0}{s + a_0}.$$

The simulation diagram is given in Figure 14.12.

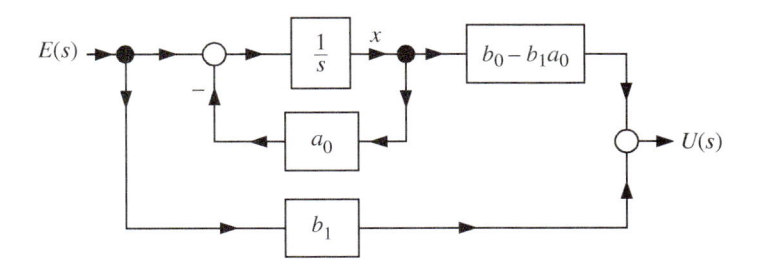

Figure 14.12. Simulation diagram for the controller given Figure 14.11.

The corresponding statespace equations are then

$$\frac{dx(t)}{dt} = -a_0 x(t) + e(t)$$
$$u(t) = (b_0 - b_1 a_0)x(t) + b_1 e(t)$$

with

$$e(t) = r(t) - c(t).$$

Since this is to be implemented on a digital computer it must be converted to a discrete-time model. To do so, let T denote the sample period and let's approximate the derivative dx/dt by

$$\left.\frac{dx}{dt}\right|_{t=kT} \approx \frac{x((k+1)T) - x(kT)}{T}.$$

Substituting this into the statespace equations along with setting $t = kT$ we obtain

$$x((k+1)T) = (1 - a_0 T)x(kT) + Te(kT)$$
$$u(kT) = (b_0 - b_1 a_0)x(kT) + b_1 e(kT)$$
$$e(kT) = r(kT) - c(kT).$$

These are the equations that are implemented using (perhaps) the C programming language. The value of $c(kT)$ comes from the output sensor and we consider the reference $r(kT)$ to be stored in the computer's memory.

Example 6 *Implementing a Transfer Function*
Consider the unity feedback control system of Figure 14.13.

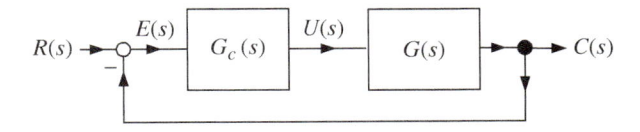

Figure 14.13. Unity feedback control system.

14 State Variables

Suppose a controller was designed and has the form

$$\frac{U(s)}{E(s)} = G_c(s) = \frac{b_{c2}s^2 + b_{c1}s + b_{c0}}{s^3 + a_{c2}s^2 + a_{c1}s + a_{c0}}.$$

A statespace realization of $G_c(s)$ is given by

$$\frac{d}{dt}\begin{bmatrix} x_1 \\ x_2 \\ x_3 \end{bmatrix} = \underbrace{\begin{bmatrix} 0 & 1 & 0 \\ 0 & 0 & 1 \\ -a_{c0} & -a_{c1} & -a_{c2} \end{bmatrix}}_{A_c} \underbrace{\begin{bmatrix} x_1 \\ x_2 \\ x_3 \end{bmatrix}}_{x} + \underbrace{\begin{bmatrix} 0 \\ 0 \\ 1 \end{bmatrix}}_{b_c} e$$

$$u = \underbrace{\begin{bmatrix} b_{c0} & b_{c1} & b_{c2} \end{bmatrix}}_{C_c} \begin{bmatrix} x_1 \\ x_2 \\ x_3 \end{bmatrix}.$$

To implement this controller in software we can use an Euler discretization approach which means we approximate the derivative of the state by

$$\begin{bmatrix} \frac{dx_1}{dt} \\ \frac{dx_2}{dt} \\ \frac{dx_3}{dt} \end{bmatrix}_{|t=kT} \approx \begin{bmatrix} \frac{x_1((k+1)T) - x_1(kT)}{T} \\ \frac{x_2((k+1)T) - x_2(kT)}{T} \\ \frac{x_3((k+1)T) - x_3(kT)}{T} \end{bmatrix}.$$

Then

$$\frac{1}{T}\left(\begin{bmatrix} x_1((k+1)T) \\ x_2((k+1)T) \\ x_3((k+1)T) \end{bmatrix} - \begin{bmatrix} x_1(kT) \\ x_2(kT) \\ x_3(kT) \end{bmatrix}\right) = \begin{bmatrix} 0 & 1 & 0 \\ 0 & 0 & 1 \\ -a_{c0} & -a_{c1} & -a_{c2} \end{bmatrix}\begin{bmatrix} x_1(kT) \\ x_2(kT) \\ x_3(kT) \end{bmatrix}$$

$$+ \begin{bmatrix} 0 \\ 0 \\ 1 \end{bmatrix} e(kT).$$

Rearranging we have

$$\begin{bmatrix} x_1((k+1)T) \\ x_2((k+1)T) \\ x_3((k+1)T) \end{bmatrix} = \underbrace{\begin{bmatrix} 1 & T & 0 \\ 0 & 1 & T \\ -Ta_{c0} & -Ta_{c1} & 1-Ta_{c2} \end{bmatrix}}_{I_{3\times 3} + TA_c} \begin{bmatrix} x_1(kT) \\ x_2(kT) \\ x_3(kT) \end{bmatrix} + T\begin{bmatrix} 0 \\ 0 \\ 1 \end{bmatrix} e(kT)$$

$$u(kT) = \begin{bmatrix} b_{c0} & b_{c1} & b_{c2} \end{bmatrix}\begin{bmatrix} x_1(kT) \\ x_2(kT) \\ x_3(kT) \end{bmatrix},$$

where

$$e(kT) = r(kT) - c(kT).$$

$c(kT)$ is the measured output brought into the computer (e.g., an A/D, optical encoder, etc.) and $r(kT)$ is stored in the computer memory. Then a programming language such as C can be used to implement this in software.

14.3 Laplace Transform of the Statespace Equations

A second-order linear statespace model has the form

$$\frac{d}{dt}\begin{bmatrix} x_1 \\ x_2 \end{bmatrix} = \underbrace{\begin{bmatrix} a_{11} & a_{12} \\ a_{21} & a_{22} \end{bmatrix}}_{A} \underbrace{\begin{bmatrix} x_1 \\ x_2 \end{bmatrix}}_{x(t)} + \underbrace{\begin{bmatrix} b_1 \\ b_2 \end{bmatrix}}_{b} u \tag{14.11}$$

$$\underbrace{\begin{bmatrix} x_1(0) \\ x_2(0) \end{bmatrix} = \begin{bmatrix} x_{01} \\ x_{02} \end{bmatrix}}_{x_0}. \tag{14.12}$$

As before we can use the Laplace transform to solve this set of equations. We have

$$\begin{bmatrix} sX_1(s) - x_1(0) \\ sX_2(s) - x_2(0) \end{bmatrix} = \begin{bmatrix} a_{11} & a_{12} \\ a_{21} & a_{22} \end{bmatrix} \begin{bmatrix} X_1(s) \\ X_2(s) \end{bmatrix} + \begin{bmatrix} b_1 \\ b_2 \end{bmatrix} U(s) \tag{14.13}$$

which upon some rearrangement gives

$$\underbrace{\left(s\begin{bmatrix} 1 & 0 \\ 0 & 1 \end{bmatrix} - \begin{bmatrix} a_{11} & a_{12} \\ a_{21} & a_{22} \end{bmatrix}\right)}_{sI-A} \underbrace{\begin{bmatrix} X_1(s) \\ X_2(s) \end{bmatrix}}_{X(s)} = \underbrace{\begin{bmatrix} x_1(0) \\ x_2(0) \end{bmatrix}}_{x_0} + \underbrace{\begin{bmatrix} b_1 \\ b_2 \end{bmatrix}}_{b} U(s) \tag{14.14}$$

or

$$\begin{bmatrix} X_1(s) \\ X_2(s) \end{bmatrix} = \left(s\begin{bmatrix} 1 & 0 \\ 0 & 1 \end{bmatrix} - \begin{bmatrix} a_{11} & a_{12} \\ a_{21} & a_{22} \end{bmatrix}\right)^{-1} \begin{bmatrix} x_1(0) \\ x_2(0) \end{bmatrix}$$

$$+ \left(s\begin{bmatrix} 1 & 0 \\ 0 & 1 \end{bmatrix} - \begin{bmatrix} a_{11} & a_{12} \\ a_{21} & a_{22} \end{bmatrix}\right)^{-1} \begin{bmatrix} b_1 \\ b_2 \end{bmatrix} U(s). \tag{14.15}$$

Computing we have

$$\underbrace{\left(s\begin{bmatrix} 1 & 0 \\ 0 & 1 \end{bmatrix} - \begin{bmatrix} a_{11} & a_{12} \\ a_{21} & a_{22} \end{bmatrix}\right)^{-1}}_{(sI-A)^{-1}} = \begin{bmatrix} s - a_{11} & -a_{12} \\ -a_{21} & s - a_{22} \end{bmatrix}^{-1}$$

$$= \frac{1}{\det(sI - A)} \begin{bmatrix} s - a_{22} & a_{12} \\ a_{21} & s - a_{11} \end{bmatrix},$$

where

$$\det(sI - A) = \det\begin{bmatrix} s - a_{11} & -a_{12} \\ -a_{21} & s - a_{22} \end{bmatrix} = s^2 + (-a_{11} - a_{22})s + a_{11}a_{22} - a_{12}a_{21}.$$

More compactly we write

$$X(s) = (sI - A)^{-1}x(0) + (sI - A)^{-1}bU(s). \tag{14.16}$$

Then

$$x(t) = \mathcal{L}^{-1}\{X(s)\} = \mathcal{L}^{-1}\{(sI - A)^{-1}x(0)\} + \mathcal{L}^{-1}\{(sI - A)^{-1}bU(s)\}. \tag{14.17}$$

Let's look at an example.

Example 7 *Inverse Laplace Transform of the Statespace Equations* [4]
Consider the statespace model

$$\frac{d}{dt}\begin{bmatrix} x_1 \\ x_2 \end{bmatrix} = \begin{bmatrix} 0 & 2 \\ -2 & -5 \end{bmatrix}\begin{bmatrix} x_1 \\ x_2 \end{bmatrix} + \begin{bmatrix} 0 \\ 1 \end{bmatrix} u.$$

Then

$$X(s) = \begin{bmatrix} X_1(s) \\ X_2(s) \end{bmatrix} = (sI - A)^{-1}x(0) + (sI - A)^{-1}bU(s).$$

Now as

$$(sI - A)^{-1} = \begin{bmatrix} s & -2 \\ 2 & s+5 \end{bmatrix}^{-1} = \frac{1}{\underbrace{(s+5)s + 4}_{\det(sI-A)}}\begin{bmatrix} s+5 & 2 \\ -2 & s \end{bmatrix} = \frac{1}{\underbrace{s^2 + 5s + 4}_{(s+4)(s+1)}}\begin{bmatrix} s+5 & 2 \\ -2 & s \end{bmatrix}$$

we have

$$\mathcal{L}^{-1}\{(sI - A)^{-1}\} = \begin{bmatrix} \mathcal{L}^{-1}\left\{\frac{s+5}{(s+4)(s+1)}\right\} & \mathcal{L}^{-1}\left\{\frac{2}{(s+4)(s+1)}\right\} \\ \mathcal{L}^{-1}\left\{-\frac{2}{(s+4)(s+1)}\right\} & \mathcal{L}^{-1}\left\{\frac{s}{(s+4)(s+1)}\right\} \end{bmatrix}.$$

Perform partial fraction expansions to obtain

$$\frac{s+5}{(s+4)(s+1)} = \frac{A}{s+4} + \frac{B}{s+1} = -\frac{1/3}{s+4} + \frac{4/3}{s+1}$$

$$\frac{2}{(s+4)(s+1)} = \frac{A}{s+4} + \frac{B}{s+1} = -\frac{2/3}{s+4} + \frac{2/3}{s+1}$$

and

$$\frac{s}{(s+4)(s+1)} = \frac{A}{s+4} + \frac{B}{s+1} = \frac{4/3}{s+4} - \frac{1/3}{s+1}.$$

Thus

$$\Phi(t) \triangleq \mathcal{L}^{-1}\{(sI - A)^{-1}\} = \begin{bmatrix} \mathcal{L}^{-1}\left\{-\frac{1/3}{s+4} + \frac{4/3}{s+1}\right\} & \mathcal{L}^{-1}\left\{-\frac{2/3}{s+4} + \frac{2/3}{s+1}\right\} \\ \mathcal{L}^{-1}\left\{\frac{2/3}{s+4} - \frac{2/3}{s+1}\right\} & \mathcal{L}^{-1}\left\{\frac{4/3}{s+4} - \frac{1/3}{s+1}\right\} \end{bmatrix}$$

$$= \begin{bmatrix} -\frac{1}{3}e^{-4t} + \frac{4}{3}e^{-t} & -\frac{2}{3}e^{-4t} + \frac{2}{3}e^{-t} \\ \frac{2}{3}e^{-4t} - \frac{2}{3}e^{-t} & \frac{4}{3}e^{-4t} - \frac{1}{3}e^{-t} \end{bmatrix}.$$

14.3 Laplace Transform of the Statespace Equations

Suppose $u(t) = 0$. Then the solution $x(t)$ is

$$x(t) = \mathcal{L}^{-1}\{(sI - A)^{-1}x(0)\} = \mathcal{L}^{-1}\{(sI - A)^{-1}\}x(0)$$
$$= \Phi(t)x(0)$$
$$= \begin{bmatrix} -\dfrac{1}{3}e^{-4t} + \dfrac{4}{3}e^{-t} & -\dfrac{2}{3}e^{-4t} + \dfrac{2}{3}e^{-t} \\ \dfrac{2}{3}e^{-4t} - \dfrac{2}{3}e^{-t} & \dfrac{4}{3}e^{-4t} - \dfrac{1}{3}e^{-t} \end{bmatrix} \begin{bmatrix} x_1(0) \\ x_2(0) \end{bmatrix}.$$

Now let the initial conditions be zero, that is, $x_1(0) = x_2(0) = 0$ and let the input be

$$u(t) = u_s(t) = \begin{cases} 1, & t \geq 0 \\ 0, & t < 0. \end{cases}$$

Then $U(s) = 1/s$ and we have

$$X(s) = (sI - A)^{-1}bU(s) = \dfrac{1}{(s+4)(s+1)}\begin{bmatrix} 2 \\ s \end{bmatrix}\dfrac{1}{s} = \begin{bmatrix} \dfrac{2}{s(s+4)(s+1)} \\ \dfrac{1}{(s+4)(s+1)} \end{bmatrix}$$

$$= \begin{bmatrix} \dfrac{1/2}{s} - \dfrac{2/3}{s+1} + \dfrac{1/6}{s+4} \\ \dfrac{1/3}{s+1} - \dfrac{1/3}{s+4} \end{bmatrix}.$$

Thus

$$x(t) = \mathcal{L}^{-1}\{(sI - A)^{-1}bU(s)\} = \mathcal{L}^{-1}\left\{\begin{bmatrix} \dfrac{1/2}{s} - \dfrac{2/3}{s+1} + \dfrac{1/6}{s+4} \\ \dfrac{1/3}{s+1} - \dfrac{1/3}{s+4} \end{bmatrix}\right\}$$

$$= \begin{bmatrix} \dfrac{1}{2}u_s(t) - \dfrac{2}{3}e^{-t} + \dfrac{1}{6}e^{-4t} \\ \dfrac{1}{3}e^{-t} - \dfrac{1}{3}e^{-4t} \end{bmatrix}.$$

Finally suppose $x_1(0) = 1, x_2(0) = 2, u(t) = u_s(t)$. The complete solution is

$$x(t) = \mathcal{L}^{-1}\{(sI - A)^{-1}x(0)\} + \mathcal{L}^{-1}\{(sI - A)^{-1}bU(s)\}$$

$$= \begin{bmatrix} -\dfrac{1}{3}e^{-4t} + \dfrac{4}{3}e^{-t} & -\dfrac{2}{3}e^{-4t} + \dfrac{2}{3}e^{-t} \\ \dfrac{2}{3}e^{-4t} - \dfrac{2}{3}e^{-t} & \dfrac{4}{3}e^{-4t} - \dfrac{1}{3}e^{-t} \end{bmatrix}\begin{bmatrix} 1 \\ 2 \end{bmatrix} + \begin{bmatrix} \dfrac{1}{2}u_s(t) - \dfrac{2}{3}e^{-t} + \dfrac{1}{6}e^{-4t} \\ \dfrac{1}{3}e^{-t} - \dfrac{1}{3}e^{-4t} \end{bmatrix}.$$

14.4 Fundamental Matrix Φ

We have considered a second-order statespace model given by

$$\frac{d}{dt}\begin{bmatrix} x_1 \\ x_2 \end{bmatrix} = \underbrace{\begin{bmatrix} a_{11} & a_{12} \\ a_{21} & a_{22} \end{bmatrix}}_{A} \underbrace{\begin{bmatrix} x_1 \\ x_2 \end{bmatrix}}_{x(t)} + \underbrace{\begin{bmatrix} b_1 \\ b_2 \end{bmatrix}}_{b} u \quad (14.18)$$

$$\underbrace{\begin{bmatrix} x_1(0) \\ x_2(0) \end{bmatrix}}_{x(0)} = \begin{bmatrix} x_{01} \\ x_{02} \end{bmatrix} \quad (14.19)$$

or, more compactly,

$$\frac{dx}{dt} = Ax + bu \quad (14.20)$$

$$x(0) = x_0. \quad (14.21)$$

With $X(s) = \mathcal{L}\{x(t)\}$ we have

$$X(s) = (sI - A)^{-1}x(0) + (sI - A)^{-1}bU(s). \quad (14.22)$$

Letting $u(t) = 0$ it follows that the solution to

$$\frac{dx}{dt} = Ax \quad (14.23)$$

$$x(0) = x_0 \quad (14.24)$$

is

$$x(t) = \mathcal{L}^{-1}\left\{(sI-A)^{-1}x(0)\right\} = \underbrace{\mathcal{L}^{-1}\{(sI-A)^{-1}\}}_{\Phi(t)}x(0) = \Phi(t)x(0). \quad (14.25)$$

$\Phi(t)$ is called the *fundamental* matrix. We now develop an explicit expression for it.

Exponential Matrix e^{At}

Recall that the Taylor series expansion for e^{at} is

$$e^{at} = 1 + at + a^2\frac{t^2}{2!} + a^3\frac{t^3}{3!} + a^4\frac{t^4}{4!} + \cdots$$

$$= \sum_{j=0}^{\infty} \frac{(at)^j}{j!}.$$

Next note that

$$\frac{d}{dt}e^{at} = \frac{d}{dt}\left(1 + at + a^2\frac{t^2}{2!} + a^3\frac{t^3}{3!} + a^4\frac{t^4}{4!} + \cdots\right)$$

$$= 0 + a + a^2 t + a^3\frac{t^2}{2!} + a^4\frac{t^3}{3!} + \cdots$$

$$= a\left(1 + at + a^2\frac{t^2}{2!} + a^3\frac{t^3}{3!} + \cdots\right)$$

$$= a\sum_{j=0}^{\infty} \frac{(at)^j}{j!}$$

$$= ae^{at}.$$

For $A \in \mathbb{R}^{n \times n}$ define the *exponential matrix* e^{At} by

$$e^{At} \triangleq I_{n \times n} + At + A^2\frac{t^2}{2!} + A^3\frac{t^3}{3!} + A^4\frac{t^4}{4!} + \cdots$$

$$= \sum_{j=0}^{\infty} A^j \frac{t^j}{j!}$$

where we recall that $0! \triangleq 1$ and $A^0 \triangleq I_{n \times n}$. Then we have

$$\frac{d}{dt}e^{At} \triangleq \frac{d}{dt}\left(I_{n \times n} + At + A^2\frac{t^2}{2!} + A^3\frac{t^3}{3!} + A^4\frac{t^4}{4!} + \cdots\right)$$

$$= 0_{n \times n} + A + A^2\frac{t}{1!} + A^3\frac{t^2}{2!} + A^4\frac{t^3}{3!} + \cdots$$

$$= A\left(I_{n \times n} + A\frac{t}{1!} + A^2\frac{t^2}{2!} + A^3\frac{t^3}{3!} + \cdots\right)$$

$$= A\sum_{j=0}^{\infty} A^j \frac{t^j}{j!}$$

$$= Ae^{At}.$$

Also note that

$$e^{At}\big|_{t=0} = I_{n \times n} + At + A^2\frac{t^2}{2!} + A^3\frac{t^3}{3!} + \cdots\bigg|_{t=0}$$

$$= I_{n \times n}.$$

Writing $x(t) = e^{At}x_0$ we have $dx/dt = Ae^{At}x_0 = Ax(t)$ and $x(0) = x_0$, that is, $x(t) = e^{At}x_0$ satisfies (14.23) and (14.24).

This shows that the exponential matrix is the fundamental matrix, i.e.,

$$\Phi(t) = \mathcal{L}^{-1}\{(sI - A)^{-1}\} = e^{At}. \tag{14.26}$$

Example 8 *Exponential Matrix*

Let
$$A = \begin{bmatrix} 0 & 1 \\ 0 & 0 \end{bmatrix}.$$

Then
$$e^{At} = \begin{bmatrix} 1 & 0 \\ 0 & 1 \end{bmatrix} + \begin{bmatrix} 0 & 1 \\ 0 & 0 \end{bmatrix}t + \underbrace{\begin{bmatrix} 0 & 1 \\ 0 & 0 \end{bmatrix}^2}_{0_{2 \times 2}}\frac{t^2}{2!} + \underbrace{\begin{bmatrix} 0 & 1 \\ 0 & 0 \end{bmatrix}^3}_{0_{2 \times 2}}\frac{t^3}{3!} + \cdots$$

$$= \begin{bmatrix} 1 & 0 \\ 0 & 1 \end{bmatrix} + \begin{bmatrix} 0 & 1 \\ 0 & 0 \end{bmatrix} t$$

$$= \begin{bmatrix} 1 & t \\ 0 & 1 \end{bmatrix}.$$

Note that

$$\frac{d}{dt}\underbrace{\begin{bmatrix} 1 & t \\ 0 & 1 \end{bmatrix}}_{e^{At}} = \begin{bmatrix} 0 & 1 \\ 0 & 0 \end{bmatrix} = \underbrace{\begin{bmatrix} 0 & 1 \\ 0 & 0 \end{bmatrix}}_{A}\underbrace{\begin{bmatrix} 1 & t \\ 0 & 1 \end{bmatrix}}_{e^{At}}.$$

Example 9 *Exponential Matrix*
Let
$$A = \begin{bmatrix} 2 & 0 \\ 0 & 3 \end{bmatrix}.$$

Then

$$e^{At} = \begin{bmatrix} 1 & 0 \\ 0 & 1 \end{bmatrix} + \begin{bmatrix} 2 & 0 \\ 0 & 3 \end{bmatrix} t + \begin{bmatrix} 2 & 0 \\ 0 & 3 \end{bmatrix}^2 \frac{t^2}{2!} + \begin{bmatrix} 2 & 0 \\ 0 & 3 \end{bmatrix}^3 \frac{t^3}{3!} + \cdots$$

$$= \begin{bmatrix} 1 & 0 \\ 0 & 1 \end{bmatrix} + \begin{bmatrix} 2 & 0 \\ 0 & 3 \end{bmatrix} t + \begin{bmatrix} 2^2 & 0 \\ 0 & 3^2 \end{bmatrix} \frac{t^2}{2!} + \begin{bmatrix} 2^3 & 0 \\ 0 & 3^3 \end{bmatrix} \frac{t^3}{3!} + \cdots$$

$$= \begin{bmatrix} 1 + 2t + \frac{(2t)^2}{2!} + \frac{(2t)^3}{3!} + \cdots & 0 \\ 0 & 1 + 3t + \frac{(3t)^2}{2!} + \frac{(3t)^3}{3!} + \cdots \end{bmatrix}$$

$$= \begin{bmatrix} e^{2t} & 0 \\ 0 & e^{3t} \end{bmatrix}.$$

Note that

$$\frac{d}{dt}\underbrace{\begin{bmatrix} e^{2t} & 0 \\ 0 & e^{3t} \end{bmatrix}}_{e^{At}} = \begin{bmatrix} 2e^{2t} & 0 \\ 0 & 3e^{3t} \end{bmatrix} = \underbrace{\begin{bmatrix} 2 & 0 \\ 0 & 3 \end{bmatrix}}_{A}\underbrace{\begin{bmatrix} e^{2t} & 0 \\ 0 & e^{3t} \end{bmatrix}}_{e^{At}}.$$

Example 10 *Exponential Matrix*
Let
$$A = \begin{bmatrix} \lambda & 1 \\ 0 & \lambda \end{bmatrix}.$$

Then
$$e^{At} = \begin{bmatrix} 1 & 0 \\ 0 & 1 \end{bmatrix} + \begin{bmatrix} \lambda & 1 \\ 0 & \lambda \end{bmatrix} t + \begin{bmatrix} \lambda & 1 \\ 0 & \lambda \end{bmatrix}^2 \frac{t^2}{2!} + \begin{bmatrix} \lambda & 1 \\ 0 & \lambda \end{bmatrix}^3 \frac{t^3}{3!} + \cdots$$

Now

$$\begin{bmatrix} \lambda & 1 \\ 0 & \lambda \end{bmatrix}^2 = \begin{bmatrix} \lambda^2 & 2\lambda \\ 0 & \lambda^2 \end{bmatrix}$$

$$\begin{bmatrix} \lambda & 1 \\ 0 & \lambda \end{bmatrix}^3 = \begin{bmatrix} \lambda^2 & 2\lambda \\ 0 & \lambda^2 \end{bmatrix}\begin{bmatrix} \lambda & 1 \\ 0 & \lambda \end{bmatrix} = \begin{bmatrix} \lambda^3 & 3\lambda^2 \\ 0 & \lambda^3 \end{bmatrix}$$

$$\begin{bmatrix} \lambda & 1 \\ 0 & \lambda \end{bmatrix}^4 = \begin{bmatrix} \lambda^3 & 3\lambda^2 \\ 0 & \lambda^3 \end{bmatrix}\begin{bmatrix} \lambda & 1 \\ 0 & \lambda \end{bmatrix} = \begin{bmatrix} \lambda^4 & 4\lambda^3 \\ 0 & \lambda^4 \end{bmatrix}.$$

14.4 Fundamental Matrix Φ

In general we have
$$\begin{bmatrix} \lambda & 1 \\ 0 & \lambda \end{bmatrix}^k = \begin{bmatrix} \lambda^k & k\lambda^{k-1} \\ 0 & \lambda^k \end{bmatrix}.$$

We may then write e^{At} as

$$e^{At} = \sum_{k=0}^{\infty} \begin{bmatrix} \lambda^k & k\lambda^{k-1} \\ 0 & \lambda^k \end{bmatrix} \frac{t^k}{k!} = \begin{bmatrix} \sum_{k=0}^{\infty} \frac{(\lambda t)^k}{k!} & \sum_{k=0}^{\infty} kt\frac{(\lambda t)^{k-1}}{k!} \\ 0 & \sum_{k=0}^{\infty} \frac{(\lambda t)^k}{k!} \end{bmatrix} = \begin{bmatrix} \sum_{k=0}^{\infty} \frac{(\lambda t)^k}{k!} & t\sum_{k=1}^{\infty} \frac{(\lambda t)^{k-1}}{(k-1)!} \\ 0 & \sum_{k=0}^{\infty} \frac{(\lambda t)^k}{k!} \end{bmatrix}$$

$$= \begin{bmatrix} \sum_{k=0}^{\infty} \frac{(\lambda t)^k}{k!} & t\sum_{m=0}^{\infty} \frac{(\lambda t)^m}{m!} \\ 0 & \sum_{k=0}^{\infty} \frac{(\lambda t)^k}{k!} \end{bmatrix}$$

$$= \begin{bmatrix} e^{\lambda t} & te^{\lambda t} \\ 0 & e^{\lambda t} \end{bmatrix}.$$

Note that

$$\frac{d}{dt}\underbrace{\begin{bmatrix} e^{\lambda t} & te^{\lambda t} \\ 0 & e^{\lambda t} \end{bmatrix}}_{e^{At}} = \begin{bmatrix} \lambda e^{\lambda t} & e^{\lambda t} + \lambda te^{\lambda t} \\ 0 & \lambda e^{\lambda t} \end{bmatrix} = \underbrace{\begin{bmatrix} \lambda & 1 \\ 0 & \lambda \end{bmatrix}}_{A} \underbrace{\begin{bmatrix} e^{\lambda t} & te^{\lambda t} \\ 0 & e^{\lambda t} \end{bmatrix}}_{e^{At}}.$$

The fundamental matrix is the exponential matrix, that is,

$$\Phi(t) = e^{At} = \mathcal{L}^{-1}\{(sI - A)^{-1}\}.$$

We have computed $\Phi(t) = e^{At}$ by using its definition as a Taylor series expansion. This was done in order to obtain familiarity with the concept of the exponential matrix. We can also compute $\Phi(t) = e^{At}$ using Laplace transforms. We do an example of this next.

Example 11 *Exponential Matrix*

Let
$$A = \begin{bmatrix} \lambda & 1 \\ 0 & \lambda \end{bmatrix}.$$

As
$$sI - A = s\begin{bmatrix} 1 & 0 \\ 0 & 1 \end{bmatrix} - \begin{bmatrix} \lambda & 1 \\ 0 & \lambda \end{bmatrix} = \begin{bmatrix} s-\lambda & -1 \\ 0 & s-\lambda \end{bmatrix},$$

it follows that

$$(sI-A)^{-1} = \begin{bmatrix} s-\lambda & -1 \\ 0 & s-\lambda \end{bmatrix}^{-1} = \frac{1}{(s-\lambda)^2}\begin{bmatrix} s-\lambda & 1 \\ 0 & s-\lambda \end{bmatrix} = \begin{bmatrix} \frac{1}{s-\lambda} & \frac{1}{(s-\lambda)^2} \\ 0 & \frac{1}{s-\lambda} \end{bmatrix}.$$

Then

$$\Phi(t) = e^{At} = \mathcal{L}^{-1}\{(sI - A)^{-1}\} = \begin{bmatrix} \mathcal{L}^{-1}\left\{\dfrac{1}{s-\lambda}\right\} & \mathcal{L}^{-1}\left\{\dfrac{1}{(s-\lambda)^2}\right\} \\ 0 & \mathcal{L}^{-1}\left\{\dfrac{1}{s-\lambda}\right\} \end{bmatrix}$$

$$= \begin{bmatrix} e^{\lambda t} & te^{\lambda t} \\ 0 & e^{\lambda t} \end{bmatrix}.$$

Recall that

$$\mathcal{L}\{e^{\lambda t}\} = \int_0^\infty e^{-st}e^{\lambda t}\,dt = \int_0^\infty e^{-(s-\lambda)t} = \frac{1}{s-\lambda}$$

for $\text{Re}\{s\} > \text{Re}\{\lambda\}$. Using this expression we then have

$$\frac{d}{ds}\int_0^\infty e^{-st}e^{\lambda t}\,dt = \frac{d}{ds}\frac{1}{s-\lambda}$$

or

$$\int_0^\infty (-t)e^{-st}e^{\lambda t}\,dt = -\frac{1}{(s-\lambda)^2}$$

or

$$\mathcal{L}\{te^{\lambda t}\} = \int_0^\infty e^{-st}te^{\lambda t}\,dt = \frac{1}{(s-\lambda)^2}.$$

Properties of the Exponential Matrix

Problem 4 at the end of this chapter asks you to show for matrices $A, B \in \mathbb{R}^{n\times n}$:

(1) $Ae^{At} = e^{At}A$.
(2) $e^{At}e^{Bt} = e^{Bt}e^{At} = e^{(A+B)t}$ if and only if $AB = BA$.
(3) $(e^{At})^{-1} = e^{-At}$.

14.5 Solution of the Statespace Equation*

Given the statespace model

$$\frac{dx}{dt} = Ax + bu \tag{14.27}$$

$$x(0) = x_0 \tag{14.28}$$

we have shown that the Laplace transform of the solution $x(t)$ is

$$X(s) = (sI - A)^{-1}x(0) + (sI - A)^{-1}bU(s). \tag{14.29}$$

We now give an explicit expression for $x(t) = \mathcal{L}^{-1}\{X(s)\}$. To start, we first do the scalar case.

Scalar Case

Consider the scalar case, that is,

$$\frac{dx}{dt} = ax + bu, \quad x(0) = x_0$$

with $x, x_0, a, b \in \mathbb{R}$. Multiply both sides by e^{-at} to obtain

$$e^{-at}\frac{dx(t)}{dt} = e^{-at}ax(t) + e^{-at}bu(t)$$

or

$$\frac{d}{dt}\left(e^{-at}x(t)\right) + ae^{-at}x(t) = e^{-at}ax(t) + e^{-at}bu(t)$$

or finally

$$\frac{d}{dt}\left(e^{-at}x(t)\right) = e^{-at}bu(t).$$

Changing the symbol t to τ in this last equation and integrating results in

$$\int_0^t \frac{d}{d\tau}\left(e^{-a\tau}x(\tau)\right) d\tau = \int_0^t e^{-a\tau}bu(\tau)d\tau$$

or

$$e^{-a\tau}x(\tau)\Big|_0^t = \int_0^t e^{-a\tau}bu(\tau)d\tau$$

or

$$e^{-at}x(t) - x_0 = \int_0^t e^{-a\tau}bu(\tau)d\tau.$$

Multiplying through by e^{at} and rearranging gives the result

$$x(t) = e^{at}x_0 + e^{at}\int_0^t e^{-a\tau}bu(\tau)d\tau = e^{at}x_0 + \underbrace{\int_0^t e^{a(t-\tau)}bu(\tau)d\tau}_{\text{Convolution}}.$$

We can confirm that this is a solution to (14.27) by direct computation. But first we need to digress and explain the Leibniz rule of differentiation.

Digression – *Leibniz's Rule for Differentiation*

Both Newton and Leibniz showed that

$$\int_{t_1}^{t_2} f(\tau)d\tau = F(t_2) - F(t_1),$$

where $F(t)$ is an antiderivative of $f(t)$, i.e., $dF(t)/dt = f(t)$. Then

$$\frac{d}{dt_2}\int_{t_1}^{t_2} f(\tau)d\tau = \frac{d}{dt_2}(F(t_2) - F(t_1)) = f(t_2)$$

$$\frac{d}{dt_1}\int_{t_1}^{t_2} f(\tau)d\tau = \frac{d}{dt_1}(F(t_2) - F(t_1)) = -f(t_1).$$

14 State Variables

Combining this with the chain rule for differentiation gives

$$\frac{d}{dt_2} \int_{t_1}^{g(t_2)} f(\tau) d\tau = \frac{d}{dt_2} \Big(F(g(t_2)) - F(t_1) \Big) = f(g(t_2)) \frac{dg(t_2)}{dt_2}.$$

Similarly

$$\frac{d}{dt_1} \int_{h(t_1)}^{t_2} f(\tau) d\tau = \frac{d}{dt_1} \Big(F(t_2) - F(h(t_1)) \Big) = -f(h(t_1)) \frac{dh(t_1)}{dt_1}.$$

This gives the *Leibniz rule*

$$\frac{d}{dt} \int_{h(t)}^{g(t)} f(\tau) d\tau = f(g(t)) \frac{dg(t)}{dt} - f(h(t)) \frac{dh(t)}{dt}.$$

Finally suppose we want to compute the derivative of

$$\frac{d}{dt} \int_{h(t)}^{g(t)} f(t, \tau) d\tau.$$

It follows immediately that

$$\frac{d}{dt} \int_{h(t)}^{g(t)} f(t, \tau) d\tau = \int_{h(t)}^{g(t)} \frac{\partial f(t, \tau)}{\partial t} d\tau + f(t, g(t)) \frac{dg(t)}{dt} - f(t, h(t)) \frac{dh(t)}{dt}.$$

End of digression

Applying Leibniz's rule to

$$x(t) = e^{at} x_0 + \int_0^t e^{a(t-\tau)} bu(\tau) d\tau = e^{at} x_0 + e^{at} \int_0^t e^{-a\tau} bu(\tau) d\tau$$

gives

$$\frac{d}{dt} x(t) = ae^{at} x_0 + \frac{d}{dt} \left(e^{at} \int_0^t e^{-a\tau} bu(\tau) d\tau \right)$$

$$= ae^{at} x_0 + ae^{at} \int_0^t e^{-a\tau} bu(\tau) d\tau + e^{at} \frac{d}{dt} \int_0^t e^{-a\tau} bu(\tau) d\tau$$

$$= ae^{at} x_0 + ae^{at} \int_0^t e^{-a\tau} bu(\tau) d\tau + e^{at} e^{-at} bu(t)$$

$$= a \underbrace{\left(e^{at} x_0 + \int_0^t e^{a(t-\tau)} bu(\tau) d\tau \right)}_{x(t)} + bu(t)$$

$$= ax(t) + bu(t).$$

Also

$$x(0) = x(t)|_{t=0} = e^{at} x_0 |_{t=0} + \int_0^t e^{a(t-\tau)} bu(\tau) d\tau \Big|_{t=0}$$

$$= x_0 + \int_0^0 e^{a(0-\tau)} bu(\tau) d\tau$$

$$= x_0.$$

Matrix Case

The matrix case is similar. Multiply both sides of (14.27) by e^{-At} to obtain

$$e^{-At}\frac{dx(t)}{dt} = e^{-At}Ax(t) + e^{-At}bu(t)$$

or

$$\frac{d}{dt}(e^{-At}x(t)) + Ae^{-At}x(t) = e^{-At}Ax(t) + e^{-At}bu(t).$$

By property (1) of the exponential matrix we have $Ae^{-At} = e^{-At}A$, and thus

$$\frac{d}{dt}(e^{-At}x(t)) = e^{-At}bu(t).$$

Replacing t by τ and integrating both sides gives

$$e^{-A\tau}x(\tau)|_0^t = \int_0^t e^{-A\tau}bu(\tau)d\tau.$$

Rearranging we have

$$x(t) = e^{At}x_0 + e^{At}\int_0^t e^{-A\tau}bu(\tau)d\tau = e^{At}x_0 + \int_0^t e^{A(t-\tau)}bu(\tau)d\tau \in \mathbb{R}^n,$$

where, as At and $-A\tau$ commute, we were able to use

$$e^{At}e^{-A\tau} = e^{At-A\tau} = e^{A(t-\tau)}.$$

Similar to the scalar case we compute

$$\frac{d}{dt}x(t) = \frac{d}{dt}e^{At}x_0 + \frac{d}{dt}\left(e^{At}\int_0^t e^{-A\tau}bu(\tau)d\tau\right)$$

$$= Ae^{At}x_0 + Ae^{At}\int_0^t e^{-A\tau}bu(\tau)d\tau + e^{At}\frac{d}{dt}\int_0^t e^{-A\tau}bu(\tau)d\tau$$

$$= Ae^{At}x_0 + Ae^{At}\int_0^t e^{-A\tau}bu(\tau)d\tau + e^{At}e^{-At}bu(t)$$

$$= A\underbrace{\left(e^{At}x_0 + Ae^{At}\int_0^t e^{-A\tau}bu(\tau)d\tau\right)}_{x(t)} + bu(t)$$

$$= Ax(t) + bu(t).$$

14.6 Discretization of a Statespace Model*

We have shown that the solution to

$$\frac{dx}{dt} = Ax + bu \tag{14.30}$$

is given by
$$x(t) = e^{At}x_0 + \int_0^t e^{A(t-\tau)}bu(\tau)d\tau \qquad (14.31)$$

with arbitrary initial condition $x(0) = x_0$. To run this statespace model on a digital computer, we must find a discrete-time model. Previously we have considered the approximation
$$\left.\frac{dx}{dt}\right|_{t=(k+1)T} \approx \frac{x\big((k+1)T\big) - x(kT)}{T},$$

where T is the sample period. Then the state equation can be discretized as
$$\frac{x\big((k+1)T\big) - x(kT)}{T} = Ax(kT) + bu(kT)$$

or
$$x\big((k+1)T\big) = (I + TA)x(kT) + Tbu(kT). \qquad (14.32)$$

However, there is a more accurate way to discretize the equations. We rewrite (14.31) as
$$x(t) = e^{A(t-t_0)}x(t_0) + \int_{t_0}^t e^{A(t-\tau)}bu(\tau)d\tau.$$

Then we let $t = (k+1)T, t_0 = kT$ and assume the input is constant over each sample period of length T, that is,
$$u(t) = u(kT) \text{ for } kT \le t < (k+1)T. \qquad (14.33)$$

This results in
$$\begin{aligned}x\big((k+1)T\big) &= e^{AT}x(kT) + \int_{kT}^{(k+1)T} e^{A((k+1)T-\tau)}bu(kT)d\tau \\ &= e^{AT}x(kT) + \left(\int_{kT}^{(k+1)T} e^{A((k+1)T-\tau)}bd\tau\right)u(kT) \\ &= e^{AT}x(kT) - \left(\int_T^0 e^{As}bds\right)u(kT), \quad \begin{cases} s = (k+1)T - \tau \\ ds = -d\tau \end{cases} \\ &= e^{AT}x(kT) + \left(\int_0^T e^{As}bds\right)u(kT).\end{aligned}$$

With
$$x_{k+1} \triangleq x\big((k+1)T\big), \ x_k \triangleq x(kT)$$
$$u_k \triangleq u(kT)$$
$$F \triangleq e^{AT} \in \mathbb{R}^{n \times n}$$
$$g \triangleq \int_0^T e^{As}bds \in \mathbb{R}^n$$

we have the discrete-time statespace model
$$x_{k+1} = Fx_k + gu_k. \qquad (14.34)$$

Typically in a computer controlled system (14.33) holds so that the discrete-time model gives the *exact* values of (14.31) at the sampling instants $0, T, 2T, \ldots$.

Note that for T small
$$F \triangleq e^{AT} = I + AT + A^2\frac{T^2}{2!} + \cdots$$
$$\approx I + AT.$$

Also for T small
$$g \triangleq \int_0^T e^{A\tau} b d\tau = \int_0^T \left(I + A\tau + A^2\frac{\tau^2}{2!} + \cdots\right) b d\tau$$
$$= b\int_0^T d\tau + Ab\int_0^T \tau d\tau + A^2 b\int_0^T \frac{\tau^2}{2!}d\tau + \cdots$$
$$= bT + Ab\frac{T^2}{2} + A^2 b\frac{T^3}{6} + \cdots$$
$$\approx bT.$$

With T small the discrete-time model (14.34) reduces to (14.32).

Problems

Problem 1 *Statespace Form of a Mass–Spring–Damper System*

In Figure 14.14, the input to the mass–spring–damper system is the position u and the output is the position y of the mass m. The right-side of the spring k_1 does not show any mass so we will let m_p denote the mass of the piston to which the spring is attached. Later we will let $m_p \to 0$.

We let z denote the position of the piston (mass m_p). If $u = z$ then the spring is relaxed. The relative velocity between the piston and the cylinder of the damper is $\dot{z} - \dot{y}$.

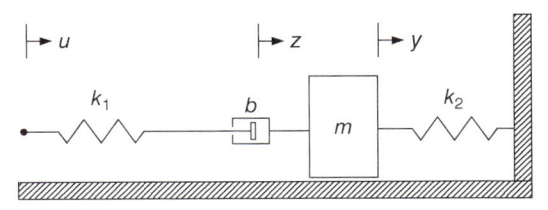

Figure 14.14. Mass–spring–damper system.

The equations of motion for m and m_p are
$$m\ddot{y} = -k_2 y + b(\dot{z} - \dot{y})$$
$$m_p\ddot{z} = -b(\dot{z} - \dot{y}) - k_1(z - u).$$

Remember that u is the input and y is the output. With zero initial conditions the Laplace transform of these equations are
$$(ms^2 + bs + k_2)Y(s) = bsZ(s)$$
$$(m_p s^2 + bs + k_1)Z(s) = bsY(s) + bk_1 U(s).$$

Eliminating $Z(s)$ gives
$$(m_p s^2 + bs + k_1)(ms^2 + bs + k_2)U(s) = b^2 s^2 Y(s) + bk_1 sU(s).$$

564   14   State Variables

Letting $m_p = 0$ and solving for $X(s)$ finally results in

$$Y(s) = \frac{bk_1 s}{(bs+k_1)(ms^2+bs+k_2)-b^2s^2} U(s) = \underbrace{\frac{bk_1 s}{bms^3+mk_1 s^2+(bk_1+bk_2)s+k_1 k_2}}_{\text{Transfer function}} U(s).$$

(a) Draw a simulation diagram for this system.

(b) Use the simulation diagram from part (a) to find a statespace representation of this system.

(c) Use the Routh–Hurwitz test to show

$$\frac{Y(s)}{U(s)} = \frac{\frac{k_1}{m} s}{s^3 + \frac{k_1}{b} s^2 + \frac{k_1+k_2}{m} s + \frac{k_1 k_2}{bm}}$$

is stable. With $U(s) = \frac{U_0}{s}$ show that $y(t) \to 0$ for any U_0. Note that $Y(s)/U(s)$ has a zero at $s = 0$.

(d) Can you explain physically why $y(\infty) = 0$ for every constant input $u(t) = U_0 u_s(t)$? That is, every constant input results in the output asymptotically going to zero. Hint: Note that $u(t) = U_0 u_s(t)$ also results in $z(\infty) = U_0$.

Problem 2 *Statespace Realization*
  Consider the transfer function

$$\frac{Y(s)}{U(s)} = \frac{2s+1}{s^2+15s+50}.$$

Find a statespace realization of this system in control canonical form. Show work which should include a simulation diagram.

Problem 3 *Statespace Realization of a PI Controller*
  Let

$$G_c(s) = \frac{U(s)}{E(s)} = K\frac{s+\alpha}{s} = K + \frac{\alpha K}{s}.$$

(a) Draw a simulation diagram for $G_c(s)$.

(b) Use the simulation diagram in part (a) to give the statespace equations for the PI Controller.

(c) With $t = kT$ and $\left.\frac{dx}{dt}\right|_{t=(k+1)T} \approx \frac{x((k+1)T)-x(kT)}{T}$ give a discrete-time version of the statespace model found in part (b).

Problem 4 *Properties of the Exponential Matrix*

(a) Show $Ae^{At} = e^{At}A$.

(b) Show $e^{At}e^{Bt} = e^{Bt}e^{At}$ if and only if $AB = BA$.

(c) Show that $(e^{At})^{-1} = e^{-At}$ for any matrix A.

Problem 5 *Solution of the Statespace Equations*

$$\frac{dx}{dt} = \begin{bmatrix} 0 & 1 \\ -1 & 0 \end{bmatrix} x, \quad x(0) = \begin{bmatrix} 0 \\ 1 \end{bmatrix}. \tag{14.35}$$

(a) With

$$A \triangleq \begin{bmatrix} 0 & 1 \\ -1 & 0 \end{bmatrix}$$

compute e^{At} using the Laplace transform method. Show work.

(b) What are the roots of $\det(sI - A) = 0$?

(c) Use your answer in part (a) to find the solution to (14.35). Show work!

(d) Does $e^{At} \to 0_{2\times 2}$ as $t \to \infty$? Just answer yes or no.

Problem 6 *Solution of the Statespace Equations*

$$\frac{dx}{dt} = \begin{bmatrix} 0 & 1 \\ 0 & 0 \end{bmatrix} x, \quad x(0) = \begin{bmatrix} 0 \\ 1 \end{bmatrix}. \tag{14.36}$$

(a) With

$$A \triangleq \begin{bmatrix} 0 & 1 \\ 0 & 0 \end{bmatrix}$$

compute e^{At} using the Laplace transform method. Show work.

(b) Use your answer in part (a) to find the solution to (14.36). Show work!

(c) Does $e^{At} \to 0_{2\times 2}$ as $t \to \infty$? Just answer yes or no.

Problem 7 *Transfer Function*

Suppose that the statespace model of a system is given by

$$\frac{dx}{dt} = \begin{bmatrix} 0 & 1 \\ 1 & 0 \end{bmatrix} x + \begin{bmatrix} 0 \\ -1 \end{bmatrix} u$$

$$y = \begin{bmatrix} 1 & 0 \end{bmatrix} x.$$

Compute the transfer function $Y(s)/U(s)$. Show work!

Problem 8 *Transfer Function*

Suppose that the statespace model of a system is given by

$$\frac{dx}{dt} = \begin{bmatrix} -15 & 1 \\ -50 & 0 \end{bmatrix} x + \begin{bmatrix} 0 \\ 2 \end{bmatrix} u$$

$$y = \begin{bmatrix} 1 & 0 \end{bmatrix} x.$$

Compute the transfer function $Y(s)/U(s)$. Show work!

Problem 9 *Solution of the Statespace Equations*

$$\frac{dx}{dt} = \begin{bmatrix} 0 & 1 \\ 6 & -1 \end{bmatrix} x, \quad x(0) = \begin{bmatrix} 0 \\ 1 \end{bmatrix} \quad (14.37)$$

(a) With

$$A \triangleq \begin{bmatrix} 0 & 1 \\ 6 & -1 \end{bmatrix}$$

compute e^{At} using the Laplace transform method. Show work!

(b) Use your answer in part (a) to find the solution to (14.37). Show work.

(c) Does $e^{At} \to 0_{2 \times 2}$ as $t \to \infty$? Just answer yes or no.

Problem 10 *Statespace Realization*
Consider the transfer function

$$\frac{Y(s)}{U(s)} = \frac{3s^2 + 2s + 1}{s^3 + 10s^2 + 15s + 20}.$$

(a) Find a statespace realization of this system in control canonical form. Show work which should include a simulation diagram.

(b) Use the Euler approximation to find a discrete-time representation of the statespace realization of part (a). Such a representation would be used to implement this system model in a software program.

Problem 11 *Statespace Realization of an Inverted Pendulum*
In Chapter 13 a transfer function model for the inverted pendulum was found to be

$$\frac{Y(s)}{U(s)} = \frac{-\kappa m g \ell}{s^2(s^2 - \alpha^2)},$$

where $Y(s) = X(s) + \left(\ell + \dfrac{J}{m\ell}\right)\theta(s)$ is the output ($X(s)$ is the cart position and $\theta(s)$ is the angle of the pendulum rod), $U(s)$ is the horizontal force on the cart, and

$$\alpha^2 = \frac{mg\ell(M+m)}{Mm\ell^2 + J(M+m)}, \quad \kappa = \frac{1}{Mm\ell^2 + J(M+m)}.$$

Compute a statespace representation of this transfer function in control canonical form. Be sure to draw the simulation diagram.

Problem 12 *Realization of a Non-Strictly Proper Controller*
Recall from Chapter 10 the output pole placement feedback controller for the inverted pendulum given in Figure 14.15 where (see Chapter 13)

$$\alpha^2 = \frac{mg\ell(M+m)}{Mm\ell^2 + J(M+m)}, \quad \kappa = \frac{1}{Mm\ell^2 + J(M+m)}.$$

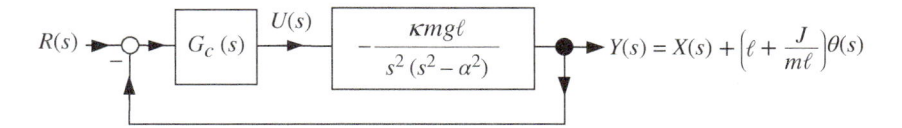

Figure 14.15. Output feedback controller for the inverted pendulum.

The minimum order controller that allows arbitrary pole placement has the form

$$G_c(s) = \frac{b_c(s)}{a_c(s)} = \frac{b_3 s^3 + b_2 s^2 + b_1 s + b_0}{s^3 + a_2 s^2 + a_1 s + a_0}.$$

(a) Find the statespace realization of $G_c(s)$ in control canonical form.
Hint: Add and subtract $b_3(s^3 + a_2 s^2 + a_1 s + a_0)$ to the numerator of $G_c(s)$ as follows

$$\frac{U(s)}{E(s)} = G_c(s)$$

$$= \frac{b_3 s^3 + b_2 s^2 + b_1 s + b_0 - b_3(s^3 + a_2 s^2 + a_1 s + a_0) + b_3(s^3 + a_2 s^2 + a_1 s + a_0)}{s^3 + a_2 s^2 + a_1 s + a_0}$$

$$= \frac{(b_2 - b_3 a_2)s^2 + (b_1 - b_3 a_1)s + (b_0 - b_3 a_0) + b_3(s^3 + a_2 s^2 + a_1 s + a_0)}{s^3 + a_2 s^2 + a_1 s + a_0}$$

$$= \underbrace{\frac{(b_2 - b_3 a_2)s^2 + (b_1 - b_3 a_1)s + (b_0 - b_3 a_0)}{s^3 + a_2 s^2 + a_1 s + a_0}}_{\text{Strictly proper}} + b_3.$$

That is,

$$U(s) = \frac{(b_2 - b_3 a_2)s^2 + (b_1 - b_3 a_1)s + (b_0 - b_3 a_0)}{s^3 + a_2 s^2 + a_1 s + a_0} E(s) + b_3 E(s)$$

is written as a strictly proper part and a direct feed through part.

(b) Give the Euler discretization of your answer in (a) which can then be use to implement $G_c(s)$ on a microprocessor.

Problem 13 *Computing the Transfer Function from the Statespace Representation*
A third-order statespace representation is given by

$$\frac{d}{dt}\begin{bmatrix} x_1 \\ x_2 \\ x_3 \end{bmatrix} = \underbrace{\begin{bmatrix} 0 & 1 & 0 \\ 0 & 0 & 1 \\ -a_0 & -a_1 & -a_2 \end{bmatrix}}_{A} \underbrace{\begin{bmatrix} x_1 \\ x_2 \\ x_3 \end{bmatrix}}_{x} + \underbrace{\begin{bmatrix} 0 \\ 0 \\ 1 \end{bmatrix}}_{b} e$$

$$u = \underbrace{\begin{bmatrix} b_0 - b_3 a_0 & b_1 - b_3 a_1 & b_2 - b_3 a_2 \end{bmatrix}}_{c} \begin{bmatrix} x_1 \\ x_2 \\ x_3 \end{bmatrix} + b_3 e.$$

Compute the transfer function $U(s)/E(s)$.

15
State Feedback

In Chapter 14 we discussed statespace models. We now show how these models can be used to design very effective controllers.

15.1 Two Examples

Let's do two examples to see how statespace models and state feedback are used in control systems.

Example 1 *Stabilization of an Inverted Pendulum*

Figure 15.1 shows the inverted pendulum on a cart. Recall the transfer function model of the inverted pendulum given by (see 13.13)

$$\theta(s) = -\frac{\kappa m \ell}{s^2 - \alpha^2} U(s), \tag{15.1}$$

where

$$\alpha^2 \triangleq \frac{mg\ell(M+m)}{Mm\ell^2 + J(M+m)}, \quad \kappa \triangleq \frac{1}{Mm\ell^2 + J(M+m)}.$$

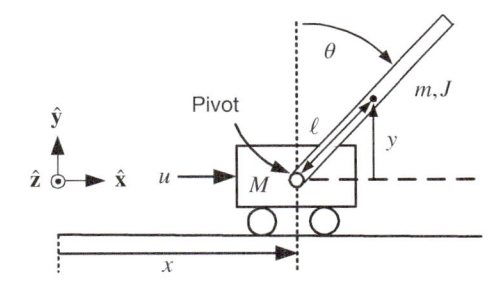

Figure 15.1. Inverted pendulum on a cart.

Let

$$x_1 = \theta$$
$$x_2 = \dot{\theta}$$

so that a statespace representation is given by

$$\frac{d}{dt}\begin{bmatrix} x_1 \\ x_2 \end{bmatrix} = \begin{bmatrix} 0 & 1 \\ \alpha^2 & 0 \end{bmatrix} \begin{bmatrix} x_1 \\ x_2 \end{bmatrix} + \begin{bmatrix} 0 \\ -\kappa m \ell \end{bmatrix} u. \tag{15.2}$$

An Introduction to System Modeling and Control, First Edition. John Chiasson.
© 2022 John Wiley & Sons, Inc. Published 2022 by John Wiley & Sons, Inc.
Companion website: www.wiley.com/go/chiasson/anintroductiontosystemmodelingandcontrol

570   15 State Feedback

Recall that the input u is the force on the cart and our objective is to keep $\theta = 0$ and $\dot{\theta} = 0$. Let's try a *state feedback* controller defined by

$$u = -\begin{bmatrix} k_1 & k_2 \end{bmatrix} \begin{bmatrix} \theta(t) \\ \dot{\theta}(t) \end{bmatrix} = -\underbrace{\begin{bmatrix} k_1 & k_2 \end{bmatrix}}_{k} \begin{bmatrix} x_1 \\ x_2 \end{bmatrix}. \tag{15.3}$$

Applying this feedback to the inverted pendulum system results in

$$\frac{d}{dt}\begin{bmatrix} x_1 \\ x_2 \end{bmatrix} = \begin{bmatrix} 0 & 1 \\ \alpha^2 & 0 \end{bmatrix}\begin{bmatrix} x_1 \\ x_2 \end{bmatrix} - \begin{bmatrix} 0 \\ -\kappa m \ell \end{bmatrix}\begin{bmatrix} k_1 & k_2 \end{bmatrix}\begin{bmatrix} x_1 \\ x_2 \end{bmatrix}$$

$$= \underbrace{\left(\begin{bmatrix} 0 & 1 \\ \alpha^2 & 0 \end{bmatrix} - \begin{bmatrix} 0 & 0 \\ -\kappa m \ell k_1 & -\kappa m \ell k_2 \end{bmatrix}\right)}_{A - bk}\begin{bmatrix} x_1 \\ x_2 \end{bmatrix}$$

$$= \begin{bmatrix} 0 & 1 \\ \alpha^2 + \kappa m \ell k_1 & \kappa m \ell k_2 \end{bmatrix}\begin{bmatrix} x_1 \\ x_2 \end{bmatrix}. \tag{15.4}$$

The solution to this system is

$$\begin{bmatrix} x_1(t) \\ x_2(t) \end{bmatrix} = e^{(A-bk)t}\begin{bmatrix} x_1(0) \\ x_2(0) \end{bmatrix}. \tag{15.5}$$

For any values of the initial conditions $x_1(0) = \theta(0), x_2(0) = \dot{\theta}(0)$, we want

$$\begin{bmatrix} x_1(t) \\ x_2(t) \end{bmatrix} \to \begin{bmatrix} 0 \\ 0 \end{bmatrix}. \tag{15.6}$$

This requires

$$e^{(A-bk)t} \to 0_{2\times 2} \text{ as } t \to \infty.$$

Let's look at how to choose k using a direct computation. Recall that we can compute $e^{(A-bk)t}$ using the Laplace transform as follows.

$$e^{(A-bk)t} = \mathcal{L}^{-1}\left\{(sI - (A-bk))^{-1}\right\} = \mathcal{L}^{-1}\left\{\left(s\begin{bmatrix} 1 & 0 \\ 0 & 1 \end{bmatrix} - \begin{bmatrix} 0 & 1 \\ \alpha^2 + \kappa m \ell k_1 & \kappa m \ell k_2 \end{bmatrix}\right)^{-1}\right\}$$

$$= \mathcal{L}^{-1}\left\{\begin{bmatrix} s & -1 \\ -(\alpha^2 + \kappa m \ell k_1) & s - \kappa m \ell k_2 \end{bmatrix}^{-1}\right\}$$

$$= \mathcal{L}^{-1}\left\{\frac{1}{\det(sI-(A-bk))}\begin{bmatrix} s - \kappa m \ell k_2 & 1 \\ \alpha^2 + \kappa m \ell k_1 & s \end{bmatrix}\right\}. \tag{15.7}$$

We compute

$$\det(sI - (A-bk)) = \det\begin{bmatrix} s & -1 \\ -(\alpha^2 + \kappa m \ell k_1) & s - \kappa m \ell k_2 \end{bmatrix} = s(s - \kappa m \ell k_2) - (\alpha^2 + \kappa m \ell k_1)$$

$$= s^2 - \kappa m \ell k_2 s - (\alpha^2 + \kappa m \ell k_1).$$

With $r_1 > 0$ and $r_2 > 0$, choose k_1 and k_2 so that

$$s^2 - \kappa m \ell k_2 s - \alpha^2 - \kappa m \ell k_1 = (s + r_1)(s + r_2) = s^2 + (r_1 + r_2)s + r_1 r_2.$$

That is, set
$$k_1 = -\frac{\alpha^2 + r_1 r_2}{\kappa m \ell}$$
$$k_2 = -\frac{r_1 + r_2}{\kappa m \ell}. \quad (15.8)$$

Then
$$e^{(A-bk)t}\bigg|_{\substack{k_1=-(r_1r_2+\alpha^2)/\kappa m\ell \\ k_2=-(r_1+r_2)/\kappa m\ell}} = \mathcal{L}^{-1}\left\{\frac{1}{(s+r_1)(s+r_2)}\begin{bmatrix} s - \kappa m\ell k_2 & 1 \\ \alpha^2 + \kappa m\ell k_1 & s \end{bmatrix}\right\}$$
$$= \mathcal{L}^{-1}\left\{\begin{bmatrix} \dfrac{s - \kappa m\ell k_2}{(s+r_1)(s+r_2)} & \dfrac{1}{(s+r_1)(s+r_2)} \\ \dfrac{\alpha^2 + \kappa m\ell k_1}{(s+r_1)(s+r_2)} & \dfrac{s}{(s+r_1)(s+r_2)} \end{bmatrix}\right\}.$$

After doing a partial fraction expansion, we have
$$e^{(A-bk)t}\bigg|_{\substack{k_1=-(r_1r_2+\alpha^2)/\kappa m\ell \\ k_2=-(r_1+r_2)/\kappa m\ell}} = \begin{bmatrix} A_{11}e^{-r_1 t} + B_{11}e^{-r_2 t} & A_{12}e^{-r_1 t} + B_{12}e^{-r_2 t} \\ A_{21}e^{-r_1 t} + B_{21}e^{-r_2 t} & A_{22}e^{-r_1 t} + B_{22}e^{-r_2 t} \end{bmatrix} \to 0_{2\times 2}. \quad (15.9)$$

The key observation here is that by our choice of k_1 and k_2 we have
$$\det(sI - (A - bk))\bigg|_{\substack{k_1=-(r_1r_2+\alpha^2)/\kappa m\ell \\ k_2=-(r_1+r_2)/\kappa m\ell}} = (s+r_1)(s+r_2).$$

This shows that every component of the 2×2 matrix $(sI - (A - bk))^{-1}$ has its poles at $-r_1, -r_2$. Consequently, the inverse Laplace transform of each component will go to 0 as $t \to \infty$. Summarizing, the row vector k must be chosen so that the roots of
$$\det(sI - (A - bk)) = 0$$
are in the *open* left half-plane. We then say that $A - bk \in \mathbb{R}^{2\times 2}$ is a *stable* matrix.

Example 2 *Trajectory Tracking by a Cart on a Track*

Recall from Chapter 13 the cart on a track system as shown in Figure 15.2. In this example we describe how to design a state feedback controller to force the cart to track a trajectory.

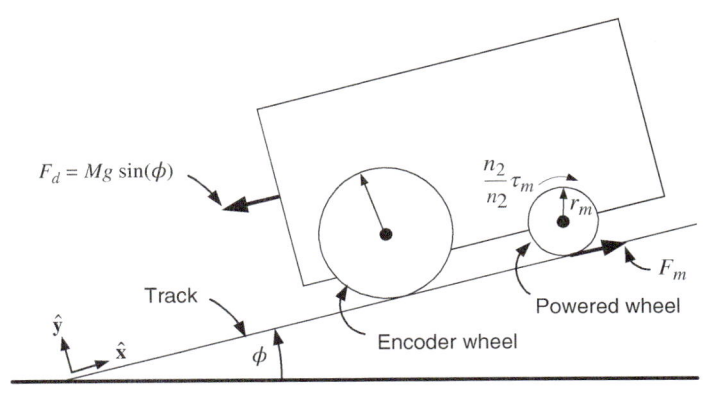

Figure 15.2. Cart on track system.

In Chapter 13 a transfer function model of this system was derived and is shown again in Figure 15.3 where

$$a_0 \triangleq \frac{K_b K_T \left(\frac{n_2}{n_1}\frac{1}{r_m}\right)^2}{RJ/r_m^2 + RM}, \quad b_0 \triangleq \frac{K_T \frac{n_2}{n_1}\frac{1}{r_m}}{RJ/r_m^2 + RM}, \quad K_D \triangleq \frac{R}{K_T \frac{n_2}{n_1}\frac{1}{r_m}}.$$

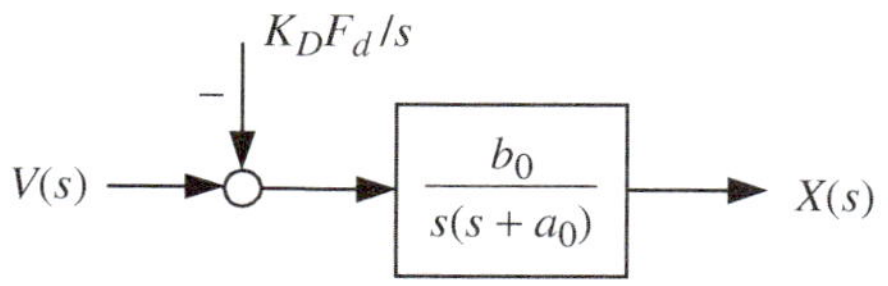

Figure 15.3. Transfer function model of the cart on a track system.

$V(t)$ is the voltage applied to the DC motor of the cart's powered wheel, x is the cart position along the track and $v = dx/dt$ is the cart's velocity. $F_d(t) = Mg\sin(\phi)$ is the disturbance force acting on the cart due to gravity as the cart moves up an incline at an angle ϕ. $K_D F_d$ is the equivalent voltage input to the motor that has the same effect on x, v as the disturbance force. A statespace model for the cart system is

$$\begin{aligned} \frac{dx}{dt} &= v \\ \frac{dv}{dt} &= -a_0 v + b_0 V(t) - \underbrace{b_0 K_D F_d}_{d}. \end{aligned} \tag{15.10}$$

The quantity $d \triangleq b_0 K_D F_d = Mg\sin(\phi)/(J/r_m^2 + M)$ is the (disturbance) acceleration of the cart body and wheels due to gravity. The objective is to design a state feedback trajectory tracking controller based on this statespace model. Specifically, we want the cart's position and speed to follow (track) a specified trajectory $(x_{ref}(t), v_{ref}(t))$. Setting the disturbance $d = 0$ for now and letting $u \triangleq V(t)$, the statespace model becomes

$$\begin{aligned} \frac{dx}{dt} &= v \\ \frac{dv}{dt} &= -a_0 v + b_0 u. \end{aligned} \tag{15.11}$$

In matrix form this is

$$\frac{d}{dt}\begin{bmatrix} x \\ v \end{bmatrix} = \begin{bmatrix} 0 & 1 \\ 0 & -a_0 \end{bmatrix}\begin{bmatrix} x \\ v \end{bmatrix} + \begin{bmatrix} 0 \\ b_0 \end{bmatrix} u.$$

Trajectory Design

We first design a trajectory for the cart to track. Denote the position reference trajectory by $x_{ref}(t)$ with $x_{ref}(0) = 0$ and $x_{ref}(t_f) = x_f$ where t_f is the final time and x_f is the final desired position. This is referred to as a *point-to-point* move. Consider the simple reference trajectory shown in Figure 15.4. To have a smooth trajectory for $x_{ref}(t)$ the speed and acceleration references are required to be continuous functions of time.

15.1 Two Examples

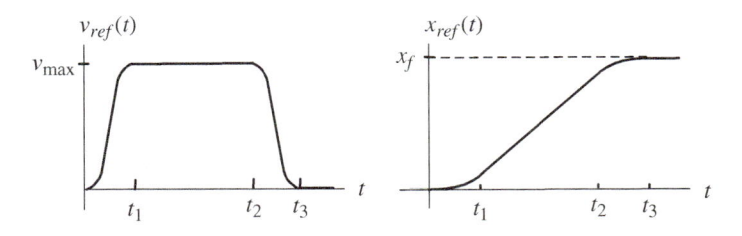

Figure 15.4. Reference speed and position profiles.

We require the reference trajectory for the speed to satisfy the following conditions.

$$v_{ref}(0) = 0 \qquad \dot{v}_{ref}(0) = 0$$
$$v_{ref}(t_1) = v_{max} \qquad \dot{v}_{ref}(t_1) = 0$$
$$v_{ref}(t) = v_{max} \qquad t_1 \leq t \leq t_2$$
$$v_{ref}(t_2) = v_{max} \qquad \dot{v}_{ref}(t_2) = 0$$
$$v_{ref}(t_3) = 0 \qquad \dot{v}_{ref}(t_3) = 0.$$

As shown on the left-side of Figure 15.4, the velocity reference $v_{ref}(t)$ is to be symmetric with respect to the midpoint of the trajectory. The requires $t_3 - t_2 = t_1$ so the time required to decelerate the motor is the same as the amount of time to accelerate with the final time given by $t_f \triangleq t_3 = t_1 + t_2$. Further, the velocity reference $v_{ref}(t)$ must satisfy

$$v_{ref}(t) = v_{ref}(t_3 - t) \text{ for } t_2 \leq t \leq t_3.$$

As the final position is to be x_f, we must have

$$\int_0^{t_3} v_{ref}(\tau) d\tau = x_f.$$

There are still many ways to define a reference trajectory and still satisfy all these conditions. Let's do this with a polynomial reference trajectory given by

$$v_{ref}(t) = c_1 t^2 + c_2 t^3 \text{ for } 0 \leq t \leq t_1.$$

This clearly satisfies $v_{ref}(0) = 0, \dot{v}_{ref}(0) = 0$. The conditions at t_1 become

$$v_{ref}(t_1) = c_1 t_1^2 + c_2 t_1^3 = v_{max}$$
$$\dot{v}_{ref}(t_1) = 2c_1 t_1 + 3c_2 t_1^2 = 0$$

or

$$\begin{bmatrix} t_1^2 & t_1^3 \\ 2t_1 & 3t_1^2 \end{bmatrix} \begin{bmatrix} c_1 \\ c_2 \end{bmatrix} = \begin{bmatrix} v_{max} \\ 0 \end{bmatrix}.$$

This has the unique solution

$$\begin{bmatrix} c_1 \\ c_2 \end{bmatrix} = \frac{1}{t_1^4} \begin{bmatrix} 3t_1^2 & -t_1^3 \\ -2t_1 & t_1^2 \end{bmatrix} \begin{bmatrix} v_{max} \\ 0 \end{bmatrix} = \begin{bmatrix} +3v_{max}/t_1^2 \\ -2v_{max}/t_1^3 \end{bmatrix}. \tag{15.12}$$

The velocity reference trajectory is then completely specified by

$$v_{ref}(t) = \begin{cases} c_1 t^2 + c_2 t^3, & 0 \le t \le t_1 \\ v_{\max}, & t_1 \le t \le t_2 \\ c_1(t_3 - t)^2 + c_2(t_3 - t)^3, & t_2 \le t \le t_3 \end{cases} \quad (15.13)$$

with c_1, c_2 given by (15.12).

We now specify the reference trajectory for the position. The distance traveled at time t_1 is

$$x_{ref}(t_1) = \int_0^{t_1} v_{ref}(\tau) d\tau = c_1 t_1^3/3 + c_2 t_1^4/4 = \frac{3v_{\max}}{t_1^2} \frac{t_1^3}{3} - \frac{2v_{\max}}{t_1^3} \frac{t_1^4}{4} = \frac{v_{\max} t_1}{2}.$$

By symmetry $\int_{t_2}^{t_3} v_{ref}(\tau) d\tau = v_{\max} t_1/2$. As the total distance traveled at time $t_f = t_3$ must be x_f, it follows that

$$x_f = \int_0^{t_3} v_{ref}(\tau) d\tau = x_{ref}(t_1) + v_{\max}(t_2 - t_1) + x_{ref}(t_1) = 2\frac{v_{\max} t_1}{2} + v_{\max}(t_2 - t_1)$$
$$= v_{\max} t_2.$$

This then puts a constraint on the choices of $v_{\max}$ and t_2. For example, if x_f and t_2 are specified, then $v_{\max} = x_f/t_2$.

The corresponding position reference is just the integral of the velocity reference. Consequently, with $v_{\max} = x_f/t_2$, we have (see Problem 6)

$$x_{ref}(t) = \begin{cases} c_1 t^3/3 + c_2 t^4/4, & 0 \le t \le t_1 \\ v_{\max} t_1/2 + v_{\max}(t - t_1), & t_1 \le t \le t_2 \\ v_{\max} t_2 - c_1(t_3 - t)^3/3 - c_2(t_3 - t)^4/4, & t_2 \le t \le t_3. \end{cases} \quad (15.14)$$

The acceleration reference $\alpha_{ref}(t)$ is just the derivative of the velocity reference and is given by

$$\alpha_{ref}(t) = \begin{cases} 2c_1 t + 3c_2 t^2, & 0 \le t \le t_1 \\ 0, & t_1 \le t \le t_2 \\ -2c_1(t_3 - t) - 3c_2(t_3 - t)^2, & t_2 \le t \le t_3. \end{cases}$$

Differentiating $\alpha_{ref}(t)$ gives the jerk reference $j_{ref}(t) \triangleq d\alpha_{ref}/dt$. $\alpha_{ref}(t)$ and $j_{ref}(t) \triangleq d\alpha_{ref}/dt$ are shown in Figure 15.5. Note that $j_{ref}(t)$ is discontinuous at $t = 0, t_1, t_2,$ and t_3.

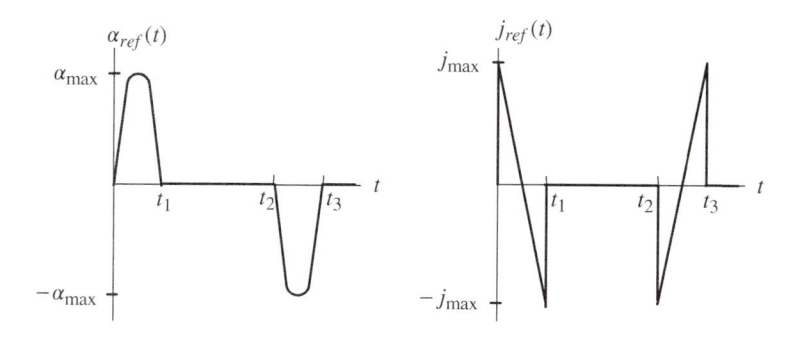

Figure 15.5. Acceleration reference and jerk reference.

By construction
$$\frac{dx_{ref}}{dt} = v_{ref}. \tag{15.15}$$

The input voltage reference u_{ref} must be chosen to satisfy
$$\underbrace{\frac{dv_{ref}}{dt}}_{a_{ref}} = -a_0 v_{ref} + b_0 u_{ref}. \tag{15.16}$$

That is, simply set
$$u_{ref}(t) \triangleq \frac{a_{ref}(t) + a_0 v_{ref}(t)}{b_0}. \tag{15.17}$$

For later reference note that for $t \geq t_f = t_3$ we have
$$\begin{bmatrix} x_{ref}(t) \\ v_{ref}(t) \\ a_{ref}(t) \\ j_{ref}(t) \end{bmatrix} = \begin{bmatrix} x_f \\ 0 \\ 0 \\ 0 \end{bmatrix} \tag{15.18}$$

and
$$u_{ref}(t) = 0. \tag{15.19}$$

Design of a State Feedback Tracking Controller

The system equations are
$$\begin{aligned} \frac{dx}{dt} &= v \\ \frac{dv}{dt} &= -a_0 v + b_0 u(t) - d, \end{aligned} \tag{15.20}$$

where $u(t)$ is the input voltage and $d \triangleq b_0 K_D F_d = Mg \sin(\phi)/(J/r_m^2 + M)$ is the (disturbance) acceleration due to gravity. The reference trajectory and reference input have been chosen to satisfy
$$\begin{aligned} \frac{dx_{ref}}{dt} &= v_{ref} \\ \frac{dv_{ref}}{dt} &= -a_0 v_{ref} + b_0 u_{ref}(t). \end{aligned} \tag{15.21}$$

Define the *error* state variables by
$$\epsilon_1(t) \triangleq x_{ref}(t) - x(t) \tag{15.22}$$
$$\epsilon_2(t) \triangleq v_{ref}(t) - v(t). \tag{15.23}$$

Subtracting (15.20) from (15.21) results in
$$\begin{aligned} \frac{d\epsilon_1}{dt} &= \epsilon_2 \\ \frac{d\epsilon_2}{dt} &= -a_0 \epsilon_2 + b_0 w + d, \end{aligned} \tag{15.24}$$

576   15 State Feedback

where
$$w(t) \triangleq u_{ref}(t) - u(t). \tag{15.25}$$

Finally we write
$$\frac{d}{dt}\begin{bmatrix}\epsilon_1\\\epsilon_2\end{bmatrix} = \underbrace{\begin{bmatrix}0 & 1\\0 & -a_0\end{bmatrix}}_{A}\begin{bmatrix}\epsilon_1\\\epsilon_2\end{bmatrix} + \underbrace{\begin{bmatrix}0\\b_0\end{bmatrix}}_{b}w + \begin{bmatrix}0\\1\end{bmatrix}d. \tag{15.26}$$

If we take $d = 0$ for now, (15.26) reduces to
$$\frac{d\epsilon}{dt} = A\epsilon + bw.$$

We want to find k such that the feedback
$$w = -\begin{bmatrix}k_1 & k_2\end{bmatrix}\begin{bmatrix}\epsilon_1\\\epsilon_2\end{bmatrix}$$

results in the closed-loop system
$$\frac{d\epsilon}{dt} = (A - bk)\epsilon$$

being stable, i.e.,
$$\epsilon(t) = e^{(A-bk)t}\begin{bmatrix}\epsilon_1(0)\\\epsilon_2(0)\end{bmatrix} \to \begin{bmatrix}0\\0\end{bmatrix}.$$

The state feedback control setup is shown in Figure 15.6.

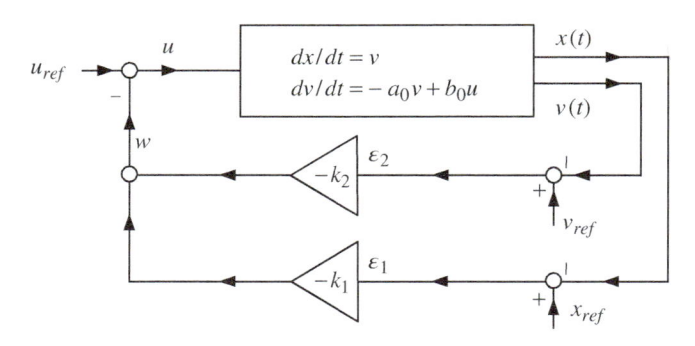

Figure 15.6. State feedback trajectory tracking controller.

We now determine k by a direct computation. Specifically we compute $e^{(A-bk)t}$ using the Laplace transform as follows.

$$e^{(A-bk)t} = \mathcal{L}^{-1}\left\{(sI - (A - bk))^{-1}\right\}$$
$$= \mathcal{L}^{-1}\left\{\left(s\begin{bmatrix}1 & 0\\0 & 1\end{bmatrix} - \left(\begin{bmatrix}0 & 1\\0 & -a_0\end{bmatrix} - \begin{bmatrix}0\\b_0\end{bmatrix}\begin{bmatrix}k_1 & k_2\end{bmatrix}\right)\right)^{-1}\right\}$$
$$= \mathcal{L}^{-1}\left\{\begin{bmatrix}s & -1\\b_0 k_1 & s + a_0 + b_0 k_2\end{bmatrix}^{-1}\right\}$$

$$= \mathcal{L}^{-1}\left\{\frac{1}{\det(sI - (A - bk))}\begin{bmatrix} s + a_0 + b_0k_2 & 1 \\ -b_0k_1 & s \end{bmatrix}\right\}.$$

Furthermore

$$\det(sI - (A - bk)) = \det\begin{bmatrix} s & -1 \\ b_0k_1 & s + a_0 + b_0k_2 \end{bmatrix} = s(s + a_0 + b_0k_2) + b_0k_1$$

$$= s^2 + (a_0 + b_0k_2)s + b_0k_1.$$

With $r_1 > 0$ and $r_2 > 0$, choose k_1 and k_2 so that

$$s^2 + (a_0 + b_0k_2)s + b_0k_1 = (s + r_1)(s + r_2) = s^2 + (r_1 + r_2)s + r_1r_2.$$

That is, set

$$k_1 = \frac{r_1 r_2}{b_0}$$
$$k_2 = \frac{r_1 + r_2 - a_0}{b_0}. \tag{15.27}$$

Then

$$e^{(A-bk)t}\bigg|_{\substack{k_1 = r_1r_2/b_0 \\ k_2 = (r_1+r_2-a_0)/b_0}} = \mathcal{L}^{-1}\left\{\frac{1}{(s+r_1)(s+r_2)}\begin{bmatrix} s + a_0 + b_0k_2 & 1 \\ -b_0k_1 & s \end{bmatrix}\right\}$$

$$= \mathcal{L}^{-1}\left\{\begin{bmatrix} \dfrac{s + a_0 + b_0k_2}{(s+r_1)(s+r_2)} & \dfrac{1}{(s+r_1)(s+r_2)} \\ \dfrac{-b_0k_1}{(s+r_1)(s+r_2)} & \dfrac{s}{(s+r_1)(s+r_2)} \end{bmatrix}\right\}.$$

After a partial fraction expansion we see that

$$e^{(A-bk)t}\bigg|_{\substack{k_1=r_1r_2/b_0 \\ k_2=(r_1+r_2-a_0)/b_0}} = \begin{bmatrix} A_{11}e^{-r_1t} + B_{11}e^{-r_2t} & A_{12}e^{-r_1t} + B_{12}e^{-r_2t} \\ A_{21}e^{-r_1t} + B_{21}e^{-r_2t} & A_{22}e^{-r_1t} + B_{22}e^{-r_2t} \end{bmatrix} \to 0_{2\times 2}.$$

Again, the key observation here is that by the choice of k_1 and k_2 we have

$$\det(sI - (A - bk))\big|_{\substack{k_1=r_1r_2/b_0 \\ k_2=(r_1+r_2-a_0)/b_0}} = (s + r_1)(s + r_2).$$

Therefore every component of the 2×2 matrix $(sI - (A - bk))^{-1}$ has its poles at $-r_1$, and $-r_2$. Consequently, the inverse Laplace transform of each component will die out.

Summarizing, the row vector k must be chosen so that the roots of

$$\det(sI - (A - bk)) = 0$$

are in the *open* left half-plane resulting in $A - bk \in \mathbb{R}^{2\times 2}$ being a *stable* matrix. In Problem 3 a you are asked to simulate this state feedback trajectory tracking system.

15.2 General State Feedback Trajectory Tracking

Let's now set up the general state feedback trajectory tracking control problem. Consider the model of a system in statespace form given by

$$\frac{dx}{dt} = Ax + bu, \quad A \in \mathbb{R}^{n \times n}, \quad b \in \mathbb{R}^n. \tag{15.28}$$

Let x_d be a state trajectory and u_d be a reference input designed to satisfy these system equations, that is,

$$\frac{dx_d}{dt} = Ax_d + bu_d. \tag{15.29}$$

Define the error state as

$$\epsilon \triangleq x_d - x$$

and the input error by

$$w \triangleq u_d - u.$$

Subtracting (15.28) from (15.29) results in

$$\frac{d}{dt}(x_d - x) = A(x_d - x) + b(u_d - u)$$

or

$$\frac{d}{dt}\epsilon = A\epsilon + bw. \tag{15.30}$$

Using the state feedback $w = -k\epsilon$ we have

$$\frac{d}{dt}\epsilon = (A - bk)\epsilon. \tag{15.31}$$

A block diagram of this feedback system is shown in Figure 15.7.

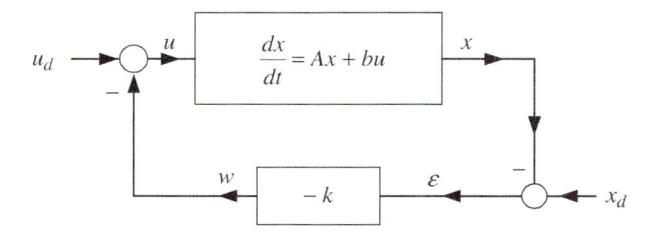

Figure 15.7. General setup for state feedback trajectory tracking.

The solution to (15.31) is

$$\begin{bmatrix} \epsilon(t) \end{bmatrix} = \begin{bmatrix} e^{(A-bk)t} \end{bmatrix} \begin{bmatrix} \epsilon(0) \end{bmatrix}.$$

The key to this feedback controller working is to be able to choose k such that

$$e^{(A-bk)t} \to 0_{n \times n}.$$

As in the aforementioned examples, this holds if and only if k can be chosen such that the roots of
$$\det(sI - (A - bk)) = 0 \tag{15.32}$$
are in the open left half-plane.

Remark If we let $r(t) \triangleq u_d(t) + kx_d(t)$ then the block diagram of Figure 15.7 is equivalent to the block diagram of Figure 15.8.

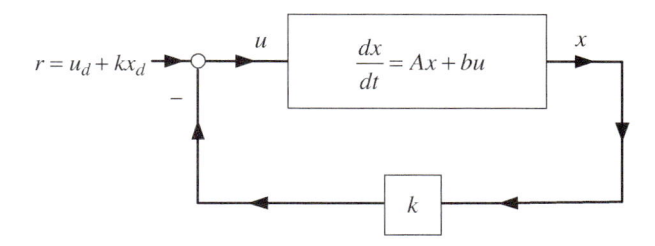

Figure 15.8. Equivalent setup for state feedback trajectory tracking.

15.3 Matrix Inverses and the Cayley–Hamilton Theorem

Before going further with state feedback we need to discuss matrix inverses and the Cayley–Hamilton theorem which are used in the remainder of this chapter.

Matrix Inverse

We review how to find the inverse of a matrix by computing its adjoint matrix and its determinant. The procedure is given without proof. Let $A \in \mathbb{R}^{3\times 3}$ be given by

$$A = \begin{bmatrix} a_{11} & a_{12} & a_{13} \\ a_{21} & a_{22} & a_{23} \\ a_{31} & a_{32} & a_{33} \end{bmatrix}. \tag{15.33}$$

Its corresponding *sign* matrix is defined by

$$S \triangleq \begin{bmatrix} 1 & -1 & 1 \\ -1 & 1 & -1 \\ 1 & -1 & 1 \end{bmatrix} \in \mathbb{R}^{3\times 3}. \tag{15.34}$$

That is, the (i,j) component s_{ij} of S is defined to be $s_{ij} = (-1)^{i+j}$. The *cofactor* matrix is defined by

$$cof(A) \triangleq \begin{bmatrix} b_{11} & b_{12} & b_{13} \\ b_{21} & b_{22} & b_{23} \\ b_{31} & b_{32} & b_{33} \end{bmatrix} \in \mathbb{R}^{3\times 3}, \tag{15.35}$$

where

$$b_{11} = \underbrace{+1}_{s_{11}} \cdot \det\begin{bmatrix} a_{22} & a_{23} \\ a_{32} & a_{33} \end{bmatrix}, \quad b_{12} = \underbrace{-1}_{s_{12}} \cdot \det\begin{bmatrix} a_{21} & a_{23} \\ a_{31} & a_{33} \end{bmatrix}, \quad b_{13} = \underbrace{+1}_{s_{13}} \cdot \det\begin{bmatrix} a_{21} & a_{22} \\ a_{31} & a_{32} \end{bmatrix}$$

$$b_{21} = \underbrace{-1}_{s_{21}} \cdot \det\begin{bmatrix} a_{12} & a_{13} \\ a_{32} & a_{33} \end{bmatrix}, \quad b_{22} = \underbrace{+1}_{s_{22}} \cdot \det\begin{bmatrix} a_{11} & a_{13} \\ a_{31} & a_{33} \end{bmatrix}, \quad b_{23} = \underbrace{-1}_{s_{23}} \cdot \det\begin{bmatrix} a_{11} & a_{12} \\ a_{31} & a_{32} \end{bmatrix}$$

$$b_{31} = \underbrace{+1}_{s_{31}} \cdot \det\begin{bmatrix} a_{12} & a_{13} \\ a_{22} & a_{23} \end{bmatrix}, \quad b_{32} = \underbrace{-1}_{s_{32}} \cdot \det\begin{bmatrix} a_{11} & a_{13} \\ a_{21} & a_{23} \end{bmatrix}, \quad b_{33} = \underbrace{+1}_{s_{33}} \cdot \det\begin{bmatrix} a_{11} & a_{12} \\ a_{21} & a_{22} \end{bmatrix}.$$

The *adjoint* matrix of A is simply the transpose of its cofactor matrix, that is,

$$adj(A) \triangleq cof^T(A) \in \mathbb{R}^{3\times 3}, \tag{15.36}$$

where $cof^T(A)$ denotes the transpose of $cof(A)$. Finally, the inverse of A is given by

$$A^{-1} = \frac{1}{\det A} adj(A). \tag{15.37}$$

The determinant A is found as follows (expanding along the first row of A).

$$\det A = \underbrace{1}_{s_{11}} \cdot a_{11} \det\begin{bmatrix} a_{22} & a_{23} \\ a_{32} & a_{33} \end{bmatrix} + \underbrace{-1}_{s_{12}} \cdot a_{12} \det\begin{bmatrix} a_{21} & a_{23} \\ a_{31} & a_{33} \end{bmatrix} + \underbrace{1}_{s_{13}} \cdot a_{13} \det\begin{bmatrix} a_{21} & a_{22} \\ a_{31} & a_{32} \end{bmatrix}.$$
$$\tag{15.38}$$

We can compute $\det A$ by expanding along *any* row or column. For example, expanding along the second column, we have

$$\det A = \underbrace{-1}_{s_{12}} \cdot a_{12} \det\begin{bmatrix} a_{21} & a_{23} \\ a_{31} & a_{33} \end{bmatrix} + \underbrace{1}_{s_{22}} \cdot a_{22} \det\begin{bmatrix} a_{11} & a_{13} \\ a_{31} & a_{33} \end{bmatrix} + \underbrace{-1}_{s_{32}} \cdot a_{32} \det\begin{bmatrix} a_{11} & a_{13} \\ a_{21} & a_{23} \end{bmatrix}.$$
$$\tag{15.39}$$

- We are assuming that $\det(A) \neq 0$. If $\det(A) = 0$ then A does *not* have an inverse.
- Though this approach was done for a 3×3 matrix, the procedure will work for any matrix $A \in \mathbb{R}^{n \times n}$.
- This approach for finding the inverse of a matrix is *not* efficient for (and should not be used to) numerically computing the inverse. The key use of this approach is for *theoretical* purposes.

Example 3 *Inverse of a Matrix*

Let
$$A = \begin{bmatrix} 0 & 1 & 0 \\ 0 & 0 & 1 \\ -\alpha_0 & -\alpha_1 & -\alpha_2 \end{bmatrix}$$

and let's compute the inverse of $sI - A$. We have

$$sI - A = s\begin{bmatrix} 1 & 0 & 0 \\ 0 & 1 & 0 \\ 0 & 0 & 1 \end{bmatrix} - \begin{bmatrix} 0 & 1 & 0 \\ 0 & 0 & 1 \\ -\alpha_0 & -\alpha_1 & -\alpha_2 \end{bmatrix} = \begin{bmatrix} s & -1 & 0 \\ 0 & s & -1 \\ \alpha_0 & \alpha_1 & s+\alpha_2 \end{bmatrix}$$

and its determinant is (expanding along the first column of $sI - A$)

$$\det(sI - A) = \underbrace{(1)}_{s_{11}} (s) \det\begin{bmatrix} s & -1 \\ \alpha_1 & s+\alpha_2 \end{bmatrix} + \underbrace{(-1)}_{s_{21}}(0) \det\begin{bmatrix} -1 & 0 \\ \alpha_1 & s+\alpha_2 \end{bmatrix}$$

15.3 Matrix Inverses and the Cayley–Hamilton Theorem

$$+ \underbrace{(1)}_{s_{31}} (\alpha_0) \det \begin{bmatrix} -1 & 0 \\ s & -1 \end{bmatrix}$$

$$= s^3 + \alpha_2 s^2 + \alpha_1 s + \alpha_0.$$

To compute the adjoint of $sI - A$ we first compute the components of the cofactor matrix of $sI - A$. These are

$$b_{11} = +1 \cdot \det \begin{bmatrix} s & -1 \\ \alpha_1 & s + \alpha_2 \end{bmatrix}, \quad b_{12} = -1 \cdot \det \begin{bmatrix} 0 & -1 \\ \alpha_0 & s + \alpha_2 \end{bmatrix}, \quad b_{13} = +1 \cdot \det \begin{bmatrix} 0 & s \\ \alpha_0 & \alpha_1 \end{bmatrix}$$

$$b_{21} = -1 \cdot \det \begin{bmatrix} -1 & 0 \\ \alpha_1 & s + \alpha_2 \end{bmatrix}, \quad b_{22} = +1 \cdot \det \begin{bmatrix} s & 0 \\ \alpha_0 & s + \alpha_2 \end{bmatrix}, \quad b_{23} = -1 \cdot \det \begin{bmatrix} s & -1 \\ \alpha_0 & \alpha_1 \end{bmatrix}$$

$$b_{31} = +1 \cdot \det \begin{bmatrix} -1 & 0 \\ s & -1 \end{bmatrix}, \quad b_{32} = -1 \cdot \det \begin{bmatrix} s & 0 \\ 0 & -1 \end{bmatrix}, \quad b_{33} = +1 \cdot \det \begin{bmatrix} s & -1 \\ 0 & s \end{bmatrix}.$$

The cofactor matrix is then

$$\text{cof}(sI - A) = \begin{bmatrix} s^2 + \alpha_2 s + \alpha_1 & -\alpha_0 & -\alpha_0 s \\ s + \alpha_2 & s^2 + \alpha_2 s & -\alpha_1 s - \alpha_0 \\ 1 & s & s^2 \end{bmatrix}.$$

The adjoint matrix is then

$$\text{adj}(sI - A) = \text{cof}^T(sI - A) = \begin{bmatrix} s^2 + \alpha_2 s + \alpha_1 & s + \alpha_2 & 1 \\ -\alpha_0 & s^2 + \alpha_2 s & s \\ -\alpha_0 s & -\alpha_1 s - \alpha_0 & s^2 \end{bmatrix}.$$

Finally, the inverse of $sI - A$ is given by

$$(sI - A)^{-1} = \frac{1}{\det(sI - A)} \text{adj}(sI - A)$$

$$= \frac{1}{s^3 + \alpha_2 s^2 + \alpha_1 s + \alpha_0} \begin{bmatrix} s^2 + \alpha_2 s + \alpha_1 & s + \alpha_2 & 1 \\ -\alpha_0 & s^2 + \alpha_2 s & s \\ -\alpha_0 s & -\alpha_1 s - \alpha_0 & s^2 \end{bmatrix}.$$

Remark We can also write $(sI - A)^{-1}$ in the form

$$(sI - A)^{-1} = \frac{1}{s^3 + \alpha_2 s^2 + \alpha_1 s + \alpha_0}$$

$$\times \left(\underbrace{\begin{bmatrix} 1 & 0 & 0 \\ 0 & 1 & 0 \\ 0 & 0 & 1 \end{bmatrix}}_{N_2} s^2 + \underbrace{\begin{bmatrix} \alpha_2 & 1 & 0 \\ 0 & \alpha_2 & 1 \\ -\alpha_0 & -\alpha_1 & 0 \end{bmatrix}}_{N_1} s + \underbrace{\begin{bmatrix} \alpha_1 & \alpha_2 & 1 \\ -\alpha_0 & 0 & 0 \\ 0 & -\alpha_0 & 0 \end{bmatrix}}_{N_0} \right)$$

$$= \frac{1}{\det(sI - A)} (N_2 s^2 + N_1 s + N_0), \quad N_i \in \mathbb{R}^{3 \times 3}.$$

Theorem 1 *Properties of Matrix Inverses and Determinants*

(1) Let $A \in \mathbb{R}^{n \times n}, B \in \mathbb{R}^{n \times n}$ be two matrices. Then

$$\det(AB) = \det(A)\det(B).$$

(2) Let $A \in \mathbb{R}^{n \times n}, B \in \mathbb{R}^{n \times n}$ be invertible, that is, $\det(A) \neq 0, \det(B) \neq 0$. Then

$$(AB)^{-1} = B^{-1}A^{-1}.$$

(3) Let $A \in \mathbb{R}^{n \times n}$ with $\det(A) \neq 0$. Then

$$\det(A^{-1}) = \frac{1}{\det A}.$$

Proof.
(1) Omitted.
(2) $(AB)(B^{-1}A^{-1}) = A(BB^{-1})A^{-1} = AIA^{-1} = AA^{-1} = I$ showing that $B^{-1}A^{-1}$ is the inverse of AB.
(3) As $AA^{-1} = I$ it follows that $(\det A)(\det A^{-1}) = \det I = 1$ or $\det(A^{-1}) = 1/\det A$. ■

Cayley–Hamilton Theorem

Let $A \in \mathbb{R}^{3 \times 3}$ be a matrix with

$$\det(sI - A) = s^3 + \alpha_2 s^2 + \alpha_1 s + \alpha_0. \tag{15.40}$$

Then

$$(sI - A)^{-1} = \frac{1}{\det(sI - A)} adj(sI - A), \tag{15.41}$$

where $adj(sI - A)$ is the adjoint of $sI - A$. The adjoint $adj(sI - A)$ may be written in the form

$$adj(sI - A) = N_2 s^2 + N_1 s + N_0, \quad N_i \in \mathbb{R}^{3 \times 3}. \tag{15.42}$$

Multiply both sides of (15.41) by $sI - A$ to obtain

$$I = (sI - A)\frac{1}{\det(sI - A)} adj(sI - A).$$

Multiplying this last expression through by $\det(sI - A)$ we have

$$\det(sI - A)I = (sI - A)adj(sI - A)$$

or

$$s^3 I + \alpha_2 s^2 I + \alpha_1 sI + \alpha_0 I = (sI - A)(N_2 s^2 + N_1 s + N_0)$$
$$= s^3 N_2 + (N_1 - AN_2)s^2 + (N_0 - AN_1)s - AN_0. \tag{15.43}$$

15.3 Matrix Inverses and the Cayley–Hamilton Theorem

Equating powers of s gives
$$I = N_2$$
$$\alpha_2 I = N_1 - AN_2$$
$$\alpha_1 I = N_0 - AN_1$$
$$\alpha_0 I = -AN_0.$$

Then
$$\alpha_0 I = -AN_0 = -A(\alpha_1 I + AN_1)$$
$$= -\alpha_1 A - A^2 N_1$$
$$= -\alpha_1 A - A^2(\alpha_2 I + AN_2)$$
$$= -\alpha_1 A - \alpha_2 A^2 - A^3.$$

Rearranging this last line gives the *Cayley–Hamilton* theorem, i.e.,
$$A^3 + \alpha_2 A^2 + \alpha_1 A + \alpha_0 I = 0_{3\times 3}.$$

In other words, in the equation $\det(sI - A) = s^3 + \alpha_2 s^2 + \alpha_1 s + \alpha_0 = 0$, we replace the scalar s^n by the matrix A^n ($s^0 = 1$ by the identity matrix I) and the scalar 0 by the zero matrix $0_{3\times 3}$ to have $A^3 + \alpha_2 A^2 + \alpha_1 A + \alpha_0 I = 0_{3\times 3}$.

Example 4 *Cayley–Hamilton Theorem*

Let
$$A = \begin{bmatrix} 0 & 1 & 0 \\ 0 & 0 & 1 \\ -\alpha_0 & -\alpha_1 & -\alpha_2 \end{bmatrix},$$

where it was shown above that
$$\det(sI - A) = s^3 + \alpha_2 s^2 + \alpha_1 s + \alpha_0.$$

We compute

$$\begin{bmatrix} 0 & 1 & 0 \\ 0 & 0 & 1 \\ -\alpha_0 & -\alpha_1 & -\alpha_2 \end{bmatrix}^3 + \alpha_2 \begin{bmatrix} 0 & 1 & 0 \\ 0 & 0 & 1 \\ -\alpha_0 & -\alpha_1 & -\alpha_2 \end{bmatrix}^2 + \alpha_1 \begin{bmatrix} 0 & 1 & 0 \\ 0 & 0 & 1 \\ -\alpha_0 & -\alpha_1 & -\alpha_2 \end{bmatrix} + \alpha_0 \begin{bmatrix} 1 & 0 & 0 \\ 0 & 1 & 0 \\ 0 & 0 & 1 \end{bmatrix}$$

$$= \begin{bmatrix} -\alpha_0 & -\alpha_1 & -\alpha_2 \\ \alpha_0\alpha_2 & \alpha_1\alpha_2 - \alpha_0 & \alpha_2^2 - \alpha_1 \\ \alpha_0\alpha_1 - \alpha_0\alpha_2^2 & \alpha_1^2 - \alpha_1\alpha_2^2 + \alpha_0\alpha_2 & -\alpha_2^3 + 2\alpha_1\alpha_2 - \alpha_0 \end{bmatrix}$$

$$+ \alpha_2 \begin{bmatrix} 0 & 0 & 1 \\ -\alpha_0 & -\alpha_1 & -\alpha_2 \\ \alpha_0\alpha_2 & \alpha_1\alpha_2 - \alpha_0 & \alpha_2^2 - \alpha_1 \end{bmatrix} + \alpha_1 \begin{bmatrix} 0 & 1 & 0 \\ 0 & 0 & 1 \\ -\alpha_0 & -\alpha_1 & -\alpha_2 \end{bmatrix} + \alpha_0 \begin{bmatrix} 1 & 0 & 0 \\ 0 & 1 & 0 \\ 0 & 0 & 1 \end{bmatrix}$$

$$= \begin{bmatrix} 0 & 0 & 0 \\ 0 & 0 & 0 \\ 0 & 0 & 0 \end{bmatrix}.$$

Definition 1 *Characteristic Polynomial and Eigenvalues*

Given a matrix $A \in \mathbb{R}^{n \times n}$, $\det(sI - A)$ is called the *characteristic polynomial* of A. The roots of

$$\det(sI - A) = 0$$

are called the *characteristic values* or *eigenvalues* of A.

15.4 Stabilization and State Feedback

Consider the third-order statespace model of a system given by

$$\frac{d}{dt}\begin{bmatrix} x_1 \\ x_2 \\ x_3 \end{bmatrix} = \underbrace{\begin{bmatrix} a_{11} & a_{12} & a_{13} \\ a_{21} & a_{22} & a_{23} \\ a_{31} & a_{32} & a_{33} \end{bmatrix}}_{A} \begin{bmatrix} x_1 \\ x_2 \\ x_3 \end{bmatrix} + \underbrace{\begin{bmatrix} b_1 \\ b_2 \\ b_3 \end{bmatrix}}_{b} u \tag{15.44}$$

with u chosen to be

$$u = -\underbrace{\begin{bmatrix} k_1 & k_2 & k_3 \end{bmatrix}}_{k} \begin{bmatrix} x_1 \\ x_2 \\ x_3 \end{bmatrix}. \tag{15.45}$$

The closed-loop system is then

$$\frac{dx}{dt} = (A - bk)x \tag{15.46}$$

with solution

$$x(t) = e^{(A-bk)t} x(0). \tag{15.47}$$

Recall that

$$\bigl(sI - (A - bk)\bigr)^{-1} = \frac{1}{\det\bigl(sI - (A - bk)\bigr)} adj\bigl(sI - (A - bk)\bigr). \tag{15.48}$$

The components of the adjoint matrix

$$adj\bigl(sI - (A - bk)\bigr) \in \mathbb{R}^{3 \times 3}$$

are *polynomials* in s. Each element of the 3×3 matrix $\bigl(sI - (A - bk)\bigr)^{-1}$ will have the same denominator given by $\det(sI - (A - bk))$. Consequently writing $e^{(A-bk)t}$ as

$$e^{(A-bk)t} = \mathcal{L}^{-1}\left\{\frac{1}{\det(sI - (A - bk))} adj\bigl(sI - (A - bk)\bigr)\right\}, \tag{15.49}$$

we see that each component of $e^{(A-bk)t}$ will go zero if and only if the roots of the characteristic polynomial

$$\det\bigl(sI - (A - bk)\bigr) = 0$$

are in the *open* left half-plane. If this is the case we say that $A - bk$ is a *stable* matrix. In other words,
$$e^{(A-bk)t} \to 0_{3\times 3}$$
if and only if $A - bk$ is a *stable* matrix.

Example 5 *Control Canonical Form*
Consider the statespace model
$$\frac{dx}{dt} = Ax + bu, \tag{15.50}$$
where
$$x = \begin{bmatrix} x_1 \\ x_2 \\ x_3 \end{bmatrix}, \quad A = \begin{bmatrix} 0 & 1 & 0 \\ 0 & 0 & 1 \\ -\alpha_0 & -\alpha_1 & -\alpha_2 \end{bmatrix}, \quad b = \begin{bmatrix} 0 \\ 0 \\ 1 \end{bmatrix} \tag{15.51}$$

and u is the scalar input. Note that the pair (A, b) are in control canonical form (see Section 14.2.1).
Now let the input be given by
$$u = -kx, \tag{15.52}$$
where
$$k \triangleq \begin{bmatrix} k_1 & k_2 & k_3 \end{bmatrix}. \tag{15.53}$$

The closed-loop system is
$$\frac{dx}{dt} = (A - bk)x$$
with solution
$$x(t) = e^{(A-bk)t}x(0).$$

We want to choose k such that
$$e^{(A-bk)t} \to 0_{3\times 3}.$$

The characteristic polynomial of $sI - (A - bk)$ is
$$\det\left(sI - (A - bk)\right) = \det \begin{bmatrix} s & -1 & 0 \\ 0 & s & -1 \\ k_1 + \alpha_0 & k_2 + \alpha_1 & s + k_3 + \alpha_2 \end{bmatrix}$$

$$= s \det \begin{bmatrix} s & -1 \\ k_2 + \alpha_1 & s + k_3 + \alpha_2 \end{bmatrix} - (-1) \det \begin{bmatrix} 0 & -1 \\ k_1 + \alpha_0 & s + k_3 + \alpha_2 \end{bmatrix}$$

$$= s\left(s(s + k_3 + \alpha_2) + k_2 + \alpha_1\right) + k_1 + \alpha_0$$

$$= s^3 + (k_3 + \alpha_2)s^2 + (k_2 + \alpha_1)s + k_1 + \alpha_0. \tag{15.54}$$

15 State Feedback

With $r_1 > 0, r_2 > 0$, and $r_3 > 0$, let's place the closed-loop poles at $-r_1, -r_2$, and $-r_3$. That is, we want

$$\det(sI - (A - bk)) = (s + r_1)(s + r_2)(s + r_3)$$
$$= s^3 + (r_1 + r_2 + r_3)s^2 + (r_1 r_2 + r_1 r_3 + r_2 r_3)s + r_1 r_2 r_3. \quad (15.55)$$

To do this simply set

$$k_3 + \alpha_2 = r_1 + r_2 + r_3$$
$$k_2 + \alpha_1 = r_1 r_2 + r_1 r_3 + r_2 r_3 \quad (15.56)$$
$$k_1 + \alpha_0 = r_1 r_2 r_3$$

or

$$k_3 = r_1 + r_2 + r_3 - \alpha_2$$
$$k_2 = r_1 r_2 + r_1 r_3 + r_2 r_3 - \alpha_1 \quad (15.57)$$
$$k_1 = r_1 r_2 r_3 - \alpha_0.$$

As A and b are in control canonical form it turns out that $(sI - (A - bk))^{-1}$ has a special form. To see this we compute

$$sI - (A - bk) = s \begin{bmatrix} 1 & 0 & 0 \\ 0 & 1 & 0 \\ 0 & 0 & 1 \end{bmatrix} - \left(\begin{bmatrix} 0 & 1 & 0 \\ 0 & 0 & 1 \\ -\alpha_0 & -\alpha_1 & -\alpha_2 \end{bmatrix} - \begin{bmatrix} 0 \\ 0 \\ 1 \end{bmatrix} \begin{bmatrix} k_1 & k_2 & k_3 \end{bmatrix} \right)$$

$$= s \begin{bmatrix} 1 & 0 & 0 \\ 0 & 1 & 0 \\ 0 & 0 & 1 \end{bmatrix} + \begin{bmatrix} 0 & -1 & 0 \\ 0 & 0 & -1 \\ \alpha_0 & \alpha_1 & \alpha_2 \end{bmatrix} + \begin{bmatrix} 0 & 0 & 0 \\ 0 & 0 & 0 \\ k_1 & k_2 & k_3 \end{bmatrix}$$

$$= s \begin{bmatrix} 1 & 0 & 0 \\ 0 & 1 & 0 \\ 0 & 0 & 1 \end{bmatrix} + \begin{bmatrix} 0 & -1 & 0 \\ 0 & 0 & -1 \\ k_1 + \alpha_0 & k_2 + \alpha_1 & k_3 + \alpha_2 \end{bmatrix}$$

$$= \begin{bmatrix} s & -1 & 0 \\ 0 & s & -1 \\ k_1 + \alpha_0 & k_2 + \alpha_1 & s + k_3 + \alpha_2 \end{bmatrix}. \quad (15.58)$$

Some tedious calculations show that

$$adj(sI - (A - bk)) = adj \begin{bmatrix} s & -1 & 0 \\ 0 & s & -1 \\ k_1 + \alpha_0 & k_2 + \alpha_1 & s + k_3 + \alpha_2 \end{bmatrix}$$

$$= \begin{bmatrix} s^2 + s(\alpha_2 + k_3) + \alpha_1 + k_2 & s + \alpha_2 + k_3 & 1 \\ -\alpha_0 - k_1 & s^2 + s(\alpha_2 + k_3) & s \\ -s\alpha_0 - sk_1 & -s(\alpha_1 + k_2) - \alpha_0 - k_1 & s^2 \end{bmatrix}.$$

15.4 Stabilization and State Feedback

Note the special form of the third column of $adj(sI - (A - bk))$. Finally, we have

$$(sI - (A - bk))^{-1} = \frac{1}{\det(sI - (A - bk))}$$
$$\times \begin{bmatrix} s^2 + s(\alpha_2 + k_3) + \alpha_1 + k_2 & s + \alpha_2 + k_3 & 1 \\ -\alpha_0 - k_1 & s^2 + s(\alpha_2 + k_3) & s \\ -s\alpha_0 - sk_1 & -s(\alpha_1 + k_2) - \alpha_0 - k_1 & s^2 \end{bmatrix}$$
$$= \frac{1}{(s+r_1)(s+r_2)(s+r_3)}$$
$$\times \begin{bmatrix} s^2 + s(\alpha_2 + k_3) + \alpha_1 + k_2 & s + \alpha_2 + k_3 & 1 \\ -\alpha_0 - k_1 & s^2 + s(\alpha_2 + k_3) & s \\ -s\alpha_0 - sk_1 & -s(\alpha_1 + k_2) - \alpha_0 - k_1 & s^2 \end{bmatrix}. \quad (15.59)$$

The inverse Laplace transform of each component (via partial fraction expansions) of this 3×3 matrix will be of the form $Ae^{-r_1 t} + Be^{-r_2 t} + Ce^{-r_3 t}$ and therefore go to 0 as $t \to \infty$. Consequently

$$e^{(A-bk)t} = \mathcal{L}^{-1}\left\{(sI - (A - bk))^{-1}\right\} \to 0_{3\times3}$$

and thus $x(t) \to 0$ for any initial condition $x(0)$.

Example 6 *Magnetic Levitation*

In Problem 9 of Chapter 13 (page 528) a third-order linear statespace model of the magnetic levitation system was found to be

$$\frac{d}{dt}\begin{bmatrix} i - i_{eq} \\ x - x_{eq} \\ v - v_{eq} \end{bmatrix} = \underbrace{\begin{bmatrix} -\frac{R}{L_0} & 0 & \frac{rL_1 \, i_{eq}}{L_{eq} \, x_{eq}^2} \\ 0 & 0 & 1 \\ -\frac{2g}{i_{eq}} & \frac{2g}{x_{eq}} & 0 \end{bmatrix}}_{A} \begin{bmatrix} i - i_{eq} \\ x - x_{eq} \\ v - v_{eq} \end{bmatrix} + \underbrace{\begin{bmatrix} \frac{1}{L_0} \\ 0 \\ 0 \end{bmatrix}}_{b} (u - u_{eq}), \quad (15.60)$$

where

$$\begin{bmatrix} i_{eq} \\ x_{eq} \\ v_{eq} \end{bmatrix} = \begin{bmatrix} x_{eq}\sqrt{\frac{2mg}{rL_1}} \\ x_{eq} \\ 0 \end{bmatrix}, \quad u_{eq} = Ri_{eq}.$$

To simplify notation let

$$\begin{bmatrix} z_1 \\ z_2 \\ z_3 \end{bmatrix} \triangleq \begin{bmatrix} i - i_{eq} \\ x - x_{eq} \\ v - v_{eq} \end{bmatrix}, \quad w = u - u_{eq} \quad (15.61)$$

and

$$A = \begin{bmatrix} a_{11} & 0 & a_{13} \\ 0 & 0 & 1 \\ a_{31} & a_{32} & 0 \end{bmatrix}, \quad b = \begin{bmatrix} b_1 \\ 0 \\ 0 \end{bmatrix}. \quad (15.62)$$

15 State Feedback

Note that A, b are *not* in control canonical form. With

$$w = -\begin{bmatrix} k_1 & k_2 & k_3 \end{bmatrix} \begin{bmatrix} z_1 \\ z_2 \\ z_3 \end{bmatrix} \tag{15.63}$$

the closed-loop system is

$$\frac{dz}{dt} = (A - bk)z.$$

Its solution is then

$$z(t) = e^{(A-bk)t} z(0),$$

where

$$e^{(A-bk)t} = \mathcal{L}^{-1}\left\{(sI - (A - bk))^{-1}\right\}.$$

We want to choose k such that $e^{(A-bk)t} \to 0_{3\times 3}$. As explained above, this means we must choose k such that the roots of

$$\det(sI - (A - bk)) = 0$$

are in the open left half-plane. We have

$$\det(sI - (A - bk)) = \det\begin{bmatrix} s + b_1 k_1 - a_{11} & b_1 k_2 & b_1 k_3 - a_{13} \\ 0 & s & -1 \\ -a_{31} & -a_{32} & s \end{bmatrix}$$

$$= (s + b_1 k_1 - a_{11})\det\begin{bmatrix} s & -1 \\ -a_{32} & s \end{bmatrix} + (-a_{31})\det\begin{bmatrix} b_1 k_2 & b_1 k_3 - a_{13} \\ s & -1 \end{bmatrix}$$

$$= (s + b_1 k_1 - a_{11})(s^2 - a_{32}) - a_{31}(-b_1 k_2 - s(b_1 k_3 - a_{13}))$$

$$= s^3 + (b_1 k_1 - a_{11})s^2 + (b_1 k_3 a_{31} - a_{13} a_{31} - a_{32})s$$

$$+ (a_{11} a_{32} - b_1 k_1 a_{32} + b_1 k_2 a_{31}). \tag{15.64}$$

With $r_1 > 0, r_2 > 0$, and $r_3 > 0$, choose k so that

$$\det(sI - (A - bk)) = (s + r_1)(s + r_2)(s + r_3)$$

$$= s^3 + (r_1 + r_2 + r_3)s^2 + (r_1 r_2 + r_1 r_3 + r_2 r_3)s + r_1 r_2 r_3. \tag{15.65}$$

That is, set

$$b_1 k_1 - a_{11} = r_1 + r_2 + r_3$$

$$b_1 k_3 a_{31} - a_{13} a_{31} - a_{32} = r_1 r_2 + r_1 r_3 + r_2 r_3$$

$$a_{11} a_{32} - b_1 k_1 a_{32} + b_1 k_2 a_{31} = r_1 r_2 r_3$$

to obtain

$$k_1 = \frac{r_1 + r_2 + r_3 + a_{11}}{b_1}$$

$$k_2 = \frac{r_1 r_2 r_3 - a_{11} a_{32} + b_1 k_1 a_{32}}{b_1 a_{31}} = \frac{r_1 r_2 r_3 + (r_1 + r_2 + r_3) a_{32}}{b_1 a_{31}} \tag{15.66}$$

$$k_3 = \frac{r_1 r_2 + r_1 r_3 + r_2 r_3 + a_{13} a_{31} + a_{32}}{b_1 a_{31}}.$$

With these values of the gains $k_1, k_2,$ and k_3 we have

$$\begin{aligned}
e^{(A-bk)t} &= \mathcal{L}^{-1}\left\{(sI - (A-bk))^{-1}\right\} \\
&= \mathcal{L}^{-1}\left\{\frac{1}{(s+r_1)(s+r_2)(s+r_3)} adj\,(sI - (A-bk))\right\} \\
&\to 0_{3\times 3}.
\end{aligned}$$

Thus

$$z(t) = \begin{bmatrix} i(t) - i_{eq} \\ x(t) - x_{eq} \\ v(t) - v_{eq} \end{bmatrix} = e^{(A-bk)t} \begin{bmatrix} i(0) - i_{eq} \\ x(0) - x_{eq} \\ v(0) - v_{eq} \end{bmatrix} \to \begin{bmatrix} 0 \\ 0 \\ 0 \end{bmatrix}.$$

As

$$w = u - u_{eq} = -kz$$

the actual voltage applied to the coil is

$$u(t) = -kz(t) + u_{eq}.$$

15.5 State Feedback and Disturbance Rejection

We now show how a state feedback controller can be designed to reject constant disturbances.

Example 7 Let's return to the controller for the cart on track system discussed earlier in Section 15.1 on page 575. There we had the statespace model

$$\begin{aligned}
\frac{dx}{dt} &= v \\
\frac{dv}{dt} &= -a_0 v + b_0 u(t) - d.
\end{aligned} \quad (15.67)$$

Here $u(t)$ is the input voltage and $d \triangleq b_0 K_D F_d = Mg\sin(\phi)/(J/r_m^2 + M)$ is the (disturbance) acceleration of the cart body and wheels due to gravity. A position and velocity reference trajectory $(x_{ref}(t), v_{ref}(t))$ and reference input $u_{ref}(t)$ were designed to satisfy

$$\begin{aligned}
\frac{dx_{ref}}{dt} &= v_{ref} \\
\frac{dv_{ref}}{dt} &= -a_0 v_{ref} + b_0 u_{ref}.
\end{aligned} \quad (15.68)$$

The *error* state variables

$$\begin{aligned}
\epsilon_1(t) &= x_{ref}(t) - x(t) \\
\epsilon_2(t) &= v_{ref}(t) - v(t)
\end{aligned}$$

satisfy (subtract (15.67) from (15.68))

$$\frac{d\epsilon_1}{dt} = \epsilon_2$$
$$\frac{d\epsilon_2}{dt} = -a_0\epsilon_2 + b_0 w + d, \tag{15.69}$$

where
$$w(t) \triangleq u_{ref}(t) - u(t). \tag{15.70}$$

In matrix form we write

$$\frac{d}{dt}\begin{bmatrix}\epsilon_1\\\epsilon_2\end{bmatrix} = \underbrace{\begin{bmatrix}0 & 1\\0 & -a_0\end{bmatrix}}_{A}\underbrace{\begin{bmatrix}\epsilon_1\\\epsilon_2\end{bmatrix}}_{\epsilon} + \underbrace{\begin{bmatrix}0\\b_0\end{bmatrix}}_{b}w + \underbrace{\begin{bmatrix}0\\1\end{bmatrix}}_{p}d. \tag{15.71}$$

As in Chapters 9 and 10, it is necessary to feed back the integral of the position error in order to have a controller that rejects constant disturbances. Specifically, the error system (15.71) is augmented with the new error state variable defined by

$$\epsilon_0(t) \triangleq \int_0^t \epsilon_1(\tau)d\tau = \int_0^t (x_{ref}(\tau) - x(\tau))d\tau. \tag{15.72}$$

The error system is now

$$\frac{d}{dt}\begin{bmatrix}\epsilon_0\\\epsilon_1\\\epsilon_2\end{bmatrix} = \underbrace{\begin{bmatrix}0 & 1 & 0\\0 & 0 & 1\\0 & 0 & -a_0\end{bmatrix}}_{A_a}\underbrace{\begin{bmatrix}\epsilon_0\\\epsilon_1\\\epsilon_2\end{bmatrix}}_{\epsilon_a} + \underbrace{\begin{bmatrix}0\\0\\b_0\end{bmatrix}}_{b_a}w + \underbrace{\begin{bmatrix}0\\0\\1\end{bmatrix}}_{p_a}d. \tag{15.73}$$

Note the pair (A_a, b_a) is *not* in control canonical form because b_0 is not equal to 1. Let $w(t)$ be the state feedback

$$w(t) = -\left(k_0 \int_0^t \epsilon_1(\tau)d\tau + k_1\epsilon_1(t) + k_2\epsilon_2(t)\right)$$
$$= -\underbrace{\begin{bmatrix}k_0 & k_1 & k_2\end{bmatrix}}_{k_a}\underbrace{\begin{bmatrix}\epsilon_0\\\epsilon_1\\\epsilon_2\end{bmatrix}}_{\epsilon_a}. \tag{15.74}$$

The setup for this state feedback control system is shown in Figure 15.9.
The closed-loop system is

$$\frac{d\epsilon_a}{dt} = A\epsilon_a - b_a k_a \epsilon_a + p_a d = (A_a - b_a k_a)\epsilon_a + p_a d. \tag{15.75}$$

Denote the Laplace transform of the error state variables by

$$E(s) = \begin{bmatrix}E_0(s)\\E_1(s)\\E_2(s)\end{bmatrix} \triangleq \begin{bmatrix}\mathcal{L}\{\epsilon_0(t)\}\\\mathcal{L}\{\epsilon_1(t)\}\\\mathcal{L}\{\epsilon_2(t)\}\end{bmatrix} = \mathcal{L}\{\epsilon_a(t)\}.$$

15.5 State Feedback and Disturbance Rejection

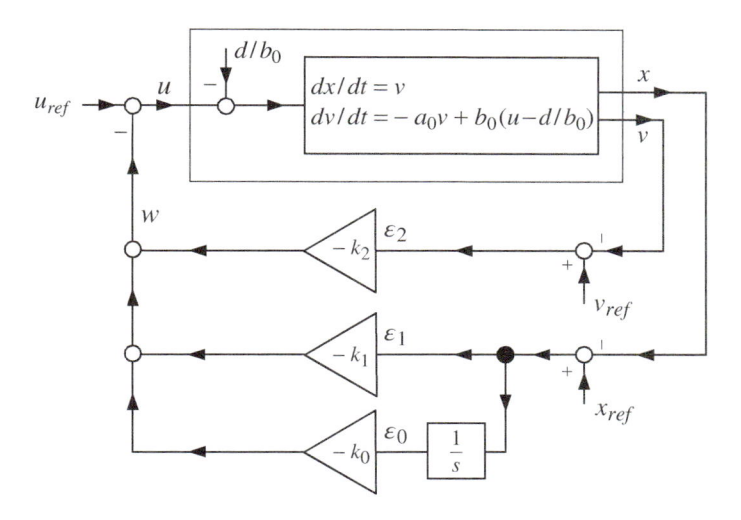

Figure 15.9. Trajectory tracking and disturbance rejection by state feedback.

Take the Laplace transform of (15.75) to obtain

$$E(s) = \left(sI - (A_a - b_a k_a)\right)^{-1} \epsilon_a(0) + \left(sI - (A_a - b_a k_a)\right)^{-1} p_a \frac{d}{s}. \tag{15.76}$$

With $r_1 > 0, r_2 > 0$, and $r_3 > 0$, we show below that k_a can be chosen such that

$$\det\left(sI - (A_a - b_a k_a)\right) = (s + r_1)(s + r_2)(s + r_3)$$

so that the poles are at $-r_1, -r_2$, and $-r_3$. First we note that $sE(s)$ given by

$$sE(s) = s\left(sI - (A_a - b_a k_a)\right)^{-1} \epsilon_a(0) + s\left(sI - (A_a - b_a k_a)\right)^{-1} p_a \frac{d}{s}$$
$$= s\left(sI - (A_a - b_a k_a)\right)^{-1} \epsilon_a(0) + \left(sI - (A_a - b_a k_a)\right)^{-1} p_a d$$

is *stable* so that by the final value theorem we have

$$\epsilon_a(\infty) = \lim_{s \to 0} sE(s) = -(A_a - b_a k_a)^{-1} p_a d.$$

It turns out that $\epsilon_1(\infty) = 0, \epsilon_2(\infty) = 0$, but $\epsilon_0(\infty) = \dfrac{d}{b_0 k_0}$, which can be found directly by substituting in the above expressions A_a, b_a, k_a, p_a (see Problem 4). However, let's do this by going back to (15.76). Rearranging (15.76) we may write

$$\underbrace{\begin{bmatrix} s & -1 & 0 \\ 0 & s & -1 \\ b_0 k_0 & b_0 k_1 & s + a_0 + b_0 k_2 \end{bmatrix}}_{sI - (A_a - b_a k_a)} \begin{bmatrix} E_0(s) \\ E_1(s) \\ E_2(s) \end{bmatrix} = \begin{bmatrix} \epsilon_0(0) \\ \epsilon_1(0) \\ \epsilon_2(0) \end{bmatrix} + \begin{bmatrix} 0 \\ 0 \\ d/s \end{bmatrix}. \tag{15.77}$$

The inverse of $sI - (A_a - b_a k_a)$ is

$$\left(sI - (A_a - b_a k_a)\right)^{-1} = \underbrace{\frac{1}{s^3 + (b_0 k_2 + a_0)s^2 + b_0 k_1 s + b_0 k_0}}_{\det\left(sI - (A_a - b_a k_a)\right)}$$

$$\times \underbrace{\begin{bmatrix} s^2 + (b_0 k_2 + a_0)s + b_0 k_1 & s + b_0 k_2 + a_0 & 1 \\ -b_0 k_0 & s^2 + (b_0 k_2 + a_0)s & s \\ -b_0 k_0 s & -(b_0 k_1 s + b_0 k_0) & s^2 \end{bmatrix}}_{\mathrm{adj}(sI - (A_a - b_a k_a))}.$$

Multiplying both sides of (15.77) on the left $\left(sI - (A_a - b_a k_a)\right)^{-1}$ we have

$$E_0(s) = \frac{\left(s^2 + (b_0 k_2 + a_0)s + b_0 k_1\right)\epsilon_0(0) + (s + b_0 k_2 + a_0)\epsilon_1(0) + \epsilon_2(0) + d/s}{s^3 + (b_0 k_2 + a_0)s^2 + b_0 k_1 s + b_0 k_0}$$

$$E_1(s) = \frac{-b_0 k_0 \epsilon_0(0) + \left(s^2 + (b_0 k_2 + a_0)s\right)\epsilon_1(0) + s\epsilon_2(0) + s(d/s)}{s^3 + (b_0 k_2 + a_0)s^2 + b_0 k_1 s + b_0 k_0} \tag{15.78}$$

$$E_2(s) = \frac{-b_0 k_0 s \epsilon_0(0) - (b_0 k_1 s + b_0 k_0)\epsilon_1(0) + s^2 \epsilon_2(0) + s^2(d/s)}{s^3 + (b_0 k_2 + a_0)s^2 + b_0 k_1 s + b_0 k_0}.$$

The characteristic polynomial of $A_a - b_a k_a$ given by

$$\det\left(sI - (A_a - b_a k_a)\right) = s^3 + (b_0 k_2 + a_0)s^2 + b_0 k_1 s + b_0 k_0$$

appears in the denominator of all three of these Laplace transforms. To make

$$\det\left(sI - (A_a - b_a k_a)\right) = (s + r_1)(s + r_2)(s + r_3)$$
$$= s^3 + (r_1 + r_2 + r_3)s^2 + (r_1 r_2 + r_1 r_3 + r_2 r_3)s + r_1 r_2 r_3$$

the gains k_0, k_1, and k_2 are chosen as

$$k_2 = \frac{r_1 + r_2 + r_3 - a_0}{b_0}$$
$$k_1 = \frac{r_1 r_2 + r_1 r_3 + r_2 r_3}{b_0} \tag{15.79}$$
$$k_0 = \frac{r_1 r_2 r_3}{b_0}.$$

With this choice of gains, the closed-loop poles are $p_1 = -r_1, p_2 = -r_2, p_3 = -r_3$. Assuming these poles are distinct, $E_1(s)$ in (15.78) becomes

$$E_1(s) = \frac{-b_0 k_0 \epsilon_0(0) + \left(s^2 + (b_0 k_2 + a_0)s\right)\epsilon_1(0) + s\epsilon_2(0) + d}{(s + r_1)(s + r_2)(s + r_3)}$$
$$= \frac{A_1}{s + r_1} + \frac{B_1}{s + r_2} + \frac{C_1}{s + r_3},$$

where A_1, B_1, and C_1 are constants. The inverse Laplace transform of $E_1(s)$ is then

$$x_{ref}(t) - x(t) = \epsilon_1(t) = A_1 e^{-r_1 t} + B_1 e^{-r_2 t} + C_1 e^{-r_3 t} \to 0.$$

Similarly, for some constants A_2, B_2, and C_2 we have

$$v_{ref}(t) - v(t) = \epsilon_2(t) = A_2 e^{-r_1 t} + B_2 e^{-r_2 t} + C_2 e^{-r_3 t} \to 0.$$

15.6 Similarity Transformations

The further the closed-loop poles are in the left half-plane, the faster $\epsilon_1(t) \to 0, \epsilon_2(t) \to 0$ and thus the faster $x(t) \to x_{ref}(t)$ and $v(t) \to v_{ref}(t)$. However, note that the further in the left half-plane the closed-loop poles are chosen, the larger r_1, r_2, r_3 and thus by (15.79) the larger the gains k_0, k_1, k_2. The difficulty with choosing large values for the feedback gains is that the resulting feedback voltage (see Figure 15.9)

$$V(t) = u(t) = u_{ref}(t) + \left(k_0 \int_0^t \epsilon_1(\tau) d\tau + k_1 \epsilon_1(t) + k_2 \epsilon_2(t) \right) \tag{15.80}$$

can be quite large and saturate the amplifier, even if the errors $\epsilon_0, \epsilon_1, \epsilon_2$ are small. It is up to the control engineer to choose the closed-loop pole locations, or equivalently, the gains. Typically, if the poles are not far enough in the left half-plane, the response is too slow. If the poles are chosen too far in the left half-plane, the amplifier will saturate. This procedure of varying the location of the closed-loop poles (thus varying the gains) is referred to as "tuning the system".

Finally, let's look at $\epsilon_0(t)$. We have

$$\lim_{t \to \infty} \epsilon_0(t) = \lim_{s \to 0} sE_0(s)$$
$$= \lim_{s \to 0} s \frac{(s^2 + (b_0 k_2 + a_0)s + b_0 k_1) \epsilon_0(0) + (s + b_0 k_2 + a_0)\epsilon_1(0) + \epsilon_2(0) + d/s}{s^3 + (b_0 k_2 + a_0)s^2 + b_0 k_1 s + b_0 k_0}$$
$$= \frac{d}{b_0 k_0}.$$

We then have

$$\lim_{t \to \infty} w(t) = \lim_{t \to \infty} -\left(k_0 \epsilon_0(t) + k_1 \epsilon_1(t) + k_2 \epsilon_2(t) \right) = -\frac{d}{b_0}.$$

However, recall that

$$w(t) \triangleq u_{ref}(t) - u(t)$$

and thus the final voltage $u(\infty) = \lim_{t \to \infty} u(t)$ into the motor is[1]

$$u(\infty) \triangleq \lim_{t \to \infty} \left(u_{ref}(t) - w(t) \right) = \lim_{t \to \infty} u_{ref}(t) - \lim_{t \to \infty} w(t) = 0 + \frac{d}{b_0}.$$

Figure 15.9 shows that $u(\infty) = d/b_0$ is exactly the voltage needed to cancel the constant disturbance d! In Problem 3b you are asked to simulate this state feedback trajectory tracking system.

15.6 Similarity Transformations

If you compare Example 5 with Example 6, Example 5 was easier to work with because the system was in *control canonical* form. Consequently, given a statespace model

$$\frac{d}{dt} \begin{bmatrix} x_1 \\ x_2 \\ x_3 \end{bmatrix} = \underbrace{\begin{bmatrix} a_{11} & a_{12} & a_{13} \\ a_{21} & a_{22} & a_{23} \\ a_{31} & a_{32} & a_{33} \end{bmatrix}}_{A} \begin{bmatrix} x_1 \\ x_2 \\ x_3 \end{bmatrix} + \underbrace{\begin{bmatrix} b_1 \\ b_2 \\ b_3 \end{bmatrix}}_{b} u \tag{15.81}$$

---

[1] See Eqs. (15.17)–(15.19) on page 575 to see that $u_{ref}(\infty) = 0$.

594   15   State Feedback

we would like to put it into the control canonical form

$$\frac{d}{dt}\begin{bmatrix} x_{c1} \\ x_{c2} \\ x_{c3} \end{bmatrix} = \underbrace{\begin{bmatrix} 0 & 1 & 0 \\ 0 & 0 & 1 \\ -\alpha_0 & -\alpha_1 & -\alpha_2 \end{bmatrix}}_{A_c}\begin{bmatrix} x_{c1} \\ x_{c2} \\ x_{c3} \end{bmatrix} + \underbrace{\begin{bmatrix} 0 \\ 0 \\ 1 \end{bmatrix}}_{b_c} u. \tag{15.82}$$

This is done using a statespace transformation of the form

$$x_c = Tx, T \in \mathbb{R}^{3 \times 3} \tag{15.83}$$

with

$$\det T \neq 0.$$

Let's first find the system equations in terms of the state x_c. Differentiating (15.83) with respect to t and using (15.81) we have

$$\frac{d}{dt}x_c = T\frac{d}{dt}x = T(Ax + bu) = TAx + Tbu. \tag{15.84}$$

As $\det T \neq 0$, T has an inverse so that

$$x = T^{-1}x_c.$$

Substituting this into (15.84) results in

$$\frac{d}{dt}x_c = \underbrace{TAT^{-1}}_{A_c}x_c + \underbrace{Tb}_{b_c}u. \tag{15.85}$$

The transformation T that takes A to TAT^{-1} is an example of a *similarity* transformation. Our goal is find T such that A_c, b_c are in the control canonical form given in (15.82). However, let's try a simpler transformation first.

Example 8 *Similarity Transformation*

Consider again the general third-order statespace model

$$\frac{d}{dt}\begin{bmatrix} x_1 \\ x_2 \\ x_3 \end{bmatrix} = \underbrace{\begin{bmatrix} a_{11} & a_{12} & a_{13} \\ a_{21} & a_{22} & a_{23} \\ a_{31} & a_{32} & a_{33} \end{bmatrix}}_{A}\begin{bmatrix} x_1 \\ x_2 \\ x_3 \end{bmatrix} + \underbrace{\begin{bmatrix} b_1 \\ b_2 \\ b_3 \end{bmatrix}}_{b} u. \tag{15.86}$$

Define

$$\mathcal{C} \triangleq \begin{bmatrix} b & Ab & A^2b \end{bmatrix} \in \mathbb{R}^{3 \times 3} \tag{15.87}$$

and suppose that $\det \mathcal{C} \neq 0$ so that $\mathcal{C}$ has an inverse. The matrix $\mathcal{C}$ is called the *controllability* matrix of the system (15.86). Consider the transformation

$$T_1 \triangleq \mathcal{C}^{-1} = \begin{bmatrix} b & Ab & A^2b \end{bmatrix}^{-1}. \tag{15.88}$$

With

$$x' \triangleq T_1 x \tag{15.89}$$

15.6 Similarity Transformations

the statespace model in the x' coordinate system is

$$\frac{d}{dt}x' = \underbrace{T_1 A T_1^{-1}}_{A'} x' + \underbrace{T_1 b}_{b'} u. \tag{15.90}$$

What do A', b' look like? We write

$$A' = T_1 A T_1^{-1}$$
$$b' = T_1 b$$

or

$$T_1^{-1} A' = A T_1^{-1}$$
$$T_1^{-1} b' = b.$$

Substituting the expression (15.88) for T_1 gives

$$\begin{bmatrix} b & Ab & A^2 b \end{bmatrix} A' = A \begin{bmatrix} b & Ab & A^2 b \end{bmatrix}$$
$$\begin{bmatrix} b & Ab & A^2 b \end{bmatrix} b' = b$$

or

$$\begin{bmatrix} b & Ab & A^2 b \end{bmatrix} A' = \begin{bmatrix} Ab & A^2 b & A^3 b \end{bmatrix}$$
$$\begin{bmatrix} b & Ab & A^2 b \end{bmatrix} b' = b. \tag{15.91}$$

Note that, by inspection, this last set of matrix equations requires A' and b' to have the form

$$A' = \begin{bmatrix} 0 & 0 & a'_{13} \\ 1 & 0 & a'_{23} \\ 0 & 1 & a'_{33} \end{bmatrix}, \quad b' = \begin{bmatrix} 1 \\ 0 \\ 0 \end{bmatrix},$$

where a'_{13}, a'_{23}, and a'_{33} must be found to satisfy $a'_{13} b + a'_{23} Ab + A^2 b = A^3 b$. We do this as follows. Let

$$\det(sI - A) = s^3 + \alpha_2 s^2 + \alpha_1 s + \alpha_0$$

be the characteristic polynomial of the matrix A. By the Cayley–Hamilton theorem we know that

$$A^3 + \alpha_2 A^2 + \alpha_1 A + \alpha_0 I = 0_{3 \times 3}.$$

Multiplying on the right by b and rearranging gives

$$A^3 b = -\alpha_0 b - \alpha_1 Ab - \alpha_2 A^2 b.$$

Thus, again by inspection of (15.91), we have

$$A' = \begin{bmatrix} 0 & 0 & -\alpha_0 \\ 1 & 0 & -\alpha_1 \\ 0 & 1 & -\alpha_2 \end{bmatrix}, \quad b' = \begin{bmatrix} 1 \\ 0 \\ 0 \end{bmatrix}. \tag{15.92}$$

The pair A', b' given in (15.92) is not quite in control canonical form. We next show how to transform A', b' to control canonical form. Note that by a direct calculation (or see Theorem 3) we have $\det(sI - A') = s^3 + \alpha_2 s^2 + \alpha_1 s + \alpha_0 = \det(sI - A)$.

596   15 State Feedback

Example 9 *Similarity Transformation*
Now consider a statespace model already in control canonical form, i.e.,

$$\frac{d}{dt}\underbrace{\begin{bmatrix} x_{c1} \\ x_{c2} \\ x_{c3} \end{bmatrix}}_{x_c} = \underbrace{\begin{bmatrix} 0 & 1 & 0 \\ 0 & 0 & 1 \\ -\alpha_0 & -\alpha_1 & -\alpha_2 \end{bmatrix}}_{A_c} \underbrace{\begin{bmatrix} x_{c1} \\ x_{c2} \\ x_{c3} \end{bmatrix}}_{x_c} + \underbrace{\begin{bmatrix} 0 \\ 0 \\ 1 \end{bmatrix}}_{b_c} u.$$

With $\mathcal{C}_c \triangleq \begin{bmatrix} b_c & A_c b_c & A_c^2 b_c \end{bmatrix} \in \mathbb{R}^{3 \times 3}$ define

$$T_2 \triangleq \mathcal{C}_c^{-1}.$$

We know that

$$\det(sI - A_c) = \det \begin{bmatrix} s & -1 & 0 \\ 0 & s & -1 \\ \alpha_0 & \alpha_1 & s + \alpha_2 \end{bmatrix} = s^3 + \alpha_2 s^2 + \alpha_1 s + \alpha_0.$$

As shown in Example 8 the coordinate transformation

$$x' \triangleq T_2 x_c$$

results in

$$\frac{d}{dt} x' = \underbrace{T_2 A_c T_2^{-1}}_{A'} x' + \underbrace{T_2 b_c}_{b'} u = \begin{bmatrix} 0 & 0 & -\alpha_0 \\ 1 & 0 & -\alpha_1 \\ 0 & 1 & -\alpha_2 \end{bmatrix} x' + \begin{bmatrix} 1 \\ 0 \\ 0 \end{bmatrix} u.$$

On the other hand, the inverse of this transformation given by

$$x_c = T_2^{-1} x' = \mathcal{C}_c x'$$

takes this system back to its original control canonical form as

$$\frac{d}{dt} x_c = T_2^{-1} \frac{d}{dt} x' = T_2^{-1}(A' x' + b' u)$$
$$= T_2^{-1}(T_2 A_c T_2^{-1} x' + T_2 b_c u)$$
$$= T_2^{-1}(T_2 A_c T_2^{-1}) T_2 x_c + T_2^{-1} T_2 b_c u$$
$$= A_c x_c + b_c u.$$

That is, if a system is in the form

$$\frac{d}{dt} x' = \begin{bmatrix} 0 & 0 & -\alpha_0 \\ 1 & 0 & -\alpha_1 \\ 0 & 1 & -\alpha_2 \end{bmatrix} x' + \begin{bmatrix} 1 \\ 0 \\ 0 \end{bmatrix} u,$$

then $x_c = T_2^{-1} x' = \mathcal{C}_c x'$ takes it into the *control canonical* form

$$\frac{d}{dt}\underbrace{\begin{bmatrix} x_{c1} \\ x_{c2} \\ x_{c3} \end{bmatrix}}_{x_c} = \underbrace{\begin{bmatrix} 0 & 1 & 0 \\ 0 & 0 & 1 \\ -\alpha_0 & -\alpha_1 & -\alpha_2 \end{bmatrix}}_{A_c} \underbrace{\begin{bmatrix} x_{c1} \\ x_{c2} \\ x_{c3} \end{bmatrix}}_{x_c} + \underbrace{\begin{bmatrix} 0 \\ 0 \\ 1 \end{bmatrix}}_{b_c} u.$$

15.6 Similarity Transformations

Theorem 2 *Transformation to Control Canonical Form*

Let a system be given by

$$\frac{d}{dt}\begin{bmatrix} x_1 \\ x_2 \\ x_3 \end{bmatrix} = \underbrace{\begin{bmatrix} a_{11} & a_{12} & a_{13} \\ a_{21} & a_{22} & a_{23} \\ a_{31} & a_{32} & a_{33} \end{bmatrix}}_{A} \begin{bmatrix} x_1 \\ x_2 \\ x_3 \end{bmatrix} + \underbrace{\begin{bmatrix} b_1 \\ b_2 \\ b_3 \end{bmatrix}}_{b} u$$

and define the *controllability matrix* to be

$$\mathcal{C} \triangleq \begin{bmatrix} b & Ab & A^2 b \end{bmatrix}.$$

Further, let

$$\det(sI - A) = s^3 + \alpha_2 s^2 + \alpha_1 s + \alpha_0.$$

If $\det \mathcal{C} \neq 0$ then the above system is transformed to the control canonical form

$$\frac{d}{dt}\underbrace{\begin{bmatrix} x_{c1} \\ x_{c2} \\ x_{c3} \end{bmatrix}}_{x_c} = \underbrace{\begin{bmatrix} 0 & 1 & 0 \\ 0 & 0 & 1 \\ -\alpha_0 & -\alpha_1 & -\alpha_2 \end{bmatrix}}_{A_c} \underbrace{\begin{bmatrix} x_{c1} \\ x_{c2} \\ x_{c3} \end{bmatrix}}_{x_c} + \underbrace{\begin{bmatrix} 0 \\ 0 \\ 1 \end{bmatrix}}_{b_c} u$$

by the transformation

$$x_c = \mathcal{C}_c \mathcal{C}^{-1} x,$$

where

$$\mathcal{C}_c \triangleq \begin{bmatrix} b_c & A_c b_c & A_c^2 b_c \end{bmatrix}.$$

Proof. By the two previous examples. ∎

Definition 2 *Similarity Transformation*

Let a linear statespace system be

$$\frac{dx}{dt} = Ax + bu, \quad A \in \mathbb{R}^{n \times n}, \quad b \in \mathbb{R}^n.$$

Using an invertible matrix $T \in \mathbb{R}^{n \times n}$, i.e., $\det(T) \neq 0$, consider the change of state variables

$$x' = Tx.$$

The statespace representation in terms of x' is then

$$\frac{dx'}{dt} = TAT^{-1} x' + Tbu.$$

The transformation of a matrix $A \in \mathbb{R}^{n \times n}$, according to

$$A \to TAT^{-1},$$

is called a *similarity* transformation.

598   15   State Feedback

Theorem 3 *Property of Similarity Transformations*

Let
$$A' = TAT^{-1}.$$

Then
$$\det(sI - A') = \det(sI - A).$$

Proof. We have $sI - A' = TsT^{-1} - TAT^{-1} = T(sI - A)T^{-1}$. Then

$$\det(sI - A') = \det(T(sI - A)T^{-1}) = \det T \det(sI - A) \det T^{-1} = \det(sI - A)$$

as $\det T \det T^{-1} = 1$. ∎

15.7   Pole Placement

Consider the system
$$\frac{dx_c}{dt} = A_c x_c + b_c u,$$

where A_c, b_c are in the control canonical form, i.e.,

$$A_c = \begin{bmatrix} 0 & 1 & 0 \\ 0 & 0 & 1 \\ -\alpha_0 & -\alpha_1 & -\alpha_2 \end{bmatrix}, \quad b_c = \begin{bmatrix} 0 \\ 0 \\ 1 \end{bmatrix}. \tag{15.93}$$

The characteristic polynomial of A_c is (expanding along the third column)

$$\det\left(s\begin{bmatrix} 1 & 0 & 0 \\ 0 & 1 & 0 \\ 0 & 0 & 1 \end{bmatrix} - \begin{bmatrix} 0 & 1 & 0 \\ 0 & 0 & 1 \\ -\alpha_0 & -\alpha_1 & -\alpha_2 \end{bmatrix}\right) = \det\begin{bmatrix} s & -1 & 0 \\ 0 & s & -1 \\ \alpha_0 & \alpha_1 & s+\alpha_2 \end{bmatrix}$$

$$= -(-1)\det\begin{bmatrix} s & -1 \\ \alpha_0 & \alpha_1 \end{bmatrix} + (s+\alpha_2)\det\begin{bmatrix} s & -1 \\ 0 & s \end{bmatrix}$$

$$= \alpha_1 s + \alpha_0 + (s+\alpha_2)s^2$$

$$= s^3 + \alpha_2 s^2 + \alpha_1 s + \alpha_0.$$

Also, with
$$k_c = \begin{bmatrix} k_{c1} & k_{c2} & k_{c3} \end{bmatrix}$$

and the feedback
$$u = -k_c x_c = -\begin{bmatrix} k_{c1} & k_{c2} & k_{c3} \end{bmatrix}\begin{bmatrix} x_{c1} \\ x_{c2} \\ x_{c3} \end{bmatrix},$$

the closed-loop system becomes
$$\frac{dx_c}{dt} = (A_c - b_c k_c)x_c.$$

15.7 Pole Placement

Explicitly, $A_c - b_c k_c$ is given by

$$A_c - b_c k_c = \begin{bmatrix} 0 & 1 & 0 \\ 0 & 0 & 1 \\ -\alpha_0 & -\alpha_1 & -\alpha_2 \end{bmatrix} - \begin{bmatrix} 0 \\ 0 \\ 1 \end{bmatrix} \begin{bmatrix} k_{c1} & k_{c2} & k_{c3} \end{bmatrix}$$

$$= \begin{bmatrix} 0 & 1 & 0 \\ 0 & 0 & 1 \\ -\alpha_0 - k_{c1} & -\alpha_1 - k_{c2} & -\alpha_2 - k_{c3} \end{bmatrix}.$$

The *closed-loop* characteristic polynomial is then

$$\det(sI - (A_c - b_c k_c)) = \det\left(s \begin{bmatrix} 1 & 0 & 0 \\ 0 & 1 & 0 \\ 0 & 0 & 1 \end{bmatrix} - \begin{bmatrix} 0 & 1 & 0 \\ 0 & 0 & 1 \\ -\alpha_0 - k_{c1} & -\alpha_1 - k_{c2} & -\alpha_2 - k_{c3} \end{bmatrix} \right)$$

$$= s^3 + (\alpha_2 + k_{c3})s^2 + s(\alpha_1 + k_{c2}) + \alpha_0 + k_{c1}.$$

Choosing

$$k_{c1} = \alpha_{d0} - \alpha_0$$
$$k_{c2} = \alpha_{d1} - \alpha_1$$
$$k_{c3} = \alpha_{d2} - \alpha_2$$

results in the closed-loop characteristic polynomial being

$$s^3 + \alpha_{d2}s^2 + \alpha_{d1}s + \alpha_{d0}.$$

For example, let $r_1 > 0, r_2 > 0, r_3 > 0$ and suppose we want

$$s^3 + \alpha_{d2}s^2 + \alpha_{d1}s + \alpha_{d0} = (s + r_1)(s + r_2)(s + r_3)$$

so the closed-loop poles are at $-r_1, -r_2, -r_3$. Then

$$(s + r_1)(s + r_2)(s + r_3) = s^3 + \underbrace{(r_1 + r_2 + r_3)}_{\alpha_{d2}} s^2 + \underbrace{(r_1 r_2 + r_1 r_3 + r_2 r_3)}_{\alpha_{d1}} s + \underbrace{r_1 r_2 r_3}_{\alpha_{d0}}$$

$$= s^3 + \alpha_{d2}s^2 + \alpha_{d1}s + \alpha_{d0}.$$

Thus if the system is in control canonical form, it is pretty easy to find the feedback gain (row) vector k to put the poles wherever it is desired. Now suppose the system is

$$\frac{d}{dt}x = Ax + bu, \quad A \in \mathbb{R}^{3\times3}, \quad b \in \mathbb{R}^3$$

which is *not necessarily* in control canonical form. Let the open-loop characteristic polynomial be written as

$$\det(sI - A) = s^3 + \alpha_2 s^2 + \alpha_1 s + \alpha_0.$$

Assume that

$$\det \mathcal{C} = \det\begin{bmatrix} b & Ab & A^2 b \end{bmatrix} \neq 0.$$

15 State Feedback

Then by the Theorem 2 we know the transformation

$$x_c = C_c C^{-1} x$$

transforms this system to the control canonical form

$$\frac{d}{dt}\underbrace{\begin{bmatrix} x_{c1} \\ x_{c2} \\ x_{c3} \end{bmatrix}}_{x_c} = \underbrace{\begin{bmatrix} 0 & 1 & 0 \\ 0 & 0 & 1 \\ -\alpha_0 & -\alpha_1 & -\alpha_2 \end{bmatrix}}_{A_c} \underbrace{\begin{bmatrix} x_{c1} \\ x_{c2} \\ x_{c3} \end{bmatrix}}_{x_c} + \underbrace{\begin{bmatrix} 0 \\ 0 \\ 1 \end{bmatrix}}_{b_c} u.$$

The feedback

$$u = -k_c x_c = -\begin{bmatrix} k_{c1} & k_{c2} & k_{c3} \end{bmatrix} \begin{bmatrix} x_{c1} \\ x_{c2} \\ x_{c3} \end{bmatrix}$$

$$= \begin{bmatrix} \alpha_{d0} - \alpha_0 & \alpha_{d1} - \alpha_1 & \alpha_{d2} - \alpha_2 \end{bmatrix} x_c$$

results in

$$\det(sI - (A_c - b_c k_c)) = s^3 + \alpha_{d2} s^2 + \alpha_{d1} s + \alpha_{d0}.$$

In the original coordinate system the feedback

$$u = -k_c x_c = \underbrace{-k_c C_c C^{-1}}_{k} x$$

results in

$$\det(sI - (A - bk)) = s^3 + \alpha_{d2} s^2 + \alpha_{d1} s + \alpha_{d0}.$$

Problem 18 gives the MATLAB code to compute $k = k_c C_c C^{-1}$.

Example 10 *Cart on the Track System*
Recall the equations of the error system for the cart on the track system given in (15.26) on page 576 and repeated here for convenience.

$$\frac{d}{dt}\begin{bmatrix} \epsilon_1 \\ \epsilon_2 \end{bmatrix} = \underbrace{\begin{bmatrix} 0 & 1 \\ 0 & -a_0 \end{bmatrix}}_{A} \begin{bmatrix} \epsilon_1 \\ \epsilon_2 \end{bmatrix} + \underbrace{\begin{bmatrix} 0 \\ b_0 \end{bmatrix}}_{b} w + \begin{bmatrix} 0 \\ 1 \end{bmatrix} d.$$

Then

$$\det(sI - A) = s(s + a_0) = s^2 + a_0 s$$

and

$$C = \begin{bmatrix} b & Ab \end{bmatrix} = \begin{bmatrix} 0 & b_0 \\ b_0 & -a_0 b_0 \end{bmatrix}, \quad C^{-1} = -\frac{1}{b_0^2}\begin{bmatrix} -a_0 b_0 & -b_0 \\ -b_0 & 0 \end{bmatrix}.$$

As

$$A_c = \begin{bmatrix} 0 & 1 \\ -\alpha_0 & -\alpha_1 \end{bmatrix} = \begin{bmatrix} 0 & 1 \\ 0 & -a_0 \end{bmatrix}$$

$$b_c = \begin{bmatrix} 0 \\ 1 \end{bmatrix}$$

we have
$$C_c = \begin{bmatrix} b_c & A_c b_c \end{bmatrix} = \begin{bmatrix} 0 & 1 \\ 1 & -a_0 \end{bmatrix}.$$

With the desired closed-loop characteristic polynomial given by
$$(s+r_1)(s+r_2) = s^2 + \underbrace{(r_1+r_2)}_{\alpha_{d1}}s + \underbrace{r_1 r_2}_{\alpha_{d0}},$$

the feedback gain k to place the poles at $-r_1, -r_2$ is given by
$$k = k_c C_c \mathcal{C}^{-1}$$
$$= \begin{bmatrix} r_1 r_2 - 0 & r_1 + r_2 - a_0 \end{bmatrix} \begin{bmatrix} 0 & 1 \\ 1 & -a_0 \end{bmatrix} \frac{1}{b_0} \begin{bmatrix} a_0 & 1 \\ 1 & 0 \end{bmatrix}$$
$$= \begin{bmatrix} \dfrac{r_1 r_2}{b_0} & \dfrac{r_1 + r_2 - a_0}{b_0} \end{bmatrix}.$$

This is the same result as given in (15.27).

Example 11 *Magnetic Levitation*

Let's now apply the pole placement procedure to the third-order linear statespace model of the magnetic levitation system in Example 6 (page 587). The system matrices A and b have the form
$$A = \begin{bmatrix} a_{11} & 0 & a_{13} \\ 0 & 0 & 1 \\ a_{31} & a_{32} & 0 \end{bmatrix}, \quad b = \begin{bmatrix} b_1 \\ 0 \\ 0 \end{bmatrix},$$

where
$$\det(sI - A) = \det \begin{bmatrix} s - a_{11} & 0 & -a_{13} \\ 0 & s & -1 \\ -a_{31} & -a_{32} & s \end{bmatrix}$$
$$= s^3 - a_{11}s^2 - (a_{13}a_{31} + a_{32})s + a_{11}a_{32}$$
$$= s^3 + \alpha_2 s^2 + \alpha_1 s + \alpha_0.$$

Then
$$\mathcal{C} = \begin{bmatrix} b & Ab & A^2 b \end{bmatrix} = \begin{bmatrix} b_1 & a_{11}b_1 & b_1 a_{11}^2 + b_1 a_{13} a_{31} \\ 0 & 0 & b_1 a_{31} \\ 0 & a_{31}b_1 & b_1 a_{11} a_{31} \end{bmatrix}$$

with
$$\mathcal{C}^{-1} = -\frac{1}{b_1^3 a_{31}^2} \begin{bmatrix} -b_1^2 a_{31}^2 & b_1^2 a_{13} a_{31}^2 & b_1^2 a_{11} a_{31} \\ 0 & b_1^2 a_{11} a_{31} & -b_1^2 a_{31} \\ 0 & -b_1^2 a_{31} & 0 \end{bmatrix} = \begin{bmatrix} \dfrac{1}{b_1} & -\dfrac{a_{13}}{b_1} & -\dfrac{a_{11}}{b_1 a_{31}} \\ 0 & -\dfrac{a_{11}}{b_1 a_{31}} & \dfrac{1}{b_1 a_{31}} \\ 0 & \dfrac{1}{b_1 a_{31}} & 0 \end{bmatrix}.$$

Further, with $\alpha_0 = a_{11}a_{32}$, $\alpha_1 = -(a_{13}a_{31} + a_{32})$, and $\alpha_2 = -a_{11}$, the control canonical form for A, b is

$$A_c = \begin{bmatrix} 0 & 1 & 0 \\ 0 & 0 & 1 \\ -\alpha_0 & -\alpha_1 & -\alpha_2 \end{bmatrix}, \quad b_c = \begin{bmatrix} 0 \\ 0 \\ 1 \end{bmatrix}.$$

The corresponding controllability matrix is

$$\mathcal{C}_c = \begin{bmatrix} b_c & A_c b_c & A_c^2 b_c \end{bmatrix} = \begin{bmatrix} 0 & 0 & 1 \\ 0 & 1 & -\alpha_2 \\ 1 & -\alpha_2 & \alpha_2^2 - \alpha_1 \end{bmatrix} = \begin{bmatrix} 0 & 0 & 1 \\ 0 & 1 & a_{11} \\ 1 & a_{11} & a_{11}^2 + a_{13}a_{31} + a_{32} \end{bmatrix}.$$

Set the desired closed-loop characteristic polynomial to be

$$s^3 + \alpha_{d2}s^2 + \alpha_{d1}s + \alpha_{d0} = s^3 + \underbrace{(r_1 + r_2 + r_3)}_{\alpha_{d2}}s^2 + \underbrace{(r_1r_2 + r_1r_3 + r_2r_3)}_{\alpha_{d1}}s + \underbrace{r_1r_2r_3}_{\alpha_{d0}}$$

$$= (s + r_1)(s + r_2)(s + r_3).$$

The gain matrix k to achieve this is

$$k = k_c \mathcal{C}_c \mathcal{C}^{-1}$$

$$= \begin{bmatrix} r_1r_2r_3 - a_{11}a_{32} & r_1r_2 + r_1r_3 + r_2r_3 + a_{13}a_{31} + a_{32} & r_1 + r_2 + r_3 + a_{11} \end{bmatrix} \times$$

$$\begin{bmatrix} 0 & 0 & 1 \\ 0 & 1 & a_{11} \\ 1 & a_{11} & a_{11}^2 + a_{13}a_{31} + a_{32} \end{bmatrix} \begin{bmatrix} \dfrac{1}{b_1} & -\dfrac{a_{13}}{b_1} & -\dfrac{a_{11}}{b_1 a_{31}} \\ 0 & -\dfrac{a_{11}}{b_1 a_{31}} & \dfrac{1}{b_1 a_{31}} \\ 0 & \dfrac{1}{b_1 a_{31}} & 0 \end{bmatrix}$$

$$= \begin{bmatrix} \dfrac{r_1 + r_2 + r_3 + a_{11}}{b_1} & \dfrac{r_1r_2r_3 + (r_1 + r_2 + r_3)a_{32}}{b_1 a_{31}} & \dfrac{r_1r_2 + r_1r_3 + r_2r_3 + a_{13}a_{31} + a_{32}}{b_1 a_{31}} \end{bmatrix}.$$

This is the same gain (row) vector k calculated in (15.66).

State Feedback Does Not Change the System Zeros

We end this section showing that feedback does not change the location of the system's zeros. Recall in Chapter 14 that a statespace realization for the transfer function

$$G(s) = \dfrac{Y(s)}{U(s)} = \dfrac{\beta_2 s^2 + \beta_1 s + \beta_0}{s^3 + \alpha_2 s^2 + \alpha_1 s + \alpha_0}$$

in control canonical form is given by

$$\dfrac{dx}{dt} = \underbrace{\begin{bmatrix} 0 & 1 & 0 \\ 0 & 0 & 1 \\ -\alpha_0 & -\alpha_1 & -\alpha_2 \end{bmatrix}}_{A} x + \underbrace{\begin{bmatrix} 0 \\ 0 \\ 1 \end{bmatrix}}_{b} u$$

$$y = \underbrace{\begin{bmatrix} \beta_0 & \beta_1 & \beta_2 \end{bmatrix}}_{c} x.$$

With the state feedback
$$u = -\underbrace{\begin{bmatrix} k_0 & k_1 & k_2 \end{bmatrix}}_{k} x + r$$

we have

$$\frac{dx}{dt} = \begin{bmatrix} 0 & 1 & 0 \\ 0 & 0 & 1 \\ -\alpha_0 & -\alpha_1 & -\alpha_2 \end{bmatrix} x - \begin{bmatrix} 0 \\ 0 \\ 1 \end{bmatrix} \begin{bmatrix} k_0 & k_1 & k_2 \end{bmatrix} x + \begin{bmatrix} 0 \\ 0 \\ 1 \end{bmatrix} r$$

$$= \begin{bmatrix} 0 & 1 & 0 \\ 0 & 0 & 1 \\ -\alpha_0 - k_0 & -\alpha_1 - k_1 & -\alpha_2 - k_2 \end{bmatrix} x + \begin{bmatrix} 0 \\ 0 \\ 1 \end{bmatrix} r$$

$$y = \underbrace{\begin{bmatrix} \beta_0 & \beta_1 & \beta_2 \end{bmatrix}}_{c} x.$$

The closed-loop transfer function is then

$$G_{CL}(s) = c(sI - (A - bk))^{-1} b$$

$$= \begin{bmatrix} \beta_0 & \beta_1 & \beta_2 \end{bmatrix} \begin{bmatrix} s & -1 & 0 \\ 0 & s & -1 \\ \alpha_0 + k_0 & \alpha_1 + k_1 & s + \alpha_2 + k_2 \end{bmatrix}^{-1} \begin{bmatrix} 0 \\ 0 \\ 1 \end{bmatrix}$$

$$= \begin{bmatrix} \beta_0 & \beta_1 & \beta_2 \end{bmatrix} \frac{\begin{bmatrix} \times & \times & 1 \\ \times & \times & s \\ \times & \times & s^2 \end{bmatrix} \begin{bmatrix} 0 \\ 0 \\ 1 \end{bmatrix}}{s^3 + (\alpha_2 + k_2)s^2 + (\alpha_1 + k_1)s + (\alpha_0 + k_0)}$$

$$= \frac{\beta_2 s^2 + \beta_1 s + \beta_0}{s^3 + (\alpha_2 + k_2)s^2 + (\alpha_1 + k_1)s + (\alpha_0 + k_0)}.$$

The closed-loop system has the same zeros as the open-loop system!

15.8 Asymptotic Tracking of Equilibrium Points

We have looked at using state feedback for trajectory tracking. A special case of this is to (asymptotically) track an equilibrium point of the system as the equilibrium point is simply a constant trajectory. To explain, consider the linear system

$$\frac{dx}{dt} = Ax + bu, \quad A \in \mathbb{R}^{n \times n}, \quad b \in \mathbb{R}^n.$$

The constant vector x_{eq} is an equilibrium point of this system if there is a constant input u_{eq} such

$$0 = \frac{dx_{eq}}{dt} = Ax_{eq} + bu_{eq}.$$

We develop a state feedback controller that forces $x(t) \to x_{eq}$. The dynamics of the error system $x(t) - x_{eq}$ is given by

$$\frac{d}{dt}(x - x_{eq}) = Ax + bu - (Ax_{eq} + bu_{eq}) = A(x - x_{eq}) + b(u - u_{eq}).$$

With the state feedback
$$u - u_{eq} = -k(x - x_{eq})$$
we obtain
$$\frac{d}{dt}(x - x_{eq}) = (A - bk)(x - x_{eq}).$$

Given the pair (A, b) is controllable, the feedback gain vector k is chosen to place the closed-loop poles in the open left half-plane with the result that
$$x(t) - x_{eq} = e^{(A-bk)t}\left(x(0) - x_{eq}\right) \to 0.$$

Note that the input u is given by
$$u(t) = -k\left(x(t) - x_{eq}\right) + u_{eq} \to u_{eq}.$$

Example 12 *Inverted Pendulum*

In Chapter 13 a linear approximate model of the inverted pendulum was found to be

$$\underbrace{\begin{bmatrix} dx/dt \\ dv/dt \\ d\theta/dt \\ d\omega/dt \end{bmatrix}}_{dz/dt} = \underbrace{\begin{bmatrix} 0 & 1 & 0 & 0 \\ 0 & 0 & -\dfrac{gm^2\ell^2}{M m\ell^2 + J(M+m)} & 0 \\ 0 & 0 & 0 & 1 \\ 0 & 0 & \dfrac{mg\ell(M+m)}{M m\ell^2 + J(M+m)} & 0 \end{bmatrix}}_{A} \underbrace{\begin{bmatrix} x \\ v \\ \theta \\ \omega \end{bmatrix}}_{z} + \underbrace{\begin{bmatrix} 0 \\ \dfrac{J+m\ell^2}{M m\ell^2 + J(M+m)} \\ 0 \\ -\dfrac{m\ell}{M m\ell^2 + J(M+m)} \end{bmatrix}}_{b} u.$$

(15.94)

It is easy to see that any equilibrium point of this linear approximate model has the form
$$z_{eq} = \begin{bmatrix} x_{eq} \\ 0 \\ 0 \\ 0 \end{bmatrix} \quad \text{with } u_{eq} = 0.$$

This equilibrium point corresponds to the cart at x_{eq} with $v = \theta = \omega = 0$. The pair (A, b) is controllable so k can be found so that the state feedback $u = -k(z - z_{eq}) + u_{eq} = -k(z - z_{eq})$ forces $z(t) \to z_{eq}$. Noting that $kz_{eq} = k_1 x_{eq}$ the right side of Figure 15.10 shows an equivalent implementation using the reference $r = k_1 x_{eq}$. Problem 11 asks the reader to write a MATLAB program to compute the feedback gain vector k and then simulate the stabilized system in SIMULINK.

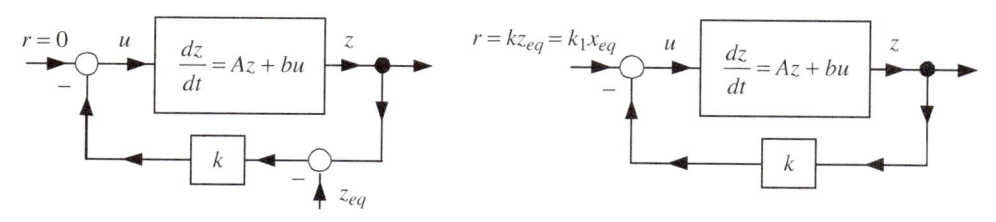

Figure 15.10. Equivalent state feedback setups for tracking step inputs.

As explained in Chapter 17 this state feedback control system for the inverted pendulum is robust. This means the controller will keep the pendulum rod upright in spite of inaccurate model parameters and small disturbances acting on either the cart or pendulum rod.

15.9 Tracking Step Inputs via State Feedback

In Chapters 9 and 10 we considered asymptotic tracking of step inputs using transfer functions. We now look at this problem using the statespace approach. We do this by considering the inverted pendulum as given in Example 12. Here the objective is to have the cart position track a step input while maintaining the pendulum upright. To the statespace model of (15.94) let's now take the cart position as output by setting

$$y = \underbrace{\begin{bmatrix} 1 & 0 & 0 & 0 \end{bmatrix}}_{c} \underbrace{\begin{bmatrix} x \\ v \\ \theta \\ \omega \end{bmatrix}}_{z}. \qquad (15.95)$$

We already have seen that the pair (A, b) is controllable and in Chapter 13 we showed that the open-loop transfer function from input $U(s)$ to output $Y(s) = X(s)$ is given by

$$\frac{X(s)}{U(s)} = G(s) = \frac{\kappa(J + m\ell^2)s^2 - \kappa mg\ell}{s^2(s^2 - \alpha^2)} = \frac{n(s)}{d(s)}, \qquad (15.96)$$

where

$$\alpha^2 = \frac{mg\ell(M+m)}{Mm\ell^2 + J(M+m)}, \quad \kappa = \frac{1}{Mm\ell^2 + J(M+m)}. \qquad (15.97)$$

For later reference we note that $G(0) \neq 0$, i.e., $n(0) = -\kappa mg\ell \neq 0$.

In Example 12 we discussed how to choose $u = -kz$ to place the closed-loop poles of the inverted pendulum at any desired location. Figure 15.11 shows the setup for the state feedback stabilization of the inverted pendulum.

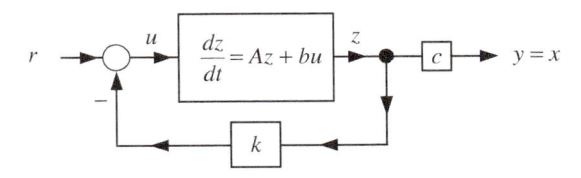

Figure 15.11. Stabilizing the inverted pendulum using state feedback.

We now show how to make this system (asymptotically) track step reference inputs $r(t) = x_{ref} u_s(t)$ for the position of the cart while keeping the pendulum upright. To proceed, we first redraw Figure 15.11 to obtain the equivalent system of Figure 15.12.

Figure 15.12. This diagram is equivalent to Figure 15.11.

The output of this system is the cart position. Figure 15.13 is a transfer function representation of Figure 15.12.

Figure 15.14 shows the cart's position error $e = x_{ref} - x$ is now put through an integrator as feedback. This control setup is suggested by the internal model principle of

$$R(s) \longrightarrow \boxed{G_k(s) = c(sI - (A - bk))^{-1}b} \longrightarrow Y(s) = X(s)$$

Figure 15.13. Transfer function representation of Figure 15.12.

Chapter 9. Specifically, if $k \in \mathbb{R}^4$, $k_0 \in \mathbb{R}$ can be chosen so that the closed-loop system is stable, then $x(t) \to x_{ref}$ as $t \to \infty$.

$$R(s) = \frac{x_{ref}}{s} \longrightarrow \bigcirc \longrightarrow \frac{1}{s} \xrightarrow{z_0} \boxed{-k_0} \longrightarrow \boxed{G_k(s)} \longrightarrow Y(s) = X(s)$$

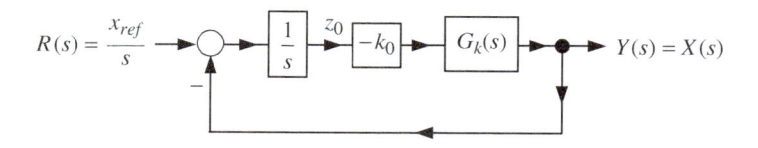

Figure 15.14. Use integrator feedback to asymptotically track step inputs.

The fact that the integrator in Figure 15.14 results in the tracking of step inputs can be shown directly as follows. Write

$$G_k(s) = c(sI - (A - bk))^{-1}b = \frac{n(s)}{d_k(s)},$$

where $d_k(s) = \det(sI - (A - bk))$ and $n(s) = \kappa(J + m\ell^2)s^2 - \kappa m g \ell$ (state feedback does not change the system zeros). The overall transfer function of the system of Figure 15.14 is

$$Y(s) = X(s) = \frac{\frac{-k_0}{s}G_k(s)}{1 + \frac{-k_0}{s}G_k(s)} R(s) = \frac{\frac{-k_0}{s}\frac{n(s)}{d_k(s)}}{1 + \frac{-k_0}{s}\frac{n(s)}{d_k(s)}} R(s) = \frac{-k_0 n(s)}{sd_k(s) - k_0 n(s)} \frac{x_{ref}}{s}.$$

It is shown in the Appendix to this chapter that k, k_0 can be found to make the closed-loop system stable. As a result $sd_k(s) - k_0 n(s)$ is a stable polynomial and therefore $sX(s)$ is stable. By the final value theorem

$$\lim_{t \to \infty} x(t) = \lim_{s \to 0} sX(s) = \lim_{s \to 0} s \left(\frac{-k_0 n(s)}{sd_k(s) - k_0 n(s)} \frac{x_{ref}}{s} \right) = \frac{-k_0 n(0)}{-k_0 n(0)} x_{ref}, \quad n(0) \neq 0$$

$$= x_{ref}.$$

To address the stability problem we redraw the block diagram of Figure 15.14 as shown in Figure 15.15.

$$r = x_{ref} u_s(t) \longrightarrow \bigcirc \longrightarrow \frac{1}{s} \xrightarrow{z_0} \boxed{-k_0} \longrightarrow \bigcirc \xrightarrow{u} \boxed{\frac{dz}{dt} = Az + bu} \xrightarrow{z} \boxed{c} \longrightarrow y = x$$

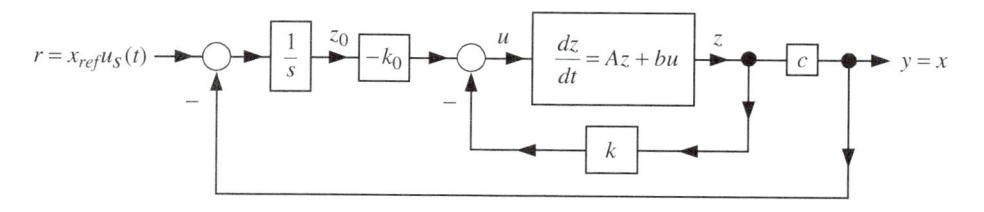

with feedback k.

Figure 15.15. Statespace tracking of step inputs.

15.9 Tracking Step Inputs via State Feedback

With z_0 the output of the integrator, the equations describing Figure 15.15 are

$$\frac{dz}{dt} = Az + bu$$

$$u = -kz - k_0 z_0$$

$$y = cz = x$$

$$\frac{dz_0}{dt} = x_{ref} - x.$$

We then can then reformulate this as the augmented system

$$\frac{d}{dt}\underbrace{\begin{bmatrix} z \\ z_0 \end{bmatrix}}_{z_a \in \mathbb{R}^5} = \underbrace{\begin{bmatrix} A & 0_{4\times 1} \\ -c & 0 \end{bmatrix}}_{A_a \in \mathbb{R}^{5\times 5}} \underbrace{\begin{bmatrix} z \\ z_0 \end{bmatrix}}_{z_a \in \mathbb{R}^5} + \underbrace{\begin{bmatrix} b \\ 0 \end{bmatrix}}_{b_a \in \mathbb{R}^5} u + \underbrace{\begin{bmatrix} 0_{4\times 1} \\ 1 \end{bmatrix}}_{p_a \in \mathbb{R}^5} x_{ref} \tag{15.98}$$

$$u = -\underbrace{\begin{bmatrix} k & k_0 \end{bmatrix}}_{k_a \in \mathbb{R}^{1\times 5}} \underbrace{\begin{bmatrix} z \\ z_0 \end{bmatrix}}_{z_a \in \mathbb{R}^5} \tag{15.99}$$

$$y = \underbrace{\begin{bmatrix} c & 0 \end{bmatrix}}_{c_a \in \mathbb{R}^{1\times 5}} \begin{bmatrix} z \\ z_0 \end{bmatrix}. \tag{15.100}$$

The pair (A, b) in the system (15.94) is controllable in $\mathbb{R}^4$ and the numerator of the open-loop transfer function $G(s)$ given in (15.96) is not zero at $s = 0$, i.e., $G(0) \neq 0$. By Theorem 4 at the end of this chapter (see the Appendix: *Disturbance Rejection in the Statespace*), these two conditions ensure that the pair of augmented matrices

$$A_a \triangleq \begin{bmatrix} A & 0_{4\times 1} \\ -c & 0 \end{bmatrix} \in \mathbb{R}^{5\times 5}, \quad b_a \triangleq \begin{bmatrix} b \\ 0 \end{bmatrix} \in \mathbb{R}^5$$

is *controllable* in $\mathbb{R}^5$. Therefore we can choose $k_a = \begin{bmatrix} k & k_0 \end{bmatrix}$ to place the closed-loop poles of $A_a - b_a k_a$ at any desired location.

This analysis guarantees $x(t) \to x_{ref}$, but do $v(t), \theta(t)$, and $\omega(t)$ go to zero? To look into this consider the compact representation of the closed-loop system (15.98) given by

$$\frac{dz_a}{dt} = A_a z_a + b_a u + p_a x_{ref}. \tag{15.101}$$

The equilibrium points z_{a_eq} and their corresponding inputs u_{eq} of (15.101) are solutions to

$$0_{5\times 1} = \frac{dz_{a\_eq}}{dt} = A_a z_{a\_eq} + b_a u_{eq} + p_a x_{ref}. \tag{15.102}$$

These solutions are given by

$$z_{a\_eq} = \begin{bmatrix} z_{eq} \\ z_{0\_eq} \end{bmatrix} = \begin{bmatrix} x_{ref} \\ 0 \\ 0 \\ 0 \\ z_{0\_eq} \end{bmatrix}, \quad u_{eq} = 0,$$

where z_{0_eq} is arbitrary. This is easily verified by substituting these expressions for z_{a_eq} and u_{eq} into (15.101) as follows:

$$\begin{bmatrix} 0 \\ 0 \\ 0 \\ 0 \\ 0 \end{bmatrix} = \underbrace{\begin{bmatrix} 0 & 1 & 0 & 0 & 0 \\ 0 & 0 & -\dfrac{gm^2\ell^2}{Mm\ell^2+J(M+m)} & 0 & 0 \\ 0 & 0 & 0 & 1 & 0 \\ 0 & 0 & \dfrac{mg\ell(M+m)}{Mm\ell^2+J(M+m)} & 0 & 0 \\ -1 & 0 & 0 & 0 & 0 \end{bmatrix}}_{A_a} \underbrace{\begin{bmatrix} x_{ref} \\ 0 \\ 0 \\ 0 \\ z_{0\_eq} \end{bmatrix}}_{z_{a\_eq}} + \underbrace{\begin{bmatrix} 0 \\ \dfrac{J+m\ell^2}{Mm\ell^2+J(M+m)} \\ 0 \\ -\dfrac{m\ell}{Mm\ell^2+J(M+m)} \\ 0 \end{bmatrix}}_{b_a} u_{eq}$$

$$+ \begin{bmatrix} 0 \\ 0 \\ 0 \\ 0 \\ 1 \end{bmatrix} x_{ref}.$$

The statespace equations for the error are found by subtracting (15.102) from (15.101) to obtain

$$\frac{d}{dt}(z_a - z_{a\_eq}) = A_a z_a + b_a u + p_a x_{ref} - (A_a z_{a\_eq} + b u_{eq} + p_a x_{ref})$$
$$= A_a(z_a - z_{a\_eq}) + b_a(u - u_{eq}).$$

As the pair (A_a, b_a) is controllable, k_a can then be chosen so that the state feedback

$$u - u_{eq} = u = -k_a(z_a - z_{a\_eq})$$

results in

$$\frac{d}{dt}(z_a - z_{a\_eq}) = (A_a - b_a k_a)(z_a - z_{a\_eq})$$

being stable. As a consequence

$$z_a(t) - z_{a\_eq} = e^{(A_a - b_a k_a)t}(z_a(0) - z_{a\_eq}) \to 0.$$

What about the value of z_{0_eq}? Well in Figure 15.15 we are feeding back

$$u = -\begin{bmatrix} k & k_0 \end{bmatrix} \begin{bmatrix} z \\ z_0 \end{bmatrix}$$

not

$$u = -\begin{bmatrix} k & k_0 \end{bmatrix} \left(\begin{bmatrix} z \\ z_0 \end{bmatrix} - \begin{bmatrix} z_{eq} \\ z_{0\_eq} \end{bmatrix} \right).$$

These are the same if we choose z_{0_eq} to satisfy

$$\begin{bmatrix} k & k_0 \end{bmatrix} \begin{bmatrix} z_{eq} \\ z_{0\_eq} \end{bmatrix} = \begin{bmatrix} k_1 & k_2 & k_3 & k_4 & k_0 \end{bmatrix} \begin{bmatrix} x_{ref} \\ 0 \\ 0 \\ 0 \\ z_{0\_eq} \end{bmatrix} = 0$$

or
$$z_{0\_eq} = -(k_1/k_0)x_{ref}.$$

Another way to explain this is to say that the feedback $u = -k_a z_a$ applied to the system (with k_a chosen so that $A_a - b_a k_a$ is stable)
$$\frac{dz_a}{dt} = A_a z_a + b_a u + p_a x_{ref}$$

results in
$$z_a(t) \to z_{a\_eq} = \begin{bmatrix} x_{ref} \\ 0 \\ 0 \\ 0 \\ -(k_1/k_0)x_{ref} \end{bmatrix}.$$

This equilibrium point satisfies $k_a z_{a\_eq} = 0$. Also see Problem 17.

As explained in Chapter 17 this control system for the inverted pendulum is robust. This means the controller will keep the pendulum rod upright in spite of inaccurate model parameters and small disturbances acting on either the cart or pendulum rod.

Pole Placement Procedure

Let
$$C_a = \begin{bmatrix} b_a & A_a b_a & A_a^2 b_a & A_a^3 b_a & A_a^4 b_a \end{bmatrix} \in \mathbb{R}^{5 \times 5}$$
$$\det(sI_{5 \times 5} - A_a) = \det \begin{bmatrix} sI_{4 \times 4} - A & 0_{4 \times 1} \\ c & s \end{bmatrix} = s^5 + \alpha_4 s^4 + \alpha_3 s^3 + \alpha_2 s^2 + \alpha_1 s + \alpha_0$$

and
$$A_{ac} = \begin{bmatrix} 0 & 1 & 0 & 0 & 0 \\ 0 & 0 & 1 & 0 & 0 \\ 0 & 0 & 0 & 1 & 0 \\ 0 & 0 & 0 & 0 & 1 \\ -\alpha_0 & -\alpha_1 & -\alpha_2 & -\alpha_3 & -\alpha_4 \end{bmatrix}, \quad b_{ac} = \begin{bmatrix} 0 \\ 0 \\ 0 \\ 0 \\ 1 \end{bmatrix}$$
$$C_{ac} = \begin{bmatrix} b_{ac} & A_{ac} b_{ac} & A_{ac}^2 b_{ac} & A_{ac}^3 b_{ac} & A_{ac}^4 b_{ac} \end{bmatrix} \in \mathbb{R}^{5 \times 5}.$$

Then setting
$$k_a = \begin{bmatrix} k & k_0 \end{bmatrix}$$
$$= \begin{bmatrix} \alpha_{d0} - \alpha_0 & \alpha_{d1} - \alpha_1 & \alpha_{d2} - \alpha_2 & \alpha_{d3} - \alpha_3 & \alpha_{d4} - \alpha_4 \end{bmatrix} C_{ac} C_a^{-1} \in \mathbb{R}^{1 \times 5}$$

results in
$$\det \begin{bmatrix} sI_{4 \times 4} - (A - bk) & bk_0 \\ c & s \end{bmatrix} = s^5 + \alpha_{d4} s^4 + \alpha_{d3} s^3 + \alpha_{d2} s^2 + \alpha_{d1} s + \alpha_{d0}.$$

This procedure for computing the feedback gain vector k_a is to be used in Problem 12 to simulate a state feedback controller that forces the inverted pendulum to track step inputs.

Step Response of an Inverted Pendulum*

A straightforward (but tedious) calculation using

$$\frac{dz_a}{dt} = (A_a - b_a k_a) z_a + p_a x_{ref}, \quad z_a(0) = 0_{5 \times 1}$$

shows that (see Problem 17)

$$X(s) = c_a \big(sI_{5\times 5} - (A_a - b_a k_a)\big)^{-1} b_a \frac{x_{ref}}{s} = -k_0 \underbrace{\frac{\kappa(J + m\ell^2) s^2 - \kappa m g \ell}{\det\big(sI_{5\times 5} - (A_a - b_a k_a)\big)}}_{\text{Closed-loop transfer function}} \frac{x_{ref}}{s}. \quad (15.103)$$

Notice that the numerator has zeros at $s = \pm\sqrt{mg\ell/(J + m\ell^2)} = \pm\sqrt{\frac{g}{\ell}}\sqrt{\frac{1}{J/(m\ell^2)+1}}$.

Recall from Chapter 8 that the step response for $x(t)$ will have *undershoot* due to the right half-plane zero. With the parameter values chosen to be those of the QUANSER system (see Problem 12 of this chapter) and using the feedback setup of Figure 15.15, let's choose $k \in \mathbb{R}^{1\times 4}, k_0 \in \mathbb{R}$ to put all five closed-loop poles at -5. Then the position $X(s)$ of the cart due to a step input of $R(s) = 0.1/s$ (closed-loop response) is given by

$$X(s) = \frac{1.0025 s^2 + 23}{(s+5)^2} \frac{0.1}{s}.$$

The two zeros of this system are $s = \pm\sqrt{23/1.0025} = \pm 4.7898$. The zero at $s = 4.7898$ in the open right half-plane indicates we will have undershoot. With the step reference input $x_{ref}(t) = 0.1 u_s(t-1)$ applied to the system, Figure 15.16 is a plot of the response $x(t)$ showing the undershoot. The cart indeed goes backwards before going forwards to end up at $x = 0.1$ m.

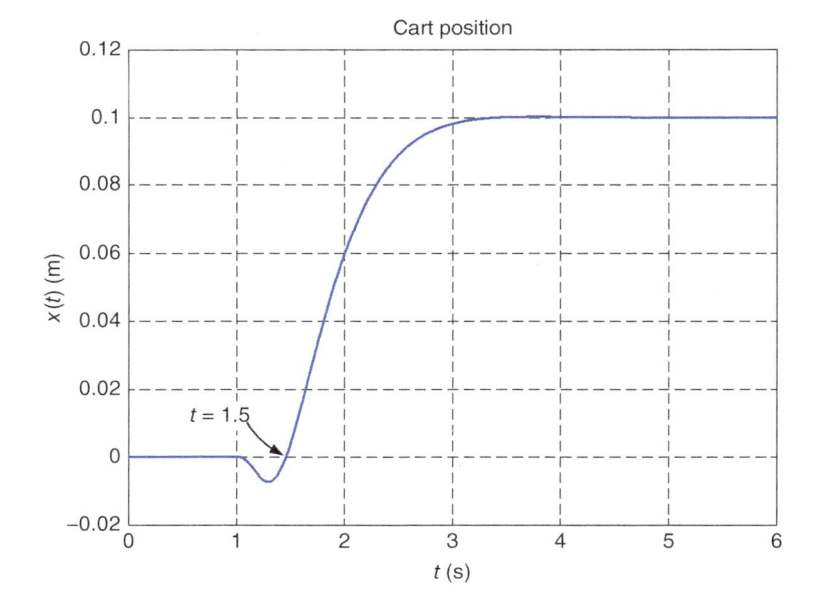

Figure 15.16. Step response $x(t)$ with the input applied at $t = 1$ second.

15.9 Tracking Step Inputs via State Feedback

Remark The closed-loop poles were chosen to be at -5. We could put them further in the left half-plane for a faster response with the trade-off being more undershoot in x (see Theorem 10 of Chapter 10).

Figure 15.17a shows the cart at $x = 0$ with the pendulum at $\theta = 0$. At $t = 1$ second the step input is applied and the cart moves backwards (the cart acceleration $u < 0$). This results in the pendulum rod rotating forward slightly (with respect to the cart) as shown in Figure 15.17b. This was necessary in order to get the pendulum pointing in the right direction for the move to the right! At $t = 1.2$ the cart's acceleration becomes positive (see Figure 15.18) so that by $t = 1.5$ the cart's position becomes positive as well (see Figures 15.16 and 15.17c). Figure 15.19 shows the corresponding response of the pendulum angle $\theta(t)$. There we see that the pendulum angle starts out positive and comes back to zero at $t = 1.75$, but then it goes *negative* as shown in Figure 15.17d. Then, as

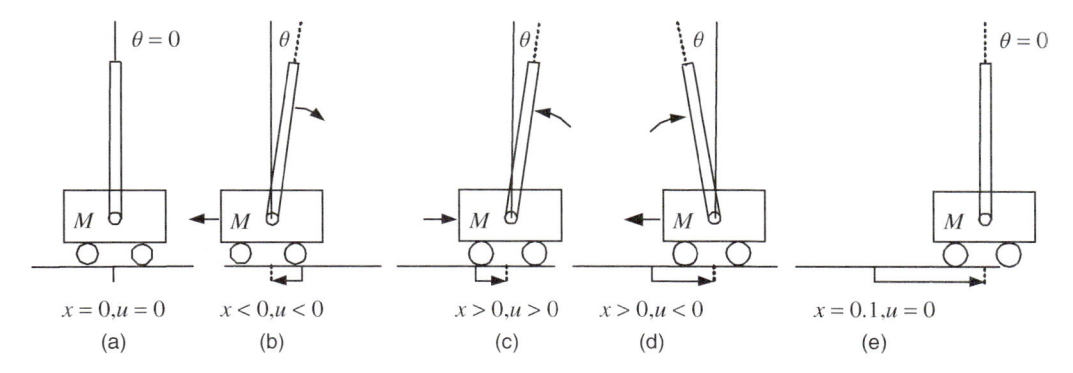

Figure 15.17. The cart and pendulum as it goes from $x = 0$ to $x = 0.1$.

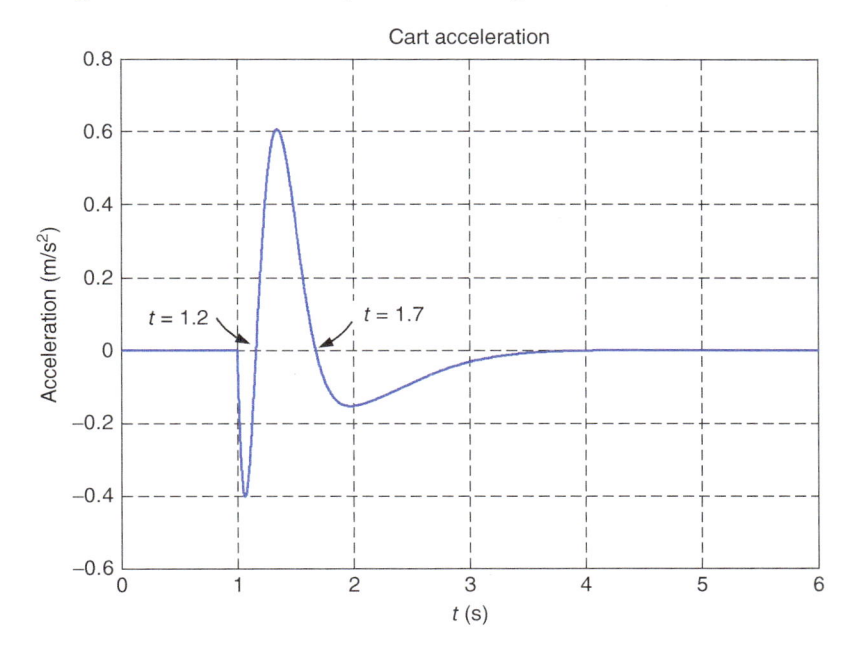

Figure 15.18. Cart acceleration.

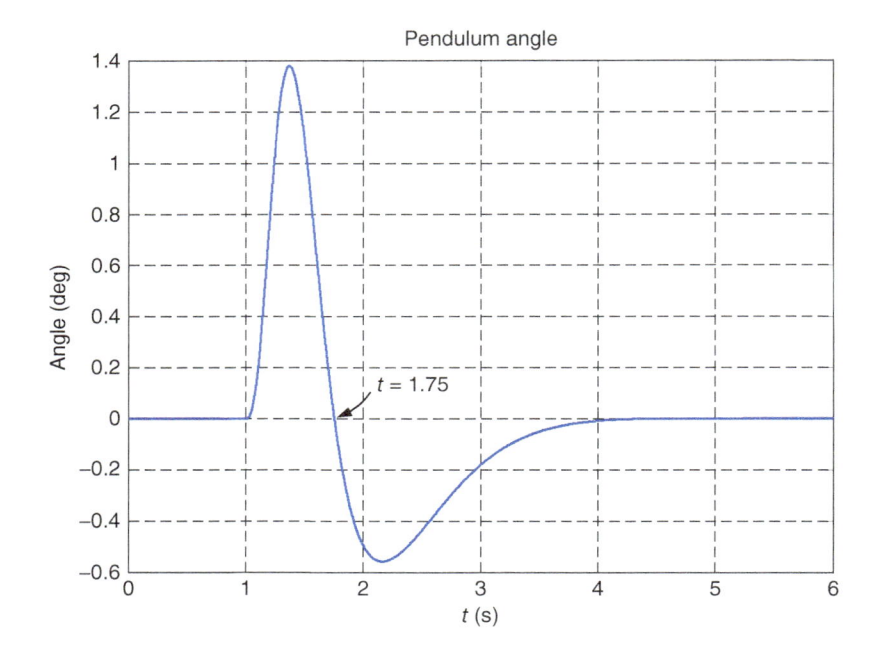

Figure 15.19. Response of the pendulum angle $\theta(t)$ to a step input in position.

Figure 15.18 shows, the cart's *acceleration* becomes negative at $t = 1.7$ resulting in the pendulum rod rotating forward to go back upright as indicated in Figure 15.17d. Finally, for $t \geq 4$, Figure 15.17e shows that $\theta = 0$, $x = 0.1$ with zero acceleration.

15.10 Inverted Pendulum on an Inclined Track*

Figure 15.20 shows the inverted pendulum with the cart's track at an incline. Note that the x direction is along the track with the y direction perpendicular to the track. As shown the angle θ is measured with respect to the y coordinate axis.

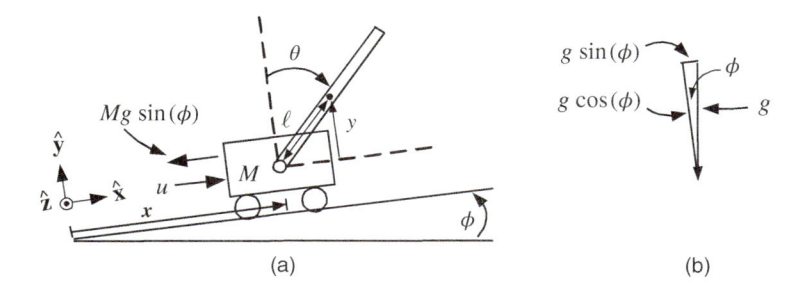

Figure 15.20. Inverted pendulum with the track raised at one end.

The (unknown) forces F_{mx} and F_{my} exerted on the rod at the pivot by the cart are taken to be along the x, y coordinate axis, respectively. Newton's equations for the translational

15.10 Inverted Pendulum on an Inclined Track* 613

motion of the pendulum rod of mass m are then

$$m\frac{d^2}{dt^2}(x+\ell\sin(\theta)) = F_{mx} - mg\sin(\phi)$$

$$m\frac{d^2}{dt^2}(\ell\cos(\theta)) = F_{my} - mg\cos(\phi).$$

The axis of rotation for the pendulum rod is the $-\hat{\mathbf{z}}$ direction through its center of mass. With J denoting the pendulum rod's moment of inertia with respect to its center of mass we have

$$\tau = J\frac{d^2\theta}{dt^2}.$$

As in the case of the flat track, the gravitational force $-mg\hat{\mathbf{y}}$ does *not* produce any *torque* on the rod as it acts through its center of mass so its moment arm is zero. The torque on the pendulum rod is only due to F_{mx} and F_{my}. The equation of motion for the rod is

$$J\frac{d^2\theta}{dt^2} = \ell F_{my}\sin(\theta) - \ell F_{mx}\cos(\theta).$$

The equation of motion for the cart is simply

$$M\frac{d^2x}{dt^2} = u(t) - F_{mx} - Mg\sin(\phi).$$

The complete set of equations for the pendulum on the cart are

$$m\frac{d^2x}{dt^2} - m\ell\sin(\theta)\left(\frac{d\theta}{dt}\right)^2 + m\ell\cos(\theta)\frac{d^2\theta}{dt^2} = F_{mx} - mg\sin(\phi) \quad (15.104)$$

$$-m\ell\cos(\theta)\left(\frac{d\theta}{dt}\right)^2 - m\ell\sin(\theta)\frac{d^2\theta}{dt^2} = F_{my} - mg\cos(\phi) \quad (15.105)$$

$$\ell F_{my}\sin(\theta) - \ell F_{mx}\cos(\theta) = J\frac{d^2\theta}{dt^2} \quad (15.106)$$

$$u(t) - F_{mx} - Mg\sin(\phi) = M\frac{d^2x}{dt^2}. \quad (15.107)$$

Linearization About an Equilibrium Point

To compute the equilibrium points of the inverted pendulum on the inclined plane we set all the derivatives to zero in Eqs. (15.104)–(15.104) and eliminate F_{mx}, F_{my} to obtain

$$\ell mg\cos(\phi)\sin(\theta) - \ell mg\sin(\phi)\cos(\theta) = \ell mg\sin(\theta - \phi) = 0$$

$$u_{eq} - (m+M)g\sin(\phi) = 0.$$

Solving gives $\theta_{eq} = \phi$ and $u_{eq} = (M+m)g\sin(\phi)$. The set of equilibrium points along their corresponding input reference are given by

$$z_{eq} = \begin{bmatrix} x_{eq} \\ v_{eq} \\ \theta_{eq} \\ \omega_{eq} \end{bmatrix} = \begin{bmatrix} x_{eq} \\ 0 \\ \phi \\ 0 \end{bmatrix}, \quad u_{eq} = (M+m)g\sin(\phi).$$

Note that x_{eq} can be set arbitrarily. Eliminating the forces F_{mx} and F_{my}, Eqs. (15.104)–(15.107) in statespace form are

$$\frac{dx}{dt} = v$$

$$\frac{dv}{dt} = \frac{m\ell(J+m\ell^2)\omega^2\sin(\theta) - gm^2\ell^2\cos(\theta)\sin(\theta-\phi) - (J+m\ell^2)(M+m)g\sin(\phi)}{JM + Jm + Mm\ell^2 + m^2\ell^2\sin^2(\theta)}$$

$$+ \frac{J+m\ell^2}{JM + Jm + Mm\ell^2 + m^2\ell^2\sin^2(\theta)} u$$

$$\frac{d\theta}{dt} = \omega \qquad (15.108)$$

$$\frac{d\omega}{dt} = \frac{mg\ell(M+m)\sin(\theta-\phi) - m^2\ell^2\omega^2\cos(\theta)\sin(\theta) + m\ell\cos(\theta)(M+m)g\sin(\phi)}{JM + Jm + Mm\ell^2 + m^2\ell^2\sin^2(\theta)}$$

$$- \frac{m\ell\cos(\theta)}{JM + Jm + Mm\ell^2 + m^2\ell^2\sin^2(\theta)} u.$$

With

$$\Delta x \triangleq x - x_{eq}$$
$$\Delta v \triangleq v - v_{eq} = v$$
$$\Delta \theta \triangleq \theta - \theta_{eq} = \theta - \phi$$
$$\Delta \omega \triangleq \omega - \omega_{eq} = \omega$$
$$\Delta u \triangleq u - u_{eq} = u - (M+m)g\sin(\phi),$$

the (nonlinear) equations of motion in terms of $\Delta x, \Delta v, \Delta \theta, \Delta \omega$, and $\Delta u = u - (M+m)g\sin(\phi)$ are

$$\frac{d\Delta x}{dt} = \Delta v$$

$$\frac{d\Delta v}{dt} = \frac{m\ell(J+m\ell^2)(\Delta\omega)^2\sin(\Delta\theta+\phi) - gm^2\ell^2\cos(\Delta\theta+\phi)\sin(\Delta\theta)}{JM + Jm + Mm\ell^2 + m^2\ell^2\sin^2(\Delta\theta+\phi)}$$

$$+ \frac{J+m\ell^2}{JM + Jm + Mm\ell^2 + m^2\ell^2\sin^2(\Delta\theta+\phi)}(\Delta u + u_{eq})$$

$$\frac{d\Delta\theta}{dt} = \Delta\omega$$

$$\frac{d\Delta\omega}{dt} = \frac{mg\ell(M+m)\sin(\Delta\theta) - m^2\ell^2(\Delta\omega)^2\cos(\Delta\theta+\phi)\sin(\Delta\theta+\phi)}{JM + Jm + Mm\ell^2 + m^2\ell^2\sin^2(\Delta\theta+\phi)}$$

$$- \frac{m\ell\cos(\Delta\theta+\phi)}{JM + Jm + Mm\ell^2 + m^2\ell^2\sin^2(\Delta\theta+\phi)}(\Delta u + u_{eq}).$$

To linearize this system of equations about z_{eq} substitute in the approximations

$$\cos(\Delta\theta+\phi) = \cos(\Delta\theta)\cos(\phi) - \sin(\Delta\theta)\sin(\phi) \approx \cos(\phi) - \Delta\theta\sin(\phi)$$

15.10 Inverted Pendulum on an Inclined Track*

$$\sin(\Delta\theta + \phi) = \sin(\Delta\theta)\cos(\phi) + \cos(\Delta\theta)\sin(\phi) \approx \Delta\theta\cos(\phi) + \sin(\phi)$$

$$\sin^2(\Delta\theta + \phi) \approx \sin^2(\phi)$$

and drop second-order or higher terms in $\Delta\omega, \Delta\theta$ to obtain

$$\frac{d}{dt}\begin{bmatrix}\Delta x \\ \Delta v \\ \Delta\theta \\ \Delta\omega\end{bmatrix} = \underbrace{\begin{bmatrix} 0 & 1 & 0 & 0 \\ 0 & 0 & -\frac{gm^2\ell^2\cos(\phi)}{den(\phi)} & 0 \\ 0 & 0 & 0 & 1 \\ 0 & 0 & \frac{mg\ell(M+m)}{den(\phi)} & 0 \end{bmatrix}}_{A_\phi}\begin{bmatrix}\Delta x \\ \Delta v \\ \Delta\theta \\ \Delta\omega\end{bmatrix} + \underbrace{\begin{bmatrix} 0 \\ \frac{J+m\ell^2}{den(\phi)} \\ 0 \\ -\frac{m\ell\cos(\phi)}{den(\phi)} \end{bmatrix}}_{b_\phi}\Delta u \qquad (15.109)$$

$$\Delta y = \underbrace{\begin{bmatrix} 1 & 0 & 0 & 0 \end{bmatrix}}_{c_\phi}\begin{bmatrix}\Delta x \\ \Delta v \\ \Delta\theta \\ \Delta\omega\end{bmatrix}. \qquad (15.110)$$

Here $den(\phi) \triangleq J(M+m) + Mm\ell^2 + m^2\ell^2\sin^2(\phi)$. The controllability matrix and its determinant are, respectively,

$$\mathcal{C} = \begin{bmatrix} b_\phi & A_\phi b_\phi & A_\phi^2 b_\phi & A_\phi^3 b_\phi \end{bmatrix}$$

$$= \begin{bmatrix} 0 & \frac{J+m\ell^2}{den(\phi)} & 0 & \frac{mg\ell m^2\ell^2\cos^2(\phi)}{den^2(\phi)} \\ \frac{J+m\ell^2}{den(\phi)} & 0 & \frac{mg\ell m^2\ell^2\cos^2(\phi)}{den^2(\phi)} & 0 \\ 0 & -\frac{m\ell\cos(\phi)}{den(\phi)} & 0 & -\frac{mg\ell(M+m)m\ell\cos(\phi)}{den^2(\phi)} \\ -\frac{m\ell\cos(\phi)}{den(\phi)} & 0 & -\frac{mg\ell(M+m)m\ell\cos(\phi)}{den^2(\phi)} & 0 \end{bmatrix}$$

$$\det\mathcal{C} = \frac{(mg\ell)^2 m\ell\cos^2(\phi)}{den^2(\phi)^4}.$$

Remark The linear statespace model (15.109) is controllable as $\det\mathcal{C} \neq 0$ for $0 \leq \phi < \pi/2$. Also, one can check the open-loop transfer function $\Delta Y(s)/\Delta u(s) = G(s) = c_\phi(sI_{4\times 4} - A_\phi)^{-1}b_\phi = n(s)/d(s)$ satisfies $n(0) \neq 0$.

State Feedback Control

To track a step reference $x_{eq}u_s(t)$ in position an integrator is added to the forward path as shown in Figure 15.21. This integrator will also provide disturbance rejection of the

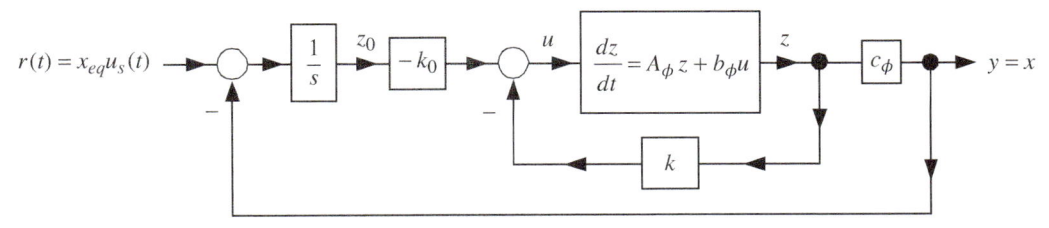

Figure 15.21. Control structure to reject the gravity force disturbance $(M+m)g\sin(\phi)$ on the cart.

gravity force $(M+m)g\sin(\phi)$ acting on the cart and pendulum system. (Note that in Figure 15.21 $z = \begin{bmatrix} x & v & \theta & \omega \end{bmatrix}^T$ not $\Delta z = \begin{bmatrix} \Delta x & \Delta v & \Delta \theta & \Delta \omega \end{bmatrix}^T$ and u is the full input not Δu.)

The augmented system is

$$\frac{d}{dt}\underbrace{\begin{bmatrix} z \\ z_0 \end{bmatrix}}_{z_a \in \mathbb{R}^5} = \underbrace{\begin{bmatrix} A_\phi & 0_{4\times 1} \\ -c_\phi & 0 \end{bmatrix}}_{A_{\phi a} \in \mathbb{R}^{5\times 5}} \underbrace{\begin{bmatrix} z \\ z_0 \end{bmatrix}}_{z_a \in \mathbb{R}^5} + \underbrace{\begin{bmatrix} b_\phi \\ 0 \end{bmatrix}}_{b_{\phi a} \in \mathbb{R}^5} u + \underbrace{\begin{bmatrix} 0_{4\times 1} \\ 1 \end{bmatrix}}_{p_a \in \mathbb{R}^5} x_{ref}$$

$$u = -\underbrace{\begin{bmatrix} k & k_0 \end{bmatrix}}_{k_a \in \mathbb{R}^{1\times 5}} \underbrace{\begin{bmatrix} z \\ z_0 \end{bmatrix}}_{z_a \in \mathbb{R}^5}$$

$$y = \underbrace{\begin{bmatrix} c_\phi & 0 \end{bmatrix}}_{c_{\phi a} \in \mathbb{R}^{1\times 5}} \begin{bmatrix} z \\ z_0 \end{bmatrix}.$$

By these remarks and the Appendix *Disturbance Rejection in the Statespace* at the end of this chapter, it follows that the pair $(A_{\phi a}, b_{\phi a})$ of the augmented system is controllable. Consequently, we can (and do) choose $k_a \triangleq \begin{bmatrix} k & k_0 \end{bmatrix}$ so that $A_{\phi a} - b_{\phi a} k_a$ is stable.[2] With x_{eq} chosen arbitrarily, an equilibrium point requires $v_{eq} = 0$, $\theta_{eq} = \phi$, and $\omega_{eq} = 0$. The feedback structure of Figure 15.21 shows $u = -k_0 z_0 - kz$ so that at equilibrium it is necessary that

$$u_{eq} = (M+m)g\sin(\phi) = -k_a z_{a\_eq} = -k_0 z_{0\_eq} - k z_{eq} = -k_0 z_{0\_eq} - k_1 x_{eq} - k_3\phi. \tag{15.111}$$

The equilibrium value for output of the integrator z_0 is then

$$z_{0\_eq} = -(k_1 x_{eq} + k_3 \phi + u_{eq})/k_0.$$

The error state $z_a - z_{a\_eq}$ satisfies

$$\frac{d}{dt}(z_a - z_{a\_eq}) = \underbrace{\begin{bmatrix} 0 & 1 & 0 & 0 & 0 \\ 0 & 0 & -\dfrac{gm^2\ell^2 \cos(\phi)}{den(\phi)} & 0 & 0 \\ 0 & 0 & 0 & 1 & 0 \\ 0 & 0 & \dfrac{mg\ell(M+m)}{den(\phi)} & 0 & 0 \\ -1 & 0 & 0 & 0 & 0 \end{bmatrix}}_{A_{\phi a}} (z_a - z_{a\_eq}) + \underbrace{\begin{bmatrix} 0 \\ \dfrac{J+m\ell^2}{den(\phi)} \\ 0 \\ -\dfrac{m\ell\cos(\phi)}{den(\phi)} \\ 0 \end{bmatrix}}_{b_{\phi a}} (u - u_{eq}).$$
(15.112)

More compactly we have

$$\frac{d}{dt}(z_a - z_{a\_eq}) = A_{\phi a}(z_a - z_{a\_eq}) + b_{\phi a}(u - u_{eq}).$$

---

[2] k_a depends on ϕ because $A_{\phi a}$ and $b_{\phi a}$ depend on ϕ. We will remedy this at the end of the section.

15.10 Inverted Pendulum on an Inclined Track*

Set

$$u - u_{eq} = -k_a(z_a - z_{a\_eq}), \quad k_a = \begin{bmatrix} k & k_0 \end{bmatrix} \in \mathbb{R}^{1\times 5}$$

to obtain

$$\frac{d}{dt}(z_a - z_{a\_eq}) = (A_{\phi a} - b_{\phi a}k_a)(z_a - z_{a\_eq}).$$

The feedback gain k_a is used to place the eigenvalues of

$$A_{\phi a} - b_{\phi a}k_a = \begin{bmatrix} A_\phi & 0_{4\times 1} \\ -c_\phi & 0 \end{bmatrix} - \begin{bmatrix} b_\phi \\ 0 \end{bmatrix}\begin{bmatrix} k & k_0 \end{bmatrix} = \begin{bmatrix} A_\phi - b_\phi k & -b_\phi k_0 \\ -c_\phi & 0 \end{bmatrix}$$

at any desired location in the open left-half plane. The feedback u to the pendulum is

$$u = -k_a(z_a - z_{a\_eq}) + u_{eq}$$

and will result in $z_a(t) \to z_{a\_eq}$. However, by (15.111), we have $u_{eq} = -k_a z_{a\_eq}$ so

$$u = -k_a(z_a - z_{a\_eq}) + u_{eq} = -k_a z_a + k_a z_{a\_eq} + u_{eq} = -k_a z_a.$$

The feedback u to the pendulum is simply given by

$$u = -k_a z_a = -k_0 z_0 - kz \tag{15.113}$$

as shown in Figure 15.21.

The Incline Angle ϕ Is Unknown

A final issue is that k_a is computed using $A_{\phi a}$ and $b_{\phi a}$, which depend on the unknown value of ϕ! To deal with this we choose the feedback gains $k_a = \begin{bmatrix} k & k_0 \end{bmatrix} \in \mathbb{R}^{1\times 5}$ to place the closed-loop poles of the *flat* track model, that is, using

$$A_a = \begin{bmatrix} 0 & 1 & 0 & 0 & 0 \\ 0 & 0 & -\dfrac{gm^2\ell^2}{J(M+m)+Mm\ell^2} & 0 & 0 \\ 0 & 0 & 0 & 1 & 0 \\ 0 & 0 & \dfrac{mg\ell(M+m)}{J(M+m)+Mm\ell^2} & 0 & 0 \\ -1 & 0 & 0 & 0 & 0 \end{bmatrix}, \quad b_a = \begin{bmatrix} 0 \\ \dfrac{J+m\ell^2}{J(M+m)+Mm\ell^2} \\ 0 \\ -\dfrac{m\ell}{J(M+m)+Mm\ell^2} \\ 0 \end{bmatrix}.$$

If applying these (flat track computed) feedback gains to the *inclined* track linear model results in

$$\begin{bmatrix} A_\phi & 0_{4\times 1} \\ -c_\phi & 0 \end{bmatrix} - \begin{bmatrix} b_\phi \\ 0 \end{bmatrix}\begin{bmatrix} k & k_0 \end{bmatrix} = \begin{bmatrix} A_\phi - b_\phi k & -b_\phi k_0 \\ -c_\phi & 0 \end{bmatrix} \tag{15.114}$$

being stable, then the controller will keep the pendulum upright, i.e. $\theta(t) \to \phi, \omega(t) \to 0$, while tracking step reference inputs, i.e. $x(t) \to x_{eq}, v(t) \to 0$. It turns out that there is a range of values $0 \leq \phi \leq \phi_{\max}$ such that (15.114) is stable. You are asked to simulate this system in Problem 19.

15.11 Feedback Linearization Control*

For both the inverted pendulum and the magnetic levitation systems we used an approximate linear model of each system around their respective equilibrium points to design their controllers. For some nonlinear systems there is a much more powerful technique that allows us to control the system and does not require its state to be close to an equilibrium point. We illustrate this approach with the magnetic levitation system.

The nonlinear statespace model of the magnetic levitation system is (see Chapter 13)

$$\frac{di}{dt} = rL_1 \frac{i}{x^2 L(x)} v - \frac{R}{L(x)} i + \frac{1}{L(x)} u \tag{15.115}$$

$$\frac{dx}{dt} = v \tag{15.116}$$

$$\frac{dv}{dt} = g - \frac{rL_1}{2m} \frac{i^2}{x^2} + \frac{f_d}{m}. \tag{15.117}$$

Here f_d is a disturbance force (perhaps due to air flow) on the steel ball acting in the x direction and $L(x) = L_0 + rL_1/x$. Define new state variables as

$$x_1 \triangleq x \tag{15.118}$$

$$x_2 \triangleq v \tag{15.119}$$

$$x_3 \triangleq g - \frac{rL_1}{2m} \frac{i^2}{x^2}. \tag{15.120}$$

We then have

$$\frac{dx_1}{dt} = x_2$$

$$\frac{dx_2}{dt} = x_3 + \frac{f_d}{m}$$

$$\frac{dx_3}{dt} = \frac{d}{dt}\left(g - \frac{rL_1}{2m} \frac{i^2}{x^2}\right).$$

x_1 is the position of the steel ball and x_2 is its velocity. The acceleration of the steel ball is $x_3 + \frac{f_d}{m}$. We calculate $\frac{dx_3}{dt}$ as follows.

$$\frac{d}{dt}\left(g - \frac{rL_1}{2m} \frac{i^2}{x^2}\right) = -\frac{rL_1}{m} \frac{i}{x^2} \frac{di}{dt} + \frac{rL_1}{m} \frac{i^2}{x^3} \frac{dx}{dt}$$

$$= -\frac{rL_1}{m} \frac{i}{x^2}\left(rL_1 \frac{i}{x^2 L(x)} v - \frac{R}{L(x)} i + \frac{1}{L(x)} u\right) + \frac{rL_1}{m} \frac{i^2}{x^3} v$$

$$= -\frac{rL_1}{m} \frac{i^2}{x^4} \frac{rL_1}{L(x)} v + \frac{rL_1}{m} \frac{i^2}{x^2} \frac{R}{L(x)} - \frac{rL_1}{m} \frac{i}{x^2} \frac{u}{L(x)} + \frac{rL_1}{m} \frac{i^2}{x^3} v$$

$$= \underbrace{-\frac{rL_1}{m} \frac{i^2}{x^4} \frac{rL_1}{L(x)} v + \frac{rL_1}{m} \frac{i^2}{x^2} \frac{R}{L(x)} + \frac{rL_1}{m} \frac{i^2}{x^3} v}_{f(i,x,v)} + \underbrace{\left(-\frac{rL_1}{m} \frac{i}{x^2 L(x)}\right)}_{g(i,x)} u.$$

15.11 Feedback Linearization Control*

In terms of the state variables x_1, x_2, and x_3, the nonlinear statespace model is

$$\frac{dx_1}{dt} = x_2$$

$$\frac{dx_2}{dt} = x_3 + \frac{f_d}{m}$$

$$\frac{dx_3}{dt} = f(i, x, v) + g(i, x)u.$$

Let $f_d = 0$ for now so this model reduces to

$$\frac{dx_1}{dt} = x_2$$

$$\frac{dx_2}{dt} = x_3$$

$$\frac{dx_3}{dt} = f(i, x, v) + g(i, x)u.$$

The point of going into this new coordinates system is that the nonlinearites are in the *same* equation as the input. Setting

$$u = \frac{-f(i, x, v) + u'}{g(i, x)}$$

we obtain the *linear* system

$$\frac{dx_1}{dt} = x_2$$

$$\frac{dx_2}{dt} = x_3$$

$$\frac{dx_3}{dt} = u'.$$

This approach is referred to as *feedback linearization* as we made the system linear by feedback rather than computing an *approximate* linear model valid only near an operating point.

Before proceeding further with this idea, let's first design a trajectory for the system. Let x_{d1} be the position reference, $x_{d2} = dx_{d1}/dt$ be the speed reference, $x_{d3} = dx_{d2}/dt$ be the acceleration reference, and $j_d = dx_{d3}/dt$ be the jerk reference so that

$$\frac{dx_{d1}}{dt} = x_{d2}$$

$$\frac{dx_{d2}}{dt} = x_{d3}$$

$$\frac{dx_{d3}}{dt} = j_d.$$

By the definitions (15.118)–(15.120) we have

$$x_d \triangleq x_{d1}, v_d \triangleq x_{d2}, \text{ and } i_d \triangleq \sqrt{\frac{2m}{rL_1}(g - x_{d3})x_{d1}^2}.$$

15 State Feedback

We want

$$\frac{dx_{d3}}{dt} = f(i_d, x_d, v_d) + g(i_d, x_d)u_d = j_d,$$

so we set the reference input voltage u_d to be

$$u_d \triangleq \frac{j_d - f(i_d, x_d, v_d)}{g(i_d, x_d)}.$$

With $\epsilon_1 = x_{d1} - x_1, \epsilon_2 = x_{d2} - x_2, \epsilon_3 = x_{d3} - x_3$ the error system is

$$\frac{d\epsilon_1}{dt} = \epsilon_2$$

$$\frac{d\epsilon_2}{dt} = \epsilon_3$$

$$\frac{d\epsilon_3}{dt} = f(i_d, x_d, v_d) + g(i_d, x_d)u_d - f(i, x, v) - g(i, x)u$$

$$= f(i_d, x_d, v_d) - f(i, x, v) + g(i_d, x_d)u_d - g(i, x)u.$$

Then choose the input u to satisfy

$$f(i_d, x_d, v_d) - f(i, x, v) + g(i_d, x_d)u_d - g(i, x)u = -\underbrace{\begin{bmatrix} k_1 & k_2 & k_3 \end{bmatrix}}_{k} \underbrace{\begin{bmatrix} \epsilon_1 \\ \epsilon_2 \\ \epsilon_3 \end{bmatrix}}_{\epsilon}.$$

That is, let the input voltage be given by (see Figure 15.22)

$$u = \frac{f(i_d, x_d, v_d) - f(i, x, v) + g(i_d, x_d)u_d + k\epsilon}{g(i, x)} = \frac{-f(i, x, v) + j_d + k\epsilon}{g(i, x)}. \quad (15.121)$$

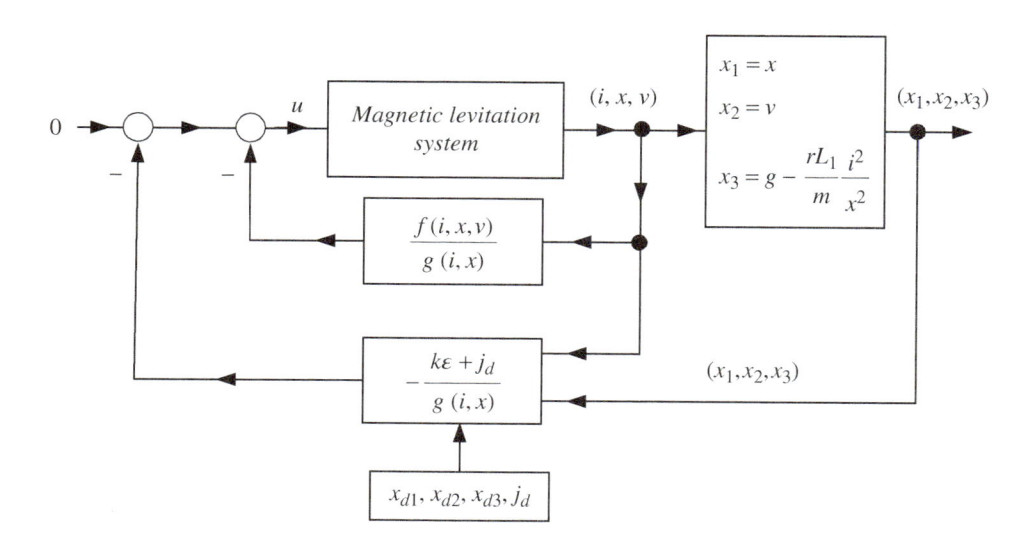

Figure 15.22. Block diagram for feedback linearization.

15.11 Feedback Linearization Control*

The error system is then

$$\frac{d\epsilon_1}{dt} = \epsilon_2$$

$$\frac{d\epsilon_2}{dt} = \epsilon_3$$

$$\frac{d\epsilon_3}{dt} = -k_1\epsilon_1 - k_2\epsilon_2 - k_3\epsilon_3.$$

In matrix form we have

$$\frac{d}{dt}\begin{bmatrix}\epsilon_1\\ \epsilon_2\\ \epsilon_3\end{bmatrix} = \begin{bmatrix}0 & 1 & 0\\ 0 & 0 & 1\\ -k_1 & -k_2 & -k_3\end{bmatrix}\begin{bmatrix}\epsilon_1\\ \epsilon_2\\ \epsilon_3\end{bmatrix}$$

with

$$\det(sI - (A_c - b_c k)) = s^3 + k_3 s^2 + k_2 s + k_1.$$

We can choose the feedback gains k_1, k_2, k_3 to place the poles of $A_c - b_c k$ at any desired location in the open left half-plane. The feedback voltage u to the amplifier is then

$$u = \frac{-f(i,x,v) + j_d + k\epsilon}{g(i,x)}.$$

In real-time we must sample the state variables i, x, v and, along with the stored variables $x_{d1} = x_d, x_{d2} = v_d, x_{d3}$ and j_d, compute u in real-time according to (15.121) as the voltage commanded to the amplifier. This nonlinear controller allows for trajectory tracking and we do *not* require (i, x, v) to be close to an equilibrium state. However, we must still make sure that the controller does not violate the voltage limit of the amplifier, that is, we must choose the feedback gains k_1, k_2, k_3 in such a way that

$$|u| = \left|\frac{-f(i,x,v) + j_d + k\epsilon}{g(i,x)}\right| \leq V_{\max}.$$

Problem 16 asks you to simulate this feedback control system.

Disturbance Rejection

Let's return to the nonlinear statespace model with the disturbance.

$$\frac{dx_1}{dt} = x_2$$

$$\frac{dx_2}{dt} = x_3 + \frac{f_d}{m}$$

$$\frac{dx_3}{dt} = f(i,x,v) + g(i,x)u.$$

The error system is then

$$\frac{d\epsilon_1}{dt} = \epsilon_2$$

$$\frac{d\epsilon_2}{dt} = \epsilon_3 - \frac{f_d}{m}$$

$$\frac{d\epsilon_3}{dt} = -f(i,x,v) + j_d - g(i,x)u.$$

15 State Feedback

We add the new error state variable

$$\epsilon_0 = \int_0^t \epsilon_1(\tau) d\tau = \int_0^t (x_{d1}(\tau) - x_1(\tau)) d\tau$$

to obtain

$$\frac{d\epsilon_0}{dt} = \epsilon_1$$

$$\frac{d\epsilon_1}{dt} = \epsilon_2$$

$$\frac{d\epsilon_2}{dt} = \epsilon_3 - \frac{f_d}{m}$$

$$\frac{d\epsilon_3}{dt} = -f(i,x,v) + j_d - g(i,x)u.$$

Set

$$u = \frac{-f(i,x,v) + j_d + k\epsilon}{g(i,x)}, \quad k\epsilon = \begin{bmatrix} k_0 & k_1 & k_2 & k_3 \end{bmatrix} \begin{bmatrix} \epsilon_0 \\ \epsilon_1 \\ \epsilon_2 \\ \epsilon_3 \end{bmatrix}$$

to obtain

$$\frac{d\epsilon_0}{dt} = \epsilon_1$$

$$\frac{d\epsilon_1}{dt} = \epsilon_2$$

$$\frac{d\epsilon_2}{dt} = \epsilon_3 - \frac{f_d}{m}$$

$$\frac{d\epsilon_3}{dt} = -k_0\epsilon_0 - k_1\epsilon_1 - k_2\epsilon_2 - k_3\epsilon_3.$$

In matrix form we have

$$\frac{d}{dt}\begin{bmatrix} \epsilon_0 \\ \epsilon_1 \\ \epsilon_2 \\ \epsilon_3 \end{bmatrix} = \begin{bmatrix} 0 & 1 & 0 & 0 \\ 0 & 0 & 1 & 0 \\ 0 & 0 & 0 & 1 \\ -k_0 & -k_1 & -k_2 & -k_3 \end{bmatrix} \begin{bmatrix} \epsilon_0 \\ \epsilon_1 \\ \epsilon_2 \\ \epsilon_3 \end{bmatrix} - \begin{bmatrix} 0 \\ 0 \\ f_d/m \\ 0 \end{bmatrix}.$$

Then

$$\begin{bmatrix} E_0(s) \\ E_1(s) \\ E_2(s) \\ E_3(s) \end{bmatrix} = \begin{bmatrix} s & -1 & 0 & 0 \\ 0 & s & -1 & 0 \\ 0 & 0 & s & -1 \\ k_0 & k_1 & k_2 & s+k_3 \end{bmatrix}^{-1} \begin{bmatrix} 0 \\ 0 \\ F_d(s)/m \\ 0 \end{bmatrix}$$

$$= \frac{1}{s^4 + k_3 s^3 + k_2 s^2 + k_1 s + k_0} \times$$

$$\begin{bmatrix} s^3 + k_3 s^2 + k_2 s + k_1 & s^2 + k_3 s + k_2 & s + k_3 & 1 \\ -k_0 & s^3 + k_3 s^2 + k_2 s & s^2 + k_3 s & s \\ -sk_0 & -k_0 - sk_1 & s^3 + k_3 s^2 & s^2 \\ -s^2 k_0 & -k_1 s^2 - k_0 s & -k_2 s^2 - k_1 s - k_0 & s^3 \end{bmatrix} \begin{bmatrix} 0 \\ 0 \\ -F_d(s)/m \\ 0 \end{bmatrix}$$

$$= -\frac{1}{s^4 + k_3 s^3 + k_2 s^2 + k_1 s + k_0} \begin{bmatrix} s + k_3 \\ s^2 + k_3 s \\ s^3 + k_3 s^2 \\ -k_2 s^2 - k_1 s - k_0 \end{bmatrix} \frac{F_d(s)}{m}.$$

Let $F_d(s) = \dfrac{F_{d0}}{s}$ and set

$$k_0 = r_1 r_2 r_3 r_4$$
$$k_1 = r_1 r_2 r_3 + r_1 r_2 r_4 + r_1 r_3 r_4 + r_2 r_3 r_4$$
$$k_1 = r_1 r_2 + r_1 r_3 + r_1 r_4 + r_2 r_3 + r_2 r_4 + r_3 r_4$$
$$k_3 = r_1 + r_2 + r_3 + r_4.$$

Then

$$E_1(s) = -\frac{s^2 + k_3 s}{(s+r_1)(s+r_2)(s+r_3)(s+r_4)} \frac{F_{d0}}{m} \frac{1}{s} = -\frac{s + k_3}{(s+r_1)(s+r_2)(s+r_3)(s+r_4)} \frac{F_{d0}}{m}.$$

With $r_1 > 0, r_2 > 0, r_3 > 0, r_4 > 0$ it follows that

$$\epsilon_1(\infty) = \lim_{t \to \infty} \epsilon_1(t) = \lim_{s \to 0} s E_1(s) = 0.$$

Remark *Feedback Linearization Control*

It turns out that the approach of feedback linearization, which was used above for the magnetic levitation system is *not* possible for the inverted pendulum [55–57]. That is, there is no transformation to a new coordinate system in which the nonlinearities can be canceled out by state feedback. However, it can be used for synchronous motors (brushless DC and stepper motors), power electronic converters, and series DC motors [7, 58, 59, 60, 70]. There is a related approach referred to as *input–output linearization*, which is applicable to induction machines [7].

Appendix: Disturbance Rejection in the Statespace

The proof of the following theorem is adapted from Chen [17].

Theorem 4 *Controllability of the Augmented System*

Consider the system $dx/dt = Ax + bu$ where $A \in \mathbb{R}^{n \times n}$ and $b \in \mathbb{R}^n$ are *controllable* in $\mathbb{R}^n$, i.e.,

$$\det \mathcal{C} = \det \begin{bmatrix} b & Ab & \cdots & A^{n-1}b \end{bmatrix} \neq 0.$$

Let the output be given by $y = cx$, $c \in \mathbb{R}^{1 \times n}$. Define the augmented system matrices

$$A_a \triangleq \begin{bmatrix} A & 0 \\ -c & 0 \end{bmatrix} \in \mathbb{R}^{(n+1) \times (n+1)}, \quad b_a \triangleq \begin{bmatrix} b \\ 0 \end{bmatrix} \in \mathbb{R}^{n+1}.$$

The corresponding *controllability matrix* is

$$\mathcal{C}_a = \begin{bmatrix} b_a & A_a b_a & \cdots & A_a^n b_a \end{bmatrix} \in \mathbb{R}^{(n+1) \times (n+1)}.$$

Then
$$\det \mathcal{C}_a \neq 0$$

if and only if the *open-loop* transfer function
$$G(s) = \frac{n(s)}{d(s)} \triangleq c(sI_{n\times n} - A)^{-1}b$$

satisfies
$$n(0) \neq 0.$$

That is, if the pair (A, b) is controllable and $n(0) \neq 0$, then one can choose $\begin{bmatrix} k & k_0 \end{bmatrix}$ to arbitrarily assign the eigenvalues of

$$\begin{bmatrix} A & 0 \\ -c & 0 \end{bmatrix} - \begin{bmatrix} b \\ 0 \end{bmatrix} \begin{bmatrix} k & k_0 \end{bmatrix} = \begin{bmatrix} A - bk & -bk_0 \\ -c & 0 \end{bmatrix}.$$

Proof. The controllability matrix of the *augmented* system is

$$\mathcal{C}_a = \begin{bmatrix} b & Ab & A^2b & \cdots & A^nb \\ 0 & -cb & -cAb & \cdots & -cA^{n-1}b \end{bmatrix} \in \mathbb{R}^{(n+1)\times(n+1)}.$$

For clarity of exposition let $n = 4$. To (slightly) simplify the presentation we multiply the last row by -1 to obtain the matrix $\mathcal{C}'_a$, which has the same rank as $\mathcal{C}_a$. We now show the matrix

$$\mathcal{C}'_a \triangleq \begin{bmatrix} b & Ab & A^2b & A^3b & A^4b \\ 0 & cb & cAb & cA^2b & cA^3b \end{bmatrix} \in \mathbb{R}^{5\times 5}$$

has full rank, i.e., it satisfies $\det \mathcal{C}'_a \neq 0$. The original (unaugmented) open-loop system has a transfer function of the form

$$G(s) = c(sI_{4\times 4} - A)^{-1}b = \frac{n(s)}{d(s)} = \frac{\beta_3 s^3 + \beta_2 s^2 + \beta_1 s + \beta_0}{s^4 + \alpha_3 s^3 + \alpha_2 s^2 + \alpha_1 s + \alpha_0}.$$

Note that $n(0) \neq 0$ is the same as $\beta_0 \neq 0$. As the pair (A, b) is given to be controllable in $\mathbb{R}^4$, we can transform the (unaugmented) system (c, A, b) to control canonical form. That is, there is a T such that

$$TAT^{-1} = A_c = \begin{bmatrix} 0 & 1 & 0 & 0 \\ 0 & 0 & 1 & 0 \\ 0 & 0 & 0 & 1 \\ -\alpha_0 & -\alpha_1 & -\alpha_2 & -\alpha_3 \end{bmatrix}, \quad b_c = Tb = \begin{bmatrix} 0 \\ 0 \\ 0 \\ 1 \end{bmatrix}$$

$$c_c = cT^{-1} = \begin{bmatrix} \beta_0 & \beta_1 & \beta_2 & \beta_3 \end{bmatrix},$$

where
$$\begin{aligned} b_c &= Tb & c_c b_c &= cb \\ A_c b_c &= TAb & c_c A_c b_c &= cAb \\ A_c^2 b_c &= TA^2 b & c_c A_c^2 b_c &= cA^2 b \\ A_c^3 b_c &= TA^3 b & c_c A_c^3 b_c &= cA^3 b \\ A_c^4 b_c &= TA^4 b & c_c A_c^4 b_c &= cA^4 b \end{aligned}$$

Appendix: Disturbance Rejection in the Statespace 625

and
$$G(s) = c(sI_{4\times 4} - A)^{-1}b = c_c(sI_{4\times 4} - A_c)^{-1}b_c.$$

Using these relationships we may write $\mathcal{C}'_a$ as

$$\mathcal{C}'_a = \begin{bmatrix} b & Ab & A^2b & A^3b & A^4b \\ 0 & cb & cAb & cA^2b & cA^3b \end{bmatrix} = \begin{bmatrix} T^{-1} & 0_{4\times 1} \\ 0_{1\times 4} & 1 \end{bmatrix}$$

$$\times \underbrace{\begin{bmatrix} b_c & A_c b_c & A_c^2 b_c & A_c^3 b_c & A_c^4 b_c \\ 0 & c_c b_c & c_c A_c b_c & c_c A_c^2 b_c & c_c A_c^3 b_c \end{bmatrix}}_{\mathcal{C}'_{ac}}.$$

A direct computation gives

$$\begin{bmatrix} b_c & A_c b_c & A_c^2 b_c & A_c^3 b_c & A_c^4 b_c \end{bmatrix}$$

$$= \begin{bmatrix} 0 & 0 & 0 & 1 & -\alpha_3 \\ 0 & 0 & 1 & -\alpha_3 & \alpha_3^2 - \alpha_2 \\ 0 & 1 & -\alpha_3 & \alpha_3^2 - \alpha_2 & \alpha_2\alpha_3 - \alpha_1 + \alpha_3(\alpha_2 - \alpha_3^2) \\ 1 & -\alpha_3 & \alpha_3^2 - \alpha_2 & \alpha_2\alpha_3 - \alpha_1 + \alpha_3(\alpha_2 - \alpha_3^2) & \alpha_2^2 - 3\alpha_2\alpha_3^2 + \alpha_3^4 + 2\alpha_1\alpha_3 - \alpha_0 \end{bmatrix}.$$

More computation gives

$$\begin{bmatrix} c_c b_c & c_c A_c b_c & c_c A_c^2 b_c & c_c A_c^3 b_c \end{bmatrix}$$

$$= \begin{bmatrix} \beta_0 & \beta_1 & \beta_2 & \beta_3 \end{bmatrix} \begin{bmatrix} 0 & 0 & 0 & 1 \\ 0 & 0 & 1 & -\alpha_3 \\ 0 & 1 & -\alpha_3 & \alpha_3^2 - \alpha_2 \\ 1 & -\alpha_3 & \alpha_3^2 - \alpha_2 & \alpha_2\alpha_3 - \alpha_1 + \alpha_3(\alpha_2 - \alpha_3^2) \end{bmatrix}$$

$$= \begin{bmatrix} \beta_3 & \beta_2 - \alpha_3\beta_3 & \beta_1 - \alpha_3\beta_2 - \beta_3(\alpha_2 - \alpha_3^2) & \beta_0 - \beta_1\alpha_3 + \beta_3(\alpha_2\alpha_3 - \alpha_1 + \alpha_3(\alpha_2 - \alpha_3^2)) - \beta_2(\alpha_2 - \alpha_3^2) \end{bmatrix}.$$

Then

$$\begin{bmatrix} b_c & A_c b_c & A_c^2 b_c & A_c^3 b_c & A_c^4 b_c \\ 0 & c_c b_c & c_c A_c b_c & c_c A_c^2 b_c & c_c A_c^3 b_c \end{bmatrix}$$

$$= \begin{bmatrix} 0 & 0 & 0 & 1 & -\alpha_3 \\ 0 & 0 & 1 & -\alpha_3 & \alpha_3^2 - \alpha_2 \\ 0 & 1 & -\alpha_3 & \alpha_3^2 - \alpha_2 & \alpha_2\alpha_3 - \alpha_1 + \alpha_3(\alpha_2 - \alpha_3^2) \\ 1 & -\alpha_3 & \alpha_3^2 - \alpha_2 & \alpha_2\alpha_3 - \alpha_1 + \alpha_3(\alpha_2 - \alpha_3^2) & \alpha_2^2 - 3\alpha_2\alpha_3^2 + \alpha_3^4 + 2\alpha_1\alpha_3 - \alpha_0 \\ 0 & \beta_3 & \beta_2 - \alpha_3\beta_3 & \beta_1 - \alpha_3\beta_2 - \beta_3(\alpha_2 - \alpha_3^2) & \beta_0 - \beta_1\alpha_3 + \beta_3(\alpha_2\alpha_3 - \alpha_1 + \alpha_3(\alpha_2 - \alpha_3^2)) - \beta_2(\alpha_2 - \alpha_3^2) \end{bmatrix}.$$

To show this has full rank if $\beta_0 \neq 0$, multiply the third row by $-\beta_3$ and add to the last row to obtain the matrix

$$\begin{bmatrix} 0 & 0 & 0 & 1 & -\alpha_3 \\ 0 & 0 & 1 & -\alpha_3 & \alpha_3^2 - \alpha_2 \\ 0 & 1 & -\alpha_3 & \alpha_3^2 - \alpha_2 & \alpha_2\alpha_3 - \alpha_1 + \alpha_3(\alpha_2 - \alpha_3^2) \\ 1 & -\alpha_3 & \alpha_3^2 - \alpha_2 & \alpha_2\alpha_3 - \alpha_1 + \alpha_3(\alpha_2 - \alpha_3^2) & \alpha_2^2 - 3\alpha_2\alpha_3^2 + \alpha_3^4 + 2\alpha_1\alpha_3 - \alpha_0 \\ 0 & 0 & \beta_2 & \beta_1 - \alpha_3\beta_2 & \beta_0 - \beta_1\alpha_3 - \beta_2(\alpha_2 - \alpha_3^2) \end{bmatrix}.$$

Next, multiplying the second row of this matrix by $-\beta_2$ and adding it to the last row gives the matrix

$$\begin{bmatrix} 0 & 0 & 0 & 1 & -\alpha_3 \\ 0 & 0 & 1 & -\alpha_3 & \alpha_3^2 - \alpha_2 \\ 0 & 1 & -\alpha_3 & \alpha_3^2 - \alpha_2 & \alpha_2\alpha_3 - \alpha_1 + \alpha_3(\alpha_2 - \alpha_3^2) \\ 1 & -\alpha_3 & \alpha_3^2 - \alpha_2 & \alpha_2\alpha_3 - \alpha_1 + \alpha_3(\alpha_2 - \alpha_3^2) & \alpha_2^2 - 3\alpha_2\alpha_3^2 + \alpha_3^4 + 2\alpha_1\alpha_3 - \alpha_0 \\ 0 & 0 & 0 & \beta_1 & \beta_0 - \beta_1\alpha_3 \end{bmatrix}$$

Finally, multiply the first row of this matrix by $-\beta_1$ and add to the last row to obtain the matrix

$$\begin{bmatrix} 0 & 0 & 0 & 1 & -\alpha_3 \\ 0 & 0 & 1 & -\alpha_3 & \alpha_3^2 - \alpha_2 \\ 0 & 1 & -\alpha_3 & \alpha_3^2 - \alpha_2 & \alpha_2\alpha_3 - \alpha_1 + \alpha_3(\alpha_2 - \alpha_3^2) \\ 1 & -\alpha_3 & \alpha_3^2 - \alpha_2 & \alpha_2\alpha_3 - \alpha_1 + \alpha_3(\alpha_2 - \alpha_3^2) & \alpha_2^2 - 3\alpha_2\alpha_3^2 + \alpha_3^4 + 2\alpha_1\alpha_3 - \alpha_0 \\ 0 & 0 & 0 & 0 & \beta_0 \end{bmatrix}.$$

This last matrix is full rank if and only if $\beta_0 \neq 0$. This is the same as saying that $n(s)$ is *not* zero at $s = 0$. With the assumption that $\beta_0 \neq 0$ we can choose $\begin{bmatrix} k & k_0 \end{bmatrix}$ so that

$$\begin{bmatrix} A - bk & -bk_0 \\ c & 0 \end{bmatrix}$$

is stable. ∎

Problems

Problem 1 *Trajectory Tracking for a Double Integrator System*
Consider the system

$$\frac{d}{dt}\begin{bmatrix} x_1 \\ x_2 \end{bmatrix} = \underbrace{\begin{bmatrix} 0 & 1 \\ 0 & 0 \end{bmatrix}}_{A}\begin{bmatrix} x_1 \\ x_2 \end{bmatrix} + \underbrace{\begin{bmatrix} 0 \\ 1 \end{bmatrix}}_{b} u.$$

We can interpret x_1 as angular or linear position, x_2 as velocity and the input u as acceleration. In such an application the physical input would be torque or force making the acceleration input either torque divided by the moment of inertia or force divided by the mass.

Remark If $y = x_1$ then, as the reader can show, $Y(s)/U(s) = 1/s^2$ and this is the reason for the terminology *double integrator* system.

Let x_{d1}, x_{d2} be the desired position trajectory and speed trajectory and u_d be the corresponding input which satisfy

$$\frac{d}{dt}\begin{bmatrix} x_{1d} \\ x_{2d} \end{bmatrix} = \underbrace{\begin{bmatrix} 0 & 1 \\ 0 & 0 \end{bmatrix}}_{A}\begin{bmatrix} x_{d1} \\ x_{d2} \end{bmatrix} + \underbrace{\begin{bmatrix} 0 \\ 1 \end{bmatrix}}_{b} u_d.$$

Define the error state as

$$\varepsilon \triangleq x_d - x$$

or
$$\varepsilon = \begin{bmatrix} \varepsilon_1 \\ \varepsilon_2 \end{bmatrix} = \begin{bmatrix} x_{d1} \\ x_{d2} \end{bmatrix} - \begin{bmatrix} x_1 \\ x_2 \end{bmatrix}.$$

The error state satisfies
$$\dot{\varepsilon} = A\varepsilon + b(u_d - u) = A\varepsilon + bw,$$

where
$$w \triangleq u_d - u.$$

We let
$$w = -k\varepsilon = -\begin{bmatrix} k_1 & k_2 \end{bmatrix}\begin{bmatrix} \varepsilon_1 \\ \varepsilon_2 \end{bmatrix} = -k_1(x_{d1} - x_1) - k_2(x_{d2} - x_2).$$

so that
$$\dot{\varepsilon} = (A - bk)\varepsilon.$$

The feedback setup is illustrated in Figure 15.23.

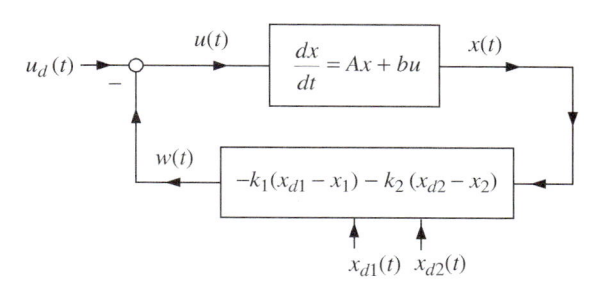

Figure 15.23. Block diagram for a state feedback trajectory tracking controller.

By direct calculation find the values of k_1, k_2 such that the roots of $\det(sI - (A - bk))$ are at $-r_1, -r_2$. In particular, if $r_1 = r_2 = 10$ what are the values of k_1, k_2?

Problem 2 *State Feedback Control of the Magnetic Levitation System*

Recall from Chapter 13 the magnetic levitation system with the amplifier configured for current command. The schematic is again shown in Figure 15.24.

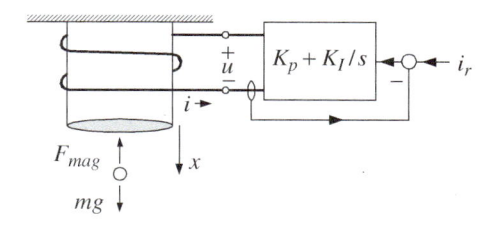

Figure 15.24. Current command amplifier for the magnetic levitation system.

With K_I, K_p chosen so that $i(t) \to i_r(t)$ fast compared with the motion of the steel ball we can consider the current $i(t)$ as the input. As shown in Chapter 13, the nonlinear

628   15 State Feedback

statespace model for this system is

$$\frac{dx}{dt} = v$$
$$\frac{dv}{dt} = g - \frac{rL_1}{2m}\frac{i^2}{x^2}.$$

(a) Let x_{eq} denote the desired distance below the electromagnet we want the steel ball to be located. The desired equilibrium point is

$$\begin{bmatrix} x_{eq} \\ v_{eq} \end{bmatrix} = \begin{bmatrix} x_{eq} \\ 0 \end{bmatrix},$$

where the reference input (current in the electromagnet) i_{eq} satisfies

$$g - \frac{rL_1}{2m}\frac{i_{eq}^2}{x_{eq}^2} = 0.$$

It was also shown in Chapter 13 that a linear statespace model valid about this equilibrium point is

$$\frac{d}{dt}\begin{bmatrix} x - x_{eq} \\ v - v_{eq} \end{bmatrix} = \underbrace{\begin{bmatrix} 0 & 1 \\ \frac{2g}{x_{eq}} & 0 \end{bmatrix}}_{A}\begin{bmatrix} x - x_{eq} \\ v - v_{eq} \end{bmatrix} + \underbrace{\begin{bmatrix} 0 \\ -\frac{2g}{i_{eq}} \end{bmatrix}}_{b}(i - i_{eq}).$$

With

$$\begin{bmatrix} z_1 \\ z_2 \end{bmatrix} \triangleq \begin{bmatrix} x - x_{eq} \\ v - v_{eq} \end{bmatrix}, \quad w \triangleq i - i_{eq}$$

the linear statespace model may be written as

$$\frac{d}{dt}\begin{bmatrix} z_1 \\ z_2 \end{bmatrix} = A\begin{bmatrix} z_1 \\ z_2 \end{bmatrix} + bw.$$

Let

$$w = -\begin{bmatrix} k_1 & k_2 \end{bmatrix}\begin{bmatrix} z_1 \\ z_2 \end{bmatrix}$$

and find the values of k_1, k_2 such that the closed-loop poles are at $-r_1, -r_2$.

(b) Simulate the state feedback current controlled magnetic levitation system of part (a). A SIMULINK block diagram for doing this is shown in Figure 15.25.
From QUANSER [34] set $L_0 = 0.4125$ H, $R = 11$ Ω, $r = 1.27 \times 10^{-2}$ m (1.27 cm), $m = 0.068$ kg, $g = 9.81$ m/s², $x_{eq} = r + 0.008$ m, $v_{eq} = 0, i_{eq} = \sqrt{\frac{2mg}{rL_1}}x_{eq}$, and $V_{max} = 26$ V. Also set $rL_1 = 6.5308 \times 10^{-4}$ Nm²/A² so $K_m = rL_1/2$. As in Problem 10 of Chapter 13 (see Figure 13.24) let the initial conditions be $x(0) = 0.014 + r, v(0) = 0$, with $i(0)$ set to satisfy $mg = K_m i^2(0)/x^2(0)$. Try using $K_p = 1000, K_I = 900$ and consider placing the two closed-loop poles for position and speed at -10.
Hint: Modify the SIMULINK block diagrams shown in Problem 10 of Chapter 13.

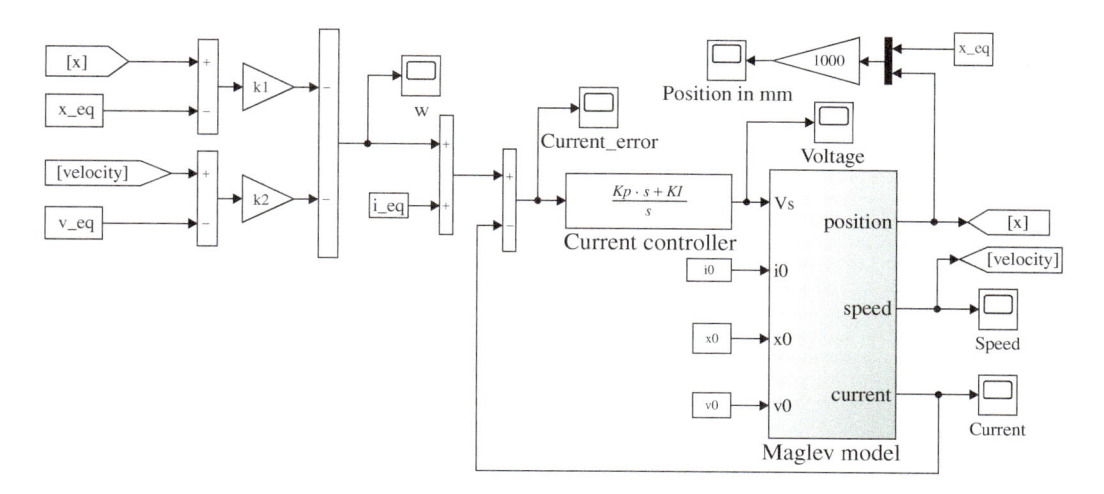

Figure 15.25. Statespace controller for the magnetic levitation system.

Remark QUANSER denotes the force constant as K_m instead of $rL_1/2$ with the value $K_m = 6.5308 \times 10^{-5}$ Nm^2/A^2. However, in the magnetic force expression QUANSER considers the position x of the steel ball to be the distance from the bottom of the magnet to the *top* of the steel ball rather than to its center of mass. Instead of setting $rL_1/2$ equal to this value of K_m, we set $rL_1/2 = 5K_m$ (or $rL_1 = 10K_m$) to compensate for this difference in the reference point for measuring x.

Problem 3 *State Feedback Tracking Control of the Cart*

Consider the cart on a track system of Example 2 (page 571). Use the QUANSER [34] parameter values given as $K_T = K_b = 7.67 \times 10^{-3}$ Nm/A, $R = 2.6$ Ω, $L = 0.18$ mH, $M = 0.57$ kg, $r_m = 6.35 \times 10^{-3}$ m, $r_g = n_2/n_1 = 3.71$ and $J = 0$ in the expressions for $a_0, b_0,$ and K_d. Set the maximum voltage as $V_{\max} = 5$ V and the simulation step size as $T = 0.001$ second. If you include the encoder in your model, use $K_{enc} = 4 \times 1024 = 4096$ counts/rev, $r_{encoder} = R_{enc} = 0.01483$ m, $g = 9.8$ m/s^2, $K_{enc} = (2\pi/N_{enc})r_{enc} = 2.275 \times 10^{-5}$ m/count. You may have already done all of this if you did Problem 12 of Chapter 13. Implement the SIMULINK simulation given in Figure 15.26) and run the simulation for the four cases (a), (b), (c), and (d). In each case, hand in your .m file, a screenshot of your SIMULINK block diagram, a plot of x and x_{ref}, and a plot of $e_1 = x_{ref} - x$. The SIMULINK block diagram for the trajectory generator is available in the student simulation files for this chapter.

(a) In the simulation implement the state feedback controller shown in Figure 15.6 of Example 2 of this chapter. Adjust the location of the closed-loop poles until you achieve good tracking without saturating the amplifier.

(b) In your simulation of part (a), remove (or zero out) the voltage reference v_{ref} and the gain k_2. The feedback is then given by $V(t) = -k_1 \left(x_{ref}(t) - x(t)\right)$, i.e., a simple proportional controller. Run the simulation and you should still get very good tracking. (If you don't, increase the gain k_1.)

(c) Go back to the state feedback trajectory tracking controller simulation in part (a) and add to the input of the cart the equivalent voltage disturbance given by $d/b_0 = K_D F_d = K_D Mg\sin(\phi)$ with $\phi = \tan^{-1}(1/2)$ or $28.65°$ (i.e., add $-d/b_0$ to the input in Figure 15.6 as shown in Figure 15.9). Run the simulation. You should see error in the position tracking.

15 State Feedback

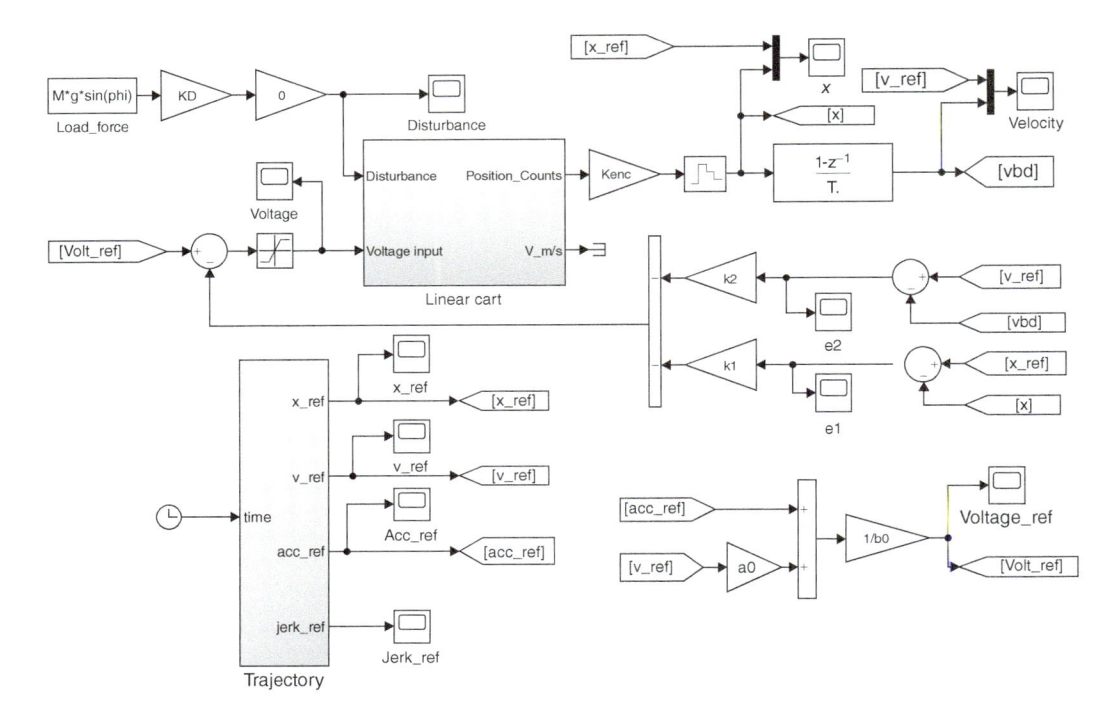

Figure 15.26. SIMULINK diagram for cart on the track state feedback controller. See Figure 13.28 of Chapter 13 for the inside of the Linear Cart block.

(d) To the simulation of part (c) add the integrator as shown in Figure 15.9 of Example 7. With appropriately chosen feedback gains k_0, k_1, and k_2, run the simulation to show that $e_1(t) = x_{ref}(t) - x(t) \to 0$. Add a scope that plots $V(t)$ and $K_D M g \sin(\phi)$ together to show that $V(t) \to K_D M g \sin(\phi)$.

Problem 4 *Disturbance Rejection for the Cart on the Track System*

The augmented cart model has system matrices (see (15.73))

$$A_a = \begin{bmatrix} 0 & 1 & 0 \\ 0 & 0 & 1 \\ 0 & 0 & -a_0 \end{bmatrix}, \quad b_a = \begin{bmatrix} 0 \\ 0 \\ b_0 \end{bmatrix}, \quad p_a = \begin{bmatrix} 0 \\ 0 \\ 1 \end{bmatrix}.$$

With $k_a = \begin{bmatrix} k_0 & k_1 & k_2 \end{bmatrix}$ show by direct computation that

$$\begin{bmatrix} \epsilon_0(\infty) \\ \epsilon_1(\infty) \\ \epsilon_2(\infty) \end{bmatrix} = -(A_a - b_a k_a)^{-1} p_a d = \begin{bmatrix} \dfrac{d}{b_0 k_0} \\ 0 \\ 0 \end{bmatrix}.$$

Problem 5 *Control Canonical Form*

Let a general fourth-order statespace system be given by

$$\frac{dx}{dt} = Ax + bu,$$

where
$$A = \begin{bmatrix} a_{11} & a_{12} & a_{13} & a_{14} \\ a_{21} & a_{22} & a_{23} & a_{24} \\ a_{31} & a_{32} & a_{33} & a_{34} \\ a_{41} & a_{42} & a_{43} & a_{44} \end{bmatrix}, \quad b = \begin{bmatrix} b_1 \\ b_2 \\ b_3 \\ b_4 \end{bmatrix}$$

and
$$\det(sI - A) = s^4 + \alpha_3 s^3 + \alpha_2 s^2 + \alpha_1 s + \alpha_0.$$

It is desired to find an invertible matrix T such that with
$$x_c \triangleq Tx,$$

the statespace model in the x_c coordinates is in control canonical form. That is,
$$\frac{dx_c}{dt} = T\frac{dx}{dt} = TAx + Tbu = \underbrace{TAT^{-1}}_{A_c} x_c + \underbrace{Tb}_{b_c} u$$

$$= \begin{bmatrix} 0 & 1 & 0 & 0 \\ 0 & 0 & 1 & 0 \\ 0 & 0 & 0 & 1 \\ -\alpha_0 & -\alpha_1 & -\alpha_2 & -\alpha_3 \end{bmatrix} x_c + \begin{bmatrix} 0 \\ 0 \\ 0 \\ 1 \end{bmatrix} u.$$

(a) Define the *controllability* matrix
$$\mathcal{C} \triangleq \begin{bmatrix} b & Ab & A^2b & A^3b \end{bmatrix} \in \mathbb{R}^{4 \times 4}$$

and assume $\det(\mathcal{C}) \neq 0$. Let
$$x' \triangleq \mathcal{C}^{-1} x.$$

Find the statespace representation in the x' coordinates. Explain your steps. (Hint: See Section 15.6.)

(b) Now let
$$x_c \triangleq \mathcal{C}_c x',$$

where
$$\mathcal{C}_c \triangleq \begin{bmatrix} b_c & A_c b_c & A_c^2 b_c & A_c^3 b_c \end{bmatrix} = \begin{bmatrix} 0 & 0 & 0 & 1 \\ 0 & 0 & 1 & -\alpha_3 \\ 0 & 1 & -\alpha_3 & \alpha_3^2 - \alpha_2 \\ 1 & -\alpha_3 & \alpha_3^2 - \alpha_2 & -\alpha_3^3 + 2\alpha_2\alpha_3 - \alpha_1 \end{bmatrix}.$$

It can be straightforwardly shown that
$$\mathcal{C}_c^{-1} = \begin{bmatrix} \alpha_1 & \alpha_2 & \alpha_3 & 1 \\ \alpha_2 & \alpha_3 & 1 & 0 \\ \alpha_3 & 1 & 0 & 0 \\ 1 & 0 & 0 & 0 \end{bmatrix}.$$

Show that the statespace transformation $x_c = Tx$ with $T \triangleq \mathcal{C}_c \mathcal{C}^{-1}$ results in the statespace representation in the x_c coordinates being in control canonical form.

15 State Feedback

Problem 6 *Trajectory Design*

Integrate the velocity reference given in (15.13) to obtain the position reference given in (15.14).

Problem 7 *State Feedback*

Consider the model of a system in statespace form given by

$$\frac{dx}{dt} = Ax + bu, \quad A \in \mathbb{R}^{n \times n}, \quad b \in \mathbb{R}^n.$$

Let x_d be a desired reference trajectory and u_d be the corresponding reference input where the pair x_d, u_d satisfy the system equations as well, that is,

$$\frac{dx_d}{dt} = Ax_d + bu_d.$$

Define the error state by $\epsilon \triangleq x_d - x$ and the error input by $w = u_d - u$.

(a) Give the state equations for ϵ.

(b) Using the state feedback $w = -k\epsilon$, write down the solution to your answer in part (a) in terms of an exponential matrix.

(c) Draw a block diagram of the complete state feedback trajectory tracking system. Be sure to label the signals and what is inside each block.

(d) Answer yes or no. $e^{(A-bk)t} \to 0_{n \times n}$ if and only if all the roots of the nth degree characteristic polynomial $\det(sI - (A - bk)) = 0$ have negative real parts.

Problem 8 *Pole Placement*

Suppose that the statespace trajectory tracking error model of a system is given by

$$\frac{d\epsilon}{dt} = \begin{bmatrix} 0 & 1 \\ 0 & 0 \end{bmatrix} \epsilon + \begin{bmatrix} 0 \\ 1 \end{bmatrix} w.$$

(a) Using the state feedback $w = -k\epsilon$, show how to choose the state feedback gain vector k so that the closed-loop poles are at -5 and -10. Show work!

(b) Using the k chosen in part (a), will $e^{(A-bk)t} \to 0_{2 \times 2}$ as $t \to \infty$. Just answer yes or no.

Problem 9 *Pole Placement*

Suppose that the statespace trajectory tracking error model of a system is given by

$$\frac{d\epsilon}{dt} = \begin{bmatrix} 0 & 1 \\ -1 & 0 \end{bmatrix} \epsilon + \begin{bmatrix} 0 \\ 1 \end{bmatrix} w.$$

(a) Using the state feedback $w = -k\epsilon$, show how to choose the state feedback gain vector k so that the closed-loop poles are at -5 and -10. Show work!

(b) Using the k chosen in part (a), will $e^{(A-bk)t} \to 0_{2 \times 2}$ as $t \to \infty$? Just answer yes or no.

Problem 10 Pole Placement
Suppose that the statespace trajectory tracking error model of a system is given by

$$\frac{d\epsilon}{dt} = \begin{bmatrix} 0 & 1 \\ -5 & -6 \end{bmatrix} \epsilon + \begin{bmatrix} 0 \\ 1 \end{bmatrix} w.$$

(a) Using the state feedback $w = -k\epsilon$, show how to choose the state feedback gain vector k so that the closed-loop poles are at -3 and -2. Show work!

(b) Using the k chosen in part (a), will $e^{(A-bk)t} \to 0_{2\times 2}$ as $t \to \infty$? Just answer yes or no.

Problem 11 Pole Placement for the Inverted Pendulum
Recall the linear statespace model for the inverted pendulum around the equilibrium point $(x_{eq}, v_{eq}, \theta_{eq}, \omega_{eq}) = (0, 0, 0, 0)$ given by

$$\underbrace{\begin{bmatrix} dx/dt \\ dv/dt \\ d\theta/dt \\ d\omega/dt \end{bmatrix}}_{dz/dt} = \underbrace{\begin{bmatrix} 0 & 1 & 0 & 0 \\ 0 & 0 & -\dfrac{gm^2\ell^2}{Mm\ell^2 + J(M+m)} & 0 \\ 0 & 0 & 0 & 1 \\ 0 & 0 & \dfrac{mg\ell(M+m)}{Mm\ell^2 + J(M+m)} & 0 \end{bmatrix}}_{A} \begin{bmatrix} x \\ v \\ \theta \\ \omega \end{bmatrix} + \underbrace{\begin{bmatrix} 0 \\ \dfrac{J+m\ell^2}{Mm\ell^2 + J(M+m)} \\ 0 \\ -\dfrac{m\ell}{Mm\ell^2 + J(M+m)} \end{bmatrix}}_{b} u.$$

The QUANSER [34] parameter values are

$M = 0.57$ kg (cart mass)

$m = 0.23$ kg (pendulum mass)

$g = 9.81$ m/s^2 (acceleration due to gravity)

$\ell = 0.6413/2$ m (pendulum rod length divided by 2)

$J = 7.88 \times 10^{-3}$ kg-m^2 (pendulum rod moment of inertia).

Use MATLAB for parts (a)–(d).

(a) Compute the controllability matrix $\mathcal{C}$ for the system (A, b).

(b) A simple computation shows that $\det(sI - A) = s^4 - \alpha^2 s^2$ with $\alpha^2 = \dfrac{mg\ell(M+m)}{Mm\ell^2 + J(M+m)}$. Use this to compute the control canonical form (A_c, b_c) for the inverted pendulum.

(c) Use your answer in part (b) to compute the controllability matrix $\mathcal{C}_c$ for the system matrices (A_c, b_c) in control canonical form.

(d) Compute the feedback gain k that places the poles at $-r_1, -r_2, -r_3, -r_4$. Use the pole placement procedure given in Section 15.7.

(e) Using Figure 15.27 as a guide, simulate this feedback control system in SIMULINK using the *nonlinear* statespace model of the inverted pendulum given in Figure 13.19 of Problem 6 in Chapter 13.

634   15 State Feedback

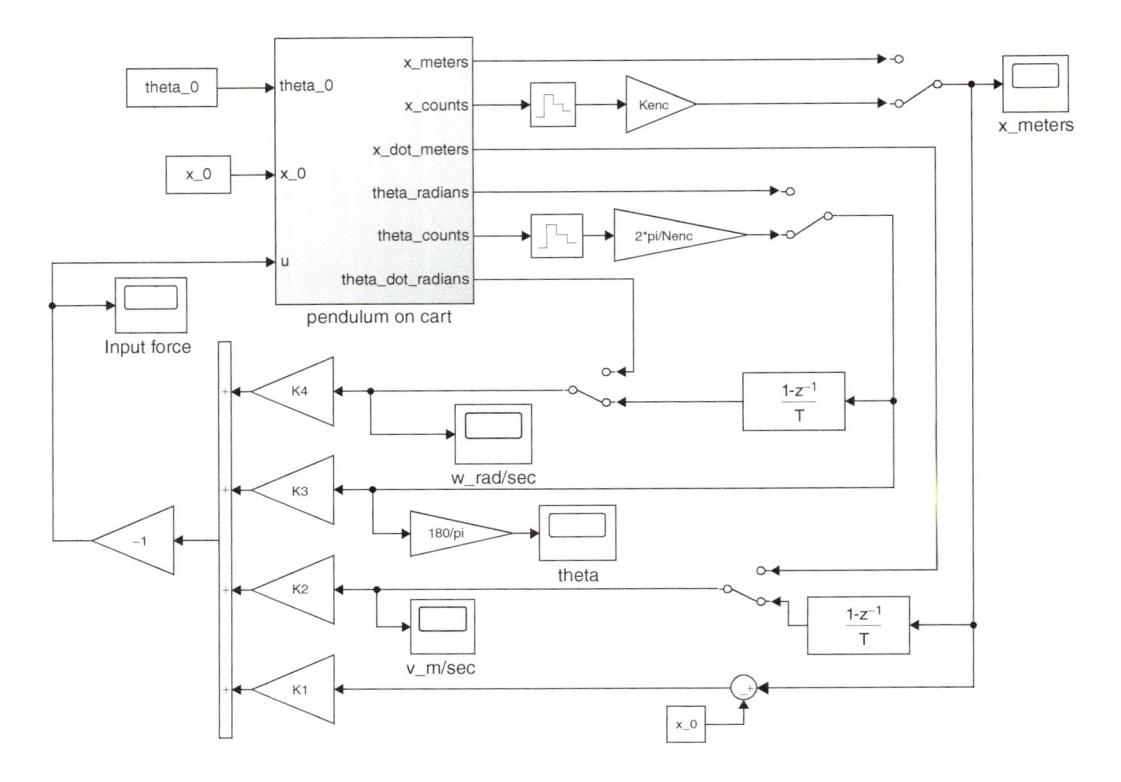

Figure 15.27. SIMULINK block diagram for state feedback control of the inverted pendulum. The inside of the Pendulum on Cart block is given in Figure 13.19 of Chapter 13.

Problem 12 *Tracking Step Inputs in the Cart Position of the Inverted Pendulum*

This problem asks you to simulate the controller of Section 15.9 that has the cart of the inverted pendulum track step inputs. Do this by following parts (a)–(f) using the same parameter values while keeping the rod vertical as in Problem 11.

(a) With A, b for the inverted pendulum given as in Problem 11 and $c = \begin{bmatrix} 1 & 0 & 0 & 0 \end{bmatrix}$, the augmented model for tracking step inputs is

$$\underbrace{\frac{d}{dt}\begin{bmatrix} z \\ z_0 \end{bmatrix}}_{z_a \in \mathbb{R}^5} = \underbrace{\begin{bmatrix} A & 0_{4\times 1} \\ -c & 0 \end{bmatrix}}_{A_a \in \mathbb{R}^{5\times 5}} \underbrace{\begin{bmatrix} z \\ z_0 \end{bmatrix}}_{z_a \in \mathbb{R}^5} + \underbrace{\begin{bmatrix} b \\ 0 \end{bmatrix}}_{b_a \in \mathbb{R}^5} u + \underbrace{\begin{bmatrix} 0_{4\times 1} \\ 1 \end{bmatrix}}_{p_a \in \mathbb{R}^5} x_{ref}.$$

Rewrite this system of equations by inserting the explicit matrices for A, b and c.

(b) Compute $\det(sI_{5\times 5} - A_a) = s^5 + \alpha_4 s^4 + \alpha_3 s^3 + \alpha_2 s^2 + \alpha_1 s + \alpha_0$ to find the values of $\alpha_4, \alpha_3, \alpha_2, \alpha_1, \alpha_0$.

(c) Using your answer to part (b) give the control canonical form A_{ac}, b_{ac} for A_a, b_a.

(d) In a MATLAB .m file code A_a, b_a and compute the controllability matrix $\mathcal{C}_a$ for the pair (A_a, b_a).

(e) Add to your MATLAB .m file the code to compute the control canonical form of A_a and b_a denoted as A_{ac} and b_{ac}, respectively. Also add code to compute the controllability matrix $\mathcal{C}_{ac}$ for the pair (A_{ac}, b_{ac}).

(f) Add to your MATLAB .m file the code to compute the feedback gain k_a that places the poles of $A_a - b_a k_a$ at $-r_1, -r_2, -r_3, -r_4, -r_5$.

(g) In SIMULINK simulate this feedback control system using the *nonlinear* statespace model of the inverted pendulum given in Figure 13.19 of Chapter 13. Let the step reference input be $x_{ref} = 0.1 u_s(t)$.

Problem 13 *Disturbance Rejection and Tracking in the Statespace*
Redraw Figure 15.9 to have the form of Figure 15.15.

Problem 14 *Inverted Pendulum Control Using the Motor Voltage Input*

The QUANSER [34] inverted pendulum system uses the cart on a track system with a pendulum rod mounted on the cart as described in Section 13.4. The physical input is not the force on the cart, but instead a voltage applied to a DC motor. This voltage produces current in the armature of the motor, which in turn produces torque to turn the powered wheel. The equation for the current i in the DC motor is

$$L\frac{di}{dt} = -Ri - K_b \omega + V,$$

where V is the input voltage, ω is the angular velocity of the motor shaft, K_b is the back emf constant, and R, L are the resistance and inductance of the armature loops, respectively. The powered wheel has radius r_m so its angular velocity is $\dot{x}/r_m$. The motor shaft is connected to the powered wheel by a set of gears with gear ratio n_2/n_1 so the angular velocity of the motor shaft is

$$\omega = \frac{n_2}{n_1}\frac{\dot{x}}{r_m}.$$

The inductance of the QUANSER DC motor is negligible and thus is taken to be zero. With $L = 0$ the motor current is given by

$$i = \frac{V - K_b \omega}{R} = \frac{1}{R}V - \frac{K_b}{Rr_m}\frac{n_2}{n_1}\dot{x}.$$

The torque put out by the motor is $\tau_m = K_T i$ where $K_T (= K_b)$ is the torque constant. The torque exerted on the powered wheel is then $(n_2/n_1)\tau_m$, which exerts a force on the teeth of the track. Denoting the reaction force of the track teeth on the powered wheel by F_m, this reaction force exerts a torque $r_m F_m$ on the powered wheel of the cart (see Figure 13.9). The moments of inertia of the powered wheel and motor shafts are negligible so $(n_2/n_1)\tau_m = r_m F_m$. Putting this altogether the force u exerted on the cart by the track is

$$u = F_m = \frac{1}{r_m}\frac{n_2}{n_1}\tau_m = \frac{1}{r_m}\frac{n_2}{n_1}K_T\left(\frac{1}{R}V - \frac{K_b}{Rr_m}\frac{n_2}{n_1}\dot{x}\right) = \frac{K_T}{Rr_m}\frac{n_2}{n_1}V - \frac{K_T K_b}{(Rr_m)^2}\left(\frac{n_2}{n_1}\right)^2 \dot{x}.$$

(a) Modify the linear statespace model of Problem 11 by substituting in this expression for u so that the input is now the voltage V. Show that the model is now

$$\begin{bmatrix} dx/dt \\ dv/dt \\ d\theta/dt \\ d\omega/dt \end{bmatrix} = \underbrace{\begin{bmatrix} 0 & 1 & 0 & 0 \\ 0 & -\dfrac{n^2 K_T^2}{r_m^2 R}\kappa(J+m\ell^2) & -\kappa g m^2 \ell^2 & 0 \\ 0 & 0 & 0 & 1 \\ 0 & \dfrac{n^2 K_T^2}{r_m^2 R}\kappa m\ell & \kappa m g \ell (M+m) & 0 \end{bmatrix}}_{A} \begin{bmatrix} x \\ v \\ \theta \\ \omega \end{bmatrix}$$

$$+ \underbrace{\begin{bmatrix} 0 \\ \dfrac{nK_T}{Rr_m}\kappa(J+m\ell^2) \\ 0 \\ -\dfrac{nK_T}{Rr_m}\kappa m\ell \end{bmatrix}}_{b} V$$

(b) Write a MATLAB program to compute the controllability matrix $\mathcal{C}$ for the system (A, b). Use the parameter values given in Problem 11 along with $R = 2.6\,\Omega$, $K_T = K_b = 0.00767$ V/(rad/s), $\dfrac{n_2}{n_1} = 3.71$ gear ratio, and $L = 0$ H.

(c) Show that

$$\det(sI - A) = s^4 + \kappa(J_p + m\ell^2)\dfrac{n^2 K_T^2}{r_m^2 R}s^3 - \kappa m g \ell(M+m)s^2 - \kappa m g \ell \dfrac{n^2 K_T^2}{r_m^2 R}s.$$

(d) Use your answer in part (c) to find the control canonical form (A_c, b_c) of the system found in part (a). Add code to your MATLAB program of part (b) to compute the controllability matrix $\mathcal{C}_c$ for the system (A_c, b_c).

(e) Add code to your MATLAB program of part (d) to compute the feedback gain k that places the poles of $A - bk$ at $-r_1, -r_2, -r_3, -r_4$. Use the pole placement procedure given in Section 15.7.

(f) In SIMULINK simulate this feedback control system using the nonlinear pendulum model given in Figure 13.17 of Problem 5 in Chapter 13.

Problem 15 *Inverted Pendulum Control Using Motor Voltage Input*
Using the model developed in Problem 14 for the voltage controlled inverted pendulum, simulate the controller described in Section 15.9 of this chapter that achieves tracking of step reference inputs of the cart position.

Problem 16 *Feedback Linearization Control of the Magnetic Levitation System*
Simulate the feedback linearization controller for the magnetic levitation system shown in Figure 15.22. Figure 13.22 of Problem 10 in Chapter 13 is a SIMULINK model of the nonlinear open-loop magnetic levitation system. As in Problem 2, use the QUANSER [34] parameter values given by $L_0 = 0.4125$ H, $R = 11\,\Omega$, $r = 1.27 \times 10^{-2}$ m (1.27 cm), $m = 0.068$ kg, $g = 9.81$ m/s², $x_{eq} = r + 0.006$ m, $i_{eq} = \sqrt{\dfrac{mg}{K_m}}x_{eq} = 0.8453$ A, and $V_{\max} = 33$ V. Finally set $rL_1 = 10K_m = 6.5308 \times 10^{-4}$ Nm²/A².

Remark QUANSER denotes the force constant as K_m instead of $rL_1/2$ with the value $K_m = 6.5308 \times 10^{-5}$ Nm^2/A^2. However, in the magnetic force expression QUANSER considers the position x of the steel ball to be the distance from the bottom of the magnet to the *top* of the steel ball rather than to its center of mass. Instead of setting $rL_1/2$ equal to this value of K_m, we set $rL_1/2 = 5K_m$ (or $rL_1 = 10K_m$) to compensate for this difference in the reference point for measuring x.

As in Problem 10 of Chapter 13 (see Figure 13.24) set the initial conditions as $x(0) = 0.014 + r, v(0) = 0, i(0) = \sqrt{\dfrac{2mg}{rL_1}x(0)}$ and set the final conditions to be $x(t_f) = x_{eq}, i(t_f) = i_{eq}, v(t_f) = 0$. The SIMULINK trajectory generator for the cart on the track system of Problem 3 can be modified for use in this problem. In the SIMULINK trajectory generator the position reference $x_{ref}(t)$ goes from 0 to x_f. So add x0 to x_ref outside the trajectory block so that the trajectory now starts at x0. In the MATLAB code for the trajectory generator set xf=x_eq-x0 so that at the final time t_f the output of the trajectory block is xf+x0=x_eq. (Better yet, ask your instructor to give you the MATLAB/SIMULINK *student* files for this problem which has the modified trajectory generator and the nonlinear open-loop magnetic levitation model.)

Problem 17 *Equilibrium Point for the Augmented Inverted Pendulum Model*

In designing a feedback scheme to keep the pendulum rod upright while the cart tracks step inputs, the following augmented system was developed.

$$\underbrace{\begin{bmatrix} dx/dt \\ dv/dt \\ d\theta/dt \\ d\omega/dt \\ dz_0/dt \end{bmatrix}}_{dz_a/dt} = \underbrace{\begin{bmatrix} 0 & 1 & 0 & 0 & 0 \\ 0 & 0 & -\kappa mgm\ell^2 & 0 & 0 \\ 0 & 0 & 0 & 1 & 0 \\ 0 & 0 & \alpha^2 & 0 & 0 \\ -1 & 0 & 0 & 0 & 0 \end{bmatrix}}_{A_a} \underbrace{\begin{bmatrix} x \\ v \\ \theta \\ \omega \\ z_0 \end{bmatrix}}_{z_a} + \underbrace{\begin{bmatrix} 0 \\ \kappa(J+m\ell^2) \\ 0 \\ -\kappa m\ell \\ 0 \end{bmatrix}}_{b_a} u + \underbrace{\begin{bmatrix} 0 \\ 0 \\ 0 \\ 0 \\ 1 \end{bmatrix}}_{p_a} x_{ref}$$

$$u = -\underbrace{\begin{bmatrix} k_1 & k_2 & k_3 & k_4 & k_0 \end{bmatrix}}_{k_a} \underbrace{\begin{bmatrix} x \\ v \\ \theta \\ \omega \\ z_0 \end{bmatrix}}_{z_a}.$$

More compactly we write

$$\frac{dz_a}{dt} = A_a z_a + b_a u + p_a x_{ref}$$
$$u = -k_a z_a$$

or

$$\frac{dz_a}{dt} = (A_a - b_a k_a) z_a + p_a x_{ref}. \tag{15.122}$$

(a) Show that equilibrium point z_{a_eq} of the closed-loop system (15.122) is

$$z_{a\_eq} = -(A_a - b_a k_a)^{-1} p_a x_{ref} = \begin{bmatrix} x_{ref} \\ 0 \\ 0 \\ 0 \\ -(k_1/k_0)x_{ref} \end{bmatrix}.$$

(b) The Laplace transform of (15.122) has the form

$$Z_a(s) = (sI_{5\times 5} - (A_a - b_a k_a))^{-1} z_a(0) + (sI_{5\times 5} - (A_a - b_a k_a))^{-1} p_a \frac{x_{ref}}{s}.$$

Show that

$$(sI_{5\times 5} - (A_a - b_a k_a))^{-1} \begin{bmatrix} 0 \\ 0 \\ 0 \\ 0 \\ 1 \end{bmatrix} \frac{x_{ref}}{s} = \frac{1}{\det(sI_{5\times 5} - (A_a - b_a k_a))}$$

$$\times \begin{bmatrix} -k_0\kappa(J + m\ell^2)s^2 + k_0\kappa mg\ell \\ -k_0\kappa(J + m\ell^2)s^3 + k_0\kappa mg\ell s \\ k_0 s^2 \kappa m\ell \\ k_0 s^3 \kappa m\ell \\ q(s) \end{bmatrix} \frac{x_{ref}}{s},$$

where

$$q(s) = s^4 + (k_2\kappa(J + m\ell^2) - k_4\kappa m\ell)s^3 - (\alpha^2 + k_3\kappa m\ell - k_1\kappa(J + m\ell^2))s^2$$
$$- k_2\kappa mg\ell s - k_1\kappa mg\ell$$

and

$$\det(sI_{5\times 5} - (A_a - b_a k_a)) = s^5 + (k_2\kappa(J + m\ell^2) - k_4\kappa m\ell)s^4$$
$$- (\alpha^2 + k_3\kappa m\ell - k_1\kappa(J + m\ell^2))s^3$$
$$- (k_2\kappa mg\ell + k_0\kappa(J + m\ell^2))s^2 - k_1\kappa mg\ell s + k_0\kappa mg\ell.$$

Hint: Use a symbolic manipulation software package such as MATHEMATICA, MAPLE, or SYM in MATLAB.

(c) Use your answer to part (b) to show

$$z_a(t) \to \begin{bmatrix} x_{ref} \\ 0 \\ 0 \\ 0 \\ -(k_1/k_0)x_{ref} \end{bmatrix}.$$

Compare this with your answer to part (a).

Remark Note that $k_a z_{a\_eq} = 0$ so that $-k_a z_a = -k_a(z_a - z_{a\_eq})$.

Problem 18 MATLAB *Code for Computing* $k = k_c C_c C^{-1}$

Run the following MATLAB program for computing $k = k_c C_c C^{-1}$.

```
% Compute k = kc*Cc*inv(C)
% Set up two arbitrary (random) A and b matrices.
A = rand(4,4);b = rand(4,1);
C = [b A*b A^2*b A^3*b]; det(C);
% Compute the control canonical form for A using Equation (15.92) on page 595.
% That is, A_prime = inv(C)*A*C and Ac = transpose(A_prime)
Ac = transpose(inv(C)*A*C));
% Control canonical form for b
bc = [0; 0; 0; 1];
% det(sI-A) = s^4 + alpha3*s^3 + alpha2*s^2 + alpha1*s + alpha0
% The coefficients of det(sI-A) are the negatives of the last row of Ac.
alpha0 = -Ac(4,1); alpha1 = -Ac(4,2); alpha2 = -Ac(4,3); alpha3 = -Ac(4,4);
% Compute the controllability matrix of (Ac,bc)
Cc = [bc Ac*bc Ac^2*bc Ac^3*bc];
det(Cc);
% Set the location of the desired closed-loop poles.
% The desired poles are -r1, -r2, -r3, -r4.
r1 = 5; r2 = r1; r3 = r2; r4 = r3;
% Set the coefficients of the desired closed-loop characteristic polynomial.
alphad3 = r1 + r2 + r3 + r4;
alphad2 = r1*r2 + r1*r3 + r1*r4 + r2*r3 + r2*r4 + r3*r4;
alphad1 = r1*r2*r3 + r1*r2*r4 + r1*r3*r4 + r2*r3*r4;
alphad0 = r1*r2*r3*r4;
% Compute the feedback gain Kc for the control canonical form.
Kc = [alphad0 - alpha0; alphad1 - alpha1; alphad2 - alpha2; alphad3 - alpha3]';
% Check the computation of K
ceig(Ac-bc*Kc);
% Compute the feedback gain K for A,b.
K = Kc*Cc*inv(C);
% Check the computation of K.
eig(A-b*K)
```

Problem 19 *Inverted Pendulum on an Inclined Track*

Modify the open-loop nonlinear SIMULINK model of the inverted pendulum to account for an inclined track at an angle ϕ with respect to the horizontal. Apply the *same* feedback controller designed in Problem 12 based on the linearized flat track model. Run the simulation and you should see that the controller still stabilizes the pendulum for angles ϕ less than about 15°. In particular, you should see that $\theta(t) \to \phi$!

Problem 20 *Tracking a Sinusoidal Reference for the Inverted Pendulum*

Consider the problem of having the cart position track a sinusoidal reference while keeping the pendulum rod upright. That is, we want $x(t) \to x_{ref}(t) = X_0 \sin(\omega_0 t + \phi_0)$ for any given X_0 and ϕ_0. By the internal model principle this requires a transfer function in the forward path of the form

$$G_c(s) = \frac{-k_{01}s - k_{00}}{s^2 + \omega_0^2}.$$

The feedback architecture is shown in Figure 15.28.

15 State Feedback

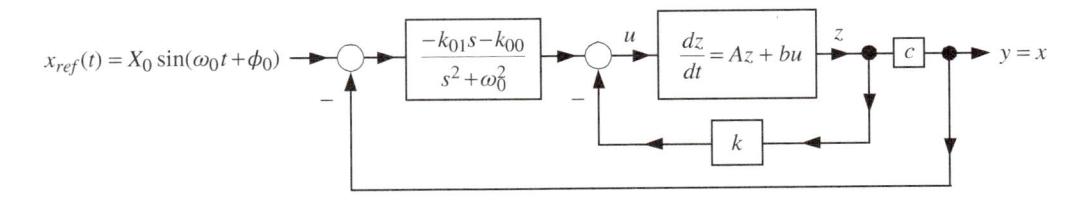

Figure 15.28. Feedback architecture for tracking a sinusoidal reference.

We can write the equations describing Figure 15.28 as

$$\frac{dz}{dt} = Az + bu$$
$$u = -kz - k_{00}z_{00} - k_{01}z_{01}$$
$$y = cz = x$$
$$\frac{d}{dt}\underbrace{\begin{bmatrix} z_{00} \\ z_{01} \end{bmatrix}}_{z_0 \in \mathbb{R}^2} = \underbrace{\begin{bmatrix} 0 & 1 \\ -\omega_0^2 & 0 \end{bmatrix}}_{A_0 \in \mathbb{R}^{2\times 2}} \begin{bmatrix} z_{00} \\ z_{01} \end{bmatrix} + \underbrace{\begin{bmatrix} 0 \\ 1 \end{bmatrix}}_{b_0 \in \mathbb{R}^2} (x_{ref}(t) - x).$$

More compactly this is written as

$$\frac{d}{dt}\underbrace{\begin{bmatrix} z \\ z_0 \end{bmatrix}}_{\in \mathbb{R}^6} = \underbrace{\begin{bmatrix} A & 0_{4\times 2} \\ 0_{1\times 4} & A_0 \\ -c & \end{bmatrix}}_{A_a \in \mathbb{R}^{6\times 6}} \underbrace{\begin{bmatrix} z \\ z_0 \end{bmatrix}}_{\in \mathbb{R}^6} + \underbrace{\begin{bmatrix} b \\ 0_{2\times 1} \end{bmatrix}}_{b_a \in \mathbb{R}^6} u + \underbrace{\begin{bmatrix} 0_{4\times 1} \\ b_0 \end{bmatrix}}_{\in \mathbb{R}^6} x_{ref}(t).$$

The open-loop transfer function of the inverted pendulum is

$$G(s) = \frac{n(s)}{d(s)} = c(sI_{4\times 4} - A)^{-1}b = \frac{\kappa(J + m\ell^2)s^2 - \kappa m g \ell}{s^2(s^2 - \alpha^2)}.$$

With

$$u = -\underbrace{\begin{bmatrix} k & k_0 \end{bmatrix}}_{k_a \in \mathbb{R}^{1\times 6}} \underbrace{\begin{bmatrix} z \\ z_0 \end{bmatrix}}_{\in \mathbb{R}^6}$$

the closed-loop dynamic system is given by

$$\frac{d}{dt}\begin{bmatrix} z \\ z_0 \end{bmatrix} = \begin{bmatrix} A & 0_{4\times 2} \\ 0_{1\times 4} & A_0 \\ -c & \end{bmatrix} \begin{bmatrix} z \\ z_0 \end{bmatrix} - \begin{bmatrix} b \\ 0 \end{bmatrix} \begin{bmatrix} k & k_0 \end{bmatrix} \begin{bmatrix} z \\ z_0 \end{bmatrix} + \begin{bmatrix} 0_{4\times 1} \\ b_0 \end{bmatrix} x_{ref}(t)$$

$$= \left(\begin{bmatrix} A & 0_{4\times 2} \\ 0_{1\times 4} & A_0 \\ -c & \end{bmatrix} - \begin{bmatrix} bk & bk_0 \\ 0_{2\times 4} & 0_{2\times 2} \end{bmatrix} \right) \begin{bmatrix} z \\ z_0 \end{bmatrix} + \begin{bmatrix} 0_{4\times 1} \\ b_0 \end{bmatrix} x_{ref}(t)$$

$$= \begin{bmatrix} A & 0_{4\times 2} \\ 0_{1\times 4} & A_0 \\ -c & \end{bmatrix} \begin{bmatrix} z \\ z_0 \end{bmatrix} + \begin{bmatrix} 0_{4\times 1} \\ b_0 \end{bmatrix} x_{ref}(t).$$

It turns out that $k \in \mathbb{R}^{1\times 4}, k_0 \in \mathbb{R}^{1\times 2}$ can be chosen to arbitrarily place the closed-loop poles. Specifically, using the fact that the pair (A, b) describing the inverted pendulum is controllable in $\mathbb{R}^4$, and the numerator $n(s)$ of the open-loop transfer function $G(s) = n(s)/d(s)$ does *not* contain the factor $s^2 + \omega_0^2$, the pair

$$A_a \triangleq \begin{bmatrix} A & 0_{4\times 2} \\ 0_{1\times 4} & A_0 \\ -c & \end{bmatrix} \in \mathbb{R}^{6\times 6}, \quad b_a \triangleq \begin{bmatrix} b \\ 0_{2\times 1} \end{bmatrix} \in \mathbb{R}^6$$

is controllable in $\mathbb{R}^6$. (The proof is similar to that given in the Appendix *Disturbance Rejection in the Statespace* of this chapter and is detailed in the Solutions Manual.) The procedure to compute the feedback $k_a = \begin{bmatrix} k & k_0 \end{bmatrix} \in \mathbb{R}^{1\times 6}$ to place the closed-loop poles is as follows. Set

$$C_a = \begin{bmatrix} b_a & A_a b_a & \cdots & A_a^4 b_a \end{bmatrix} \in \mathbb{R}^{6\times 6}$$

$$\det(sI_{6\times 6} - A_a) = \det \begin{bmatrix} sI_{4\times 4} - A & 0_{4\times 2} \\ 0_{1\times 4} & sI_{2\times 2} - A_0 \\ c & \end{bmatrix}$$

$$= s^6 + \alpha_5 s^5 + \alpha_4 s^4 + \alpha_3 s^3 + \alpha_2 s^2 + \alpha_1 s + \alpha_0$$

$$A_{ac} = \begin{bmatrix} 0 & 1 & 0 & 0 & 0 & 0 \\ 0 & 0 & 1 & 0 & 0 & 0 \\ 0 & 0 & 0 & 1 & 0 & 0 \\ 0 & 0 & 0 & 0 & 1 & 0 \\ 0 & 0 & 0 & 0 & 0 & 1 \\ -\alpha_0 & -\alpha_1 & -\alpha_2 & -\alpha_3 & -\alpha_4 & -\alpha_5 \end{bmatrix}, \quad b_{ac} = \begin{bmatrix} 0 \\ 0 \\ 0 \\ 0 \\ 0 \\ 1 \end{bmatrix}$$

$$C_{ac} = \begin{bmatrix} b_{ac} & A_{ac} b_{ac} & \cdots & A_{ac}^5 b_{ac} \end{bmatrix} \in \mathbb{R}^{6\times 6}.$$

Letting

$$k_a = \begin{bmatrix} k & k_0 \end{bmatrix}$$
$$= \begin{bmatrix} \alpha_{d0} - \alpha_0 & \alpha_{d1} - \alpha_1 & \alpha_{d2} - \alpha_2 & \alpha_{d3} - \alpha_3 & \alpha_{d4} - \alpha_4 & \alpha_{d5} - \alpha_5 \end{bmatrix} C_{ac} C_a^{-1} \in \mathbb{R}^{1\times 6}$$

results in

$$\det \begin{bmatrix} sI_{4\times 4} - (A - bk) & bk_0 \\ 0_{1\times 4} & sI_{2\times 2} - (A_0 - bk_0) \\ c & \end{bmatrix}$$
$$= s^6 + \alpha_{d5} s^5 + \alpha_{d4} s^4 + \alpha_{d3} s^3 + \alpha_{d2} s^2 + \alpha_{d1} s + \alpha_{d0}.$$

Simulate this feedback control system using the parameter values given in Problem 11 with $x_{ref}(t) = 0.25 \sin(2\pi t)$ and the closed-loop poles all at -5.

16

State Estimators and Parameter Identification

16.1 State Estimators

A problem with state feedback is that one needs the full state measurement, i.e., values of all the state variables. If we do not have a measurement of the full state, then we need to estimate the state variables that are not directly measured. Let's see how to do this by starting with a concrete example.

Example 1 *Speed Estimator for the Cart*

Consider the cart on the track system from Chapters 6 and 15 where an optical encoder was used to measure position. With T the sample period the speed at time jT is computed by numerically differentiating the position measurement according to

$$v_{bd}(jT) \triangleq \frac{x(jT) - x((j-1)T)}{T}.$$

This is the *backward difference* speed estimate and it is noisy due to the finite resolution of the encoder. Another approach is to estimate the rotor speed using an *observer* to obtain a hopefully smoother (less noisy) estimate of speed. In this example we will assume the cart is on a flat track. From Example 2 of Chapter 15 (page 568) the equations describing the cart on the track are

$$\frac{d}{dt}x(t) = v$$
$$\frac{d}{dt}v(t) = -a_0 v(t) + b_0 u(t). \tag{16.1}$$

In matrix form we have

$$\frac{d}{dt}\begin{bmatrix} x \\ v \end{bmatrix} = \begin{bmatrix} 0 & 1 \\ 0 & -a_0 \end{bmatrix}\begin{bmatrix} x \\ v \end{bmatrix} + \begin{bmatrix} 0 \\ b_0 \end{bmatrix} u \tag{16.2}$$

$$y = \begin{bmatrix} 1 & 0 \end{bmatrix}\begin{bmatrix} x \\ v \end{bmatrix}. \tag{16.3}$$

The system (16.2), (16.3) is of the form

$$\frac{dz}{dt} = Az + bu$$
$$y = cz$$

An Introduction to System Modeling and Control, First Edition. John Chiasson.
© 2022 John Wiley & Sons, Inc. Published 2022 by John Wiley & Sons, Inc.
Companion website: www.wiley.com/go/chiasson/anintroductiontosystemmodelingandcontrol

with the obvious definitions for $A, b, c, z,$ and u. A speed *observer* is then defined by

$$\frac{d}{dt}\begin{bmatrix}\hat{x}\\\hat{v}\end{bmatrix} = \begin{bmatrix}0 & 1\\0 & -a_0\end{bmatrix}\begin{bmatrix}\hat{x}\\\hat{v}\end{bmatrix} + \begin{bmatrix}0\\b_0\end{bmatrix}u + \underbrace{\begin{bmatrix}\ell_1\\\ell_2\end{bmatrix}}_{\ell}(y-\hat{y}) \tag{16.4}$$

$$\hat{y} = \begin{bmatrix}1 & 0\end{bmatrix}\begin{bmatrix}\hat{x}\\\hat{v}\end{bmatrix}. \tag{16.5}$$

That is, we sample the output $y = x$ to bring it into the computer and then numerically integrate the set of Eqs. (16.4) and (16.5) in real-time. If the observer gain vector ℓ is zero the observer is simply a real-time simulation of the cart system. However the correction term $\ell(y-\hat{y})$ in (16.4) is the key to having the observer actually work. Specifically, we now show that the gains ℓ_1 and ℓ_2 can be chosen such that $\hat{v} \to v$. To do so, define the estimation errors to be $\varepsilon_1 = x - \hat{x}$ and $\varepsilon_2 = v - \hat{v}$. Upon subtracting (16.4) from (16.2) we obtain

$$\frac{d}{dt}\begin{bmatrix}\varepsilon_1\\\varepsilon_2\end{bmatrix} = \begin{bmatrix}0 & 1\\0 & -a_0\end{bmatrix}\begin{bmatrix}\varepsilon_1\\\varepsilon_2\end{bmatrix} - \begin{bmatrix}\ell_1\\\ell_2\end{bmatrix}(y-\hat{y})$$

$$= \begin{bmatrix}0 & 1\\0 & -a_0\end{bmatrix}\begin{bmatrix}\varepsilon_1\\\varepsilon_2\end{bmatrix} - \begin{bmatrix}\ell_1\\\ell_2\end{bmatrix}\begin{bmatrix}1 & 0\end{bmatrix}\begin{bmatrix}\varepsilon_1\\\varepsilon_2\end{bmatrix}$$

$$= \begin{bmatrix}-\ell_1 & 1\\-\ell_2 & -a_0\end{bmatrix}\begin{bmatrix}\varepsilon_1\\\varepsilon_2\end{bmatrix}. \tag{16.6}$$

More compactly, the error system is given by

$$\frac{d\varepsilon}{dt} = (A - \ell c)\varepsilon. \tag{16.7}$$

If we can choose the column vector ℓ such that $A - \ell c$ is stable, then

$$\begin{bmatrix}\varepsilon_1(t)\\\varepsilon_2(t)\end{bmatrix} = e^{(A-\ell c)t}\begin{bmatrix}\varepsilon_1(0)\\\varepsilon_2(0)\end{bmatrix} \to \begin{bmatrix}0\\0\end{bmatrix}. \tag{16.8}$$

A block diagram for the combined state feedback trajectory tracking controller and speed observer is shown in Figure 16.1.

Figure 16.1 shows the addition of a state observer to the block diagram of Figure 15.6. The gains k_1, k_2 for the state feedback in Figure 16.1 are chosen as given in Eq. (15.27). We now choose the gains ℓ_1, ℓ_2 for the state observer.

A direct way to determine the observer gains ℓ_1, ℓ_2 is to simply compute

$$\det(sI - (A - \ell c)) = \det\begin{bmatrix}s+\ell_1 & -1\\s+\ell_2 & s+a_0\end{bmatrix} = s^2 + (\ell_1 + a_0 + 1)s + (\ell_2 + \ell_1 a_0).$$

If we want the poles at $-r_1, -r_2$ then we set

$$s^2 + (\ell_1 + a_0 + 1)s + (\ell_2 + \ell_1 a_0) = (s+r_1)(s+r_2) = s^2 + (r_1+r_2)s + r_1 r_2$$

which requires
$$\ell_1 = -(a_0 + 1) + r_1 + r_2$$
$$\ell_2 = -\ell_1 a_0 + r_1 r_2.$$

Problem 5a asks you to simulate the combined state feedback and state observer system of Figure 16.1.

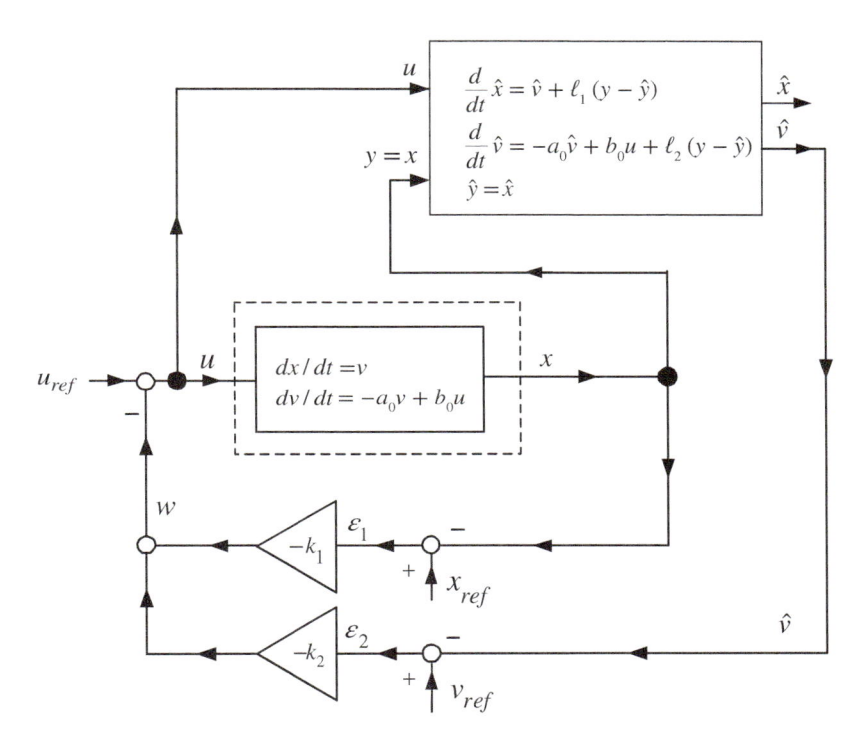

Figure 16.1. Combined state feedback trajectory tracking controller and speed observer. The position $x(t)$ is sampled. The observer equations are then numerically integrated in real time and the solution $\hat{x}, \hat{v}$ are used as the estimates of x, v, respectively.

Example 2 *Position Cannot Be Estimated from Speed*

Suppose a tachometer is available to measure speed, but there is no position measurement. Can the position be estimated from the speed measurements? Well, the answer is no. An easy way to understand this is to realize that $x(t) = x(0) + \int_0^t v(\tau)d\tau$, but that
$$\frac{d}{dt}\left(c + \int_0^t v(\tau)d\tau\right) = v(t)$$
for *any* constant c. That is, there are an infinite number of position signals with the same speed signal. Consequently, the speed signal does not contain enough information to obtain an estimate for the position. From the point of view of observer theory, again consider the cart on the track system. We have
$$\frac{d}{dt}\begin{bmatrix} x \\ v \end{bmatrix} = \begin{bmatrix} 0 & 1 \\ 0 & -a_0 \end{bmatrix}\begin{bmatrix} x \\ v \end{bmatrix} + \begin{bmatrix} 0 \\ b_0 \end{bmatrix}u$$
$$y = \begin{bmatrix} 0 & 1 \end{bmatrix}\begin{bmatrix} x \\ v \end{bmatrix}.$$

The output y is now the speed v. Let's try an observer given by

$$\frac{d}{dt}\begin{bmatrix}\hat{x}\\\hat{v}\end{bmatrix}=\begin{bmatrix}0&1\\0&-a_0\end{bmatrix}\begin{bmatrix}\hat{x}\\\hat{v}\end{bmatrix}+\begin{bmatrix}0\\b_0\end{bmatrix}u+\begin{bmatrix}\ell_1\\\ell_2\end{bmatrix}(y-\hat{y})$$

$$\hat{y}=\begin{bmatrix}0&1\end{bmatrix}\begin{bmatrix}\hat{x}\\\hat{v}\end{bmatrix}.$$

With $\varepsilon_1 = x - \hat{x}, \varepsilon_2 = v - \hat{v}$ the error system is

$$\frac{d}{dt}\begin{bmatrix}\varepsilon_1\\\varepsilon_2\end{bmatrix}=\begin{bmatrix}0&1\\0&-a_0\end{bmatrix}\begin{bmatrix}\varepsilon_1\\\varepsilon_2\end{bmatrix}-\begin{bmatrix}\ell_1\\\ell_2\end{bmatrix}(y-\hat{y})$$

$$=\left(\begin{bmatrix}0&1\\0&-a_0\end{bmatrix}-\begin{bmatrix}\ell_1\\\ell_2\end{bmatrix}\begin{bmatrix}0&1\end{bmatrix}\right)\begin{bmatrix}\varepsilon_1\\\varepsilon_2\end{bmatrix}$$

$$=\begin{bmatrix}0&-\ell_1\\0&-a_0-\ell_2\end{bmatrix}\begin{bmatrix}\varepsilon_1\\\varepsilon_2\end{bmatrix}.$$

The characteristic polynomial of $A - \ell c$ is

$$\det(sI-(A-\ell c))=\det\begin{bmatrix}s&\ell_1\\0&s+a_0+\ell_2\end{bmatrix}=s^2+(a_0+\ell_2)s,$$

which cannot be made stable for any choice of the observer gains ℓ_1, ℓ_2.

Example 3 *Speed and Disturbance Estimator for the Cart*

Let's again look at the cart on the track system still using an optical encoder to measure position, but now the cart is considered to be on an incline so it has a (gravitational force) disturbance acting on it. The disturbance must also be estimated in order to obtain an unbiased estimate of the speed. To proceed we model this disturbance on the cart as a state variable and consider it to be constant over the time we are estimating its value. (This method will also work if the disturbance is "slowly varying".) The equations describing the cart on the track are then

$$\frac{d}{dt}x(t)=v$$
$$\frac{d}{dt}v(t)=-a_0v(t)+b_0u(t)-d \qquad (16.9)$$
$$\frac{d}{dt}d(t)=0,$$

where $d = b_0 K_D F_d = Mg\sin(\phi)/(J/r_m^2 + M)$ is the (acceleration) disturbance on the cart due to gravity (see Section 15.5). In matrix form we write

$$\frac{d}{dt}\begin{bmatrix}x\\v\\d\end{bmatrix}=\begin{bmatrix}0&1&0\\0&-a_0&-1\\0&0&0\end{bmatrix}\begin{bmatrix}x\\v\\d\end{bmatrix}+\begin{bmatrix}0\\b_0\\0\end{bmatrix}u \qquad (16.10)$$

$$y=\begin{bmatrix}1&0&0\end{bmatrix}\begin{bmatrix}x\\v\\d\end{bmatrix}. \qquad (16.11)$$

The system (16.10), (16.11) is of the form

$$\frac{dz}{dt} = Az + bu$$
$$y = cz$$

with the obvious definitions for A, b, c, z, and u. A speed and disturbance *observer* is defined by

$$\frac{d}{dt}\begin{bmatrix}\hat{x}\\\hat{v}\\\hat{d}\end{bmatrix} = \begin{bmatrix}0 & 1 & 0\\0 & -a_0 & -1\\0 & 0 & 0\end{bmatrix}\begin{bmatrix}\hat{x}\\\hat{v}\\\hat{d}\end{bmatrix} + \begin{bmatrix}0\\b_0\\0\end{bmatrix}u + \underbrace{\begin{bmatrix}\ell_1\\\ell_2\\\ell_3\end{bmatrix}}_{\ell}(y - \hat{y}) \quad (16.12)$$

$$\hat{y} = \begin{bmatrix}1 & 0 & 0\end{bmatrix}\begin{bmatrix}\hat{x}\\\hat{v}\\\hat{d}\end{bmatrix}. \quad (16.13)$$

Figure 16.2 shows the addition of this state observer to the block diagram of Figure 15.6 of Chapter 15. The gains k_0, k_1, k_2 for the state feedback in Figure 16.2 are chosen as given in Eq. 15.79. The output $y = x$ is sampled to bring it into the computer. The set of Eqs. (16.12), (16.13) is integrated in real-time with the solutions $\hat{v}(t)$ and $\hat{d}(t)$ used as the estimates of the speed and disturbance. As we now show, the correction term $\ell(y - \hat{y})$ in (16.12) is the key to making the estimates $\hat{v}(t)$ and $\hat{d}(t)$ converge to $v(t)$ and $d(t)$, respectively. Specifically, we show the gains ℓ_1, ℓ_2, and ℓ_3 can be chosen to force $\hat{v}(t) \to v(t)$ and $\hat{d}(t) \to d$.

To proceed, define the estimation errors to be $\varepsilon_1 = x - \hat{x}, \varepsilon_2 = v - \hat{v}$, and $\varepsilon_3 = d - \hat{d}$. Subtracting (16.12) from (16.10) results in

$$\frac{d}{dt}\begin{bmatrix}\varepsilon_1\\\varepsilon_2\\\varepsilon_3\end{bmatrix} = \begin{bmatrix}0 & 1 & 0\\0 & -a_0 & -1\\0 & 0 & 0\end{bmatrix}\begin{bmatrix}\varepsilon_1\\\varepsilon_2\\\varepsilon_3\end{bmatrix} - \begin{bmatrix}\ell_1\\\ell_2\\\ell_3\end{bmatrix}(y - \hat{y})$$

$$= \begin{bmatrix}0 & 1 & 0\\0 & -a_0 & -1\\0 & 0 & 0\end{bmatrix}\begin{bmatrix}\varepsilon_1\\\varepsilon_2\\\varepsilon_3\end{bmatrix} - \begin{bmatrix}\ell_1\\\ell_2\\\ell_3\end{bmatrix}\begin{bmatrix}1 & 0 & 0\end{bmatrix}\begin{bmatrix}\varepsilon_1\\\varepsilon_2\\\varepsilon_3\end{bmatrix}$$

$$= \begin{bmatrix}-\ell_1 & 1 & 0\\-\ell_2 & -a_0 & -1\\-\ell_3 & 0 & 0\end{bmatrix}\begin{bmatrix}\varepsilon_1\\\varepsilon_2\\\varepsilon_3\end{bmatrix}. \quad (16.14)$$

This error system is written compactly as

$$\frac{d\varepsilon}{dt} = (A - \ell c)\varepsilon. \quad (16.15)$$

With the column vector ℓ chosen so that $A - \ell c$ is stable, it follows that

$$\begin{bmatrix}\varepsilon_1(t)\\\varepsilon_2(t)\\\varepsilon_3(t)\end{bmatrix} = e^{(A-\ell c)t}\begin{bmatrix}\varepsilon_1(0)\\\varepsilon_2(0)\\\varepsilon_3(0)\end{bmatrix} \to \begin{bmatrix}0\\0\\0\end{bmatrix}. \quad (16.16)$$

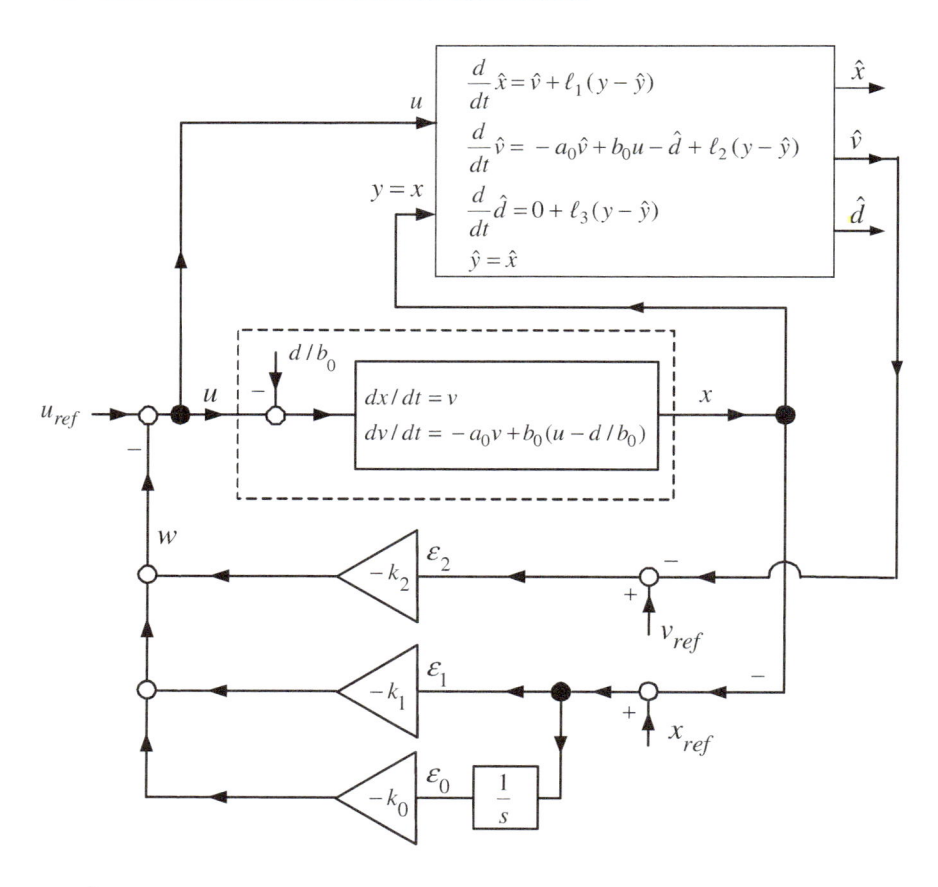

Figure 16.2. Combined state feedback trajectory tracking controller and speed observer. The position $x(t)$ is sampled. The observer equations are then numerically integrated in real-time and the solutions $\hat{x}, \hat{v}, \hat{d}$ are used as the estimates of x, v, d, respectively.

To find the gains ℓ_1, ℓ_2, and ℓ_3 that make $A - \ell c$ is stable we compute

$$\det(sI - (A - \ell c)) = \det \begin{bmatrix} s + \ell_1 & -1 & 0 \\ \ell_2 & s + a_0 & 1 \\ \ell_3 & 0 & s \end{bmatrix} = s^3 + (\ell_1 + a_0)s^2 + (\ell_2 + a_0\ell_1)s - \ell_3.$$

To place the observer poles at $-r_1, -r_2$, and $-r_3$ set

$$s^3 + (\ell_1 + a_0)s^2 + (\ell_2 + a_0\ell_1)s - \ell_3 = (s + r_1)(s + r_2)(s + r_3)$$
$$= s^3 + (r_1 + r_2 + r_3)s^2 + (r_1r_2 + r_1r_3 + r_2r_3)s + r_1r_2r_3.$$

The observer gains are then

$$\ell_1 = -a_0 + r_1 + r_2 + r_3$$
$$\ell_2 = -a_0\ell_1 + r_1r_2 + r_1r_3 + r_2r_3$$
$$\ell_3 = -r_1r_2r_3.$$

Problem 5b asks you to simulate the combined state feedback and state observer system of Figure 16.2.

General Procedure for State Estimation

The open-loop statespace model is

$$\frac{dx}{dt} = Ax + bu, \quad A \in \mathbb{R}^{n \times n}, \quad b \in \mathbb{R}^n$$

$$y = cx, \quad c \in \mathbb{R}^{1 \times n}.$$

Given $y(t), u(t), A, b$, and c can we estimate $x(t)$ in a practical manner? Let's first just try a real-time simulation of our system model. That is, using the values of the known input $u(t)$, we integrate the system model equations

$$\frac{d\hat{x}}{dt} = A\hat{x} + bu$$

in real-time to obtain $\hat{x}$. Figure 16.3 is a block diagram of this real-time simulator.

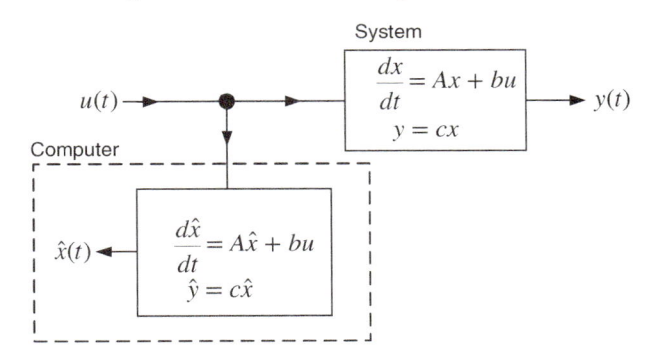

Figure 16.3. Simulation of the system.

The question is how good will $\hat{x}$ be as an estimate of x? Remember that we do not know the initial state $x(0)$. We have

$$\frac{dx}{dt} = Ax + bu$$

$$\frac{d\hat{x}}{dt} = A\hat{x} + bu$$

with A, b, and u assumed to be known. The estimation error $\hat{x} - x$ satisfies

$$\frac{d}{dt}(x - \hat{x}) = A(x - \hat{x}).$$

The solution is

$$x(t) - \hat{x}(t) = e^{At}\bigl(x(0) - \hat{x}(0)\bigr).$$

If A is stable, then $x(t) - \hat{x}(t) \to 0$ for any (unknown) initial state $x(0)$. However, if A is unstable we are out of luck. Further, even if A is stable, the rate that $\hat{x}(t) - x(t)$ goes to zero is fixed by the eigenvalues of A. To get around these issues we use the measurement $y(t)$ in the estimator. To develop a general procedure for state estimation we go back to

16 State Estimators and Parameter Identification

the open-loop statespace model given as

$$\frac{dx}{dt} = Ax + bu, \quad A \in \mathbb{R}^{n \times n}, \quad b \in \mathbb{R}^n \tag{16.17}$$

$$y = cx, \quad c \in \mathbb{R}^{1 \times n}.$$

Define a state estimator by

$$\frac{d\hat{x}}{dt} = A\hat{x} + bu + \ell(y - \hat{y}) \tag{16.18}$$

$$\hat{y} = c\hat{x}$$

or

$$\left[\frac{d\hat{x}}{dt}\right] = \begin{bmatrix} A \end{bmatrix}\begin{bmatrix} \hat{x} \end{bmatrix} + \begin{bmatrix} b \end{bmatrix} u + \begin{bmatrix} \ell \end{bmatrix}\begin{bmatrix} c \end{bmatrix}\begin{bmatrix} x - \hat{x} \end{bmatrix} \tag{16.19}$$

$$\hat{y} = \begin{bmatrix} c \end{bmatrix}\begin{bmatrix} \hat{x} \end{bmatrix}.$$

Figure 16.4 is a block diagram of the system and observer. Again, the idea here is that we know the value of the input $u(t)$ and the sampled output $y(t)$, and we use them to integrate (in real-time) the observer Eq. (16.19). The solution $\hat{x}(t)$ is then used as our estimate of the state $x(t)$.

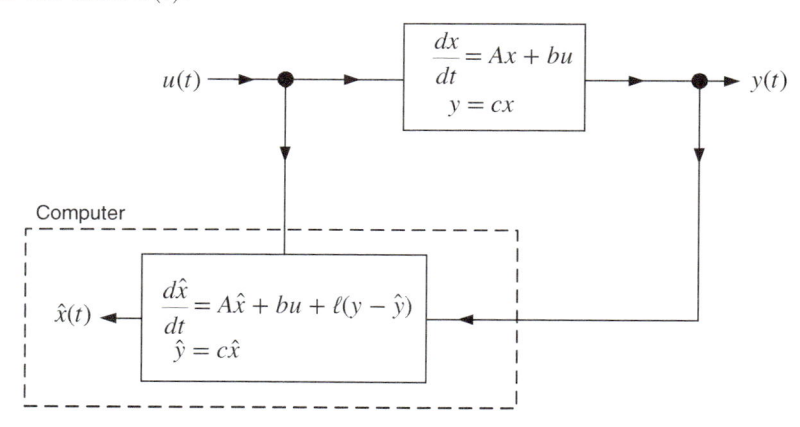

Figure 16.4. State estimator.

To show that this setup works we subtract (16.18) from (16.17) to see that the estimation error satisfies

$$\frac{d}{dt}(x - \hat{x}) = A(x - \hat{x}) - \ell c(x - \hat{x}) = (A - \ell c)(x - \hat{x}).$$

With $\varepsilon \triangleq x - \hat{x}$ we have

$$\left[\frac{d\varepsilon}{dt}\right] = \left(\begin{bmatrix} A \end{bmatrix} - \begin{bmatrix} \ell \end{bmatrix}\begin{bmatrix} c \end{bmatrix}\right)\begin{bmatrix} \varepsilon \end{bmatrix}.$$

The solution is

$$\begin{bmatrix} \varepsilon(t) \end{bmatrix} = \begin{bmatrix} e^{(A-\ell c)t} \end{bmatrix} \begin{bmatrix} \varepsilon(0) \end{bmatrix}.$$

If $A - \ell c$ is stable then $\varepsilon(t) = e^{(A-\ell c)t}\varepsilon(0) \to 0_{n\times 1}$ for arbitrary $\varepsilon(0) = x(0) - \hat{x}(0)$.

The key problem is to find $\ell \in \mathbb{R}^n$ such that $A - \ell c$ is stable. We show how to do this by going through some examples.

Example 4 *System Matrices in Observer Canonical Form*
Let

$$A_o = \begin{bmatrix} 0 & 0 & -\beta_0 \\ 1 & 0 & -\beta_1 \\ 0 & 1 & -\beta_2 \end{bmatrix}, \quad c_o = \begin{bmatrix} 0 & 0 & 1 \end{bmatrix}$$

We say that the pair (c_o, A_o) is in *observer canonical* form.

We compute

$$\det(sI - A_o) = \det\begin{bmatrix} s & 0 & \beta_0 \\ -1 & s & \beta_1 \\ 0 & -1 & s+\beta_2 \end{bmatrix} = s^3 + \beta_2 s^2 + \beta_1 s + \beta_0.$$

With

$$\ell = \begin{bmatrix} \ell_0 \\ \ell_1 \\ \ell_2 \end{bmatrix}$$

we have

$$\det(sI - (A_o - \ell c_o)) = \det\left(\begin{bmatrix} s & 0 & 0 \\ 0 & s & 0 \\ 0 & 0 & s \end{bmatrix} - \left(\begin{bmatrix} 0 & 0 & -\beta_0 \\ 1 & 0 & -\beta_1 \\ 0 & 1 & -\beta_2 \end{bmatrix} - \begin{bmatrix} \ell_0 \\ \ell_1 \\ \ell_2 \end{bmatrix}\begin{bmatrix} 0 & 0 & 1 \end{bmatrix}\right)\right)$$

$$= \det\left(\begin{bmatrix} s & 0 & 0 \\ 0 & s & 0 \\ 0 & 0 & s \end{bmatrix} - \left(\begin{bmatrix} 0 & 0 & -\beta_0 \\ 1 & 0 & -\beta_1 \\ 0 & 1 & -\beta_2 \end{bmatrix} - \begin{bmatrix} 0 & 0 & \ell_0 \\ 0 & 0 & \ell_1 \\ 0 & 0 & \ell_2 \end{bmatrix}\right)\right)$$

$$= \det\left(\begin{bmatrix} s & 0 & 0 \\ 0 & s & 0 \\ 0 & 0 & s \end{bmatrix} - \begin{bmatrix} 0 & 0 & -\beta_0 - \ell_0 \\ 1 & 0 & -\beta_1 - \ell_1 \\ 0 & 1 & -\beta_2 - \ell_2 \end{bmatrix}\right)$$

$$= \det\left(\begin{bmatrix} s & 0 & \beta_0 + \ell_0 \\ 1 & s & \beta_1 + \ell_1 \\ 0 & 1 & s+\beta_2 + \ell_2 \end{bmatrix}\right)$$

$$= s^3 + (\beta_2 + \ell_2)s^2 + (\beta_1 + \ell_1)s + \beta_0 + \ell_0.$$

With the desired closed-loop characteristic polynomial given by $s^3 + \beta_{d2}s^2 + \beta_{d1}s + \beta_{d0}$, simply choose

$$\ell = \begin{bmatrix} \ell_0 \\ \ell_1 \\ \ell_2 \end{bmatrix} = \begin{bmatrix} \beta_{d0} - \beta_0 \\ \beta_{d1} - \beta_1 \\ \beta_{d2} - \beta_2 \end{bmatrix}$$

to have
$$\det(sI - (A_o - \ell c_o)) = s^3 + \beta_{d2}s^2 + \beta_{d1}s + \beta_{d0}.$$

Digression *Digression on Matrix Multiplication*
Note that
$$rT \triangleq \begin{bmatrix} r_1 & r_2 & r_3 \end{bmatrix} \begin{bmatrix} t_{11} & t_{12} & t_{13} \\ t_{21} & t_{22} & t_{23} \\ t_{31} & t_{32} & t_{33} \end{bmatrix}$$
$$= r_1 \begin{bmatrix} t_{11} & t_{12} & t_{13} \end{bmatrix} + r_2 \begin{bmatrix} t_{21} & t_{22} & t_{23} \end{bmatrix} + r_3 \begin{bmatrix} t_{31} & t_{32} & t_{33} \end{bmatrix}.$$

That is, we can view the multiplication of the square matrix T on the left by the row vector r as simply a linear combination of the *rows* of T.

Similarly, we have
$$RT \triangleq \begin{bmatrix} r_{11} & r_{12} & r_{13} \\ r_{21} & r_{22} & r_{23} \\ r_{31} & r_{32} & r_{33} \end{bmatrix} \begin{bmatrix} t_{11} & t_{12} & t_{13} \\ t_{21} & t_{22} & t_{23} \\ t_{31} & t_{32} & t_{33} \end{bmatrix}$$

$$= \begin{bmatrix} r_{11}\begin{bmatrix}t_{11} & t_{12} & t_{13}\end{bmatrix} + r_{12}\begin{bmatrix}t_{21} & t_{22} & t_{23}\end{bmatrix} + r_{13}\begin{bmatrix}t_{31} & t_{32} & t_{33}\end{bmatrix} \\ r_{21}\begin{bmatrix}t_{11} & t_{12} & t_{13}\end{bmatrix} + r_{22}\begin{bmatrix}t_{21} & t_{22} & t_{23}\end{bmatrix} + r_{23}\begin{bmatrix}t_{31} & t_{32} & t_{33}\end{bmatrix} \\ r_{31}\begin{bmatrix}t_{11} & t_{12} & t_{13}\end{bmatrix} + r_{32}\begin{bmatrix}t_{21} & t_{22} & t_{23}\end{bmatrix} + r_{33}\begin{bmatrix}t_{31} & t_{32} & t_{33}\end{bmatrix} \end{bmatrix}.$$

Each *row* of RT is a linear combination of the *rows* of T.

End of Digression

Example 5 *System Matrices Not in Observer Canonical Form*
We are given the system
$$\frac{dx}{dt} = Ax + bu, \quad A \in \mathbb{R}^{3\times 3}, \quad b \in \mathbb{R}^3$$
$$y = cx, \quad c \in \mathbb{R}^{1\times 3}$$

with $\det(sI - A) = s^3 + \beta_2 s^2 + \beta_1 s + \beta_0$. Suppose the pair (c, A) is not in observer canonical form. The previous example indicates that we could easily compute the gain vector ℓ if we could transform the pair (c, A) into observer canonical form. To do this, define the *observability* matrix $\mathcal{O}$ as
$$\mathcal{O} \triangleq \begin{bmatrix} c \\ cA \\ cA^2 \end{bmatrix}.$$

Suppose $\mathcal{O}$ is invertible, i.e., $\det \mathcal{O} \neq 0$. Define the statespace transformation
$$x' \triangleq \mathcal{O}x.$$

16.1 State Estimators

In terms of x' the statespace model becomes

$$\frac{dx'}{dt} = \underbrace{\mathcal{O}A\mathcal{O}^{-1}}_{A'}x' + \underbrace{\mathcal{O}b}_{b'}u$$

$$y = \underbrace{c\mathcal{O}^{-1}}_{c'}x'.$$

That is,

$$A' = \mathcal{O}A\mathcal{O}^{-1}$$
$$c' = c\mathcal{O}^{-1}$$

or

$$A'\mathcal{O} = \mathcal{O}A$$
$$c'\mathcal{O} = c.$$

The pair (c', A') have a special form as we now show. Write

$$\underbrace{\begin{bmatrix} a'_{11} & a'_{12} & a'_{13} \\ a'_{21} & a'_{22} & a'_{23} \\ a'_{31} & a'_{32} & a'_{33} \end{bmatrix}}_{A'} \underbrace{\begin{bmatrix} c \\ cA \\ cA^2 \end{bmatrix}}_{\mathcal{O}} = \underbrace{\begin{bmatrix} cA \\ cA^2 \\ cA^3 \end{bmatrix}}_{\mathcal{O}A} \quad \text{and} \quad \underbrace{\begin{bmatrix} c'_1 & c'_2 & c'_3 \end{bmatrix}}_{c'} \underbrace{\begin{bmatrix} c \\ cA \\ cA^2 \end{bmatrix}}_{\mathcal{O}} = c.$$

With reference to the above digression on matrix multiplication this can be rewritten as

$$\underbrace{\begin{bmatrix} a'_{11}c + a'_{12}cA + a'_{13}cA^2 \\ a'_{21}c + a'_{22}cA + a'_{23}cA^2 \\ a'_{31}c + a'_{32}cA + a'_{33}cA^2 \end{bmatrix}}_{A'\mathcal{O}} = \underbrace{\begin{bmatrix} cA \\ cA^2 \\ cA^3 \end{bmatrix}}_{\mathcal{O}A} \quad \text{and} \quad \underbrace{c'_1 c + c'_2 cA + c'_3 cA^2}_{c'\mathcal{O}} = c. \qquad (16.20)$$

By inspection it follows that A' and c' must have the form

$$A' = \begin{bmatrix} 0 & 1 & 0 \\ 0 & 0 & 1 \\ a'_{31} & a'_{32} & a'_{33} \end{bmatrix}, \quad c' = \begin{bmatrix} 1 & 0 & 0 \end{bmatrix},$$

where $a'_{31}, a'_{32}, a'_{33}$ must be found to satisfy

$$a'_{31}c + a'_{32}cA + a'_{33}cA^2 = cA^3.$$

To do this we use the characteristic polynomial of A which is $\det(sI - A) = s^3 + \beta_2 s^2 + \beta_1 s + \beta_0$. By the Cayley–Hamilton theorem it follows that

$$A^3 + \beta_2 A^2 + \beta_1 A + \beta_0 I = 0_{3\times 3}.$$

Multiplying this on the left by c and rearranging gives

$$-\beta_0 c - \beta_1 cA - \beta_2 cA^2 = cA^3.$$

That is, $a'_{31} = -\beta_0, a'_{32} = -\beta_1$, and $a'_{33} = -\beta_2$ so that we may now write

$$A' = \begin{bmatrix} 0 & 1 & 0 \\ 0 & 0 & 1 \\ -\beta_0 & -\beta_1 & -\beta_2 \end{bmatrix}, \quad c' = \begin{bmatrix} 1 & 0 & 0 \end{bmatrix}. \tag{16.21}$$

This is *not* observer canonical form, but we are almost there. Continuing, consider

$$\frac{dx_o}{dt} = A_o x_o + b_o u, \quad A_o \in \mathbb{R}^{3\times 3}, \ b_o \in \mathbb{R}^3$$

$$y = c_o x_o, \quad c_o \in \mathbb{R}^{1\times 3}$$

with (c_o, A_o) already in observer canonical form, i.e.,

$$A_o = \begin{bmatrix} 0 & 0 & -\beta_0 \\ 1 & 0 & -\beta_1 \\ 0 & 1 & -\beta_2 \end{bmatrix}, \quad c_o = \begin{bmatrix} 0 & 0 & 1 \end{bmatrix}. \tag{16.22}$$

Define

$$\mathcal{O}_o \triangleq \begin{bmatrix} c_o \\ c_o A_o \\ c_o A_o^2 \end{bmatrix}.$$

We just showed that the transformation $x' = \mathcal{O}_o x_o$ results in

$$\frac{dx'}{dt} = \underbrace{\mathcal{O}_o A_o \mathcal{O}_o^{-1}}_{A'} x' + \mathcal{O}_o b_o u$$

$$y = \underbrace{c_o \mathcal{O}_o^{-1}}_{c'} x'$$

with

$$A' = \mathcal{O}_o A_o \mathcal{O}_o^{-1} = \begin{bmatrix} 0 & 1 & 0 \\ 0 & 0 & 1 \\ -\beta_0 & -\beta_1 & -\beta_2 \end{bmatrix}, \quad b' = \mathcal{O}_o b_o = \begin{bmatrix} b'_1 \\ b'_2 \\ b'_3 \end{bmatrix}$$

$$c' = c_o \mathcal{O}_o^{-1} = \begin{bmatrix} 1 & 0 & 0 \end{bmatrix}. \tag{16.23}$$

Thus the inverse transformation

$$x_o = \mathcal{O}_o^{-1} x'$$

will take a system in the form (16.23) to the observer canonical form (16.22).

Theorem 1 *Transformation to Observer Canonical Form*

Consider the linear statespace system given by

$$\frac{dx}{dt} = Ax + bu, \quad A \in \mathbb{R}^{3\times 3}, \ b \in \mathbb{R}^3 \tag{16.24}$$

$$y = cx, \quad c \in \mathbb{R}^{1\times 3}. \tag{16.25}$$

16.1 State Estimators 655

Let
$$\det(sI - A) = s^3 + \beta_2 s^2 + \beta_1 s + \beta_0$$

$$\mathcal{O} \triangleq \begin{bmatrix} c \\ cA \\ cA^2 \end{bmatrix} \text{ with } \det \mathcal{O} \neq 0.$$

Define
$$c_o = \begin{bmatrix} 0 & 0 & 1 \end{bmatrix}, \quad A_o = \begin{bmatrix} 0 & 0 & -\beta_0 \\ 1 & 0 & -\beta_1 \\ 0 & 1 & -\beta_2 \end{bmatrix}, \quad \mathcal{O}_o \triangleq \begin{bmatrix} c_o \\ c_o A_o \\ c_o A_o^2 \end{bmatrix}.$$

Then the transformation
$$x_o = \mathcal{O}_o^{-1} \mathcal{O} x$$

takes the system (16.24) and (16.25) to the observer canonical form given by

$$\frac{dx_o}{dt} = \underbrace{\begin{bmatrix} 0 & 0 & -\beta_0 \\ 1 & 0 & -\beta_1 \\ 0 & 1 & -\beta_2 \end{bmatrix}}_{A_o} x_o + \underbrace{\begin{bmatrix} b_{o1} \\ b_{o2} \\ b_{o3} \end{bmatrix}}_{b_o} u$$

$$y = \underbrace{\begin{bmatrix} 0 & 0 & 1 \end{bmatrix}}_{c_o} x_o.$$

Proof. This follows from the previous two examples. ∎

Theorem 2 *Placement of the Observer Poles*

Let a linear time invariant system be given by
$$\frac{dx}{dt} = Ax + bu, \quad A \in \mathbb{R}^{3 \times 3}, \quad b \in \mathbb{R}^3$$

$$y = cx, \quad c \in \mathbb{R}^{1 \times 3}$$

with
$$\det(sI - A) = s^3 + \beta_2 s^2 + \beta_1 s + \beta_0$$

and
$$\mathcal{O} \triangleq \begin{bmatrix} c \\ cA \\ cA^2 \end{bmatrix} \text{ with } \det \mathcal{O} \neq 0.$$

Define
$$A_o = \begin{bmatrix} 0 & 0 & -\beta_0 \\ 1 & 0 & -\beta_1 \\ 0 & 1 & -\beta_2 \end{bmatrix}, \quad c_o = \begin{bmatrix} 0 & 0 & 1 \end{bmatrix}, \quad \mathcal{O}_o \triangleq \begin{bmatrix} c_o \\ c_o A_o \\ c_o A_o^2 \end{bmatrix}.$$

16 State Estimators and Parameter Identification

Let the desired closed-loop characteristic polynomial be

$$s^3 + \beta_{d2}s^2 + \beta_{d1}s + \beta_{d0}.$$

Then choosing

$$\ell = \begin{bmatrix} \ell_0 \\ \ell_1 \\ \ell_2 \end{bmatrix} = \begin{bmatrix} c \\ cA \\ cA^2 \end{bmatrix}^{-1} \begin{bmatrix} c_o \\ c_o A_o \\ c_o A_o^2 \end{bmatrix} \begin{bmatrix} \beta_{d0} - \beta_0 \\ \beta_{d1} - \beta_1 \\ \beta_{d2} - \beta_2 \end{bmatrix} = \mathcal{O}^{-1}\mathcal{O}_o(\beta_d - \beta)$$

results in

$$\det(sI - (A - \ell c)) = s^3 + \beta_{d2}s^2 + \beta_{d1}s + \beta_{d0}.$$

Proof. The observer is

$$\frac{d\hat{x}}{dt} = A\hat{x} + bu + \ell(y - \hat{y})$$

$$\hat{y} = c\hat{x}.$$

Using the transformation $x_o = \mathcal{O}_o^{-1}\mathcal{O}x$ this becomes

$$\frac{d\hat{x}_o}{dt} = \underbrace{(\mathcal{O}_o^{-1}\mathcal{O})A(\mathcal{O}_o^{-1}\mathcal{O})^{-1}}_{A_o}\hat{x}_o + \underbrace{(\mathcal{O}_o^{-1}\mathcal{O})bu}_{b_o} + \underbrace{(\mathcal{O}_o^{-1}\mathcal{O})\ell}_{\ell_o}(y - \hat{y})$$

$$\hat{y} = \underbrace{c(\mathcal{O}_o^{-1}\mathcal{O})^{-1}}_{c_o}\hat{x}_o.$$

The pair (c_o, A_o) is in observer canonical form. By Example 4 taking

$$\ell_o = \beta_d - \beta = \begin{bmatrix} \beta_{d0} - \beta_0 \\ \beta_{d1} - \beta_1 \\ \beta_{d2} - \beta_2 \end{bmatrix}$$

results in $\det(sI - (A_o - \ell_o c_o)) = \det(sI - (A - \ell c)) = s^3 + \beta_2 s^2 + \beta_1 s + \beta_0$. ∎

Example 6 *State Estimation*

Suppose that the system model is given by

$$\frac{dx}{dt} = \underbrace{\begin{bmatrix} 0 & 1 \\ -5 & -6 \end{bmatrix}}_{A} x + \underbrace{\begin{bmatrix} 0 \\ 1 \end{bmatrix}}_{b} u$$

$$y = \underbrace{\begin{bmatrix} 0 & 1 \end{bmatrix}}_{c} x.$$

Based on the measurement y we design an observer to estimate the full state. The observability matrix is

$$\mathcal{O} = \begin{bmatrix} c \\ cA \end{bmatrix} = \begin{bmatrix} 0 & 1 \\ -5 & -6 \end{bmatrix}.$$

The open-loop characteristic polynomial is

$$\det(sI - A) = s^2 + \beta_1 s + \beta_0 = \det\begin{bmatrix} s & -1 \\ 5 & s+6 \end{bmatrix} = s^2 + 6s + 5.$$

The observer (state estimator) is given by

$$\frac{d\hat{x}}{dt} = \begin{bmatrix} 0 & 1 \\ -5 & -6 \end{bmatrix}\hat{x} + \begin{bmatrix} 0 \\ 1 \end{bmatrix}u + \begin{bmatrix} \ell_1 \\ \ell_2 \end{bmatrix}(y - \hat{y})$$

$$\hat{y} = \begin{bmatrix} 0 & 1 \end{bmatrix}\hat{x}.$$

Using Theorem 2 we now compute the gain vector ℓ that will place the two closed-loop poles at -5. First note that the desired characteristic polynomial of the observer is

$$s^2 + \beta_{d1}s + \beta_{d0}s = (s+5)(s+5) = s^2 + 10s + 25.$$

The observer canonical form and the observability matrix for the pair (c, A) are

$$A_o = \begin{bmatrix} 0 & -\beta_0 \\ 1 & -\beta_1 \end{bmatrix} = \begin{bmatrix} 0 & -5 \\ 1 & -6 \end{bmatrix}, \quad c_o = \begin{bmatrix} 0 & 1 \end{bmatrix}$$

$$\mathcal{O}_o = \begin{bmatrix} c_o \\ c_o A_o \end{bmatrix} = \begin{bmatrix} 0 & 1 \\ 1 & -6 \end{bmatrix}.$$

By Theorem 2 we have

$$\ell = \mathcal{O}^{-1}\mathcal{O}_o \begin{bmatrix} \beta_{d0} - \beta_0 \\ \beta_{d1} - \beta_1 \end{bmatrix} = \begin{bmatrix} 0 & 1 \\ -5 & -6 \end{bmatrix}^{-1}\begin{bmatrix} 0 & 1 \\ 1 & -6 \end{bmatrix}\begin{bmatrix} 25 - 5 \\ 10 - 6 \end{bmatrix} = \begin{bmatrix} -4 \\ 4 \end{bmatrix}.$$

As a check we compute

$$\det(sI - (A - \ell c)) = \det\left(\begin{bmatrix} s & 0 \\ 0 & s \end{bmatrix} - \left(\begin{bmatrix} 0 & 1 \\ -5 & -6 \end{bmatrix} - \begin{bmatrix} -4 \\ 4 \end{bmatrix}\begin{bmatrix} 0 & 1 \end{bmatrix}\right)\right)$$

$$= \det\left(\begin{bmatrix} s & 0 \\ 0 & s \end{bmatrix} - \begin{bmatrix} 0 & 5 \\ -5 & -10 \end{bmatrix}\right)$$

$$= \det\begin{bmatrix} s & -5 \\ 5 & s+10 \end{bmatrix}$$

$$= s^2 + 10s + 25.$$

The observer is then

$$\frac{d\hat{x}}{dt} = \begin{bmatrix} 0 & 1 \\ -5 & -6 \end{bmatrix}\hat{x} + \begin{bmatrix} 0 \\ 1 \end{bmatrix}u + \begin{bmatrix} -4 \\ 4 \end{bmatrix}(y - \hat{y})$$

$$\hat{y} = \begin{bmatrix} 0 & 1 \end{bmatrix}\hat{x}.$$

Separation Principle

Consider the statespace model given by

$$\frac{d}{dt}x = Ax + bu, \quad A \in \mathbb{R}^{n \times n}, \quad b \in \mathbb{R}^n$$

$$y = cx, \quad c \in \mathbb{R}^{1 \times n}.$$

(16.26)

16 State Estimators and Parameter Identification

The reference trajectory x_d and the reference input r are designed to satisfy

$$\frac{d}{dt}x_d = Ax_d + br. \qquad (16.27)$$

We are given that the pair (c, A) is observable and thus assume $\ell \in \mathbb{R}^n$ is chosen so that $A - \ell c \in \mathbb{R}^{n \times n}$ is stable. We are also given that the pair (A, b) is controllable so we also assume a row vector $k \in \mathbb{R}^{1 \times n}$ is chosen so that $A - bk \in \mathbb{R}^{n \times n}$ is stable. The input to the physical system is

$$u = k(x_d - \hat{x}) + r.$$

Figure 16.5 illustrates the feedback setup.

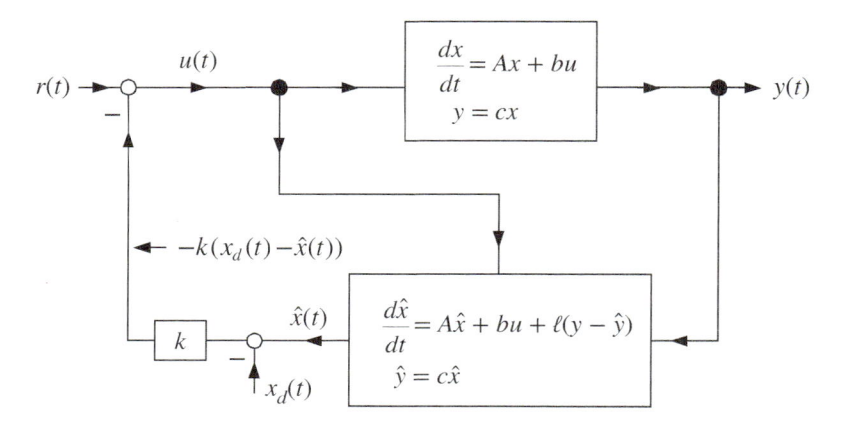

Figure 16.5. Trajectory tracking with a state estimator.

With $r_d(t) \triangleq r(t) + kx_d$ Figure 16.5 is equivalent to Figure 16.6.

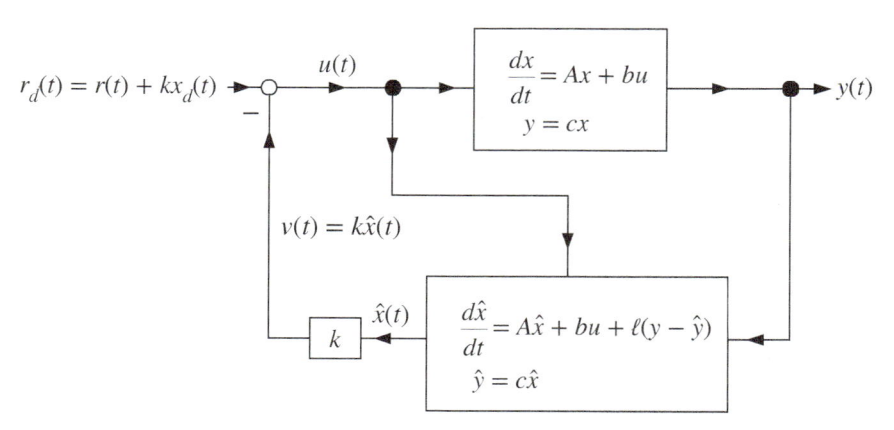

Figure 16.6. Equivalent setup to Figure 16.5.

A statespace model for the system of Figure 16.6 is given by

$$\frac{d}{dt}x = Ax - bk\hat{x} + br_d$$

$$\frac{d}{dt}\hat{x} = (A - \ell c - bk)\hat{x} + \ell cx + br_d \qquad (16.28)$$

$$y = \begin{bmatrix} c & 0_{1\times n} \end{bmatrix} \begin{bmatrix} x \\ \hat{x} \end{bmatrix}.$$

In matrix form this is

$$\frac{d}{dt}\begin{bmatrix} x \\ \hat{x} \end{bmatrix} = \begin{bmatrix} A & -bk \\ \ell c & A - \ell c - bk \end{bmatrix}\begin{bmatrix} x \\ \hat{x} \end{bmatrix} + \begin{bmatrix} b \\ b \end{bmatrix} r_d \qquad (16.29)$$

$$y = \begin{bmatrix} c & 0_{1\times n} \end{bmatrix} \begin{bmatrix} x \\ \hat{x} \end{bmatrix}. \qquad (16.30)$$

With $e \triangleq x - \hat{x}$ define the statespace (similarity) transformation

$$\begin{bmatrix} x \\ e \end{bmatrix} = \underbrace{\begin{bmatrix} I_{n\times n} & 0_{n\times n} \\ I_{n\times n} & -I_{n\times n} \end{bmatrix}}_{T}\begin{bmatrix} x \\ \hat{x} \end{bmatrix},$$

where

$$T^{-1} = \begin{bmatrix} I_{n\times n} & 0_{n\times n} \\ I_{n\times n} & -I_{n\times n} \end{bmatrix} = T.$$

Then applying the transformation T to the system (16.29) and (16.30), we have

$$\frac{d}{dt}\begin{bmatrix} x \\ e \end{bmatrix} = T\begin{bmatrix} A & -bk \\ \ell c & A - \ell c - bk \end{bmatrix}T^{-1}\begin{bmatrix} x \\ e \end{bmatrix} + T\begin{bmatrix} b \\ b \end{bmatrix}r_d$$

$$y = \begin{bmatrix} c & 0_{1\times n} \end{bmatrix}T^{-1}\begin{bmatrix} x \\ e \end{bmatrix}$$

or

$$\frac{d}{dt}\begin{bmatrix} x \\ e \end{bmatrix} = \begin{bmatrix} A - bk & -bk \\ 0_{n\times n} & A - \ell c \end{bmatrix}\begin{bmatrix} x \\ e \end{bmatrix} + \begin{bmatrix} b \\ 0_{n\times 1} \end{bmatrix}r_d \qquad (16.31)$$

$$y = \begin{bmatrix} c & 0_{1\times n} \end{bmatrix}\begin{bmatrix} x \\ e \end{bmatrix}. \qquad (16.32)$$

As $\begin{bmatrix} A - bk & -bk \\ 0_{n\times n} & A - \ell c \end{bmatrix}$ is in block diagonal form it follows that

$$\det\begin{bmatrix} sI - (A - bk) & bk \\ 0_{n\times n} & sI - (A - \ell c) \end{bmatrix} = \det(sI - (A - bk))\det(sI - (A - \ell c)). \qquad (16.33)$$

The eigenvalues of the system matrix (16.31) are the union of the eigenvalues of $A - bk$ and the eigenvalues of $A - \ell c$. Next we show that the trajectory tracking error $\varepsilon \triangleq x_d - x$ goes to zero. Substituting $r_d = r + kx_d$ into (16.31) gives

$$\frac{d}{dt}x = (A - bk)x - bke + b(r + kx_d).$$

Using the expression for dx_d/dt given in (16.27), it is seen that the trajectory tracking $x_d - x$ satisfies

$$\frac{d}{dt}(x_d - x) = Ax_d + br - \big((A - bk)x - bke + b(r + kx_d)\big) = (A - bk)(x_d - x) + bke.$$

Putting this all together the statespace system for the error states $[\varepsilon\ e]^T = [x_d - x, x - \hat{x}]^T$ is

$$\frac{d}{dt}\begin{bmatrix}\varepsilon\\e\end{bmatrix} = \begin{bmatrix}A - bk & bk\\0_{n\times n} & A - \ell c\end{bmatrix}\begin{bmatrix}\varepsilon\\e\end{bmatrix}. \tag{16.34}$$

From (16.33) we see this error system is stable with the consequence that both $\varepsilon(t) \triangleq x_d(t) - x(t) \to 0$ and $e(t) = x(t) - \hat{x}(t) \to 0$. The gain vector $k \in \mathbb{R}^{1\times n}$ is used to place the poles of $A - bk$ of the trajectory tracking error while *separately* the gain vector $\ell \in \mathbb{R}^n$ is used to place the poles of $A - \ell c$ of the state estimation error. This is referred to as the *separation principle*.

16.2 State Feedback and State Estimation in the Laplace Domain*

We now look at state feedback using state estimation in the Laplace domain. Let the statespace model of the open-loop system be

$$\frac{dx}{dt} = Ax + bu$$
$$y = cx$$

with $A \in \mathbb{R}^{n\times n}, b \in \mathbb{R}^n, c \in \mathbb{R}^{1\times n}$. We assume (A, b) is controllable and (c, A) is observable. We choose $k \in \mathbb{R}^{1\times n}$ so that $A - bk$ is stable and $\ell \in \mathbb{R}^n$ so that $A - \ell c$ is stable. The state trajectory x_d and reference input r are designed to satisfy

$$\frac{dx_d}{dt} = Ax_d + br.$$

With reference to Figure 16.6 let

$$r_d \triangleq kx_d + r$$

so that the input u may be written as $u = -k\hat{x} + r_d$.

The state observer is

$$\frac{d\hat{x}}{dt} = A\hat{x} + bu + \ell(y - \hat{y}) = (A - \ell c)\hat{x} + \ell y + bu$$
$$\hat{y} = c\hat{x}.$$

With $\hat{x}(0) = 0$ the Laplace transform of $\hat{x}(t)$ is

$$\hat{X}(s) = \big(sI - (A - \ell c)\big)^{-1}\ell Y(s) + \big(sI - (A - \ell c)\big)^{-1}bU(s).$$

With the definitions

$$G(s) \triangleq c(sI - A)^{-1}b = \frac{b(s)}{a(s)}$$

16.2 State Feedback and State Estimation in the Laplace Domain*

$$G_{c1}(s) \triangleq k(sI - (A - \ell c))^{-1}b = \frac{n(s)}{\delta(s)}$$

$$G_{c2}(s) \triangleq k(sI - (A - \ell c))^{-1}\ell = \frac{m(s)}{\delta(s)}$$

we may write

$$V(s) \triangleq k\hat{X}(s) = \frac{n(s)}{\delta(s)}U(s) + \frac{m(s)}{\delta(s)}Y(s).$$

In the s domain the block diagram of Figure 16.6 becomes that of Figure 16.7.

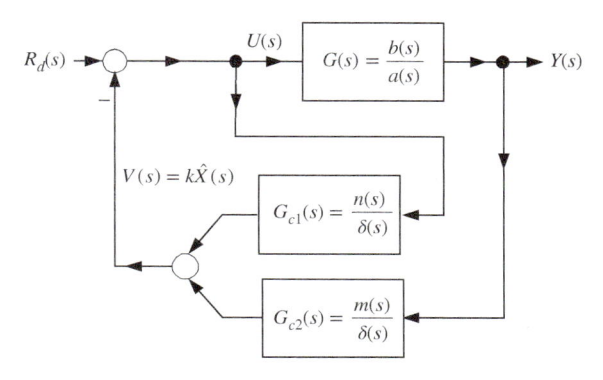

Figure 16.7. Transfer function representation of Figure 16.6.

$G(s), G_{c1}(s), G_{c2}(s)$ in Figure 16.7 are each of order n so the complete system appears to be of order $3n$. That is, in going from the statespace block diagram of Figure 16.6 to the transfer function block diagram of Figure 16.7, it seems we have gone from a $2n$ order system to a $3n$ order system. However, with $n = 3$, a statespace implementation of the feedback $V(s) = G_{c1}(s)U(s) + G_{c2}(s)Y(s)$ is

$$\frac{d}{dt}\begin{bmatrix} z_1 \\ z_2 \\ z_3 \end{bmatrix} = \begin{bmatrix} -\delta_2 & 1 & 0 \\ -\delta_1 & 0 & 1 \\ -\delta_0 & 0 & 0 \end{bmatrix}\begin{bmatrix} z_1 \\ z_2 \\ z_3 \end{bmatrix} + \begin{bmatrix} n_2 \\ n_1 \\ n_0 \end{bmatrix}u + \begin{bmatrix} m_2 \\ m_1 \\ m_0 \end{bmatrix}y$$

$$v = \begin{bmatrix} 1 & 0 & 0 \end{bmatrix}\begin{bmatrix} z_1 \\ z_2 \\ z_3 \end{bmatrix}.$$

This is verified by the computation

$$V(s) = \begin{bmatrix} 1 & 0 & 0 \end{bmatrix}\begin{bmatrix} s+\delta_3 & -1 & 0 \\ \delta_2 & s & -1 \\ \delta_1 & 0 & s \end{bmatrix}^{-1}\left(\begin{bmatrix} n_2 \\ n_1 \\ n_0 \end{bmatrix}U(s) + \begin{bmatrix} m_2 \\ m_1 \\ m_0 \end{bmatrix}Y(s)\right)$$

$$= \frac{\begin{bmatrix} s^2 & s & 1 \end{bmatrix}}{s^3 + \delta_2 s^2 + \delta_1 s + \delta_0}\left(\begin{bmatrix} n_2 \\ n_1 \\ n_0 \end{bmatrix}U(s) + \begin{bmatrix} m_2 \\ m_1 \\ m_0 \end{bmatrix}Y(s)\right)$$

$$= G_{c1}(s)U(s) + G_{c2}(s)Y(s).$$

That is, $G_{c1}(s)$ and $G_{c2}(s)$ are implemented with the same n integrators. So, with this implementation, Figure 16.7 represents a $2n$ order system as well. A simple rearrangement of Figure 16.7 gives the block diagram of Figure 16.8.

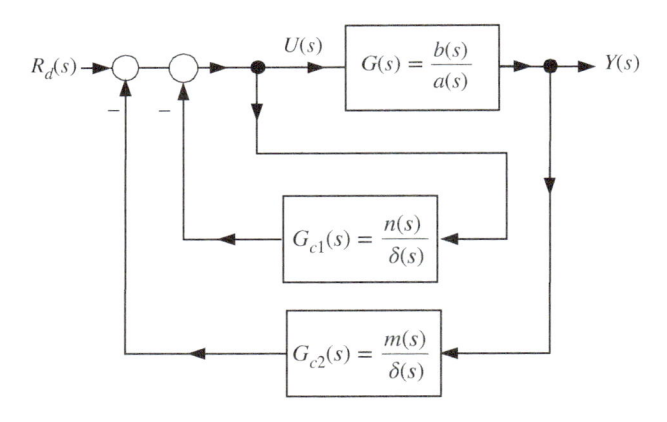

Figure 16.8. Block diagram equivalent to Figure 16.7.

Finally, rearrange Figure 16.8 to obtain the block diagram of Figure 16.9.

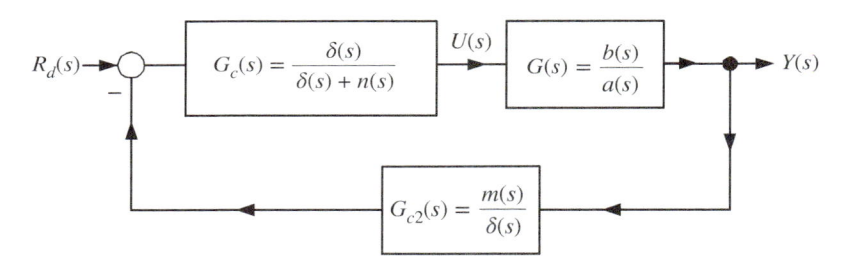

Figure 16.9. Block diagram equivalent to Figure 16.8.

With reference to Figure 16.9, the closed-loop transfer function is

$$Y(s) = \frac{\dfrac{\delta(s)}{\delta(s)+n(s)}\dfrac{b(s)}{a(s)}}{1+\dfrac{m(s)}{\delta(s)}\dfrac{\delta(s)}{\delta(s)+n(s)}\dfrac{b(s)}{a(s)}} R_d(s) = \frac{\delta^2(s)b(s)}{\delta(s)\bigl(\delta(s)+n(s)\bigr)a(s)+m(s)\delta(s)b(s)} R_d(s)$$

$$= \frac{\delta(s)b(s)}{\bigl(\delta(s)+n(s)\bigr)a(s)+m(s)b(s)} R_d(s).$$

As ℓ is chosen so that $\delta(s) = \det(sI - (A - \ell c))$ is stable the cancellation in the last step would be a *stable* pole–zero cancellation. However, as explained above, $G_{c1}(s)$ and $G_{c2}(s)$ are implemented with the same n integrators so this cancellation is just an artifact of the transfer function block diagram representation that was used and does not actually occur. With $a(s) \triangleq \det(sI - (A - bk))$ the separation principle allows us to write

$$\bigl(\delta(s)+n(s)\bigr)a(s) + m(s)b(s) = \det(sI - (A - \ell c))\det(sI - (A - bk)) = \delta(s)\alpha(s).$$

16.3 Multi-Output Observer Design for the Inverted Pendulum*

The closed-loop transfer function then becomes

$$Y(s) = \frac{\delta(s)b(s)}{(\delta(s)+n(s))\,a(s)+m(s)b(s)}R_d(s) = \frac{\delta(s)b(s)}{\delta(s)\alpha(s)}R_d(s) = \frac{b(s)}{\alpha(s)}R_d(s), \quad (16.35)$$

where a stable pole–zero cancellation occurred. We end up with the same closed-loop transfer function as if full state feedback was used!

Remark In Chapters 9 and 10 a transfer function model of an open-loop system was placed in a unity feedback configuration to design a controller for it. This section indicates that it would be better to start with the control structure of Figure 16.9. This is because starting with this control structure the end result is (16.35) showing the closed-loop system has only the zeros of the open-loop model and the closed-loop poles can be placed at any desired location in the open left-half plane. Specifically, the designer freely choooses $\delta(s)$ and $\alpha(s)$ to be nth order stable polynomials. Then, with $n(s)$ and $m(s)$ polynomials of degree $n-1$ or less, the controllers $G_c(s)$ and $G_{c1}(s)$ are specified to have the forms $G_c(s) = \dfrac{\delta(s)}{\delta(s)+n(s)}$ and $G_c(s) = \dfrac{m(s)}{\delta(s)}$, respectively. As shown in Kailath [61] (see pages 208 and 276–277 of [61]), the polynomials $n(s)$ and $m(s)$ can be found such that $(\delta(s)+n(s))\,a(s) + m(s)b(s) = \delta(s)\alpha(s)$. This then results in the expression for $Y(s)$ given in (16.35).

16.3 Multi-Output Observer Design for the Inverted Pendulum*

The observer design procedure in this chapter has been done assuming a single output. In the case of the inverted pendulum with both the cart position x and the pendulum rod angle θ measured, we have two outputs. Rather than combine them into a single output, e.g., as $y = x + \left(\ell + \dfrac{J}{m\ell}\right)\theta$, we now present an approach for the state estimate based on the *vector* output measurement given by

$$\underbrace{\begin{bmatrix} y_1 \\ y_2 \end{bmatrix}}_{y} = \begin{bmatrix} x \\ \theta \end{bmatrix}.$$

The starting point for the observer design is the linear model of the inverted pendulum about the equilibrium point $(x_{eq}, v_{eq}, \theta_{eq}, \omega_{eq}) = (0,0,0,0)$. That is, the model given by

$$\underbrace{\begin{bmatrix} dx/dt \\ dv/dt \\ d\theta/dt \\ d\omega/dt \end{bmatrix}}_{dz/dt} = \underbrace{\begin{bmatrix} 0 & 1 & 0 & 0 \\ 0 & 0 & -\kappa g m^2 \ell^2 & 0 \\ 0 & 0 & 0 & 1 \\ 0 & 0 & \alpha^2 & 0 \end{bmatrix}}_{A} \underbrace{\begin{bmatrix} x \\ v \\ \theta \\ \omega \end{bmatrix}}_{z} + \underbrace{\begin{bmatrix} 0 \\ \kappa(J+m\ell^2) \\ 0 \\ -\kappa m\ell \end{bmatrix}}_{b} u$$

$$\underbrace{\begin{bmatrix} y_1 \\ y_2 \end{bmatrix}}_{y} = \underbrace{\begin{bmatrix} 1 & 0 & 0 & 0 \\ 0 & 0 & 1 & 0 \end{bmatrix}}_{C} \underbrace{\begin{bmatrix} x \\ v \\ \theta \\ \omega \end{bmatrix}}_{z},$$

where $\kappa = \dfrac{1}{Mm\ell^2 + J(M+m)}, \alpha^2 = \dfrac{mg\ell(M+m)}{Mm\ell^2 + J(M+m)}$. The gain matrix L for this vector output is

$$L = \begin{bmatrix} l_{11} & l_{12} \\ l_{21} & l_{22} \\ l_{31} & l_{32} \\ l_{41} & l_{42} \end{bmatrix} \in \mathbb{R}^{4\times 2}.$$

The observer for this system has the form

$$\frac{d\hat{z}}{dt} = A\hat{z} + bu + L(y - \hat{y})$$

$$\hat{y} = C\hat{z}.$$

The system model is

$$\frac{dz}{dt} = Az + bu$$

$$y = Cz.$$

The state estimation error $z - \hat{z}$ satisfies

$$\frac{d}{dt}(z - \hat{z}) = A(z - \hat{z}) - LC(z - \hat{z}) = (A - LC)(z - \hat{z}).$$

With $L \in \mathbb{R}^{4\times 2}$ chosen so the eigenvalues of $A - LC$ are in the open left-half plane, it follows that

$$z(t) - \hat{z}(t) = e^{(A-LC)t}(z(0) - \hat{z}(0)) \to 0_{4\times 1}.$$

The key to making this work is being able to choose the gain matrix $L \in \mathbb{R}^{4\times 2}$ to place the poles of $A - LC$. To do so we first compute

$$A - LC = \begin{bmatrix} 0 & 1 & 0 & 0 \\ 0 & 0 & -\kappa g m^2 \ell^2 & 0 \\ 0 & 0 & 0 & 1 \\ 0 & 0 & \alpha^2 & 0 \end{bmatrix} - \begin{bmatrix} l_{11} & l_{12} \\ l_{21} & l_{22} \\ l_{31} & l_{32} \\ l_{41} & l_{42} \end{bmatrix} \begin{bmatrix} 1 & 0 & 0 & 0 \\ 0 & 0 & 1 & 0 \end{bmatrix}$$

$$= \begin{bmatrix} -l_{11} & 1 & -l_{12} & 0 \\ -l_{21} & 0 & -g\kappa m^2 \ell^2 - l_{22} & 0 \\ -l_{31} & 0 & -l_{32} & 1 \\ -l_{41} & 0 & \alpha^2 - l_{42} & 0 \end{bmatrix}.$$

Next choose $l_{12} = 0, l_{22} = -g\kappa m^2 \ell^2, l_{31} = 0, l_{41} = 0$ so that $A - LC$ becomes

$$A - LC = \begin{bmatrix} -l_{11} & 1 & 0 & 0 \\ -l_{21} & 0 & 0 & 0 \\ 0 & 0 & -l_{32} & 1 \\ 0 & 0 & \alpha^2 - l_{42} & 0 \end{bmatrix}.$$

Notice that $A - LC$ is now *block diagonal* consisting of two 2×2 matrices on the main diagonal. As a result

$$\det(sI - (A - LC)) = \det \begin{bmatrix} s + l_{11} & -1 & 0 & 0 \\ l_{21} & s & 0 & 0 \\ 0 & 0 & s + l_{32} & -1 \\ 0 & 0 & l_{42} - \alpha^2 & s \end{bmatrix}$$

$$= \det\begin{bmatrix} s+l_{11} & -1 \\ l_{21} & s \end{bmatrix} \det\begin{bmatrix} s+l_{32} & -1 \\ l_{42}-\alpha^2 & s \end{bmatrix}$$
$$= (s^2 + l_{11}s + l_{21})(s^2 + l_{32}s + l_{42} - \alpha^2).$$

Finally, choosing $l_{11} = r_1 + r_2, l_{21} = r_1 r_2, l_{32} = r_3 + r_4, l_{42} = \alpha^2 + r_3 r_4$, it follows that

$$\det(sI - (A - LC)) = (s+r_1)(s+r_2)(s+r_3)(s+r_4).$$

Summarizing, the observer gain matrix given by

$$L = \begin{bmatrix} r_1 + r_2 & 0 \\ r_1 r_2 & -g\kappa m^2 \ell^2 \\ 0 & r_3 + r_4 \\ 0 & \alpha^2 + r_3 r_4 \end{bmatrix}$$

places the poles of $A - LC$ at $-r_1, -r_2, -r_3, -r_4$. This state estimator is implemented using the observer equation

$$\frac{d\hat{z}}{dt} = (A - LC)\hat{z} + bu + Ly,$$

where

$$y(t) = \begin{bmatrix} x(t) \\ \theta(t) \end{bmatrix} \in \mathbb{R}^2, \quad \hat{z}(t) = \begin{bmatrix} \hat{x}(t) \\ \hat{v}(t) \\ \hat{\theta}(t) \\ \hat{\omega}(t) \end{bmatrix}.$$

Problem 9 asks you to add this simulation to a state feedback controller of the inverted pendulum.

Remark This is an example of a *multi-output* observer. In this particular example we were able to pick the gain matrix L to place the closed-loop poles anywhere we choose. Note that we did not change to a new coordinate system in order to do this. This was possible because the pair (C, A) happened to be in multi-output observer form, i.e., we were lucky! See Section 6.4.6 of Kailath [61] for the theory of transforming multi-input multi-output (MIMO) control systems into their canonical forms.

16.4 Properties of Matrix Transpose and Inverse

For both the section on *Duality* and the section on *Parameter Identification* we need to use some more matrix results about transpose and inverse. We present them now.

Theorem 3 $(AB)^T = B^T A^T$
For any two matrices we have

$$(AB)^T = B^T A^T.$$

Proof.
$$(AB)_{ij} = \sum_{\ell=1}^{n} a_{i\ell} b_{\ell j} \implies ((AB)^T)_{ij} = \sum_{\ell=1}^{n} a_{j\ell} b_{\ell i}$$

$$(B^T A^T)_{ij} = \sum_{\ell=1}^{n} (B^T)_{i\ell} (A^T)_{\ell j} = \sum_{\ell=1}^{n} b_{\ell i} a_{j\ell} = ((AB)^T)_{ij}.$$

∎

Example 7 *Transpose of the Product of Two Matrices*
Let
$$A = \begin{bmatrix} 1 & 2 \\ 3 & 4 \end{bmatrix}, \; B = \begin{bmatrix} 5 & 6 \\ 7 & 8 \end{bmatrix}.$$

Then
$$(AB)^T = \left(\begin{bmatrix} 1 & 2 \\ 3 & 4 \end{bmatrix}\begin{bmatrix} 5 & 6 \\ 7 & 8 \end{bmatrix}\right)^T = \left(\begin{bmatrix} 19 & 22 \\ 43 & 50 \end{bmatrix}\right)^T = \begin{bmatrix} 19 & 43 \\ 22 & 50 \end{bmatrix}$$

while
$$B^T A^T = \left(\begin{bmatrix} 5 & 6 \\ 7 & 8 \end{bmatrix}\right)^T \left(\begin{bmatrix} 1 & 2 \\ 3 & 4 \end{bmatrix}\right)^T = \begin{bmatrix} 5 & 7 \\ 6 & 8 \end{bmatrix}\begin{bmatrix} 1 & 3 \\ 2 & 4 \end{bmatrix} = \begin{bmatrix} 19 & 43 \\ 22 & 50 \end{bmatrix}.$$

Example 8 *Transpose of the Product of Two Matrices*
Let
$$A = \begin{bmatrix} 1 & 2 \end{bmatrix}, \; B = \begin{bmatrix} 5 & 6 \\ 7 & 8 \end{bmatrix}$$

then
$$(AB)^T = \left(\begin{bmatrix} 1 & 2 \end{bmatrix}\begin{bmatrix} 5 & 6 \\ 7 & 8 \end{bmatrix}\right)^T = (\begin{bmatrix} 19 & 22 \end{bmatrix})^T = \begin{bmatrix} 19 \\ 22 \end{bmatrix}$$

and
$$B^T A^T = \left(\begin{bmatrix} 5 & 6 \\ 7 & 8 \end{bmatrix}\right)^T (\begin{bmatrix} 1 & 2 \end{bmatrix})^T = \begin{bmatrix} 5 & 7 \\ 6 & 8 \end{bmatrix}\begin{bmatrix} 1 \\ 2 \end{bmatrix} = \begin{bmatrix} 19 \\ 22 \end{bmatrix}.$$

Theorem 4 $(A + B)^T = A^T + B^T$
For any two matrices of the same dimensions we have
$$(A + B)^T = A^T + B^T.$$

Proof. Hopefully this is obvious. ∎

Theorem 5 $(A^T)^{-1} = (A^{-1})^T$
Let A be an invertible matrix, i.e., $\det A \neq 0$. Then
$$(A^T)^{-1} = (A^{-1})^T.$$

Proof. Writing $(A^T)^{-1}$ means that
$$(A^T)^{-1} A^T = I.$$

16.4 Properties of Matrix Transpose and Inverse

Take the transpose of both sides to obtain

$$((A^T)^{-1}A^T)^T = I^T = I.$$

By Theorem 3 we may rewrite this as

$$(A^T)^T((A^T)^{-1})^T = I$$

or

$$A((A^T)^{-1})^T = I.$$

This means that

$$((A^T)^{-1})^T = A^{-1}.$$

Finally, take the transpose of both sides of this last equation to obtain

$$(A^T)^{-1} = (A^{-1})^T.$$

Let's do an example.

Example 9 *Transpose and Inverse Commute*
Let
$$A = \begin{bmatrix} 1 & 2 \\ 3 & 4 \end{bmatrix}.$$

Then

$$(A^T)^{-1} = \left(\begin{bmatrix} 1 & 2 \\ 3 & 4 \end{bmatrix}^T\right)^{-1} = \left(\begin{bmatrix} 1 & 3 \\ 2 & 4 \end{bmatrix}\right)^{-1} = -\frac{1}{2}\begin{bmatrix} 4 & -3 \\ -2 & 1 \end{bmatrix}$$

while

$$(A^{-1})^T = \left(\begin{bmatrix} 1 & 2 \\ 3 & 4 \end{bmatrix}^{-1}\right)^T = \left(-\frac{1}{2}\begin{bmatrix} 4 & -2 \\ -3 & 1 \end{bmatrix}\right)^T = -\frac{1}{2}\begin{bmatrix} 4 & -3 \\ -2 & 1 \end{bmatrix}.$$

Theorem 6 $(AB)^{-1} = B^{-1}A^{-1}$
For any two square invertible matrices A, B we have

$$(AB)^{-1} = B^{-1}A^{-1}.$$

Proof. By the definition of inverse $(AB)^{-1}$ satisfies

$$AB(AB)^{-1} = I.$$

The computation
$$(AB)(B^{-1}A^{-1}) = ABB^{-1}A^{-1} = AA^{-1} = I$$

shows that $B^{-1}A^{-1}$ is the inverse of AB.

16.5 Duality*

The observer pole placement problem is to find $\ell \in \mathbb{R}^n$ to place the poles of

$$A - \ell c = \underbrace{\begin{bmatrix} & & \\ & A & \\ & & \end{bmatrix}}_{\in \mathbb{R}^{n \times n}} - \underbrace{\begin{bmatrix} \\ \ell \\ \\ \end{bmatrix}}_{\in \mathbb{R}^n} \underbrace{[\; c \;]}_{\in \mathbb{R}^{1 \times n}}.$$

Taking the transpose we have

$$(A - \ell c)^T = A^T - c^T \ell^T = \underbrace{\begin{bmatrix} & & \\ & A^T & \\ & & \end{bmatrix}}_{\in \mathbb{R}^{n \times n}} - \underbrace{\begin{bmatrix} \\ c^T \\ \\ \end{bmatrix}}_{\in \mathbb{R}^n} \underbrace{[\; \ell^T \;]}_{\in \mathbb{R}^{1 \times n}}.$$

Make the definitions[1]

$$A_d \triangleq A^T, \; b_d \triangleq c^T$$
$$k \triangleq \ell^T,$$

where the pair $(A_d, b_d) = (A^T, c^T)$ is referred to as the *dual* of (c, A). With

$$\det(sI - A) = s^3 + \beta_2 s^2 + \beta_1 s + \beta_0,$$

the observer canonical form for (c, A) is

$$A_o = \begin{bmatrix} 0 & 0 & -\beta_0 \\ 1 & 0 & -\beta_1 \\ 0 & 1 & -\beta_2 \end{bmatrix}$$

$$c_o = \begin{bmatrix} 0 & 0 & 1 \end{bmatrix}.$$

The control canonical form for dual system (A_d, b_d) is

$$A_{dc} = A_o^T = \begin{bmatrix} 0 & 1 & 0 \\ 0 & 0 & 1 \\ -\beta_0 & -\beta_1 & -\beta_2 \end{bmatrix}, \; b_{dc} = c_o^T = \begin{bmatrix} 0 \\ 0 \\ 1 \end{bmatrix}.$$

With $k = \ell^T$ it follows that

$$\det(sI - (A - \ell c)) = \det(sI - (A^T - c^T \ell^T)) = \det(sI - (A_d - b_d k)).$$

This shows that placing the closed-loop poles of the observer system $A - \ell c$ is equivalent to placing the closed-loop poles of the dual system $A_d - b_d k$. That is, we need only find $k \in \mathbb{R}^{1 \times n}$ to assign poles of $A_d - b_d k$ and then $\ell = k^T \in \mathbb{R}^n$ will place the poles of $A - \ell c$ at the same locations. Continuing, the controllability matrix of the dual system is

$$\mathcal{C}_d = \begin{bmatrix} b_d & A_d b_d & A_d^2 b_d \end{bmatrix} = \begin{bmatrix} c^T & A^T c^T & (A^T)^2 c^T \end{bmatrix} = \begin{bmatrix} c \\ cA \\ cA^2 \end{bmatrix}^T = \mathcal{O}^T.$$

---

[1] The subscript d stands for "dual."

Thus the dual system is controllable if and only if the original system is observable. Further, the controllability matrix for the control canonical form (A_{dc}, b_{dc}) of the dual system is

$$\mathcal{C}_{dc} = \begin{bmatrix} b_{dc} & A_{dc}b_{dc} & A_{dc}^2 b_{dc} \end{bmatrix} = \begin{bmatrix} c_o^T & A_o^T c_o^T & (A_o^T)^2 c_o^T \end{bmatrix} = \begin{bmatrix} c_o \\ c_o A_o \\ c_o A_o^2 \end{bmatrix}^T = \mathcal{O}_o^T.$$

This is simply the transpose of observability matrix of (c_o, A_o).

Then, by results of Section 15.7, we have

$$k = \begin{bmatrix} \beta_{do} - \beta_0 & \beta_{d1} - \beta_1 & \beta_{d2} - \beta_2 \end{bmatrix} \mathcal{C}_{dc} \mathcal{C}_d^{-1}$$

$$= \begin{bmatrix} \beta_{do} - \beta_0 & \beta_{d1} - \beta_1 & \beta_{d2} - \beta_2 \end{bmatrix} \begin{bmatrix} c_o \\ c_o A_o \\ c_o A_o^2 \end{bmatrix}^T \left(\begin{bmatrix} c \\ cA \\ cA^2 \end{bmatrix}^T \right)^{-1}$$

$$= \begin{bmatrix} \beta_{do} - \beta_0 & \beta_{d1} - \beta_1 & \beta_{d2} - \beta_2 \end{bmatrix} \begin{bmatrix} c_o \\ c_o A_o \\ c_o A_o^2 \end{bmatrix}^T \left(\begin{bmatrix} c \\ cA \\ cA^2 \end{bmatrix}^{-1} \right)^T$$

results in

$$\det(sI - (A_d - b_d k)) = \det(sI - (A^T - c^T \ell^T)) = s^3 + \beta_{d2} s^2 + \beta_{d1} s + \beta_{d0}.$$

Equivalently, we have

$$\ell = k^T = \left(\begin{bmatrix} \beta_{do} - \beta_0 & \beta_{d1} - \beta_1 & \beta_{d2} - \beta_2 \end{bmatrix} \mathcal{C}_{dc} \mathcal{C}_d^{-1} \right)^T$$

$$= \begin{bmatrix} c \\ cA \\ cA^2 \end{bmatrix}^{-1} \begin{bmatrix} c_o \\ c_o A_o \\ c_o A_o^2 \end{bmatrix} \begin{bmatrix} \beta_{do} - \beta_0 \\ \beta_{d1} - \beta_1 \\ \beta_{d2} - \beta_2 \end{bmatrix}$$

$$= \mathcal{O}^{-1} \mathcal{O}_o \begin{bmatrix} \beta_{do} - \beta_0 \\ \beta_{d1} - \beta_1 \\ \beta_{d2} - \beta_2 \end{bmatrix}$$

and

$$\det(sI - (A - \ell c)) = \det(sI - (A^T - c^T \ell^T)) = s^3 + \beta_{d2} s^2 + \beta_{d1} s + \beta_{d0}.$$

16.6 Parameter Identification

In this section we want to present a method to estimate the parameters of the open loop model of the cart. In order to do this we first must do some more matrix theory.

Symmetric and Positive Definite Matrices

Definition 1 *Symmetric Matrix*
A matrix Q is *symmetric* if
$$Q^T = Q.$$

16 State Estimators and Parameter Identification

Definition 2 *Positive Semidefinite Matrix*
A symmetric matrix $Q \in \mathbb{R}^{m \times m}$ is *positive semidefinite* if for all $x \in \mathbb{R}^m$,
$$x^T Q x \geq 0.$$

Definition 3 *Positive Definite Matrix*
A symmetric matrix $Q \in \mathbb{R}^{m \times m}$ is *positive definite* if for all $x \in \mathbb{R}^m$
$$x^T Q x \geq 0$$
and
$$x^T Q x = 0$$
if and only if x is the zero vector, i.e., $x = 0_{n \times 1}$.

Example 10 *Positive Definite Matrix*
Let
$$Q_1 = \begin{bmatrix} 1 & 0 \\ 0 & 2 \end{bmatrix}$$
and note that $Q_1 = Q_1^T$. Then
$$x^T Q_1 x = \begin{bmatrix} x_1 & x_2 \end{bmatrix}^T \begin{bmatrix} 1 & 0 \\ 0 & 2 \end{bmatrix} \begin{bmatrix} x_1 \\ x_2 \end{bmatrix} = x_1^2 + 2x_2^2 \geq 0 \text{ for all } x \in \mathbb{R}^2$$
and the only way it can equal zero is if $x_1 = 0$ and $x_2 = 0$. That is, Q_1 is positive definite.

Example 11 *Positive Semidefinite Matrix*
Let
$$Q_2 = \begin{bmatrix} 0 & 0 \\ 0 & 2 \end{bmatrix}$$
and note that $Q_2 = Q_2^T$. Then
$$x^T Q_2 x = \begin{bmatrix} x_1 & x_2 \end{bmatrix}^T \begin{bmatrix} 0 & 0 \\ 0 & 2 \end{bmatrix} \begin{bmatrix} x_1 \\ x_2 \end{bmatrix} = 2x_2^2 \geq 0 \text{ for all } x \in \mathbb{R}^2.$$

Thus Q_2 is positive semidefinite. However, in this example, $x = \begin{bmatrix} 1 \\ 0 \end{bmatrix}$ results in $x^T Q_2 x = 0$, that is, Q_2 is *not* positive definite.

Example 12 *Indefinite Matrix*
Let
$$Q_3 = \begin{bmatrix} -1 & 0 \\ 0 & 2 \end{bmatrix}$$
and note that $Q_3 = Q_3^T$. Then
$$x^T Q_3 x = \begin{bmatrix} x_1 & x_2 \end{bmatrix}^T \begin{bmatrix} -1 & 0 \\ 0 & 2 \end{bmatrix} \begin{bmatrix} x_1 \\ x_2 \end{bmatrix} = -x_1^2 + 2x_2^2.$$

In this example, $x^T Q_3 x$ can be positive if $x = \begin{bmatrix} 0 \\ 1 \end{bmatrix}$, or negative if $x = \begin{bmatrix} 1 \\ 0 \end{bmatrix}$. Consequently, Q_3 is neither positive definite nor positive semidefinite.

16.6 Parameter Identification

Least-Squares Identification

Recall the model of the cart on the track system from Chapter 13 given by

$$\frac{dx}{dt} = v$$
$$\frac{d^2x}{dt^2} = -a\frac{dx}{dt} + bV(t). \tag{16.36}$$

In order to design a controller based on these equations the values of the parameters a, b need to be found. This can be done by an experiment in which a voltage $V(t)$ is applied to the motor and it, along with the cart position $x(t)$, are recorded. This data is then used to determine the parameters. To understand how this is accomplished we rewrite the second equation of (16.36) as

$$\begin{bmatrix} -\dfrac{dx(t)}{dt} & V(t) \end{bmatrix} \begin{bmatrix} a \\ b \end{bmatrix} = \dfrac{d^2x(t)}{dt^2}. \tag{16.37}$$

This is a linear equation in the unknown parameters a, b. The coefficients of this linear equation are found from the measured/calculated data $V(t), x(t), dx(t)/dt, d^2x(t)/dt^2$. Let

$$\frac{dx(nT)}{dt} \triangleq \left.\frac{dx(t)}{dt}\right|_{t=nT}$$

$$\frac{d^2x(nT)}{dt^2} \triangleq \left.\frac{d^2x(t)}{dt^2}\right|_{t=nT}$$

$$V(nT) \triangleq V(t)|_{t=nT}$$

be the speed, acceleration, and voltage, respectively, at time nT. At each time $t = nT$, (16.37) should hold, that is,

$$\begin{bmatrix} -\dfrac{dx(nT)}{dt} & \cdot V(nT) \end{bmatrix} \begin{bmatrix} a \\ b \end{bmatrix} = \dfrac{d^2x(nT)}{dt^2}. \tag{16.38}$$

With

$$W(nT) \triangleq \begin{bmatrix} -\dfrac{dx(nT)}{dt} & V(nT) \end{bmatrix} \in \mathbb{R}^{1\times 2}$$

$$y(nT) \triangleq \dfrac{d^2x(nT)}{dt^2} \in \mathbb{R}$$

$$K \triangleq \begin{bmatrix} a \\ b \end{bmatrix} \in \mathbb{R}^2,$$

(16.38) may be written as

$$W(nT)K = y(nT). \tag{16.39}$$

Here W is referred to as the *regressor* matrix. The desire here is to find the constant vector K that satisfies this for all n! To do so, multiply both sides of (16.39) by $W^T(nT)$ to obtain

$$W^T(nT)W(nT)K = W^T(nT)y(nT), \tag{16.40}$$

where

$$W^T(nT)W(nT) = \begin{bmatrix} -\dfrac{dx(nT)}{dt} \\ V(nT) \end{bmatrix} \begin{bmatrix} -\dfrac{dx(nT)}{dt} & V(nT) \end{bmatrix}$$

$$= \begin{bmatrix} \left(\dfrac{dx(nT)}{dt}\right)^2 & -\dfrac{dx(nT)}{dt}V(nT) \\ -\dfrac{dx(nT)}{dt}V(nT) & V^2(nT) \end{bmatrix}$$

(16.41)

and

$$W^T(nT)y(nT) = \begin{bmatrix} -\dfrac{dx(nT)}{dt} \\ V(nT) \end{bmatrix} \dfrac{d^2x(nT)}{dt^2} = \begin{bmatrix} -\dfrac{dx(nT)}{dt}\dfrac{d^2x(nT)}{dt^2} \\ V(nT)\dfrac{d^2x(nT)}{dt^2} \end{bmatrix}.$$

Then (16.40) is

$$\begin{bmatrix} \left(\dfrac{dx(nT)}{dt}\right)^2 & -\dfrac{dx(nT)}{dt}V(nT) \\ -\dfrac{dx(nT)}{dt}V(nT) & V^2(nT) \end{bmatrix} \begin{bmatrix} a \\ b \end{bmatrix} = \begin{bmatrix} -\dfrac{dx(nT)}{dt}\dfrac{d^2x(nT)}{dt^2} \\ V(nT)\dfrac{d^2x(nT)}{dt^2} \end{bmatrix}. \quad (16.42)$$

Note that $W^T(nT)W(nT) \in \mathbb{R}^{2\times 2}$, that is, it is a square matrix. It would then be nice if it was possible to multiply both sides of (16.40) (or (16.42)) by $\left(W^T(nT)W(nT)\right)^{-1}$ to solve for K. However, $W^T(nT)W(nT)$ is *never* invertible! To see this, we simply compute

$$\det \begin{bmatrix} \left(\dfrac{dx(nT)}{dt}\right)^2 & -\dfrac{dx(nT)}{dt}V(nT) \\ -\dfrac{dx(nT)}{dt}V(nT) & V^2(nT) \end{bmatrix} = \left(\dfrac{dx(nT)}{dt}\right)^2 V^2(nT) - \left(\dfrac{dx(nT)}{dt}V(nT)\right)^2 \equiv 0.$$

Something else has to be done. As (16.40) must hold for *all* n with K a fixed parameter vector we sum up it for N time instants to obtain

$$\left(\sum_{n=1}^{N} W^T(nT)W(nT)\right)K = \sum_{n=1}^{N} W^T(nT)y(nT). \quad (16.43)$$

Define

$$R_W \triangleq \sum_{n=1}^{N} W^T(nT)W(nT) \in \mathbb{R}^{2\times 2}$$

$$R_{Wy} \triangleq \sum_{n=1}^{N} W^T(nT)y(nT) \in \mathbb{R}^{2}$$

(16.44)

so that (16.43) may be rewritten as

$$R_W K = R_{Wy}. \quad (16.45)$$

Now suppose the matrix sum $R_W \triangleq \sum_{n=1}^{N} W^T(nT)W(nT)$ is invertible. Then we may multiply both sides of (16.45) by $R_W^{-1} \triangleq \left(\sum_{n=1}^{N} W^T(nT)W(nT)\right)^{-1}$ to obtain K as

$$K = R_W^{-1} R_{Wy}. \tag{16.46}$$

This solution is called the *least-squares* solution. The key to this method is making sure R_W is invertible. This is dependent on choosing an input voltage that results in R_W being invertible, as any arbitrary input voltage will not work. For example, suppose $V(t) \equiv 0$ so that $x(t) \equiv 0$ and thus $dx(t)/dt \equiv 0$. In this case, $W(nT) \equiv 0$ for all n so that $R_W \triangleq \sum_{n=1}^{N} W^T(nT)W(nT) \equiv 0$ and is therefore *not* invertible. It is up to the control engineer to find an input $V(t)$ so that R_W has an inverse.

Remark Recall from Chapter 8 that the parameters a and b can be found by applying a step input and measuring the peak time and peak overshoot. However, the least-squares approach uses *all* of the data rather than a single time point. That is, the model parameters determined using the least-squares approach are the best fit for all time, not just a single point in time. Further, the method based on the peak time and peak overshoot is only valid for second order systems whose closed loop transfer function can be written in the form $C(s)/R(s) = \omega_n^2/(s^2 + 2\zeta\omega_n s + \omega_n^2)$ while it turns out that the least-squares approach is applicable to *any* linear or nonlinear system whose regressor is linear in the parameters.

Least-Squares Approximation

The analysis to derive $K = R_W^{-1} R_{Wy}$ was based on the equation

$$\underbrace{\begin{bmatrix} -\dfrac{dx(nT)}{dt} & V(nT) \end{bmatrix}}_{W(nT)} \underbrace{\begin{bmatrix} a \\ b \end{bmatrix}}_{K} = \underbrace{\dfrac{d^2 x(nT)}{dt^2}}_{y(nT)}$$

being true for all n. However, in the "real-world" that engineers work this is never true. The model (16.36) is not an exact description of the cart. For example, the motor inductance was assumed to be zero, $V(t)$ and $x(t)$ cannot be measured perfectly, and the derivatives $dx(t)/dt, d^2x/dt^2$ cannot be computed exactly. Thus there will not be a single fixed parameter vector K that satisfies $W(nT)K = y(nT)$ for all n.

One can still run an experiment and collect the data $x(t), V(t)$ to compute R_W, R_{Wy} and thus determine

$$\hat{K} = R_W^{-1} R_{Wy}.$$

The key question is then "How well does $\hat{K} = R_W^{-1} R_{Wy}$ satisfy $W(nT)K = y(nT)$ for all n?" To answer this question recall the definitions

$$W(nT) \triangleq \begin{bmatrix} -\dfrac{dx}{dt}(nT) & V(nT) \end{bmatrix} \in \mathbb{R}^{1 \times 2}$$

$$y(nT) \triangleq \dfrac{d^2 x}{dt^2}(nT) \in \mathbb{R}.$$

Then the error between the output $y(nT)$ and its predicted value $\hat{y}(nT) = W(nT)K$ is

$$y(nT) - \underbrace{W(nT)K}_{\hat{y}(nT)} \in \mathbb{R}.$$

16 State Estimators and Parameter Identification

The problem is to find the value of K that makes this difference as small as possible for *all* n. Specifically, we want to find the value of K that minimizes

$$E^2(K) \triangleq \sum_{n=1}^{N} \left(y(nT) - W(nT)K\right)^T \left(y(nT) - W(nT)K\right)$$

$$= \sum_{n=1}^{N} \left(y(nT) - \hat{y}(nT)\right)^T \left(y(nT) - \hat{y}(nT)\right)$$

$$= \sum_{n=1}^{N} e^2(nT), \qquad (16.47)$$

where

$$e(nT) \triangleq y(nT) - W(nT)K = y(nT) - \hat{y}(nT) \in \mathbb{R}.$$

In the jargon of identification theory $y(nT)$ is considered the *output* while $\hat{y}(nT) = W(nT)K$ is the *predicted output* based on K as the estimate of the parameters. Consequently $e(nT)$ is the *error* and

$$E^2(K) \triangleq \sum_{n=1}^{N} \left(y(nT) - W(nT)K\right)^T \left(y(nT) - W(nT)K\right)$$

is the total *squared error*. If a parameter vector K can be found that minimizes $E^2(K)$, then it is referred to as the *least-squares* estimate.

It is now shown that there is a unique solution and it equals $R_W^{-1} R_{Wy}$. To do so, expand this expression for $E^2(K)$ to obtain (recall $(AB)^T = B^T A^T$)

$$E^2(K) = \sum_{n=1}^{N} \left(y^T(nT) - K^T W^T(nT)\right)\left(y(nT) - W(nT)K\right)$$

$$= \sum_{n=1}^{N} \left(y^T(nT)y(nT) - y^T(nT)W(nT)K - K^T W^T(nT)y(nT) + K^T W^T(nT)W(nT)K\right)$$

$$= \sum_{n=1}^{N} y^T(nT)y(nT) - \left(\sum_{n=1}^{N} y^T(nT)W(nT)\right)K - K^T\left(\sum_{n=1}^{N} W^T(nT)y(nT)\right)$$

$$+ K^T \left(\sum_{n=1}^{N} W^T(nT)W(nT)\right)K.$$

Adding to the definitions given in (16.44), we define

$$R_{yW} \triangleq \sum_{n=1}^{N} y^T(nT)W(nT) \in \mathbb{R}^{1 \times 2}$$

$$R_y \triangleq \sum_{n=1}^{N} y^T(nT)y(nT) \in \mathbb{R}.$$

Note that $R_{yW} = R_{Wy}^T$. For a fixed N, the matrices R_y, R_{yW}, R_{Wy}, and R_W are constant and known from data measurements. $E^2(K)$ is written compactly as

$$E^2(K) = R_y - R_{yW}K - K^T R_{Wy} + K^T R_W K$$
$$= R_y - R_{yW} R_W^{-1} R_{Wy} + (K - R_W^{-1} R_{Wy})^T R_W (K - R_W^{-1} R_{Wy}). \quad (16.48)$$

To verify (16.48) we first show that R_W is a *symmetric positive semidefinite* matrix. We have

$$R_W \triangleq \sum_{n=1}^{N} W^T(nT) W(nT) = \sum_{n=1}^{N} \begin{bmatrix} -\frac{dx}{dt}(nT) \\ V(nT) \end{bmatrix} \begin{bmatrix} -\frac{dx}{dt}(nT) & V(nT) \end{bmatrix} \in \mathbb{R}^{2 \times 2}$$

is symmetric as

$$R_W^T \triangleq \left(\sum_{n=1}^{N} W^T(nT) W(nT) \right)^T = \sum_{n=1}^{N} \left(W^T(nT) W(nT) \right)^T = \sum_{n=1}^{N} W^T(nT) \left(W^T(nT) \right)^T$$
$$= \sum_{n=1}^{N} W^T(nT) W(nT)$$
$$= R_W.$$

R_W is positive semidefinite as

$$x^T R_W x \triangleq \sum_{n=1}^{N} x^T W^T(nT) W(nT) x = \sum_{n=1}^{N} \left(W(nT) x \right)^T W(nT) x = \sum_{n=1}^{N} \| W(nT) x \|^2 \geq 0.$$

Under the assumption that R_W has an inverse we verify (16.48) as follows.

$$(K - R_W^{-1} R_{Wy})^T R_W (K - R_W^{-1} R_{Wy})$$
$$= (K^T - R_{Wy}^T (R_W^{-1})^T) R_W (K - R_W^{-1} R_{Wy})$$
$$= (K^T - R_{yW} R_W^{-1}) R_W (K - R_W^{-1} R_{Wy})$$
$$= K^T R_W (K - R_W^{-1} R_{Wy}) - R_{yW} R_W^{-1} R_W (K - R_W^{-1} R_{Wy})$$
$$= K^T R_W K - K^T R_W R_W^{-1} R_{Wy} - R_{yW} R_W^{-1} R_W K + R_{yW} R_W^{-1} R_W R_W^{-1} R_{Wy}$$
$$= K^T R_W K - K^T R_{Wy} - R_{yW} K + R_{yW} R_W^{-1} R_{Wy}.$$

Thus

$$R_y - R_{yW} R_W^{-1} R_{Wy} + (K - R_W^{-1} R_{Wy})^T R_W (K - R_W^{-1} R_{Wy})$$
$$= R_y - R_{yW} R_W^{-1} R_{Wy} + K^T R_W K - K^T R_{Wy} - R_{yW} K + R_{yW} R_W^{-1} R_{Wy}$$
$$= R_y + K^T R_W K - K^T R_{Wy} - R_{yW} K$$
$$= E^2(K).$$

The control engineer will design the experiment, i.e., the specification of $V(t)$, so that R_W is invertible (see Problem 15). It turns out that if a symmetric, positive semidefinite matrix is also *invertible*, then it must be *positive definite*. Consequently R_W is positive definite. Looking at

$$E^2(K) = R_y - R_{yW} R_W^{-1} R_{Wy} + \underbrace{(K - R_W^{-1} R_{Wy})^T}_{\in \mathbb{R}^{1\times 2}} \underbrace{R_W}_{\in \mathbb{R}^{2\times 2}} \underbrace{(K - R_W^{-1} R_{Wy})}_{\in \mathbb{R}^{2}}$$

we want to choose K so that $E^2(K)$ is as small as possible. As R_W is positive definite we have for $K \in \mathbb{R}^2$

$$(K - R_W^{-1} R_{Wy})^T R_W (K - R_W^{-1} R_{Wy}) \geq 0.$$

This term equals zero if and only if $K - R_W^{-1} R_{Wy} = \begin{bmatrix} 0 \\ 0 \end{bmatrix}$. That is, by inspection of (16.48), $E^2(K)$ is minimized for $K = \hat{K} = R_W^{-1} R_{Wy}$! Choosing $K = \hat{K} = R_W^{-1} R_{Wy}$ results in the squared error having its least value, that is, the *least-squares* error.

Error Index

How good is the least-squares estimate? Well, the exact value of K is not known so the error between the "exact" value of the parameter vector and its estimate, i.e., $K - \hat{K}$ is unknown. However, an indication of how good the estimate $\hat{K}$ is can be found by comparing it with a given (and thus known) value of K. Specifically, if $K = 0$, then the squared error is $E(0) = R_y$ as seen by putting $K = 0$ in (16.48). Using the least-squares estimate $\hat{K}$, that is, setting $K = \hat{K} \triangleq R_W^{-1} R_{Wy}$ in (16.48), the error is

$$E^2(K)|_{K=\hat{K} \triangleq R_W^{-1} R_{Wy}} = R_y - R_{yW} R_W^{-1} R_{Wy}.$$

This is called the *residual* error, that is, it is the total squared error after using the value of K which minimizes the squared error. As R_W is positive definite, it turns out that its inverse R_W^{-1} must also be positive definite. Further, as $R_{yW} \in \mathbb{R}^{1\times 5}$, $R_{Wy} \in \mathbb{R}^{5\times 1}$, and $R_{yW} = R_{Wy}^T$, it follows that $R_{yW} R_W^{-1} R_{Wy} \geq 0$ so that $E^2(\hat{K}) = R_y - R_{yW} R_W^{-1} R_{Wy} \leq R_y = E^2(0)$. As a result, the quantity

$$\frac{E^2(\hat{K})}{E^2(0)} = \frac{R_y - R_{yW} R_W^{-1} R_{Wy}}{R_y} \leq 1.$$

$E^2(\hat{K})/E^2(0)$ is a measure of the minimum squared error relative to the squared error obtained from taking the parameter vector K to be the zero vector. By taking the square root, a measure of the relative error rather than squared error is obtained. This motivates the definition of the so-called *error index* as

$$\text{Error index} \triangleq \sqrt{\frac{E^2(\hat{K})}{E^2(0)}} = \sqrt{\frac{R_y - R_{yW} R_W^{-1} R_{Wy}}{R_y}} \leq 1.$$

Note that if the error index is close to 1, then our estimate is not much better than taking all the parameter values equal to zero! Thus, the error index must be much less than one for the estimate to be of any value. If the error index is close to one, then we would suspect that the original parametric model of the system is incorrect.

How does the error index help in determining a good estimate for K? Well, in using the formula $\hat{K} = R_W^{-1} R_{Wy}$ to compute $\hat{K}$, it is necessary to measure the data $x(t), V(t)$ and then calculate $dx(t)/dt, d^2x/dt^2$ to compute R_W, R_{Wy}. However, for example, to calculate dx/dt using the backward difference approximation defined by

$$\frac{d\hat{x}}{dt}(nT) = \frac{x(nT) - x((n-1)T)}{T}$$

requires specifying a sample period T and then filtering to remove the noise. What kind of filter should one use? What order of filter is needed? What should the cutoff frequency of the filter be? Well, one can "head in the right direction" by using the error index! That is, one could consider a particular filter with two different cutoff frequencies. Then $\hat{K}$ is computed twice, once having the noisy speed filtered using the first cutoff frequency and the second time with the other cutoff frequency. The $\hat{K}$ with the smaller error index would be the one to use. Using this approach one can "home in" on a good choice for the cutoff frequency.

Problems

Problem 1 *State Estimation*

Let a statespace model be given by

$$\frac{dx}{dt} = \underbrace{\begin{bmatrix} 0 & 1 \\ -1 & 0 \end{bmatrix}}_{A} x + \underbrace{\begin{bmatrix} 0 \\ 1 \end{bmatrix}}_{b} u$$

$$y = \underbrace{\begin{bmatrix} 1 & 0 \end{bmatrix}}_{c} x.$$

(a) Compute the observability matrix and the open-loop characteristic polynomial, i.e., $\det(sI - A)$. Is the system observable?

(b) Write down the equations for a state estimator.

(c) Compute the gain vector ℓ that will place the two closed-loop poles of your state estimator from part (b) at -5. Use Theorem 2.

(d) Using the ℓ chosen in part (c), will $\hat{x}(t) \to x(t)$ as $t \to \infty$? Just answer yes or no.

Problem 2 *State Estimation*

Let a statespace model be given by

$$\frac{dx}{dt} = \underbrace{\begin{bmatrix} 0 & 1 \\ -1 & 0 \end{bmatrix}}_{A} x + \underbrace{\begin{bmatrix} 0 \\ 1 \end{bmatrix}}_{b} u$$

$$y = \underbrace{\begin{bmatrix} 0 & 1 \end{bmatrix}}_{c} x.$$

(a) Compute the observability matrix and the open-loop characteristic polynomial, i.e., $\det(sI - A)$. Is the system observable?

(b) Write down the equations for a state estimator.

(c) Compute the gain vector ℓ that will place the two closed-loop poles of your state estimator from part (b) at -5. Use Theorem 2.

(d) Using the ℓ chosen in part (c), will $e^{(A-\ell c)t} \to 0_{2\times 2}$ as $t \to \infty$? Just answer yes or no.

Problem 3 *State Estimation*
Let the statespace model be given by

$$\frac{dx}{dt} = \underbrace{\begin{bmatrix} 0 & 1 \\ 0 & -6 \end{bmatrix}}_{A} x + \underbrace{\begin{bmatrix} 0 \\ 1 \end{bmatrix}}_{b} u$$

$$y = \underbrace{\begin{bmatrix} 0 & 1 \end{bmatrix}}_{c} x.$$

(a) Compute the observability matrix. Is the system observable?

(b) Let $\ell = \begin{bmatrix} \ell_1 & \ell_2 \end{bmatrix}^T$. Can ℓ_1, ℓ_2 be chosen so that $A - \ell c$ is stable?

(c) Can a state estimator be designed to have $\hat{x} \to x$? Explain briefly.

Problem 4 *State Estimation*
Let the statespace model be given by

$$\frac{dx}{dt} = \underbrace{\begin{bmatrix} -2 & 1 \\ 0 & -6 \end{bmatrix}}_{A} x + \underbrace{\begin{bmatrix} 0 \\ 1 \end{bmatrix}}_{b} u$$

$$y = \underbrace{\begin{bmatrix} 0 & 1 \end{bmatrix}}_{c} x.$$

(a) Compute the observability matrix. Is the system observable?

(b) Let $\ell = \begin{bmatrix} \ell_1 & \ell_2 \end{bmatrix}^T$. Can ℓ_1, ℓ_2 be chosen so that $A - \ell c$ is stable? Explain. Hint: The answer is yes! Explain.

(c) Can a state estimator be designed to have $\hat{x} \to x$. Explain briefly.

(d) Can ℓ be chosen to arbitrarily place the poles of $A - \ell c$? Explain.

Problem 5 *State Estimators for the Cart on the Track System*

(a) *Speed estimator* Starting with the simulation of the state feedback controller for the cart on the track system of Problem 3a in Chapter 15, add the state estimator given in Figure 16.1 to that simulation and use it for state feedback.

(b) *Speed and disturbance estimator* Starting with the simulation of the state feedback controller for the cart on the track system of Problem 3(c) in Chapter 15, add the state estimator given in Figure 16.2 to that simulation and use it for state feedback.

Problem 6 *State Estimator for the Inverted Pendulum with Output* $y = x + \left(\ell + \frac{J}{m\ell}\right)\theta$
The linear statespace model of the inverted pendulum around the equilibrium point

$(x_{eq}, v_{eq}, \theta_{eq}, \omega_{eq}) = (0, 0, 0, 0)$ is

$$\underbrace{\begin{bmatrix} dx/dt \\ dv/dt \\ d\theta/dt \\ d\omega/dt \end{bmatrix}}_{dz/dt} = \underbrace{\begin{bmatrix} 0 & 1 & 0 & 0 \\ 0 & 0 & -\dfrac{gm^2\ell^2}{M m\ell^2 + J(M+m)} & 0 \\ 0 & 0 & 0 & 1 \\ 0 & 0 & \dfrac{mg\ell(M+m)}{M m\ell^2 + J(M+m)} & 0 \end{bmatrix}}_{A} \begin{bmatrix} x \\ v \\ \theta \\ \omega \end{bmatrix} + \underbrace{\begin{bmatrix} 0 \\ \dfrac{J+m\ell^2}{M m\ell^2 + J(M+m)} \\ 0 \\ \dfrac{m\ell}{M m\ell^2 + J(M+m)} \end{bmatrix}}_{b} u.$$

Let the output equation be

$$y = \underbrace{\begin{bmatrix} 1 & 0 & \ell + \dfrac{J}{m\ell} & 0 \end{bmatrix}}_{c} \underbrace{\begin{bmatrix} x \\ v \\ \theta \\ \omega \end{bmatrix}}_{z}.$$

(a) Compute the observability matrix $\mathcal{O}$ for the system (c, A).

(b) Compute the open-loop characteristic polynomial, i.e., $\det(sI - A) = s^4 + \beta_3 s^3 + \beta_2 s^2 + \beta_1 s + \beta_0$.

(c) Use your answer in part (b) to compute the observability canonical form (c_o, A_o) for the inverted pendulum.

(d) Use your answer in part (c) to compute the observability matrix $\mathcal{O}_o$ for the pair (c_o, A_o).

(e) Give the observer gain ℓ that places the poles at $-\gamma_1, -\gamma_2, -\gamma_3, -\gamma_4$. Use the pole placement procedure given in Theorem 2 of this chapter.

(f) In the simulation of Problem 11 of Chapter 15 use this state estimator to implement the state feedback.

Remark It will be shown in Chapter 17 that this particular controller, i.e., state feedback combined with a state estimator based on the output $y = x + \left(\ell + \dfrac{J}{m\ell}\right)\theta$, results in the closed-loop system being *very* sensitive to small disturbances. This means that small disturbance inputs can cause the pendulum rod to swing so far from its equilibrium angle of $\theta = 0$ that the linear model of the inverted pendulum is no longer a valid approximation to the nonlinear model. Consequently this controller will most likely not be able to return the pendulum rod to the upright position.

Problem 7 *State Estimator for the Inverted Pendulum with Output $y = x$*
The linear statespace model for the inverted pendulum around the equilibrium point $(x_{eq}, v_{eq}, \theta_{eq}, \omega_{eq}) = (0, 0, 0, 0)$ is

16 State Estimators and Parameter Identification

$$\underbrace{\begin{bmatrix} dx/dt \\ dv/dt \\ d\theta/dt \\ d\omega/dt \end{bmatrix}}_{dz/dt} = \underbrace{\begin{bmatrix} 0 & 1 & 0 & 0 \\ 0 & 0 & -\dfrac{gm^2\ell^2}{M m\ell^2 + J(M+m)} & 0 \\ 0 & 0 & 0 & 1 \\ 0 & 0 & \dfrac{mg\ell(M+m)}{M m\ell^2 + J(M+m)} & 0 \end{bmatrix}}_{A} \begin{bmatrix} x \\ v \\ \theta \\ \omega \end{bmatrix} + \underbrace{\begin{bmatrix} 0 \\ \dfrac{J+m\ell^2}{M m\ell^2 + J(M+m)} \\ 0 \\ -\dfrac{m\ell}{M m\ell^2 + J(M+m)} \end{bmatrix}}_{b} u.$$

Let the output equation be

$$y = \underbrace{\begin{bmatrix} 1 & 0 & 0 & 0 \end{bmatrix}}_{c} \underbrace{\begin{bmatrix} x \\ v \\ \theta \\ \omega \end{bmatrix}}_{z}.$$

(a) Compute the observability matrix $\mathcal{O}$ for the system (c, A).

(b) Compute the open-loop characteristic polynomial, i.e., $\det(sI - A) = s^4 + \beta_3 s^3 + \beta_2 s^2 + \beta_1 s + \beta_0$.

(c) Use your answer in part (b) to compute the observability canonical form (c_o, A_o) for the inverted pendulum.

(d) Use your answer in part (c) to compute the observability matrix $\mathcal{O}_o$ for the pair (c_o, A_o).

(e) Give the observer gain ℓ that places the poles at $-\gamma_1, -\gamma_2, -\gamma_3, -\gamma_4$. Use the pole placement procedure given in the text.

(f) In the simulation of Problem 11 of Chapter 15 add this observer to implement the state feedback.

Remark It will be shown in Chapter 17 that this particular controller (state feedback using a state estimator based on the cart position x) results in the closed-loop system being *extremely* sensitive to small disturbances. Specifically, small disturbance inputs can cause the pendulum angle to swing so far from its equilibrium angle $\theta = 0$ that the linear model is no longer a valid approximation of the nonlinear model. Though one can make a working simulation of this system, there is no chance this controller will ever work in practice!

Problem 8 *State Estimator for the Inverted Pendulum with Output $y = \theta$*

The linear statespace model for the inverted pendulum around the equilibrium point $(x_{eq}, v_{eq}, \theta_{eq}, \omega_{eq}) = (0, 0, 0, 0, 0)$ is

$$\underbrace{\begin{bmatrix} dx/dt \\ dv/dt \\ d\theta/dt \\ d\omega/dt \end{bmatrix}}_{dz/dt} = \underbrace{\begin{bmatrix} 0 & 1 & 0 & 0 \\ 0 & 0 & -\dfrac{gm^2\ell^2}{M m\ell^2 + J(M+m)} & 0 \\ 0 & 0 & 0 & 1 \\ 0 & 0 & \dfrac{mg\ell(M+m)}{M m\ell^2 + J(M+m)} & 0 \end{bmatrix}}_{A} \begin{bmatrix} x \\ v \\ \theta \\ \omega \end{bmatrix} + \underbrace{\begin{bmatrix} 0 \\ \dfrac{J+m\ell^2}{M m\ell^2 + J(M+m)} \\ 0 \\ -\dfrac{m\ell}{M m\ell^2 + J(M+m)} \end{bmatrix}}_{b} u.$$

Let the output equation be

$$y = \underbrace{\begin{bmatrix} 0 & 0 & 1 & 0 \end{bmatrix}}_{c} \underbrace{\begin{bmatrix} x \\ v \\ \theta \\ \omega \end{bmatrix}}_{z}.$$

(a) Compute the observability matrix $\mathcal{O}$ for the system (c, A).

(b) With $y = \theta$ can an observer be designed to estimate the full state? Explain why or why not.

Problem 9 *Multi-Output Observer for the Inverted Pendulum Tracking Step Inputs in x*

Add the multi-output observer of Section 16.3 to the inverted pendulum controller of Problem 12 of Chapter 15. It is recommended you use the SIMULINK State-Space block to implement this observer as follows. The state estimator equation to be implemented is

$$\frac{d\hat{z}}{dt} = (A - LC)\hat{z} + bu + Ly = (A - LC)\hat{z} + B\mathbf{u} \quad (16.49)$$

$$\hat{z} = I_{4\times 4}\hat{z} + D\mathbf{u}, \quad (16.50)$$

where

$$B \triangleq \begin{bmatrix} b & L \end{bmatrix} = \begin{bmatrix} 0 & l_{11} & l_{12} \\ \kappa(J + m\ell^2) & l_{21} & l_{22} \\ 0 & l_{31} & l_{32} \\ -\kappa m\ell & l_{41} & l_{42} \end{bmatrix}, \quad \mathbf{u} \triangleq \begin{bmatrix} u \\ x \\ \theta \end{bmatrix}, \quad D \triangleq \begin{bmatrix} 0 & 0 & 0 \\ 0 & 0 & 0 \\ 0 & 0 & 0 \\ 0 & 0 & 0 \end{bmatrix}.$$

The SIMULINK block called Vector Concatenate combines the input u to the cart and the measurements x, θ into a single vector $\mathbf{u} \in \mathbb{R}^3$. The vector $\mathbf{u}$ is the quantity going into the State-Space block of Figure 16.10. Double clicking on the State-Space block opens up the dialog box shown in Figure 16.11 on the next page.

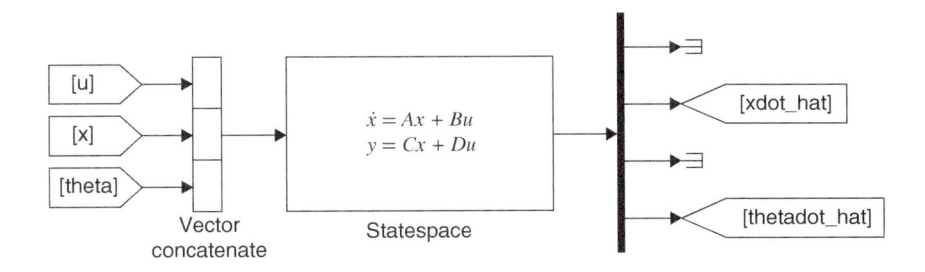

Figure 16.10. The SIMULINK state-space block.

With reference to Eqs. (16.49) and (16.50), in the dialog box set $A - LC \to A$, $B \to B$, $I_{4\times 4} \to C$, and $D \to D$. Along with the SIMULINK block diagram of Figure 16.10, you will need to add the following MATLAB code to your simulation from Problem 12 of Chapter 15.

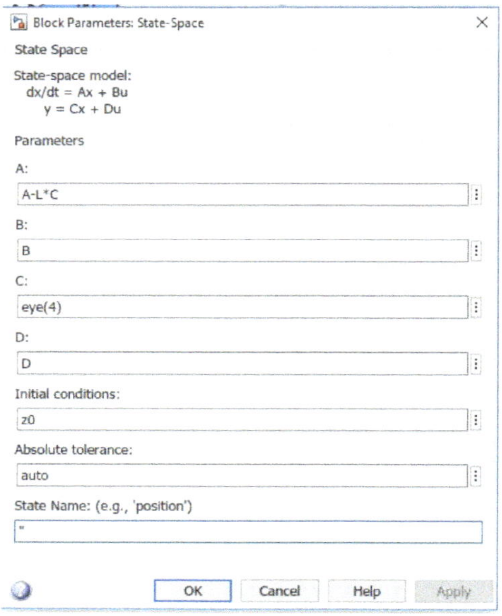

682   16   State Estimators and Parameter Identification

Figure 16.11. Dialog box for `State-Space` block.

```
% Observer
A = [0 1 0 0; 0 0 -kappa*m*g*m*Lp^2 0; 0 0 0 1; 0 0 kappa*m*g*Lp*(M+m) 0];
b = [0 kappa*(Jp+m*Lp^2) 0 -kappa*m*Lp]';
z0 = [x_0 0 theta_0 0]; ro1 = 10; ro2 = ro1; ro3 = ro1; ro4 = ro1;
C = [1 0 0 0; 0 0 1 0];
l_11 = ro1+ro2; l_12 = 0; l_21 = ro1*ro2; l_22 = -g*kappa*(m*Lp)^2;
l_31 = 0; l_32 = ro3+ro4; l_41 = 0; l_42 = alpha_sq + ro3*ro4;
L = [l_11 l_12; l_21 l_22; l_31 l_32; l_41 l_42];
eig(A-L*C)
B = horzcat([b,L]); D = [0 0 0; 0 0 0; 0 0 0; 0 0 0];
```

Problem 10 *Velocity Observer for the Magnetic Levitation System*

In Chapter 13 the nonlinear equations modeling the magnetically levitated steel ball are given by

$$\frac{di}{dt} = rL_1 \frac{i}{x^2 L(x)} v - \frac{R}{L(x)} i + \frac{1}{L(x)} u$$

$$\frac{dx}{dt} = v$$

$$\frac{dv}{dt} = g - \frac{rL_1}{2m} \frac{i^2}{x^2},$$

where $L(x) = L_0 + rL_1/x$. In the physical setup of this system the current and position are available measurements, but the velocity is not. This problem looks at designing a velocity estimator. Consider

$$\frac{dx}{dt} = v$$

$$\frac{dv}{dt} = g - \frac{rL_1}{2m} \frac{i^2}{x^2},$$

where i^2/x^2 is known as i, x are measured. Define an observer by

$$\frac{d\hat{x}}{dt} = \hat{v} + \ell_1(x - \hat{x})$$

$$\frac{d\hat{v}}{dt} = g - \frac{rL_1}{2m}\frac{i^2}{x^2} + \ell_2(x - \hat{x}).$$

(a) Let $\epsilon_1 \triangleq x - \hat{x}$ and $\epsilon_2 \triangleq v - \hat{v}$. Give the system of differential equations satisfied by these error state variables.

(b) Show how to choose the gains ℓ_1 and ℓ_2 so that the observer has its poles at $-r_1$ and $-r_2$.

(c) Add this observer to the simulation of magnetic levitation system of Problem 2 of Chapter 15.

Problem 11 *State Estimator for the Inverted Pendulum with $y = x + \left(\ell + \frac{J}{m\ell}\right)\theta$*

In Problem 6 of this chapter you developed an observer for the inverted pendulum using the measurement $y = x + \left(\ell + \frac{J}{m\ell}\right)\theta$. Add this observer to the state feedback controller given in Problem 12 of Chapter 15 that provides tracking of step inputs in the cart position.

Problem 12 *State Estimator for the Inverted Pendulum with $y = x$*

In Problem 7 of this chapter you developed an observer for the inverted pendulum using the measurement $y = x$. Add this observer to the state feedback controller given in Problem 12 of Chapter 15 that provides tracking of step inputs in the cart position.

Problem 13 *Multi-Output Observer for the Voltage Controlled Inverted Pendulum*

In Problem 14 of Chapter 15 the following linear model of the voltage controlled QUANSER inverted pendulum system was developed

$$\begin{bmatrix} dx/dt \\ dv/dt \\ d\theta/dt \\ d\omega/dt \end{bmatrix} = \underbrace{\begin{bmatrix} 0 & 1 & 0 & 0 \\ 0 & -\frac{n^2 K_T^2}{r_m^2 R}\kappa(J + m\ell^2) & -\kappa g m^2 \ell^2 & 0 \\ 0 & 0 & 0 & 1 \\ 0 & \frac{n^2 K_T^2}{r_m^2 R}\kappa m\ell & \kappa m g \ell(M + m) & 0 \end{bmatrix}}_{A} \begin{bmatrix} x \\ v \\ \theta \\ \omega \end{bmatrix}$$

$$+ \underbrace{\begin{bmatrix} 0 \\ \frac{n K_T}{R r_m}\kappa(J + m\ell^2) \\ 0 \\ -\frac{n K_T}{R r_m}\kappa m\ell \end{bmatrix}}_{b} V.$$

16 State Estimators and Parameter Identification

With x and θ measured, along with the obvious definition for the a_{ij}, b_i, we may write

$$\begin{bmatrix} dx/dt \\ dv/dt \\ d\theta/dt \\ d\omega/dt \end{bmatrix} = \underbrace{\begin{bmatrix} 0 & 1 & 0 & 0 \\ 0 & a_{22} & a_{23} & 0 \\ 0 & 0 & 0 & 1 \\ 0 & a_{42} & a_{43} & 0 \end{bmatrix}}_{A} \begin{bmatrix} x \\ v \\ \theta \\ \omega \end{bmatrix} + \underbrace{\begin{bmatrix} 0 \\ b_2 \\ 0 \\ b_4 \end{bmatrix}}_{b} V, \; y = \underbrace{\begin{bmatrix} 1 & 0 & 0 & 0 \\ 0 & 0 & 1 & 0 \end{bmatrix}}_{C} \underbrace{\begin{bmatrix} x \\ v \\ \theta \\ \omega \end{bmatrix}}_{z}.$$

Then

$$A - LC = \begin{bmatrix} 0 & 1 & 0 & 0 \\ 0 & a_{22} & a_{23} & 0 \\ 0 & 0 & 0 & 1 \\ 0 & a_{42} & a_{43} & 0 \end{bmatrix} - \begin{bmatrix} l_{11} & l_{12} \\ l_{21} & l_{22} \\ l_{31} & l_{32} \\ l_{41} & l_{42} \end{bmatrix} \begin{bmatrix} 1 & 0 & 0 & 0 \\ 0 & 0 & 1 & 0 \end{bmatrix}$$

$$= \begin{bmatrix} -l_{11} & 1 & -l_{12} & 0 \\ -l_{21} & a_{22} & a_{23} - l_{22} & 0 \\ -l_{31} & 0 & -l_{32} & 1 \\ -l_{41} & a_{42} & a_{43} - l_{42} & 0 \end{bmatrix}.$$

Choosing $l_{12} = 0, l_{22} = a_{23}, l_{31} = 0$, and $l_{41} = 0$, it is seen that $A - LC$ becomes

$$A - LC = \begin{bmatrix} -l_{11} & 1 & 0 & 0 \\ -l_{21} & a_{22} & 0 & 0 \\ 0 & 0 & -l_{32} & 1 \\ 0 & a_{42} & a_{43} - l_{42} & 0 \end{bmatrix}.$$

Unlike the example in Section 16.3, $A - LC$ is *not* in block diagonal form due to $a_{42} \neq 0$. However,

$$\det(sI - (A - LC)) = \det \begin{bmatrix} s + l_{11} & -1 & 0 & 0 \\ l_{21} & s - a_{22} & 0 & 0 \\ 0 & 0 & s + l_{32} & 1 \\ 0 & 0 & -a_{42} & a_{43} - l_{42} & s \end{bmatrix}$$

$$= (s^2 + (l_{11} - a_{22})s + l_{21} - a_{22}l_{11})(s^2 + l_{32}s + l_{42} - a_{43}).$$

That is, the determinant does not depend on a_{42}. Choosing $l_{11} = r_1 + r_2 + a_{22}, l_{21} = a_{22}l_{11} + r_1r_2, l_{32} = r_3 + r_4$, and $l_{42} = a_{43} + r_3r_4$, it follows that

$$\det(sI - (A - LC)) = (s + r_1)(s + r_2)(s + r_3)(s + r_4).$$

In summary, the observer gains are set as

$$\begin{bmatrix} l_{11} & l_{12} \\ l_{21} & l_{22} \\ l_{31} & l_{32} \\ l_{41} & l_{42} \end{bmatrix} = \begin{bmatrix} r_1 + r_2 + a_{22} & 0 \\ a_{22}l_{11} + r_1r_2 & a_{23} \\ 0 & r_3 + r_4 \\ 0 & a_{43} + r_3r_4 \end{bmatrix}.$$

Add this observer to the simulation of Problem 14 of Chapter 15. See Problem 9 of this chapter for a straightforward procedure to implement this observer in SIMULINK.

Problem 14 *Canonical Form for the Multi-Output Observer*

As shown in Problem 13 the observer model the voltage controlled pendulum on a cart is

$$\frac{d\hat{z}}{dt} = (A - LC)\hat{z} + bu + Ly. \tag{16.51}$$

Define a change of coordinates by

$$\hat{z}_o \triangleq T_o \hat{z},$$

where

$$T_o \triangleq \begin{bmatrix} 1 & 0 & 0 & 0 \\ -a_{22} & 1 & 0 & 0 \\ 0 & 0 & 1 & 0 \\ -a_{42} & 0 & 0 & 1 \end{bmatrix}, \quad T_o^{-1} = \begin{bmatrix} 1 & 0 & 0 & 0 \\ a_{22} & 1 & 0 & 0 \\ 0 & 0 & 1 & 0 \\ a_{42} & 0 & 0 & 1 \end{bmatrix}.$$

We now show this transformation puts the state estimator (16.51) into multi-output observer canonical form. Multiplying (16.51) on the left by T_o and replacing $\hat{z}$ by $T_o^{-1}\hat{z}_o$ the state estimator (16.51) becomes

$$\frac{d\hat{z}_o}{dt} = \underbrace{T_o(A - LC)T_o^{-1}}_{A_o - L_o C_o}\hat{z}_o + \underbrace{T_o b}_{b_o} u + \underbrace{T_o L}_{L_o} y \tag{16.52}$$

$$\hat{y} = \underbrace{CT_o^{-1}}_{C_o}\hat{z}_o. \tag{16.53}$$

Explicitly we have

$$A_o \triangleq T_o A T_o^{-1} = \begin{bmatrix} 1 & 0 & 0 & 0 \\ -a_{22} & 1 & 0 & 0 \\ 0 & 0 & 1 & 0 \\ -a_{42} & 0 & 0 & 1 \end{bmatrix} \begin{bmatrix} 0 & 1 & 0 & 0 \\ 0 & a_{22} & a_{23} & 0 \\ 0 & 0 & 0 & 1 \\ 0 & a_{42} & a_{43} & 0 \end{bmatrix} \begin{bmatrix} 1 & 0 & 0 & 0 \\ a_{22} & 1 & 0 & 0 \\ 0 & 0 & 1 & 0 \\ a_{42} & 0 & 0 & 1 \end{bmatrix}$$

$$= \begin{bmatrix} a_{22} & 1 & 0 & 0 \\ 0 & 0 & a_{23} & 0 \\ a_{42} & 0 & 0 & 1 \\ 0 & 0 & a_{43} & 0 \end{bmatrix}$$

$$C_o \triangleq CT_o^{-1} = \begin{bmatrix} 1 & 0 & 0 & 0 \\ 0 & 0 & 1 & 0 \end{bmatrix} \begin{bmatrix} 1 & 0 & 0 & 0 \\ a_{22} & 1 & 0 & 0 \\ 0 & 0 & 1 & 0 \\ a_{42} & 0 & 0 & 1 \end{bmatrix} = \begin{bmatrix} 1 & 0 & 0 & 0 \\ 0 & 0 & 1 & 0 \end{bmatrix}$$

and $T_o(A - LC)T_o^{-1} = T_o A T_o^{-1} - T_o LCT_o^{-1} = A_o - L_o C_o$. Then

$$A_o - L_o C_o = \begin{bmatrix} a_{22} & 1 & 0 & 0 \\ 0 & 0 & a_{23} & 0 \\ a_{42} & 0 & 0 & 1 \\ 0 & 0 & a_{43} & 0 \end{bmatrix} - \begin{bmatrix} l_{o11} & l_{o12} \\ l_{o21} & l_{o22} \\ l_{o31} & l_{o32} \\ l_{o41} & l_{o42} \end{bmatrix} \begin{bmatrix} 1 & 0 & 0 & 0 \\ 0 & 0 & 1 & 0 \end{bmatrix}$$

$$= \begin{bmatrix} a_{22} - l_{o11} & 1 & -l_{o12} & 0 \\ -l_{o21} & 0 & a_{23} - l_{o22} & 0 \\ a_{42} - l_{o31} & 0 & -l_{o32} & 1 \\ -l_{o41} & 0 & a_{43} - l_{o42} & 0 \end{bmatrix}. \tag{16.54}$$

We say that (C_o, A_o) is in observer canonical form because, as (16.54) shows, we can set the first and third columns of $A_o - L_o C_o$ to anything we like using L_o. Further the second and fourth columns are zero except for a "1" in the "right spots".
In (16.54) set $l_{o12} = 0, l_{o22} = a_{23}, l_{o31} = a_{42}, l_{o41} = 0$ so that $A_o - L_o C_o$ becomes

$$A_o - L_o C_o = \begin{bmatrix} a_{22} - l_{o11} & 1 & 0 & 0 \\ -l_{o21} & 0 & 0 & 0 \\ 0 & 0 & -l_{o32} & 1 \\ 0 & 0 & a_{43} - l_{o42} & 0 \end{bmatrix}.$$

Note that $A_o - L_o C_o$ is now block diagonal. The characteristic polynomial is then

$$\det(sI - (A_o - L_o C_o)) = \det \begin{bmatrix} s + l_{o11} - a_{22} & -1 & 0 & 0 \\ l_{o21} & s & 0 & 0 \\ 0 & 0 & s + l_{o32} & -1 \\ 0 & 0 & l_{o42} - a_{43} & s \end{bmatrix}$$

$$= (s^2 + (l_{o11} - a_{22})s + l_{o21})(s^2 + l_{o32}s + l_{o42} - a_{43}).$$

Setting $l_{o11} = r_1 + r_2 + a_{22}, l_{o21} = r_1 r_2, l_{o32} = r_3 + r_4$, and $l_{o42} = a_{43} + r_3 r_4$ this becomes

$$\det(sI - (A_o - L_o C_o)) = (s + r_1)(s + r_2)(s + r_3)(s + r_4).$$

In summary setting

$$L_o = \begin{bmatrix} l_{o11} & l_{o12} \\ l_{o21} & l_{o22} \\ l_{o31} & l_{o32} \\ l_{o41} & l_{o42} \end{bmatrix} = \begin{bmatrix} r_1 + r_2 + a_{22} & 0 \\ r_1 r_2 & a_{23} \\ a_{42} & r_3 + r_4 \\ 0 & a_{43} + r_3 r_4 \end{bmatrix}$$

places the poles of results in $A_o - L_o C_o$ at $-r_1, -r_2, -r_3, -r_4$. Finally, as $L_o = T_o L$ the gain matrix in the original coordinate system is

$$L = T_o^{-1} L_o = \begin{bmatrix} 1 & 0 & 0 & 0 \\ a_{22} & 1 & 0 & 0 \\ 0 & 0 & 1 & 0 \\ a_{42} & 0 & 0 & 1 \end{bmatrix} \begin{bmatrix} r_1 + r_2 + a_{22} & 0 \\ r_1 r_2 & a_{23} \\ a_{42} & r_3 + r_4 \\ 0 & a_{43} + r_3 r_4 \end{bmatrix}$$

$$= \begin{bmatrix} r_1 + r_2 + a_{22} & 0 \\ (r_1 + a_{22})(r_2 + a_{22}) & a_{23} \\ a_{42} & r_3 + r_4 \\ a_{42}(r_1 + r_2 + a_{22}) & a_{43} + r_3 r_4 \end{bmatrix}.$$

Add this observer to the simulation of Problem 14 of Chapter 15. See Problem 9 of this chapter for a straightforward procedure to implement this observer in SIMULINK.

Problem 15 *Parameter Identification of the Cart on the Track System*

Figure 16.12 is a SIMULINK block diagram for collecting (simulated) data to identify the parameters of the linear cart model of Chapter 15. The cart has the transfer function model $G(s) = \dfrac{b}{s(s+a)}$ where a, b are to be estimated from measurements of the voltage V and the position x. Set $a = 10, b = 2$ in the simulation of this system and set the saturation limits to $\pm V_{\max} = \pm 5$ V. The voltage input is chosen to be a chirp signal given by $V(t) = 4 \sin(\phi(t))$

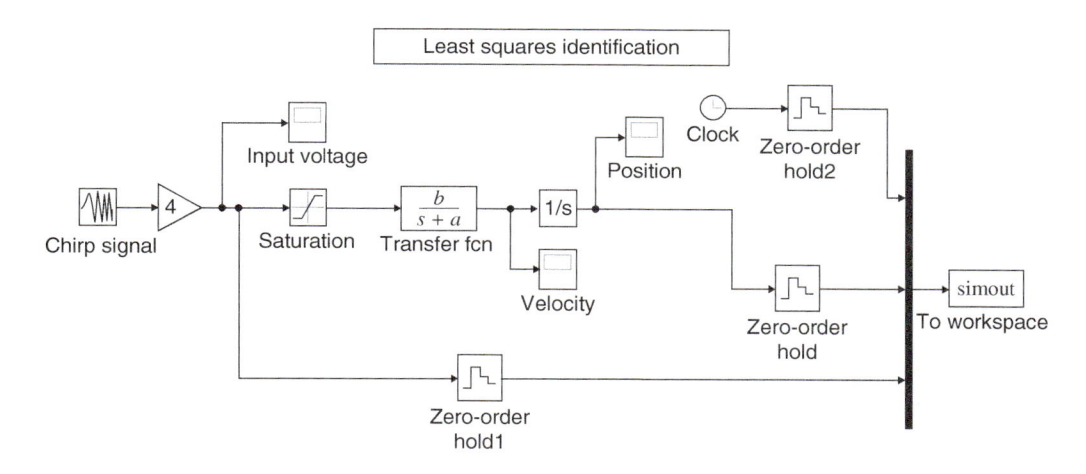

Figure 16.12. SIMULINK block diagram for collecting data.

where $\phi(t) = \dfrac{2\pi f_2 - 2\pi f_1}{T_f} \dfrac{t^2}{2} + 2\pi f_1 t$ and $\dot{\phi}(t) = \dfrac{2\pi f_2 - 2\pi f_1}{T_f} t + 2\pi f_1$. That is, $V(t)$ is a sinusoidal voltage whose instantaneous frequency $\dot{\phi}(t)$ increases linearly from $\dot{\phi}(0) = 2\pi f_1$ to $\dot{\phi}(T_f) = 2\pi f_2$.

Figure 16.13 is the dialog box for the chirp signal block. Set $f_1 = 0.04$, $f_2 = 4$, and $T_f = 10$.

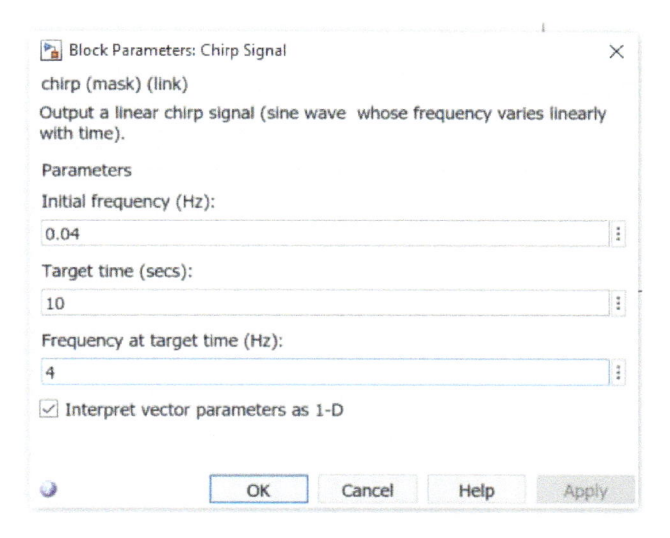

Figure 16.13. Dialog box for the chirp signal. The *Initial Frequency* and the *Frequency at target time* are in hertz.

The (simulated) data is collected using a zero-order hold block to simulate an analog-to-digital converter. The dialog box for the zero-order hold block is given in Figure 16.14 with the sample period T set to 0.001 seconds.

16 State Estimators and Parameter Identification

Figure 16.14. Dialog box for the `zero-order hold` block.

The data is stored using the `To Workspace` block. The dialog box for the `To Workspace` block is shown in Figure 16.15.

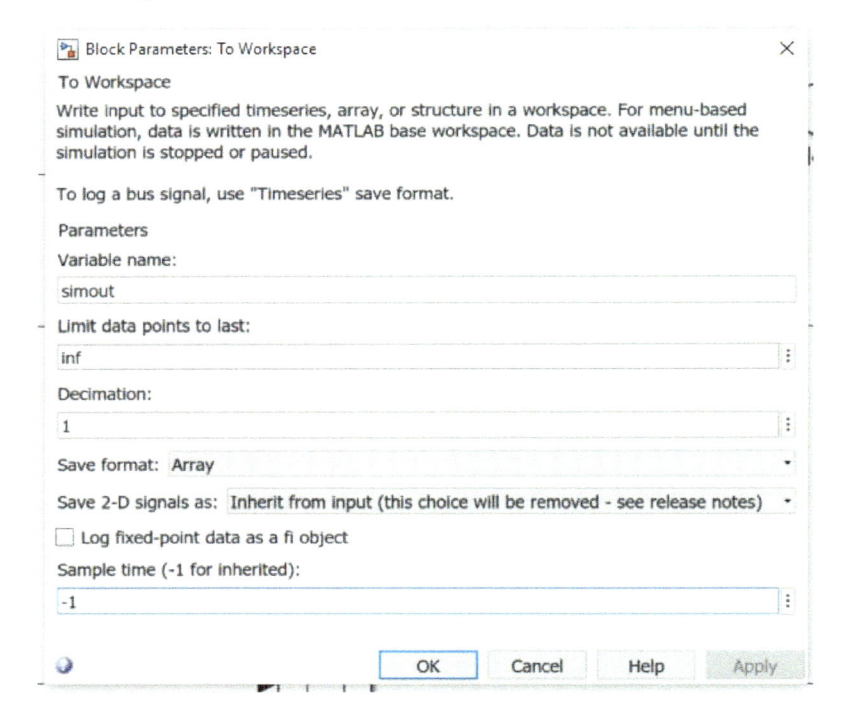

Figure 16.15. Dialog box for the `To Workspace` block.

(a) Implement this SIMULINK simulation and then write a program to use the time t, position x, and voltage V stored in `simout` to compute $\dot{x}, \ddot{x}$. The following code should be useful. It is standard to filter the measured signals with a low-pass filter. The low-pass filtered signals have a delay due to the filter itself. This makes it important to filter all signals used in the calculation of the parameters with the same filter so they all have the same delay.

```
t = simout(:,1); x = simout(:,2); V = simout(:,3);
derv = [1 -1]/T;
```

```
% Compute the velocity by differentiating position
xdot = filter(derv,1, x);
% Compute the acceleration by differentiating velocity
acc = filter(derv,1,xdot);
% Put all signals through a low-pass Butterworth filter to remove
% high-frequency noise.
fs = 1/T;
% Sampling frequency in Hz
fsc = (1/2)*fs;
% Nyquist frequency in Hz defined as 1/2 of the sampling freq.
wn = 10/fsc;
% wn is the cutoff freq (= 10 Hz) divided by the Nyquist freq.
[bf,af] = butter(2,wn);
% Second-order Butterworth filter with a cutoff freq of 10 Hz.
xdot_f = filter(bf,af,xdot);
acc_f = filter(bf,af,acc);
Vf_f = filter(bf,af,V);
```

Plot xdot_f and xdot vs. time on the same plot.

Plot acc_f and acc vs. time on the same plot.

Plot V_f and V vs. time on the same plot.

In each of these plots you will see a small delay between the signal and its filtered version.

(b) Add to your program of part (a) the calculations of R_y, R_W, and R_{Wy}. Then use them to compute the estimates $\hat{a}, \hat{b}$ of a, b according to

$$\begin{bmatrix} \hat{a} \\ \hat{b} \end{bmatrix} = R_W^{-1} R_{Wy}.$$

Compare your estimates $\hat{a}, \hat{b}$ with the values used in your simulation by computing the fractional error $(\hat{a} - a)/a$ and $(\hat{b} - b)/b$.

(c) Add to your program of part (b) the calculation of the *residual error*

$$E^2(K)\Big|_{K=\hat{K} \triangleq R_W^{-1} R_{Wy}} = R_y - R_{yW} R_W^{-1} R_{Wy}$$

and the *residual error index*

$$\text{Error index} = \sqrt{\frac{R_y - R_{yW} R_W^{-1} R_{Wy}}{R_y}}.$$

Problem 16 *Parameter Identification of the Cart on the Track System with an Encoder*

This problem uses the SIMULINK model of an optical encoder as given in Problem 10 of Chapter 13. Figure 16.16 is a SIMULINK block diagram for collecting (simulated) data to identify the parameters of the linear cart system with the position measured using an optical encoder. The key reason for including the encoder is that it is a chief source of the noise in the computation of the velocity $\dot{x}$ and acceleration $\ddot{x}$ of the cart position. The cart has the transfer function model $G(s) = \dfrac{b}{s(s+a)}$ where a, b are to be estimated from measurements of the voltage V and the position x. You can set $a = 10, b = 2$ in the simulation of the cart and set the saturation limits to $\pm V_{\max} = \pm 5$ V. The conversion from encoder counts to the cart position in meters is given by $K_{enc} = r_{enc}\dfrac{2\pi}{N_{enc}} = 0.01483\dfrac{2\pi}{4 \cdot 1024} = 2.2749 \times 10^{-5}$ where r_{enc} is the radius of the encoder wheel and N_{enc} is the number of counts the encoder puts out per revolution.

The voltage input is chosen to be a chirp signal given by $V(t) = 4\sin(\phi(t))$ where $\phi(t) = \dfrac{2\pi f_2 - 2\pi f_1}{T_f}\dfrac{t^2}{2} + 2\pi f_1 t$ and $\dot{\phi}(t) = \dfrac{2\pi f_2 - 2\pi f_1}{T_f}t + 2\pi f_1$. That is, $V(t)$ is a sinusoidal voltage whose instantaneous frequency $\dot{\phi}(t)$ increases linearly from $\dot{\phi}(0) = 2\pi f_1$ to $\dot{\phi}(T_f) = 2\pi f_2$. Set $f_1 = 0.04, f_2 = 4$, and $T_f = 10$. The (simulated) data is collected using a `zero-order hold` block to simulate an analog-to-digital converter. Set the sample period T be 0.001 seconds. The data is then stored using the `To Workspace` block.

Rather than do the differentiation and filtering in a MATLAB program, they can be done in the SIMULINK simulation as shown in Figure 16.16.

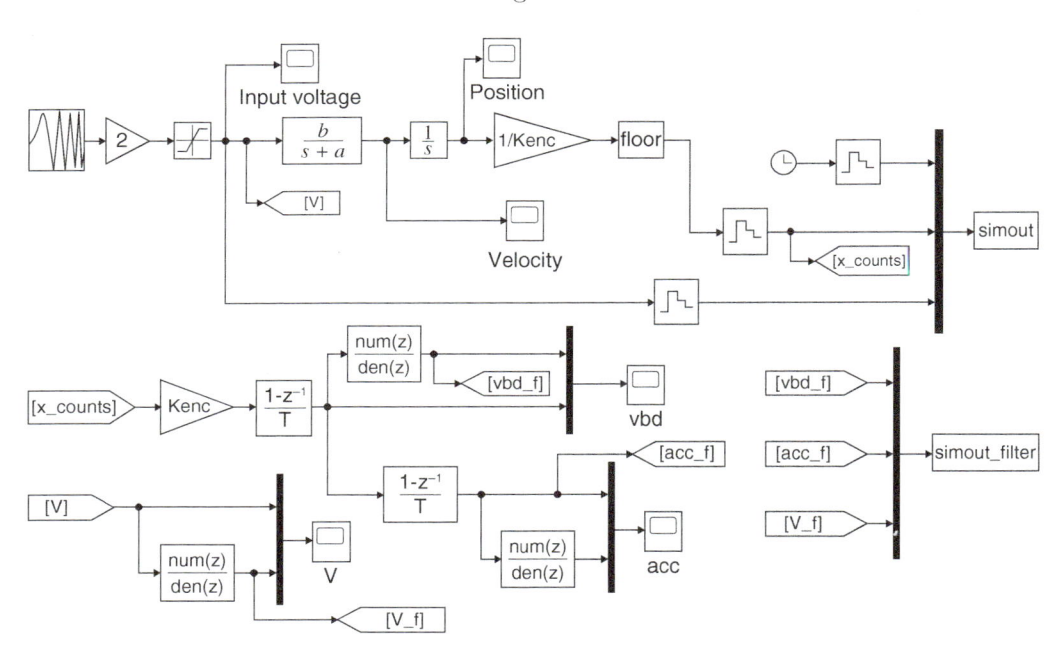

Figure 16.16. SIMULINK block diagram for collecting data.

There are three identical low-pass Butterworth filters in Figure 16.16 shown as $\dfrac{num(z)}{den(z)}$ `discrete filter` blocks. The dialog box for these `discrete filter` blocks is shown in Figure 16.17. The filter parameters `af` and `bf` given in the dialog box of Figure 16.17 are computed as given in part (a) of Problem 15.

There are two differentiators shown as $\dfrac{1-z^{-1}}{T}$ blocks in Figure 16.16. The dialog box for the differentiation filters is shown in Figure 16.18. The sample period T is set to 0.001 seconds.

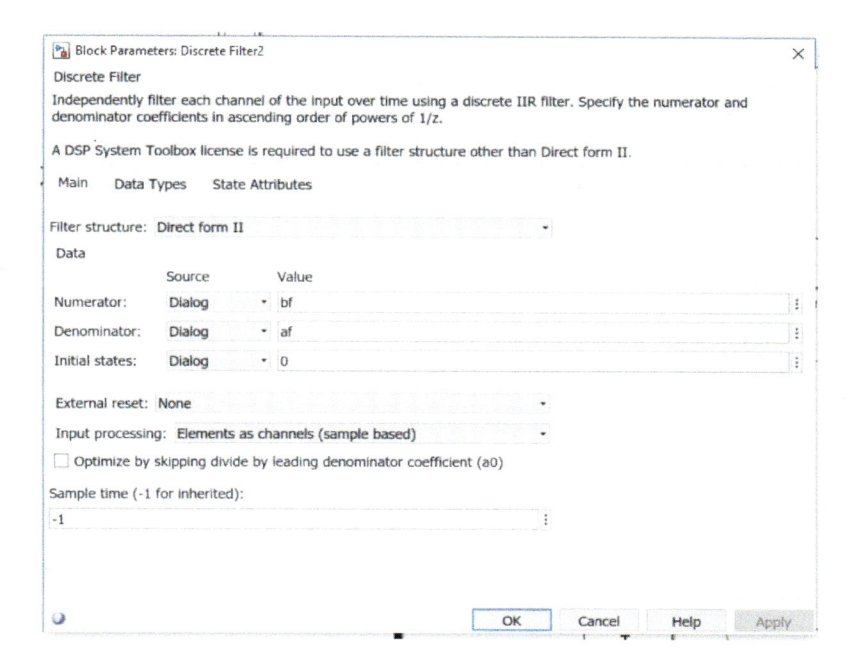

Figure 16.17. Dialog box for the `discrete filter` block.

Figure 16.18. Dialog box for the differentiation filter.

(a) Implement this SIMULINK simulation. Either use the program from part (a) of Problem 15 or the `simout_filter` from the simulation to obtain $V_f, \dot{x}_f$, and $\ddot{x}_f$, which are low-pass filtered versions of $V, \dot{x}$, and $\ddot{x}$, respectively. The low-pass filtered signals have a delay due to the filter itself. So it is important to filter all signals used in the calculation of the parameters with the same filter so they all have the same delay.

Plot `xdot_f` and `xdot` vs. time on the same plot.

Plot `acc_f` and `acc` vs. time on the same plot.

Plot `V_f` and `V` vs. time on the same plot.

In each of these plots you will see a small delay between the signal and its filtered version.

(b) Add to your program of part (a) the calculations of R_y, R_W, R_{Wy}, and

$$\begin{bmatrix} \hat{a} \\ \hat{b} \end{bmatrix} = R_W^{-1} R_{Wy}.$$

Compare your estimates $\hat{a}, \hat{b}$ with the values used in your simulation by computing the fractional errors $(\hat{a} - a)/a$ and $(\hat{b} - b)/b$.

(c) Add to your program of part (b) the calculation of the *residual error*

$$E^2(K)\big|_{K=\hat{K} \triangleq R_W^{-1} R_{Wy}} = R_y - R_{yW} R_W^{-1} R_{Wy}$$

and the *residual error index*

$$\text{Error index} = \sqrt{\frac{R_y - R_{yW} R_W^{-1} R_{Wy}}{R_y}}.$$

17

Robustness and Sensitivity of Feedback

Chapters 8, 9, 11, and 12 of this text make up "classical control theory". These methods of Bode [47], Nyquist [48], and Evans [53] (root locus) were developed in the 1930s and 1940s. Most applications of these techniques were to single-input single-output (SISO) control loops. The statespace approach appeared in the 1960s and 1970s and is usually referred to as "modern control theory". A great advantage of the statespace method was that it could easily deal with multi-input multi-output systems. However, design for robustness (i.e., good gain and phase margins) was not so manageable in the statespace. The 1980s and 1990s saw the return of the frequency domain as the Nyquist theory gives a way to measure robustness.

An alternative way to view classical control has been given by Bruce Francis [62] as follows.

> The subject of this first course is classical control. That means the systems to be controlled are single-input, single-output and stable except possibly for a single pole at the origin, and design is done in the frequency domain, usually with Bode diagrams.

We make a slight addition to this view by letting the open-loop system (system to be controlled) have up to two poles at the origin along with the inclusion of root locus as a usual design method. Controllers for such systems can be designed using the methods of output pole placement (Chapter 10), Bode/Nyquist (Chapter 11), root locus (Chapter 12), or the statespace (Chapters 15 and 16). These systems constitute the majority of the examples in this text (as well as other elementary texts) and the designs result in good stability margins. However, if the system to be controlled has poles in the *open* right-half plane then things are quite different. Though we can always stabilize such a system (e.g., using output pole placement), it can be very difficult or even impossible to obtain adequate stability margins. For example, the output feedback controller designed for the inverted pendulum in Chapter 10 (page 295) was shown to have small stability margins in Section 11.9. In this chapter it is shown that controllers designed for systems with poles in the open right-half plane are fundamentally limited in terms of the amount of robustness that can be achieved. We demonstrate this by looking at the stability margins of four different controllers designed to stabilize the inverted pendulum (see Woodyatt et al. [63], Stein [36], Goodwin et al. [6], and Doyle et al. [49]). The starting point is the model of the inverted pendulum derived in Chapter 13, which is valid for θ and $\dot{\theta}$ both close to zero. Given as a

---

An Introduction to System Modeling and Control, First Edition. John Chiasson.
© 2022 John Wiley & Sons, Inc. Published 2022 by John Wiley & Sons, Inc.
Companion website: www.wiley.com/go/chiasson/anintroductiontosystemmodelingandcontrol

transfer function the model is

$$X(s) = \underbrace{\frac{\kappa(J+m\ell^2)s^2 - \kappa mg\ell}{s^2(s^2-\alpha^2)}}_{G_X(s)}U(s)$$

$$+ \frac{-\kappa g(m\ell)^2(s\theta(0)+\dot{\theta}(0)) + (s^2-\alpha^2)(sx(0)+\dot{x}(0))}{s^2(s^2-\alpha^2)} \tag{17.1}$$

$$\theta(s) = -\underbrace{\frac{\kappa m\ell}{s^2-\alpha^2}}_{G_\theta(s)}U(s) + \frac{s\theta(0)+\dot{\theta}(0)}{s^2-\alpha^2}, \tag{17.2}$$

where $\alpha^2 = \dfrac{mg\ell(M+m)}{Mm\ell^2+J(M+m)}$, $\kappa = \dfrac{1}{Mm\ell^2+J(M+m)}$.

17.1 Inverted Pendulum with Output x

Let's now look at the control of the inverted pendulum using just the measurement of the cart position x as feedback. Using the pole placement method of Chapter 10 a controller can be designed that keeps the *simulated* pendulum rod upright (see Problem 1 of this chapter). As pointed out on page 256 of Goodwin et al. [6], this controller is essentially doomed to failure when implemented on an actual cart and pendulum system. To explain, Figure 17.1 shows a unity feedback control architecture where $G_c(s)$ is designed to make the closed-loop system stable. In Figure 17.1 $p(s,x(0),\dot{x}(0),\theta(0),\dot{\theta}(0))$ denotes the numerator of the initial condition term in (17.1). Note that the transfer function $G_X(s) \triangleq X(s)/U(s)$ has a pole and a zero in the open right half-plane which are, respectively,

$$p = \sqrt{\frac{mg\ell(M+m)}{Mm\ell^2+J(M+m)}}, \quad z = \sqrt{\frac{mg\ell}{J+m\ell^2}}. \tag{17.3}$$

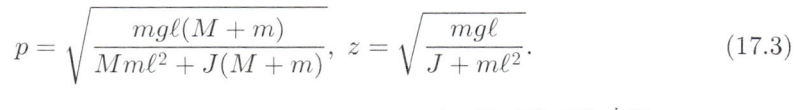

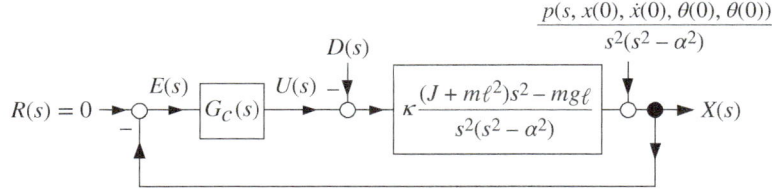

Figure 17.1. Output pole placement controller.

To be specific, consider the QUANSER inverted pendulum on a cart. It has a pendulum rod of length $\ell_Q = 0.6413$ m whose center of mass is at the midpoint of the rod. As in Chapter 13, we take the axis of rotation of the rod be in the $-\hat{\mathbf{z}}$ direction going through its center of mass (Figure 13.1). The moment of inertia of the rod about this axis of rotation is $J = m\ell^2/3$ where $\ell \triangleq \ell_Q/2$. The pole and zero in (17.3) simplify to

$$p = \sqrt{\frac{mg\ell(M+m)}{Mm\ell^2+(m\ell^2/3)(M+m)}} = \sqrt{\frac{3}{4}\frac{g}{\ell}}\sqrt{\frac{M+m}{M+m/4}} \tag{17.4}$$

$$z = \sqrt{\frac{mg\ell}{m\ell^2/3+m\ell^2}} = \sqrt{\frac{3}{4}\frac{g}{\ell}}. \tag{17.5}$$

17.1 Inverted Pendulum with Output x

The QUANSER values $M = 0.57$ and $m = 0.23$ result in $\sqrt{\dfrac{M+m}{M+m/4}} = \sqrt{\dfrac{0.57+0.23}{0.57+0.23/4}} =$ 1.129 showing that the right half-plane pole is only slightly greater than the right half-plane zero. With these values of the parameters, the open-loop transfer function of the inverted pendulum evaluates to

$$G_X(s) = \frac{1.59s^2 - 36.57}{s^2(s^2 - 29.25)} = 1.59\frac{(s-4.7958)(s+4.7958)}{s^2(s-5.408)(s+5.408)}. \tag{17.6}$$

By the pole placement method a controller of the form

$$G_c(s) = \frac{b_3 s^3 + b_2 s^2 + b_1 s + b_0}{s^3 + a_2 s^2 + a_1 s + a_0}$$

will allow arbitrary placement of the closed-loop poles. Choosing the seven closed-loop poles to all be at -5 results in

$$G_c(s) = \frac{4262s^3 + 22\,579s^2 - 2991s - 2137}{s^3 + 35s^2 - 6240s - 30\,583}$$
$$= 4262\frac{(s+5.4086)(s-0.3668)(s+0.2533)}{(s+96.4235)(s-66.2137)(s+4.7902)}. \tag{17.7}$$

Problem 1 asks you to simulate this controller where you will see that it keeps the pendulum rod vertical. We now use Nyquist theory to show that this controller has no chance of working in practice, that is, if you try implement this controller on the actual inverted pendulum it will not be able to keep the rod vertical! The Nyquist contour is given in Figure 17.2a where the two poles of $G_c(s)G_X(s)$ at the origin and the two in the open right half-plane are indicated. This shows that $P = 2$ as $G_c(s)G_X(s)$ has two poles inside the

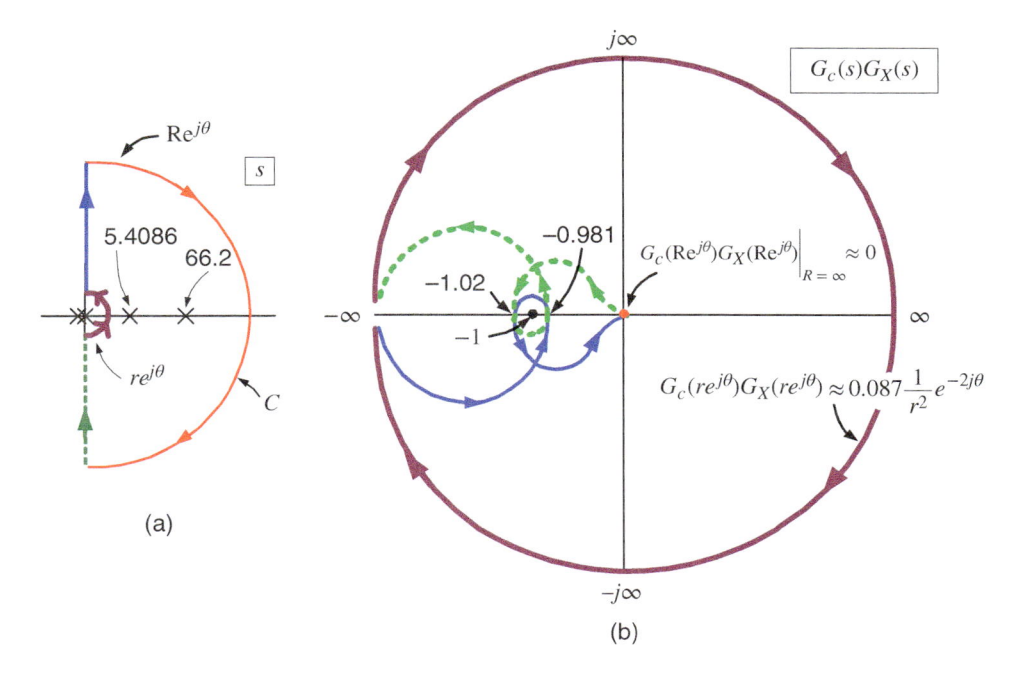

Figure 17.2. (a) Nyquist contour. (b) Nyquist plot.

696   17   Robustness and Sensitivity of Feedback

Nyquist contour. The corresponding Nyquist plot of $G_c(s)G_X(s)$ is shown in Figure 17.2b. (For r small $G_c(re^{j\theta})G_X(re^{j\theta}) \approx (2137/30583)(36.57/29.25)e^{-j2\theta}/r^2 = 0.087e^{-j2\theta}/r^2$.)

From the Nyquist plot it is seen that $N = -2$ so the closed-loop system is stable. (Of course we knew this already as the seven closed-loop poles are all at -5!) However, the gain margin is very small with $\dfrac{KG_c(s)G_X(s)}{1 + KG_c(s)G_X(s)}$ stable only for $-1.02 < -1/K < -0.981$ or $1/1.02 = 0.980 < K < 1.019 = 1/0.981$! The phase margin is only of the order $\tan^{-1}(0.02/1) = 0.02$ radians or 1.15 degrees! That is, if the polar plot is rotated by this amount, the number of encirclements of $-1 + j0$ changes. One might argue that with $G_c(s)$ accurately implemented using a microprocessor then we should be okay. But the parameter values of transfer function $G_X(s)$ may not be known accurately enough that the controller $G_c(s)$ designed based on this model will be able to stabilize the actual inverted pendulum. In other words, this controller is *not* robust to small uncertainties in the model of the inverted pendulum.

The situation is even worse than just model uncertainty! Even if the linear model $G_X(s)$ is very accurate around the (pendulum up) operating point, this controller will *not* work in practice. To show this we need to discuss the *sensitivity* of a control system. We start by magnifying the part of the Nyquist plot around the $-1 + j0$ point as shown in Figure 17.3, but showing only the part of $G_c(j\omega)G_X(j\omega)$ for $\omega > 0$. As seen in this figure the quantity

$$1 + G_c(j\omega)G_X(j\omega)$$

is a vector (complex number) from $-1 + j0$ to $G_c(j\omega)G_X(j\omega)$. For some frequencies the length of this vector is very small (about 0.02). The Bode diagram for $G_c(j\omega)G_X(j\omega)$ is given in Figure 17.4 and shows for the frequency range $0.8 \leq \omega \leq 12$ that $G_c(j\omega)G_X(j\omega)$ is close to $-1 + j0$ or, equivalently, $|1 + G_c(j\omega)G_X(j\omega)|$ is quite small.

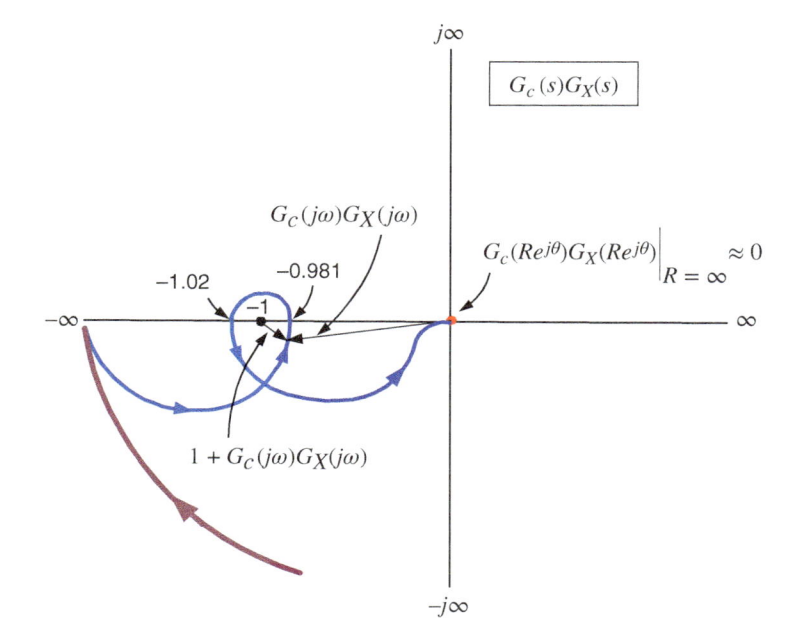

Figure 17.3. Nyquist plot of $G_c(s)G_X(s)$.

17.1 Inverted Pendulum with Output x

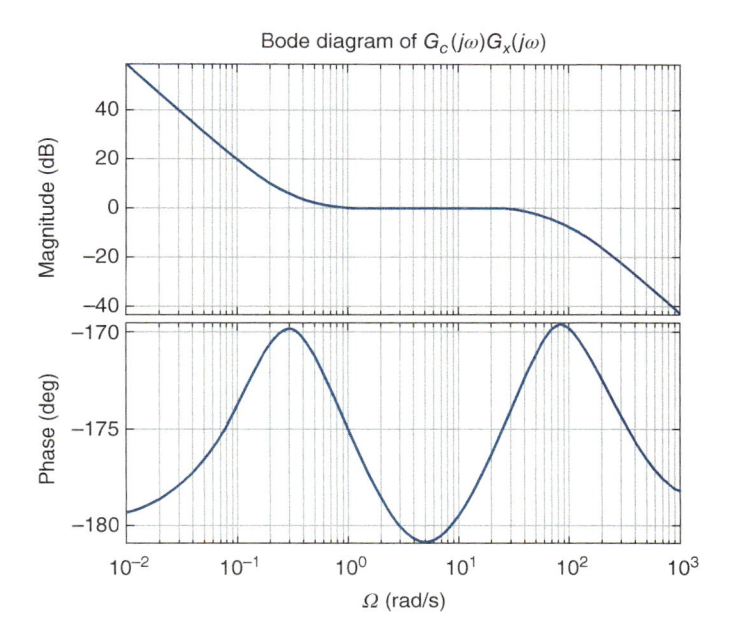

Figure 17.4. $L(j\omega) = G_c(j\omega)G_X(j\omega)$. For $0.8 \leq \omega \leq 12$, $|L(j\omega)|$ is close to 1 and $\angle L(j\omega)$ is close to $180°$

The *sensitivity* function is defined to be

$$S(j\omega) \triangleq \frac{1}{1 + G_c(j\omega)G_X(j\omega)}.$$

The Bode diagram of the sensitivity is shown in Figure 17.5.
From the Bode diagram of $S(j\omega)$ it is found that

$$\begin{aligned}
|S(j\omega)| &> 1 \text{ (0 dB)} &&\text{for } \omega > 0.25 \\
|S(j\omega)| &> 10 \text{ (20 dB)} &&\text{for } 1 < \omega < 25 \\
\max_{\omega} |S(j\omega)| &= 66.8 \text{ (36.5 dB)} &&\text{for } \omega = 5.
\end{aligned} \qquad (17.8)$$

Let's now use the sensitivity function to explain why trying to control the pendulum by only feeding back the cart position x won't work in practice. The closed-loop transfer function is

$$X(s) = \underbrace{\frac{G_c(s)G_X(s)}{1 + G_c(s)G_X(s)}}_{G_{CL}(s)} X_{ref}(s). \qquad (17.9)$$

In the frequency range $0.8 < \omega < 12$ we have $|S(j\omega)| = \left|\frac{1}{1 + G_c(j\omega)G_X(j\omega)}\right| > 10$ so $|1 + G_c(j\omega)G_X(j\omega)| < 0.1$ or, equivalently, $0.9 < |G_c(j\omega)G_X(j\omega)| < 1.1$. As a consequence

$$|G_{CL}(j\omega)| = \left|\frac{G_c(j\omega)G_X(j\omega)}{1 + G_c(j\omega)G_X(j\omega)}\right| > (10)(0.9) = 9.$$

17 Robustness and Sensitivity of Feedback

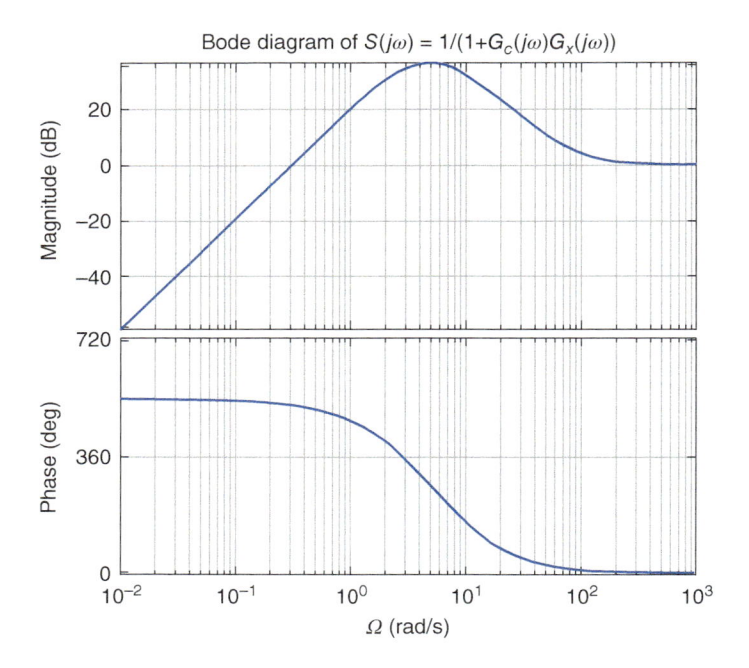

Figure 17.5. Bode diagram of $S(j\omega) = \dfrac{1}{1+G_c(j\omega)G_X(j\omega)}$.

The track of the QUANSER pendulum system is 1 m long. With the initial cart position at the center of the track we apply the reference input $r(t) = R_0 \sin(\omega t)$ with $R_0 = 0.1$ and $\omega = 0.8$ (0.13 Hz). Then

$$x(t) \to x_{ss}(t) = 0.1 \underbrace{|G_{CL}(j0.8)|}_{9} \sin\bigl(0.8t + \angle G_{CL}(j0.8)\bigr) = 0.9 \sin\bigl(0.8t + \angle G_{CL}(j0.8)\bigr).$$

With the amplitude of the reference input equal to 0.1 m, the sinusoidal steady-state position response has an amplitude of 0.9 m. That is, the cart hits the end of the track! In the frequency range where $|S(j\omega)|$ is large, the steady-state output response will be large. In other words, a small amplitude input can cause a large output response if the input contains these frequencies.

Even if the track was longer one should still not expect the pendulum to stay upright! To explain we look at the response of $\theta(t)$. The transfer function from $\theta(s)$ to $X(s)$ is

$$X(s) = G_X(s)U(s) = G_X(s)\frac{1}{G_\theta(s)}\theta(s) = \frac{\dfrac{\kappa(J+m\ell^2)s^2 - \kappa m g\ell}{s^2(s^2-\alpha^2)}}{-\dfrac{\kappa m \ell}{s^2-\alpha^2}}\theta(s)$$

$$= -\frac{(J+m\ell^2)s^2 - mg\ell}{m\ell s^2}\theta(s). \tag{17.10}$$

17.1 Inverted Pendulum with Output x

Solving for $\theta(s)$ and using $X(s) = \dfrac{G_c(s)G_X(s)}{1+G_c(s)G_X(s)}X_{ref}(s)$, the closed-loop transfer function from $X_{ref}(s)$ to $\theta(s)$ is (see Problem 2)

$$\theta(s) = \underbrace{-\dfrac{m\ell s^2}{(J+m\ell^2)s^2 - mg\ell}}_{G_{\theta X}(s)}X(s) = G_{\theta X}(s)\underbrace{\dfrac{G_c(s)G_X(s)}{1+G_c(s)G_X(s)}}_{G_{\theta X_{ref}}(s)}X_{ref}(s). \qquad (17.11)$$

The Bode magnitude diagram of $G_{\theta X_{ref}}(s)$ is given in Figure 17.6.

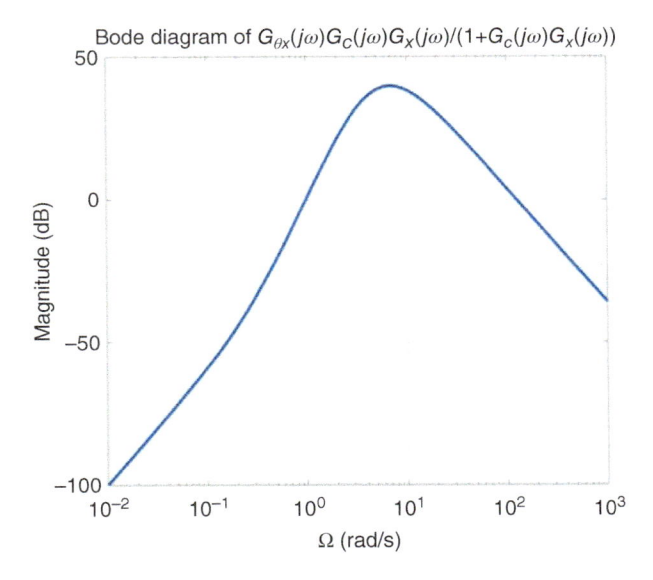

Figure 17.6. Bode diagram of $G_{\theta X_{ref}}(j\omega) = \dfrac{G_{\theta X}(j\omega)G_c(j\omega)G_X(j\omega)}{1+G_c(j\omega)G_X(j\omega)}$.

For $1.9 < \omega < 39$ the Bode diagram shows that $\left|G_{\theta X_{ref}}(j\omega)\right| > 10$ and $\max_{\omega}\left|G_{\theta X_{ref}}(j\omega)\right| = 96.7$ (39.7 dB) with the maximum value achieved at $\omega = 7$. With $X_{ref}(t) = R_0\sin(\omega t)$ we have

$$\theta(t) \to \theta_{ss}(t) = R_0\left|G_{\theta X_{ref}}(j\omega)\right|\sin\!\left(\omega t + \angle G_{\theta X_{ref}}(j\omega)\right).$$

For example, with $R_0 = 0.02$ m and $\omega = 7$ (1.1 Hz) it follows that

$$\theta_{ss}(t) = R_0\underbrace{\left|G_{\theta X_{ref}}(j7)\right|}_{96.7}\sin\!\left(7t + \angle G_{\theta X_{ref}}(j7)\right) = 1.93\sin\!\left(7t + \angle G_{\theta X_{ref}}(j7)\right).$$

That is, $\theta_{ss}(t)$ is oscillating between ± 1.93 rad or $\pm 111°$! The linearized model of the inverted pendulum from which $G_X(s)$ and $G_\theta(s)$ were found are only valid for a variation in θ of perhaps as much as $\pm 20°$ (± 0.35 rad). If $\theta(t)$ varies too far from its equilibrium angle 0 then the transfer function $G_X(s)$ no longer accurately represents the nonlinear model of the pendulum. As a consequence the controller $G_c(s)$ designed using the approximate linear model is no longer guaranteed to (and in all likelihood won't) keep the pendulum rod vertical. Though the discussion here used a pure sinusoidal reference input, typically x_{ref} is a *step* input that contains a whole range of frequencies where $|S(j\omega)|$ is large.

Even if there is no reference input there will be small disturbances between the cart and the track. The block diagram of Figure 17.1 indicates a disturbance $D(s)$ between the track and the cart. With a flat track we don't expect to have a constant disturbance. However, to give an idea of the kind of disturbances that can arise, Figure 17.7 shows a close up of the QUANSER cart used to carry the pendulum rod [34]. The small front gear on the right is powered by a DC motor to propel the cart backwards and forwards along the track. (The large back gear is connected to an optical encoder which is used to compute the cart position x along the track.) The interaction of the teeth of these gears with the teeth of the track can result in small (in magnitude) disturbances forces acting on the cart. We let $D(s)$ denote (the Laplace transform of) any such disturbance.

Figure 17.7. Close up of the QUANSER cart used to carry the pendulum bar [34]. The small front gear is powered by a DC motor to propel the cart backwards and forwards along the track. Source: Quanser – Real Time Control Experiments for Education and Research, www.quanser.com.

The corresponding effect of the disturbance $D(s)$ on the cart position and pendulum angle are given by

$$X(s) = -\underbrace{\frac{G_X(s)}{1 + G_c(s)G_X(s)}}_{G_{XD}(s)} D(s) \tag{17.12}$$

$$\theta_D(s) = -\underbrace{\frac{m\ell s^2}{(J + m\ell^2)s^2 - mg\ell}}_{G_{\theta X}(s)} X(s) = -\underbrace{\frac{G_{\theta X} G_X(s)}{1 + G_c(s)G_X(s)}}_{G_{\theta D}(s)} D(s). \tag{17.13}$$

Both transfer functions $G_{XD}(s) = X(s)/D(s)$ and $G_{\theta D}(s) = \theta(s)/D(s)$ are stable by design (see Problem 3) and they both have the sensitivity $S(s) \triangleq \dfrac{1}{1 + G_c(s)G_X(s)}$ as a factor. Figure 17.8 shows the Bode diagram for $G_{XD}(s)$ where it is seen for $0 \leq \omega \leq 1$ ($0 \leq f \leq 0.16$) that $|G_{XD}(j\omega)| = 14$ (23 dB).

17.1 Inverted Pendulum with Output x 701

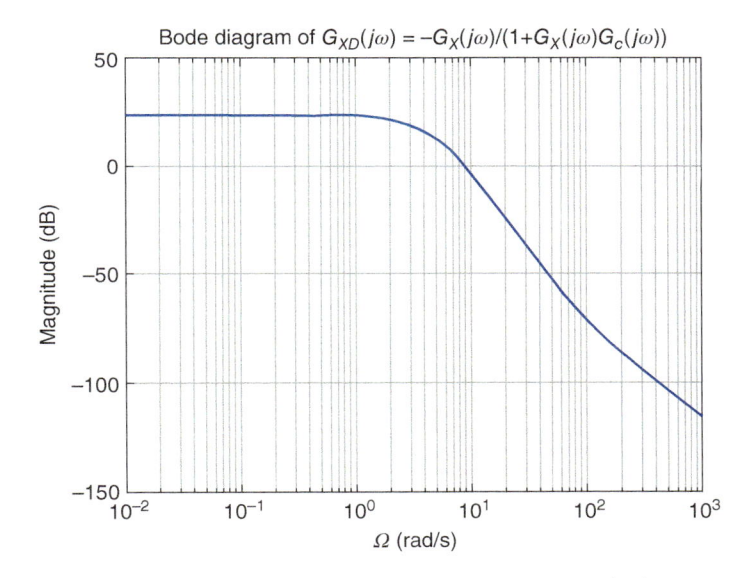

Figure 17.8. Bode diagram of $G_{XD}(j\omega) = -\dfrac{G_X(j\omega)}{1 + G_c(j\omega)G_X(j\omega)}$.

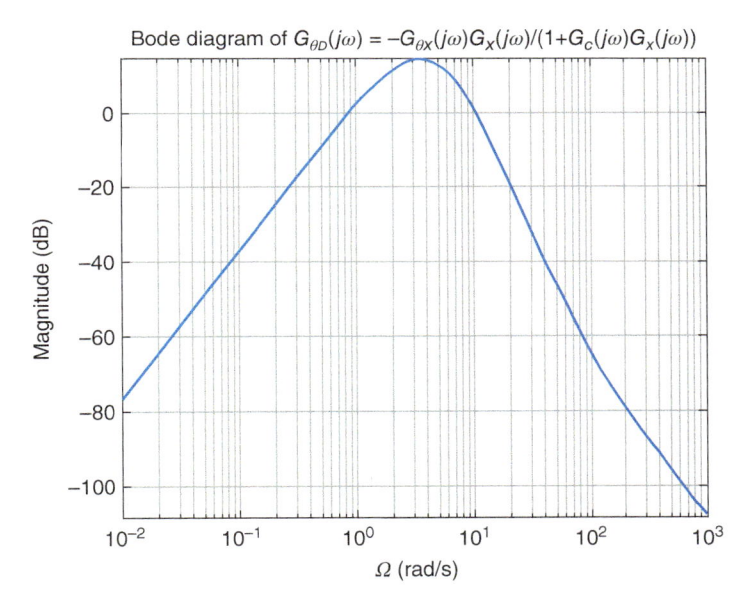

Figure 17.9. Bode diagram of $G_{\theta D}(j\omega) = -\dfrac{G_{\Theta X}G_X(j\omega)}{1 + G_c(j\omega)G_X(j\omega)}$.

Figure 17.9 shows the Bode diagram for $G_{\theta D}(s)$ where it is seen for $2.8 < \omega < 4.6$ ($0.45 < f < 0.73$) that $|G_{\theta D}(j\omega)| > 5$ (14 dB).

For example, the disturbance $d(t) = 0.05\sin(2.8t)$ will result in $\theta_{ss}(t) = |G_{\theta D}(j2.8)|(0.05)\sin(2.8t + \angle G_{\theta D}(j2.8))$ where $|G_{\theta D}(j2.8)|(0.05) > 5(0.05) = 0.25$ radians or $14.8°$. A small disturbance can cause the pendulum angle to swing so far from $\theta = 0$ that the controller based on the linearized model is no longer able to bring it back upright.

17 Robustness and Sensitivity of Feedback

Right Half-Plane Poles and Zeros and the Control Problem

The above robustness and sensitivity problems of controlling an inverted pendulum are *not* due to the particular controller we chose, i.e., $G_c(s)$ being chosen to put all the closed-loop poles at -5. These problems will emerge for *any* controller $G_c(s)$ that stabilizes the inverted pendulum based on cart position feedback [35, 63, 64]. That is, any controller based on the model

$$G_X(s) = \frac{X(s)}{U(s)} = \frac{1.59s^2 - 36.57}{s^2(s^2 - 29.25)} = 1.59\frac{(s - 4.7958)(s + 4.7958)}{s^2(s - 5.408)(s + 5.408)}$$

will not work in practice. This open-loop model has a pole at 5.408 and a zero at 4.7958, which are both in the open right-half plane. Using this information, we show that *any* stabilizing controller results in the closed-loop system being so sensitive to modeling errors, disturbance inputs, reference inputs, etc. that it will not work in practice. Before we can explain all this we need more background starting with Bode's sensitivity integral.

Theorem 1 *Bode's Integral Theorem* [47]
Let

$$S(j\omega) = \frac{1}{1 + G_c(j\omega)G(j\omega)}.$$

Suppose the following conditions hold.

(a) $S(j\omega)$ is stable.
(b) $G_c(j\omega)G(j\omega)$ has no poles in the open right half-plane.
(c) $G_c(j\omega)G(j\omega) = \frac{b_c(j\omega)b(j\omega)}{a_c(j\omega)a(j\omega)}$ has relative degree of at least 2, i.e.,

$$\deg\{a_c(s)a(s)\} - \deg\{b_c(s)b(s)\} \geq 2.$$

Then

$$\int_0^\infty \log_{10}|S(j\omega)|d\omega = 0. \tag{17.14}$$

Proof. See [47] for the original work. Perhaps the simplest proof is given in [65]. ∎

We first work out an example to illustrate this theorem.

Example 1 *Double Integrator* $G(s) = 1/s^2$
In Chapter 11 we stabilized the system $G(s) = 1/s^2$ with the lead controller $G_c(s) = 10\frac{s + 0.1}{s + 10}$. The closed-loop transfer function is

$$C(s) = \underbrace{\frac{G_c(s)G(s)}{1 + G_c(s)G(s)}}_{G_{CL}(s)} R(s) = \frac{10\frac{s+0.1}{s+10}\frac{1}{s^2}}{1 + 10\frac{s+0.1}{s+10}\frac{1}{s^2}}R(s) = \frac{10(s+0.1)}{s^3 + 10s^2 + 10s + 1}R(s).$$

The sensitivity is given by

$$S(s) = \frac{1}{1+G_c(s)G(s)} = \frac{1}{1+10\frac{s+0.1}{s+10}\frac{1}{s^2}} = \frac{s^2(s+10)}{s^3+10s^2+10s+1},$$

which is stable. $G_c(s)G(s)$ has no poles in the *open* right half-plane and it has relative degree 2.

Thus the conditions of the Theorem 1 are satisfied resulting in

$$\int_0^\infty \log_{10}\big|S(j\omega)\big|d\omega = \int_0^\infty \log_{10}\left|\frac{1}{1+10\frac{j\omega+0.1}{j\omega+10}\frac{1}{(j\omega)^2}}\right|d\omega = 0.$$

A plot of $\log_{10}\big|S(j\omega)\big|$ for $0.001 < \omega < 100$ is given in Figure 17.10. Note that the horizontal axis is a linear scale in ω not $\log_{10}(\omega)$ as it is for Bode diagrams. Figure 17.11 on the next page zooms into this plot where it is seen to be positive for $\omega > 2$. For $\omega = 2$ it shows that $\log_{10}\big|S(j2)\big| = 0$ ($\big|S(j2)\big| = 1$). In this example the Bode sensitivity integral tells us that the (negative) area under the curve $\log_{10}\big|S(j\omega)\big|$ for $\omega < 2$ has the same magnitude as the area under this curve for $\omega > 2$. Also, in this example, the magnitude of the sensitivity is small for *all* ω. In fact, $\max_{\omega \geq 0}\big|S(j\omega)\big| = 1.0036$ (0.031 dB). The key takeaway of this

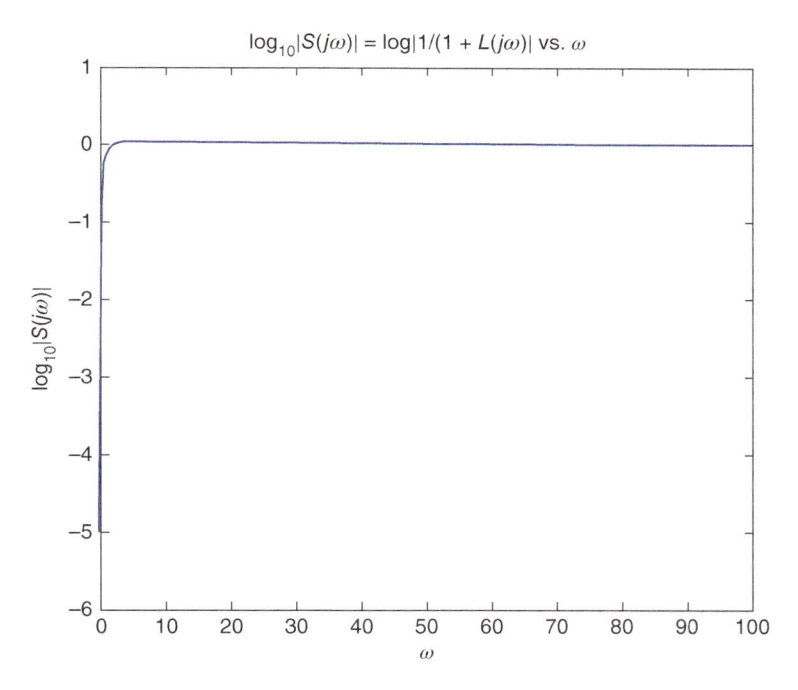

Figure 17.10. Plot of $\log_{10}\big|S(j\omega)\big|$ where $S(j\omega) = \dfrac{1}{1+10\dfrac{j\omega+0.1}{j\omega+10}\dfrac{1}{(j\omega)^2}}$.

17 Robustness and Sensitivity of Feedback

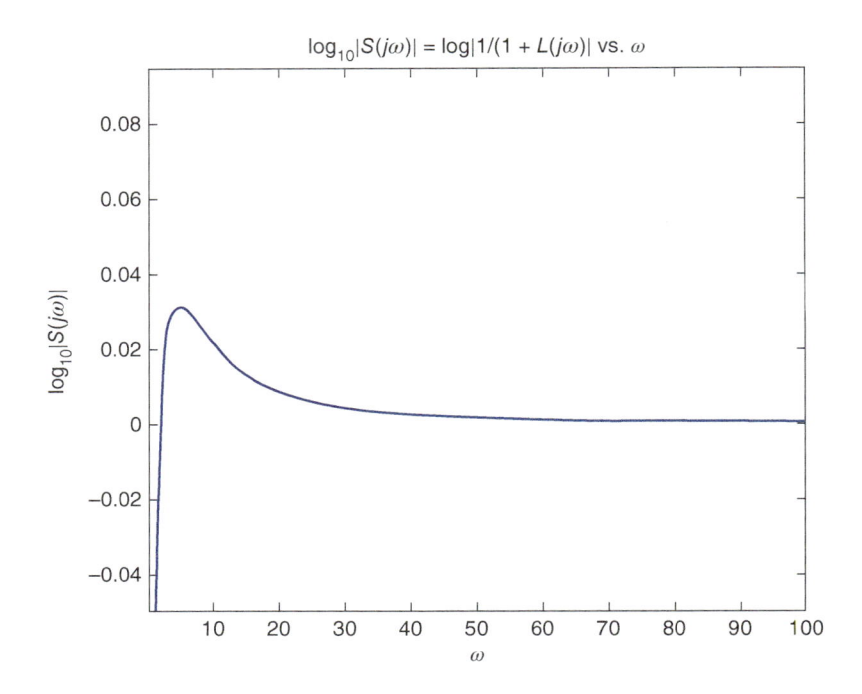

Figure 17.11. Plot of $\log_{10}|S(j\omega)|$ for $\omega > 2$. The horizontal axis is ω (not $\log_{10}\omega$).

example is that $10\dfrac{s+0.1}{s+10}\dfrac{1}{s^2}$ has no poles in the *open* right-half plane, which corresponds to the sensitivity being small for all ω.

Let's return to the inverted pendulum whose transfer function has a pole and a zero in the open right-half plane. With $G_c(s)$ the controller that places the closed-loop poles at -5 we have

$$G_c(s)G_X(s) = 4262\frac{(s+5.4086)(s-0.3668)(s+0.2533)}{(s+96.4235)(s-66.2137)(s+4.7902)}1.59\frac{(s-4.7958)(s+4.7958)}{s^2(s-5.408)(s+5.408)}.$$

The relative degree of $G_c(s)G_X(s)$ is two and $S(s) = \dfrac{1}{1+G_c(s)G_X(s)}$ is stable by our choice $G_c(s)$. However $G_c(s)G_X(s)$ has a right half-plane pole at $s = 5.408$ from the pendulum model and another right half-plane pole at 66.2137 from the controller. Bode's sensitivity integral was generalized by Freudenberg and Looze [35] to handle such a case.

Theorem 2 *Generalization of Bode's Integral theorem* Freudenberg and Looze [35]
Let
$$S(j\omega) = \frac{1}{1+G_c(j\omega)G(j\omega)}.$$

Suppose the following conditions hold.

(a) $S(j\omega)$ is stable.
(b) $G_c(j\omega)G(j\omega)$ has n_p poles in the open right half-plane at $p_1, p_2, \ldots, p_{n_p}$.

(c) $G_c(j\omega)G(j\omega) = \dfrac{b_c(j\omega)b(j\omega)}{a_c(j\omega)a(j\omega)}$ has relative degree of at least 2, i.e.,

$$\deg\{a_c(s)a(s)\} - \deg\{b_c(s)b(s)\} \geq 2.$$

Then

$$\int_0^\infty \log_{10}|S(j\omega)|d\omega = \pi \log_{10}(e) \sum_{i=1}^{n_p} \mathrm{Re}\{p_i\}. \tag{17.15}$$

Note $\log_{10}(e) = 0.434$.

Proof. See [35] or [65]. ∎

Remarks As $\log_{10}|x| = \log_{10}(e)\ln|x|$ this result may also be written as $\int_0^\infty \ln|S(j\omega)|d\omega = \pi \sum_{i=1}^{n_p} \mathrm{Re}\{p_i\}$.

The poles come in complex conjugate pairs so $\sum_{i=1}^{n_p} \mathrm{Re}\{p_i\} = \sum_{i=1}^{n_p} p_i$.

Note that the open-loop system $G_c(s)G(s)$ may have poles on the $j\omega$ axis, but they make no contribution to the right-hand side of (17.15).

Example 2 *Inverted Pendulum $G_c(s)G_X(s)$*

Let's now go back to the unity feedback pole placement controller for the inverted pendulum where

$$G_c(s)G_X(s) = 4262\dfrac{(s+5.4086)(s-0.3668)(s+0.2533)}{(s+96.4235)(s-66.2137)(s+4.7902)}1.59\dfrac{(s-4.7958)(s+4.7958)}{s^2(s-5.408)(s+5.408)}.$$

By the Theorem 2 we have

$$\int_0^\infty \log_{10}|S(j\omega)|\,d\omega = \pi\log_{10}(e)(5.408 + 66.2137) = 97.72. \tag{17.16}$$

Figure 17.12 on the next page is the plot of $\log_{10}|S(j\omega)|$ vs. ω where $\log_{10}|S(j\omega)| > 0$ for $\omega > 1$.[1] This plot shows, as already pointed out earlier, that $\max_\omega |S(j\omega)| = 66.8$ ($\max_\omega \log_{10}|S(j\omega)| = 1.8$) with the maximum value achieved at $\omega = 6$. Typically it is desired to have $\max_{\omega \geq 0}|S(j\omega)| \leq 2$ or, equivalently, $\max_{\omega \geq 0}\log_{10}|S(j\omega)| \leq \log_{10}(2) = 0.69$ (see page 97 of [49]).

Some important points are to be made here.

(1) As $G_c(j\omega)$ is proper and $G_X(j\omega)$ is strictly proper it follows that $\lim_{\omega \to \infty}|G_c(j\omega)G_X(j\omega)| = 0$ and so

$$\lim_{\omega \to \infty} \log_{10}|S(j\omega)| = \lim_{\omega \to \infty} \log_{10}\left|\dfrac{1}{1+G_c(j\omega)G_X(j\omega)}\right| = 0.$$

(2) As $\lim_{\omega \to 0}|G_X(j\omega)| = \infty$ it follows that

$$\lim_{\omega \to 0} \log_{10}|S(j\omega)| = \lim_{\omega \to 0} \log_{10}\left|\dfrac{1}{1+G_c(j\omega)G_X(j\omega)}\right| = -\infty.$$

---

[1]This same information is shown in the top of Figure 17.5, which is a plot of $20\log_{10}|S(j\omega)|$ vs. $\log_{10}\omega$.

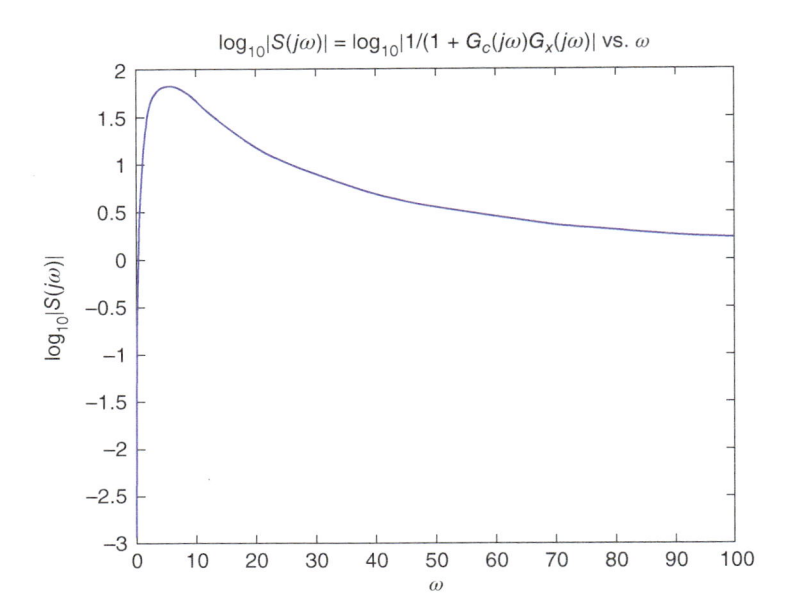

Figure 17.12. Plot of $\log|S(j\omega)|$ for the inverted pendulum with the closed-loop poles all at -5.

(3) The theorem tells us that $G_c(j\omega)G_X(j\omega)$ having right half-plane poles means that $|S(j\omega)|$ has more positive area than negative area. This positive area must be at lower frequencies as $\lim_{\omega \to \infty} \log_{10}|S(j\omega)| = 0$ by point (1) above.

(4) No matter what stabilizing controller $G_c(s)$ is used, the right-hand side of (17.16) must be greater or equal to $\pi \log_{10}(e)(5.408) = 7.38$. This is a fundamental limitation in designing a controller to reduce $|S(j\omega)|$ and is a consequence of $G_X(s)$ having a pole in the open right-half plane.

(5) Equations (17.4) and (17.5) show that $p > z$. It turns out that this implies that any stabilizing controller $G_c(s)$ *must* have a right half-plane pole.[2] Further, this right half-plane pole of $G_c(s)$ must be to the *right* of the zero of $G_X(s)$ [67].[3] Consequently, for *any* controller it follows that

$$\int_0^\infty \log_{10}|S(j\omega)|d\omega \geq \pi \log_{10}(e)(5.4086 + 4.7958) = 13.92.$$

One must expect that a small disturbance will cause the pendulum rod angle to swing far from $\theta = 0$ with the result that the linear controller will not be able to return it back to 0.

Poisson Integral for Sensitivity

Theorem 2 tells us to expect the sensitivity to be large if the system model has poles in the open right-half plane. The *Poisson integral for sensitivity* is a constraint on the sensitivity in terms of the poles and zeros of $G_c(s)G_X(s)$ that are in the open right half-plane.

---

[2] This follows from the *parity interlacing property*, which implies the controller $G_c(s)$ must be unstable [46, 49, 66].

[3] This is based on the real-axis root locus for $1 + KG_c(s)G_X(s)$. See Problem 6.

17.1 Inverted Pendulum with Output x

In particular, if a right half-plane pole is close to right half-plane zero, it shows that *any* controller in the setup of Figure 17.1 will result in the sensitivity being large. To explain, recall $G_c(s)G_X(s)$ is given by

$$G_c(s)G_X(s) = 4262\frac{(s+5.4086)(s-0.3668)(s+0.2533)}{(s+96.4235)(s-66.2137)(s+4.7902)}1.59\frac{(s-4.7958)(s+4.7958)}{s^2(s-5.408)(s+5.408)}.$$

$G_c(s)G_X(s)$ has two poles in the open right half-plane (RHP): one at $p_1 = 5.408$ from $G_X(s)$ and a second at $p_2 = 66.2137$ from $G_c(s)$. It also has two zeros in the open RHP: one at $z_1 = 4.7958$ from $G_X(s)$ and a second at $z_2 = 0.3668$ from $G_c(s)$. There is a *Poisson integral* for each of the zeros in the open RHP. Specifically (see [68, 69])

$$\int_0^\infty \log_{10}|S(j\omega)|\frac{z_1}{z_1^2+\omega^2}d\omega = -\pi\log_{10}(e)\left(\log_{10}\left|\frac{z_1-p_1}{z_1+p_1}\right| + \log_{10}\left|\frac{z_1-p_2}{z_1+p_2}\right|\right) \quad (17.17)$$

and

$$\int_0^\infty \log_{10}|S(j\omega)|\frac{z_2}{z_2^2+\omega^2}d\omega = -\pi\log_{10}(e)\left(\log_{10}\left|\frac{z_2-p_1}{z_2+p_1}\right| + \log_{10}\left|\frac{z_2-p_2}{z_2+p_2}\right|\right). \quad (17.18)$$

As z_1 and p_1 are quite close together, the first integral is the most illuminating at showing why we should expect the sensitivity to be large regardless of the controller. Evaluating the first Poisson integral gives

$$\int_0^\infty \log_{10}|S(j\omega)|\frac{4.7958}{(4.7958)^2+\omega^2}d\omega$$

$$= -\pi\log_{10}(e)\left(\log_{10}\left|\frac{4.7958-5.408}{4.7958+5.408}\right| + \log_{10}\left|\frac{4.7958-66.2137}{4.7958+66.2137}\right|\right) \quad (17.19)$$

$$= -\pi\log_{10}(e)(-1.222 - 0.063)$$

$$= 1.753.$$

Multiplying both sides by $z_1 = 4.7958$ we rewrite this last equation as

$$\int_0^\infty \log_{10}|S(j\omega)|\frac{1}{1+(\omega/4.7958)^2}d\omega = (1.753)(4.7958) = 8.41. \quad (17.20)$$

$S(j\omega)$ in (17.20) is (of course) the same sensitivity function as in (17.16). Multiplying (17.17) by z_1 it follows that, no matter what stabilizing controller is designed, we must have

$$\int_0^\infty \log_{10}|S(j\omega)|\frac{1}{1+(\omega/4.7958)^2}d\omega \geq -z_1\pi\log_{10}(e)\log_{10}\left|\frac{z_1-p_1}{z_1+p_1}\right|$$

$$= -(4.7958)\pi\log_{10}(e)(-1.222)$$

$$= 7.996. \quad (17.21)$$

Further, as $\frac{1}{1+(\omega/4.7958)^2}$ drops off rapidly toward 0 for $\omega > z_1 = 4.7958$, we see that $|S(j\omega)|$ must be sufficiently large at the lower frequencies in order to satisfy (17.21). No feedback controller can make this Poisson integral less than 7.996 leading us to expect that $\max_{|\omega|\leq 4.7958}|S(j\omega)|$ will be large for *any* controller.

The takeaway from this example is that a different scheme[4] is needed to control the inverted pendulum. We consider this in the next sections. For more discussion of the Poisson integral constraint see Section 9.2 of Goodwin et al. [6].

---

[4]That is, different from just feeding back the measured cart position in a unity feedback configuration.

H_∞ Norm

Looking back at Figure 17.1 the transfer function from the reference input $R(s)$ to the error $E(s)$ is

$$E(s) = \frac{1}{1+G_c(s)G(s)}R(s).$$

The quantity

$$\max_{-\infty<\omega<\infty}|S(j\omega)| = \max_{-\infty<\omega<\infty}\left|\frac{1}{1+G_c(j\omega)G(j\omega)}\right|$$

is the upper bound on the maximum amplitude of the error response due to a sinusoidal input. We have seen that it is this upper bound that must not be too large if we are to be able to have the controller work in practice. More specifically, recall (17.11) for the transfer function for angle of the pendulum rod given by

$$\theta(s) = G_{\theta X}(s) \underbrace{\frac{G_c(s)G_X(s)}{1+G_c(s)G_X(s)}}_{G_{\theta X_{ref}}(s)} X_{ref}(s).$$

It was shown that for any controller $G_c(s)$ that stabilized the closed-loop system, the quantity

$$\max_{-\infty<\omega<\infty}\left|G_{\theta X_{ref}}(j\omega)\right|$$

was large. It is so large that, in practice, the pendulum rod would not remain upright during step changes in the cart's position. This discussion motivates the following definition.

Definition 1 H_∞ Norm

Let $G(j\omega)$ be a stable and strictly proper transfer function. The H_∞ norm of $G(j\omega)$ is defined by

$$\max_{-\infty<\omega<\infty}|G(j\omega)|.$$

Remark If the closed-loop system has a small H_∞ norm then we should expect to have the controller work in practice. In other words, not only must the feedback controller stabilize the closed-loop system, but it must not have too large of an H_∞ norm so that it works when implemented. Designing H_∞ controllers (controllers that stabilize the closed-loop system and minimize the H_∞ norm taking into account model uncertainty) is involved and we refer you to [49] for an elementary introduction to the approach.

17.2 Inverted Pendulum with Output $y(t) = x(t) + \left(\ell + \frac{J}{m\ell}\right)\theta(t)$

Let's now put a sensor to measure the pendulum rod angle and combine that measurement with the cart position. Specifically we take the output to be

$$y(t) = x(t) + \left(\ell + \frac{J}{m\ell}\right)\theta(t).$$

17.2 Inverted Pendulum with Output $y(t) = x(t) + (\ell + \frac{J}{m\ell})\theta(t)$

Then using (17.1) and (17.2) we have

$$Y(s) = X(s) + \left(\ell + \frac{J}{m\ell}\right)\theta(s) = \frac{\kappa(J + m\ell^2)s^2 - \kappa m g \ell}{s^2(s^2 - \alpha^2)}U(s) - \kappa\left(\ell + \frac{J}{m\ell}\right)\frac{m\ell s^2}{s^2(s^2 - \alpha^2)}U(s)$$

$$+ \frac{\left(\ell + (J/(m\ell))\right)s^2 - \kappa g(m\ell)^2\right)(s\theta(0) + \dot{\theta}(0)) + (s^2 - \alpha^2)\left(sx(0) + \dot{x}(0)\right)}{s^2(s^2 - \alpha^2)}$$

$$= \underbrace{-\frac{\kappa m g \ell}{s^2(s^2 - \alpha^2)}}_{G_Y(s)} U(s) + \frac{p_Y(s)}{s^2(s^2 - \alpha^2)}$$

with the obvious definition for $p_Y(s)$. Note that the transfer function $G_Y(s)$ no longer has any zeros. This is the reason for choosing this output! Figure 17.13 shows the unity feedback control configuration using this new output of the inverted pendulum.

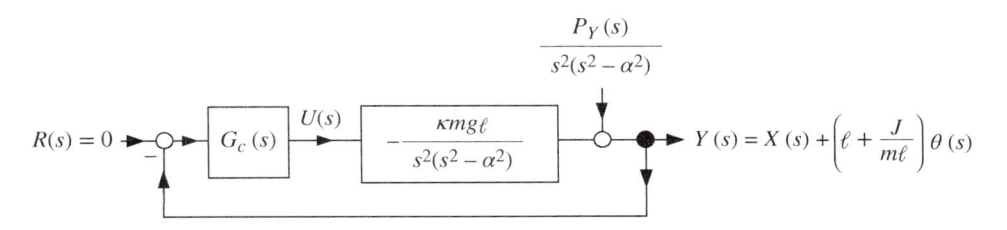

Figure 17.13. Output pole placement controller.

Using the QUANSER parameter values ($\kappa m g \ell = 36.57, \alpha^2 = 29.26$), the open-loop transfer function of the inverted pendulum is

$$G_Y(s) = \frac{-36.57}{s^2(s^2 - 29.26)} = \frac{-36.57}{s^2(s + 5.409)(s - 5.409)}.$$

In order to arbitrarily place the closed-loop poles with a minimum order controller we choose $G_c(s)$ to be of the form

$$G_c(s) = \frac{b_c(s)}{a_c(s)} = \frac{b_3 s^3 + b_2 s^2 + b_1 s + b_0}{s^3 + a_2 s^2 + a_1 s + a_0}.$$

In Example 5 of Chapter 10 the details of this design were presented. There it was chosen to place closed-loop poles at -5 and resulted in the controller

$$G_c(s) = -\frac{1041.6s^3 + 6113.7s^2 + 2990.8s + 2136.3}{s^3 + 35s^2 + 554.3s + 5399}$$

$$= -1041.6 \frac{(s + 5.409)(s^2 + 0.46s + 0.3792)}{(s + 20.835)(s^2 + 14.17s + 259.13)}. \qquad (17.22)$$

Note that $G_c(s)$ has no RHP poles or zeros.

We first look at what Nyquist theory can tell us about this controller design. Figure 17.14a is the Nyquist contour and Figure 17.14b is the corresponding Nyquist plot for $G_c(s)G_Y(s)$.

710   17 Robustness and Sensitivity of Feedback

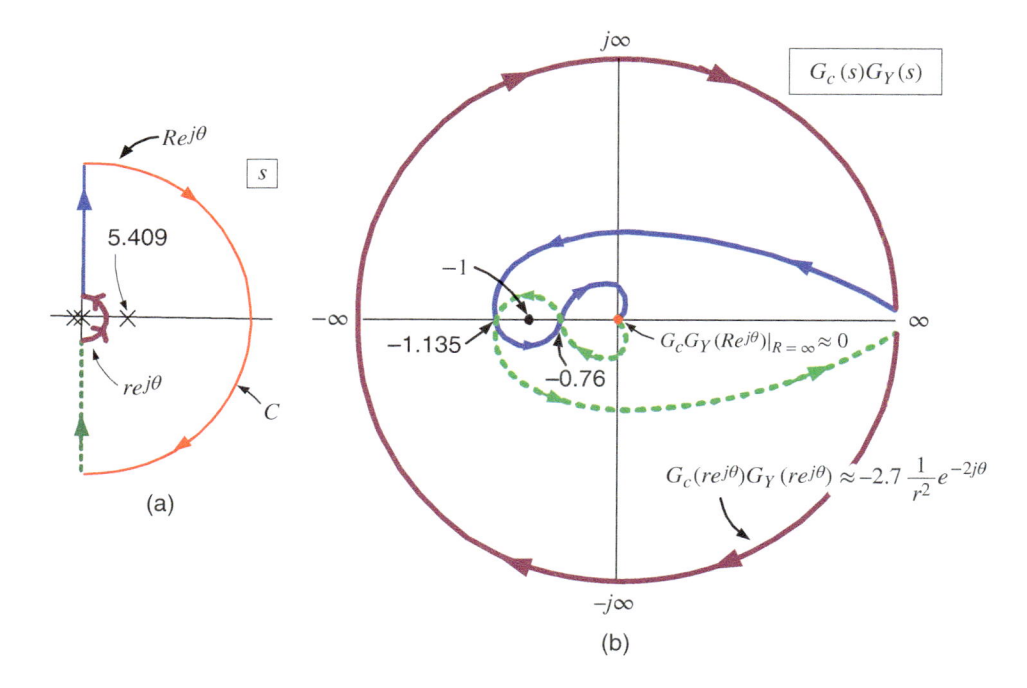

Figure 17.14. (a) Nyquist contour. (b) Nyquist plot.

From the Nyquist plot it is seen that the closed-loop transfer function

$$\frac{KG_c(s)G_Y(s)}{1 + KG_c(s)G_Y(s)}$$

is stable for $-1.135 < -\frac{1}{K} < -0.78$ or $1/1.135 = 0.881 < K < 1.282 = 1/0.78$. Though this is a much larger gain margin compared the control system when the output was just x, it is still quite small! The phase margin turns out to be about $7°$ (small, but much larger than the phase margin of $1.15°$ using only x as the feedback). Let's look at the sensitivity of this controller. $G_c(s)G_Y(s)$ is given by

$$G_c(s)G_Y(s) = -\frac{1041.6s^3 + 6113.7s^2 + 2990.8s + 2136.3}{s^3 + 35s^2 + 554.3s + 5399} \frac{-36.57}{s^2(s^2 - 29.26)}.$$

The sensitivity function is

$$S(s) = \frac{1}{1 + G_c(s)G_Y(s)}.$$

Theorem 2 tell us that

$$\int_0^\infty \log_{10}\left|S(j\omega)\right| d\omega = \pi \log_{10}(e)(5.409) = \pi \log_{10}(e)(5.409) = 7.38.$$

A plot of $\log_{10}|S(j\omega)|$ is given in Figure 17.15, which shows $\max_{\omega \geq 0} \log_{10}|S(j\omega)| = 0.94$ or $\max_{\omega \geq 0} |S(j\omega)| = 8.7$. This maximum sensitivity is 7 times smaller compared with the sensitivity with only x as the feedback. However $\max_{\omega \geq 0} |S(j\omega)| = 8.7$ is still large making it questionable if this controller will actually work in practice.

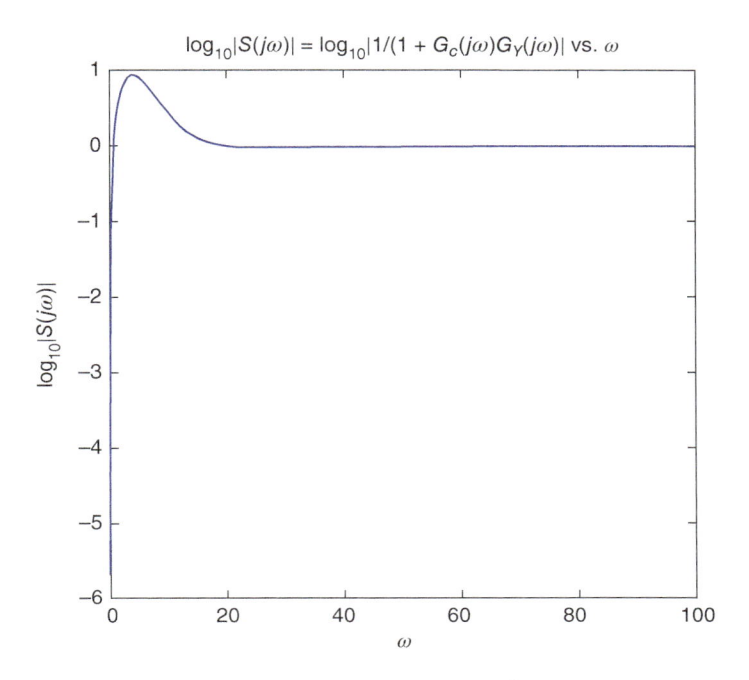

Figure 17.15. $\log_{10}\bigl|S(j\omega)\bigr| = \log_{10}\dfrac{1}{1+G_c(j\omega)G_Y(j\omega)}$ vs. ω.

17.3 Inverted Pendulum with State Feedback

We now consider the sensitivity of a control system for the inverted pendulum that uses state feedback. In Chapter 13 a linear statespace model of the inverted pendulum was found to be

$$\underbrace{\begin{bmatrix} dx/dt \\ dv/dt \\ d\theta/dt \\ d\omega/dt \end{bmatrix}}_{dz/dt} = \underbrace{\begin{bmatrix} 0 & 1 & 0 & 0 \\ 0 & 0 & -\kappa g m^2 \ell^2 & 0 \\ 0 & 0 & 0 & 1 \\ 0 & 0 & \kappa m g \ell (M+m) & 0 \end{bmatrix}}_{A} \underbrace{\begin{bmatrix} x \\ v \\ \theta \\ \omega \end{bmatrix}}_{z} + \underbrace{\begin{bmatrix} 0 \\ \kappa(J+m\ell^2) \\ 0 \\ -\kappa m \ell \end{bmatrix}}_{b} u, \qquad (17.23)$$

where $\kappa = \dfrac{1}{Mm\ell^2 + J(M+m)}$. This model is valid as long as both θ and $\dot\theta$ are not too far from 0. In Chapter 15 it was shown that one can choose $k \in \mathbb{R}^{1\times 4}$ such that the feedback

$$u(t) = -\underbrace{\begin{bmatrix} k_1 & k_2 & k_3 & k_4 \end{bmatrix}}_{k} \underbrace{\begin{bmatrix} x \\ v \\ \theta \\ \omega \end{bmatrix}}_{z}$$

will force $z(t) = [x(t), v(t), \theta(t), \omega(t)]^T \to [0,0,0,0]^T$. A block diagram of such a state feedback controller is shown in Figure 17.16a. An equivalent transfer function block diagram of the state feedback controller is shown in Figure 17.16b where $y \triangleq kz$. Using the QUANSER

712   17 Robustness and Sensitivity of Feedback

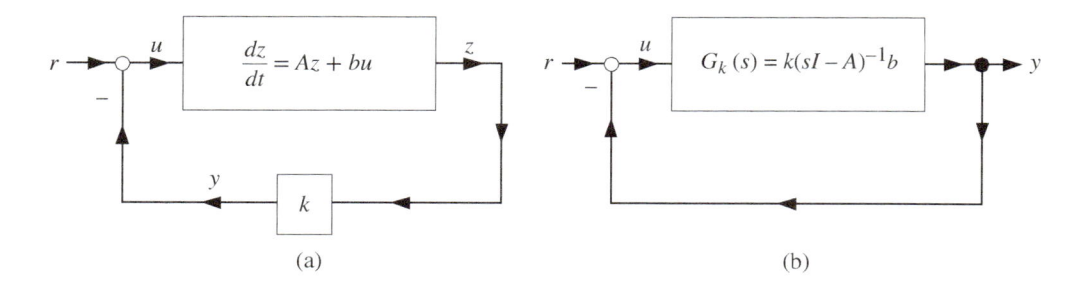

Figure 17.16. (a) State feedback controller. (b) Equivalent transfer function representation.

parameter values and placing the closed-loop poles at -5 results in

$$k = \begin{bmatrix} -29.0255 & -23.2204 & -56.4917 & -12.0901 \end{bmatrix}$$

$$G_k(s) \triangleq k(sI-A)^{-1}b = \frac{20s^3 + 179.3s^2 + 500s + 625}{s^4 - 29.3s^2} = \frac{20(s+5.41)(s^2+3.55s+5.777)}{s^2(s+5.409)(s-5.409)}.$$

$G_k(s)$ has a single pole at 5.409 in the *open* right-half plane, but no right half-plane zeros.

The Nyquist contour and corresponding Nyquist plot of $G_k(s)$ are shown in Figure 17.17. (For r small $G_k(re^{j\theta}) \approx -(625/29.3)e^{-j2\theta}/r^2 = -21.33e^{-j2\theta}/r^2$.)

The pole at 5.409 of $G_k(s)$ is inside the Nyquist contour so $P=1$. The Nyquist plot goes around $-1+j0$ once in the counter-clockwise direction making $N=-1$. Thus $Z = N + P = 0$ showing the closed-loop system is stable (as we already knew!). Next, if we multiply $G_k(s)$ by the scalar gain K_{gain}, the Nyquist plot shows that $\dfrac{K_{gain}G_k(s)}{1+K_{gain}G_k(s)}$ is

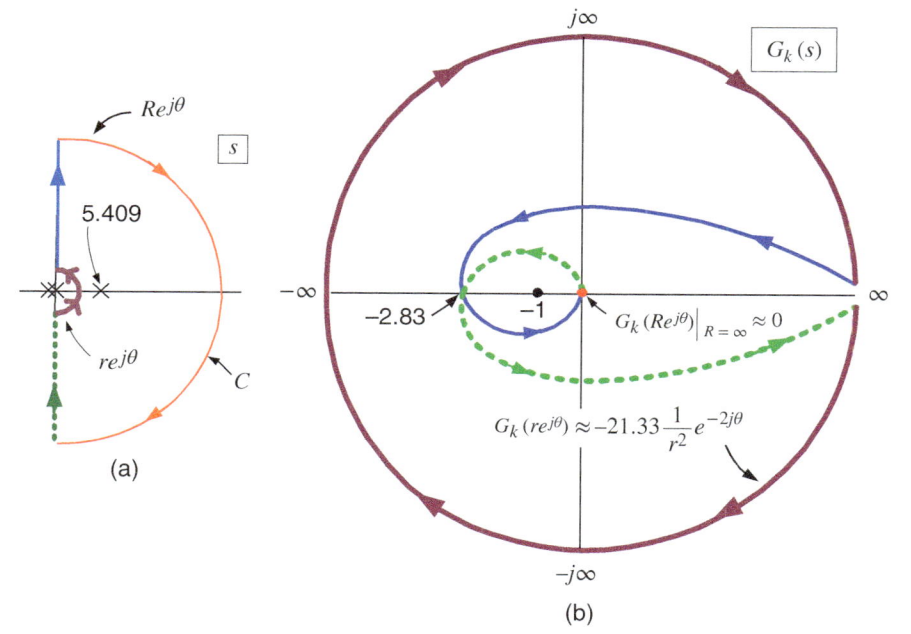

Figure 17.17. Nyquist contour and plot for $G_k(s) = k(sI-A)^{-1}b = \dfrac{20s^3+176.9s^2+500s+625}{s^4-26.92s^2}.$

stable for $-2.83K_{gain} < -1$ or $K_{gain} > 1/2.83 = 0.353$. The Nyquist plot is not drawn to scale so the phase margin cannot be read from it. However, using MATLAB it is found that the phase margin is 64°. We shouldn't have any trouble getting this controller to work in practice!

The sensitivity function for the control system shown in Figure 17.16b is

$$S_k(j\omega) = \frac{1}{1 + G_k(j\omega)}.$$

As shown above $G_k(s)$ has one right half-plane pole and no right half-plane zeros. However, $G_k(s)$ has relative degree 1 and so Theorem 2 does not apply. We need the following generalization of it.

Theorem 3 $\int_0^\infty \log_{10}|S(j\omega)| \, d\omega = \log_{10}(e)\left(-\eta\frac{\pi}{2} + \pi \sum_{i=1}^{N_p} \text{Re}\{p_i\}\right)$

Define

$$S(j\omega) \triangleq \frac{1}{1 + G_c(j\omega)G(j\omega)}$$

$$\eta \triangleq \lim_{s \to \infty} sG_c(s)G(s)$$

and suppose the following conditions hold.

(a) $S(j\omega)$ is stable.
(b) $G_c(j\omega)G(j\omega)$ has n_p poles in the open right half-plane at $p_1, p_2, \ldots, p_{n_p}$.
(c) $G_c(j\omega)G(j\omega) = \frac{b_c(j\omega)b(j\omega)}{a_c(j\omega)a(j\omega)}$ has relative degree 1 or higher.

Then

$$\int_0^\infty \log_{10}|S(j\omega)| \, d\omega = \log_{10}(e)\left(-\eta\frac{\pi}{2} + \pi \sum_{i=1}^{N_p} \text{Re}\{p_i\}\right). \quad (17.24)$$

Note $\log_{10}(e) = 0.434$.

Proof. See [65]. ∎

Example 3 *Inverted Pendulum Controlled by State Feedback*

We apply the Theorem 3 to the transfer function model $G_k(s)$ of the inverted pendulum. From above we have

$$G_k(s) = \frac{20s^3 + 179.3s^2 + 500s + 625}{s^4 - 29.3s^2} = \frac{20(s + 5.41)(s^2 + 3.55s + 5.777)}{s^2(s + 5.409)(s - 5.409)},$$

which has relative degree 1 and a single pole in the open right half-plane at 5.409. Also $\eta \triangleq \lim_{s \to \infty} sG_k(s) = 20$ so by Theorem 3 it follows that

$$\int_0^\infty \log_{10}|S_k(j\omega)| \, d\omega = \log_{10}(e)\left(-20\frac{\pi}{2} + \pi(5.409)\right) = -6.264.$$

The integral of the logarithm of the sensitivity is *negative* and thus we do not expect $|S_k(j\omega)|$ to be large. In fact, $\max_{\omega \geq 0} |S_k(j\omega)| \leq 1$! However, this is the sensitivity of the output

$$y = kz = k_1 x + k_2 v + k_3 \theta + k_4 \omega$$

and we are really interested in the sensitivity of the output θ. We don't want small disturbances in the input or small non zero initial conditions to cause θ to vary far from its equilibrium value of 0. That is, θ and $\dot{\theta}$ must be kept small for all time so that this controller based on the inverted pendulum's linear statespace model is able to ensure closed-loop stability of the actual (nonlinear) system. Let's look at the response with the output being θ. Our setup is

$$\frac{dz}{dt} = Az + bu$$
$$u = -kz + r$$
$$y_\theta = \underbrace{\begin{bmatrix} 0 & 0 & 1 & 0 \end{bmatrix}}_{c_\theta} \underbrace{\begin{bmatrix} x \\ v \\ \theta \\ \omega \end{bmatrix}}_{z}$$

or, equivalently,

$$\frac{dz}{dt} = (A - bk)z + br$$
$$y_\theta = c_\theta z.$$

The transfer function for $\theta(s)$ is then

$$\frac{\theta(s)}{R(s)} = \frac{Y_\theta(s)}{R(s)} = c_\theta \left(sI - (A - bk) \right)^{-1} b.$$

In Figure 17.18 the reference input r is a force as is the feedback $-kz$. In this context r better models a force disturbance input (perhaps due to the cart wheels interacting with the track). With the closed-loop poles all at -5 it turns out that (see Problem 5)

$$G_\theta(s) = \frac{\theta(s)}{R(s)} = c_\theta \left(sI - (A - bk) \right)^{-1} b = -\frac{m\kappa\ell s^2}{(s+5)^4},$$

Figure 17.18. $G_\theta(s) = c_\theta \left(sI - (A - bk) \right)^{-1} b.$

where $m\kappa\ell = 3.728$. A Bode plot of $G_\theta(j\omega)$ reveals that

$$\max_{\omega \geq 0} |G_\theta(j\omega)| \leq 0.037.$$

Consequently, if there is any additive disturbance $d(t)$ acting at the input $u(t)$ (acting on the cart body) this controller will keep $\theta(t)$ close to 0. In particular, if there is a step disturbance $r(t) = D_0 u_s(t)$ in the input we see that

$$\lim_{t \to \infty} \theta(t) = \lim_{s \to 0} s G_\theta(s) \frac{D_0}{s} = \lim_{s \to 0} -\frac{m\kappa\ell s^2}{(s+5)^4} D_0 = 0.$$

The pendulum rod angle is still kept at $\theta = 0$!

17.4 Inverted Pendulum with an Integrator and State Feedback

In Chapter 15 the control architecture of Figure 17.19 was used to show that the pendulum rod could be kept upright while having the cart position track step inputs. We now consider the sensitivity of this feedback scheme.

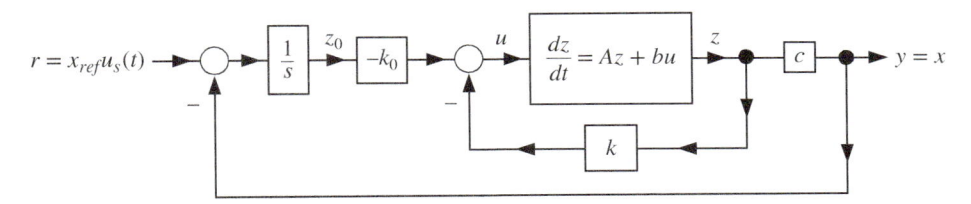

Figure 17.19. Statespace tracking of step inputs.

With A, b as given in (17.23) and $c = \begin{bmatrix} 1 & 0 & 0 & 0 \end{bmatrix}$, define

$$G_k(s) \triangleq c(sI - (A - bk))^{-1} b.$$

To do a Nyquist stability analysis of this control system consider the equivalent block diagram of Figure 17.19 given in Figure 17.20.

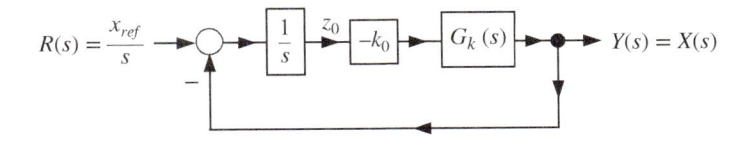

Figure 17.20. Equivalent block diagram of Figure 17.19.

It was shown in Chapter 15 that $k_a = \begin{bmatrix} k & k_0 \end{bmatrix} = \begin{bmatrix} k_1 & k_2 & k_3 & k_4 & k_0 \end{bmatrix}$ can be chosen to place the closed-loop poles at any desired set of values. Choosing the closed-loop poles to be at -5 results in

$$k_a = \begin{bmatrix} k & k_0 \end{bmatrix} = \begin{bmatrix} -85.4514 & -37.9043 & -111.4406 & -22.9103 & 85.4514 \end{bmatrix}$$

and

$$G_k(s) = c(sI - (A - bk))^{-1}b = \frac{1.59s^2 - 36.57}{s^4 + 25s^3 + 250s^2 + 1386s + 3125}$$
$$= 1.59\frac{(s + 4.7958)(s - 4.7958)}{(s + 13.5367)(s + 1.2474)(s^2 + 10.216s + 9.0599)}.$$

With $G_{k_0}(s) \triangleq -\frac{k_0}{s}$ the sensitivity function is

$$S(s) = \frac{1}{1 + G_{k_0}(s)G_k(s)} = \frac{1}{1 + \frac{-k_0}{s}G_k(s)}.$$

A plot of $\log_{10}|S(j\omega)|$ vs. ω is given in Figure 17.21. It is seen that $\max_\omega |S(j\omega)| = 1.6092$ (0.2066 dB) with the maximum value achieved at $\omega = 2.88$.

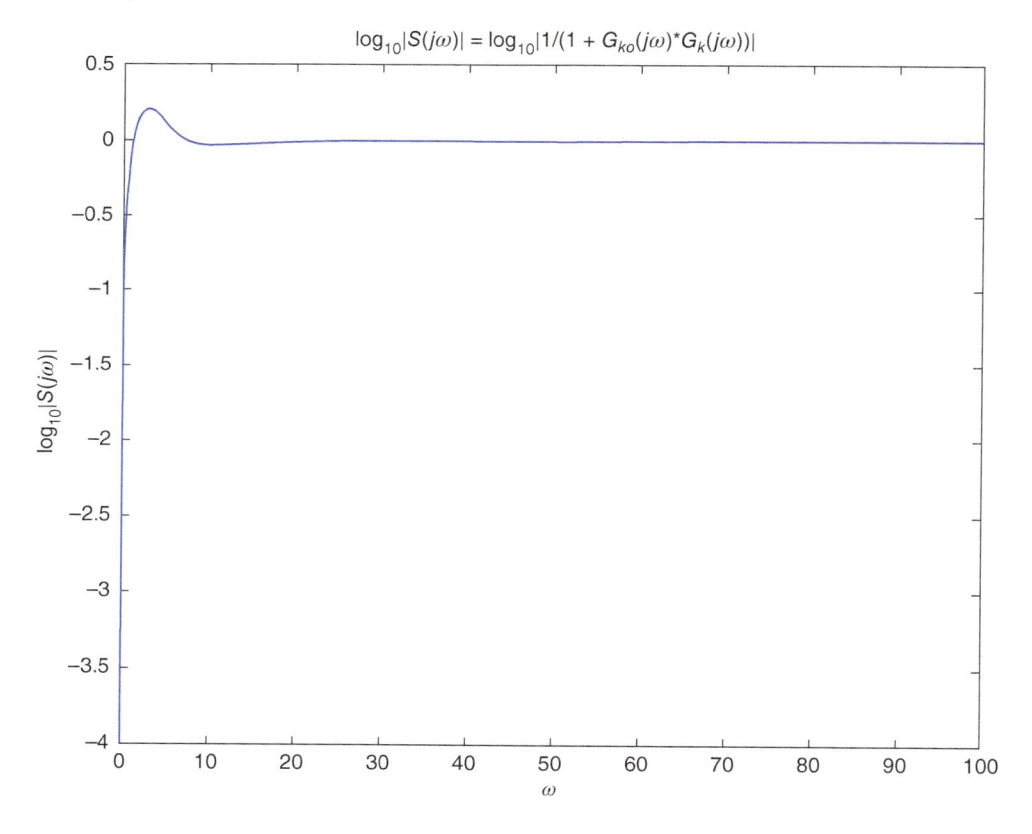

Figure 17.21. Plot of $\log_{10}|S(j\omega)|$ vs. ω.

Let's compute the closed-loop transfer function from $R(s)$ to $\theta(s)$. From Section 15.9 the statespace setup is

$$\frac{d}{dt}\underbrace{\begin{bmatrix} z \\ z_0 \end{bmatrix}}_{z_a \in \mathbb{R}^5} = \underbrace{\begin{bmatrix} A & 0_{4\times 1} \\ -c & 0 \end{bmatrix}}_{A_a \in \mathbb{R}^{5\times 5}} \underbrace{\begin{bmatrix} z \\ z_0 \end{bmatrix}}_{z_a \in \mathbb{R}^5} + \underbrace{\begin{bmatrix} b \\ 0 \end{bmatrix}}_{b_a \in \mathbb{R}^5} u$$

$$u = -\underbrace{\begin{bmatrix} k & k_0 \end{bmatrix}}_{k_a \in \mathbb{R}^{1\times 5}} \underbrace{\begin{bmatrix} z \\ z_0 \end{bmatrix}}_{z_a \in \mathbb{R}^5} + r,$$

where $r(t) = x_{ref} u_s(t)$ and $z_a = \begin{bmatrix} x & v & \theta & \omega & z_0 \end{bmatrix}^T$. Taking $y_\theta \triangleq \theta$ as the output we write

$$y_\theta = \theta = \underbrace{\begin{bmatrix} 0 & 0 & 1 & 0 & 0 \end{bmatrix}}_{c_{\theta a} \in \mathbb{R}^{1\times 5}} \begin{bmatrix} z \\ z_0 \end{bmatrix}.$$

Then

$$G_\theta(s) = \frac{Y_\theta(s)}{R(s)} = c_{\theta a}\left(sI_{5\times 5} - (A_a - b_a k_a)\right)^{-1} b_a = \frac{-m\kappa\ell s^3}{(s+5)^5},$$

with $m\kappa\ell = 3.728$. We can make a Bode plot of $G_\theta(s)$ to find $\max_{\omega \geq 0} |G_\theta(j\omega)| \leq 0.028$. This shows the pendulum rod angle θ is *not* sensitive to step reference inputs.

17.5 Inverted Pendulum with State Feedback via State Estimation

In Example 12 of Chapter 15 state feedback control of the inverted pendulum was considered. In Problem 6 of Chapter 16 a state estimator using the output $y = x + \left(\ell + \dfrac{J}{m\ell}\right)\theta$ was used to estimate the state for the feedback control. The setup is shown in Figure 17.22 where x_{ref} is the desired cart position.[5] In this section we see if state feedback control using a state estimator (this approach is still output feedback) helps with the robustness and sensitivity of the closed-loop system.[6]

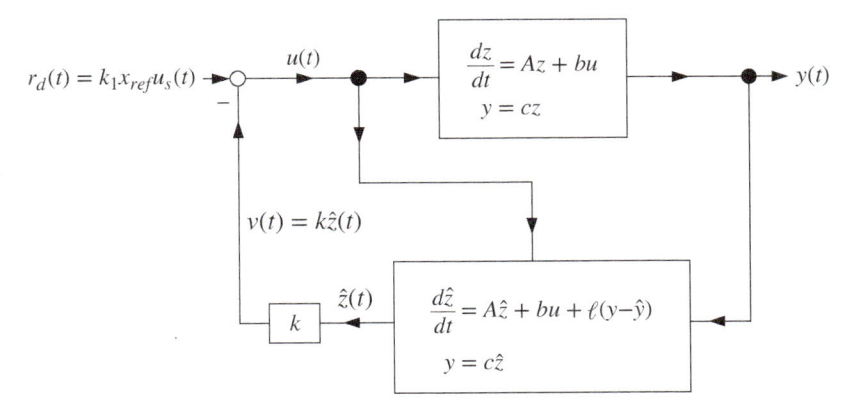

Figure 17.22. State feedback control of the inverted pendulum using a state estimator.

---

[5] Note the abuse of notation in that ℓ is used to denote the half length of the pendulum rod as well as the observer gain vector.
[6] Spoiler alert. It won't!

17 Robustness and Sensitivity of Feedback

The statespace model of the pendulum around the upright position is

$$\underbrace{\begin{bmatrix} dx/dt \\ dv/dt \\ d\theta/dt \\ d\omega/dt \end{bmatrix}}_{dz/dt} = \underbrace{\begin{bmatrix} 0 & 1 & 0 & 0 \\ 0 & 0 & -\dfrac{gm^2\ell^2}{M m\ell^2 + J(M+m)} & 0 \\ 0 & 0 & 0 & 1 \\ 0 & 0 & \dfrac{mg\ell(M+m)}{M m\ell^2 + J(M+m)} & 0 \end{bmatrix}}_{A} \underbrace{\begin{bmatrix} x \\ v \\ \theta \\ \omega \end{bmatrix}}_{z} + \underbrace{\begin{bmatrix} 0 \\ \dfrac{J+m\ell^2}{M m\ell^2 + J(M+m)} \\ 0 \\ -\dfrac{m\ell}{M m\ell^2 + J(M+m)} \end{bmatrix}}_{b} u$$

$$y = \underbrace{\begin{bmatrix} 1 & 0 & \ell + \dfrac{J}{m\ell} & 0 \end{bmatrix}}_{c} \underbrace{\begin{bmatrix} x \\ v \\ \theta \\ \omega \end{bmatrix}}_{z}.$$

(17.25)

In Section 16.2 it was shown that Figure 17.22 is equivalent to Figure 17.23 where

$$G(s) = \frac{b(s)}{a(s)} = c(sI - A)^{-1}b, \quad G_{c1}(s) = \frac{n(s)}{\delta(s)} = k(sI - (A - \ell c))^{-1}b,$$

$$G_{c2}(s) = \frac{m(s)}{\delta(s)} = k(sI - (A - \ell c))^{-1}\ell.$$

A disturbance $D(s)$ has also been included acting at the input.

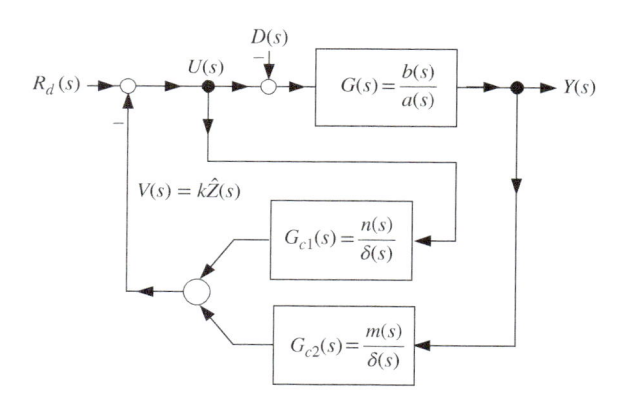

Figure 17.23. Block diagram equivalent to Figure 17.22.

Using the QUANSER parameter values and choosing k to place all four poles of $A - bk$ at -5 and ℓ to place the four poles of $A - \ell c$ at -10 results in (see Problem 6 of Chapter 16)

$$k = \begin{bmatrix} -17.090280684076106 & -13.672224547260877 & -55.391420765269693 & -11.209832050489917 \end{bmatrix}$$

$$\ell = \begin{bmatrix} -185.308353999079 & -608.778931086277 & 527.039326655450 & 2896.001382027047 \end{bmatrix}^T.$$

In Problem 7 you are asked to show that the transfer functions of Figure 17.23 are

$$G(s) = c(sI - A)^{-1}b = \frac{b(s)}{a(s)} = \frac{-36.5705}{s^4 - 29.2564s^2}$$

17.5 Inverted Pendulum with State Feedback via State Estimation

$$G_{c1}(s) = k(sI - (A - \ell c))^{-1}b = \frac{n(s)}{\delta(s)} = \frac{20s^3 + 979.2564s^2 + 2.0255 + 236830}{s^4 + 40s^3 + 600s^2 + 4000s + 10000}$$

$$G_{c2}(s) = k(sI - (A - \ell c))^{-1}\ell = \frac{m(s)}{\delta(s)} = \frac{-50167s^3 - 303420s^2 - 205080s - 170900}{s^4 + 40s^3 + 600s^2 + 4000s + 10000}.$$

Some block diagram reduction shows that Figure 17.23 is equivalent to that of Figure 17.24 where

$$G_c(s) = \frac{\delta(s)}{\delta(s) + n(s)} = \frac{s^4 + 40s^3 + 600s^2 + 4000s + 10\,000}{s^4 + 60s^3 + 1579.3s^2 + 4000s + 246\,830}$$

$$= \frac{s^4 + 40s^3 + 600s^2 + 4000s + 10\,000}{(s^2 + 63.3642s + 1642.2)(s^2 - 3.364s + 150.31)}.$$

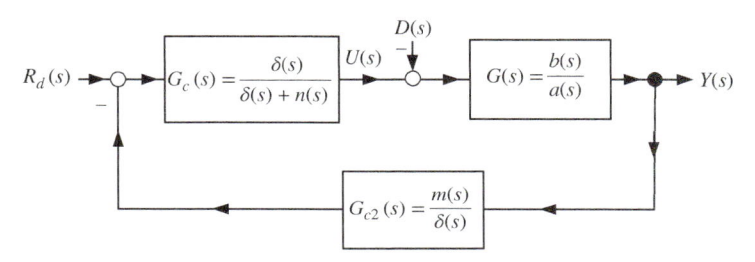

Figure 17.24. Block diagram equivalent to Figure 17.23.

The transfer function from $D(s)$ to $Y(s)$ is

$$\frac{Y(s)}{D(s)} = -\frac{G(s)}{1 + G_c(s)G_{c2}(s)G(s)} = -\frac{\frac{b(s)}{a(s)}}{1 + \frac{m(s)}{\delta(s) + n(s)}\frac{b(s)}{a(s)}}.$$

The sensitivity function is

$$S(s) \triangleq \frac{1}{1 + \frac{m(s)}{\delta(s) + n(s)}\frac{b(s)}{a(s)}} = \frac{1}{1 + \frac{(s + 5.408\,854)(s^2 + 0.639\,34s + 0.629\,82)}{(s^2 + 63.3642s + 1642.2)(s^2 - 3.364s + 150.31)} \times \frac{36.5705}{s^2(s^2 - 29.2564)}}$$

and, by Theorem 2, it follows that

$$\int_0^\infty \log_{10}|S(j\omega)|\,d\omega = \pi\log_{10}(e)(3.364 + \sqrt{29.2564}) = 16.6.$$

This indicates that the output $y = x + \left(\ell + \frac{J}{m\ell}\right)\theta$ is very sensitive to input disturbances. The point here is that using an observer to estimate the state is still an output feedback controller and Theorem 2 is applicable. Consequently, for this example, the fact that $G_c(s)G_{c2}(s) = \frac{m(s)}{\delta(s) + n(s)}$ and $G(s) = \frac{b(s)}{a(s)}$ have right half-plane poles tells us to expect the sensitivity to be large.

This state feedback with an observer approach resulted in the transfer function $G_c(s)G_{c2}(s)$ having two poles in the open right-half plane at $1.682 \pm j12.144$ (the roots of $s^2 - 3.364s + 150.31 = 0$) along with the pole of $G(s)$ in the open right-half plane. Compare this with the output pole placement approach of Section 17.2 that fed back $y(t) = x + (\ell + J/(m\ell))\theta(t)$ using the controller given in (17.22) whose poles and zeros are all in the open left half-plane.

Problems

Problem 1 *Output Position Feedback for the Inverted Pendulum*

The objective of this problem is to design an output pole placement controller for the inverted pendulum feeding back just the cart position x Use the parameter values of the QUANSER [34] inverted pendulum system given by

$M = 0.57$ kg (cart mass)

$m = 0.23$ kg (pendulum mass)

$g = 9.81$ m/s^2 (acceleration due to gravity)

$\ell = 0.6412/2 = 0.3206$ m (pendulum rod length divided by 2)

$J = m\ell^2/3 = 7.88 \times 10^{-3}$ kg-m^2 (pendulum rod moment of inertia about its center of mass).

(a) With the output taken to be x, Eq. (17.1) shows $X(s)$ is given by

$$X(s) = \kappa \underbrace{\frac{(J+m\ell^2)s^2 - mg\ell}{s^2(s^2 - \alpha^2)}}_{\text{Transfer function}} U(s)$$

$$+ \frac{s^3 x(0) + s^2 \dot{x}(0) - (\alpha^2 x(0) + m\ell\beta_0\theta(0))s - m\ell\beta_0\dot{\theta}(0) - \alpha^2 \dot{x}(0)}{s^2(s^2 - \alpha^2)}.$$

Let $p\left(s, x(0), \dot{x}(0), \theta(0), \dot{\theta}(0)\right)$ denote the numerator of the initial condition term and consider the control system of Figure 17.25.

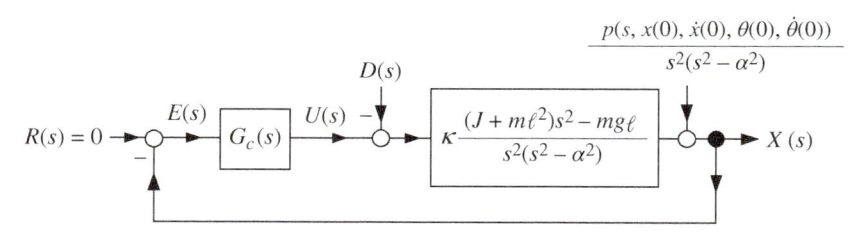

Figure 17.25. Pole placement controller using cart position for feedback.

Design the lowest order unity feedback controller that stabilizes the closed-loop system of Figure 17.25 and allows arbitrary pole placement.

(b) Figure 17.26 is a SIMULINK block diagram of a unity feedback control system using cart position as feedback. Simulate the linear statespace pendulum model along with

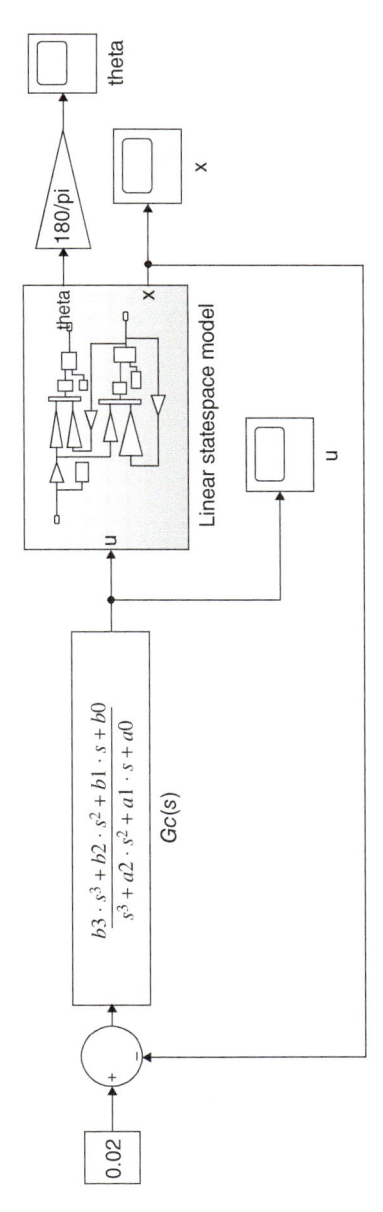

Figure 17.26. SIMULINK simulation to control the inverted pendulum using cart position feedback.

722 17 Robustness and Sensitivity of Feedback

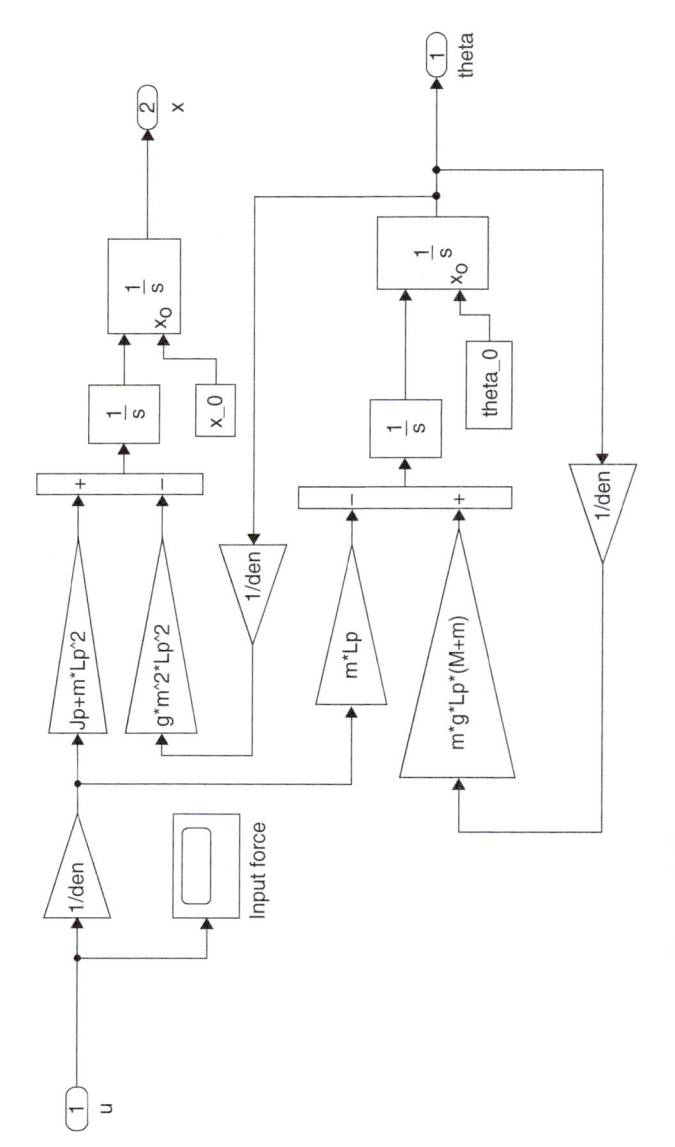

Figure 17.27. Inside the Linear Statespace Model block of Figure 17.26.

the stabilizing controller designed in part (a). Set the initial conditions as $\theta(0) = 0, \dot{\theta}(0) = 0, x(0) = 0, \dot{x}(0) = 0$. The pendulum angle $\theta(t)$ should not vary by more than about $\pm 20°$ so that the linear model remains a valid representation of the nonlinear model. The variable den in Figure 17.27 is $Mm\ell^2 + J(M+m) = 1/\kappa$. Further, in Figure 17.27, Lp corresponds to ℓ and Jp corresponds to J. The reason for using the statespace model to simulate the pendulum is so that both $x(t)$ and $\theta(t)$ can be plotted. Use the Euler integration algorithm with a step size of $T = 0.001$ second.

With the reference input $r = 0.02$ m (2 cm), the simulation should show that θ swings out past $50°$!

Problem 2 $G_{\theta X}(s) = \theta(s)/X_{ref}(s)$
Show that

$$G_\theta(s) = G_{\theta X}(s) \frac{G_c(s)G_X(s)}{1 + G_c(s)G_X(s)} X_{ref}(s)$$

$$= -\frac{\kappa m\ell s^2(b_3 s^3 + b_2 s^2 + b_1 s + b_0)}{(s^3 + a_2 s^2 + a_1 s + a_0)s^2(s^2 - \alpha^2)} X_{ref}(s)$$
$$+ (b_3 s^3 + b_2 s^2 + b_1 s + b_0)(\kappa(J + m\ell^2)s^2 - \kappa mg\ell)$$

$$= -\frac{\kappa m\ell s^2(b_3 s^3 + b_2 s^2 + b_1 s + b_0)}{(s+5)^7} X_{ref}(s),$$

where $G_c(s)$ was chosen to place the seven closed-loop poles at -5, i.e., $G_c(s)$ is given by (17.7).

Problem 3 $G_{XD}(s)$ and $G_{\theta D}(s)$
Let

$$G_{\theta X} = -\frac{\kappa m\ell s^2}{\kappa(J + m\ell^2)s^2 - \kappa mg\ell}, \quad G_X(s) = \frac{\kappa(J + m\ell^2)s^2 - \kappa mg\ell}{s^2(s^2 - \alpha^2)},$$

$$G_c(s) = \frac{b_3 s^3 + b_2 s^2 + b_1 s + b_0}{s^3 + a_2 s^2 + a_1 s + a_0}.$$

(a) Show that

$$G_{\theta D}(s) = \frac{G_{\theta X} G_X(s)}{1 + G_c(s)G_X(s)}$$

$$= \frac{-\kappa m\ell(s^3 + a_2 s^2 + a_1 s + a_0)s^2}{(s^3 + a_2 s^2 + a_1 s + a_0)s^2(s^2 - \alpha^2)}$$
$$+(b_3 s^3 + b_2 s^2 + b_1 s + b_0)(\kappa(J + m\ell^2)s^2 - \kappa mg\ell)$$

$$= \frac{-\kappa m\ell(s^3 + a_2 s^2 + a_1 s + a_0)s^2}{(s+5)^7},$$

where $G_c(s)$ was chosen to place the seven closed-loop poles at -5, i.e., $G_c(s)$ is given by (17.7).

(b) Show that

$$G_{XD}(s) = -\frac{G_X(s)}{1 + G_c(s)G_X(s)}$$

$$= -\frac{(s^3 + a_2s^2 + a_1s + a_0)(\kappa(J + m\ell^2)s^2 - \kappa mg\ell)}{(s^3 + a_2s^2 + a_1s + a_0)s^2(s^2 - \alpha^2) + (b_3s^3 + b_2s^2 + b_1s + b_0)(\kappa(J + m\ell^2)s^2 - \kappa mg\ell)}$$

$$= -\frac{(s^3 + a_2s^2 + a_1s + a_0)(\kappa(J + m\ell^2)s^2 - \kappa mg\ell)}{(s + 5)^7},$$

where $G_c(s)$ was chosen to place the seven closed-loop poles at -5, i.e., $G_c(s)$ is given by (17.7).

Problem 4 *Poisson Integral for the Inverted Pendulum System*

In the Poisson integral for the inverted pendulum given in (17.17) the quantity

$$-\pi \log_{10}(e) \log_{10}\left|\frac{p_1 - z_1}{p_1 + z_1}\right| = -\pi \log_{10}(e) \log_{10}\left|\frac{4.7958 - 5.408}{4.7958 + 5.408}\right| = 1.67$$

is large due to the closeness in value of p_1 and z_1. Using Eqs. (17.4) and (17.5) where $p_1 = p$ and $z_1 = z$, show that as $m \to \infty$ (or $M \to 0$) this quantity reduces to $-\pi \log_{10}(e) \log_{10}|1/3| = 0.651$. Then show, similar to (17.20), that

$$\int_0^\infty \log_{10}|S(j\omega)| \frac{1}{1 + \omega^2/z_1^2} d\omega \geq 3.122.$$

The point here is that one would still expect the sensitivity to be so large that *any* unity feedback controller using just position feedback won't be able to keep the rod upright.

Problem 5 $G_\theta(s) = c_\theta(sI - (A - bk))^{-1}b$

Using state feedback and taking the output to be the pendulum angle θ, the system equations are

$$\underbrace{\begin{bmatrix} dx/dt \\ dv/dt \\ d\theta/dt \\ d\omega/dt \end{bmatrix}}_{dz/dt} = \underbrace{\begin{bmatrix} 0 & 1 & 0 & 0 \\ 0 & 0 & -\kappa g m^2 \ell^2 & 0 \\ 0 & 0 & 0 & 1 \\ 0 & 0 & \kappa mg\ell(M+m) & 0 \end{bmatrix}}_{A} \underbrace{\begin{bmatrix} x \\ v \\ \theta \\ \omega \end{bmatrix}}_{z} + \underbrace{\begin{bmatrix} 0 \\ \kappa(J + m\ell^2) \\ 0 \\ -\kappa m\ell \end{bmatrix}}_{b} u,$$

with

$$y = \underbrace{\begin{bmatrix} 0 & 0 & 1 & 0 \end{bmatrix}}_{c_\theta} \begin{bmatrix} x \\ v \\ \theta \\ \omega \end{bmatrix}, \quad u(t) = -\begin{bmatrix} k_1 & k_2 & k_3 & k_4 \end{bmatrix} \underbrace{\begin{bmatrix} x \\ v \\ \theta \\ \omega \end{bmatrix}}_{z} + r$$

and $\kappa = \dfrac{1}{Mm\ell^2 + J(M+m)}$. The transfer function from $R(s)$ to $\theta(s)$ is

$$\theta(s)/R(s) = G_\theta(s) = c_\theta(sI - (A - bk))^{-1}b.$$

(a) Some computation shows that

$$\det(sI - (A - bk))$$
$$= s^4 + ((m\ell^2 + J)\kappa k_2 - \kappa m\ell k_4)s^3 + ((J + m\ell^2)\kappa k_1 - \kappa m\ell k_3 - (M+m)\kappa mg\ell)s^2$$
$$- \kappa mg\ell k_2 s - \kappa mg\ell k_1.$$

Show how to choose $k_1, k_2, k_3,$ and k_4 so that

$$\det(sI - (A - bk)) = (s + r_1)(s + r_2)(s + r_3)(s + r_4)$$
$$= s^4 + (r_1 + r_2 + r_3 + r_4)s^3 + (r_1 r_2 + r_1 r_3 + r_1 r_4 + r_2 r_3 + r_2 r_4 + r_3 r_4)s^2$$
$$+ (r_1 r_2 r_3 + r_1 r_2 r_4 + r_1 r_3 r_4 + r_2 r_3 r_4)s + r_1 r_2 r_3 r_4.$$

(b) Some more computation shows that

$$c_\theta \operatorname{adj}(sI - (A - bk)) = \begin{bmatrix} 0 & 0 & 1 & 0 \end{bmatrix} \operatorname{adj}(sI - (A - bk))$$
$$= \begin{bmatrix} \kappa m\ell k_1 s & \kappa m\ell k_1 + \kappa m\ell k_2 s & s^3 + ((J + m\ell^2)k_2 - m\ell k_4)\kappa s^2 \\ + (J + m\ell^2)\kappa k_1 s s^2 + (J + m\ell^2)\kappa k_2 s + (J + m\ell^2)\kappa k_1 \end{bmatrix}.$$

The notation "adj" stands for the *adjoint* of a matrix. See Chapter 15 (page 569) for an explanation of the adjoint of a matrix as well as the inverse and determinant of a matrix.

(c) Use parts (a) and (b) to show that

$$G_\theta(s) = c_\theta(sI - (A - bk))^{-1} b = \frac{-m\ell\kappa s^2}{(s+r_1)(s+r_2)(s+r_3)(s+r_4)}.$$

Problem 6 *Inverted Pendulum – Root Locus of* $1 + KG_c(s)G_X(s)$

The unity feedback control structure of Figure 17.1 with $G_c(s)G_X(s)$ given by

$$G_c(s)G_X(s) = 4262 \frac{(s + 5.4086)(s - 0.3668)(s + 0.2533)}{(s + 96.4235)(s - 66.2137)(s + 4.7902)} 1.59 \frac{(s - 4.7958)(s + 4.7958)}{s^2(s - 5.408)(s + 5.408)}$$

results in the seven closed-loop poles placed at -5. The real-axis root loci for $1 + KG_c(s)G_X(s)$ is shown in Figure 17.28 on next page. Breakaway points at $0, 1.97,$ and 12.8 are also indicated. Note that the pole $p = 5.408$ of $G_X(s)$ and the pole $p_c = 66.2317$ of $G_c(s)$ come together and breakaway from the real axis at 12.8. As K is increased these poles migrate to the left half-plane. For $K = 1$ all the poles are at -5 by design of $G_c(s)$. The pole $p = 5.408$ of $G_X(s)$ is to the right of the zero $z = 4.7958$ of $G_X(s)$. As $p > z$ it turns out that any stabilizing controller $G_c(s)$ must have a pole p_c to the right of the zero $z = 4.7958$. In this example $p_c = 66.2317$ is the unstable pole of $G_c(s)$.

To show why any stabilizing controller must have a pole to right of z we use a root locus argument. By contradiction, suppose that there is no pole p_c of $G_c(s)$ to the right of z. The pole-zero plot of $G_X(s)$ along with a pole p_c of $G_c(s)$ closest to z is shown in Figure 17.29 on the next page. As p_c (stable or not) is to the left of z it follows that the interval from $z = 4.7958$ to $p = 5.408$ is part of the real-axis root locus. There can be no breakaway point in this interval because there is only a single pole that starts at p (unlike the case shown

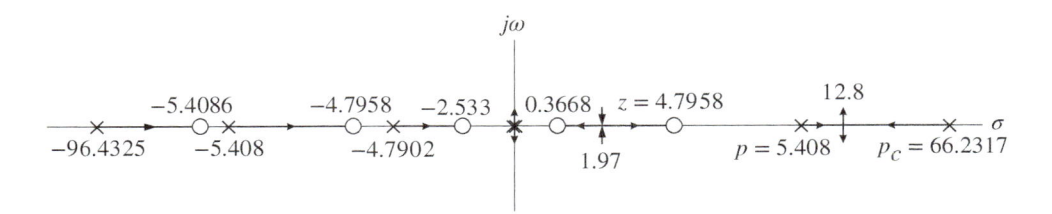

Figure 17.28. Real axis root locus for $1 + KG_c(s)G_X(s)$.

in Figure 17.28). This means there is a closed-loop pole in this interval for all K. This contradicts $G_c(s)$ being a stabilizing controller and so the assumption $p_c < z$ cannot hold.

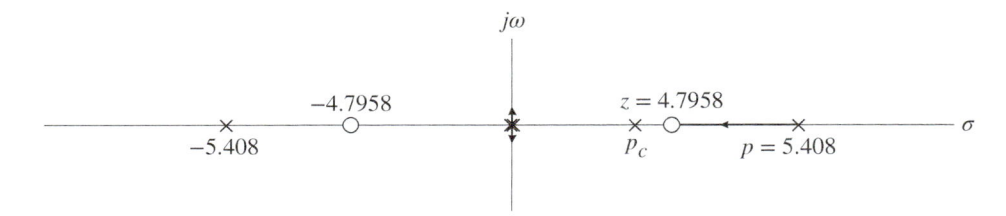

Figure 17.29. Pole–zero plot of $G_X(s)$ along with a presumed right half-plane pole p_c of $G_c(s)$ with $p_c < z$.

Plot the root locus of $1 + KG_c(s)G_X(s)$.

Problem 7 *Output Feedback via State Estimation for the Inverted Pendulum*

Consider the statespace model of the inverted pendulum given in (17.25) with the QUANSER parameter values as given in Problem 1. Write a MATLAB program to compute the state feedback vector k that places all the poles of $A - bk$ at -5. Also compute the state estimator vector ℓ that places all the poles of $A - \ell c$ at -10. Then append to your MATLAB program the code to compute $G(s), G_{c1}(s), G_{c2}(s)$, and $G_c(s)$ as given in Section 17.5.

Hint: The following MATLAB code shows how to compute $c(sI - A)^{-1}b$ from the A, b, c matrices.

```
A = [1 2; 3 4]; b = [5; 6]; c = [7 8];
sympref('FloatingPointOutput',true);
syms s;
G = simplify(c*inv((s*eye(2)-A))*b)
G_num = c*adjoint((s*eye(2)-A))*b
G_den = det((s*eye(2)-A))
```

References

[1] K. Ogata, *Modern Control Engineering*. Englewood Cliffs, NJ: Prentice-Hall, 2002.

[2] B. C. Kuo, *Automatic Control Systems*. Englewood Cliffs, NJ: Prentice-Hall, 1987.

[3] G. F. Franklin, J. D. Powell and A. Emami-Naeini, *Feedback Control of Dynamic Systems*, 6th ed. Prentice-Hall, 2009.

[4] C. L. Philips and J. M. Parr, *Feedback Control Systems*. Englewood Cliffs, NJ: Prentice-Hall, 2011.

[5] L. Qiu and K. Zhou, *Introduction to Feedback Control*. Prentice-Hall, 2010.

[6] G. C. Goodwin, S. F. Graebe and M. E. Salgado, *Control System Design*. Upper Saddle River, NJ: Prentice-Hall, 2001.

[7] J. Chiasson, *Modeling and High-Performance Control of Electric Machines*. John Wiley & Sons, 2005.

[8] Licensed under CC BY-SA 3.0 via Wikimedia Commons (2010). *User: ComputerGeezer and Geof, Venturiflow*. Available: https://commons.wikimedia.org/wiki/File:VenturiFlow.png.

[9] S. Karp, *How it works flight controls*, October 12, 2013. Available: https://www.youtube.com/watch?v=AiTk5r-4coc.

[10] W. Barie, *Design and Implementation of a Nonlinear State-Space Controller for a Magnetic Levitation System*. University of Pittsburgh, 1994.

[11] M. Bodson, "Explaining the Routh–Hurwitz criterion," *IEEE Control Systems*, vol. 40, no. 1, pp. 45–51, Feb 2020.

[12] C. L. Philips and R. D. Harbor, *Feedback Control Systems*. Englewood Cliffs, NJ: Prentice-Hall, 1988.

[13] R. W. Beard, T. W. McLain and C. Peterson, *Introduction to Feedback Control Using Design Studies*. Independently published, 2016.

[14] K. R. Symon, *Mechanics*, 2nd ed. Addison-Wesley, 1960.

[15] W. J. Palm III, *System Dynamics*, 2nd ed. McGraw-Hill, 2010.

[16] D. Halliday and R. Resnick, *Physics Parts 1 and 2*. New York: John Wiley & Sons, 1978.

[17] C. T. Chen, *Linear System Theory and Design*, 4th ed. Oxford Press, 2013.

[18] D. Halliday and R. Resnick, *Physics Volume II*. New York: John Wiley & Sons, 1962.

[19] U. Haber-Schaim, J. H. Dodge, R. Gardner and E. Shore, *PSSC Physics*, 7th ed. Dubuque, IA: Kendall/Hunt, 1991.

[20] S. J. Chapman, *Electric Machinery Fundamentals*. New York: McGraw-Hill, 1985.

[21] L. W. Matsch and J. D. Morgan, *Electromagnetic and Electromechanical Machines*, 3rd ed. New York: John Wiley & Sons, 1986.

[22] C. W. deSilva, *Control Sensors and Actuators*. Englewood Cliffs, NJ: Prentice-Hall, 1989.

[23] C. W. deSilva, *Mechatronics: An Integrative Approach*. Boca Raton, FL: CRC Press, 2004.

[24] P. C. Krause and O. Wasynczuk, *Electromechanical Motion Devices*. New York: McGraw-Hill, 1989.

[25] G. R. Slemon and A. Straughen, *Electric Machines*. Reading, MA: Addison-Wesley, 1980.

[26] M. Vidyasagar, "On undershoot and nonminimum phase zeros," *IEEE Transactions on Automatic Control*, vol. 31, no. 5, p. 440, May 1986.

[27] B. Francis and M. Wonham, "The internal model principle for linear multivariable regulators," *Applied Mathematical Optimization*, vol. 2, no. 2, pp. 170–194, June 1975.

[28] B. A. Francis and M. W. Wonham, "The internal model principle of control theory," *Automatica*, vol. 12, no. 5, pp. 457–465, Sept 1975.

[29] E. J. Davison, "The robust control of the servomechanism problem for linear time invariant multivariable systems," *IEEE Transactions on Automatic Control*, vol. 21, no. 1, pp. 25–34, Feb 1976.

[30] M. Bodson, *Foundations of Control Engineering*. Independently published, 2020.

[31] B. Messner and D. Tilbury (1998), *Control tutorials for Matlab and Simulink*. Available: http://ctms.engin.umich.edu.

[32] Wikipedia, *PID controller*, 2019 [Online]. Available: https://en.wikipedia.org/wiki/PIDcontroller. (accessed 25 March 2019).

[33] G. F. Franklin, J. D. Powell and A. Emami-Naeini, *Feedback Control of Dynamic Systems*. Reading, MA: Addison-Wesley, 1986.

[34] Quanser, *Real Time Control Experiments for Education and Research*. Available: www.quanser.com.

[35] J. S. Freudenberg and D. P. Looze, "Right half-plane poles and zeros and design tradeoffs in feedback systems," *IEEE Transactions on Automatic Control*, vol. 30, no. 6, pp. 555–565, June 1985.

[36] G. Stein, "Respect the unstable," *IEEE Control Systems Magazine*, vol. 23, no. 4, pp. 12–25, Aug 2003.

[37] S. Darbha and S. P. Bhattacharyya, "On the synthesis of controllers for a non-overshooting step response," *IEEE Transactions on Automatic Control*, vol. 48, no. 5, pp. 797–799, May 2003.

[38] J. W. Howze and S. P. Bhattacharya, "Robust tracking, error feedback, and two-degree-of-freedom controllers," *IEEE Transactions on Automatic Control*, vol. 42, no. 7, 980–983, pp. July 1997.

[39] W. A. Wolovich, *Automatic Control Systems: Basic Analysis and Design*. Saunders College Publishing, 1994.

[40] K. J. Åström and R. Murray, *Feedback Systems: An Introduction for Engineers and Scientists*. Princeton University Press, 2008.

[41] Aerotech Inc., *Harmonic cancellation: optimize periodic trajectories, reject periodic disturbances*. Available: https://www.aerotech.com/harmonic-cancellation-optimize-periodic-trajectories-reject-periodic-disturbances/.

[42] Aerotech, Inc., *Advanced control techniques*, 2010. Available: https://www.youtube.com/watch?v=5A-R-JKisPY.

[43] M. Bodson, A. Sacks and P. Khosla, "Harmonic generation in adaptive feedforward cancellation schemes," in *Proceedings of the 31st Conference on Decision and Control*, Tucson, AZ, December 1992.

[44] X. Guo and M. Bodson, "Equivalence between adaptive feedforward cancellation and disturbance rejection using the internal model principle," *Adaptive Control and Signal Processing*, vol. 24, no. 3, pp. 211–218, Apr 2009.

[45] L. H. Keel and S. P. Bhattacharya, "Robust, Fragile, or optimal," *IEEE Transactions on Automatic Control*, vol. 42, no. 8, pp. 1098–1105, Aug 1997.

[46] P. Dorato, *Analytical Feedback System Design An Interpolation Approach*. Pacific Grove, CA: Brooks/Cole, 1999.

[47] H. W. Bode, *Network Analysis and Feedback Amplifier Design*. Princeton, NJ: D. Van Nordstand, 1945.

[48] H. Nyquist, "Regeneration theory," *Bell System Technical Journal*, vol. 11, pp. 126–147, Jan 1932.

[49] J. C. Doyle, B. A. Francis and A. R. Tannenbaum, *Feedback Control Theory*. Macmillan, 1992.

[50] A. V. Oppenheim and A. S. Willsky, *Signals & Systems*, 2nd ed. Englewood Cliffs, NJ: Prentice-Hall, 1997.

[51] E. W. Kamen and B. S. Heck, *Fundamentals of Signals and Systems using the Web and Matlab*, 2nd ed. Upper Saddler River, NJ: Prentice-Hall, 2000.

[52] M. Viola, Inverse Fourier Transform of $\frac{1}{\alpha + j\omega}$. Stack Exchange. https://math.stackexchange.com/questions/1531134/inverse-fourier-transform-of-frac1ajw, Nov 2015.

[53] W. R. Evans, "Control system synthesis by root locus method," *Transactions of the AIEE*, vol. 69, pp. 66–69, Jan 1950.

[54] W. Barie, "Design and implementation of a nonlinear state-space controller for a magnetic levitation system," Master's thesis, University of Pittsburgh, 1994.

[55] A. Isidori, *Nonlinear Control Systems*, 2nd ed. Berlin: Springer-Verlag, 1989.

[56] R. Marino and P. Tomei, *Nonlinear Control Design - Geometric, Adaptive and Robust*. Englewood Cliffs, NJ: Prentice-Hall, 1995.

[57] H. Nijmeijer and A. J. van der Schaft, *Nonlinear Dynamical Control Systems*. Springer-Verlag, 1990.

[58] M. Bodson, J. Chiasson, R. Novotnak and R. Rekowski, "High performance nonlinear control of a permanent magnet stepper motor," *IEEE Transactions on Control Systems Technology*, vol. 1, no. 1, pp. 5–14, Mar 1993.

[59] M. Bodson and J. Chiasson, "Differential-geometric methods for control of electric motors," *International Journal of Robust and Nonlinear Control*, vol. 8, pp. 923–954, Sept 1998.

[60] S. Mehta, "Control of a series DC motor by feedback linearization," Master's thesis, University of Pittsburgh, 1996.

[61] T. Kailath, *Linear Systems*. Englewood Cliffs, NJ: Prentice-Hall, 1980.

[62] B. Francis, *Classical control*, 2015. Available: http://www.scg.utoronto.ca/~francis/main.pdf.

[63] A. R. Woodyatt, R. H. Middleton and J. S. Freudenberg, "Fundamental contraints for the inverted pendulum," The University of Newcastle, Tech. Rep. EE9716, 1997.

[64] S. Skogestad and I. Postlethwaite, *Multivariable Feedback Control - Analysis and Design*. John Wiley & Sons, 2005.

[65] B.-F. Wu and E. A. Jonckherre, "A simplified approach to Bode's theorem for continuous-time and discrete-time systems," *IEEE Transactions on Automatic Control*, vol. 37, no. 7, pp. 175–182, Nov 1992.

[66] M. Vidyasagar, *Control System Synthesis: A Factorization Approach*. MIT Press, 1985.

[67] J. B. Hoagg and D. S. Bernstein, "Nonminimum-phase zeros: much to do about nothing," *IEEE Control Systems Magazine*, vol. 27, no. 3, pp. 45–57, June 2007.

[68] A. Emami-Naeini and D. de Roover, "Bode's sensitivity integral constraints: The waterbed effect revisited," January 2019. arXiv:1902.11302v1.

[69] S. Boyd and C. A. Desoer, "Subharmonic functions and performance bounds in linear time-invariant feedback systems," *IMA Journal of Mathematical Control and Information*, vol. 2, pp. 153–170, June 1985.

[70] J. Chiasson, "The Differential-Geometric Approach to Nonlinear Control," independently published, 2021.

Index

Aerotech Inc.
    harmonic cancellation, 345

Bandwidth, 409
Block diagram
    DC motor, 185
    example, 191, 192, 194
Block diagram reduction, 187
Bode diagram
    complex pair of poles, 377
    peak and resonant values for a
        second-order system, 378
Bode diagrams, 361
    examples, 365
Brushless DC motor, 175

Calculus
    Leibniz formula, 559
Cart on a Track System, 516
    block diagram, 519
    electrical equations, 518
    mechanical equations, 517
Cayley-Hamilton theorem, 579
Characteristic polynomial
    definition, 310, 584
Characteristic value
    definition, 584
Closed-loop system
    characteristic polynomial, 310
Collocated
    mass-spring-damper, 351
    satellite model, 119
Commutation
    multiloop DC motor, 171
    single-loop DC motor, 153
Complex numbers, 26
Control canonical form, 541
Convolution, 44
Current command
    DC motor, 195

Cylindrical coordinates
    definition, 152

DC motor
    armature flux, 160
    back emf, 159
    block diagram, 185
    brushless DC motor, 175
    commutation
        multiloop motor, 171
        single-loop motor, 153
    current command, 195
    energy conversion, 164
    flexible shaft, 355, 356, 443
    mathematical model, 163
    multiloop DC motor, 170
    schematic diagram, 164
    self-inductance, 161
    simulation
        open-loop, 180
    single-loop, 151
    tachometer, 168
        single-loop motor, 169
    torque, 151
    voltage and current limits, 165
Derivative feedback
    practical implementation, 249
Differential equations, 49
    phasor method of solution, 53
    sinusoidal steady-state response, 50
    unstable response with a sinusoidal
        input, 52, 54
Disturbance rejection, 244
    PI controller, 245
    proportional controller, 244
    sinusoidal disturbance, 256, 288
Duality, 668

Eigenvalue
    definition, 584

An Introduction to System Modeling and Control, First Edition. John Chiasson.
© 2022 John Wiley & Sons, Inc. Published 2022 by John Wiley & Sons, Inc.
Companion website: www.wiley.com/go/chiasson/anintroductiontosystemmodelingandcontrol

Electromotive force (emf)
  definition, 155
Equilibrium point
  definition, 507
  for a system with an input, 507
Euler's formula, 40
Exponential function, 39
Exponential matrix, 554
  properties, 558

Faraday's law, 155, 175
  linear DC machine, 158
  sign convention for the induced emf, 157
  single-loop DC motor, 159
  surface element vector, 156
Final value theorem, 57
First-order systems
  time constant
    definition, 205
  transient response, 205
Fourier transform, 409
  step function, 411
Frequency response
  Bode diagram, 361
  definition, 56
Friction, 91, 129
  kinetic friction, 130
  static friction, 129
  viscous friction, 91
Fundamental matrix, 554

Gain and phase margins, 402
Gears, 120
  algebraic relationships, 121
  dynamic relationships, 121

H-infinity controllers, 409, 708
H-infinity norm, 708
Harmonic cancellation, 345

I-PD controller, 283
Internal model principle, 256
  example, 256, 282, 283
  harmonic cancellation, 345
  phase-locked loops, 258
Internal stability, 308
Inverted pendulum, 497
  inclined track

  convergence to an equilibrium point, 617
  equilibrium points, 613
  linear approximate model, 500
  mathematical model, 497
  multi-output observer, 663, 681, 683, 685
  nested loop feedback structure, 503
  output feedback
    Nyquist plot, 426
    robustness, 426
  pole placement, 295, 352, 633
  simulation, 633
    statespace model, 720
  statespace model
    linear, 501
    nonlinear, 500
  step response of cart position, 610
  tracking a sinusoidal reference, 639
  tracking of step reference inputs, 604, 605
  transfer function, 501

Lag and lead compensation, 414
Lag compensation, 414
  reduction of steady-state error, 415
Laplace transforms
  asymptotic response of f(t), 39
  complex exponential, 18
  convolution, 44
  definition, 17
  differential equation, 23
  exponential function, 18
  final value theorem, 57
  integration, 44
  Matlab, 25, 28
  open left half-plane, 38
  partial fraction expansion, 24
    multiple poles, 30
  poles and partial fractions, 36
  poles and zeros, 35
  properties, 20
  stable rational transfer functions
    definition, 62
  step input, 17
  table, 45
  transfer function
    definition, 50
Lead compensation, 414

as an approximation to a PD
    controller, 281, 416
Leibniz formula, 559
Leibniz rule for differentiation, 559
Linearization of nonlinear
    systems, 506
  example, 508, 509

Magnetic field
  DC motor
    air-gap magnetic field, 151
  definition, 149
Magnetic force
  definition, 149
Magnetic levitation
  conservation of energy, 511
  feedback linearization control, 618
    simulation, 636
  linear statespace model, 514
  mathematical model, 510
  nonlinear statespace model
    current input, 513
    equilibrium point, 514
    linearization of the current input
      model, 514
    linearization of the voltage input
      model, 528
    voltage input, 512, 513
  simulation, 529
  transfer function, 515
Mass-spring-damper systems, 89
  equilibrium point, 92
    equations of motion with respect to
      the equilibrium point, 92,
      101–103
  mechanical work, 89, 90
  simulation, 102
  transfer function, 91, 93–95
Matlab
  bode command, 427
  feedback gain for pole placement, 638
  filter
    Butterworh, 688
  multinomial expansions, 328
  nyquist command, 427, 439
  root locus, 455
    code, 224
Matrices
  adjoint, 580, 725

Cayley-Hamilton theorem, 582
cofactor, 579
determinant, 580
inverse, 579
Inverse of a 2x2, 545
inverse of a product, 667
matrix multiplication as a linear
    combination of rows, 652
positive definite, 670
positive semidefinite, 670
similarity transformations, 593
symmetric, 669
transpose, 665
transpose and inverse compute, 666
transpose of a product, 666
transpose of a sum, 666
Mechanical work and kinetic energy, 89
Minimum phase system, 220
Missile
  tracking and disturbance rejection, 279
Model uncertainty, 265
Moment of inertia
  cylinder, 112
  definition, 111
  sphere, 135
Multi-output observer
  inverted pendulum, 663, 681, 683, 685
Multinomial expansions
  using Matlab, 328

Natural logarithm, 41
Newton's law of rotational motion, 112
Non-minimum phase, 435
Non-minimum phase system, 221, 349,
    350
Noncollocated
  mass-spring-damper, 350
  satellite model, 119
Notch filter
  DC motor with flexible shaft, 356, 445
  satellite with solar panels, 484, 495
Nyquist theory, 382
  gain and phase margins, 402
  lag and lead compensation, 414
  lag compensation to reduce
    steady-state error, 415
Nyquist contour, 389
Nyquist polar plots, 390
Nyquist plot, 436

Nyquist theory (*continued*)
  Nyquist stability test, 390
  principle of the argument, 382
  robustness, 407

Observers, 643
  cart on a track, 643
    on an inclined track, 646
  duality, 668
  general procedure, 649
  in canonical form, 651
  Laplace domain, 660
  observability matrix, 652
  placement of the observer poles, 655
  position cannot be estimated from speed, 645
  separation principle, 657
  transformation to observer canonical form, 654
Open left half-plane
  definition, 38
Optical encoder model, 165
Output pole placement, see Pole placement, 285
Overshoot
  stable system with real poles and one RHP zero, 332, 333
  type 2 systems, 329, 330

Parameter estimation
  second-order system
    from the step response, 210, 226
Parameter identification, 669
  error index, 676
  least-squares approximation, 673
  least-squares identification, 671
Parity interlacing property, 706
Phase-locked loops
  internal model principle, 258
Phasor method, 52
PI controller
  interpretation, 246
  rejecting a constant load torque, 245
  speed control of a DC motor, 217
  summary, 247
  tracking a ramp input, 243
PI-D controller, 250, 258
PID controller, 248
  applied to a servomechanism, 249

I-PD controller, 283
PI-D controller, 258
practical implementation of derivative feedback, 249
Pitch control
  via 2 DOF control, 316
  via PID, 258
Poisson integral
  output feedback control of the inverted pendulum, 706
Pole placement
  choice of pole location, 318
  first-order system
    tracking a step input and rejecting a constant disturbance, 340
  second-order system
    tracking a step input and rejecting a constant disturbance, 285, 287
  second-order system tracking a sinusoid, 288
  via state feedback, 598
  via the Diophantine equation, 326
Pole-zero cancellation, 311, 312
Poles and zeros
  definition, 35
Polynomial
  stable polynomial
    definition, 63
Principle of the argument, 382

Realization, 539
  discrete-time, 544, 549, 566
  not strictly proper transfer function, 566, 567
Right half-plane poles
  parity interlacing property, 706
Right half-plane zeros
  eliminating overshoot summary, 306
  example, 289
  fundamental limitations, 306, 706
  second-order systems, 221
  two right half-plane zeros, 304
Rigid body rotational dynamics, 111
  gears, 120
  kinetic energy, 111
  moment of inertia
    definition, 111
  motorized rolling cylinder, 133
  Newton's law of rotational motion, 112

rack and pinion example, 115, 116
rolling cylinder, 127
    no slip condition, 128
rotational mass-spring-damper, 118, 139
    damper, 119, 140
    torsional spring, 119, 140
torque
    definition, 113
Robustness, 306, 706
    inverted pendulum, 426
    Nyquist theory, 407
Rolling cylinder, 127
    down an incline, 129
    flat surface, 128
    up an incline, 133
Rolling mill, 124, 183
Root locus, 447
    angle of departure, 463
    asymptotes and their intercept, 457
    breakaway points, 482
    effect of open-loop poles on the root locus, 480
    effect of open-loop zeros on the root locus, 481
    proportional control, 224
    root locus rules, 449
    satellite with solar panels
        non collocated, 484
Routh-Hurwitz stability test, 65
    main result, 68
    Routh array, 67
    second-order polynomial, 71
    secondary result, 68
    special cases
        row of zeros in the Routh array, 72
        zero in the first column, but not a row of zeros, 75
    third-order polynomial, 71

Satellite positioning system
    tracking and disturbance rejection, 276
Satellite with solar panels
    collocated case, 321
    differential equation model, 119
        simulation, 140
    non collocated case, 353, 484
    transfer function model, 119
        simulation, 140

Second-order systems
    effects of zeros, 219
    peak overshoot, 210
    peak time, 210
    rise time, 214
    settling time, 212
    step response, 205
        by partial fraction expansion, 210, 224
        with a zero, 217
    transient response, 207
    with a left half-plane zero, 220
    with a right half-plane zero, 221, 229
    with a zero, 217
Sensitivity, 268, 697
    Bode integral, 702
    Bode integral for relative degree one systems, 713
    Bode integral with RHP poles, 704
    definition, 271
    double integrator control, 702
    output feedback control of the inverted pendulum, 697, 705, 710
    output feedback control via state estimation, 717
    Poisson integral, 706
    reduction by feedback, 268
    state feedback control of the inverted pendulum, 713, 715
Separation principle, 657
Sevomechanism
    modeling, 233
Similarity transformations, 593
Simulation, 96
    cart on the track, 532
    DC motor, 201
        identification of the model, 227
    in Simulink, 96
    inverted pendulum
        nested loops, 522
        output position feedback, 720
        pole placement, 352
    magnetic levitation, 529
        3rd order model, 531
        feedback linearization control, 636
    parameter identification
        with optical encoder, 689
        without optical encoder, 686

736 Index

Simulation (*continued*)
    pitch control, 258, 316
    speed control of a DC motor, 230
Simulation diagram, 542
Simulink
    cart on the track, 532
    DC motor, 105, 180, 201
        identification of the model, 227
    inverted pendulum
        nested loops, 522
        output position feedback, 720
        pole placement, 352
    magnetic levitation, 529
        3rd order model, 531
    mass-spring-damper system, 96
    optical encoder, 180
    parameter identification
        with optical encoder, 689
        without optical encoder, 686
    pitch control, 258, 316
    speed control of a DC motor, 230
    stable differential equation, 107
    unstable differential equation, 109
Stable transfer function
    definition, 65
    necessary condition for stability, 67
State
    definition, 538, 539
State estimators, 643
    cart on a track, 643
        on an inclined track, 646
    duality, 668
    general procedure, 649
    in observer canonical form, 651
    Laplace domain, 660
    observability matrix, 652
    placement of the observer poles, 655
    position cannot be estimated from speed, 645
    separation principle, 657
    transformation to observer canonical form, 654
State feedback, 569
    augmented system for disturbance rejection, 605, 623
    augmented system for tracking step reference inputs, 605, 623
    control canonical form, 585
    disturbance rejection, 589

    feedback linearization, 618
        magnetic levitation, 618
    inverted pendulum
        tracking step inputs in horizontal position, 605
    magnetic levitation, 587
    pole placement, 598
        cart on the track system, 600
        magnetic levitation, 601
    stabilization, 584
    trajectory design, 571
    trajectory tracking, 578
State variables, 537
    definition, 538, 539
Statespace
    control canonical form, 541
    discrete-time, 548, 561
    exponential matrix, 554
    fundamental matrix, 554
    Laplace transform, 551
    simulation diagram, 542
    solution of the statespace equations, 558
Statespace form, 537
    cart on a track, 519
    DC motor, 538
    inverted pendulum, 500
    magnetic levitation, 512
    spring-mass-damper system, 538
Sylvester resultant matrix, 326
System responses
    first-order system, 203
        time constant, 205
    second-order system, 205
        peak time and peak overshoot, 210
        rise time, 214
        settling time, 212
        with a zero, 217
    step response
        by partial fraction expansion, 210, 224
    third-order system, 222

Tachometer
    DC motor, 168
        single-loop motor, 169
Tension
    definition, 125
    rope or cable, 125

Third-order systems
   one real pole and two
      complex-conjugate poles, 222
Torque
   DC motor, 151
   definition, 113
   sign convention, 114
Torsional damper, 119, 140
Torsional spring, 119, 140
Tracking
   PI controller, 243
   sinusoidal input, 288
   stable reference signals, 278
   step and ramp inputs, 239
Tracking and disturbance rejection, 239
   general theory, 252
   I-PD controller, 283
   PI-D controller, 250
   PID controller, 248
   type number
      definition, 252
   using the internal model principle, 256

Trajectory design
   for state feedback, 571
Transfer function
   definition, 50
   order, 50
   realization, 539
   stable rational transfer functions, 62
Two degrees of freedom controllers, 298, 303
   input filter, 300
   pitch control, 316
   satellite with solar panels
      collocated case, 321
      non collocated case, 353
Type number
   definition, 252

Undershoot, 336
   magnitude versus settling time, 338
   odd number of real right half-plane zeros, 222
Unstable pole-zero cancellation, 311, 312